The Routledge Handbook of Agricultural Economics

This Handbook offers an up-to-date collection of research on agricultural economics. Drawing together scholarship from experts at the top of their profession and from around the world, this collection provides new insights into the area of agricultural economics.

The Routledge Handbook of Agricultural Economics explores a broad variety of topics including welfare economics, econometrics, agribusiness, and consumer economics. This wide range reflects the way in which agricultural economics encompasses a large sector of any economy, and the chapters present both an introduction to the subjects as well as the methodology, statistical background, and operations research techniques needed to solve practical economic problems. In addition, food economics is given a special focus in the Handbook due to the recent emphasis on health and feeding the world population a quality diet. Furthermore, through examining these diverse topics, the authors seek to provide some indication of the direction of research in these areas and where future research endeavors may be productive.

Acting as a comprehensive, up-to-date, and definitive work of reference, this Handbook will be of use to researchers, faculty, and graduate students looking to deepen their understanding of agricultural economics, agribusiness, and applied economics, and the interrelationship of those areas.

Gail L. Cramer is Professor Emeritus of Agricultural Economics at Louisiana State University (LSU) and LSU Agricultural Center, USA. He has held teaching and research positions for 48 years at Montana State University, University of Arkansas and Louisiana State University. His academic specialty is in agricultural marketing, policy, transportation, and international trade.

Krishna P. Paudel is the Gilbert Durbin Endowed Professor of Environmental and Resource Economics at Louisiana State University (LSU) and LSU Agricultural Center, USA. He holds a PhD from the University of Georgia and has been a faculty member at Auburn University and Louisiana State University.

Andrew Schmitz is the Ben Hill Griffin, Jr., Eminent Scholar and Professor of Food and Resource Economics, University of Florida; Research Professor, University of California, Berkeley; and Adjunct Professor, University of Saskatchewan.

The Routledge Handbook of Agricultural Economics

*Edited by Gail L. Cramer, Krishna P. Paudel,
and Andrew Schmitz*

LONDON AND NEW YORK

First published 2019
by Routledge
2 Park Square, Milton Park, Abingdon, Oxon OX14 4RN

and by Routledge
711 Third Avenue, New York, NY 10017

Routledge is an imprint of the Taylor & Francis Group, an informa business

© 2019 selection and editorial matter, Gail L. Cramer; individual chapters, the contributors

The right of Gail L. Cramer to be identified as the author of the editorial material, and of the authors for their individual chapters, has been asserted in accordance with sections 77 and 78 of the Copyright, Designs and Patents Act 1988.

All rights reserved. No part of this book may be reprinted or reproduced or utilised in any form or by any electronic, mechanical, or other means, now known or hereafter invented, including photocopying and recording, or in any information storage or retrieval system, without permission in writing from the publishers.

Trademark notice: Product or corporate names may be trademarks or registered trademarks, and are used only for identification and explanation without intent to infringe.

British Library Cataloguing-in-Publication Data
A catalogue record for this book is available from the British Library

Library of Congress Cataloging-in-Publication Data
Names: Cramer, Gail L., editor. | Paudel, Krishna P., editor. | Schmitz, Andrew, editor.
Title: The Routledge handbook of agricultural economics / edited by Gail L. Cramer, Krishna P. Paudel, & Andrew Schmitz.
Description: New York, NY : Routledge, 2018. | Includes bibliographical references and index.
Identifiers: LCCN 2018005264 (print) | LCCN 2018007743 (ebook) |
 ISBN 9781315623351 (Ebook) | ISBN 9781138654235 (hardback : alk. paper)
Subjects: LCSH: Agriculture—Economic aspects—Handbooks, manuals, etc.
Classification: LCC HD1415 (ebook) | LCC HD1415 .R615 2018 (print) |
 DDC 338.1—dc23
LC record available at https://lccn.loc.gov/2018005264

ISBN: 978-1-138-65423-5 (hbk)
ISBN: 978-1-315-62335-1 (ebk)

Typeset in Bembo
by Apex CoVantage, LLC

Contents

Editor biographies	ix
List of contributors	xi
Acknowledgments	xiv

1	An introduction to *The Routledge Handbook of Agricultural Economics* Gail L. Cramer	1
2	Broadening the validity of welfare economics for quantitative public policy analysis Richard E. Just and Andrew Schmitz	10

PART 1
Food economics   35

3	World food: new developments Shenggen Fan	37
4	Industrial organization of the food industry: the role of buyer power Ian Sheldon	50
5	Economic ramifications of obesity: a selective literature review Oral Capps, Jr., Ariun Ishdorj, Senarath Dharmasena, and Marco A. Palma	70
6	Behavioral economics in food and agriculture Wen Li and David R. Just	84
7	Product quality and reputation in food and agriculture Anthony R. Delmond, Jill J. McCluskey, and Jason A. Winfree	96

Contents

8   Food marketing in the United States   108
    Stephen Martinez and Krishna P. Paudel

9   Food from the water – fisheries and aquaculture   134
    Frank Asche, James L. Anderson, and Taryn M. Garlock

10 The economics of antibiotics use in agriculture   159
    Jutta Roosen and David A. Hennessy

11 Advancements in the economics of food security   175
    P. Lynn Kennedy

PART 2
Environmental and resource economics   189

12 How does climate change affect agriculture?   191
    *Panit Arunanondchai, Chengcheng Fei, Anthony Fisher,
    Bruce A. McCarl, Weiwei Wang, and Yingqian Yang*

13 Habitat conservation in agricultural landscapes   211
    Chad Lawley and Charles Towe

14 Water quality trading   232
    James S. Shortle and Aaron Cook

15 Natural capital and ecosystem services   254
    Marcello Hernández-Blanco and Robert Costanza

16 Water management and economics   269
    Louis Sears and C.-Y. Cynthia Lin Lawell

17 Water supply and dams in agriculture   285
    Biswo N. Poudel, Yang Xie, and David Zilberman

PART 3
Trade, policy, and development   299

18 A synopsis of trade theories and their applications   301
    Stephen Devadoss and Jeff Luckstead

19 International agricultural trade: a road map   327
    Andrew Schmitz and Ian Sheldon

20	Use of subsidies and taxes and the reform of agricultural policy *G. Cornelis Van Kooten, David Orden, and Andrew Schmitz*	355
21	The political economy of agricultural and food policies *Johan Swinnen*	381
22	Macroeconomic issues in agricultural economics *Michael Reed and Sayed Saghaian*	399
23	Models of economic growth: application to agriculture and structural transformation *Terry L. Roe and Munisamy Gopinath*	412
24	Rural development – theory and practice *Thomas G. Johnson and J. Matthew Fannin*	427

PART 4
Methods   443

25	Econometrics for the future *Hector O. Zapata, R. Carter Hill, and Thomas B. Fomby*	445
26	On the evolution of agricultural econometrics *David A. Bessler and Marco A. Palma*	467
27	New empirical models in consumer demand *Timothy J. Richards and Celine Bonnet*	488
28	A survey of semiparametric regression methods used in the environmental Kuznets curve analyses *Krishna P. Paudel and Mahesh Pandit*	512
29	Reconstructing deterministic economic dynamics from volatile time series data *Ray Huffaker, Ernst Berg, and Maurizio Canavari*	533
30	Agricultural development impact evaluation *J. Edward Taylor*	548
31	Estimating production functions *Daniela Puggioni and Spiro E. Stefanou*	581

Contents

PART 5
Production   601

32 Role of risk and uncertainty in agriculture   603
    Jean-Paul Chavas

33 Investing in people in the twenty-first century: education, health, and migration   616
    Wallace E. Huffman

34 The economics of biofuels   637
    Farzad Taheripour, Hao Cui, and Wallace E. Tyner

35 Farm management: recent applications   658
    Patricia Duffy and Sam Funk

36 Economics of agricultural biotechnology   670
    David Zilberman, Justus Wesseler, Andrew Schmitz, and Ben Gordon

PART 6
Marketing   687

37 The financial economics of agriculture and farm management   689
    Charles B. Moss, Jaclyn D. Kropp, and Maria Bampasidou

38 Futures markets and hedging   713
    T. Randall Fortenbery

39 Cooperative extension system: value to society   728
    Russell Tronstad and Mike Woods

40 Theory of cooperatives: recent developments   748
    Michael L. Cook and Jasper Grashuis

41 Agribusiness economics and management   760
    Michael Boland and Metin Çakır

List of reviewers   779
Index   781

Editor biographies

Gail L. Cramer

Gail L. Cramer is Emeritus Professor at Louisiana State University in Agricultural Economics. He was Department Head of the Agricultural Economics and Agribusiness Department from 2000 to 2015 at LSU. In his tenure, the Department graduated about 100 PhD students, 200 master's students, and 870 undergraduates.

He was Professor and L.C. Carter Endowed Chair of Rice and Soybean Marketing and Policy at the University of Arkansas from 1987 to 2000. From 1967 to 1987, he was Assistant, Associate, and Professor at Montana State University. He taught agricultural economics and economics courses and his research areas were grain marketing, agricultural policy, and international trade. He is the author of several books and over 220 research papers. His introductory *Agricultural Economics* book (co-authored by Clarence Jensen) was used throughout the world and translated into Spanish, Chinese, Malaysian, and Russian. His contributions in research have been in increased efficiency in wheat and rice marketing, increased understanding of the world food problem, and a better understanding of world rice marketing and price formation. Also, he trained many Chinese students in research methods and basic econometrics in the early 1990s. He held visiting professorships at Harvard University, the University of California at Berkeley, The Ohio State University, Winrock International, and the University of Illinois. He was selected for three teaching awards at Montana State University, where he taught large agricultural economics and economics classes. He earned a Bachelor of Science degree from Washington State University in 1963, a Master of Science degree from Michigan State University in 1964, and a PhD from Oregon State University in 1968. Major research awards received include an Outstanding Research Achievement award from the University of Arkansas and the U.S. Rice Industry; a Lifetime Achievement award from SAEA; a Distinguished Alumni Achievement Award from Washington State University; the International Research Award for Gamma Sigma Delta; the Edwin Nourse PhD award from the National Council of Agricultural Cooperatives, a Fellows award from IAMA; and three awards from the American Agricultural Economics Association. He was editor of *The Review of Agricultural Economics* from 1999–2001.

Editor biographies

Krishna P. Paudel

Krishna P. Paudel is the Gilbert Durbin Endowed Professor of Environmental and Resource Economics at Louisiana State University. He holds a PhD from the University of Georgia. He has been a faculty member at Auburn University and Louisiana State University. He has been a Visiting Scholar at Vanderbilt University and a Visiting Assistant Professor at NASA's Global Hydrology and Climate Change Center. His research in environmental and resource economics and agricultural economics has appeared in major agricultural economics and environmental economics journals. He has worked in the economics of agricultural, developmental, and environmental issues for more than 20 years. His environmental and resource economics work has been in areas that intersect environment/resource economics and agricultural economics, specifically in water quality and water quantity areas. He is currently working on water quality and water quantity issues in Louisiana and other parts of the world. He served as the editor of the *Journal of Agricultural and Applied Economics* between 2013 and 2016.

Andrew Schmitz

Professor Andrew Schmitz is the Ben Hill Griffin, Jr., Eminent Scholar and Professor of Food and Resource Economics, University of Florida; Research Professor, University of California, Berkeley; and Adjunct Professor, University of Saskatchewan. He received awards for both his master's thesis (University of Saskatchewan) and his PhD dissertation (University of Wisconsin–Madison). He is a Fellow of the American Agricultural Economics Association. His publications have won six major research awards and three research of enduring quality awards from the American Agricultural Economics Association and two major awards from the Canadian Agricultural Economics Association. He was awarded the Who's Who in America in 2004, the Southern Agricultural Economics Association Lifetime Achievement Award in 2003, and the Earned Doctor of Letters degree from the University of Saskatchewan in 1999. He has been a consultant to hundreds of private and public organizations. Some of his recent books include *Government Policy and Farmland Markets* (Iowa State Press); *The Welfare Economics of Public Policy* (Elgar Publishing); *Agricultural Policy, Agribusiness, and Rent-Seeking Behavior* (University of Toronto Press); *Agricultural Policy, Agribusiness, and Rent-Seeking Behavior, Second Edition* (University of Toronto Press); *Sugar and Related Sweetener Markets* (CABI Publishing); and *The Economics of Alternative Energy Sources and Globalization* (Bentham Publishing).

Contributors

JAMES L. ANDERSON
University of Florida

PANIT ARUNANONDCHAI
Texas A&M University

FRANK ASCHE
University of Florida

MARIA BAMPASIDOU
Louisiana State University (LSU) and LSU Agricultural Center

ERNST BERG
University of Bonn

DAVID A. BESSLER
Texas A&M University

MICHAEL BOLAND
University of Minnesota

CELINE BONNET
Toulouse School of Economics

METIN ÇAKIR
University of Minnesota

MAURIZIO CANAVARI
University of Bologna

ORAL CAPPS, JR.
Texas A&M University

JEAN-PAUL CHAVAS
University of Wisconsin

AARON COOK
Penn State University

MICHAEL L. COOK
University of Missouri

ROBERT COSTANZA
Australian National University

GAIL L. CRAMER
Louisiana State University (LSU) and LSU Agricultural Center

HAO CUI
Purdue University

ANTHONY R. DELMOND
Washington State University

STEPHEN DEVADOSS
Texas Tech University

SENARATH DHARMASENA
Texas A&M University

PATRICIA DUFFY
Auburn University

SHENGGEN FAN
International Food Policy Research Institute (IFPRI)

J. MATTHEW FANNIN
Louisiana State University (LSU) and LSU Agricultural Center

CHENGCHENG FEI
Texas A&M University

ANTHONY FISHER
University of California

THOMAS B. FOMBY
Southern Methodist University

T. RANDALL FORTENBERY
Washington State University

Contributors

SAM FUNK
AgServe, LLC

TARYN M. GARLOCK
University of Florida

MUNISAMY GOPINATH
Economic Research Service, USDA

BEN GORDON
University of California – Berkeley

JASPER GRASHUIS
University of Missouri

DAVID A. HENNESSY
Michigan State University

MARCELLO HERNÁNDEZ-BLANCO
Australian National University

R. CARTER HILL
Louisiana State University

RAY HUFFAKER
University of Florida

WALLACE E. HUFFMAN
Iowa State University

ARIUN ISHDORJ
Texas A&M University

THOMAS G. JOHNSON
University of Missouri-Columbia

DAVID R. JUST
Cornell University

RICHARD E. JUST
University of Maryland

P. LYNN KENNEDY
Louisiana State University (LSU) and LSU Agricultural Center

G. CORNELIS VAN KOOTEN
University of Victoria

JACLYN D. KROPP
University of Florida

C.-Y. CYNTHIA LIN LAWELL
Cornell University

CHAD LAWLEY
University of Manitoba

WEN LI
Cornell University

JEFF LUCKSTEAD
University of Arkansas

STEPHEN MARTINEZ
Economics Research Service, USDA

BRUCE A. McCARL
Texas A&M University

JILL J. McCLUSKEY
Washington State University

CHARLES B. MOSS
University of Florida

DAVID ORDEN
Virginia Tech University

MARCO A. PALMA
Texas A&M University

MAHESH PANDIT
Comerica Bank, Dallas

KRISHNA P. PAUDEL
Louisiana State University (LSU) and LSU Agricultural Center

BISWO N. POUDEL
Kathmandu University

DANIELA PUGGIONI
Bank of Mexico, Mexico City

MICHAEL REED
University of Kentucky

TIMOTHY J. RICHARDS
Arizona State University

TERRY L. ROE
University of Minnesota

JUTTA ROOSEN
Technical University of Munich

SAYED SAGHAIAN
University of Kentucky

ANDREW SCHMITZ
University of Florida

LOUIS SEARS
Cornell University

Contributors

IAN SHELDON
Ohio State University

JAMES S. SHORTLE
Penn State University

SPIRO E. STEFANOU
University of Florida

JOHAN SWINNEN
University of Leuven

FARZAD TAHERIPOUR
Purdue University

J. EDWARD TAYLOR
University of California

CHARLES TOWE
University of Connecticut

RUSSELL TRONSTAD
University of Arizona

WALLACE E. TYNER
Purdue University

WEIWEI WANG
University of Illinois at Urbana-Champaign

JUSTUS WESSELER
Wageningen University

JASON A. WINFREE
University of Idaho

MIKE WOODS
Oklahoma State University

YANG XIE
University of California-Riverside

YINGQIAN YANG
Texas A&M University

HECTOR O. ZAPATA
Louisiana State University

DAVID ZILBERMAN
University of California – Berkeley

Acknowledgments

This Handbook was organized so as to take advantage of the differing expertise of the co-editors. The task was divided into three parts, with each co-editor assuming responsibility for approximately one-third of the manuscripts.

Thanks are extended to the authors of the individual manuscripts. These authors are eminent economists at their respective universities or the USDA.

Permission has been granted by each of the authors to publish the manuscripts in this Handbook. The authors have each stated that there is no proprietary information included in their manuscripts that required separate permission to publish. Any quotation from this Handbook should be cited as follows: Cramer, Gail L., Krishna P. Paudel, and Andrew Schmitz, 2018, *The Routledge Handbook of Agricultural Economics*, Routledge (Taylor and Francis), U.K.

In addition, thanks to the many reviewers of each of the manuscripts published in this Handbook. Reviewers' names are listed after Chapter 41 of the Handbook.

The co-editors appreciate the diligent work of Anna Priddy, who read and copyedited most manuscripts. Others who have provided assistance are Fan Yang, Carol Fountain, Carole Schmitz, Claudene Chagini, Marilyn Cramer, and Suniti Bhattarai. The coeditors would also like to thank the Routledge publishing team specifically Anna Cuthbert (editorial assistant), Kelly Cracknell (production editor), Holly Smithson (project manager from Apex CoVantage), and other editors who we have contributed to this handbook.

The cooperation of the LSU AgCenter, Louisiana State University, and the University of Florida is greatly appreciated. Special thanks to Dr. William B. Richardson, Vice President of Agriculture and Dean of College of Agriculture, and Dr. Michael E. Salassi, Department Head, for providing an excellent working environment to complete this Handbook.

Any error and omissions are the responsibility of the co-editors.

1
An introduction to *The Routledge Handbook of Agricultural Economics*

Gail L. Cramer

An academic handbook is a reference book of the literature in a subject or area of study. Routledge Handbooks are

> prestige reference works providing an overview of a whole subject area or sub-discipline and which survey the state of a field including emerging and cutting edge areas. The aim is to produce a comprehensive, up-to-date, definitive work of reference which can be cited as an authoritative source on the subject. (https://www.routledgehandbooks.com/collection/Business_Economics)

It is used primarily by researchers, faculty, and graduate students to continue to deepen their understanding of areas of interest and the interrelationship of those areas. There are all kinds of Handbooks. This one summarizes the research in agricultural economics, agribusiness, and applied economics. It provides some indication of the direction of the relevant research and where future research endeavors may be productive. In addition, the reference list at the end of each chapter summarizes relevant journal articles and other publications in the field.

This Handbook covers the entire field of Agricultural Economics. Therefore, it integrates microeconomics, macroeconomics, production and consumption, finance, international trade, natural resources and the environment, marketing, management, business, econometrics, and other sub-disciplines.

Agricultural Economics started as a profession by applying farm management and economic and business principles to farms. The current subject covers food from farm to fork and all the important resources influenced by agriculture such as land, water, clean air, environment, animals, rural development, education, forestry, trade, restaurants, oil, gas, agribusinesses, and farms. Therefore, Agricultural Economics has been broadened to include applied economics and business. The title of the profession is Agricultural Economics and Applied Economics. Applied Economics includes all the fields where agricultural economics overlaps with other disciplines. Also, undergraduate degrees are conferred in Agribusiness, Agrifinance, Natural Resources, Food Economics, International Trade or Business, and Economic or Community Development.

Agricultural Economics continues to grow nationally and around the world because the faculty and students are trained to solve practical business and economic problems. Agricultural

Economics programs include classwork in agriculture, business and economics, statistics, natural and social sciences, mathematics, language, and research methods. Students can use their research methods and relevant classwork to provide alternative solutions to problems. This background and training allows them to be hired by any organization or business, not just by agribusinesses. Some business schools have developed substitutes for agricultural economists through specific training in operation research techniques, MBAs, and other management and finance programs. In addition, many economics departments have developed more research-oriented master's and PhD programs, and in recent years, they have added more econometrics and statistics courses to their degree requirements.

The historical Agricultural Economics Handbooks are ageing, although a "major work" was published by Routledge Press in 2011 in four volumes edited by Gail L. Cramer. These volumes reviewed the entire literature in Agricultural and Applied Economics in 76 papers and specifically analyzed "critical concepts in economics."[1] These volumes were reviewed in the *American Journal of Agricultural Economics* by Dr. Patricia Duffy at Auburn.[2] As stated by Dr. Duffy,

> I found this set of volumes a highly worthwhile addition to my reference shelf. In addition to the collected articles, the work contains a thoughtful introduction by the editor, giving a broad overview of our discipline that I enjoyed reading.

Three other surveys of the field of Agricultural Economics have been published. One survey was completed by the AAEA (Agricultural and Applied Economics Association) between 1977 and 1992 by Lee Martin. A second was edited by Bruce Gardner and Gordon Raiser and was published in 2001. For the 100th anniversary of AAEA, the organization reviewed the literature again, producing *The Centennial Issue of the American Journal of Agricultural Economics*, Vol 92, Issue 2, April 2010. This issue covered the field of Agricultural Economics as production economics, marketing, agricultural policy, international trade, agricultural development, natural resources and conservation, environmental economics, food and consumer economics, rural development, agribusiness, and economic developments. This was the first official recognition of the broad field of Agricultural and Applied Economics. Up to the time of Walter Armbruster's presidency of the AAEA, the AAEA regarded anything out of the areas of production, marketing and trade, econometrics, statistics, and possibly policy to be outside of its core. Other studies were published in *AJAE* (*American Journal of Agricultural Economics*), but economic development, international trade, consumer economics, rural and regional economics, and environmental economics were low priority at annual meetings and in the journals. Bruce Gardner advocated for publishing an agricultural policy journal and the cooperative extension economists expressed their desire for a popular extension outlet such as the *Choices* magazine.

Organization of this Handbook

This Handbook covers the entire field of Agricultural Economics and will be referenced as the primary source of current data, references, and literature.

The Routledge Handbook of Agricultural Economics begins with the topic of welfare theory. Alternative welfare theories are examined and applied to federal agricultural policy. Next, the important subjects of world food and food security are covered. Food and food safety are a daily concern in all countries. Food recalls are becoming more common. Next, natural resources and the environment, including climate change, are covered. Water is given priority because of its significance in food production and its impact on land availability. International trade theories and growth models are explained. Trade usually improves both the production and distribution of good and services. Influences on the marketing system coming from the general economy or macroeconomics is discussed. Advancements in econometrics and modeling are presented and used in estimating demand and supply functions. The marketing system is developed with agribusiness, futures, options, and specific industry studies. The research important to farmers is distributed in the U.S. by the Cooperative Extension Service at Land Grant Universities. The last few chapters emphasize production economics and farm management.

These 41 chapters are well written and represent the entire field of Agricultural Economics at an advanced level. Seniors and graduate students should have the background to read and understand these papers.

Climate change is given significant coverage because of the impact weather has on the world and the question of whether man can reduce the impact of weather changes. It is known that glaciers are melting, sea levels are rising, air and water are being polluted, and the environment is decaying and/or changing. Hypoxia and other factors affecting the planet can be studied and research may find alternative ways to adjust to these influences or reduce their impact on human utility and comfort. The planet is ageing and, with that reality, science must find new ways to improve the human condition.

Progress has been made in combating world hunger. Even as recently as 50 years ago, extreme food shortages were evident in China, India, Russia, Africa, and South and Central America, while elsewhere food was plentiful. With the aid of the CGIAR (formerly known as Consultative Group for International Agricultural Research) and their associated research organizations, as well as United States Agency for International Development (USAID) and changes in free markets, the world food situation has improved substantially. Food deficits occur every day in the United States as well as all other countries of the world. It is possible for the world to produce enough food to feed the world's population, but we lack the economic and distribution systems to close the food gap. Current resources and policies have proven inadequate to solve this huge problem, which affects about 814 million people worldwide.

The papers in this Handbook were written by outstanding professors in each academic area. These professors teach doctoral courses at the highest levels and are aware of the training that researchers must have in order to make additional academic progress. In fact, some of the articles were written with their PhD students as co-authors.

The future of the study of food, agricultural economics, agribusiness, and applied economics looks very bright, not only at the bachelor degree level, but also at the graduate level. The primary reason for this is the methodology and training of the students. The students are taught logic through their use of economic theory. In addition, they are taught how to use the scientific method in course research papers. With the assistance of basic economic, mathematical, and statistics courses with linear and nonlinear modeling, they can use deductive reasoning to analyze many practical problems.

Students who have been trained in the classes, have a command of language, and can solve problems will always be able find employment. With new course work on entrepreneurial concepts, they can start their own firms. Other attractive reasons to major in Applied Economics are the versatility of employment (many types of positions) and the flexibility of employment (ability to change). Economics and business students are in demand by all types of governmental agencies and private enterprises.

Continuing on for a master's degree is encouraged. A typical MS (Master of Science) or MBA (Master of Business Administration) degree requires about two years. This consists of course work and/or the completion of a thesis. A PhD degree is primarily a research degree. The degree requires about two years of course work and one to two years of research and the completion of a PhD dissertation or three research essays. There are positions for PhDs in government, academia, and private firms. Most of the government positions in the U.S. are with Economic Research Service and Foreign Agricultural Service (USDA), USAID, World Bank, International Monetary Fund, and the Departments of the Interior, Treasury, and Commerce. Full-time teaching positions are usually with smaller colleges. Private companies hire economists to model commodities or to use big data to forecast in order to increase the precision of business decisions. Saving a few cents per pound on a unit train of sugar or flour can amount to a substantial sum of money.

The value of the current university training of agricultural economists has been proven over the years. The basic problem the agricultural economics profession faces is how to broaden its base to include the solving of economics and business problems in other industries. The name "Applied Economics" really fits as the needed title, but there is a reluctance to use this terminology because it would create two "economics departments," and universities might want to combine them. There is little desire to combine Economics and Agricultural Economics departments because of their different missions and different methodologies. Over time, very few departments have found it useful to merge. Most departments that have merged have either separated or operate as two individual departments within the department.

Applied research

Applied Economics programs were started to attract bright incoming students and PhDs to these programs. The graduates receive compensation approximately equivalent to that of engineering and computer science graduates and it is rewarding work.

Agricultural Economics fits well with Applied Economics because agriculture is a major user of land, water, equipment, and other capital items such as health care, roads and highways, energy, and education. Agricultural Economics therefore covers many areas of the general economy because those areas overlap in a large way.

The overlapping of agriculture with the general economy is highlighted in Davis and Goldberg's book *A Concept of Agribusiness*, as shown by input–output tables.

Following are examples of applications of cooperation between disciplines in order to solve difficult problems. For example, Midwest farmers affect the quality of water in the Gulf of Mexico in an enormous way. The fertilizer applied to corn fields in Minnesota, Iowa, Ohio, Illinois, and other environs, primarily nitrogen (N) and phosphorous (P), leaches through the farm soils into the underground water supply that eventually drains into the Mississippi River. This flow continues to the Gulf of Mexico and causes the "dead zone" or hypoxia in the Gulf that extends over 100 miles of coast line and kills millions of fish. This situation is being researched by several academic departments, but agricultural economics has been at the forefront because this profession deals with the interrelationships between production inputs.

Many practical research problems result from the hypoxia problem. Some of these involve Midwest corn production, such as what is the optimal amount of N and P to apply to maximize profits for the farmers, or what additional results can be achieved by applying less N and P to lower the N and P in the Mississippi River. Other interesting research would include how much N and P can be eliminated by establishing "no corn growing easements" along the Mississippi River? Also, would it be feasible to take the N and P out of the river water at some location up river? Other research could involve the fish kill. Could the dead fish be used for fertilizer or for some other use?

Another interesting study is to determine if the water could be mixed to eliminate the dead zone. Chemical engineers at Louisiana State University are working on the physical side of the issue. They are attempting to find ways to mix the river water to add oxygen to the dead zone.

This would be accomplished by taking the oxygen-rich upper layer of water and mixing it with the oxygen-dead layer below. These water layers can be mixed to provide enough oxygen for fish. What needs to be determined is the type of machine to use to mix the two water levels. Experiments are being implemented to determine the type of propeller, the shape of the propeller blades, and how to provide energy for the machine. Working with the physical scientists, the agricultural economists can deal with the costs and return to an investor or society and suggest if should be done or if it makes economic sense. This is a relatively small situation, but the solution to this problem requires many disciplines working together to find an answer.

Biologists are also working on solving the hypoxia problem biologically. A biological solution could be to create many diversion streams from the Mississippi River into the River Delta areas. These streams would carry heavily loaded N and P water through marsh lands to reduce the N and P naturally, while letting the "clean water" continue to flow to the Gulf. This solution is very expensive and one that would take a large investment in or to build the necessary diversion dams and streams. The hypoxia problem did not exist before 1927, when the U.S. government channeled the Mississippi River. Before 1927, the river flowed wherever nature took it. It would flow south and deposit rich soil all over the Delta. Farmers farmed those acres. Now, with the river kept within its channeled banks, all of the water flows to the same area of the Gulf, and the Gulf waves keep it close to the Louisiana and Texas coasts. The water has large amounts of N and P and it pollutes the water next to the coast line. The hypoxia zone is 8,776 square miles in size.

Another problem that involves international trade policy is free trade in rice. Both Japan the South Korea have been unwilling to negotiate their rice tariffs. Japan has the largest tariff and wants to keep out U.S. short grain rice produced in California. Currently, Japan purchases small quantities of short grain rice from the U.S. that the Japanese government diverts to livestock feed rather than permitting that rice to be sold for human consumption. Japan and South Korea are never going to purchase American rice until it is available in their countries for human consumption. In several consumer blind surveys, consumers in Japan could not tell the difference between Japanese-produced rice and American rice. Therefore, if better "trade deals" were made, American rice could be marketed in restaurants and institutions in Japan and South Korea where rice would be accepted as breakfast, lunch, or dinner. More free trade will benefit American farmers and Japanese consumers. Japanese consumers will get rice at a lower price and American rice farmers will receive a higher price. This is a win–win situation.

Total free trade normally would increase total trade, reduce the prices to importing countries, and increase the prices in exporting countries. However, lower prices for rice would change the importing consumers' demand for other goods that are substitutes or complements. The agricultural economist's position is to build "economic models" to explain and predict the new production and consumption patterns. This is just another example of what the profession does.

Funding agricultural economics education

Because of the practical nature of work and because it affects actual business relationships, the demand for PhD agricultural economists will continue to increase as will the salaries in the profession. It is impossible to build "models" and predict in Agricultural Economics without understanding the underlying "commodity cycles" that exist in agriculture. To fully understand one commodity system such as wheat, rice, or feed grains takes about five years of training. The Economic Research Service of the USDA trains PhDs in commodity analyst positions. The Economic Research Service hires these economists, who often have an agricultural background. Most agricultural economists, once they are trained in a commodity area, spend the rest of their research career in that commodity area or a closely related area. This is due to the fact that it is extremely costly to change research or teaching areas at universities or other organizations.

University programs offer opportunities for teaching and research. Any agricultural economist should be able to teach the freshman and sophomore courses, but if the Agricultural Economics professor is to be able to teach the specialized courses in production, marketing, policy, operations research, or econometrics, specific training is required. The training is provided in graduate schools. The student needs to carefully evaluate prospective graduate programs, as areas of study differ between programs.

In recent years, the funding of higher education in the U.S. has become a difficult process. The sizes and scopes of programs offered at colleges have increased immensely. The public cost of these schools has reached the limit of many states' ability to fund them. Nearly every high school graduate is planning on some post–high school training to allow him or her to be prepared for a better paying job, resulting in benefits to society in many ways. The benefits to society include such things as leadership and service to the communities in which the graduates live and work. This in itself justifies the input of a public subsidy. The large payoff from education has attracted many more students to the universities. Universities have been forced to respond by requiring professors to teach more and larger classes, move to the semester system, reduce the number of credit hours required for graduation, increase class sizes, and use more multiple choice and true/false tests (to accommodate machine grading).

The result has been a "watering down" of educational quality. Professors no longer have the time for research at many colleges. With that, comes the reduction of the funding that had been provided through grants to research projects. If the research innovations of the past are to be continued, it is time for the federal government to start a new program similar to the Land Grant College System developed in 1862 during the Lincoln administration. Under this program, land grants were awarded to each state providing funding for the establishment of Land Grant Colleges.

Additional Research Institutions need to be established at which outstanding research faculty currently at universities, as well as newly trained PhDs, may carry out their research projects. Examples of the type of Research Institutions are M.D. Anderson in Houston, Texas; Oak Ridge National Laboratory in Oak Ridge, Tennessee; and CERN in France and Switzerland. If the United States is going to be the leader of learning, innovation, discovery, and economic development, new institutions are necessary. These "super research" universities could advance work on disease, artificial intelligence, food production systems, space travel and colonization, human replacement organs, and sea exploration, to name just a few areas of needed research. The organizations that currently exist do excellent research in their areas of expertise, but the public Research Universities could provide the opportunity for many PhD research assistants and associates to collaborate as well as professors. Also, grants could be written to help fund and guide research. Super Universities could provide the forum for renewed vigor in a system that has been drained by budget cuts and low morale.

Notes

1 Cramer, Gail L., ed. 2011. *Agricultural Economics: Critical Concepts in Economics*, Volume I–IV. (Volume I: Production Economics; Volume II: Resource and Environmental Economics; Volume III: Agribusiness, Marketing, and Consumption Economics; and Volume IV: Agricultural Policy, International Trade, and Development Economics.)
2 Duffy, P. 2011. "Agricultural Economics, Critical Concepts in Economics." Book Review. *American Journal of Agricultural Economics*, 93, 1229.

Additional references

Annou, M.M., E. J. Wailes, and G. L. Cramer. 2001. *Economic Analysis of Liberty Link Rice: Preliminary Results.* USDA Rice Situation and Outlook Yearbook, January.

Arthur, H. B., and G. L. Cramer. 1976. Brighter Forecast for the World's Food Supply. *Harvard Business Review.*

Babcock, M. W., G. L. Cramer, and W. A. Nelson. 1985. The Impact of Transportation Rates on the Location of the Wheat Flour Milling Industry. *Agribusiness* 1(1): 61–71.

Bierlen, R., E. J. Wailes, and G. L. Cramer. 1998. Unilateral Reforms, Trade Blocs, and Law of One Price: Mercour Rice Markets. *Agribusiness* 14(3): 183–198.

Bierlen, R., E. J. Wailes, and G. L. Cramer. 1997. *The Mercosur Rice Economy*, Vol. 954. Arkansas Agricultural Experiment Station.

Kolstad, C.D. 2000. *Environmental Economics.* Oxford: Oxford University Press, p. 5.

Copeland, M. D., and G. L. Cramer. 1973. An Efficient Organization of the Montana Wheat Marketing System, Bulletin No. 667.

Cramer, G. L. 1993. Mexico's Rice Market; Current Status and Prospects for US Trade (No. 04; HD9066, C7.).

Cramer, G. L. 1996. Domestic Reforms and Regional Integration: Can Argentina and Uruguay Increase Non-Mercosur Rice Exports? *Agribusiness* 12(5): 473–484.

Cramer, G. L. 2002. Perspectives on Our Profession. *Journal of Agricultural and Applied Economics* 34(2): 239–241.

Cramer, G. L. 2005. Rural Development Makes Louisiana a Better Place. *Louisiana Agriculture* 48(4): 7.

Cramer, G. L. 2011. The Microtheory of Innovative Entrepreneurship. *American Journal of Agricultural Economics* 93(5):1410–1412.

Cramer, G. L., J. M. Hansen, and E. J. Wailes. 1999. Impact of Rice Tariffication on Japan and the World Rice Market. *American Journal of Agricultural Economics* 81(5): 1149–1156.

Cramer, G. L., C. W. Jensen, and D. Southgate Jr. 2001. *Agricultural Economics and Agribusiness.* New York: John Wiley and Sons.

Cramer, G. L., E. J. Wailes, and S. Shui. 1993. Impacts of Liberalizing Trade in the World Rice Market. *American Journal of Agricultural Economics* 75(1): 219–226.

Cramer, G. L., E. J. Wailes, B. Jiang, and L. Hoffman. 1999. Market Efficient Tests of US Rough Rice Futures Market. *Research Series-Arkansas Agricultural Experiment Station*: 389–392.

Cramer, G. L., and A. Maurer. 1987. The Agricultural Chemical Industry in Ecuador, Final USAID Report.

Cramer, G. L., and A. Maurer. 1987. The Agricultural Machinery Industry in Ecuador, Final USAID Report.

Cramer, G. L., and E. J. Wailes. 1993. *Grain Marketing*, Second Edition, Boulder, CO: Westview Press, p. 460.

Cramer, G. L., and T. M. Billings. 1979. Feeding and Hedging Strategies for Montana Cattle Feeders, Bul. published by the Chicago Mercantile Exchange.

Cramer, G. L., E. J. Wailes, B. Gardner, and B. Lin. 1990. Regulation in the U.S. Rice Industry, 1965–1989. *American Journal of Agricultural Economics* 72(4): 1056–1065.

Cramer, G. L., K. B. Young, and E. J. Wailes. 2003. *Rice Marketing, Rice: Origin, History, Technology, and Production*, edited by C. Wayne Smith and Robert H. Dilday, Wiley Series in Crop Science, Chapter 38, pp. 473–488.

Davis, J. H., and R. A. Goldberg. 1957. *Concept of Agribusiness.* Boston, MA: Research Division, Harvard Business School.

Fan, S., G. L. Cramer, and E. J. Wailes. 1994. Food Demand in Rural China: Evidence from Rural Household Survey. *Agricultural Economics* 11(1): 61–69.

Fan, S., E. J. Wailes, and G. L. Cramer. 1995. Household Demand in Rural China: A Two-Stage LES-AIDS Model. *American Journal of Agricultural Economics* 77(1): 54–62.

Fan, S., E. Wailes, and G. L Cramer. 1994. Impact of Eliminating Government Interventions on China's Rice Sector. *Agricultural Economics* 11(1): 71–81.

Fortenbery, T. R., G. L. Cramer, and B. R. Beattie. 1982. Highways and Railroads in Montana: Problems and Opportunities. Special Report, Montana Agricultural Station Bulletin.

Gao, X. M., E. J. Wailes, and G. L. Cramer. 1995. A Microeconometric Model Analysis of US Consumer Demand for Alcoholic Beverages. *Applied Economics* 27(1): 59–69.

Gao, X. M., E. J. Wailes, and G. L. Cramer. 1995. Double-Hurdle Model with Bivariate Normal Errors: An Application to US Rice Demand. *Journal of Agricultural and Applied Economics* 27(2): 363–376.

Gao, X. M., E. J. Wailes, and G. L. Cramer 1996. A Two-Stage Rural Household Demand Analysis: Microdata Evidence from Jiangsu Province, China. *American Journal of Agricultural Economics* 78(3):604–613.

Gao, X. M., E. J. Wailes, and G. L. Cramer. 1996. Partial Rationing and Chinese Urban Household Food Demand Analysis. *Journal of Comparative Economics* 22(1): 43–62.

Gao, X. M., E. J. Wailes, and G. L. Cramer. 1997. A Microeconometric Analysis of Consumer Taste Determination and Taste Change for Beef. *American Journal of Agricultural Economics* 79(2): 573–582.

Gardner, B. L., and G. C. Rausser. 2002. *Handbook of Agricultural Economics*, Vol. 2. Amsterdam: Elsevier.

Garoian, L., and G. L. Cramer. 1968. Merger Component of Growth of Agricultural Cooperatives. *American Journal of Agricultural Economics* 50(5): 1472–1482.

Johnson, G. L. 1986. *Research Methodology for Economists*. New York: Palgrave Macmillan Publishing, pp. 12–14.

Kimble, W. J., G. L Cramer, and V. W. House. 1976. The Quality of Education in Rural Montana, Montana Agri. Exp. Sta. Bul. No. 685.

Koo, W. W., and G. L. Cramer. 1977. Shipment Patterns of Montana Wheat and Barley Under Alternative Rail and Truck-Barge Rate Structures. Bulletin-Montana Agricultural Experiment Station (USA).

Larson, D. E. 2008. Biofuel Production Technologies: Status, Prospects and Implications for Trade and Development, United Conference on Trade and Development. DITC/TED/2007/10, www.UNCTAD.org/en/docs/ditcted200710_en.Pdf. Accessed September 1, 2017.

Martin, L. R. 1981. *A Survey of Agricultural Economics Literature: Traditional Fields of Agricultural Economics, 1940s to 1970s*, Vol. 3. Minneapolis: University of Minnesota Press.

Maurer, A., and G. L. Cramer. 1987. An Economic Review of the Fertilizer Industry in Ecuador, Final USAID Report.

Maurer, A., and G. L. Cramer. 1987. An Economic Review of the Improved Seed Industry in Ecuador, Final USAID Report.

Michael, B., and A. Jay. National Food and Agribusiness Management Education Commission, National Food and Agribusiness Management Education Commission, July 16, 2004. www.farmfoundation.org/news/articlefiles/1016-2isc-executivesummary-final73104_2__2_.pdf. Accessed December 17, 2017.

Mittelhammer, R. 2009. Applied Economics – Without Apology. *American Journal of Agricultural Economics* 91(5): 1161–1174.

Mjelde, J. W. 1982. Optimal Decision Rules for Marketing and Storage of Wheat and Corn. Master's thesis, Montana State University-Bozeman, College of Agriculture.

Just, R. E., D. L. Heuth, and A. Schmitz. 2004. *The Welfare Economics of Public Policy*. Cheltenham, UK: Edward Elgar Publishing Limited.

Runge, C. F. 2006. Agricultural Economics: A Brief Intellectual History. No. 13649. University of Minnesota, Department of Applied Economics.

Salassi, M. E., M. Fannin, K. Guidry, M. Dunn, and G. L. Cramer. 2009. *Economic Importance of Agricultural Research and Extension to Louisiana*. Louisiana: Louisiana State University Agricultural Center.

Shumway, C. R. 1995. Recent Duality Contributions in Production Economics. *Journal of Agricultural and Resource Economics*: 178–194.

Smith, R. K., E. J. Wailes, and G. L. Cramer. 1990. The Market Structure of the US Rice Industry. Fayetteville, AR: Agricultural Experiment Station, Division of Agriculture, University of Arkansas.

Tweeten, L. G., and G. L. Cramer. 1994. 1995 Farm Bill Options, the Ohio State Univ. Agri. Experiment Station.

Wailes, E. J., and G. L. Cramer. 1991. Japan's Rice Market: Policies and Prospects for Trade Liberalization. Arkansas University, Fayetteville, Ark. USA.

Wailes, E. J., K. B. Young, and G. L. Cramer. 1993. Rice and Food Security in Japan: An American Perspective. In *Japanese and American Agriculture: Tradition and Progress in Conflict*. Westview Press, Boulder, CO, pp. 337–393.

Wailes, E. J., G. L. Cramer, E. C. Chavez, and J. Hansen. 1997. Arkansas Global Rice Model – International Baseline Projections for 1997–2010. Special Report 177. Arkansas Agricultural Experiment Station, Fayetteville, AR.

Wailes, E. J., G. L. Cramer, E. Chavez, and J. Hansen. 1998. Arkansas Global Rice Model: International Baseline Projections for 1998–2010, Special Report 189, 72 pp. Arkansas Agricultural Experiment Station, AR.

Wang, J., X. M. Gao, E. J. Wailes, and G. L. Cramer. 1996. US Consumer Demand for Alcoholic Beverages: Cross-Section Estimation of Demographic and Economic Effects. *Review of Agricultural Economics* 18(3): 477–489.

Wang, J., E. J. Wailes, and G. L. Cramer. 1996. A Shadow-Price Frontier Measurement of Profit Efficiency in Chinese Agriculture. *American Journal of Agricultural Economics* 78(1): 146–156.

Wheeler, R. D., G. L. Cramer, K. B. Youngand, and E. Ospina. 1981. *The World Livestock Product, Feedstuff, and Food Grain System*. Morrilton, AR: Winrock International.

WHO (World Health Organization), United Nations.

Woolverton, M. W., G. L. Cramer, and T. M. Hammonds. 1985. Agribusiness: What Is It All About? *Agribusiness* 1(1).

World Health Organization. 2009. World Health Statistics 2009. World Health Organization.

Young, K. B., and G. L. Cramer. 1986. The Impact of Livestock Feed Demand in Centrally Planned Countries on Grain and Oilseed Imports. *Agricultural Systems* 21(1): 69–82.

Young, K. B., P. Amir, and G. L. Cramer. 1990. Implications of Dairy Development in Indonesia. *Agribusiness* 6(6): 559–574.

Young, K. B., E. J. Wailes, and G. L. Cramer. 1994. Economic Analysis of Rice Bran Oil Processing and Potential Use in the United States. Bulletin/Arkansas Agricultural Experiment Station, Division of Agriculture, University of Arkansas (USA).

Young, K. B., E. J. Wailes, G. L. Cramer, and N. T. Khiem. 2002. Vietnam's Rice Economy: Developments and Prospects. Fayetteville, AR: Arkansas Agricultural Experiment Station.

Young, K. B., G. L. Cramer, and E. J. Wailes. 1998. An Economic Assessment of the Myanmar Rice Sector: Current Developments and Prospects. Fayetteville, AR: Arkansas Agricultural Experiment Station.

Young, K. B., and G. L. Cramer. 1886. Economic Analysis of the Antigua/Barbada Livestock Sector, Vol. 1: Policy Analysis, USAID Final Report, Winrock International, p. 101.

2
Broadening the validity of welfare economics for quantitative public policy analysis

Richard E. Just and Andrew Schmitz

1 Introduction[1]

For well over a century, the principles of welfare economics, initially popularized by Alfred Marshall (1890), have dominated the quantitative economic analysis of public policy. Some of the basic constructs, however, have come under scrutiny and criticism along the way, which has led to waves of widespread reservations about its applications (Just, Hueth, and Schmitz 2004). Generalizations of simplistic original definitions of consumer surplus generated criticisms of early interpretations as money measures of utility change (Samuelson 1942). Development of willingness-to-pay and willingness-to-accept concepts subsequently resolved mathematical ambiguities. However, these concepts introduced competing definitions and competing views about whether actual compensation is required for the validity of implications compared to a potential compensation criterion and its paradoxical possibilities (Hicks 1941, 1943, 1956; Scitovsky 1941; Coleman 1980).

Regardless of these controversies, the standard concepts of welfare economics, with their various refinements, have continued to be the major tools used for quantitative analysis of the benefits and costs of public policy because few, if any, alternatives exist. This is the reason a significant sub-discipline of economics, with its own academic society, has sprung up in the new millennium – the Society of Benefit Cost Analysis.

The tools and concepts of economic welfare analysis have been generalized to broaden assumptions and the scope of application, and thus expand the set of issues that can be addressed and potentially quantified for purposes of public policy analysis. Alternative means of incorporating non-market values have been developed, such as travel cost and random utility models, hedonic and experimental methods, and contingent valuation approaches. Related advances have broadened the application of welfare economics to incorporate use and existence values, altruism, and value of life. However, these methods have raised further controversy about the use of stated versus revealed preferences or blended versions of the two (Bockstael and McConnell 2007; Azevedo, Herriges, and Kling 2003).

Other possibilities for examining distributional concerns of public policy have also been developed. Through all of these controversies and refinements, the tools of welfare economics have continued to offer the most widespread means of quantitative evaluation of public policy. More recently, perhaps the greatest challenge to the continued use of welfare economics for public policy analysis has grown out of the path-breaking work initiated by Kahneman and Tversky (1979) that demonstrated failures of rationality in decision making. Consistent rational optimization by economic agents is a basic assumption at the heart of classical welfare economics. With increasing frequency since the introduction of Kahneman and Tversky's work, the literature has generated experimental studies that identify both specific and random behavioral anomalies. Studies that identify specific behavioral patterns often suggest potential policy changes as candidates for improving economic welfare but without a comprehensive quantitative analysis that accounts for indirect effects, alternatives, and possibilities for social optimization.

Several recent generalizations permit adaptation of the principles of welfare economics in order to facilitate its application to quantitative analysis of public policy that accounts for both repetitive and random behavioral anomalies. In this chapter, we review these generalizations and argue that behavioral anomalies need not block economists from comprehensive quantitative analysis of public policy. Rather, the generalizations of economic welfare principles should be exploited to reinvigorate the more normative role that economists once aspired to fill.

This review considers the joint implications of three relatively recent bodies of literature related to welfare measurement: the behavioral welfare economics literature, the debate about structural versus reduced form practices in econometrics, and the focus on finding sufficient statistics by which to characterize behavior. We argue that the collective implications of these new directions, as well as some related general equilibrium considerations, can be incorporated into robust concepts of economic welfare measurement. This provides justification for a resurgence of the terminology and concepts of traditional welfare economics, although under a much broader framework of understanding.

2 Conceptual generalizations in welfare measurement

The new behavioral welfare economics literature (Bernheim and Rangel 2009) has made major advances in showing that economic welfare concepts can provide valid measures of willingness to pay (WTP) when a variety of the controversial assumptions about behavior are relaxed.[2] The debate about structural versus reduced form econometrics (Keane 2010) reveals that (1) reduced form approaches require unacknowledged implicit assumptions and are not as assumption-free as often suggested, and (2) alternative underlying frameworks may admit competing implications. This debate is relevant for welfare analysis because required parameters that are estimable with structural approaches cannot be derived from reduced form estimates without imposing sufficient structure, implicitly or explicitly, to satisfy related identification conditions. The controversy stems from a lack of consensus about the robustness and empirical validity of identification conditions. The literature on sufficient statistics (Chetty 2009) is important because it attempts to bridge the gap, allowing estimation of parameters necessary for welfare analysis without some of the specific assumptions of structural estimation.

From this literature, sufficient statistics for welfare analysis are those that determine traditional graphical concepts of consumer and producer surplus, although under much broader assumptions. The potential value of these tools in this broader behavioral and general equilibrium context lies in the ability to convey understanding to semi-lay users of policy analysis, such as Congressional staffers and applied economists. By comparison, large tables of general equilibrium results based on complex mathematical models, possibly incorporating a variety of heterogeneous behavioral criteria, may appear as black boxes from which little intuitive understanding or confidence is generated.

This review assumes that normative policy analysis is a major purpose of empirical microeconomic analysis and that at least a sub-discipline of economists will continue to try to fulfill demands of the policy process for aggregate and distributional economic welfare analysis with the best tools available (Harberger 1964; Baumol, Starr, and Wilson 2000; Just, Hueth, and Schmitz 2004; Schmitz and Zerbe 2009). This is evidenced by the continued use of surplus measures and deadweight loss concepts by applied economists and benefit–cost practitioners despite successive criticisms and rehabilitations of consumer surplus (as reviewed in earlier stages of development by Currie, Murphy, and Schmitz 1971 and Just, Hueth, and Schmitz 2004). One reason for this continued use is that graphical surplus concepts motivate intuitive understanding, which has made them highly desirable for the presentation of results.

Surplus concepts are defined throughout as the triangle-like area behind supply or demand with the understanding that they measure WTP accurately if based on Hicksian compensated relationships (Hicks 1941, 1943) or represent Willig (1976) approximations if based on uncompensated relationships (without supposition of linearity of the hypotenuse). Also, Scitovsky's (1941) compensation principle is assumed to provide the basis for additive aggregation of WTP that, if compensation is paid, is equivalent to employing a Benthamite social welfare function, as some prefer to describe it (Heckman and Vytlacil 2005, 2007). Together with development of WTP valuation, methods for non-market goods, resource stocks, and the value of life, these concepts are the accepted foundations of policy analysis among environmental and resource policy economists. We review here generalizations of the behavioral hypotheses under which they have approximate or accurate relevance.

To broaden the validity of economic welfare concepts, users should bear in mind several general and self-evident principles that describe the context of their application.

> **Principle 1.** Projects and policies of significance affect large groups of individuals.
> **Principle 2.** Evaluation of economic welfare effects of projects and policies requires consideration of all affected individuals.
> **Principle 3.** Heterogeneity in characteristics, conditions, behavior, and perceptions is a widely recognized and critical concern, which implies that analyzing the distribution of welfare effects is an important dimension of project and policy analysis in addition to efficiency analysis (Ben-David et al. 1999; Long and Soubeyran 1997; Kirman 2006; Sen 2004).

Some examples illustrate that generalizations are possible by integrating the three bodies of literature on behavioral welfare economics, structural versus reduced form estimation, and sufficient statistics. The producer case is useful as a complementary starting point because these three literatures have focused primarily on consumer issues, whereas many resource and environmental policy issues impact producer behavior.

For purposes of comparison of later generalizations, we first outline the standard producer short-run profit maximization model subject only to a technology specification:

$$\pi(p,r) \equiv \max_x \{pq(x) - rx - k \mid x \geq 0\}, \qquad (1)$$

where output q is non-negative, continuous, concave, and positively monotonic in the input vector x, respective prices are p and r, and k is fixed cost.[3] This model has provided the basis for most producer welfare evaluation. First-order conditions yield factor demands $\hat{x}(p,r) \equiv -\pi_r$ and supply $\hat{q}(p,r) \equiv \pi_p \equiv q(\hat{x})$.[4] Short-run profit is measured by the producer surplus triangle:[5]

$$S(p,w) \equiv p\hat{q} - r\hat{x} = \int_0^p \hat{q}(t,r)\,dt. \tag{2}$$

Marginal WTP follows $dS = \hat{q}\,dp - \hat{x}\,dr$, according to slopes of supply and demand, whereas WTP for a discrete change is measured by the change in the producer surplus triangle S for any combination of price and non-price (technology) changes.[6]

Virtually all applications of this framework have overlooked basic alternative behavioral theories of the firm (McGuire 1964; McCloskey 1982) as well as more sophisticated behavioral departures from profit maximization under risk aversion (von Neumann and Morgenstern 1944; Kahneman and Tversky 1979). However, WTP is readily captured by producer surplus triangles under a number of alternative behavioral theories when the simple model of (1) misrepresents the underlying structure (Just, Hueth, and Schmitz 2004, pp. 93–97).

3 Behavioral welfare economics

Behavioral economics studies observed human behavior without assuming the pure optimization of profits or utility of consumption imposed in neoclassical economics. While most studies in this field use experimental and survey-based research merely to identify a qualitative role of factors not included in neoclassical models, many such factors can be incorporated to generalize standard optimization models to measure quantitative behavior as necessary for economic welfare analysis.

3.1 Behavioral welfare economics for producers

Suppose an entrepreneur operates with self-imposed behavioral constraints that depart from profit maximization for religious, family, moral, or psychological reasons, as represented by a vector condition, $b(q(x),x) \leq 0$, that is continuous, positively monotonic, and convex in x. The problem in (1) thus becomes

$$\max_x \{pq(x) - rx - k \mid b(q(x),x) \leq 0, x \geq 0\}. \tag{3}$$

Maximization of the associated Lagrangian, $L = pq(x) - rx - k - \lambda b(q(x),x)$, with Lagrangian multiplier vector λ yields input demands and output supply also denoted by $\hat{x}(p,r)$ and $\hat{q}(p,r) \equiv q(\hat{x})$. Measurement of WTP follows (2) because first-order conditions require $\lambda b(\hat{q},\hat{x}) = 0$ where neither p nor w appears in the added constraints. This WTP is a willingness to pay given the self-imposed behavioral constraints of the decision maker. Mathematically, adding such behavioral constraints is equivalent to altering the technology. Thus, equivalent welfare implications for surplus concepts remain intact even though the structural model of decisions can differ drastically.

Alternatively, suppose a firm's expenditures cannot exceed y due to available working capital, credit availability, and behavioral attitudes toward debt. The problem in (1) becomes

$$\max_x \{pq(x) - rx - k \mid rx + k \leq y, x \geq 0\}. \tag{4}$$

Maximization of the associated Lagrangian, $L = pq(x) - rx - k - \lambda(rx + k - y)$, yields factor demands and supply of the form $\hat{x}(p,r,y-k)$ and $\hat{q}(p,r,y-k) \equiv q(\hat{x})$. Again, WTP measurement follows (2), meaning that the change in producer surplus is a proper measure of WTP for changes in all prices as well as the expenditure constraint. However, the presence of input prices as a multiplier of x in the constraint implies that $dS = \hat{q}\,dp - (1+\lambda)\hat{x}\,dr$. Thus, marginal WTP for output price changes follows the slope of supply, but slopes of factor demands underestimate marginal WTP for input price changes by a factor $1 + \lambda$.[7]

A further result finds applicability of surplus triangles under risk averse behavior. Suppose the problem in (1) is replaced by

$$V(\overline{p},\overline{r},\alpha,k) \equiv \max_x \{E[U(pq(x) - rx + w - k)] \mid x \geq 0\}, \qquad (5)$$

where U is a positive monotonic and concave utility function; prices are stochastic; x choices are based on mean prices, $\overline{p}$ and $\overline{r}$, and other distributional parameters such as variances that determine risk are in α; and w represents wealth. Input demands are $\hat{x}(\overline{p},\overline{r},\alpha,k-w) \equiv -V_{\overline{r}}/V_k$ and supply is $\hat{q}(\overline{p},\overline{r},\alpha,k-w) \equiv V_{\overline{p}}/V_k \equiv q(\hat{x})$. In this case, Hicksian concepts parallel to the consumer case yield a compensated supply defined implicitly by

$$V(\overline{p},\overline{r}_0,\alpha_0,k_0 - w_0 + C) \equiv V(\overline{p}_0,\overline{r}_0,\alpha_0,k_0 - w_0) \equiv V_0, \qquad (6)$$

where C is the compensating variation of a change from $\overline{p}_0$ to $\overline{p}$.

Total differentiation of (6) with respect to $\overline{p}$ and C generates the compensated output supply, $\tilde{q}(\overline{p},\overline{r}_0,\alpha_0,V_0)$, as a function of the mean output price, for which the producer surplus triangle is $S(\overline{p},\overline{r}_0,\alpha_0,V_0) = \int_0^{\overline{p}} \tilde{q}(t,\overline{r}_0,\alpha_0,V_0)\,dt$. This surplus triangle measures WTP for circumstances $(\overline{p},\overline{r}_0,\alpha_0,k_0-w_0)$ relative to the shutdown case where expected utility is $U(w-k)$. Accordingly, WTP for any combination of changes in mean prices, risk as represented by α, or technology is captured by the change in this producer surplus triangle; thus, compensating variation is $S(\overline{p}_1,\overline{r}_1,\alpha_1,V_0) - S(\overline{p}_0,\overline{r}_0,\alpha_0,V_0)$. Equivalent variation is derived similarly by holding (6) constant at $V_1 \equiv V(\overline{p}_1,\overline{r}_1,\alpha_1,k_1-w_1)$ rather than V_0. Similar results also reveal that marginal WTP for changes in mean prices follows $dS = \tilde{q}\,dp - \tilde{x}\,dr$, where $\tilde{q}$ and $\tilde{x}$ are compensated supply and demands. As in the consumer case, marginal WTP is approximated by ordinary supply and demand following $dS = \hat{q}\,dp - \hat{x}\,dr$ (see Just, Hueth, and Schmitz 2004, pp. 518–526, for more detail and for measurement when the technology is stochastic).

3.2 Project and policy evaluation

The above illustrates how WTP can be measured with the same welfare concepts under a variety of behavioral assumptions. Historically, studies of producer behavior have represented economic choice under a common decision criterion, merely conditioning on heterogeneity in circumstances such as endowments, capital, and education. In reality, individuals may differ in behavioral criteria as well. One of the most documented variations in behavior among agents is risk aversion among producers, especially farmers (e.g., Cox, Smith, and Walker 1985).

Against Principles 1–3, which describe the circumstances of practical policy evaluation, the behavioral models in (3)–(5) illustrate some critical points. First, the underlying structural models that determine observed supply and demand behavior may differ drastically among individuals. Within model (3) alone, the specific structural equations that represent alternative behavioral constraints may differ dramatically. The data-intensive analysis necessary to identify heterogeneity in *behavioral criteria* as well as heterogeneity in characteristics in a broad population may be so demanding as to be impractical.

> **Principle 4**. Limited data and research budgets may make empirical identification of the full structure of behavior and perceptions for all individuals, or even all individual groups, infeasible for practical purposes.

Second, the models in (3)–(5) illustrate that elasticities of output supplies and input demands can have the same welfare significance across many underlying structural models.[8] Thus, simple identification of marginal output supply or marginal input demand may capture marginal WTP for a variety of underlying behavioral criteria, even though full structural models of all contributing agents are not identified. This is the concept of sufficiency of a statistic for welfare measurement – a statistic that captures information required for welfare evaluation regardless of the underlying structure (Chetty 2009).

> **Principle 5**. Statistics can be identified that are minimally sufficient to capture welfare significance across a variety of behavioral criteria. In particular, supply and demand elasticities suffice for marginal WTP measurement for price changes under a variety of behaviors.

The models in (1) and (3)–(5) illustrate how structure under each behavioral criterion can be distinctively different, depending on religious or moral criteria, family constraints, psychological factors, perceptions, credit attitudes and availability, and risk preferences. While analysis of a general model that includes all of these models as special cases can, in principle, permit identification of which behavioral variation describes a given agent, such detailed analysis is rarely feasible due to data requirements and research budget constraints.

Another lesson from the variants in (3)–(5) is that the marginal approach of measuring WTP for price changes does not suffice for measuring the WTP of non-price changes such as changes in risk, technology, and credit availability, which represent the effects of many public policies. These examples illustrate how WTP for discrete non-price changes is captured by changes in surplus triangles rather than marginal behavior at equilibrium. These are important points because much of the formal justification for welfare measurement in the literature focuses entirely on marginal WTP using the slope of supplies and demands (Harberger 1964).

> **Principle 6**. Surplus triangles can serve as sufficient statistics for WTP measurement of discrete non-price impacts under a variety of behaviors.[9]

Another important implication of the model in (5) has to do with the role of human perceptions. Imperfect perceptions, if ignored, are falsely attributed to preferences or construed as anomalous behavior. Expected prices and perceived risk of variation from these that guide agent choice presumably represent perceptions of the agent making the choices. Thus, public information that affects perceptions interacts with the agent's risk preferences, causing welfare consequences. Correct representation of both preferences and perceptions is therefore critical for the assessment of welfare gains from improved information provided by government actions (Just 2008). The subjective nature of this framework also offers explanations for a host of actions that are often interpreted as anomalies in producer behavior, such as failure to adopt new technology (Feder, Just, and Zilberman 1983).

3.3 Aggregation under heterogeneity

For producer behavior, additive aggregation of sufficient statistics can readily accommodate aggregate welfare analysis over a variety of heterogeneous behavioral criteria. For example, marginal supply and demand behavior readily aggregate additively over heterogeneous agents to represent marginal group behavior. Also, individual surplus triangles based on heterogeneity of behavioral criteria readily aggregate over agents to represent corresponding group surplus triangles (Just, Hueth, and Schmitz 2004, pp. 298–305). Thus, when necessary structural identification for all individuals is infeasible, estimation of aggregate supply captures WTP for a population with a broader class of heterogeneous behavioral criteria than typically considered. Supply elasticities serve as sufficient statistics for marginal welfare analysis of price changes, while estimated supply relationships serve as sufficient statistics for welfare analysis of discrete changes of both price and non-price factors.[10] By comparison, imposing theoretical relationships associated with any single behavioral criterion would be inappropriate when data are generated under heterogeneity of behavioral criteria.

> **Principle 7.** When WTP is measured in a common monetary unit for all individuals, sufficient statistics for welfare measurement permit convenience in both aggregation over agents (for efficiency analysis) and comparison of individuals or groups (for distributional analysis) when both behavioral criteria and characteristics are heterogeneous within fairly broad classes of behavioral criteria.

Dynamic generalizations of the producer problem also readily follow whereby short-run behavior is conditioned on physical capital by vintage and longer-run behavior includes physical capital procurement. When conditioned appropriately on capital stock and time horizon, the surplus triangles again measure WTP (Just, Hueth, and Schmitz 2004, pp. 85–93 and 618–629) and, by analogy with (1)–(6), apply under these alternative behavioral criteria. At the aggregate level, discrete choices of physical capital investment are also easily accommodated along lines for the consumer case, discussed later.

3.4 Behavioral welfare economics for consumers

Traditionally, skepticism over the use of consumer surplus has been greater than for producer surplus. This skepticism was largely due to the lack of theoretical underpinnings before (1) the work of Hicks (1941, 1943) and Willig (1976), (2) the development of methodology for measuring Hicksian surplus from ordinary demand estimates (Hause 1975; Hausman 1981; Vartia 1983), and (3) the development of flexible integrable demand systems (Blackorby, Primont, and Russell 1978; Deaton and Muelbauer 1980). Since then, however, the path-breaking work of Kahneman and Tversky (1979) and others has documented a number of behavioral patterns that are anomalous in the context of the standard utility maximization model.

Many economists and psychologists have regarded these anomalies as invalidation of the standard consumer utility model. Because the standard model has served as the basis for traditional consumer welfare evaluation, many have also taken these results as a rejection of welfare economics and, more broadly, of the normative potential of the revealed preference paradigm (Ariely, Loewenstein, and Prelec 2003).[11] Some behavioral economists, pressed with the ensuing vacuum of possibilities for normative analysis, have suggested alternative measures of "true utility" in contrast to the "decision utility" implied by observed choice (Kahneman 1994; Kahneman and Tversky 1979). In fact, Kahneman (1994) distinguishes four types of utility: remembered utility, anticipated utility, choice utility, and experienced utility. Some have explored neuroeconomics as an alternative means of normative measurement (Camerer, Loewenstein, and Prelec 2005). But many applied welfare economists remain skeptical of normative measurements that are not validated by choice behavior (Bernheim 2009a,b).

More recently, assumptions under which traditional concepts of welfare analysis have validity for consumers have been relaxed by considering behavioral modifications, much as in the producer examples earlier. For example, Kőszegi and Rabin (2008a,b) have suggested that anomalies represent mistakes in decision making due to errors in beliefs. They argue that modeling the possibility of mistakes in perceptions is far more sensible than abandoning assumptions that allow normative analysis respecting people's values.

More generally, Bernheim and Rangel (2009) have introduced the concept of ancillary conditions to explain the anomalous behavior of non-standard decision makers. They argue that seemingly anomalous behavior that follows documented patterns can be modeled by adding ancillary conditions to normative consumer choice for purposes of obtaining a model that permits positive application. This approach can represent documented repetitive anomalies while the bulk of other behavior is captured by optimization.

In this context, each "different self," among the "multi-self" Pareto optima suggested by the psychological literature, is activated by a different ancillary condition (Bernheim and Rangel 2007). Choices often construed as anomalous (or inconsistent preferences) are explained by imperfect subjective perceptions due to poor learning, memory, and reasoning; overweighting of recent or familiar experiences; etc. The framework of Bernheim and Rangel thus retains the libertarian principle by which individual choices manifest revealed preferences after controlling for ancillary conditions.

Possible ancillary conditions include timing of choice, labeling of choices, and prior exposure to anchoring experiences found to cause documented variations in behavior. Adding these phenomena to the consumer choice model can yield positive explanations for hyperbolic discounting, status quo bias, experiential bias, etc. Ancillary conditions can also differentiate earned versus unearned income to account for differences in satisfaction from spending, as documented by Lowenstein and Issacharoff (1994) and Zink et al. (2004).

While Bernheim and Rangel develop more general results using a general choice correspondence rather than a utility function, some of the most practical and salient advantages can be illustrated by a simple variation of the standard consumer utility maximization problem,

$$V(p,m) = \max_q \{U(q) \mid pq = m, q \geq 0\}, \tag{7}$$

where U is a continuous, quasi-concave, and positive monotonic utility function, q is a vector of consumer choices, p is a corresponding price vector, and m represents fixed income. First-order conditions for maximizing the Lagrangian $L = U(q) + \lambda(m - pq)$ when $q > 0$ yield ordinary demands $\hat{q}(p,m) = -V_p/V_m$ following Roy's identity. Thus, the WTP (compensating variation), C, for a change in prices from p_0 to p is defined implicitly by

$$V(p, m_0 - C) = V(p_0, m_0) \equiv V_0. \tag{8}$$

Total differentiation of (8) with respect to p and C generates compensated demands $\bar{q}(p, m_0, V_0) = -V_p/V_C$. The WTP of a discrete change from p_0 to p_1 is measured accurately by the change in Hicksian consumer surplus, C, and approximately by the uncompensated consumer surplus, ΔS, measured by

$$C = \int_L dm - \bar{q}(p, V_0) dp \cong \int_L dm - \hat{q}(p, m_0) dp = \Delta S, \tag{9}$$

where L is any path of integration from p_0 to p_1 (Willig 1976). Any further change in income merely adds to C, as implied by (8).

Alternatively, now suppose that a consumer makes choices with self-imposed or experientially imposed behavioral constraints for religious, family, moral, or psychological reasons. Adding a vector of behavioral constraints, $b(q) \leq 0$ (assuming continuity, positive monotonicity, and convexity in q), the problem in (7) becomes

$$V(p,m) = \max_q \{U(q) \mid pq = m, b(q,m) \leq 0, q \geq 0\}. \tag{10}$$

This adds another term in the Lagrangian that alters the structure of ordinary demands. But behavior is still represented as $\hat{q}(p,m) \equiv -V_p/V_m$. Furthermore, the mathematics in (8)–(9) remains intact. Thus, demand elasticities (compensated or ordinary) that determine (9), although altered by the behavioral deviations, retain their welfare significance. Consumer welfare concepts (e.g., consumer surplus) retain the same welfare significance even though behavioral deviations invalidate the standard utility maximization model.[12]

Another example is the case where consumers have heterogeneous and endogenous perceptions of the quality of goods. Initially, perceived quality may be determined by media reports, advertising, and experience of friends, but perceptions are modified with personal experience. Where the perceived quality of goods is represented by a vector z, the consumer problem becomes

$$V(p,m,z) = \max_q \{U(q,z) \mid pq = m, q \geq 0\}. \tag{11}$$

This generates a structural model conditioned on z where demands are denoted as $\hat{q}(p,m,z) \equiv -V_p/V_m$. However, the results in (8)–(9) remain valid when conditioned on z. Thus, again, the structural equations describing behavior are altered, possibly becoming dynamically dependent on prior consumption through z, but demand elasticities remain sufficient statistics for measuring WTP.

Consumers may also have different perceptions of how well certain goods can be transformed into non-market goods for home consumption in a household production framework:

$$V(p,m) = \max_{q,x} \{U(q,t(x)) \mid p(q+x) = m, q \geq 0\}, \tag{12}$$

where t represents the home production technology (with standard properties) for a vector of non-market goods using input vector x. Consumers may differ in their technology $t(\cdot)$ or perceptions of it. But, again, results (8)–(9) remain intact, where demands include both q and x (Bockstael and McConnell 1983). As in the producer case, measuring WTP for discrete changes in household technology (a non-price change) can be accomplished by integrating demands with respect to the behavioral changes they induce.

3.5 Aggregation and discrete choice under heterogeneity

As in the producer case, if WTP is measured in common monetary units for all consumers, WTP among heterogeneous consumers or consumer groups accommodates both aggregate efficiency and distributional analysis based on sufficient statistics. That is, both heterogeneous marginal behavior and heterogeneous surplus measures aggregate to market demands and surplus triangles (Just, Hueth, and Schmitz 2004, pp. 306–310).[13] Results also generalize to the case of household supply of services such as labor and the sales of home produced goods (Just,

Hueth, and Schmitz 2004, pp. 221–237). Estimation of aggregate supply captures WTP in these cases for a population with heterogeneous technology as well as behavioral criteria of the types illustrated in (7)–(12).

To illustrate, suppose the model in (12) becomes

$$V_{ij}(p,m) + \varepsilon_{ij} = \max_{q,x} \{U(q, t_j(x)) \mid p(q+x) + k_j = m_i, q \geq 0\},\qquad(13)$$

where V_{ij} denotes the indirect utility of individual i if technology t_j is chosen, k_j denotes the cost of technology t_j, and ε_{ij} represents unobservable differences in perceptions or efficiencies, considered random by the econometrician.[14] For example, j might represent one of J possible choices for an appliance (or automobile) to use as the household technology to produce heat (or transportation). Similarly, the j subscript could be added to the quality variables in (11) to represent discrete choices among goods of differing qualities.

Assuming the ε_{ij} have extreme value (or normal) distributions over individuals, standard multinomial logit (or probit) choice equations apply for individuals (McFadden 1973, 1976). More generally, if the distributions of ε_{ij} are smooth, integration to the aggregate level produces smooth market relationships that have the standard welfare implications for marginal and discrete changes (Small and Rosen 1981; Hanemann 1984; Chetty 2009).

These consumer models also generalize to the dynamic case where appliances, automobiles, and houses are used for home production of comfort, transportation, and convenience; short-run behavior is conditioned on existing durables, possibly by vintage; and longer-run behavior includes durable purchases. When conditioned appropriately on durable stocks and time horizons, surplus triangles again measure WTP (Just, Hueth, and Schmitz 2004, pp. 629–635) and, by analogy with (10)–(13), apply whether or not these behavioral modifications apply.

Thus, consumer surplus measures generalize welfare analysis under a variety of behavioral variations that invalidate the standard utility maximization model. These generalizations can accommodate many types of documented behavior often characterized as anomalous. For example, choices can depend on timing, prior consumption, and stimuli at the time of choice, which would enter the problem as exogenous variables either through behavioral constraints or the perceived quality of goods, much as $y - k$ adds an exogenous factor in (4).

In conclusion, if behavior is coherent and can be rationalized by including documented determinants of behavior, then it can be incorporated into ancillary conditions that modify the standard choice problem (Bernheim and Rangel 2009). Examples include various causes of the coherent arbitrariness observed by Ariely, Loewenstein, and Prelec (2003) and hyperbolic discounting popularized by Laibson (1997) and O'Donoghue and Rabin (1999). As long as the mathematics in (8)–(9) is preserved, demand elasticities serve as sufficient statistics for welfare analysis even though differing structures of behavioral criteria apply. Thus, Principles 1–7 apply to consumer as well as producer welfare measurement, with the exception that welfare measurement for discrete or non-price changes, as well as changes involving non-market goods, require Mler's (1974) weak complementarity condition (because utility depends on a vector of goods that may drop out of the consumption bundle, depending on circumstances).

4 Mistakes in decision making and random behavior

Ancillary conditions in (10)–(12) reflect behavior that affects WTP in documented repetitive ways. Such behavior when conditions (including information, subjective assessments, and experiential history) are repeated can be defined as rational behavior even though failure to understand preferences or observe all relevant conditions may make some decisions appear to be irrational or involve errors. Aside from data availability, which is a pervasive problem of econometrics, this can happen in two important ways: first, a decision maker may not observe all information at the decision time, in which case a suboptimal decision may be later regretted when full information is realized (rationality with learning); or, second, a decision maker may not have the capacity to evaluate all available information (bounded rationality).

Cases of addictive behavior are interesting as possible examples of errors under rationality with learning. They are arguably described by dynamic models accounting for previous consumption, discounting, and incomplete information about future implications (Becker and Murphy 1988; Gruber and Kőszegi 2001). However, some behavioral economists argue that such models are inadequate or that neuroeconomics can add understanding (Camerer, Loewenstein, and Prelec 2004; Bickel et al. 2007). In any case, if coherent behavioral patterns of addiction can be identified, then consumer choice models can be generalized to include them while preserving WTP measurement (Bernheim and Rangel 2004); however, if addictive behavior causes external effects on society, then standard Pigouvian corrections may be needed.

The basic approach of Bernheim and Rangel is to correctly model all conditions affecting decision making necessary to rationalize behavior as recurrent. On the other hand, econometric practice typically treats random and non-recurrent errors as econometric errors. Some errors associated with bounded rationality may represent minor ancillary conditions that cannot be identified separately for practical purposes. However, Bernheim and Rangel (2009) show convergence to welfare measurement accuracy as such errors become small.

Alternatively, Kőszegi and Rabin (2008a,b) argue that mistakes should be modeled explicitly in a framework of rationality with learning. To identify errors in belief, they suggest eliciting subjective problem-relevant beliefs, then their impact on choices can be estimated to rationalize behavior given beliefs. Explicit modeling of mistakes permits better preference identification because estimated behavioral equations are conditioned on perceptions.

Eliciting heterogeneous problem-relevant perceptions presents a significant data burden because elicitations are required of the same agents at the same times that generate the revealed preference data. Just, Calvin, and Quiggin (1999) used this approach to explain heterogeneous producer crop insurance choices based on elicited risk perceptions. This use permitted econometric distinction of preferences and perceptions as well as empirical separation of adverse selection incentives from risk aversion incentives.

Of course, elicitation raises concerns about hypothetical, strategic, and temporal bias (Whitehead and Blomquist 2006). But the science of framing questions that minimizes such bias has become quite refined, and slightly biased perception measurement is likely superior to treating mistakes as random variation when significant. Alternatively, mistakes in quality perceptions can sometimes be modeled as a function of exposure to advertising, media reports, or other major events, which makes elicitation unnecessary (Foster and Just 1989).

In conclusion, if mistakes can be measured, then welfare analysis can make use of them to measure WTP according to implied preferences given perceptions. This permits evaluation of policies that reduce mistakes, such as the measurement of welfare loss associated with false advertising in the Becker and Murphy (1993) framework where advertising complements the advertised good. Also, measurement of mistakes permits measuring the welfare effects of public policy that provides or withholds information on prices or product quality (Foster and Just 1989).

> **Principle 8.** If mistakes in behavior are due to incorrect perceptions that can be measured, then behavior can be better identified by modeling perceptions and experience, and the welfare cost of mistakes and policies that reduce mistakes can be measured.

5 When should policy makers disregard revealed preferences?

Existence of externalities, public goods, common-property resources, and exhaustible resources call for policy makers to consider government intervention or internalization to correct market failures. Methodologies to measure welfare effects according to underlying preferences of decision makers are well developed for these cases (Just, Hueth, and Schmitz 2004). More generally, however, documented behavior that departs from the standard model in (7) has led many to reject the normative interpretation of observed choice as a guide for social planning (Camerer et al. 2003; Sunstein and Thaler 2003; Camerer, Loewenstein, and Prelec 2005; Gul and Pesendorfer 2007).

However, if agent sovereignty is the only justification for normative analysis, as argued by Sugden (2004), then loss of quantitative interpretation of WTP leaves economists with only the very limited qualitative scope of Pareto comparability (Just, Hueth, and Schmitz 2004, pp. 29–31). The Pareto principle enjoys widespread consensus but limited applicability for major policy issues that almost always involve gainers and losers. The compensation principle has provided a much more widely used approach that is defensible if interpreted correctly for measuring aggregate efficiency based on aggregate WTP, while facilitating distributional analysis based on the WTP of individuals or aggregations of WTP within individual groups.

> **Principle 9.** Non-WTP–based social criteria fail to provide a sufficient means of quantifying both efficiency and distributional implications of policies involving both gainers and losers.

In the extreme, some behavioral constraints in (10) may completely determine behavior. Some might be tempted to conclude that such information does not inform the policy process and should be ignored. However, examples in (10)–(12) show how WTP should be measured and included in policy evaluation because the cost of meeting behavioral constraints may change.

In summary, the philosophical validity of measuring WTP given variation in behavior and perceptions depends on evaluating WTP in the behavioral context in which they arise. Compensating an agent based on an action that would not be taken seems indefensible. Also, policy evaluation should consider how variations in behavior or perceptions are altered by public action. Thus, in the WTP paradigm, the only economic justification for a policy maker to evaluate goods differently than implied by revealed preferences of the agents making the choices (aside from correcting typical problems of market failure) is where the policy maker has more information than the agent. Even then, use of correct information appears to be justified only to guide individuals to the choices representing their ex post optima after gaining the information. This is usually best done by providing the information to agents in a form that will be used. Then the information can be evaluated according to individual revealed preferences rather than overruling them. Without knowing the preferences of individuals and how changes in information affect choice, the optimal actions of a policy maker, aside from correcting market failure, appear to amount essentially to providing accurate information, perhaps in a highly processed form that facilitates ready understanding and correct use by individuals.

> **Principle 10.** When individual behavior is subject to measurable mistakes, a social planner should not evaluate policies as though mistakes are not made, but rather according to preferences revealed under mistakes in perceptions, which calls for measuring WTP according to models that explain and allow for reducing mistakes.

6 Econometric support for behavioral welfare economics

At the same time documentation of behavioral anomalies led to skepticism of standard economic welfare analysis, structural estimation was undergoing replacement by reduced form (RF) estimation, often based on little, if any, theory in order to avoid parametric assumptions. Structural estimation requires specification of multidimensional interactions of individual or group behavior, technology, and stochastic phenomena to find how exogenous forces directly affecting each interact to determine observed outcomes. In contrast, the motivation for RF econometrics as characterized by Keane (2010, p. 4) is that "if we can just find 'natural experiments' or 'clever instruments,' we can learn interesting things about behavior without making strong *a priori* assumptions, and without using 'too much' economic theory," whereas "results of structural econometric analysis cannot be trusted because they hinge on 'too many assumptions'."

Principle 11. Estimating full structural models of decision making behavior involves explicit structural assumptions that complicate econometric analysis and call for structural inference.

While RF estimation has long been considered an alternative to structural estimation, experimental economics has spawned growing interest. Initially, university students and later field respondents were truly randomized into treatment and control groups to motivate simple treatment-effect econometrics (see the review by Kagel and Roth 1995). However, these experiments lacked real world relevance because the respondents either lacked real-world experience or did not bear the realistic consequences of their responses (Harrison, List, and Towe 2007). Nevertheless, the attractive simplicity of treatment-effect econometrics in lab and field experimentation motivated attempts to view naturally occurring data as quasi-experiments. Early studies made substantial claims of robustness, often based on a cavalier choice of a single instrument claimed to be uncorrelated with excluded data on differences in characteristics, which is typically an unrealistic possibility (Angrist and Krueger 2001). In response, a number of more general estimation techniques were developed to add covariates to standard treatment-effect models, such as through propensity score matching and difference-in-difference-in-difference approaches (Gruber 1994; Heckman, Ichimura, and Todd 1998; Imbens 2004).

More generally, instrumental variables (IV) estimation of atheoretic reduced forms became popular using small and arbitrary sets of instruments as a way of avoiding parametric assumptions. However, when instruments are poorly correlated with the regressors they replace, Nelson and Startz (1990a,b) have shown that IV estimators (1) reflect the amount of feedback rather than the true coefficient, (2) can be even worse than Ordinary Least Squares (OLS), (3) can be significant even when the true coefficient is zero, and (4) even asymptotic distributions can be poor.[15] The discovery of these problems in IV estimation and quasi-experimental economics closely parallels Working's (1927) discovery long ago of exogenous variables that he called "extraneous and complicating factors" in the estimation of simultaneous structural demand and supply systems. Implicitly, the choice of instruments is an assumption on the underlying structure, whereas avoiding such assumptions was the main reason for abandoning the structure of simultaneous equations models (Keane 2010).[16]

Principle 12. Treatment-effect econometrics and atheoretic RF estimation is not assumption-free, but rather imposes significant implicit assumptions that approximate those of structural modeling when extended to enable welfare analysis.

From the standpoint of welfare economics, while atheoretic nonparametric and semiparametric approaches have seemingly relaxed the parametric specifications otherwise required by structural methods, this has often come with a sacrifice of potential use for welfare analysis. Comparison of field and lab experiments has identified both similarities and differences explained by background conditions such as economic, social, moral, and ethical factors (Harrison, List, and Towe 2007). From the standpoint of welfare analysis, these studies identify *qualitative* factors that are critical to generalizing normative models for robust welfare analysis but are of limited direct use for *quantitative* evaluation of changes in policy parameters. Without a theory, estimated parameters cannot be translated into WTP estimates necessary for welfare analysis (Blundell 2010; Heckman and Urzúa 2010; Rust 2010).

Principle 13. Treatment-effect econometrics documents behavioral deviations from standard models that should be used to generalize models of welfare analysis, but atheoretic approaches are insufficient to support social optimization, broad-based counterfactual policy analysis, or analysis of indirect consequences of policy actions.

Structural models are often used to identify supply and demand elasticities from which WTP calculations can be made. With aggregation, these elasticities permit estimation of market equilibrium responses to prices and other factors as necessary for counterfactual policy analysis. With heterogeneity, the estimation of counterfactual distributional effects requires individual or group elasticities (Heckman and Vytlacil 2007; Heckman and Urzúa 2010). However, necessary elasticity estimates can also be derived from theory-based RF equations, which is a basic property of identified structures (Greene 2007).

In contrast to atheoretic RF estimation, we define theory-based RF estimation as estimation of partially reduced forms that can correspond to one or several plausible theoretical structures that produce statistics sufficient for welfare analysis without estimating full structures, including all of their primitives. The challenge is to develop specifications that appropriately represent a variety of behavioral structures. In contrast to treatment-effect analysis, which tends to narrow the set of policy issues to comparison of only a few historical alternatives, carefully specified theory-based RF equations can permit analysis of a wide range of counterfactual policies as well as forward-looking optimization without major additional assumptions (Heckman and Urzúa 2010).

This calls for a return to a structural basis for econometric practice but with greater breadth. Keane (2010) has argued that *all* econometric work relies on assumptions, but whereas behavior must be explicit with a structural approach, the experimentalist leaves assumptions implicit and often unrecognized. He argues that the difference is not in the number of assumptions, but in the extent to which they are made explicit. For examples of strong implicit assumptions, see Heckman (1997), Keane (2010), and Rosenzweig and Wolpin (2000). Heckman and Urzúa (2010) and Heckman and Vytlacil (2005, 2007) conclude that the experimental advantage disappears when treatment and control groups are dissimilar because theory must be used to control for differences; implicit assumptions vulnerable to errors are required just as in structural approaches. However, a major advantage of structural analysis is the possibility of explicit validation of assumptions (Keane 2010).

Principle 14. Use of instrumental variables estimators does not avoid the need to conceptualize the underlying structure. Rather, the choice of instruments represents an underlying structure whether reflected implicitly or imposed explicitly.

7 Sufficient statistics for welfare analysis

Structural modeling requires a great deal of data, institutional understanding, and modeling effort, particularly when supplemented with proper model validation (Keane 2010; Rust 2010). Data requirements can become impractical for modeling heterogeneous behavioral criteria (not simply effects of heterogeneous characteristics). Heckman and Urzúa (2010) suggest finding a middle ground between the IV estimation of treatment-effect models and full structural analysis by appealing to Marschak's Maxim – estimators should be selected on the basis of their ability to answer well-posed economic problems with minimal assumptions (Marschak 1953). They suggest that many well-posed questions can be answered by estimating certain combinations of structural parameters without full structural estimation, as was suggested by Marschak (1953).

Heckman and Vytlacil (2005, 2007) present efforts to bridge this gap with nonparametric and semiparametric methods in the case of discrete choice by developing the concept of a marginal treatment effect (MTE). This concept measures WTP for a marginal expansion of a program at the point where the marginal individual switches between discrete choice groups. Within this class of problems, they are able to avoid a great deal of parametric specificity while allowing heterogeneity in responses. But these methods allow only heterogeneity in responses, not choices.

Chetty (2009) focuses more generally on finding sufficient statistics that bridge the gap between atheoretic RF approaches and full structural approaches, which fits our definition of theory-based RF estimation. He suggests preserving the feasible simplicity and presumed robustness of RF analysis subject to retaining the ability of structural analysis to make precise statements about welfare, which yields his concept of sufficient statistics for marginal welfare analysis. This can avoid the complicated demands of finding all of the deep primitives of structural analysis. By focusing on high-level elasticities, welfare consequences can be evaluated with more robustness because many combinations of underlying structures that match the elasticities can have the same welfare consequences, as shown in Section 3. Chetty illustrated several specific problems where structural studies permit welfare analysis while RF studies do not, but yet other studies using sufficient statistics were adequate. The remaining challenge is to find ways to estimate sufficient statistics that apply under a variety of behavioral differences.

8 General equilibrium welfare analysis

A major shortcoming of the widespread emphasis on treatment-effect and atheoretic econometrics has been a neglect of general equilibrium (GE) concepts of policy impacts. The treatment effect on an individual or group has typically been applied as a partial equilibrium (PE) concept, assuming that the untreated are unaffected by equilibrium responses to changes in behavior of the treated. Such analysis fails to inform the policy process of indirect consequences of policy actions, and tends to be useful only for projects so small that related markets and conditions are virtually unaffected (Just, Hueth, and Schmitz 2008; Schmitz, Haynes, and Schmitz 2016).

In one of the few studies that has extended treatment-effect econometrics to the GE case using the MTE concept, Heckman, Lochner, and Taber (1998) find for their application that (1) PE analysis produces the intuitive result (e.g., tuition subsidies increase college education), (2) GE analysis produces counterintuitive but explainable effects (e.g., a glut of college graduates depresses salaries, resulting in a subsequent decline in college education from what it would have been otherwise), and (3) GE effects are small compared to PE effects. Obtaining these results with a treatment-effect model in their case, however, requires adding significant structure involving a full dynamic adjustment model.

> **Principle 15.** Policy impacts based on fixed economic circumstances are partial equilibrium measures that are possibly highly misleading as an assessment of ultimate policy implications upon consideration of indirect consequences of policy actions.
>
> **Principle 16.** Understanding indirect consequences of policy actions requires general equilibrium analysis of feedback effects of adjustments in the rest of the economy.

By definition, GE analysis tends to be structural because it must sort out interactions among markets. However, sufficient statistics can also be found for GE purposes that simplify both estimation and intuitive understanding for semi-lay users. Just, Hueth, and Schmitz (2004) show that actions of heterogeneous individuals generated by the models in (1) and (7) can be aggregated into market supply and demand relationships across all markets using the results in (2) and (9) to capture both PE and GE welfare measurements. By analogy, the behavioral variations in (4)–(6) and (10)–(13) can also be included in these aggregates without altering the interpretation associated with (2), (8), and (9). Bernheim and Rangel (2009, pp. 78–79) also show that the concept of general equilibrium is easily modified to include behavioral competitive equilibrium by building on Fon and Otani (1979).

To represent these concepts graphically, suppose in Figure 2.1 that a distortion is introduced in a particular market where initially demand is q^d and supply is q^s and all other prices are represented by $o(0)$. According to PE analysis, the introduction of a tax in the amount of δ_0 causes equilibrium to shift from quantity q_0 to $q^\star$ at producer price $p^s(\delta_0)$, exclusive of the tax and consumer price $p^d(\delta_0)$, inclusive of the tax. PE analysis falsely suggests that welfare declines by area $b + c + e$ for consumers and area $g + h + j$ for producers, with tax revenue equal to area $b + g$ and a deadweight loss of area $c + e + h + j$.

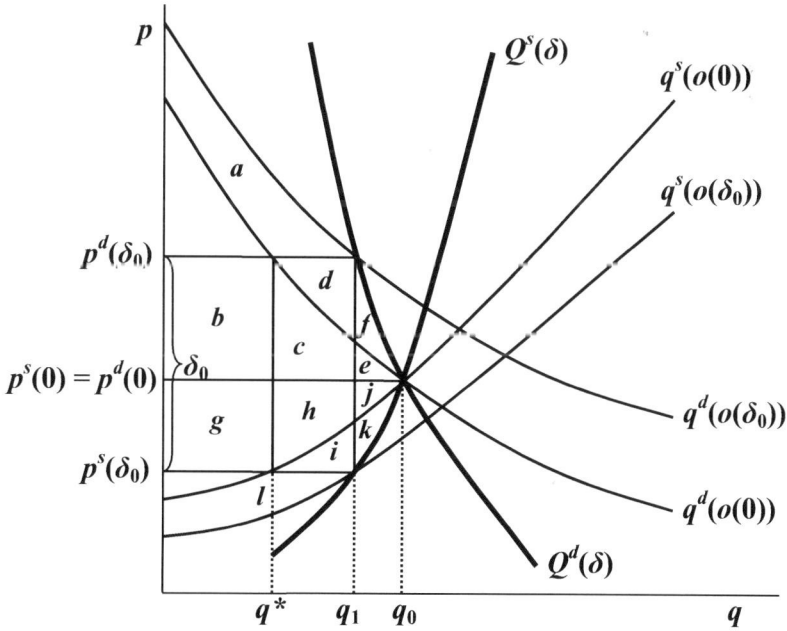

Figure 2.1 Partial and general equilibrium welfare effects of a tax

With GE analysis, however, increasing the consumer price increases demand for substitutes (decreases demand for complements), which bids up (down) their prices in $o(\delta_0)$. Consequently, the demand for q is higher after GE adjustment. Similarly, reducing producer price reduces demand for factor inputs, bidding down their prices. Consequently, the ordinary supply shifts outward so that GE with the tax δ_0 occurs at q_1.

Just, Hueth, and Schmitz (2004) show for a competitive economy with heterogeneous producers and consumers that varying the distortion δ traces out GE relationships Q^d and Q^s that reflect the marginal value to consumers and marginal cost to producers in this particular market as a function of the tax δ (a subsidy if negative). Furthermore, they show that the area behind these curves measures the aggregate change in private welfare aggregated over all producers and consumers throughout the economy. Private valuation of effects transmits through GE adjustments across all markets to the point where a distortion is introduced or altered.[17] This is essentially the same result presented by Harberger (1964) in a model where all distortions are represented as taxes.

The simple graphical analysis thus summarizes aggregate private GE welfare effects as a loss of area $b + c + d + e + f + g + h + i + j + k$.[18] Of this aggregate loss, the losses incurred by buyers and sellers involved directly in this market are area $b + c + e - a$ and area $g + h + j - l$, respectively. In other words, PE welfare measures are applicable if conditioned properly on GE price changes. Private welfare effects that occur throughout the rest of the economy are the difference in aggregate private effects and the effects on participants involved in this market, area $a + d + f + i + k + l$.

Just, Hueth, and Schmitz (2004) show that these conclusions hold whether or not other markets are distorted by (1) fixed per unit taxes and subsidies, (2) quotas or caps if quota rents are appropriated privately, and (3) exogenous government buying and selling in fixed amounts. These results hold because none of these distortions prevents full GE marginal valuation to private individuals from being transmitted to the market where the distortion δ is altered.

This simple graphical result demonstrates how the concept of sufficient statistics generalizes to the GE level and identifies the sufficient statistics required for both aggregate and distributional GE welfare analysis. The sufficient statistics required for aggregate efficiency analysis of private welfare are the direct elasticities of the GE relationships Q^d and Q^s. If these elasticities can be identified at the aggregate level, then the distribution of behavioral criteria among the various cases in (1)–(13) is inconsequential except as it affects the distribution of effects among individuals. The sufficient statistics for distributional welfare analysis are the direct partial equilibrium supply and demand elasticities of individuals (or groups) as well as the cross elasticities that determine shifts in individual demands and supplies in response to changes in the GE price vector. If these concepts can be measured for individuals or behavioral groups, then distributional effects among individuals or associated with behavioral groups can be measured.

Thus, sufficient statistics can permit explanation of welfare effects in a GE framework that facilitates intuitive understanding, much like the classical Marshallian graphical concepts of welfare economics, but under much broader assumptions, including heterogeneity of decision criteria as well as characteristics. These tools can convey powerful understanding to semi-lay users of policy analysis of both direct and indirect consequences of policy actions. By comparison, complex mathematics and large tables of numerical results can appear as black boxes. If sufficient intuitive understanding of indirect consequences cannot be conveyed to semi-lay users of policy analysis, then results will likely fail to gain traction in the policy process.

For example, Figure 2.1 offers simple intuition for why the GE effects of a policy can be much smaller than implied by PE intuition and analysis, depending on elasticities. This understanding can be critical for avoiding over-reactions by policy makers. The quantity effects of a move to q_1 are much smaller than a move to $q^\star$ because adjustments are spread across many markets and, consequently, the true deadweight loss, area $e + f + j + k$, is much smaller than the area $c + e + h + j$ suggested by PE analysis.

Principle 17. Sufficient statistics can be defined for general equilibrium welfare analysis of both efficiency and distribution in which typical partial equilibrium surplus measures facilitate distributional analysis if conditioned on general equilibrium prices.

Much of the GE welfare literature relies heavily on Harberger's (1964) arguments, as though welfare consequences of policies can be decomposed into tax-equivalent effects (Chetty 2009). A further example reveals that this is not always the case while underscoring the robustness of sufficient statistics in a graphical framework akin to Figure 2.1. Assume in Figure 2.2 that initial equilibrium occurs at p_0 and q_0 along initial supply q^s and demand q^d where other prices are o_0. Suppose public research produces an improvement in technology that increases supply to $q^{s\star}$ if other prices remain at o_0. Then PE analysis suggests a new equilibrium at $p^\star$ and $q^\star$, which implies a relatively mild price decline and major welfare gains for producers (not shown explicitly).

In GE, however, the consumer price reduction causes reduced (increased) demand for substitutes (complements), bidding down (up) their prices represented in o_1. Consequently, demand declines after GE adjustment as represented by q^d evaluated at o_1. GE adjustments also likely occur in factor input markets. If the new technology is primarily output increasing, the new equilibrium along $q^d(o_1)$ may use fewer inputs, implying a factor price reduction and ultimate supply at $q^s(o_1)$ to the right of $q^{s\star}(o_0)$ and a new GE at p_1 and q_1. If the new technology is primarily cost reducing, output expansion may require more inputs, implying a factor price increase and ultimate supply to the left of $q^{s\star}(o_0)$ (not shown).

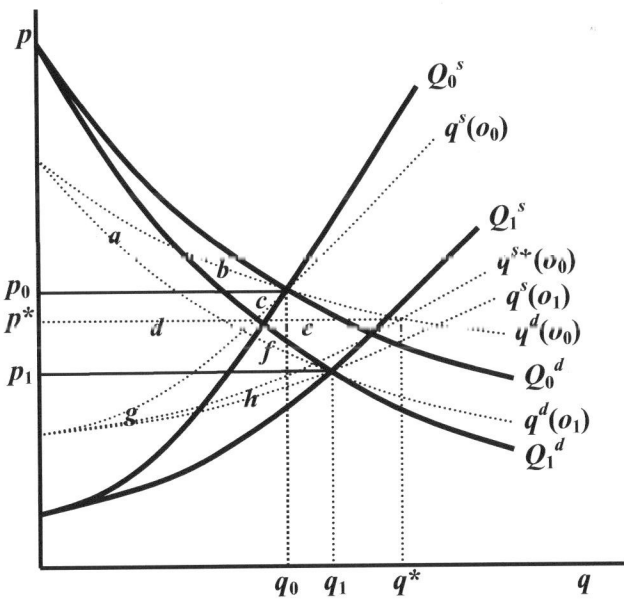

Figure 2.2 Partial and general equilibrium effects of a technology improvement

Just and Hueth (1979) show that the GE relationship Q^s can be defined that traces out GE responses of quantity supplied as the price in this market is changed, accounting for all consequent GE price adjustments on the input side of the market. Similarly, the GE relationship Q^d can be defined that traces out GE responses of quantity demanded as the price in this market is changed, accounting for all consequent GE price adjustments on the output side of the market. Moreover, aggregate GE welfare effects of change are captured by the surplus triangles these relationships define. The GE effect on aggregate private welfare is an increase from area $a + b + c + d + g$ to area $a + d + f + g + h$ for a net gain of area $f + g - b - c$.

As for distributional effects, the welfare effects on buyers and sellers involved in this market are again captured by the changes in PE surplus triangles defined by the dotted line relationships but evaluated at GE prices before and after the change. Furthermore, the gain for buyers in this market combined with other parties on the output side of this market is area $d + f - b$, and the gain for sellers in this market combined with other parties on the input side of the market is area $h - d - f$.

Because the latter gain may be negative, depending on elasticities, the producers, who would be the expected gainers from improved technology according to PE intuition, can be the losers as a result of indirect consequences of actions. Thus, this single understandable graphical presentation can capture the explanation of how indirect consequences of a policy may be counterintuitive, that is, qualitatively the opposite of what policy makers using PE intuition might anticipate. Of course, a complete welfare analysis of such a policy would also require accounting of the welfare costs of the public research required to generate the new technology.

This simple graphical explanation of GE effects suffices for measuring private welfare effects when distortions in other parts of the economy are fixed per unit taxes and subsidies, the rents on any quotas or caps are appropriated privately, and exogenous government buying and selling is determined by a price-insensitive political process. For net social welfare implications, this analysis must be supplemented by the aggregate GE change in tax revenues less subsidies in the rest of the economy.[19, 20]

9 Alternative methodological approaches

Finding GE price vectors that apply before and after implementation of a policy can be difficult for counterfactual policies. Even when prices are observed both before and after a change for ex post analysis, the task of controlling for changes in exogenous factors affecting related markets is difficult. For this purpose, some have used the Computable General Equilibrium (CGE) approach developed by Scarf (1973) (see Shoven and Whalley 1992 for a review) and made accessible by Rutherford (1999), while others have tried to approximate GE by developing a multimarket equilibrium model extensive enough to include econometrically identifiable effects.[21] Typically, CGE models impose heavy-handed assumptions (constant returns to scale, linearly homogenous preferences, constant elasticity of substitution, zero profits) and exclude behavioral variations that are now well documented. Required calibration with non-market goods and externalities requires further heavy assumptions, with unknown and potentially serious consequences (Klaiber and Smith 2009).

The alternative of modeling less than the full economy but approximating major GE interactions with theory-based RF equations in a multimarket framework can relax these assumptions to improve reliability and permit use of sufficient statistics (Just, Hueth, and Schmitz 2004, pp. 365–368; Klaiber and Smith 2009), particularly when individuals may be following a variety of behavioral criteria. This avoids the need for quantification associated with distortions related to market power and externalities in markets with inconsequential effects, which tend to be ignored in practical applications of the CGE approach anyway.

In summary, the approach of estimating sufficient statistics in GE tends to bring practices full circle to a resurgence of the terminology and surplus concepts of classical welfare economics, although under a much broader framework of understanding.

Principle 18. Robust economic welfare analysis should, insofar as possible, accommodate a variety of behavioral criteria while avoiding unverifiable assumptions as much as possible.

While Principle 18 offers a challenging research agenda, the principles developed in this review suggest a sizeable conceptual step is underway toward empirical feasibility.

10 Conclusions

Many observed types of behavior are anomalous compared to standard representations of consistent preferences. However, many modifications of behavior that capture so-called anomalies can be accommodated in revisions of optimization models, particularly when random errors are admitted. This can preserve aspects of traditional models that have seemed to capture the bulk of economic behavior empirically, while adding generalizations that are well documented in the behavioral literature. More importantly from a welfare standpoint, this can preserve welfare interpretations of revealed preferences that can inform the policy process.

By adding the variety of behavioral possibilities, structural representations of behavioral criteria as well as socioeconomic characteristics may vary among individuals. Full structural identification of heterogeneous behavioral criteria for a group of policy concerns is an overwhelming, if not infeasible, task in many cases. Conversely, atheoretic reduced form and traditional treatment-effect models typically capture inadequate information for welfare analysis and social optimization. Newer marginal treatment-effect models offer greater possibilities, especially for welfare analysis of policies with voluntary participation, but are not well-suited to many other policy issues, particularly those with heterogeneous choice.

The concept of sufficient statistics has been developed as a means of bridging this gap and capturing the key parameters that facilitate social optimization without requiring full structural identification. The most critical sufficient statistics that support welfare analysis are the elasticities of supply and demand by the individuals and groups they represent. This calls for a new focus on estimating theoretically based reduced form models.

This review demonstrates by reference to results elsewhere that the concepts of sufficient statistics can be extended to a useful general equilibrium context. This appears to offer the best possibilities for robust welfare analysis because it can include a variety of behavioral variations documented in the literature. Approximating general equilibrium analysis by multimarket equilibrium analysis of all markets with econometrically identifiable effects appears to be the most attractive approach in this context. Moreover, this approach facilitates use of a graphical single-market framework to convey intuitive understanding to semi-lay users of policy analysis of both direct and indirect consequences, which preserves much of the simplicity of classical welfare economics but under much broader assumptions.

Notes

1 This chapter incorporates major parts of Just (2011). We thank the president and editors of the *Annual Review of Resource Economics* for permission to do so.
2 For convenience, we refer to WTP generically so as to include willingness-to-accept concepts.

3 The single-product case enables simplicity, but generalizations are straightforward (Just, Hueth, and Schmitz 2004).
4 Subscripts that represent variables denote differentiation throughout.
5 Kuhn-Tucker conditions are assumed so that $q(0) = 0$ and $x = 0$ if output price is below some threshold.
6 Econometric identification of supply near the axis can be poor if some observable data are not close to the axis. However, alternative approaches (e.g., flexible profit functions) reduce this weakness by the assumptions they impose. For analysis of an observed change, this problem is mitigated by using the change in observed profit. Alternatively, measurement of WTP for price changes avoids some of this weakness by sequentially evaluating price changes by their respective supplies and demands (Just, Hueth, and Schmitz 2004).
7 These conclusions assume WTP is measured by ex post compensation, in which case the change in consumer surplus associated with an essential input demand underestimates WTP by a factor $1 + \lambda$. With ex ante compensation, these conclusions are reversed so $dS = \hat{q}\,dp/(1+\hat{\lambda}) - \hat{x}\,dr$, where changes in producer surplus overestimate WTP by a factor $1/(1+\lambda)$, but slopes of demands measure marginal WTP accurately.
8 Of course, not all behavioral criteria are easily grouped for welfare analysis. For example, if a firm manager maximizes sales subject to a minimum profit constraint, then WTP measures the change in revenue. Consequently, the change in the producer surplus triangle is proportional to WTP but is underestimated by a factor involving the Lagrangian multiplier of the profit constraint (Just, Hueth, and Schmitz 2004).
9 When the "variety of behaviors" for producers becomes more general than the examples earlier, by allowing preferences to depend on a vector of outcomes rather than a single aggregate, such as constrained profit or certainty equivalent profit, the conclusion regarding discrete non-price changes in Principle 6 requires an assumption similar to Mäler's (1974) weak complementarity condition. This is the almost universal approach of the literature on consumer welfare analysis of non-market goods (Just, Hueth, and Schmitz 2004).
10 Again, the qualification of endnote 9 may apply depending on the generality of behaviors.
11 Curiously, while producer anomalies (i.e., departures from profit maximization) such as those considered in the previous section have been widely observed, economists have been far less concerned with them than for the consumer case.
12 The mathematics here is similar to the case of quantity restrictions (Randall and Stoll 1980).
13 Chetty (2009) presents equivalent results based on Harberger (1964) regarding marginal analysis.
14 Regardless of how unobservables enter (13), they can be represented as additive where the parameters of the distribution of ε_{ij} possibly depend on p and m. That is, define $\varepsilon_{ij} \equiv V_{ij}^*(p,m,\varepsilon_{ij}) - V_{ij}(p,m)$ such that $E(\varepsilon_{ij}) = 0$. The critical matter for empirical modeling depends on how restrictive the distributional assumptions must be to accommodate standard choice models used for logit or probit analysis. Thus, limitations are defined by econometric possibilities rather than the conceptual requirements of welfare analysis.
15 For a more complete review of the many issues involved in the correct choice of instruments and the evolution of treatment-effect econometrics as it relates to welfare measurement, see Just (2011).
16 A lesson of traditional structural econometric methods (e.g., two-stage least squares) is that the best choice of instruments for jointly endogenous variables is the full set of exogenous variables in the structural system that explains them, except in very small samples (Greene 2007). Determination of this full set of valid instruments is closely akin to specifying the full system structure. In contrast, many other variables may exist in the economy that can pass typical exogeneity tests but may yet be poor instruments with spurious correlations that cause bias.
17 For practical GE analysis, the consumer model may require expansion to include labor and household production. But this is a straightforward generalization following Just, Hueth, and Schmitz (2004, pp. 216–237).
18 As with surplus concepts for individuals, these results apply accurately (approximately) if responses are compensated (uncompensated) and readily disaggregate to the heterogeneous agents they represent.
19 For more general distortions that change the difference in marginal valuations of buyers and sellers in the rest of the economy or where the political process seeks only domestic welfare in an open economy, see Just, Hueth, and Schmitz (2004), pp. 361–366.

20 Additionally, when public finances are derived by taxation in a way that causes deadweight loss, further consideration of that social cost is necessary. Many studies have estimated this marginal cost with a wide range of results. See Just (2011) for a brief review.
21 One such model used in this way is the locational sorting model where households can benefit from local changes, but changes can induce locational choices in response (Tiebout 1956; Klaiber and Smith 2009).

References

Angrist, J.D., and A.B. Krueger. 2001. "Instrumental Variables and the Search for Identification: From Supply and Demand to Natural Experiments." *Journal of Economic Perspectives* 15:69–85.

Ariely, D., G. Loewenstein, and D. Prelec. 2003. "Coherent Arbitrariness: Stable Demand Curves Without Stable Preferences." *Quarterly Journal of Economics* 118:73–105.

Azevedo, C.D., J.A. Herriges, and C.L. Kling. 2003. "Combining Revealed and Stated Preferences: Consistency Tests and Their Implications." *American Journal of Agricultural Economics* 85:525–537.

Baumol, W.J., C.V. Starr, and C.A. Wilson. 2000. *Welfare Economics*. Cheltenham, UK: Edward Elgar.

Becker, G.S., and K.M. Murphy. 1988. "A Theory of Rational Addiction." *Journal of Public Economics* 104:675–700.

———. 1993. "A Simple Theory of Advertising as a Good or Bad." *Quarterly Journal of Economics* 108:941–964.

Ben-David, S., D.S. Brookshire, S. Burness, M. McKee, and C. Schmidt. 1999. "Heterogeneity, Irreversible Production Choices, and Efficiency in Emission Permit Markets." *Journal of Environmental Economics and Management* 38:176–194.

Bernheim, B.D. 2009a. "Behavioral Welfare Economics." *Journal of European Economics Association* 7:267–319.

———. 2009b. "Neuroeconomics: A Sober but Hopeful Appraisal." *American Economic Journal of Microeconomics* 1:1–41.

Bernheim, B.D., and A. Rangel. 2004. "Addiction and Cue-Triggered Decision Processes." *American Economic Review* 94:1558–1590.

———. 2007. "Toward Choice Theoretic Foundations for Behavioral Welfare Economics." *American Economic Review* 97(2):464–470.

———. 2009. "Beyond Revealed Preference: Choice Theoretic Foundations for Behavioral Welfare Economics." *Quarterly Journal of Economics* 124:51–104.

Bickel, W.K., M.L. Miller, R.Yi, B.P. Kowal, D.M. Lindquist, and J.A. Pitcock. 2007. "Behavioral and Neuroeconomics of Drug Addiction: Competing Neural Systems and Temporal Discounting Processes." *Drug Alcohol Dependence* 90:S85–91.

Blackorby, C., D. Primont, and R.R. Russell. 1978. *Duality, Separability, and Functional Structure: Theory and Economic Applications*. New York: Elsevier North-Holland.

Blundell, R. 2010. "Comments on: Michael P. Keane's Structural vs. Atheoretic Approaches to Econometrics." *Journal of Econometrics* 156:25–26.

Bockstael, N.E., and K.E. McConnell. 1983. "Welfare Measurement in the Household Production Framework." *American Economic Review* 73:806–814.

———. 2007. *Environmental and Resource Valuation with Revealed Preferences: A Theoretical Guide to Empirical Models*. Dordrecht, Netherlands: Springer.

Camerer, C., S. Issacharoff, G. Loewenstein, T. O'Donoghue, and M. Rabin. 2003. "Regulation for Conservatives: Behavior Economics and the Case for 'Asymmetric Paternalism'." *University of Pennsylvania Law Review* 151:1211–1254.

Camerer, C., G. Loewenstein, and D. Prelec. 2004. "Neuroeconomics: Why Economics Needs Brains." *Scandinavian Journal of Economics* 106:555–579.

———. 2005. "Neuroeconomics: How Neuroscience Can Inform Economics." *Journal of Economics Literature* 43:9–64.

Chetty, R. 2009. "Sufficient Statistics for Welfare Analysis: A Bridge Between Structural and Reduced Form Methods." *Annual Review of Economics* 1:451–488.

Coleman, J. 1980. "Efficiency, Utility, and Wealth Maximization." *Hofstra Law Review* 8:509–551.

Cox, J.C., V.L. Smith, and J.M. Walker. 1985. "Experimental Development of Sealed-Bid Auction Theory: Calibrating Controls for Risk Aversion." *American Economic Review* 75:160–165.

Currie, J.M., J.A. Murphy, and A. Schmitz. 1971. "The Concept of Economic Surplus and Its Use in Economic Analysis." *Economics Journal* 81:741–799.

Deaton, A.S., and J. Muellbauer. 1980. *Economics and Consumer Behavior*. Cambridge: Cambridge University Press.

Feder, G., R.E. Just, and D. Zilberman. 1983. "Adoption of Agricultural Innovations in Developing Countries: A Survey." *Economic Development and Cultural Change* 33(2):255–298.

Fon, V., and Y. Otani. 1979. "Classical Welfare Theorems with Non-Transitive and Non-Complete Preferences." *Journal of Economics Theory* 20:409–418.

Foster, W., and R.E. Just. 1989. "Measuring the Welfare Effects of Product Contamination with Consumer Uncertainty." *Journal of Environmental Economics and Management* 17:266–283.

Greene, W.H. 2007. *Econometric Analysis* (6th ed.). New York: Prentice Hall.

Gruber, J. 1994. "The Incidence of Mandated Maternity Benefits." *American Economic Review* 84:622–642.

Gruber, J., and B. Kőszegi. 2001. "Is Addiction 'Rational'? Theory and Evidence." *Quarterly Journal of Economics* 116:1261–1303.

Gul, F., and W. Pesendorfer. 2007. "Welfare Without Happiness." *American Economic Review* 97:471–476.

Hanemann, M. 1984. "Discrete/Continuous Models of Consumer Demand." *Econometrica* 52:541–561.

Harberger, A.C. 1964. "The Measurement of Waste." *American Economic Review* 54:58–76.

Harrison, G.W., J.A. List, and C.E. Towe. 2007. "Naturally Occurring Preferences and Exogenous Laboratory Experiments: A Case Study of Risk Aversion." *Econometrica* 75:433–458.

Hause, J. 1975. "The Theory of Welfare Cost Measurement." *Journal of Political Economics* 83:1154–1178.

Hausman, J. 1981. "Exact Consumer's Surplus and Deadweight Loss." *American Economic Review* 71:662–676.

Heckman, J. 1997. "Instrumental Variables: A Study of Implicit Behavioral Assumptions Used in Making Program Evaluations." *Journal of Human Resources* 32:441–462.

Heckman, J.J., H. Ichimura, and P. Todd. 1998. "Matching as an Econometric Evaluation Estimator." *Review of Economics Studies* 65:261–294.

Heckman, J.J., L.J. Lochner, and C. Taber. 1998. "General-Equilibrium Treatment Effects: A Study of Tuition Policy." *American Economic Review* 88:381–386.

Heckman, J.J., and S. Urzúa. 2010. "Comparing IV with Structural Models: What Simple IV Can and Cannot Identify." *Econometrics* 156:27–37.

Heckman, J.J., and E.J. Vytlacil. 2005. "Structural Estimation, Treatment Effects, and Econometric Policy Evaluation." *Econometrica* 73:669–738.

———. 2007. "Econometric Evaluation of Social Programs, Part II: Using the Marginal Treatment Effect to Organize Alternative Economic Estimators to Evaluate Social Programs and to Forecast Their Effects in New Environments." In J. Heckman and E. Leamer, eds., *Handbook of Econometrics* (Vol. 6). Amsterdam: Elsevier, pp. 4875–5144.

Hicks, J.R. 1941. "The Rehabilitation of Consumer(s Surplus." *Review of Economic Studies* 8:108–116.

———. 1943. "The Four Consumer's Surpluses." *Review of Economic Studies* 11:31–41.

———. 1956. *A Revision of Demand Theory*. Oxford: Clarendon Press.

Imbens, G. 2004. "Nonparametric Estimation of Average Treatment Effects Under Exogeneity: A Review." *Review of Economics and Statistics* 86:4–29.

Just, R.E. 2008. "Distinguishing Preference and Perceptions for Meaningful Policy Analysis." *American Journal of Agricultural Economics* 90:1165–1175.

———. 2011. "Behavior, Robustness, and Sufficient Statistics in Welfare Measurement." *Annual Review of Resource Economics* 3:37–70.

Just, R.E., L. Calvin, and J. Quiggin. 1999. "Adverse Selection in Crop Insurance: Actuarial and Asymmetric Information Incentives." *American Journal of Agricultural Economics* 81:834–849.

Just, R.E., and D.L. Hueth. 1979. "Welfare Measures in a Multimarket Framework." *American Economic Review* 69:947–954.

Just, R.E., D.L. Hueth, and A. Schmitz. 2004. *The Welfare Economics of Public Policy*. Northampton, MA: Edward Elgar.

———. 2008. *Applied Welfare Economics*. Cheltenham: Edward Elgar.

Kagel, J.H., and A.E. Roth. 1995. *The Handbook of Experimental Economics*. Princeton, NJ: Princeton University Press.

Kahneman, D. 1994. "New Challenges to the Rationality Assumption." *Journal of Institutional and Theoretical Economics* 150:18–36.

Kahneman, D., and A. Tversky. 1979. "Prospect Theory: An Analysis of Decision Under Risk." *Econometrica* 47:263–292.

Keane, M. 2010. "Structural vs. Atheoretic Approaches to Econometrics." *Econometrics* 156:3–20.

Kirman, A. 2006. "Heterogeneity in Economics." *Journal of Economic Interaction and Coordination* 1:89–117.

Klaiber, H.A., and V.K. Smith. 2009. "General Equilibrium Benefit Analysis for Social Programs." Unpublished Paper, Pennsylvania State University.

Kőszegi, B., and M. Rabin. 2008a. "Choices, Situations, and Happiness." *Journal of Public Economics* 92:1821–1832.

Kőszegi, B., and M. Rabin. 2008b. "Revealed Mistakes and Revealed Preferences." In A. Caplin and A. Schotter, eds., *The Foundations of Positive and Normative Economics*. New York: Oxford University Press, pp. 193–209.

Laibson, D. 1997. "Golden Eggs and Hyperbolic Discounting." *Quarterly Journal of Economics* 112:443–477.

Long, N.V., and A. Soubeyran. 1997. "Cost Heterogeneity, Industry Concentration, and Strategic Trade Policies." *Journal of International Economics* 43:207–220.

Lowenstein, G., and S. Issacharoff. 1994. "Source-Dependence in the Valuation of Objects." *Journal of Behavioral Decision Making* 7:157–168.

Mäler, K.-G. 1974. *Environmental Economics: A Theoretical Inquiry*. Baltimore: Johns Hopkins Press.

Marschak, J. 1953. "Economic Measurements for Policy and Prediction." In W. Hood and T. Koopmans, eds., *Studies in Econometric Method*. New York: Wiley, pp. 1–26.

Marshall, A. 1890. *Principles of Economics*. London: Palgrave Macmillan.

McCloskey, D.N. 1982. *The Applied Theory of Price*. New York: Macmillan.

McFadden, D. 1973. "Conditional Logit Analysis of Qualitative Choice Behavior." In P. Zarembke, ed., *Frontiers of Econometrics*. New York: Academic Press, pp. 105–142.

———. 1976. "Quantal Choice Analysis." *Annals of Economic and Social Measurement* 5:363–390.

McGuire, J.W. 1964. *Theories of Business Behavior*. New York: Prentice-Hall.

Nelson, C.R., and R. Startz. 1990a. "Some Further Results on the Exact Small Sample Properties of the Instrumental Variables Estimator." *Econometrica* 58:967–976.

———. 1990b. "The Distribution of the Instrumental Variables Estimator and Its T-Ratio When the Instrument Is a Poor One." *Journal of Business* 63:5125–5140.

O'Donoghue, T., and M. Rabin. 1999. "Doing It Now or Later." *American Economic Review* 89:103–124.

Randall, A., and J.R. Stoll. 1980. "Consumer's Surplus in Commodity Space." *American Economic Review* 71:449–457.

Rosenzweig, M.R., and K.I. Wolpin. 2000. "Natural 'Natural Experiments' in Economics." *Journal of Economic Literature* 38:827–874.

Rust, J. 2010. "Comments on Structural vs. Atheoretic Approaches to Econometrics." *Econometrics* 156:21–24.

Rutherford, T.F. 1999. "Applied General Equilibrium Modeling with MPSGE as a GAMS Subsystem: An Overview of the Modeling Framework and Syntax." *Computational Economics* 14:1–46.

Samuelson, P.A. 1942. "Constancy of the Marginal Utility of Income." In O. Lange, F. McIntyre, and T. Yntema, eds., *Studies in Mathematical Economics and Econometrics*. Chicago: University of Chicago Press, pp. 75–91.

Scarf, H. 1973. *The Computation of Economic Equilibrium*. New Haven, CT: Yale University Press.

Schmitz, A., D. Haynes, and T.G. Schmitz. 2016. "The Not-So-Simple Economics of Production Quota Buyouts." *Journal of Agricultural and Applied Economics* 48(2):119–147.
Schmitz, A., and R.O. Zerbe. 2009. *Applied Benefit–Cost Analysis*. Cheltenham: Edward Elgar.
Scitovsky, T. 1941. "A Note on Welfare Propositions in Economics." *Review of Economic Studies* 9:77–88.
Sen, A. 2004. "Economic Methodology: Heterogeneity and Relevance." *Social Research: An International Quarterly* 71:583–614.
Shoven, J., and J. Whalley. 1992. *Applying General Equilibrium*. New York: Cambridge University Press.
Small, K.A., and H.S. Rosen. 1981. "Applied Welfare Economics with Discrete Choice Models." *Econometrica* 49:105–130.
Sugden, R. 2004. "The Opportunity Criterion: Consumer Sovereignty Without the Assumption of Coherent Preferences." *American Economic Review* 94:1014–1033.
Sunstein, C., and R. Thaler. 2003. "Libertarian Paternalism." *American Economic Review* 93:175–179.
Tiebout, C.M. 1956. "A Pure Theory of Local Expenditures." *Journal of Political Economics* 64:416–424.
Vartia, Y.O. 1983. "Efficient Methods of Measuring Welfare Change and Compensated Income in Terms of Ordinary Demand Functions." *Econometrica* 51:79–98.
von Neumann, J., and O. Morgenstern. 1944. *Theory of Games and Economic Behavior*. Princeton, NJ: Princeton University Press.
Whitehead, J.C., and G.C. Blomquist. 2006. "The Use of Contingent Valuation in Benefit–Cost Analysis." In A. Alberini and J. Kahn, eds., *Handbook on Contingent Valuation*. Cheltenham, UK: Edward Elgar Publishing, pp. 92–115.
Willig, R.D. 1976. "Consumer's Surplus Without Apology." *American Economic Review* 66:589–597.
Working, E.J. 1927. "What Do Statistical Demand Curves Show?" *Quarterly Journal of Economics* 41:212–235.
Zink, C.F., G. Pagnoni, M.E., Martin-Skurski, J.C. Chappelow, and G.S. Berns. 2004. "Human Striatal Response to Monetary Reward Depends on Saliency." *Neuron* 42:509–517.

Part 1
Food economics

3
World food
New developments

Shenggen Fan

1 Introduction

The world food sector has been fundamentally transformed over the last several decades. Much progress has been made in increasing food production as well as in reducing hunger and undernutrition. Nevertheless, challenges in feeding the future global population remain enormous, as the world faces persistent hunger and malnutrition as well as emerging challenges such as demographic shifts and urbanization, climate change, continued conflict, and uncertainty in the global political landscape. Innovations in technologies, policies, and institutions, as well as reforms in global and local governance, are urgently needed to reshape the global food system to produce sufficient nutritious and affordable food to meet changing food needs and to end hunger and malnutrition sustainably.

From 1960 to 2015, agricultural production more than tripled, largely due to enhanced productivity through Green Revolution technologies and the expanded use of land, water, and other natural resources. During the same period, food demand increased 2.2 percent annually and dietary preferences and patterns of consumption evolved. With a projected world population of almost 10 million by 2050, food demand would increase by 50 percent from 2013 levels. This poses several key questions: what will food supply and demand look like in 2050, and what is needed for the world to be able to feed its population?

2 Supply and demand projections: IMPACT model[1]

We can explore these questions by using International Food Policy Research Institute's (IFPRI's) International Model for Policy Analysis of Agricultural Commodities and Trade (IMPACT), which is a partial equilibrium, multimarket economic model that simulates national and international agricultural markets, solving for production, demand, and prices that equate supply and demand globally. The model links various modules on climate, water, crops, land use, nutrition, and health, as well as welfare analysis. Some of the information flows among these component modules are one-way (for instance, from the climate and water modules to crop simulation models) and some capture feedback loops (for example, water demand from the core market model and water supply from the water module to be reconciled to estimate water stress impact on crop yields). The core model includes 159 countries, 320 regions referred to as food production units, and 62 agricultural commodity markets. The model provides various climate change scenario analyses rather than forecasting and covers only agricultural commodities. The integrated projections and implications of physical, biophysical, and socioeconomic trends allow for various in-depth analyses on key issues of interest to policy makers (IFPRI 2017; Robinson et al. 2015b).

2.1 Supply projections

According to the latest IMPACT model supply projections, global food production will grow by approximately 60 percent by 2050 (relative to 2010 levels).[2] Production will grow more rapidly in developing countries at 71 percent compared to developed countries at 29 percent. Africa and the Middle East will more than double their food production, South Asia will increase its production by 91 percent, and Latin America and the Caribbean by 72 percent. Meat production will rise globally by 66 percent and by 78 percent in developing countries. Production of fruits and vegetables, pulses, and oilseeds will grow more rapidly, by more than 80 percent globally. Cereals and roots and tuber production will grow more slowly, by around 40 percent globally, though it will double in sub-Saharan Africa (IFPRI 2017).

2.2 Demand projections

Total food demand will continue to increase over the coming decades, and the composition of diets will continue to evolve with changes in income and preferences (Figure 3.1). Staple food demand, including cereals, roots, and tubers, will grow by around 40 percent, and meat demand will grow by over 60 percent. Demand for fruits and vegetables will grow at a more rapid rate, though it is starting from lower levels. Despite population growth and climate change, per capita consumption is estimated to increase by 9 percent worldwide to reach more than 3,000 kilocalories per day. Per capita consumption of fruits and vegetables in developing countries is expected to surpass that of developed countries by 2050, which could have important nutrition and health benefits (IFPRI 2017).

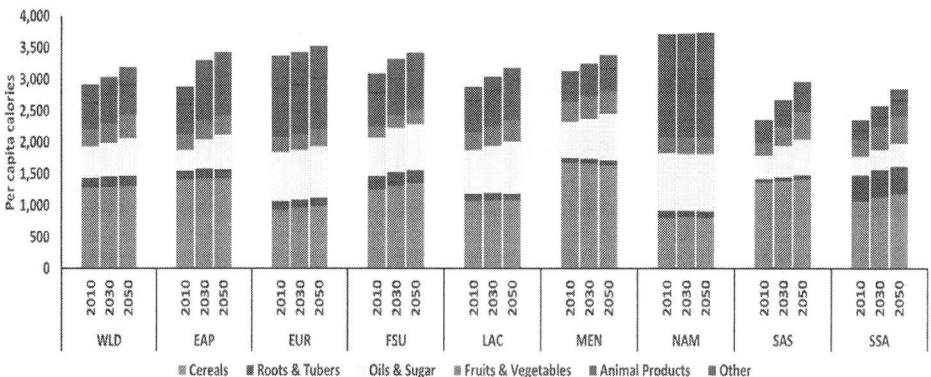

Figure 3.1 Changing composition of diets per capita[3]

Source: IFPRI, IMPACT model version 3.2, November 2015.

Note: WLD = World; EAP = East Asia and Pacific; EUR = Europe; FSU = Former Soviet Union; LAC = Latin America and Caribbean; MEN = Middle East and North Africa; NAM = North America; SAS = South Asia; SSA = Sub-Saharan Africa

World food

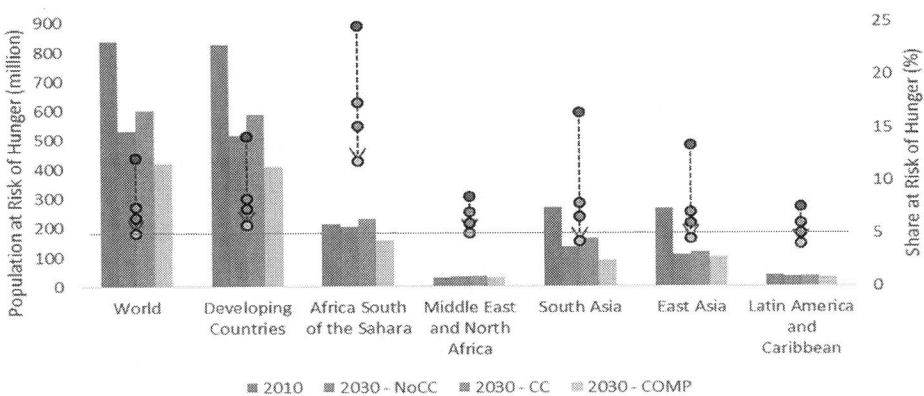

Figure 3.2 Hunger in 2030 by climate and investment scenario
Source: Rosegrant et al. (2017), based on IMPACT model version 3.3.

Nevertheless, disparities in food access and the impacts of climate change imply that 592 million people will remain at risk of hunger by 2030 (Figure 3.2). By 2050, 477 million people will be at risk of hunger, 461 million of whom will live in developing countries. Those living in Africa south of the Sahara alone will account for 189 million.

3 Current and emerging challenges

3.1 Triple burden of malnutrition

Tremendous progress has been made in reducing global hunger, from 19 percent to 11 percent from 1990 to 2016 (FAO 2017). Childhood undernutrition has also declined in developing countries, where the number of stunted children under 5 years old decreased from 254 million in 1990 to 155 million in 2016 (UNICEF, WHO, and World Bank 2017). Much of this reduction occurred in rural areas, whereas the proportion of stunted children rose from 23 to 31 percent in cities between 1985 and 2011 (Paciorek et al. 2013). However, 815 million people are still estimated to be undernourished globally in 2016 (FAO, IFAD, UNICEF, WFP and WHO 2017). More than 2 billion suffer from a lack of micronutrients, the so-called hidden hunger. At the same time, the number of those considered overweight or obese has increased globally among children and adults. The number of overweight children rose by more than 50 percent from 1990 to 2011, and, as of 2015, 73 percent of all overweight children live in Africa and Asia (Ruel, Garrett, and Yosef 2017; UNICEF, WHO, and World Bank 2017). The rise in overweight and obese adults has also been drastic – 2 million out of 5 million adults worldwide were overweight or obese in 2016 (IFPRI 2016).

This rising co-existence of energy deficiencies from undernourishment, micronutrient deficiencies, and over-nutrition in the form of obesity or being overweight is also referred to as the "triple burden of malnutrition" (Fan 2017). Undernourishment is the result of insufficient caloric intake and contributes to negative health outcomes with the prevalence of being underweight, wasting, or stunting, particularly among children. Micronutrient malnutrition refers to a deficient intake of vitamins and minerals that are necessary for good health and results from poor dietary composition and disease (Gomez et al. 2013). Obesity and being overweight arise from excessive dietary energy intake and are associated with increases in the risk of noncommunicable diseases, such as diabetes and cardiovascular disease (Gomez et al. 2013).

While these burdens are growing globally, it is particularly a challenge for low- and middle-income countries. In Africa, 8 percent of adults over age 20 are obese, and 13 countries face serious levels of under age 5 stunting, anemia in reproductive age women, and adult overweight and obesity (Haddad et al. 2016). Additionally, 2 billion people suffer from hidden hunger, especially in Africa south of the Sahara (von Grebmer et al. 2014). Middle-income countries such as Brazil, China, India, Indonesia, and Mexico have made progress in reducing the number of the chronically hungry, yet 363 million people who fit this definition – accounting for almost half of the world's hungry – live in these countries (Fan and Cousin 2015). Malnutrition imposes high economic costs in these countries. For example, micronutrient deficiencies cost India up to 3 percent of its annual GDP, and noncommunicable diseases linked with being overweight or obese accounted for 13 percent of all health care expenditures in 2008 in Mexico (Stein and Qaim 2007). Rising inequalities across wealth, gender, and education hinder human capital development, which jeopardizes food security and nutrition (Fan and Cousin 2015). Rapid urbanization, which is also concentrated in low- and middle-income countries, is changing consumer preferences away from traditional cereal-based diets to protein-rich diets (OECD and FAO 2014). While social protection has often provided greater access to food in these countries, assistance is often not complemented with nutrition education and is subject to poor targeting and leakages.

3.2 Urbanization

For the first time in history, more than half of the world population lives in urban areas, and by 2050, 2.5 billion people will either be born in or move to urban areas, increasing the urban population to two-thirds of the world. Africa and Asia account for 90 percent of this growth, and currently have urban populations of 40 percent and 47 percent, respectively. Furthermore, China, India, and Nigeria alone are expected to add 900 million urban dwellers by 2050 (United

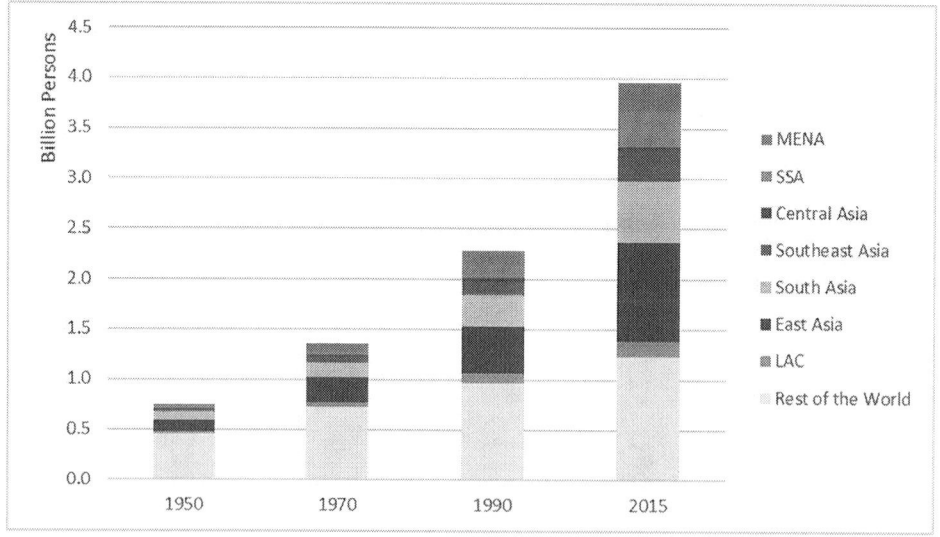

Figure 3.3 Growth of urban population in major developing regions
Source: Food and Agriculture Organization of the United Nations, FAOSTAT (2016).

Nations 2014). Rapid growth in urban populations (Figure 3.3) entails new and unique challenges and opportunities with regards to poverty, food security, and nutrition.

Urbanization is bringing significant changes to urban food systems and agriculture. Food supplies in urban areas are more diverse than in rural areas, driven by increased urban demand for more and better food. Urban food diversity is also enhanced by infrastructure and population densities in urban centers that facilitate distribution, transportation, and technologies that allow suppliers to reach more consumers at a lower cost (Ruel, Garrett, and Yosef 2017). Supermarkets are growing in developing countries and emerging economies, accounting for 30 to 50 percent of the food markets in Southeast Asia, Central America, Argentina, Chile, and Mexico by mid-2000s. Yet the urban poor depend on informal markets and street vendors to purchase food (Weatherspoon and Reardon 2003). The informal economy is key for urban food security due to their proximity to low-income housing areas, and it improves food availability by providing low prices and serving as an income source for the poor (Roever 2014). A survey of over 6,000 low-income neighborhood households in 11 African cities found that 70 percent of urban households depend on informal markets and street vendors for food (Resnick 2017). Urban agriculture – growing crops or raising livestock in urban or peri-urban areas – provides an additional source of food, employment, and income for urban residents. While there is a wide range in the scale of urban agriculture, from 11 percent household participation in Indonesia to 70 percent in Nicaragua and Vietnam, much of urban agriculture is utilized for household consumption (Zezza and Tasciotti 2010).

Despite the availability of diverse foods in many urban areas for wealthier consumers, the urban poor face obstacles in accessing high quality, diverse, safe, and affordable food, thereby increasing their risk of poor health and nutrition. While urban dwellers are more likely to meet protein and energy requirements than are rural dwellers, urban consumers are also more likely to have imbalanced diets with energy-dense processed foods and higher intake in calories, saturated fats, refined sugars, and salt. While urban diets also tend to have more fruits and vegetables, these diets are still low in micronutrients such as iron, zinc, and vitamin A (Ruel, Garrett, and Yosef 2017). This shift in diets in urban areas arises from factors including the urban food environment (such as the availability of energy-rich, nutrient-poor processed food and convenient fast food), changes in food habits linked to increased income, changes in employment, and changes in norms and attitudes towards traditional food.

3.3 Climate change

Climate change threatens agricultural production systems in many ways. It changes temperatures as well as amounts of rainfall, distribution, and intensity in many areas, which can negatively impact growing seasons, plant growth, and crop yields. For example, increased night-time temperatures can reduce rice yields by up to 10 percent for each 1°C change in minimum dry season temperatures (Thornton and Lipper 2014). The IFPRI IMPACT model projects significant climate impacts on rain-fed crop yields. Between 2005 and 2050, maize yields are projected to increase by 124 percent without climate change but would increase by only 74 percent with climate change. Similarly, rice yields are projected to increase by 43 percent without climate change and by 25 percent under climate change (Robinson et al. 2015a). Climate change is projected to increase global food production by 9 percentage points less than would be the case without climate change, and is also expected to place 71 million additional people at risk of hunger by 2050 (IFPRI 2017). Many of the climate impacts on agriculture will occur through water (Thornton and Lipper 2014). The increasing variability in rainfall is expected to continue, as are occurrences of floods and droughts. The impact on freshwater systems will likely outweigh any benefits from the overall increase in precipitation from global warming. Global demand for water withdrawals for agriculture is projected to increase 11 percent by 2050, further adding to the water security challenge (Bruinsma 2009). As of 2010, almost 40 percent of the world's grain production took place in water-stressed regions, yet if we continue business as usual, by 2050 we will depend on these regions for 49 percent of world grain production (Veolia Water and IFPRI 2013).

The four dimensions of food security (availability, access, utilization, and stability) are also directly affected by climate change. In terms of food availability, research indicates that crop yields are more negatively affected in tropical areas than in areas of higher latitudes, and the impact will become more severe as the degree of climate change progresses. This impact on crop productivity will be especially pronounced in areas that currently face high burdens of hunger. For example, maize yields could decline on average by 5 percent in Africa and 16 percent in South Asia compared to current yields (Knox et al. 2012). Food access largely depends on household- and individual-level income. Climate change could alter the ability of households and individuals to produce certain products, as well as the prices of various inputs and resources necessary for agricultural production, thereby threatening income levels connected to food access. Food utilization, or the attainment of adequate nutrients from food, may be directly impacted by climate change through increased levels of carbon dioxide. Some studies have found that elevated levels of carbon dioxide in the atmosphere were associated with decreases in the concentration of iron and zinc; for example, wheat grains showed 9 percent less zinc and 5 percent less iron (Myers et al. 2014). Climate change will also affect the availability of clean drinking water, and extreme weather events causing floods or droughts make good health care and dietary practices challenging. Diet quality may also be impacted from ecological shifts and diseases in crops or food. The stability of the food system as a whole may be jeopardized under climate change, as climate is a key determinant of future price trends. Food security of the poor is heavily dependent on staple food prices, and food market volatility from supply-side or demand-side shocks from climate change will be a significant challenge (Wheeler and von Braun 2013).

At the same time, agriculture is a main contributor to climate change. The agriculture sector contributes between 19 and 29 percent of greenhouse gas (GHG) emissions, with nearly three-quarters of emissions occurring in developing countries. This share could grow to as much as 80 percent by 2050 (Smith et al. 2008). Total emissions from livestock from 1995 to 2005 were between 5.6 and 7.5 million metric tons of carbon dioxide equivalent, the major source of emissions being from feed production and land use for animal feed and pastures. The developing world accounts for 70 percent of emissions from ruminants and 53 percent from monogastrics, and these figures will continue to grow to meet growing food demand (Thornton and Lipper 2014). Globally, agricultural emissions alone almost reach the full 2-degree target emission allowance in 2050 under business-as-usual scenarios. This indicates that the agriculture and food systems need a drastic change in order to meet future food needs sustainably (Bajzelj et al. 2014).

3.4 Conflict

Currently, 1.5 billion people in the world live in fragile, conflict-affected areas, and these people are nearly twice as likely to be malnourished than those in other developing countries. Food insecurity is often a direct result consequence of conflict. Conflict destroys agricultural assets and infrastructure, reduces food availability, and increases risk in accessing food markets due to destruction of physical infrastructure, which also drives up local food prices (Breisinger et al. 2014b). Studies have shown the impact of past conflicts on food security. The Rwandan genocide impacted child stunting, the exposure to war in Burundi decreased children's height-for-age, and even exposure to conflict in utero or during early life had negative implications for height-for-age (Breisinger et al. 2014a).

At the same time, food insecurity can fuel conflicts, especially in an environment of unstable political regimes, a youth bulge, slow economic development and growth, and inequality (Brinkman and Hendrix 2011). Rising food prices in particular have increased risk of political and social unrest. The 2007–2008 global food crisis led to riots in 28 countries, and food insecurity played a large role in sparking the Arab awakening (Maystadt et al. 2014). For example, Yemen has experienced various conflicts over the past decade. In addition to Al Qaeda-linked activities, the country suffered multiple economic shocks, including the 2007–2008 global food crisis and food price spikes in 2011, and has dealt with an influx of internally displaced people, all contributing to food insecurity and conflicts (Breisinger et al. 2014b).

3.5 Uncertainty in global political landscape

Despite commitments to sustainable development and food security in 2016, the outlook for the future global development landscape is uncertain, namely in the prospects for economic growth and changing global political paradigms. Economic prospects vary greatly across countries and regions. While Asia and other major emerging economies indicate robust growth, Africa, south of the Sahara, is experiencing a slowdown, which threatens to reverse gains there in poverty reduction and food security (IMF 2016). Furthermore, several sub-Saharan countries will undergo transitions in political leadership, and political uncertainties in Latin America will lead to uncertainties in economic and social stability. New political regimes in Asia, North America, and Western Europe are likely to have a shift in their approach to trade and development, as the changes in advanced economies adds further to uncertain domestic and global growth. Rising income inequality within countries in the context of rapid globalization also has uncertain implications for global trade and immigration (World Bank Group 2017).

4 Discussion: future strategies to fill the gaps

4.1 Innovations in technologies

As the world population continues to grow and face competing demands for natural resources in agriculture, urbanization, industry, and energy, food production growth to meet increasing demands will need to come from greater productivity than from land expansion. To achieve this, accelerated investments in agricultural research and development (R&D) on technologies to produce more with less will be crucial. Technologies should focus not only on increasing yield and productivity, but also on achieving multiple wins, including climate adaptation, GHG mitigation, and nutrition. Biofortification, the process of increasing vitamin and mineral density in crops through plant breeding, transgenic techniques, and agronomic practices, improves human health and nutrition in a cost-effective way that can reach underserved, rural populations (Bouis and Saltzman 2017). Studies also have found that improved crop varieties or types, such as drought-tolerant or heat-tolerant crops, can increase crop yields while reducing yield variability, increase climate resilience, and increase soil carbon storage. The use of crop covers increases soil fertility and yields due to nitrogen fixing in soils, improves water holding capacity, and has high mitigation potential from increased soil carbon sequestration (Bryan et al. 2011).

Research based on the IMPACT model projects that the food-insecure population in developing countries by 2050 could be reduced by 12 percent with successful development and adoption of nitrogen-use efficiency (NUE) technologies, by 9 percent with wider adoption of no-till technology, and by 8 percent if heat-tolerant varieties and precision agriculture (PA) are adopted more widely (Rosegrant et al. 2014).[4] It will be important to target technologies based on the needs and contexts of regions and countries. For example, heat-tolerant and drought-tolerant varieties in North America and South Asia; drought tolerance in Latin America and the Caribbean, the Middle East, North Africa, and sub-Saharan Africa; and greater crop protection in Eastern Europe, South Asia, and sub-Saharan Africa would be particularly beneficial. Furthermore, NUE is critical for resource efficiency in most developing regions, particularly in South Asia, East Asia and the Pacific, and sub-Saharan Africa. These technologies will also be important to address abiotic stress expected from climate change. To sustainably meet future needs under climate change, increases in crop productivity through enhanced investment in agricultural R&D, and resource-conserving management and increased investment in irrigation will be key (Rosegrant et al. 2014).

4.2 Innovations in policies

Fiscal policy can also contribute to multiple wins in meeting future food needs under climate change by ensuring that food prices reflect the full cost of production to the environment. Recent research using the IMPACT model in collaboration with Oxford University projected the potential impact of GHG taxation of food commodities, based on their contribution to GHG emissions, on climate and health outcomes. Average taxes in the model were highest for animal-sourced foods, such as beef, lamb, and pork and poultry, intermediate for vegetable oils, milk and eggs, and rice, while taxes were low on other crops such as fruits, vegetables, grains, and legumes. This resulted in increases in prices and reductions in consumption, which were high for beef, vegetable oils, milk, and lamb, and intermediate for poultry. Under a tax scenario that was optimized for each region to maximize health benefits, the analysis projected 510,000 avoided deaths due to changes in diet and reduced global GHG emissions by 8.6 percent. Levying such targeted GHG emission pricing to reflect the environmental impact of food commodities can promote health while mitigating climate change. Health benefits from reductions in obesity related to emissions-based taxes on foods, particularly animal-sourced foods, would outweigh potential health losses from increases in the numbers of people who are underweight (Springmann et al. 2017). This would also allow tax revenues to be redirected to support the production of more climate-smart, nutritious foods.

In addition, tailored policies and programs to populations uniquely impacted by current trends in food security and nutrition, namely rural small-scale farmers and the urban poor, are important. In order to enhance climate adaptation in smallholder agriculture, policies can promote land rights and efficient land markets (including new arrangements in land rental markets), and improve risk-management, mitigation, and adaptation strategies through insurance and information services. Social protection systems can build on successes, such as Ethiopia's Productive Safety Net Program, to provide both safety nets and agricultural support to help secure basic livelihoods while building resilience to shocks. At the same time, measures to increase the access of the urban poor to healthy and nutritious foods and to promote healthy choices will be important.

4.3 Innovations in institutions

Under rapid urbanization, institutions have a key role in encouraging inclusive value chains. Institutions can support the "quiet revolution" taking place in traditional value chains, as seen in the expansion and modernization of farms, mills, and markets in Asia's rice value chain

(Reardon et al. 2014). Institutions can also enhance vertical and horizontal coordination by promoting efficiency-building competition among different farming models such as cooperatives and family farms, and also by improving farm-to-market synchronization. Urban and rural policy makers can coordinate to support the flow of products into cities and also to fully harness the opportunities available from growing urbanization to integrate smallholders, traders, and others into the urban markets along the full food value chain.

Furthermore, institutions are crucial for climate-smart agriculture, particularly in serving as a center for information, innovation, investment, and insurance. To encourage adoption of climate-smart agricultural practices, local institutions will facilitate access to many resources and to information for stakeholders. Institutions providing insurance and related information for climate-smart agriculture will need to be more inclusive in incorporating local perspectives while tailoring the products and practices for smallholders. As many climate-smart agricultural practices will need to occur at a large spatial scale and over a longer time frame, institutions will be key for coordination and continued support through social safety nets. Additionally, institutions can promote partnerships for climate-smart adaption and support various climate-friendly financial arrangements.

4.4 Reforms in global and local governance

To achieve the end of hunger and malnutrition, along with the other Sustainable Development Goals (SDGs), global governance of development efforts must be reformed for greater stability of investments and the inclusion of developing countries. In particular, emerging economies, such as those of Brazil, China, and India, have the opportunity to play a greater role in global governance. These countries have significant potential to contribute to global food security by not only making progress in alleviating domestic hunger and malnutrition, but also by increasing trade, investments, and exchanges in technology. Particularly in the current global landscape of uncertainty around advanced economies' openness to trade and investment in developing countries, new entities led by the emerging economics, such as the New Development Bank (NDB), could significantly expand their role in the global food system.

In the globalized international arena, promoting a global trade regime that is open, transparent, and fair will continue to have important benefits for food security. Trade policies should seek to reduce transport and transaction costs and increase productivity. Harmful trade policies, such as import tariffs and export bans, can hurt the poor by impacting their ability to access or afford food and also hinder the efficiency of agricultural markets. Open, transparent, and fair trade can fill domestic gaps with appropriate imports and enable domestic production in countries to be geared toward what is most resource-efficient. Furthermore, trade can entail environmental benefits through avoided domestic environmental costs by importing crops and reducing environmental costs from food production based on domestic resource availability and efficiency (Martinez-Melendez and Bennett 2016). Trade can also facilitate the creation of global and regional grain reserves, especially in poor, food-importing countries, to help stabilize grain prices. Increasing the transfer of technology, technical assistance, and investments through South–South cooperation, including joint ventures, cooperation contracts, and public–private partnerships, could also be beneficial.

Global governance of food security and nutrition also needs enhanced engagement with multiple stakeholders, namely civil society and the private sector. Civil society organizations have been key in advocating for consumers and producers, especially smallholder farmers, while the private sector is a major player in the food processing industry. The Committee on World Food Security (CFS) is well placed to strengthen the involvement of multi-stakeholders by building on its current arrangements. The role of CFS can be further expanded to coordinate at the national and regional levels, promote accountability, and develop a global strategic framework. In this regard, it will be important to provide CFS the authority to adopt strategies and policy guidelines with appropriate accountability as well as enforcement mechanisms.

At the same time, local governance of both rural and urban areas will be just as important as urbanization continues. Horizontal coordination across sectors and vertical coordination among different tiers of government is needed. In particular, governance of the informal economy will have a growing importance in food security and nutrition. Urban households, particularly the urban poor, depend largely on the informal sector for food security as well as employment. Small-scale farmers in rural areas also benefit from the informal economy, as the low barrier to entry allows them to participate in the rural–urban agricultural value chain. Effectively incorporating the informal economy in national policies and governing the informal sector is an increasingly important challenge, as legislation in many African countries penalizes informal economy participants, and reported violence against members of the informal sector has been rising over the last two decades (Resnick 2017).

5 Conclusions

As the world faces challenges in the triple burden of malnutrition (undernutrition, food insecurity, overweight and obesity), urbanization, climate change, conflict, and uncertainty in the decades ahead, research will be critical to continue to provide insights into how to address these challenges and reshape the global food system to achieve multiple development goals.

Areas for future research

There is a need for more data and research to understand current opportunities and challenges to inform effective policy design and implementation. As much of the current data on food security are either outdated or incomplete, comprehensive, precise, and timely data on the state of poverty, food insecurity, and malnutrition are needed. Evidence on the factors that influence food choices, current nutrition gaps and dietary patterns, and various impacts of food environments will also be required. It will also be important for such data to be disaggregated (for example by gender, age, and income) to be able to better inform policies that are designed and tailored to particular needs.

Research on the enabling environments for improved food security and nutrition will also be necessary. Enabling environments for food involve social, policy, institutional, and spatial dimensions across individual, national, regional, and global levels. A better understanding of what drives malnutrition and food insecurity across this wide span, for example the impact of supermarkets on dietary choices, can aid in addressing the underlying drivers of the challenges in food security. Research in this regard will also need to go beyond the public sector to shed light on the important role of the private sector, particularly in creating incentives and improving the access to and affordability of healthy foods. Potential areas for the private sector to take a lead in addressing food security and nutrition requires further research.

Notes

1 Similar projections are found in the OECD–FAO Agricultural Outlook 2016–2025.
2 The scenarios used in the IMPACT model draw from the fifth assessment report by the Intergovernmental Panel on Climate Change (IPCC) and are defined by two major components: Shared Socioeconomic Pathways (SSPs), representing various socioeconomic challenges to climate mitigation and adaption, and Representative Concentration Pathways (RCPs), representing climates based on various greenhouse gas

emission levels. The projections in this chapter assume population and income changes based on SSP2 (middle-of-the-road scenario following historical trends) and emissions based on RCP8.5 (the most rapid climate change scenario).
3 Projections are based on SSP@, no climate change scenario.
4 Among a variety of high-tech and low-tech, traditional, conventional, and advanced practices with proven potential for yield improvement, a select group of technologies were evaluated under the IMPACT model: no-till, integrated soil fertility management (ISFM), precision agriculture (PA), organic agriculture (OA), nitrogen-use efficiency (NUE), water harvesting, drip irrigation, sprinkler irrigation, improved varieties (drought-tolerant characters), improved varieties (heat-tolerant characters), and crop protection.

References

Bajželj, B., K.S. Richards, J.M. Allwood, P. Smith, J.S. Dennis, E. Curmi, and C.A. Gilligan. 2014. "Importance of Food-Demand Management for Climate Mitigation." *Nature Climate Change* 4(10):924–929.

Bouis, H.E., and A. Saltzman. 2017. "Improving Nutrition Through Biofortification: A Review of Evidence from Harvest Plus, 2003 Through 2016." *Global Food Security* 12:49–58.

Breisinger, C., O. Ecker, J. Maystadt, J. Trinh Tan, P. Al-Riffai, K. Bouzar, A. Sma, and M. Abdelgadir. 2014a. *How to Build Resilience to Conflict: The Role of Food Security*. Washington, DC: International Food Policy Research Institute (IFPRI). doi:10.2499/9780896295667.

———. 2014b. "Food Security Policies for Building Resilience to Conflict." In S. Fan, R. Pandya-Lorch, and S. Yosef, eds., *Resilience for Food and Nutrition Security*. Washington, DC: International Food Policy Research Institute (IFPRI), pp. 37–44.

Brinkman, H-J., and C.S. Hendrix. 2011. *Food Insecurity and Violent Conflict: Causes, Consequences, and Addressing the Challenges*. Rome: World Food Programme (WFP).

Bruinsma, J. 2009. *The Resources Outlook to 2050 – By How Much Do Land, Water, and Crop Yields Need to Increase by 2050?* Rome: Food and Agriculture Organization of the United Nations (FAO).

Bryan, E., C. Ringler, B. Okoba, J. Koo, M. Herrero, and S. Silvestri. 2011. *Agricultural Management for Climate Change Adaptation, Greenhouse Gas Mitigation, and Agricultural Productivity*. Washington, DC: International Food Policy Research Institute (IFPRI).

Fan, S. 2017. "Food Security and Nutrition in an Urbanizing World." In *2017 Global Food Policy Report*. Washington, DC: International Food Policy Research Institute (IFPRI), pp. 6–13.

Fan, S., and E. Cousin. 2015. "Reaching the Missing Middle: Overcoming Hunger and Malnutrition in Middle Income Countries." In *2014–2015 Global Food Policy Report*. Washington, DC: International Food Policy Research Institute (IFPRI), pp. 13–17.

FAO (Food and Agriculture Organization of the United Nations). 2017. *FAOSTAT*. Rome: FAO. www.fao.org/faostat/. FAO, IFAD, UNICEF, WFP and WHO. 2017. *The State of Food Security and Nutrition in the World 2017. Building Resilience for Peace and Food Security*. Rome: FAO. Accessed 2 March 2017.

Gómez, M.I., C.B. Barrett, T. Raney, P. Pinstrup-Andersen, J. Meerman, A. Croppenstedt, B. Carisma, and B. Thompson. 2013. "Post green Revolution Food Systems and the Triple Burden of Malnutrition." *Food Policy* 42:129–138.

Haddad, L.J., M. Ag Bendech, K. Bhatia, K. Eriksen, I. Jallow, and N. Ledlie. 2016. "Africa's Progress toward Meeting Current Nutrition Targets." In N. Covic and S.L Hendriks, eds., *Achieving a Nutrition Revolution for Africa: The Road to Healthier Diets and Optimal Nutrition*. Washington, DC: International Food Policy Research Institute (IFPRI), pp. 12–27. doi:10.2499/9780896295933_03.

International Food Policy Research Institute (IFPRI). 2016. *Global Nutrition Report 2016: From Promise to Impact: Ending Malnutrition by 2030*. Washington, DC: International Food Policy Research Institute (IFPRI).

———. 2017. *2017 Global Food Policy Report, Annex*. Washington, DC: International Food Policy Research Institute (IFPRI).

International Monetary Fund (IMF). 2016. *World Economic Outlook October 2016: Subdued Demand: Symptoms and Remedies*. Washington, DC: IMF.

Knox, J., T. Hess, A. Daccache, and T. Wheeler. 2012. "Climate Change Impacts on Crop Productivity in Africa and South Asia." *Environmental Research Letters* 7(3):041001.

Martinez-Melendez, L.A., and E.M. Bennett. 2016. "Trade in the U.S. and Mexico Helps Reduce Environmental Costs of Agriculture." *Environmental Research Letters* 11(5):055004.

Maystadt, J-F., J-F. Trinh Tan, and C. Breisinger. 2014. "Does Food Security Matter for Transition in Arab Countries?" *Food Policy* 46:106–115.

Myers, S.S., A. Zanobetti, I. Kloog, P. Huybers, A.D. Leakey, A.J. Bloom, E. Carlisle, L.H. Dietterich, G. Fitzgerald, T. Hasegawa, and N.M Holbrook. 2014. "Increasing CO_2 Threatens Human Nutrition." *Nature* 510(7503):139–142.

OECD and FAO. 2014. "Overview of the OECD–FAO 2014 Outlook 2014–2023." In *OECD–FAO Agricultural Outlook 2014*. Paris: Organisation for Economic Co-operation and Development (OECD), pp. 21–61. doi:10.1787/agr_outlook-2014-4-en.

Paciorek, C.J., G.A. Stevens, M.M. Finucane, and M. Ezzati. 2013. "Children's Height and Weight in Rural and Urban Populations in Low-Income and Middle-Income Countries: A Systematic Analysis of Population-Representative Data." *Lancet Global Health* 1:e300–e309. www.ncbi.nlm.nih.gov/pubmed/25104494. Accessed on 2 March 2017.

Reardon, T., K.Z. Chen, B. Minten, L. Adriano, T.A. Dao, J. Wang, and S.D. Gupta. 2014. "The Quiet Revolution in Asia's Rice Value Chains." *Annals of the New York Academy of Sciences* 1331(1):106–118.

Resnick, D. 2017. "Informal Food Markets in Africa's Cities." In *2017 Global Food Policy Report*. Washington, DC: International Food Policy Research Institute (IFPRI), pp. 50–57.

Robinson, S., D. Mason-D'Croz, S. Islam, N. Cenacchi, B. Creamer, A. Gueneau, G. Hareau, U. Kleinwechter, K.A. Mottaleb, S. Nedumaran, and R. Robertson. 2015a. *Climate Change Adaptation in Agriculture: Ex Ante Analysis of Promising and Alternative Crop Technologies Using DSSAT and IMPACT*. Washington, DC: International Food Policy Research Institute (IFPRI).

Robinson, S., D. Mason-D'Croz, T. Sulser, S. Islam, R. Robertson, T. Zhu, A. Gueneau, G. Pitois, and M.W. Rosegrant. 2015b. *The International Model for Policy Analysis of Agricultural Commodities and Trade (IMPACT): Model Description for Version 3*. Washington, DC: International Food Policy Research Institute (IFPRI).

Roever, S. 2014. *Informal Economy Monitoring Study Sector Report: Street Vendors*. Cambridge, MA: Women in Informal Employment Globalizing and Organizing (WIEGO).

Rosegrant, M.W., J. Koo, N. Cenacchi, C. Ringler, R.D. Robertson, M. Fisher, C.M. Cox, K. Garrett, N.D Perez, and P. Sabbagh. 2014. *Food Security in a World of Natural Resource Scarcity: The Role of Agricultural Technologies*. Washington, DC: International Food Policy Research Institute (IFPRI).

Rosegrant, M.W., T.B. Sulser, D. Mason-D'Croz, N. Cenacchi, A. Nin-Pratt, S. Dunston, T. Zhu, C. Ringler, K. Wiebe, S. Robinson, D. Willenbockel, H. Xie, H.-Y. Kwon, T. Johnson, F. Wimmer, R. Schaldach, G.C. Nelson, B. Willaarts, and others. 2017. "Quantitative Foresight Modeling to Inform the CGIAR Research Portfolio." International Food Policy Research Institute (IFPRI), Washington, DC, pp. 14–17. www.ifpri.org/publication/foresight-modeling-agricultural-research.

Ruel, M., J. Garrett, and S. Yosef. 2017. "Growing Cities, New Challenges." In *2017 Global Food Policy Report*. Washington, DC: International Food Policy Research Institute (IFPRI), pp. 24–33.

Smith, P., D. Martino, Z., Cai, D. Gwary, H. Janzen, P. Kumar, B. McCarl, S. Ogle, F. O'Mara, C. Rice, B. Scholes, O. Sirotenko, M. Howden, T. McAllister, G. Pan, V. Romanenkov, U. Schneider, S. Towprayoon, M. Wattenbach, J. Smith. 2008. "Greenhouse Gas Mitigation in Agriculture." *Philosophical Transactions of the Royal Society B: Biological Sciences* 363(1492):789–813.

Springmann, M., D. Mason-D'Croz, S. Robinson, K. Wiebe, H.C.J. Godfray, M. Rayner, and P. Scarborough. 2017. "Mitigation Potential and Global Health Impacts from Emissions Pricing of Food Commodities." *Nature Climate Change* 7:69–74.

Stein, A.J., and M. Qaim. 2007. "The Human and Economic Cost of Hidden Hunger." *Food and Nutrition Bulletin* 28(2):125–134.

Thornton, P.K., and L. Lipper. 2014. *How Does Climate Change Alter Agricultural Strategies to Support Food Security?* Washington, DC: International Food Policy Research Institute (IFPRI).

UNICEF, WHO, and World Bank. 2016. *Joint Child Malnutrition Estimates, September 2016 Edition*. Washington, DC: World Health Organization (WHO).

United Nations. 2014. *World Urbanization Prospects: The 2014 Revision*. New York: United Nations.

Veolia Water North America and the International Food Policy Research Institute (IFPRI). 2013. *Finding the Blue Path for a Sustainable Economy*. Chicago, IL and Washington, DC: VWNA and IFPI. http://icma.org/Documents/Document/Document/302537. Accessed on 2 March 2017.

von Grebmer, K., A. Saltzman, E. Birol, D. Wiesmann, N. Prasai, S. Yin, Y. Yohannes, P. Menon, J. Thompson, and A. Sonntag. 2014. *2014 Global Hunger Index: The Challenge of Hidden Hunger*. Bonn, Washington, DC, and Dublin: Welthungerhilfe, International Food Policy Research Institute, and Concern Worldwide. doi:10.2499/9780896299580.

Weatherspoon, D., and T. Reardon. 2003. "The Rise of Supermarkets in Africa: Implications for Agrifood Systems and the Rural Poor." *Development Policy Review* 21(3):333–355.

Wheeler, T., and J. von Braun. 2013. "Climate Change Impacts on Global Food Security." *Science* 341(6145):508–513.

World Bank Group. 2017. *Global Economic Prospects: Weak Investment in Uncertain Times*. Washington, DC: World Bank.

Zezza, A., and L. Tasciotti. 2010. "Urban Agriculture, Poverty, and Food Security: Empirical Evidence from a Sample of Developing Countries." *Food Policy* 35:265–273.

4
Industrial organization of the food industry
The role of buyer power

Ian Sheldon

1 Introduction

The food industry consists of a system of vertically interrelated markets: agricultural producers sell raw agricultural commodities downstream to food processors, who in turn sell food products to the food retailing sector, followed by the latter marketing final food products and other retail services to consumers (Figure 4.1). Industrial organization in the food industry is the study of its market structure and the competitive relationships within the system. The literature can be broken down into three distinct phases. First, stimulated by Clodius and Mueller (1961), early applied industrial organization studies focused predominantly on describing and analyzing the consolidation of specific stages of the food industry through application of the structure–conduct–performance (SCP) approach (Connor et al. 1985; Marion 1986). Second, starting with Just and Chern (1980), researchers have adopted methods from the so-called new empirical industrial organization (NEIO) to estimate the extent of market power in the food industry (Sheldon and Sperling 2003). Third, since the publication of Sexton and Lavoie's (2001) earlier handbook chapter, and a review article by McCorriston (2002), there has been a distinct shift away from analyzing market structure and competition between firms at a specific stage (horizontal) of the food industry to examining coordination between different stages (vertical) of the food industry, and its implications for economic welfare and anti-trust policy (Sheldon 2017).

In the case of U.S.-oriented research, the focus has moved to careful analysis of the link between processors and farmers, emphasizing a specific issue: the possibility that concerns about food processor buyer power are overstated because orthodox models fail to capture the economic logic for observed vertical coordination between suppliers of raw agricultural products and food processors (Sexton 2013). Contemporaneously, in EU-oriented research, the emphasis has been on vertical coordination between food processors and retailers in the EU food industry, given concerns about consolidation in the retailing sector and its potential impact on price transmission (McCorriston 2014).

This shift in emphasis has been driven by a key conceptual problem: NEIO analysis of the food industry has found little empirical evidence for exploitation of horizontal market power in either the food processing or retailing sectors. This may of course be due to technical issues with the methodology itself or the lack of necessary data, but it is more likely due to the fact

Industrial organization of the food industry

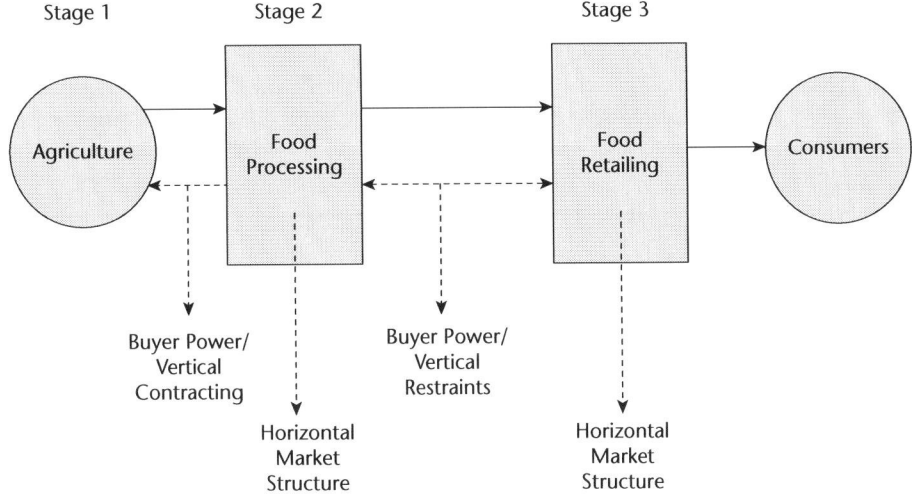

Figure 4.1 Market structure: the food industry

that vertical coordination between downstream food processors and suppliers of raw agricultural commodities is often characterized by extensive use of contracts designed to internalize the inefficiencies of using spot markets, and vertical coordination between upstream food processors and downstream food retailers through vertical restraints such as slotting allowances designed to resolve the problem of double-marginalization created by successive oligopoly. Neither characteristic is adequately captured in typical NEIO models focusing on single homogenous products transformed/sold through a series of vertically interrelated stages; the approach ignores the impact of multi-product/multi-brand food processing and food retailing on vertical coordination, and the potential feedback to suppliers of agricultural inputs.

2 Market structure of the food industry

Although it is well understood that high levels of seller concentration do not necessarily imply abuse of market power (Tirole 2015), either within any given stage or in coordination between stages, it is still useful to obtain a sense of the current market structure of the food industry in developed countries such as the United States (U.S.) and the European Union (EU). Apart from being a set of stylized facts in virtually all applied work in the field, the description of market structure has been very central to the public debate concerning consolidation in the U.S. and EU food industries, such as the United States Department of Justice (2012), the United Kingdom Competition Commission (2008), and the Organisation for Economic Co-operation and Development [OECD] (2014).

2.1 Food processing

Crespi, Saitone, and Sexton (2012) provide a detailed description of the market structure in U.S. food processing, drawing on data from the 2007 Census of Manufacturers, which measures the degree of seller concentration by the share of the largest four firms in a specific industry's sales (CR4). For a sample of food processing industries defined at the six-digit level of the North American Industrial Classification System (NAICS), high levels of seller concentration can be found in several industries, including dog and cat food manufacturing (71 percent), wet corn milling (84 percent), soybean processing (82 percent), breakfast cereal manufacturing (80 percent), and cane sugar refining (95 percent), which compare with an average CR4 of 50 percent across a sample of 47 NAICS industries. In addition, a majority of these 47 industries have exhibited an increase in seller concentration over a ten-year period since 1997.

Interestingly, the CR4s for animal slaughtering and for poultry slaughtering were 59 and 46 percent, respectively, which do not deviate that much from the average for food processing as a whole. However, these indices of market structure hide the extent to which these particular industries are characterized by seller concentration in the purchasing of raw agricultural inputs. Using disaggregated data on seller concentration from the United States Department of Agriculture (USDA) Grain Inspection, Packers, and Stockyards Administration (GIPSA), Crespi, Saitone, and Sexton (2012) find that for selected meat packing industries, the CR4 in 2007 was 80, 70, 65, 57 and 51 percent, respectively, for steers and heifers, sheep and lambs, hogs, broilers, and turkeys, and over the 1980 to 2010 period, the CR4 in these meat packing industries increased on average by 69 percent, albeit the rate of increase slowing after 1995 as the rate of mergers in these industries decelerated.

Data on the market structure of food processing in the European Union are less recent in terms of the period of time, with McCorriston (2014) reporting the same data by Cotterill (1999) from the mid-1990s. Average three-firm seller concentration ratios (CR3) in the food processing sector are reported for a sample of countries in the European Union, with the average CR3 across these countries being 64 percent. This suggests that even in the mid-1990s there were high levels of seller concentration in the EU food processing sector, although the level varied across countries. As McCorriston (2014) points out, there was also considerable variation in seller concentration across specific industries. For example, the CR3 for breakfast cereals ranged from 92 percent in Ireland to 65 percent in the United Kingdom, while baby food production ranged from 98 percent in Ireland to 54 percent in Spain. Even though these figures are rather dated, they do confirm that the EU food processing sector was already highly concentrated by the mid-1990s, and seller concentration is unlikely to have fallen given the extent of domestic and EU cross-border merger activity in the sector over the past 20 years (McCorriston 2014).

2.2 Food retailing

Leading food retailers in the United States and the European Union have become dominant in terms of market share at the consumer end of their food industries. In the case of the United States, Richards and Pofahl (2010) report a CR4 for food retailing of 48.8 percent in 2008, the largest four retailers consisting of Walmart (21.3 percent), Kroger (12.6 percent), Safeway (8.1 percent), and Costco (6.8 percent), with concentration having increased from a CR4 of 16.8 percent in 1992 (Sexton 2013). Of course, the United States as a whole is not really the appropriate market definition for food retailing, relevant markets being much more localized given the spatial distribution of consumers and associated transport costs (Sexton 2013). As a consequence, seller concentration levels in food retailing are much higher at the city level, average food retailing CR4s being 79 percent in 2006 for 229 metropolitan statistical areas (Sexton 2013).

There have been several significant changes in the U.S. food retailing landscape: in the 1980s regional and local supermarket chains were dominant, while in the 1990s large grocery retail chains merged or bought out other regional food retailers as large warehouse clubs and large discount general merchandise stores expanded into food retailing. Between 1996 and 1999, there were 385 mergers in the sector as incumbent firms were forced to compete with supercenter/discounter firms such as Walmart and Costco (Richards and Patterson 2003). These developments have not been driven by economies of scale in food retailing (Richards and Hamilton 2013). Instead, Ellickson (2007) argues that supermarkets have had to compete for customers through offering a wider variety of products in their stores, requiring fixed investment in distribution. In order to gain a larger market share, more stores have to be built, escalating fixed costs and discouraging the entry of other firms. Using store-level data for 1998, Ellickson (2007) finds that U.S. food retailing has a two-tiered structure: a small number of firms capturing the majority of sales, competing with an expanding fringe of stores offering a narrower variety of products.

Recent evidence suggests that U.S. food retailing may be undergoing further structural change. Traditional food retailers are finding it harder to follow a strategy of product variety expansion, given competitive pressure on their margins from discounters such as Walmart and Aldi (Hausmann and Leibtag 2007; Basker and Noel 2009) and the increasing incentive to focus on a narrower range of high-quality fresh unpackaged foods. In addition, online food retailing is expected to grow more rapidly compared to traditional food retailing (Richards, Hamilton, and Empen 2015).

In the case of the European Union, the average level of seller concentration in food retailing in 2004–2005 was very similar to the national level of concentration in the United States, with an average 5-firm sales concentration ratio (CR5) across the EU-15 being just above 50 percent. However, as reported by Buceviciute, Dierx, and Ilzcovitz (2009), levels of seller concentration vary considerably around this average; take, for example, the CR5s for Italy (32 percent), France (52 percent), the United Kingdom (60 percent), Denmark (75 percent), Sweden (76 percent), and Finland (84 percent). In contrast, newer EU members from Eastern Europe typically have lower levels of seller concentration in food retailing, including CR5s for Bulgaria (14 percent), Poland (22 percent), and Romania (21 percent), although these levels of seller concentration have grown significantly in a short period of time (McCorriston 2014). While merger activity has been significantly greater on average in the EU food processing sector over the past two decades, like the United States, there was a spike in food retail mergers in the mid to late-1990s (McCorriston 2014).

3 "New" industrial organization analysis of the food industry

3.1 Basic model

A useful starting point for analyzing the industrial organization of the food industry is the model developed by Sexton (2000), Sexton and Lavoie (2001), Sexton and Zhang (2001), and Sexton et al. (2007). The model consists of a vertical structure where a processing sector purchases raw product from an upstream agricultural sector at the price p^f, and then, undertaking some processing activities, sells a homogeneous food product downstream to a retailing sector at the wholesale price of p^w. In turn, the retailing sector sells the food product to consumers at the retail price of p^r. All transactions occur at arms' length through linear pricing.

Inverse demand at retail for the food product is defined as $p^r = D(Q)$, where Q is the total amount of the food product sold at retail. The upstream agricultural sector is assumed to be perfectly competitive, the inverse supply curve of the raw agricultural product being given as $p^f = S(Q^f)$, where Q^f is the amount of raw product.

In terms of technology, both food processing and retailing are assumed to produce via a one-to-one fixed proportions technology where $Q^f = Q^w = Q^r = Q$ (Sexton and Lavoie 2001). A representative food processor purchasing q^f of the raw agricultural product incurs variable costs $C^w = p^f q^f + c^w q^f$, where c^w are constant per unit processing costs. A representative food retailer purchasing q^w of the processed food product incurs variable costs $C^r = p^w q^w + c^r q^w$, where c^r are constant per unit food retailing costs.

A retailing firm's first-order condition for profit maximization is

$$p^r\left(1 - \frac{\lambda^r}{\eta^r}\right) = \left(p^w + c^r\right). \tag{1}$$

In equilibrium, mark-up of the retail price p^r over marginal retailing cost (the Lerner index) is conditioned on a conjectural elasticity, λ^r, capturing the extent of retailer exploitation of oligopoly market power, and the absolute value of the price-elasticity of demand at retail, η^r. The index of retail market power, $\lambda^r \in [0,1]$, ranges from perfect competition to monopoly.

Given the inverse derived demand function for the food processor at wholesale, $p^w = D(Q \mid \lambda^r, c^r)$, a processing firm's first-order condition for profit maximization is

$$p^w\left(1 - \frac{\lambda^w}{\eta^w}\right) = p^f\left(1 + \frac{\theta^f}{\varepsilon^f}\right) + c^w. \tag{2}$$

In equilibrium, mark-up of the wholesale price, p^w, over marginal wholesaling cost is conditioned on the wholesale market conjectural elasticity, λ^w, capturing the extent of any processor exploitation of oligopoly market power, and the price-elasticity of derived demand for the processed good, η^w, while mark-down of the raw product price, p^f, below its marginal value product is conditioned on a conjectural elasticity capturing processor exploitation of oligopsony power, θ^f, and the price-elasticity of supply from the agricultural sector, ε^f. The indices of food processor market power, $\theta^f \in [0,1]$ ($\lambda^w \in [0,1]$), range from competition to monopsony (monopoly) in the raw agricultural input (processed food) market(s).

The case of processor–oligopsony and successive processor–retailer oligopoly is illustrated in Figure 4.2. To keep the analysis simple, processing and retailing costs have been normalized, $c^w = 0$, $c^r = 0$; $p^f(Q)$ is linear inverse farm supply, PMC^w is the perceived marginal cost of the raw product at wholesale, $p^r(Q)$ is linear inverse demand at retail, and PMR^r and PMR^w are perceived marginal revenue at retail and wholesale, respectively. Equilibrium output under this market structure is Q_o, and p_1^f, p_1^w, p_1^r are the farm-gate, wholesale, and retail prices respectively, with the associated deadweight loss from oligopsony and successive oligopoly of areas ABD and DEF, respectively. This compares to the perfectly competitive output of Q_c and prices $p^c = p_2^f = p_2^w = p_2^r$.

3.2 Empirical analysis

The model outlined is subject to two well-merited criticisms in the context of vertical market structure: (1) transactions between stages of the food industry occur at arms' length, ruling out principal–agent-type contracts to resolve vertical inefficiencies, and (2) the relative bargaining power of processors and retailers is not modeled. Notwithstanding these criticisms, the model has the advantage that it can be used for empirical analysis, assuming estimates of the price and conjectural elasticities are available (Sexton 2000; Sexton et al. 2007). In particular, the basic structure is very similar to what formed the modeling basis of applied research of industry behavior in the 1980s under the rubric of the NEIO, characterized by efforts to estimate the extent of imperfect competition in single industries.

Industrial organization of the food industry

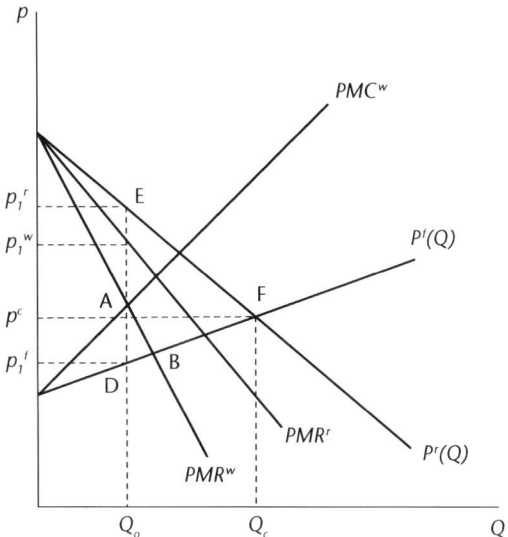

Figure 4.2 Vertical market structure

The NEIO grew out of dissatisfaction with the SCP paradigm pioneered by Bain (1951), whereby a one-way causal relationship was posited from market structure (seller concentration) to conduct to performance in a specific industry. Typically in cross-sectional empirical research, average accounting profits in a specific industry were regressed on seller concentration in that industry, a positive coefficient on market structure being taken to mean that higher seller concentration facilitated collusion among firms (Geroski 1988).

Up to the mid-1980s, most studies of the food industry typically followed the SCP approach (Connor et al. 1985), a positive correlation being found consistently between seller concentration and industry profits (Sexton and Lavoie 2001). However, starting with Just and Chern (1980) (tomato processing), and subsequently Schroeter (1988) (beef packing), emphasis began to shift to NEIO-type studies drawing on the methodologies for estimating market conduct described in articles by Appelbaum (1982) and Bresnahan (1982).

Development of the NEIO methodology grew out of significant criticisms of the SCP approach (Schmalensee 1989): Demsetz (1973) questioned the view that profits in concentrated industries reflect collusive behavior as opposed to differential efficiency, while Fisher and McGowan (1983) criticized SCP studies for using accounting measures of profitability to infer market power. Perhaps most seriously, Clarke and Davies (1982), in extending Cowling and Waterson's (1976) efforts to provide a theoretical underpinning to SCP models, showed that all variables in such models were logically endogenous, making it impossible to infer any direction of causality between market structure and performance. In contrast, the NEIO drew on models of imperfectly competitive profit-maximizing firms to guide specification, estimation, and testing of structural time series econometric models of industry behavior (Bresnahan 1989).

In detailed surveys, Sexton (2000), Sexton and Lavoie (2001), and Sheldon and Sperling (2003) made the following observations concerning extant NEIO studies in the agricultural economics literature: (1) the majority of studies concerned the U.S. food processing sector;

(2) the estimates of market power varied widely across industries, conjectural elasticities for food processing typically being 0.25 or less; (3) the value of the Lerner index depended not only on the value of the conjectural elasticity parameter, but also on the estimated price elasticity; (4) the U.S. meat marketing system had been subject to the most analysis; and (5) analysis of food retailing had received virtually no attention in terms of this methodology.

As noted by Sexton (2000), Sexton and Lavoie (2001), Sheldon and Sperling (2003), Kaiser and Suzuki (2006), and Perloff, Karp, and Golan (2007), key issues arose with the use of the NEIO methodology. These issues included, *inter alia*, the lack of a dynamic oligopoly framework in the majority of studies; the modeling of the food processing technology (fixed vs. variable proportions); poorly defined product markets; specific *ex ante* choices of functional forms for the demand and supply functions and processing technology; and a failure to account for economies of scale.

These and other criticisms of the NEIO methodology may partially explain the relative decline in its popularity among applied industrial organization (IO) researchers examining the food industry. However, as noted in a recent survey by Sheldon (2017), of the NEIO-type studies undertaken since 2000, several have made efforts to address some of these issues, while others have adopted new or adapted NEIO-type methodologies. Nevertheless, with a few exceptions, the majority of the studies have found only modest departures from perfect competition in the industries studied.

In their earlier survey, Sheldon and Sperling (2003) noted that virtually all NEIO models failed to account for the multi-product nature of food processing and retailing. At the time of writing, Nevo's (1998, 2001) methodology was an exception in focusing on a differentiated food product (breakfast cereal). In the intervening period, Nevo's (1998) analytical approach, and variations on it, has become the "workhorse" for examining horizontal competition in food processing and retailing and also vertical relationships between food processors and retailers. While discussion of the latter is delayed until later in this chapter, it is useful at this point to describe the approach and its early application to food processing.

The NEIO methodology pioneered by Applebaum (1982) and Bresnahan (1982) focused on homogeneous product industries, extension to differentiated products being limited by the difficulty of estimating demand and then solving out for the conjectural elasticities. To get around this problem, Nevo (1998) proposed the following approach. First, aggregate demand for varieties of a differentiated product is estimated using the discrete demand methodology suggested by Berry, Levinsohn, and Pakes (1995). Second, assuming a Bertrand–Nash equilibrium in prices, the profit-maximizing conditions of the firms selling these varieties are derived, implying price–cost margins for each variety. Third, a pricing equation is derived where price is a function of marginal cost and the markup. Fourth, using estimates of the parameters for the demand equation, price–cost margins are calculated without observing marginal costs. Separating the margins into three specific sources (single-product differentiation, multi-product firm pricing, and potential price collusion), Nevo (2001) applied this methodology to the U.S. ready-to-eat breakfast cereals industry, his results indicating that despite high price–cost margins, prices in the industry were consistent with multi-product firms playing a Bertrand–Nash game in prices rather than colluding over price.

What is important about this methodology is that it provides a robust framework for analyzing the behavior of multi-product firms in the current food industry and is likely a better approach than the typical NEIO model historically applied to food processing. Nevertheless, it is perhaps not unreasonable for many earlier NEIO studies to have focused on estimating market power in a homogeneous product setting such as beef and poultry processing.

4 Vertical market control

A typical vertical market structure is successive oligopoly between two stages, upstream processing and downstream retailing (Figure 4.2), with the potential for a vertical externality (Tirole 1988). The best known example of such an externality is the problem of double-marginalization: at the processing stage, processors mark up their wholesale price p_1^w over the price of the raw input p_1^f, which is passed through at arms' length under linear pricing to retailers who then mark up the retail price p_1^r over the wholesale price p_1^w. The vertical externality arises because retailers fail to take account of the processors' marginal profits when setting the retail price, resulting in downstream consumption of the product produced upstream being too low at Q_o and aggregate profit of the vertical system being lower than it would be under vertical integration.

Alternatively, it is argued that the processors have an incentive to impose a vertical restraint on retailers, thereby eliminating the vertical externality (Katz 1989). The standard example of such a restraint is nonlinear pricing, specifically a two-part tariff; that is, processors implement a contract consisting of a wholesale price equal to their marginal costs p_1^f and a franchise fee g equal to the vertical system's profit. Such a contract allows processors to realize the profits of downstream vertical integration by making the retailing sector the residual claimant on any marginal profit in the system; that is, they are given the correct incentive to set the oligopolistic price at retail (in other words, processors are the principal and retailers the agent). Alternatively, processors are able to charge the wholesale oligopoly price p_1^w, and impose resale-price maintenance, whereby the retail price p_2^r is set equal to p_1^w, with processors capturing all the vertical profits through g. Elimination of double marginalization is unambiguously welfare increasing, with retail consumers facing a lower price, p_2^r.

This conventional approach to analyzing vertical market control focuses on processors as the principal and retailers as the agent, ignoring the potential for each to exercise buyer power over their upstream suppliers (Dobson and Waterson 1997). Two types of downstream market power can be identified, contingent on whether or not upstream suppliers have market power (Chen 2007). If there is competition among suppliers upstream, the normal selling price would be competitive, and downstream buyer power is because of oligopsony. On the other hand, if the upstream market is oligopolistic, and the normal selling price would be in excess of the competitive price, downstream buyer power arises from the exercise of countervailing power.

5 Processor buyer power

Focusing on the first definition of buyer power: if this were a true characterization of the relationship between suppliers of raw agricultural commodities and downstream food processors, market transactions would all be through spot markets, and there should be some empirical evidence for the exertion of oligopsony power. As already noted, NEIO studies have found only modest departures from perfectly competitive pricing, and in the case of the U.S. meat packing industry, there is considerable empirical support for the view that the actual buying behavior of downstream food processors is driven not by exertion of market power, but rather through realization of economies of scale and the need to operate processing plants at full capacity (MacDonald and Ollinger 2000). In addition, food processing firms are meeting increased demands for quality from food retailers, who in turn are responding to consumers' willingness to pay for a broad range of attributes in the foods they consume (Sexton 2013). As a consequence, food processors have an incentive to enter into vertical contracts that ensure sufficient plant throughput of uniform and high quality inputs, thereby maintaining processing plant-level profitability (MacDonald and McBride 2009).

The latter point has been extensively documented in the literature – since the 1980s, there has been a significant decline in the use of spot markets across a range of agricultural commodities and a significant increase in vertical coordination through contractual arrangements between upstream suppliers of raw agricultural commodities and downstream food processing. MacDonald (2015), using USDA's Agricultural Resource Management (ARMS) dataset for 2013, reports that by 2011 40 percent of the value of U.S. agricultural commodity production was governed by contracts compared to 28 percent in 1991, although this share did fall to 35 percent by 2013. The extent of contracting does vary quite a bit across commodity-type, with MacDonald (2015) reporting that by 2013, the share of production under contract ranged from 22 percent for all crops to 52 percent for all livestock, but within all crops, the share varied from 13 percent for wheat to 57 percent for peanuts, and within all livestock, the share varied from 32 percent for cattle to 74 and 84 percent for hogs and poultry, respectively.

Much of the public concern about contracting has involved the livestock sector, where vertical coordination is typically conducted through production contracts. For example, in the broiler industry, producers sink costs into specialized housing and equipment and provide labor, while poultry processing firms provide key inputs such as chicks, feed, veterinary services, and management guidance (Knoeber and Thurman 1995). Producers deliver finished birds to processing plants, and they are paid on the basis of how well they performed in transforming chicks and feed into broiler meat. MacDonald (2015) reports that in 2013 such contracts accounted for US$58 billion worth of U.S. agricultural production, with US$48 billion in poultry and hogs.

Importantly, production contracts typically bind upstream producers to either a specific or limited number of downstream processors for multiple production periods due to investment in specific assets such as production units and because of scale economies in downstream processing. As Wu (2006) notes, relationship-specific investments in upstream production facilities create quasi-rents, the difference between the profit a producer can make within a contractual relationship, and the next best use of those assets. Bargaining over and appropriation of quasi-rents by downstream food processors, and, hence, the risk to the upstream supplier of a "hold-up," is at the heart of concerns over buyer power, which should be treated as distinct from the exertion of textbook oligopsony power (MacDonald 2015).

Crespi, Saitone, and Sexton (2012) and Sexton (2013) argue strongly that downstream processors do not have an incentive to exert oligopsony power over suppliers of raw agricultural commodities. This follows from the fact that the processors have themselves sunk investment into large-scale, location-specific processing plants committed to supplying a specific product to the downstream food retailing sector. As a consequence, driving their input prices below their marginal value product is short-sighted, generating not only the traditional deadweight loss triangles, but also running the risk of pushing upstream suppliers out of business as the rate of return on their specific investment falls below the competitive level. Downstream food processors will in fact seek contracts in order to guarantee a stable supply of high-quality raw agricultural commodities and minimize transactions costs. As Crespi, Saitone, and Sexton (2012) indicate, the connection between the transaction cost–benefits of vertical coordination and firm/relationship-specific assets was originally made by Williamson (1986) and has been noted in the agricultural context by several researchers, including, *inter alia*, Goodhue (2000) and Wu (2006, 2014).

Following Crespi, Saitone, and Sexton (2012), suppose the per unit surplus to a transaction between a downstream food processor i and an upstream supplier j is $S_{ij} = p_i^w - c_i - c_j - t_{ij}$, where p_i^w is the wholesale price charged to downstream food retailers, c_i and c_j are downstream and upstream variable production costs respectively, and t_{ij} are the downstream processing firm's transactions costs, which include contract monitoring, enforcement, and other agency costs that cannot be contracted on. Given this, the downstream processor will offer the upstream supplier a nonlinear price consisting of, $p_{ij}^u = c_j + \delta S_{ij}$, that is, a two-part tariff that covers the upstream supplier's variable costs plus a share of the vertical market surplus $0 \leq \delta \leq 1$. As Sexton (2013) notes, if upstream suppliers sell to competitive downstream firms, $\delta = 1$, in which case upstream firms also earn a return on their fixed investment c_j^f. In principle, in the contracting case, the downstream processor only has to offer a contract that covers the upstream variable costs c_j, with $\delta = 0$, assuming that the upstream supplier has no alternative outlet. This has the structure of a standard principal–agent problem, where the downstream food processor (the principal) has the ability to make take-it-or-leave-it contracts to the upstream supplier of the raw agricultural commodity (the agent). Essentially, the contract as described satisfies the agent's participation constraint.

In a dynamic setting, such a contract is likely to be inefficient due to the fact that if the upstream firms make no return on their fixed investment c_j^f in the long run, they will be forced to exit the industry. In turn, this would result in suboptimal use of capacity by the downstream food processor or the transaction costs of seeking alternative suppliers of high-quality raw agricultural commodities. Crespi, Saitone, and Sexton (2012) and Sexton (2013) argue that as a consequence, in a long-run setting, upstream suppliers will be offered a portion of the available vertical surplus, $0 < \delta^* \leq 1$, such that $p_{ij}^u \geq c_j^f$. In other words, downstream processing firms have an incentive to internalize the vertical externality that would be generated by exploitation of their buyer power due to oligopsony.

Wu (2006, 2014) suggests that this characterization of long-run vertical coordination is rooted in the notion of relational contracts (Levin 2003). Wu (2014) defines relational contracts as incomplete contracts that govern contract performance via informal incentives, which are self-enforced via a repeated game. In a simple contracting setting, suppose an upstream supplier j and downstream processor i sign a contract to trade a unit of the raw agricultural commodity, and during the production stage, the upstream supplier can choose to either invest or not invest in some action that will ensure a higher quality of the agricultural commodity and thereby raise the wholesale price, p_i^w, the downstream food processor can charge. Based on this, the downstream processor can choose either to pay or not pay the upstream supplier a bonus. Assuming the investment and bonus actions are observable but not contractible, in a one-shot game the equilibrium is one where the upstream supplier does not invest and the downstream processor offers no bonus.

If this game is repeated indefinitely, however, and there is a sufficiently high discount rate, the upstream supplier has an incentive to invest as the downstream food processor will credibly promise to pay bonus payments; that is, a relational contract is established. Wu (2006) also shows that in this setting the downstream food processor has an incentive to invest in relationship-specific assets to increase their payoff at wholesale. This is a relationship-specific asset due to the fact that the additional payoff to the downstream food processor only exists when it cooperates with the existing upstream supplier. Of course, parties to such a contract are not wholly immune to opportunism; for example, the downstream processing firm may heavily discount the future, reverting to the short-run optimum of not offering a bonus. Crespi, Saitone, and Sexton (2012) argue that this will only occur in cases of extreme buyer financial distress; that is, their discount rate would be low. However, typical models of contracting, by assuming the principal offers a take-it-or-leave-it contract, ignore any potential for bargaining in the contracting process.

6 Retailer buyer power

6.1 Countervailing power

Countervailing buying power, long ignored since the term was introduced into the economics literature by Galbraith (1952), has generated considerable interest among economists since the mid-1990s. The key argument is that if a small number of large retailers extract lower wholesale prices through countervailing power, the benefits of lower retail prices will outweigh any negative effect of increased consolidation in retailing. However, this result depends quite crucially on the nature of competition downstream and contracts between upstream and downstream firms.

Both von Ungern-Sternberg (1996) and Dobson and Waterson (1997) assume a structure where a single upstream firm sells a homogenous good to downstream retailers in the context of a two-stage game: at the first stage, wholesale prices are the outcome of a Nash bargaining game between the upstream processor with each retailer; at the second stage, retail prices are solved as a Nash equilibrium. Their results show that retailer consolidation may have two opposing effects on prices downstream: on the one hand, it results in an increase in their bargaining power, driving down the price that the upstream firm can charge, while on the other hand, consolidation downstream may result in firms increasing their mark-ups over marginal cost, pushing up downstream prices. Which effect dominates depends on how intense competition is at the downstream stage. von Ungern-Sternberg (1996) finds that if downstream firms play Cournot–Nash, retail prices increase, while Dobson and Waterson (1997) show that if retailers play Bertrand–Nash, retail prices will fall if other retail services are perfect substitutes.

Chen (2003) assumes a vertical market structure of a single upstream firm selling a homogeneous good to a dominant downstream firm competing with a competitive fringe, with wholesale prices set via nonlinear contracts, and joint profits shared through Nash bargaining. As the bargaining power of the dominant retailer increases, the upstream firm reacts to its declining share of these joint profits by lowering the wholesale price it charges fringe retailers in order to boost sales, and in turn fringe retailers lower their prices to consumers.

In contrast, Erutku (2005) assumes a market structure where an upstream firm sells to both a large national retailer and smaller retailers competing in spatially separate local markets, with the large retailer bargaining with the upstream supplier over discounts in the wholesale price as opposed to their joint profits. The key result is that the wholesale price charged to local retailers initially increases with the large retailer's bargaining power, pushing up local retail prices. With further increases in bargaining power, the upstream supplier is forced to lower its wholesale price to local retailers, retail prices falling for both the national and local retailers.

In a similar market structure, Inderst and Valetti (2011) assume that retailers incur fixed costs of seeking an alternative supplier if they reject the wholesale price; for example, retailers may invest in production and marketing of a private label good as opposed to carrying the national brand. The large retailer, by spreading fixed costs over more outlets, forces the upstream supplier to lower its wholesale price, which is then passed on to consumers, taking away market share from smaller retailers and raising the fixed costs of seeking an alternative supplier. As a consequence, the upstream firm raises its wholesale price to smaller retailers. In the retail price-setting game, the impact on smaller retailers' prices depends on two separate effects: the price response of smaller retailers to the large retailer lowering its retail price, and the retail price response of smaller retailers to an increase in their wholesale price. If prices are strategic substitutes, smaller retailers end up paying higher wholesale prices and raising retail prices – termed a "waterbed effect." However, if prices are strategic substitutes, the latter effect is countered by strong competition at retail.

6.2 Vertical restraints

Vertical restraints cover a wide range of activities, including exclusive dealing and exclusive territories, which are contractual provisions restricting a retailer to carrying only one processor's brand, and the geographical area of sales for that brand, and full-line forcing, which relates to a retailer having to carry the complete range of a processor's products. The standard argument in favor of such vertical restraints is that without them intensive competition between downstream retailers can result either in an inefficient level of pre-retail services (Matthewson and Winter 1984; Rey and Tirole 1986) or excessive post-sale quality differentiation (Bolton and Bonnano 1988). However, other arguments suggest that the efficiency enhancing effects of such restraints ignore their potential effect on firm behavior at the upstream processing and downstream retailing stages (Rey and Stiglitz 1988; Innes and Hamilton 2009).

Consider a result due to Bonnano and Vickers (1988): two upstream processors sell a differentiated product to a downstream retailing sector consisting of two firms, and at both stages firms compete in price. Also assume that there is exclusive dealing; that is, each upstream processor delegates just one of the retailers to sell their branded product, charging them a wholesale price equal to marginal cost, franchise fees being set equal to zero. Under such a contract, neither retailer can credibly raise price beyond marginal cost (i.e., the Bertrand–Nash equilibrium). Alternatively, suppose each upstream processor sets its wholesale price above marginal cost, which they are able to do because of exclusive dealing. This allows retailers downstream to credibly raise their prices, with upstream processors capturing the additional retailing profits via franchise fees. In other words, exclusive dealing in conjunction with a two-part tariff while removing the vertical externality of inter-brand competition can actually reduce consumer welfare.

The above result is very sensitive to the assumption that the upstream processor(s) is able to make take-it-or-leave-it offers to the downstream retailer(s), ignoring the possibility that downstream retailers are able to bargain in their favor over setting contract terms due to their countervailing power. Following Shaffer (1991), suppose an upstream processing sector sells homogeneous products to a downstream retailing duopoly differentiated by characteristics such as location, range of goods and services, etc. Each upstream processor sets a wholesale price, and then each retailer chooses an upstream processor as their supplier and sets a retail price. Given competition between processors for retail shelf-space, with linear pricing no upstream processor can set a wholesale price above marginal cost as they will be undercut by other processors, and neither retailer can raise the price above the marginal cost because they will be undercut by the other retailer. Alternatively, a two-part tariff can be offered where the wholesale price is marked up above the marginal cost along with a negative franchise fee paid by the upstream processor(s) to the downstream retailer(s). The negative franchise fee compensates downstream retailers for the higher wholesale price, but at the same time, in paying the higher wholesale price, competition is lessened at retail, feeding back into higher retail profits. This result highlights again that countervailing power at retail is not necessarily in the interest of consumer welfare.

The extensive use of negative franchise fees, or slotting allowances as they are more commonly known, has been at the heart of the debate about increased buyer power by the food retailing sector. Innes and Hamilton (2013) report that their use is pervasive in U.S. food retailing, fees taking various forms, including among others, new product introduction fees and pay-to-stay fees on existing stocks. Other fees include firms producing established products that pay "facing allowances" for better shelf-positioning, end-aisle displays requiring "street money" from upstream firms, and contributions by upstream processors to "market development funds" (Shaffer 2005).

Since Shaffer's (1991) seminal article, several authors, drawing on different market structure assumptions, have all shown that slotting allowances may facilitate market control and are therefore welfare-reducing. Shaffer (2005) describes a set-up where a dominant upstream processor and competitive fringe compete for retail shelf space by selling differentiated products A and B, respectively. The dominant firm makes a take-it-or-leave-it two-part tariff consisting of a wholesale price and slotting fee, while retailers either accept the contract and carry the dominant firm's product A or reject it and carry product B supplied by the competitive fringe. This is designed to capture the idea that smaller processors in the competitive fringe are unable to get their product B onto retailer shelves due to the exclusionary nature of slotting allowances. Shaffer (2005) finds that the dominant firm is more likely to induce exclusion of the competitive fringe when products A and B are more substitutable.

Innes and Hamilton (2006, 2009) model how slotting allowances can be used in a multi-product retailing environment to facilitate cross-market control by a dominant upstream processor of a single product. In Innes and Hamilton (2006), they assume a similar horizontal market structure upstream to Shaffer (2005), but downstream there is a duopoly selling the processed products A and B in a spatially differentiated, multi-product retail market. Here the dominant upstream processor offers a contract to retailers that consists of a monopoly wholesale price for product A and a requirement that slotting allowances be imposed on the competitive fringe product B. The effect of slotting allowances is to reduce supply of the competitive fringe and raise the monopoly profits of the dominant upstream processor, despite retailer competition, where the additional monopoly rents are essentially extracted from fringe consumers. In other words, slotting allowances act like auction prices paid by fringe processors to be the retailer's second brand.

A similar market structure at the processing and retailing stages is also assumed in Innes and Hamilton (2009), but in this case the requirement that slotting allowances be imposed on the competitive fringe is ruled out – either because of anti-trust rules or because the behavior of retailers cannot be monitored by the dominant upstream firm. Instead, the dominant upstream processor chooses a vertical contract combining a wholesale price below marginal cost with a franchise fee paid by the retailers and minimum resale price maintenance for its own product A. In turn, this will elicit a demand by downstream retailers for payment of slotting allowances by upstream suppliers in exchange for wholesale prices of product B being set above the marginal cost. If minimum resale price maintenance is deemed illegal by anti-trust rules, the upstream dominant firm will increase its wholesale price for product A due to the slotting allowance having raised the wholesale price of product B. In other words, slotting allowances charged to the competitive fringe results in higher retail prices for both goods.

A problem with all of these results is that other than the possibility of scarce shelf-space, retailers are not actually exercising any countervailing power; rather, it is a dominant upstream processor using vertical restraints as a means to exert market power. An exception to this is seen in a recent article by Innes and Hamilton (2013), in which they assume it is retailers that exercise their countervailing power through slotting allowances in a manner more similar to Shaffer's (1991) original result. A retailing duopoly is assumed to compete for consumers spatially, as well as through the variety of products on their shelves. Each retailer selects a number of varieties to stock and offers a nonlinear price to the upstream supplier of each variety, consisting of a wholesale price and a slotting allowance, and then retailers compete in price. In the absence of slotting allowances, a retailer choosing more varieties will prompt the other retailers to lower prices in order to make up for lost consumer traffic, which in turn will result in fewer varieties actually being offered in equilibrium. With slotting allowances, retailers are able to commit to paying higher wholesale prices, as in Shaffer (1991), and this frees up retailers to increase the number of varieties they are able to offer consumers. Consequently, there is a welfare tradeoff: higher retail prices versus greater product variety.

6.3 Private labels

The growth of private labels and their market penetration has become an important feature of how food retailers in both the United States and the European Union compete horizontally with each other, as well as how they compete vertically with national brand food processors. McCorriston (2014) reports that private labels account for 15 percent of retail sales in North America and 23 percent in the European Union, the level of penetration varying widely across EU member states and product categories.

Most of the theoretical research on private labels has focused on the vertical dimension of private label products and whether or not competition between them and national brands results in higher or lower consumer prices. Bergès-Sennou, Bontems, and Réquillart (2004) note that the typical model in the literature has the following structure: that is, an upstream food processor sells a high-quality national brand to a downstream food retailer, the problem of double-marginalization arising. If the food retailer introduces a lower-quality private label product into the market, this limits the market power of the upstream processor, driving down the wholesale price of the national brand; consumer surplus increases due to removal of double-marginalization, assuming the cost of producing the private label product is not too high and that it is sold at marginal cost by a competitive fringe of upstream food processors.

A key assumption affecting results in the literature relates to the production costs for private label products and competing national brands. Mills (1995) assumed that these costs are the same, his results driven by the quality level of the private label product. Specifically, if the private label reaches a certain quality level, the firm producing the national brand has an incentive to lower its wholesale price in order to deter the retailer selling the private label product, but as private label quality increases, the firm producing the national brand accommodates the private label product, lowering the wholesale price. Consumers benefit from the introduction of the private label product, as it drives down the retail price of the national brand.

In contrast, Bontems, Monier-Dilhan, and Réquillart (1999) assume that the marginal costs of producing private label products differ from the national brand, and costs increase in quality. If the private label product is of low quality, the producer of the national brand cannot prevent entry of the private label product, the wholesale price of the national brand falls, and the retailer sells both products. As private label quality increases, even though it gets more competitive with the national brand, it induces a cost increase, which may result in the wholesale price of the national brand increasing. If the national brand price rises enough, though, it may exit the market, especially if consumers have a low willingness to pay for quality. For intermediate levels of private label quality, the firm producing the national brand can set a limit on the wholesale price to deter sales of the private label product, the increasing cost of producing the private label making limit pricing easier. For high levels of private label quality, the national brand recovers its monopoly position, as the private label product is not competitive.

6.4 Evaluating vertical coordination

Actually assessing vertical coordination empirically is difficult due to the lack of data on wholesale prices, and as yet only a limited number of articles have managed to generate any meaningful results, notably Villas-Boas (2007), Rennhoff (2008), Bonnet and Dubois (2010), and Bonnet and Bouamra-Mechemache (2016), who examine the U.S. yogurt, U.S. ketchup,

French bottled water, and French fluid milk markets, respectively. Extending the workhorse empirical methodology pioneered by Berry, Levinsohn, and Pakes (1995) and Nevo (2001) to a vertical market setting, and drawing on earlier work by, *inter alia*, Sudhir (2001), Villas-Boas and Zhao (2005), and Hellerstein (2008), the approach of these studies is essentially to use alternative models of vertical coordination between upstream processors and downstream retailers in order to establish whether or not vertical contracting relationships are being used to avoid the double-marginalization problem.

Using demand estimates from a discrete choice demand formulation, and allowing for branded and private label yogurt products, Villas-Boas (2007) computes the price–cost margins for U.S. yogurt processors and retailers implied by different models of vertical coordination and then compares these with price–cost margins from direct estimates of cost. The results indicate that double-marginalization is being avoided in the sector, processors are pricing at marginal cost, and profit maximizing prices are being set by retailers. Using a similar methodology, and again allowing for branded and private label bottled water, Bonnet and Dubois (2010) compute price–cost margins for both linear pricing and two-part tariff contracts with or without resale price maintenance; their results indicate that processors in the French bottled water sector use nonlinear pricing with resale price maintenance, even though the latter vertical restraint is actually illegal in France.

While this empirical analysis of vertical coordination is encouraging, it does not shed much light on the impact of slotting allowances or the exercise of buying power by food retailers. For example, Villas-Boas' (2007) results are consistent with nonlinear pricing by processors via two-part tariffs, except that in the U.S. yogurt market positive franchise fees are either very small or nonexistent. She interprets this to imply that food retailers have bargaining power, driving down the wholesale price of yogurt, but without detailed data on fixed fees (positive or negative), she is unable to formally identify what type of vertical contract is driving her results or how retailers are exercising their countervailing power. Rennhoff (2008), using a similar methodology, develops a vertical market model with two-part tariffs, where processors do not hold total bargaining power, thereby allowing for negative franchise fees. Importantly, he is able to infer the level of the fees by tying them to in-store promotion decisions by retailers in the U.S. ketchup sector, his estimates suggesting an average weekly fee of US$1.60 per bottle of ketchup. Overall, Rennhoff (2008) concludes that negative franchise fees result in a loss of aggregate economic welfare, an increase in retailer profits being outweighed by a decrease in processor profits and consumer surplus.

By contrast, Richards and Hamilton (2013) have cast doubt on the viability of retailers paying high wholesale prices and accepting slotting allowances, focusing on the network effects that arise from two-sided demand for shelf space in supermarkets. In their model, food retailers are multi-product platforms connecting food processors with consumers. Based on earlier empirical research (Richards and Hamilton 2006), they argue that because consumers value a variety of food products, they also value supermarket platforms that attract a range of branded food products. At the same time, food processors benefit from supplying supermarket platforms that attract a large number of consumers. As a consequence, both consumers and food processors are willing to pay for access to the platform, jointly determining retail and wholesale margins.

Key to how the total margin is shared between processor and retailer are both direct and indirect effects. The former depends on the relative value consumers and food processors place on the size of the platform (Armstrong 2006), while the latter depends on indirect network effects on the pricing power of food processors and retailers. In a multi-product framework, an increase in the number of varieties drives down total margins due to increased price competition

(Hamilton and Richards 2009), but at the same time, increased variety increases consumer demand within a retail location, generating indirect network effects, thereby raising retail margins at the expense of wholesale margins. Basically, there is a network externality whereby all food processors benefit from a new variety being added to a retailer's range. However, while consumers are able to internalize this externality due to their utility increasing in product variety, retailers extract increased rents from consumers' willingness to pay for more variety, while food processors fail to recognize the externality they generate from adding an extra product at the margin. As a consequence, retailers internalize indirect network effects by increasing their retail margins. Richards and Hamilton (2013) use demand estimates from a Berry, Levinsohn, and Pakes (1995) type model to solve for the retail choices concerning variety and price; their empirical results for six supermarkets in Visalia, California indicate that as variety increases retailers have an incentive to raise retail margins, while food processors accept lower wholesale margins in competing for the higher retail margins. Richards and Hamilton (2013) argue that their results provide an explanation for the emergence of Walmart as a supermarket platform in the United States.

The possibility that retailers may be able to exert bargaining power over processors is also explicitly tested for by Bonnet and Bouamra-Mechemache (2016), who draw on the methodology of Draganska, Klapper, and Villas-Boas (2010) to model wholesale price as the outcome of a Nash bargaining game between processors and retailers in the French fluid milk market. Their approach involves using estimates from a discrete choice demand model to recover retail margins as per Nevo (2001), from which they are able to estimate the relative bargaining power of processors versus retailers, given exogenous cost variables for fluid milk products. From the estimates of retail margins and bargaining power parameters, Bonnet and Bouamra-Mechemache (2016) are able to infer wholesale margins and hence total fluid milk margins. Their results indicate that processor margins exceed retailer margins for organic milk as compared to conventional milk, and that retailer bargaining power is still lower even when they sell private label milk. In the case of conventional milk, retailers have greater bargaining power, although it varies by retailer and brand pair.

Turning to the impact of private labels, the empirical research indicates that they are associated with national brand prices rising. For example, Cotterill, Putsis, and Dhar (2000), using cross-sectional U.S. data for 143 product categories in 57 geographic markets, found that private label and national brand prices tend to be higher when the retail market is concentrated and share of the national brand is high. Ward et al. (2002), using monthly time series data for 34 products in the United States, found that an increase in the market share of private labels is associated with an increase (or no change) in the price of national brands, a fall in the price (or no change) of private label products, and a negative impact on average prices or no change. Finally, Bontemps, Orozco, and Réquillart (2008), using a sample of 218 French products over the 1998–2001 period, find that there is a positive relationship between private label product development and national brand prices, the effect being stronger when the private label product is a closer substitute in terms of quality for the national brand.

Unlike the existing empirical literature on vertical restraints, which is constrained by the lack of data on two-part tariffs, the empirical results for private labels seem quite consistent. Nevertheless, as McCorriston (2014) points out, the empirical research on private label products is of a reduced form, such that it is difficult to choose between alternative theoretical explanations that might be consistent with the data. Second, the focus of both theoretical and empirical research is entirely on the effect of private label introduction on national brand prices, with nothing said about how it affects horizontal competition across retailers.

7 Concluding comments

In the period since Sexton and Lavoie's (2001) handbook chapter, there has been a significant shift in emphasis away from analyzing horizontal competition in processing and retailing to analyzing the role of buyer power in the food industry. Two overall conclusions can be drawn from recent developments in the literature.

First, there is now an extensive body of contract theory focusing on agency costs and incentives, but it has yet to rigorously incorporate downstream processer buyer power, and data constraints make it difficult to conduct any robust empirical analysis of contracting. Developing the latter is critical if public debate is to go beyond concentrated food processing market structures being seen as evidence of exploitation of buyer power by food processors.

Second, the body of research on vertical markets is now very rich in different models of the impact of vertical restraints and how to capture countervailing power of retailers. Importantly, these models emphasize that the impact of buyer power on consumers is also conditioned on the nature of horizontal interaction at each stage of the food industry. However, as is often the case in industrial organization analysis, different modeling results are very sensitive to key underlying assumptions. As a consequence, these models are potentially hard to evaluate, particularly because of a lack of data on wholesale prices. Nevertheless, some progress has been made empirically in pinning down the structure of vertical contracts, through adoption/adaptation of the multi-product approach originally pioneered by Berry, Levinsohn, and Pakes (1995). Continued progress in this area is important for the application of appropriate anti-trust interventions in food retailing.

References

Appelbaum, E. 1982. "The Estimation of the Degree of Oligopoly Power." *Journal of Econometrics* 19(2/3):287–299.

Armstrong, M. 2006. "Competition in Two-Sided Markets." *Rand Journal of Economics* 37(3):668–691.

Bain, J.S. 1950. "Workable Competition in Oligopoly: Theoretical Considerations and Some Empirical Evidence." *The American Economic Review* 40(2):35–47.

Bain, J.S. 1951. "Relation of Profit Rate to Industry Concentration: American Manufacturing." *Quarterly Journal of Economics* 65(3):293–324.

Basker, E., and M. Noel. 2009. "The Evolving Food Chain: Competitive Effects of Wal-Mart's Entry Into the Supermarket Industry." *Journal of Economics and Management Strategy* 18(4):977–1009.

Bergès-Sennou, F., P. Bontems, and V. Réquillart. 2004. "Economics of Private Labels: A Survey of Literature." *Journal of Agricultural and Food Industrial Organization* 2(3):1037–1062.

Berry, S., J. Levinsohn, and A. Pakes. 1995. "Automobile Prices in Market Equilibrium." *Econometrica* 63(4):841–890.

Bolton, P., and G. Bonnano. 1988. "Vertical Restraints in a Model of Vertical Differentiation." *Quarterly Journal of Economics* 103(3):555–570.

Bonnano, G., and J. Vickers. 1988. "Vertical Separation." *Journal of Industrial Economics* 36(3):257–265.

Bonnet, C., and Z. Bouamra-Mechemache. 2016. "Organic Label, Bargaining Power, and Profit-Sharing in the French Fluid Milk Market." *American Journal of Agricultural Economics* 98(1):113–133.

Bonnet, C., and P. Dubois. 2010. "Inference on Vertical Contracts Between Manufacturers and Retailers Allowing for Nonlinear Pricing and Resale Price Maintenance." *Rand Journal of Economics* 41(1):139–164.

Bontemps, C., V. Orozco, and V. Réquillart. 2008. "Private Labels, National Brands and Food Prices." *Review of Industrial Organization* 33(1):1–22.

Bontems, P., S. Monier-Dilhan, and V. Réquillart. 1999. "Strategic Effects of Private Labels." *European Review of Agricultural Economics* 26(2):147–165.

Bresnahan, T.F. 1982. "The Oligopoly Solution Concept Is Identified." *Economics Letters* 10(1/2):87–92.

———. 1989. "Empirical Studies of Industries with Market Power." In R. Schmalensee and R. Willig, eds., *Handbook of Industrial Organization.* Amsterdam, Netherlands: North-Holland, pp. 1011–1057.

Buceviciute, L., A. Dierx, and F. Ilzkovitz. 2009. "The Functioning of the Food Supply Chain and Its Effect on Food Prices in the European Union." Occasional Paper 47. European Commission, Brussels, Belgium.

Chen, Z. 2003. "Dominant Retailers and Countervailing Hypothesis." *Rand Journal of Economics* 34(4):612–625.

———. 2007. "Buyer Power: Economic Theory and Antitrust Policy." *Research in Law and Economics* 22:17–40.

Clarke, R., and S.W. Davies. 1982. "Market Structure and Price-Cost Margins." *Economica* 49(195):277–287.

Clodius, R.L., and W.F. Mueller. 1961. "Market Structure Analysis as an Orientation for Research in Agricultural Economics." *Journal of Farm Economics* 43(3):515–553.

Connor, J.M., R.T. Rogers, B.W. Marion, and W.F. Mueller. 1985. *The Food Manufacturing Industries: Structure, Strategies, Performance, and Policies.* Lexington, MA: Lexington Books.

Cotterill, R.W. 1999. "Continuing Concentration in Food Industries Globally: Strategic Challenges to an Unstable Status Quo." Research Report 49. Food Policy Research Center, University of Connecticut, Storrs, CT.

Cotterill, R.W., W.P. Putsis, and R. Dhar. 2000. "Assessing the Competitive Interaction Between Private Labels and National Brands." *Journal of Business* 73(1):109–137.

Cowling, K., and M. Waterson. 1976. "Price-Cost Margins and Market Structure." *Economica* 43(171):267–274.

Crespi, J.M., T.L. Saitone, and R.J. Sexton. 2012. "Competition in the U.S. Farm Products Markets: Do Long-Run Incentives Trump Short-Run Market Power?" *Applied Economic Perspectives and Policy* 34(4):669–695.

Demsetz, H. 1973. "Industry Structure, Market Rivalry, and Public Policy." *Journal of Law and Economics* 16(1):1–9.

Dobson, P.W., and M. Waterson. 1997. "Countervailing Power and Consumer Prices." *Economic Journal* 107(441):418–430.

Draganska, M., D. Klapper, and S.B. Villas-Boas. 2010. "A Larger Slice of a Larger Pie? An Empirical Investigation of Bargaining Power in the Distribution Channel." *Marketing Science* 29(1):57–74.

Ellickson, P.B. 2007. "Does Sutton Apply to Supermarkets?" *Rand Journal of Economics* 38(1):43–59.

Erutku, C. 2005. "Buying Power and Strategic Interactions." *Canadian Journal of Economics* 38(4):1160–1172.

Fisher, F.M., and J.J. McGowan. 1983. "On the Misuse of Accounting Rates of Return to Infer Monopoly Profits." *American Economic Review* 73(1):82–97.

Galbraith, J.K. 1952. *American Capitalism: The Concept of Countervailing Power.* New York: Houghton Mifflin.

Geroski, P.A. 1988. "In Pursuit of Monopoly Power: Recent Quantitative Work in Industrial Economics." *Journal of Applied Econometrics* 3(2):107–123.

Goodhue, R.E. 2000. "Broiler Production Contracts as a Multi-Agent Problem: Common Risk, Incentives and Heterogeneity." *American Journal of Agricultural Economics* 82(2):606–622.

Hamilton, S.F., and T.J. Richards. 2009. "Product Differentiation, Store Differentiation, and Assortment Depth." *Management Science* 55(8):1368–1376.

Hausmann, J., and E. Leibtag. 2007. "Consumer Benefits from Increased Competition in Shopping Outlets: Measuring the Effect of Wal-Mart." *Journal of Applied Econometrics* 22(7):1157–1177.

Hellerstein, R. 2008. "Who Bears the Cost of a Change in the Exchange Rate? Pass-Through Accounting for the Case of Beer." *Journal of International Economics* 76(1):14–32.

Inderst, R., and T.M. Valletti. 2011. "Buyer Power and the 'Waterbed Effect'." *Journal of Industrial Economics* 59(1):1–20.

Innes, R., and S.F. Hamilton. 2006. "Naked Slotting Fees for Vertical Control of Multi-Product Retail Markets." *International Journal of Industrial Organization* 24(2):303–318.

———. 2009. "Vertical Restraints and Horizontal Control." *Rand Journal of Economics* 40(1):120–143.

———. 2013. "Slotting Allowances Under Supermarket Oligopoly." *American Journal of Agricultural Economics* 95(5):1216–1222.

Just, R.E., and W.S. Chern. 1980. "Tomatoes, Technology, and Oligopsony." *Bell Journal of Economics* 11(2):584–602.

Kaiser, H.M., and N. Suzuki. 2006. *New Empirical Industrial Organization and the Food System*. New York: Peter Lang.

Katz, M. 1989. "Vertical Contractual Relations." In R. Schmalensee and R. Willig, eds., *Handbook of Industrial Organization*. Amsterdam, Netherlands: North-Holland, pp. 655–721.

Knoeber, C.R., and W.N. Thurman. 1995. "'Don't Count Your Chickens. . .' Risk and Risk Shifting in the Broiler Industry." *American Journal of Agricultural Economics* 77(3):486–496.

Levin, J. 2003. "Relational Incentive Contracts." *American Economic Review* 93(3):835–857.

MacDonald, J.M. 2015. "Trends in Agricultural Contracts." *Choices* 30(3).

MacDonald, J.M., and W. McBride. 2009. "The Transformation of the U.S. Livestock Agriculture: Scale, Efficiency, and Risks." Economic Information Bulletin 43, USDA-ERS, Washington, DC.

MacDonald, J.M., and M.E. Ollinger. 2000. "Scale Economies and Consolidation in Hog Slaughter." *American Journal of Agricultural Economics* 82(2):334–346.

Marion, B.W. 1986. *The Organization and Performance of the U.S. Food System*. Lexington, MA: Lexington Books.

Matthewson, G., and R. Winter. 1984. "An Economic Theory of Vertical Restraints." *Rand Journal of Economics* 15(1):27–38.

McCorriston, S. 2002. "Why Should Imperfect Competition Matter to Agricultural Economists?" *European Review of Agricultural Economics* 29(3):349–372.

———. 2014. "Background Note." In *Competition Issues in the Food Chain Industry*. Paris, France: OECD, pp. 9–45.

Mills, D.E. 1995. "Why Retailers Sell Private Labels." *Journal of Economics and Management Strategy* 4(3):509–528.

Nevo, A. 1998. "Identification of the Oligopoly Solution Concept in a Differentiated-Products Industry." *Economics Letters* 59(3):391–395.

———. 2001. "Measuring Market Power in the Ready-To-Eat Breakfast Cereal Industry." *Econometrica* 69(2):307–342.

Organisation for Economic Co-operation and Development [OECD]. 2014. *Competition Issues in the Food Chain Industry*. Paris, France: OECD.

Perloff, J.M., L.S. Karp, and A. Golan. 2007. *Estimating Market Power and Strategies*. Cambridge: Cambridge University Press.

Rennhoff, A. 2008. "Paying for Shelf Space: An Investigation of Merchandising Allowances in the Grocery Industry." *Journal of Agricultural and Food Industrial Organization* 6(1):1–40.

Rey, P., and J.E. Stiglitz. 1988. "Vertical Restraints and Producers' Competition." *European Economic Review* 32(2/3):561–568.

Rey, P., and J. Tirole. 1986. "The Logic of Vertical Restraints." *American Economic Review* 76(5):921–939.

Richards, T.J., and S.F. Hamilton. 2006. "Rivalry and Price and Variety Among Supermarket Retailers." *American Journal of Agricultural Economics* 88:710–726.

———. 2013. "Network Externalities in Supermarket Retailing." *European Review of Agricultural Economics* 40(1):1–22.

Richards, T.J., S.F. Hamilton, and J. Empen. 2015. "Attribute Search in Online Retailing." Selected paper presented at the AAEA Conference, San Francisco, CA.

Richards, T.J., and P.M. Patterson. 2003. "Competition in Fresh Produce Markets: An Empirical Analysis of Marketing Channel Performance." Contractors and Cooperators Report 1, USDA-ERS, Washington, DC.

Richards, T.J., and G. Pofahl. 2010. "Pricing Power By Supermarket Retailers: A Ghost in the Machine." *Choices* 25(2).

Schmalensee, R. 1989. "Inter-Industry Studies of Structure and Performance." In R. Schmalensee and R. Willig, eds., *Handbook of Industrial Organization*. Amsterdam, Netherlands: North-Holland, pp. 951–1009.

Schroeter, J.R. 1988. "Estimating the Degree of Market Power in the Beef Packing Industry." *Review of Economics and Statistics* 70(1):158–162.

Sexton, R.J. 2000. "Industrialization and Consolidation in the U.S. Food Sector: Implications for Competition and Welfare." *American Journal of Agricultural Economics* 82(5):1087–1104.

———. 2013. "Market Power, Misconceptions, and Modern Agricultural Markets." *American Journal of Agricultural Economics* 95(2):209–219.

Sexton, R.J., and N. Lavoie. 2001. "Food Processing and Distribution: An Industrial Organization Approach." In B. Gardner and G. Rausser, eds., *Handbook of Agricultural Economics*. Amsterdam, Netherlands: North-Holland, pp. 475–535.

Sexton, R.J., I.M. Sheldon, S. McCorriston, and H. Wang. 2007. "Agricultural Trade Liberalization and Economic Development: The Role of Downstream Market Power." *Agricultural Economics* 36(2):253–270.

Sexton, R.J., and M. Zhang. 2001. "An Assessment of the Impact of Food Industry Market Power on U.S. Consumers." *Agribusiness: An International Journal* 17(1):59–79.

Shaffer, G. 1991. "Slotting Allowances and Resale Price Maintenance: A Comparison of Facilitating Practices." *Rand Journal of Economics* 22(1):120–135.

———. 2005. "Slotting Allowances and Optimal Product Variety." *Advances in Economic Analysis and Policy* 5(1):1083–1118.

Sheldon, I.M. 2017. "The Competitiveness of Agricultural Product and Input Markets: A Review and Synthesis of Recent Research." *Journal of Agricultural and Applied Economics* 49(1):1–44.

Sheldon, I.M., and R. Sperling. 2003. "Estimating the Extent of Imperfect Competition in the Food Industry: What Have We Learned?" *Journal of Agricultural Economics* 54(1):89–109.

Sudhir, K. 2001. "Structural Analysis of Manufacturer Pricing in the Presence of a Strategic Retailer." *Marketing Science* 20(3):244–264.

Tirole, J. 1988. *Theory of Industrial Organization*, Cambridge, MA: MIT Press.

———. 2015. "Market Failures and Public Policy." *American Economic Review* 105(6):1665–1682.

United Kingdom Competition Commission. 2008. *The Supply of Groceries in the UK Market Investigation*. London, England: Competition Commission.

United States Department of Justice. 2012. *Voices from the Workshops on Agriculture and Antitrust Enforcement in our 21st Century Economy and Thoughts on the Way Forward*. Washington, DC: U.S. Department of Justice.

Villas-Boas, S.B. 2007. "Vertical Relationships Between Manufacturers and Retailers: Inference with Limited Data." *Review of Economic Studies* 74(2):625–652.

Villas-Boas, S.B., and Y. Zhao. 2005. "Retailer, Manufacturers, and Individual Consumers: Modeling the Supply Side in the Ketchup Marketplace." *Journal of Marketing Research* 42(1):83–95.

von Ungern-Sternberg, T. 1996. "Countervailing Power Revisited." *International Journal of Industrial Organization* 14(4):507–520.

Ward, M.B., J.P. Shimshack, J.M. Perloff, and J.M. Harris. 2002. "Effects of Private-Label Invasion in Food Industries." *American Journal of Agricultural Economics* 84(4):961–973.

Williamson, O.E. 1986. "Vertical Integration and Related Variations on a Transactions-Cost Economics Theme." In J. Stiglitz and F. Matthewson, eds., *New Developments in the Analysis of Market Structure*. London, England: Palgrave Macmillan, pp. 149–174.

Wu, S.Y. 2006. "Contract Theory and Agricultural Policy Analysis: A Discussion and Survey of Recent Developments." *Australian Journal of Agricultural and Resource Economics* 50(4):490–509.

———. 2014. "Adapting Contract Theory to Fit Contract Farming." *American Journal of Agricultural Economics* 96(5):1241–1256.

5
Economic ramifications of obesity
A selective literature review

Oral Capps, Jr., Ariun Ishdorj, Senarath Dharmasena, and Marco A. Palma

1 Introduction

Obesity is a pervasive problem not only domestically but also globally. As of 2016, one-fifth of the global population and one-third of the U.S. population were obese. Until the early 1980s, about one in six adults was obese. As reported by the World Health Organization (WHO), the top ten most obese industrialized countries are (in order of obesity ranking) the United States, New Zealand, Canada, Israel, the United Kingdom, Greece, Lithuania, Poland, Hungary, and France, respectively.

Globally, men and women face markedly different risks of obesity. In general, obesity is more prevalent among women than men. According to the Behavioral Risk Factor Surveillance System of the Centers for Disease Control and Prevention (CDC), in 2014, obesity prevalence varied widely across U.S. states and territories. No state had a prevalence of obesity less than 20 percent in 2014, with the Midwest having the highest prevalence of obesity (30.7 percent), followed by the South (30.6 percent), the Northeast (27.3 percent), and the West (25.7 percent). Additionally, racial variations were evident in the prevalence of obesity. Non-Hispanic blacks had the highest prevalence of self-reported obesity (38.1 percent), followed by Hispanics (31.3 percent), and non-Hispanic whites (27.1 percent) in 2014.

The coverage of the obesity epidemic has been quite extensive across medical, sociological, psychological, political, and economic disciplines for at least 30 years. Simply put, our objective is to provide a selective review of the literature relevant to agricultural economics on this topic. In this review, we address various issues: (1) the metrics of obesity; (2) the causes and the consequences of obesity; (3) the role of government in attempting to reduce obesity rates; and (4) insights from behavioral economics in combating the incidence of obesity.

2 Metrics of obesity

The primary sources of obesity statistics are the CDC and the WHO. Typically, adults are classified as obese if their body mass index (BMI) or Quetelet index exceeds 30kg/m^2. Put simply, the calculation of BMI takes into account the height and weight of any individual. Formally, BMI is defined as weight in kilograms divided by the square of height in meters, or alternatively as weight in pounds*703/height in inches². BMI is universally expressed in units of kg/m². Commonly accepted BMI ranges are as follows: underweight: under 18.5 kg/m²; normal weight: 18.5 to 25 kg/m²; overweight: 25 to 30 kg/m²; and obese: over 30 kg/m².

3 Causes and consequences of obesity

Obesity, currently classified as a disease by the CDC, is attributed to caloric imbalance, where more calories are consumed than expended. The actual causes of obesity are far more complex. Obesity results from a combination of causes and contributing factors that can be genetic, behavioral, economic, environmental, social, and even political. Additional contributing factors include education and marketing/promotion. This section reviews the literature on some of the underlying causes and consequences of obesity, using only peer-reviewed research.

3.1 Causes of obesity

Unhealthy diet and inadequate physical activity, as well as other individual, biological, and genetic factors, have been identified as primary causes of increases in obesity in the existing literature (Philipson and Posner 2003; Baum and Ruhm 2009). However, these factors alone might not provide a sufficient explanation, since many economic and social factors outside of the control of individuals also can affect obesity (Philipson and Posner 2003; Lakdawalla, Philipson, and Bhattacharya 2005). To illustrate, Finkelstein, Ruhm, and Kosa (2005) argue that owing to technological innovations in food processing, the price of calorie-dense prepackaged or prepared foods has fallen relative to the price of less calorie-dense foods. As such, owing to economic forces, individuals have shifted their consumption of foods toward cheaper, calorie-dense alternatives.

Additionally, the literature supports the notion that changes in technology have contributed to the rise in obesity (Lakdawalla and Philipson 2002; Philipson and Posner 2003, 2008). That is, improved food processing technologies have made food more affordable and readily available than ever before, resulting in increased food consumption (Cutler, Glaeser, and Shapiro 2003). At the same time, labor-saving technologies in workplaces have made jobs more sedentary, requiring less caloric expenditure (Philipson 2001; Lakdawalla and Philipson 2002; Philipson and Posner 2003; Wansink and Huckabee 2005). Finkelstein, Ruhm, and Kosa (2005) argued that the decline in manual labor began well before the rapid rise in obesity; hence, technological progress may be only partially responsible for the increase in obesity.

Improvements in technology allowed for mass production of food and the widespread distribution of prepared foods to consumers (Cutler, Glaeser, and Shapiro 2003). Cutler, Glaeser, and Shapiro (2003) found that reductions in time costs of food preparation led to reductions in food prices and increases in calorie intake from prepared snack foods. Additionally, technological change brought about a notable reduction in time spent on household production processes and helped increase labor force participation rates of women. While studies by Cutler, Glaeser, and Shapiro (2003) and Loureiro and Nayga (2007) found that the participation rate of women in the labor force had no significant effect on the incidence of obesity in working women, Anderson, Butcher, and Levine (2003) found that increased hours worked per week by mothers were associated with a significant increase in children's weight.

In general, the consumption of energy-dense and nutrient-empty foods such as added fats and sugars, salty snacks, refined grain products, sweets, beverages, and fast foods is linked to obesity. A number of studies looked at the relationship between the availability of fast food restaurants, food prices, and obesity (Rashad, Grossman, and Chou 2006; Chou, Grossman, and Saffer 2004; Chou, Rashad, and Grossman 2008; Currie et al. 2010; Dunn 2010) as well as between food-away-from-home expenditures and obesity (Chou, Grossman, and Saffer 2004; Rashad, Grossman, and Chou 2006; Drichoutis, Nayga, and Lazaridis 2012). Currie et al. (2010) found that proximity to fast food restaurants had a significant effect on the risk of obesity, while proximity to non-fast food restaurants had no effect. Chou, Grossman, and Saffer (2004) examined the effect of restaurant density and the relative prices of fast food and full-service restaurants and food-consumed-at-home on obesity and found that lower fast food and full-service restaurants prices were associated with higher-weight outcomes. Moreover, Chou, Grossman, and Saffer (2004) concluded that technological changes and economies of scale that resulted in a reduction in prices in fast food restaurants led to increases in the demand for food-away-from-home. Rashad, Grossman, and Chou (2006) reported that increases in the per capita number of restaurants resulted in increased obesity.

Furthermore, the literature supports the idea that variations in the neighborhood environment may be responsible for increases in obesity rates by affecting diet and exercise. Existing research examining the relationship between urban sprawl and obesity based on cross-sectional data found that living in a sprawling county or metropolitan area was associated with higher rates of obesity (Ewing et al. 2003). But no relationship was found based on longitudinal data (Ewing, Brownson, and Berrigan 2006). Using a sample of individuals who had recently moved, Plantinga and Bernell (2007) examined the relationship between patterns of urban land development and obesity. They found that an individual's BMI is a significant factor in determining the choice of a dense or sprawling location. As well, they found that individuals who move to more dense population locations not only lose weight but also that the greater the change in population density, the greater the weight loss.

3.2 Consequences of obesity

Obesity is a major public health issue and a known cause of many chronic health conditions, such as diabetes, high blood pressure, asthma, cholesterol, and cardiovascular disease; as such, obesity is a primary driver of health care spending (Mokdad et al. 2001; Flegal et al. 2002; Bray 2004; Grundy 2004; Hu 2008; Dixon 2010). Obesity and its associated chronic health conditions raise medical expenditures, negatively affect the health care system, and result in productivity losses due to disability, illness, and premature mortality (Quesenberry, Bette, and Alice 1998; Finkelstein, Fiebelkorn, and Wang 2003; Andreyeva, Roland, and Ringel 2004). Compared to normal weight or overweight individuals, obese individuals have much higher mortality rates and higher risks of disability (Allison et al. 1999; Calle et al. 1999; Engeland et al. 2003; Fontaine et al. 2003; Peeters et al. 2003; Flegal et al. 2005; Sturm 2002; Sturm, Ringel, and Andreyeva 2004).

Many studies have estimated the effect of obesity on national health care costs (Finkelstein, Fiebelkorn, and Wang 2003; Thorpe et al. 2004; Finkelstein et al. 2009; Trasande et al. 2009; Cawley and Meyerhoefer 2012). According to Cawley and Meyerhoefer (2012), annual medical costs were estimated to be US$2,741 on average for men and women taken together, US$3,613 for women as a group, and US$1,152 for men as a group. Per capita medical spending for obese individuals was higher by US$1,429, roughly 42 percent higher, than for normal weight individuals (Finkelstein et al. 2009). For both men and women, the costs of obesity were much higher than the costs of being overweight. The annual medical costs of being obese were US$4,879 for an obese woman and US$2,646 for an obese man, whereas the annual costs of being overweight were US$524 and US$432 for women and men, respectively (Dor et al. 2010).

Cawley and Meyerhoefer (2012) estimated the medical care costs of obesity-related illness in adults in the United States to be US$209.7 billion in 2005. As such, roughly 21 percent of U.S. national health expenditures were spent treating obesity-related illnesses in 2005. Finkelstein et al. (2009) estimated the costs of obesity among full-time employees in the United States to be US$85.7 billion. Overweight and obese individuals were more likely to be absent from work due to health-related issues than non-obese and non-overweight individuals. Obese men miss up to two additional days of work and obese women miss up to five more work days annually compared to normal-weight men and women, respectively (Finkelstein, Ruhm, and Kosa 2005). Using the 2006 Medical Expenditure Panel Survey and the 2008 National Health Wellness Survey, Finkelstein et al. (2010) quantified per capita and aggregate medical expenditures and the value of lost productivity, including absenteeism. They estimated the annual cost of obesity among full-time employees to be US$73.1 billion; roughly one-fifth of this annual cost resulted from increases in absenteeism.

The effects of obesity on wages are well documented in the extant literature (Averett and Korenman 1996; Baum and Ford 2004; Cawley 2004; Morris 2007; Cawley, Han, and Norton 2009; Han, Norton, and Stearns 2009). In addition, there exists an extensive literature on the effect of obesity on employment (Morris 2007; Norton and Han 2008; Cawley, Han, and Norton 2009; Han, Norton, and Stearns 2009). Obese individuals have a substantially lower probability of being employed than do healthy-weight individuals (Morris 2007; Han, Norton, and Stearns 2009). Additionally, obesity has a negative impact on wages and leads to an increase in the cost of life insurance and other personal expenses. Cawley (2004) found that overweight and obese females of various ethnic groups tended to earn less compared to healthy-weight females. Furthermore, Dor et al. (2010) estimated annual wage losses of US$750 for obese men and US$1,855 for obese women. One explanation for lower wages among overweight and obese adults is that they are more likely to suffer from chronic diseases, thus leading to more expensive medical bills paid by employers (Cawley 2004; Yang and Hall 2008).

Ricci and Chee (2005) estimated that costs associated with reduced productivity were US$358 per obese worker per year. Goetzel et al. (2010) estimated the costs per obese worker to be US$54, while Gates et al. (2008) estimated these costs to be US$575. Thompson et al. (1998) estimated that an additional US$2.6 billion was spent on life insurance as a result of being overweight or obese. Compared to normal-weight individuals, overweight and obese individuals incurred an additional US$14 and US$111, respectively, in annual life insurance costs (Dor et al. 2010).

4 Role of government

U.S. government intervention efforts at federal, state, and city levels are focused on a multitude of methods to combat the U.S. obesity epidemic. These methods include the imposition of taxes at various stages of the supply chain (manufacturer and consumer level taxes) and quantity reduction efforts through regulations such as banning the consumption of certain oversized food and beverage products containing added sugar designed to lower calorie consumption.

Various studies in the literature have shown that the consumption of sugar-sweetened beverages (SSBs) has contributed to rising obesity rates in the United States (Ruyter et al. 2012; Ebbeling et al. 2012; Qi et al. 2012; Kaiser et al. 2013). The most widely proposed (and used) government intervention has been the use of excise or sales taxes on SSBs (Jacobson and Brownell 2000; Brownell et al. 2009; Chaloupka, Powell, and Chriqui 2009; Hahn 2009) and taxes on snack foods (Kuchler et al. 2005; Chouinard et al. 2007). According to Dharmasena, Davis, and Capps (2014), the empirical literature on evaluating taxes on SSBs can be partitioned into three groups. The first group of studies focused only on the direct effects of taxes on SSBs while ignoring the possible substitution effects as a result of a tax (Jacobson and Brownell 2000; Brownell et al. 2009; Andreyeva, Chaloupka, and Brownell 2011). The second group of studies incorporated possible substitution effects between unhealthy beverages that are high in added sugars (added calories) and healthy beverages that are naturally high in calories such as 100 percent fruit juices and milk (Finkelstein et al. 2010; Smith, Lin, and Lee 2010; Zhen et al. 2010; Lin et al. 2011; Dharmasena and Capps 2012). The third group of studies (Fletcher, Frisvold, and Tefft 2010; Sturm et al. 2010; Dharmasena, Davis, and Capps 2014) looked at differences across states in terms of soft drink tax rates and then estimated differences in caloric intake and weight associated with different tax rates.

In Table 5.1, various studies in the literature dealing with taxing beverages are compared in terms of data used, the outcomes of the tax implemented on caloric intake, and the reductions in body weight. According to these studies, it is clear that the reduction in body weight (and obesity) as a result of a tax on SSBs is quite small. To add to this finding, Finkelstein et al. (2013) estimated the effect of a 20 percent tax on SSBs on body weight when substitutions to high-calorie non-beverage items were considered. They found that the average per capita weight loss was on the order of 1.6 pounds in the first year and a cumulative weight loss of 2.9 pounds.

Furthermore, several studies investigated the impacts of SSB taxes on the supply side of the economy by calculating lost revenue to manufacturers (Andreyeva, Chaloupka, and Brownell 2011; Dharmasena, Davis, and Capps 2014). Bottom line, when taking into account substitution possibilities attributed to a tax on SSBs, the extant literature reveals that this intervention does not offer much to combat obesity.

As a result of these excise or sales taxes, the consumption of taxed food and beverages is expected to decline, which could have negative consequences on revenue generated by agri-businesses as well as on available employment opportunities. However, due to cross-price effects associated with taxed and non-taxed beverages, the consumption of non-taxed food and beverage products is expected to rise, which could increase calorie consumption, thus exacerbating obesity-related issues, especially if consumers substitute other high-calorie food and beverage products for the aforementioned taxed products (Dharmasena and Capps 2012).

Table 5.1 Comparison of studies on taxing SSBs in combating obesity in the United States

Study	Data	Tax work
Dharmasena and Capps (2012)	Monthly time series January 1998–December 2003 derived from Nielsen Homescan data	20% tax on SSBs (isotonics, regular soft drinks, fruit drinks) results in a reduction in body weight between 1.54 pounds per person and 2.55 pounds per person per year
Smith, Lin and Lee (2010)	Monthly time series January 1998–December 2007 derived from Nielsen Homescan data and 2003–2006 NHANES data	20% price increase induced by a tax on SSBs leads to a reduction in per capita body weight by 3.8 pounds per year
Zhen et al. (2010)	Cross-sectional data 2004–2006 from Nielsen Homescan data	One-half cent tax on SSBs results in moderate reduction in consumption of these beverages
Fletcher, Frisvold, and Tefft (2010)	Cross-sectional data 1999–2006 NHANES	Soft drink tax designed to reduce consumption is offset by the intake of high-calorie milk

Economists, politicians, nutritionists, and journalists have questioned the impact of U.S. farm policies in making so-called calorie-dense foods cheaper compared to their healthy counterparts. Alston, Rickard, and Okrent (2010) showed that U.S. farm policies have had modest and mixed effects on the prices and quantities of farm commodities, with very small effects on the prices paid by consumers for food and beverage products. As a result, it was concluded that the effects of U.S. farm policy on dietary patterns and obesity were marginal at best. Similar effects also have been shown by Alston, Sumner, and Vosti (2008), Beghin and Jensen (2008), and Okrent (2010). Importantly, these findings suggest that agricultural policy designed to assist domestic producers does not contribute to the obesity epidemic in the United States.

Another perspective on the role of government in reducing the obesity epidemic pertains to nutrition education programs designed to assist in decreasing the consumption of food and beverage products high in calories while increasing the consumption of fruits, vegetables, whole grains, and dietary fiber. The Dietary Guidelines for Americans (DGA), put together by the U.S. Department of Agriculture and the U.S. Department of Health and Human Services, provide valuable information concerning nutrition education for U.S. consumers in choosing healthy food and beverage products to mitigate obesity. A few studies in the existing literature have investigated the effect of the U.S. dietary guidelines on calorie intake and body weight. These studies generated mixed outcomes in regard to the effectiveness of the U.S. dietary guidelines on the intake of calories (Dharmasena, Capps, and Clauson 2011), whole grains (Mancino and Kuchler 2012), and dietary fiber (Senia and Dharmasena 2016).

The U.S. Supplemental Nutrition Assistance Program (SNAP), the Women's, Infants, and Children Program (WIC), and the National School Lunch Program (NSLP) were implemented to provide food assistance and nutrition interventions principally to low-income individuals. Several research studies in the literature investigated the impact of these programs on obesity (Wolf and Colditz 1998; Casey et al. 2001; Jyoti et al. 2005; Casey et al. 2006; Dubois et al. 2006; Yen et al. 2008; Finkelstein et al. 2009; Nord and Golla 2009; Gundersen, Kreider, and Pepper 2011; Cawley and Meyerhoefer 2012; Tiehen, Jolliffe, and Gundersen 2012; Dharmasena, Bessler, and Capps 2016). The extant literature has provided mixed outcomes. For example, some studies found that participation in food assistance programs led to increases in obesity rates in the United States (Cawley and Meyerhoefer 2012; Finkelstein et al. 2009; Wolf and Coldtiz 1998). Casey et al. (2001, 2006), Dubois et al. (2006), and Jyoti et al. (2005) found evidence to support a positive relationship between obesity and food insecurity, especially in children. According to Dharmasena, Bessler, and Capps (2016), obesity and participation in SNAP are *indirectly* related to race, income, poverty, food insecurity, and unemployment. Furthermore, obesity and food insecurity are *directly* related to income, food taxes, race, and ethnicity.

Moreover, Wilde and Nord (2005) suggested that federally sponsored promotion programs, known as commodity checkoff programs, result in increased consumption of beef, pork, and dairy products. As such, commodity checkoff programs may lead to increases in incidences of obesity. Hence, inconsistencies are evident with respect to government involvement in dealing with the obesity problem in the United States.

5 Insights from behavioral economics

Neoclassical economics assumes that rational agents make decisions based on full information and act according to their own self-interests. In reality, individuals rarely possess full information or unconstrained time and cognitive resources to make decisions. Incorporating psychological aspects into economics to explain human behavior has gained popularity following the works of Simon (1955), Kahneman and Tversky (1979), and Thaler (1980), among others. A key

question often asked about the use of behavioral economics principles is whether policy makers should "paternalistically" intervene to influence agent decisions. An emerging movement known as *libertarian paternalism* (Thaler and Sunstein 2003) or *asymmetric paternalism* (Camerer et al. 2003) poses that behavioral economics can be used for enacting policies designed to "nudge" individuals while still preserving their freedom of choice (Bhargava and Loewenstein 2015; Wansink 2015).

The general idea is that the cost of decisions can be reduced by simplifying choices or aligning "default choices" to make them beneficial to the agent's self-interest. For example, most restaurant menu options have one or two side dishes that are included in the price of the meal. In this case, most patrons will perceive not eating the side dishes as a loss since the cost of the meal is fixed. What if the "default" options were healthy alternatives, or if customers only had to buy the main entrée at a lower price and each side dish were charged separately? Just and Wansink (2011) suggest that there may be a reduction in calories consumed since individuals in a fixed-price context may eat more to get their money's worth. Note that this setup does not limit individual choices and the relative prices can be set to keep the total cost of the meal at the same level. This setup incorporates behavioral "nudges" and financial incentives. Although profit maximization of restaurant owners is not accounted for in this illustration, it is conceivable that a reduction in price of the main entrée may result in additional customers, thus increasing overall sales.

Behavioral economics can provide useful answers about the potential outcomes to obesity intervention programs designed to influence food choices and physical activities (Just 2006; Just and Payne 2009; Galizzi 2014; Liu et al. 2014; List and Samek 2015). However, obesity is a complex issue and the optimal weight of an individual can be affected by psychological, environmental, social, and cultural aspects related to appearance, self-esteem, and social norms (Levy 2002). Most obese individuals do not normally receive weight-related counseling (Bleich, Pickett-Blakely, and Cooper 2011). Simple measures such as medical preventative visits or counseling can reduce calorie intake and promote exercising (Loureiro and Nayga 2007; Bleich, Pickett-Blakely, and Cooper 2011).

Should policy be used to try to reduce obesity? The main arguments to do so are that the actions of an individual affect others and that the public cost of obesity is enormous (Yaniv, Rosin, and Tobol 2009). Males and females trying to lose weight use similar weight-loss strategies (Bish et al. 2005), but it is likely that self-image and identity factors yield asymmetric results in intervention programs by gender. Intervention programs should consider customizing activities by gender, age, and targeted goals (i.e., programs targeting extremely obese individuals need to set realistic goals to account for potential social and peer influence). In reality, weight loss is difficult to achieve, but sustaining attained weight losses over time may be even harder (Dragone 2009; Rosin 2012).

As previously discussed, one of the approaches to reduce obesity is to implement *direct policy interventions* such as imposing taxes on unhealthy food or adding subsidies for healthy food or exercise equipment. While some studies find that taxing unhealthy food may reduce consumption (Zhen et al. 2010), the full effects of such policies need to be carefully evaluated (Streletskaya et al. 2014). For example, Yaniv, Rosin, and Tobol (2009) showed that due to spillover income and substitution effects of food intake and leisure time for exercising, taxing unhealthy food or subsidizing healthy food may actually increase obesity.

Other policies may be categorized as improving transparency in information or promoting healthy food (Nayga 2008; Rusmevichientong et al. 2014). These policies implicitly assume that better-informed individuals make better food choices (Burton et al. 2006; Øvrum et al. 2012; Shimokawa 2013; Liu et al. 2014). But as pointed out by Wansink, van Ittersum, and

Painter (2004), this assumption may not hold. Prior studies have documented that the type of information presented to consumers and how it is presented to consumers influence their food choices (Burton et al. 2006; Drichoutis et al. 2008; Kim et al. 2012; Liaukonyte et al. 2013; Puhl, Peterson, and Luedicke 2013; Banterle and Cavaliere 2014; Becker et al. 2015; Zhu, Lopez, and Liu 2015;). Adding healthy options may effectively increase healthy choices in some cases (McCluskey, Mittelhammer, and Asiseh 2012), but healthy nutritional labels may decrease consumption due to the perceptions of tradeoffs between healthy food and taste (Berning, Chouinard, and McCluskey 2010). Low-fat nutrition labels may even result in the opposite intended effect by reducing consumption guilt or changing the perception of serving size (Wansink and Chandon 2006).

Commitment devices can be effective in changing food consumption behavior (Zhang and Rashad 2008; Chandon and Wansink 2012; Fan and Yanhong 2013; Guthrie, Mancino, and Lin 2015; Wansink 2015). Pioneering work by Wansink (2007) and his colleagues shows promising venues that can be effective in changing "mindless eating" behavior. Small changes in the food environment, such as choosing healthier alternatives (Just, Mancino, and Wansink 2007; Wansink and Chandon 2014), can produce positive results. As Wansink and Chandon (2014) cleverly put it: "It is easier to change our food environment than to change our mind" (p. 413).

The implementation of commitment devices is needed to evaluate the outcomes of financial and nonfinancial incentives and to assess their feasibility for public policy (List and Samek 2015). Economic and behavioral incentives may work to some extent, but programs also need to carefully consider that some individuals may lack the financial resources to purchase healthier food. The use of preventative measures targeting children and adolescents is likely to yield high returns (Boumtje et al. 2005). Targeting adolescents and children should be a priority since they are a vulnerable, at-risk group still forming life-long habits (Li et al. 2016).

6 Concluding remarks

Obesity is a complex, multifaceted health problem/disease in the United States involving medical, sociological, psychological, political, and economic dimensions. The pronounced focus on obesity prevention by government intervention programs has provided mixed outcomes in dealing with the widespread problem of obesity. Because many variables affecting and affected by obesity interact in a complex food–nutrition–consumer–producer–government interface, finding permanent solution(s) to the pervasive problem of obesity is as complex as the problem itself. Perhaps more concerted efforts through customized government intervention programs would provide a better outcome to curb obesity in the United States.

Alternatively, there are areas in which behavioral economics can provide valuable insights in dealing with the seemingly ubiquitous issue of obesity. These general areas can be grouped into the following categories: (1) isolate the effects of existing policies in a more controlled environment; (2) examine the role and feasibility of monetary and non-monetary incentives in changing outcomes of obesity through public policy; (3) evaluate the potential demand for commitment devices and the associated outcomes for obesity; (4) investigate the persistence of effective short-term policies over the longer term; and (5) identify potential asymmetries in obesity outcome gains by different at-risks groups to customize policy interventions (perhaps with a greater emphasis on children and adolescents). One potential negative externality of behavioral interventions is that the same behavioral principles can be used by firms to promote the consumption of their products, whether or not they are healthy. This begs the question as to what role, if any, policies should have in protecting consumers in what Bhargava and Loewenstein (2015) call "behavioral exploitation" (p. 398). That said, without question, the use of the guiding principles of behavioral economics in conjunction with other emerging disciplines such as neuroeconomics will likely continue to gain the attention of agricultural economists and policy makers as ways to combat obesity.

References

Allison, D.B., R. Zannolli, and K.M. Narayan. 1999. "The Direct Health Care Costs of Obesity in the United States." *American Journal of Public Health* 89(8):1194–1199.

Alston, J.M., B.J. Rickard, and A.M. Okrent. 2010. "Farm Policy and Obesity in the United States." *Choices* 25(3):470–479.

Alston, J.M., D.A. Sumner, and S.A. Vosti. 2008. "Farm Subsidies and Obesity in the United States: National Evidence and International Comparisons." *Food Policy* 33:470–479.

Anderson, P.M., K.F. Butcher, and P.B. Levine. 2003. "Maternal Employment and Overweight Children." *Journal of Health Economics* 22(3):477–504.

Andreyeva, T., F.J. Chaloupka, and K.D. Brownell. 2011. "Estimating the Potential of Taxes on Sugar-Sweetened Beverages to Reduce Consumption and Generate Revenue." *Preventive Medicine* 52:413–416.

Andreyeva, T., S. Roland, and J.S. Ringel. 2004. "Moderate and Severe Obesity Have Large Differences in Health Care Costs." *Obesity Research* 12:1936–1943.

Averett, S., and S. Korenman. 1996. "The Economic Reality of the Beauty Myth." *Journal of Human Resources* 31(2):304–330.

Banterle, A., and A. Cavaliere. 2014. "Is There a Relationship Between Product Attributes, Nutrition Labels, and Excess Weight? Evidence from an Italian Region." *Food Policy* 49(1):241–249.

Baum, C.L., and W.F. Ford. 2004. "The Wage Effects of Obesity: A Longitudinal Study." *Health Economics* 13(9):885–899.

Baum, C.L., and C.J. Ruhm. 2009. "Age, Socioeconomic Status, and Obesity Growth." *Journal of Health Economics* 28(3):635–648.

Becker, M.W., N.M. Bello, R.P. Sundar, C. Peltier, and L. Bix. 2015. "Front of Pack Labels Enhance Attention to Nutrition Information in Novel and Commercial Brands." *Food Policy* 56:76–86.

Beghin, J.C., and H.H. Jensen. 2008. "Farm Policies and Added Sugars in US Diets." *Food Policy* 33(6):480–448.

Berning, J.P., H.H. Chouinard, and J.J. McCluskey. 2010. "Do Positive Nutrition Shelf Labels Affect Consumer Behavior? Findings from a Field Experiment with Scanner Data." *American Journal of Agricultural Economics* 93(2):364–369.

Bhargava, S., and G. Loewenstein. 2015. "Behavioral Economics and Public Policy 102: Beyond Nudging." *American Economic Review* 105(5):396–401.

Bish, C.L., H.M. Blanck, M.K. Serdula, M. Marcus, H.W. Kohl, and L.K. Khan. 2005. "Diet and Physical Activity Behaviors Among Americans Trying to Lose Weight: 2000 Behavioral Risk Factor Surveillance System." *Obesity Research* 13(3):596–607.

Bleich, S.N., O. Pickett-Blakely, and L.A. Cooper. 2011. "Physician Practice Patterns of Obesity Diagnosis and Weight-Related Counseling." *Patient Education and Counseling* 82(1):123–129.

Boumtje, P.I., C.L. Huang, J. Lee, and B. Lin. 2005. "Dietary Habits, Demographics, and the Development of Overweight and Obesity Among Children in the United States." *Food Policy* 30(2):115–128.

Bray, G.A. 2004. "Medical Consequences of Obesity." *Journal of Clinical Endocrinology and Metabolism* 89(6):2583–2589.

Brownell, Kelly D., T. Farley, W.C. Willett, B.M. Popkin, F.J. Chaloupka, J.W. Thompson, D.S. Ludwig. 2009. "The Public Health and Economic Benefits of Taxing Sugar-Sweetened Beverages," *The New England Journal of Medicine*, Health Policy Report, 1–7.

Burton, S., E.H. Creyer, J. Kees, and K. Huggins. 2006. "Attacking the Obesity Epidemic: The Potential Health Benefits of Providing Nutrition Information in Restaurants." *American Journal of Public Health* 96(9):1669–1675.

Calle, E.E., M.J. Thun, J.M. Petrelli, C. Rodriguez, and C.W. Heat. 1999. "Body-Mass Index and Mortality in a Prospective Cohort of U.S. Adults." *New England Journal of Medicine* 341(15):1097–1105.

Camerer, C., S. Issacharoff, G. Loewenstein, T. O'Donoghue, and M. Rabin. 2003. "Regulation for Conservatives: Behavioral Economics and the Case for Asymmetric Paternalism." *University of Pennsylvania Law Review* 151(3):1211–1254.

Casey, P., P. Simpson, J. Gossett, M. Bogle, C. Champagne, C. Connell, D. Harsha, B. McCabe-Sellers, and J. Robbins. 2006. "The Association of Child and Household Food Insecurity with Childhood Weight Status." *Pediatrics* 118:1406–1413.

Casey, P., K. Szeto, S. Lensing, M. Bogle, and J. Weber. 2001. "Children in Food-Insufficient, Low-Income Families: Prevalence, Health and Nutrition Status." *Archives of Pediatrics and Adolescent Medicine* 155:508–514.

Cawley, J. 2004. "Impact of Obesity on Wages." *Journal of Human Resources* 39(2):451–474.

Cawley, J., E. Han, and E.C. Norton. 2009. "Obesity and Labor Market Outcomes Among Legal Immigrants to the United States from Developing Countries." *Economics and Human Biology* 7(2):153–164.

Cawley, J., and C.D. Meyerhoefer. 2012. "The Medical Cost of Obesity: An Instrumental Variables Approach." *Journal of Health Economics* 31:219–230.

Chaloupka, F.J., L.M. Powell, and J.F. Chriqui. 2009. "Sugar Sweetened Beverage Taxes and Public Health," Research Brief, Robert Wood Johnson Foundation and School of Public Health, University of Minnesota.

Chandon, P., and B. Wansink. 2012. "Does Food Marketing Need to Make Us Fat? A Review and Solutions." *Nutrition Reviews* 70(10):571–593.

Chou, S.Y., M. Grossman, and H. Saffer. 2004. "An Economic Analysis of Adult Obesity: Results from the Behavioral Risk Factor Surveillance System." *Journal of Health Economics* 23:565–587.

Chou, S.Y., I. Rashad, and M. Grossman. 2008. "Fast-Food Restaurant Advertising on Television and Its Influence on Childhood Obesity." *Journal of Law and Economics* 51(4):599–618.

Chouinard, H.H., D.E. David, J.T. LaFrance, and J.M. Perloff. 2007. "Fat Taxes: Big Money for Small Change." *Forum for Health Economics & Policy (Obesity)* 10(2):1–28.

Currie, J., S. Dellavigna, E. Moretti, and V. Pathania. 2010. "The Effect of Fast Food Restaurants on Obesity and Weight Gain." *Journal of Economics and Political Economy* 2(3):32–63.

Cutler, D.M., E.L. Glaeser, and J.M. Shapiro. 2003. "Why Have Americans Become More Obese?" *Journal of Economic Perspectives* 17(3):93–118.

Dharmasena, S., D.A. Bessler., and O. Capps Jr. 2016. "Food Environment in the United States as a Complex Economic System." *Food Policy* 61:163–175.

Dharmasena, S., and O. Capps Jr. 2012. "Intended and Unintended Consequences of a Proposed National Tax on Sugar-Sweetened Beverages to Combat the U.S. Obesity Problem." *Health Economics* 2:669–694.

Dharmasena, S., O. Capps Jr., and A. Clauson. 2011. "Ascertaining the Impact of the 2000 USDA Dietary Guidelines for Americans in the Intake of Calories, Caffeine, Calcium, and Vitamin C from At-Home Consumption of Non-Alcoholic Beverages." *Journal of Agricultural and Applied Economics* 43(1):13–27.

Dharmasena, S., G.C. Davis, and O. Capps Jr. 2014. "Partial Versus General Equilibrium Caloric and Revenue Effects Associated with a Sugar-Sweetened Beverage Tax." *Journal of Agricultural and Resource Economics* 39(2):157–174.

Dixon, J.B. 2010. "The Effect of Obesity on Health Outcomes." *Molecular and Cellular Endocrinology* 316(2):104–108.

Dor, A., C. Ferguson, C. Langwith, and E. Tan. 2010. *A Heavy Burden: The Individual Costs of Being Overweight and Obese in the United States*. Washington, DC: George Washington University.

Dragone, D. 2009. "A Rational Eating Model of Binges, Diets and Obesity." *Journal of Health Economics* 28(4):799–804.

Drichoutis, A.C., P. Lazaridis, R.M. Nayga, M. Kapsokefalou, and G. Chryssochoidis. 2008. "A Theoretical and Empirical Investigation of Nutritional Label Use." *The European Journal of Health Economics* 9(3):293–304.

Drichoutis, A.C., R.M. Nayga, and P. Lazaridis. 2012. "Food Away from Home Expenditures and Obesity Among Older Europeans: Are There Gender Differences?" *Empirical Economics* 42:1051–1078.

Dubois, L., A. Farmer, M. Girard, and M. Porcherie. 2006. "Family Food Insufficiency Is Related to Overweight Among Preschoolers." *Social Science and Medicine* 63:1503–1516.

Dunn, R.A. 2010. "The Effect of Fast-Food Availability on Obesity: An Analysis By Gender, Race, and Residential Location." *American Journal of Agricultural Economics* 92(4):1149–1164.

Ebbeling, C.B., H.A. Feldman, V.R. Chomitz, T.A. Antonelli, S.L. Gortmaker, S.K. Osganian, D.S. Ludwig. 2012. "A Randomized Trial of Sugar-Sweetened Beverages and Adolescent Body Weight." *New England Journal of Medicine* 367:1407–1417.

Engeland, A., S. Tretli, and T. Bjorge. 2003. "Height, Body Mass Index, and Prostate Cancer: A Follow-Up of 950 000 Norwegian Men." *British Journal of Cancer* 89(7):1237–1242.

Ewing, R., R. Brownson, and D. Berrigan. 2006. Relationship Between Urban Sprawl and Weight of United States Youth." *American Journal of Preventative Medicine* 31(6):464–474.

Ewing, R., T. Schmid, R. Killingsworth, A. Zlot, and S. Raudenbush. 2003. "Relationship Between Urban Sprawl and Physical Activity, Obesity and Morbidity." *American Journal of Health Promotion* 18:47–57.

Fan, M., and J. Yanhong. 2013. "Obesity and Self-control: Food Consumption, Physical Activity, and Weight-Loss Intention." *Applied Economic Perspectives and Policy* 36(1):125–145.

Finkelstein, E.A., M. DiBonaventura, S.M. Burgess, and B.C. Hale. 2010. "The Costs of Obesity in the Workplace." *Journal of Occupational and Environmental Medicine* 52(10):971–976.

Finkelstein, E.A., I.C. Fiebelkorn, and G. Wang. 2003. "National Medical Expenditures Attributable to Overweight and Obesity: How Much, and Who's Paying?" *Health Affairs* W3:219–226.

Finkelstein, E.A., C.J. Ruhm, and K.A. Kosa. 2005. "Economic Causes and Consequences of Obesity." *Annual Review of Public Health* 26:239–257.

Finkelstein, E.A., J.G. Trogdon, J.W. Cohen, and W. Dietz. 2009. "Annual Medical Spending Attributable to Obesity: Payer- and Service-Specific Estimates." *Health Affairs* 28(5):822–831.

Finkelstein, E.A., C. Zhen, M. Bilger, J. Nonnemaker, A.M. Fareoqui, and J.E. Todd. 2013. "Implications of a Sugar-Sweetened Beverage (SSB) Tax When Substitutions to Non-Beverage Items Are Considered." *Journal of Health Economics* 32(1):219–239.

Flegal, K.M., M.D. Carrol, C.L. Ogden, and C.L. Johnson. 2002. "Prevalence and Trends in Obesity Among U.S. Adults, 1999–2000." *Journal of the American Medical Association* 288(14):1723–1727.

Flegal, K.M., B.I. Graubard, D.F. Williamson, and M.H. Gail. 2005. "Excess Deaths Associated with Underweight, Overweight, and Obesity." *Journal of the American Medical Association* 293(15):1861–1867.

Fletcher, J.M., D.E. Frisvold, and N. Tefft. 2010. "The Effects of Soft Drink Taxes on Child and Adolescent Consumption and Weight Outcomes." *Journal of Public Economics* 94(11/12):967–974.

Fontaine, K.R., D.T. Redden, W. Chenxi, A.O. Westfall, and D.B. Allison. 2003. "Years of Life Lost Due to Obesity." *Journal of the American Medical Association* 289(2):187–193.

Galizzi, M.M. 2014. "What Is Really Behavioral in Behavioral Health Policy? And Does It Work?" *Applied Economic Perspectives and Policy* 36(1):25–60.

Gates, D.M., P. Succop, B.J. Brehm, G.L. Gillespie, and B.D. Sommers. 2008. "Obesity and Presenteeism: The Impact of Body Mass Index on Workplace Productivity." *Journal of Occupational and Environmental Medicine* 50(1):39–45.

Goetzel, R.Z., T.B. Gibson, M.E. Short, B.C. Chu, J. Waddell, J. Bowen, S.C. Lemon, I.D. Fernandez, R.J. Ozminkowski, M.G. Wilson, and D.M. DeJoy. 2010. "A Multi-Worksite Analysis of the Relationships among Body Mass Index, Medical Utilization, and Worker Productivity." *Journal of Occupational and Environmental Medicine* 52(Suppl 1):S52–S58.

Grundy, S.M. 2004. "Obesity, Metabolic Syndrome, and Cardiovascular Disease." *Journal of Clinical Endocrinology and Metabolism* 89(6):2595–2600.

Gundersen, C., B. Kreider, and J. Pepper. 2011. "The Economics of Food Security in the United States." *Applied Economic Perspectives and Policy* 33(3):281–303.

Guthrie, J., L. Mancino, and C-T.J. Lin. 2015. "Nudging Consumers Toward Better Food Choices: Policy Approaches to Changing Food Consumption Behaviors." *Psychology and Marketing* 32(5):501–511.

Hahn, R. 2009. "The Potential Economic Impact of a U.S. Excise Tax on Selected Beverages." Report prepared for the American Beverage Association, Washington, DC: Georgetown University, Center for Business and Public Policy.

Han, E., E.C. Norton, and S.C. Stearns. 2009. "Weight and Wages: Fat versus Lean Paychecks." *Health Economics* 18(5):535–548.

Hu, F. 2008. *Obesity Epidemiology: Methods and Applications.* New York: Oxford University Press, Inc.

Jacobson, M.F., and K.D. Brownell. 2000. "Small Taxes on Soft Drinks and Snack Foods to Promote Health." *American Journal of Public Health* 90(6):854–857.

Just, D.R. 2006. "Behavioral Economics, Food Assistance, and Obesity." *Agricultural and Resource Economics Review* 35(2):209–220.

Just, D.R., L. Mancino, and B. Wansink. 2007. *Could Behavioral Economics Help Improve Diet Quality for Nutrition Assistance Program Participants?* Washington, DC: USDUA/ERS.

Just, D.R., and C.R. Payne. 2009. "Obesity: Can Behavioral Economics Help?" *Annals of Behavioral Medicine* 38(1):47–55.

Just, D.R., and B. Wansink. 2011. "The Flat-Rate Pricing Paradox: Conflicting Effects of 'All-You-Can-Eat' Buffet Pricing." *Review of Economics and Statistics* 93(1):193–200.

Jyoti, D., E. Frongillo, and S. Jones. 2005. "Food Insecurity affects School Children's Academic Performance, Weight Gain, and Social Skills." *Journal of Nutrition* 135:2831–2839.

Kahneman, D., and A. Tversky. 1979. "Prospect Theory: An Analysis of Decision Under Risk." *Econometrica* 47(2):263–292.

Kaiser, K.A., J.M. Shikany, K.D. Keating, and D.B. Allison. 2013. "Will Reducing Sugar-Sweetened Beverage Consumption Reduce Obesity? Evidence Supporting Conjecture is Strong, But Evidence When Testing Effect Is Weak." *Obesity Reviews* 14(8):620–633. doi:10.1111/obr.12048.

Kim, H., L.A. House, G. Rampersaud, and Z. Gao. 2012. "Front-of-Package Nutritional Labels and Consumer Beverage Perceptions." *Applied Economic Perspectives and Policy* 34(4):599–614.

Kuchler, F., A. Tegene, and J.M. Harris. 2005. "Taxing snack Foods: Manipulating Diet Quality or Financing Information Programs." *Review of Agricultural Economics* 27(1):4–20.

Lakdawalla, D., and T. Philipson. 2002. "The Growth of Obesity and Technological Change: A Theoretical and Empirical Examination." NBER Working Paper 8946, National Bureau of Economic Research, Washington, DC.

Lakdawalla, D., T. Philipson, and J. Bhattacharya. 2005. "Welfare-Enhancing Technological Change and the Growth of Obesity." *American Economic Review* 95(2):253–257.

Levy, A. 2002. "Rational Eating: Can It Lead to Overweightness or Underweightness?" *Journal of Health Economics* 21 (5):887–899.

Li, Y., M.A. Palma, S.D. Towne, J.L. Warren, and M.G. Ory. 2016. "Peer Effects on Childhood Obesity from an Intervention Program." *Health Behavior and Policy Review* 3(4):323–335.

Liaukonyte, J., N.A. Streletskaya, H.M. Kaiser, and B.J. Rickard. 2013. "Consumer Response to 'Contains' and 'Free of' Labeling: Evidence from Lab Experiments." *Applied Economic Perspectives and Policy* 98(1):41–53.

Lin, B.H., T.A. Smith, J.Y. Lee, and K.D. Hall. 2011. "Measuring Weight Outcomes for Obesity Intervention Strategies: The Case of a Sugar-Sweetened Beverage Tax." *Economics and Human Biology* 9(4):329–41. doi:10.1016/j.ehb.2011.08.007.

List, J.A., and A.S. Samek. 2015. "The Behavioralist as Nutritionist: Leveraging Behavioral Economics to Improve Child Food Choice and Consumption." *Journal of Health Economics* 39:135–146.

Liu, P.J., J. Wisdom, C.A. Roberto, L.J. Liu, and P.A. Ubel. 2014. "Using Behavioral Economics to Design More Effective Food Policies to Address Obesity." *Applied Economic Perspectives and Policy* 36(1):6–24.

Loureiro, M.L., and R.M. Nayga, Jr. 2007. "Physician's Advice Affects Adoption of Desirable Dietary Behaviors." *Applied Economic Perspectives and Policy* 29(2):318–330.

Mancino, L., and F. Kuchler. 2012. "Demand for Whole-Grain Bread Before and After the Release of the Dietary Guidelines." *Applied Economic Perspectives and Policy* 34(1):76–101.

McCluskey, J.J., R.C. Mittelhammer, and F. Asiseh. 2012. "From Default to Choice: Adding Healthy Options to Kids' Menus." *American Journal of Agricultural Economics* 94(2):338–343.

Mokdad, A.H., B.A. Bowman, E.S. Ford, F. Vinicor, J.S. Marks, and J.P. Koplan. 2001. "The Continuing Epidemic of Obesity and Diabetes in the United States." *Journal of the American Medical Association* 286(10):1195–1200.

Morris, S. 2007. "The Impact of Obesity on Employment." *Labour Economics* 14(3):413–433.

Nayga, R.M. 2008. "Nutrition, Obesity, and Health: Policies and Economic Research Challenges." *European Review of Agricultural Economics* 35(3):281–302.

Nord, M., and A.M. Golla. 2009. "Does SNAP Decrease Food Insecurity? Untangling the Self-Selection Effect." Economic Research Report No 85, U.S. Department of Agriculture, Economic Research Service.

Norton, E.C., and E. Han. 2008. "Genetic Information, Obesity, and Labor Market Outcomes." *Health Economics* 17(9):1089–1104.

Okrent, A.M. 2010. *The Effects of Farm Commodity and Retail Food Policies on Obesity and Economic Welfare in the United States*. PhD dissertation, University of California, Davis.

Øvrum, A., F. Alfnes, V.L. Almli, and K. Rickertsen. 2012. "Health Information and Diet Choices: Results from a Cheese Experiment." *Food Policy* 37(5):520–529.

Peeters, A., J.J. Barendregt, F. Willekens, J.P. Mackenbach, and A. Al Mamun. 2003. "Obesity in Adulthood and Its Consequences for Life Expectancy: A Life-Table Analysis." *Annals of Internal Medicine* 138(1):24–32.

Philipson, T. 2001. "The World-Wide Growth in Obesity: An Economic Research Agenda." *Health Economics* 10:1–7.

Philipson, T.J., and R.A. Posner. 2003. "The Long-Run Growth in Obesity as a Function of Technological Change." *Perspectives in Biology and Medicine* 46(S3):S87–S107.

———. 2008. "Is the Obesity Epidemic a Public Health Problem? A Decade of Research on the Economics of Obesity." NBER Working Paper 14010, National Bureau of Economic Research, Washington, DC.

Plantinga, A.J., and S. Bernell. 2007. "The Association between Urban Sprawl and Obesity: Is it a Two-Way Street?" *Journal of Regional Science* 45(3):473–492.

Puhl, R., J.L. Peterson, and J. Luedicke. 2013. "Fighting Obesity or Obese Persons [Quest] Public Perceptions of Obesity-Related Health Messages." *International Journal of Obesity* 37(6):774–782.

Qi, Q., A.Y. Chu, J.H. Kang, M.K. Jensen, G.C. Curhan, L.R. Pasquale, P.M. Ridker, D.J. Hunter, W.C. Willett, E.B. Rimm, D.I. Chasman, F.B. Hu, L. Qi. 2012. "Sugar-Sweetened Beverages and Genetic Risk of Obesity." *New England Journal of Medicine* 367:1387–1397.

Quesenberry, C.P., C. Bette, and J. Alice. 1998. "Obesity, Health Services Use, and Healthcare Costs Among Members of a Health Maintenance Organization." *Archives of Internal Medicine* 158(5):466–472.

Rashad, I., M. Grossman, and S.Y. Chou. 2006. "The Supersize of America: An Economic Estimation of Body Mass Index and Obesity in Adults." *Eastern Economic Journal* 32:133–148.

Ricci, J.A., and E. Chee. 2005. "Lost Productive Time Associated with Excess Weight in the U.S. Workforce." *Journal of Occupational and Environmental Medicine* 47(12):1227–1234.

Rosin, O. 2012. "Weight-Loss Dieting Behavior: An Economic Analysis." *Health Economics* 21(7):825–838.

Rusmevichientong, P., N.A. Streletskaya, W. Amatyakul, and H.M. Kaiser. 2014. "The Impact of Food Advertisements on Changing Eating Behaviors: An Experimental Study." *Food Policy* 44:59–67.

Ruyter, J.C., M.R. Olthof, J.C. Seidell, and M.B. Katan. 2012. "A Trial of Sugar-Free or Sugar-Sweetened Beverages and Body Weight in Children." *New England Journal of Medicine*. 367:1397–1407.

Senia, M.C., and S. Dharmasena. 2016. "Ascertaining the Role of Socio-Economic-Demographic and Government Food Policy Related Factors on the Per Capita Intake of Dietary Fiber Derived from Consumption of Various Foods in the United States." http://purl.umn.edu/235757. Accessed 11 January 2017.

Shimokawa, S. 2013. "When Does Dietary Knowledge Matter to Obesity and Overweight Prevention?" *Food Policy* 38:35–46.

Simon, H.A. 1955. "A Behavioral Model of Rational Choice." *Quarterly Journal of Economics* 69(1):99–118.

Smith, T.A., B.H. Lin, and J.Y. Lee. 2010. *Taxing Caloric Sweetened Beverages: Potential Effects on Beverage Consumption, Calorie Intake and Obesity*. Washington, DC: USDA/ERS.

Streletskaya, N.A., P. Rusmevichientong, W. Amatyakul, and H.M. Kaiser. 2014. "Taxes, Subsidies, and Advertising Efficacy in Changing Eating Behavior: An Experimental Study." *Applied Economic Perspectives and Policy* 36(1):146–174.

Sturm, R. 2002. "The Effects of Obesity, Smoking, and Drinking on Medical Problems and Costs." *Health Affairs* 21(2):245–253.

Sturm, R., L.M. Powell, J.F. Chriqui, and F.J. Chaloupka. 2010. "Soda Taxes, Soft Drink Consumption, and Children's Body Mass Index." *Health Affairs* 29:1052–1058.

Sturm, R., J.S. Ringel, and T. Andreyeva. 2004. "Increasing Obesity Rates and Disability Trends." *Health Affairs* 23(2):199–205.

Thaler, R. 1980. "Toward a Positive Theory of Consumer Choice." *Journal of Economic Behavior and Organization* 1(1):39–60.

Thaler, R.H., and C.R. Sunstein. 2003. "Libertarian Paternalism." *American Economic Review* 93(2):175–179.

Thompson, D., J. Edelsberg, K.L. Kinsey, and G. Oster. 1998. "Estimated Economic Costs of Obesity to U.S. Business." *American Journal of Health Promotion* 13(2), 120–127.

Thorpe, K.E., C.S. Florence, D.H. Howard, and J. Peter. 2004. "The Impact of Obesity on Rising Medical Spending." *Health Affairs* 23(6):4480–4486.

Tiehen, L., D. Jolliffe, and C. Gundersen. 2012. *Alleviating Poverty in the United States: The Critical Role of SNAP Benefits*. Washington, DC: USDA/ERS.

Trasande, L., Liu, Y., Fryer, G., Weitzman, M. 2009. "Effects of Childhood Obesity on Hospital Care and Costs, 1999–2005." *Health Affairs* 28(4):w751–w760.

Wansink, B. 2007. *Mindless Eating: Why We Eat More Than We Think*. Bantam.

———. 2015. "Change Their Choice! Changing Behavior Using the CAN Approach and Activism Research." *Psychology and Marketing* 32 (5):486–500.

Wansink, B., and P. Chandon. 2006. "Can 'Low-Fat' Nutrition Labels Lead to Obesity?" *Journal of Marketing Research* 43(4):605–617.

———. 2014. "Slim by Design: Redirecting the Accidental Drivers of Mindless Overeating." *Journal of Consumer Psychology* 24(3):413–431.

Wansink, B., and M. Huckabee. 2005. "De-Marketing Obesity." *California Management Review* 47(4):6–18.

Wansink, B., K. van Ittersum, and J.E. Painter. 2004. "How Diet and Health Labels Influence Taste and Satiation." *Journal of Food Science* 69(9):S340–S346.

Wilde, P., and M. Nord. 2005. "The Effect of Food Stamps in Food Security: A Panel Data Approach." *Review of Agricultural Economics* 27(3):425–432.

Wolf, A.M., and G.A. Colditz. 1998. "Current Estimates of the Economic Cost of Obesity in the United States." *Obesity Research* 6(2):97–106.

Yang, Z., and A.G. Hall. 2008. "The Financial Burden of Overweight and Obesity Among Elderly Americans: The Dynamics of Weight, Longevity, and Health Care Cost." *Health Services Research Journal* 43(3):849–868.

Yaniv, G., O. Rosin, and Y. Tobol. 2009. "Junk-Food, Home Cooking, Physical Activity, and Obesity: The Effect of the Fat Tax and the Thin Subsidy." *Journal of Public Economics* 93(5/6):823–830.

Yen, S.T., M. Andrews, Z. Chen, and D.B. Eastwood. 2008. "Food Stamp Program Participation and Food Insecurity: An Instrumental Variables Approach." *American Journal of Agricultural Economics* 90(1):117–132.

Zhang, L.E.I., and I. Rashad. 2008. "Obesity and Time Preference: The Health Consequences of Discounting the Future." *Journal of Biosocial Science* 40(1):97–113.

Zhen, C., M.K. Wohlgenant, S. Karns, and P. Kaufman. 2010. "Habit Formation and Demand for Sugar-Sweetened Beverages." *American Journal of Agricultural Economics* 93(1):175–195.

Zhu, C., R.A. Lopez, and X. Liu. 2015. "Information Cost and Consumer Choices of Healthy Foods." *American Journal of Agricultural Economics* 98(1):41–55.

6
Behavioral economics in food and agriculture

Wen Li and David R. Just

1 Introduction

Over the past two decades, there has been a tremendous growth in the use of behavioral models and methods to address common topics in agricultural and applied economics. The initial forays into behavioral economics occurred due to a confluence of events. The long history of work in contingent valuation naturally led to questions about the commonly observed phenomenon known as the endowment effect, whereby individuals tend to place a greater value on an object they own than an identical object that they do not own. Indeed, it was the environmental economist Jack Knetsch who initially observed the endowment effect (Knetsch and Sinden 1984), leading to his famous work with Kahneman and Thaler (Kahneman, Knetsch, and Thaler 1991).

Over roughly the same period, the field of economics underwent radical changes in its approach to decision under risk. Theories of decision under risk have been a staple of the agricultural production literature from the outset. Agricultural economists' approaches to risk have roughly mirrored that of the general literature, progressing from maximizing expected profit (Brownlee and Gainer 1949), to maximizing expected utility of profit (Sandmo 1971), to models that now take account of behavioral anomalies – often based upon prospect theory (Collins, Musser, and Mason 1991). Agricultural economists have been active in this literature throughout, often providing some of the first practical applications of new theories of risky decision making.

One additional factor that led to the heavy infiltration of behavioral economics among agricultural economists is the tie to experimental research. Behavioral economics has a special relationship with experimental research, as it is often the only internally valid and credible way to isolate a behavioral anomaly (Roe and Just 2009). As the experimental economics literature has grown, agricultural economics programs have invested disproportionately in experimental labs and faculty who specialize in experimental methods. However, this is not a new phenomenon. Indeed, experiments in consumer choice (Stanton 2001, pp. 129–131) and production have a long history in agricultural economics (Stanton 2001, p. 4). Levitt and List (2009), for instance, credit this early work as laying the foundation upon which Vernon Smith and others built.

In this chapter, we will summarize the major contributions of the large and growing intersection of behavioral economics and food and agricultural economics. This work can be broken into two distinct literatures: consumer behavior and production decisions. We will first discuss the literature on food and consumer behavior, including discussions of both the research and current contributions to policy. We will then turn our attention to behavioral research, addressing production decisions and the potential for future policy contributions.

2 General consumption behaviors

Traditional research related to food consumption ordinarily uses standard economic theory to estimate price and income effects. More recently, researchers have applied behavioral economic concepts to understand an individual's consumption behaviors. The majority of this work is empirical, with some minor contributions to theory and modeling. On average, each individual makes more than 221 food-related decisions per day, although most individuals appear to be unaware of this behavior (Wansink and Sobal 2007). Several have argued that the sheer volume of decisions makes rational and informed decisions difficult, if not impossible.

General models of behavioral time discounting (Laibson 1997) suggest that individuals display time inconsistent preferences. The implication is that while individuals project a willingness to do difficult but good things in future periods (e.g., go on a diet, exercise, save money), when that time arrives their preferences will change and they will continue with their poor habits. If individuals are aware of this, they would prefer to use commitment devices to ensure better behavior in the future (O'Donoghue and Rabin 1999). Experimental evidence provides some support for this in a field setting. While the study by Just et al. (2008) found that college students who preselected their lunch items did not necessarily make healthier food choices than students who ordered their food while standing in line, two other studies with children found a positive impact (Miller et al. 2016; Hanks, Just, and Wansink 2013). In all three studies, students chose more vegetables, and in the latter two, students chose more fruit. However, in the first study, fewer students chose dessert items when ordering in line.

Much of the behavioral work on food selection has been squarely centered on addressing policy tradeoffs between providing consumers greater health information versus alternative measures. While health information can have some impact on household behaviors, often the effects are relatively modest. Mancino and Kinsey (2004) found that providing accurate health and diet information to consumers decreases their calorie intake by a small amount each meal. Using the data from the National Health Interview Survey, Variyam and Cawley (2006) found that nutrition labeling does not have a positive effect in the general population, although some sub-populations appear to be impacted. Indeed, response to nutrition information can differ substantially by race and education levels (Gould and Lin 1994; Elbel et al. 2009). The relatively modest impacts of health information have led both to calls for more direct paternalistic interventions and for the examination of behavioral measures.

Given the volume of food decisions, a prominent theory supposes that individuals make most food decisions with little cognitive investment, using heuristics and rules of thumb (Just and Payne 2009). Relying on such rules of thumb and heuristics often gives great power to the environment in which the decision is made over eventual food behaviors. For example, Wansink (1996) found that food package size influences product consumption (i.e., the larger the product package size, the greater the volume that will be consumed). Similar impacts can be found by changing the visibility, convenience, or suggested social norms within a situation (Just and Gabrielyan 2016). Milliman (1986) found that restaurant background music can influence customers' meal duration. Compared to fast background music, slow background music influences customers to spend more money on average and to lengthen their meal time by about 11 minutes. The people with whom one eats the meal can also affect one's eating behavior. De Castro (1994) found that eating meals with friends or family significantly increases the amount of food consumed and the time taken for the meal when compared to eating alone. The size of the eating group can also affect the duration of meals. For example, a group of two will spend more time eating than a group of four (Clendenen, Herman, and Polivy 1994).

One of the primary applications of behavioral economics in food consumption has been to school lunches. School lunches in many ways offer an ideal setting to test behavioral concepts because of the degree of control offered to a researcher and the relative ease in obtaining student-level observations. Some work in this direction has demonstrated how direct incentives and health information impact school lunch choices. List and Samek (2015) found that health information alone in the form of verbal educational messages, such as those detailing the benefits of eating fruit, does not persuade students to select healthier food items. This result is perhaps unsurprising given the vast literature on nutrition education. List and Samek find that verbal educational messages are only effective when combined with an incentive (e.g., rewarding students who choose fruit items for lunch).

Similarly, Just and Price (2013) found that providing rewards such as a quarter ($0.25) to students increases the percentage of students who eat at least one serving of fruits or vegetables by 38.5 percent, while a nickel ($0.05) reward increases fruit or vegetable selection by only 15.4 percent. This must be placed in contrast to simple manipulations of the environment that are relatively inexpensive. For example, the study by Greene et al. (2017) found an increase in fruit consumption of 23 percent when implementing strategies associated with the Smarter Lunchrooms program. This program is a collection of strategies that cost small fractions of a cent per serving (such as placing fruit in a visible location). Interestingly, comparing these two approaches suggests that paying individuals to eat fruit may not be the most economically efficient means of encouraging fruit consumption. Similar work engaged by a number of researchers demonstrates a clear ability to change consumption using relatively inexpensive environmental interventions (Just and Wansink 2010; Hakim and Meissen 2013; Siegel et al. 2015; Adams et al. 2016).[1]

The Smarter Lunchrooms program has been incorporated into several state and national policies, including the Healthier U.S. School Challenge, encouraging schools to adopt the program as part of their overall wellness approach. Other work has examined the potential to use similar techniques to influence SNAP and WIC participants in grocery stores (Payne et al. 2015). Such efforts are in their infancy, but hold promise that after some trial and error, programs could help encourage healthier choices without restricting choice sets or spurring backlash among program participants.

3 Agricultural production and behavioral economics

While food choice in many ways makes for an ideal setting to conduct behavioral experiments, agricultural production presents a severe challenge. While many food decisions can be cast as random experiments where many consumers make nearly identical decisions, each farm will face very unique production possibilities and constraints. Nonetheless, there has long been intuition, if not evidence, that farmer behaviors do not quite fall in line with the neoclassical models of production theory. Many of these early hints came from the literature on decisions under risk.

3.1 Risk in agricultural production

After the boom and bust in commodity prices surrounding the onset of World War I, there was a general push to stabilize agricultural prices with the explicit argument that farmers should be protected from price risk (Camp 1924). During that era, economists had a hard time squaring such arguments with common economic models at the time. Up to that point, farms and other producers had primarily been modeled as maximizing profit – or, in the case of risk, expected profit. Stabilizing prices was primarily argued based upon the notion that stable prices would help alleviate credit constraints (Drummond 1948). It was not until the era of Sandmo (1971) and Turnovsky (1973) that economists could reconcile the desire to reduce risk with profit motives, via the introduction of expected utility theory.[2]

While expected utility theory is built upon rational axioms (Savage 1954), at its heart, expected utility theory is a behavioral model. It was originally proposed to reconcile the St. Petersburg paradox – that individuals are unwilling to pay an infinite amount for gambles with infinite expected value, offering larger and larger payouts with vanishing probabilities. The notion that one will value their thousandth dollar less than their first dollar is essentially a behaviorally rational explanation of why such gambles offer only finite value to the potential player.

While the introduction of expected utility theory and diminishing marginal utility of wealth was able to explain general motivations to avoid risk, it is still difficult to reconcile several puzzling behaviors. Expected utility theory commonly assumes that decision makers display diminishing marginal utility of wealth over the support of wealth considered under all relevant choices. If this is the case, farmers should be willing to reduce their expected profits by reasonable amounts in order to reduce the variation of profits. Nonetheless, while farmers do appear to seek a general reduction of risk, for example, through crop diversification (Pope and Prescott 1980), they do not tend to use futures markets to the extent necessary to ensure output prices (Shapiro and Brorsen 1988). This seems a particular conundrum because doing so would not decrease profits on average. Moreover, Goodwin and Schroeder (1994) find that risk attitude is not related to hedging behavior in any intuitive way.

Additionally perplexing is farmers' general reluctance to purchase subsidized crop insurance. Introduced in the 1930s, federally subsidized crop insurance has been offered in several different formats over the years. Barnett and Skees (1995) note that for many years, despite the fact that federal crop insurance offered a significant reduction in downside yield risk and an increase in expected profits, adoption rates were relatively low. Alternatively, expected utility theory suggests that anyone who is risk averse should be willing to purchase such insurance. Alternative explanations suggest that farmers are purchasing purely to obtain the subsidies and that the scale of risk is not salient. Du, Feng, and Hennessy (2016) demonstrate that coverage choices are inconsistent with expected utility theory.

Finally, Just and Peterson (2003, 2010) point out a puzzle that is at once more esoteric, but perhaps more fundamental. They derive a tool to calibrate the minimum amount of diminishing marginal utility that would explain particular observed risk-averse behavior. When applying this tool to estimated stochastic profit functions from the agricultural production literature, they find that observed behavior suggests that farmers' utility functions must be negatively sloped over some portion of wealth within the practical support of profit. This suggests that estimates of stochastic profit functions are severely misspecified, farmers are not maximizing expected utility of profit, or perhaps both.

3.2 Using behavioral models to address the puzzles

Each of these puzzles could be reconciled by the use of behavioral models of risk that account for additional motivations when dealing with risk. Perhaps the most commonly employed alternative model of risk is prospect theory (Kahneman and Tversky 1979; Tversky and Kahneman 1992). This model supposes that the decision maker displays loss aversion – displaying risk-averse behavior over potential gains, but risk-loving behavior over losses. In addition to loss aversion, prospect theory also presumes that individuals display probability weighting. That is, individual decision makers treat small probabilities as if they are larger and treat large probabilities as if they are smaller. Interestingly, loss aversion provides one coherent explanation that could potentially explain each of the three behavioral puzzles, although the plausibility of the explanations is still the subject of substantial debate.

Mattos, Garcia, and Pennings (2008) demonstrate that prospect theory predicts much lower ratios of hedge positions in the futures market to value of commodity being hedged. In this case, probability weighting does a majority of the work. While hedge ratios for a loss-averse decision maker with no probability weights behaves nearly identically to a simple risk-averse decision maker (with some exceptions based upon whether loss or gains are calculated historically or contemporaneously), adding probability weights can drive hedge ratios down to ranges that are much closer to those observed in practice.

A study by Pennings and Garcia (2001) uses survey methods to determine how futures contract behavior relates to various measures of risk attitude. Tantalizingly, they find risk attitudes that conform to loss aversion (risk-loving for low prices and risk-averse for higher prices), and they find that risk attitudes are related to hedging activity. However, their work is designed only to find associations between general risk-aversion measures and behavior, and thus cannot provide strong support for this behavioral model.

Du, Feng, and Hennessy (2016) suggest cumulative prospect theory as a potential solution to the low observed adoption rates for subsidized crop insurance as well. Indeed, Babcock (2015) picks up this thread by calibrating a prospect theoretic model to the crop insurance decisions of three representative farms (a Nebraska corn farm, a Kansas wheat farm, and a Texas cotton farm). In implementing such a model, it is important to determine the reference point and whether one should consider insurance decisions jointly with production decisions or consider them separately.

Thaler's (1999) theory of mental accounting (which incorporates prospect theory) makes ambiguous predictions because he notes that sometimes individuals will group transactions together when determining gains or losses, and at other times they will separate such transactions. Whether the decision maker integrates or segregates transactions can have substantive impacts on whether they consider themselves better or worse off after a series of transactions. Nonetheless, there is reasonable effort that individuals are guided in whether to integrate or segregate by the way in which the transactions are framed (called hedonic framing), rather than strategically deciding whether they want to integrate or segregate transactions (called hedonic editing) (Thaler 1999). Hedonic framing is influenced by the timing and order of decisions as well as the way in which the decisions are pitched (see Just 2013 for a full review). In this case, Babcock (2015) finds that farmers behave in a way that is consistent with prospect theory in which insurance decisions are segregated from the production decisions. This may indicate that crop insurance is sold in a way that emphasizes the impact of the insurance given historical production performance.

Prospect theory has also been proposed as a solution to the problem of excessive risk aversion, as was observed by Just and Peterson (2003, 2010) in the context of agricultural production.

Rabin (2000) noted the problem of excessive risk aversion in small gambles, developing a key theorem to help identify when such behavior suggests expected utility theory is not a possible explanation for observed behavior. Rabin's calibration theorem (which addresses only dichotomous choice for simple gambles) and the general calibration tool developed by Just and Peterson (2010) for continuous distributions and continuous choice rely on the notion of finding the potential utility function with the minimum change in slope over the support of a risky choice. This yields a function that consists of two connected line segments. Essentially, if one makes a choice that displays excessive risk aversion in a small gamble, it will imply even greater excessive risk aversion in larger gambles. In extreme cases, the behavior in the smaller gamble can imply negatively sloped portions of the utility function, which would seem to suggest the behavior is entirely inconsistent with expected utility theory.

Just and Peterson (2010) note that whenever one is making a continuous choice involving risk, the decision maker is trading off small amounts of risk on the margin. Thus, expected utility theory is extremely sensitive to the impact of small changes in continuous choices. Because empirical approaches have not considered this phenomenon, it is possible to estimate utility functions and production functions that look entirely consistent with expected utility theory, despite the fact that such functions could not predict the observed behavior. Indeed, the authors have not yet discovered an empirical application in agriculture that does not suffer from this excessive risk-aversion problem.

By creating additional degrees of freedom in the minimization problem, prospect theory can allow for excessive risk aversion in small gambles without implying excessive risk aversion in larger gambles. It is tempting, therefore, to suggest that prospect theory could explain the apparent misspecification problem that is apparently general to the agricultural production literature. Indeed, Bocqueho, Jacquet, and Reynaud (2013) find support for prospect theoretic behavior among French farmers using a field experiment. In the United States, Collins, Musser, and Mason (1991) also find some evidence of prospect theoretic production decisions. Nonetheless, it is not likely that prospect theory could account for all of the excessive risk aversion. While prospect theory frees global risk aversion from the tyranny of local risk aversion near the reference point, it does not add any new degrees of freedom when we move away from the reference point. The same problems can arise when we thus compare two similar gambles that only have outcomes in the gain domain. Accounting for all of the excessive risk aversion would render the concept of a reference point vacuous in terms of predictive theory. Furthermore, Neilson (2001) notes that prospect theory can actually exacerbate the problem given the assumed shape of probability weighting functions.

3.3. Identifying behavioral risk models

Several other positive approaches to behavioral risk in agricultural production have been published, although few that have had any long lasting impact on the profession. Buschena and Zilberman (2000) use experimental evidence to show that many of the behavioral models of risk are consistent with expected utility theory with a heteroscedastic error term that is correlated with how similar the risky alternatives considered are. In effect, the behavioral anomalies are only likely to occur where expected utility was similar to begin with. Buschena and Zilberman (1994) review the importance of the non-expected utility approach for addressing agricultural production issues. Buschena (2003) later speculates that the most likely behavioral anomaly to affect agricultural production in important ways are violations of the independence axiom. Most generalized expected utility theories address such violations, which arise when individuals do not value gambles as a linear combination of the value of the individual potential outcomes of the gamble.

Just and Lybbert (2009), using experimental data, suggest that perhaps farmers respond to changes in risk rather than changes in outcomes. This result is suggested by the observation that participants in a lab-in-the-field experiment displayed overall risk-averse choices, but that on the margin they preferred more risk. Such behavior, if general, would be inconsistent with any of the well-known theories of risky behavior. Chambers (2015) later points out that their experiment was ill-suited to the task at hand and may result from an inappropriate measure of risk.

Lee, Bellemare, and Just (2015) test this notion of risk-taking at the margin explicitly in the context of Sandmo in a set of three experiments. Within these experiments, subjects were given production functions and a description of the distribution of output prices. They also find that risk preferences on the margin are different from overall risk preferences; however, the pattern appears to be entirely different from that of Just and Lybbert. In essence, participants decide to produce more when risk is present than when there is no risk present, suggesting risk-loving behavior overall. Alternatively, participants choose to produce less when they face greater price risk, suggesting risk-averse behavior on the margin. This behavior was observed both with American students as participants and with Peruvian agricultural household heads. The results create an additional interesting puzzle that may be important in determining both how farmers respond to risk and how markets with evolving price risk function.

While there have been many forays into the application of behavioral risk models in agricultural production, their use is still not widespread. One of the primary reasons for this is the substantial challenge of econometric identification of production and risk preferences generally (Just and Just 2011). Each of the prominent behavioral models can be thought of as an expected utility model, with degrees of freedom added to account for behavioral anomalies. These additional degrees of freedom only add to the difficulty of identifying behavioral models of risk in applications to agricultural production. Indeed, Just and Just (2016) find that it is generally impossible to test entire classes of behavioral risk models against one another in an applied setting due to the additional degrees of freedom. Specifically, any behavior that can be described by an augmented utility function can also be described by an augmented probability weighting function unless we introduce restrictions based upon assertion of theory. At some level, we must use our intuition to restrict the types of models we will consider before the data can express a preference for one over the other. Without this realization, the economics literature addressing risk and production is prone to muddy results that will often produce conflicting conclusions regarding behavior. Indeed, in order to come to any agreed-upon approach, we must embrace some common set of assumptions or restrictions for imposing the prominent models in question.

3.4 Ambiguity in agricultural production

A more recent and somewhat narrower literature has examined the decisions of agricultural producers in response to ambiguity. Where risk is characterized by decisions in which the decision maker can assign probabilities to all possible outcomes of all possible choices, ambiguity is characterized by situations in which the decision maker may not be able to assign a single set of probabilities or may not be able to identify all possible outcomes (Knight 1921). Like risk, there is experimental evidence that individuals tend to be ambiguity-averse (Ellsberg 1961). Bougherara et al. (2017) find experimental evidence that farmers are indeed ambiguity-averse, meaning that they will choose well-defined outcomes over undefined outcomes – even if there is some suggestion that the undefined outcomes could dominate.

Ambiguity aversion potentially plays a central role in the behaviors that are so key to agricultural economics. However, the relative difficulty of applying the concept of ambiguity aversion empirically has been a challenge to advancing this literature. The potential importance is highlighted by Barham et al. (2014), who examine the adoption of genetically modified (GM) crops by Midwestern farmers. They use laboratory experimental measures of both risk aversion and ambiguity aversion to look for potential behavioral impacts on the timing of adoption of GM crops. Despite the long and prominent literature examining the impact of risk aversion on adoption, they find only a modest relationship. Alternatively, they find a sizable impact of ambiguity aversion on early GM corn adoption – a relationship that is absent for GM soy. Barham et al. argue that this is due to the prominence of Bt corn, which resists pests, potentially reducing ambiguity surrounding the potential for pest damage to crops. Alternatively, the GM soy available in the United States is primarily developed to resist herbicides, reducing labor and other costs that are perhaps much less impacted by ambiguity. This suggests that many of the issues we model as risk may in fact be better modeled as ambiguity – and that the distinction is important when examining producer decisions.

3.5 Potential for application to agricultural policy

Agricultural policy applications of behavioral relationships have been extremely limited and disconnected from the academic research on behavioral production decisions. While research on behavioral production has focused almost entirely on risk and ambiguity, applications to policy have mainly focused on participation in government programs. The United States Department of Agriculture (USDA) under the Obama administration began to show significant interest in using behavioral tools specifically to influence environmental and stewardship decisions of farmers. Such applications are the primary focus of the Center for Behavioral and Experimental Agri-Environmental Research (CBEAR), established in 2014. One of the first prominent studies along these lines examines the potential to nudge bidding in a competitive auction for incentives to reduce farm runoff (Wu, Palm-Forster, and Messer 2017). A field experiment was conducted on farmers in the Chesapeake, Delaware, and Galveston Bay watersheds. Farmers would bid by proposing cost shares for potential projects, with a variety of potential projects available. Bids were entered on a sliding scale, although the researchers randomized the start position of the slider. Importantly, they find that starting the slider at 100 percent marginally (but significantly) increased the cost share of the bids, leading to somewhat greater benefits for dollars spent for the program.

While behavioral design of agricultural policy is not widespread, efforts from groups like CBEAR show there is potential for improved efficiency. Similar efforts have been used to examine whether language-framing programs to encourage adaptation to climate change may be deterring use (Ellis et al. 2016). Farmers in general make up a broad and diverse set of interests, encompassing several different political and economic motives. The use of A–B behavioral tests when introducing new programs or materials could enhance communications, encourage greater participation, and avoid potential waste. While the efforts of CBEAR to introduce such a paradigm are modest in scope, the potential importance is tremendous.

4 Conclusions

Advances in the use of behavioral economics in food and agricultural economics has largely mirrored the rise of behavioral economics within the broader discipline. Behavioral and experimental economists are found in each of the research active departments across the United States and in much of Europe. While behavioral research has been broadly applied to many of the key areas of research with agricultural and applied economics, there has been substantially greater emphasis on consumer choice both in research and policy applications. This is likely due at least in part to the greater access researchers might have to food selection and consumption decisions for field experimentation. This has led both to a large encyclopedic literature discovering behavioral anomalies in food choice and a growing body of policy applications designed to take advantage of such anomalies to encourage greater health.

The use of experiments among farmers is much more limited. This is because it can be difficult to find venues to recruit farmers and because providing realistic incentives for farmers is prohibitive (particularly in developed countries). The large-scale collection of farm production data through surveys, censuses, and tax records helped propel agricultural economics to the leading edge of early econometric applications. However, such measurements are poorly equipped to reveal behavioral effects because they provide little opportunity to identify deviations from rational behavior.

This is not to say that behavioral work in production agriculture is not relevant. There is substantive evidence both that behavioral models provide more parsimonious explanations of some producer behaviors, and that such behavioral models can form the basis for more effective policy implementation. As behavioral work among developing country farmers has made substantive contributions (Mullainathan 2005), the work in developed countries is both promising and necessary. Behavioral work in general business strategy and management suffers from the same barriers to widespread data collection, yet these literatures have found venues for experimentation through school advisory boards (Bogan and Just 2009), executive education (Fehr and List 2004), and other significant gatherings. In paving a path toward future application, it may be important to begin to adapt techniques from the general business literature to agricultural production ventures.

Notes

1 An annotated bibliography of similar results can be found here: www.smarterlunchrooms.org/sites/default/files/documents/SLM%2060%20point%20Scorecard%20Lit%20Review.pdf.
2 Interestingly, expected utility theory was used a decade earlier to argue that farmers were risk-loving (Johnson 1962).

References

Adams, M.A., M. Bruening, P. Ohri-Vachaspati, and J.C. Hurley. 2016. "Location of School Lunch Salad Bars and Fruit and Vegetable Consumption in Middle Schools: A Cross-Sectional Plate Waste Study." *Journal of the Academy of Nutrition and Dietetics* 116(3):407–416.

Babcock, B.A. 2015. "Using Cumulative Prospect Theory to Explain Anomalous Crop Insurance Coverage Choice." *American Journal of Agricultural Economics* 97(5):1371–1384.

Barham, B.L., J.P. Chavas, D. Fitz, V.R. Salas, and L. Schechter. 2014. "The Roles of Risk and Ambiguity in Technology Adoption." *Journal of Economic Behavior & Organization* 97:204–218.

Barnett, B.J., and J.R. Skees. 1995. "Region and Crop Specific Models of the Demand for Federal Crop Insurance." *Journal of Insurance Issues* 19:47–65.

Bocqueho, G., F. Jacquet, and A. Reynaud. 2013. "Expected Utility or Prospect Theory Maximisers? Assessing Farmers' Risk Behaviour from Field-Experiment Data." *European Review of Agricultural Economics* 41(1):135–172.

Bogan, V., and D. Just. 2009. "What Drives Merger Decision Making Behavior? Don't Seek, Don't Find, and Don't Change Your Mind." *Journal of Economic Behavior & Organization* 72(3):930–943.

Bougherara, D., X. Gassmann, L. Piet, and A. Reynaud. 2017. "Structural Estimation of Farmers' Risk and Ambiguity Preferences: A Field Experiment." *European Review of Agricultural Economics* 44(5):782–808. doi:10.1093/erae/jbx011.

Brownlee, O.H., and W. Gainer. 1949. "Farmers' Price Anticipations and the Role of Uncertainty in Farm Planning." *Journal of Farm Economics* 31(2):266–275.

Buschena, D.E. 2003. "Expected Utility Violations: Implications for Agricultural and Natural Resource Economics." *American Journal of Agricultural Economics* 85(5):1242–1248.

Buschena, D.E., and D. Zilberman. 1994. "What Do We Know about Decision Making under Risk and Where Do We Go from Here?" *Journal of Agricultural and Resource Economics* 19:425–445.

———. 2000. "Generalized Expected Utility, Heteroscedastic Error, and Path Dependence in Risky Choice." *Journal of Risk and Uncertainty* 20(1):67–88.

Camp, W.R. 1924. "The Organization of Agriculture in Relation to the Problem of Price Stabilization: I." *Journal of Political Economy* 32(3):282–314.

Chambers, R.G. 2015. "On Marginal-Risk Behavior." *American Journal of Agricultural Economics* 98(2):406–421.

Clendenen, V.I., C.P. Herman, and J. Polivy. 1994. "Social Facilitation of Eating Among Friends and Strangers." *Appetite* 23(1):1–13.

Collins, A., W.N. Musser, and R. Mason. 1991. "Prospect Theory and Risk Preferences of Oregon Seed Producers." *American Journal of Agricultural Economics* 73(2):429–435.

De Castro, J.M. 1994. "Family and Friends Produce Greater Social Facilitation of Food Intake Than Other Companions." *Physiology & Behavior* 56(3):445–455.

Drummond, W.M. 1948. "Objectives and Methods of Government Pricing of Farm Products." *Journal of Farm Economics* 30(4):665–679.

Du, X., H. Feng, and D.A. Hennessy. 2016. "Rationality of Choices in Subsidized Crop Insurance Markets." *American Journal of Agricultural Economics* 99(3):732–756.

Elbel, B., R. Kersh, V.L. Brescoll, and L.B. Dixon. 2009. "Calorie Labeling and Food Choices: A First Look at the Effects on Low-Income People in New York City." *Health Affairs* 28(6):w1110–w1121.

Ellis, S.F., J.R. Fooks, K.D. Messer, and M.J. Miller. 2016. "The Effects of Climate Change Information on Charitable Giving for Water Quality Protection: A Field Experiment." *Agricultural and Resource Economics Review* 45(2):319–337.

Ellsberg, D. 1961. "Risk, Ambiguity, and the Savage Axioms." *The Quarterly Journal of Economics* 75(4):643–669.

Fehr, E., and J.A. List. 2004. "The Hidden Costs and Returns of Incentives – Trust and Trustworthiness Among CEOs." *Journal of the European Economic Association* 2(5):743–771.

Goodwin, B.K., and T.C. Schroeder. 1994. "Human Capital, Producer Education Programs, and the Adoption of Forward-Pricing Methods." *American Journal of Agricultural Economics* 76(4):936–947.

Gould, B.W., and H.C. Lin. 1994. "Nutrition Information and Household Dietary Fat Intake." *Journal of Agricultural and Resource Economics* 19(2):349–365.

Greene, K.N., G. Gabrielyan, D.R. Just, and B. Wansink. 2017. "Fruit-Promoting Smarter Lunchrooms Interventions: Results from a Cluster RCT." *American Journal of Preventive Medicine* 52(4):451–458.

Hakim, S.M., and G. Meissen. 2013. "Increasing Consumption of Fruits and Vegetables in the School Cafeteria: The Influence of Active Choice." *Journal of Health Care for the Poor and Underserved* 24(2):145–157.

Hanks, A.S., D.R. Just, and B. Wansink. 2013. "Preordering School Lunch Encourages Better Food Choices By Children." *JAMA Pediatrics* 167(7):673–674.

Johnson, P.R. 1962. "Do Farmers Hold a Preference for Risk?" *Journal of Farm Economics* 44(1):200–207.

Just, D.R. 2013. *Introduction to Behavioral Economics*. New York: John Wiley.

Just, D.R., and G. Gabrielyan. 2016. "Food and Consumer Behavior: Why the Details Matter." *Agricultural Economics* 47(S1):73–83.

Just, R.E., and D.R. Just. 2011. "Global Identification of Risk Preferences with Revealed Preference Data." *Journal of Econometrics* 162(1):6–17.

Just, D.R., and R.E. Just. 2016. "Empirical Identification of Behavioral Choice Models Under Risk." *American Journal of Agricultural Economics* 98(4):1181–1194.

Just, D.R., and T.J. Lybbert. 2009. "Risk Averters That Love Risk? Marginal Risk Aversion in Comparison to a Reference Gamble." *American Journal of Agricultural Economics* 91(3):612–626.

Just, D.R., and C.R. Payne. 2009. "Obesity: Can Behavioral Economics Help?" *Annals of Behavioral Medicine* 38(1):47–55.

Just, D.R., and H.H. Peterson. 2003. "Diminishing Marginal Utility of Wealth and Calibration of Risk in Agriculture." *American Journal of Agricultural Economics* 85(5):1234–1241.

———. 2010. "Is Expected Utility Theory Applicable? A Revealed Preference Test." *American Journal of Agricultural Economics* 92(1):16–27.

Just, D.R., and J. Price. 2013. "Using Incentives to Encourage Healthy Eating in Children." *Journal of Human Resources* 48(4):855–872.

Just, D., and B. Wansink. 2010. "Better School Meals on a Budget: Using Behavioral Economics and Food Psychology to Improve Meal Selection." *Choices* 24(3):19–24.

Just, D.R., B. Wansink, L. Mancino, and J. Guthrie. 2008. "Behavioral Economic Concepts to Encourage Healthy Eating in School Cafeterias." ERR-68. United States Department of Agriculture, Economic Research Service, Washington, DC.

Kahneman, D., J.L. Knetsch, and R.H. Thaler. 1991. "Anomalies: The Endowment Effect, Loss Aversion, and Status Quo Bias." *The Journal of Economic Perspectives* 5(1):193–206.

Kahneman, D., and A. Tversky. 1979. "Prospect Theory: An Analysis of Decision Under Risk." *Econometrica* 47(2):263–291.

Knetsch, J.L., and J.A. Sinden. 1984. "Willingness to Pay and Compensation Demanded: Experimental Evidence of an Unexpected Disparity in Measures of Value." *The Quarterly Journal of Economics* 99(3):507–521.

Knight, F.H. 1921. *Risk, Uncertainty, and Profit*. Boston, MA: Houghton-Mifflin.

Laibson, D. 1997. "Golden Eggs and Hyperbolic Discounting." *The Quarterly Journal of Economics* 112(2):443–478.

Lee, Y.N., M.F. Bellemare, and D.R. Just. 2015. "Was Sandmo Right? Experimental Evidence on Attitudes to Price Uncertainty." Working Paper, University of Minnesota, Minneapolis, MN.

Levitt, S.D., and J.A. List. 2009. "Field Experiments in Economics: The Past, the Present, and the Future." *European Economic Review* 53(1):1–18.

List, J.A., and A.S. Samek. 2015. "The Behavioralist as Nutritionist: Leveraging Behavioral Economics to Improve Child Food Choice and Consumption." *Journal of Health Economics* 39:135–146.

Mancino, L., and J. Kinsey. 2004. "Diet Quality and Calories Consumed: The Impact of Being Hungrier, Busier, and Eating Out." Working Paper 04–02, University of Minnesota, Minneapolis, MN.

Mattos, F., P. Garcia, and J.M. Pennings. 2008. "Probability Weighting and Loss Aversion in Futures Hedging." *Journal of Financial Markets* 11(4):433–452.

Miller, G.F., S. Gupta, J.D. Kropp, K.A. Grogan, and A. Mathews. 2016. "The Effects of Pre-Ordering and Behavioral Nudges on National School Lunch Program Participants' Food Item Selection." *Journal of Economic Psychology* 55:4–16.

Milliman, R.E. 1986. "The Influence of Background Music on the Behavior of Restaurant Patrons." *Journal of Consumer Research* 13(2):286–289.

Mullainathan, S. 2005. *Development Economics Through the Lens of Psychology*. Washington, DC: World Bank.

Neilson, W.S. 2001. "Calibration Results for Rank-Dependent Expected Utility." *Economics Bulletin* 4(10):1–5.

O'Donoghue, T., and M. Rabin. 1999. "Doing It Now or Later." *American Economic Review* 89(1):103–124.

Payne, C.R., M. Niculescu, D.R. Just, and M.P. Kelly. 2015. "Shopper Marketing Nutrition Interventions: Social Norms on Grocery Carts Increase Produce Spending without Increasing Shopper Budgets." *Preventive Medicine Reports* 2:287–291.

Pennings, J.M., and P. Garcia. 2001. "Measuring Producers' Risk Preferences: A Global Risk-Attitude Construct." *American Journal of Agricultural Economics* 83(4):993–1009.

Pope, R.D., and R. Prescott. 1980. "Diversification in Relation to Farm Size and Other Socioeconomic Characteristics." *American Journal of Agricultural Economics* 62(3):554–559.

Rabin, M. 2000. "Risk Aversion and Expected-Utility Theory: A Calibration Theorem." *Econometrica* 68(5):1281–1292.

Roe, B.E., and D.R. Just. 2009. "Internal and External Validity in Economics Research: Tradeoffs Between Experiments, Field Experiments, Natural Experiments, and Field Data." *American Journal of Agricultural Economics* 91(5):1266–1271.

Sandmo, A. 1971. "On the Theory of the Competitive Firm Under Price Uncertainty." *The American Economic Review* 61(1):65–73.

Savage, L.J. 1954. *The Foundations of Statistics*. New York: John Wiley.

Shapiro, B.I., and B.W. Brorsen. 1988. "Factors Affecting Farmers' Hedging Decisions." *North Central Journal of Agricultural Economics* 10(2):145–153.

Siegel, R.M., A. Anneken, C. Duffy, K. Simmons, M. Hudgens, M.K. Lockhart, and J. Shelly. 2015. "Emoticon Use Increases Plain Milk and Vegetable Purchase in a School Cafeteria Without Adversely Affecting Total Milk Purchase." *Clinical Therapeutics* 37(9):1938–1943.

Stanton, B.F. 2001. *Agricultural Economics at Cornell: A History, 1900–1990*. Ithaca, NY: Cornell University.

Thaler, R.H. 1999. "Mental Accounting Matters." *Journal of Behavioral Decision Making* 12(3):183.

Turnovsky, S.J. 1973. "Production Flexibility, Price Uncertainty, and the Behavior of the Competitive Firm." *International Economic Review* 14(2):395–413.

Tversky, A., and D. Kahneman. 1992. "Advances in Prospect Theory: Cumulative Representation of Uncertainty." *Journal of Risk and Uncertainty* 5(4):297–323.

Variyam, J.N., and J. Cawley. 2006. "Nutrition Labels and Obesity." NBER w11956. National Bureau of Economic Research, Washington, DC.

Wansink, B. 1996. "Can Package Size Accelerate Usage Volume?" *The Journal of Marketing* 60(3):1–14.

Wansink, B., and J. Sobal. 2007. "Mindless Eating: The 200 Daily Food Decisions We Overlook." *Environment and Behavior* 39(1):106–123.

Wu, S., L. Palm-Forster, and K.D. Messer. 2017. "Heterogeneous Agents and Information Nudges in Non-Point Source Water Pollution Management." Working Paper, University of Delaware, Newark, DE.

7
Product quality and reputation in food and agriculture

Anthony R. Delmond, Jill J. McCluskey, and Jason A. Winfree

1 Introduction

The agricultural industry and food consumers have a long history of product quality concerns. For example, the first few problems in the Rhind Papyrus showed a very early use of fractions to create food quality standards for bread and beer (Clagett 1999). However, while food standards have a long history, they are not always adequate in creating optimal food quality. Standards can be difficult to implement and enforce given the multitude of food quality dimensions and the lack of quality information for consumers at the time of purchase. Therefore, consumers often rely on the opinion of other consumers regarding the reputation of the food.

The foundations of economic research on product quality and reputation are firmly rooted in the literature of industrial organization and game theory. While abstract analytical models explaining how firms and consumers use and react to quality abound, difficulties in the process of definition, determination, and quantification of quality have presented an early hindrance to researchers attempting to adapt the theoretical models for real-world applications. Without the ability to identify data for quality, it is nearly impossible to empirically examine reputations for quality. New data sources, techniques, and creative interpretations allow for empirical applications to move forward, but the progress has been gradual. This is especially true for food quality and reputation; therefore, general economics research can supplement our knowledge of food quality and the effect on consumers.

Agricultural goods once were treated as fairly homogenous in the market. However, differentiation of products along quality lines has grown increasingly important in the food and agricultural industries. In the current age of biotechnological advances, production has spiked and the reliance of firms on product specificity and quality to stake claims to some market share has become the norm rather than the exception. With the proliferation of product specificity (e.g., "grass-fed Wagyu beef" and "Mendocino County Organic Merlot") – especially when quality exhibits some degree of *ex ante* or *ex post* uncertainty – comes an increasing need to examine the strategic decisions of various agents in the market. As economics expands further into behavioral explanations of the consumer decision making process, *perception* of quality is increasingly becoming an important component of economic models. Both consumers and producers stand to benefit from a more detailed understanding of how their economic counterpart invests in, perceives, and exploits information about production quality.

This chapter follows the evolution of economic research on product differentiation, specifically along the quality axis, and continues through the major developments and economic applications of the various types of reputation. The discussion begins with a general overview of the theoretical underpinnings of product quality in Section 2. This concept is disaggregated into (1) exogenous product quality in which product quality is externally determined and (2) endogenous product quality in which firms consciously choose product quality to maximize their objectives. Section 3 continues with an overview of the economics of reputation, examining individual and collective reputations. Policy implications are covered in Section 4. Section 5 briefly discusses a common methodological approach used in empirical studies of quality and reputation before going through some important applications. Section 6 concludes with some suggestions for future directions for the field.

2 Product quality

Products can be differentiated along either the horizontal or the vertical dimension.[1] Horizontally differentiated products differ by characteristics in which consumers disagree over the optimal level. An example is the level of spiciness in food. The optimal level of spice for a given consumer can range from totally bland to extremely hot. Vertical differentiation, on the other hand, implies an objective ordering on which all consumers can agree, with all consumers preferring more of the vertical quality attribute to less. An example is food safety. *Ceteris paribus*, all consumers should prefer safer food. Consumers may have a different willingness to pay for a vertical product attribute. Different levels of willingness to pay for a vertical attribute usually depend upon differences in income or the intensity of other preferences, such as sensitivity to pathogens in the food safety case.

Regarding horizontal product differentiation, Hotelling's (1929) location model projects product heterogeneity onto a linear scale where prices are reactionary and firms choose the distance between their competing products' attributes to maximize profits. Salop (1979) further develops this model, removing the endpoints of Hotelling's linear "city," introducing a circular product space, and incorporating monopolistic competition explicitly. The classic findings of Salop's (1979) model indicate that under certain circumstances the market may overprovide variety (in his case, firms) compared to socially optimal levels.

Seminal works by Mussa and Rosen (1978), Gabszewicz and Thisse (1979), and Shaked and Sutton (1982) examine vertical product differentiation. Mussa and Rosen (1978) analyze product quality decisions for firms that can produce multiple goods of various qualities. Their model has heterogeneous consumers, and product quality depends upon the market structure. They argue that monopolists will generally sell a lower quality product to a given consumer when compared to competitive markets. Gabszewicz and Thisse's (1979) model describes two firms, each offering a good of a different quality: one high and the other low. Consumer preferences are homogeneous, but incomes vary. They show that income differences are the driving force behind product choice when quality is observable and consumer preferences are identically ranked from high-quality to low-quality goods. Shaked and Sutton (1982) derive an equilibrium using a game theoretic approach. In their three-stage sequential-move game, firms first choose whether to enter, then set quality, and lastly set price. They find that, conditional on a two-product upper bound on the number of surviving products in the market and their additional assumptions, the only perfect equilibrium has two firms entering the market and producing distinct products.

Although it has become the norm in economic analysis to distinguish between horizontal and vertical differentiation, rarely does product differentiation fall exclusively along one dimension. Many of the current topics in agriculture involve food attributes that do not fall neatly into the classification of either vertical or horizontal differentiation. For example, many consumers have a preference for organic food, but not all consumers agree whether "organic" implies that the food is of higher quality. Some consumers may see organic food as simply a different kind of food. Other consumers have a preference for food that is not genetically modified. In the case of genetically modified food, it is not obvious if modified or non-genetically modified food is of higher quality. While some consumers warn of dangers from genetically modified food, some genetic modifications can lead to better nutrition.

One can think of situations in which there is both horizontal and vertical differentiation. As an example, wines are produced in different regions with different terroirs and from different grape varieties that appeal to the heterogeneous preferences of consumers, but all consumers would prefer a wine that is not "corked." Neven and Thisse (1989) present the first model of double differentiation using a technically complex model. Duvaleix-Tréguer et al. (2012) make a simplifying assumption in the modeling of consumer preferences for the vertical attribute. They assume that two groups of consumers are each located at one point on the vertical axis of quality rather than distributed uniformly on this axis. They point out that this assumption allows them to avoid the technical complexity associated with Neven and Thisse's (1989) model without inducing significant effects on the robustness of the qualitative results. Gabszewicz and Wauthy (2012) nest the two approaches into a duopoly (and triopoly with entry) model that allows for varying consumer preferences for horizontally differentiated products (i.e., varying natural market size across firms). They find that in equilibrium, prices are increasing in the level of symmetry between the two markets, and a horizontally differentiated market is always less competitive (in price terms) than a vertically differentiated market.

All of these approaches to product differentiation explicitly treat quality as observable by the consumer. However, since quality is not always observable *ex ante*, the mere existence of heterogeneous levels of product quality is only part of the story. Models incorporating information asymmetries are required to examine further the economics behind products of heterogeneous quality. If a firm's reputation influences demand for its product, then the incentives for production and quality will depend upon the characteristics of the market. A key factor is whether the product quality is exogenous to the firm (i.e., outside the firm's control), endogenous (i.e., influenced directly or indirectly by the firm's actions), or a combination of the two.

2.1 Exogenous product quality

Akerlof (1970) describes a market in which only sellers know the product quality. This asymmetric information creates a scenario where sellers with high quality goods do not want to enter the market because buyers will underestimate the quality. In this example, product quality is exogenous and the seller cannot change the product quality. This can describe agricultural products when the quality of food is determined by factors such as weather or other exogenous factors. Akerlof's solution to this asymmetric information problem is to issue a guarantee or a warranty for the product. While a warranty may be infeasible when selling food directly to consumers, it is possible to offer guarantees or incentives through contracts in the food supply chain. Food suppliers and processors often have contracts that are dependent upon food quality.

Leland (1979) also models a market with asymmetric information and considers minimum quality standards and licensing to fix the information problem. While consumers can potentially benefit from some type of standard, Leland points out that it can also be used to exert market power, which can be harmful to consumers. This seems to describe the benefits and costs of quality standards found in agricultural marketing orders. Some marketing orders create quality standards for commodities. Given that the United States Department of Agriculture (USDA) is concerned with food quality, they have been in favor of such quality standards. However, the United States Department of Justice is more concerned with market power issues, so they have been against such standards.

Martin (2017), treating quality as exogenously drawn from a distribution, takes a behavioral approach, arguing that some consumers rationally choose to ignore quality. Martin extends the classic signaling game for quality by allowing the buyer to choose an information structure that stochastically generates posterior beliefs prior to the purchase decision. Attention is costly for consumers, and if the cost of information gets too high, some consumers will choose to buy a product of uncertain quality. Firms set prices knowing the marginal cost of consumers' attention.

2.2 Endogenous product quality

When product quality is endogenous, firms can invest in technology to directly or indirectly shape consumers' perception of their products. Much of the literature on firm reputation has focused on forms that can endogenously determine their product quality at a cost. Dorfman and Steiner (1954) distinguish endogenous quality decisions as those that influence the demand curve for a product at some variable cost. They find that the production of higher quality products depends upon consumer preferences, price elasticity of demand, and the relationship between average costs and changes in quality.

McCluskey (2000) uses a repeated game approach to analyze firms' incentives to provide quality with different degrees of asymmetric information. She shows that with experience goods, high quality products can be provided in repeat-purchase relationships as long as the firm cares enough about the future. With credence goods, she shows that for authentic high-quality credence goods to be available in the market there is a minimum necessary level of monitoring (i.e., probability of getting caught cheating) that depends upon the price premium for the high-quality product, the difference in production costs between the high-quality product and a conventional product, and the discount rate.

Kranton (2003) explains that when firms face an endogenous quality choice in production, the institutional structure under which they operate and compete determines their incentives. She finds that the reputation mechanism enforced by consumers is itself insufficient to incentivize firms to produce high quality. Her argument is premised upon collusive pricing acting as a deterrent to firms switching from high to low quality. If firms have the ability to punish those that deviate to produce lower quality through the price mechanism, then a high-quality equilibrium can be maintained. However, if no credible punishment strategy exists, then every firm should have an incentive to deviate and there will be no high-quality product in the competitive equilibrium. Klein and Leffler (1981) argue that enforcement by consumers can maintain high quality, but as their argument relies on reputation, it is discussed in Section 3.

3 Reputation

When product quality is unobservable prior to purchase, as in the case of experience goods (Nelson 1970), consumers rely on firm and product reputation to inform their purchase decisions. Consumers can either construct those reputations individually from personal past experience, or, as Nelson describes, they can use "guided sampling" of the opinions of other consumers. Information asymmetries, in which sellers know the quality of their products but consumers do not, influence the incentives of those sellers to invest in quality. Those incentives often are at the expense of consumers. While food is often cited as an example of a product whose quality is unknown at the time of purchase, many, if not most, products involve some degree of quality uncertainty prior to consumption. It is in these circumstances that reputations matter to consumers making purchase decisions. For many businesses, reputation is a main driver of demand for their products. A firm producing experience goods has several options available to build its reputation. It can treat its reputation as an asset, independently building its own brand through investment in quality (Shapiro 1982). Alternatively, a firm can rely partly on the reputation of some group of which it is a member (Tirole 1996).

The level of aggregation of reputations can vary from individual agents to individual firms to small or large regions. In many instances, businesses utilize some combination of reputational factors in their marketing schemes to exploit all potential avenues for maximizing revenue. The depth of aggregation employed by firms is determined, at least in part, by consumers' ability to identify and distinguish between products at each level. For example, if consumers are unable to construct a preference ordering between identical varieties of apples from two different counties in Washington, then drawing a distinction, *ceteris paribus*, between the reputations of the two regions is futile. In that case, reputation aggregated at the county level would not be a fruitful enterprise.

Development of firm reputation, collective reputation, or a combination of both depends upon the characteristics of the market. Markets with firm reputation only offer products that are tractable to the firm, and the firm's location is irrelevant. An example is Starbucks. Other markets exhibit only collective reputation. In these markets, the product is non-traceable to the firm, and there is a recognizable region of production. An example of collective-only reputation is Washington apples. When consumers buy apples at the grocery store, they do not know which farmer grew the apples. Other markets feature products with dual reputations (i.e., both firm and collective reputations). In this situation, the product is traceable to the firm, and the location is relevant. This is typically the case for wine, such as those from Leonetti Cellars in Walla Walla, Washington.

Some of the empirical research discussed in this section disentangles the effects of different levels of reputation aggregation to determine the sources of their marginal benefits. Much of this research relates to consumers' willingness to incur the search costs associated with independently determining quality and verifying reputations.

3.1 Firm reputation

Shapiro (1982) offers a model in which revenues depend upon current reputation, which is a function of past quality. He shows that imperfect information will lead to a lower investment in quality when compared with perfect information. As we will go on to discuss, many applications of reputation in agricultural and food settings will use some variant or extension of Shapiro's (1982) model. Shapiro (1983) shows that one solution to the imperfect information problem is a quality standard. Another possible solution is to improve consumer information and the speed with which consumers can update their information.

Treating reputations as non-enforceable contracts between producers and consumers, Klein and Leffler (1981) explain that consumers can punish sellers who revert to a lower quality product independent of government enforcement simply by terminating their business relationship. Additionally, the higher the sunk costs incurred in brand-name capital investment, the larger the potential losses to the firm from switching to a lower-than-expected quality. This approach assumes that markets can effectively supply a range of quality without any explicit regulation, and consumers can rely on price as an indicator of quality. Reputation itself incentivizes firms to meet consumers' expected levels of quality as long as the firms' expected payoff for a single period of deviation is sufficiently low. Klein and Leffler's result does not necessarily hold for credence goods – goods whose quality is unknown both before and after purchase. In an agricultural context, genetically modified organisms (GMOs) in the 1990s provide a good example of credence goods. Consumers had a more limited comprehension of the long-run health effects of genetically modified foods and agricultural products before regulatory capacity and risk research caught up with the agricultural biotechnology (Jasanoff 1997; Gaskell et al. 1999).

As Darby and Karni (1973) show, competitive markets are insufficient to prevent fraud by established firms because consumers cannot verify the quality of the services rendered. When Rogerson (1983) takes those credence goods described by Darby and Karni (1973) and proposes a model to include reputation and word-of-mouth advertising, he finds that there exist incentives for firms to produce higher quality goods because high-quality firms have more satisfied and repeat customers. The less impaired the consumers' ability to differentiate product quality, either *ex ante* or *ex post*, the higher the share of high-quality products in the market.

3.2 Collective reputation

There may be cases where consumers rely on a reputation but do not distinguish between individual firms. For example, consumers probably do not know the individual Florida orange grower or the individual Walla Walla onion grower. In these cases, consumers rely on collective reputations. Tirole (1996) first modeled collective reputation and showed that individuals will want to leave groups with a low reputation. If individuals or firms cannot leave a group, their incentive to maintain the collective reputation is stronger if the group currently has a high reputation.

Winfree and McCluskey (2005) model a collective reputation and show that firms have an incentive to under-produce quality and free ride on the collective reputation. Because firms do not receive the full benefit of an investment into quality, they will not invest the socially optimal amount into it. Fleckinger (2007) also examines collective reputation and shows that alleviating the free riding problem by limiting the number of firms can also create inefficiencies due to market power.

While minimum quality standards are a potential solution to firms free-riding on a collective reputation, it is not always feasible. For example, the jurisdiction of a standard may not be the same as the location that shares the collective reputation. Or firms may try to cheat on the standard. For example, Olmstead and Rhode (2003) give the example of low-quality cotton growers who produced their crop in one region but sold their crop in a different region known for high-quality cotton in the early twentieth century. The high-quality growers were unable to stop the free riding on their reputation, so they were unable to sustain higher prices for their cotton. Severson (2014) and Bushak (2014) give more recent examples. Under these conditions, Winfree (2016) argues that rather than increasing the standard, which may be bad for incumbent firms, the focus should be on enlarging the jurisdiction or increasing firm traceability. Saak (2012) examines the scenario where firms can monitor and punish other firms at a cost. Under this model, social norms may play an important role in certain groups, such as agricultural cooperatives. Social norms may be an easier and less costly way of maintaining quality as opposed to a quality standard.

While collective reputations may be geographic, they may also be industry specific. Berning, Costanigro, and McCullough (2017) argue that the craft beer brewing industry utilizes collective reputation. Given the number of craft brewers, they often rely on collective names. Rather than promoting their individual breweries, many brewers might promote the style of beer or the geographic location for the production of the beer. Some of the geographic promotions by the breweries can also coincide with tourism efforts in the area.

Often, a collective reputation is shared by a country. Donnenfeld and Mayer (1987) and Chiang and Masson (1988) make this assumption in an international trade setting. Donnenfeld and Mayer argue that exporting countries might impose a voluntary restraint to ensure that average quality is maintained, which helps all exporting firms in that country. Similarly, Chiang and Masson argue that limiting the number of firms will alleviate free riding on quality for exporting countries. McQuade, Salant, and Winfree (2016) also assume that each country has a reputation and suggest a further partitioning of geographic areas as a solution to the collective reputation problem. For example, while France as a country has a reputation for wine, it also has geographical areas (regions) within the country that are noted for their reputation for particular types of wines. This may be a preferable solution, as opposed to limiting the number of firms or output, because of market power concerns. If consumers can identify the region, this helps alleviate free riding on the reputation since there are fewer firms in the region. However, the effectiveness of this type of strategy hinges upon consumers being able to identify and remember the quality of products from more refined regions.

When a collective reputation stems from a geographic location, geographic protections are sometimes used. One reason that regions use geographical names is to promote the reputation of that area. Another reason is to protect the reputation by not allowing other regions to free ride on that reputation. For example, the Champagne region in France wants to be the only region that produces champagne and the Parma region in Italy wants to be the only region that produces parmesan cheese. These protections help protect the region's brand and reputation, but the possible downside is that it could create inefficiencies from market power. The magnitude of the benefits and costs of these policies may largely depend on the knowledge of consumers.

3.3 Dual reputation

Although most of the literature focuses on either firm or collective reputations, in many instances they are co-existent. A prominent example is the wine industry. Wine is one of a class of products in which each firm wants to develop its own reputation for quality and each firm wants its region to produce the highest quality. Geographical indications (GIs), which are common in Europe, are another example. Costanigro, Bond, and McCluskey (2012) analyze markets where reputations are dual. In the dual reputation strategic game, each firm also considers the marginal value of its own and collective reputation, but positive externalities accruing to other firms are not internalized. Menapace and Moschini (2012) use a dual reputation approach. They begin with Shapiro's (1983) model of firms utilizing trademarks to indicate firm reputation and extend it to include the aggregate reputation of geographical indications. They show the complementary nature of these dual reputations. Firms' costs may diminish because the amount of individual investment in a collective reputation is below that necessary to sustain an individual reputation. Their model maintains that certification is necessary to guarantee a certain level of quality to consumers and to avoid firms within a geographical indication free riding on the quality investments of other members.

4 Empirical work

Much empirical work has been undertaken in food and agricultural economics using the theoretical underpinnings of the reputation models described earlier. Hedonic regression analysis is a frequent approach used to isolate specific product attributes and examine those components' (e.g., reputation) effects on price. Though many papers have expounded upon and improved the process, the seminal paper developing the hedonic model was that of Rosen (1974). For a discussion of hedonic price analysis in food markets, see Costanigro and McCluskey (2011). Hedonic analysis is a powerful methodological tool, and it features prominently in many of the empirical papers discussed here.

One bountiful area of empirical research, owing largely to its unique and broad-ranging data sources, is wine, which is often treated as an experience good for which the consumer only knows the true quality after consumption. In terms of reputation construction, wine offers the unique advantage of comprising both individual firm reputations and collective regional reputations. Firms have a choice of whether to utilize collective reputations in their branding, but most research indicates that consumers use regional information in their purchase decisions for wine (Atkin and Johnson 2010; Menival and Charters 2014). Atkin and Johnson (2010), examining consumer use of geographic information using data collected in an online survey, find that while brand is the element used by the largest percentage of respondents (76.7 percent), the second and third most utilized elements were those comprising a collective reputation: country (55.7 percent) and region (55.2 percent).

Many economic studies of the wine industry treat expert reviews as a signal of quality and some aggregate lagged expert reviews to form individual or collective reputations for firms or regions, respectively. Schamel and Anderson (2003) perform a hedonic regression using vintage rating for quality, external winery rating as firm reputation, and regional dummy variables for collective regional reputations, among other regressors, to determine the value of quality and reputation on price differences of wine in Australia and New Zealand. They find that regional quality differentiation in Australia is more striking than in New Zealand, and the firm reputation variable also exhibits smaller and less significant parameter estimates for the former compared to the latter.

Modeling reputation as high-quality or low-quality dummy variables based on one standard deviation above or below the regional average, Schamel (2009) shows that as regional reputations gain credence, prices rely more on collective reputation than individual firm reputation. Costanigro, McCluskey, and Goemans (2010) take the empirical individual versus collective reputation comparison a step farther, deducing that for higher price quantiles of wine, consumers rely more on individual firm reputation than collective reputation. In other words, as the price premiums for quality increase, consumers are more willing to spend time researching individual firms than just relying on the collective reputation of the region. The reluctance of consumers to bear search costs is inversely related to the price.

Using data on Bordeaux wine, Landon and Smith (1997, 1998)[2] construct a novel set of aggregate reputation variables. These include several specifications at the firm and regional levels using quality ratings from a wine magazine and a set of collective reputation dummy variables at the quality-classification level using an official ranking system dating back to a request by Napoleon III in 1855. Their estimation results show that both firm and collective reputation are key determinants of price, and reputation is actually more significant for price determination than true improvements in product quality.

Examining wine producers in a coalition with minimum quality standards, Castriota and Delmastro (2015) demonstrate that collective reputation and group size exhibit an inverted-U relationship (i.e., as the number of producers in a coalition increases, that coalition's reputation increases to a maximum and decreases thereafter). Institutional signals such as classification systems, however, are not significant after controlling for other variables. Their research suggests that there is an optimal size for a group sharing a collective reputation. If the group is too small, it is difficult to promote or establish a successful marketing campaign; if the group is too large, there is less of an incentive to produce high quality and more incentive to shirk or free ride.

Very few researchers have considered how new reputations are formed and can be influenced. Rickard, McCluskey, and Patterson (2015) designed a laboratory experiment and auction to estimate consumer responses to information tying U.S. wine regions to high-reputation wine production regions in France. Their results show that discussing the similarities of new U.S. wine regions to the high-reputation French regions increases bids for the wines produced in burgeoning U.S. wine regions. The effect of "reputation tapping," or linking by comparison of a product to a high-reputation region, had a stronger effect for products from emerging regions compared to more established regions.

5 Policy implications

Policy implications can vary widely depending upon the assumptions of each model or estimation results. However, many of the policy implications and prescriptions involve trying to maintain a certain level of quality, either through minimum quality standards, warrantees, certifications, licenses, or other restrictions on certain products, and either consumer-imposed, firm-imposed, or imposed by some external regulating body.

Several of the papers mentioned earlier examine these policy implications and their resulting welfare effects. Shapiro (1983) discusses the welfare effects of minimum quality standards on consumers' uncertainty over product quality. Menapace and Moschini (2012) describe a certification scheme, whereby membership in a collective requires some internally verifiable level of quality. This allows firms sharing a regional reputation to cooperate and self-regulate in order to maintain the value of that collective reputation. In the case of an unregulated regional reputation, firms have a financial incentive to shirk on quality investment, relying on other members to keep the average level of quality high (Winfree and McCluskey 2005). Firms are powerless to compel their offending compatriots to toe the quality line. Certification and minimum quality standards for membership are among the most useful ways to force those sharing a collective reputation to oblige in maintaining that reputation's value.

Part of the problem with these policy prescriptions is that they defy the basic policy recommendations of industrial organization. For example, it can be difficult to build a distinct reputation in a competitive market, so quality problems may be worse as competition increases. Therefore, remedies such as standards or certifications can be seen as barriers to entry, which can lead to firms with more individual market power and the concomitant decreases in social welfare.

Leland (1979) identified this problem early on. Suppose a group of doctors are trying to identify an optimal standard for becoming a doctor. While there is certainly a valid argument that not everyone should be able to become a doctor, incumbent members will decide to set the standard above the socially optimal level so that competition is lessened and there is more demand for their services. This intuition can be used for goods or services.

Third-party verification, such as expert reviews and labeling, can be an effective way to reduce information asymmetries, especially with food. There is a long history of research on food labeling. Caswell and Padberg (1992) discuss the use of food labels to mitigate the imperfect information problem in food safety, arguing that quality signaling through labeling promotes market incentives with relatively limited government involvement. Caswell and Mojduszka (1996) argue that food labeling can transform a credence good product attribute into a search good attribute. Labeling communicates product differentiation by featuring product attributes that can be found in specific niche markets. A concern is that so much information is offered on labels that the important information becomes lost in the crowd. Another issue is that labels may contain misleading claims that exploit consumer knowledge gaps, with the example of gluten-free water (McFadden 2017).

6 Future research

Given the previous difficulties in obtaining data for empirical studies, there is potential for future empirical analyses to shed light on these quality and reputation issues in the agricultural industry. The research has been growing, but there is considerable room both for empirical verification of existing theoretical abstractions and for new theory derived directly from patterns observed in the data. The expansion of potential data sources, including scanner data and data from laboratory and field experiments, can be used to improve our understanding of how information might influence signals of quality concerning food quality reputations at all levels of specificity.

As food purchases change, the types of quality concerns may change as well. For example, with Amazon selling food, new concerns about food quality and delivery may arise. Similarly, Alibaba is a growing online seller of food in China. Online platforms will need to make some quality assurances to build their reputation so that consumers feel comfortable purchasing food in this manner. Online sales may also provide data for more empirical analyses.

Finally, many theoretical and empirical studies of reputation have analyzed models that are applied variations of Shapiro's (1982) model. Although attempts have been made to more rigorously formalize the interactions and dynamics of firm and collective reputations (Benavente 2013), there does not exist any other comprehensive model to which all reputation analysis easily conforms. It may be useful, therefore, to construct a generalizable model that bridges the gap between the game theoretical and dynamic optimization approaches. This would be a good first step in moving from the ad hoc to a broad-scope formalized theory of reputation.

Notes

1 Chamberlin (1962) described product differentiation slightly differently. In his dichotomous conception, differentiation occurs either by product characteristics, including quality, design, trademarks, and product peculiarities, or by the conditions surrounding the product's sale; for example, seller's location or seller's reputation in the sales process (i.e., fairness and courtesy).
2 Landon and Smith (1997) examine firm reputation and quality, and Landon and Smith (1998) extend the model to include an empirical investigation of the effects of collective reputations.

References

Akerlof, G.A. 1970. "The Market for 'Lemons': Quality Uncertainty and the Market Mechanism." *Quarterly Journal of Economics* 84(3):488–500.
Atkin, R., and R. Johnson. 2010. "Appellation as an Indicator of Quality." *International Journal of Wine Business Research* 22(1):42–61.
Benavente, D. 2013. *The Economics of Geographical Indications*. Geneva, Switzerland: Graduate Institute Publications.
Berning, J., M. Costanigro, and M.P. McCullough. 2017. "Can the Craft Beer Industry Tap Into Collective Reputation?" *Choices* 32(3):1–6.
Bushak, L. 2014. "Mozzarella Arrests: Factory Shuttered for Making Bad Cheese That Contained 20 Times More Bacteria Than Permitted." *Medical Daily*. www.medicaldaily.com/mozzarella-arrests-factory-shutteredmaking-bad-cheese-contained-20-times-more-bacteria-permitted. Accessed 21 April 2014.
Castriota, S., and M. Delmastro. 2015. "The Economics of Collective Reputation: Evidence from the Wine Industry." *American Journal of Agricultural Economics* 97(2):469–489.

Caswell, J.A., and E.M. Mojduszka. 1996. "Using Informational Labeling to Influence the Market for Quality in Food Products." *American Journal of Agricultural Economics* 78:1248–1253.

Caswell, J.A., and D.I. Padberg. 1992. "Toward a More Comprehensive Theory for Food Labels." *American Journal of Agricultural Economics* 74:460–468.

Chamberlin, E.H. 1962. *The Theory of Monopolistic Competition: A Re-orientation of the Theory of Value* (8th ed.). Cambridge, MA: Harvard University Press.

Chiang, S.-C., and R.T. Masson. 1988. "Domestic Industrial Structure and Export Quality." *International Economic Review* 29(2):261–270.

Clagett, M. 1999. *Ancient Egyptian Science: A Source Book*. Philadelphia, PA: American Philosophical Society.

Costanigro, M., C.A. Bond, and J.J. McCluskey. 2012. "Reputation Leaders, Quality Laggards: Incentive Structure in Markets with Both Private and Collective Reputations." *Journal of Agricultural Economics* 63(2):245–264.

Costanigro, M., and J.J. McCluskey. 2011. "Hedonic Price Analysis in Food Markets." In Lusk, J., J. Roosen, and J.F. Shogren, eds., *The Oxford Handbook of the Economics of Food Consumption and Policy*. New York: Oxford University Press, pp. 152–180.

Costanigro, M., J.J. McCluskey, and C. Goemans. 2010. "The Economics of Nested Names: Name Specificity, Reputations, and Price Premia." *American Journal of Agricultural Economics* 92(5):1339–1350.

Darby, M.R., and E. Karni. 1973. "Free Competition and the Optimal Amount of Fraud." *Journal of Law & Economics* 16(1):67–88.

Donnenfeld, S., and W. Mayer. 1987. "The Quality of Export Products and Optimal Trade Policy." *International Economic Review* 28(1):159–174.

Dorfman, R., and P.O. Steiner. 1954. "Optimal Advertising and Optimal Quality." *American Economic Review* 44(5):826–836.

Duvaleix-Tréguer, A. Hammondi, L. Rouached, and L.G. Soler. 2012. "Firms' Responses to Nutritional Policies." *European Review of Agricultural Economics* 39(5):843–877.

Fleckinger, P. 2007. *Collective Reputation and Market Structure: Regulating the Quality vs. Quantity Trade-off*. Discussion Paper, Ecole Polytechnique, Paris, France.

Gabszewicz, J.J., and J.-F. Thisse. 1979. "Price Competition, Quality and Income Disparities." *Journal of Economic Theory* 20(3):340–359.

Gabszewicz, J.J., and X.Y. Wauthy. 2012. "Nesting Horizontal and Vertical Differentiation." *Regional Science and Urban Economics* 42(6):998–1002.

Gaskell, G., M.W. Bauer, J. Durant, and N.C. Allum. 1999. "Worlds Apart? The Reception of Genetically Modified Foods in Europe and the United States." *Science* 285(5426):384–387.

Hoch, S.J., and Banerji, S. 1993. "When Do Private Labels Succeed?" *Sloan Management Review* 34(4):57.

Hotelling, H. 1929. "Stability in Competition." *Economic Journal* 39(153):41–57.

Jasanoff, S. 1997. "Product, Process, or Programme: Three Cultures and the Regulation of Biotechnology." In M.W. Bauer, ed., *Resistance to New Technology: Nuclear Power, Information Technology, and Biotechnology*. Cambridge: Cambridge University Press, pp. 311–331.

Klein, B., and K.B. Leffler. 1981. "The Role of Market Forces in Assuring Contractual Performance." *Journal of Political Economy* 89(4):615–641.

Kranton, R.E. 2003. "Competition and the Incentive to Produce High Quality." *Economica* 70(279):385–404.

Landon, W., and C.E. Smith. 1997. "The Use of Quality and Reputation Indicators By Consumers: The Case of Bordeaux Wine." *Journal of Consumer Policy* 20(3):289–323.

———. 1998. "Quality Expectations, Reputation, and Price." *Southern Economic Journal* 64(3):628–647.

Leland, H.E. 1979. "Quacks, Lemons, and Licensing: A Theory of Minimum Quality Standards." *Journal of Political Economy* 87(6):1328–1346.

Martin, D. 2017. "Strategic Pricing with Rational Inattention to Quality." *Games and Economic Behavior* 104:131–145.

McCluskey, J.J. 2000. "A Game Theoretic Approach to Organic Foods: An Analysis of Asymmetric Information and Policy." *Agricultural and Resource Economics Review* 29(1):1–9.

McFadden, B. 2017. "Gluten-Free Water Shows Absurdity of Trend in Labeling What's Absent." *The Conversation.com*, August 28.

McQuade, T., S.W. Salant, and J. Winfree, 2016. "Markets with Untraceable Goods of Unknown Quality: Beyond the Small-Country Case." *Journal of International Economics* 100:112–119.

Menapace, L., and G. Moschini. 2012. "Quality Certification By Geographical Indications, Trademarks and Firm Reputation." *European Review of Agricultural Economics* 39(4):539–566.

Menival, D., and S. Charters. 2014. "The Impact of Geographic Reputation on the Value Created in Champagne." *Australian Journal of Agricultural and Resource Economics* 58(2):171–184.

Mussa, M., and S. Rosen. 1978. "Monopoly and Product Quality." *Journal of Economic Theory* 18:301–317.

Nelson, P. 1970. "Information and Consumer Behavior." *Journal of Political Economy* 78(2):311–329.

Neven, D. and Thisse, J.-F. 1989. "Choix des Produits: Concurrence en Qualité et en Variété." *Annals d'Économie et de Statistique* 15/16:85–112.

Olmstead, A.L., and P.W. Rhode. 2003. "Hog-Round Marketing, Seed Quality, and Government Policy: Institutional Change in U.S. Cotton Production, 1920–1960." *The Journal of Economic History* 63(2):447–488.

Rickard, B., J.J. McCluskey, and R. Patterson. 2015. "Reputation Tapping." *European Review of Agricultural Economics* 42(4):675–701.

Rogerson, W.P. 1983. "Reputation and Product Quality." *Bell Journal of Economics* 14(2):508–516.

Rosen, S. 1974. "Hedonic Prices and Implicit Markets: Product Differentiation in Pure Competition." *Journal of Political Economy* 82(1):34–55.

Saak, A.E. 2012. "Collective Reputation, Social Norms, and Participation." *American Journal of Agricultural Economics* 94(3):763–785.

Salop, S.C. 1979. "Monopolistic Competition with Outside Goods." *Bell Journal of Economics* 10(1):141–156.

Schamel, G. 2009. "Dynamic Analysis of Brand and Regional Reputation: The Case of Wine." *Journal of Wine Economics* 4:62–80.

Schamel, G., and K. Anderson. 2003. "Wine Quality and Varietal, Regional, and Winery Reputations: Hedonic Prices for Australia and New Zealand." *Economic Record* 79(246):357–369.

Severson, K. 2014. "Vidalia Onions: A Crop with an Image to Uphold." *The New York Times* April 7.

Shaked, A., and J. Sutton. 1982. "Relaxing Price Competition Through Product Differentiation." *Review of Economic Studies* 49(1):3–13.

Shapiro, C. 1982. "Consumer Information, Product Quality, and Seller Reputation." *Bell Journal of Economics* 13(1):20–35.

———. 1983. "Premiums for High Quality Products as Returns to Reputations." *Quarterly Journal of Economics* 98(4):659–679.

Tirole, J. 1996. "A Theory of Collective Reputations (with Applications to the Persistence of Corruption and to Firm Quality)." *Review of Economic Studies* 63(1):1–22.

Winfree, J. 2016. "Partial Adherence to Voluntary Quality Standards for Experience Goods." *Journal of Agricultural and Food Industrial Organization* 14(1):81–89.

Winfree, J.A., and J.J. McCluskey. 2005. "Collective Reputation and Quality." *American Journal of Agricultural Economics* 87(1):206–213.

8
Food marketing in the United States

Stephen Martinez[1] and Krishna P. Paudel

1 Introduction

The U.S. food marketing system is comprised of five broad stages of economic activity: production, processing and manufacturing, wholesaling, retailing, and consumption (see Figure 8.1). The food manufacturing and distribution stages serve as the bridge between production and consumption, coordinating the delivery of farm products in the form, place, and time preferred by consumers. Vertical coordination between stages of the system is achieved through a variety of methods, including spot markets, contracts, alliances, and vertical integration, which varies by specific food industry. In 2015, the food and fiber marketing system employed more than 17.7 million full and part-time workers, which accounted for 12.0 percent of total U.S. employment (U.S. Department of Commerce, Bureau of Economic Analysis). According to calculations based on U.S. Department of Commerce, U.S. Census Bureau data, the food marketing system also accounted for 11.7 percent of the value of all merchandise exported by the United States in 2016. The activities and services provided by food manufacturers and distributors accounted for 84.4 percent of consumer food expenditures in 2015, while the farm share component accounted for the remainder (U.S. Department of Agriculture, Economic Research Service).

The economic contribution of the U.S. food and fiber marketing system is measured by its value added to the gross domestic product (GDP) of the United States. The net value added by each major stage of the system is shown in Table 8.1. Among the seven stages identified, the value added by food services and drinking establishments is the greatest (2.1 percent of GDP in 2016), followed by food and beverage processing and tobacco manufacturing (1.4 percent). Farming itself contributed less than 1 percent of the system's value added in 2007 and 2016. Food and beverage stores played rather minor roles. The entire food system, as defined in Table 8.1, accounted for 5.4 percent of U.S. GDP in 2016. Food services and drinking establishments as a category increased its share of value added from 1.8 percent in 1997 to 2.1 percent in 2016, while farming fell from 1.0 to 0.7 percent.

The objective of this chapter is to provide an overview of developments in the U.S. food marketing system, emphasizing food supply chain industries involved in post-farm activities.

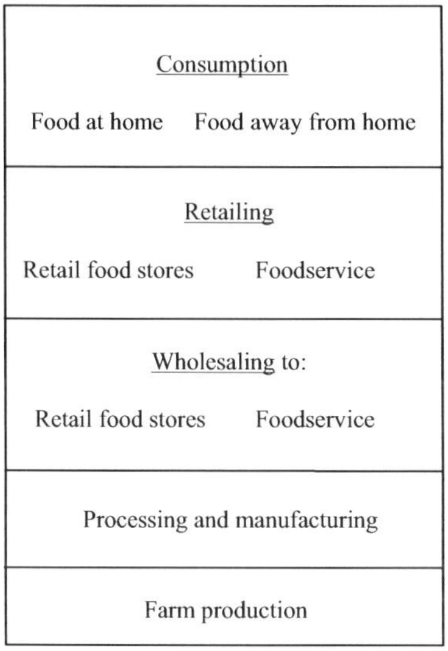

Consumption
Food at home   Food away from home

Retailing
Retail food stores   Foodservice

Wholesaling to:
Retail food stores   Foodservice

Processing and manufacturing

Farm production

The U.S. food marketing system is generally composed of five broad stages. Over 25,000 food and beverage processors purchase output from more than 2.1 million farms. Food processors handle over 90 percent of the value of U.S. farm production, with the remainder reaching consumers in unprocessed form (Connor and Schiek 1997).

Processed and packaged products are then sold to 30,500 wholesalers, 84,000 food retailers, and 443,000 foodservice companies for distribution to over 116 million households. Grocery wholesalers deliver products to retail food stores, such as supermarkets, for food eaten mainly at home. Specialized foodservice wholesalers also distribute to foodservice outlets that serve meals and snacks for immediate consumption on site (food away from home). The foodservice and food retail segments compete for the consumers' food dollar, but the wholesalers that supply them do not compete directly with each other. Foodservice distributors serve more locations with smaller order sizes, as the number of items typically required by a foodservice establishment is considerably smaller than those offered through grocery stores.

Figure 8.1 Economic stages of the U.S. food marketing system

Sources: U.S. Department of Commerce, U.S. Census Bureau (2012); U.S. Department of Agriculture, National Agricultural Statistics Service (2014); Lofquist et al. (2012).

Table 8.1 The contribution of the U.S. food system to GDP – 1997, 2007, and 2016[1]

Stage of the food system	1997	2007	2016
	Value added as a percent of GDP		
Farms	1.0	0.8	0.7
Forestry, fishing, and related activities	0.2	0.3	0.2
Food and beverage processing and tobacco manufacturing	1.6	1.3	1.4
Textile mills and textile product mills	0.3	0.1	0.1
Apparel and leather and allied products	0.3	0.1	0.1
Food and beverage stores	1.0	0.9	0.9
Food services and drinking places	1.8	1.9	2.1
Total food system/U.S. GDP	6.3	5.3	5.4

1 Excludes contribution of grocery wholesale trade and food transportation activities, which cannot be identified.
Source: Bureau of Economic Analysis.

The chapter is organized into three major sections: (1) evolving food distribution channels, including the growing diversity of retail outlets selling food. The influx of retailers not traditionally involved in selling food has been one of the most important developments in the U.S. food system; (2) the growth of locally sourced and produced foods; and (3) changes in market structure related to consolidation, mergers and acquisitions, and methods of vertical coordination.

2 Competition in food retailing

As food companies strive to grow or maintain market share in a slowly growing domestic food economy, distribution channels for marketing food products in the United States are changing. The food industry has seen an influx of companies not traditionally involved in food sales, led by supercenters, along with continued growth in convenience stores and the foodservice segment.

2.1 Growth of nontraditional grocery retailers

Nontraditional food retail outlets, including supercenters, warehouse clubs, drugstores, and dollar stores, increased their share of grocery sales from 8.8 percent in 1994 to 39.6 percent in 2016 (Table 8.2). Although gross margins for grocery items are lower than that of other retail products, by adding groceries to their product mix, nontraditional outlets can draw customers who purchase other nonfood items (Clifford 2011). Most sales growth was due to supercenters and wholesale club stores, which together took in 27.8 percent of grocery sales in 2016, up from 6.8 percent in 1994. Supercenters offer a wide variety of food and nonfood merchandise at lower prices than traditional stores. Wholesale clubs compete by catering to businesses and consumer groups with a limited variety of products and a grocery line dedicated to large-size packages and bulk sales (Inmar Willard Bishop Analytic 2017).

Drugstores and dollar stores have also emerged as formidable competitors to traditional food retailers. Drugstores compete in both the food and drug industries. By focusing on costs, dollar stores appeal to bargain and low-income shoppers.

2.1.1 Growth of supercenters

Supercenters have led the growth in grocery sales by nontraditional outlets (Table 8.2). Primarily through its supercenters, Walmart has emerged as a prominent player in the food industry. Within 12 years after opening its first supercenter in 1988, Walmart became the nation's leading grocery retailer.

Walmart's tremendous growth is largely due to efficiencies in managing its supply chain to lower costs (Irwin and Clark 2006). By offering food at lower prices with very low profit, Walmart can attract customers who also buy the store's more profitable general merchandise. The company is also expanding its offerings of private label food items. Given its immense size, Walmart is changing the traditional relationship between manufacturer and retailer (Greenhouse 2004). The company is the biggest customer of many of the nation's leading food processors, and its share continues to slowly increase. For example, Walmart is the largest customer of Dean Foods, General Mills, Kellogg Company, Campbell Soup, Tyson Foods, and Pepsico, accounting

Table 8.2 Share of grocery sales by store format[1]

	1994	2003	2007	2011	2016
			Percent		
Nontraditional formats					
Wholesale club	4.8	6.9	7.9	8.5	9.0
Supercenter (e.g., Walmart, Meijer)	2.0	11.3	15.0	17.2	18.8
Dollar stores (e.g., Dollar General, Dollar Tree)	Na	1.4	1.7	2.2	2.8
Drug stores (e.g., Walgreens, CVS)	Na	4.4	5.0	5.5	5.1
Mass (e.g., traditional Walmart and Target stores)	Na	6.6	4.8	4.4	3.5
Military commissaries	Na	0.6	0.6	0.5	0.4
Other (mini club, deep discount drug store)	2.0	Na	Na	Na	Na
Total nontraditional formats	8.8	31.3	35.0	38.2	39.5
Convenience stores	10.1	12.4	16.1	15.1	15.9
Total traditional formats[2]	81.0	56.3	48.9	46.7	44.6

Na=not available.

1 Sales of edible and nonedible grocery items, health and beauty items, greeting cards and magazines, alcohol, tobacco, and some seasonable items. Excludes clothing, gasoline, prescription drugs, toys, and other general merchandise.
2 Includes traditional supermarkets, fresh formats, limited assortment stores, super warehouses, and other (small grocery stores). These formats accounted for 94.6 percent of grocery sales in 1980.

Sources: Inmar Willard Bishop Analytics 2017; Sowka 2008, 2012; Little 2004; Rogers 2000.

for 13 to 20 percent of their net sales in 2016. In 2004, the share of these companies' sales going to Walmart ranged from 11 to 15 percent.

2.1.2 Other nontraditional retail outlets

Other types of nontraditional food outlets have recently emerged as competitors in the food industry (Table 8.2). Drug stores, such as Walgreens Boots Alliance and CVS Health, accounted for 5.1 percent of all food and nonfood grocery sales in 2016. At least 20 percent of drug store sales are generated from food and nonfood grocery items, general merchandise, and seasonal items (Inmar Willard Bishop Analytics 2017). Each store offers a small assortment of food and beverages, but the large number of stores operated by each company (e.g., CVS with 9,659 stores, Walgreens with 8,050 stores, and Rite Aid with 4,561) adds up to significant food sales. Customers are attracted by the pharmacy and the convenience of food product offerings, which earn higher margins for the company than drug sales.

Dollar stores are driving sales growth by opening new stores and adding grocery items to attract more customers. Grocery products account for between 20–66 percent of total sales (Inmar Willard Bishop Analytics 2017). The growth strategy for dollar stores is to open many small stores, which make them convenient for quick shopping trips that combine to generate significant sales (Wells 6 Feb 2017). Dollar General alone added more than 900 stores in 2016, for a total of more than 13,000 locations, up from 11,000 in 2014 (Giammona 2017; Webber 2014). The chain generates roughly 75 percent of its revenue from consumable items such as food, soap, and paper towels. In response, manufacturers are adapting their products for dollar store customers who prefer smaller packages sold at lower prices (Wong 2014).

2.2 Convenience stores

Convenience stores are small, higher-margin stores that offer an edited selection of staple groceries, nonfoods, and ready-to-eat foods. The share of grocery sales accounted for by convenience stores increased from 10.1 percent in 1994 to 15.9 percent in 2016 as more stores added fresh and prepared foods to their product assortment (Table 8.2). As with drug stores and dollar stores, their small size and many locations appeal to consumer shopping preferences for convenience and quick shopping trips. There is a notable difference between the top-quartile and bottom-quartile convenience store companies, with nearly a ten-fold difference in store operating profit between them (National Association of Convenience Stores 2017). In terms of foodservice, the top quartile had over three times more prepared food sales than the bottom quartile.

2.3 Foodservice gains

Grocery retailers also face formidable competition from the foodservice channel as consumers slowly increase their share of food expenditures from foodservice outlets. In 2014, the food-away-from home market (limited or quick service restaurants, full service restaurants, hotels and motels, schools, and other institutions that prepare and serve food) captured 50.1 percent of total food expenditures, compared with 44.5 percent in 1994 (U.S. Department of Agriculture; Economic Research Service). To continue building market share, foodservice companies are offering new products and services that cater to the lifestyles of today's health-conscious, time-pressed, and demanding consumers.

Many foodservice companies are focusing on "cleaner" and more natural ingredients (American Institute of Food Distribution, Inc. 2016). Papa John's, Caribou Coffee, Pizza Hut, Subway, Panera Bread, Taco Bell, McDonald's, Denny's, Burger King, Starbucks, and KFC have committed to removing artificial flavors, preservatives, and/or antibiotics from some menu items. Panera Bread plans to source all of its egg-based products from cage-free eggs by 2020. McDonald's also announced it would shift to cage-free eggs in the U.S.

Foodservice companies are addressing consumer demand for convenience with digital integration, expanded delivery services, and new snack offerings (American Institute of Food Distribution, Inc. 2016). Companies are installing ordering kiosks, incorporating mobile payment options, and improving online ordering. McDonald's and Wendy's began introducing self-service kiosks, where consumers can place orders to their exact specifications. Panera Bread expanded delivery service to 10 percent of its restaurants in 2016. Others, such as Taco Bell and Chili's, are planning or testing delivery services. Companies introducing mobile and online ordering options include Panera Bread, Wendy's, Subway, Taco Bell, Chili's, and Dunkin Donuts. New snack-sized menu options offered at Arby's, Taco Bell, McDonald's, and others appeal to customers who prefer smaller portion sizes consumed throughout the day.

Some limited-service chains are catering to consumer preferences for "casual indulgence" with more "upscale" items. Some of the fastest growing chains are those offering specialty products and gourmet ingredients. Starbucks, the world's leading retailer of specialty coffee, had the second-largest increase in units from 2014 to 2015 among the top 50 chains (Duncan 2016). The company now ranks as the second largest limited service chain, behind only McDonald's. Panera Bread also ranks among the top ten chains serving sandwiches made from all-natural specialty breads.

2.4 Strategic responses by traditional grocery outlets

Nontraditional outlets are important because of the efficiencies they can trigger in other retailers, in addition to the options they provide consumers (Hausman and Leibtag 2005). They can provide the impetus for traditional operators to close inefficient operations. For example, several A&P stores closed in 2011 due to increased competition from Walmart Supercenters (Springer 2011). In 2015, A&P filed for Chapter 11 bankruptcy protection and sold or closed its stores after operating for 156 years.

Conventional outlets are in direct competition with every retail channel and are responding with a number of competitive strategies to retain market share. Establishing a competitive position requires an understanding of what factors influence costs, which attributes are preferred by consumers, and how cost factors and attributes vary by customer segment (Besanko et al. 1996; Porter 1990).

Traditional grocery stores are differentiating from the competition with expanded product offerings, new store layouts, and innovative in-store technologies. As supermarkets compete with the foodservice sector, their annual sales of fresh prepared foods grew by an annual rate of 10.4 percent between 2006 and 2014, making supermarkets one of the highest performing segments in the food industry (Robards 2016). In 2014, supermarket foodservice sales grew 5.9 percent compared to an overall restaurant industry growth of 3.8 percent (Smith 2017). Companies are offering a variety of precooked, precut, or premarinated meats for warming or cooking. These foods offer consumers the convenience of buying a meal while purchasing groceries or filling their gas tanks. Many retailers have launched initiatives to compete with meal kit delivery services such as Hello Fresh and Blue Apron, including Kroger's Prep + Pared, that come with the ingredients necessary to prepare a meal for two in about 20 minutes; Giant Food Stores, with its Fresh Meal Kits; Publix through its Aprons brand; Whole Foods, which began selling meal kits made by Purple Carrot and Salted; and Supervalu's meal solutions under the Quick & Easy Meals brand (Progressive Grocer 2017; Wells 29 Sept 2017; Riemenschneider 2017). Albertsons recently acquired Plated, which is considered to be the second largest meal-kit provider in the U.S. behind Blue Apron (Hamstra 2017). A recent Harris Poll found that 25 percent of U.S. adults purchased a meal kit in 2016, primarily to save time with planning and preparing meals (Johnsen 2017).

Significant numbers of supermarkets are now offering in-store amenities such as café seating, wi-fi, and adult beverage sampling to complement their prepared foods programs (Robards 2016). Hy-Vee opened full-service restaurant operations, referred to as Market Grille, located next to select Hy-Vee stores. The restaurants and community gathering places offer a slightly more upscale dining experience and trendy offerings like craft beer, breakfast skillets, and flatbreads.

New store designs offer consumers both upgraded and expanded product offerings. Building on Kroger's acquisition of Fred Meyer, a large supercenter selling groceries and merchandise, Kroger Marketplace stores offer full-service grocery, pharmacy, and health and beauty care departments, as well as an expanded perishable section and general merchandise area that includes apparel, home goods, and toys. The stores are about three-quarters the size of a Walmart supercenter. New England supermarket chain Big Y Foods operates the Fresh Acres Market, a concept that mixes an open-air-style farmers' market with upscale fare. Some Big Y supermarkets have been renovated to include a new Living Well Eating Smart area with natural, organic, and gluten-free foods, a ready-to-eat chicken wing bar and stir-fry station, and a cafe area with a "coffee house" atmosphere and ambiance.

In response to preferences for quick shopping trips, grocers are operating more convenience-focused stores (Wells 10 April 2017). Giant Eagle's GetGo gas stations are being remodeled and rebranded as GetGo Café + Market and carry groceries as well as a variety of prepared foods. Hy-Vee has also invested in its convenience stores, while H-E-B plans to operate a new convenience store concept that will carry more groceries and prepared foods than a typical convenience store (The American Institute of Food Distribution 2016). Kroger's, the nation's largest grocery retail chain, introduced its new convenience store concept, referred to as Fresh Eats MKT, that features made-to-order foodservice from Fresh Eats Kitchen and includes mobile/online ordering and indoor/outdoor seating (Springer 2017). They also offer a larger assortment of fresh products than what typical convenience stores carry, including fresh produce, meat, bakery items, and dairy.

Fresh format stores (e.g., Whole Foods, The Fresh Market, Sprouts Farmers Market) have seen expansive growth. These stores emphasize perishables and offer assortments that differ from those of other traditional retailers, most notably in the areas of ethnic, natural, and organics (Inmar Willard Bishop Analytics 2017). In 2015, fresh format stores represented the largest increase in store count among all formats (Willard Bishop 2016).

In response to the growing health-conscious consumer population, some supermarkets are opening their own organic and natural food stores and producing their own corporate-brand organics. Companies offering their own corporate-brand organic products include Kroger (Simple Truth Organic), Giant Food (Nature's Promise), and Shaw's (Wild Harvest). In 2016, Simple Truth Organic reached over $1 billion in sales (Watrous 2017). Publix' GreenWise Markets features organic, natural, and specialty products, including organic produce, meats with no added hormones, and more healthful prepared foods, along with conventional grocery items. Ahold created a chain of smaller stores in cities, called Bfresh, where most space is devoted to produce, which also accounts for most sales. The expansion of natural and organic food at conventional supermarkets has usurped market share from niche grocers, such as Whole Foods and Fresh Market (Peterson 2017).

Many grocery retailers across the United States have committed to cage-free egg policies, including Price Chopper, Bi-Lo, Winn Dixie, ShopRite, Lowes Foods, and Tops Markets, with most setting a target year of 2025 for 100 percent cage-free egg offerings (Progressive Grocer April 2016). Publix was the last of the major grocery operators to make such a commitment, with a goal of carrying 100 percent cage-free eggs by 2026 (Progressive Grocer July 2016). To satisfy consumer demand and retail requirements, a number of major manufacturers have also committed to transitioning to cage-free eggs, including Nestlé, General Mills, Mondelēz, Kellogg, ConAgra, and Hormel.

3 The emergence of direct sales/local foods

Consumer preferences have expanded to encompass where the product was produced, and they are demanding more local products in the places they shop. Local foods appear to be an increasingly important component of the retail landscape through direct-to-consumer (e.g., farmers' markets, roadside stands, u-pick) and intermediated marketing channels (e.g., direct to retail food stores, restaurants, institutions, or to regional food aggregators) (Low et al. 2015). A 2016 online survey of 1,298 professional chef members of the American Culinary Federation by the National Restaurant Association found that hyper-local sourcing (e.g., restaurant gardens, onsite beer brewing, house-made items) ranked as the leading concept trend. Locally grown produce and locally sourced meats and seafood ranked fifth and sixth, respectively (National Restaurant Association 2017).

While the scale and logistical challenges of sourcing local food may seem at odds with the size of large retailers, many are also increasing their supply of these items as a differentiation strategy (Wells 26 April 2017). To compete with traditional grocers' service and product offerings, in 2010, Walmart pledged to double its sales of locally grown produce. By 2015, the company had exceeded that goal with 10 percent of its total produce sales from local sources, which it intends to double again by 2025. Meijer spends more than $100 million each year on local produce grown within the six states where their stores are located. Southeastern Grocers, the nation's 11th largest retail grocery chain in 2017, estimates that 30 percent of its in-season produce is derived from local sources and announced a policy that gives priority to produce grown in the southeastern U.S.

The number of farmers engaged in direct marketing grew from about 340 in 1970 to over 3,000 in 2001 and dramatically increased after the passing of the U.S. Public Law 94–463 (PL 94–463), the farmer-to-consumer Direct Marketing Act of 1976 (Brown 2002). Farm operations with direct-to-consumer (DTC) sales of food for human consumption increased from 116,733 to 144,530 between 2002 and 2012 (Low et al. 2015). Direct marketing has been enjoying per capita spending of $4.17 and an average growth in sales of 1.63 percent per year since 2007 (Boys and Blank 2016). Uematsu and Mishra (2011) indicated that the economic incentives available for both producers and consumers have contributed to the recent trend in the increased use of direct marketing strategies by U.S. farmers. Angelo et al. (2016) posit that vertical integration and consolidation in conventional food supply chains make it difficult for small and mid-sized farmers to access markets that could provide a significant return on investment.

The existing literature on direct marketing has mainly focused on factors affecting consumer decisions to purchase directly from farmers through channels such as pick-your-own, catalogue, and internet sales operations, consumer cooperatives agriculture (CSA), and locally branded meats (see Brown et al. 2006; Monson et al. 2008; Kohls and Uhl 2014; Buhr 2004). Wolf (1997) finds that the majority of consumers in San Luis Obispo County, California, preferred produce at farmers markets rather than supermarkets because of taste (freshness), quality, and price. Brown et al. (2006) use an ordinary least square (OLS) regression model and test the relationship between county-level direct marketing sales and economic and location characteristics in West Virginia. They find that higher median housing value, population density among the young, and distance to Washington, DC, increases the likelihood of purchasing at direct markets. The study also finds that the spillover effect of political awareness via regulations across industry can further empower consumers to make more informed choices about food consumption. Gallons et al. (1997) posit that consumers in Delaware choose direct market outlets because of the variety of locally grown products available. However, the findings are only relevant for states with a large proportion of small part-time farmers close to metropolitan areas.

Uematsu and Mishra (2011) and Monson et al. (2008) document that there are relatively fewer studies that examine determinants of producer decisions to engage in direct marketing. By synthesizing the literature, we find that farmers located in the Northeast are more likely to engage in direct marketing than farmers who are located in the Midwest, Pacific, and Mountain regions in the United States. This finding is consistent with Boys and Blank (2016), who indicate that consumers in the U.S. states of Maine, New Hampshire, and Vermont purchase notably more from farmers than do consumers elsewhere in the country. They also found that the Southeast part of the United States lags behind the national average with respect to direct purchase of farm products. By noting these regional differences, policy makers will be better able to channel programs that support the marketing of local foods in the United States.

Another important parameter to control for is age, which is considered to be one of the most important parameters in the configuration of the customer base network of a farmers' choice of location and direct marketing outlets. The hypothesis is that farmers draw naturally from the neighborhoods nearby where they are located and most of the customers in farmers' markets, for instance, are Caucasian (if race is reported) (Capstick 1982). Among factors that influence the decision to adopt direct marketing strategies regardless of earning performance is whether farmers are organic producers or not. The argument is that farmers who adopt organic markets are required to have a third-party certification of compliance with the United States Department of Agriculture (USDA)'s National Organic Standard (NOS) if sales are greater or equal to $5,000 (Monson et al. 2008). Thus, if farmers are organic producers and sales are less than $5,000, farmers have no incentive to seek the USDA's National Organic Standard compliance, and therefore regardless of earnings potential they will adopt direct marketing outlets. Gender is another factor to consider, because female farmers are more inclined to adopt greater direct marketing channel outlets. The argument is that the ratio of women to men in high-value markets, including direct marketing channels, is higher than in commodity markets (Monson et al. 2008). Education level also increases the likelihood of engaging in direct marketing. Farmers who a have higher earning potential in off-farm activities are more inclined to invest in high-value products and sell in direct marketing outlets given that farming is intended as a hobby or retirement activity (Monson et al. 2008). Farmers who depend on their location in a county and off-farm activity are not motivated by earning performance when choosing to sell in direct marketing outlets.

In the urban economics literature, retail locations have been characterized by common preferences of clustering among retailers to create an opportunity for place marketing activities (Teller and Reutterer 2008). Thus, consumers, producers, and markets all affect the structure of direct marketing locations. Jarosz (2008) enumerates a list of county infrastructural parameters influencing market channel outlet, such as (1) distances between producer and consumer, (2) farm size and scale, (3) whether organic or holistic, (4) agribusiness-orientation, (5) alliance cooperatives, and (6) commitment to the social (social justice/equity concerns), economic activism, and environmental considerations of food production.

These infrastructural parameters illustrate what Guthman et al. (2006) explain as a region or country's agricultural geography and history, which play important roles in production and farm location. For instance, Sage (2012) found that in Washington State the western district contains 65 percent of all the farmer's markets, but also has the lowest per capita number of markets of any of the districts. This indicates a potential pull factor in engaging in direct marketing in term of attractiveness (Hay and Smith 1980). In this literature, the distance or the radius of trade are the most influential factor that affects marketing outlet strategies, which leads to their agglomeration in specific counties. Such agglomeration enables the use of common infrastructure and environment and provides access to consumers regardless of the purchasing power, let alone retailers' total sales (Teller and Reutterer 2008). For instance, Govindasamy et al. (1998) find that farmers in New Jersey travel 1–70 miles to access direct markets. Metcalf (1999) reports that in Dane Country, Wisconsin, farmers' markets drew farmers and customers from a radius of 240 miles.

4 Modeling approach to study direct marketing

Different econometrics approaches have been used to study either the impact of direct marketing on sales or the choice of direct marketing channel by producers. Park, Paudel, and Sene (2017) use an endogenous multinomial treatment model developed by Deb and Tribedi (2006a, 2006b) to understand the effect of the direct market channel on sales volume. They find a reduction in sales volume if producers sell directly to consumers or directly to both consumers and retailers. They found no effect on sales volume if producers only sell directly to retailers. Park (2015) posits that producers who plan to sell more in local outlets should expect sales to decline (Also see, Park et al. 2014; Low and Vogel 2011; Park 2009; Darby et al. 2008; Schneider and Francis 2005).

Moreover, Park (2015) examines the impact of direct marketing on the entire distribution of farm sales using an unconditional quantile regression model. The unconditional quantile regression model results are valid even when different sets of explanatory variables are considered, thereby making it more robust than alternative control measures. Park finds that demographic variables, such as operator's experience, education, and gender, to be significant in the model. Results also indicated that smaller operations' farm sales' volume is affected more by direct marketing compared to larger operations.

Ahearn and Sterns (2013) examine the impact of direct marketing on profitability and rate of return on assets in the Southeast USA. Using a logistic regression, they find that direct marketing has a negative impact on profits but a positive impact on rate of return.

Outside of the United States, Wang et al. (2014) study the impact of direct marketing on farm income using robust econometric models (propensity score matching and OLS on average treatment effect) that control for selectivity bias. They find that direct marketing substantially increases total farm income among Vietnamese farmers.

Park, Mishra, and Wozniak (2014) use a multinomial logit model to identify variables that impact the choice of direct marketing outlets. Producers who are able to control input costs are more likely to choose direct to consumer or direct to retail marketing channels.

Uematsu and Mishra (2011) estimate a zero-inflated negative binomial model to identify variables that affect up to seven different direct marketing strategies adopted by U.S. farmers. In the second stage, they use a quantile regression to identify the predicted value of intensity of direct marketing strategies on total gross farm income at five different quantile levels (0.10, 0.25, 0.50, 0.75, 0.90). The results show that the intensity of adoption has no significant impact on gross cash farm income and that participation in farmers markets is negatively correlated with gross cash farm income at all five quantiles estimated.

Detre et al. (2011) use a double hurdle model to understand the relationship between direct marketing and farm income. In the first step, they model whether a farmer participates in a direct marketing strategy or not, and in the second stage, they identify factors affecting gross sales from direct marketing. Unlike the other studies, Detre et al. find a positive effect on gross farm value from using a direct marketing strategy. Some of the significant variables affecting the gross sales volume in the model are organic production and regional location of farmers.

5 Structural changes in food distribution and processing

One widely used characteristic of market structure is market concentration, or the degree to which economic activity is concentrated in the hands of a few large firms. Many factors contribute to increasing concentration in the U.S. food system. Among the pressures to concentrate are technological improvements with large fixed costs that tend to favor larger processing and marketing participants with better access to capital. Firms may rely on economies of scale and scope to lower average costs relative to rivals producing smaller quantities of the same product. Economies of scale allow larger volumes to be produced at lower per unit production costs. Economies of scope exist if a firm achieves cost savings as it increases the variety of goods it produces. Economies of scale and scope not only affect the size of firms, but also influence business strategies, such as decisions on merging with another firm or whether market expansion can be achieved through long-term cost reductions (Besanko et al. 1996). For example, Walmart's demand for low-cost products may influence merger decisions by manufacturers needing to achieve economies of scale and increased efficiency in order to supply Walmart (Hopkins 2003).

Other factors that can increase concentration include the exit of firms unable to compete with larger, more efficient firms, the cost of new technology, consolidation to offset market power as concentration increases at other stages of the food marketing system, and slow overall demand growth. When demand is decreasing, new facilities may be discouraged. Consolidation may also be encouraged by other economies resulting from size, such as increased access to capital for research and advertising, volume-based price reductions on production inputs, or price premiums for large volumes of specific outputs.

5.1 Food distribution

Concentration ratios published by the U.S. Census Bureau provide useful summary indicators of the importance of large firms in the food system, as classified under the six-digit North American Industrial Classification System (NAICS). At the national level, food wholesaling is the most concentrated of the food distribution industries, followed by grocery stores and foodservice (Table 8.3). Among grocery stores, consolidation led to sharp increases in national concentration from 1997 to 2002 and has since slowed or fallen slightly. From 2007 to 2012, concentration across all stages of distribution slowed or declined.

Grocery store concentration ratios exclude supercenters, which the Census does not classify as food stores because food does not account for the majority of their sales. Figure 8.2 shows concentration ratios estimated by the U.S. Department of Agriculture's Economic Research Service that include sales at Walmart and Target supercenters. These ratios likely offer a better

Table 8.3 Percent of U.S. wholesale, retail, and foodservice sales by the 4, 8, 20, and 50 largest food companies, Census years 1997–2012[1]

		1997	2002	2007	2012	Percent change		
			Percent			1997–2002	2002–2007	2007–2012
General-line grocery merchants	4	40.9	40	39.5	38.2	−2.2	−1.3	−3.3
	8	49.9	52.6	52.6	53.6	5.4	0.0	1.9
	20	65	69	71.8	73.7	6.2	4.1	2.6
	50	79.8	83.5	85.9	87.7	4.6	2.9	2.1
Grocery stores	4	19.9	31	30.7	29.8	55.8	−1.0	−2.9
	8	32.9	43.4	43.6	42.8	31.9	0.5	−1.8
	20	46.6	54.5	56.2	56.2	17.0	3.1	0.0
	50	59.0	65.3	67.5	67.9	10.7	3.4	0.6
Foodservice	4	6.4	5.7	6.3	6.3	−10.9	10.5	0.0
	8	9.6	10	10.8	9.9	4.2	8.0	−8.3
	20	14.3	14.9	15.9	14.3	4.2	6.7	−10.1
	50	19.8	20.1	21.3	19.3	1.5	6.0	−9.4

1 2012 is the latest Census year available. The next Economic Census is scheduled to be conducted for 2017. The data for that year are unlikely to be available before 2019 or 2020.
Source: U.S. Census Bureau, Economic Census of the United States.

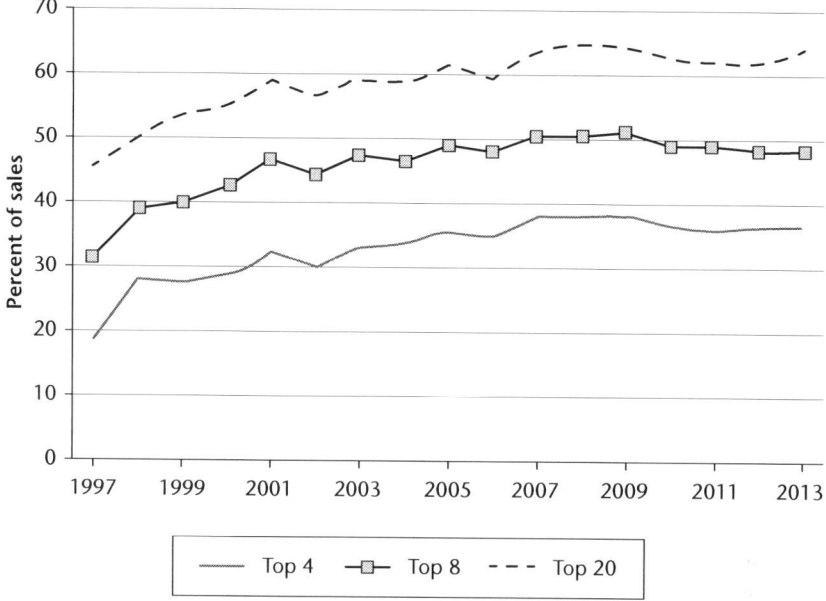

Figure 8.2 Top 4, 8 and 20 firms' share of U.S. grocery retail sales, 1997–2013

reflection of the largest food retailers irrespective of whether food is their primary business. Shares held by the largest 4, 8, and 20 supermarket and supercenter retailers have remained steady since the Great Recession, which lasted from December of 2007 to June of 2009. However, the longer-term trend shows an increasing concentration of sales among the nation's largest grocery retailers. One contributing factor to such increases has been the steady growth of Walmart supercenters.

Mergers and acquisitions can have profound effects on the number and size of companies. Mergers may increase market power and result in higher prices for the product being offered. On the other hand, mergers may result in a more efficient, lower-cost entity, and with sufficient competition from low-cost rivals like Walmart, lead to lower prices.

Since 2012, several important mergers and acquisitions have occurred in retailing. In 2013, Kroger purchased Harris Teeter for $2.5 billion, and AB Acquisition, the third largest grocery retail chain, purchased five retail banners from Supervalu, including Albertsons, Acme, Jewel-Osco, Shaw's, and Star Market stores. In 2014, AB Acquisition also agreed to acquire Safeway for $9 billion. Dutch retailer Royal Ahold agreed to purchase Belgian retailer Delhaize Group in 2015 for $10.4 billion. The combined company, referred to as Ahold Delhaize, was the fourth largest U.S. grocery chain in 2017.

Past studies have examined the effect of the mergers and acquisitions on operating efficiencies and food prices. Bjornson and Sykuta (2002) analyze the effect of acquisitions by Albertsons, Kroger, and Safeway on financial performance of the firms over the 1993–1999 period. They find evidence of improved cost control and buying efficiencies for Kroger and Safeway, while Albertsons showed signs of inefficiencies from its acquisitions. They also find that gross margin increased for each firm, but they were unable to distinguish how much of the increase was due to greater market power or improved operating efficiencies and product offerings. Sharkey and Stiegert (2006) examine the effect of retail consolidation on food prices in major metropolitan areas from 1993 to 2003. They find that if supermarkets did experience efficiency gains from mergers, cost savings were not passed on to consumers.

Rising concentration in the grocery retail segment is of concern to manufacturers because it makes them dependent on fewer retailers for maintaining their market share. Growing retail concentration could indicate a shift in bargaining power from manufacturers to grocery retailers. If food processors must sell their products to a few large distributors, the bargaining power of distributors may increase. Davis (2010) finds that food prices were negatively related to supermarket chains' share of total U.S. food sales from 1997 to 1999, which was a period in which supermarket merger activity peaked. This supports the hypothesis that supermarket chains enjoyed economies of scale or benefited from an improved post-merger bargaining position with suppliers to lower their costs and prices to consumers.

The fastest growing segment of food distribution during the 1997 to 2012 period was the foodservice sector (Table 8.4). An expanding foodservice sector and continual growth in the number of foodservice companies and establishments have kept concentration levels in check. Slow and negative inflation-adjusted growth in annual sales at traditional food stores was likely due in part to increased competition from nontraditional food retailers and foodservice. Over the 2007–2012 period comprising the Great Recession, each of the three food distribution channels experienced declining or negative real sales growth.

The number of grocery wholesaling companies and establishments declined throughout the 1997–2012 period. Increasing concentration has resulted in more direct negotiations between manufacturers and large retailers offering a broad assortment of items, reducing the power and influence of traditional wholesalers. Self-distribution is the preferred method of vertical coordination for large grocery chain stores. A decline in the number of wholesalers leaves smaller independent grocers (those with 1–10 stores operating under similar policies and programs) with fewer options for obtaining merchandise.

Table 8.4 Number of food distribution establishments and companies and changes in real sales, Census years, 1997–2012

	Grocery wholesalers	Food stores[1]	Foodservice
Number of establishments			
1997	41,760	118,915	486,906
2002	38,646	119,847	504,641
2007	37,979	114,599	571,621
2012	37,935	114,954	598,656
Number of companies			
1997	35,088	85,757	365,588
2002	32,559	87,303	377,717
2007	31,356	83,890	425,844
2012	30,481	83,960	443,238
Real sales (percent change)			
1997–2002	3.4	1.8	12.3
2002–2007	7.1	2.4	16.3
2007–2012	1.7	–0.3	2.6

1 Includes grocery stores and specialty food stores.

Sources: U.S. Census Bureau, Economic Census of the United States and Bureau of Labor Statistics.

In addition to value added, employment is another way of demonstrating the relative importance of industry segments within the U.S. food system. Workers in food processing and grocery wholesaling tend to be full-time, while a large share of retail food workers are part-timers. To adjust for part-time employment, estimates of employment on a full-time equivalent basis is displayed in Table 8.5 for 2008 and 2015. Food system employment increased from 13.4 million workers in 2008 to 14.7 million in 2015. Within the food system, agriculture and processing employment have increased at a rate equal to the U.S. economy, while food stores and foodservice establishments generated jobs at a relatively faster pace. The biggest change in employment was an increase of 1.1 million foodservice jobs, or a 0.8 percentage point increase in the percentage of private sector employment accounted for by foodservice.

Bureau of Labor Statistics data indicate that labor productivity (i.e., real output per employee hour worked) in food store retailing, foodservice, and grocery wholesaling rose from 1997 to 2007 (Figure 8.3). This followed general declines during the early to mid-1990s. It is likely that consolidation in the retail grocery sector and cost-cutting moves in response to strong price competition from the nontraditional outlets contributed to increases in food store labor productivity. Since the Great Recession of 2007–2009, changes in labor productivity at each stage of the distribution system have slowed.

Increases in labor productivity tend to be greater in food processing than foodservice, as replacing labor with other inputs has historically proved more difficult in food distribution compared to processing (Connor and Schiek 1997 and Figure 8.3). While food processing has

Table 8.5 Full-time equivalent employees in the U.S. food system, 2008 and 2015[1]

Stage of the food system	Year	
	2008	2015
	Millions	
Farms	0.6	0.7
Forestry, fishing, and related activities	0.5	0.5
Food and beverage processing and tobacco manufacturing	1.6	1.7
Textile mills and textile product mills	0.3	0.2
Apparel and leather and allied products	0.2	0.2
Food and beverage stores	2.5	2.7
Food services and drinking establishments	7.7	8.8
Total food system	13.4	14.7
	Percent[2]	
Farms	0.6	0.6
Forestry, fishing, and related activities	0.4	0.5
Food and beverage processing and tobacco manufacturing	1.5	1.5
Textile mills and textile product mills	0.3	0.2
Apparel and leather and allied products	0.2	0.1
Food and beverage stores	2.3	2.4
Food services and drinking establishments	7.1	7.9
Total food system/U.S. private total	12.4	13.1

1 Excludes contribution of grocery wholesaling and food transportation workers, which cannot be distinguished from all such workers.
2 Share of private sector employment.

Source: Bureau of Economic Analysis.

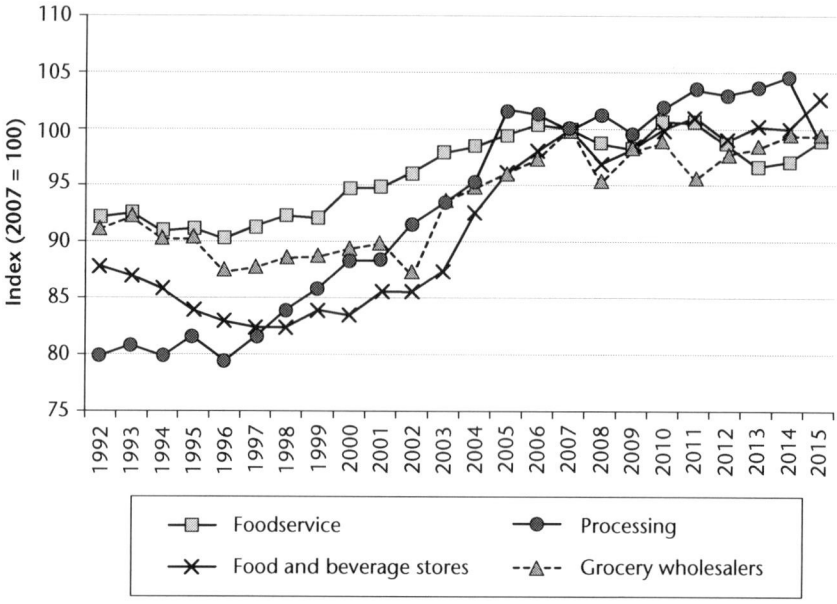

Figure 8.3 Labor productivity index, 1992–2015 (output per employee hour)

achieved growth through changes in labor productivity, growth in foodservice has required additional employees. However, the foodservice industry continues to look for ways to reduce labor costs and increase efficiencies.

5.2 Food and beverage processing

Concentration of ownership at the food and beverage manufacturing stage is of special concern to farmers because manufacturers are the primary purchasers of agricultural products. Over the 1997 to 2012 period, national concentration in the food processing industry peaked in 2002 (Table 8.6). According to the Census Bureau, the top 50 food processors accounted for 51 percent of U.S. food processing sales in 2007 and 2012, down from 53 percent in 2002. The beverage industry is much more highly concentrated. In 2007 and 2012, the top 50 beverage processors accounted for 83 percent of sales, up from 79 percent in 1997. Notable increases occurred in share of sales accounted for by the top 4 and 8 processors from 2007 to 2012, which rose by 10 and 8 percentage points, respectively.

Every manufacturing establishment, or plant, is owned by at least one company. Large manufacturing companies usually own several operating establishments and often operate plants in several industries (Connor and Schiek 1997). If ownership is consolidating, the number of companies will fall faster than the number of plants. When firm divestitures outweigh mergers, the number of companies could increase, even though the number of plants is steady or decreasing.

Over the 1997 to 2012 period, the number of food manufacturing companies and plants peaked in 2002, before settling at around 21,400 companies and 25,600 plants (Table 8.6). All plant size categories as measured by the number of employees per plant contributed to the reduction in food manufacturing plants from 2002 to 2012 (Table 8.7).

Table 8.6 Number of food and beverage processing plants and companies, and sales concentration, Census years 1997–2012

Census year	Plants	Companies	Top 4, 8, 20, and 50 share of shipment value			
			4	8	20	50
	Number	Number	Percent			
Food						
1997	26,302	21,958	14	22	35	51
2002	27,915	23,344	17	25	40	53
2007	25,616	21,355	15	23	38	51
2012	25,619	21,464	16	24	38	51
Beverage						
1997	2,622	2,169	41	52	66	79
2002	2,908	2,444	40	53	69	82
2007	3,717	3,160	39	52	71	83
2012	4,958	4,357	49	60	71	83

Sources: U.S. Census Bureau, Economic Census of the United States.

Table 8.7 Size distribution of food processing plants, 2002 and 2012

Size (employees)	2002				2012			
	Plants		Value of shipments		Plants		Value of shipments	
	Number	%	Million $	%	Number	%	Million $	%
Food								
Less than 20	19,179	68.7	19,973	4.4	17,458	68.1	33,335	4.5
20–49	3,548	12.7	34,769	7.6	3,256	12.7	61,042	8.3
50–99	1,930	6.9	48,896	10.7	1,761	6.9	79,667	10.8
100–249	1,804	6.5	104,694	22.8	1,803	7.0	179,179	24.2
250–999	1,265	4.5	173,131	37.7	1,180	4.6	269,294	36.4
1000 or more	189	0.7	77,324	16.9	161	0.6	116,755	15.8
Total	27,915	100.0	458,787	100.0	25,619	100.0	739,272	100.0
Beverage[1]								
Less than 20	2,038	67.4	2,067	2.0	3,858	76.1	4,979	3.5
20–49	391	12.9	3,347	3.2	593	11.7	6,875	4.8
50–99	215	7.1	6,632	6.3	256	5.0	11,254	7.9
100–249	237	7.8	22,428	21.2	246	4.9	33,173	23.3
250–999	131	4.3	31,079	29.4	114	2.2	52,739	37.1
1000 or more	13	0.4	40,161	38.0	5	0.1	33,215	23.4
Total	3,025	100	105,714	100.0	5,072	100.0	142,235	100.0

1 Includes tobacco manufacturing.
Source: U.S. Census Bureau, Manufacturing: Economic Census of the United States.

Small plants (fewer than 20 employees) accounted for 68–69 percent of all plants, while their share of shipment value was below 5 percent. Many of the smallest plants operate for only a few weeks of the year, oftentimes with unpaid family labor.

Growth in manufacturing operations has come from the beverage manufacturing industry (Table 8.6). This industry has seen continual increases in plants and companies, led by wineries (+1,953 plants and +1,890 companies from 1997 to 2012) and breweries (+345 plants, +343 companies). The increase in wineries comes amidst a steady increase in U.S. wine consumption, from 1.74 gallons per person in 1993 to 2.83 gallons per person in 2015, with consumption growth exceeding that of beer (Notte 2017; Hisano 2017). Small plants account for most of the increase in plant numbers (Table 8.7). In 2012, 76 percent of beverage plants were small, up from 67 percent in 2002, while the percentage of plants in all other size categories decreased. The share of shipment value accounted for by the largest plants (1000 employees or more) fell by nearly 15 percentage points. As in the food industry, small-size plants account for a small share (less than 4 percent) of beverage processing shipments.

In general, consolidation and investment in optimum-sized processing plants are expected to lead to productivity increases, assuming that the acquiring firms are more efficient operators (Ollinger et al. 2006). Food processing labor productivity (output per employee hour) increased significantly from 1996 to 2005 but has since slowed (Figure 8.4). One possible explanation for the reduction in labor productivity growth is that food processors' quality control efforts and quality assurance programs likely lead to additional costs and input use (Connor and Schiek 1997). Over the past several decades, increases in food processing labor productivity have been less than in the rest of manufacturing, which dipped by 7 percentage points during the Great Recession and has since slowed.

The Bureau of Labor Statistics also provides a multifactor productivity index for food, beverage, and tobacco manufacturing industries combined. Generally, changes in multifactor productivity, which compares output growth with changes in the quantities of all required inputs, are

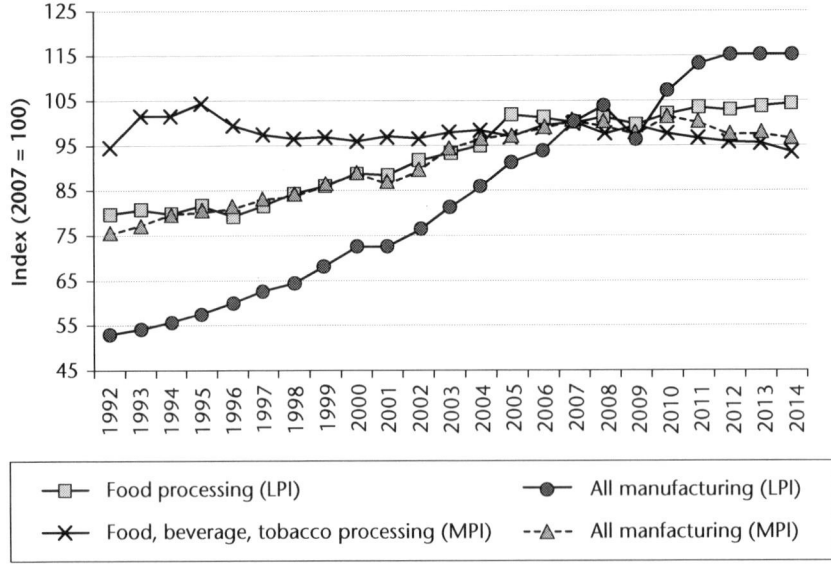

Figure 8.4 Labor productivity index (LPI) and multifactor productivity index (MPI), 1992–2014

smaller than single-factor productivity changes. Unlike labor productivity, multifactor productivity in the food processing industry has declined since reaching its peak in 1995.

5.2.1 Specific food and beverage processing industries

Consolidation can vary widely across specific food and beverage manufacturing industries, as defined by the NAICS industry classifications. It should be noted, however, that the NAICS system is less than ideal for measuring concentration because some industries are too narrowly defined, while other definitions are overly broad. For example, products from the beet sugar and cane sugar refining industries are near-perfect substitutes, which suggests that combining these industries for analysis would provide a more accurate assessment of concentration. On the other hand, the animal slaughtering industry provides an imperfect measure of concentration because output from cattle and hog slaughter plants is combined. Despite these weaknesses, Census industry classifications are the only publicly available source of food processing sales' concentration.

In 2012 (the latest Census year available), the average national market 4-firm concentration ratio (CR4), or percent of shipment value accounted for by the four largest firms, across 37 food and beverage manufacturing industries, was 50.3 percent. This was nearly unchanged from 50.8 in 2002. Four industries had relatively low levels of national concentration (i.e., CR4 less than 30), while 11 had high levels (i.e., CR4 greater than 70) (Table 8.8).

The brewing industry was the most highly concentrated, with the top four companies accounting for 88 percent of sales. Since 2012, there have been several notable mergers in the brewing industry. In 2016, Molson Coors Brewing completed its purchase of the remainder of its U.S. joint venture (formed in 2008) with SABMiller, MillerCoors. In 2015, Anheuser-Busch (AB) InBev announced a final agreement to purchase SABMiller.

Table 8.8 National 4-firm concentration ratios[1] (CR4) for food and beverage manufacturers, 2012

CR4 < 30 (low)[2]		CR4 > 70 (high)	
Industry	CR4	Industry	CR4
Fruit and vegetable canning	20.4	Breweries	87.8
Seafood product preparation and packaging	22.9	Wet corn milling	86.4
Perishable prepared food	29.6	Breakfast cereal manufacturing	79.2
Cheese manufacturing	29.9	Soybean and other oilseed processing	78.5
		Beet sugar manufacturing	77.5
		Creamery butter manufacturing	74.6
		Flavoring syrup and concentrate manufacturing	74.6
		Specialty canning	74.4
		Other snack food manufacturing[3]	73.9
		Malt manufacturing	71.2
		Bottled water manufacturing	70.6

1 CR4 = the percent of shipment value accounted for by the four largest firms.
2 Excludes animal feed and retail bakeries, which are local market industries, and "all other miscellaneous food." CR4 is meaningless for the miscellaneous food product category.
3 Excludes roasted nuts and peanut butter.
Source: U.S. Census Bureau, 2012 Economic Census of the United States.

AB InBev and Molson Coors now account for 90 percent of U.S. brewing industry sales, with the remainder going to craft breweries, which have benefited from consumers' interest in more localized and diversified beers (Koch 2017; Doering 3 May 2017). The total number of breweries in the U.S. more than doubled between 2012 and 2016. Much of the growth is attributed to craft breweries, which accounted for 99 percent of all U.S. breweries in 2016 (Brewers Association 2017). As craft beer continues to grow, it is one of the last areas for large non-craft beverage companies to expand by purchasing smaller craft breweries.

Over the 2002–2012 period, nearly half of all food and beverage manufacturing industries became more concentrated. Fourteen had sizeable increases in concentration, as measured by changes in CR4s exceeding 10 percent (Table 8.9). Industries with processing operations that serve local markets also had fewer firms. Firms serving local markets have historically exited at a faster rate than other food processors, which suggests that merger activity has been more intense in these industries (Connor and Schiek 1997).

Table 8.9 Change in concentration, sales (volume) growth, and number of plants and firms, select industries, 2002–2012[1]

Industry	CR4			Real sales (volume) growth	Change in number of plants	Change in number of firms
	2002	2012	Percent change	Percent		
11–16 percent increase in CR4:						
Seafood product preparation and packaging	20.5	22.9	11.7	−18.8	−20.3	−22.7
Coffee and tea manufacturing	51.2	57.5	12.3	58.4	59.8	61.6
Bottled water manufacturing	62.6	70.6	12.8	127.1	17.6	5.0
Spice and extract manufacturing	28.7	32.4	12.9	9.9	21.9	21.5
Fats and oils refining and blending	48.2	55.3	14.7	55.3	4.1	17.1
Frozen fruit, juice, and vegetable manufacturing	39.4	45.5	15.5	−10.4	−6.4	−4.6
16–30 percent increase in CR4:						
Perishable prepared food manufacturing	23.6	29.6	25.4	27.5	15.1	14.1
Wet corn milling	68.7	86.4	25.8	9.8	4.9	−6.1
Retail bakeries (L)	3.7	4.7	27.0	−26.1	−13.0	−10.6
Creamery butter manufacturing	57.5	74.6	29.7	19.6	5.7	−12.1
>30 percent increase in CR4:						
Rendering and meat byproduct processing	34.1	44.5	30.5	61.5	−7.8	−5.7
Soft drink manufacturing (L)	51.9	68.3	31.6	−12.0	−10.9	−13.6
Meat processed from carcasses	24.2	32.8	35.5	14.1	0.8	0.9
Ice manufacturing (L)	42.9	61.7	43.8	−22.3	−10.4	−27.7

CR4 = four-firm concentration ratio.

L = local market industry.

[1] Industries listed are those with a change in the 4-firm concentration ratio (CR4) exceeding 10 percent from 2002 to 2012. Excludes soybean processing and other oilseed processing; cane sugar; and dry pasta, dough, and flour mixes manufactured from purchased flour; which were segregated into separate categories in 2002. The CR4 was not provided for flavoring syrup and concentrate in 2002 to avoid disclosing individual company data. In 1997, CR4 for this industry was 81.

Source: U.S. Census Bureau, 2002 and 2012 Economic Census of the United States.

Falling demand (reduction in real sales volume) was associated with the reduction in plants and firms in five of the industries, including soft drinks (Table 8.9). Declining consumption of soft drinks is one factor influencing consolidation. U.S. bottled water consumption surpassed soft drinks for the first time in 2016 to become the largest beverage category by volume, increasing from 11.8 billion gallons in 2015 to 12.8 billion gallons (Loria March 2017a).

Among the industries with increases in concentration exceeding 10 percent, bottled water had the largest increase in demand growth (real sales volume growth), more than doubling over the 2002–2012 period. The shift coincides with consumer preferences for healthier refreshments and reduced sugar consumption. Local governments have also contributed to the decline in soda consumption by levying taxes on sugary beverages (Silver et al. 2017; Doering 4 May 2017; Loria April 2017b).

An increase in the concentration of bottled water production can be traced to a small number of companies. Diversification by firms into multiple product markets may affect the costs of entry when products are related through economies of scope in production, distribution, or marketing. Pepsi-Cola and Coca-Cola, leveraged by their extensive bottling and distribution networks, entered the bottled water industry and quickly ranked among the top four firms. The companies also relied on their own brand-building capabilities to introduce and promote their own brands of bottled water.

The U.S. Department of Agriculture provides information on market structure that enables more precise meat slaughter classifications than U.S. Census Bureau data does (Table 8.10). Long-term trends indicate that fewer and larger hog slaughter plants account for an increasing share of annual slaughter. However, the share of slaughter by the four largest hog slaughter firms has remained in the low- to mid-60s range since 2004, after increasing by nearly 20 percentage points since 1994. In 2015, JBS USA, a subsidiary of Brazilian company JBS S.A. (the world's largest processor of fresh beef and pork), purchased Cargill's hog and pork business. At the time, Cargill was the fourth largest U.S. pork slaughter firm, behind Smithfield, Tyson, and JBS USA (Meyer 2015).

In the cattle slaughter sector, relatively high concentration is coupled with continued consolidation into fewer and larger plants. The 4-firm concentration ratio for steer and heifer slaughter increased by 6 percentage points from 2004 to 2012 and has since remained at around 85 percent. JBS S.A. entered the U.S. meat market in 2007 with its purchase of Swift & Company (the third largest U.S. beef processor), making it the world's largest processor of fresh beef. In

Table 8.10 Hog and cattle slaughter plants, percent of animals slaughtered in large plants, and slaughter concentration, 1994, 2004, 2012, and 2015

	Number of federally inspected slaughter plants				Percent slaughtered in large plants[1]				Share of slaughter by the 4 largest firms[2]			
	1994	2004	2012	2015	1994	2004	2012	2015	1994	2004	2012	2015
Hogs	830	664	600	613	62	78	87	87	45	64	64	66
Cattle	882	689	627	641	46	52	55	57	82	79	85	85

1 Large hog slaughter plants are those slaughtering at least 2 million head annually. Large cattle slaughter plants are those slaughtering at least 1 million head annually. The definition of large plants is not directly comparable for hogs and cattle because a 1-million-head cattle plant produces much more meat than a 2-million-head hog plant.
2 Concentration figures are for steer and heifer slaughter.

Source: U.S. Department of Agriculture, National Agricultural Statistics Service (selected issues) and U.S. Department of Agriculture, Grain Inspection, Packers and Stockyards Administration (selected issues).

2008, JBS S.A. purchased the beef operations of Smithfield Foods, which was the fifth largest beef processor in the United States. By 2012, JBS USA ranked as the second largest U.S. beef processor based on processing capacity.

5.2.2 Vertical coordination

Vertical coordination refers to the methods by which goods and services may be exchanged between different stages of production, as among farmers, wholesalers, processors, and retailers. Broadly, exchange may be coordinated through:

- Spot markets (cash), where many buyers and sellers interact on a frequent basis and establish daily cash prices;
- Contracts, in which specific buyers and sellers reach more formal and longer-term agreements that specify product characteristics, terms and duration of exchange, and product volumes;
- Vertical integration, in which units at different stages of production are owned by the same firm and product flows are coordinated through administrative means.

Over the course of the twentieth century, capital-intensive technological improvements increased the production of agricultural goods, while the advantage of scale efficiencies also led to increased market concentration at the farm level. The number of farms accounting for half the value of all sales of several major U.S. commodities fell by at least 50 percent from 1987 to 2012 (Adjemian et al. 2016). Marketing contracts and vertical integration are substitutes for spot market sales and are the preferred methods of vertical coordination between large manufacturing firms and large agricultural producers.

A related factor contributing to contracting and vertical integration is heightened product heterogeneity as consumers demand more differentiated food products (Adjemian et al. 2016). In addition to traditional preferences of taste, appearance, and convenience, consumer preferences have expanded to include food attributes related to healthfulness, environmental concerns, animal welfare, location of production, and fair trade. Meeting these evolving demands requires a level of coordination that is typically not possible in a traditional, competitive spot market in which goods are homogeneous. Instead, food processors may enter into contractual relationships with producers who can consistently deliver products in a timely manner and with the required attributes. Increased coordination between producers and processors afforded by contracts can reduce production costs and transmit more information about consumer demand than traditional spot markets.

In highly concentrated industries, farmers often have only a small number of potential buyers. Consolidation in manufacturing, coupled with contracting and vertical integration, give rise to policy concerns related to volatile and thinly traded spot markets where prices are subject to manipulation by manufacturing firms; proprietary nature of contract information; and the ability of smaller, independent producers to compete with contract producers.

Beef and pork slaughter has been of special interest to policy officials, given the historically high and growing rates of concentration and alternative methods of vertical coordination. In 2013, 63.5 percent of all hogs were sold through marketing contracts, and 29.2 percent were owned and slaughtered by the same packer (USDA/GIPSA 2013). Only 3.3 percent of hogs were purchased through negotiated spot market transactions, compared with 6.8 percent in 2009. For fed cattle, the percentage of slaughter that was contracted or packer-owned increased from 49.6 percent in 2009 to 70.6 percent in 2013.

Food marketing in the United States

Table 8.11 Share of commodity production under contract, by commodity

Commodity	1991–93	1996–97	2001–02	2005	2008	2013
	Share of production under contract (percent)					
All crops	25	23	28	30	27	22
Corn	11	13	15	20	26	17
Soybeans	10	13	9	18	25	19
Wheat	6	9	7	8	23	13
Rice	20	26	39	27	45	51
Peanuts	48	34	28	65	73	57
Fruit	NA	42	42	49	38	50
Vegetables	NA	28	28	41	39	29

NA = data not available for commodity detail.

Sources: MacDonald and Korb 2011; MacDonald 2015.

The poultry industry is highly vertically integrated and, consequently, the use of spot markets for poultry is virtually nonexistent. Poultry processors, or "integrators," typically contract with producers for grower services to raise chicks to slaughter size and weight. The integrator owns the birds and slaughters and further processes the poultry.

An analysis of trends in contract use across crops finds that the share of crop production under contract varies by commodity (Table 8.11). In 2013, contracts covered 57 percent of peanut production, but they are much less prevalent in corn (17 percent of production), soybeans (19 percent), and wheat (13 percent). There is no broad trend in the use of agricultural contracts in the 2000s; rather, there are substantial episodic shifts in specific commodity classes. Use of contracting in corn, soybeans, and wheat rose by 11 to 16 percentage points between 2001 and 2008. Contract shares then fell by 6 to 10 percentage points between 2008 and 2013. Corn, soybean, and wheat producers who use contracts tend to be larger producers that use marketing contracts to cover a substantial share of production. While contracts do not account for large shares of corn, soybeans, and wheat production, contract production of these field crops amounts to half of the value of all crop production under contract (MacDonald 2015).

In peanuts, contract use expanded sharply after the cessation of federal marketing quota programs in the early 2000s. Marketing quotas provided price stability for producers, and contracts provided a way to manage emerging price risks after the end of the programs, while also tying payments more closely to product attributes (MacDonald 2015). Contracts covered 65 percent of peanut production in 2005, compared to 28 percent in 2001–2002, before the program changes.

6 Conclusions

This chapter on the U.S. food and beverage marketing system (food manufacturing, grocery wholesaling, grocery retailing, and foodservice) focuses on key marketing and structural changes and adjustments in companies' competitive strategies to maintain or gain market share. It also presents comparisons between the various stages of the food marketing system. This focus provides insights on the food system's contributions to meeting emerging consumer demands and the implications for changes in industry structure.

A relatively recent development in the U.S. food marketing landscape is growing consumer demand for local foods at direct-to-consumer outlets and intermediated marketing channels. Recent implementation of the Food Safety Modernization Act (FSMA) may provide incentives to increase direct marketing among farmers with produce sales below $25,000 (last three-year average), because they are exempt from FSMA regulations. The FSMA also has less stringent requirements for farmers with food sales of less than $500,000 and for those that sell to local establishments (restaurants or retail food stores) or within a 275 miles radius (USFDA 2015).

The U.S. food marketing system represents an adaptive, multidimensional system, as firms compete for market share with important implications for food prices, product offerings, and services. It represents one of the most significant components of the U.S. economy and affects the social and economic well-being of nearly all Americans. The interconnections between the U.S. food system, health, environment, and quality of life means that policies affecting one component of the system may have spillover effects on other components. The challenge for agricultural economists will be informing decision makers in food and agricultural practices and policies by viewing marketing as a system of economic activities in ways that improve overall efficiency.

Note

1 The views expressed here are those of the authors and cannot be attributed to ERS or the U.S. Department of Agriculture.

References

Adjemian, M.K., B.W. Brorsen, T.L. Saitone, and R.J. Sexton. 2016. "Thin Markets Raise Concerns, But Many Are Capable of Paying Producers Fair Prices." *Amber Waves*. Washington, DC: U.S. Department of Agriculture, Economic Research Service, March.

Ahearn, M., and J. Sterns. 2013. "Direct-to-Consumer Sales of Farm Products: Producers and Supply Chains in the Southeast." *Journal of Agricultural and Applied Economics* 45(3):497–508.

American Institute of Food Distribution, Inc. 2016. *Food Industry Review: 2016 Edition*. Upper Saddle River, NJ.

Angelo, B.E., B.B. Jablonski, and D. Thilmany. 2016. "Meta-analysis of US Intermediated Food Markets: Measuring What Matters." *British Food Journal* 118(5):1146–1162.

Besanko, D., D. Dranove, and M. Shanley. 1996. *Economics of Strategy*. New York: John Wiley and Sons.

Bjornson, B., and M.E. Sykuta. 2002. "Growth By Acquisition and the Performance of Large Food Retailers." *Agribusiness* 18:263–281.

Boys, K.A., and S.C. Blank. 2016. "The Evolution of Local Foods: A Retrospective and Prospective Consideration." In J. Stanton and M. Lang, eds., *The Meaning of Local Foods: A Food Marketing Perspective*. Philadelphia, PA: The Institute of Food Product Marketing.

Brewers Association. 2017. *Number of Breweries*. Boulder, CO. Philadelphia, PA: The Institute of Food Product Marketing.

Brown, A. 2002. "Farmers' Market Research 1940–2000: An Inventory and Review." *American Journal of Alternative Agriculture* 17(04):167–176.

Brown, C., J.E. Gandee, and G. D'Souza. 2006. "West Virginia Farm Direct Marketing: A County Level Analysis." *Journal of Agricultural and Applied Economics* 38(03):575–584.

Buhr, B.L. 2004. "Case Studies of Direct Marketing Value-added Pork Products in a Commodity Market." *Review of Agricultural Economics* 26(2):266–279.

Capstick, D.F. 1982. *A Study of Direct Marketing of Farm Produce in Arkansas*. Bulletin-Arkansas, Agricultural Experiment Station (USA).

Clifford, S. 2011. "Big Retailers Fill More Aisles with Groceries." *The New York Times* January 16.

Connor, J.M., and W.A. Schiek. 1997. *Food Processing: An Industrial Powerhouse in Transition* (2nd ed.). New York: John Wiley and Sons.

Darby, K., M.T. Batte, S. Ernst, and B. Roe. 2008. "Decomposing Local: A Conjoint Analysis of Locally Produced Foods." *American Journal of Agricultural Economics* 90(2):476–486.

Davis, D.E. 2010. "Prices, Promotions, and Supermarket Mergers." *Journal of Agricultural & Food Industrial Organization* 8(1):Article 8.

Deb, P., and Trivedi, P. K. 2006a. "Maximum Simulated Likelihood Estimation of a Negative Binomial Regression Model with Multinomial Endogenous Treatment." *Stata Journal*, 6(2), 246–255.

Deb, P., and P.K. Trivedi. 2006b. "Specification and Simulated Likelihood Estimation of a Non-Normal Treatment-Outcome Model with Selection: Application to Health Care Utilization." *The Econometrics Journal*, 9(2), 307–331.

Detre, J.D., T.B. Mark, A.K. Mishra, and A. Adhikari. 2011. "Linkage Between Direct Marketing and Farm Income: A Double-hurdle Approach." *Agribusiness* 27(1):19–33.

Doering, C. 2017. "Molson Coors Revenue Surges Following Miller Coors Deal." *Food Dive* May 3.

———. 2017. "'A Money Grab': Dr. Pepper, Coke Execs Blast Soda Taxes as Bad for Business." *Food Dive* May 4.

Duncan, N. 2016. "The QSR 50." *QSR Magazine* August.

Gallons, J., U.C. Toensmeyer, J.R. Bacon, and C.L. German. 1997. "An Analysis of Consumer Characteristics Concerning Direct Marketing of Fresh Produce in Delaware: A Case Study." *Journal of Food Distribution Research* 28(1):98–106.

Giammona, C. 2017. "Why the Retail Crisis Could Be Coming to American Groceries." *Bloomberg* May 4.

Govindasamy, R., M. Zurbriggen, J. Italia, A.O. Adelaja, P. Nitzsche, and R. VanVranken. 1998. "Farmers Markets: Consumer Trends, Preferences, and Characteristics." http://ageconsearch.umn.edu/bitstream/36722/2/pa980798.pdf. Accessed 20 March 2017.

Greenhouse, S. 2004. "Wal-Mart, a Nation Unto Itself." *The New York Times* April 17.

Guthman, J., A.W. Morris, and P. Allen. 2006. "Squaring Farm Security and Food Security in Two Types of Alternative Food Institutions." *Rural Sociology* 71(4):662.

Hamstra, M. 2017. "Blue Apron Trims Staff By 6%." *Supermarket News* October 19.

Hausman, J., and E. Leibtag. 2005. "Consumer Benefits from Increased Competition in Shopping Outlets: Measuring the Effect of Wal-Mart." Working Paper No. 11809, National Bureau of Economic Research, Washington, DC.

Hay, A.M., and R.H. Smith. 1980. "Consumer Welfare in Periodic Market Systems." *Transactions of the Institute of British Geographers* 5:29–44.

Hisano, A. 2017. "Reinventing the American Wine Industry: Marketing Strategies and the Construction of Wine Culture." Working Paper 17–099, Harvard Business School, Harvard University.

Hopkins, J. 2003. "U.S. Economy Follows the Wal-Mart Way." *USA Today* February 3.

Inmar Willard Bishop Analytics. 2017. *2017 Future of Food Retailing*. Winston-Salem, NC.

Irwin, E.G., and J. Clark. 2006. "The Local Costs and Benefits of Wal-Mart." Unpublished, Ohio State University.

Jarosz, L. 2008. "The City in the Country: Growing Alternative Food Networks in Metropolitan Areas." *Journal of Rural Studies* 24(3):231–244.

Johnsen, M. 2017. "Grocers Can Set Themselves Apart by Tapping Meal Kit Opportunity." *Own Brands Now* April 10.

Koch, J. 2017. "Is It Last Call for Craft Beer?" *The New York Times* April 7.

Kohls, R.L., and J.N. Uhl. 2014. *Marketing of Agricultural Products* (10 ed.). Prentice-Hall Inc.

Little, P. 2004. *Channel Blurring Redefines the Grocery Market*. Long Grove, IL: Willard Bishop Consulting, June.

Lofquist, D., T. Lugaila, M. O'Connell, and S. Feliz. 2012. *Households and Families: 2010*. Washington, DC: U.S. Department of Commerce, U.S. Census Bureau, 2010 Census Briefs C2010BR-14, April.

Loria, K. 2017a. "Bottled Water Officially Edges Soda as America's Most Popular Drink." *Food Dive* March 15.

———. 2017b. "Study: Berkeley Soda Tax Reduced Sales of Sugary Drinks 10%." *Food Dive* April 21.

Low, S.A., A. Adalja, E. Beaulieu, N. Key, S. Martinez, A. Melton, A. Perez, K. Ralston, H. Stewart, S. Suttles, S. Vogel, and B.B.R. Jablonski. 2015. *Trends in U.S. Local and Regional Food Systems*. Washington, DC: U.S. Department of Agriculture, ERS AP-068, February.

Low, S.A., and S.J. Vogel. 2011. *Direct and Intermediated Marketing of Local Foods in the United States*. USDA Economics Research Service, Economic Research Report Number 118. www.ers.usda.gov/webdocs/publications/err128/8276_err128_2_.pdf. Accessed 1 March 2017.

MacDonald, J.M. 2015. "Trends in Agricultural Contracts." *Choices* 30(3):1–6.

MacDonald, J.M., and P. Korb. 2011. *Agricultural Contracting Update: Contracts in 2008*. Washington, DC: U.S. Department of Agriculture, ERS EIB-72, February.

Metcalf, J. 1999. "Dane County Farmers' Market offers Diversity, Quality." *The Vegetable Growers News*, Sparta Michigan January:62–63.

Meyer, S. 2015. "Cargill–JBS Deal Changes Pork Industry Landscape." *National Hog Farmer* July 6.

Monson, J., D. Mainville, and N. Kuminoff. 2008. "The Decision to Direct Market: An Analysis of Small Fruit and Specialty-product Markets in Virginia." *Journal of Food Distribution Research* 39(2):1–11.

National Association of Convenience Stores. 2017. "Convenience Stores Hit Record In-Store Sales in 2016." Press Release, Alexandria, VA, April 5.

National Restaurant Association. 2017. *What's Hot: 2017 Culinary Forecast*. Washington, DC.

Notte, J. 2017. "Opinion: How Can Craft Beer Companies Survive? Use Ratings." *MarketWatch* April 9.

Ollinger, M., S.V. Nguyen, D. Blayney, B. Chambers, and K. Nelson. 2006. *Food Industry Mergers and Acquisitions Lead to Higher Labor Productivity*. Washington, DC: U.S. Department of Agriculture, ERS ERR-27, October.

Park, T. 2015. "Direct Marketing and the Structure of Farm Sales: An Unconditional Quantile Regression Approach." *Journal of Agricultural and Resource Economics* 40(2):266–284.

Park, T., A.K. Mishra, and S.J. Wozniak. 2014. "Do Farm Operators Benefit from Direct to Consumer Marketing Strategies?" *Agricultural Economics* 45(2):213–224.

Park, T.A. 2009. "Assessing the Returns from Organic Marketing Channels." *Journal of Agricultural and Resource Economics* 34(3):483–497.

Park, T.A., K.P. Paudel, and S.O. Sene. 2017. "Sales Impacts of Direct Marketing Choices: Treatment Effects with Multinomial Selectivity." *European Review of Agricultural Economics* (forthcoming).

Peterson, H. 2017. "Whole Foods Is Facing Its Worst Nightmare After an Unexpected Threat Stole Millions of Customers." *Business Insider* March 27.

Porter, M.E. 1990. *The Competitive Advantage of Nations*. New York: The Free Press.

Progressive Grocer. 2016. "More Retailers Convert to Cage-Free." April 15.

———. 2016. "Publix Adopts Cage-Free Egg Timeline." July 15.

———. 2017. "Kroger Enters Meal Kit Business." May 5.

Riemenschneider, P. 2017. "Supervalu Launches Meal Kits." *The Packer* November 20.

Robards, J. 2016. "FMI and Technomic Top 10 Trends in Supermarket Fresh Prepared Foods." *LinkedIn* February 8.

Rogers, R.T. 2000. "The U.S. Food Marketing System." In F.J. Francis, ed., *Wiley Encyclopedia of Food Science and Technology*. New York: John Wiley and Sons, pp. 2701–2724.

Sage, J.L. 2012. *A Geographic Exploration of the Social and Economic Sustainability of Farmers' Markets and the Rural Communities That Make Them Work*. PhD dissertation, Washington State University.

Schneider, M.L., and C.A. Francis. 2005. "Marketing Locally Produced Foods: Consumer and Farmer Opinions in Washington County, Nebraska." *Renewable Agriculture and Food Systems* 20(04):252–260.

Sharkey, T., and K. Stiegert. 2006. "Impacts of Nontraditional Food Retailing Supercenters on Food Price Changes." *Food System Research Group Monograph Series #20*. University of Wisconsin: Dept. of Agr. and App. Econ.

Silver, L.D., S.W. Ng, S. Ryan-Ibarra, L.S. Taillie, M. Induni, D.R. Miles, J.M. Poti, and B.M. Popkin. 2017. "Changes in Prices, Sales, Consumer Spending, and Beverage Consumption One Year After a Tax on Sugar-Sweetened Beverages in Berkeley, California, US: A Before-and-After Study." *PLoS Med* 14(4).

Smith, K. 2017. "Prepared Foods Will Continue to Drive Growth." *The Shelby Report* April 28.

Sowka, H. 2008. *The Future of Food Retailing*. Long Grove, IL: Willard Bishop.

Springer, J. 2011. "A&P to Close 32 More Stores." *Supermarket News* February 21.

———. 2017. "Fresh Eats Heralds 'New Way to Shop,' Kroger Officials Say." *Supermarket News* May 10.

Teller, C., and T. Reutterer. 2008. "The Evolving Concept of Retail Attractiveness: What Makes Retail Agglomerations Attractive When Customers Shop at Them?" *Journal of Retailing and Consumer Services* 15(3):127–143.

U.S. Department of Agriculture, Grain Inspection, Packers and Stockyards Administration (GIPSA). "Selected Issues." *Packers and Stockyards Program Annual Report*. Washington, DC.

U.S. Department of Agriculture, National Agricultural Statistics Service (NASS). 2014. *2012 United States Census of Agriculture*. Washington, DC, May.

———. "20 Selected Issues." *Livestock Slaughter Annual Summary*. Washington, DC.

U.S. Department of Commerce, U.S. Census Bureau. "Census Years 1997–2012." *Economic Census*. Washington, DC.

U.S. Food and Drug Administration (USFDA). 2015. "Final Rule: Standards for the Growing, Harvesting, Packing, and Holding of Produce for Human Consumption." www.federalregister.gov/articles/2015/11/27/2015-28159/standards-for-the-growing-harvesting-packing-and-holding-of-produce-for-human-consumption#sec-112–114. Accessed 1 September 2017.

Uematsu, H., and A.K. Mishra. 2011. "Use of Direct Marketing Strategies by Farmers and Their Impact on Farm Business Income." *Agricultural and Resource Economics Review* 40(1):1–19.

Wang, H., P. Moustier, and N.T.T. Loc. 2014. "Economic Impact of Direct Marketing and Contracts: The Case of Safe Vegetable Chains in Northern Vietnam." *Food Policy* 47:13–23.

Watrous, M. 2017. "Simple Truth a Bright Spot in Disappointing Year for Kroger." *FoodBusinessNews.net* March 3.

Webber, L. 2014. "Winners and Losers in Five Years of Top 75." *Supermarket News* January 21.

Wells, J. 2017. "Dollar Disruptors: How Discount Stores Are Shaking Up the Grocery World." *Food Dive* February 6.

———. 2017. "C-Stores are Selling More Snacks, Drinks and Prepared Foods Than Ever." *Food Dive* April 10.

———. 2017. "Growing Pains: Why Supermarkets Are Struggling to Source Local Products." *Food Dive* April 26.

———. 2017. "Report: Walmart Will Sell Meal Kits on Its Website Starting in December." *Food Dive* September 29.

Willard Bishop, an Inmar Analytics Company. 2016. *The Future of Food Retailing*. Winston-Salem, NC.

Wolf, M.M. 1997. "A Target Consumer Profile and Positioning for Promotion of the Direct Marketing of Fresh Produce: A Case Study." *Journal of Food Distribution Research* 28(3):11–17.

Wong, V. 2014. "Dollar Stores Want to Be Grocery Stores, But Cheaper." *Bloomberg Businessweek* July 15.

9
Food from the water – fisheries and aquaculture

Frank Asche, James L. Anderson, and Taryn M. Garlock

1 Introduction

In recent decades, seafood is the segment of the food production and market system where the most dramatic changes have taken place.[1] In 1970, global seafood production was primarily based on traditional capture fisheries. Although there was increasing evidence of over-exploitation, production was still rapidly increasing due to improved fishing technologies and targeting of new stocks. Traditional fisheries harvest had to take place where the fish were located. Therefore, most fish had to be canned, smoked, frozen, or preserved in some form before it reached the consumer, as the fishing grounds were generally located at substantial distances from most large markets. While there certainly existed locally supplied markets with fresh seafood supplied by local fishers, most supply chains were long with many independent agents in markets that cleared at each level, and with few opportunities to exploit economies of scale and scope. In the 1970s, several events took place that are still having a major impact on the transformation of the seafood market.

Many fish stocks were indeed overfished. This became obvious as fishing vessels had to steam farther, use better technology, and increase effort, but still caught less fish in many traditionally important fisheries. This was of course the result when exploiting a renewable natural resource with basically no, or ineffective, management. Such fisheries provide good examples of the tragedy of the commons playing out.[2] As most fish stocks can be found on or along the continental shelf relatively close to the coastline, extending an Exclusive Economic Zone (EEZ) to 200 nautical miles provided the part of the solution to the overfishing challenge in the ocean. Fish stocks within the 200-mile national jurisdiction could be managed by the associated coastal country. After some tense episodes like the Fish War between the UK and Iceland when Iceland extended its EEZ unilaterally, the 200-mile zone became global practice in 1976, although it did not formally become an international law until 1982 with the UN Convention on the Law of the Sea (UNCLOS). The 200-mile EEZ allowed a country, or a group of countries, to impose a total quota, mostly known as the total allowable catch (TAC), that protected a fish stock from overfishing. As the 200-mile EEZs excluded a number of countries' fishing fleets from their traditional fishing grounds, the so-called distant water fleets, this also created a substantial incentive for increased international trade of seafood.

The fish stocks that were overfished saw a reduction in supply, and correspondingly price increased. This created incentives to find alternative sources. This provided an additional motivation to increase international seafood trade, as imported fish could substitute for what could not be locally produced. It also provided strong incentives to develop new technologies for producing seafood. Aquaculture is an old production technology that was practiced in ancient China, Egypt, and Rome (Lockwood 2017). However, in the 1970s a "blue revolution" occurred as knowledge from the agro-sciences increasingly was transferred to the culture of aquatic animals, causing productivity growth that reduced production cost and increased competitiveness for aquaculture products (Asche 2008). This proved immensely successful, as aquaculture has been the world's fastest growing food production technology since the 1970s (Smith et al. 2010; FAO 2016). The growth in aquaculture production was further helped by the fact that the growth in the global landings of wild fish leveled off in the mid-1980s and has been varying around 90 million metric tons (mmt) since then. By 2014, aquaculture provided over one half of the seafood for human consumption (FAO 2016).[3]

Initial efforts to address overfishing have often come in the form of restrictive gear and vessels regulations, size limits, seasonal closures, and various input controls. However, in many countries, particularly in the developed world, the overfishing challenge was largely solved by the introduction of total allowable catch (TAC) regulations. Yet, TACs do little to solve the economic incentives associated with the tragedy of the commons (Munro and Scott 1985; Wilen 2000; Curtis and Squires 2007). Rather, it created a "race to fish" or "Olympic fishing" as fishers competed to get their share of the resource rent. Because the stock was protected with a TAC and therefore provided a higher catch than what is associated with open access, the over-capacity in a TAC-regulated fishery could be substantially higher than under open access (Homans and Wilen 1997). The negative consequences of the race to fish were often addressed by a number of additional restrictions, such as size limitations, entry restrictions, and gear limitations, which the fishers largely found ways to circumvent (Munro and Scott 1985; Dupont 1991). The problem started to be solved when incentive-compatible regulations, mostly in the form of rights-based systems like Individual Fishing Quotas (IFQs) or catch shares, were introduced. However, these remain controversial since it is a "closing" of the commons, or the exclusion of members of the coastal community, which creates a number of social grievances (Copes 1986).

The race to fish also reduced the value of the fishery as poorer quality was landed or the catch was harvested in a short period of time, which did not allow many higher paying markets to be served, as shown first by Homans and Wilen (2005). Aquaculture benefited by this because the higher control of the production process allowed them to supply higher quality products and to more closely meet consumer needs (Anderson and Bettencourt 1993). In fisheries where management improved, the degree of control the fishers have with the harvesting process increased, enabling them to optimize along more margins (Smith 2012). This is important in a seafood market that, on one hand, became more globalized because of increased seafood trade and, on the other hand, became more segmented as increased control of the production process allowed more specific consumer preferences to be targeted (Roheim et al. 2007, 2011; Tveteras et al. 2012). This segmentation also created demand for credence attributes such as sustainably harvested fish and the use of ecolabels to provide this information (Wessells 2002).

In this chapter, we will discuss in more detail how the seafood market has changed since the 1970s. We will start by showing the development in seafood production, highlighting the importance of production technology and trade. In Section 3, we will discuss the impacts on the supply of wild fish, first from overfishing and then from improved regulations. In Section 4, the factors leading to productivity growth and increased production in aquaculture will be discussed together with the new environmental challenges the industry caused. In Section 5, market and trade will be investigated with particular focus on globalization and the competition between farmed and wild fish.

2 Seafood production

It may be useful to start by putting seafood into a larger perspective, particularly as a source for animal protein. In Figure 9.1, global animal production is shown by main category in 1980 and 2013. With a total animal production of 465.5 mmt, seafood makes up 34.6 percent in 2013 with a production of 161.3 mmt. Of this, 91.9 mmt is from fisheries and 69.3 mmt is from aquaculture. Pork is the largest terrestrial meat category with a production of 113 mmt, followed by poultry (109.7 mmt), beef (67.7 mmt), and mutton (13.9 mmt). However, there have been substantial changes in composition over these three decades.

This is illustrated in Figure 9.2, where the annual growth rates are shown for the period. Aquaculture has by far the largest growth rate at 8.3 percent. Poultry, the fastest growing terrestrial meat, has an annual growth rate of 4.6 percent. As total meat production has an annual growth rate of 2.9 percent, all other meat sources are losing production share. This is relatively moderate for pork, mutton, and wild fish, as they all have annual growth rates over 2 percent. For beef it is more dramatic, as an annual growth rate of 1.1 percent has led the production share to decline from 25.8 percent in 1980 to 14.5 percent in 2013.

Seafood is produced with two main technologies: aquaculture and fisheries. The development in production since 1970 is shown in Figure 9.3. There was a dramatic shift in the late 1980s. The landings of wild fish stagnated and stabilized around 90 mmt (the mean landings in the period 1986–2015 is 88.7 mmt). Fish farming was quite limited in 1970, with a harvested

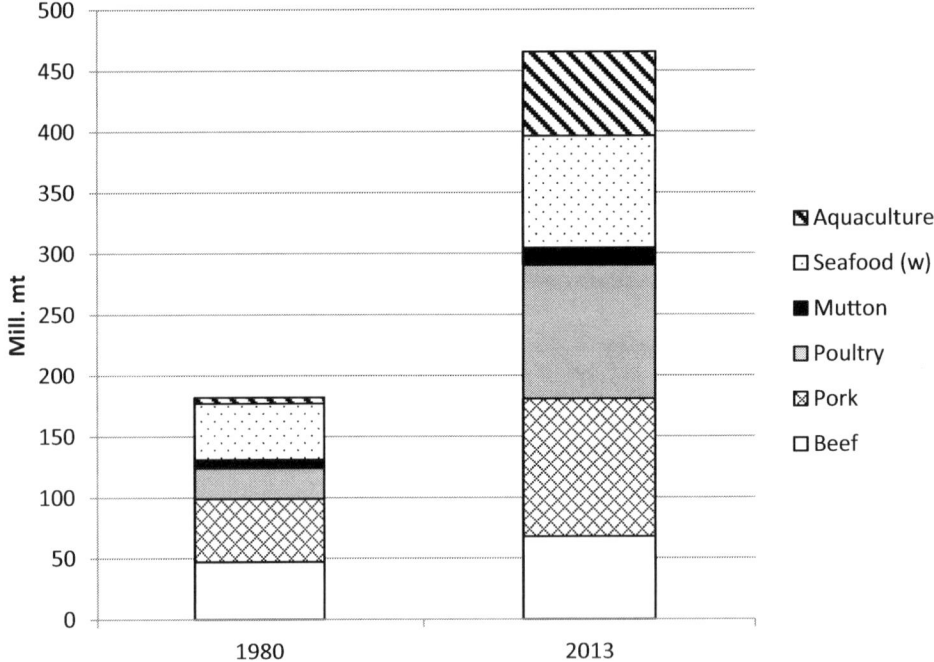

Figure 9.1 Global production of animal protein by main category

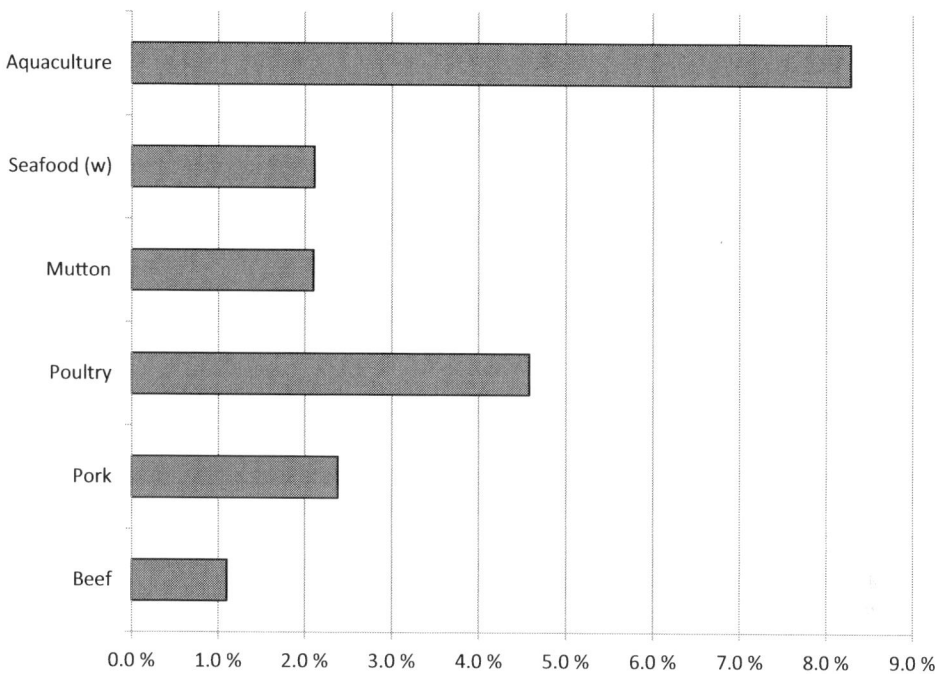

Figure 9.2 Growth rates in global animal protein production by category

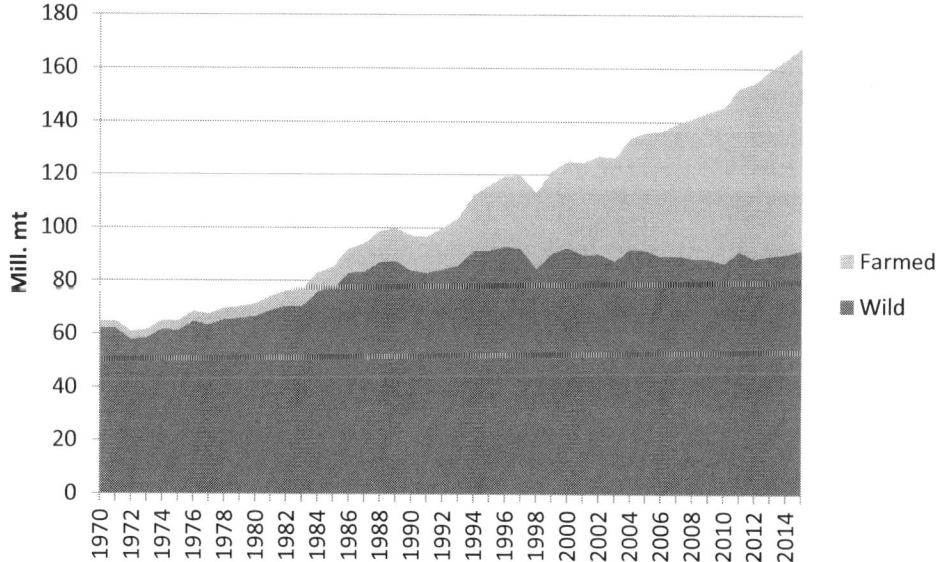

Figure 9.3 Global seafood production by production technology

quantity of about 2.6 mmt. Since then there has been a virtual explosion in aquaculture production; in 2015, farmed seafood made up 45.2 percent of the total seafood supply, with a production of 75.9 mmt. Hence, aquaculture is essentially the only reason why global seafood supply has continued to increase since 1990. Recently, farmed seafood became larger than wild fisheries as a source for food; it made up over 50 percent of seafood for human consumption in 2014, since about 20 mmt of the landed fish goes to reduction to fishmeal and other nonfood uses (FAO 2016). Total production reached 167.7 mmt in 2015 (FAO 2017). This production growth is strong enough to increase not only the total supply of seafood, but also the global per capita consumption, as per capita consumption of seafood passed 20kg in 2014, the highest on record (FAO 2016).

During the last 30 years, the world's seafood markets have also changed profoundly. Improved logistics and distribution as well as lower transportation costs have created global markets for a number of species, where there earlier were regional or local markets. Seafood is regarded as an industrial product by the World Trade Organization (WTO) and therefore is not included among agricultural products. Hence, trade barriers have not been a major obstacle to trade, particularly for product forms with a limited degree of processing. This has made seafood one of the most traded groups of food products. In 2013, 36 percent of seafood production was traded and as much as 78 percent of the production is estimated to be exposed to trade competition (FAO 2016; Tveteras et al. 2012). However, increased production and trade have not come without controversy, as indicated by the prevalence of aquaculture species in antidumping/countervailing duty conflicts, concerns related to seafood quality, fraud and adulteration, and issues related to food security in developing countries with a high dependence on fisheries and often ineffective fisheries governance. Not surprisingly, the two most valuable farmed species, salmon and shrimp, show up most often in these cases, although a number of other species have also been involved in antidumping cases.[4]

Increased seafood trade has been facilitated by technological innovations and globalization (Anderson 2003; Anderson et al. 2010). Transportation and logistics have improved significantly. Substantial reductions in transportation costs by surface and air have promoted the international trade of fresh seafood. Lower transportation costs have given new producers access to the global market. Improved logistics have also created economies of scale and scope on all levels of the supply chain, particularly in the retail sector where supermarkets have replaced fishmongers and markets in a number of places. Progress in storage and preservation has continued, allowing a wider range of seafood products to be traded. Freezing technology has improved to such an extent in recent years that many product forms can be frozen twice, allowing products to be processed in locations with competitive advantages in processing fish rather than in locations close to where the fish is caught. Lastly, the improved control in the harvesting process in fisheries and throughout the production process in aquaculture has enabled producers to better target the needs of the modern consumer and to further innovate in the supply chains.

These various factors tend to reinforce each other, even though the strength of each differs by market and species. Increased trade has profoundly affected seafood markets; an increasing number of markets have gone from regional to global and more species from widely different places have become substitutes (Asche et al. 2001; Tveteras et al. 2012). Moreover, a growing share of producers have access to the global market as global transportation systems improve and can take advantage of new market opportunities, increasing trade competition in export as well as import markets. For those consumers with the ability to pay, these trends increase the available supply of seafood in the short run. Hence, the share of imports of developed countries – the European Union, Japan, and the U.S. in particular – remains high. Economic growth in many developing countries also increases demand (Delgado et al. 2003; Kobayashi et al. 2015). As a result, there is a declining import share for developed countries despite growth in total values of seafood exports from developing to developed countries. Economic growth in China has caused it to become the number one importer by quantity used both for domestic consumption and value-added re-export. It is also the number one exporter.

Trade with seafood in general, and thereby also for aquaculture, creates food security concerns, as it is perceived to move large volumes of fish of high nutritional value from poor (i.e., developing) to rich (i.e., developed) countries. In 2011, developing countries accounted for only 26.3 percent of the value of global imports of seafood but accounted for 52.6 percent of the value of global exports of seafood, creating a substantial seafood trade surplus. From a food security perspective, this could be interpreted as a substantial problem, as it might mean that poor countries may be deprived of sorely needed proteins (Swartz et al. 2010). On the other hand, this could be interpreted as contributing to poverty alleviation due to the increased earnings and purchasing power resulting from export growth. Béné et al. (2009) provide an overview of the literature on these different perspectives on seafood trade. Thus, while the increase in trade flows and aggregate economic growth is indisputable, the effect on poverty reduction, via economic growth, of those trade flows is contentious (Roheim 2004; Asche et al. 2016).

Asche et al. (2015) show how the Food and Agricultural Organization of the United Nations' (FAO) data on global seafood trade can be used to separate exports from and imports to both developing and developed countries for the period 1976–2011.[5] The international trade of seafood, as measured in total real value exported, has grown substantially over the past four decades. In 1976, the total traded value was US$30.9 billion. This increased to US$129.2 billion in 2011, which is more than a fourfold increase. The developing countries' share of seafood exports rose steadily from 36.5 percent in 1976 to 52.6 percent in 2011. Exports from developing countries, where most of the aquaculture production takes place, have grown faster than the total increase in exports.

Global seafood imports for developing and developed countries tell a different story, as developing countries' share of the imports by value is much lower. In 1976, developing countries imported only 12.2 percent of the total value of seafood imports. While that share steadily increased throughout the period 1976–2011, it was no more than 26.3 percent of the total value of seafood imports in 2011. This asymmetry in exports and import shares between developing and developed countries provides an indication that exporting seafood is detrimental for the food security of developing countries. However, according to Asche et al. (2015), the focus solely on values can be misleading, as one must also account for quantities and their impact on unit value.

While developing countries make up only 26.3 percent of the imports in 2011 when measured in value, they made up 45.2 percent of the imports measured in quantity. In other words, the seafood trade deficit for developing countries is much smaller when measured in quantity than in dollar value. Asche et al. (2015) show that the much larger value share of the exports than imports means that developing countries in aggregate are very well compensated for the quantities they give up. This export revenue can be spent on imports of other goods, including other foods, and accordingly contribute to food security as well as other economic activity.

3 Fisheries

Fisheries are our last large and economically significant hunting industry, and for a long time, it was perceived as inexhaustible. For instance, Professor Thomas Huxley in his inaugural address at the London fisheries exhibition in 1883 stated:

> I believe, then, that the cod fishery, the herring fishery, the pilchard fishery, the mackerel fishery, and probably all the great sea fisheries, are inexhaustible; that is to say, that nothing we do seriously affects the number of the fish. And any attempt to regulate these fisheries seems consequently, from the nature of the case, to be useless.
>
> *(Huxley [1883] 1998)*

While this is a perspective that can still be found among a surprising number of fishers, it was already close to being obsolete, as new technology in the form of motorized vessels had started to increase fishing power. Leading fisheries biologists started to worry about a decline in fish stocks in the Baltic Sea and the North Atlantic, leading to the formation of the International Council for the Exploration of the Sea (ICES) in 1902 (Rozwadowski 2002; Eggert 2010a,b).[6] The council's objective was to work on practical fisheries problems and to serve as a multidisciplinary forum including all disciplines related to marine sciences. Sea mammals were among the most vulnerable species. For instance, severely depleted stocks led to the formation of the North Pacific Fur Seal Commission and a treaty that prohibited pelagic sealing in 1911 (Wilen 1976).[7]

Until the middle of the twentieth century, fishing technology was not sufficiently powerful to pose a threat to most fish stocks, and management efforts were largely isolated affairs due to special circumstances. This changed as engines and steel hulls became cheaper, and new technologies like sonar and stronger materials allowing larger nets were introduced (Squires and Vestergaard 2013), as described e.g., by Gordon and Hannesson (2015) and Kvamsdal (2016). As fishing effort increased and stocks were fished down, the first bioeconomic models were developed to explain the interaction between fishers and a fish stock. This is known as the Gordon–Schaefer model after Gordon (1954), who explained fisher behavior, and Schaefer (1957), who provided the most commonly used biological growth function. Although a static equilibrium model, this is an extremely powerful model in explaining the basic mechanisms leading to most known fisheries outcomes in a single species setting.

The basic model is illustrated in Figures 9.4 and 9.5.[8] The size of the stock biomass is shown on the x-axis and the crescent shaped curve is the Schaefer growth function, with the growth rate of the fish stock shown on the y-axis. Under equilibrium conditions, the growth rate of fish stock is equal to the harvest rate, which is also shown the y-axis. The growth is zero when the stock biomass is zero (or sufficiently small) as there is no reproduction, and it is zero at the point K, which is known as the natural carrying capacity. Beyond this point, lack of food and habitat constraints will prevent the stock from growing further, as the growth rate is negative. The crescent shape of the Schaefer growth function indicates there will be positive surplus production between zero and K. Without any harvest, any shock that reduces the stock size below K will be corrected, as there is positive growth until the stock again reaches the natural carrying capacity. At small stock sizes, the growth function is increasing as the stock exploits good growth conditions until it reaches a peak, after which scarcer food, habitat, or predation slows the increase in growth rate and the stock reaches the natural carrying capacity at K.

In Figure 9.4, there are also depicted three harvest functions for different levels of effort. The intersection of the harvest function and the growth curve represents the equilibrium

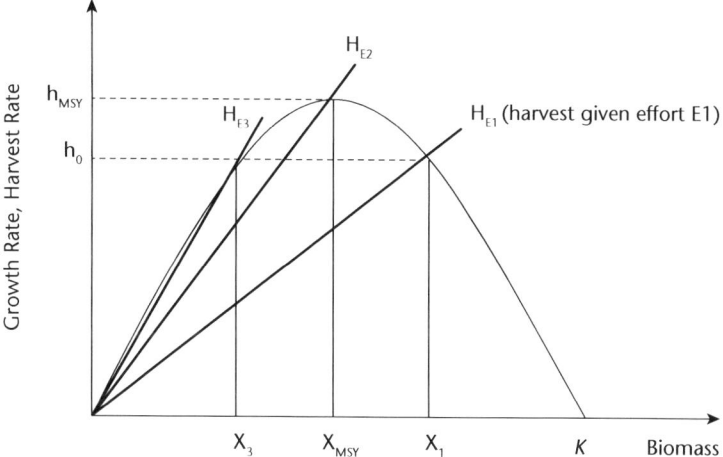

Figure 9.4 Bioeconomic model, harvest, biomass, and effort

harvest and fish biomass (stock) for the given level of effort. The effort is zero at the natural carrying capacity, and as effort is increased, harvest increases and the stock biomass is reduced. The harvest function of effort E_1 depicts a relatively inefficient technology or ineffective fishing skill. If the fishery is harvested with this level of effort E_1, biomass is reduced to a level below K, and the sustainable growth rate becomes positive. As long as the harvest level associated with E_1 is below the growth function, growth will be higher than the harvest, bringing the stock back towards the carrying capacity. The point where the harvest function crosses the growth function at stock size X_1 and harvest level H_0, one reaches equilibrium as the harvest is equal to the stock growth. At effort level E_2, the equilibrium harvest is the highest point on the growth curve. This is known as the Maximum Sustainable Yield (MSY) and corresponds to the biomass at MSY, X_{MSY}. This is an important point, and many fisheries management systems are required to set quotas so that the harvest is MSY. If the fishing effort is maintained higher than the MSY harvest rate (or above the MSY growth rate), the stock will be driven to extinction in this simple model.

As the effort level is increased to E_3, the equilibrium harvest rate (growth rate) will decline and the fish stock is driven below X_{MSY}. Note, effort levels E_1 and E_3 result in the same equilibrium harvest with different fish stock levels. This effect is known as the stock externality, since harvesting H_0 at a fish stock level of X_3, results in a waste of effort resources.

Figure 9.5 characterizes the yield curve (Panel B), Revenue and costs (Panel A) and Supply (Panel C). The relationship between equilibrium effort and harvest levels from Figure 9.4 are illustrated in Figure 9.5, Panel B. This is known as the yield curve. When this yield function is multiplied by a price p, it results in a simple total revenue function in Figure 9.5, Panel A. Assuming a constant harvest and cost per unit of effort, the total cost relationship is a simple linear function of effort level. In a competitive open access equilibrium, economic profits are zero or total cost (TC) is equal to total revenue (TR).[9]

In Figure 9.5, Panel C, the supply relationship for an open access fishery is shown. It is derived for the equilibrium of total revenue for various prices and total cost. In contrast to the typical supply relationship in food systems, the open access supply relationship is backward-bending when equilibrium harvest exceeds MSY. If the fishery is managed to maximize resource rents,

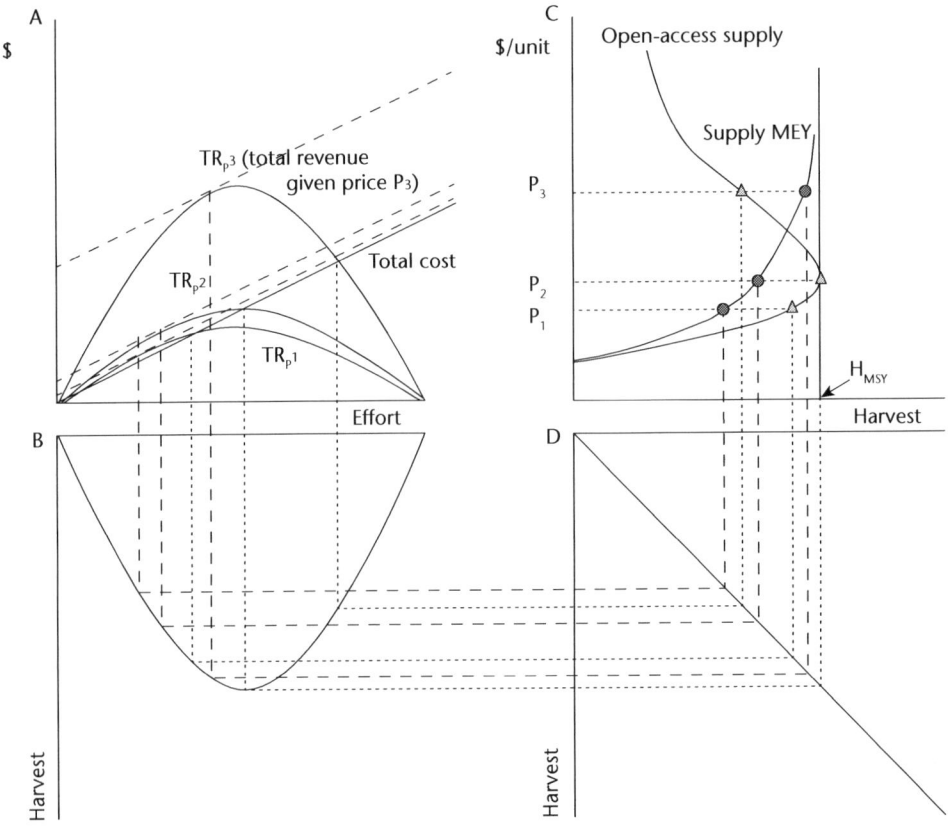

Figure 9.5 Relationship between the effort, harvest, and supply

the harvest in this simple model would correspond to a point where marginal revenue equals marginal costs. This is shown in Panel A at the point where slope of TC is equal to the slope of TR. This is known as the Maximum Economic Yield (MEY). Following this through to the supply relationship (Panel C) it can be seen that the MEY supply relationship is strictly upward sloping with price and does not exceed the MSY harvest level. Note MEY is not a constant as is the MSY, as it changes with the price of fish as well as input factor prices, and it is rarely considered a target in most publicly managed fisheries. In this static model, MEY is always associated with a stock size larger than MSY. However, in a simple dynamic model, it is conceptually possible to be economically optimal to harvest a fish stock to total depletion (extinction) if the discount rate is sufficiently higher than the intrinsic growth rate of profit in the fishery (Clark 1973). This has never been a relevant option in practice of managing wild fisheries, although it is in a fish farming context.

Meaningful fisheries management of open access fisheries often starts only when stocks were severely depleted and economic rents have been dissipated. The focus of most early management systems was to protect the stocks using command and control methods such as gear, vessel and crew constraints, area and seasonal closures, and size limits. Often, setting a total allowable catch (TAC) has been the primary means to protect the fish stock. Appropriately set, a TAC can be used to rebuild a fish stock and to maintain the stock at a target level, which in most management systems is MSY. When a TAC is the primary regulation, it is typically enforced by opening the fishery at a given time, and then closing it when the managers estimate that the TAC has been harvested. These approaches are often referred to as regulated open access. The fishers know that when the TAC is taken, the fishery will be closed. Hence, this form of management gives fishers incentives to catch as much fish as possible before the fishery is closed, creating a race-to-fish and subsequently so-called derby or Olympic fisheries. This in itself tends to shorten the period when the fishery is open. Moreover, as larger, more powerful fishing vessels catch more fish, it gives incentives to build larger vessels, and at the same time smaller vessels lose out. When the management system succeeds at increasing the harvest and fish stock, the increase in TAC will attract more fishers as this increases the revenue one competes for (Homans and Wilen 1997). Moving up the revenue schedule in Figure 9.5, the largest revenue, and therefore greatest effort, will be associated with MSY (assuming the TAC is never set to exceed MSY). As this fishing effort is generally not even close to the level of effort that maximizes economic rent, a fishery operating under these conditions will have excess capacity, as the fish could have been harvested with a substantially smaller fleet with a higher resource rent.

Initially, fishery managers' response to rapid increases in effort associated with the race to fish was additional restrictions. These came in two main forms. First, access was restricted in that one needed a license to participate in a fishery; with a restricted number of licenses, this limited the ability of new vessels to enter the fishery, creating so-called restricted access fisheries. As those fishers with a license still had incentives to increase their capacity (until total revenue is equal to total cost), additional regulations tended to limit vessel size, which gear could be used and how many fishing days were available for a vessel. In many fleets, the vessels were organized in different groups to maintain the small-scale fishermen, but still allowing a race-to-fish within each vessel group. In summary, these command-and-control measures attempt to manage the fishery by limiting harvest through decreasing fishing efficiency and/or constraining fishing time, but without creating effective incentives to change behavior. It is also of interest to note that when these allowable-catch, command and control, and limited-entry regulations are partly successful, the gains are capitalized in the license value, as shown for Alaska salmon by Valderrama and Anderson (2010). The license value will be approximately the expected net present value of the resource rent that the management system generates.

Given that fisheries management is often not seriously implemented until the stock was critically depleted, the starting point is in many cases an open access fishery with excess capacity. As rebuilding stocks usually requires fishing less, rebuilding fish stocks is a serious challenge, since that requires even less effort and generally less employment in the fishing activity as well as in processing plants. Moreover, in a fishery with over-capacity and virtually no profit, it is very difficult to exit since the over-capacity also limits the value of the capital. Throughout much of the world, weak and failing fisheries received subsidies. These came in two main forms; operational subsidies to supplement income such as price supports and to reduce costs such as fuel and investment subsidies, and capacity reducing subsidies in the form of buy-back programs.[10,11] While the subsidies were well-intended, they still just fueled the overfishing problem by explicitly or implicitly reducing the fishing cost.

Almost immediately after fisheries management systems were introduced, most economists realized that open access and the lack of property rights meant that command-and-control management tools did little to address the real issues, but rather addressed the symptoms of poor management (Wilen 2006). The main problem creating the race-to-fish is that the fishers' economic incentives are to maximize profits, and as long as there is any resource rent left, this appears as excess profitability or returns in the fishery. Fishers, as any other economic agent, will be willing to incur additional cost to compete for the excess profit until it is zero. To address this challenge, one must fix these incentives. One way to do this is with a revenue tax so that the rent is taxed away.[12] While this is theoretically feasible, it is sufficiently controversial, unrealistic, non-adaptive, and impractical that it has not been implemented to any extent. The most successful approach to provide better incentives has been the creation of various forms of property or tenure rights (Costello et al. 2007). These come in many forms such as individual quotas, known as Individual Fishing Quotas (IFQs), Individual Transferable Quotas (ITQs), catch-shares and Territorial Use Rights Fisheries (TURFs) or Vessel Day Schemes. These approaches change the fishers' incentives to maximize the profits for their quota rather than to maximize their share of the catch as under different forms of open access/race-to-fish. In fisheries where IFQs were introduced, the power of the correct incentives has been astounding. First, with the race-to-fish removed, the harvesting season was expanded (Birkenbach et al. 2017). In several fisheries, the race-to-fish in itself had contributed to reduce rent by reducing the price the fishers received. Homans and Wilen (2005) show this for the Alaska halibut fishery, in which halibut fishers were forced to freeze their fish when they landed the whole quota in a few days. When the season was extended, they could serve the higher paying fresh market.

While well-designed rights can fix the incentive problem, they do not directly address over-capacity. However, this is facilitated by making the IFQs tradable, or Individual Transferable Quotas (ITQs). An IFQ will have a higher value for better fishers, either because they have lower harvesting cost or because they can create more value out of the catch. When the IFQ is tradable, fishers with a higher valuation of the quota will then buy quota from those with a lower valuation. Arnarson (1990) shows that under fairly reasonable assumptions, the total value of the quotas in an ITQ fishery will be equal to the net present value of the resource rent. Hence, while in any race-to-fish fishery the resource rent is dissipated with excessive effort and over-capacity, in a well-designed ITQ fishery the expected rent is monetized. That also means that an ITQ fishery will be operating at MEY if the TAC is set to maximize value of the quota.

Still, IFQs, and in particular ITQs, are controversial. For such management systems to work, access to the fishery is limited. Moreover, as profitability in a race-to-fish type fishery is low, quotas tend to be provided for the active fishers for free, so-called grandfathering. However, while the quota value tends to be low when the ITQs are introduced, the value tends to increase over time as over-capacity is bought out and markets are better served. These values are often substantial. Wilen's rule of thumb is that the rent constitutes half of the landed value, although this will vary between fisheries.[13] There are also a number of other perceived or real challenges with rights-based systems such as ITQs, including that it gives larger fishing operations or wealthier fishers with access to capital markets an unbeatable advantage and that it endangers coastal communities as fishers are bought out (e.g., Copes 1986). While substantial parts of this criticism do not hold up against the evidence where it can be tested, it has been sufficiently influential to introduce precautions in most IFQ systems that limit transferability.[14] Virtually all systems have a limit on how much quota a single company can own. Moreover, in most systems there are different vessel groups by size to protect the smaller vessels, and there are often regional and gear preferences. While these measures prevent full rationalization, they also protect inefficiencies and at a substantial cost. For instance, Kroetz et al. (2015) show that a third of the potential resource rent is dissipated for U.S. sablefish and halibut to maintain three different size-groups of vessels. Moreover, it is the medium-sized vessels that are most profitable, indicating that the trade restrictions protect not only the smallest vessels but also the largest. There are also Territorial Use Rights Fisheries (TURF) systems, which assign rights to a community or group, thereby giving the community the power to both economically manage the fishery and to address cultural traditions or equity concerns.

There are of course a number of extensions to these basic models that are beyond the scope of a short chapter. Smith (1969) set up a model where the fish resource was treated as (natural) capital, making the optimization problem dynamic. Clark and Munro (1975) show how there is a modified golden rule for a renewable natural resource that accounts for the stock effect, so that the optimal harvest rate is equal to the interest rate adjusted for the marginal stock effect. Clark (1973) uses a version of this to show that for some stocks with very low growth rates, it may be economically optimal to drive them to extinction. Virtually all analytically solvable bioeconomic models (and management systems) are single species. There exist some papers, such as Sumaila (1997), that specify two-species predator prey models, but the insights derived from such models are limited beyond the obvious conclusion that it may be optimal to limit exploitation of a prey species if the predator is more valuable.

The use of aquaculture to enhance "wild" harvest in capture fisheries is being employed around the world. For example, released hatchery fish are an essential part of almost all salmon fisheries (Knapp et al. 2007); fish fattening operations are now used in bluefin tuna fishing; and habitat creation/restoration and stocking is common in oyster fisheries and many other fisheries in Asia and elsewhere (Taylor et al. 2017). This creates some interesting and complex ecological and economic issues. Unfortunately, there has been only limited work on the economics of commercial fisheries enhancement, such as the hatchery-based salmon fisheries in the Pacific. Anderson (1985a) analytically evaluates the bioeconomics of enhancement with a focus on how harvest will influence the state of wild stocks and hatchery-based stocks.

As most fish stocks are patchy, a spatial dimension was introduced in bioeconomic models (Sanchirico and Wilen 1999). Marine Protected Areas have become a conservation tool, and there is a large literature base investigating the economic impact of this tool, and was another important area for spatial considerations (Holland and Sutinen 2000; Smith and Wilen 2003; Haynie and Layton 2010). Rotation schemes for opening and closing fishing areas have become a valuable management approach, especially for relatively sedentary species such as clams and scallops (Valderrama and Anderson 2007). Spatial considerations were also a major factor in introducing behavioral economic considerations into fisher decisions (Squires 1987; Pascoe et al. 2007; Nøstbakken 2008; Schnier and Felthoven 2011), but also in other economic decisions such as gear choice (Eggert and Tveteras 2004), which species to target (Zhang and Smith 2011; Abbott et al. 2015), and other externalities (Haynie et al. 2009; Huang and Smith 2014). With the over-capacity and potential rents (and rent seeking), these are important areas for research (Weninger 1998; Pascoe et al. 2012). The development of quota markets and how they work has also become an area of research (Newell 2005; Costello and Deacon 2007; Nøstbakken 2012) as well as the interaction between regulations and markets (Nielsen 2005; Scheld and Anderson 2014).

4 Aquaculture

Aquaculture is distinguished from other aquatic production such as fishing by the degree of human intervention and control that is possible (Anderson 2002; Asche and Bjørndal 2011), and can be defined as the human cultivation of organisms in water.[15] Hence, aquaculture is more similar to terrestrial animal production than to traditional capture fisheries, even though it is surprisingly hard to draw the precise line between aquaculture and a fishery (Klinger et al. 2013). Typically, aquaculture starts with human intervention in a part of the production cycle, often providing protection and/or feed in some part of the organism's life cycle. This may, for instance, be improvements in mussel beds, a hatchery to increase survival rates for fingerlings, or a fenced-in area to limit predation.[16] Aquaculture with limited intervention, which also tends to be small scale, is often known as extensive aquaculture. As one obtains more control with the production cycle, the aquaculture operation intensifies, and one says that the production cycle is closed when it does not depend on wild stocks for reproduction. More intensive operations can also be larger scale. The production process in aquaculture is determined by biological, technological, economic, and environmental factors. However, the key factor with respect to competitiveness is that the production process can be brought under human control, as the control enables innovation and systematic research and development (R&D). This has been essential for the rapid technological development, productivity growth, and cost reduction that have fueled aquaculture production growth since the early 1970s, as shown in Figure 9.3.

Salmon, the aquaculture species with the second highest global production value in 2015 and also one of the most intensive production processes, illustrates well the importance of productivity growth in aquaculture. Figure 9.6 shows real production cost and export price for salmon in Norway, as well as global production of Atlantic salmon.[17] Both price and cost have a clear downward trend, and the gap between them is consistently small, even though there is some evidence of cycles. The average price in 2003 was about a quarter of the price in 1985, and the reduction in production cost was of the same magnitude. The important message here is that there is a close relationship between the development of productivity and production cost and falling export prices. Productivity gains are therefore able to explain a great deal of the decline in farmed salmon prices, as the price has been moving down with the production cost, keeping the profit margin relatively constant. This is to be expected in a competitive industry, since high profitability is the market's signal to increase production. As the cost reduction has been translated into lower prices, it is also clear that the productivity gains have been passed on to consumers. The main effect for the producers is that they become larger and earn a higher profit due to the higher quantities produced.

The reduction in production costs is due to three main factors. First, fish farmers have become more efficient, so they produce more salmon with the same input. This is normally referred to as the fish farmers' productivity growth. Second, improved input factors (e.g., better feed and feeding technology and improved genetic attributes due to salmon breeding) make the

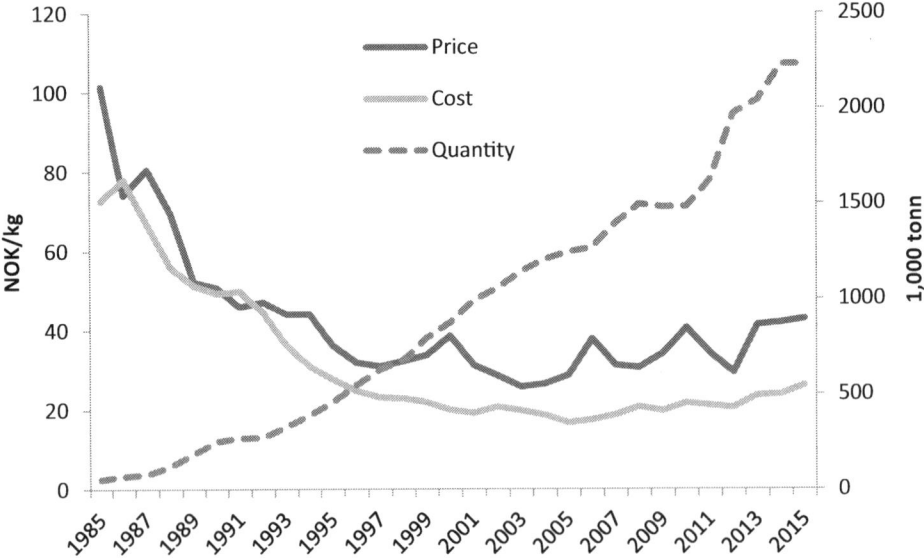

Figure 9.6 Global Atlantic salmon production and Norwegian export price and cost

production process less costly. This is due to technological change for fish farmers and productivity growth for the fish farm suppliers. This distinction is often missed, and the productivity growth for the farmers, as well as for their suppliers, is somewhat imprecisely referred to as productivity growth for the whole industry. Third, scale has increased substantially, as a Norwegian farm in 2015 produces approximately 20 times as much fish as a farm in 1980. In addition, while the focus is on the production process, productivity gains in the distribution chain to the retail outlet are equally important.

The most important input in salmon farming is the salmon feed, which represented around 50 percent of operating costs in 2015. Other important inputs are smolts (10 percent cost share), capital (6 percent), and labor (8 percent). The share of feed has been increasing (from about 25 percent in the mid-1980s), making the production process more feed intensive. Because feed is the factor most closely related to production volume, this development indicates better exploitation of the capital and labor employed at each farm. This can be explained to a large extent by increased production on each farm. Several studies using data from the 1980s found that substitution was possible among feed, capital, and labor. For instance, hand feeding was at the time more efficient than machine feeding. However, with the increased cost share of feed, these substitution possibilities have been reduced. Guttormsen (2002) suggested that they had largely disappeared in the 1990s. This implies that salmon production now, after investments in capital equipment have been made, can be characterized as a technology with a close to fixed relative factor share in the production process. The production process then becomes one of converting a cheaper feed into a more desirable product for the consumers. So, even if the substitution possibilities between capital, labor, and feed are limited, the farmers can substitute between different types of feed.

In most markets, price is the most important argument with respect to which products a retailer will stock and a consumer will buy. Total production cost will then be the main factor explaining which aquaculture products will be produced. Total production cost means the total cost of bringing the product to the consumer, which then includes not only production, but also transportation and processing costs. The fact that the aquaculture producer has greater control over the quality of the product and when to harvest it also gives an advantage compared with other seafood suppliers in the supply chain. The control allows more predictability and better capacity utilization. It also enables exploitation of economies of scale and scope in distribution and marketing. Hence, using the terminology of Smith (2012), aquaculture also optimizes over margins in the supply chain to a larger extent than fisheries.

Figure 9.3 suggests that global fisheries production did not flatten out until the mid-1980s. However, for a number of important species such as cod, the landings peaked substantially earlier due to overfishing. This created an economic opportunity, and the creation of modern aquaculture in the 1970s is to a substantial extent a response to this economic opportunity. Initially, farmers targeted the same market as the wild fishers. Assuming the two products to be perfect substitutes, there is one market supplied by two sources of production with very different characteristics. Anderson (1985b) set up the first model with market interactions between farmed and wild species. This is an extension of the bioeconomic model and is graphed in Figure 9.7. Due to the growth function, the supply schedule from the fishery is backward bending, while the supply schedule from aquaculture is of the traditionally upward sloping type. Assume that we start in a situation with a depleted open access fishery with price P_0 and quantity Q_0. If a relatively inefficient aquaculture technology is introduced with the supply schedule S_1, price is reduced to P_1, and total quantity is increased to Q_1. However, note that due to the backward bending supply schedule, supply from the fishery actually increases. This is due to the fact that the lower price reduces the rent and therefore attracts less effort. Hence, the introduction of

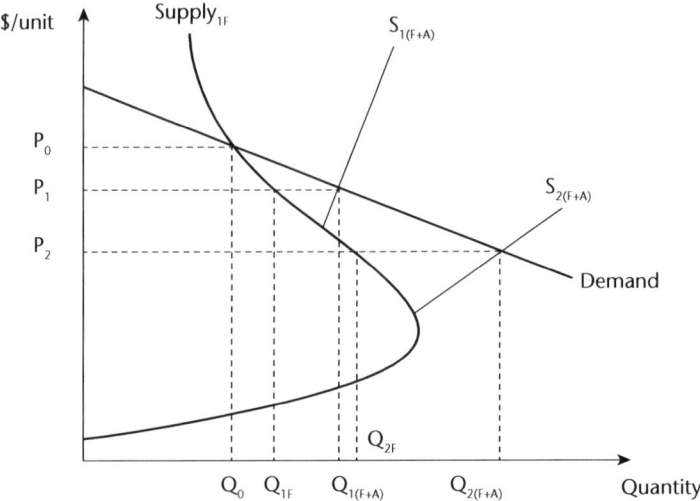

Figure 9.7 Market interaction between a fish stock and aquaculture

aquaculture reduces overfishing and helps improve the fish stock.[18] Productivity growth in aquaculture will shift the supply schedule from aquaculture downward. For instance, at S_2, the supply from the fishery is at the MSY level, and a further increase in aquaculture productivity reduces supply from the fishery and increases the fish stock. In fact, the price reduction because of the increased aquaculture production works in a similar manner as a tax on the sales price for the fishers, and increased aquaculture production that reduces the price more acts in a similar way as a higher tax. Hence, in principle the "right" production level from aquaculture can "fix" the overfishing problem. For various forms of fisheries management, the supply response will be different. For instance, the supply from the fishery with a TAC will not change as long as the price is high enough, and otherwise it will follow the bioeconomic supply schedule. Valderrama and Anderson (2010) show that in a limited entry fishery, reduced prices first reduce license value, and when the price is sufficiently low it will start reducing effort. In a regulated open access fishery, first excess-capacity will be reduced before over-capacity.

While the development of new technology has significantly increased the production potential of aquaculture and has the potential to create positive environmental externalities by reducing fishing pressure, the increased production has also raised questions about the negative environmental impacts and thereby the sustainability of aquaculture. This is certainly an issue because aquaculture, like other biological production processes, interacts with the surrounding environment. Moreover, for some species there is a global supply network because fishmeal and fish oil are used in the feed. The environmental challenges appear as three distinct but important issues: increased fishing pressure on species harvested for aquafeed, local environmental carrying capacity, and competition for locations.

The environmental issues in intensive aquaculture must be seen in relation to the introduction of a new technology that uses the environment as an input. The greater the production at any site and the more intensive the process, the greater the potential for environmental damage. However, the greater degree of control within the production process in intensive aquaculture also makes it easier to address these issues. Like all new technologies, there will be unexpected side effects, and there will be a time lag from when an issue arises until it can be addressed. First, the impact and the causes must be properly identified. Second, the solution to the problems will require modifications of existing technology or perhaps entirely new technology. In both cases, pollution reduction implies some form of induced innovation. Tveterås (2002) argues that industry growth has a positive effect on pollution in line with the environmental Kuznets curve, which refers to an empirical observation that pollution tends to increase with economic growth up to a certain point, after which growth will reduce pollution. This gives the pollution profile over time the shape of an inverted U. However, to what extent this will occur is largely a function of the management system (Chu et al. 2010; Nielsen 2012; Kumar and Engle 2016).

There are a number of examples of poor environmental practices in aquaculture (as in agriculture). However, that does not make the production method inherently unsustainable; there are a number of examples of sustainable aquaculture. Still, intensive and particularly large-scale intensive aquaculture has a larger potential to produce detrimental environmental effects than do other technologies. The higher the degree of control with the production process does, on the other hand, give these farmers a better opportunity to control the negative effects of their production. Hence, whether aquaculture is carried out in a sustainable manner is independent of the level of intensity. Therefore, the real issue with aquaculture and sustainability is whether institutions are developed to internalize externalities and create the incentives such that it is in the fish farmers' best interest to use sustainable practices. This will be an issue primarily of local regulations and governance but may also be influenced by consumer initiatives and ecolabels.

Regnier and Schubert (2017) provide the most extensive model of interactions between aquaculture and fisheries so far, and include market interactions, the use of marine ingredients in the feed, and its impact on wild edible species by reducing availability of wild species. They conclude that "when biological interactions are moderate, the introduction of aquaculture is beneficial in the long run: it improves consumer utility and alleviates the pressure on the edible fish stock (p. 186)." However, they note that when the interactions are strong, this conclusion is reversed. The conditions for when the interaction is strong are in most cases highly unlikely. For instance, there is a global market for fishmeal (Asche et al. 2013), implying that production responds to a global price determination process. Moreover, farmers stay competitive by reducing cost and substitute away from marine ingredients in the feed when they get more expensive (Kristofersson and Anderson 2006). For most fed species, the share of marine ingredients have been strongly reduced over time (Tacon and Metian 2008), and salmon in Norway now provides more marine ingredients than it uses (Ytrestøl et al. 2015).

5 Seafood markets and trade

Seafood markets have changed tremendously during recent decades. The shift in production technology helps explain this, but increased trade is just as important in creating global markets for the main groups of seafood species. At the same time, increased consumer concerns with respect to sustainability, health, and communities have created increased opportunities for market segmentation.

Seafood has been a traded commodity for thousands of years. From early on, the quantity traded was limited, mainly because seafood is perishable, and conserving fish (e.g., by producing dried fish) was time consuming, costly, and often inefficient. However, improved storage and preservation technologies and cheaper transportation have steadily increased fish trade. Still, there has been a virtual revolution in seafood trade over the last 30 years. After adjustment for inflation, from 1976 to 2013 world seafood trade value increased threefold, from US$28.3 billion to US$86.4 billion, and also the increase in traded quantity is approximately threefold. This makes seafood the most traded major food category (Anderson 2003; Anderson et al. 2010).

A number of factors have facilitated and/or caused the increased trade in seafood. The limited production growth from fisheries and the introduction of 200-mile EEZs created economic opportunity, and the fact that seafood is treated as an industrial product in the General Agreement on Tariffs and Trade (GATT) and the World Trade Organization (WTO) has led to lower barriers to trade. Improved transportation and logistics have led to substantial reductions in transportation costs and improved trade of new product forms, such as fresh seafood, and have given new producers access to the global market. Improved logistics has allowed economies of scale and scope on all levels in the supply chain, and particularly in the retail sector, where the supermarket has replaced fishmongers and fish markets in a number of places. Progress in storage and preservation has allowed a wider range of seafood products to be traded; of particular importance is improved freezing technology, which has allowed many product forms to be frozen twice. This is exploited by processing the products in locations with competitive advantages in processing fish rather than locations close to where the fish is caught.

As one would expect, ability to pay is important, and there is a tendency that trade flows from poorer to wealthier countries (Delgado et al. 2003; Kobayashi et al. 2015). Developing countries make up about 50 percent of export value, but just over 20 percent of import value (Asche et al. 2015).[19] The USA is the largest seafood importer, and over 90 percent of U.S. seafood consumption is imported. Japan is the second largest seafood importer, although the EU as a group is also larger than the USA. Combined, these three regions make up over 70 percent of global seafood imports by value. However, it is worthwhile to note that this share is declining, and particularly Southeast Asia is becoming a more important import region.

The geographical extent of seafood markets was traditionally limited by the perishability of the product. Until a hundred years ago, there were only a few heavily processed product forms that were shipped over longer distances. For other product forms, the market was, at best, regional, and often very local. As transportation, conservation, and logistics improved, and fueled by increased demand that could not be met from regional fisheries, the sources for the fish became increasingly global. The whitefish market is a good example (Asche, Roll, and Trollvik 2009). In 1980 it included primarily North Atlantic species, such as cod, saithe, and haddock. By 1990, Alaska pollock and Pacific cod were established as a major part of the market, linking the North Atlantic and North Pacific fisheries. During the 1990s, species such as Nile perch, Argentinean and Namibian hake, and hoki from New Zealand, and after the turn of the century farmed species such as pangasius and tilapia, made the market truly global (Bronnmann et al. 2016).

A similar development has taken place for most main species groups, creating global markets with a common price determination process. This is the case for shrimp (Vinuya 2007; Asche et al. 2012), salmon (Asche, Bremnes, and Wessells 1999), and tuna (Bose and McIlgorm 1996; Jeon et al. 2007). Moreover, when farmed fish becomes a large enough part of supply, production cost in aquaculture determines the price (Asche, Bremnes, and Wessells 1999). This is because farmers adjust production (primarily as some go bankrupt) when prices are low, while in most fisheries it is just the over-capacity that is adjusted.

Most of the main trends in the seafood market lead to more trade and more integrated markets. However, during the last decades there have also been an increasing number of factors that contribute to market segmentation, particularly in the European Union and the United States. The most important concern is the environmental impact of the fishing or aquaculture activity, although quality or safety signals can also segment the market by allowing producers to obtain a margin, and in some cases to obtain a completely independent price determination process.

The overfishing problems that led to the flattening out of global landings of fish (Figure 9.3) led a number of people to conclude that governments and particularly fisheries managers had not done their job. As a consequence, several non-governmental organizations (NGOs) have become advocates of market-based tools to help in management of fisheries, as they believe that consumers do not accept the mismanagement of fish stocks (Gudmundsson and Wessells 2000). Ecolabels are a market-based tool since they allow the consumers to choose seafood only from well-managed fisheries. Certification, labeling, and meeting specific standards have the effect of segmenting the market into those products where the standard is met and those where it is not. Consumers who respond to the label will then increase demand for labeled product and reduce demand for unlabeled product, creating a premium for labeled products. Meeting the standards requires that producers provide information that otherwise would not be provided and carry out costly additions to the production processes they otherwise would not undertake. This makes some producers unable or unwilling to meet the standards and therefore further segments the market and reduces trade or changes trade patterns.

A number of studies have investigated general ecolabels for seafood with two main approaches, survey data and hedonic price analysis. Survey data has the advantage that they can be obtained even before labeled products exist in a market, and one can also segment consumers. A number of studies show that consumers have a positive willingness to pay (WTP) for ecolabeled products, but that this varies by country and species (Wessells et al. 1999; Salladarré et al. 2010; Fonner and Sylvia 2015). However, it has been questioned whether this intention transfers to actual consumer behavior, particularly since there is often limited knowledge about ecolabels as well as issues among consumers (Grünert et al. 2014). More recently, several studies have used hedonic price functions to estimate price premiums associated with ecolabels on actual market data (Roheim et al. 2011). While most of these studies find a price premium, they do not provide conclusive evidence as they do not account for the quantity effect. There are only two studies that do provide mixed evidence; Teisl et al. (2002) show that a dolphin-safe label increases demand for labeled tuna cans, while Hallstein and Villa-Boas (2013) show no evidence of changes in demand for green or red labeled fish but reduced demand for yellow labeled fish in a San Francisco area retail chain. There is also a rapid increase in the number of ecolabels, as there are separate labels for aquaculture (e.g., the Aquaculture Stewardship Council (ASC)) that can correct the negative perception of aquaculture partly or fully (Bronnmann and Asche 2017), as well as other more general best practice standards such as Global Aquaculture Alliance (GAA) and GLOBALGAP or organic labeling (Ankamah-Yeboah et al. 2016).

While some standards seem justified, the myriad of requirements that differ among countries creates barriers for many producers. This is particularly true for producers in developing countries, where limited infrastructure makes it very hard to document the production process even when it is compliant. Sampson et al. (2015) show how hard this can be in that Fisheries Improvement Projects (FIPs), which are intended to help developing countries, did not produce much improvement in fisheries management. Myriad labels can also reduce impact as they create confusion among consumers (Roheim 2009). There is also evidence that wild fish is preferred to farmed (Uchida et al. 2014), but that this can be addressed with ecolabels for farmed fish (Bronnmann and Asche 2017).

6 The ocean ecosystem

There is a growing consensus that our oceans are in poor and declining health, plagued by a myriad of issues ranging from pollution and coastal development to climate change and ocean acidification, all of which impact the future of fisheries. Pollution, in forms of untreated sewage and industrial and agricultural runoff, has contributed to the rapid expansion of hypoxic zones characterized by low productivity and reductions in carrying capacity (Diaz and Rosenberg

2008). Increased plastic and microplastic pollution has received equal, if not more, attention and is expected to continue increasing. While there is little evidence linking plastics to stock population levels, it poses direct threats to human health by entering our seafood supply (Lusher et al. 2017).

The impact of climate change on fishery production remains undetermined, but shifts in migration patterns and jurisdictional issues will likely be the greatest impact (United Nations 2017). Changes in fish migration patterns may create international jurisdiction issues and may even cause problems within the current U.S. governance structure (i.e., regional fishery councils) as effort is subject to spillover effects (Cunningham et al. 2016). Greater emphasis on adaptive management and capacity (i.e., economic development and alternative livelihoods) is needed to cope with the uncertainty surrounding climate change and its impact on fisheries (Allison et al. 2009; Melnychuk et al. 2013).

7 Concluding remarks

Per capita seafood production reached a new record high, surpassing 20kg in 2014, following an increasing trend for decades. This makes seafood increasingly important for food and livelihoods (Smith et al. 2010; Anderson et al. 2015; Golden et al. 2016). As seafood has a number of positive health effects, its importance in the human diet and public health is hard to understate (Mozaffarian and Rimm 2006).

Approximately 71 percent of the earth's surface is covered with water. However, human ability to penetrate the water surface and thereby fully utilize ocean resources has been limited. When it comes to food production, until relatively recently, we have been constrained by what the aquatic ecosystems provide. Fishing for a long time has been the planet's largest hunting industry, which is suggestive of the immense resources held by the oceans. However, with the rapid development of modern aquaculture, the productivity and use of aquatic resources are in a remarkable transformation. Aquaculture has for three decades been the world's fastest growing food production technology, suggesting tremendous potential. Aquaculture is already as important as wild capture fisheries in providing seafood for human consumption, and it will be very surprising if aquaculture does not increasingly dominate the global seafood supply just as domesticated animals dominate the meat and poultry industry.

Notes

1 By seafood, we will mean all fish, crustaceans, and other marine animals harvested from oceans, lakes, and rivers, but not marine mammals and plants.
2 It is not accidental that the first economic model of the tragedy of the commons and how to prevent it was set in the context of a fishery (Gordon 1954).
3 Fisheries landings are still larger in total as about 25 mmt are landed for other purposes, with reduction to fishmeal and fishoil as the most important.
4 Kinnucan and Myrland (2002) and Keithly and Poudel (2008) provide two case studies for Atlantic salmon and shrimp respectively.
5 The product categories "aquatic plants," "inedible," and "sponges, corals, shells" from the FAO seafood trade statistics are excluded to focus on the trade of seafood products.
6 It is of interest to note that the Dane Jens Warming developed a bioeconomic model that explained the tragedy of the commons as early as 1911, although as this was in Danish it was overlooked by the international academic community (Andersen 1983; Eggert 2010a, b).
7 Although other easily accessible species were vulnerable, including fisheries in fresh water lakes and rivers, and management systems some for such species (e.g., salmon) date back centuries, overfishing in large lakes is largely a modern phenomenon with the same drivers as in marine fisheries, with Eggert et al. (2015) providing a recent example.

8 A more detailed but easily accessible version of the model can be found in most introductory texts in natural resource economics. A complete treatment can be found in Clark (1990).
9 Economic profits are zero in all competitive industries, as all factors are paid their opportunity cost. In general, accounting profit will be positive in such industries, as the accounting profit is the return on the capital to the investors.
10 In buy-back programs, fish managers reduce capacity by buying the vessel out of the fishery, often with stipulations that the vessel is scrapped to avoid contributing to over-capacity in other fisheries. The basic objective is to increase profitability and contribute to the livelihoods of the remaining fishers since effort is reduced (Curtis and Squires 2007). However, this will only be successful if the one can also restrict the remaining fishers race-to-fish, which is hard under command-and-control management.
11 There are also researchers that will regard management cost and even infrastructure and research costs as subsidies, but there is no consensus on these definitions. While it may be clear that a specific fishing dock can be a subsidy, it is less clear that a road to a community is a fishing subsidy.
12 This was also first suggested by Warming (Eggert 2010a).
13 For instance, cod and pollock are typically harvested with the same vessels and gears, but as cod prices are substantially higher than saithe prices, the resource rent will be higher for pollock.
14 For instance, Abbott et al. (2010) show that while the crab rationalization program in Alaska reduced the number of fishers, it did not reduce man-hours to a significant extent. Hence, the longer seasons did not reduce total employment, but it redistributed it to more stable jobs for fewer fishers.
15 However, plant production is normally excluded, and that will be done in this chapter too.
16 And this is also where the difficulty comes in (Klinger et al. 2013). Danish mussel fishers tend to spend more resources improving the beds than Dutch mussel farmers. Who, then, is the farmer? And the feed conversion ratio in the New England lobster fishery is about four to one, as four times as much bait fish by weight is expended than lobster being landed, and the traps also provide protection. In a number of Alaska salmon fisheries, most of the landed fish are from hatcheries, etc.
17 Norway is by far the largest salmon producing country, with a production share around 60 percent, and Norwegian prices and cost are representative of global prices and costs as there is an integrated market (Asche and Bjørndal 2011).
18 Jensen et al. (2014) show that in a similar fashion, if fisheries management is improved, the increased supply is detrimental to aquaculture production.
19 Imports to developing countries are much higher by quantity, with a share of over 40 percent.

References

Abbott, J.K., B. Garber-Younts, and J.E. Wilen. 2010. "Employment and Remuneration Effects of ITQs in the Bering/Aleutean Islands Crab Fisheries." *Marine Resource Economics* 25:333–354.

Abbott, J.K., A.C. Haynie, and M.N. Reimer. 2015. "Hidden Flexibility: Institutions, Incentives, and the Margins of Selectivity in Fishing." *Land Economics* 91(1):169–195.

Allison, E.H., A.L. Perry, M.C. Badjeck, W.N. Adger, K. Brown, et al. 2009. "Vulnerability of National Economies to the Impacts of Climate Change on Fisheries." *Fish and Fisheries* 10:173–196. doi:10.1111/j.1467–2979.2008.00310.x

Andersen, P. 1983. "'On Rent of Fishing Grounds': A Translation of Jens Warming's 1911 Article, with an Introduction." *History of Political Economy* 15(3):391–396.

Anderson, J.L. 1985a. "Private Aquaculture and Commercial Fisheries: Bioeconomics of Salmon Ranching." *Journal of Environmental Economics and Management* 12(4):353–370.

———. 1985b. "Market Interactions Between Aquaculture and the Common-Property Commercial Fishery." *Marine Resource Economics* 2:1–24.

———. 2002. "Aquaculture and the Future: Why Fisheries Economists Should Care." *Marine Resource Economics* 17(2):133–151.

———. 2003. *The International Seafood Trade*. Cambridge: Woodhead Publishing.

Anderson, J.L., C.M. Anderson, J. Chu, J. Meredith, F. Asche, G. Sylvia, M.D. Smith, D. Anggraeni, R. Arthur, A. Guttormsen, J.K. McCluney, T. Ward, W. Akpalu, H. Eggert, J. Flores, M.A. Freeman, D.S. Holland, G. Knapp, M. Kobayashi, S. Larkin, K. MacLauchlin, K. Schnier, M. Soboil, S. Tveteras, H. Uchida, and D. Valderrama. 2015. "The Fishery Performance Indicators: A Management Tool for Triple Bottom Line Outcomes." *PLOS ONE* 4.

Anderson, J.L., F. Asche, and S. Tveterås. 2010. "World Fish Markets." (pp. 113-122) In R.O. Grafton, R. Hilborn, D. Squires, M. Tait, and M. Williams, eds., *Handbook of Marine Fisheries Conservation and Management*. Oxford: Oxford University Press.

Anderson, J.L., and S.U. Bettencourt. 1993. "A Conjoint Approach to Model Product Preference: The New England Market for Fresh and Frozen Salmon." *Marine Resource Economics* 8(1):31–49.

Ankamah-Yeboah, I., M. Nielsen, and R. Nielsen. 2016. "Price Premium of Organic Salmon in Denmark." *Ecological Economics* 122:54–60.

Arnarson, R. 1990. "Minimum Information Management in Fisheries." *Canadian Journal of Economics* 23:630–653.

Asche, F. 2008. "Farming the Sea." *Marine Resource Economics* 23(4):527–547.

Asche, F., M. Bellemare, C. Roheim, M.D. Smith, and S. Tveteras. 2015. "Fair Enough? Food Security and the International Trade of Seafood." *World Development* 67:151–160.

Asche, F., L. Bennear, A. Oglend, and M.D. Smith. 2012. "U.S. Shrimp Market Integration." *Marine Resource Economics* 27(2):181–192.

Asche, F., and T. Bjørndal. 2011. *The Economics of Salmon Aquaculture*. Chichester: Wiley-Blackwell.

Asche, F., T. Bjørndal, and J.A. Young. 2001. "Market Interactions for Aquaculture Products." *Aquaculture Economics and Management* 5:303–318.

Asche, F., H. Bremnes, and C.R. Wessells. 1999. "Product Aggregation, Market Integration and Relationships Between Prices: An Application to World Salmon Markets." *American Journal of Agricultural Economics* 81(3):568–581.

Asche, F., A. Oglend, and S. Tveteras. 2013. "Regime Shifts in the Fish Meal/Soybean Meal Price Ratio." *Journal of Agricultural Economics* 64(1):97–111.

Asche, F., C.A. Roheim, and M.D. Smith. 2016. "Trade Intervention: Not a Silver Bullet to Address Environmental Externalities in Global Aquaculture." *Marine Policy* 69:194–201.

Asche, F., K.H. Roll, and T. Trollvik. 2009. "New Aquaculture Species – The Whitefish Market." *Aquaculture Economics and Management* 13(2):76–93.

Béné, C., R. Lawton, and E.H. Allison. 2009. "Trade Matters in the Fight Against Poverty: Narratives, Perceptions and, (Lack of) Evidence in the Case of Fish Trade in Africa." *World Development* 38:933–954.

Birkenbach, A.M., D. Kaczan, and M.D. Smith. 2017. Catch Shares Slow the Race to Fish. *Nature* 544(7649):223–226.

Bose, S., and A. McIlgorm. 1996. "Substitutability Among Species in the Japanese Tuna Market: A Cointegration Analysis." *Marine Resource Economics* 11(3):143–155.

Bronnmann, J., I. Ankamah Yeboah, and M. Nielsen. 2016. "Market Integration Between Farmed and Wild Fish: Evidence from the Whitefish Market in Germany." *Marine Resource Economics* 31(4):421–432.

Bronnmann, J., and F. Asche. 2017. "Sustainable Seafood from Aquaculture and Wild Fisheries: Insights from a Discrete Choice Experiment in Germany." *Ecological Economics* 142:113–119.

Chu, J., J.L. Anderson, F. Asche, and L. Tudur. 2010. "Stakeholders' Perceptions of Aquaculture and Implications for Its Future: A Comparison of the U.S.A. and Norway." *Marine Resource Economics* 25:61–76.

Clark, C.W. 1973. "Profit Maximization and the Extinction of Animal Species." *Journal Political Economics* 81:950–961.

Clark, C.W. 1990. *Mathematical Bioeconomics: The Optimal Management of Renewable Resources*. New York: Wiley-Interscience.

Clark, C.W., and G.R. Munro. 1975. "The Economics of Fishing and Modern Capital Theory: A Simplified Approach." *Journal of Environmental Economics and Management*: 92–106.

Copes, P. 1986. "A Critical-Review of the Individual Quota as a Device in Fisheries Management." *Land Economics* 62(3):278–291.

Costello, C., and R. Deacon. 2007. "The Efficiency Gains from Fully Delineating Rights in an ITQ Fishery." *Marine Resource Economics* 22:347–361.

Costello, C., S. Gaines, and J. Lynham. 2008. "Can Catch Shares Prevent Fisheries Collapse?" *Science* 321(5896):1678–1681.

Cunningham, S., L.S. Bennear, and M.D. Smith. 2016. "Spillovers in Regional Fisheries Management: Do Catch Shares Cause Leakage?" *Land Economics* 92:344–362.

Curtis, R., and D. Squires. 2007. *Fisheries Buybacks.* Ames, Iowa: Blackwell.

Delgado, C.L., N. Wada, M.W. Rosengrant, S. Meijer, and M. Ahmed. 2003. *Fish to 2020: Supply and Demand in Changing Global Markets.* Washington, DC: IFPRI.

Diaz, R.J., and R. Rosenberg. 2008. "Spreading Dead Zones and Consequences for Marine Ecosystems." *Science* 321:926–929.

Dupont, D.P. 1991. "Testing for Input Substitution." *American Journal of Agricultural Economics* 73(Feb):155–164.

Eggert, H. 2010a. "Jens Warming on Open Access, the Pigovian Tax, and Property Rights." *History of Political Economy* 42(3):469–481.

Eggert, H. 2010b. "The Danish Right to Eel Weir. A Translation of Jens Warming's 1911 Article." *History of Political Economy* 42(3):469–481.

Eggert, H., M. Greaker, and A. Kidane. 2015. "Trade and Resources: Welfare Effects of the Lake Victoria Fisheries Boom." *Fisheries Research* 167:156–163.

Eggert, H., and R. Tveteras. 2004. "Stochastic Production and Heterogeneous Risk Preferences: Commercial Fishers' Gear Choices." *American Journal of Agricultural Economics* 86:199–212.

FAO. 2016. *The State of World Fisheries and Aquaculture 2016.* Rome: FAO, 200 pp.

FAO. 2017. "FishstatPlus." Universal software for fishery statistical time series. Rome. Accessed July 20, 2017.

Fonner, R., and G. Sylvia. 2015. "Willingness to Pay for Multiple Seafood Labels in a Niche Market." *Marine Resource Economics* 30:51–70.

Golden, C.D., E.H. Allison, W.W.L. Cheung, M.M. Dey, B.S. Halpern, D.J. McCauley, M. Smith, B. Vaitla, D. Zeller, and S.S. Myer. 2016. "Nutrition: Fall in Fish Catch Threatens Human Health." *Nature* 534:317–320.

Gordon, H.S. 1954. "The Economic Theory of a Common Property Resource: The Fishery." *Journal of Political Economy* 62:124–142.

Gordon, D.V., and R. Hannesson. 2015. "The Norwegian Winter Herring Fishery: A Story of Technological Progress and Stock Collapse." *Land Economics* 91(2):362–385.

Grünert, K.G., S. Hieke, and J. Wills. 2014. "Sustainability Labels on Food Products: Consumer Motivation, Understanding and Use." *Food Policy* 44:177–189.

Gudmundsson, E., and C.R. Wessells. 2000. "Ecolabeling Seafood for Sustainable Production: Implications for Fisheries Management." *Marine Resource Economics* 15(2):97–113.

Guttormsen, A.G. 2002. "Input Factor Substitutability in Salmon Aquaculture." *Marine Resource Economics* 17:91–102.

Hallstein, E., and S.B. Villas-Boas. 2013. "Can Household Consumers Save the Wild Fish? Lessons from a Sustainable Seafood Advisory." *Journal of Environmental Economics and Management* 66(1):52–71.

Haynie, A., and D.F. Layton. 2010. "An Expected Profit Model for Monetizing Fishing Location Choices." *Journal of Environmental Economics and Management* 59:165–176.

Haynie, A.C., R.L. Hicks, and K.E. Schnier. 2009. "Common Property, Information, and Cooperation: Commercial Fishing in the Bering Sea." *Ecological Economics* 69(2):406–413.

Holland, D.S., and J.G. Sutinen. 2000. "Location Choice in New England Trawl Fisheries: Old Habits Die Hard." *Land Economics* 76(1):133–149.

Homans, F.R., and J.E. Wilen. 1997. "A Model of Regulated Open Access Resource Use." *Journal of Environmental Economics and Management* 32:1–21.

Homans, F.R., and J.E. Wilen. 2005. "Markets and Rent Dissipation in Regulated Open Access Fisheries." *Journal of Environmental Economics and Management* 49:381–404.

Huang, L., and M.D. Smith. 2014. "The Dynamic Efficiency Costs of Common-Pool Resource Exploitation." *The American Economic Review* 104(12):4071–4103.

Huxley, Thomas Henry. [1883] 1998. Inaugural Address to the Fisheries Exhibition. London. Available at The Huxley File, created by C. Blinderman and D. Joyce. http://aleph0.clarku.edu/huxley/SM5/fish.html. Accessed 3 December 2010.

Jensen, F., M. Nielsen, and R. Nielsen. 2014. "Increased Competition for Aquaculture from Fisheries: Does Improved Fisheries Management Limit Aquaculture Growth?" *Fisheries Research* 159:25–33.

Jeon, Y., C. Reid, and D. Squires. 2007. "Is There a Global Market for Tunas? Policy Implications for Tropical Tuna Fisheries." *Ocean Development and International Law* 39(1):32–50.

Keithly, W.R. Jr., and P. Poudel. 2008. "The Southeast U.S. Shrimp Industry: Issues Related to Trade and Antidumping Duties." *Marine Resource Economics* 23:459–483.

Kinnucan, H.W., and Ø. Myrland. 2002. "The Relative Impact of the Norway-EU Salmon Agreement: A Mid-Term Assessment." *Journal of Agricultural Economics* 53:195–220.

Klinger, D., M. Turnipseed, J.L. Anderson, F. Asche, L. Crowder, A.G. Guttormsen, B.S. Halpern, M.I. O'Connor, R. Sagarin, K.A. Selkoe, G. Shester, M.D. Smith, and P. Tyedmers. 2013. "Moving Beyond the Fished or Farmed Dictomy." *Marine Policy* 38:369–374.

Knapp, G., C.A. Roheim, and J.L. Anderson. 2007. *The Great Salmon Run: Competition Between Wild and Farmed Salmon.* Washington, DC: Traffic North America, World Wildlife Fund, 302 pp.

Kobayashi, M., S. Msangi, M. Batka, S. Vannuccini, M.M. Dey, and J.L. Anderson. 2015. "Fish to 2030: The Role and Opportunity for Aquaculture." *Aquaculture Economics & Management* 193:282–300.

Kumar, G., and C. Engle. 2016. "Technological Advances That Led to Growth of Shrimp, Salmon, and Tilapia Farming." *Reviews in Fisheries Science & Aquaculture* 24(2):136–152.

Kristofersson, D., and J.L. Anderson. 2006. "Is There a Relationship Between Fisheries and Farming? Interdependence of Fisheries, Animal Production and Aquaculture." *Marine Policy* 30(6):721–725.

Kroetz, K., J.N. Sanchirico, and D.K. Lew. 2015. "Efficiency Costs of Social Objectives in Tradable Permit Programs." *Journal of the Association of Environmental and Resource Economists* 2(3):339–366.

Kvamsdal, S. 2016. "Technical Change as a Stocastic Trend in a Fisheries Model." *Marine Resource Economics* 31:403–419.

Lockwood, G.S. 2017. *Aquaculture: Will It Rise to It's Potential to Feed the World?* Carmel Valley, CA: George Lockwood.

Lusher, A., P. Hollman, and J. Mendoza-Hall. 2017. "Microplastics in Fisheries and Aquaculture: Status of Knowledge on Their Occurrence and Implications for Aquatic Organisms and Food Safety." FAO Fisheries and Aquaculture Technical Paper No. 615.

Melnychuk, M.C., J.A. Banobi, and R. Hilborn. 2013. "The Adaptive Capacity of Fishery Management Systems for Confronting Climate Change Impacts on Marine Populations." *Reviews in Fish Biology and Fisheries* 24:561–575.

Mozaffarian, D., and E. Rimm. 2006. "Fish Intake, Contaminants, and Human Health: Evaluating the Risks and the Benefits." *JAMA* 296(15):1885–1899. doi:10.1001/jama.296.15.1885.

Munro, G.R., and A.D. Scott. 1985. "The Economics of Fisheries Management." In A.V. Kneese, and J.L. Sweeny, eds., *Handbook of Natural Resource and Energy Economics.* Amsterdam: North Holland.

Newell, R., J.N. Sanchirico, and S. Kerr. 2005. "Fishing Quota Markets." *Journal of Environmental Economics and Management* 49(3):437–462.

Nielsen, M. 2005. "Price Formation and Market Integration on the European First-hand Market for Whitefish." *Marine Resource Economics* 20:185–202.

Nielsen, R. 2012. "Introducing Individual Transferable Quotas on Nitrogen in Danish Freshwater Aquaculture: Production and Profitability Gains." *Ecological Economics* 75:83–90.

Nøstbakken, L. 2008. "Fisheries Law Enforcement: A Survey of the Economic Literature." *Marine Policy* 32(3):293–300.

———. 2012. "Investment Drivers in a Fishery with Tradable Quotas." *Land Economics* 88:400–424.

Pascoe, S., L. Coglan, A.E. Punt, and C. Dichmont. 2012. Impacts of Vessel Capacity Reduction Programmes on Efficiency in Fisheries. *Journal of Agricultural Economics* 63:425–443.

Pascoe, S., P. Koundouri, and T. Bjørndal. 2007. "Estimating Targeting Ability in Multi-Species Fisheries: A Primal Multi-Output Distance Function Approach." *Land Economics* 83(3):382–397.

Regnier, E., and K. Schubert. 2017. "To What Extent Is Aquaculture Socially Beneficial? A Theoretical Analysis." *American Journal of Agricultural Economics* 99(1):186–206.

Roheim, C. 2004. "Trade Liberalization in Fish Products: Impacts on Sustainability of International Markets and Fish Resources." In A. Aksoy and J. Beghin, eds., *Global Agricultural Trade and Developing Countries.* Washington, DC: The World Bank, pp. 275–295.

Roheim, C.A. 2009. "An Evaluation of Sustainable Seafood Guides: Implications for Environmental Groups and the Seafood Industry." *Marine Resource Economics* 24:301–310.

Roheim, C.A., L. Gardiner, and F. Asche. 2007. "Value of Brands and Other Attributes: Hedonic Analysis of Retail Frozen Fish in the UK." *Marine Resource Economics* 22:239–253.

Roheim, C., F. Asche, and J. Insignares Santos. 2011. "The Elusive Price Premium for Ecolabelled Products: Evidence from Seafood in the UK Market." *Journal of Agricultural Economics* 62(3):655–668.

Rozwadowski, H.M. 2002. *The Sea Knows No Boundaries*. Seattle: University of Washington Press and the International Council for the Exploration of the Sea.

Salladarré, F., P. Guillotreau, Y. Perraudeau, and M.-C. Monfort. 2010. "The Demand for Seafood Eco-Labels in France." *Journal of Agricultural and Food Industrial Organization* 8:1–24.

Sampson, G.S., J.N. Sanchirico, C.A. Roheim, S.R. Bush, J.E. Taylor, E.H. Allison, J.L. Anderson, N.C. Ban, R. Fujita, S. Jupiter, and J.R. Wilson. 2015. "Secure Sustainable Seafood from Developing Countries." *Science* 348(6234):504–506.

Sanchirico, J.N., and J.E. Wilen. 1999. "Bioeconomics of Spatial Exploitation in a Patchy Environment." *Journal of Environmental Economics and Management* 37(2):129–150.

Schaefer, M.B. 1957. "Some Aspects of the Dynamics of Populations Important to the Management of Commercial Marine Fisheries." *Bulletin of the Inter-American Tropical Tuna Commission* 1:25–56.

Scheld, A., and C. Anderson. 2014. "Market Effects of Catch Share Management: The Case of New England Multispecies Groundfish." *ICES Journal of Marine Science* 71(7):1835–1845.

Schnier, K., and R. Felthoven. 2011. "Accounting for Spatial Heterogeneity and Autocorrelation in Spatial Discrete Choice Models: Implications for Behavioral Predictions." *Land Economics* 87(3):382–402.

Smith, M.D. 2012. "The New Fisheries Economics: Incentives Across Many Margins." *Annual Review of Resource Economics* 4:379–402.

Smith, M.D., C.A. Roheim, L.B. Crowder, B.S. Halpern, M. Turnipseed, J.L. Anderson, F. Asche, L. Bourillón, A.G. Guttormsen, A. Kahn, L.A. Liguori, A. McNevin, M. O'Connor, D. Squires, P. Tyedemers, C. Brownstein, K. Carden, D.H. Klinger, R. Sagarin, and K.A. Selkoe. 2010. "Sustainability and Global Seafood." *Science* 327:784–786.

Smith, M.D., and J.E. Wilen. 2003. "Economic Impacts of Marine Reserves: The Importance of Spatial Behavior." *Journal of Environmental Economics and Management* 46(2):183–206.

Smith, V.L. 1969. "On Models of Commercial Fishing." *Journal of Political Economy* 77:181–198.

Squires, D. 1987. "Public Regulation and the Structure of Production in Multiproduct Indus- tries: An Application to the New England Otter Trawl." *Rand Journal of Economics* 18(2):232–247.

Squires, D., and N. Vestergaard. 2013. "Technical Change and the Commons." *The Review of Economics and Statistics* 95(5):1769–1787.

Sumaila, U.R. 1997. "Strategic Dynamic Interaction: The Case of Barents Sea Fisheries." *Marine Resource Economics* 12(2):77–94.

Swartz, W., U.R. Sumaila, R. Watson, and D. Pauly. 2010. "Sourcing Seafood for the Three Major Markets: The EU, Japan and the USA." *Marine Policy* 34:1366–1373.

Tacon, A.G.J., and M. Metian. 2008. "Global Overview on the Use of Fish Meal and Fish Oil in Industrially Compounded Aquafeeds: Trends and Future Prospects." *Aquaculture* 285:146–158.

Taylor, T.D., R.C. Chick, K. Lorenzen, A.-L. Agnalt, K. Leber, H.L. Blankenship, G.V. Haegen, and N.R. Loneragen. 2017. "Fisheries Enhancement and Restoration in a Changing World." *Fisheries Research* 186:407–412.

Teisl, M.F., B. Roe, and R.L. Hicks. 2002. "Can Eco-Labels Tune a Market? Evidence from Dolphin-Safe Labeling." *Journal of Environmental Economics and Management* 43(3):339–359.

Tveterås, S. 2002. "Norwegian Salmon Aquaculture and Sustainability: The Relationship Between Environmental Quality and Industry Growth." *Marine Resource Economics* 17:121–132.

Tveteras, S., F. Asche, M.F. Bellemare, M.D. Smith, A.G. Guttormsen, A. Lem, K. Lien, and S. Vannuccini. 2012. "Fish Is Food – The FAO's Fish Price Index." *PLoS ONE* 7(5):1–10.

Uchida, H., Y. Onozaka, T. Morita, and S. Managi. 2014. "Demand for Ecolabeled Seafood in the Japanese Market: A Conjoint Analysis of the Impact of Information and Interaction with Other Labels." *Food Policy* 44:68–76.

United Nations. 2017. Oceans and the Law of the Sea: Report of the Secretary-General, A/72/70. http://undocs.org/A/72/70. Accessed 11 December 2017.

Valderrama, D., and J.L. Anderson. 2007. "Improving Utilization of the Atlantic Sea Scallop Resource: An Analysis of Rotational Management of Fishing Grounds." *Land Economics* 83(1):86–103.

———. 2010. Market Interactions Between Aquaculture and Common-Property Fisheries: Recent Evidence from the Bristol Bay Sockeye Salmon Fishery in Alaska." *Journal of Environmental Economics and Management* 59:115–128.

Vinuya, F.D. 2007. "Testing for Market Integration and the Law of One Price in World Shrimp Markets." *Aquaculture Economics and Management* 11(3):243–265.

Weninger, Q. 1998. "Assessing Efficiency Gains from Individual Transferable Quotas: An Application to the Mid-Atlantic Surf clam and Ocean Quahog Fishery." *American Journal of Agricultural Economics* 80(4):750–764.

Wessells, C.R. 2002. "Markets for Seafood Attributes." *Marine Resource Economics* 17:153–162.

Wessells, C., R. Johnston, and H. Donath. 1999. "Assessing Consumer Preferences for Ecolabeled Seafood: The Influence of Species, Certifier and Household Attributes." *American Journal of Agricultural Economics* 81:1084–1089.

Wilen, J.E. 1976. "Common Property Resources and the Dynamics of Overexploitation: The Case of the North Pacific Fur Seal." Department of Economics Research Paper 3, University of British Columbia, Vancouver.

———. 2000. "Renewable Resource Economists and Policy: Have We Made a Difference?" *Journal of Environmental Economics and Management* 37:129–150.

———. 2006. "Why Fisheries Management Fails: Treating Symptoms Rather Than the Cause." *Bulletin of Marine Science* 78(3):529–546.

Ytrestøl, T., T.S. Aas, and T. Åsgård. 2015. "Utilisation of Feed Resources in Production of Atlantic Salmon (Salmo salar) in Norway." *Aquaculture* 448:365–374.

Zhang, J., and Martin D. Smith. 2011. "Estimation of a Generalized Fishery Model: A Two-Stage Approach." *Review of Economics and Statistics* 93(2):690–699.

10

The economics of antibiotics use in agriculture

Jutta Roosen and David A. Hennessy

1 Introduction

Worldwide growth of intensive farm animal production units has raised several concerns about the public health implication of disease management in farm animals. An issue that has received considerable attention is how antibiotic use in agriculture relates to the development of antibiotic resistance in human medicine. This chapter discusses the economics of antibiotic use in agriculture, including the implications for animal stock management, and the impacts beyond the farm. It covers the use of antimicrobials in general. However, to better converse in layperson terms, we assume that antibiotics and antimicrobials are synonymous. To approach the main economic issues surrounding antibiotic use in animal agriculture, Section 2 reviews the question of resistance management from the perspective of a common property resource. The farm economic perspective is analyzed in Section 3, while Section 4 reviews studies on the consumers' willingness to pay (WTP) and industry initiatives. Section 5 concludes the chapter with a summary and a discussion of topics that we believe merit future research. However, before turning to an economic analysis, we first provide an overview of the extent of antibiotic use in agriculture and develop the rationale for concern about public health implications in the remainder of this introduction.

Antibiotics find different usages in agriculture. While they can be applied in tree fruit crops (Roosen and Hennessy 2001), the predominant mass of applications in the sector occur in animal agriculture. In animal production, the drug can have three specific uses: as a (1) prophylactic treatment (control and prevention) of disease events; (2) treatment after an existing bacterial infection; and (3) growth promotant. While these categories seem precisely defined in theory, assessing the use intended by a farm operator in terms of these categories may be difficult in practice. All three uses have been of interest since the technology's earliest commercial use. The history of antibiotics commenced with Alexander Fleming's discovery of penicillin's antimicrobial effect in the late 1920s. A decade passed before other United Kingdom-based scientists used this knowledge to develop the first antibiotics and provide proof of concept in human subjects. Under pressure to manage the health of troops during World War II, much of the wartime research and development took place in the United States, where commercial quantities of the microbial disease-fighting drug soon became available. The growth promotant effect of antibiotics in non-ruminant farm animals had been noted during the 1940s, and the U.S. Food and Drug Administration approved their use as feed additives in 1951. European countries passed national regulations on antibiotics as a medicine and as feed additives in the 1950s and 1960s (Castanon 2007). A common European legislation was formed in 1970. In all instances, the antibiotics had to be registered for veterinary use and for use as growth promotant.

Recent decades have seen an increasing realization that antibiotic use in agriculture poses significant public health concerns. Antimicrobial drug residues in food products are of some concern, and the Codex Alimentarius as well as different national regulations stipulate maximum residue levels for specific drugs. The main public health concern, however, surrounds the development of resistant bacteria and other microbials, which, in turn, can pose threats to human health. The Centers for Disease Control in the United States estimated that more than two million individuals fell sick to antibiotic-resistant bacteria strains and 23,000 patients died in 2013 (Centers for Disease Control and Prevention 2013). For the European Union (EU), the European Centre for Disease Prevention and Control estimated that 25,000 people died in 2007 as a result of resistant strains (ECDC 2009). Worldwide estimates amount to 700,000 deaths annually (O'Neill 2016). As early as 1965, the implications of antibiotic use in animal agriculture for the possible spread of resistance among bacteria were described by Anderson and Lewis (1965). Several government reports have addressed the issue over the decades, and an overview of these reports can be found in National Research Council (NRC 1999, pp. 16–17) and United States (U.S.) Food and Drug Administration (US FDA, 2012).

An appreciation of the public health concerns posed by antibiotics requires some deliberation on ways through which resistance in bacteria can emerge and spread. Consumers can come in contact with resistant bacteria that remain in or on food products leaving the farm. Resistance can spread by selection pressure toward resistant bacteria that survive drug applications and through mitochondrial exchange of genetic information within and across bacterial species. Hence, the intake of resistant microbials into the human body may either cause illness directly, due to resistant bacteria, or indirectly, because the ingested resistant bacteria passes the resistance on to other pathogens and, hence, spreads resistance. Resistance may also spread through manufacturing waste disposal whereby antibiotics enter the environment directly, or alternatively via manure and other forms of animal waste (O'Neill 2015).

The implications of agricultural use of antibiotics for public health led the World Health Organization in 1997 and the EU in 1998 to declare the use of antibiotics in agriculture a public health issue (Castanon 2007). In consequence, the use of antibiotics as a growth promotant was phased out in the EU in 2006. However, continued widespread use of antibiotics for therapeutic and blanket uses remains a concern. To empower public agents to address this issue, EU Regulation 2016/429 on Transmissible Animal Diseases stipulates that microbials that have developed resistance to antimicrobials should be treated as a transmissible disease. Hence, these bacteria and other microbials qualify for specific measures of control in farm operations under the EU's existing "One Health" action plan. In contrast to the mandatory approach pursued in the EU, the United States has taken a voluntary approach of engagement with the pharmaceutical industry in which companies are strongly encouraged to withdraw certain antibiotics for use as growth promotants (US FDA 2013). By 2017, all manufacturers removed the label claim for growth promotants or increasing feed efficiency so that all antibiotics medically important to humans are effectively withdrawn from the market for use as growth promotants (US FDA 2017).

While it is known that antibiotic use in agriculture is widespread, data on the exact extent and purpose of farm use remain surprisingly sketchy. The World Organisation for Animal Health has recently evaluated the legislation and monitoring of antimicrobial use in 130 countries.

Legislation is lacking and data are missing in 110 of the surveyed countries. In many developing and emerging countries, antimicrobials are freely available and often adulterated, so that low-dose applications risk rapid resistance development (World Organisation for Animal Health 2015). However, monitoring is also incomplete in the developed countries. A report by the Economic Research Service in the United States (Sneeringer et al. 2015) suggests that between 2004 and 2009, the share of finished hogs that received antibiotics as growth promotants fell from 52 to 40 percent. The respective numbers for nursery hogs fell from 29 to 23 percent. At the same time, the farms' uncertainty about whether they used antibiotics in feed or drinking water has increased from 7 to 22 percent and from 5 to 20 percent, respectively. Lack of knowledge is a problem that prevails, in particular, among integrated farms that receive feed mixes from their integrator. For broilers in the United States, the share of farms reporting that they do not apply antibiotics as growth promotants rose from 42 percent in 2006 to 48 percent in 2011. Antibiotics are in particular an issue in large-scale beef feedlots, where they are applied to about 49 percent of cattle. Antibiotics are also applied for preventing infections during the dry period in 90 percent of dairy operations.

For animal agricultural sectors in European countries, it is also difficult to obtain a detailed picture of how antibiotics are used. Regulatory oversight by the European Surveillance of Veterinary Antimicrobial Consumption has established statistics on the amount of antibiotics sold in the EU. Between 2011 and 2013, sales (in metric tons) have declined by 10.5 percent on average. Upon correction for the food-animal population using a measure called population correction unit (PCU), use in mg/PCU has declined by 7.9 percent. More detailed statistics regarding animal species and use are not available at the EU level, and figures are difficult to compare across countries because agricultural sectors differ in regard to species composition (Smith et al. 2016a).

With the intention of becoming better informed and better placed to implement policies, regulators have used different instruments. These include approval processes for drugs, bans on specific uses, monitoring requirements for antibiotic use and resistance development, and, more recently, labeling intended to support market-based solutions (Sneeringer et al. 2015). We discuss these in context and in more detail in the subsequent sections of this chapter. But first, we explain why a market failure is to be expected and why interventions to limit and direct use are needed. This is done in Section 2, which provides an overview of the literature on the social planner's problem of managing resistance in a multipurpose drug setting, in which there are human medicine and veterinary uses.

2 Common property resource perspective on antibiotic resistance management

Bacteria can become resistant to antibiotics through selection pressure in the manner that pests can become resistant to pesticides. The development of resistance can be viewed as an externality that emerges with the application of antibiotics. Early treatments of this externality issue are Brown and Layton (1996), Secchi (2000), and Laxminarayan and Brown (2001). McNamara and Miller (2002) discussed the implications of antibiotic use in agriculture. The problem of resistance or susceptibility management is set up as one of optimal control. Here we adopt the presentation by Brown and Layton (1996) because it facilitates clarity in understanding the social planner's problem of weighing the benefits and costs of antibiotics for human against those for agricultural uses.

The benefit to a human patient from treatment is denoted as B^h and the cost as C^h. The cost of treatment is considered to be a function of susceptibility S, as reduced susceptibility requires a higher dose or treatment by a more costly antimicrobial. A patient will seek treatment by a given dose of antibiotics, a^h, as long as $B^h > C^h(S)$. On the farm, the benefit is denoted as $\max_{a^f} B^f(a^f) - C^f(a^f, S)$, where a^f denotes the amount of antibiotics administered to animals on the farm. The farm's maximization problem would yield the optimal doses to apply to animals as that which equates marginal benefits with marginal costs. We will deal with this problem in more detail in Section 3.

While patients and farmers have to manage their own individual disease problem, the social planner has the additional task of managing how susceptibility is depleted (i.e., the development of resistance). This depletion can be described by the equation of motion $\dfrac{dS}{dt} = \dot{S} = -(a^h + \gamma a^f)$. Hence, the pool of susceptibility is diminished by the amount of antibiotics applied in the human population of size N^h and the farm animal population of size N^f, where the factor $\gamma > 0$ accounts for the rate of depletion arising from animal applications. In consequence, the social planner's problem, recognizing the size of the human population N^h and the animal population N^f, becomes

$$\max_{a^h, a^f} \int_0^\infty e^{-\rho t}\left(B^h\left(N^h, a^h\right) - C^h\left(a^h, S\right) + B^f\left(N^f, a^f\right) - C^f\left(a^f, S\right)\right) dt$$

$$\text{s.t.} \quad \dfrac{dS}{dt} = \dot{S} = -\left(a^h + \gamma a^f\right)$$

where the parameter ρ denotes the social discount rate. Building the current-value Hamiltonian, $\tilde{H} = \left[B^h(N^h, a^h) - C^h(a^h, S)\right] + \left[B^f(N^f, a^f) - C^f(a^f, S)\right] - \lambda(a^h + \gamma a^f)$ yields the first order conditions where in human medicine: $MB_a^h = MC_a^h + \lambda$ in agricultural production: $MB_a^f = MC_a^f + \gamma\lambda$ and optimal depletion of susceptibility: $\dot{\lambda} = \delta\lambda + MC_S^h + MC_S^f$

Here, MB and MC denote the marginal benefit and marginal cost for human (h) and farm (f) use with respect to the subscript dose (a) or susceptibility (S).

The first-order conditions demonstrate that varying the amount of antibiotics used (a^h and a^f) can yield the marginal benefit of curing an infection. At the same time, it creates a marginal *private* cost of treatment (C) plus a marginal *public* cost of reduced effectiveness of the drug at later stages in time. This latter effect is captured in the shadow value to the equation of motion λ. The simple optimal control problem is one way of characterizing the negative externality of resistance development such that interventions intended to manage resistance can improve social welfare. In its simplest form, the common resource problem will yield an exploitation path that will deplete susceptibility and develop resistance too quickly because it ignores the public cost of depletion. In comparison with individuals, the optimizing social planner will acknowledge the higher cost of future treatments (both for humans and animals) and the changes in benefits arising from higher resistance. To solve this resistance management issue, the social planner needs insights into the rate γ with which resistance spreads between human and animal pathogens.

The basic optimal control model above has been extended in multiple ways to acknowledge the complexity of antibiotic resistance management. An issue addressed by Laxminarayan and Brown (2001) is the optimal use of multiple antibiotic drugs in human disease treatments for managing resistance. They showed that the optimal timing of antibiotic use depends on the difference in the rates at which resistance to these drugs develops and the difference in their costs. Secchi (2000) developed a similar optimal control model that acknowledges different scenarios for alternative drugs becoming available through investment in technology development.

Overall, she found that investments into alternative technologies would limit the increase in the number of people treated with an existing therapy. However, when discovery is uncertain then the number of people treated should be restricted through time. Indeed, the rate of over-prescription of 40 percent, as estimated in Fleming-Dutra et al. (2016) for the United States, gives rise to concerns in light of Secchi's conclusions. The issue of drug development in monopoly situations was also analyzed by Fischer and Laxminarayan (2005), who conclude that, depending on the number and quality of available backstop technologies, the monopolist may under- or over-exploit an exhaustible resource when compared to the social planner.

Discussions to this point have been crude in regard to characterizing the nature of farm-level demand for antibiotics. An understanding of the origins of demand is important if policies are to be developed that seek to substitute for antibiotics as inputs. In the following section, we turn to a representation of the farmer's input choice problem. In particular, we consider the nature of antibiotics in farm animal production as a damage-control agent, yielding the optimal use of antibiotics in agriculture. Thus, we leave the social-planner problem aside for the moment and return to it in the conclusion.

3 Antibiotics as production inputs in agriculture

In comparison to the representation above, where antibiotics are seen as directly productive inputs, they can and probably should be viewed as a damage control input (Lichtenberg and Zilberman 1986), an interpretation that has implications for how these inputs are viewed for both microeconomic analysis and econometric estimation. The matter is important because it implies that demand may be discontinuous and may even disappear when marginal value of product is high (Fox and Weersink 1995; Hennessy 2018).

As mentioned in the Introduction, the drugs have potentially three uses in animal agriculture. They can be applied as a prophylactic prior to specific information suggesting that a disease problem exists. They can be applied as therapy in response to evidence that a disease event exists, and they can also be applied for purposes other than disease control. For example, they have been routinely included in the feed mixes of non-ruminants, where it is known that (likely for reasons related to the microbiology of gut bacteria) feed conversion efficiency is often higher if the feed contains antibiotics (Sneeringer et al. 2015).

A small body of research has sought to address connections between therapeutic application and sub-therapeutic application with intent to control and prevent disease. If antibiotics are not used to prevent biotic disease, then more disease will likely occur and any subsequent antibiotic use may be more intensive. The microeconomic foundations of the tradeoff have been pursued in Hennessy (2018), where the choice between blanket early administration and selective information-conditioned administration later is viewed as a real option. There it is shown that inelastic demand for the input implies that a ban on sub-therapeutic use will likely decrease mean use, and it is argued that inelastic demand is to be expected. A review of data available from countries that have limited sub-therapeutic use suggests that the overall load is likely to decrease as a result of use restrictions (Jensen 2016; Jensen and Hayes 2014).

The third use – for reasons other than disease management – is the most controversial, as it is viewed as the one with the lowest private marginal value of product. The extent of improved performance due to use for non-medical purposes was thought to be very large at one time, but different studies have arrived at widely varying results. Teillant, Brower, and Laxminarayan (2015) provided a detailed overview of the cost implications of discontinuing the use of antibiotic growth promoters. Upon consideration of a literature founded largely on experimental studies, as well as the Danish and Swedish experience in restricting use, they came to the conclusion that the effect may be very limited but is likely to vary with region and the farmers' biosecurity management skills. Similar conclusions were reached by Sneeringer et al. (2015).

Measurement of the effect on production performance is problematic for several reasons. One is that observed behavior does not allow one to distinguish between application in a sub-therapy program and application motivated by reasons other than disease management. In addition, comparisons of market outcomes for groups that do and do not use antibiotics cannot be interpreted directly, as they are subject to the concern that growing operations make the antibiotic decision based on the extent of the problems faced. True randomized control trials are rare and very costly to conduct.

Employing methods to control for selection bias in the farmers' responses provided to the USDA Agricultural Resource Management Survey of 2009 on feeder-to-finish pig operations, Key and McBride (2014) found that antibiotics increased mean productivity and decreased variance of productivity, but these effects were not large. Rojo-Gimeno et al. (2016) conducted treatment-and-control experiments on farrow-to-finish pig farms in northern Belgium and found minimal net benefits to antibiotic use. Most likely, the productivity effects have diminished over time as more modern on-farm facilities seek to relieve animals from the weather and other stressors and are generally easier to clean and better secured against disease entry. Productivity gains from use to address non-disease issues are likely greater in environments with weak infrastructure, such as those in developing countries.

3.1 Simple model, with implications for production structure

The model presented below is an adaptation of a technology-driven motive for risk management developed by Froot, Scharfstein, and Stein (1993). A farm has available a fixed proportions technology with representative effective input amount $e \geq 0$ and revenue $R(e)$, a positive, differentiable, increasing, and concave function. The amount of actual input I is comprised of two parts, effective input $e \geq 0$ and a random component ω, where $I = e + \omega$. Difference $\omega = I - e$ arises due to unforeseen events, including disease, as well as weather and infrastructure breakdowns (water, air conditioning, nutrition, and labor quality) that leave animals more vulnerable to biotic debilitation. In turn, the random component can be controlled through the use of antibiotics. The cost of effective input e is a positive, differentiable, increasing, and convex function $c(e + \omega)$, where the random component is comprised of two parts, $\omega = \mu(a) + \sigma(a)\varepsilon$. Here, the amount of antibiotics used is $a \geq 0$, with direct unit cost $w > 0$. As we are now dealing only with the farm problem, we drop the superscript f introduced in Section 2. Functions $\mu(a)$ and $\sigma(a)$ are both well-defined, positive, continuously differentiable, and decreasing functions. Variable $\varepsilon \in [-1,1]$ is random with distribution function $F(\varepsilon)$ and mean value $\mathbb{E}[\varepsilon] = 0$.

The time sequence is as follows. The antibiotic choice is made at time point 0. The extent of production problems, as reflected by ω, emerges at time point 1. At time point 2, the effective input level is chosen to maximize the expectation of revenue less control costs. The time point 2 problem is $\max_e R(e) - c(e + \omega)$, with first-order condition $R_e(e) = c_e(e + \omega)$ and ω-conditioned optimal choice $e^*(\omega)$. The valuation function then becomes

$$P(\omega) \equiv R(e^*(\omega)) - c(e^*(\omega) + \omega), \qquad (1)$$

where $R_e(e) = c_e(e + \omega)$ ensures that

$$P_\omega(\omega) = R_e(e^*(\omega))e^*_\omega(\omega) - c_e(e^*(\omega) + \omega)[e^*_\omega(\omega) + 1] = -c_e(e^*(\omega) + \omega) < 0. \qquad (2)$$

The first-order condition also allows for simplification of the expression for curvature: $P_{\omega\omega}(\omega) = R_{ee}(e^*(\omega))[e^*_\omega(\omega)]^2 - c_{ee}(e^*(\omega) + \omega)[e^*_\omega(\omega) + 1]^2 < 0$.

In addition, $R_e(e^*(\omega)) = c_e(e^*(\omega) + \omega)$ ensures that

$$e^*_\omega(\omega) = \frac{c_{ee}(e^*(\omega) + \omega)}{R_{ee}(e^*(\omega)) - c_{ee}(e^*(\omega) + \omega)} \in [-1, 0], \tag{3}$$

leading to

$$P_{\omega\omega}(\omega) = -R_{ee}(e^*(\omega))e^*_\omega(\omega) < 0. \tag{4}$$

The time point 0 decision is then to take the antibiotic action that maximizes the expected value of $P(\omega)$; that is,

$$\max_a \int P(\mu(a) + \sigma(a)\varepsilon)dF(\varepsilon) - wa. \tag{5}$$

Noting that $\mathbb{E}[\varepsilon] = 0$, we may write $\int P_\omega(\mu(a) + \sigma(a)\varepsilon)\varepsilon dF(\varepsilon) = \text{Cov}[P_\omega(\mu(a) + \sigma(a)\varepsilon), \varepsilon]$, Cov[.,.] being covariance. Given (4), the two arguments in the covariance functional vary in opposing directions as ε changes value and so the covariance expression has negative value. The optimality condition in (5) is

$$\overbrace{\mu_a(a)\mathbb{E}[P_\omega(\mu(a) + \sigma(a)\varepsilon)]}^{\text{Input effectiveness motive}} + \overbrace{\sigma_a(a)\text{Cov}[P_\omega(\mu(a) + \sigma(a)\varepsilon), \varepsilon]}^{\text{Control motive}} = w. \tag{6}$$

In this decomposition, we have ascribed two motives for antibiotic use. Both left-hand expressions in (6) have positive value. Our model is a very much reduced form and a more developed model would articulate several channels through which each effect might occur. Our basic point here is that some of the production effects of antibiotics can be viewed as directly increasing productivity in all states of nature, while others can be viewed as increasing productivity by controlling bad outcomes. We now proceed to discuss each effect in more detail.

The input effectiveness motivation regards directly increasing marginal value product. For instance, as antibiotics improve feed conversion efficiency, they increase the marginal value product of feed and so should promote more rapid finishing (Hennessy, Zhang, and Bai 2018). This should, in turn, increase the rate of return on capital investments in feedlot premises, thus deepening the incentive to make capital investments in animal feeding. Bear in mind that the holding of animals involves daily maintenance costs, such that a 1 percent rate of increase in daily production translates to a greater rate of increase in daily profit obtained. Thus, we argue that the increase in feed conversion efficiency can precipitate further changes in production format.

The control motive for use refers to why $c(e + \omega)$ is convex. It appeals to the input's damage-control attribute. When disease occurs, or a disruption occurs in some other input that leaves an animal stressed, there is need for intervention. Intervention is costly, as it consumes scarce resources, especially labor and entrepreneur time. Resources differ in their capacity to adapt to changing circumstances. In the main, the intelligence and understanding that attend labor provide the labor input with the flexibility to adapt to special needs. When some or all of an animal flock are stressed, then labor often identifies the cause, isolates it, and provides relief. Capital can replace the menial component of labor through automated feeding, bedding, and manure-handling systems. Labor costs have risen over time, while technological advances now enable mechanized substitutes for many forms of menial labor. Less readily replaced is the capacity to adapt in the presence of stressors. Application of antibiotics provides one way of alleviating the stressors, thus removing a major impediment toward replacing labor with capital.

The costs of capital investments are, however, scale sensitive with large fixed costs, so larger production units are best suited to adopt such technologies. Scale is not the only requirement for successful capital investment. A substitute for the good husbandry component of disease control needs to be found, for disruptions to address stressed herds would reduce throughput and the return on capital investment. Chandler (1990, p. 24) made the point that

> if the realized volume of flow fell below capacity, then actual costs per unit rose rapidly. They did so because fixed costs remained much higher and "sunk costs" (the original capital investment) were also much higher than in the more labor-intensive industries.

Quite how antibiotics could promote capital investments and scaled-up production is explained in Hennessy and Wang (2012, pp. 85–88). Their points are that capital investment and antibiotic use are complementary and that both will tend to be adopted when labor wage rates become sufficiently high.

3.2 Regulatory approaches at the farm level

New regulatory efforts try to improve documentation of antibiotic use, raise awareness of appropriate use, and change the labeling practice for veterinary drugs (Jensen 2016; European Commission 2016). In this regard, a study by Ortega et al. (2015) on how Chinese aquaculture farmers value antibiotic-use reduction is quite informative. Drug management ranks in the lower half of issues of importance for managing their operation. In addition, Sneeringer et al. (2015) report that many farmers in the United States, in particular those producing under contract, are not aware if their feed contains antibiotics. A better understanding of how integrators manage antibiotic applications would be necessary to uncover opportunities for use limitations. Additionally, the European experience highlights opportunities through the development of drug alternatives, as was successfully implemented in Norwegian salmon farms, where antibiotic treatments now occur in less than 1 percent of the production units (Smith et al. 2016b).

A second important step is the control and documentation of antibiotic use for veterinary purposes. The introduced documentation requirements have enabled the development of coherent insights into European veterinary antibiotic use (Smith et al. 2016a, p. 66). For example, data collected by these means have revealed that antibiotic use in mg per PCU is highly variable across animal species, production locations, and agricultural production systems. Among intense users, it varies between 400 mg per PCU in Cyprus and less than 100 mg per PCU in the Netherlands. This variation points to possibilities for significant curtailment that can be attained by improving farmers' and veterinarians' awareness and documentation tasks.

Indeed, new documentation and disease management requirements in animal production units in Germany point toward such an effect. Farmers are required to record their antibiotic treatments, including the number of animals treated, into a database (HIT database). From these figures, average treatment days are calculated and benchmarked against the German median and third quartile values, as calculated over a half-year period. Farmers whose average treatment days are above the median are required to consult their veterinarian to check if their antibiotic use can be further reduced. Those with an average treatment-day indicator in the fourth quartile

Table 10.1 Changes in average treatment days in food-animals in Germany after installing, reporting, and control requirements, 2014–2015

Animal species and use	Median			Third quartile		
	Jul – Dec 2014	Jul – Dec 2015	Δ	Jul – Dec 2014	Jul – Dec 2015	Δ
Fattening calves up to 8 months	0.00	0.00	.	5.06	2.71	−46%
Fattening cattle older than 8 months	0.00	0.00	.	0.02	0.00	−100%
Piglets ≤30 kg body weight	4.79	3.49	−27%	26.19	13.57	−48%
Fattening pigs of body weight >30 kg	1.20	0.55	−54%	9.49	4.64	−51%
Broiler	19.56	11.86	−39%	35.03	22.02	−37%
Turkey	23.03	18.36	−20%	47.49	32.34	−32%

Source: BVL 2015; BVL 2016.

have to submit a written plan of their intended antibiotic control measures. The plans have to be written in cooperation with the farms' veterinarians.

Table 10.1 reports results for the benchmarks from the earliest (second half of 2014) to the latest available term (second half of 2015). The table shows, for example, that the median treatment days for turkeys (last row) in a 182-day period have decreased from 23 average treatment days to 18 average treatment days. This reduction corresponds to a 20 percent decrease, as indicated in the column headed Δ. It is clear that the average number of days of treatment has declined among monogastric animals. The third-quartile benchmark experienced a stronger reduction than the median in piglets and turkeys. For fattening pigs and broilers, shifts in the distribution were similar for both quartile levels. The data seem to suggest that monitoring and control requirements can be quite effective. Further research should investigate the mechanisms by which these impacts are attained. The size of the effect suggests that ample room remains for efficiently replacing antibiotics with improved hygiene and disease management. Also, the involvement of veterinarians in preparing the farms' antibiotic control measures, as explained above, may orient farmers towards developing better animal health management skills.

4 Studies on consumer valuations and industry initiatives to limit antibiotic use

Given the widespread concern about antibiotic resistance as a public health issue, consumers are increasingly worried about antibiotic use in animal production. The food industry has responded with new initiatives to reduce antibiotics use in animal production and has sought to communicate its program efforts to consumers. Most of these initiatives function as a voluntary industry pledge, where companies promise to end non-curative, and sometimes even all, antibiotic use. Lacking evidence on consumer valuations in the market context, some consumer studies have attempted to estimate the consumers' valuation of the "raised without antibiotics" characteristic using online surveys or lab experiments. This section first reviews studies on the consumers' WTP before turning to a review of industry efforts and studies on market regulations.

4.1 Consumers' WTP for food produced without antibiotics

A literature overview by Clark et al. (2016) reviewed 80 studies dealing with animal welfare and production disease in intense animal farming systems, of which 21 dealt with antibiotic use. They concluded that consumers condone therapeutic antibiotic use for the treatment of injury and disease, but that sub-therapeutic antibiotic use is frowned upon by many consumers and citizens because it is viewed as unnatural. In addition, the public tends to associate non-use of antibiotics with animal welfare-friendly production systems.

In line with this finding, individual studies report much concern about the health implications of antibiotic residues in food. For example, in their survey, Dahlhausen, Rungie, and Roosen (2015) found that the concern about residues is highest for antibiotics (4.26 on a 5-point scale, with 5 = very high), followed by pesticides (4.25) and hormones (4.16). These results are comparable to those of Hwang, Roe, and Teisl (2005) in the United States. Also, Lusk's Food Demand Survey persistently finds high concern regarding antibiotic use among U.S. consumers (Lusk 2017). However, when looking at the likelihood of purchasing chicken in a sample of Delaware consumers, Bernard, Pesek, and Pan (2007) found that the use of antibiotics impacts purchase intentions less than does irradiation or genetic modification of either animals or feed. Wolf et al. (2016) analyzed a hypothetical referendum situation for a ban on therapeutic and preventive antibiotic use in dairy cattle. Using a sample of 2,001 U.S. food shoppers in an online survey, they found that 66 percent of the respondents would vote for an antibiotic ban for purposes other than treating a disease. Older and female consumers, as well as consumers with higher incomes, higher food expenditures, more exposure to animal welfare, media views, and greater concerns about animal welfare, are more likely to vote for a ban on antibiotic treatments in dairy production.

Such a ban, however, would come at a cost, and judging the efficiency of antibiotic regulations requires economic welfare measures. In this regard, consumer studies assess WTP as a measure of the compensating variation for limiting antibiotic use. Beyond the usual concern that survey-based estimates are prone to overestimate the WTP that drives real purchasing decisions, the question is even more complicated for antibiotic use. In addition to the normal private good aspect, there is a relevant public good dimension in questions concerning resistance management. Hence, even under the best circumstances, it is not possible to measure the full economic value in real purchase decisions. Indeed, Lusk, Norwood, and Pruitt (2006) studied the consumers' WTP for antibiotic-friendly pork by considering these two aspects. First, they used a choice experiment on food demand for private consumption and then conducted a choice experiment on donation-giving to measure the externality aspect of antibiotic use. They found substantive welfare effects that amount to an incremental WTP for pork of US$1.86 per pound, corresponding to 77 percent of market value. However, the confidence interval is large and depends on the knowledge that consumers have about sub-therapeutic use of antibiotics in agriculture.

In a similar approach, Dahlhausen, Rungie and Roosen (2015) used an online sample of 802 respondents in Germany to measure consumers' WTP for credence attributes in animal-based food products. Their concern is mainly to identify a coherent preference structure over product attributes that are preferred for altruistic (e.g., support farmers, respect for animals) or egoistic reasons (e.g., personal health concerns). Using a structural choice–modeling approach to compare preferences among pork chops, eggs, and egg noodles for "local production," "organic," "free of antibiotics," and "enhanced animal welfare" attributes, they found a unique preference proportion for the health- and quality-motivated aspects of products labeled "free of antibiotics," which correlates with labels for organic and enhanced animal welfare. Also, in an experimental study of four Eastern U.S. states, Bernard and Bernard (2009) showed that the WTP premium for organic milk coincides with that for milk produced without the use of antibiotics. This suggests that organic and non-use of antibiotics act as surrogate attributes signaling quality.

Beyond measures of WTP for verified non-use of antibiotics per se, a number of studies have further investigated consumer preferences over alternative certification bodies and how these preferences depend on product origin. In general, it is found that government certification is trusted more than private industry initiatives (McKendree et al. 2013). In addition, products labeled by government and third-party certifiers from the United States receive a positive WTP, while WTP is zero when certification is given by a Chinese entity (Ortega, Wang, and Olynk Widmar 2014). Finally, Ortega, Wang, and Olynk-Widmar (2015) reported that the WTP for cessation of antibiotic use depends on the origin of the product. While consumers were found to be willing to pay significant amounts for U.S. shrimp, products from Asia (China or Thailand) did not receive a statistically significant premium.

4.2 Industry initiatives and market assessments of bans on antibiotic use in agriculture

While online surveys and lab studies reveal that consumers state a positive willingness to pay for products raised without antibiotics, translating these stated preferences to market premia at commercial scales poses many challenges. For instance, differentiating and labeling such products are rather complex issues. Given the three uses of antibiotics in animal agriculture, concern mostly exists regarding the use of antibiotics as growth promotants. Animal welfare advocates have ethical concerns that banning the use of antibiotics altogether will leave sick animals in unnecessary pain. The issue of prophylactic use is challenging, as cases of therapy and applications of control and prevention are difficult to differentiate, and sometimes, particularly in large animal production units, control and prevention may prevent a large number of subsequent therapeutic administrations.

In the United States, Perdue Farms and Tyson Foods introduced a label "Raised without Antibiotics" in 2007. The label was initially endorsed by the U.S. Department of Agriculture Food Safety Inspection Service (FSIS) regulators. FSIS rescinded the endorsement of the claim after it became apparent that the label was based on false claims leading to consumer confusion and claims of unfair competition. Bowman et al. (2016) presented a case study of the events and discussed the resulting market impacts. Despite such difficulties in implementing this label on meat products, similar efforts are being pursued by other companies. The food industry has taken action following the US FDA Guidance for Industry in 2012, which was intended to promote the judicious use of medically important antimicrobial drugs in food-producing animals (US FDA 2012). For example, in 2015, Subway announced that it intended to serve chicken meat only from animals that have never been treated with antibiotics of importance in human medicine (Subway 2015). McDonald's made a similar pledge for its U.S. business and claims to have made good on its pledge by 2016 (McDonald's 2016).

Saitone, Sexton, and Sumner (2015) analyzed the impact of intermediaries implementing restrictive process standards and conducted a simulation study of quick-service restaurants or retailers requiring pork raised without antibiotics. They considered three consumer groups: those who (1) either do not care or are not aware of such a procurement policy, (2) do not buy pork products without the policy but would do so if the process standard was implemented, and (3) do buy pork in both situations and have a positive WTP for pork without antibiotics. Only a small share of pork products is sold to food services, primarily bacon, processed ham, and breakfast sausage products. Thus, the supply share of meat from animals raised without antibiotics has to be about double the demand share. In different scenarios, the authors estimated that 6 to 11 percent of demand growth would need to come from adding new consumers who would buy the restricted meat, but not the conventional meat, in order to restore the consumer surplus. They concluded that these policies may also be motivated by the quest for a better company image and to preempt future regulations (Segerson 1999). Trade and marketing implications of voluntary standards can be favorable for industry incumbents if they have a comparative advantage in accommodating the standard (Hobbs 2010; Bech-Larsen and Aschemann-Witzel 2012). In the case of banning antibiotics as growth promotants, the cost burden will depend, in many cases, on the contract arrangements within the industry.

Market-level outcomes of a ban on sub-therapeutic antibiotic use have been studied by, for example, Wade and Barkley (1992) and McNamara and Miller (2002). The NRC (1999) used an estimated consumer cost of a ban on sub-therapeutic use of US$4.84 to US$9.72 per capita. This estimate was obtained based on the assumption that consumer demand for animal products would not change in response to the ban. Others, such as Wade and Barkley (1992) and the study by Saitone, Sexton, and Sumner (2015) discussed earlier, assumed a response in demand.

Regarding international trade, Wilson, Otsuki, and Majumdsar (2003) studied the costs resulting from residue limits that deviate from the Codex recommendations for international beef trade by imposing a stricter standard. Based on trade data from 1995 to 2000, they concluded that, if all countries were to apply Codex standards of 0.6 ppm for tetracycline residues, then the international beef trade would increase significantly, where the exporting countries such as Argentina, Brazil, and South Africa would be the primary beneficiaries.

5 Conclusions

Due to accumulating evidence on the extent and origins of resistance against antibiotics, and on the implications for human health, public efforts to curtail antibiotic use in agriculture and to engage the industry in doing so are being established and extended in many countries. In particular, the use as growth promotants of antibiotics that are important for human treatment is considered to be critical. Given high concerns in these countries, Sweden and Denmark have been the first to discontinue these types of uses and serve as case studies in international discussions. Reassured by experiences in those countries, the EU banned antibiotics as growth promotants in 2006. However, experience shows that, as of 2016, total use remains widespread in high-income countries and even more so in many middle- and low-income countries, so supplementary monitoring requirements are being installed.

The review in this chapter has shown that antibiotics can be understood as a damage-control input, which leads to an input effectiveness motive and a control motive for use. Our basic point is to stress that demand for antibiotics can be viewed as having two components, one in which productivity increases in all states of nature and one in which productivity increases by controlling disruptions that arise from bad outcomes. Further research should attempt to quantify both aspects as well as further elucidate the relationship between antibiotics and other inputs in agriculture. The implications of antibiotics for labor, genetics, and fixed capital inputs merit particular attention because these inputs are most salient in characterizing the distinctions between traditional-format and large-scale animal production. Studies that seek to expand upon the input effectiveness and control motives should use a more articulated production technology. Control is best understood as the management of information, and so the decision structure needs to be sufficiently rich to include a decision vector that is contingent on state outcomes.

In regard to empirics, data that allow for a clear understanding of when antibiotics are used and why they are used are needed. Standard data layouts that delineate aggregate use on farms of different sizes and use different sets of input bundles may not be adequate for the purpose of inferring how antibiotics are used. A more epidemiological approach may be more useful, in which detailed production logs show animal age, condition when the decisions were made, who made the decisions, and what their incentives were. Economic approaches would be important in such a line of inquiry since, ultimately, use is driven by incentives. As with antibiotics in human medicine, for example, when used to satisfy a patient seeking treatment of a viral infection, it is conceivable that some motives for antibiotic use in animals are grounded in psychology. Finer detail on incentives may provide further insights. For example, if production employees make decisions on the administration of antibiotics and receive bonuses in proportion to gross production or pigs weaned per sow, they may have no incentive to control antibiotic applications.

Diverse groups have stakes in the regulations of antibiotics, including agriculture, human health, environmentalists, and animal-welfare advocates. Indeed, the externalities to be considered when regulating antibiotics are likely to go beyond the issue of resistance and public health implications. For example, feed efficiency, as determined by animal productivity and influenced by the use of antibiotics, may have an impact on managing environmental externalities such as nitrogen in groundwater or climate change. In addition, the magnitude of the public health risks is obscured by our incomplete understanding of the dynamic and stochastic processes that underlie resistance development. It is inevitable, therefore, that regulatory processes will be ponderous and contentious. Economic considerations loom large in the optimal management of resistance. As indicated in this chapter, economic information needed to assist in resistance management include demand elasticities of antibiotics in relationship to own price and the prices of other inputs. The extent of the substitution relationship between disease prevention and disease treatment also needs to be quantified in empirical studies; U.S. National Animal Health Monitoring System (NAHMS) data may possibly be useful in this regard. The Veterinary Feed Directive has required veterinary oversight of antibiotics use in animal agriculture, where antibiotics manufacturers and veterinary associations have joined with the federal government in a commitment to eliminate the nontherapeutic use of antibiotics. Veterinarian behavior in terms of prescription data may also be an interesting route for future research. Finally, data that clarifies the role of buildings and infrastructure investments on disease would also be useful when seeking to reduce the demand for antibiotics.

References

Anderson, E.S., and M.J. Lewis. 1965. "Drug Resistance and Its Transfer in Salmonella Typhimurium." *Nature* 206(4984):579–583.

Bech-Larsen, T., and J. Aschemann-Witzel. 2012. "A Macromarketing Perspective on Food Safety Regulation: The Danish Ban on Trans-Fatty Acids." *Journal of Macromarketing* 32(2):208–219.

Bernard, J.C., and D.J. Bernard. 2009. "What Is It About Organic Milk? An Experimental Analysis." *American Journal of Agricultural Economics* 91(3):826–836.

Bernard, J.C., J.D. Pesek, and X. Pan. 2007. "Consumer Likelihood to Purchase Chickens with Novel Production Attributes." *Journal of Agricultural and Applied Economics* 39(3):581–596.

Bowman, M., K.K. Marshall, F. Kuchler, and L. Lynch. 2016. "Raised Without Antibiotics: Lessons from Voluntary Labeling of Antibiotic Use Practices in The Broiler Industry." *American Journal of Agricultural Economics* 98(2):622–642.

Brown, G., and D.F. Layton. 1996. "Resistance Economics: Social Cost and the Evolution of Antibiotic Resistance." *Environment and Development Economics* 1(3):349–355.

BVL. 2015. Bekanntmachung des Medians und des dritten Quartils der vom 1. Juli 2014 bis 31. Dezember 2014 erfassten bundesweiten betrieblichen Therapiehäufigkeiten für Mastrinder, Mastschweine, Masthühner und Mastputen nach § 58c Absatz 4 des Arzneimittelgesetzes. Bundesanzeiger (March 31, 2015):1.

BVL. 2016. Bekanntmachung des Medians und des dritten Quartils der vom 1. Juli 2015 bis 31. Dezember 2015 erfassten bundesweiten betrieblichen Therapiehäufigkeiten für Mastrinder, Mastschweine, Masthühner und Mastputen nach § 58c Absatz 4 des Arzneimittelgesetzes. Bundesanzeiger (March 31, 2016):1.

Castanon, J.I.R. 2007. "History of the Use of Antibiotic as Growth Promoters in European Poultry Feeds." *Poultry Science* 86(11):2466–2471.

Centers for Disease Control and Prevention (CDC). 2013. "Antibiotic Resistance Threats in the United States, 2013. U.S. Department of Health and Human Services." www.cdc.gov/drugresistance/threat-report-2013/. Accessed December 26, 2016.

Chandler, A.D. 1990. *Scale and Scope: The Dynamics of Industrial Capitalism*. Cambridge, MA: Belknap/Harvard University Press.

Clark, B., G.B. Stewart, L.A. Panzone, I. Kyriazakis, and L.J. Frewer. 2016. "A Systematic Review of Public Attitudes, Perceptions and Behaviours Towards Production Diseases Associated with Farm Animal Welfare." *Journal of Agricultural and Environmental Ethics* 29(3):455–478.

Dahlhausen, J.L., C. Rungie, and J. Roosen. 2015. "The Value of Ethical Concern." *Paper Presented at the 143rd Joint EAAE/AAEA Seminar*, March 25–27, Naples, Italy.

ECDC (European Center for Disease Prevention and Control). 2009. The Bacterial Challenge: Time to React. http://ecdc.europa.eu/en/publications/Publications/0909_TER_The_Bacterial_Challenge_Time_to_React.pdf. Accessed December 26, 2016.

European Commission. 2016. Antimicrobial Resistance (AMR) Evaluation of the 2011–2016 Action Plan. Factsheet. http://ec.europa.eu/dgs/health_food-safety/amr/docs/amr_evaluation_2011-16_factsheet.pdf. Accessed February 12, 2017.

Fischer, C., and R. Laxminarayan. 2005. "Sequential Development and Exploitation of an Exhaustible Resource: Do Monopoly Rights Promote Conservation?" *Journal of Environmental Economics and Management* 49(3):500–515.

Fleming-Dutra, K.E., A.L. Hersh, D.J. Shapiro, M. Bartoces, E.A. Enns, T.M. File, J.A. Finkelstein, J.S. Gerber, D.Y. Hyun, J.A. Linder, R. Lynfield, D.J. Margolis, L.S. May, D. Merenstein, J.P. Metlay, J.G. Newland, J.F. Piccirillo, R.M. Roberts, G.V. Sanchez, K.J. Suda, A. Thomas, T.M. Woo, R.M. Zetts, and L.A. Hicks. 2016. "Prevalence of Inappropriate Antibiotic Prescriptions Among US Ambulatory Care Visits, 2010–2011." *Journal of the American Medical Association* 315(17):1864.

Fox, G., and A. Weersink. 1995. "Damage Control and Increasing Returns." *American Journal of Agricultural Economics* 77(1):33–39.

Froot, K.A., D.S. Scharfstein, and J.C. Stein. 1993. "Risk Management: Coordinating Corporate Investment and Financing Policies." *Journal of Finance* 48(5):1629–1658.

Hennessy, D.A. 2018. "Managing Derived Demand for Antibiotics in Animal Agriculture." *Selected Paper Presented at AAEA Annual Meetings*, Washington, DC, August 5–7, 2018.

Hennessy, D.A., and T. Wang. 2012. Animal Disease and the Industrialization of Agriculture. In D. Zilberman, J. Otte, D. Roland-Holst, and D. Pfeiffer, eds., *Health and Animal Agriculture in Developing Countries*. New York: Springer, pp. 77–100.

Hennessy, D.A., J. Zhang, and N. Bai. 2018. "Animal Health Inputs, CAFOs, the Structure of Protein Production and Contract Information Flows That Reduce Supply Chain Frictions." *Food Policy*. https://www.sciencedirect.com/science/article/pii/S0306919217307261

Hobbs, J.E. 2010. "Public and Private standards for Food Safety and Quality: International Trade Implications." *The Estey Centre Journal of International Law and Trade Policy* 11(1):136–152.

Hwang, Y.-J., B. Roe, and M.F. Teisl. 2005. "An Empirical Analysis of United States Consumers' Concerns About Eight Food Production and Processing Technologies." *AgBioForum* 8(1):40–49.

Jensen, H.H. 2016. "Reducing Antibiotic Use in Animal Production Systems." *Agricultural Policy Review* 2016(2):Article 3.

Jensen, H.H., and D.J. Hayes. 2014. "Impact of Denmark's Ban on Antimicrobials for Growth Promotion." *Current Opinion in Microbiology* 19(June):30–36.

Key, N., and W.D. McBride. 2014. "Sub-therapeutic Antibiotics and the Efficiency of U.S. Hog Farms." *American Journal of Agricultural Economics* 96(3):831–850.

Laxminarayan, R., and G. Brown. 2001. "Economics of Antibiotic Resistance: A Theory of Optimal Use." *Journal of Environmental Economics and Management* 42(2):183–206.

Lichtenberg, E., and D. Zilberman. 1986. "The Econometrics of Damage Control: Why Specification Matters." *American Journal of Agricultural Economics* 68(2):261–273.

Lusk, J.L. 2017. Food Demand Survey. www.agecon.okstate.edu/agecon_research.asp. Accessed January 6, 2017.

Lusk, J.L., F.B. Norwood, and J.R. Pruitt. 2006. "Consumer Demand for a Ban on Antibiotic Drug Use in Pork Production." *American Journal of Agricultural Economics* 88(4):1015–1033.

McDonald's. 2016. About Our Food: Our Commitment to Quality. www.mcdonalds.com/us/en-us/about-our-food/our-food-philosophy/commitment-to-quality.html. Accessed December 26, 2016.

McKendree, M., N. Wydmar, D. Ortega, and K. Foster. 2013. "Consumer Preferences for Verified Pork Practices in the Production of Ham Products." *Journal of Agricultural and Resource Economics* 38(3):397–417.

McNamara, P.E., and G.Y. Miller. 2002. "Pigs, People and Pathogens: A Social Welfare Framework for the Analysis of Animal Antibiotic Use Policy." *American Journal of Agricultural Economics* 84(5):1293–1300.

National Research Council (NRC) Committee on Drug Use in Food Animals. 1999. *The Use of Drugs in Food Animals: Risks and Benefits*. Washington, DC: National Academy Press.

O'Neill, J. 2015. Antimicrobials in Agriculture and the Environment: Reducing Unnecessary Use and Waste. Review on Antimicrobial Resistance commissioned by the UK Government and the Wellcome Trust. https://amr-review.org/sites/default/files/Antimicrobials in agriculture and the environment – Reducing unnecessary use and waste.pdf. Accessed October 10, 2017.

O'Neill, J. 2016. Tackling Drug-Resistance Infections Globally: Final Report and Recommendations. Review on Antimicrobial Resistance commissioned by the UK Government and the Wellcome Trust. https://amr-review.org/sites/default/files/160525_Final paper_with cover.pdf. Accessed February 6, 2017.

Ortega, D.L., H.H. Wang, and N.J. Olynk Widmar. 2014. "Aquaculture Imports from Asia: An Analysis of U.S. Consumer Demand for Select Food Quality Attributes." *Agricultural Economics* 45(5):625–634.

Ortega, D.L., H.H. Wang, and N.J. Olynk Widmar. 2015. "Effects of Media Headlines on Consumer Preferences for Food Safety, Quality and Environmental Attributes." *Australian Journal of Agricultural and Resource Economics* 59(3):433–445.

Rojo-Gimeno, C., M. Postma, J. Dewulf, H. Hogeveen, L. Lauwers, and E. Wauters. 2016. "Farm-Economic Analysis of Reducing Antimicrobial Use Whilst Adopting Improved Management Strategies on Farrow-to-Finish Pig Farms." *Preventive Veterinary Medicine* 129:74–87

Roosen, J., and D.A. Hennessy. 2001. "An Equilibrium Analysis of Antibiotics Use and Replanting Decision in Apple Production." *Journal of Agricultural and Resource Economics* 26(2):539–553.

Saitone, T.L., R.J. Sexton, and D.A. Sumner. 2015. "What Happens When Food Marketers Require Restrictive Farming Practices?" *American Journal of Agricultural Economics* 97(4):1021–1043.

Secchi, S. 2000. *Economic Issues in Resistance Management*. Ames, IA: Dissertation. Iowa State University.

Segerson, K. 1999. "Mandatory Versus Voluntary Approaches to Food Safety." *Agribusiness* 15(1):53–70.

Smith, E., C.A. Lichten, J. Taylor, C. MacLure, L. Lepetit, E. Harte, A. Martin, I. Ghiga, E. Pitchforth, J. Sussex, E. Dujso, and J. Littmann. 2016a. *Evaluation of the EU Action Plan against the rising threats from antimicrobial resistance. Final Report*. Brussels: European Commission.

———. 2016b. *Evaluation of the EU Action Plan against the rising threats from antimicrobial resistance. Final Report – Appendices*. Brussels: European Commission.

Sneeringer, S., J.M. MacDonald, N. Key, W. McBride, and K. Matthews. 2015. Economics of Antibiotic Use in U.S. Livestock Production. Economic Research Service Report Number 200, U.S. Department of Agriculture, November.

Subway. 2015. SUBWAY Restaurants Elevates Current Antibiotic-Free Policy U.S. *Press Release*. www.subway.com/PressReleases/AntibioticFreeRelease10.20.15.pdf. Accessed December 26, 2016.

Teillant, A., C.H. Brower, and R. Laxminarayan. 2015. "Economics of Antibiotic Growth Promoters in Livestock." *Annual Review of Resource Economics* 7(1):349–374.

US FDA (US Food and Drug Administration). 2012. "The Judicious Use of Medically Important Antimicrobial Drugs in Food-Producing Animals." Guidance for Industry #209. *Federal Register* (July 6, 2012):1–19.

US FDA (US Food and Drug Administration). 2013. New Animal Drugs and New Animal Drug Combination Products Administered in or on Medicated Feed or Drinking Water of Food- Producing Animals: Recommendations for Drug Sponsors for Voluntarily Aligning Product Use Conditions with GFI #209. Guidance for Industry #213. *Federal Register* (September 12, 2013):1–18.

US FDA (US Food and Drug Administration). 2017. FDA Announces Implementation of GFI #213, Outlines Continuing Efforts to Address Antimicrobial Resistance. Release February 17. www.fda.gov/AnimalVeterinary/NewsEvents/CVMUpdates/ucm535154.htm. Accessed October 8, 2017.

Wade, M.A., and A.P. Barkley. 1992. "The Economic Impacts of a Ban on Subtherapeutic Antibiotics in Swine Production." *Agribusiness* 8(2):93–108.

Wilson, J.S., T. Otsuki, and B. Majumdsar. 2003. "Balancing Food Safety and Risk: Do Drug Residue Limits Affect International Trade in Beef?" *The Journal of International Trade & Economic Development* 12(4):377–402.

Wolf, C.A., G.T. Tonsor, M.G.S. McKendree, D.U. Thomson, and J.C. Swanson. 2016. "Public and Farmer Perceptions of Dairy Cattle Welfare in the United States." *Journal of Dairy Science* 99(7):5892–5903.

World Organisation for Animal Health (OIE). 2015. Risks Associated with the Use of Antimicrobials in Animals Worldwide. Editorial. www.oie.int/en/for-the-media/editorials/detail/article/risks-associated-with-the-use-of-antimicrobials-in-animals-worldwide/. Accessed February 7, 2017.

11
Advancements in the economics of food security

P. Lynn Kennedy

1 Introduction

Food security is a broad concept that covers the entire scope and mission of agriculture. Over two centuries ago, Malthus (1798) recognized the potential for population growth to exceed that of food production. Food production and availability have long been considered an important dimension of food security (Schmitz and Kennedy 2015), but it is also recognized that poverty and famine can result from a lack of entitlements (Sen 1981). Entitlements and the access dimension of food security are important, as starvation and malnutrition have been shown to occur even in areas of ample food supply (President's Commission 1969).

The 1996 World Food Summit defined food security as the situation where "all people, at all times, have physical, social and economic access to sufficient, safe and nutritious food that meets their dietary needs and food preferences for an active and healthy life" (FAO 1996, p. 3). The summit also established four dimensions of food security: physical availability, economic access, adequate utilization, and stability.

Although availability and access are necessary conditions for food security, they are insufficient to prevent malnutrition (Smith 1998). As a result, more recent definitions distinguish between food security and nutrition security (FAO, IFAD, and WFP 2012). The definition of nutrition security recognizes that nutrition insecurity can be shown as undernutrition, overnutrition, and through micronutrient deficiencies (Pinstrup-Andersen 2007). The multiple dimensions and variables associated with food security have also resulted in the existence of many ways to measure food security (Anderson 1990; Magana-Lemus and Lara-Alvarez 2015). The development and use of appropriate measures of food security is particularly important as a means of identifying and solving problems associated with chronic poverty and malnutrition in developing countries and emerging markets (Henneberry and Carrasco 2015).

It is unsurprising that the current state of research in the area of food and nutrition security is broad, given the four dimensions and multiple attributes of food security. Population growth and economic expansion are two factors influencing the global demand for food (Krivonos, Morrison, and Canigiani 2015). Forecasts by the Food and Agriculture Organization of the United Nations (FAO) estimate that world population will increase from 7.2 billion in 2015 to 9.5 billion in 2050. Combined with economic growth, this will necessitate a 60 percent increase in the production of food (Alexandratos and Bruinsma 2012). Despite this, recent growth in income and production has been accompanied by decreased food insecurity; these gains have also brought concerns over the increased incidence of obesity and associated diseases (Hawkes 2015).

The relationship between international trade and food security is strongly debated, with some suggesting that free trade results in food insecurity (Madeley 2000), while others suggest that free trade promotes food security (Griswold 1999). The 1943 UN conference on Food and Agriculture in its final declaration highlighted the link between poverty and malnutrition and suggested that international trade, as a vehicle to affect poverty, can serve to influence food security (Shaw 2007). Macroeconomic and trade policies, through their potential to increase household income, can alleviate hunger and malnutrition (Diaz-Bonilla 2015).

This chapter seeks to provide an overview of the current state of research in the economics of food security. Given the multiple dimensions and diverse attributes of food security, the issues and research will be categorized according to the four dimensions of food security: availability, access, utilization, and stability. The breadth of current work reflects the complexity of this topic as producers, policy makers, and researchers seek solutions that will allow all members of society to live an active and healthy life.

2 Availability

The dimension of food availability involves the availability of sufficient quantities of food of appropriate quality, supplied through domestic production or imports, including food aid (FAO 2006). This section reviews factors influencing food availability, including technology, genetically modified products, research and development, and resource constraints.

2.1 Technology

Productivity gains derived through technological developments, including the adoption of semi-dwarf wheat and rice hybrids and the Haber–Bosch process for fertilizer production, allowed food availability to keep pace with a growing population in the twentieth century. Now, with gains from crop-breeding decelerating, genetic engineering has become the leading potential vehicle for technological advancement in agriculture.

Koo and Taylor (2015) examine whether the world food supply will be sufficient to meet the growing demand for food by the year 2050. They find that corn, soybean, rice, and wheat yield increases ranged between 50 and 65 percent from 1980 to 2011. Koo and Taylor (2015) question whether productivity gains can continue to meet food consumption needs as world population is expected to increase another 30 percent by 2050. Their analysis finds that global food supply will be insufficient to meet food demand by the year 2050, and they anticipate severe food shortages in Africa and significant food-price inflation. Meyers and Kalaitzandonakes (2015) come to a similar conclusion in their comparison of world population projections with grain and oilseed supply forecasts, highlighting the dominance of anticipated population growth in Africa.

Wailes, Durand-Morat, and Diagne (2015) examine the issue of food shortages in Africa using the Arkansas Global Rice Model. West Africa accounts for 25 percent of global rice imports; national rice development strategies have targeted the doubling of 2008 production levels by 2018. Should West African self-sufficiency in rice be achieved, the resulting elimination of rice imports would reduce global rice prices. They postulate that the level of land expansion and intensification necessary to achieve self-sufficiency would result in decreased quality levels, and, in turn, discounted domestic rice prices relative to that of imported product. This would reduce benefits to both producers and consumers.

Schroeder and Meyers (2016) conduct a broad analysis of food insecurity and malnutrition for the region of Europe and Central Asia. They use macro, sectoral, and household data to examine production, consumption, and trade of agricultural and food products. Components of food security considered in their analysis include trends in economic growth and poverty, constraints to agricultural productivity, and food security-related policies. Although food insecurity and poverty have been reduced in the study area, progress can still be achieved with respect to micronutrient intake, quality of diet, obesity rates, and dependence on food imports. The authors outline public actions that can contribute to the further reduction of food insecurity and malnutrition.

Technological developments in crop production can be achieved via many avenues. Among these, the adoption of more productive hybrid varieties and advancements in mechanization technologies can be substitutes or complements. Schmitz, Kennedy, and Salassi (2017) examine the adoption of high-yielding sugarcane varieties that would not have been feasible without the accompanying introduction of a new generation of mechanical sugarcane harvesters. Their work highlights the need for coordinated strategies in the development and adoption of new technologies to best provide long-term food security.

2.2 Genetically modified products

The adoption of genetically modified crop varieties promises much in the way of increased food availability, with the potential to boost agricultural productivity and reduce resource use. However, the benefits to society are not always positive and vary among interest groups. In their 2015 study, Lakkakula, Haynes, and Schmitz determine the economic impacts of genetic engineering for food security based on the introduction of a yield-increasing, genetically modified rice variety. Although their analysis shows a net gain to society, these technological gains have resulted in economic losses to producers. Kennedy, Lewis, and Schmitz (2017) use the case of genetically modified sugar beets to examine the impact of genetically modified crops on food security. Their analysis shows that supply-induced food security gains can be negated if consumers are averse to genetically modified food products.

Global population is expected to double in the second half of the twenty-first century. Given that arable farmland is limited, future production growth will need to result from yield growth as opposed to acreage expansion. Von Witzke and Noleppa (2016) build upon the traditional returns to research literature; they incorporate the environmental benefits of yield growth. They find that the environmental benefits of yield growth greatly exceed the direct producer and consumer gains from expanded production. Given this, von Witzke and Noleppa (2016) postulate that the restriction of analyses to only price and quantity effects seriously underestimates the societal benefits of modern agriculture.

2.3 Research and development

While many are uncertain as to the source of future gains in agricultural productivity, aquaculture promises much to alleviate food insecurity. According to Anderson et al. (2017), the current rate of growth in aquacultural production exceeds that of all other meat production sectors. Those gains are expected to continue as aquaculture has not yet fully borrowed from technological developments in the traditional meat sectors. Advances in aquaculture technology contribute to the availability dimension of food security through increased food production and to the access dimension as farmed seafood increases the incomes of rural households.

Kristkova, van Kijk, and van Meijl (2017) use a computable general equilibrium model to determine the impact of public agricultural research and development (R&D) on agricultural productivity and food security using dynamic accumulation of R&D stock. While one might expect the food security gains from R&D to manifest themselves mainly through the availability dimension, Kristkova, van Kijk, and van Meijl (2017) find that R&D investment brings gains especially through the access dimension of food security, particularly in the case of Sub-Saharan Africa.

2.4 Resource constraints

The availability of an adequate food supply is often limited due to resource constraints. Throughout the world, water is often a limiting factor of production; agricultural production limitations based on the water supply is a critical consideration for the Kingdom of Saudi Arabia (KSA). Pieters and Swinnen (2016) formulate a water–energy–food nexus framework to analyze the interaction between water scarcity, energy abundance, and food supply and demand. Strategies are proposed to meet the KSA food security objectives. Pieters and Swinnen (2016) suggest that the Kingdom of Saudi Arabia would benefit from a broader food security strategy in which food stocks and subsidies are augmented by in-kind and cash transfers.

The availability dimension of food security often suggests research, development, and technological gains. However, the existence of input and product markets along with other infrastructure serve to facilitate food production. Beghin and Teshome (2017) examine the impact of various impediments to food security, including limited land and transportation and telecommunication infrastructures. They conclude that improved infrastructures and better access to land would enhance food security, both through increased food availability and enhanced farm household income.

3 Access

Food access is defined as access by individuals to adequate resources (entitlements) for acquiring appropriate foods for a nutritious diet (FAO 2006). There are a variety of avenues by which consumers' access to food can be limited. Recent literature has identified (1) barriers that impede access, (2) the means to measure access, and (3) the ways to improve access. The impact of food security measurement, income, and supply chain efficiency are examined in this section.

3.1 Measurement

One barrier to achieving food security has been the lack of a widely accepted measure that is easily constructed using available data, provides temporal and spatial comparability, and accounts for the various components of food security. Antle, Adhikari, and Price (2015) review various measures of food security and conclude that previous measures focused primarily on a single component of food security. They propose an Income-Based Food Security (IBFS) indicator that measures the ability of a household to purchase a basket of goods that meets its nutritional requirements. The IBFS indicator is compatible with any framework that generates predictions regarding per capita or farm household income.

Thome et al. (2017) use a demand-based modeling framework to analyze food security in Ethiopia. Their approach considers the impact of population growth, income growth, food prices, and real exchange rates on food consumption. Models utilizing multiple components of food security allow for the consideration of a more relevant set of scenarios and can provide a much richer set of information for policy analysis.

Beyond measurement of food security, others have examined food aid policy to determine how to most efficiently allocate food aid or best design food security programs. Miljkovic (2015) proposes a nonnormative approach to examine the quality, effectiveness, and efficiency of food security. On the demand side, programs should be evaluated based on their impact on the quality of life of an at-risk population. On the supply side, the motives of food aid donors are discussed in the context of the deservingness heuristic. The model identifies three problems in measuring food security-related welfare: different expectations, different endowments, and different perceived needs. Determining the behavioral causes of food aid from both the demand and supply sides can help predict whether the needy will receive food aid at all.

3.2 Income

Population growth was the primary driver behind demand growth in the twentieth century. Although global population is expected to grow to over 9 billion people by 2050, population growth is rapidly decelerating. Fukase and Martin (2017) show that per capita income growth in developing countries is likely to be a bigger factor in global food demand than population growth in the next 30 years. Their analysis argues that economic convergence between the rich and the poor will have a greater impact on food demand than across-the-board growth, because lower-income households tend to spend a greater share of their income on food. The relative rates of income growth between the rich and the poor will have significant impacts on the balance between world food supply and demand.

Food aid is an important source of nourishment in times of emergency or shortage. The provision of food aid gives access to food in the short run, but does it result in food security in the long run? To examine the effect of donor investments in a market channel that rewards product quality, Moss, Oehmke, and Lyambabaje (2016) estimate Working's Model: the relationship between the share of income spent on food and income. Their empirical results are consistent with Working's Model; that is, the share of income spent on food declines as income increases. However, this result does not affect whether the household benefits from the improved market channel; increased household income is shown to improve food security.

Just as Sen (1981) argued that famine is not always an availability issue, off-farm labor opportunities are an important factor in determining rural food security. Seidu and Onel (2016) examine the local and non-local off-farm incomes for at-home food consumption expenditures to determine the food security implications of off-farm labor reallocation decisions. They find that local off-farm income has a positive effect on per capita food consumption expenditures of farm households, while private remittances from non-local off-farm income have the opposite effect. Seidu and Onel (2016) suggest that the development and implementation of programs to enhance off-farm labor opportunities would increase food security. These results are consistent with those of Mishra and Khanal (2017), who find that Bangladeshi households participating in non-farm income-generating activities experience higher levels of household income and diet diversity.

As a sub-component of off-farm labor, *ganyu* labor is off-farm, informal, piece-work labor that requires no formal education or training. According to Sitienei, Mishra, and Khanal (2016), *ganyu* laborers tend to be less-educated males from large households with lower cropland and farm size. Their research shows a positive relationship between *ganyu* labor participation and the number of meals consumed per day, thus enhancing food security via the access dimension.

Similar to off-farm labor, remittances extend the earning potential of the household. Regmi and Paudel (2016) examine food security in Bangladesh using the Food Consumption Score (FCS) and Household Hunger Scale (HHS). Their results indicate that remittances, off-farm income, and literacy are important determinants of food security.

3.3 Supply chain efficiency

Access to food is restricted not only by means, but also by an inefficient supply chain. Clark and Hobbs (2015) explore the structure, goals, and objectives of international food assistance organizations. They examine how Ready-to-Use Therapeutic Foods and Ready-to-Use Supplementary Foods have become an important tool for achieving food assistance goals. Food aid has shifted from a focus on acute malnutrition to include chronic undernutrition. Concurrently, an earlier dependence on transoceanic food aid has given way to an increased focus on the development of local and regional partnerships.

Vulnerability of the supply chain, particularly with respect to food aid shipments, has become an important issue in recent years. Kerr (2015) examines the cost-effectiveness of efforts to protect World Food Program shipments to Somalia from seaborne pirates. The analysis examines previous strategies to provide insight as to how to avoid the high costs that have been incurred to protect Somalian food aid shipments.

4 Utilization

The utilization dimension of food security is defined as the utilization of food through adequate diet, clean water, sanitation, and health care to reach a state of nutritional well-being where all physiological needs are met (FAO 2006). This section reviews factors influencing food utilization, including adequate diet and food waste.

4.1 Adequate diet

Beyond assuring adequate caloric intake, food security involves the attainment of key dietary targets. To this end, Msangi and Batka (2015b) use a global multimarket agricultural model to determine how progress toward dietary intake targets compares across regions. They look particularly at the Bottom Billion, which is comprised of the lower sixth of global nutritional attainment. Their findings show that most populations in the Bottom Billion are deficient in carbohydrate, protein, and fiber intake; the majority of the Bottom Billion are located in Africa and Asia. Msangi and Batka (2015b) suggest that food security policies must account for changing socioeconomic patterns that influence food intake and dietary trends. Their evaluation of those populations suffering the greatest nutritional deficits sets the stage for a more meaningful policy discussion aimed at improving food security among this vulnerable group.

Msangi and Batka (2015a) build on the framework of nutrient disaggregation to determine how the nutritional status of the Bottom Billion can be enhanced. Their model determines the effect of various policy interventions on the gap between nutrition baseline trends and target nutrition intake levels. They show that policies which increase household income or enhance agricultural productivity are the most effective pathways toward improving nutritional outcomes.

Inadequate nutrition is often associated with famine and starvation, but it can also be associated with the epidemic of obesity experienced in many locales across the world. Anfinson et al. (2016) use a general linear equation to determine which factors influence adolescent obesity. Their framework distinguishes between the nutritional characteristics of the foods consumed and the socioeconomic and demographic factors. Their analysis shows that foods consumed at home and away from home have a similar positive effect on obesity as measured by the Body Mass Index (BMI). Anfinson et al. (2016) find that income growth and rapid urbanization may contribute to the increased incidence of adolescent obesity in China.

Food decisions are also examined by Useche and Twyman (2016). They determine the relationship between household and regional characteristics and the joint demand for the various components of a household's diet. Wealth, education, demographic structure, market access, geographic location, and past shocks are important influences on household food intake patterns. Obesogenic foods are found to be complements to food consumption, but milk is found to be a strong substitute for these foods. Results also show a high risk in female-headed households of inadequate consumption of nutrient-rich foods, and households with children show a propensity toward high consumption of drinks with added sugars. They recommend that nutritional information and education be available to allow for informed consumer decisions. Policy makers should also create means by which healthy foods would be made available during periods of economic and climatic crisis to avoid the resulting post-crisis increases in health issues.

Nutrient availability and access are important dimensions of food security. The challenge of concurrently providing sufficient nutrient availability and access will only intensify as the population increases and resources become increasingly scarce. McFadden and Schmitz (2017) examine the need for dietary guidelines to improve nutrition security. They also review the history of dietary guidelines in the United States, examine compliance, and conclude with implications for future dietary guidelines.

4.2 Food waste

Just as the adage says that a penny saved is a penny earned, so too does the prevention of food waste increase food availability. Just and Swigert (2016) address the potential to reduce food waste by focusing on behavioral nudges. Their work seeks to identify interventions that will be effective in reducing food waste, which is prevalent in both developed and developing countries. They conclude that food waste can be reduced through basic behavioral interventions that can be employed in both institutional and household settings.

While Just and Swigert (2016) address food waste at the consumer level, Minten et al. (2016) seek to measure and propose means to reduce food waste in the food value chain prior to consumption. The potato sectors of various Asian countries are examined to determine the total quantity of potatoes wasted in the post-harvest period and the off-season storage period. They suggest that increased use of cold storage can reduce food waste in the value chain.

Koester (2017) provides the contra view of this issue and questions whether the prevention of food waste is economically viable. He raises several questions regarding food waste. First, is the measurement of food waste based on sound economic principles? Second, does the elimination of food waste truly have the potential to reduce world hunger? Third, is there a moral problem in analyzing food waste while ignoring the waste of other goods or the consumption of luxury goods? Koester (2017) postulates that the cost of food waste avoidance must be compared to the value of the food waste reduction.

5 Stability

To be food secure, a population, a household, or an individual must have access to adequate food at all times. They should not risk losing access to food as a consequence of sudden shocks or cyclical events. The concept of stability can therefore refer to both the availability and access dimensions of food security (FAO 2006). This section reviews factors influencing the stability dimension of food security, primarily focusing on price volatility, production instability, climate change, and agricultural policy incentives.

5.1 Price volatility and production instability

Price volatility and production instability have direct impacts on food security. The existence of instability in the food system through the various dimensions of food security implies a probability of fatality or adverse health effects. Zilberman and Jin (2016) employ a risk management framework to examine food security. They determine cost-minimizing policies to achieve food security through availability, access, and vulnerability while accounting for the impact of randomness and uncertainty on the system. They show that ignoring key sources of variability can have dire consequences for at-risk groups. Programs to remedy food insecurity in emergency situations must find the appropriate balance between an enhanced overall food supply and the provision of emergency food aid and health care to vulnerable peoples.

Price volatility is a component of food security as price fluctuations affect production and impact buying power and access. The impact of food prices on poor and vulnerable households is examined by Ivanic and Martin (2015). Their analysis focuses on individual households living near the World Bank poverty standard of US$1.25 per day of purchasing power. Findings indicate that exogenous increases in food prices increase poverty in the short run given that many poor households are net purchasers of grain. In the long run, wages to unskilled off-farm labor fully adjust to the price shock and food output increases. Although price increases can exacerbate short-run poverty, higher food prices can contribute to long-run poverty alleviation.

Mitchell, Kayombo, and Cochrane (2017) determine the impact of the global food crisis of 2007–2008 on Tanzania's real food prices and the corresponding cost of the typical food basket. They find that the 2007/08 global food crisis did not cause food prices in Tanzania to increase because domestic factors are more important in the determination of food prices and food costs. They find that impacts of the global food crisis on food prices and costs in developing countries have been overestimated and, as a result, the global policy response may have been inappropriate.

Production instability and price volatility can be mitigated through food storage. Schmitz and Kennedy (2016) review models of commodity price stabilization brought about through storage programs by the government or the private sector (firms). The objectives of the government and the private sector may be quite different, as the government's role is to provide food from its stockholdings in times of emergency while the private sector manages its stockholdings with an objective of profit-maximization. They conclude that storage is not a means to reduce food insecurity in the long term. Similarly, the accumulation of large stocks of food grains, such as wheat, can be quite costly.

5.2 Climate change

The four dimensions of food security (availability, access, utilization, and stability) are influenced by a variety of interconnected factors. Mitchell et al. (2015) identify key pathways between climate, water, food, and conflict. Fluctuations among these correlated elements can all serve to undermine food security. Their analysis postulates that the alleviation of food insecurity through each of the four dimensions can serve to eliminate some of the underlying sources of conflict.

Drivers by which climate change can affect food security include temperature levels and variability, precipitation patterns, extreme events, and carbon dioxide and ozone levels. Chen, McCarl, and Thayer (2017) suggest that the impacts of climate change on food production systems will vary across regions, causing some regions to experience increased food security and other regions to experience food insecurity. In addition, crop and animal enterprises within an individual region may be affected differently. Chen, McCarl, and Thayer (2017) maintain that adapting to the potential impacts of climate change is necessary to promote continued food security, and research to determine how to best deal with this potential food security threat is of vital importance.

According to Letson (2017), from the perspective of the state of Florida, climate change will impact food security through invasive alien species, sea level rise flooding, and storm surges. For Florida's agricultural sector to maintain their provision of food security, Letson (2017) suggests that increased innovation by the agricultural sector will be necessary to keep pace with climate change.

Adequate production and appropriate consumption of fruits and vegetables is a good first step to prevent malnutrition. Ebert (2017) considers the impact of climate change on fruit and vegetable cropping systems. To promote resilient and sustainable agronomic systems, he stresses the need for integrated disease and pest management and the use of diverse production systems.

5.3 Agricultural policies and incentives

Domestic and trade policies can serve to enhance the stability dimension of food security, while ill-conceived policies can exacerbate problems for at-risk groups. Baylis, Fulton, and Reynolds (2016) analyze export restriction policies in Vietnam and India to determine their impact on price fluctuations using a political–economic framework. Given the downward pressure of these policies on domestic grain prices, the elimination of export restrictions could enhance food security through increased production and increased farm-household incomes. Efforts to reform these policies are unlikely given the political and economic power of elite local interest groups.

The agricultural policies of developed countries impact domestic production, influence domestic and international commodity prices, and affect global food security. Smith and Glauber (2017) consider U.S. agricultural policy and food aid programs and conclude that U.S. agricultural subsidy programs have little impact on production, prices, or food-insecure households. They determine that the Renewable Fuels Standard program has increased the prices of food grains, feed grains, and oilseeds, while negatively impacting food security among at-risk households both domestically and in developing countries.

In analyzing the role of agricultural trade in food security, Martin (2017a, b) states that restrictive trade policies, such as export bans or import restrictions, serve as beggar thy neighbor price insulation strategies and increase food price volatility, while free trade policies allow land-abundant counties to provide exports to land-poor countries, thereby increasing efficiency. Trade liberalization increases dietary diversity, increases access to food, diversifies supply sources, and reduces food price volatility.

Multilateral trade agreements have been successful in reducing many barriers to agricultural and food product trade. However, barriers to trade in biotechnology-related agricultural products have remained high relative to their non-biotech counterparts. Viju, Smyth, and Kerr (2017) examine three preferential trade agreements and their negotiations with respect to biotechnology products. Given the potential, and perhaps necessity, of biotechnology to keep pace with a growing population, trade agreements must address barriers to biotechnology trade.

Policies designed to promote the safety of food products can have adverse consequences with respect to food security. Zhang and Seale (2017) simulate the U.S. Food Safety Modernization Act (FSMA) to determine its implications for quantity demanded, revenue, and profit for various sized farms. Their analysis of domestic and foreign tomato producers shows that small and very small farms will experience significant losses in profit because of the FSMA. While the overall objective of the FSMA may be achieved and contribute to the availability component of food security, decreased farm profits will adversely impact access via reduced income among certain rural households.

6 Conclusions

As the world population continues to grow, how will food production keep pace with food demand? How can poverty be alleviated so at-risk households can enjoy access to a nutritionally adequate food supply? What policies can guarantee that the diet of at-risk households sufficiently provides the appropriate levels and combinations of nutrients? What are the most appropriate mechanisms to ensure price, income, and production stability within the food system? These questions and many others like them shed light on the complexity of the global food system and the fragile nature of food and nutrition security, particularly in the case of at-risk households.

Food security is a complex issue. Since the time of Malthus (1798), the problem has been refined to its current definition and dimensions. As the dynamics of the global economy and food system change, so too must models be sufficiently flexible to account for changes in technology, population, income, nutritional needs, and other variables. Modeling frameworks must continue to be developed to provide the appropriate information to producers and policy makers so that "all people, at all times, have physical, social and economic access to sufficient, safe and nutritious food that meets their dietary needs and food preferences for an active and healthy life" (FAO 1996, p. 3).

Acknowledgments

Examples and studies cited in this work draw on the three *Food Security* volumes edited by Schmitz, Kennedy, and Schmitz (2015, 2016, and 2017) in the series "Frontiers of Economics and Globalization." I am indebted to Dr. Andrew Schmitz and Dr. Troy Schmitz for their collaborations on these volumes. I also appreciate the very useful comments and suggestions of Dr. Will Martin.

References

Anderson, S. 1990. "Core Indicators of Nutritional State for Difficult-to-Sample Populations." *Journal of Nutrition* 120:1559–1600.

Anderson, J.L., F. Asche, T. Garlock, and J. and Chu. 2017. "Aquaculture: Its Role in the Future of Food." In A. Schmitz, P.L. Kennedy, and T.G. Schmitz, eds., *World Agricultural Resources and Food Security (Vol. 17). Frontiers of Economics and Globalization.* Bingley: Emerald Group Publishing Limited.

Alexandratos, N., and J. Bruinsma. 2012. "World Agriculture toward 2030/2050: The 2012 Revision." ESA Working Paper No. 12–03, FAO, Rome.

Anfinson, C., T.I. Wahl, J.L. Seale Jr., and J. Bai. 2016. "Factors Affecting Adolescent Obesity in Urban China." In A. Schmitz, P.L. Kennedy, and T.G. Schmitz, eds., *Food Security in a Food Abundant World: An Individual Country Perspective (Vol. 16). Frontiers of Economics and Globalization.* Bingley, UK: Emerald Group Publishing Limited.

Antle, J., R. Adhikari, and S. Price. 2015. "An Income-Based Food Security Indicator for Agricultural Technology Impact Assessment." In A. Schmitz, P.L. Kennedy, and T.G. Schmitz, eds., *Food Security in an Uncertain World: An International Perspective (Vol. 15). Frontiers of Economics and Globalization.* Bingley: Emerald Group Publishing Limited.

Baylis, K., M.E. Fulton, and T. Reynolds. 2016. "The Political Economy of Export Restrictions: The Case of Vietnam and India." In A. Schmitz, P.L. Kennedy, and T.G. Schmitz, eds., *Food Security in a Food Abundant World: An Individual Country Perspective (Vol. 16). Frontiers of Economics and Globalization*. Bingley: Emerald Group Publishing Limited.

Beghin, J., and Y. Teshome. 2017. "The Coffee-Food Security Interface for Subsistence Households in Jimma Zone Ethiopia." In A. Schmitz, P.L. Kennedy, and T.G. Schmitz, eds., *World Agricultural Resources and Food Security (Vol. 17). Frontiers of Economics and Globalization*. Bingley: Emerald Group Publishing Limited.

Chen, J., B.A. McCarl, and A. Thayer. 2017. "Climate Change and Food Security: Threats and Adaptation." In A. Schmitz, P.L. Kennedy, and T.G. Schmitz, eds., *World Agricultural Resources and Food Security (Vol. 17). Frontiers of Economics and Globalization*. Bingley: Emerald Group Publishing Limited.

Clark, L.F., and J.E. Hobbs. 2015. "Innovations in International Food Assistance Strategies and Therapeutic Food Supply Chains." In A. Schmitz, P.L. Kennedy, and T.G. Schmitz, eds., *Food Security in an Uncertain World: An International Perspective (Vol. 15). Frontiers of Economics and Globalization*. Bingley: Emerald Group Publishing Limited.

Diaz-Bonilla, E. 2015. "Macroeconomic Policies and Food Security." In A. Schmitz, P.L. Kennedy, and T.G. Schmitz, eds., *Food Security in an Uncertain World: An International Perspective (Vol. 15). Frontiers of Economics and Globalization*. Bingley: Emerald Group Publishing Limited.

Ebert, A.W. 2017. "Vegetable Production, Diseases, and Climate Change." In A. Schmitz, P.L. Kennedy, and T.G. Schmitz, eds., *World Agricultural Resources and Food Security (Vol. 17). Frontiers of Economics and Globalization*. Bingley: Emerald Group Publishing Limited.

Food and Agriculture Organization of the United Nations (FAO). 1996. *Declaration on World Food Security and World Food Summit Plan of Action*. Rome: FAO.

Food and Agriculture Organization of the United Nations (FAO). 2006. *Food Security*. Rome: FAO.

Food and Agriculture Organization of the United Nations, International Fund for Agricultural Development, and World Food Program (FAO, IFAD, and WFP). 2012. *The State of Food Insecurity in the World, 2012*. Rome: FAO.

Fukase, E., and W. Martin. 2017. "Economic Growth, Convergence and World Food Demand and Supply." IFPRI Working Paper, International Food Policy Research Institute: Washington, DC.

Griswold, D.T. 1999. "Bringing Economic Sanity to Agricultural Trade." Cato Institute, Washington, DC.

Hawkes, C. 2015. "Diet, Chronic Disease, and the Food System: Making the Links, Pushing for Change." In *Advancing Health and Well-Being in Food Systems: Strategic Opportunities for Funders*. Toronto: Global Alliance for the Future of Food.

Henneberry, S.R., and C.D. Carrasco. 2015. "Food Security Issues: Concepts and the Role of Emerging Markets." In A. Schmitz, P.L. Kennedy, and T.G. Schmitz, eds., *Food Security in an Uncertain World: An International Perspective (Vol. 15). Frontiers of Economics and Globalization*. Bingley: Emerald Group Publishing Limited.

Ivanic, M., and W. Martin. 2015. "Managing High and Volatile Food Prices in Developing Countries since 2000." In A. Schmitz, P.L. Kennedy, and T.G. Schmitz, eds., *Food Security in an Uncertain World: An International Perspective (Vol. 15). Frontiers of Economics and Globalization*. Bingley: Emerald Group Publishing Limited.

Just, D.R., and J.M. Swigert. 2016. "The Role of Nudges in Reducing Food Waste." In A. Schmitz, P.L. Kennedy, and T.G. Schmitz, eds., *Food Security in a Food Abundant World: An Individual Country Perspective (Vol. 16). Frontiers of Economics and Globalization*. Bingley: Emerald Group Publishing Limited.

Kennedy, P.L., K.E. Lewis, and A. Schmitz. 2017. "Food Security through Biotechnology: The Case of Genetically Modified Sugar Beets in the United States." In A. Schmitz, P.L. Kennedy, and T.G. Schmitz, eds., *World Agricultural Resources and Food Security (Vol. 17). Frontiers of Economics and Globalization*. Bingley: Emerald Group Publishing Limited.

Kerr, W.A. 2015. "Food Security and Anti-Piracy Strategies: The Economics of Protecting World Food Program Shipments." In A. Schmitz, P.L. Kennedy, and T.G. Schmitz, eds., *Food Security in an Uncertain World: An International Perspective (Vol. 15). Frontiers of Economics and Globalization*. Bingley: Emerald Group Publishing Limited.

Koester, U. 2017. "Food Loss and Waste as an Economic and Policy Problem." In A. Schmitz, P.L. Kennedy, and T.G. Schmitz, eds., *World Agricultural Resources and Food Security (Vol. 17). Frontiers of Economics and Globalization*. Bingley: Emerald Group Publishing Limited.

Koo, W.W., and R. Taylor. 2015. "Who Will Feed the Growing Population in the Developing Nations? Implications of Bioenergy Production." In A. Schmitz, P.L. Kennedy, and T.G. Schmitz, eds., *Food Security in an Uncertain World: An International Perspective (Vol. 15). Frontiers of Economics and Globalization*. Bingley: Emerald Group Publishing Limited.

Kristkova, Z.S., M. van Kijk, and H. van Meijl. 2017. "Assessing the Impact of Agricultural R&D Investments on Long-term Projections of Food Security." In A. Schmitz, P.L. Kennedy, and T.G. Schmitz, eds., *World Agricultural Resources and Food Security (Vol. 17). Frontiers of Economics and Globalization*. Bingley: Emerald Group Publishing Limited.

Krivonos, E., J. Morrison, and E. Canigiani. 2015. "Trade and Food Security: Links, Processes, and Prospects." In A. Schmitz, P.L. Kennedy, and T.G. Schmitz, eds., *Food Security in an Uncertain World: An International Perspective (Vol. 15). Frontiers of Economics and Globalization*. Bingley: Emerald Group Publishing Limited.

Lakkakula, P., D.J. Haynes, and T.G. Schmitz. 2015. "Genetic Engineering and Food Security: A Welfare Economics Perspective." In A. Schmitz, P.L. Kennedy, and T.G. Schmitz, eds., *Food Security in an Uncertain World: An International Perspective (Vol. 15). Frontiers of Economics and Globalization*. Bingley: Emerald Group Publishing Limited.

Letson, D. 2017. "Climate Change and Food Security: Florida's Agriculture in the Coming Decades." In A. Schmitz, P.L. Kennedy, and T.G. Schmitz, eds., *World Agricultural Resources and Food Security (Vol. 17). Frontiers of Economics and Globalization*. Bingley: Emerald Group Publishing Limited.

Madeley, J. 2000. "Trade and Hunger." GRAIN, Barcelona, Spain. Magana-Lemus, D., and J. Lara-Alvarez. 2015. "Food Security Measurement: An Empirical Approach." In A. Schmitz, P.L. Kennedy, and T.G. Schmitz, eds., *Food Security in an Uncertain World: An International Perspective (Vol. 15). Frontiers of Economics and Globalization*. Bingley: Emerald Group Publishing Limited.

Malthus, T.R. 1798. *An Essay on the Principle of Population*. London: J. Johnson Publishing.

Martin, W. 2017a. "Agricultural Trade and Food Security." ADBI Working Paper Series, Asian Development Bank Institute, Tokyo, Japan.

———. 2017b. "The Research Agenda for International Agricultural Trade and Development." IFPRI Working Paper, International Food Policy Research Institute, Washington, DC.

McFadden, B., and T. Schmitz. 2017. "The Nexus of Dietary Guidelines and Food Security." In A. Schmitz, P.L. Kennedy, and T.G. Schmitz, eds., *World Agricultural Resources and Food Security (Vol. 17). Frontiers of Economics and Globalization*. Bingley: Emerald Group Publishing Limited.

Meyers, W.H., and N. Kalaitzandonakes. 2015. "World Population, Food Growth, and Food Security Challenges." In A. Schmitz, P.L. Kennedy, and T.G. Schmitz, eds., *Food Security in an Uncertain World: An International Perspective (Vol. 15). Frontiers of Economics and Globalization*. Bingley: Emerald Group Publishing Limited.

Miljkovic, D. 2015. "Dual Nature and the Human Face of Food (In)security." In A. Schmitz, P.L. Kennedy, and T.G. Schmitz, eds., *Food Security in an Uncertain World: An International Perspective (Vol. 15). Frontiers of Economics and Globalization*. Bingley: Emerald Group Publishing Limited.

Minten, B., T. Reardon, S.D. Gupta, D. Hu, and K.A.S. Murshid. 2016. "Wastage in Food Value Chains in Developing Countries: Evidence from the Potato Sector in Asia." In A. Schmitz, P.L. Kennedy, and T.G. Schmitz, eds., *Food Security in a Food Abundant World: An Individual Country Perspective (Vol. 16). Frontiers of Economics and Globalization*. Bingley: Emerald Group Publishing Limited.

Mishra, A.K., and A.R. Khanal. 2017. "Assessing Food Security in Rural Bangladesh: The Role of Non-Farm Economy." In A. Schmitz, P.L. Kennedy, and T.G. Schmitz, eds., *World Agricultural Resources and Food Security (Vol. 17). Frontiers of Economics and Globalization*. Bingley: Emerald Group Publishing Limited.

Mitchell, D., D. Hudson, R. Post, P. Bell, and R.B. Williams. 2015. "Food Security and Conflict." In A. Schmitz, P.L. Kennedy, and T.G. Schmitz, eds., *Food Security in an Uncertain World: An International Perspective (Vol. 15). Frontiers of Economics and Globalization*. Bingley: Emerald Group Publishing Limited.

Mitchell, D., A. Kayombo, and N. Cochrane. 2017. "Food Costs during the Food Crisis: The Case of Tanzania." In A. Schmitz, P.L. Kennedy, and T.G. Schmitz, eds., *World Agricultural Resources and Food Security (Vol. 17). Frontiers of Economics and Globalization*. Bingley: Emerald Group Publishing Limited.

Moss, C.B., J.F. Oehmke, and A. Lyambabaje. 2016. "Food Security, Subsistence Agriculture, and Working's Model." In A. Schmitz, P.L. Kennedy, and T.G. Schmitz, eds., *Food Security in a Food Abundant World: An Individual Country Perspective (Vol. 16). Frontiers of Economics and Globalization*. Bingley: Emerald Group Publishing Limited.

Msangi, S., and M. Batka. 2015a. "Dietary Change and Global Drivers of Change: How Can We Improve the Nutritional Status of the Bottom Billion?" In A. Schmitz, P.L. Kennedy, and T.G. Schmitz, eds., *Food Security in an Uncertain World: An International Perspective (Vol. 15). Frontiers of Economics and Globalization*. Bingley: Emerald Group Publishing Limited.

———. 2015b. "Major Trends in Diets and Nutrition: A Global Perspective to 2050." In A. Schmitz, P.L. Kennedy, and T.G. Schmitz, eds., *Food Security in an Uncertain World: An International Perspective (Vol. 15). Frontiers of Economics and Globalization*. Bingley: Emerald Group Publishing Limited.

Pieters, H., and J. Swinnen. 2016. "Food Security Policy at the Extreme of the Water-Energy-Food Nexus: The Kingdom of Saudi Arabia." In A. Schmitz, P.L. Kennedy, and T.G. Schmitz, eds., *Food Security in a Food Abundant World: An Individual Country Perspective (Vol. 16). Frontiers of Economics and Globalization*. Bingley: Emerald Group Publishing Limited.

Pinstrup-Andersen, P. 2007. "Agricultural Research and Policy for Better Health and Nutrition in Developing Countries: A Food Systems Approach." *Agricultural Economics* 37:187–198.

President's Commission. 1969. *Poverty amid Plenty: The American Paradox*. Washington, DC: Government Printing Office.

Regmi, M., and K.P. Paudel. 2016. "Impact of Remittances on Food Security in Bangladesh." In A. Schmitz, P.L. Kennedy, and T.G. Schmitz, eds., *Food Security in a Food Abundant World: An Individual Country Perspective (Vol. 16). Frontiers of Economics and Globalization*. Bingley: Emerald Group Publishing Limited.

Schmitz, A., and P.L. Kennedy. 2015. "Food Security: Starvation in the Midst of Plenty." In A. Schmitz, P.L. Kennedy, and T.G. Schmitz, eds., *Food Security in an Uncertain World: An International Perspective (Vol. 15). Frontiers of Economics and Globalization*. Bingley: Emerald Group Publishing Limited.

———. 2016. "Food Security and the Role of Food Storage." In A. Schmitz, P.L. Kennedy, and T.G. Schmitz, eds., *Food Security in a Food Abundant World: An Individual Country Perspective (Vol. 16). Frontiers of Economics and Globalization*. Bingley: Emerald Group Publishing Limited.

Schmitz, A., P.L. Kennedy, and M. Salassi. 2017. "Sugarcane Yields and Production: Florida and Louisiana." In A. Schmitz, P.L. Kennedy, and T.G. Schmitz, eds., *World Agricultural Resources and Food Security (Vol. 17). Frontiers of Economics and Globalization*. Bingley: Emerald Group Publishing Limited.

Schmitz, A., P.L. Kennedy, and T.G. Schmitz (editors). 2015. *Food Security in an Uncertain World: An International Perspective (Vol. 15). Frontiers of Economics and Globalization*. Bingley: Emerald Group Publishing Limited.

———. 2016. *Food Security in a Food Abundant World: An Individual Country Perspective (Vol. 16). Frontiers of Economics and Globalization*. Bingley: Emerald Group Publishing Limited.

———. 2017. *World Agricultural Resources and Food Security (Vol. 17). Frontiers of Economics and Globalization*. Bingley: Emerald Group Publishing Limited.

Schroeder, K.G., and W.H. Meyers. 2016. "The Status and Challenges of Food Security in Europe and Central Asia." In A. Schmitz, P.L. Kennedy, and T.G. Schmitz, eds., *Food Security in a Food Abundant World: An Individual Country Perspective (Vol. 16). Frontiers of Economics and Globalization*. Bingley: Emerald Group Publishing Limited.

Seidu, A., and G. Onel. 2016. "Off-Farm Labor Allocation, Income, and Food Consumption among Rural Farm Households in Transitional Albania." In A. Schmitz, P.L. Kennedy, and T.G. Schmitz, eds., *Food Security in a Food Abundant World: An Individual Country Perspective (Vol. 16). Frontiers of Economics and Globalization*. Bingley: Emerald Group Publishing Limited.

Sen, A. 1981. *Poverty and Famines: An Essay on Entitlement and Deprivation*. Oxford: Oxford University Press.

Shaw, D.J. 2007. *World Food Security: A History Since 1945*. New York: Palgrave Macmillan.

Sitienei, I., A.K. Mishra, and A.R. Khanal. 2016. "Informal 'Ganyu' Labor Supply, and Food Security: The Case of Malawi." In A. Schmitz, P.L. Kennedy, and T.G. Schmitz, eds., *Food Security in a Food Abundant World: An Individual Country Perspective (Vol. 16). Frontiers of Economics and Globalization.* Bingley: Emerald Group Publishing Limited.

Smith, L.C. 1998. "Can FAO's Measure of Chronic Undernourishment Be Strengthened?" *Food Policy* 23(5):425–445.

Smith, V.H., and J.W. Glauber. 2017. "U.S. Agricultural Policy: Impacts on Domestic and International Food Security." In A. Schmitz, P.L. Kennedy, and T.G. Schmitz, eds., *World Agricultural Resources and Food Security (Vol. 17). Frontiers of Economics and Globalization.* Bingley: Emerald Group Publishing Limited.

Thome, K., B. Meade, S. Rosen, and J.C. Beghin. 2017. "Assessing Food Security in Ethiopia with USDA ERS's Demand-based Food Security Model." In A. Schmitz, P.L. Kennedy, and T.G. Schmitz, eds., *World Agricultural Resources and Food Security (Vol. 17). Frontiers of Economics and Globalization.* Bingley: Emerald Group Publishing Limited.

Useche, P., and J. Twyman. 2016. "Sugar, Fat, or Protein: Are All Food Insecure Households Eating the Same? The Case of Small Rice Producers in Peru." In A. Schmitz, P.L. Kennedy, and T.G. Schmitz, eds., *Food Security in a Food Abundant World: An Individual Country Perspective (Vol. 16). Frontiers of Economics and Globalization.* Bingley: Emerald Group Publishing Limited.

Viju, C., S.J. Smyth, and W.A. Kerr. 2017. "Agricultural Biotechnology and Food Security: Can the CETA, TPP, and TTIP Become Venues to Facilitate Trade in GM Products?" In A. Schmitz, P.L. Kennedy, and T.G. Schmitz, eds., *World Agricultural Resources and Food Security (Vol. 17). Frontiers of Economics and Globalization.* Bingley: Emerald Group Publishing Limited.

Von Witzke, H., and S. Noleppa. 2016. "The High Value to Society of Modern Agriculture: Global Food Security, Climate Protection, and Preservation of the Environment – Evidence from the European Union." In A. Schmitz, P.L. Kennedy, and T.G. Schmitz, eds., *Food Security in a Food Abundant World: An Individual Country Perspective (Vol. 16). Frontiers of Economics and Globalization.* Bingley: Emerald Group Publishing Limited.

Wailes, E.J., A. Durand-Morat, and M. Diagne. 2015. "Regional and National Rice Development Strategies for Food Security in West Africa." In A. Schmitz, P.L. Kennedy, and T.G. Schmitz, eds., *Food Security in an Uncertain World: An International Perspective (Vol. 15). Frontiers of Economics and Globalization.* Bingley: Emerald Group Publishing Limited.

Zhang, L., and J. Seale, Jr. 2017. "Food Security and the Food Safety Modernization Act." In A. Schmitz, P.L. Kennedy, and T.G. Schmitz, eds., *World Agricultural Resources and Food Security (Vol. 17). Frontiers of Economics and Globalization.* Bingley: Emerald Group Publishing Limited.

Zilberman, D., and Y. Jin. 2016. "A Probabilistic Approach to Food Security." In A. Schmitz, P.L. Kennedy, and T.G. Schmitz, eds., *Food Security in a Food Abundant World: An Individual Country Perspective (Vol. 16). Frontiers of Economics and Globalization.* Bingley: Emerald Group Publishing Limited.

Part 2
Environmental and resource economics

12
How does climate change affect agriculture?

Panit Arunanondchai, Chengcheng Fei, Anthony Fisher, Bruce A. McCarl, Weiwei Wang, and Yingqian Yang

1 Introduction

Climate change, according to the Intergovernmental Panel on Climate Change (IPCC), refers to a change in the state of the climate that can be identified by changes in the mean and/or the variability of its properties and that persists for an extended period, typically decades or longer (IPCC 2013). Climate change has been observed to have widespread impacts on human and natural systems (IPCC 2014a), including agriculture. Of course, it is not normally possible to attribute all extreme events, such as heat waves, hurricanes, and droughts, to climate change, but a recent study of extreme weather events in 2012 has detected the "finger prints" of climate change in half of them, including "Superstorm Sandy" (Peterson et al. 2013).

The essential questions of this chapter are: (1) how do the effects of and possible scientific responses to climate change affect agriculture? and (2) how might agricultural economists contribute to understanding and reducing the magnitude of the problem? We will address these items, raising issues and providing a supporting literature review (partially because of the vastness of the literature). In doing this, we will review recent climate and climate change driver developments, climate change effects on agriculture, possible adaptation strategies, and possible agricultural actions to mitigate/limit the future extent of climate change.

1.1 Recent history of climate change and drivers

The IPCC has devoted substantial effort to documenting the extent of climate change, in which the most recent report (IPCC 2013) indicates that:

- Between 1880 and 2012, the combined land and ocean surface temperature increased by 0.85°C with a range of 0.65 to 1.06°C. Additionally, it exhibited substantial decadal and inter-annual variability.
- Precipitation over the mid-latitude land areas of the Northern Hemisphere has increased since 1901.

- Between 1901 and 2010, global mean sea level rose by 0.19 m with a range of 0.17 to 0.21 m. The rate of sea level rise since the mid-twentieth century has been larger than the mean rate during the previous two millennia.
- Increases in extreme climate related events have been observed since about 1950. These include heat waves, extreme precipitation events, droughts, floods, cyclones, and wildfires.
- Frequency of cold days and nights has decreased and that of warm days and nights has increased.

Natural and anthropogenic processes that alter the earth's energy budget have been identified as drivers of climate change (IPCC 2014b). Anthropogenic greenhouse gas (GHG) concentrations and emissions have greatly increased, with CO_2 rising from 275 parts per million (ppm) in the atmosphere in 1850 to over 400 ppm and CO_2 equivalent to over 485 ppm (NOAA 2016), yielding concentrations much higher than in the last half million years (IPCC 2014b).

1.2 Projected climate change

Climate change is projected to continue, although the extent of it depends upon GHG concentrations in the atmosphere and mitigation efforts to limit net emissions. The future projections are summarized in the graph in Figure 12.1 from Knutti and Sedláček (2012), as modified by McCarl (2015).

In Figure 12.1, IPCC (2014a) defines two eras of future climate change. Era 1 is the period between now and 2040 and is called the period of committed climate change; Era 2 is the period between 2040 and 2100 and is called the era of climate options (IPCC 2014a). Also, in the graph, the continuous black line represents the amount of temperature change to date, while the other lines show climate change evolution under different degrees of GHG mitigation and are called representative concentration pathways (RCPs).

Note in Era 1, there is a commitment to about a 1°C change without very large effects from the different mitigation scenarios. However, in Era 2, there is a temperature change spanning

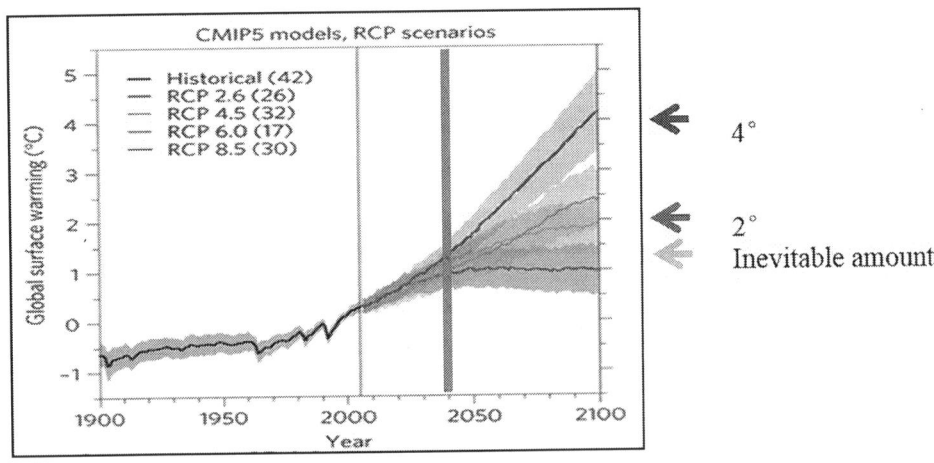

Figure 12.1 IPCC AR5 graph of future temperature change under alternative GHG scenarios, modified by McCarl (2015)

between 2 and 4°C depending on different mitigation efforts (where RCP8.5 has much less mitigation effort than RCP4.5). Agriculture will need to operate in both of these Eras, and thus will be affected by climate change and may participate in reaching the lower RCP levels through mitigation actions.

Before proceeding, we should note that IPCC projections of future climate change indicators such as emissions, temperature change, and sea level rise tend to be conservative, with a history of under-predicting these indicators. We briefly note a couple of problems with the most recent projections.

One is that they omit the potential increase in global mean temperature (GMT) due to Arctic permafrost melting and the release of carbon dioxide and methane. This omission may be defensible on the grounds that the cumulative release is uncertain at this time, but it is already occurring on a small scale (United Nations Environment Program 2012). The concern is that this is a positive feedback loop: the greater the release of these greenhouse gases, the greater the increase in temperature, and so on. Moreover, the situation could worsen dramatically, because there are much larger stores of methane in undersea structures (methane clathrates, methane trapped in crystal structures of water found under sediment on ocean floors), already releasing an estimated 19 million tons annually from the East Siberian Arctic Shelf (Shakhova et al. 2013). Under the right circumstances, releases could be much greater comparable to land-based emissions (Whiteman, Hope, and Wadhams 2013).

Also omitted are effects on sea level rise from melting of the two major potential sources: the Greenland Ice Sheet and the West Antarctic Ice Sheet. In this regard, there are two recent and concerning discoveries – that the formerly stable ice sheet in northeast Greenland is now melting rapidly (Khan et al. 2014), along with previously known melting in southern Greenland, and that early-stage collapse of the West Antarctic Ice Sheet has begun and is irreversible (Joughin, Smith, and Medley 2014). Such findings make more likely and bring nearer in time a massive sea level rise of as much as 15 meters (Poore, Williams, and Tracey 2000) and a resulting inundation of coastal and low-lying areas, including prime agricultural land. The latter is projected to have a major impact in the decades beyond 2100, but this raises another issue: what is the appropriate time horizon for projections?

Temperature increase and sea level rise will continue well beyond 2100, especially under high emissions like those under RCP8.5. A more appropriate end-point for projections, when a natural equilibrium may be reached (Cline 1992; Fisher and Le 2014), is probably around 2300, when the increase in GMT will have reached about 10°C, with attendant catastrophic impacts, according to an extended RCP8.5 scenario developed by the German Climate Computing Center (2014).

An obvious question here is: why should we care about events far in the future which will be heavily discounted in a benefit–cost analysis? Discussions of discounting in the context of climate change policy are often set in the framework of the Ramsey equation, which states that the consumption or goods discount rate is equal to the sum of two components: the pure (social) rate of time preference and the product of the elasticity of the marginal utility of consumption and the rate of growth in per capita consumption. Although not all would agree, Stern (2007), Heal (2008), and others have argued that the social rate of time preference should be set at or very close to zero, implying a low consumption or goods discount rate. Furthermore, once uncertainty about the course of future consumption growth, affected by, among other things, uncertainty about potential impacts of climate change, is introduced into the Ramsey framework, there are strong theoretical arguments, presented and summarized by Arrow et al. (2014), for a declining discount rate, i.e., a discount rate schedule in which the rate applied to benefits and costs occurring in the future declines over time. The point is that, in thinking about the potential impacts of climate change, the future, even the distant future, matters more than in a typical benefit–cost analysis. Interestingly, as they note, this is in fact the practice in evaluating projects and regulations in France and the United Kingdom, though not in the U.S., where the Office of Management and Budget (OMB) requires use of a constant exponential rate.

We are not predicting that the kinds of catastrophic impacts in the more distant future, beyond the end of the present century, as discussed by Stern (2007), Fisher and Le (2014), and others, will happen, for several reasons, including technical change in the energy sector and adoption of policies to more dramatically reduce emissions. Rather, unless economists and decision makers have a clear understanding of the potential impacts of unconstrained emissions, analyses and policies that would lead to a timely development of energy alternatives will be problematic.

1.3 Broad ways climate change affects agriculture

The agriculture sector is highly dependent upon weather and climate. Thus, it is vulnerable to climate change (called effects). The extent of this vulnerability depends on adaptive actions taken by humans to moderate the effects (called adaptation). Agriculture is also a significant source of anthropogenic GHG emissions (about 24 percent globally when including deforestation) and may participate in emissions reduction by altering management (called mitigation). Thus, the ways the climate change issue affects agriculture depends on the simultaneous extent of three forces:

- The effects of climate change brought on by altered temperatures, precipitation, extreme events, sea level rise, pest responses, and other climate-related forces (as reviewed in Porter et al. 2014; Hatfield et al. 2014).
- The adaptation actions to alter agricultural processes in response to climate change to reduce damages and/or exploit opportunities (as reviewed in McCarl, Thayer, and Jones 2016).
- The mitigation actions to limit the future extent of climate change by altering agricultural management (Smith et al. 2008; Smith et al. 2014).

Thus, when one examines how climate change will impact agriculture, one needs to consider:

- What is the extent of the climate change effects?
- What are the impacts on agriculture of adaptation and mitigation actions?
- How do these forces jointly interact and what is the total impact on productivity, cost of production, income, food costs, food security, and many other agricultural attributes?

This chapter will provide coverage across all of these items.

2 Effects of climate change on agriculture

Climate change can significantly affect crops, livestock, farmland values, household welfare, and food security. Studies have shown that its overall effect on agriculture can be complicated. For instance, while changes in temperature and precipitation can be beneficial for some crops in some places (areas limited by cold or improper amounts of rainfall), they can be detrimental for other crops in other places (places that are hot and dry). Nonlinear climate effects are also expected as low or high extremes in heat or precipitation are detrimental.

In this section, a number of effects of climate change are highlighted with major pieces of research identified. The main items examined are shown in the following subsections: (1) crops; (2) livestock; (3) land values; and (4) incomes, consumer welfare, prices, food security, and poverty.

2.1 Crops

Crop yields are strongly affected by temperature, precipitation, and extremes thereof, along with carbon dioxide (CO_2), and are also affected by sea level rise and pests. Econometric and simulation approaches have been used to examine these issues.

Crop simulation was the original technique used to study crop yield sensitivity. Two studies employing this method are Adams et al. (1990) and Reilly et al. (2002). Collectively these studies found: (1) crop yield effects can be positive or negative and vary by region, with southern areas generally having larger and more negative effects; (2) carbon dioxide is an important factor in yields of some crops, and its consideration can reverse the nature of overall effects; and (3) effects can be overstated if one does not consider changes in planting dates, varieties, and other farm-level adaptations.

More recently, a number of econometric studies have examined effects using historical yields. For example McCarl, Villavicencio, and Wu (2008) developed an econometric estimate of annual average climate change effects on yield distributions using data on major agricultural crops across the U.S. They considered effects of temperature, precipitation, variance of intra-annual temperature, a constructed index of rainfall intensity, and the Palmer Drought Severity Index (PDSI), but ignored nonlinearities. Their empirical results show that stationarity of crop yield distributions does not hold for either the mean or the variance. The effects were found to differ by crop and location.

Schlenker, Hanemann, and Fisher (2006) studied extremes and found that an increase in very hot days (measured by growing season days with temperatures above 34°C) strongly affects agricultural land values in the U.S. They projected that an increase in such hot days under a business as usual scenario, such as the subsequent IPCC RCP8.5, would have a large and mostly statistically significant negative impact on U.S. farmland value. Land value changes were found to be unevenly distributed, with most counties harmed and a few, in a northern tier along the Canadian border, helped by a longer growing season.

More generally, changes in the frequency and severity of extreme events, such as droughts and floods, have been found to adversely affect crop production. A few studies of such issues and items focused on are climate variability (Porter and Semenov 2005); shifts in El Nino event frequency (Chen, McCarl and Adams 2001); incidence of very hot days (Schlenker, Hanemann and Fisher 2006); variance of intra-annual temperature, rainfall intensity, and drought frequency (McCarl et al. 2008); and hurricanes (Chen and McCarl 2009).

Schlenker and Roberts (2009) studied the nonlinear impact of climate change and found that crop yields are stable or slowly increase with temperature up to 29°C for corn and 30°C for soybeans, but they then fall sharply with additional increases (see Figure 12.2).

One additional factor that alters crop yield is carbon dioxide (CO_2), which stimulates growth for crops that fix carbon via the Calvin cycle (which are know as C3 crops like cotton, wheat, and rice) but has small growth effects on crops that fix carbon via the Hatch-Slack pathway (the so-called C4 crops such as sugarcane, corn, sorghum, and millet), except under drought (Atta-vanich and McCarl 2014).

This is a difficult item to include in models as it has progressed with time and can be confused with technological progress. Attavanich and McCarl (2014) merged a dataset of experimental results over alternative CO_2 with a historical dataset to econometrically investigate the

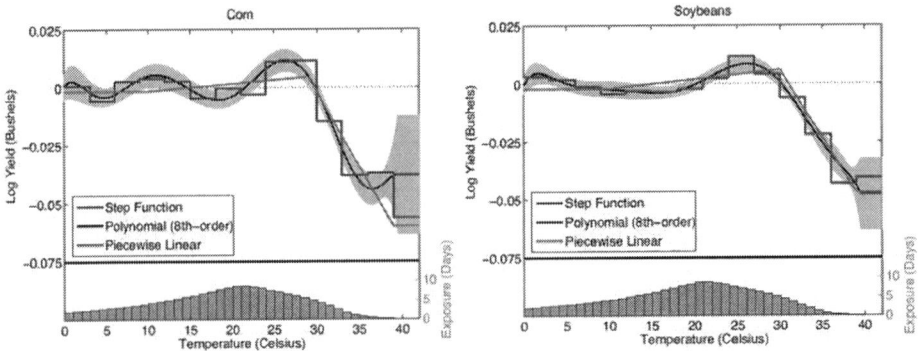

Figure 12.2 Nonlinear impact of temperature on U.S. corn and soybean crop yields
Source: Schlenker and Roberts (2009; Figure 12.1, p. 15595).

relationship between climate and yield variability. They found that ignoring atmospheric CO_2 leads to an overestimate of technical progress and found the expected results that yields of C3 crops positively respond while yields of C4 crops do not except under drought stress. They did find nonlinear effects of temperature and precipitation.

Crop yields are affected in the short run, but there is also the likelihood that longer run technical progress is altered. Additionally, climate change also has been found to influence technological progress and have regionally specific effects on total returns to research investments, with southern areas seeing reductions and some northern areas increases (Feng, McCarl, and Havlik 2011; Villavicencio et al. 2013).

In terms of pests, several studies have investigated climate influences on pests and pesticide costs. Juroszek and von Tiedemann (2013) provided a review of the biophysical effects. Chen and McCarl (2001) looked at cost effects under the assumption that pest incidence increases would be matched by pesticide cost increases and examined climate effects on cost. They found that pesticide expenditures rise with increased temperatures and precipitation for the majority of crops. Shakhramanyan, Schneider, and McCarl (2013) extended the work, looking at individual compounds and their resultant environmental costs, finding that climate change induced the increased use of pesticides and associated runoff and, in turn, environmental and health costs.

Sea level is also a factor where producing lands can be inundated, particularly in low-lying areas of countries like Bangladesh, Japan, Taiwan, Egypt, Myanmar, and Vietnam. Low-lying areas are vulnerable to sea level rise that causes land loss and salt water intrusion affecting water supplies. The realization of these threats can cause substantial regional crop productivity losses (Chen, McCarl, and Chang 2012).

2.2 Livestock

Climate change can alter livestock growth, reproduction rates, death rates, feed supplies, nutritional content, and disease and pest incidence. Some specific findings are:

- Heat waves have been found to affect animal performance, animal mortality, illness, feed intake, feed conversion rates, rates of gain, milk production, conception rates, and appetite loss (St-Pierre, Cobanov, and Schnitkey 2003; Gaughan et al. 2009; Mader et al. 2009; Mader 2014).

- A longer animal feeding period will be needed to obtain the same volume of animal products (Mader et al. 2009).
- Feed availability and feed quality are altered through effects on crop, pasture, and forage growth and nutritional quality as discussed earlier, plus grasses have altered nutritional value and digestibility (Craine et al. 2010).
- Increased rainfall intensity has been observed to increase range and pasture land soil degradation (Howden, Crimp, and Stokes 2008).
- Pest and disease incidence has been found to increase, altering animal health, disease outbreaks, and fecundity (Gale et al. 2009; Mu et al. 2014).

2.3 Regions and land values

Most of the early studies predicted adverse effects of global climate change on U.S. agriculture (Adams et al. 1990; Adams et al. 1995). Mendelsohn, Nordhaus, and Shaw (1994) used a reduced-form hedonic model with the value of farmland to estimate the impact of climate change and found economic benefits for agriculture. This is possible because their model allows adaptation as conditions change (see Figure 12.3).

Schlenker, Hanemann, and Fisher (2005) argue that the hedonic model approach is vulnerable to misspecification problems. They argue in particular that the inadequate treatment of irrigation might bias the results. Hence, they incorporated irrigation into the hedonic model. They find that the economic effects of climate change on agriculture need to be assessed differently in primarily rain-fed and primarily irrigated areas. In a semi-arid area, climate change is likely to affect agriculture in two distinct ways. One pathway is the direct effect of climate on crop growth. Another potential pathway is through the supply of water for irrigation. The main point is that local precipitation, as used by Mendelsohn et al. (1994), is not the right variable for irrigated areas, as water supplies are drawn from sometimes distant geographic areas

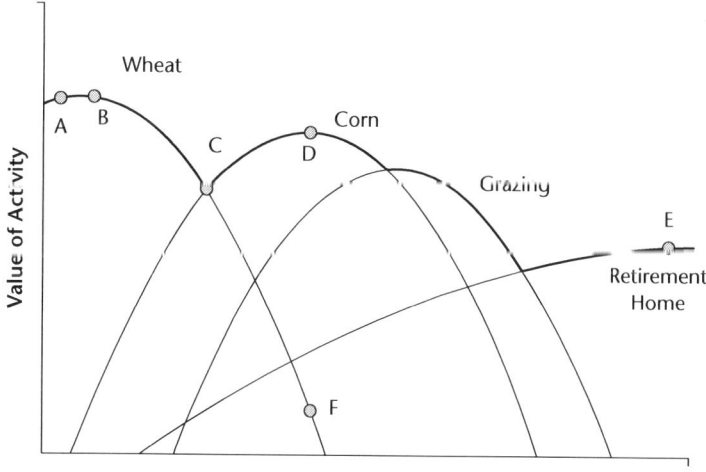

Figure 12.3 Hypothetical values of output in four different sectors as a function of temperature
Source: Mendelsohn, Nordhaus, and Shaw (1994; Figure 12.1, p. 754).

with differing precipitation and from aquifers. Due to this, they limit their analysis to rainfed areas, constituting of about two-thirds of U.S. counties, where water supplies are measured by local precipitation. Their results suggest that the economic effects of projected climate change on farmland values are unambiguously negative. In a subsequent study of the effects of irrigation water availability on farmland values in California, they find that water availability strongly capitalizes into farmland values and that the anticipated decrease in availability can be expected to have a large and significant negative impact on farmland values (Schlenker, Hanemann, and Fisher 2007).

2.4 Incomes, income distribution, and food security

Welfare, food security, and incomes can also be affected by climate change. For instance, climate change may adversely affect agricultural productivity, raising food prices and farm incomes but lowering food supplies and enhancing food insecurity (Butt et al. 2005). Moreover, it may affect poor people's livelihood through its effects on health, access to water and natural resources, and infrastructure. As stated in previous subsections, several empirical studies provide significant evidence that changes in climate can substantially affect agricultural output. However, there is considerable heterogeneity in outcomes based upon geographical location with higher latitude regions and larger countries, like the U.S., possibly increasing welfare (Malcolm et al. 2012). Furthermore, collective work shows the opposite effects of productivity on consumers and producers, with producers gaining and consumers losing when productivity is negatively affected and the opposite when it is positively affected (Reilly et al. 2002).

In a global setting, Ahmed, Diffenbaugh, and Hertel (2009) find that the most significant climate change impacts on poverty are likely to occur among urban wage laborers due to food price increases and that self-employed rural households are less affected because they benefit from higher prices. They also find that climate change exacerbates poverty vulnerability in many nations and that climate extremes exert stress on low-income populations. In terms of food security, Baldos and Hertel (2014) find that climate change is still a key driver of nutritional outcomes, particularly in regions where chronic malnutrition is prevalent.

3 Adaptation

Adaptation actions can be taken to reduce the negative impacts of climate change or to exploit opportunities. There is an inevitable need for agriculture to adapt given the future projected temperature path. In particular, as noted in Figure 12.1, in the next 25 years (Era 1), society faces about one degree of temperature increase regardless of mitigation efforts (see the explanation in McCarl 2015).

Adaptation can be natural or the result of human action. Natural adaptations are those undertaken by ecosystems, like changes in bird migrations or the mix of tree species at a site. Human adaptations are commonly classified as autonomous or planned adaptations (IPCC 2014a). Autonomous adaptation actions refer to those occurring "without directed intervention of public agency," adaptations (IPCC 2014a) which are generally private actions taken by decision makers in their own best interests. Planned adaptations refer to actions undertaken by public groups resolving public good, like market failures (McCarl, Thayer, and Jones 2016). Here we discuss a number of aspects of adaptation, including (1) observed adaptation actions, (2) benefits and costs, (3) constraints and economic barriers, and (4) adaptation deficits and maladaptation.

3.1 Observed adaptation actions

A number of studies have examined the types of agricultural adaptations to climate change that have been undertaken. Here we list some of the adaptations observed, grouped under crops and livestock.

3.1.1 Cropping

The most common agricultural adaptation involves crop selection and crop timing. In terms of crop timing, Sacks and Kucharik (2011) found that farmers are planting soybeans and corn earlier than before. The EPA indicates that "the average length of the growing season in the contiguous 48 states has increased by nearly two weeks since the beginning of the 20th century" (EPA 2013).

Park, McCarl, and Wu (2016) found in a U.S.-based econometric study that temperature and precipitation influence farmers' crop selection and that farmers alter crop mix in response to climate. In the coldest areas, barley, soybeans, and spring wheat dominate. As the temperature increases, soybeans and winter wheat enter and become dominant crops; then, with more temperature increase, we see the entry of upland cotton and sorghum, and finally rice is planted in the hottest areas.

Reilly et al. (2002) and Cho (2015) found that many crops in U.S. have shifted to higher latitudes and elevations as temperatures increased. Attavanich et al. (2013) found such a crop mix shifts as well and observed that this changes demand on the grain transportation system and mode usage.

Employing irrigation, managing water, and building infrastructure are also adaptations used in cropping (Howden et al. 2007), as are an increase in pesticide use (Chen and McCarl 2001) and a movement of land from cropping to pasture/range (Mu, McCarl, and Wein 2013).

3.1.2 Livestock

Climate change also causes producers to make shifts in the livestock species they raise, plus the breeds used and the management of the animals. In terms of species choice, Seo and Mendelsohn (2008) found that in Africa species incidence shifts with climate. They found farmers in warmer places chose goats and sheep instead of beef cattle, and farmers in wetter places shifted from cattle and sheep to goats and chickens. On breeds, Zhang, Hagerman, and McCarl (2013) found that more heat-tolerant cattle (*Bos indicus*) were selected to adapt to hotter conditions and as temperatures increase in Texas.

In terms of management, strategies like adjusting stocking rate, varying season of grazing, and altering pest management have been used (Joyce et al. 2013). Mu et al. (2013) found that cattle stocking rates are decreased, with reductions in precipitation and an increasing summer temperature–humidity index (THI). Gaughan et al. (2009) observed (1) adjustments in livestock and feed supply mix, including altered forages, (2) alteration of facilities like provision of shade, misting, and water, and (3) relocation of livestock among different regions. Howden et al. (2007) found that integrated crop and livestock production systems are employed to adapt to climate change.

3.2 Adaptation benefits and costs

Howden et al. (2007) argue adaption can increase economic benefits and reduce the impact due to climate change, in turn increasing profit and social welfare, improving food security, and providing valuable time for mitigation to work. A number of studies have addressed such benefits quantitatively:

- Shifts in crop mixes, livestock species, and livestock breeds were found to increase yield and decrease the costs of drought and heat waves (Adams et al. 1999; Howden et al. 2007; IPCC 2014a).
- Butt, McCarl, and Kergna (2006) found that crop management adaption (such as crop mixes and drought-resistant varieties) and increased international trade could substantially improve food security problems in Mali, as well as increase social welfare.
- Lobell et al. (2008) argued crop adaptation to climate change is necessary to enhance food security, while IPCC (2014a) indicates adaption could contribute to current and future economic growth plus distribute risk.
- Aisabokhae, McCarl, and Zhang (2011) found adaptation in crop planting dates and time to maturity were the most valuable of the variety of adaptations they studied.
- Crop variety shifts have been found to allow producers to adapt climate change under increased heat and drought frequency, due to crop yield change and altered pest populations (Yu and Babcock 2010; Malcolm et al. 2012).
- Mendelsohn and Dinar (1999) found farmers in developing countries could reduce the potential damages from climate change by up to one half through adaptation activities.
- Increases in technical progress have been found to be strategies to adapt to climate change, such as sea level induced losses in planted rice areas and yield losses (Chen et al. 2012).
- Integrated crop and livestock production systems have been found to be a valuable adaptation option that greatly reduces income fluctuation (Howden et al. 2007). A few studies have examined the cost of adaptation, considering fixed capital formation, infrastructure development, research and development, extension, transitional assistance, and trade policy. As part of the UNFCCC (2007) study, McCarl (2007) developed an estimate that global agricultural public funding needs were about US$14 billion per year. Parry et al. (2009) estimated a similar amount.

3.3 Adaptation constraints and economic barriers

A number of constraints impede adaptation implementation. According to IPCC (2014a), the main constraints include: (1) knowledge, awareness, and technology availability, (2) physical and biological limits, (3) economic and financial resources, (4) human capabilities, (5) social and cultural considerations, and (6) governance and institutions. Economic considerations pose another set of barriers. These include transaction costs, information and adjustment costs, market failure, and missing markets, etc. (IPCC 2014a).

3.4 Deficits, residual damages, and maladaptation

Adaptation is not only a concern that addresses future climate change but can also involve alterations to accommodate the current climate; in addition, there is the possibility that adaptation by one party can worsen the situation of another. Finally, adaptation may be unable to accommodate some climate change effects. This raises the issues of adaptation deficits, maladaptation, and residual damages.

Decision makers may face an adaptation deficit when current systems are not fully adapted to the current climate (Burton and May 2004). For example, a region facing more intense precipitation may not be well adapted to current heavy precipitation events and when pursuing, say, flood control actions, might correct both current and future issues.

Adaptation actions by one party may cause maladaptation (Barnett and O'Neill 2010). Namely, actions by one party may increase climate change vulnerability for: (1) the party doing the adaptation through perhaps poor implementation, poor project choice, or diverted resources; (2) parties somewhere else (for example, flood control in one place may worsen it elsewhere); and (3) parties in the future (exploitation of depletable groundwater now may worsen adaptation possibilities in the future). Note maladaptation actions may be rational if the benefits to the adapting party outweigh the losses to the maladapted policy in real terms.

Finally, in spite of adaptation measures, residual damages are likely to exist. Adaptation cannot totally overcome all climate change effects. For example, sea level rise and the loss of coastal property or species extinction may be practically irreversible.

4 Mitigation

Mitigation refers to a human intervention to reduce the extent of climate change by limiting climate change drivers. This mainly involves limiting greenhouse gas (GHG) emissions but can also involve geoengineering.

4.1 Greenhouse gas emissions sources

The IPCC (2014b) states that current GHG emissions mainly arise from: electricity and heat production; agricultural, forestry, and other land use; buildings; transport; and industry. Energy use is the largest contributor. Emissions from electricity and heat generation contributed 75 percent of the energy total, followed by 16 percent for fuel production and transmission, and 8 percent for petroleum refining (IEA 2012). The agricultural sector accounts for an estimated 56 percent of 2005 non-CO_2 emissions (EPA 2011; IPCC 2014b) and 24 percent of total GHG emissions (IPCC 2014b).

4.2 The mitigation imperative

The need for GHG-based mitigation can be highlighted by referring to Figure 12.1. There, the different RCP (essentially representing alternative atmospheric GHG concentrations) scenarios show mitigation makes a huge difference in the realized extent of climate change by the end of the century.

4.3 Potential agricultural role

Agriculture can play a role in GHG mitigation. McCarl and Schneider (2001) indicate there are four ways agriculture can be involved. First, agriculture is a GHG emitter releasing significant amounts of methane, nitrous oxide, and CO_2 (Smith et al. 2008; Smith et al. 2014) and can act to reduce those. This can involve reductions in emissions from livestock, equipment, nitrogen fertilizer, rice, and deforestation, just to mention the large ones. Second, agriculture can increase carbon stores in the ecosystem (sequestration) by using less intensive tillage, afforestation, avoiding deforestation, and the converting of crop lands to grass, among other items. Third, agriculture could provide substitutes for GHG emission intensive products by producing biomass to replace fossil energy or using wood in construction instead of concrete block and steel. Fourth, agriculture could do little direct action but still face higher fossil fuel prices (reflective of emissions control in the energy industry), which in turn stimulates agricultural emissions reductions through substitution for fossil fuels.

4.4 Policies, measures, and instruments

Climate change is largely an economic externality brought about by market failure. Pricing and regulatory approaches can be used in an attempt to reduce the magnitude of the externality. Economists generally favor sending a price signal to emitters to cut back emissions that is reflective of the externality costs. Several means may be used to send such a signal:

- Fuel taxes: one could tax fossil fuel use. Such taxes have low transaction costs and yield revenues that can be used to finance other mitigation policies (Schneider and McCarl 2005). Increased fossil fuel costs have been projected to reduce on-farm fuel usage plus increase bioenergy production (USDA 1999; Schneider and McCarl 2005).
- Subsidies: one can lower existing subsidies to fossil fuel production/use or increase subsidies for alternative practices reducing emissions.
- Cap-and-trade: one can reduce GHG emissions below current levels with cap-and-trade schemes that allocate permits to emitters at a level below current emissions and allow them to be traded. Require those who want to emit more to acquire additional permits (Tietenberg 2010) and allow high reduction cost emitters to buy emission rights from low reduction cost emitters. The initial allocation can involve schemes like an auction (Cramton and Kerr 2002).

Regulatory approaches towards GHG emission reductions include:

- Clean Power Plan: the EPA in response to a U.S. Supreme Court ruling on controlling GHGs announced the Clean Power Plan, which is aimed at reducing U.S. electric power carbon emissions to 32 percent below 2005 levels by 2030. Implementation is up to U.S. states, and the plan mentions the use of market-based mechanisms, including interstate cap-and-trade programs, carbon taxes, and tradable renewable certificates. The use of biomass is explicitly addressed, and states may use qualified biomass resources as a component of their state plan (EPA 2015). However, note that as of fall 2017 the Clean Power Plan will not be implemented.
- Paris Climate Agreement: the key elements of the Paris Agreement are: (1) limiting the global temperature increase to 1.5–2°C; (2) countries must employ transparent standardized accounting, monitoring, and verification; (3) developed countries agree to provide a $100 billion per year fund for adaptation and mitigation; (4) subsequent adaption of increasingly ambitious reduction targets. Some other notable features are:
    - As of December 2015, Intended Nationally Determined Contributions (INDCs) under the Paris Agreement represent 96 percent of global emissions as opposed to 14 percent under the Kyoto Protocol (UNFCCC 2015).
    - The 2015 U.S. commitment under the Paris Agreement involves reducing GHG emissions 26–28 percent below 2005 level by 2025, plus a commitment of $800 million per year to the fund (US Whitehouse Press Office 2015) although as of 2017 the US announced its intentions to withdraw from the agreement.
    - Specific mechanisms are up to individual countries.

- California Air Resources Board (CARB): the CARB program provides a limited number of tradable emission permits and sets up a trading system. It contains the Low Carbon Fuel Standard (CARB 2007), which is intended to reduce the carbon intensity of transportation fuel by at least 10 percent by 2020. For agriculture, there are protocols for livestock, forestry, and rice (CARB 2011).
- Renewable Fuel Standard: an element of the Energy Independence and Security Act of 2007 that in implementation requires certain volumes of renewable fuel, with the fuel in classes depending on their associated GHG emissions (EPA 2016).

4.5 Economic appraisal of the effects of agricultural mitigation actions

A number of economic studies have examined climate change mitigation in an agricultural setting; here we review some of the important findings.

- Agricultural emission reductions and offsets, e.g., agricultural soil carbon sequestration, can contribute net emission reductions at relatively low carbon prices (McCarl and Schneider 2001; Murray et al. 2005).
- Agricultural soil-based carbon sequestration can be competitive at low carbon prices (Murray et al. 2005). Studies have shown that the realized sequestration acreage is substantially lower than the total estimated potential, and the differences narrow as carbon price increases (McCarl and Schneider 2001).
- Agricultural mitigation efforts are competitive with food production. McCarl and Schneider (2001) find that there is a substantial decrease in food production and an increase in food prices with higher carbon prices.
- Among large potential agricultural mitigation contributions is the replacement of power plant coal-fired electricity generation with biomass-fueled electricity, but only at carbon prices above $50 per ton. Afforestation is another large potential contribution. At low carbon prices, changes in tillage and forest management dominate, but these have limited potential (McCarl and Schneider 2001).
- Generally the sequestration possibilities like tillage, land use change, afforestation, and forest management only accumulate additional carbon for a limited amount of time and then need to be maintained (Lee, McCarl, and Gillig 2005; Kim, McCarl, and Murray 2008).
- There are co-benefits from agricultural mitigation actions where, for example, sequestration activities can also create reductions in soil erosion, alter soil organic matter, increase soil water-holding capacity, increase wildlife populations, and lower fertilizer use and chemical runoff (Murray et al. 2005). However, evaluation of such co-benefits for a mitigation possibility implies you should also do this for a wide variety of competing possibilities and may impose too large a burden (Elbakidze and McCarl 2007).
- Competition with food production may reduce production, increase prices, and cause emissions to increase elsewhere – called leakage (Murray, McCarl, and Lee 2004).
- Mitigation activity stimulated by carbon price increases generally improves producers' welfare and decreases consumers' welfare (McCarl and Schneider 2001).

4.6 Challenges in implementing agricultural mitigation

Many mitigation actions can be identified; however, implementing them in agriculture introduces challenges. Here we discuss a few.

One key challenge involves coverage. Agriculture has generally been regarded as a sector that will not be capped in emissions but rather will be a producer of offsets (Murray et al. 2005). Rules that have emerged have generally treated agriculture on an item by item basis without coverage of many of the opportunities and with various rules concerning prior use of the practice. Some studies have shown that partial coverage, say, ignoring the non-CO_2 possibilities, causes a substantial price increase in the cost of controlling emissions (Fawcett and Sands 2006).

A second challenge involves heterogeneity and management interdependencies (McCarl and Schneider 2001). Agriculture is spatially diverse with respect to soils, climate conditions, and land management history. These three factors combined results in heterogeneous GHG mitigation potential, making uniform national policies a challenge (Antle et al. 2001), but transaction cost can be a major difficulty in localizing policies (Murray et al. 2005).

Third, interdependencies of crop and livestock management actions can interact to affect the costs and the potentials for agricultural GHG mitigations in four ways: (1) many agricultural mitigation strategies compete with traditional agricultural production, but some are complementary (i.e., cutting back on livestock herd size reduces methane emissions and those from feed production); (2) mitigation strategies impact one another (using land for bioenergy reduces its availability for afforestation); (3) mitigation strategies can have different effects across different GHGs (adding fertilizer increases the negative contribution through soil carbon sequestration but positively increases nitrous oxide emissions); and (4) management decisions aimed at GHG emission abatement may also impact other environmental properties, such as soil erosion and nutrient leaching. As a result, it is important to simultaneously consider the suite of agricultural greenhouse gas mitigation strategies (Murray et al. 2005).

5 Towards societal action

Society broadly observed is pursing both mitigation and adaptation. There are some elements of designing such an approach that merit discussion.

5.1 Mitigation and adaptation incentive program design

Programs are emerging that will fund mitigation and adaptation projects. In deciding what to fund, there are a number of issues that affect eventual project success. These include additionality, permanence, uncertainty, and leakage/maladaptation, as well as transactions cost (see the review of these in a mitigation context in Murray et al. 2005 and in an adaptation context in McCarl et al. 2016). A brief overview of the listed characteristics is given below:

- Additionality: in both contexts, it is desirable to fund projects that would not have been implemented in the absence of funding. In an adaptation context, there is an added meaning that the funding applied to adaptation be additional funds, not repurposed traditional development funds (see discussion in McCarl et al. 2016).
- Permanence: in the mitigation context, one worries about the persistence of the carbon offset, particularly for sequestered carbon, which cannot be guaranteed to be permanent and may be discounted (Kim et al. 2008). Similarly, in adaptation one worries about how long the adaptation will be effective and how its benefits evolve over time.
- Uncertainty: in both cases, one worries about the degree of uncertainty in benefits and may discount the benefits to reach a level one is confident in (see treatment in Kim and McCarl 2009).

- Leakage: mainly a mitigation context issue and arises when a project alters traditional commodities in the market place, raising their price and potentially stimulating increased emissions in other regions (Murray et al. 2004).
- Maladaptation: refers to cases where actions by one party reduce the adaptation of people elsewhere or in the future (see discussion in McCarl et al. 2016).
- Transactions costs refer to the costs of conveying the money to those that will implement the mitigation or adaptation projects plus monitor performance. As such, the evaluation at the project level may be misleading on costs and a wider set needs to be included (see mitigation-related discussion in Murray et al. 2005).

5.2 Mix of adaptation and mitigation

There is a need to simultaneously implement adaptation and mitigation. Mitigation will not prevent much climate change before mid-century, but requires substantial near-term effort to limit end of century climate change (IPCC 2007; 2014a; 2014b; NAS 2010). Also, in some regions, policy action to reduce emissions is progressing at a slow pace while emissions growth continues worldwide. As a consequence, adaptation is needed to reduce current impacts while mitigation is needed to affect the future.

However, the optimal combination of adaptation and mitigation is a challenging problem. First, one should consider interactions between adaptation strategies and mitigation potential. For example, shifting adaptation in the form of intensification may increase sequestration loss and fertilization-related emissions. Second, climate change impacts on agriculture will affect not only agricultural productivity, but also sequestration levels (Rosenzweig and Tubiello 2007).

Several studies have investigated the optimal mix of adaptation and mitigation. For example, de Bruin et al. (2009) found that adaptation dominated mitigation initially, while mitigation was predominant in later periods, as did Wang and McCarl (2013). On the other hand, Bosello et al. (2010) found that mitigation started immediately, while adaptation was delayed.

The discount rate is a key factor in all investment decisions, including the mitigation/adaptation mix (Nordhaus 2007). Most of the benefits from current mitigation policy efforts would take the form of avoided damages many years from now, whereas many of the costs of that policy would be borne in the nearer term (Goulder and Williams 2012). Adaptation tends to be more immediate. Thus, a high discount rate tends to reduce mitigation and favor adaptation (Wang and McCarl 2013).

The discount rate issue has varied substantially in studies. For example, the *Stern Review* (2007) applied a low discount rate of 1.4 percent and advocated substantial mitigation. Others argued that the *Review*'s rate was inappropriately low (Mendelsohn 2008), and Nordhaus (2007) found considerably less mitigation to be appropriate when using a higher discount rate.

6 Concluding comments

Climate change is certainly an agricultural issue portending negative effects in some regions and positive effects in others, including implications for crops, livestock, pests, and their variability, with a changing environment for all. Society will inevitably need to adapt and pursue mitigation to reduce the future implications and extent of climate change. Agricultural economists will need to be well engaged in the issue, looking at and evaluating vulnerability and appropriate levels of adaptation and mitigation, as well as designing effective policy approaches and project evaluation procedures.

References

Adams, R.M., R.A. Fleming, C.-C. Chang, B.A. McCarl, and C. Rosenzweig. 1995. "A Reassessment of the Economic Effects of Global Climate Change on U.S. Agriculture." *Climatic Change* 30(2):147–167.

Adams, R.M., B. McCarl, K. Segerson, C. Rosenzweig, K. Bryant, B. Dixon, R. Conner, R. Evenson, and D. Ojima. 1999. "Economic Effects of Climate Change on US Agriculture." In R. Mendelsohn and J.E. Neumann, eds., *The Impact of Climate Change on the US Economy*. Cambridge: Cambridge University Press, pp. 18–54.

Adams, R.M., C. Rosenzweig, R.M. Peart, J.T. Ritchie, B.A. McCarl, J.D. Glyer, R.B. Curry, J.W. Jones, K.J. Boote, and L.H. Allen. 1990. "Global Climate Change and US Agriculture." *Nature* 345(6272):219–224.

Ahmed, S.A., N.S. Diffenbaugh, and T.W. Hertel. 2009. "Climate Volatility Deepens Poverty Vulnerability in Developing Countries." *Environmental Research Letters* 4(3):034004.

Aisabokhae, R., B.A. McCarl, and Y.W. Zhang. 2011. "Agricultural Adaptation: Needs, Findings and Effects." In R. Mendelsohn and A. Dinar, eds., *Handbook on Climate Change and Agriculture*. Northampton, MA: Edward Elgar, pp. 327–341.

Antle, J.M., S.M. Capalbo, S. Mooney, E.T. Elliott, and K.H. Paustian. 2001. "Economic Analysis of Agricultural Soil Carbon Sequestration: An Integrated Assessment Approach." *Journal of Agricultural and Resource Economics* 26(2):344–367.

Arrow, K.J., M.L. Cropper, C. Gollier, B. Groom, G.M. Heal, R.G. Newell, W.D. Nordhaus, R.S. Pindyck, W.A. Pizer, P.R. Portney, T. Sterner, R.S.J. Tol, and M.L. Weitzman. 2014. "Should Governments Use a Declining Discount Rate in Project Analysis?" *Review of Environmental Economics and Policy* 8(2):145–163.

Attavanich, W., and B.A. McCarl. 2014. "How is CO2 Affecting Yields and Technological Progress? A Statistical Analysis." *Climatic Change* 124(4):747–762.

Attavanich, W., B.A. McCarl, Z. Ahmedov, S.W. Fuller, and D.V. Vedenov. 2013. "Effects of Climate Change on US Grain Transport." *Nature Climate Change* 3(7):638–643.

Baldos, U.L.C., and T.W. Hertel. 2014. "Global Food Security in 2050: The Role of Agricultural Productivity and Climate Change." *Australian Journal of Agricultural and Resource Economics* 58(4):554–570.

Barnett, J., and S. O'Neill. 2010. "Maladaptation." *Global Environmental Change* 20(2):211–213.

Bosello, F., C. Carraro, and E. De Cian. 2010. "Climate Policy and the Optimal Balance Between Mitigation, Adaptation and Unavoided Damage." *Climate Change Economics* 1(2):71–92.

de Bruin, K.C., R.B. Dellink, and R.S.J. Tol. 2009. "AD-DICE: An Implementation of Adaptation in the DICE Model." *Climatic Change* 95(1–2):63–81.

Burton, I., and E. May. 2004. "The Adaptation Deficit in Water Resource Management." *IDS Bulletin* 35(3):31–37.

Butt, T.A., B.A. McCarl, J. Angerer, P.T. Dyke, and J.W. Stuth. 2005. "The Economic and Food Security Implications of Climate Change in Mali." *Climatic Change* 68(3):355–378.

Butt, T.A., B.A. McCarl, and A.O. Kergna. 2006. "Policies for Reducing Agricultural Sector Vulnerability to Climate Change in Mali." *Climate Policy* 5(6):583–598.

California Air Resources Board (CARB). 2007. "Low Carbon Fuel Standard." www.arb.ca.gov/fuels/lcfs/lcfs.htm. Accessed July 27, 2016.

———. 2011. *Compliance Offset Program*. www.arb.ca.gov/cc/capandtrade/offsets/offsets.htm. Accessed July 27, 2016.

Chen, C.C., and B.A. McCarl. 2001. "An Investigation of the Relationship Between Pesticide Usage and Climate Change." *Climatic Change* 50(4):475–487.

———. 2009. "Hurricanes and Possible Intensity Increases: Effects on and Reactions from US Agriculture." *Journal of Agricultural and Applied Economics* 41:125–144.

Chen, C.C., B.A. McCarl, and R.M. Adams. 2001. "Economic Implications of Potential ENSO Frequency and Strength Shifts." *Climatic Change* 49(1):147–159.

Chen, C.C., B.A. McCarl, and C.C. Chang. 2012. "Climate Change, Sea Level Rise and Rice: Global Market Implications." *Climatic Change* 110(3–4):543–560.

Cho, S. 2015. *Three Essays on Climate Change Adaptation and Impacts: Econometric Investigations*. PhD dissertation, Texas A&M University.

Cline, W.R. 1992. *The Economics of Global Warming*. Washington, DC: Institute for International Economics.

Craine, J.M., A.J. Elmore, K.C. Olson, and D. Tolleson. 2010. "Climate Change and Cattle Nutritional Stress." *Global Change Biology* 16(10):2901–2911.

Cramton, P., and S. Kerr. 2002. "Tradeable Carbon Permit Auctions: How and Why to Auction Not Grandfather." *Energy Policy* 30(4):333–345.

Elbakidze, L., and B.A. McCarl. 2007. "Sequestration Offsets Versus Direct Emission Reductions: Consideration of Environmental Co-Effects." *Ecological Economics* 60(3):564–571.

EPA. 2011. *Draft: Global Anthropogenic Non-CO_2 Greenhouse Gas Emissions: 1990–2030*. https://www3.epa.gov/climatechange/Downloads/EPAactivities/EPA_NonCO2_Projections_2011_draft.pdf. Accessed July 27, 2016.

———. 2013. *Climate Change Indicators in the United States: Length of Growing Season*. www.epa.gov/climatechange/indicators. Accessed July 27, 2016.

———. 2015. *Fact Sheet: Overview of the Clean Power Plan*. www.epa.gov/cleanpowerplan/fact-sheet-overview-clean-power-plan. Accessed July 27, 2016.

———. 2016. *The Renewable Fuel Standard*. www.epa.gov/renewable-fuel-standard-program. Accessed April 10, 2018.

Fawcett, A.A., and R.D. Sands. 2006. "Non-CO_2 Greenhouse Gases in the Second Generation Model." *The Energy Journal* SI2006(01):305–322.

Feng, S., B.A. McCarl, and P. Havlik. 2011. "Crop Yield Growth and Its Implication for the International Effects of US Bioenergy and Climate Policies." Paper presented at the 2011 Agricultural & Applied Economics Association (AAEA) Annual Meeting, Pittsburgh, Pennsylvania.

Fisher, A.C., and P.V. Le. 2014. "Climate Policy: Science, Economics, and Extremes." *Review of Environmental Economics and Policy* 8(2):307–327.

Gale, P., T. Drew, L.P. Phipps, G. David, and M. Wooldridge. 2009. "The Effect of Climate Change on the Occurrence and Prevalence of Livestock Diseases in Great Britain: A Review." *Journal of Applied Microbiology* 106(5):1409–1423.

Gaughan, J., N. Lacetera, S.E. Valtorta, H.H. Khalifa, L. Hahn, and T. Mader. 2009. "Response of Domestic Animals to Climate Challenges." In K.L. Ebi, I. Burton, and G.R. McGregor, eds., *Biometeorology for Adaptation to Climate Variability and Change*. Dordrecht: Springer Netherlands, pp. 131–170. http://link.springer.com/10.1007/978-1-4020-8921-3_7. Accessed July 27, 2016.

German Climate Computing Center (DKRZ). 2014. "Laboratory for Climate Researchers." www.dkrz.de/Klimaforschung-en/konsortial-en/ipcc-ar5/ergebnisse/Mitteltemperatur-en. Accessed July 27, 2016.

Goulder, L.H., and R.C. Williams. 2012. "The Choice of Discount Rate for Climate Change Policy Evaluation." *Climate Change Economics* 3(4):1250024.

Hatfield, J., G. Takle, P. Holden, R.C. Izaurralde, T. Mader, E. Marshall, and D. Liverman. 2014. "Agriculture. Climate Change Impacts in the United States." In J.M. Melillo, T.C Richmond, and G.W. Yohe, eds., *The Third National Climate Assessment*. U.S.: Global Change Research Program, pp. 150–174. Washington, DC: Government Printing Office.

Heal, G. 2008. "Climate Economics: A Meta-Review and Some Suggestions for Future Research." *Review of Environmental Economics and Policy* 3(1):4–21.

Howden, S.M., S.J. Crimp, and C.J. Stokes. 2008. "Climate Change and Australian Livestock Systems: Impacts, Research and Policy Issues." *Australian Journal of Experimental Agriculture* 48(7):780.

Howden, S.M., J.-F. Soussana, F.N. Tubiello, N. Chhetri, M. Dunlop, and H. Meinke. 2007. "Adapting Agriculture to Climate Change." *Proceedings of the National Academy of Sciences* 104(50):19691–19696.

International Energy Agency (IEA). 2012. "CO2 Emission from Fuel Combustion." In *OECD*, Pairs, France. www.oecd-ilibrary.org/docserver/download/6112181e.pdf?expires=1469659529&id=id&accname=ocid76019061&checksum=6B7655E61C4F389860D7B0FFFD5E28C6. Accessed July 27, 2016.

IPCC. 2007. *Climate Change 2007: Mitigation of Climate Change*. Cambridge: Cambridge University Press.

———. 2013. *Climate Change 2013: The Physical Science Basis*. Cambridge and New York: Cambridge University Press.

———. 2014a. *Climate Change 2014: Impacts, Adaptation, and Vulnerability*. Cambridge and New York: Cambridge University Press.

———. 2014b. *Climate Change 2014: Synthesis Report*. Contribution of Working Groups I, II to the Fifth Assessment Report of the Intergovernmental Panel on Climate Change, Geneva, Switzerland.

Joughin, I., B.E. Smith, and B. Medley. 2014. "Marine Ice Sheet Collapse Potentially Under Way for the Thwaites Glacier Basin, West Antarctica." *Science* 344(6185):735–738.

Joyce, L.A., D.D. Briske, J.R. Brown, H.W. Polley, B.A. McCarl, and D.W. Bailey. 2013. "Climate Change and North American Rangelands: Assessment of Mitigation and Adaptation Strategies." *Rangeland Ecology & Management* 66(5):512–528.

Juroszek, P., and A. Von Tiedemann. 2013. "Plant Pathogens, Insect Pests and Weeds in a Changing Global Climate: A Review of Approaches, Challenges, Research Gaps, Key Studies and Concepts." *Journal of Agricultural Science* 151(2):163–188.

Khan, S.A., K.H. Kjær, M. Bevis, J.L. Bamber, J. Wahr, K.K. Kjeldsen, Anders A. Bjørk, Niels J. Korsgaard, L.A. Stearns, M.R. van den Broeke, L. Liu, N.K. Larsen, and I.S. Muresan. 2014. "Sustained Mass Loss of The Northeast Greenland Ice Sheet Triggered by Regional Warming." *Nature Climate Change* 4(4):292–299.

Kim, M.-K., and B.A. McCarl. 2009. "Uncertainty Discounting for Land-Based Carbon Sequestration." *Journal of Agricultural and Applied Economics* 41(1):1–11.

Kim, M.-K., B.A. McCarl, and B.C. Murray. 2008. "Permanence Discounting for Land-based Carbon Sequestration." *Ecological Economics* 64(4):763–769.

Knutti, R., and J. Sedláček. 2012. "Robustness and Uncertainties in the New CMIP5 Climate Model Projections." *Nature Climate Change* 3(4):369–373.

Lee, H.-C., B.A. McCarl, and D. Gillig. 2005. "The Dynamic Competitiveness of U.S. Agricultural and Forest Carbon Sequestration." *Canadian Journal of Agricultural Economics* 53(4):343–357.

Lobell, D.B., M.B. Burke, C. Tebaldi, M.D. Mastrandrea, W.P. Falcon, and R.L. Naylor. 2008. "Prioritizing Climate Change Adaptation Needs for Food Security in 2030." *Science* 319(5863):607–610.

Mader, T.L. 2014. "Animal Welfare Concerns for Cattle Exposed to Adverse Environmental Conditions." *Journal of Animal Science* 92(12):5319–5324.

Mader, T.L., K.L. Frank, J.A. Harrington, G.L. Hahn, and J.A. Nienaber. 2009. "Potential Climate Change Effects on Warm-Season Livestock Production in the Great Plains." *Climatic Change* 97(3–4):529–541.

Malcolm, S., E. Marshall, M. Aillery, P. Heisey, M. Livingston, and K. Day-Rubenstein. 2012. "Agricultural Adaptation to a Changing Climate: Economic and Environmental Impacts Vary by U.S. Region." In *U.S. Department of Agriculture, Economic Research Service. Economic Research Report No. 136*. July.

McCarl, B.A. 2007. "Adaptation Options for Agriculture, Forestry and Fisheries. A Report to the UNFCCC Secretariat Financial and Technical Support Division." https://unfccc.int/files/cooperation_and_support/financial_mechanism/application/pdf/mccarl.pdf. Accessed July 27, 2016.

———. 2015. "Elaborations on Climate Adaptation in US Agriculture." *Choices* 30(2).

McCarl, B.A., and U.A. Schneider. 2001. "Climate Change: Greenhouse Gas Mitigation in U.S. Agriculture and Forestry." *Science* 294(5551):2481–2482.

McCarl, B.A., A. Thayer, and J.P. Jones. 2016. "The Challenge of Climate Change Adaptation for Agriculture: An Economically Oriented Review." *Journal of Agricultural and Applied Economics* 48(4):321–344. doi:10.1017/aae.2016.27.

McCarl, B.A., X. Villavicencio, and X. Wu. 2008. "Climate Change and Future Analysis: Is Stationarity Dying?" *American Journal of Agricultural Economics* 90(5):1241–1247.

Mendelsohn, R. 2008. "Is the Stern Review an Economic Analysis?" *Review of Environmental Economics and Policy* 2(1):45–60.

Mendelsohn, R., and A. Dinar. 1999. "Climate Change, Agriculture, and Developing Countries: Does Adaptation Matter?" *The World Bank Research Observer* 14(2):277–293.

Mendelsohn, R., W.D. Nordhaus, and D. Shaw. 1994. "The Impact of Global Warming on Agriculture: A Ricardian Analysis." *The American Economic Review* 84(4):753–771.

Mu, J.E., B.A. McCarl, and A.M. Wein. 2013. "Adaptation to Climate Change: Changes in Farmland Use and Stocking Rate in the U.S." *Mitigation and Adaptation Strategies for Global Change* 18(6):713–730.

Mu, J., B.A. McCarl, X. Wu, and M.P. Ward. 2014. "Climate Change and the Risk of Highly Pathogenic Avian Influenza Outbreaks in Birds." *British Journal of Environment and Climate Change* 4(2):166–185.

Murray, B., B. Sohngen, A. Sommer, B. Depro, K. Jones, B.A. McCarl, D. Gillig, B. de Angelo, and K. Andrasko. 2005. "Greenhouse Gas Mitigation Potential in US Forestry and Agriculture." In *EPA Report 430-R-05-006*, November.

Murray, B.C., B.A. McCarl, and H.-C. Lee. 2004. "Estimating Leakage from Forest Carbon Sequestration Programs." *Land Economics* 80(1):109.

NAS. 2010. *America's Climate Choices: Advancing the Science of Climate Change*. Chaired by P.A. Matson. Washington, DC: National Academy of Sciences.

NOAA. 2016. *The NOAA Annual Greenhouse Gas Index (AGGI)*. Earth System Research Laboratory. http://esrl.noaa.gov/gmd/aggi/aggi.html. Accessed July 27, 2016.

Nordhaus, W.D. 2007. "A Review of the Stern Review on the Economics of Climate Change." *Journal of Economic Literature* XLV:686–702.

Park, J., B.A. McCarl, and X. Wu. 2016. *The Effects of Climate on Crop Mix and Climate Change Adaptation*. Texas A&M University, unpublished.

Parry, M., N. Arnell, P. Berry, D. Dodman, and S. Fankhauser (editors). 2009. *Assessing the Costs of Adaption to Climate Change: A Review of the UNFCCC and Other Recent Estimates*. London: International Institute for Environment and Development (IIED).

Peterson, T.C., R.R. Heim, R. Hirsch, D.P. Kaiser, H. Brooks, N.S. Diffenbaugh, R.M. Dole, J.P. Giovannettone, K. Guirguis, T.R. Karl, R.W. Katz, K. Kunkel, D. Lettenmaier, G.J. McCabe, C.J. Paciorek, K.R. Ryberg, S. Schubert, V.B.S. Silva, B.C. Stewart, A.V. Vecchia, G. Villarini, R.S. Vose, J. Walsh, M. Wehner, D. Wolock, K. Wolter, C.A. Woodhouse, and D. Wuebbles. 2013. "Monitoring and Understanding Changes in Heat Waves, Cold Waves, Floods, and Droughts in the United States: State of Knowledge." *Bulletin of the American Meteorological Society* 94(6):821–834.

Poore, R.Z., R.S.J. Williams, and C. Tracey. 2000. "Sea Level and Climate: U.S. Geological Survey Fact Sheet 002–000." http://pubs.usgs.gov/fs/fs2-00/. Accessed July 27, 2016.

Porter, J.R., and M.A. Semenov. 2005. "Crop Responses to Climatic Variation." *Philosophical Transactions of the Royal Society B: Biological Sciences* 360(1463):2021–2035.

Porter, J.R., L. Xie, A.J. Challinor, K. Cochrane, S.M. Howden, M.M. Iqbal, D.B. Lobell, and M.I. Travasso. 2014. "Food Security and Food Production Systems." *IPCC 2014 a Climate Change 2014: Impacts, Adaptation, and Vulnerability. Part A: Global and Sectoral Aspects*. Cambridge and New York: Cambridge University Press, pp. 485–533.

Reilly, J.M., J. Hrubovcak, J. Graham, D.G. Abler, R. Darwin, S.E. Hollinger, R.C. Izaurralde, S. Jagtap, J.W. Jones, J. Kimble, B.A. McCarl, L.O. Mearns, D.S. Ojima, E.A. Paul, K. Paustian, S.J. Riha, N.J. Rosenberg, C. Rosenzweig, and F.N. Tubiello. 2002. "Changing Climate and Changing Agriculture: Report of the Agricultural Sector Assessment Team, US National Assessment." *USGCRP National Assessment of Climate Variability*. Cambridge University Press.

Rosenzweig, C., and F.N. Tubiello. 2007. "Adaptation and Mitigation Strategies in Agriculture: An Analysis of Potential Synergies." *Mitigation and Adaptation Strategies for Global Change* 12(5):855–873.

Sacks, W.J., and C.J. Kucharik. 2011. "Crop Management and Phenology Trends in the U.S. Corn Belt: Impacts on Yields, Evapotranspiration and Energy Balance." *Agricultural and Forest Meteorology* 151(7):882–894.

Schlenker, W., W.M. Hanemann, and A.C. Fisher. 2005. "Will U.S. Agriculture Really Benefit from Global Warming? Accounting for Irrigation in the Hedonic Approach." *American Economic Review* 95(1):395–406.

———. 2006. "The Impact of Global Warming on U.S. Agriculture: An Econometric Analysis of Optimal Growing Conditions." *Review of Economics and Statistics* 88(1):113–125.

———. 2007. "Water Availability, Degree Days, and the Potential Impact of Climate Change on Irrigated Agriculture in California." *Climatic Change* 81(1):19–38.

Schlenker, W., and M.J. Roberts. 2009. "Nonlinear Temperature Effects Indicate Severe Damages to U.S. Crop Yields Under Climate Change." *Proceedings of the National Academy of Sciences* 106(37):15594–15598.

Schneider, U.A., and B.A. McCarl. 2005. "Implications of a Carbon Based Energy Tax for US Agriculture." *Agricultural and Resource Economics Review* 34:265–279.

Seo, S.N., and R. Mendelsohn. 2008. "Measuring Impacts and Adaptations to Climate Change: A Structural Ricardian Model of African Livestock Management." *Agricultural Economics* 38(2):151–165.

Shakhova, N., I. Semiletov, I. Leifer, V. Sergienko, A. Salyuk, D. Kosmach, D. Chernykh, C. Stubbs, D. Nicolsky, V. Tumskoy, and Ö. Gustafsson. 2013. "Ebullition and Storm-Induced Methane Release from the East Siberian Arctic Shelf." *Nature Geoscience* 7(1):64–70.

Shakhramanyan, N.G., U.A. Schneider, and B.A. McCarl. 2013. "US Agricultural Sector Analysis on Pesticide Externalities – the Impact of Climate Change and a Pigovian Tax." *Climatic Change* 117(4):711–723.

Smith, P., M Bustamante, H. Ahammad, H. Clark, H. Dong, E.A. Elsiddig, H. Haberl, R. Harper, J. House, M. Jafari, O. Masera, C. Mbow, N.H. Ravindranath, C.W. Rice, R. Abad, A. Romanovskaya, F. Sperling, and F.N. Tubiello. 2014. "Agriculture, Forestry and Other Land Use (AFOLU)." *IPCC, 2014: Climate Change 2014: Mitigation of Climate Change*. New York: Cambridge Press.

Smith, P., D. Martino, Z. Cai, D. Gwary, H. Janzen, P. Kumar, B. McCarl, S. Ogle, F. O'Mara, C. Rice, B. Scholes, O. Sirotenko, M. Howden, T. McAllister, G. Pan, V. Romanenkov, U. Schneider, S. Towprayoon, M. Wattenbach, and J. Smith. 2008. "Greenhouse Gas Mitigation in Agriculture." *Philosophical Transactions of the Royal Society B: Biological Sciences* 363(1492):789–813.

Stern, N. 2007. *The Economics of Climate Change: The Stern Review*. Cambridge and New York: Cambridge University Press.

St-Pierre, N.R., B. Cobanov, and G. Schnitkey. 2003. "Economic Losses from Heat Stress by US Livestock Industries." *Journal of Dairy Science* 86:E52–E77.

Tietenberg, T. 2010. "Cap-and-Trade: The Evolution of an Economic Idea." *Agricultural and Resource Economics Review* 39(3):359–367.

United Nations Environment Program. 2012. *Policy Implications of Warming Permafrost*. Nairobi, Kenya: United Nations Environment Program.

United Nations Framework Convention on Climate Change (UNFCCC). 2007. *Investment and Financial Flows to Address Climate Change*. http://unfccc.int/resource/docs/publications/financial_flows.pdf. Accessed July 27, 2016.

———. 2015. *Paris Climate Agreement*. http://unfccc.int/paris_agreement/items/9485.php. Accessed July 27, 2016.

United States Department of Agriculture (USDA). 1999. *Economic Analysis of U.S. Agriculture and the Kyoto Protocol*. Office of The Chief Economist, Global Change Program Office.

US Whitehouse Press Office. 2015. *Fact Sheet: U.S. Reports its 2025 Emissions Target to the UNFCCC*. www.whitehouse.gov/the-press-office/2015/03/31/fact-sheet-us-reports-its-2025-emissions-target-unfccc. Accessed July 27, 2016.

Villavicencio, X., B.A. McCarl, X. Wu, and W.E. Huffman. 2013. "Climate Change Influences on Agricultural Research Productivity." *Climatic Change* 119(3–4):815–824.

Wang, W., and B.A. McCarl. 2013. "Temporal Investment in Climate Change Adaptation and Mitigation." *Climate Change Economics* 04(02):1350009.

Whiteman, G., C. Hope, and P. Wadhams. 2013. "Climate Science: Vast Costs of Arctic Change." *Nature* 499(7459):401–403.

Yu, T., and B.A. Babcock. 2010. "Are U.S. Corn and Soybeans Becoming More Drought Tolerant?" *American Journal of Agricultural Economics* 92(5):1310–1323.

Zhang, Y.W., A.D. Hagerman, and B.A. McCarl. 2013. "Influence of Climate Factors on Spatial Distribution of Texas Cattle Breeds." *Climatic Change* 118(2):183–195.

13
Habitat conservation in agricultural landscapes

Chad Lawley and Charles Towe

1 Introduction

Continued habitat conversion to more intensive uses, such as agricultural production and residential and commercial development, is a pervasive feature of agricultural landscapes. Conservation groups – including governments, private organizations, or partnerships of both – are advocates for the flora and fauna that have existed for millennia in these landscapes. The scarcity of productive land resources and competing interests for these lands brings a role for economists. The challenges regarding land use decisions, the government's role, and private citizens' rights have been significant for decades; the rising specter of an evolving meteorological climate that is shifting habitat and altering productivity and a fiscal climate that demands more effective conservation are daunting.

This chapter focuses on habitat conservation in agricultural landscapes in Canada and the United States. Land in agricultural landscapes tends to be privately held and opportunities to extract value from maintenance of natural habitat is limited. For example, in Canada and the United States, waterfowl are considered public property rather than the property of the landowner on whose land they temporarily reside (Porter and van Kooten 1993). This creates a conflict between public interests that seek to increase provision of habitat and private interests that seek to maximize the productive value of private land.[1] One option to reconcile this conflict is direct regulation of land use, for instance in the form of restrictions on habitat conversion such as wetland drainage and conversion of grasslands to annual crops. There is, however, substantial landowner resistance to these types of direct regulations due to their limitations on the use of land and their expected impact on land values. Historically, jurisdictions have had difficulty – both in economic and political terms – imposing and enforcing direct regulations on the use of agricultural land (Lichtenberg and Smith-Ramirez 2011; Lawley and Towe 2014).

More flexible approaches to providing financial incentives to landowners have emerged in the absence of direct restrictions. In the context of agricultural land use, two prominent examples include fee simple acquisition of farmland and conservation easements on habitat within working farmland. Whereas fee simple acquisition involves direct transfer of ownership of a parcel, a conservation easement is an agreement between a landowner and conservation agency in which the landowner agrees to conserve existing or restored habitat in exchange for compensation from the conservation agency. Parker (2004) presents an economic model describing the conservation agency's choice between conservation easements and fee simple acquisition. Conservation easements remove conversion rights from the property but permit the landowner to continue current productive uses. Assuming landowners have a comparative advantage in farming, the conservation easement creates a setting wherein gains due to specialization can be captured. Monitoring and enforcement of easement restrictions increases transaction costs. Fee simple conservation, in contrast, transfers full ownership of the property to the conservation agency. This reduces monitoring and enforcement costs, and the conservation agency assumes responsibility to manage and maintain the property.

Both private and government agencies protect habitat in Canada and the United States. Historically, private conservation agencies tended to protect habitat through long-term fee simple acquisition and subsequent management of the acquired property. Private conservation agencies now employ a mix of fee simple acquisitions and conservation easements. Similar to fee simple acquisition, conservation easements taken on by private conservation agencies have tended to contract for permanent restrictions on the use of conserved habitat. Government conservation, by contrast, has tended to involve temporary easements. In this chapter, we review the economics of fee simple acquisitions and conservation easements within agricultural landscapes in Canada and the United States. Our review does not cover the large literature on agricultural preservation programs that seek to decelerate the development of farmland for urban development as reviewed in McConnell and Walls (2005). Furthermore, our review does not cover the extensive literature on habitat conservation in the developing world as reviewed in Alix-Garcia and Wolff (2014). Finally, we focus on compensation schemes and do not directly address endangered species legislation that targets species preservation, such as the Endangered Species Act in the U.S. or the Species at Risk Act in Canada, as discussed in Adamowicz (2016). Innes and Frisvold (2009) provide a comprehensive survey of the literature documenting the economics of endangered species, including the role of landowner compensation.

We start with a discussion of the history of habitat conservation programs within the two countries and a discussion of the evolution of conservation objectives over time. Next, we discuss several issues of design and implementation that are common to both fee simple acquisition and conservation easements. The literature examining conservation program design and implementation has developed theoretical frameworks to analyze many of the important features of the programs, including optimal targeting, contract length, and incentives for agglomeration of conservation activity. Simulation exercises focused on targeting conservation resources are common in this literature. Following discussion of program design and implementation, we discuss economic evaluations of habitat conservation programs, which are comprised primarily of econometric studies examining slippage, additionality, land market spillovers, and spatial interactions in habitat conservation.

We then provide some direction for future research utilizing the fast-evolving universe of large datasets, such as those derived from satellite data sources that are being processed at ever more granular scales both in space and time. For instance, new data sources from private micro-satellites and nano-satellites are coming online, and society has a phenomenal amount of untapped handheld computing and data-gathering capacity at its fingertips. Furthermore, and perhaps a result of these forces, there is the seemingly unprecedented impetus to cross-collaborate with the remote sensing community, private companies, non-governmental groups, environmental organizations, and social scientists. We conclude with a discussion of the potential for newly available data not yet the focus of our profession and the potential these sources have to refine evaluation of habitat conservation programs.

2 Brief history of habitat conservation in agricultural landscapes in Canada and the United States

Efforts to conserve natural areas in Canada and the United States date back to early efforts to establish national parks, forests, and wilderness areas, starting in 1872 with the creation of Yellowstone National Park in the United States and in 1885 with the creation of Banff National Park in Canada. These early initiatives to protect natural areas were motivated primarily by a desire to protect scenic locations and to encourage tourism and recreation (Newmark 1985). Attempts to protect land in agricultural landscapes first emerged in the United States in the 1930s. The 1936 Soil Conservation and Domestic Allotment Act provided financial support for farmers to switch to soil-conserving crops such as grasses, legumes, and cover crops. This Act is consistent with several measures in agricultural policy to control the supply of agricultural commodities with the intention of increasing producer surplus. The Agriculture Act of 1956 introduced a soil bank program, which paid farmers to remove land from production under a ten-year contract. The environmental benefits of early conservation programs in agriculture were limited to efforts to reduce soil erosion, with the objective of improving on-farm productivity (Cain and Lovejoy 2004).

The first efforts to directly target conservation emerged in the United States with the 1985 Farm Bill, which included the Conservation Reserve Program (CRP) and cross-compliance programs, with the stated objective of conserving highly erosive and/or biologically sensitive land to reduce off-farm threats to the environment (Cain and Lovejoy 2004). The cross-compliance measures linked receipt of farm program benefits to the preservation of existing natural grasslands and wetlands. Conservation programming under the mandate of the United States Department of Agriculture (USDA) continued to evolve in the 1990s as the CRP started to explicitly target environmental benefits using an environmental benefits index that placed greater weight on maintenance and restoration of wildlife habitat. The Wetland Reserve Program (WRP) was first introduced in the Food, Agriculture, Conservation, and Trade Act of 1990, with the goal of enrolling one million acres in wetland restoration projects (Parks and Kramer 1995). The Wildlife Habitat Incentive Program (WHIP) was introduced in 1996 to assist landowners wanting to increase habitat on their lands and received a substantial increase in funding in 2002 (Cain and Lovejoy 2004) but was repealed in 2014.

Canada is a small country in an international trade context, and supply restrictions have minimal potential impact on commodity prices. It is therefore not surprising that agricultural land retirement or set-aside programs have taken on less importance in Canada relative to the United States. The Canadian federal government introduced the Permanent Cover Program (PCP) in 1989, with the objective of reducing soil erosion and government program expenditures on marginal land. The PCP targeted marginal annual cropland to convert to perennial forages or tree cover for 10- or 21-year contracts. In exchange for conversion to permanent cover, landowners were provided a one-time payment that included a fixed seeding payment to establish perennial cover and an additional payment based on the market value of similar land and the length of the contract. The PCP enrolled more than 1.2 million acres, and just under two-thirds of the contracts were for 21 years (Vaisey, Weins, and Wetlaufer 1996). While there has been substantial research into the economics of the CRP, there has been very little economic analysis of the PCP and no overall evaluation of the program.[2]

Protection of waterfowl habitat has played an important role in habitat conservation in the agricultural landscapes of Canada and the U.S. This has evolved to include both government and private conservation agency participation. Refuge lands were first secured in 1929 by the Bureau of the Biological Survey, followed by the introduction of duck stamps and taxes on arms and munitions in the mid-1930s to fund land acquisitions (Porter and van Kooten 1993). The North American Waterfowl Management Plan (NAWMP) emerged in the 1980s as a response to declines in waterfowl populations. The Canadian and U.S. governments signed the NAWMP in 1986 with the goal of restoring wetlands and their associated upland habitat to meet habitat objectives and population goals for several waterfowl species (Nichols, Johnson, and Williams 1995). The ultimate goal of the NAWMP is to restore populations to their mid-1970s levels (Porter and van Kooten 1993). Initially, the NAWMP committed US$1 billion over 15 years (with the United States responsible for 75 percent) to encourage landowners and farmers to set aside agricultural land for wetlands and upland habitat (van Kooten 1993a). Public–private joint partnerships were formed out of the NAWMP, the largest of which is the Prairie Habitat Joint Venture (PHJV) in the western Canadian Prairie Provinces. The PHJV initially secured U.S. funds through the U.S. Fish and Wildlife Service (FWS) and the private conservation agency Ducks Unlimited (Porter and van Kooten 1993).[3]

Government land acquisition in the United States and Canada dates back to the late 1800s. Private conservation groups were slower to develop. Ducks Unlimited was established in 1937 and Ducks Unlimited Canada was established in 1938, both as responses to declining waterfowl populations that were exacerbated by extensive wetland drainage and the 1930s Prairie drought (Hansen 1982; Wetherell 2016). More recently, private land trusts have emerged as an important source of conservation investment. The number of land trusts in the United States increased dramatically between 1990 and 2000 (Merenlender et al. 2004). Land trusts use a combination of private and public grant dollars to conserve land. For instance, in the United States, several federal programs such as the U.S. Forest Service Forest Legacy Program, the U.S. FWS National Fish and Wildlife Foundation, and the U.S. Farm Bill all provide grants for the protection of habitat through conservation easements (Hohman et al. 2014; Merenlender et al. 2004). In Canada, the federal government introduced the Natural Areas Conservation Program (NACP) in 2007 as an attempt to increase private investment – through land trusts and conservation groups such as Nature Conservancy Canada and Ducks Unlimited Canada – in conservation activities, including conservation easements and land acquisition (Benidickson 2011).

3 Issues in design and implementation of habitat conservation programs

By their nature, the success of conservation programs in agricultural landscapes is contingent on several important design and implementation challenges related to private ownership of agricultural land. Two of these concerns are at the forefront of much of the literature on habitat conservation programming in Canada and the U.S. First, financial incentive programs to encourage conservation on private land will suffer from issues of asymmetric information, where the conservation agency has less information about the landowners' private opportunity costs due to conservation. This gives rise to information rents that can be captured by landowners (Kirwan, Lubowski, and Roberts 2005; Ferraro and Pattanayak 2006). This concern motivated the literature on dynamic targeting, which incorporates both the costs of conservation and the risk of future land use changes.

Second, private agricultural land ownership limits the ability of conservation agencies to coordinate conservation activity. Ecologists recognize the importance of the size of protected habitat, the number of reserves, the proximity and connectivity of reserve sites to other reserve sites, reserve shape, and buffer zones (Williams, ReVelle, and Levin 2005). Consistent with this, economists have evaluated the potential for spatial interactions in habitat conservation arising from threshold effects and the potential for increasing or decreasing the marginal benefits of conservation. This concern is addressed in the literature both on spatial targeting of habitat conservation and addressing agglomeration bonuses in habitat conservation compensation schemes.

In this section, we start with a discussion of the extensive conservation targeting literature, incorporating both static and dynamic considerations. We cover the more recent literature assessing optimal contract lengths and the use of auctions and agglomeration bonuses encouraging spatially contiguous habitat conservation. We conclude with a discussion of a novel thread of the literature that examines several alternative habitat conservation funding mechanisms that attempt to link landowner compensation more directly to the value of the conservation provided.

3.1 Targeting

A large literature on conservation targeting strategies dates back at least to the mid-1990s. Much of this research was first developed in the context of the CRP in the United States, with a focus on the benefits and costs of several different conservation targeting strategies. An alternative strand of the conservation targeting literature was developed in the context of "systematic conservation planning," with an emphasis on fee simple land acquisition by land trusts and government conservation agencies. There is a large literature on systematic conservation planning that explores the importance of incorporating ecological benefits, land costs, future risk of conversion, and the spatial arrangement of conserved parcels (Polasky and Segerson 2009).

Babcock et al. (1997) consider direct acquisition of environmental amenities – including reduced water and soil erosion, improved surface water quality, and improved wildlife habitat – through the CRP, allowing for site-specific heterogeneity in both agricultural productivity and environmental benefits. Three targeting strategies are considered: (1) maximize environmental benefits, (2) minimize acquisition costs, and (3) maximize net benefits. They show that the efficiency losses due to the cost minimization and the benefit maximization strategies depend on the degree of heterogeneity in farmland productivity and environmental benefits as well as the correlation between the two. For instance, wind erosion and CRP rental rates are negatively correlated, which implies that targeting based on cost will effectively reduce wind erosion. Alternatively, wildlife habitat is uniformly distributed across CRP land, and as a consequence there is little efficiency loss associated with any one of the three targeting strategies considered.

Wu, Zilberman, and Babcock (2001) develop an analytical framework to assess the incidence of alternative conservation targeting strategies, with a focus on environmental and distributional effects. The framework they introduce considers the effects of slippage in the context of the CRP, which in this case refers to the conversion of non crop to cropland acres, in response to removal of cropland acres. In their model, if output demand is not perfectly elastic, then output prices may increase in response to the removal of cropland acres, leading to slippage. Once again, the variability of and correlation between agricultural productivity and environmental benefits are important factors in the effectiveness of the targeting strategies. Wu, Zilberman, and Babcock (2001) consider welfare impacts on consumers, resource owners, input providers, and environmentalists. Consumers and input providers (labor) prefer benefit targeting because it has the smallest impact on output prices and overall resource use. Producers favor cost targeting strategies that cause the largest reduction in production and the largest increase in output price. Finally, environmentalists prefer benefit–cost targeting, although in the case of negatively sloped output demand benefit–cost targeting is shown to no longer maximize total environmental benefit for a given budget.

Systematic conservation planning (or reserve site selection) emerged as a theme in the economics literature starting in the late 1990s. Whereas the optimal targeting literature focused on temporary conservation easement programs that paid landowners to change behavior, the systematic conservation planning literature focused on fee simple land acquisition by governments, conservation agencies, and land trusts, with the objective of maintaining the status quo land use. Ando et al. (1998) use U.S. county-level land prices and species distributions to show that accounting for heterogeneity in land costs can significantly improve the efficiency of conservation investments. Polasky, Camm, and Garber-Yonts (2001) use species distributions and estimated land values for a similar exercise in the state of Oregon. The original formulation of benefits and costs in Ando et al. (1998) has been refined over time to incorporate important biological and economic aspects of reserve site selection. Notable among these studies, Wu and Boggess (1999) consider the impact of cumulative effects of environmental benefits on the allocation of conservation funds across a landscape and demonstrate that ignoring cumulative effects can result in conservation funds being spread too thin across the landscape.

Much of the early systematic conservation planning literature assumed that the conservation agency can dictate where conservation takes place. In many cases, the conservation agency cannot unilaterally allocate conservation across the landscape and instead must attempt to influence the choices of many private landowners who have different opportunity costs of conservation. To the extent that the landowner's opportunity cost is private information, conservation agencies will be operating in a situation of incomplete information.

Whereas the early literature uses aggregate land value estimates to account for conservation acquisition costs, a subsequent strand of the literature estimates landowner opportunity cost using econometric models of land use choices. The econometric land use analyses are derived from the crop choice and land use studies developed by Lichtenberg (1989), Stavins and Jaffe (1990), and Plantinga (1996). In this framework, land use choices are a function of the returns to alternative land uses as well as physical characteristics of parcels, including soil quality and proximity to urban centers. The early literature relied on county-level measures of land use changes and returns to alternative land uses. Applications of these models to targeting conservation activities used parcel-level observations of land use changes between broad categories of land use such as urban, crop, and pasture. Parcel-level land use change is combined with parcel-level measures of soil quality and proximity to urban centers, as well as county-level measures of returns to alternative land uses. In the United States, for instance, this literature has made extensive use of National Resource Inventory satellite data, which is based on randomly selected satellite images of land use categories updated every five years (Lewis and Plantinga 2007).

Lewis, Plantinga, and Wu (2009) examine reserve site selection in a voluntary incentive program to reduce habitat fragmentation, in which landowners hold private information about their opportunity costs. The theoretical framework considers a simple targeting strategy that converts all or none of the agricultural land on a parcel of land over to forest. The empirical analysis uses an econometric model of land use change to determine the subsidies needed to incentivize landowners to convert, and a geographic information system (GIS)-based simulation compares the targeting strategy to a uniform payment strategy and identifies large efficiency gains from the targeting strategy. They find that optimal spatial targeting should focus on regions with more existing forest and with more available forest land in order to increase the extent of core forest that can better support wildlife.

Lewis et al. (2011) build on the previous empirical work assessing the determinants of land use change and evaluate the efficiency of voluntary programs for land use change. As in Lewis, Plantinga, and Wu (2009), Lewis et al. (2011) use the parameters of an econometric model of land use change to estimate landowners' willingness to accept conservation payments – an assumption of the model is that landowner willingness to accept is private information held by landowners. One objective of the conservation program they consider is to protect spatially contiguous land parcels in an attempt to increase the size of reserve sites. They show that an information-revealing mechanism such as an auction combined with spatial agglomeration bonuses can substantially improve the efficiency of conservation programs.

3.2 Dynamic targeting

Costello and Polasky (2004) explore important dynamic considerations in reserve site selection, relaxing the assumption that the choice of which parcels to protect is a one-time choice, as in, for example, the research in Ando et al. (1998) and Polasky et al. (2001). Costello and Polasky (2004) account for annual budget constraints and future development of parcels. They consider a parcel-level model with three potential states: "developed," "reserved," and "unreserved." The objective is to maximize the number of species at the end of the conservation agency's planning horizon. Costello and Polasky (2004) show that the solution to the dynamic problem can be quite complex, especially with a potentially large number of reserve sites. Interestingly, they show that a static heuristic that accounts for the probability of conversion and species distributions across sites results in relatively small efficiency losses.

Whereas much of the prior economic literature on conservation targeting focused on land costs, Newburn, Berck, and Merenlender (2006) account for both land costs and the probability of land use conversion, thereby addressing two conflicting results from the literature (Newburn et al. 2005). The first result, as presented in Ando et al. (1998), suggests that benefit–cost targeting taking land values into account will avoid purchases of very high-cost land. Alternatively, Abbitt, Scott, and Wilcove (2000) take conversion threat into account and suggest that sites that are more vulnerable to development should be targeted first. Since land prices and vulnerability to development are positively correlated, these two approaches will generate conflicting results. Newburn, Berck, and Merenlender (2006) use parameters estimated from a hedonic model to estimate land costs and use a spatially explicit land use change model to estimate the risk of conversion. These two estimates (of acquisition cost and conversion risk) are combined in a dynamic model comparing an expected benefit–cost strategy that incorporates both land costs and vulnerability to conversion to (1) benefit–cost targeting, which ignores vulnerability and (2) expected benefit targeting, which ignores land costs. They find that ignoring vulnerability biases the program towards purchasing parcels far from development, on lower quality land, and in areas with strict zoning regulations.

More recent work incorporates uncertainty about future costs and the benefits of conservation. Climate change is an important source of uncertainty for both future land values and the geographic distribution of habitat and species (Ando and Mallory 2012). On the cost side, Lawley (2014) presents evidence that the implicit costs of wetland and other natural habitat acreage in Manitoba have increased over the course of the last couple of decades; this has increased private landowners' opportunity costs of conservation. Specific to the Prairie Pothole region of western Canada, Withey and van Kooten (2011; 2013) present evidence that the number of wetlands in western Canada will decrease due to climate change, and that the impact will vary across the landscape, with most of the loss occurring in Saskatchewan. The future pattern of loss (avoiding loss is the benefit of protecting habitat) is important for targeting conservation investment today. Ando and Mallory (2012), Mallory and Ando (2014), and Shah et al. (2017) propose the use of modern portfolio theory to allocate reserve sites across the U.S. Prairie Pothole region under future climate uncertainty. Shah and Ando (2015) extend the use of modern portfolio theory to incorporate downside risk and apply their model to conserving bird populations in the Eastern U.S.

3.3 Contract length

Ando and Chen (2011) evaluate optimal lengths for conservation contracts in a model in which the landowner chooses between accepting a conservation contract and not accepting a conservation contract. The landowner, in this case a farmer, faces a tradeoff between farming returns and a lump-sum conservation payment for the duration of the contract. At the end of each contract, the farmer has the option to renew at the same contract payment and length. Ando and Chen (2011) abstract away from the non-monetary private conservation benefits the farmer might enjoy and assume risk neutrality. They focus on the evolution of ecological benefits over time and the resulting impact on optimal contract length. Higher payments induce more farmer participation in the conservation program, whereas longer contract lengths reduce the number of farmers willing to enroll. Using a logistic ecological benefit function, the authors find that program benefits increase with contract length for ecological benefits that accumulate quickly, whereas if benefits accumulate slowly, conservation program benefits first increase with contract length and then decrease with contract length.

Shah and Ando (2016) incorporate uncertainty in both conservation and conversion returns into a real options framework. They use this framework to examine the effectiveness of permanent and temporary policy incentives to prevent land conversion to more intensive agricultural or urban uses. Shah and Ando (2016) estimate the temporary and permanent one-time payments that are sufficient to induce private landowners to forgo conversion opportunities. The permanent lump-sum payment plus the ongoing (perpetual) benefits of the land in the conservation use must exceed the value of the land with the conversion option. The temporary lump-sum payment must cover the expected difference in profits between the conservation use and the conversion use for the length of the contract. The authors find that actions to reduce the uncertainty in lump-sum payments reduce the size of lump-sum payments needed to enroll landowners, whereas an increase in the uncertainty of returns to conversion increases the size of temporary payments necessary to enroll landowners.

3.4 Auctions and agglomerations bonuses

A large literature examines the use of reverse auctions to conserve habitat and other ecological goods and services. Boxall et al. (2017) and Hellerstein (2017) provide comprehensive surveys of this literature relevant to the NAWMP and CRP, respectively. Studies based on field pilot programs include Brown et al. (2011), who examine uniform price auctions for conservation easements in the Prairie Pothole region of western Canada based on an early field program conducted by Ducks Unlimited Canada, and Hill et al. (2011), who examine a subsequent field project involving auctions for wetland restoration in Saskatchewan. Palm-Forster et al. (2016) note that the transaction costs of participation in conservation auctions can be high and, consequently, reduce the number of participants. This leads to fewer bids, which increases the likelihood that bids with higher cost–benefit ratios are accepted and may lead to lower program benefits if the budget is not exhausted. In addition, Palm-Forster et al. (2016) show that a spatially targeted uniform payment can be more cost effective than reverse auctions if the transaction costs of a reverse auction are sufficiently high.

Protecting spatially contiguous habitat is one of the criteria used by public and private conservation agencies when allocating scarce conservation dollars. For example, Oregon's Conservation Reserve Enhancement Program (CREP) offers a one-time bonus if a sufficient quantity of neighboring land is enrolled in the program (Grout 2009; Hanley et al. 2012). The Nature Conservancy focuses conservation efforts within priority regions, has minimum acreage requirements, and targets new habitat based in part on the protected status of adjacent habitat (Kiesecker et al. 2007). Recent research has also pointed out challenges associated with conservation planning in regions dominated by privately held land where conservation agencies cannot unilaterally dictate the location of conserved parcels (Banerjee and Shogren 2012; Polasky et al. 2014). This is particularly problematic in regions where many private landowners have little incentive to coordinate their conservation activity (Parkhurst and Shogren 2007).

A direct approach to influence the spatial configuration of conserved habitat is to provide agglomeration bonuses; this has received substantial attention in the experimental economics literature. Parkhurst et al. (2002) develop a theoretical model of agglomeration bonuses that involve a two-part payment. The first part of the payment is for enrolling in the conservation program, and the second is a bonus if neighboring landowners have also enrolled land in the conservation program. The agglomeration bonus provides landowners with incentives to enroll land that is adjacent to other enrolled land, creating a positive network externality (Parkhurst et al. 2002). Earlier studies present evidence that agglomeration bonuses improve the spatial configuration of conserved land (Parkhurst and Shogren 2007; Warziniack, Shogren, and Parkhurst 2007). Later studies present evidence that agglomeration bonuses are sensitive to coordination failures (Banerjee, Kwasnica, and Shortle 2012; Banerjee et al. 2014).

Banerjee et al. (2017) use experiments to evaluate the impact of transaction costs (e.g., including the cost of attending meetings and interacting with program administrators, the cost of assessing whether or not to sign a contract, and the cost of hiring outside advisors) on the performance of agglomeration bonus schemes. Banerjee et al. (2017) represent spatial interactions using a circular local network, where each participant has a neighbor on the left and a neighbor on the right, and all participants are indirectly connected. This mimics agricultural landscapes where physical distances can be large, and interactions are more frequent between immediate neighbors. Strategic uncertainty arises due to the unknown actions of neighbors and the potential for multiple equilibria. Their evidence suggests that higher transaction costs reduce program participation. Banerjee et al. (2017) also allow participants to communicate with each other, which has the effect of reducing strategic uncertainty.

Fooks et al. (2016) evaluate the effects of spatial targeting and network bonuses on the performance of reverse auctions for environmental services. Spatial targeting is shown to increase the purchase of contiguous attributes but can decrease an auction's efficiency since landowners with advantageously situated parcels gain market power. Network bonuses cause a small increase in contiguous habitat, but only one-third of the bonus payments are capitalized into bids. Fooks et al. (2016) show that while there are complementarities between spatial targeting and network bonuses, the added cost of the network bonuses results in little improvement in social benefits from using the two mechanisms together.

3.5 Funding mechanisms

Several funding sources are accessed to raise funds for habitat conservation in agricultural landscapes in Canada and the U.S. For instance, the U.S. CRP is funded through government revenue, whereas the U.S. FWS funds wetland protection through the Duck Stamp program. Private conservation groups, such as Ducks Unlimited (DU) and The Nature Conservancy (TNC), raise funds from private citizens, and, in some cases, funds are secured through government grants. The NAWMP is an interesting example of a partnership between government and private conservation groups. Funding in support of the NAWMP in Canada comes from several sources; Canadian federal and provincial sources contribute approximately 31 percent of funds; the U.S. federal government contributes 26 percent; Canadian contributions from private entities (primarily private conservation agencies) account for 21 percent; and U.S. non-federal sources account for 23 percent (NAWMP 2016). Thus, of the CA$2,218 million in total contributions between 1986 and 2016, only a quarter of NAWMP funding in Canada is derived directly from the conservation agencies administering programs on the ground. Moreover, approximately one-half of total funding for Canadian wetland and waterfowl conservation is derived from U.S. sources, reflecting the transboundary aspects of waterfowl conservation in Canada and the U.S. It is notable that a large share of total funding is derived from government sources.

Despite the significant variation in approaches to securing funds for habitat conservation, less attention has been devoted to funding mechanisms in the economics literature. A notable exception is a strand of the literature investigating approaches to increasing private contributions to conservation. Government and private philanthropy (through wildlife conservation groups or land trusts) both have limitations; government programs tend to target compensation payments according to the supply side opportunity costs of landowners rather than on the basis of benefits, while private philanthropy is subject to free-rider incentives that will tend to underprovide conservation (Swallow et al. 2008). In light of these limitations, Swallow et al. (2008) propose developing mechanisms enabling conservation entrepreneurs to raise funds based on demand for conservation actions. Swallow (2013) proposes linking payments from beneficiaries of public goods more directly to the providers of the public good, specifically using Lindahl (1958)-inspired auctions that elicit private consumer values for the public good. Swallow, Anderson, and Uchida (2018) evaluate two mechanisms based on field experiments in Jamestown, Rhode Island, that allowed local community residents to contribute to a public good, namely provision of grassland habitat for birds.

4 Evaluation of habitat conservation programs

As just discussed, the economics literature has focused on issues of design and implementation of conservation programs, with less attention paid to evaluation. Several studies outside of economics point to the improved environmental outcomes of North American conservation programming, specifically dealing with the CRP and the NAWMP (Ryan, Burger, and Kurzejeski 1998 provide an early review). These studies draw conclusions about the effectiveness of conservation programming based on associations between population trends and the variation in implementation of the conservation program. For instance, Herkert (2009) uses 21 years of bird population trend data collected *after* the establishment of the CRP to assess the impact of land set-asides on bird populations. Based on variation in the influence of set-aside lands on each bird species, Herkert (2009) predicts that those birds that benefit the most from land set-asides will show the greatest response to the CRP. The study finds that bird species uncommon in set-aside fields showed no population benefits from set-asides, whereas bird species that are common in set-aside fields were also species for which long-term population declines have reversed or lessened following the establishment of the CRP. Subsequent studies also point to the wildlife population benefits of the CRP (Hiller et al. 2015).

Although Herkert (2009) finds an association between CRP and bird population trends, the analysis is by no means causal. Subsequent research on land sparing versus land sharing – where land sparing refers to the extensive margin of land conservation through land set-asides and land sharing refers to the intensive margin through, for example, pesticide applications – notes one of many potential confounding factors not accounted for by Herkert (Fischer et al. 2014; Quinn et al. 2017). Furthermore, this literature does not account for land market feedbacks as a potential consequence of the CRP. Lubowski, Plantinga, and Stavins (2008) estimate the parameters from a national study of land use determinants, including net returns to alternative land uses, government program payments, and CRP rental rates. They find that the CRP offset the effects of government support programs that increased returns to cropland. Interestingly, they also find that the CRP reduced the quantity of acreage in pasture because cropland that would have been converted to pasture in the absence of the CRP was instead enrolled in the CRP. The authors note that their analysis does not capture reduced grassland acreage due to the potential impact of CRP on cropland returns (through a price effect as identified in the slippage literature).

Comprehensive program evaluations of joint ventures in the NAWMP have been slow to develop (Anderson and Padding 2015). Among the few studies conducted to date, the PHJV on the Canadian side of the Prairie Pothole region has been evaluated using simulation models based on the methodology developed in Devries et al. (2004). This approach traces out the counterfactual trend in waterfowl populations without the activities of the PHJV, based on changes in upland habitat/wetlands and waterfowl productivity models (PHJV 2009). Using this methodology, the authors of the PHJV (2009) report find that, as of 2006, the PHJV program decreased the hatched nest deficit by approximately 24 percent.

In the most recent economic assessment of the performance of the NAWMP, Wong, van Kooten, and Clarke (2012) estimate that between 1986 and 2008 the PHJV spent between $107 and $204 to increase the waterfowl population by one duck. This is consistent with earlier research suggesting that conservation spending was used to protect wetlands and habitat of lower quality and at lower risk of conversion (van Kooten and Schmitz 1992; van Kooten 1993a). The objective of Wong, van Kooten, and Clarke (2012) is to assess the impact of the agricultural activity on waterfowl populations accounting for spatial correlation in population numbers. To do so, the authors rely on relatively coarse strata-level data as defined by the U.S. FWS. As the authors acknowledge, future analysis of spatial interdependencies would benefit from data at a finer spatial resolution. This is a point we will return to shortly.

4.1 Additionality

Programs that compensate landowners to maintain the status quo are especially prone to problems related to additionality. For instance, the perpetual conservation easements used by conservation agencies in the Prairie Pothole regions of the United States and Canada pay landowners to maintain the status quo with little information on landowners' intentions to convert that land in the future. Early evidence from a NAWMP Canadian pilot program suggested that the wetlands and associated uplands protected by license agreements were on wetlands unlikely to be converted to cropland (van Kooten and Schmitz 1992). van Kooten and Schmitz (1992) are among the first to empirically identify the potential for low additionality in a habitat conservation program.

A primary challenge in identifying additionality in conservation programs on existing natural land is that vulnerability to conversion is unobserved. Furthermore, the time of conversion may be far into the future and so the impact of conservation undertaken today may not "pay off" or be realized for many years. Short-term evaluations of conservation expenditures and conversion rates on comparable land will, therefore, understate the true extent of additionality. Lawley and Towe (2014) devise a new approach to assess additionality in a perpetual conservation easement program. Recall that a conservation easement is an agreement between a landowner and a conservation agency where the landowner forfeits the right to convert existing habitat in exchange for a one-time payment. The conservation easement is perpetual and follows the land title. If retaining the right to convert existing habitat has value, a conservation easement will reduce the farmland prices of encumbered parcels. If retaining the right to convert does not have value, then an easement on that parcel is protecting habitat that is at very low risk of conversion in the future.

Lawley and Towe (2014) assess the impact of habitat conservation easements on farmland transaction prices as a means of quantifying the value of retaining the option to convert on those parcels. If the option to convert has zero or very low impact on land prices, this suggests that additionality on the parcel is low. In a well-functioning conservation easement program, the capitalized costs of the easements will be greater than zero and less than the one-time payment from the conservation agency.[4] Using data on conservation easements in the Prairie Pothole region of Manitoba held by Ducks Unlimited Canada and the Manitoba Habitat Heritage Corporation, Lawley and Towe (2014) present evidence that the conservation agencies in Manitoba targeted habitat at risk of conversion and that landowners on average are receiving a premium of approximately 16 percent.

4.2 Slippage

The definition of slippage in the context of the CRP has changed over time as the focus of the program has changed. For instance, in the early economics literature, the term slippage applied to situations where average yields increase in response to land retirement of more marginally productive farmland (Rausser, Zilberman, and Just 1984; Hoag, Babcock, and Foster 1993). More recently, slippage refers to the conversion of non-cropland to cropland acreage due to the CRP (Wu 2000). Slippage is an unintended consequence of the CRP and is a concern because it counteracts the environmental objective of the program, which is to remove acreage from crop production.

Wu (2000) outlines two potential reasons that slippage might occur in a large program such as the CRP. First, if the cropland acreage removed from production is sufficiently large, then output supply will fall, and, in a large country such as the United States, crop prices will increase. Farmers respond to the price increase by bringing more acreage into crop production. In other words, this price effect can occur if the demand for crop output is less than perfectly elastic. Second, economies of scale and fixed inputs might cause substitution effects such that farmers that enroll cropland in the CRP have an incentive to convert non-cropland to cropland acreage. Using cross-sectional data, Wu (2000) estimates slippage in the CRP of more than 20 percent.

A subsequent debate suggests that the empirical approach used in Wu (2000) is unable to identify slippage due to a price effect (price changes were uniform across the study region) and that evidence about the role of substitution effects is inconclusive (Roberts and Bucholtz 2005, 2006; Wu 2005). The debate revolves around the extent to which CRP acreage enrollment is endogenous and, if so, the suitability of using highly erodible land as an instrumental variable. Several subsequent studies attempt to identify slippage in agricultural and forestry conservation programs. Lichtenberg and Smith-Ramirez (2011) find that cost sharing for agricultural conservation practices tends to increase the share of acreage allocated to cropland due to a substitution effect. Alix-Garcia, Shapiro, and Sims (2012) evaluate a forest conservation program in Mexico and present evidence that both the substitution and price effects identified in Wu (2000) contributed to increased deforestation on unprotected land. Uchida (2014) examines in-farm substitution in the CRP and finds that the average CRP participant converts 14 percent of non-cropland to cropland acreage after enrollment.

4.3 Land market spillovers

There are several potential land market spillovers due to habitat conservation, particularly as a consequence of fee simple purchases that have the potential to increase the development potential of nearby land. Conserved land parcels might, therefore, increase the vulnerability of neighboring parcels to conversion as well as make neighboring parcels more expensive to protect in the future. Furthermore, both conservation agencies and developers can profit from land price speculation if they know where conservation will occur in the future. For instance, Costello and Polasky (2004) point out that a conservation agency might purchase a large tract of land with the intention of selling a portion of the land now neighboring newly conserved land at a premium. At the same time, developers might purchase land in areas where they suspect future conservation will occur.

Chamblee et al. (2011) use an event study approach to assess the land price spillovers of easement and fee simple conservation in Buncombe County, North Carolina. They combine transaction data documenting land sales between 1996 and 2008 with land title data documenting conservation interests in land either through fee simple ownership or conservation easement. This dataset provides spatial and temporal variation in conservation activity. Price spillovers might result from the amenity value of nearby conserved space, from a reduction in the supply of developable land, or through an increase in property tax rates due to newly conserved land. The first two are expected to increase nearby land values, whereas the last is expected to reduce land values.

The event study approach used in Chamblee et al. (2011) estimates the pre-conservation values of land near conserved parcels as well as the post-conservation impact of neighboring easement and fee conserved land. Their results suggest that conservation easements are located in regions with lower land values relative to fee simple conservation, implying that conservation agencies might be targeting land more at risk of development with fee simple purchases. They find evidence of substantial price premiums due to neighboring conservation: vacant land that is immediately adjacent to conserved land increases in value by approximately 46 percent.

A common criticism of protected areas is that they erode the local property tax base to the detriment of the provision of local services. Liu et al. (2013) evaluate the property value impacts of U.S. National Wildlife Refuges in the eastern United States near urban centers. They find that wildlife refuges increase nearby property values. To the extent that increased property values lead to increases in property taxes, Liu et al. (2013) consider this a benefit of protected areas. They estimate that the average capitalized value of increased property values is US$11 million per refuge and estimate that this generates an additional US$1 million in local annual property taxes, therefore counteracting reduced property taxes on land allocated to the refuge.

An important issue in conservation programs is their impact on property values. Wu and Lin (2010) specify a model of the CRP and land values and find that the value of developed land consists of (1) the net returns to agriculture or the CRP, (2) growth premium, (3) irreversibility premium due to the loss of option value once a parcel is developed, and (4) the amenity and accessibility premium associated with the convenience of developed parcels. The CRP increases the net return to agriculture or the CRP, but reduces the growth premium, the option value (due to lower development), and the accessibility value of farmland. Results of the econometric model suggest that the CRP increased farmland values by US$18 to US$25 per acre, which works out to increases of approximately 1.3 to 1.8 percent. Increases are greatest in the Mountain, Southern Plain, and Northern Plain regions, which account for more than 60 percent of CRP acreage.

Jacobs, Thurman, and Marra (2014) evaluate the impact of CRP Conservation Priority Area (CPA) designation of the Prairie Pothole region on submitted bids in CRP auctions. CPA designation exogenously increases environmental benefits index (EBI) points and can lead landowners to adjust their bids. Jacobs, Thurman, and Marra (2014) use variation across crop reporting districts and find that high-rent landowners bid further below their maximum rental rate than low-rent landowners, but their response to the exogenous increase in EBI (and therefore increase in the maximum allowable bid) is greater.

4.4 Spatial interactions and temporal spillovers

Albers and Ando (2003) initiate a literature examining spatial variation in land trust activity. They develop a model of the optimal number of conservation agents in a region as a function of spatial externalities in conservation and diversity in conservation objectives. Thinking of habitat conservation as a public good, private provision will typically under-provide the public good, although the free-rider problem might be mitigated if providers specialize.[5] Furthermore, using insights from industrial organization, Albers and Ando (2003) suggest that the land trust "industry" might be monopolistically competitive, so that the number of land trusts in a state is not at a socially optimal level. They consider two competing factors in the determination of the optimal number of land trusts: (1) niche diversification and specialization versus (2) coordination of conservation decisions. In an empirical analysis of the number of land trusts in a state, Albers and Ando (2003) find little evidence of widespread inefficiency in the U.S. land trust sector and note that land trust activity tends to be higher in U.S. states with more government-protected land.

Following up on Albers and Ando (2003), Albers, Ando, and Batz (2008) present a theoretical model of spatial interactions in land conservation, focusing on the impact of government conservation investment on the investment choices of private conservation agencies. Specifically, they consider the potential for crowding in and crowding out of private investment. Albers, Ando, and Chen (2008) follow up on the theoretical work of Albers, Ando, and Batz (2008) with an econometric analysis of spatial attraction and repulsion between government and privately protected land using a township-level cross-sectional dataset for the states of California, Illinois, and Massachusetts. Several hypotheses are formulated, once again related to whether the marginal benefits of conservation are constant, increasing, or decreasing. Albers, Ando, and Chen (2008) find that publicly protected areas repel private reserves in Illinois and Massachusetts, whereas they attract private reserves in California. The results suggest that private conservation agencies in California are attempting to create spatially concentrated reserves as opposed to fragmented ones. In contrast, in Illinois, much of the public-conserved land is concentrated in one region and private conservation agencies are behaving as if there are decreasing marginal benefits of spatially concentrated conservation in that region and have chosen to protect land elsewhere in the state.

Parker and Thurman (2011) investigate the role of crowding in and crowding out of U.S. federal land conservation programs on private land trust activity using a county-level panel dataset for the years 1990 and 2000. The panel approach in Parker and Thurman (2011) allows the researchers to examine the relationship between growth in government-conserved land and private land trust acreage. The authors find that federal land conservation programs affect private conservation investment; federal conservation has a small crowding out effect on private land trusts that focus on preserving open space and a crowding in effect on land held in trust by the Nature Conservancy, which selects sites on the basis of biodiversity benefits.

Lawley and Yang (2015) build on Albers, Ando, and Chen (2008) and Parker and Thurman (2011) with a parcel-level analysis of spatial interactions in habitat conservation easements in the

Prairie Pothole region of Manitoba. Whereas the prior studies use aggregate data to examine crowding-in and crowding-out of conservation activity, Lawley and Yang (2015) use parcel-level data, which permits analysis of the determinants of the spatial configuration of conserved habitat. Positive or negative spatial spillovers might occur in habitat conservation for several reasons. First, as highlighted in the previous literature, conservation agencies might target spatially contiguous habitat due to increasing marginal benefits of conservation associated with thresholds and increasing returns. Alternatively, conservation agencies might attempt to diversify the location of conserved habitat, for example, in response to uncertainty about the spatial consequences of future climatic change as studied in Ando and Hannah (2011) and Ando and Mallory (2012). Second, it is possible that neighbors learn from their neighbors' experience with conservation easements and, assuming the experience of their neighbors is positive, are more likely to enter into conservation easements themselves. This would generate a positive spatial spillover. Lawley and Yang (2015) find that the addition of a neighboring easement roughly doubles the likelihood of an easement within ten years, reflecting a combination of conservation agency targeting and interactions among neighboring landowners. They also find evidence of a small crowding-out effect of government-protected land, which is strongest when a substantial number of neighboring parcels are protected by the government.

Finally, Roberts and Lubowski (2007) evaluate the enduring impacts of the CRP and find that approximately 62 percent of land exiting CRP contracts was subsequently converted to crop production. The land not converted to crop production remained in pasture, range, or forest. Roberts and Lubowski (2007) interpret this finding as evidence that enrollment in the CRP generates environmental benefits (due to land use in pasture, range, or forest) that persist beyond the contract term – a form of temporal spillover. Fixed costs in converting land back to crops combined with uncertain future land returns might explain the persistence of pasture, range, and forest cover. The CRP requires that potential CRP land be in crops in at least two of the five years prior to CRP enrollment. An alternative interpretation of the results presented in Roberts and Lubowski (2007) is that land exiting the CRP was on the cropland, non-cropland margin prior to enrollment and that the additionality on this land is low.

5 Future directions

To date, much of the research has rightfully focused on aspects of program design, including targeting strategies and contract structures, or on the behavioral interactions of the program in practice – additionality, slippage, and spillovers. What remains understudied in practice is the outcome or success of the conservation activity. Habitat is not conserved simply for the sake of acreage counting, without regard to the improvement of habitat and species that depend on that habitat to survive. Advances in satellite data processing and analysis include but are not limited to in-season and cross-season land uses, land covers, vegetation health, crop choice, and even atmospheric composition. While early researchers were hindered by a lack of data on both the spatial and temporal front, we are past the infancy of large data collection and into issues of distribution and computational resources. In this section, we will direct readers to sources and provide some pointers to acquisition and usage of these data, but the direct use of data is left to the creativity and the able minds of researchers in this field.

On the distribution front, governmental agencies are working on platforms for distribution of primarily government-generated or -funded data. Many of these datasets are available with standardized metadata on the GeoPlatform[6] launched in 2010, with over 160,000 geospatial datasets already accessible. Older repositories exist at NASA under the Giovanni platform, the EnviroAtlas and Water Quality Data Portal by the U.S. EPA, and the United

States Geological Survey (USGS) Earth explorer at USGS. Some of the data available will need processing from its primitive state to be useful to social scientists, and errors in processing and classification must be considered carefully. It is our suggestion that social scientists use professionally processed data or collaborate with the community of remote sensing academics and professionals to produce the highest quality data for research input. Discussions and collaboration with government agencies producing data and private conservation groups allocating funds are not only necessary but essentially required to justify continued conservation expenditures. As governments increasingly pull back in resource expenditures for primary data collections, private organizations are attaining the skill and permission to advance the field. Two prime examples of this are Google Earth Engine and Planet Networks.[7] The latter group is flying low-altitude small satellites to collect extremely granular land use and cover data with the goal to map the earth daily regarding land cover, land use, and field-level analysis. About the former, we are all familiar with Google, a company that interacts with us on a range of services including maps, phones, video, and internet searches. To the suite of Google endeavors, they have added Google Earth Engine and are sharing their immense computational power to analyze massive amounts of data in human accessible timescales.[8] Several example case studies now exist using the platform, including a global forest change analysis done by Hansen et al. (2013) at the University of Maryland and a partnership between several environmental groups, the University of California, Santa Barbara (UCSB), and Global Fishing Watch aimed at using vessel-level data in near real time to analyze various environmental outcomes from overfishing to marine reserve management and conservation effectiveness.[9]

These satellite-based data sources are excluding perhaps the largest and most elusive type of data with the potential to impact social science: that of the handheld smart phones ubiquitous in society today. The device itself collects vast amounts of data for marketers and content providers, but the users of these devices can be trained and are a likely volunteer pool to collect ground level data. Application development, training, and direction of users using smartphone apps and the device's GIS geolocation abilities make for seemingly endless opportunities for data.

6 Conclusions

There is no one size fits all program design for habitat conservation in agricultural landscapes. Research into the implications of program design and behavioral responses of landowners and their neighbors is as important now as it has been at any time in the past. While much progress has been made in the "practice of" and "research in" habitat conservation, we are approaching a crossroads where large datasets (in time and space) are becoming reliably accessible in both distribution and processing. These data and their many advantages are an opportunity for the social science community to address questions of regional, national, and international importance regarding the efficacy of historical programmatic activity and, perhaps more importantly, the issues that are on the forefront of the environmental agenda, including how climate change will impact the need and location of habitat for a multitude of species and how we can effectively stretch conservation budgets to more effectively address policy goals.

Notes

1 Swallow et al. (2008) highlight restrictions on the ability of providers of "public goods" and "fugitive resources" to capture returns from beneficiaries.

2 Observers suggest that long-term contracts were popular because in the long-run forage production is more sustainable than annual crop production on those marginal lands, calling into question the additionality the PCP achieved (van Kooten 1993b). Further, the buyout provisions provided landowners with an early withdrawal option.
3 van Kooten and Schmitz (1992) and van Kooten (1993a) suggest that early pilot projects of the NAWMP simply protected wetlands and upland habitat that would not be used for annual crops in the absence of the program. van Kooten and Schmitz (1992) identify inadequate financial compensation paid to landowners and an overreliance on moral suasion as the reasons that the NAWMP was not a success early on.
4 As pointed out in Lawley and Towe (2014), it is possible that some "warm glow" landowners are enrolling habitat with an opportunity cost in excess of the one-time easement payment.
5 Niche diversification will increase the number of land trusts in order to fulfill several different roles, which in turn increases donations to conservation activities.
6 www.geoplatform.gov, last accessed July 13, 2017.
7 www.planet.com/, last accessed July 13, 2017.
8 Google Earth Engine Team, 2015. Google Earth Engine: A planetary-scale geospatial analysis platform. https://earthengine.google.com, last accessed July 13, 2017.
9 http://sfg.msi.ucsb.edu/research/global-fishing-watch, last accessed July 13, 2017.

References

Abbitt, R.J., J.M. Scott, and D.S. Wilcove. 2000. "The Geography of Vulnerability: Incorporating Species Geography and Human Development Patterns into Conservation Planning." *Biological Conservation* 96(2):169–175.

Adamowicz, W.L. 2016. "Economic Analysis and Species at Risk: Lessons Learned and Future Challenges." *Canadian Journal of Agricultural Economics* 64(1):21–32.

Albers, H.J., and A.W. Ando. 2003. "Could State-Level Variation in the Number of Land Trusts Make Economic Sense?" *Land Economics* 79(3):311–327.

Albers, H.J., A.W. Ando, and M. Batz. 2008. "Patterns of Multi-Agent Land Conservation: Crowding In/Out, Agglomeration, and Policy." *Resource and Energy Economics* 30(4):492–508.

Albers, H.J., A.W. Ando, and X. Chen. 2008. "Spatial-Econometric Analysis of Attraction and Repulsion of Private Conservation by Public Reserves." *Journal of Environmental Economics and Management* 56(1):33–49.

Alix-Garcia, J.M., E.N. Shapiro, and K.R. Sims. 2012. "Forest Conservation and Slippage: Evidence from Mexico's National Payments for Ecosystem Services Program." *Land Economics* 88(4):613–638.

Alix-Garcia, J., and H. Wolff. 2014. "Payment for Ecosystem Services from Forests." *Annual Review of Resource Economics* 6(1):361–380.

Anderson, M.G., and P.I. Padding. 2015. "The North American Approach to Waterfowl Management: Synergy of Hunting and Habitat Conservation." *International Journal of Environmental Studies* 72(5):810–829.

Ando, A., J. Camm, S. Polasky, and A. Solow. 1998. "Species Distributions, Land Values, and Efficient Conservation." *Science* 279(5359):2126–2128.

Ando, A.W., and X. Chen. 2011. "Optimal Contract Lengths for Voluntary Ecosystem Service Provision with Varied Dynamic Benefit Functions." *Conservation Letters* 4(3):207–218.

Ando, A.W., and L. Hannah. 2011. "Lessons from Finance for New Land Conservation Strategies Given Climate-Change Uncertainty." *Conservation Biology* 25(2):412–414.

Ando, A.W., and M.L. Mallory. 2012. "Optimal Portfolio Design to Reduce Climate-Related Conservation Uncertainty in the Prairie Pothole Region." *Proceedings of the National Academy of Sciences* 109(17):6484–6489.

Babcock, B.A., P.G. Lakshminarayan, J. Wu, and D. Zilberman. 1997. "Targeting Tools for the Purchase of Environmental Amenities." *Land Economics* 73(1):325–339.

Banerjee, S., T.N. Cason, F.P. Devries, and N. Hanley. 2017. "Transaction Costs, Communications, and Spatial Coordination in Payment for Ecosystem Services Schemes." *Journal of Environmental Economics and Management* 83:68–89.

Banerjee, S., F.P. Devries, N. Hanley, and D.P. van Soest. 2014. "The Impact of Information Provision on Agglomeration Bonus Performance: An Experimental Study on Local Networks." *American Journal of Agricultural Economics* 96(4):1009–1029.

Banerjee, S., A.M. Kwasnica, and J.S. Shortle. 2012. "Agglomeration Bonus in Small and Large Local Networks: A Laboratory Examination of Spatial Coordination." *Ecological Economics* 84:142–152.

Banerjee, P., and J.F. Shogren. 2012. "Material Interests, Moral Reputation, and Crowding Out Species Protection on Private Land." *Journal of Environmental Economics and Management* 63(1):137–149.

Benidickson, J. 2011. *Legal Framework for Protected Areas: Canada.* Gland, Switzerland: IUCN.

Brown, L.K., E. Troutt, C. Edwards, B. Gray, and W. Hu. 2011. "A Uniform Price Auction for Conservation Easements in the Canadian Prairies." *Environmental and Resource Economics* 50(1):49–60.

Boxall, P.C., O. Perger, K. Packman, and M. Weber. 2017. "An Experimental Examination of Target Based Conservation Auctions." *Land Use Policy* 63:592–600.

Cain, Z., and S. Lovejoy. 2004. "History and Outlook for Farm Bill Conservation Programs." *Choices* 19(4):37–42.

Chamblee, J.F., P.F. Colwell, C.A., Dehring, and C.A. Depken. 2011. "The Effect of Conservation Activity on Surrounding Land Prices." *Land Economics* 87(3):453–472.

Costello, C., and S. Polasky. 2004. "Dynamic Reserve Site Selection." *Resource and Energy Economics* 26(2):157–174.

Devries, J.H., K.L. Guyn, R.G. Clark, M.G. Anderson, D. Caswell, S.K. Davis, D.G. McMaster, T. Sopuck, and D. Kay. 2004. *Prairie Habitat Joint Venture (PHJV) Waterfowl Habitat Goals.* Edmonton, Canada: Environment Canada.

Ferraro, P.J., and S.K. Pattanayak. 2006. "Money for Nothing? A Call for Empirical Evaluation of Biodiversity Conservation Investments." *PLoS Biol* 4(4):e105.

Fischer, J., D.J. Abson, V. Butsic, M.J. Chappell, J. Ekroos, J. Hanspach, T. Kuemmerle, H.G. Smith, and H. Wehrden. 2014. "Land Sparing Versus Land Sharing: Moving Forward." *Conservation Letters* 7(3):149–157.

Fooks, J.R., N. Higgins, K.D. Messer, J.M. Duke, D. Hellerstein, and L. Lynch. 2016. "Conserving Spatially Explicit Benefits in Ecosystem Service Markets: Experimental Tests of Network Bonuses and Spatial Targeting." *American Journal of Agricultural Economics* 98(2):468–488.

Grout, C.A. 2009. "Incentives for Spatially Coordinated Land Conservation: A Conditional Agglomeration Bonus." *Western Economic Forum* 8(2):21–29.

Hanley, N., S. Banerjee, G.D. Lennox, and P.R. Armsworth. 2012. "How Should We Incentivize Private Landowners to 'Produce' More Biodiversity?" *Oxford Review of Economic Policy* 28(1):93–113.

Hansen, M.C., P.V. Potapov, R. Moore, M. Hancher, S.A. Turubanova, A. Tyukavina, D. Thau, S.V. Stehman, S.J. Goetz, T.R. Loveland, A. Kommareddy, A. Egorov, L. Chini, C.O. Justice, and J.R.G. Townshend. 2013. "High-Resolution Global Maps of 21st-Century Forest Cover Change." *Science* 342(6160):850–853.

Hansen, P.L. 1982. "Issues, Conflicts and Strategies in Wetland Management for Waterfowl." *Canadian Water Resources Journal* 7(4):1–21.

Hellerstein, D.M. 2017. "The US Conservation Reserve Program: The Evolution of an Enrollment Mechanism." *Land Use Policy* 63:601–610.

Herkert, J.R. 2009. "Response of Bird Populations to Farmland Set-Aside Programs." *Conservation Biology* 23(4):1036–1040.

Hill, M.R., D.G. McMaster, T. Harrison, A. Hershmiller, and T. Plews. 2011. "A Reverse Auction for Wetland Restoration in the Assiniboine River Watershed, Saskatchewan." *Canadian Journal of Agricultural Economics* 59(2):245–258.

Hiller, T.L., J.S. Taylor, J.J. Lusk, L.A. Powell, and A.J. Tyre. 2015. "Evidence That the Conservation Reserve Program Slowed Population Declines of Pheasants on a Changing Landscape in Nebraska, USA." *Wildlife Society Bulletin* 39(3):529–535.

Hoag, D.L., B.A. Babcock, and W.E. Foster. 1993. "Field-Level Measurement of Land Productivity and Program Slippage." *American Journal of Agricultural Economics* 75(1):181–189.

Hohman, W.L., E.B. Lindstrom, B.S. Rashford, and J.H. Devries. 2014. "Opportunities and Challenges to Waterfowl Habitat Conservation on Private Land." *Wildfowl* 4:368–406.

Innes, R., and G. Frisvold. 2009. "The Economics of Endangered Species." *Annual Review of Resource Economics* 1(1):485–512.

Jacobs, K.L., W.N. Thurman, and M.C. Marra. 2014. "The Effect of Conservation Priority Areas on Bidding Behavior in the Conservation Reserve Program." *Land Economics* 90:1–25.

Kiesecker, J.M., T. Comendant, T. Grandmason, E. Gray, C. Hall, R. Hilsenbeck, P. Kareiva, L. Lozier, P. Naehu, A. Rissman, and M.R. Shaw. 2007. "Conservation Easements in Context: A Quantitative Analysis of Their Use By the Nature Conservancy." *Frontiers in Ecology and the Environment* 5(3):125–130.

Kirwan, B., R.N. Lubowski, and M.J. Roberts. 2005. "How Cost-Effective Are Land Retirement Auctions? Estimating the Difference Between Payments and Willingness to Accept in the Conservation Reserve Program." *American Journal of Agricultural Economics* 87(5):1239–1247.

Lawley, C. 2014. "Changes in Implicit Prices of Prairie Pothole Habitat." *Canadian Journal of Agricultural Economics* 62(2):171–190.

Lawley, C., and C. Towe. 2014. "Capitalized Costs of Habitat Conservation Easements." *American Journal of Agricultural Economics* 96(3):657–672.

Lawley, C., and W. Yang. 2015. "Spatial Interactions in Habitat Conservation: Evidence from Prairie Pothole Easements." *Journal of Environmental Economics and Management* 71:71–89.

Lewis, D.J., and A.J. Plantinga. 2007. "Policies for Habitat Fragmentation: Combining Econometrics with GIS-Based Landscape Simulations." *Land Economics* 83(2):109–127.

Lewis, D.J., A.J. Plantinga, E. Nelson, and S. Polasky. 2011. "The Efficiency of Voluntary Incentive Policies for Preventing Biodiversity Loss." *Resource and Energy Economics* 33(1):192–211.

Lewis, D.J., A.J. Plantinga, and J. Wu. 2009. "Targeting Incentives to Reduce Habitat Fragmentation." *American Journal of Agricultural Economics* 91(4):1080–1096.

Lichtenberg, E. 1989. "Land Quality, Irrigation Development, and Cropping Patterns in the Northern High Plains." *American Journal of Agricultural Economics* 71(1):187–194.

Lichtenberg, E., and R. Smith-Ramírez. 2011. "Slippage in Conservation Cost Sharing." *American Journal of Agricultural Economics* 93(1):113–129.

Lindahl, E. 1958. "Just Taxation – a Positive Solution." In R.A. Musgrave and A.T. Peacock, eds., *Classics in the Theory of Public Finance*. London: Palgrave Macmillan, pp. 168–176.

Liu, X., L.O. Taylor, T.L. Hamilton, and P.E. Grigelis. 2013. "Amenity Values of Proximity to National Wildlife Refuges: An Analysis of Urban Residential Property Values." *Ecological Economics* 94:37–43.

Lubowski, R.N., A.J. Plantinga, and R.N. Stavins. 2008. "What Drives Land-Use Change in the United States? A National Analysis of Landowner Decisions." *Land Economics* 84(4):529–550.

Mallory, M.L., and A.W. Ando. 2014. "Implementing Efficient Conservation Portfolio Design." *Resource and Energy Economics* 38:1–18.

McConnell, V., and M.A. Walls. 2005. *The Value of Open Space: Evidence from Studies of Nonmarket Benefits*. Washington, DC: Resources for the Future.

Merenlender, A.M., L. Huntsinger, G. Guthey, and S.K. Fairfax. 2004. "Land Trusts and Conservation Easements: Who Is Conserving What for Whom?" *Conservation Biology* 18(1):65–76.

Newburn, D., P. Berck, and A. Merenlender. 2006. "Habitat and Open Space at Risk of Land-Use Conversion: Targeting Strategies for Land Conservation." *American Journal of Agricultural Economics* 88(1):28–42.

Newburn, D., S. Reed, P. Berck, and A. Merenlender. 2005. "Economics and Land-Use Change in Prioritizing Private Land Conservation." *Conservation Biology* 19(5):1411–1420.

Newmark, W.D. 1985. "Legal and Biotic Boundaries of Western North American National Parks: A Problem of Congruence." *Biological Conservation* 33(3):197–208.

Nichols, J.D., F.A. Johnson, and B.K. Williams. 1995. "Managing North American Waterfowl in the Face of Uncertainty." *Annual Review of Ecology and Systematics* 26(1):177–199.

North American Waterfowl Management Plan (NAWMP). 2016. "Habitat Matters: 2016 Canadian NAWMP Report," September 2016. http://nawmp.wetlandnetwork.ca/Media/Content/files/Hab%20Mat2016EN.pdf. Accessed April 10, 2018.

Palm-Forster, L.H., S.M. Swinton, F. Lupi, and R.S. Shupp. 2016. "Too Burdensome to Bid: Transaction Costs and Pay-for-Performance Conservation." *American Journal of Agricultural Economics* 98(5):1314–1333.

Parker, D.P. 2004. "Land Trusts and the Choice to Conserve Land with Full Ownership or Conservation Easements." *Natural Resources Journal* 44:483–518.

Parker, D.P., and W.N. Thurman. 2011. "Crowding Out Open Space: The Effects of Federal Land Programs on Private Land Trust Conservation." *Land Economics* 87(2):202–222.

Parkhurst, G.M., and J.F. Shogren. 2007. "Spatial Incentives to Coordinate Contiguous Habitat." *Ecological Economics* 64(2):344–355.

Parkhurst, G.M., J.F. Shogren, C. Bastian, P. Kivi, D. Donner, and R.B. Smith. 2002. "Agglomeration Bonus: An Incentive Mechanism to Reunite Fragmented Habitat for Biodiversity Conservation." *Ecological Economics* 41(2):305–328.

Parks, P.J., and R.A. Kramer. 1995. "A Policy Simulation of the Wetlands Reserve Program." *Journal of Environmental Economics and Management* 28(2):223–240.

Plantinga, A.J. 1996. "The Effect of Agricultural Policies on Land Use and Environmental Quality." *American Journal of Agricultural Economics* 78(4):1082–1091.

Polasky, S., J.D. Camm, and B. Garber-Yonts. 2001. "Selecting Biological Reserves Cost-Effectively: An Application to Terrestrial Vertebrate Conservation in Oregon." *Land Economics* 77(1):68–78.

Polasky, S., D.J. Lewis, A.J. Plantinga, and E. Nelson. 2014. "Implementing the Optimal Provision of Ecosystem Services." *Proceedings of the National Academy of Sciences* 111(17):6248–6253.

Polasky, S., and K. Segerson. 2009. "Integrating Ecology and Economics in the Study of Ecosystem Services: Some Lessons Learned." *Annual Review of Resource Economics* 1:409–434.

Porter, R.M., and G.C. van Kooten. 1993. "Wetlands Preservation on the Canadian Prairies: The Problem of the Public Duck." *Canadian Journal of Agricultural Economics* 41(4):401–410.

Prairie Habitat Joint Venture. 2009. *NAWMP Continental Assessment PHJV Triennial Report*. Edmonton, Canada: Environment Canada.

Quinn, J.E., T. Awada, F. Trindade, L. Fulginiti, and R. Perrin. 2017. "Combining Habitat Loss and Agricultural Intensification Improves Our Understanding of Drivers of Change in Avian Abundance in a North American Cropland Anthrome." *Ecology and Evolution* 7(3):803–814.

Rausser, G.C., D. Zilberman, and R.E. Just. 1984. "The Distributional Effects of Land Controls in Agriculture." *Western Journal of Agricultural Economics* 9(2):215–232.

Roberts, M.J., and S. Bucholtz. 2005. "Slippage in the Conservation Reserve Program or Spurious Correlation? A Comment." *American Journal of Agricultural Economics* 87(1):244–250.

———. 2006. "Slippage in the Conservation Reserve Program or Spurious Correlation? A Rejoinder." *American Journal of Agricultural Economics* 88(2):512–514.

Roberts, M.J., and R.N. Lubowski. 2007. "Enduring Impacts of Land Retirement Policies: Evidence from the Conservation Reserve Program." *Land Economics* 83(4):516–538.

Ryan, M.R., L.W. Burger, and E.W. Kurzejeski. 1998. "The Impact of CRP on Avian Wildlife: A Review." *Journal of Production Agriculture* 11(1):61–66.

Shah, P., and A.W. Ando. 2015. "Downside Versus Symmetric Measures of Uncertainty in Natural Resource Portfolio Design to Manage Climate Change Uncertainty." *Land Economics* 91(4):664–687.

———. 2016. "Permanent and Temporary Policy Incentives for Conservation under Stochastic Returns from Competing Land Uses." *American Journal of Agricultural Economics* 98(4):1074–1094.

Shah, P., M.L. Mallory, A.W. Ando, and G.R. Guntenspergen. 2017. "Fine-Resolution Conservation Planning with Limited Climate-Change Information." *Conservation Biology* 31(2):278–289.

Stavins, R.N., and A.B. Jaffe. 1990. "Unintended Impacts of Public Investments on Private Decisions: The Depletion of Forested Wetlands." *The American Economic Review* 80:337–352.

Swallow, S.K. 2013. "Demand-side Value for Ecosystem Services and Implications for Innovative Markets: Experimental Perspectives on the Possibility of Private Markets for Public Goods." *Agricultural and Resource Economics Review* 42(1):33–56.

Swallow, S.K., C.M. Anderson, and E. Uchida. 2018. "The Bobolink Project: Selling Public Goods from Ecosystem Services Using Provision Point Mechanisms." *Ecological Economics* 143:236–252.

Swallow, S.K., E.C. Smith, E. Uchida, and C.M. Anderson. 2008. "Ecosystem Services Beyond Valuation, Regulation, and Philanthropy: Integrating Consumer Values Into the Economy." *Choices* 23(2):47–52.

Uchida, S. 2014. "Indirect Land Use Effect of Conservation: Disaggregate Slippage in the U.S. Conservation Reserve Program." WP 14–05, Department of Agricultural and Resource Economics, University of Maryland, College Park, MD.

Vaisey, J., T. Weins, and R. Wetlaufer. 1996. "The Permanent Cover Program: Is Twice Enough?" In T.L. Napier, S.M. Napier, and J. Tvrdon, eds., *Soil and Water Conservation Policies and Programs: Successes and Failures*. Boca Raton, FL: CRC Press, pp. 515–533.

van Kooten, G.C. 1993a. "Preservation of Waterfowl Habitat in Western Canada: Is the North American Waterfowl Management Plan a Success." *Natural Resources Journal* 33:759–775.

———. 1993b. *Land Resource Economics and Sustainable Development: Economic Policies and the Common Good*. Vancouver, BC: UBC Press.

van Kooten, G.C., and A. Schmitz. 1992. "Preserving Waterfowl Habitat on the Canadian Prairies: Economic Incentives Versus Moral Suasion." *American Journal of Agricultural Economics* 74(1):79–89.

Warziniack, T., J.F. Shogren, and G. Parkhurst. 2007. "Creating Contiguous Forest Habitat: An Experimental Examination on Incentives and Communication." *Journal of Forest Economics* 13(2):191–207.

Wetherell, D.G. 2016. *Wildlife, Land, and People: A Century of Change in Prairie Canada*. Montreal, QC: McGill-Queen's University Press.

Williams, J.C., C.S. ReVelle, and S.A. Levin. 2005. "Spatial Attributes and Reserve Design Models: A Review." *Environmental Modeling and Assessment* 10(3):163–181.

Withey, P., and G.C. van Kooten. 2011. "The Effect of Climate Change on Optimal Wetlands and Waterfowl Management in Western Canada." *Ecological Economics* 70(4):798–805.

———. 2013. "The Effect of Climate Change on Wetlands and Waterfowl in Western Canada: Incorporating Cropping Decisions Into a Bioeconomic Model." *Natural Resource Modeling* 26(3):305–330.

Wong, L., G.C. van Kooten, and J.A. Clarke. 2012. "The Impact of Agriculture on Waterfowl Abundance: Evidence from Panel Data." *Journal of Agricultural and Resource Economics* 37(2):321–334.

Wu, J. 2000. "Slippage Effects of the Conservation Reserve Program." *American Journal of Agricultural Economics* 82(4):979–992.

———. 2005. "Slippage Effects of the Conservation Reserve Program: Reply." *American Journal of Agricultural Economics* 87(1):251–254.

Wu, J., and W.G. Boggess. 1999. "The Optimal Allocation of Conservation Funds." *Journal of Environmental Economics and Management* 38(3):302–321.

Wu, J., and H. Lin. 2010. "The Effect of the Conservation Reserve Program on Land Values." *Land Economics* 86(1):1–21.

Wu, J., D. Zilberman, and B.A. Babcock. 2001. "Environmental and Distributional Impacts of Conservation Targeting Strategies." *Journal of Environmental Economics and Management* 41(3):333–350.

14
Water quality trading

James S. Shortle and Aaron Cook

1 Introduction

Water quality trading (WQT) is generally described as a market-based mechanism that enables individual water pollution sources (e.g., municipal or industrial wastewater dischargers) to purchase environmentally equivalent pollution reductions from other pollution sources (e.g., other municipal or industrial dischargers, sources of agricultural runoff) (Selman et al. 2009). The concept of WQT was introduced by John Dales (1968), who proposed trading as a mechanism for efficiently managing water pollution in Ontario, Canada.[1] Dales proposed the creation of tradable rights to pollute water, where a right would specify an allowed volume of emissions over some time period (e.g., nitrogen emissions per year). A pollution control authority would control the total level of pollution by limiting the total supply of pollution rights available. Dales imagined a market in pollution rights emerging in which polluters with relatively low abatement costs would reduce their emissions and sell these unused rights to higher abatement cost polluters, resulting in an equilibrium allocation of emissions reductions across polluters that would minimize the total costs of pollution abatement. Dales shares credit with Tom Crocker (1966) as the inventor of the concept of pollution permit trading.

Since Dales' and Crocker's proposal, the most prominent and economically significant developments in pollution trading have occurred in the domain of air emissions control (Goulder 2013). Active water quality markets to date are small in number and in economic and geographic scope, but there is substantial interest in WQT as a means to achieve unmet water quality goals at a lower cost than the highly inefficient regulatory approaches that currently dominate water quality protection regimes (Shortle 2013, 2017). In this chapter, we examine the theory and practice of water quality markets, with an emphasis on markets for controlling nutrient pollution. Nutrient pollution control is a high priority for water quality policy makers in most developed countries and remains the target of most WQT programs developed to date. The relevance of WQT to agriculture is almost entirely due to the focus on nutrients in WQT programs and agriculture's large role as a nutrient source.

2 WQT theory: from simple to complex

Formal theoretical economic research on pollution trading has focused on the capacity of markets to achieve cost-minimizing allocations of emissions across multiple sources. While social cost minimization is not the sole objective of policy makers, the expectation that markets can perform well by this criterion is the fundamental reason for policy makers' interest in WQT (Johnstone and Tietenberg 2004).

This section has two main objectives. One is to explore the requirements of efficient allocations relevant to WQT. Specifically, if the purpose of a market is to minimize the social costs of water pollution control, what criteria must the market equilibrium satisfy in order to do so? The second objective is to explore the design of markets that would enable an efficient allocation to be implemented in equilibrium. We explore these issues using a suite of models that illustrate the physical characteristics of the water pollution processes that are of fundamental importance to efficiency criteria and efficient market design.

2.1 The basic cap-and-trade model

We begin with the simplest economic, institutional, and physical environment. A lake is contaminated by a single pollutant discharged from n sources (e.g., farms, municipalities, industrial plants). The emissions from each source can be metered accurately and continuously at little or no cost. Furthermore, emissions have no random determinants and are therefore completely under the control of the individual sources. The discharge from any source i is denoted e_i. Water circulation in the lake mixes the emissions to produce a uniform ambient concentration across the lake. The concentration is determined by the total pollution load (L), which is the sum of the individual emissions ($L = \sum_i^n e_i$).

Environmental damage costs increase with the pollution concentration. We assume the pollutant is assimilated rapidly and does not accumulate, meaning that the concentration in a period (e.g., a month or year as appropriate for management) is determined solely by that period's pollution load. With these assumptions, we write the environmental damage cost function as $D(\sum_i^n e_i)$, where D is increasing, continuous, and convex. Polluters have perfect information about the technologies for controlling their own emissions and choose them so as to minimize their individual costs. Inputs affecting pollution levels are all variable (i.e., capital adjustment costs are zero) and there are no input or product market imperfections or policy distortions that cause divergences between the private and social costs of pollution abatement. For any polluter i, their profit (or utility or other appropriate welfare value measure), conditional on their discharge level e_i, is denoted $\pi_i(e_i)$. The profit functions are continuous and concave. Given that polluting is not inherently profitable, π increases with discharges, reaches a maximum, and then declines. The level of e_i at this maximum represents the discharges that occur when polluter i's resources are configured to maximize profits without regard for the environmental costs. Reductions below this level incur costs.

Given this physical and economic environment, a particular Pareto optimal allocation maximizes the net social benefit of emissions, $NSB = \sum_i^n \pi_i(e_i) - D(\sum_i^n e_i)$. The first-order necessary conditions for an optimal allocation require that the marginal profit from emissions equals the marginal damage cost for each source, $\pi_i' = D'$. Alternatively, the condition can be interpreted as requiring that the marginal cost of pollution abatement (i.e., the profits forgone at the margin from increases in pollution control) equal the marginal benefit of the pollution reduction. We denote the optimal discharge for any source i as e_i^* and the optimal aggregate pollution load (L) as $L^* = \sum_i^{n^*} e_i^*$.

Due to the assumption of uniform mixing, the marginal damage cost of emissions from any source is equal for all polluters regardless of the aggregate pollution level. In consequence, given that the marginal profits (or marginal abatement costs) for each source equals the common marginal damage costs, it is also the case that marginal profits are equalized across sources. The equalization of marginal profits (or abatement costs) is required for the minimization of pollution control costs and is known as the equimarginal principle. An efficient or cost-minimizing allocation for any maximum load target L_{max} maximizes polluters' profits, thus minimizing their control costs, subject to the aggregate emissions being less than or equal to the target. The equimarginal principle is the first-order necessary condition for emissions in the efficient allocation.[2] The intuition is that if the marginal abatement cost for one source exceeds that of another, a reallocation of abatement from the relatively high-cost source to the relatively low-cost source will reduce overall abatement costs. Eliminating all differences in marginal abatement costs will therefore minimize total abatement costs.

With these conditions in hand, the next step is to construct a market. For the market to be efficient, it must implement the equimarginal principle in equilibrium. For overall optimality, it must implement the equimarginal principle and the optimal aggregate pollution level. Research on market design typically assumes that the target load is determined by environmental regulators according to criteria that may not be economic. The focus is, as noted previously, on cost-effectiveness in the allocation of emissions for a politically determined target load rather than complete optimality. If the market is designed efficiently and the target is optimal, then complete optimality is also achieved.

Basic elements of market design are the definition of the commodity traded in the market, the specification of trading rules governing exchanges of the commodity between market participants, an initial distribution of the commodity to market participants, and a cap on the supply of the commodity so as to assure that the pollution target is achieved. With the goal of regulating the pollution load L, the fact that emissions relate directly (and linearly) to the load, are observable at minimal cost, and are completely under the control of the individual polluters, make emissions a natural choice for the tradable commodity. Note also that the efficiency conditions described above pertain specifically to the costs of managing emissions. Regulating emissions directly through discharge permits, rather than indirectly through restrictions on technology, allows polluters the flexibility to choose least-cost pollution control practices as required for efficiency. In the markets originally envisioned by Dales and Crocker, dischargers would be required to hold permits in order to pollute. In the simplest case, a permit would indicate allowable emissions over a specified time period.

For the market to function and for efficiency to be achieved, permits must be tradeable and flow from relatively low-cost sources to relatively high-cost sources until the equimarginal principle is satisfied in equilibrium. Should any restrictions be imposed on exchanges? We should ask because rules governing exchanges are common in WQT programs in the form of trading ratios that establish exchange rates for permits (or the pollution reduction credits we will introduce subsequently) between buyers and sellers based on location, source type, or other factors. For the purposes of environmental protection, trading rules must assure that trading does not have adverse environmental consequences. The fundamental issue here is one of environmental substitution; a trade substitutes emissions reductions by one source for those of another. Under the assumptions about the physical environment in this case, alternative sources are perfect substitutes in terms of their effects on water quality (a unit reduction by one source will have an equivalent environmental impact as a unit reduction by another). This physical environment is described in the economic literature as one with uniform (or perfect) mixing. A trading ratio that requires a unit reduction in emissions by one source to offset a unit increase by another would assure in this environment that trades do not degrade water quality. In this world, the marginal rate of discharge substitution for a given level of environmental quality (MRES), i.e., the rate at which emissions can be exchanged at the margin to maintain a given environmental quality level, between alternative sources is constant and equal to 1. A trading ratio that required permit suppliers' discharge reductions to more than offset demanders' discharge increases would result in an aggregate reduction in emissions and an improvement in water quality. But such a rule would be at odds with the economic objective of abatement cost-minimization.

For the purposes of social cost-minimization, trading rules must facilitate meeting the conditions for abatement cost-minimization given the water quality objective. The issue here is again one of substitution, but in this case substitution in cost. The marginal rate of discharge substitution for a given level of aggregate abatement cost (MRCS), i.e., the rate that emissions can be substituted between two sources to maintain a given aggregate abatement cost, is given by the ratio of their marginal abatement costs. In a least-cost allocation, equality of the marginal abatement costs implies an MRCS of one. Indeed, the conditions for cost-minimization require MRES = 1 = MRCS. For this to be realized, the trade ratio must equal one. To illustrate, suppose we have two sources with identical marginal abatement costs. An efficient allocation would have equal abatement for the two sources. One-to-one trading could achieve this outcome, but a trade ratio other than one would not permit a simultaneous equality of marginal abatement costs and abatement levels. An allocation that achieved the target would necessarily have higher costs than the least-cost allocation.[3]

Having selected the commodity and trading rule, a cap is required on the total supply of permits to assure that the emissions do not exceed the target. Let $\tilde{e}_i$ denote the emissions allowed by source i's permits; source i's actual emissions, e_i, cannot exceed this amount. Accordingly, the target L_{max} will be satisfied if the aggregate supply of permits satisfies the rule $\sum_i \tilde{e}_i \leq L_{max}$. Finally, to implement the market, permits must be distributed by the regulatory authority. Possible mechanisms are auctions or free distribution.

We have now set up a basic cap-and-trade (CAT) market. The next question is what to expect from trading activity, which takes us from trading rules to the behavior of permit holders. A common assumption in theoretical research is that the permit market is perfectly competitive. Given equilibrium market prices, polluters would buy or sell permits to maximize their profits. Suppose that p is the market price of permits. The price will be positive if the load target and corresponding permit supply are binding, creating economic scarcity. Permits can be either bought or sold at this price. A polluter's profit after buying or selling permits is the net revenues from permit sales plus the profits received from its primary economic activities contingent on its actual emissions. Consider polluter i and let $\tilde{x}_i^s$ be the permits sold and $\tilde{x}_i^b$ be those purchased. The profit of the polluter is $\pi_i(e_i) + p(\tilde{e}_i^s - e_i^b)$. The polluter may have an initial endowment of permits. Denote this non-negative endowment $\tilde{e}_i^0$. Given that emissions cannot exceed the polluter's permit holdings, and that a profit maximizer will not hold an unused inventory of permits,[4] it must be the case that $\tilde{e}_i^s + e_i = \tilde{e}_i^b + \tilde{e}_i^0$. Given that the buying and selling prices are the same, there would be no value to the firm to buy permits in order to sell them. In consequence, firms will be either buyers or sellers in the market. Which is the case will depend on the initial endowment, the permit price, and the marginal costs of abatement.

If a polluter has no initial endowment and profits from positive emissions, it will be a buyer. If a polluter has a positive initial endowment, it may be a buyer or seller. For any emissions level, e, a polluter's maximum willingness to pay for an additional permit is the marginal profit it would obtain. The polluter will prefer to buy permits to cover all emissions for which marginal profit exceeds the permit price p and will prefer to reduce e wherever the marginal profit is less than the permit price. In this way, a polluter's economically optimal emissions occur where the marginal profit equals the permit price. Given that all polluters face the same price, the result will be equalization of marginal abatement costs and thus an efficient allocation of abatement across sources for the aggregate pollution target.

The key conclusion from this exercise is that in the very simple economic, institutional, and physical environment that we have assumed, a market can be designed to allocate emissions efficiently across multiple sources. The perfectly competitive market we have assumed coordinates emissions reductions by establishing a uniform price for permits that translates into a uniform price for emissions. The same outcome could be achieved with a uniform tax on emissions that was set equal to the market equilibrium price p^*. But setting the tax rate would require the regulator to know the aggregate marginal abatement cost curve so as to determine the tax that would result in the target load, whereas achieving the target efficiently with the market requires no such knowledge. Alternatively, the efficient outcome could be implemented in theory by firm-specific emissions standards if the regulator again has requisite knowledge of individual costs. In the simple world we have assumed, the regulator need only set up the market and set the aggregate permit supply cap to obtain the efficient solution. This informational efficiency was the foundation of the case for markets made by Dales (1968) and Crocker (1966).[5]

But the world is not so simple. Complexity can be explored along several dimensions. The physical environment is greatly simplified by the assumptions of a single flow pollutant that can be controlled deterministically, is observable at little or no cost, affects a single confined water body, and is uniformly mixed. The economic environment is greatly simplified by implicit assumptions that economic and institutional conditions necessary for a perfectly competitive market equilibrium are met (e.g., nonattenuated property rights, perfect information, zero transactions costs, price taking behavior by polluters in the permit market and in the markets for goods and inputs in which they operate), and that there are no regulatory or fiscal distortions that cause the private and social costs of pollution abatement to diverge. We have assumed implicitly a simple institutional environment in which there exists a single agency with authority over pollution control, and that this agency has the authority and capacities needed to implement, monitor, and enforce the permit market without legal or political constraints that could harm the efficiency of the market.

Substantial and interrelated economic, institutional, and physical complexity exists in the regulation of water pollution from agriculture, and thus in the design of WQT markets for agriculture. We will take up these complexities and their implications for water quality trading to varying degrees in subsequent sections. Particular emphasis is given to complexities in the physical environment, as these pose the greatest challenges to achieving hoped-for efficiency gains in the regulation of water pollution.

2.2 Nonuniform mixing and multiple receptors

The equimarginal principle for an efficient allocation of emissions as expressed above requires that emissions be perfect environmental substitutes. Greenhouse gas emissions are a standard example of uniformly mixed pollutants as these gases spread through the atmosphere from their point of discharge and mix to determine global atmospheric concentrations. Water pollutants of interest for water quality trading programs typically do not mix uniformly.

Consider nutrient pollution, a major target of existing and proposed water quality trading programs. Nutrient pollution problems involve multiple pollutants, with two being of particular concern: nitrogen (N) and phosphorus (P). These nutrients are necessary to healthy aquatic ecosystems but can cause significant environmental harm when concentrations are increased by nutrient-laden runoff from fields and barnyards, industrial and municipal wastewater emissions, and wet or dry deposition of nitrogen compounds emitted into the atmosphere with fossil fuel combustion. Nutrient pollution problems are pervasive and in developed countries often associated with intensive agricultural production (OECD 2012). They are a major concern in U.S. water quality policy making, considered by the US Environmental Protection Agency (USEPA) to become one of the costliest, most difficult environmental problems we face (USEPA 2009). High profile examples of nutrient pollution are found in the Baltic Sea, Chesapeake Bay, Lake Erie, and the Gulf of Mexico.

Nutrients discharged upstream in watersheds are consumed and transformed as they move downstream, making the amount transferred from upstream to downstream locations less than the amount originally released. The delivered fraction to a particular downstream location will vary with the location of the upstream discharge, and generally will be higher for sources that are nearby as compared to sources further away. For example, within the Mississippi Basin, delivered fractions to the mouth vary for nitrogen and phosphorous from less than 10 percent in upper portions of the basin to 100 percent in lower portions (Alexander et al. 2007). This implies that emissions at the source are not perfect environmental substitutes in determining loads to the mouth of the Basin and should not trade one to one in an efficient allocation.

Constructing an efficient market to address spatial variations in delivered fractions is not a major challenge provided that they are known and other assumptions of the basic model are maintained. Returning to the lake, let the fraction of emissions delivered by source i to the lake be denoted $\beta_i (0 \leq \beta_i \leq 1)$. The delivered load that must be managed by the market is now $\sum_i^n \beta_i e_i$. An efficient allocation for a specific load limit L_{max} will maximize polluters' profits subject to the constraint $\sum_i^n \beta_i e_i \leq L_{max}$. With some rearrangement, the first order necessary condition for emissions is $\pi_i' / \beta_i = \pi_j' / \beta_j$ for any two polluters i and j. This is a more general expression of the equimarginal principle, requiring the marginal profits (or marginal abatement costs) of delivered emissions be equal. The natural form of a CAT market for this case is one in which the permits are defined in terms of contributions to the delivered load and the cap placed on the total delivered load. While emissions at the source are not perfect environmental substitutes in this physical environment, delivered emissions are. To illustrate, suppose delivery factors are such that one pound of nitrogen discharges from source A leads to a half-pound of nitrogen delivered to the lake, while one pound of nitrogen discharges from source B leads to a quarter pound of nitrogen delivered to the lake. Given these fixed delivery factors, proper trading rules would dictate that emissions between source A and B would exchange at a two-to-one ratio, i.e., if source A wanted to increase emissions by one pound, source B would have to decrease emissions by two pounds in order offset source A's delivered emissions. A perfectly competitive market in delivered emissions would establish a uniform price for delivered emissions that leads to the implementation of the required equimarginal rule. Following Krupnick, Oates, and Van de Verg (1983), we will refer to this type of market as an ambient permit system (APS). Systems in which permits restrict allowable emissions at the source as opposed to the pollution at receptors are referred to as emissions permit systems (EPS). Other things being equal, the higher a source's delivery ratio (β), the lower its marginal cost of reducing delivered pollution. Accordingly, if marginal costs and delivery ratios are not positively correlated, an APS will tend to allocate higher emission reductions to high delivery ratio sources.

So far, we have considered the management of pollution from multiple sources for a single location. The next step towards the reality of water pollution processes is to allow for emissions from any source to have adverse water quality impacts at multiple locations, generally referred to as receptors in economic research on markets. In U.S. water quality management, water quality standards are defined for specific geographically delineated water resources based on designated uses. Along streams, for example, standards are set to support designated uses for defined stream reaches. In this case, receptors are stream segments. Pollutants moving downstream from the point of discharge may flow through and affect water quality in multiple reaches.

The addition of multiple receptors adds a significant increase in complexity. Suppose there are m receptors in a defined basin. The delivered load to receptor k is L_k, which is determined by upstream emissions. Delivery ratios in the basin must now be differentiated for source and receptor. The delivered fraction of emissions from any source i that reaches receptor k is β_{ik}, which is between zero and 1. Given that water flows downstream, delivery ratios to any receptor k for sources downstream of the receptor will be zero. Ratios for sources upstream of a receptor will vary between zero and 1. The load target for receptor k is denoted (L_{max}^k). Given variations in the characteristics and uses of receiving waters and other factors influencing water quality goals and the effects of pollutants on water quality within receptors, the targets would generally vary across receptors. The analogue to the load targets in U.S. water quality management is the Total Maximum Daily Load (TMDL). A TMDL is the maximum load of a pollutant that can be allowed to enter a water body (on a daily basis in the term but typically implemented annually in practice) and still achieve the water quality standards set for the resource within a margin of error. TMDLs are required by the Clean Water Act (CWA) for waters that do not meet water quality standards based on their designated uses.

An efficient allocation of emissions given the set of targets maximizes polluters' profits subject to m constraints of the form $\sum_{i}^{n} \beta_{ik} e_i \leq L_{max}^k$. Let ρ_k^* denote the shadow price of the load constraint for the kth receptor in the efficient solution. The first order necessary condition for the ith source can then be expressed as $\pi_i' = \sum_{n}^{m} \rho_k^* \beta_{ik}$. Unlike previous cases, the n conditions of this form, one for each polluter, cannot be manipulated to obtain a single equimarginal-type rule.

Extending the APS concept to this case would entail the creation of permits defining allowable delivered pollution and an aggregate load cap for each receptor. Polluters would be required to hold a portfolio of permits for all the receptors they affect. A polluter buying permits to allow an increase in its pollution at a particular receptor would be required to have or buy permits to cover the coinciding increase in pollution in all others it affects. Unlike the single-price EPS and APS, there would be m permit prices, one for each receptor market, in the multi-receptor APS system. In an efficient equilibrium, the market prices would correspond to the efficient shadow prices introduced earlier. Accordingly, the opportunity cost of emissions to an individual polluter is a linear aggregation of the prices in the multiple markets weighted by the polluter's delivery ratios.

In a seminal paper on the theory of trading, Montgomery (1972) compared the cost-effectiveness of APS and EPS. He finds that EPS only guarantees a least-cost trading equilibrium for very specific initial permit allocations, namely those for which the pollution constraint is binding at each receptor. In general, such allocations may not exist. The APS, on the other hand, can achieve the m pollution targets at minimum cost, regardless of the initial permit distribution. The intuition of this result can be understood simply by noting that an APS deploys m instruments (markets/prices), while an EPS only deploys one. APS therefore retains more flexibility in allocating emissions reductions where they are least costly.

While appealing in theory, the multiple markets and multiple prices required by a multiple receptor APS likely imply significant complexity and high transaction costs for market participants. This has led to investigation of alternatives. One approach is the pollution offset system (POS) proposed by Krupnick, Oates, and Van de Verg (1983). The mechanism resembles

Montgomery's EPS in that emissions trade in a single market, but whereas Montgomery included the restriction that no trade could *increase* the pollution level at any receptor, the POS would merely require that no trade produce a *violation* of the pollution constraint at any receptor. This simple modification significantly broadens trading possibilities and makes the cost-minimizing emissions allocation achievable for any initial permit distribution. To ensure that emissions satisfy all pollution constraints, the POS would use a physical pollution model to determine the effects of any emissions reallocation on the ambient pollution levels at all receptors. The regulator would approve trades as long as the resulting spatial distribution of ambient pollution did not violate the pollution target at any receptor. Furthermore, for any trade that caused pollution levels at any receptor to fall below the targeted levels, the agency would distribute additional permits until the pollution constraint was binding at all receptors.

While the POS manages pollution at multiple receptors with a single market, the trading ratios between sources are not specified *ex ante*. Instead, proposed trades are simulated in a physical pollution model and are allowed provided that pollution standards are maintained at all receptors. If a trade violates the standard at any receptor, the rates of substitution between the hypothetical trading partners' emissions are adjusted until the pollution constraints are satisfied. This potentially cumbersome process of calculating allowable trade ratios for each proposed trade both increases administrative and transaction costs and limits firms' capacity to plan. Additionally, the regulator's role of distributing additional permits to eliminate "slack" at any receptor is problematic for two reasons. First, it is not clear who ought to receive these newly created permits. Second, issuing further rights to pollute beyond the level specified at the start of the program might be perceived as inconsistent with the protection of environmental quality.

An alternative approach that has received significant attention as a tractable way to address imperfect substitution in environments with multiple receptors is the use of fixed location-based trade ratios (Hung and Shaw 2005; Farrow et al. 2005). To illustrate, Hung and Shaw's trading-ratio system (TRS) divides a waterway into zones, specifies delivery factors between zones (taking advantage of the fact that pollution moves from upstream to downstream zones and never the reverse), and allows polluters to trade emissions at rates that correspond with these factors. Environmental quality standards are first defined at the level of the zone, which then imply zone-level emissions caps based on the quality standards and emissions delivery structure. When satisfied, these zone-level emissions caps are equivalent to the zone-level pollution criteria. In this way, all polluters can participate in one market, trading a well-defined commodity (emissions) within or across zones at pre-specified rates. Constructing zone-level allowances in this manner also guarantees that the environmental constraints will be binding in each zone, thereby satisfying the condition in Montgomery's EPS that ensures a least cost outcome in an emissions trading market.

2.3 Dynamics

Complex substitution relationships relevant to the definition of efficient outcomes and market design exist over time as well as space. These may be due to various dynamical processes, including capital adjustment costs, lags between the time that pollutants are discharged and the time they enter the waters in which they cause problems, and persistent pollutants that accumulate in the environment over time to cause problem over multiple periods. In such cases, tradeoffs relevant to control costs and environmental outcomes exist between actions taken today versus actions taken in the future. With inter-temporal tradeoffs, the appropriate measures of benefits and costs are present values. An economically optimal allocation will be a path of emissions that maximizes the present value of net benefits. An efficient allocation for a given path will minimize (maximize) the present value of the costs (profits) of achieving the path.

The presence of dynamic processes can lead to significant complexity in efficiency conditions and market design. A problem with simple lags in pollutant delivery serves to illustrate. Returning to the nutrient example, the delivery of nutrients from agricultural land to water systems may be subject to considerable variability and delay due to meteorological and hydrological factors. These delays, sometimes on the order of decades (Bouraoui and Grizzetti 2011; Behrendt et al. 2002), suggest that *when* a pollutant's impact will be felt is as important for optimality as *where*.

To illustrate how delays influence the economics of pollution management, consider n sources whose emissions contribute to the pollution concentration at a single site with heterogeneous delay, and let l be an n-vector of non-negative integers whose ith element denotes firm i's lag length. In this case $l_i = 0$ would indicate that firm i's emissions affect ambient pollution at a receptor immediately, $l_i = 1$ would indicate a one-period delay, and so on. For simplicity, suppose that the n sources differ only with respect to the lag length between emissions discharge and delivery to a single receptor. The present case bears similarities with the spatial one in that efficient allocation must account for heterogeneous impacts among the sources. The difference comes in that here the heterogeneity is in the timing of the pollution effects, rather than their magnitude.

Efficient management of lagged pollutants will entail tradeoffs over time. For example, two sources with different lag lengths may have their pollutants delivered in the same time period, but with costs incurred during different periods. Other things being equal, social discounting will give preference to the shorter lagged source that incurs costs later in time. The appropriate concept of efficiency in this situation is dynamic efficiency. An allocation is dynamic-efficient if it minimizes the present value of the costs of achieving the load target in multiple future periods. Under the assumptions above, an efficient allocation will satisfy a type of equimarginal principle requiring that at any time t emissions from each firm (delivered in period t but discharged in $t - l_i$) should be chosen to equate the present value of marginal abatement costs across all firms.

In theory, an efficient solution for this simple case with lags would be to create permits that specify the allowable deliveries of pollutants to the receptor at particular points in time. Caps would be set for each period, and a polluter discharging at time t with lag time l would be required to have permits for period $t + l$ (and every other period in which their emissions are delivered). For efficiency, polluters would be allowed to trade over time and space. Given well-functioning futures markets, an efficient solution would be achieved in theory.

But this case introduces significant complexity. Like Montgomery's APS system, this would be a market solution with multiple markets (but for multiple time periods rather than multiple locations). Polluters would be required to hold a portfolio of permits and incur the transactions cost of participating in multiple markets over multiple time periods. A polluter seeking to increase its emission rate in the current period would be required to purchase permits for each of the subsequent time periods in which the increased emissions would be delivered. As with spatial heterogeneity, plausible markets would be second best. Given that uncertainty about the future creates profound challenges for the design of even second-best markets, the temporal substitution problem may be substantially more difficult to address with plausible second-best markets than the spatial substitution problem. Investigation into this topic is limited and is ripe for research. Anastasiadis and Kerr (2013) and Shortle et al. (2016) have numerically compared management strategies that do and do not optimally address lags in the context of specific nutrient problems in New Zealand and the U.S. Results of both studies conclude that simple versus more complex management structures may work well in practice, but results are parameter dependent and would not necessarily hold for other problems (Fleming et al. 1995; Hart 2003; Iho 2010; Iho and Laukkanen 2012).

Another example of dynamic complexity relevant to the characterization of efficient allocations and market design is the persistence and nonlinear dynamics of nutrients (particularly P) in water bodies. P dynamics have received significant attention in the study of the "lake problem," in which nonlinear dynamic processes within shallow lake systems produce positive feedbacks in nutrient concentrations. Consequences include multiple locally optimal, long-run outcomes, with the state of the world (e.g., ambient pollution levels) determining which of the various optima should be pursued (Brock and Starrett 2003; Mäler et al. 2003). Efficiency conditions for optimal management are far more complex than in other cases considered, with obvious implications for the complexity of first-best market designs.

2.4 Multiple pollutants

Water quality problems are generally determined by multiple stressors. In our ongoing nutrient pollution example, we have discussed nutrient pollution as a function of multiple pollutants (N and P in particular), but our analysis of efficiency requirements and markets to implement them have been for single pollutant models.

Multiple pollutants introduce a new substitution issue for efficiency requirements and markets. This is the substitution between the pollutant types. When managing two pollutants that jointly produce environmental damage, an economically optimal solution may call for the control of both pollutants (an interior solution) or just one (a corner solution). Kuosmanen and Laukkanen (2011) examine this issue assuming that two pollutants, x and y, determine water quality (z). Environmental damages are an increasing, convex function D of z. An economically optimal allocation maximizes the social net benefit of pollution given by aggregate polluters' profits less environmental damages. They demonstrate that if the social net benefit function is not positive semi-definite at the optimum, a corner solution will prevail, and only one pollutant should be targeted for abatement. In this case, a market for only one pollutant would be required. Otherwise, a market may be required for each. If one or both pollutants must be regulated but the relationship between z and the individual pollutants is linear, then a CAT market in the water quality indicator z can be used to obtain an efficient solution (Zylicz 1994). If the relationship is nonlinear, a market in each pollutant is required for an efficient outcome.

2.5 Unobservable emissions and stochastic processes

The most consequential assumptions we have made about the physical environment are that emissions can be accurately observed at little or no cost and are completely under the control of the polluter. When these assumptions hold, economically appropriate concepts of, and conditions for, efficient allocations are defined in terms of emissions. Permits are optimally defined in terms of emissions or contributions to ambient concentrations that map to emissions. These assumptions are generally reasonable for municipal and industrial point sources of water pollution, but do not apply to nonpoint sources of pollution. The defining characteristic of nonpoint source water pollution is that source-specific emissions cannot be observed accurately and at low cost due to the diffuse and complex pathways pollutants follow from individual sources to receptors. Further, the formation and transport of nonpoint pollutants is typically driven by stochastic weather processes that in turn make nonpoint pollution flows stochastic. Water pollution from agricultural activities is predominantly of the nonpoint type, and for many water quality problems with significant nonpoint causes, agriculture is often the predominant nonpoint source.

The nonpoint problem is well-illustrated by N from agriculture. Farmers apply N to cropland in fertilizer or manure to boost crop output. The N applied to land is partially utilized by crops, but rainfall will cause some to leach into groundwater or to move in surface runoff to streams or lakes. As N is highly labile, some may also escape into the atmosphere. Nitrogen moving through groundwater can move back into surface waters. Nitrogen in surface waters may escape into the atmosphere as gases but subsequently return to land or water, often slowly, over the period of years or decades. Nitrogen escaping from soils into the atmosphere may move in nonreactive forms (forms that do not stimulate plant growth) resulting from the biological process of denitrification, or as ammonia gas, a reactive form. The fate of the N applied is therefore subject to a variety of stochastic processes, and the movement of N from fields and farms to water resources cannot be routinely metered so as to accurately determine contributions to ambient concentrations in managed water resources.

The economic models developed in previous sections to characterize efficient allocations and to frame permit markets are models of point source pollution management. These models do not apply to nonpoint problems because they require observable emissions. Markets for nonpoint management are, however, of considerable interest to policy makers and an area of active market development.

The introduction of unobservable and highly stochastic emissions leads directly to two interrelated issues. One is the nature of the tradeable property – if permits in observable emissions cannot be traded, then what should be traded to manage nonpoint source allocations? A second is the structure of the cap. With stochastic and unobservable emissions, a deterministic cap that requires emissions or ambient concentrations to be below a target level with a 100 percent probability, i.e., the kind of caps we have discussed earlier may be infeasible or entirely uneconomic. The uncertainty about emissions and ambient concentrations implies the need for caps that encompass stochastic variability. These may take a variety of forms (Horan and Shortle 2011; Shortle and Horan 2001, 2013). The simplest would be to place a restriction on average concentrations, but this would not manage harmful spikes and may impose costly restrictions when strong pollution controls are not needed. Another is a safety-first approach in which emissions or ambient concentrations are required to be below a target level with some probability. The TMDL concept introduced earlier is compatible with this approach. A third approach and one with significant economic appeal is a cap on the expected environmental damage costs.[6]

The optimal structure of the permit will depend on the type of the cap. For example, if the cap is placed on the total annual average discharge in a uniformly mixed physical environment, then permits can be defined in terms of annual average emissions. Things get very complex if the variability of pollution is to be regulated as well. In the simplest case, if the cap is a safety-first type limiting the probability of the total annual discharge exceeding a target in a uniformly mixed physical environment, and if the dischargers are large in number and their emissions independently distributed, then the Central Limit Theorem allows a safety-first cap to be written in terms of the annual average emissions and their variances (Beavis and Walker 1983). If the mean and variance of individual emissions are not strictly proportional, permits would be multi-attribute assets establishing caps on means and variances of emissions. If small numbers and/or correlated emissions invalidate the Central Limit Theorem, the structure of the probabilistic cap can become much more complex. Correlated emissions, for example, would require consideration of covariances in emissions in the cap design and in permit specification, essentially creating third-party effects in trades. Further, the optimal feasible set for a safety-first type constraint when emissions are correlated may not be convex (Beavis and Walker 1983).

A key point is that the management of nonpoint pollution is inherently a risk management problem. The distribution of nonpoint source emissions, rather than actual emissions, is what must be managed. Economic research on the design of markets for nonpoint sources has given scant attention to first-best designs for this problem. Instead, the focus has been on second-best market designs that better inform real world policy development. Where we started earlier thinking about the structure of the cap to inform our choice of the tradeable property right, water quality policy makers begin with the latter. Policy makers have generally addressed the commodity definition problem posed by unobservable and stochastic emissions by defining property rights in terms of modeled (predicted) annual average emissions.[7] In markets that integrate point and nonpoint sources, policy makers have recognized that more or less deterministic point source emissions and modeled annual average nonpoint source discharge are not perfect substitutes. Analogous to trade ratios to address heterogeneity in spatial impacts, market designers have used point/nonpoint uncertainty trade ratios to address this imperfect substitution. These are commonly expressed at the ratio of the nonpoint reduction required to offset a forgone point source reduction.

Existing water quality markets and common guidance on market design have point/nonpoint trade ratios that require more than one unit of expected nonpoint pollution reduction to offset a forgone reduction of point source pollution, often by a substantial margin (e.g., ratios commonly exceed 2:1). The argument is that a margin of safety requires the greater mean nonpoint reduction to offset the uncertainty in the nonpoint reduction. This argument fails to appropriately frame the risk management problem, as it implicitly assumes that overall nonpoint risk is independent of the trade. Economic research indicates that optimal trade ratios may be close to one or even less than one (Horan 2001; Horan and Shortle 2005, 2011, 2017; Malik, Letson, and Crutchfield 1983; Shortle 1987, 1990). In addition to correct framing of the risk tradeoffs, an essential element of the economic research is to recognize the endogeneity of risk. Markets that trade expected nonpoint emissions provide direct incentives for controlling mean emissions but no direct incentives for controlling the variance of emissions. Optimal trade ratios for means must consider the variance responses that result from cost-minimizing choices to trade means. If the result is that if the variance moves in the same direction as the mean, then nonpoint trades provide a double dividend of reduced mean emissions and reduced water quality risk. Trade ratios in excess of one would then suboptimally penalize risk-reducing trades with nonpoint sources.

2.6 Emissions reduction credit trading and baseline and credit trading

Before concluding this section on theory, we will discuss an alternative to CAT permit markets, and in doing so take a step towards understanding the actual practice of water quality trading. CAT is the economists' ideal for organizing markets for efficient regulation of polluting emissions and the best platform for thinking about how markets ought to be designed in theory. But CAT is one in a menu of alternative trading mechanisms. Early air emissions trading schemes were emissions reduction credit (ERC) trading mechanisms. ERC trading is essentially a modification of traditional emissions regulations that allows source-specific emissions limits to be met with ERCs acquired from other sources. ERCs are created by pollution sources when they over-comply with regulatory requirements (i.e., emissions are reduced below the regulatory limit). Rules for credit generation can vary, but in the simplest case the credit earned by over-compliance is equal to the reduction in emissions relative to the regulatory limit. Thus, in ERC trading, ERCs flow from sources that over-comply with regulatory requirements to sources that under-comply. Assuming uniform mixing, one to one trading is environmentally neutral and can produce abatement cost-savings.

ERC and CAT trading programs are sometimes discussed as though they are equivalent mechanisms. The differences between them can be profound, depending on the details of implementation (Dewees 2001; Shortle 2012). A fundamental difference is that CAT programs by definition entail explicit caps on aggregate emissions, whereas ERC programs do not.

With very few exceptions, WQT schemes follow the ERC concept. That this is the case in the U.S., the location of all but a few of the extant WQT programs, can be explained by the necessity that WQT programs comply with the CWA. Under the CWA, point sources of pollution are required to possess nontradable discharge permits under the National Pollution Discharge Elimination System (NPDES). The permits specify a technology-based effluent limit (TBEL) and may also specify a more stringent water quality-based effluent limit (WQBEL). Traditionally, point sources were required to meet the limits with their own emissions, but the USEPA issued new policies in the early 2000s that enabled point sources to utilize ERCs generated by overcompliance by other sources to meet TBELs if applicable. The use of ERCs involves no permit trades; instead, the "offsite" effluent reduction is essentially treated as a technology for meeting the permit requirement and is written into the permit.[8] Importantly, the institutional restriction limiting the use of credits to meeting TBELs limits the potential efficiency gains from trading. For maximizing the economic benefits of ERC trading, a source should be able to utilize credits so as to minimize the joint costs of discharge reductions and credit purchases.

A major objective of USEPA trading policy developments in the early 2000s was to reduce the costs of meeting increasingly stringent water quality standards by allowing regulated point sources to purchase cheaper pollution reductions from other point sources or from nonpoint sources. The inclusion of nonpoint sources in trading poses conceptual challenges as noted earlier – most importantly, what to trade when trading with nonpoint sources given that individual nonpoint emissions cannot be metered and are stochastic. The solution, as noted earlier, has been to use estimated nonpoint pollution reductions as a substitute for actual nonpoint discharge reductions in the calculation of nonpoint credits, and in almost all programs to use point/nonpoint trading ratios to address imperfect substitution between point and nonpoint sources. But another issue emerges here. What is the discharge level used to compute nonpoint credits? Addressing this question brings additional conceptual challenges and a set of institutional challenges.

To be conceptually equivalent to a point source credit, a nonpoint's credit should be computed relative to a regulatory limit on its emissions, but such limits do not exist due to differences in policy approaches to the management of point and nonpoint sources. The CWA federalized the control of point sources of pollution and implemented controls through the NPDES permit system. The CWA delegated the authority for nonpoint pollution to the states. To the extent that states have undertaken the regulation of nonpoint sources, they have done so largely through the regulation of farming or other observable land use practices that influence nonpoint pollution loads, such as erosion and sediment controls and nutrient management practices. The focus on practices is understandable given the complexity of metering nonpoint emissions, but absent federal or state discharge limits, there is no regulatory basis analogous to WQBELs for defining nonpoint credits.

The standard solution is a "baseline and credit trading (BCT)" system where the regulator specifies a "baseline" for nonpoint sources and issues "credits" for reductions relative to this baseline. Baselines may be implemented in various ways.[9] In the agricultural context, one would be to define the baseline as the level of the predicted annual average discharge resulting from actual farming practices within a defined time frame and to issue credits for any reduction relative to this level. Given that agricultural nonpoint reductions are generally less costly than point source reductions, this approach maximizes the opportunities for costs savings, though it may incentivize a reduction in the use of conservation practices if farmers anticipate the design. This baseline may also be at odds with applicable state laws if farmers were required by law to meet certain conservation or stewardships standards but had not done so. An alternative that would eliminate such adverse incentives is a baseline based on compliance with set conservation or stewardship standards. This system would be consistent with prior legal requirements, fairer in its treatment of point relative to nonpoint sources, and fairer in the treatment of farmers who are "good environmental actors" compared to those that are not. Pollution reductions through implementation of control practices could produce no credits until those conservation standards were met. This approach conforms to USEPA policy guidance, but necessarily reduces the potential gains from trade by restricting the supply of low-cost agricultural abatement to offset higher cost point sources.

The BCT system raises some issues that do not arise in CAT permit systems. One is the "additionality" of the credited pollution reductions. Credits earned are essentially predictions of future pollution reductions that would not otherwise occur without the actions that produce them. If those actions would have been undertaken in whole or in part in any case, then the credited amount is in part or whole non-additional, and the reduction of the total pollutant load is overstated. This perverse outcome is at odds with the pollution control objectives of a credit-trading program. At its core, the problem arises because of the need to establish some baseline against which to calculate "surplus" pollution reduction and to predict alternative futures.

The risk of non-additional reductions is affected by the baseline choice. To illustrate, suppose the regulator essentially sets no baseline, allowing polluters to generate credits for any and all actions they take to manage nutrient runoff. Under this baseline choice, farms with existing nutrient management infrastructure would be credited with reductions even though those reductions would have occurred in the absence of the crediting program (and would therefore be non-additional). Buyers of these credits would then increase emissions, leading to a net emissions increase. Alternatively, if the regulator sets a strict baseline, requiring all polluters to adopt stringent management practices before becoming eligible to earn credits, the reductions that firms undertake to meet the baseline would not be credited as such and therefore would not lead to higher emissions from credit buyers. In this way, a strict baseline could potentially generate many "additional" credits, where total emissions are reduced on net.

Another issue that arises in the emergent BCT model is referred to as leakage. Emissions by regulated point sources in these programs are individually capped by their NPDES permit. Agricultural nonpoint sources are not. Leakage occurs when a source implements pollution control practices to produce credits but compensates for this action by adjusting production in other ways that are unregulated and unobserved in the credit approval process. The result is that total emissions from the source are reduced by less than what the credit implies. To take an example, suppose a farm removes land from cultivation to install a riparian buffer and simultaneously brings other highly erodible land into cultivation. A regulator may be able to award nutrient reduction credits to this farm for this buffer installation but may not be able to account for the increased nutrient and sediment loads associated with the newly cultivated land without undertaking a more thorough (and costly) assessment of the farm's entire production process. The lack of a comprehensive monitoring system for uncapped sources increases the risk of leakage when these sources become involved in credit trades.

In this review of credit trading, we see the relevance of institutional complexity in the design of trading programs. De novo, an economically optimal trading program would restrict aggregate emissions and would entail rules that allowed the pattern of trading to minimize abatement costs. The emergent BCT trading model seeks to enable trading in a political and institutional environment that restricts design choices in ways that limit economic gains from trade and may pose hidden environmental risks. An additional institutional complexity in U.S. agricultural BCT trading programs is the treatment of government-subsidized pollution reductions. The predominant mechanism for reducing agricultural nonpoint pollution is federal and state cost-sharing of best management practice adoption. The largest program is the U.S. Department of Agriculture Environmental Quality Incentives Program. A policy issue for states developing trading programs is whether to allow farmers who use cost-shared practices to produce credits they can sell. To do so is called double dipping. At first glance, allowing double dipping would seem to reward farmers twice for the same action, implying one payment would be redundant. But this may not be the case if the combined mechanisms lead to actions that the individual mechanisms would not. Some research shows that considering limitations of trading and cost-sharing programs, coordination of these programs can produce better environmental outcome at lower costs than the individual mechanisms (Breetz and Fisher-Vanden 2007; Horan, Shortle, and Abler 2004).

3 Water quality trading programs

Our discussion to this point has focused on the theory of trading. We turn our attention now to a short review of the practice of WQT. Several dozen WQT initiatives have been developed in some form since the mid-1980s.[10] These initiatives include planning exercises and pilot and active trading programs. Agriculture is addressed in many of these initiatives, and almost all are within the U.S. Significant non-U.S. implementations are found in Australia, Canada, and New Zealand. Substantial interest is found elsewhere, most notably among Nordic countries for nutrient pollution control in the Baltic.

Most of the WQT development in the U.S. has occurred since the mid-1990s, prompted by the interest of state water quality authorities in cost-effective approaches to TMDL compliance, encouraged and supported by USEPA policy guidance and USEPA and USDA technical and financial support for trading projects. Most programs are to manage nutrients, especially phosphorous, but programs for selenium, temperature, and dissolved solids have also been developed.

The programs take several forms. One form is a one-time offset agreement in which a point source contracts for pollution reductions with one or more other point or nonpoint sources (outside of an organized multi-source trading system) in order to meet facility-specific standards. In two prominent examples, the Minnesota Pollution Control Agency allowed industrial point sources on the Minnesota River (Rahr Malting Company in 1997 and the Southern Minnesota Beet Sugar Cooperative in 1999) to utilize agricultural and other nonpoint source nutrient pollution reductions to help meet their permit requirements. One-time offsets are a positive development for improving the efficiency of U.S. water pollution control within the context of the legal structure created by the CWA, but they do not provide for the kind of systematic, low-transactions cost trading that is required to fully realize the expected efficiency gains of trading.

Several states have undertaken or enabled the implementation of programs that facilitate trading between multiple sources contributing to the pollution of specific water bodies. These programs are typically implemented as mechanisms to help achieve a specific water quality standard (e.g., a TMDL) (Fisher-Vanden and Olmstead 2013; Morgan and Wolverton 2005). Ideally, for effective and efficient management, trading programs will encompass both point and nonpoint sources when the two are substitutes in determining water quality outcomes. Some that have developed enable trading between both types, but others limit trading to point sources. Significant point–point examples (year developed/pollutant) are the Connecticut Nitrogen Credit Exchange (2002/N), Minnesota River Basin Trading (2006/P), and Virginia's Chesapeake Bay Watershed Nutrient Credit Exchange (2011/N–P). Significant examples of programs that enable trading between point and nonpoint sources are the Greater Miami River Watershed Trading Pilot (2006/N–P), Neuse River Basin Total Nitrogen Trading (2002/N), and the Pennsylvania Nutrient Credit Trading (2006/N–P). These programs are generally intended to allow high cost point sources to achieve abatement cost savings by acquiring credits from low cost agricultural nonpoint sources. Absent requisite regulatory drivers (i.e., regulatory limits on agricultural emissions that could be met through trading), it is no surprise that there are no active programs designed exclusively for agricultural (or other) nonpoint–nonpoint trades in the U.S.[11]

Significant trading programs outside of the U.S. are the Hunter River Salinity Trading Program (2004/salinity), a point–point trading program in New South Wales, Australia; the South Nation River Watershed Trading (2000/P), a point–nonpoint program in Ontario, Canada; and the Lake Taupo nitrogen trading program (2009/N) in Waikato, New Zealand. The Lake Taupo program is unique as it is the only trading program devoted exclusively to agricultural sources.

With the exception of the Hunter River and Lake Taupo programs (both true CAT programs), all active WQT initiatives trade ERCs. Specific rules governing participation, credit generation, and trades vary widely, though there are some general features. Nutrient trading programs typically adjust for spatial variations in delivery when defining credits and spatial trade ratios. Trading with nonpoint sources is commonly discouraged economically through the use of point/nonpoint (or uncertainty) trade ratios that penalize the use of nonpoint credits to offset point source emissions. Agricultural nonpoint credits are based on predictions of annual average steady state reductions, eliminating complex dynamic and stochastic processes.

Coordination mechanisms vary widely in form and their market orientation (Woodward, Kaiser, and Wicks 2002). Some examples illustrate. Though counted as a trading program, the Connecticut Nutrient Credit Exchange (CNCE) is better described as a fixed rate effluent charge system. Dischargers are assigned effluent limits but are allowed to exceed them. Sources that over-comply are awarded ERCs for which they receive payments from the agency managing the program. Sources that under-comply are debited and are required to make payments to the agency for the shortage. The payment rate for credits and debits is not set by the market but by the managing agency. Dischargers do not exchange credits, and the "supply" and "demand" for credits are not required to balance. The South Nation River program provides P ERCs to point sources that are generated by agricultural nonpoint reductions. Unlike the CNCE, credits supplied and demand must balance, but like the CNCE, the price per credit is set by a trading authority rather than by the market. The trading authority, also the credit supplier, uses trading revenues to finance agricultural pollution control projects analogous to traditional public financing of agricultural conservation through cost sharing programs. Individual farmers do not sell credits or participate in the market.

Bilateral exchange and clearinghouse type mechanisms are found in the more market-like programs, in which exchanges of credits occur between buyers and sellers and prices are determined by the market participants (Fisher-Vanden and Olmstead 2013; Morgan and Wolverton 2005; Shortle 2013; Shortle and Horan 2013; Woodward, Kaiser, and Wicks 2002). Among the more interesting of these from the perspective of market design are the Pennsylvania Nutrient Trading Program (PNTP), the Hunter River Salinity Trading (HRST) program, and the Greater Miami River Watershed Trading Pilot (GMRWTP). The PNTP was initially developed as a bilateral exchange program in which buyers and sellers would negotiate terms subject to trading rules. The state of Pennsylvania subsequently developed a formal double auction market to connect buyers and sellers and set prices. Most trades, however, occur outside of the formal market. The HRST operates an online trading tool to connect buyers and sellers and set prices. The GMRWTP operates periodic reverse auctions in which farmers, in cooperation with their local participating Soil and Water Conservation District, sell credits to a credit trading authority. The authority has been funded by participating point sources and by grants from federal agencies as a trading demonstration project.

The most noted feature of WQT programs developed to date is the lack of trading activity. Most have had few if any trades, and there are no major success stories such as those found for air emissions trading or from the application of markets to fisheries management. This leads some analysts to be pessimistic about the potential to realize the expected cost-savings. However, it is important not to give too much weight to poor performing programs. Program assessments indicate that in many instances these programs suffer from limited investment in the development of successful platforms, design flaws, and an absence of economic fundamentals needed to drive trades. It is also important not to develop expectations for WQT based on air emissions trading. The physical, economic, and institutional environments for air emissions management are far more conducive to fluid, high volume, efficient markets than water quality environments (Fisher-Vanden and Olmstead 2013). We should note that the CWA is itself a significant institutional barrier to efficient markets in the U.S. (Fisher-Vanden and Olmstead 2013; Stephenson and Shabman 2011). Limiting trading to only WQBEL compliance and the exclusion of agriculture from an integrated capped point–nonpoint management system are fundamental limitations (Shortle and Horan 2013).

4 Concluding comments

Given the importance of controlling agricultural nonpoint pollution to significant progress on unmet water quality goals and to the overall efficiency of water quality protection, and given policy makers' evident interest in WQT, a key question is whether markets can be designed to effectively and efficiently reduce agricultural nonpoint pollution. Although there are no "great" successes in agricultural WQT, several experiments suggest that the mechanism has potential as an element of better water quality policy (Shortle 2013). The Greater Miami River, Lake Taupo, and South Nation River cases are especially encouraging for the potential of WQT programs that include agriculture. While not the kind of market or trading envisioned by economists, the South Nation River program has been effective in using novel incentives to increase agricultural nonpoint pollution control activity and reduce the costs of point source compliance (Shortle 2013). The Greater Miami River program has demonstrated that carefully designed market mechanisms can be effective in inducing farmers to participate in a water quality market and adopt pollution control practices (Newburn and Woodward 2012; Shortle 2013). The Lake Taupo program is the only true cap-and-trade program including agricultural sources and is also uniquely limited to agricultural sources. Assessments indicate that trading activity, while modest, is advancing the goal of efficient achievement of the water quality goals (Duhon, McDonald, and Kerr 2015). Like other WQT programs, Lake Taupo's includes design elements that facilitate trading and public acceptance, which are not normally envisioned in textbook EPS and APS markets.

Experience is an essential guide to the design of policy instruments for complex environmental problems. The experience with WQT to date is, however, too limited to provide significant guidance about what does and does not work in various settings. The same can be said about theory. While there is a substantial body of research on the design of trading systems, the theory is of limited relevance to WQT because of the focus on generic point source problems. Theoretical research on nonpoint trading is limited to a few articles. Additional theoretical research is needed that addresses the complexity of the water quality management problem emerging from multiple stressors, the spatial and temporal processes of water pollution, stochastic components, and various forms of uncertainty. Although the guidance from a robust theory remains to be developed, the complexity of the problem ultimately implies that plausible markets will be second best or worse.

Given the limits of theory, and the fact that the best choice among a set of suboptimal choices is inherently an empirical problem, *ex ante* empirical assessments of WQT markets are also essential to good policy guidance. Some studies assume that market equilibria will be cost-minimizing for the market design and explore the effects of design parameters (e.g., uncertainty trade ratios, BCT baseline definitions, participation rules) on the efficiency of the market using simulation models (e.g., Ghosh, Ribaudo, and Shortle 2011; Horan, Shortle, and Abler 2002; Horan et al. 2002; Horan and Shortle 2005, 2017; Ribaudo, Heimlich, and Peters 2005; Ribaudo, Savage, and Talberth 2014). Some of these studies use models that embrace characteristics of the agricultural nonpoint pollution problem, but most do not, instead applying point source paradigms. The assumption that even well-designed markets will be efficient is a strong one given that WQT markets will often be thin, will involve heterogeneous participants acting under various forms of uncertainty, and will entail high transactions costs (DeBoe and Stephenson 2016; Fang, Easter, and Brezonik 2005; Fisher-Vanden and Olmstead 2013; Nguyen et al. 2013; Woodward, Kaiser, and Wicks 2002). This has led to a few studies focused on individual and aggregate behavior under alternative market structures using experimental economic and agent-based models (e.g., Nguyen et al. 2013; Suter, Spraggon, and Poe 2013). Within this set, Nguyen et al. (2013) is of particular interest for a robust consideration of alternative types of transactions costs (e.g., search, contracting, information) and their impacts on trading volumes and gains from trade under alternative market structures. Limited case studies indicate "noneconomic factors," such as participants' experience with and trust of institutional actors and public engagement in market development, can be important to the success of agricultural markets (e.g., Breetz et al. 2005; O'Grady 2011; Newburn and Woodward 2012).

Like theory, empirical research to guide market design is in short supply. Particularly important is research that embraces the scope of the complex physical and institutional environments of nonpoint pollution and explores market design within a mixed instrument environment (Fisher-Vanden and Olmstead 2013; Horan and Shortle 2011; Shortle 2013, 2017; Shortle and Horan 2013, 2017).

Notes

1. Dales did not use the now standard label WQT to describe trading the rights to pollute water. Effluent trading was the more common label until recently.
2. We do not address the issue formally here, but there is an additional requirement for cost-minimization and net benefit maximization, and this is that the set of polluters is cost-minimizing for the pollution level. It is for this reason that we have a superscript ⋆ on the number of polluters in the summation of emissions to the optimal total load L^*. To illustrate, suppose that all sources have identical cost functions. A least-cost allocation will impose the same total cost as well as marginal cost on each polluter. The total cost for all polluters will equal the cost per polluter times the number of polluters. If a lower total cost can be achieved by increasing (decreasing) the number of polluters while decreasing (increasing) the cost per polluter so as to satisfy the equimarginal principle and achieve the target, then the efficient allocation will require an optimal number of polluters as well as an optimal level of pollution per polluter. In the context of a market, the question would be what is the optimal size of the market in terms of the number of participants given efficient allocations among those who are in the market? With differential costs, the optimal structure will include not only an optimal number of sources, but an optimal set of cost functions (Spulber 1985).

3 Trade ratios that would require that trade produce net discharge reductions have environmental appeal in that trades that occur would produce environmental quality improvements. Two economic issues emerge. One is whether the benefits of the environmental gains exceed the costs. A second is that such a trade ratio acts as a tax on trading and would be expected to depress trading activity and limit the desired cost-saving from trading.
4 In a dynamic model, a firm may hold or bank permits for future use. This will not be the case in a static model such as this, as permits have no future value.
5 The informational efficiency of markets relative to emissions taxes for achieving specific load targets at least cost in the simple world modeled does not imply that markets are economically preferred to taxes. If the objective of the environmental authority is to maximize the benefits of emissions less the environmental damage costs, but polluter's benefits (or abatement costs) are private, implying *ex ante* uncertainty about their individual and aggregate responses to taxes, it may prefer a tax to a market or a market to a tax. Other things being equal, a single price on emissions, established by market or administratively determined as a tax, will allocate emissions efficiently across sources when the regulator has asymmetric information about polluters' costs. However, the realized economic and environmental outcomes, and therefore the *ex post* efficiency of the instruments, will generally differ under the two approaches when designed with asymmetric information. Maintaining the simplifying assumptions about the economic and physical environment, the relative ranking is an empirical question dependent on the shapes of the aggregate marginal emissions abatement benefit and cost curves and their joint distribution (Weitzman 1974; Stavins 1996; Malcomson 1978; Borisova, Shortle, and Abler 2005). In a nutshell, if the *ex post* optimal pollution level does not vary much with under alternative states of the world, quantity controls have lower expected costs than taxes. The reverse is true if the *ex post* optimal pollution level varies widely across states of the world.
6 The first-best cap and trade models we have presented that apply to point sources achieve water quality targets at least cost. If those targets are optimally selected from an economic perspective to minimize the social costs of pollution and its control, then the markets are not only cost-effective for the targets but Pareto optimal. Of the various probabilistic caps for a stochastic pollutant, only a market with an expected damage cost cap can implement *ex ante* Pareto optimal allocation (Shortle and Horan 2001, 2013).
7 Numerous mathematical models have been developed by researchers to manage nonpoint emissions. The models use data on soils, topography, land use practices, weather, and other determinants of nonpoint emissions. These models exhibit significant predictive uncertainty because they are subject to significant errors related to spatial resolution, the quality of input data on the physical characteristics of landscapes, weather, and land use practices, parameter errors, and structural model error (Arabi et al. 2012; Fishbach et al. 2015; Robertson et al. 2009; Veith et al. 2010).
8 The term permit trading is used extensively in the literature to refer to the trading of emissions allowances, and we have used that term in presenting the theory of emission trading. But trading need not entail permit trading. This is illustrated by the ERC trading we describe in this section. Under the CWA, discharge permits are not tradable. ERCs can be utilized as a "technology" to meet specific WQBEL requirements but not to meet TBEL requirements.
9 State regulation of agricultural nonpoint sources has been very modest in comparison to that of point sources, with the result that agricultural nonpoint pollution is often a leading cause of remaining water quality problems and a comparatively low-cost source of pollution reductions. Policy approaches to agricultural nonpoint pollution control have to a large extent given farmers a presumptive right to pollute that does not exist for point sources. This political and institutional fact means that water quality markets that integrate point and agricultural nonpoint sources are generally designed to achieve point source costs savings through the purchase of credits from agricultural nonpoint sources.
10 Economically focused case studies and assessments of WQT include Breetz et al. 2005; Duhon, McDonald, and Kerr 2015; Duhon, Young, and Kerr 2011, 2015; Fisher-Vanden and Olmstead 2013; Morgan and Wolverton 2005; Newburn and Woodward 2012; Ribaudo and Gottlieb 2011; Shabman and Stephenson 2007; Shortle 2012, 2013; Shortle and Horan 2008, 2013; Stephenson and Shabman 2011; Woodward, Kaiser, and Wicks 2002).
11 The now inactive California's Grassland Area Farmer trading program enabled trading between irrigation districts to meet a cap on selenium in the late 1990s. Although agricultural runoff was the source of the selenium, collection systems converted the nonpoint sources into point sources.

References

Alexander, R.B., R.A. Smith, G.E. Schwarz, E.W. Boyer, J.V. Nolan, and J.W. Brakebill. 2007. "Differences in Phosphorus and Nitrogen Delivery to the Gulf of Mexico from the Mississippi River Basin." *Environmental Science and Technology* 42(3):822–830.

Anastasiadis, S., and S. Kerr. 2013. "Mitigation and Heterogeneity in Management Practices on New Zealand Dairy Farms." Working Paper 13–11, Motu Economic and Public Policy Research, Wellington, New Zealand.

Arabi, M., D.W. Meals, and D.L.K. Hoag. 2012. "Watershed Modelling: National institute of Food and Agriculture – Conservation Effects Assessment Project." In D.L. Osmond, D.W. Meals, D.L.K. Hoag, and M. Arabi, eds., *How to Build Better Agricultural Conservation Programs to Protect Water Quality: National Institute of Food and Agriculture Conservation Effects Assessment Project*. Ankeny, IA: Soil Conservation Society of America, pp. 84–119.

Beavis, B., and M. Walker. 1983. "Achieving Environmental Standards with Stochastic Emissions." *Journal of Environmental Economics and Management* 10(2):103–111.

Behrendt, H., M. Kornmilch, D. Opitz, O. Schmoll, and G. Scholz. 2002. "Estimation of the Nutrient Inputs Into River Systems – Experiences from German Rivers." *Regional Environmental Change* 3(1):107–117.

Borisova, T., J. Shortle, R.D. Horan, and D. Abler. 2005. "Value of Information for Water Quality Management." *Water Resources Research* 41(6):WR06004.

Bouraoui, F., and B. Grizzetti. 2011. "Long-Term Change of Nutrient Concentrations of Rivers Discharging in European Seas." *Science of the Total Environment* 409(23):4899–4916.

Breetz, H.L., and K. Fisher-Vanden. 2007. "Does Cost-Share Replicate Water Quality Trading Projects? Implications for a Possible Partnership." *Review of Agricultural Economics* 29(2):201–215.

Breetz, H.L., K. Fisher-Vanden, H. Jacobs, and C. Schary. 2005. "Trust and Communication: Mechanisms for Increasing Farmers' Participation in Water Quality Trading." *Land Economics* 81(2):170–190.

Brock, W.A., and D. Starrett. 2003. "Managing Systems with Non-Convex Positive Feedback." *Environmental and Resource Economics* 26(4):575–602.

Crocker, T.D. 1966. "The Structuring of Atmospheric Pollution Control Systems." In H. Wolozin, ed., *The Economics of Air Pollution*. New York: W.W. Norton, pp. 61–86.

Dales, J. 1968. *Pollution, Property and Prices*. Toronto: University of Toronto Press.

DeBoe, G., and K. Stephenson. 2016. "Transactions Costs of Expanding Nutrient Trading to Agricultural Working Lands: A Virginia Case Study." *Ecological Economics* 130:176–185.

Dewees, D.N. 2001. "Emissions Trading: ERCs or Allowances?" *Land Economics* 77(4):513–526.

Duhon, M., H. McDonald, and S. Kerr. 2015. "Nitrogen Trading in Lake Taupo: An Analysis and Evaluation of an Innovative Water Management Policy." doi:10.2139/ssrn.2653472.

Duhon, M., J. Young, and S. Kerr. 2011. "Nitrogen Trading in Lake Taupo: An Analysis and Evaluation of an Innovative Water Management Strategy." *New Zealand Agricultural and Resource Economics Society Conference*. New Zealand: Nelson.

Fang, F., K.W. Easter, P.L. Brezonik. 2005. "Point-Nonpoint Source Water Quality Trading: A Case Study in The Minnesota River Basin." *JAWRA Journal of the American Water Resources Association* 41(3):645–657.

Farrow, R.S., M.T. Schultz, P. Celikkol, and G.L. Van Houtven. 2005. "Pollution Trading in Water Quality Limited Areas: Use of Benefits Assessment and Cost-Effective Trading Ratios." *Land Economics* 81(2):191–205.

Fischbach, J.R., R.J. Lempert, E. Molina-Perez, A.A. Tariq, M.L. Finucane, and F. Hoss. 2015. *Managing Water Quality in the Face of Uncertainty*. Santa Monica, CA: RAND Corporation.

Fisher-Vanden, K., and S. Olmstead. 2013. "Moving Pollution Trading from Air to Water: Potential, Problems, and Prognosis. *The Journal of Economic Perspectives* 27(1):147–171.

Fleming, R.A., R.M. Adams, and C.S. Kim. 1995. "Regulating Groundwater Pollution: Effects of Geophysical Response Assumptions on Economic Efficiency." *Water Resources Research* 31(4):1069–1076.

Ghosh, G., M. Ribaudo, and J. Shortle. 2011. "Baseline Requirements Can Hinder Trades in Water Quality Trading Programs: Evidence from the Conestoga Watershed." *Journal of Environmental Management* 92:2076–2084.

Goulder, L.H. 2013. "Markets for Pollution Allowances: What Are the (New) Lessons?" *Journal of Economic Perspectives* 27(1):87–102.

Hart, R. 2003. "Dynamic Pollution Control – Time Lags and Optimal Restoration of Marine Ecosystems." *Ecological Economics* 47(1):79–93.

Horan, R.D. 2001. "Differences in Social and Public Risk Perceptions and Conflicting Impacts on Point/Nonpoint Trading Ratios." *American Journal of Agricultural Economics* 83(4):934–941.

Horan, R.D., and J.S. Shortle. 2005. "When Two Wrongs Make a Right: Second Best Point Nonpoint Trading." *American Journal of Agricultural Economics* 87(2):340–352.

———. 2011. "Economic and Ecological Rules for Water Quality Trading." *Journal of the American Water Resources Association* 47(1):59–69.

———. 2017. "Endogenous Risk and Point-Nonpoint Uncertainty Trading Ratios." *American Journal of Agricultural Economics* 99(2):427–446.

Horan, R., D. Abler, J. Shortle, and J. Carmichael. 2002. "Probabilistic, Cost-Effective Point/Nonpoint Management in the Susquehanna River Basin." *Journal of the American Water Resource Association* 38:467–477.

Horan, R., J. Shortle, and D. Abler. 2002. "Nutrient Point-Nonpoint Trading in the Susquehanna River Basin." *Water Resources Research* 38(5):WR000853.

Horan, R.D., J.S. Shortle, and D.G. Abler. 2004. "Point-Nonpoint Trading Programs and Agri-Environmental Policies." *Agricultural and Resource Economics Review* 33(1):61–78.

Hung, M.F., and D. Shaw. 2005. "A Trading-Ratio System for Trading Water Pollution Discharge Permits." *Journal of Environmental Economics and Management* 49(1):83–102.

Iho, A. 2010. "Spatially Optimal Steady-State Phosphorus Policies in Crop Production." *European Review of Agricultural Economics* 37(2):187–208.

Iho, A., and M. Laukkanen. 2012. "Precision Phosphorus Management and Agricultural Phosphorus Loading." *Ecological Economics* 77:91–102.

Johnstone, N., and T. Tietenberg. 2004. "Ex Post Evaluation of Tradeable Permits." *Tradeable Permits*. Paris: OECD Publishing. pp. 9–44.

Kuosmanen, T., and M. Laukkanen. 2011. "(In) Efficient Environmental Policy with Interacting Pollutants." *Environmental and Resource Economics* 48(4):629–649.

Krupnick, A.J., W.E. Oates, and E. Van De Verg. 1983. "On Marketable Air-Pollution Permits: The Case for a System of Pollution Offsets." *Journal of Environmental Economics and Management* 10(3):233–247.

Malcomson, J.M. 1978. "Prices vs. Quantities: A Critical Note on the Use of Approximations." *The Review of Economic Studies* 45(1):203–207.

Mäler, K.G., A. Xepapadeas, and A. De Zeeuw. 2003. "The Economics of Shallow Lakes." *Environmental and Resource Economics* 26:603–624.

Malik, A.S., D. Letson, and S.R. Crutchfield. 1993. "Point/nonpoint Source Trading of Pollution Abatement: Choosing the Right Trading Ratio." *American Journal of Agricultural Economics*, 75(4):959–967.

McCann, L., B. Colby, K.W. Easter, A. Kasterine, and K.V. Kuperan. 2005. "Transaction Cost Measurement for Evaluating Environmental Policies." *Ecological Economics* 52(4):527–542.

Montgomery, W.D. 1972. "Markets in Licenses and Efficient Pollution Control Programs." *Journal of Economic Theory* 5(3):395–418.

Morgan, C., and A. Wolverton. 2005. "Water Quality Trading in the United States." National Center for Environmental Economics Working Paper (05–07), USEPA, Washington, DC.

Newburn, D.A., and R.T. Woodward. 2012. "An Ex Post Evaluation of Ohio's Great Miami Water Quality Trading Program." *Journal of the American Water Resources Association* 48(1):156–169.

Nguyen, N.P., J.S. Shortle, P.M. Reed, and T.T. Nguyen. 2013. "Water Quality Trading with Asymmetric Information, Uncertainty and Transaction Costs: A Stochastic Agent-Based Model." *Resource and Energy Economics* 35(1):60–90.

O'Grady, D. 2011. "Sociopolitical Conditions for Successful Water Quality Trading in the South River Nation Watershed, Ontario, Canada." *Journal of the American Water Resources Association* 47(1):39–51.

Organisation for Economic Co-operation and Development (OECD). 2012. *Water Quality and Agriculture: Meeting the Challenge*. Paris: OECD Publishing.

Ribaudo, M.O., and J. Gottlieb. 2011. "Point-Nonpoint Trading – Can It Work?" *Journal of the American Water Resources Association* 47(1):5–14.

Ribaudo, M.O., R. Heimlich, and M. Peters. 2005. "Nitrogen Sources and Gulf Hypoxia: Potential for Environmental Credit Trading." *Ecological Economics* 52(2):159–168.

Ribaudo, M., J. Savage, and J. Talberth. 2014. "Encouraging Reductions in Nonpoint Source Pollution through Point-Nonpoint Trading: The Roles of Baseline Choice and Practice Subsidies." *Applied Economic Perspectives and Policy* 36(3):560–576.

Robertson, D.M., G.E. Schwarz, D.A. Saad, and R.B. Alexander. 2009. "Incorporating Uncertainty into the Ranking of SPARROW Model Nutrient Yields from Mississippi/Atchafalaya River Basin Watersheds." *Journal of the American Water Resources Association* 45(2):534–549.

Selman, M., S. Greenhalgh, E. Branosky, C. Jones, and J. Guiling. 2009. "Water Quality Trading Programs: An International Overview." World Resources Institute, Washington, DC.

Shabman, L., and K. Stephenson. 2007. "Achieving Nutrient Water Quality Goals: Bringing Market-Like Principles to Water Quality Management." *Journal of the American Water Resources Association* 43(4):1076–1089.

Shortle, J. 2012. *Water Quality Trading in Agriculture*. Paris: OECD Publishing.

———. 2013. "Economics and Environmental Markets: Lessons from Water-Quality Trading." *Agricultural and Resource Economics Review* 42(1):57–74.

———. 2017. "Policy Nook: Economic Incentives for Water Quality Protection." *Water Economics and Policy* 3(2):1771004.

Shortle, J., D. Abler, Z. Kaufman, and K.Y. Zipp. 2016. "Simple vs. Complex: Implications of Lags in Pollution Delivery for Efficient Load Allocation and Design of Water-Quality Trading Programs." *Agricultural and Resource Economics Review* 45(2):367–393.

Shortle, J., and R. Horan. 2001. "The Economics of Nonpoint Pollution." *Journal of Economic Surveys* 15:255–290.

———. 2008. "The Economics of Water Quality Trading." *International Review of Environmental and Resource Economics* 2(2):101–133.

———. 2013. "Policy Instruments for Water Quality Protection." *Annual Review of Resource Economics* 5(1):111–138.

Shortle, J.S. 1987. "The Allocative Implications of Comparisons Between the Marginal Costs of Point and Nonpoint Source Pollution Abatement." *Northeastern Journal of Agricultural and Resource Economics* 17:17–23.

———. 1990. "The Allocative Efficiency Implications of Abatement Cost Comparisons." *Water Resources Research* 26:793–797.

Spulber, D.F. 1985. "Effluent Regulation and Long-Run Optimality." *Journal of Environmental Economics and Management* 12(2):103–116.

Stavins, R.N. 1996. "Correlated Uncertainty and Policy Instrument Choice." *Journal of Environmental Economics and Management* 30(2):218–232.

Stephenson, K., and L. Shabman. 2011. "Rhetoric and Reality of Water Quality Trading and the Potential for Market-Like Reform." *Journal of the American Water Resources Association* 47(1):15–28.

Suter, J.F., J.M. Spraggon, and G.L. Poe. 2013. "Thin and Lumpy: An Experimental Investigation of Water Quality Trading." *Water Resources and Economics* 1:36–60.

U.S. Environmental Protection Agency (USEPA). 2009. "An Urgent Call to Action: Report of the State-EPA Nutrient Innovations Task Group." USEPA, Washington, DC.

Veith, T.L., M.W. Van Liew, D.D. Bosch, and J.G. Arnold. 2010. "Parameter Sensitivity and Uncertainty in SWAT: A Comparison Across Five USDA-ARS Watersheds." *Transactions of the ASABE* 53(5):1477–1485.

Weitzman, M.L. 1974. "Prices vs. Quantities." *The Review of Economic Studies* 41(4):477–491.

Woodward, R.T., R.A. Kaiser, and A.M.B. Wicks. 2002. "The Structure and Practice of Water Quality Trading Markets." *Journal of the American Water Resources Association* 38(4):967–979.

Zylicz, T. 1994. "Improving Environment Through Permit Trading: The Limits to a Market Approach." In E.C. van Ierland, ed., *International Environmental Economics: Theories, Models and Applications to Climate Change, International Trade and Acidification*. Amsterdam: Elsevier, pp. 283–306.

15
Natural capital and ecosystem services

Marcello Hernández-Blanco and Robert Costanza

1 A short history of natural capital and ecosystem services

Gomez and De Groot (2010) state that the concept of natural capital was introduced for the first time in 1973 by Schumacher in his book entitled *Small Is Beautiful: A Study of Economics As If People Mattered* (Gómez-Baggethun and De Groot 2010, p. 108). The term "nature's services" appeared for the first time in the literature in a paper published in *Science* by Walter Westman, titled "How much are nature's services worth?"(Westman 1977). "Ecosystem services" as synonymous to "nature's services" was mentioned for the first time in Ehrlich and Ehrlich (1981), and more systematically in Ehrlich and Mooney (1983).

In 1988, Pearce made one of the earliest introductions to the concept of natural capital, stating that "sustainability requires at least a constant stock of natural capital, construed as the set of all environmental assets" (Pearce 1988). Pearce's goal was to stimulate discussion and research around the topic of sustainability within the field of neoclassical economics. As Akerman states, the concept was then redefined by Costanza and Daly, who brought ecosystem thinking into economic analysis, implying a theoretical change in the understanding of how both ecological and economic systems worked, opening the path for the emerging field of ecological economics (Akerman 2003, p. 443). A more detailed history of ecosystem services focused on its economics roots is provided by Gómez-Baggethun, de Groot, Lomas, and Montes (2010) and L. C. Braat and de Groot (2012), who summarize the history of the concept from the perspective of ecology, economics, and ecological economics.

The year 1997 was a turning point in research and the conceptualization of natural capital and ecosystem services. First, the book *Nature's Services: Societal Dependence on Natural Ecosystems* (Daily 1997) was published, the product of a meeting in October 1995 of Pew Scholars in Conservation and the Environment in New Hampshire, which included scholars such as Jane Lubchenco, Stephen Carpenter, Paul Ehrlich, Gretchen Daily, Hal Mooney, Robert Costanza, and others. Second, during this meeting Robert Costanza proposed the idea to synthesize all the information being assembled and develop a global assessment of the value of ecosystem services. This was done through a workshop called "The Total Value of the World's Ecosystem Services and Natural Capital," held on 17–21 June 1996 with the financial support of the U.S. National Science Foundation (NSF)-funded National Center for Ecological Analysis and

Natural capital and ecosystem services

Synthesis (NCEAS) and with the participation of 13 scholars from a range of disciplines. The results were published in *Nature* (Costanza et al. 1997). They provided a "meta-analysis" of all existing studies on 17 ecosystem services across 16 biomes that were valued in the range of US$16–54 trillion per year, with an average of US$33 trillion per year, a value significantly higher than gross domestic product (GDP) at the time. These two publications sparked an explosion of research and policy interest in ecosystem services, helping to visualize the dependence that humans have on healthy ecosystems and therefore the importance of protecting natural capital for human well-being.

2 Classifying resources: basic principles for natural capital definition

Before analyzing the concept of capital (and specifically natural capital), we need to consider some basic definitions that are implicit in it. First, it is important to make a distinction between types of scarce resources, stock-flow and fund-service. On the one hand, in Daly and Farley (2004), Georgescu-Roegen defines a stock-flow resource as one that is materially transformed into what it produces, can be used at any rate desired, can be stockpiled, and is used up instead of worn out (e.g., goods such as timber, water, minerals, and fish). On the other hand, a fund-service resource is defined as one that cannot be materially transformed into what it produces, can only be used at a given rate, cannot be stockpiled, and is worn out instead of used up (e.g., services such carbon sequestration, erosion control, pollination, and water retention) (Daly and Farley 2004, p. 71).

Second, the classification of resources under the principles of excludability and rivalry is key because it is directly related to the concepts of stock-flow and fund-service. An excludable resource is one which its owner can use while simultaneously denying its use to others (the opposite is a non-excludable resource). A rival resource is one that, when consumed or used by one person, reduces the amount available for everyone else, and a non-rival resource is one in which the use by one person does not affect its use by another. In general terms, most stock-flow resources are rival, while fund-service resources are non-rival (Daly and Farley 2004, p. 73) (Figure 15.1).

These definitions frame both the consumption possibilities of resources as well as their governance, which at the end determines their sustainability, key to maintaining the well-being of current and future generations.

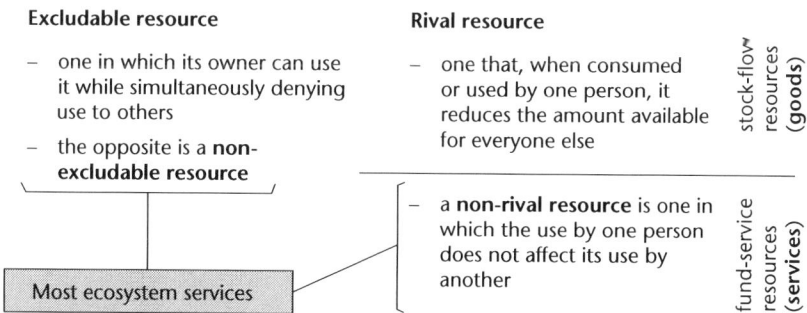

Figure 15.1 Types of scarce resources. Most ecosystem services are non-excludable and non-rival, which pose a challenge for their sustainable management

3 Natural capital concept

Capital can be defined as a "stock of materials or information that exists at a point in time" (Costanza et al. 1997, p. 254) or, moreover, as "a stock of something that yields a flow of useful goods or services" (Costanza et al. 2014, p. 119).

Classical economics identifies three economic factors of production: land, labor, and human-made capital. Neoclassical economics tends to focuses primarily on labor and human-made capital in its production functions, omitting land. Corresponding to these three traditional economic factors of production, three types of capital can be defined as natural, human, and manufactured or built capital (Costanza and Daly 1992, p. 38, and T. Prugh et al. 1995, p. 53). Moreover, Ekins (2003) proposes a disaggregation of the capital stock, adding a fourth type of capital: social capital (Ekins et al. 2003, p. 166). Costanza (2014) states that these four types of capital are necessary to support the economy and its goal of providing human well-being, and describes each one of them as follows:

- Natural capital: the natural environment and its biodiversity; it is the planet's stock of natural resources, the ecosystems that provide benefits to people (i.e., ecosystem services).
- Social capital: the web of interpersonal connections, social networks, cultural heritage, traditional knowledge, and trust, and the institutional arrangements, rules, norms, and values that facilitate human interactions and cooperation between people.
- Human capital: human beings and their attributes, including physical and mental health, knowledge, and other capacities that enable people to be productive members of society.
- Built capital: buildings, machinery, transportation infrastructure, and all other human artifacts and services (Costanza et al. 2014, pp. 129–130).

Following the definition of capital cited earlier, natural capital can be defined as "a stock of natural resources (i.e., ecosystems) that yield a flow of goods and services (i.e., ecosystem services)," such as the case of a mangrove forest that provides food and water filtration to communities. Costanza and Daly explain the flow of goods and services as the "natural income" and the stock that yields the flow as the "natural capital" (Costanza and Daly 1992, p. 38). Sustainability (more on this later) is therefore centered in the wise use of income; depleting the stocks is called capital consumption (T. Prugh et al. 1995, p. 51) and is the reason for ecosystems' loss and degradation.

Berkes and Folke (1992) state that natural capital and built capital are fundamentally complementary; it is not possible to create built capital without support from natural capital. Furthermore, it is important to note that natural capital (i.e., ecosystems) cannot provide benefits to people without its interaction with the other three types of capital. Ecosystem services (defined in the next section) do not flow directly from natural capital to human well-being (Costanza et al. 2014, p. 153). Therefore, "ecosystem services refer to the relative contribution of natural capital to the production of various human benefits, in combination with the three other forms of capital" (Figure 15.2) (Costanza 2012, p. 103).

Perceiving natural capital in insolation from the other forms of capital produces a bias in its management. Often, management of natural capital is the responsibility of the ministries of the environment and does not include other ministries, such as industry, agriculture, or finance. In the private sector, natural capital management is commonly the responsibility of the corporate sustainability department and does not come up in boardrooms (Guerry et al. 2015, p. 7350).

Natural capital and ecosystem services

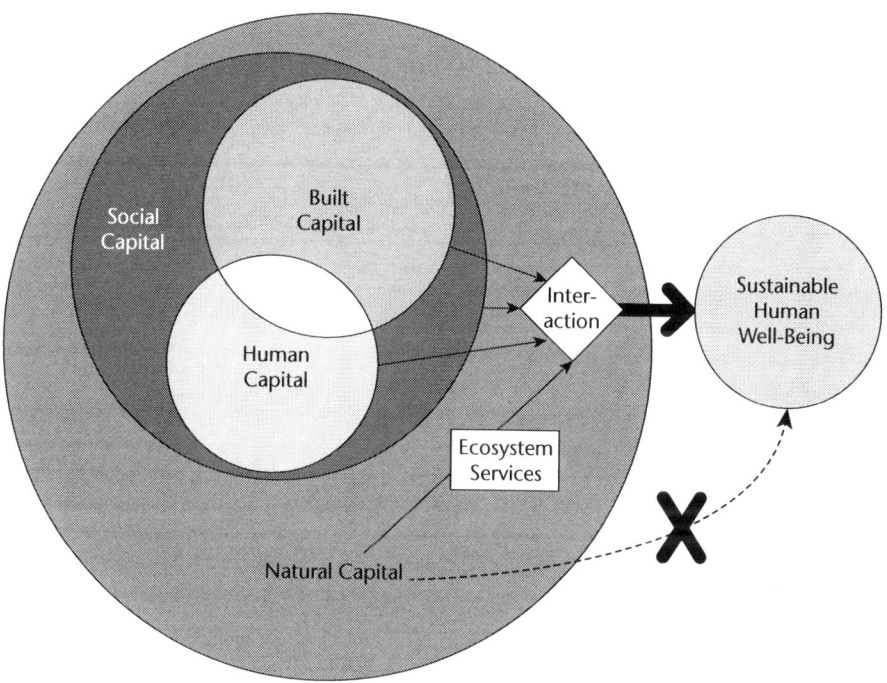

Figure 15.2 Interaction between social, built, human, and natural capital to contribute to well-being

3.1 Types of natural capital

According to Costanza and Daly (1992), there are two broad types of natural capital. Renewable natural capital, such as ecosystems, are active and self-maintaining using solar energy; they are analogous to machines and subject to entropic depreciation. Nonrenewable natural capital, such as mineral deposits and fossil fuels, are more passive and generally do not produce services until extracted. They are analogous to inventories and therefore are subject to liquidation (Costanza and Daly 1992, p. 38).

Prugh et al. (1995) describes a third category of natural capital, a hybrid that can be called cultivated natural capital, which includes agricultural and aquacultural systems, as well as planted forests, among other things. The main characteristic of this type of natural capital is that its components are not man-made, but they are not completely natural either (T. Prugh et al. 1995, p. 52).

4 Ecosystem services concept

Ecosystem services are defined as "the benefits that people obtain from ecosystems" (MEA 2005, p. v). A more complete definition of ecosystem services is "the benefits people derive from functioning ecosystems, the ecological characteristics, functions, or processes that directly or indirectly contribute to human well-being" (Costanza et al. 2011, p. 1).

Although these definitions of ecosystem services are very straightforward, they have been the subject of debate for two decades, and some clarification is therefore needed. First, it is important to distinguish between ecosystem processes and functions, on the one hand, and ecosystem services, on the other. Ecosystem processes and functions refer to biophysical relationships that exist regardless of whether humans benefit. The opposite is the case with ecosystem services, which only exist if they contribute to human well-being (Braat 2013).

This human-dependent definition of ecosystem services has led some to argue (Thompson and Barton 1994; McCauley 2006) that the concept represents an anthropocentric, utilitarian, or instrumental view of nature: that nature only exists to service humans. Nevertheless, the goal of the concept of ecosystem services is not to be anthropocentric, but rather to recognize the dependence of humans on nature for their well-being and their survival, and to visualize *Homo sapiens* as an integral part of the current biosphere. Moreover, instead of implying that humans are what matter most, or are the only thing that matters, the concept of ecosystem services implies that the whole system matters, both to humans and to the other species we are interdependent with.

4.1 Types of ecosystem services

Pearce (1998) classifies the goods and services that flow from natural capital into four categories: (1) supply of natural resource inputs to the economic production process (e.g., water, genetic diversity, and soil quality), (2) assimilation of waste products and residuals from the economic process, (3) source of direct human welfare though aesthetic and spiritual appreciation of nature, and (4) support systems–biogeochemical cycles and general ecosystem functioning (Pearce 1988).

These four categories were used almost two decades later in the Millennium Ecosystem Assessment under the names of provisioning, regulating, cultural, and supporting services:

- Provisioning services, such as timber, water, fiber, and food. A clear example of how these services interact with the other three types of capital is fishing activity, where fish provided to people as food requires fishing boats (built capital), fishermen (human capital), and fishing communities (social capital).
- Regulating services, such as pollination, flood control, water regulation, pest control, climate control, water purification, and air quality maintenance. For example, storm protection provided by wetlands (natural capital) to infrastructure such as hotels and houses on the coast (built capital), protecting its residents and other members of the community. Contrary to provisioning services, these services are not marketed.
- Cultural services that provide spiritual, recreational, and aesthetic benefits. A recreational benefit requires natural capital, such as a waterfall; built capital like a trail or a road; human capital that appreciates the waterfall; and social capital, such as friends and family and the institutions that make the waterfall accessible.
- Supporting services, such as photosynthesis, nutrient cycling, and soil formation. These types of services do not require interaction with human, social, and built capital; they affect human well-being indirectly by maintaining key processes that are necessary for the other three types of services. Using this description of supporting services, some scholars have argued that instead of ecosystem services they are ecosystem functions. Although this is true, supporting services can be used as a proxy to evaluate services in the other categories if more direct measures are not available (Costanza et al. 2011; MEA 2005)

Costanza et al. (1997) identified 17 ecosystem services. Other key reports and initiatives, such as the Millennium Ecosystem Assessment (already mentioned earlier), The Economics of Ecosystems and Biodiversity (TEEB), and, more recently, the Common International Classification of Ecosystem Services (CICES), have established classifications of ecosystem services in order to frame and enable discussions, assessments, modeling, and valuation. Table 15.1 compares these four ecosystem services classification systems, making evident that they are broadly similar.

Table 15.1 Comparison of four of the main ecosystem services classification systems used worldwide and their differences and similarities (Costanza et al. 2017)

	Costanza et al. 1997 (a)	Millennium Ecosystem Assessment, 2005	TEEB, 2010	CICES 4.3 (v. 2013) (b)
Provisioning	Food production (13)	Food	Food	Biomass – nutrition
	Water supply (5)	Fresh water	Water	Water
	Raw materials (4)	Fiber etc.	Raw materials	Biomass – fiber, energy, & other materials
		Ornamental resources	Ornamental resources	
	Genetic resources (15)	Genetic resources	Genetic resources	
		Biochemicals and natural medicines	Medicinal resources	
	X	X	X	Biomass – mechanical energy
Regulating & habitat	Gas regulation (1)	Air quality regulation	Air purification	Mediation of gas & air flows
	Climate regulation (2)	Climate regulation	Climate regulation	Atmospheric comp. & climate regulation
	Disturbance regulation (storm protection & flood control) (3)	Natural hazard regulation	Disturbance prevention or moderation	Mediation of air and liquid flows
	Water regulation (4) (e.g., natural irrigation & drought prevention)	Water regulation	Regulation of water flows	Mediation of liquid flows
	Waste treatment (9)	Water purification and waste treatment	Waste treatment (esp. water purification)	Mediation of waste, toxics, and other nuisances
	Erosion control & sediment retention (8)	Erosion regulation	Erosion prevention	Mediation of mass-flows
	Soil formation (7)	Soil formation (*supporting service*)	Maintaining soil fertility	Maintenance of soil formation and composition
	Pollination (10)	Pollination	Pollination	Life cycle maintenance (incl. pollination)
	Biological control (11)	Regulation of pests & human diseases	Biological control	Maintenance of pest and disease control

(*Continued*)

Table 15.1 (Continued)

	Costanza et al. 1997 (a)	Millennium Ecosystem Assessment, 2005	TEEB, 2010	CICES 4.3 (v. 2013) (b)
Supporting & habitat	Nutrient cycling (8)	Nutrient cycling & photosynthesis, primary production "Biodiversity"	X	X
	Refugia (12) (nursery, migration habitat)		Lifecycle maintenance (esp. nursery) Gene pool protection	Life cycle maintenance, Habitat, and gene pool protection
Cultural	Recreation (16), incl. eco-tourism & outdoor activities	Recreation & eco-tourism Aesthetic values Cultural diversity	Recreation & eco-tourism Aesthetic information Inspiration for culture, art, & design	Physical and experiential interactions
	Cultural (17) (incl. aesthetic, artistic, spiritual, education, & science)	Spirit. & religious val.	Spiritual experience	Spiritual and/or emblematic interactions
		Knowledge systems Educational values	Information for cognitive development	Intellectual and representative interactions

5 Ecosystem services valuation

5.1 The concept of value

Ecosystem services can be valued through different methods depending on the service, but before explaining these methods, it is important to understand the concept of value in this context.

A good start to better comprehend the theory behind the valuation of goods and services is the distinction that Adam Smith made in the eighteenth century between exchange value and use value, wherein he used the diamond–water paradox to explain it. Diamonds have a high exchange value and people are willing to pay a great price depending on the quality of the diamond, but diamonds have low use value because they are mainly useful as jewelry (among other uses that were implemented after Smith). Water, on the other hand, has a low exchange value, which means that people would pay very low prices to consume it, but the use value of water is high since it is a resource we need in order to survive.

Smith used this paradox to dismiss the use value as a basis for exchange value, and he instead formulated a cost of production theory of value based on wages, profit, and rent as the source of exchange value. He suggested a labor theory of exchange value, using the beaver–deer example: if it takes twice the labor to kill a beaver than to kill a deer, then one beaver will be sold for as much as two deer. Therefore, when labor is the only scarce factor, services and goods will be "priced" based on the ratio of labor used (Farber et al. 2002). It is worth noting that this point of view of value excluded completely natural capital, perhaps because at the time it was not a scarce resource.

In the twentieth century, the "marginal" revolution in value theory originated through the convergence of related streams of economic thought. Menger stated that the intensity of desire for one additional unit declines with successive units of the good. Exchanging the term "desire for one additional unit" with the term "marginal utility" results in the economic principle of diminishing marginal utility. The marginal utility theory of value is of great importance in the valuation of ecosystem services, because it can be used to measure use values instead of just exchange values, in monetary units (Farber et al. 2002).

The exchange value of goods and services is determined by the willingness to pay (WTP) to obtain them or the willingness to accept (WTA) compensation for losing them. WTP and WTA can be based on marginal changes in the availability of these goods and services, or on larger changes, including their complete absence. Exchange-based values of goods and services are determined by the prices at which they are exchanged. Overall, economists set the value of a good based on want satisfaction and pleasure, meaning that things only have value if they are desired, which is a problematic point of view in valuing natural capital, as explained later. Furthermore, as the good becomes scarcer, the desire increases, and therefore so does its value (Farber et al. 2002).

5.2 Valuation

As stated earlier, ecosystem services are the benefits people derive from ecosystems; they are provided by natural capital in combination with built, social, and human capital. The value of ecosystem services is therefore the relative contribution of ecosystems to well-being (Turner et al. 2016). This contribution can be expressed in various units (any units of the four types of capitals), where monetary units are often the most used and convenient since most people understand values in these units. Nevertheless, other units, such as time, energy, and land, can also be used. The selection will depend on which units help to better communicate to different stakeholders in a given decision making context (Costanza, et al. 2014). Valuation allows a more efficient use of limited funds by identifying where environmental protection and restoration is economically most significant, supporting the determination of the amount of compensation that should be paid for the degradation and/or loss of ecosystem services, and improving the financial mechanisms (e.g., incentives) for the conservation and sustainable use of natural capital (e.g., Payment for Ecosystem Services) (De Groot et al. 2012).

The value of ecosystem services can also be estimated by determining the cost to replicate them by artificial means (Costanza et al. 1997), for example, how much it would cost a farmer to pollinate his crops artificially. It is useful to attempt to calculate the impact in human well-being from changes in quantity or quality of natural capital that can occur due to different development decisions (Costanza et al. 1997). Valuation is therefore a tool for evaluating the tradeoffs required to achieve a shared goal, where in the past and in the present these tradeoffs have been addressed mainly through marketed goods and services (e.g., fuel or food) using commodity prices, leaving outside the equation other goods and services that currently do not have a price but that contribute equally or even more greatly to well-being (Turner et al. 2016).

Valuing ecosystem services has been criticized as unwise or even impossible because we supposedly cannot put a value on "intangibles" like human life and nature. In reality, we implicitly value these things on a daily basis through, for example, measures to protect human life, such as construction standards for housing and public infrastructure that will require spending more money in order to preserve human lives (Costanza et al. 1997). Therefore, the overall goal is not to put a price tag on nature for exchange purposes, but to visualize the effect of a change in ecosystem services provision to human well-being in terms of a rate of tradeoff against other things people value (Turner et al. 2003).

5.3 Valuation methods

After the identification, quantification, and mapping of ecosystem services for a particular area or scale, there are different types of methods used to conduct a Total Economic Valuation (TEV). These can be divided into revealed preference, stated preference, and non-preference–based methods. Revealed preference methods to estimate the benefits from ecosystems are based on market prices, which limits the use of these methods to only a few ecosystem services that are traded in markets (mostly provisioning services) (Turner et al. 2016). Revealed preference methods analyze the choices of people in real world settings and infer the value from those observed choices (Costanza et al. 2011). Non-preference methods recognize the limits of an individual's information about ecosystem services' connection to their well-being and use modeling and other techniques to estimate these connections.

Stated preference methods try to construct pseudo markets through the use of surveys in which people are asked to state their willingness to pay for ecosystem services that are not traded in current markets. These methods therefore rely on the response of people to hypothetical scenarios (Costanza et al. 2011). Stated preference approaches have limitations because people surveyed often do not completely understand or are not aware of the relation between healthy ecosystems and human well-being, because they do not feel comfortable in stating tradeoffs for ecosystems in monetary units, and finally because the willingness to pay can be significantly different to the real payment when it comes to that point (Turner et al. 2016).

Table 15.2 summarizes the different methods for ecosystem services valuation using conventional economic valuation and non-monetizing valuation (from Turner et al. 2016, which is an adaptation from Farber et al. 2006).

Table 15.2 List of methods for ecosystem services valuation

Conventional economic valuation	Revealed-preference approaches	*Travel cost*: valuations of site-based amenities are implied by the costs people incur to enjoy them (e.g., cleaner recreational lakes)
		Market methods: valuations are directly obtained from what people must be willing to pay for the service or good (e.g., timber harvest)
		Hedonic methods: the value of a service is implied by what people will be willing to pay for the service through purchases in related markets, such as housing markets (e.g., open-space amenities)
		Production approaches: service values are assigned from the impacts of those services on economic outputs (e.g., increased shrimp yields from increased area of wetlands)
	Stated-preference approaches	*Contingent valuation*: people are directly asked their willingness to pay or accept compensation for some change in ecological service (e.g., willingness to pay for cleaner air)
		Conjoint analysis: people are asked to choose or rank different service scenarios or ecological conditions that differ in the mix of those conditions (e.g., choosing between wetlands scenarios with differing levels of flood protection and fishery yields)
	Cost-based approaches	*Replacement cost*: the loss of a natural system service is evaluated in terms of what it would cost to replace that service (e.g., tertiary treatment values of wetlands if the cost of replacement is less than the value society places on tertiary treatment)
		Avoidance cost: a service is valued on the basis of costs avoided, or of the extent to which it allows the avoidance of costly averting behaviors, including mitigation (e.g., clean water reduces costly incidents of diarrhea)
Non-monetizing valuation	–	*Individual index-based methods*, including rating or ranking choice models, expert opinion
		Group-based methods, including voting mechanisms, focus groups, citizen juries, stakeholder analysis

Due to the nature of the service, each ecosystem service can be valuated through one or more particular methods. For each service, the amenability to economic valuation and the transferability across sites will vary from low to high. Table 15.3 summarizes the set of methods that are appropriate to valuate each ecosystem service (Turner et al. 2016).

Due to constraints in time and budget, it is often not possible to conduct original/primary studies to value ecosystem services (Wilson and Hoehn 2006; Plummer 2009), which has led to a wider use of secondary data (Richardson, Loomis, Kroeger, and Casey 2015) for this purpose through valuation techniques such as value/benefit transfer. Although this technique has limitations, it is sometimes the only option to inform policy decisions that require a first approximation to natural capital valuation (Richardson et al. 2015).

In simple terms, value transfer consists in "applying economic value estimates from one location to a similar site in another location" (Plummer 2009, p. 38). The site where primary data

Table 15.3 Valuation methods for each ecosystem service (Farber et al. 2006)

Ecosystem services		Amenability to economic valuation	Most appropriate method for valuation	Transferability across sites
Provisioning service	Water supply	High	AC, RC, M, TC	Medium
	Food	High	M, P	High
	Raw material	High	M, P	High
	Genetic resources	Low	M, AC	Low
	Medicinal resources	High	AC, RC, P	High
	Ornamental resources	High	AC, RC, H	Medium
Regulating services	Gas regulation	Medium	CV, AC, RC	High
	Climate regulation	Low	CV	High
	Disturbance regulation	High	AC	Medium
	Biological regulation	Medium	AC, P	High
	Water regulation	High	M, AC, RC, H, P, CV	Medium
	Soil retention	Medium	AC, RC, H	Medium
	Waste regulation	High	RC, AC, CV	Medium High
	Nutrient regulation	Medium	AC, CV	Medium
Cultural services	Recreation	High	TC, CV, ranking	Low
	Aesthetics	High	H, CV, TC, ranking	Low
	Science and education	Low	Ranking	High
	Spiritual and historic	Low	CV, ranking	Low

AC = avoided cost, CV = contingent valuation, H = hedonic pricing, M = market pricing, P = production approach, RC = replacement cost, TC = travel cost.

was collected and processed is called the study site, and the site to which this data (i.e., ecosystem services values) is going to be applied is called the policy site (because the values are commonly used for policy decisions such as land use change or the establishment of financial mechanisms) (Plummer 2009). The transfer can be spatial (across different sites, national, or international) or temporal (where the study site and the policy sites are different moments in time) (Navrud and Bergland 2004).

Other authors have proposed the following definitions of value transfer, all of them sharing the core elements of the technique:

- "Transfer of original ecosystem service value estimates from an existing 'study site' or multiple study sites to an unstudied 'policy site' with similar characteristics that is being evaluated" (Richardson et al. 2015).
- "Transposition of monetary environmental values estimated at one site (study site) through market-based or non-market-based economic valuation techniques to another site (policy site)" (Brouwer 2000).

Although the valuation technique is often referred as benefit transfer, Navrud states that the method can also be related to the transfer of damage estimates, and thus a more accurate term would be value transfer (Navrud and Bergland 2004), which will be used henceforth.

The aggregation of these methods through a value transfer make the technique useful in academic and policy settings in which ecosystem services values are not required with a high level of accuracy but need to be accurate enough to support a project or policy, and are not suitable when more accurate values are required, in cases such as the calculation of compensation payments for environmental damages (polluter pays principle) (Navrud and Ready 2007).

5.4 Difficulties in valuing ecosystem services

Valuing natural capital is far from a perfect science but is without any doubt a needed one. Turner et al. (2003) identified the following main difficulties when conducting these assessments:

- Marginality: the data used in ecosystem services are "marginal" values rather than aggregated global values; this is because what it is calculated is the value of ecosystem services' degradation or loss.
- Double counting: this problem can often occur because many ecosystem services are not complementary, which means that the provision of one is precluded by others.
- Typological issues: these are related to the design and strategy of the valuation assessments, where it is important to distinguish between valuations of the *in situ* ecosystem stock and estimates of the value of the flow of goods and services from a given stock.
- Spatial and temporal transfer: these difficulties are specifically for the aggregation method of basic value (or benefit) transfer, including the requirement of good quality studies of similar situations, the potential change of characteristics between time periods, and a failure to assess novel impacts (i.e., thresholds or resilience).
- Distribution of benefits and costs: developing countries invest high local costs to natural capital conservation that yield large global benefits, in contrast to developed countries that tend to incur relatively low local costs that produce lower global benefits.

6 Natural capital and sustainability

A key point is the understanding of the relation between sustainability and the maintenance of capital stocks. Ekins et al. (2003) explains that if sustainability depends on the maintenance of the capital stock, then there are two possibilities: (1) maintaining the total stock of capital, allowing substitutions between its components, or (2) whether certain components of capital, mainly natural capital, are non-substitutable. Ekins elaborates on these two possibilities by framing them under two types of sustainability: (1) weak sustainability, which considers that natural capital can be replaced completely by built capital under the perception that welfare is not dependent on a specific form of capital, and (2) strong sustainability, which considers complete substitution of natural capital by built capital to be impossible since natural capital provides a unique contribution to welfare, and ultimately it is the inputs for built capital and the basis of critical life support systems (Ekins et al. 2003, p. 167).

The concept of natural capital, as well as its research and policy implications, becomes relevant more than ever in the current national and global economic growth strategy. In the past (mainly before the industrial revolution), we lived in what some scholars call an empty world, empty of humans and their artifacts, full of natural resources. Now, we live in a full world, full of humans and their artifacts, with an increasingly reduced natural environment. In the former world the limiting factor was built capital, while natural capital and social capital were abundant; in the latter world, quite the contrary abounds.

In order to recognize natural capital as a limiting factor, and therefore its need of conservation and sustainable consumption, a different vision of the interaction between the economic and ecological systems is needed. Fenech et al. (2003, p. 5) propose that, instead of looking at the ecological system as part of the economic system, we need to consider the economy as part of the ecosystem.

The consideration of the economy as part of the ecosystem acknowledges the limits of growth of the economy since the ecosystem is finite. Costanza and Daly state that growth is related to throughput increase, which is destructive of natural capital, with the negative consequence of having higher costs in the medium and long term than the benefits gained in the short term (Costanza and Daly 1992, p. 43). This cost–benefit analysis for natural capital is often ignored by economic interests, undervaluing natural capital and only recognizing its value when it is lost (Ehrlich et al. 2012 p. 70). Development, on the contrary, means an increase of efficiency and quality improvement, and therefore does not reduce natural capital (Costanza and Daly 1992, p. 43).

From the natural capital perspective, development under this framework would mean that natural income must be sustainable, which should be at least the case for renewable natural capital. Since nonrenewable natural capital is reduced with use, income can be constant only if the total natural capital (renewable natural capital plus nonrenewable natural capital) is maintained constant, which implies a certain level of reinvestment of the nonrenewable natural capital consumed into the renewable natural capital (Costanza and Daly 1992, p. 43). This is relevant, especially for low income countries, since they have a higher dependency on natural capital both for growth and development (Pearce 1988).

References

Akerman, M. 2003. "What Does 'Natural Capital' Do? The Role of Metaphor in Economic Understanding of the Environment." *Environmental Values* 12(4):431–448.

Berkes, F., and C. Folke. 1992. "A Systems Perspective on the Interrelations Between Natural, Human-Made and Cultural Capital." *Ecological Economics* 5(1):1–8.

Braat, L. 2013. "The Value of the Ecosystem Services Concept in Economic and Biodiversity Policy." In S. Jacobs, N. Dendoncker, and H. Keune, eds., *Ecosystem Services—Global Issues, Local Practices* (pp. 97–103). San Diego, CA: Elsevier.

Braat, L.C., and R. de Groot. 2012. "The Ecosystem Services Agenda: Bridging the Worlds of Natural Science and Economics, Conservation and Development, and Public and Private Policy." *Ecosystem Services* 1(1):4–15. doi:10.1016/j.ecoser.2012.07.011.

Brouwer, R. 2000. "Environmental Value Transfer: State of the Art and Future Prospects." *Ecological Economics* 32(1):137–152.

Costanza, R. 2012. "The Value of Natural and Social Capital in Our Current Full World and in a Sustainable and Desirable Future." In *Sustainability Science*. New York: Springer, pp. 99–109. http://link.springer.com.virtual.anu.edu.au/chapter/10.1007/978-1-4614-3188-6_5. Accessed November 1, 2017.

Costanza, R., J.H. Cumberland, H. Daly, R. Goodland, R.B. Norgaard, I. Kubiszewski, and C. Franco. 2014. *An Introduction to Ecological Economics*. Boca Raton, FL: CRC Press. https://books-google-com-au.virtual.anu.edu.au/books?hl=en&lr=&id=w29YBQAAQBAJ&oi=fnd&pg=PP1&dq=An+Introduction+to+Ecological+Economics&ots=R93WJNitQ0&sig=iHqycAvHzcpJvVDeGt5_VnuTPVk. Accessed November 1, 2017.

Costanza, R., R. d'Arge, R. De Groot, S. Farber, M. Grasso, B. Hannon, et al. 1997. "The Value of the World's Ecosystem Services and Natural Capital." *Nature* 387(6630):253–260.

Costanza, R., and H.E. Daly. 1992. "Natural Capital and Sustainable Development." *Conservation Biology* 6(1):37–46.

Costanza, R., R. de Groot, P. Sutton, S. van der Ploeg, S.J. Anderson, I. Kubiszewski, and R.K. Turner. 2014. "Changes in the Global Value of Ecosystem Services." *Global Environmental Change* 26:152–158.

Costanza, R., R. de Groot, L. Braat, I. Kubiszewski, L. Fioramonti, P. Sutton, S. Farber, and M. Grasso. 2017. "Twenty Years of Ecosystem Services: How Far Have We Come and How Far Do We Still Need to Go?" *Ecosystem Services* 28:1–16.

Costanza, R., I. Kubiszewski, D.E. Ervin, R. Bluffstone, D. Brown, H. Chang, et al. 2011. "Valuing Ecological Systems and Services." http://pdxscholar.library.pdx.edu/iss_pub/71/?utm_source=pdxscholar.library.pdx.edu%2Fiss_pub%2F71&utm_medium=PDF&utm_campaign=PDFCoverPages. Accessed November 1, 2017.

Daily, G. 1997. *Nature's Services: Societal Dependence on Natural Ecosystems*. Washington DC: Island Press.

Daly, H.E., and J. Farley. 2004. *Ecological Economics: Principles and Applications*. Washington DC: Island Press. https://books-google-com-au.virtual.anu.edu.au/books?hl=en&lr=&id=20R9_6rC-LoC&oi=fnd&pg=PR5&dq=ecological+economics+principles+and+applications+2004&ots=ylSDF9HQTV&sig=l0fSFvDK60s5LLjJc0IVXZIr2RM. Accessed November 1, 2017.

De Groot, R., L. Brander, S. van der Ploeg, R. Costanza, F. Bernard, L. Braat, et al. 2012. "Global Estimates of the Value of Ecosystems and Their Services in Monetary Units." *Ecosystem Services* 1(1):50–61.

Ehrlich, P., and A. Erlich. 1981. *Extinction: The Causes and Consequences of the Disappearance of Species*. New York: Random House.

Ehrlich, P.R., P.M. Kareiva, and G.C. Daily. 2012. "Securing Natural Capital and Expanding Equity to Rescale Civilization." *Nature* 486(7401):68.

Ehrlich, P.R., and H.A. Mooney. 1983. "Extinction, Substitution, and Ecosystem Services." *BioScience* 33(4):248–254. doi:10.2307/1309037.

Ekins, P., S. Simon, L. Deutsch, C. Folke, and R. De Groot. 2003. "A Framework for the Practical Application of the Concepts of Critical Natural Capital and Strong Sustainability." *Ecological Economics* 44(2):165–185.

Farber, S., R. Costanza, D.L. Childers, J. Erickson, K. Gross, M. Grove, C.S. Hopkinson, J. Kahn, S. Pincetl, A. Troy, P. Warren, and M. Wilson. 2006. "Linking Ecology and Economics for Ecosystem Management: A Services-Based Approach with Illustrations from LTER Sites." *BioScience* 56:117–129.

Farber, S.C., R. Costanza, and M.A. Wilson. 2002. "Economic and Ecological Concepts for Valuing Ecosystem Services." *Ecological Economics* 41(3):375–392.

Fenech, A., J. Foster, K. Hamilton, and R. Hansell. 2003. "Natural Capital in Ecology and Economics: An Overview." *Environmental Monitoring and Assessment* 86(1):3–17.

Gagnon Thompson, S.C., and M.A. Barton. 1994. "Ecocentric and Anthropocentric Attitudes Toward the Environment." *Journal of Environmental Psychology* 14(2):149–157. doi:10.1016/S0272-4944(05)80168-9.

Gómez-Baggethun, E., and R. De Groot. 2010. "Natural Capital and Ecosystem Services: The Ecological Foundation of Human Society." In *Ecosystem Services*, pp. 105–121. http://pubs.rsc.org.virtual.anu.edu.au/en/content/chapter/bk9781849730181-00105/978-1-84973-018-. Accessed November 1, 2017.

Gómez-Baggethun, E., R. de Groot, P.L. Lomas, And C. Montes. 2010. "The History of Ecosystem Services in Economic Theory and Practice: From Early Notions to Markets and Payment Schemes." *Ecological Economics* 69(6):1209–1218. doi:10.1016/j.ecolecon.2009.11.007.

Guerry, A.D., S. Polasky, J. Lubchenco, R. Chaplin-Kramer, G.C. Daily, R. Griffin, and B. Vira. 2015. "Natural Capital and Ecosystem Services Informing Decisions: From Promise to Practice." *Proceedings of the National Academy of Sciences* 112(24):7348–7355.

McCauley, D.J. 2006. "Selling Out on Nature." *Nature* 443(7107):27–28. doi:10.1038/443027a.

MEA, M.E.A. 2005. *Ecosystems and Human Well-Being: Synthesis*. Washington, DC: Island.

Navrud, S., and R. Ready. 2007. "Lessons Learned for Environmental Value Transfer." In *Environmental Value Transfer: Issues and Methods*. New York: Springer, pp. 283–290. http://link.springer.com/content/pdf/10.1007/1-4020-5405-X_15.pdf. Accessed November 1, 2017.

Pearce, D. 1988. "Economics, Equity and Sustainable Development." *Futures* 20(6):598–605.

Pearce, D. 1998. "Auditing the Earth: The Value of the World's Ecosystem Services and Natural Capital." *Environment: Science and Policy for Sustainable Development* 40(2):23–28.

Plummer, M.L. 2009. "Assessing Benefit Transfer for the Valuation of Ecosystem Services." *Frontiers in Ecology and the Environment* 7(1):38–45.

Prugh, T., R. Costanza, John H. Cumberland, Herman Daily, Robert Goodland, and Richard B. Norgaard. 1995. *Natural Capital and Human Economic Survival*. Solomons, MD: International Ecological Economics Society Press.

Richardson, L., J. Loomis, T. Kroeger, and F. Casey. 2015. "The Role of Benefit Transfer in Ecosystem Service Valuation." *Ecological Economics* 115:51–58. doi:10.1016/j.ecolecon.2014.02.018.

Stale, Navrud, and Olvar Bergland. 2004. "Value Transfer and Environmental Policy." *The International Yearbook of Environmental and Resource Economics 2004/2005*, 189.

TEEB, 2010. Mainstreaming the Economics of Nature: A Synthesis of the Approach, Conclusions and Recommendations of TEEB Earthscan, London and Washington (2010).

Turner, K.G., S. Anderson, M. Gonzales-Chang, R. Costanza, S. Courville, T. Dalgaard, et al. 2016. "A Review of Methods, Data, and Models to Assess Changes in the Value of Ecosystem Services from Land Degradation and Restoration." *Ecological Modelling* 319:190–207.

Turner, R.K., J. Paavola, P. Cooper, S. Farber, V. Jessamy, and S. Georgiou. 2003. "Valuing Nature: Lessons Learned and Future Research Directions." *Ecological Economics* 46(3):493–510.

Westman, W.E. 1977. "How Much Are Nature's Services Worth?" *Science* 197(4307):960–964.

Wilson, M.A., and J.P. Hoehn. 2006. "Valuing Environmental Goods and Services Using Benefit Transfer: The State-of-the-Art and Science." *Ecological Economics* 60(2):335–342.

16
Water management and economics

Louis Sears and C.-Y. Cynthia Lin Lawell

1 Introduction

The sustainable management of groundwater resources for use in agriculture is a critical issue worldwide. Many of the world's most productive agricultural basins depend on groundwater and have experienced declines in water table levels. The food consumers eat, the farmers who produce that food, and the local economies supporting that production are all affected by the availability of groundwater (Lin Lawell 2016). Worldwide, about 70 percent of groundwater withdrawn is used for agriculture, and, in some countries, the percent of groundwater extracted for irrigation can be as high as 90 percent (National Groundwater Association 2016).

Increasing competition for water for cities and for environmental needs, as well as concerns about future climate variability and more frequent droughts, have caused policy makers to look for ways to decrease the consumptive use of water (Lin Lawell 2016). Approximately 25 percent of global crops are being grown in water-stressed areas (Siebert et al. 2013).

In this chapter, we discuss the economics of sustainable agricultural groundwater management, including the importance of dynamic management; the importance of spatial management; the possible perverse consequences of incentive-based agricultural groundwater conservation programs; property rights; the groundwater–energy nexus; and the effects of climate change.

Throughout this chapter, we also discuss the application of the economics of sustainable agricultural groundwater management to agricultural groundwater management in Kansas and California. California is experiencing its third-worst drought in 106 years (Howitt and Lund 2014). While California Governor Jerry Brown officially ended the drought state of emergency in all California counties except Fresno, Kings, Tulare, and Tuolumne in April 2017, the hydrologic effects of the drought will take years to recover (USGS 2017). From 1960 to the present, there has been significant deterioration in the groundwater level in the Central Valley of California, making current levels of groundwater use unsustainable (Famiglietti 2014). Groundwater management is particularly important in California as the state produces almost 70 percent of the nation's top 25 fruit, nut, and vegetable crops (Howitt and Lund 2014). Most crops in California come from two areas: the Central Valley, including the Sacramento and San Joaquin valleys; and the coastal region, including the Salinas Valley, often known as America's "salad bowl." Farmers in both areas rely heavily on groundwater (York and Sumner 2015). Understanding the economics of sustainable agricultural groundwater management is particularly timely and important for California as legislation allowing regulation of groundwater is being implemented there gradually over the next several years (York and Sumner 2015).

2 Surface water

The sources of water can be categorized into two types: surface water and groundwater. Surface water includes lakes, streams, and oceans. Surface water is a renewable resource and is provided by the earth's hydrologic cycle (Hartwick and Olewiler 1998).

The relevant notion of efficiency for surface water is allocative efficiency. Allocative efficiency arises when natural resources are allocated to their more valuable uses. The efficient allocation and price for surface water is that for which the marginal value of water is equalized among all groups of users and set equal to the marginal cost of supplying water (Hartwick and Olewiler 1998).

The condition for allocative efficiency for surface water, that the marginal value of water should be equalized among all groups of users and set equal to the marginal cost of supplying water, can be generalized along several dimensions. First, if there are environmental uses for surface water, then the efficient allocation and price for surface water would equalize the marginal value of water for environmental uses with the marginal value of water for each of the other groups of users.

Second, if there are environmental externalities associated with supplying surface water, then the costs of these environmental externalities should be included in the social marginal cost of supplying water used to determine the efficient price for water.

Third, since the marginal costs of supplying water vary over time and by region, the marginal costs used to determine the efficient surface water price should be allowed to vary over time and by region.

In addition to surface water, the other main source of water is groundwater. Groundwater is water that is held in underground aquifers. When managing water, water managers should account for both sources of water. Mani, Tsai, and Paudel (2016) find that a conjunctive-use framework for managing surface water and groundwater resources can raise groundwater levels. Tsur and Graham-Tomasi (1991) find that when utilized with a stochastic source of surface water for irrigation, groundwater may serve to mitigate fluctuations in the supply of water, and the benefit corresponding to this service, known as the buffer value of groundwater, is positive.

The economics of managing groundwater for agricultural use is the focus of the remainder of this chapter.

3 Dynamic management

Aquifers are recharged through the percolation of rain and snow (Hartwick and Olewiler 1998). If an aquifer receives very little recharge, then it is at least partially a nonrenewable resource and therefore should be managed dynamically and carefully for long-term sustainable use (Lin Lawell 2018c).

The idea behind dynamic management is that water managers need to account for the future when making current decisions. In particular, water managers may wish to extract less groundwater today in order to save more for tomorrow (Gisser and Sanchez 1980; Feinerman and Knapp 1983).

There are two main reasons why groundwater needs to be managed dynamically, particularly if the aquifer receives very little recharge. First, groundwater extraction today decreases the amount of groundwater available tomorrow. Second, groundwater extraction today increases the cost of extraction tomorrow because removal of water today increases the "lift-height" needed to lift the remaining stock to the surface tomorrow, thereby increasing the pumping cost (Timmins 2002; Sears, Bertone Oehninger, Lim, and Lin Lawell 2018; Sears, Lim and Lin Lawell 2018b). Thus, because the extraction of groundwater both decreases the future amount of groundwater available and increases the future cost of extracting groundwater, sustainable agricultural groundwater extraction may entail extracting less groundwater today in order to avoid future supply shocks (Sears, Bertone Oehninger, Lim, and Lin Lawell 2018; Sears, Lim and Lin Lawell 2018b).

The appropriate notion of efficiency for a nonrenewable resource is that of dynamic efficiency. The dynamically efficient outcome is one that maximizes the present discounted value of the entire stream of net benefits to society. Because the extraction of groundwater both decreases the future amount of groundwater available and increases the future cost of extracting groundwater, the dynamically efficient price for groundwater is higher than the marginal cost of supplying water. Thus, while the (statically) efficient price for surface water is its marginal cost, the dynamically efficient price for groundwater is higher than marginal cost.

Dynamic management may be important in Kansas, for example, where the portion of the High Plains Aquifer that lies beneath western Kansas receives very little recharge (Lin Lawell 2018c). Thus, groundwater in Kansas is at least partially a nonrenewable resource and therefore should be managed dynamically (Lin Lawell 2018c).

Dynamic management may similarly be important in California, where recharge rates are low as well. Comparing aquifer systems found in irrigated agricultural regions in the U.S., aquifers in the Central Valley have recharge rates of between 420–580 mm per year, which is within the range found in the High Plains, an aquifer which receives little recharge (Lin Lawell 2018c). This is higher than recharge rates in the Pacific Northwest and is lower than recharge rates in the Alluvium aquifer system (McMahon et al. 2011). Thus, groundwater in California is at least partially a nonrenewable resource and should be managed dynamically (Sears, Bertone Oehninger, Lim, and Lin Lawell 2018).

4 Spatial management

In addition to dynamic considerations, sustainable agricultural groundwater management needs to account for spatial considerations as well. Spatial considerations arise because groundwater users face a common pool resource problem: because farmers are sharing the aquifer with other farmers, other farmers' pumping affects their extraction cost and the amount of water they have available to pump. Consequently, groundwater pumping by one user raises the extraction cost and lowers the total amount that is available to other nearby users (Pfeiffer and Lin 2012; Lin and Pfeiffer 2015; Lin Lawell 2016). Spatial externalities resulting from the inability to completely capture the groundwater to which property rights are assigned can lead to over-extraction (Sears, Lin Lawell, and Lim 2018; Sears, Lim, and Lin Lawell 2018a).

Theoretically, spatial externalities are potentially important causes of welfare loss (Dasgupta and Heal 1979; Eswaran and Lewis 1984; Negri 1989; Provencher and Burt 1993; Brozovic, Sunding, and Zilberman 2006; Rubio and Casino 2003; Msangi 2004; Saak and Peterson 2007). Owing in large part to spatial externalities, the issue of managing water resource use across political boundaries is particularly important (Dinar and Dinar 2016).

If spatial externalities in groundwater use are significant, they allow insight into the causes of resource over-exploitation. If they are not significant or are very small in magnitude, a simpler model of groundwater user behavior, where each user essentially owns his own stock, is sufficient. Both outcomes would give guidance to policy makers, although it is important to note that the results are highly dependent on the hydrological conditions of the aquifer (Lin and Pfeiffer 2015; Lin Lawell 2016). To make optimal spatial management more politically feasible, Pitafi and Roumasset (2009) devise an inter-temporal compensation plan that renders switching from the status quo to optimal spatial management Pareto-improving.

Pfeiffer and Lin (2012) empirically examine whether the amount of water one farmer extracts depends on how much water his neighbor extracts. Their econometric model is spatially explicit, taking advantage of detailed spatial data on groundwater pumping from the portion of western Kansas that overlies the High Plains Aquifer system. Their study is the first study to empirically measure economic relationships between groundwater users.

According to their results, Pfeiffer and Lin (2012) find evidence of a behavioral response to this movement in the agricultural region of western Kansas overlying the High Plains Aquifer. Using an instrumental variable and spatial weight matrices to overcome estimation difficulties resulting from simultaneity and spatial correlation, they find that on average, the spatial externality causes over-extraction that accounts for about 2.5 percent of total pumping. These farmers would apply 2.5 percent less water in the absence of spatial externalities (Pfeiffer and Lin 2012; Pfeiffer and Lin 2015; Lin Lawell 2016).

Strengthening the evidence of the behavioral response to the spatial externalities caused by the movement of groundwater is the empirical result that a farmer who owns multiple wells does not respond to pumping at his own wells in the same manner as he responds to pumping at neighboring wells owned by others. In fact, the response to pumping at his own wells is to marginally decrease pumping, thus trading off the decrease in water levels between spatial areas and internalizing the externality that exists between his own wells (Pfeiffer and Lin 2012; Pfeiffer and Lin 2015; Lin Lawell 2016).

Sears, Lim, and Lin Lawell (2018a) present a dynamic game framework for analyzing spatial groundwater management. In particular, they characterize the Markov perfect equilibrium resulting from non-cooperative behavior and compare it with the socially optimal coordinated solution. In order to analyze the benefits from internalizing spatial externalities in California, they calibrate our dynamic game framework to California and conduct a numerical analysis to calculate the deadweight loss arising from non-cooperative behavior. Results show that the benefits from coordinated management in California are particularly high under conditions of extreme drought, and also when the possibility of extreme rainfall situations are high.

Spatial externalities in groundwater may arise not only between neighboring farmers, but also between neighboring groundwater management jurisdictions as well. Sears, Lin Lawell, and Lim (2018) present a model of inter-jurisdictional spatial externalities in groundwater management. They find that groundwater managers each managing a subset of the plots of land that overlie an aquifer and each behaving non-cooperatively with respect to other groundwater managers will over-extract water relative to the socially optimal coordinated solution if there is spatial movement of water between the patches that are managed by different groundwater managers. Moreover, transactions costs and the difficulty of observing and verifying aquifer boundaries, groundwater levels, and groundwater extraction may preclude individual groundwater managers from coordinating with each other to achieve an efficient outcome (Lin 2010; Lin Lawell 2018a; Lin Lawell 2018b).

In order to internalize any inter-jurisdictional spatial externalities, the jurisdictions of local groundwater managers should be large enough to internalize all externalities, so that there are no trans-boundary issues between jurisdictions. This means that local groundwater managers should each cover an entire groundwater basin, and also that a groundwater basin should not be managed by multiple groundwater managers (Sears, Lin Lawell, and Lim 2018).

In 2015, the California Department of Water Resources developed a Strategic Plan to implement its 2014 Sustainable Groundwater Management Act (California Department of Water Resources 2015). Under California's 2014 Sustainable Groundwater Management Act and 2015 Strategic Plan for implementing it, each groundwater basin is to be managed at the local level by locally controlled Groundwater Sustainability Agencies (GSAs).

In order to internalize any inter-jurisdictional spatial externalities in California, local agencies should each cover an entire groundwater basin, and a groundwater basin should not be managed by multiple Groundwater Sustainability Agencies (Sears, Lin Lawell, and Lim 2018). However, Sears, Lin Lawell, and Lim (2018) find that although California's 2014 Sustainable Groundwater Management Act and 2015 Strategic Plan for implementing it may have specified the efficient allocation of regulatory responsibility between central and local tiers of government, the jurisdictions for the local agencies may not internalize all the spatial externalities. As a consequence, the local agencies may behave non-cooperatively, leading to over-extraction relative to the socially optimal coordinated solution (Sears, Lin Lawell, and Lim, 2018).

Inter-jurisdictional jurisdictional externalities may also arise if an aquifer is shared across state or country borders. In these cases, transactions costs may be particularly acute and it may be especially difficult to observe and verify aquifer boundaries, groundwater levels, and groundwater extraction; as a consequence, it may be highly unlikely that individual groundwater managers are able to coordinate with each other to achieve an efficient outcome (Lin 2010; Lin Lawell 2018a; Lin Lawell 2018b).

For example, trans-boundary issues may arise between California and Nevada. The Basin and Range aquifers are located in an area that comprises most of Nevada and the southern California desert, and many of the basins are hydraulically connected (Sears, Lim, and Lin Lawell 2018a). Thus, groundwater managers in California and Nevada may behave non-cooperatively with each other, leading to over-extraction relative to the socially optimal coordinated solution. When inter-jurisdictional externalities arise across state borders, it may be efficient for both the state and federal government to be allocated regulatory authority (Lin 2010; Lin Lawell 2018a; Lin Lawell 2018b).

Aquifer heterogeneity can affect the extent of the spatial externality. Aquifers vary in rock composition, which determines the extent to which the water resource is shared. Portions of an aquifer where water moves rapidly, or those with high hydraulic conductivity, as well as those that receive less yearly recharge, face a more costly common-pool problem and therefore receive higher benefits from coordinated management (Edwards 2016). Edwards (2016) uses the introduction of management districts in Kansas to test the effect of underlying aquifer heterogeneity on changes in agricultural land value, farm size, and crop choice. A landowner in a county with hydraulic conductivity one standard deviation higher sees a relative land value increase of 5–8 percent when coordinated management is implemented. Counties with lower recharge also see relative increases in land value. Changes in farm size and percentage of cropland in corn are also consistent with the proposition that the effect of coordinated management is unequal and depends on the properties of the physical system (Edwards 2016; Lin Lawell 2016; Sears, Lim, and Lin Lawell 2017; Sears, Lim, and Lin Lawell 2018a).

Another aspect of spatial management is the possible need for spatially differentiated groundwater pumping regulations. One reason it may be important to have spatially differentiated groundwater pumping regulations is that groundwater pumping from aquifers can reduce the flow of surface water in nearby streams through a process known as stream depletion. Although the marginal damage of groundwater use on stream flows depends crucially on the location of pumping relative to streams, current regulations are generally uniform over space (Kuwayama and Brozovic 2013). Kuwayama and Brozovic (2013) use a population dataset from irrigation wells in the Nebraska portion of the Republican River Basin to analyze whether adopting spatially differentiated groundwater pumping regulations leads to significant reductions in farmer abatement costs and costs from damage to streams. They find that regulators can generate most of the potential savings in total social costs without accounting for spatial heterogeneity. However, if regulators need to increase the protection of streams significantly from current levels, spatially differentiated policies will yield sizable cost savings (Kuwayama and Brozovic 2013; Lin Lawell 2016; Sears, Lin Lawell, and Lim 2018; Sears, Lim, and Lin Lawell 2018a).

5 Perverse incentives from policy

When designing groundwater management policies, it is important to consider any possible perverse consequences from the policy. For example, incentive-based water conservation programs are extremely popular policies for water management. Farmers can receive a subsidy for upgrading their irrigation systems; less groundwater is "wasted" through runoff, evaporation, or drift; marginal lands can be profitably retired; and farmers can choose whether to participate. However, such policies can have perverse consequences (Pfeiffer and Lin 2010; Lin 2013; Pfeiffer and Lin 2014a; Pfeiffer and Lin 2014b; Lin Lawell 2016; Sears, Bertone Oehninger, Lim, and Lin Lawell 2018; Sears, Lim, and Lin Lawell 2017; Sears, Lim, and Lin Lawell 2018b).

In many places, policy makers have attempted to decrease rates of groundwater extraction through incentive-based water conservation programs. Between 1998 and 2005, the state of Kansas spent nearly $6 million on incentive programs, such as the Irrigation Water Conservation Fund and the Environmental Quality Incentives Program, to fund the adoption of more efficient irrigation systems. Such programs paid up to 75 percent of the cost of purchasing and installing new or upgraded irrigation technology, and much of the money was used for conversions to dropped nozzle systems (NRCS 2004). These policies were implemented under the auspices of groundwater conservation, in response to declining aquifer levels occurring in some portions of the state due to extensive groundwater pumping for irrigation (Committee 2001; Pfeiffer and Lin 2014a).

In California, the State Water Efficiency and Enhancement Program (SWEEP) provides financial assistance in the form of grants to implement irrigation systems that reduce greenhouse gases and save water on California agricultural operations, including evapotranspiration-based irrigation scheduling to optimize water efficiency for crops; and micro-irrigation or drip systems (California DWR and CFDA 2017). San Luis Canal Company in the San Joaquin Valley offered $250 per acre to encourage the transition to pressurized irrigation systems (CEC 2015a; Sears, Bertone Oehninger, Lim, and Lin Lawell 2018; Sears, Lim and Lin Lawell 2017; Sears, Lim, and Lin Lawell 2018b).

Similarly, though funding for this order was not passed, under the Water and Energy Saving Technologies Executive Order B-29-15, the California Energy Commission, Department of Water Resources, and State Water Resources Control Board were to provide funding for innovative technologies, including rebates for conversion from high-pressure to low-pressure drip irrigation systems (CEC 2015b; Sears, Bertone Oehninger, Lim, and Lin Lawell 2018; Sears, Lim, and Lin Lawell 2017; Sears, Lim, and Lin Lawell 2018b).

However, although they are extremely popular, policies that encourage the adoption of more efficient irrigation technology may not have the intended effect. Irrigation is said to be "productivity enhancing"; it allows the production of higher value crops on previously marginal land. Thus, a policy of subsidizing more efficient irrigation technology can induce a shift away from dry-land crops to irrigated crops. It may also induce the planting of more water-intensive crops on already irrigated land, as by definition more efficient irrigation increases the amount of water the crop receives per unit extracted (Pfeiffer and Lin 2014a; Lin Lawell 2016).

Similarly, land and water conservation and retirement programs may not necessarily reduce groundwater extraction, although they are billed as such. An example of a land retirement program is the Conservation Reserve Program (CRP) created by the federal government in 1985 to provide technical and financial assistance to eligible farmers and ranchers to address soil, water, and related natural resource concerns on their lands in an environmentally beneficial and cost-effective manner (USDA 2014). These programs include payments to landowners to retire, leave fallow, or plant non-irrigated crops on their land. Such programs operate on an offer-based contract between the landowner and the coordinating government agency. The contractual relationship is subject to asymmetric information, and adverse selection may arise because the landowner has better information about the opportunity cost of supplying the environmental amenity than does the conservation agent. As a consequence, farmers may enroll their least productive, least intensively farmed lands in the programs while receiving payments higher than their opportunity costs, thus accruing rents. It is quite unlikely that an irrigated parcel, which requires considerable investment in a system of irrigation (which, in turn, enhances the productivity of the parcel), will be among a farmer's plots with the lowest opportunity cost and thus enrolled in the program. Instead, farmers may opt to enroll non-irrigated plots in the CRP program, which does not have any effect on the amount of irrigation water extracted (Pfeiffer and Lin 2009; Pfeiffer and Lin 2010; Lin 2013; Lin Lawell 2016).

In a recent study that has been featured in such media outlets as the *New York Times* (Wines 2013), the *Washington Post* (Howitt and Lund 2014), *Bloomberg View* (Ferraro 2016), and *AgMag Blog* (Cox 2013), Pfeiffer and Lin (2014a) focus on incentive-based groundwater conservation policies in Kansas and find that measures taken by the state of Kansas to subsidize a shift toward more efficient irrigation systems have not been effective in reducing groundwater extraction. The subsidized shift toward more efficient irrigation systems has in fact increased extraction through a shift in cropping patterns. Better irrigation systems allow more water-intensive crops to be produced at a higher marginal profit. The farmer has an incentive to both increase irrigated acreage and produce more water-intensive crops (Pfeiffer and Lin 2014a; Lin Lawell 2016). Similarly, land and water conservation and retirement programs are not effective in reducing groundwater pumping, which occurs, by definition, on irrigated and, thus, very productive land (Pfeiffer and Lin 2009; Pfeiffer and Lin 2010; Lin 2013; Lin Lawell 2016).

In California, SWEEP grant funds cannot be used to expand existing agricultural operations or to convert additional new acreage to farmland (California DWR and CFDA 2017), which may limit how much a farmer can respond to the increased irrigation efficiency resulting from SWEEP grant funds to increase irrigated acreage. However, by lowering the marginal cost of irrigation, SWEEP grant funds may encourage farmers to continue irrigating more marginal lands. Furthermore, this increased efficiency may allow farmers to continue growing more water-intensive crops, even as groundwater becomes scarcer. Thus, SWEEP funds could make farmers in water-stressed locations less sensitive to existing price signals as groundwater becomes scarce, thereby slowing their adjustment to depleting groundwater stocks over the long term (Sears, Bertone Oehninger, Lim, and Lin Lawell 2018; Sears, Lim, and Lin Lawell 2017; Sears, Lim, and Lin Lawell 2018b).

The California Department of Agriculture and the California Department of Water recently introduced a pilot program within SWEEP that incentivizes joint action by farmers and larger water suppliers to implement more efficient irrigation technology in return for an agreement to halt the use of groundwater for agricultural purposes (California DWR and CFDA 2016).

However, farmers and water suppliers who rely relatively little on groundwater as a source may use this program most. In this case, while irrigation may become more efficient, this may have little effect on groundwater use, the target of the policy. As a result, the costs of the program may unfortunately exceed its benefits (Sears, Bertone Oehninger, Lim, and Lin Lawell 2018; Sears, Lim, and Lin Lawell 2017; Sears, Lim, and Lin Lawell 2018b).

While heavily irrigated, California's cropland still includes almost one million acres of dry land farming, or non-irrigated land used for planting crops. Dry land farming constitutes about 9 percent of total cropland and 3.5 percent of total farmland in California. Another half a million acres of cropland is currently left to pasture but could be converted to cropland without improvements. In addition, farmland in California includes about 13 million acres of rangeland and pasture, only about half a million of which is irrigated (USDA 2012). Thus, a possible perverse consequence of California's SWEEP grant funds is that farmers may choose to convert more marginal land that is currently used for rangeland and dry land farming to more productive irrigated cropland as part of any efficiency gains from new irrigation technology purchased with state incentives, and this possible increase in irrigated acreage may lead to an increase in groundwater consumption (Sears, Bertone Oehninger, Lim, and Lin Lawell 2018; Sears, Lim, and Lin Lawell 2017; Sears, Lim, and Lin Lawell 2018b).

Land retirement programs at the federal and state level have had limited effectiveness in California and may also have perverse consequences. The largest federal land retirement program, the Conservation Reserve Program, provides rental payments to landowners who retire their land and follow conservation practices for a contracted period of time, usually ten years. While this program has retired 35 million acres of land nationally, it had only enrolled about 138,000 acres in California as of 2007, well below its share in total farmed acres (Champetier de Ribes and Sumner 2007). This is due in large part to the relatively high value of agricultural land, particularly irrigated farmland, in California (Sears, Bertone Oehninger, Lim, and Lin Lawell 2018; Sears, Lim, and Lin Lawell 2017; Sears, Lim, and Lin Lawell 2018b).

The most important state-level land retirement program in California is the Central Valley Project Improvement Act Land Retirement Program, which purchases land and water rights from owners (Land Retirement Technical Committee 1999). Between 1992 and 2011, the program retired about 9,000 acres as part of a planned 100,000-acre retirement (California DWR 2016).

The modest effect of land retirement programs on groundwater extraction in California is evidence of a design flaw in land retirement programs. In areas of high value agricultural production like California, farmers will demand much higher payments to voluntarily abandon crop production. Since California's most water-stressed regions coincide with areas of high value irrigated agricultural production, land retirement programs in these areas may be limited in their effectiveness, or very costly. In addition, the relatively low levels of spending by the Conservation Reserve Program in California suggest that the land that has been enrolled in the program is likely low-value land. Thus, just as in Kansas, land conservation programs may be ineffective in reducing groundwater extraction in California (Sears, Bertone Oehninger, Lim, and Lin Lawell 2018; Sears, Lim, and Lin Lawell 2017; Sears, Lim, and Lin Lawell 2018b).

The result that increases in irrigation efficiency may increase water consumption is an example of a rebound effect, or "Jevons' Paradox," which arises when the invention of a technology that enhances the efficiency of using a natural resource does not necessarily lead to less consumption of that resource (Jevons 1865). Jevons (1865) found this to be true with the use of coal in a wide range of industries (Lin 2013). In the case of agricultural groundwater, irrigation technology that increases irrigation efficiency does not necessarily lead to less consumption of groundwater (Lin 2013; Lin Lawell 2016). In particular, if demand is elastic enough, the higher efficiency technology operates at a lower marginal cost, and the higher efficiency technology increases revenue, then irrigation efficiency will increase applied water (Pfeiffer and Lin 2014a; Lin Lawell 2016).

Thus, when designing policies, policy makers need to be wary of any potential unintended consequences. Incentive-based groundwater conservation programs are a prime example of a well-intentioned policy gone awry (Lin Lawell 2016).

6 Property rights

An important component of sustainable agricultural groundwater management is complete, measured, enforceable, and enforced property rights that consider the physical properties of the resource (Lin Lawell 2016).

A variety of property rights doctrines and institutions governing groundwater have evolved in the western United States. Many more institutions, both formal and informal, are in place in other locations around the world (Lin Lawell 2018c).

The absolute ownership doctrine, which is the groundwater rights doctrine in Texas, gives owners of land the absolute right to extract water from their parcels. The correlative rights doctrine, which is the groundwater rights doctrine in Nebraska and Oklahoma, allows a property right to a portion of the aquifer related to the size of the land parcel owned (Lin Lawell 2018c).

The prior appropriation doctrine, which is the groundwater rights doctrine in Colorado, Kansas, New Mexico, South Dakota, and Wyoming, allots water rights based on historical use, with priority going to those who claimed their right first. Often, rights holders under the prior appropriation doctrine are allowed a maximum level of extraction per year (Sax and Abrams 1986). Leonard and Libecap (2017) analyze the economic determinants and effects of prior appropriation water rights that were voluntarily implemented across a vast area of the U.S. West, replacing common-law riparian water rights (Lin Lawell 2018c).

Current water rights in Kansas follow the prior appropriation doctrine. Before 1945, Kansas applied the common law of absolute ownership doctrine to groundwater. Water rights were not quantified in any way (Peck 2007). In 1945, following multiple conflicts between water users and several major water cases that reached the Kansas Supreme Court, the "Arid Region Doctrine of Appropriation" was adopted, which permitted water extraction based on the principle of "first in time, first in right" (Peck 1995).

In Kansas, the earliest appropriators of water maintain the first rights to continue to use water in times of shortage or conflict. The water right comes with an abandonment clause; if the water is not used for beneficial purposes for longer than the prescribed time period, it is subject to revocation (Peck 2003). To obtain a new water right, an application stating the location of the proposed point of diversion, the maximum flow rate, the quantity desired, the intended use, and the intended place of use must be submitted to and approved by the Department of Water Resources (*Kansas Handbook of Water Rights* 2006). Since 1945, Kansas has issued more than 40,000 groundwater appropriation permits (Peck 1995). The permits specify an amount of water that can be extracted each year and are constant over time (Lin Lawell 2018c).

Through the 1970s, the period of intensive agricultural development in Kansas, groundwater-pumping permits were granted to nearly anyone who requested them. Some permits are as old as 1945, but the majority (about 75 percent) were allocated between 1963 and 1981 (Lin Lawell 2018c). In 1972, owing to concerns that the aquifer was over-appropriated, Kansas created five groundwater management districts (GMDs). The GMDs regulate well spacing and prohibit new water extraction within a designated radius of existing wells, which varies by GMD (Lin Lawell 2018c).

The adoption of the prior appropriation doctrine, together with the development of GMDs to regulate new appropriations of water rights, arguably eliminated uncontrolled entry and the resulting over-exploitation commonly associated with common property resources in Kansas. Restricting water rights can reduce groundwater extraction in Kansas even when *ex post* the water rights are not binding (Li and Zhao 2018). However, appropriation contracts distort the incentive to optimize dynamically over the life of the resource, because the farmer is essentially guaranteed his appropriated amount of water until the resource becomes so scarce that it is no longer economical to pump (Lin Lawell 2018c).

California has historically relied on a system of two forms of groundwater property rights. First, overlying property rights allow owners of land to beneficially use a reasonable share of any groundwater basin lying below the surface of the land. Second, any surplus groundwater from the basin may then be beneficially used or sold by individuals or businesses that do not own land directly overlying the basin through an appropriative right. This system of dual rights arose from a 1903 California Supreme Court decision in the case of *Katz v. Walkinshaw*, which put an end to a period of "absolute ownership" rights, which guaranteed landowners the right to unlimited use of water underneath their properties (California State Water Resources Control Board 2011; Sears and Lin Lawell 2018).

The system of dual rights in California is designed to operate under both instances of surplus groundwater, when inflow exceeds the use of overlying users, and overdraft, when the groundwater table begins to decline due to extraction exceeding inflows. Appropriative groundwater rights are subordinate to overlying rights, and in times of overdraft a "first in line" system requires that more recent appropriative users cease their extraction (Sears and Lin Lawell 2018).

In practice, though, this relies on California's court system adjudicating property rights during periods of overdraft. The court's response to these periods has varied widely over time. Prior to 1949, appropriative right holders could obtain a "prescriptive right" that was senior to overlying rights by demonstrating that they had extracted from an overdrafted basin for at least five consecutive years (Lambert 1984).

The California State Supreme Court moderated this position in 1949 by creating a system of "mutual prescription" in which users of an overdrafted basin were allocated extraction in proportion to their prior use, and total extraction was to be within a "safe yield" (California DWR 2003). This created an incentive for overdrafted basin users to expand their groundwater use during times of overdraft, in order to receive a more favorable court allocation. Mutual prescription was modified in 1975 so that it could not infringe on public water agencies' rights to groundwater (California DWR 2003). In addition, the state legislature later moderated this by allowing the adjudicated allocation to be based also on supplemental water used in lieu of groundwater during an overdraft period (Lambert 1984; California Water Code 1005.1–4).

As part of determining allocation, the California State Water Resources Control Board has monitored groundwater use in four counties in Southern California since the 1950s through the California Groundwater Recordation Program (California Water Code 4999 et. seq.). The program allows the California State Water Resources Control Board to determine both the extraction shares of users and when periods of overdraft occur (Sears and Lin Lawell 2018).

7 Groundwater–energy nexus

Energy is an important input needed to extract groundwater for irrigation. Dumler et al. (2009) estimate that the energy cost of extracting irrigation water represents approximately 10 percent of the costs for growing corn in western Kansas, which is a slightly greater share of costs than land rent. Of the acres irrigated from groundwater wells in Kansas, about 50 percent are supplied by pumps powered with natural gas, 25 percent are supplied by pumps powered with diesel fuel, and 22 percent are supplied by pumps powered with electricity (USDA 2004).

In California, most of the energy for irrigation comes from electricity, though a substantial amount comes from diesel (CEC 2005). Of the total on-farm energy expense for pumping irrigation water in California in 2013, 86 percent was for electricity; 13 percent was for diesel and biodiesel; and the remaining less than 1 percent was for liquefied petroleum gas, gas propane, butane, natural gas, gasoline, and ethanol (USDA 2013).

Pfeiffer and Lin (2014c) report that energy prices do have an effect on groundwater extraction, causing water use to decrease along both the intensive and extensive margins. Increasing energy prices would affect crop selection decisions, crop acreage allocation decisions, and the demand for water by farmers. This finding is particularly important in the face of possible increases in energy prices in the future, which may cause farmers to respond by decreasing their water use. Their results also suggest that policies that reduce energy prices would cause groundwater extraction to increase, therefore posing a potential concern to conservationists who are worried about declining water table levels in many of the world's most productive agricultural basins that depend on groundwater.

8 Climate change

Climate change has the potential to impact groundwater availability in several ways. First, changes in climate may indirectly impact groundwater availability by causing changes in agricultural land use and changes in agricultural practices that then result in changes in water availability. For example, climate change may cause farmers to change the crops they plant or the amount of water they apply, both of which have implications for water availability (Bertone Oehninger, Lin Lawell, and Springborn 2018a).

Second, climate change may affect water availability directly. For example, changing climates may result in melting snowcaps and/or changes in precipitation that would affect the availability of water for agriculture (Bertone Oehninger, Lin Lawell, and Springborn 2018a).

Climate change is characterized by uncertainty and the possibility of catastrophic damages with small but non-negligible probabilities (Weitzman 2014). Tsur and Zemel (1995) study the optimal exploitation of renewable groundwater resources when extraction affects the probability of the occurrence of an event that renders the resource obsolete. They find that under uncertainty, when the event occurrence level is unknown, the expected loss due to the event occurrence is so high that it does not pay to extract in excess of recharge, even though under certainty doing so would be beneficial (Tsur and Zemel 1995).

Bertone Oehninger, Lin Lawell, and Springborn (2018a) analyze the effects of changes in temperature, precipitation, and humidity on groundwater extraction for agriculture using an econometric model of a farmer's irrigation water pumping decision that accounts for both the intensive and extensive margins. They find that changes in climate variables influence crop selection decisions, crop acreage allocation decisions, technology adoption, and the demand for water by farmers. Bertone Oehninger, Lin Lawell, and Springborn (2018b) find that such changes in behavior could affect land use and agricultural biodiversity.

9 Conclusion

Sustainable agricultural groundwater management policies need to account for dynamic and spatial considerations that arise with groundwater, as well as for any possible perverse consequences from the policy. Important components of sustainable agricultural groundwater management are complete, measured, enforceable, and enforced property rights that consider the physical properties of the resource, as well as carefully designed policies that internalize any externalities, whether they are caused by the physical movement of water, by environmental damages or benefits, or by other causes (Lin Lawell 2016). Groundwater management policies should also consider any tradeoffs involved between water quantity and water quality, as it is possible for groundwater management policies to lower quantity while improving quality or vice versa (Roseta-Palma 2002).

Incentive-based groundwater conservation programs are a prime example of a well-intentioned policy that may have perverse consequences, meaning that they may actually increase rather than decrease groundwater extraction. When designing policies and regulation, policy makers need to be aware of the full range of implications of their policy, including any potential perverse consequences.

The water management and economics discussed in this chapter – including the importance of dynamic management; the importance of spatial management; the possible perverse consequences of incentive-based agricultural groundwater conservation programs; property rights; the groundwater–energy nexus; and the effects of climate change – have important implications for the design of policies for sustainable agricultural groundwater management worldwide.

An important direction for future research is the design and evaluation of sustainable agricultural groundwater management policies worldwide that synthesize and incorporate the economics of sustainable agricultural groundwater management discussed in this chapter.

Acknowledgments

We benefited from the excellent research of Lisa Pfeiffer, Ernst Bertone Oehninger, David Lim, and Ellen Bruno. We thank Emmanuel Asinas, Chris Bowman, Mark Carlson, Colin Carter, Ariel Dinar, Jim Downing, Dietrich Earnhart, Paul Ferraro, Roman Hernandez, Richard Howitt, Richard Kravitz, Alan Krupnick, Yusuke Kuwayama, Jay Lund, Julie McNamara, Krishna P. Paudel, H. Michael Ross, Jim Roumasset, Jim Sanchirico, Kurt Schwabe, Rich Sexton, Cindy Simmons, Mike Springborn, Dale Squires, Dan Sumner, David Sunding, Ed Taylor, Kristina Victor, Jim Wilen, and David Zilberman for invaluable comments, insight, support, and encouragement. We also benefited from comments from participants at the conference on "Water Pricing for a Dry Future: Policy Ideas from Abroad and Their Relevance to California," and at our Honorable Mention Bacon Lectureship at the University of California Center Sacramento. We received funding from the Giannini Foundation of Agricultural Economics and from the 2015–2016 Bacon Public Lectureship and White Paper Competition. All errors are our own.

References

Bertone Oehninger, E., C.-Y.C. Lin Lawell, and M.R. Springborn. 2018a. "The Effects of Climate Change on Agricultural Groundwater Extraction." Working Paper, Cornell University.
———. 2018b. "The Effects of Climate Change on Crop Choice and Agricultural Variety." Working Paper, Cornell University.
Brozovic, N., D.L. Sunding, and D. Zilberman. 2006. "Optimal Management of Groundwater Over Space and Time." *Frontiers in Water Resources Economics* 29:109–135.
California Department of Water Resources [DWR]. 2003. *DWR Bulletin 118: Appendices.* https://www.water.ca.gov/LegacyFiles/pubs/groundwater/bulletin_118/california's_groundwater__bulletin_118_-_update_2003_/bulletin118-appendices.pdf. Accessed April 10, 2018.

———. 2015. *Sustainable Groundwater Management Program: Draft Strategic Plan.* http://water.ca.gov/groundwater/sgm/pdfs/DWR_GSP_DraftStrategicPlanMarch2015.pdf. Accessed April 10, 2018.

———. 2016. *Other Resource Management Strategies: A Resource Management Strategy of the California Water Plan.* www.water.ca.gov/waterplan/docs/rms/2016/31_Other_RMS_July2016.pdf. Accessed April 10, 2018.

California Department of Water Resources [DWR] and California Department of Food and Agriculture [CFDA]. 2016. *Draft Agricultural Water Use Efficiency and State Water Efficiency and Enhancement Program: DWR/CDFA Joint Request for Proposals.* www.water.ca.gov/wuegrants/Docs/DWR%20CDFA%20Joint%20RFP%2008.25.2016-DRAFT.pdf. Accessed April 10, 2018.

———. 2017. *2017 State Water Efficiency and Enhancement Program.* www.cdfa.ca.gov/oefi/sweep/docs/2017-SWEEP_ApplicationGuidelines.pdf. Accessed April 10, 2018.

California Energy Commission [CEC]. 2005. *2005 Integrated Energy Policy Report.* www.energy.ca.gov/2005publications/CEC-100-2005-007/CEC-100-2005-007-CMF.PDF. Accessed April 10, 2018.

———. 2015a. *California's Drought: Opportunities to Save Water and Energy.* www.energy.ca.gov/drought/rebate/2015-07-10_Opportunities_To_Save_Water.pdf. Accessed November 18, 2016.

———. 2015b. *California's Water Energy Technology Program: Investing in Innovative Drought Solutions.* www.energy.ca.gov/2015publications/CEC-500-2015-025/CEC-500-2015-025-FS-REV2.pdf. Accessed April 10, 2018.

California State Water Resources Control Board. 2011. *The History of the Water Boards: The Early Years of Water Rights.* https://www.waterboards.ca.gov/about_us/water_boards_structure/history_water_rights.html. Accessed April 10, 2018.

California Water Code Section 1005, 1–4. Definitions and Interpretations of Divisions. http://leginfo.legislature.ca.gov/faces/codes_displayText.xhtml?lawCode=WAT&division=2.&title=&part=1.&chapter=1.&article=. Accessed April 10, 2018.

California Water Code Section 4999 et. seq. Recordation of Water Extractions and Diversions. http://leginfo.legislature.ca.gov/faces/codes_displayText.xhtml?lawCode=WAT&division=2.&title=&part=5.&chapter=&article=. Accessed April 10, 2018.

Champetier de Ribes, A., and D.A. Sumner. 2007. "Agricultural Conservation and the 2007 Farm Bill: A California Perspective." *University of California Agricultural Issues Center (AIC) Farm Bill Brief Number 6.* http://aic.ucdavis.edu/research/farmbill07/AIC_FBIB_6Conservation.pdf. Accessed April 10, 2018.

Committee, O.A.M.A. 2001. *Discussion and Recommendations for Long-Term Management of the Ogallala Aquifer in Kansas.* Technical Report. Topeka, KS.

Cox, C. 2013. "Programs to Reduce AG's Water Use Must Be Strengthened, Not Cut." *AgMag BLOG,* Environmental Working Group, May 28, 2013. www.ewg.org/agmag/2013/05/programs-reduce-ag-s-water-use-must-be-strengthened-not-cut. Accessed April 10, 2018.

Dasgupta, P., and G.M. Heal. 1979. *Economic Theory and Exhaustible Resources.* Cambridge: Cambridge University Press.

Dinar, S., and A. Dinar. 2016. *International Water Scarcity and Variability: Managing Resource Use Across Political Boundaries.* Berkeley: University of California Press.

Dumler, T.J., D.M. O'Brien, B.L. Olson, and K.L. Martin. 2009. "Center-Pivot-Irrigated Corn Cost–Return Budget in Western Kansas." *Farm Management Guide MF-585,* Kansas State University.

Edwards, E.C. 2016. "What Lies Beneath? Aquifer Heterogeneity and the Economics of Groundwater Management." *Journal of the Association of Environmental and Resource Economists* 3(2):453–491.

Eswaran, M., and T. Lewis. 1984. "Appropriability and the Extraction of a Common Property Resource." *Economica* 51(204):393–400.

Famiglietti, J. 2014. "Epic California Drought and Groundwater: Where Do We Go from Here?" *National Geographic* February 4, 2014. http://voices.nationalgeographic.com/2014/02/04/epic-california-drought-and-groundwater-where-do-we-go-from-here/. Accessed April 10, 2018.

Feinerman, E., and K.C. Knapp. 1983. "Benefits from Groundwater Management: Magnitude, Sensitivity, and Distribution." *American Journal of Agricultural Economics* 65(4):703–710.

Ferraro, P.J. 2016. "When Conservation Efforts End Up Using More Water." *Bloomberg View* August 2, 2016. www.bloomberg.com/view/articles/2016-08-02/when-conservation-efforts-end-up-using-more-water. Accessed April 10, 2018.

Gisser, M., and D.A. Sanchez 1980. Competition Versus Optimal Control in Groundwater Pumping. *Water Resources Research* 16(4):638–642.

Hartwick, J.M., and N.D. Olewiler. 1998. *The Economics of Natural Resource Use*. New York: Addison-Wesley.

Howitt, R., and J. Lund. 2014. "Five Myths About California's Drought." *Washington Post* August 29, 2014.

Jevons, W.S. 1865. *The Coal Question*. London: Palgrave Macmillan and Co.

Kansas Handbook of Water Rights. 2006. http://cdm16884.contentdm.oclc.org/cdm/ref/collection/p16884coll5/id/84. Accessed April 10, 2018.

Kuwayama, Y., and N. Brozovic. 2013. "The Regulation of a Spatially Heterogeneous Externality: Tradable Groundwater Permits to Protect Streams." *Journal of Environmental Economics and Management* 66(2):364–382.

Lambert, D. 1984. "District Management for California's Groundwater." *Ecology Law Quarterly* 11(3):373–400.

Land Retirement Technical Committee. 1999. *Task 3 Land Retirement. Final Report*. Sacramento (CA): The San Joaquin Valley Drainage Implementation Program and the University of California Salinity/Drainage Program. www.water.ca.gov/pubs/groundwater/land_retirement_final_report__san_joaquin_valley_drainage_implementation_program/05-landretirement.pdf. Accessed April 10, 2018.

Leonard, B., and G.D. Libecap. 2017. "Collective Action By Contract: Prior Appropriation And The Development Of Irrigation In The Western United States." NBER Working Paper No. 22185. www.nber.org/papers/w22185.pdf. Accessed April 10, 2018.

Li, H., and J. Zhao. 2018. "Rebound Effects of New Irrigation Technologies: The Role of Water Rights." *American Journal of Agricultural Economics*.

Lin, C.-Y.C. 2010. "How Should Standards Be Set and Met?: On the Allocation of Regulatory Power in a Federal System." *B.E. Journal of Economic Analysis and Policy: Topics* 10(1): Article 51.

———. 2013. "Paradox on the Plains: As Water Efficiency Increases, So Can Water Use." *California Water Blog*. http://californiawaterblog.com/2013/08/13/paradox-on-the-plains-as-water-efficiency-increases-so-can-water-use/. Accessed April 10, 2018.

Lin, C.-Y.C., and L. Pfeiffer. 2015. "Strategic Behavior and Regulation Over Time and Space." In K. Burnett, R. Howitt, J.A. Roumasset, and C.A. Wada, eds., *Routledge Handbook of Water Economics and Institutions*. New York: Routledge, pp. 79–90.

Lin Lawell, C.-Y.C. 2016. "The Management of Groundwater: Irrigation Efficiency, Policy, Institutions, and Externalities." *Annual Review of Resource Economics* 8:247–259.

———. 2018a. "A Theory of Regulatory Federalism." Working Paper, Cornell University.

———. 2018b. "On the Optimal Allocation of Regulatory Power for Public Goods Provision." Working Paper, Cornell University.

———. 2018c. "Property Rights and Groundwater Management in the High Plains Aquifer." Working Paper, Cornell University.

Mani, A., F.T.-C. Tsai, and K.P. Paudel. 2016. "Mixed Integer Linear Fractional Programming for Conjunctive Use of Surface Water and Groundwater." *Journal of Water Resources Planning and Management* 142(11).

McMahon, P.B., L.N. Plummer, J.K Bohlke, S.D. Shapiro, and S.R. Hinkle. 2011. "A Comparison of Recharge Rates in Aquifers of the United States Based on Groundwater-Age Data." *Hydrogeology Journal* 19(4):779–800.

Msangi, S. 2004. *Managing Groundwater in the Presence of Asymmetry: Three Essays*. PhD dissertation, University of California at Davis.

National Groundwater Association. 2016. *Facts About Global Groundwater Usage*. www.ngwa.org/Fundamentals/Documents/global-groundwater-use-fact-sheet.pdf. Accessed April 10, 2018.

Negri, D.H. 1989. "Common Property Aquifer as a Differential Game." *Water Resources Research* 25(1):9–15.

NRCS. 2004. *Farm bill 2002: Environmental Quality Incentives Program Fact Sheet*. U.S. Department of Agriculture. Washington, DC. www.nrcs.usda.gov/Internet/FSE_DOCUMENTS/nrcs144p2_024349.pdf. Accessed April 10, 2018.

Peck, J.C. 1995. "The Kansas Water Appropriation Act: A Fifty-Year Perspective." *Kansas Law Review* 43:735–756.

———. 2003. "Property Rights in Groundwater: Some Lessons from the Kansas Experience." *Kansas Journal of Law and Public Policy* XII(12):493–520.

———. 2007. "Groundwater Management in the High Plains Aquifer in the USA: Legal Problems and Innovations." In M. Geordano and K.G. Villholth, eds., *The Agricultural Groundwater Revolution: Opportunities and Threats to Development*. Wallingford: CAB International, pp. 296–319.

Pfeiffer, L., and C.-Y.C. Lin. 2009. "Incentive-Based Groundwater Conservation Programs: Perverse Consequences?" *Agricultural and Resource Economics Update* 12(6):1–4.

———. 2010. "The Effect of Irrigation Technology on Groundwater Use." *Choices* 25(3).

———. 2012. "Groundwater Pumping and Spatial Externalities in Agriculture." *Journal of Environmental Economics and Management* 64(1):16–30.

———. 2014a. "Does Efficient Irrigation Technology Lead to Reduced Groundwater Extraction? Empirical Evidence." *Journal of Environmental Economics and Management* 67(2):189–208.

———. 2014b. "Perverse Consequences of Incentive-Based Groundwater Conservation Programs." *Global Water Forum*, Discussion paper 1415.

———. 2014c. "The Effects of Energy Prices on Agricultural Groundwater Extraction from the High Plains Aquifer." *American Journal of Agricultural Economics* 96(5):1349–1362.

Pitafi, B.A., and J.A. Roumasset. 2009. "Pareto-Improving Water Management Over Space and Time: The Honolulu Case." *American Journal of Agricultural Economics* 91(1):138–153.

Provencher, B., and O. Burt. 1993. "The Externalities Associated with the Common Property Exploitation of Groundwater." *Journal of Environmental Economics and Management* 24(2):139–158.

Roseta-Palma, C. 2002. "Groundwater Management When Water Quality Is Endogenous." *Journal of Environmental Economics and Management* 44(1):93–105.

Rubio, S.J., and B. Casino. 2003. "Strategic Behavior and Efficiency in the Common Property Extraction of Groundwater." *Environmental and Resource Economics* 26(1):73–87.

Saak, A.E., and J.M. Peterson. 2007. "Groundwater Use Under Incomplete Information." *Journal of Environmental Economics and Management* 54(2):214–228.

Sax, J.L., and R.H. Abrams. 1986. *Legal Control of Water Resources*. St. Paul: West Publishing Company.

Sears, L., E. Bertone Oehninger, D. Lim, and C.-Y.C. Lin Lawell. 2018. "The Economics of Sustainable Agricultural Groundwater Management: Recent Findings." Working Paper, Cornell University.

Sears, L., D. Lim, and C.-Y.C. Lin Lawell. 2017. "Agricultural Groundwater Management in California: Possible Perverse Consequences?" *Agricultural and Resource Economics Update* 20(3):1–3.

———. 2018a. "Spatial Groundwater Management: A Dynamic Game Framework and Application to California." Working Paper, Cornell University.

———. 2018b. "The Economics of Agricultural Groundwater Management Institutions: The Case of California." Water Economics and Policy.

Sears, L., and C.-Y.C. Lin Lawell. 2018. "Dual Rights to Groundwater: Theory and Application to California." Working Paper, Cornell University.

Sears, L., C.-Y.C. Lin Lawell, and D. Lim. 2018. "Interjurisdictional Spatial Externalities in Groundwater Management." Working Paper, Cornell University.

Siebert, S., V. Henrich, K. Frenken, and J. Burke. 2013. *Global Map of Irrigation Areas version 5*. Reminisce Friedrich-Wilhelms-University, Bonn, Germany/Food and Agriculture Organization of the United Nations, Rome, Italy. www.fao.org/nr/water/aquastat/irrigationmap/index10.stm. Accessed April 10, 2018.

Timmins, C. 2002. "Measuring the Dynamic Efficiency Costs of Regulators' Preferences: Municipal Water Utilities in the Arid West." *Econometrica* 70(2):603–629. Tsur, Y., and T. Graham-Tomasi. 1991. "The Buffer Value of Groundwater with Stochastic Surface Water Supplies." *Journal of Environmental Economics and Management* 21(3):201–224.

Tsur, Y., and A. Zemel. (1995). "Uncertainty and Irreversibility in Groundwater Resource Management." *Journal of Environmental Economics and Management* 2(2):149–161.

U.S. Department of Agriculture [USDA]. 2004. *2003 Farm and Ranch Irrigation Survey Volume 3, Special Studies Part 1*. National Agricultural Statistics Service. http://usda.mannlib.cornell.edu/usda/AgCensusImages/2002/02/06/Complete%20Report.pdf. Accessed April 10, 2018.

———. 2012. *2012 Census of Agriculture. Volume 1, Chapter 1: State-Level Data: California*. https://www.agcensus.usda.gov/Publications/2012/Full_Report/Volume_1,_Chapter_1_State_Level/California/. Accessed April 10, 2018.

———. 2013. *2013 Farm and Ranch Irrigation Survey*. National Agricultural Statistics Service. https://www.agcensus.usda.gov/Publications/2012/Online_Resources/Farm_and_Ranch_Irrigation_Survey/. Accessed April 10, 2018.

———. 2014. *Conservation Reserve Program*. Farm Service Agency. www.fsa.usda.gov/FSA/webapp?area=home&subject=copr&topic=crp. Accessed April 10, 2018.

U.S. Geological Survey [USGS]. 2017. *California Drought*. https://ca.water.usgs.gov/data/drought/. Accessed August 16, 2017.

Weitzman, M.L. 2014. "Fat Tails and the Social Cost of Carbon." *American Economic Review* 104(5):544–546.

Wines, M. 2013. "Wells Dry, Fertile Plains Turn to Dust." *New York Times*, May 19, 2013.

York, T., and D.A. Sumner. 2015. "Why Food Prices Are Drought-Resistant." *Wall Street Journal* April 12, 2015.

17
Water supply and dams in agriculture

Biswo N. Poudel, Yang Xie, and David Zilberman

1 Introduction

Water is crucial for agriculture, but it needs to be allocated and managed effectively to address variability over space and time. The design and management of water infrastructure has traditionally been a role of governments in major societies (Water Technology Net 2016).

Dams and aqueducts have been major sources of the supply of water for agriculture, which accounts for 70 percent of total water use (van der Zaag and Gupta 2008). The role of dams in agriculture has been primarily to address seasonal water variation and to assure the success of multiple cropping (Brown and Lall 2006). Dams are also crucial in reducing the risk of flood and allocating water among seasons and between regions (IEA 2011). Dams supply 86 percent of the renewable energy in the world. However, increased urban water demand and emphasis on environmental amenities have raised concerns about the construction of new dams, and this has exacerbated the challenges of optimal water resource management. This chapter addresses some of the challenges of water infrastructure, mostly in the context of agriculture.

There has been a long, multidisciplinary policy debate about the value and viability of dams. The next section provides background on some of the debate and major issues. Afterwards, we present conceptual modeling on the design and implications of water storage systems. We also discuss farmers' response to dams, followed by an overview of empirical analysis and a conclusion.

2 Background on dams

Dam technology dates back to 1300 BC. There are as many as 50,000 large dams in the world today, compared to about 5,700 in 1950 (Scudder 2012). Dams supply water for the irrigation of 30–40 percent of the 271 million hectares of irrigated lands (FAO 2015, 2017). Irrigated agriculture produces about 40 percent of the food and fiber in the world. On a per-unit land basis, irrigated land produces more than five times the economic value of non-irrigated land (Schoengold and Zilberman 2007). This is reflected in land values, such as in California, where irrigated acreage is worth three times that of non-irrigated acreage (USDA-NASS 2012).

Dams and water projects have contributed significantly to increases in both agricultural productivity and manufacturing. Dams are an economic marvel. Almost all large dams cost a significant portion of a nation's gross domestic product (GDP). For example, the proposed Budhigandaki project in Nepal is estimated to cost more than 5 percent of Nepal's annual GDP.

Dams and water projects were a major engine of growth in the western United States during the twentieth century, as they provided an inexpensive energy source. This led, for instance, to the establishment of the aeronautical industry, which needed significant electricity for aluminum production. In addition, much of the research and engineering introduced in developed countries spread throughout the world as the damming of rivers became a major public investment (Reisner 1993). The World Commission on Dams (WCD 2000) noted that dams have made a net positive contribution to human development, but their use also has some major drawbacks.

The World Wildlife Fund (Kraljevic, Meng, and Schelle 2013) noted the potential "seven sins" of dam construction: (1) the choice of the wrong river on which to build the dam, (2) neglect of downstream flows, (3) neglect of biodiversity, (4) a reliance on bad economics (e.g., underestimated costs and construction delays (Bacon and Besant-Jones 1998; Ansar et al. 2014)), (5) a failure to acquire a *social license* to operate, (6) mishandling risks and unintended impacts, and (7) giving too much weight to policy makers' bias in construction decisions. Furthermore, optimal sites for large dam construction are finite, and reservoirs are losing 1 percent of their total capacity each year due to sedimentation deposit, especially in China and the South Asian countries (McCully 1996).

Most dams reallocate resources from riparian local users to many non-riparian stakeholders, which is often politically controversial, with displacement and resettlement a significant social issue. On average, 13,000 people have been displaced for every large dam (World Bank 1996). About 40 to 80 million people have been displaced by dams worldwide, and the lifestyle of the remaining populations have been altered, sometimes negatively (Attwood 2005). Major studies have documented flaws in the design and management of dams, including underestimation of the associated environmental costs (Stone 2011), improper sedimentation management (Poudel 2010), greenhouse gas emissions (Rudd et al. 1993; Louis et al. 2000; Tremblay, Lambert, and Gagnon 2004; Barros et al. 2011), overestimation of energy production (WCD 2000), and economic inefficiencies (Duflo and Pande 2007; Ansar et al. 2014).

Many of the criticisms of dams have singled out large dams, and there has been a lively debate on whether large dams are desirable. Large reservoirs are often correlated with increased economy of scale in terms of both benefits and costs. Large reservoirs support many farmers, attract industries to the region, and support knowledge sharing among the farmers (Lipton, Litchfield, and Faures 2003). While small dams give little aid in coping with serious droughts, large dams increase the welfare of the population (Attwood 2005).

The evidence in favor of large dams over small dams, however, is inconclusive. Hussain (2007) concluded that there is no systematic pattern of increasing or decreasing poverty associated with the size of irrigation projects. While Blanc and Strobl (2014) found that small dams tend to have a higher internal rate of return, their calculation was limited because it did not factor in both the positive (e.g., hydropower and recreational) and negative (e.g., ecological) impacts of large dams. Dillon (2011) showed that while both small and large irrigation systems have similar impacts on agricultural production, small irrigation systems tend to have higher productivity and income impact per hectare, while large systems have a higher consumption effect because they attract more people and create more non-farm employment opportunities.

One of the main arguments against large dams is the significant amount of greenhouse gas emission they create. Barros et al. (2011) estimated 4 percent of global carbon emissions from inland water could be associated with reservoirs, and Louis et al. (2000) estimated almost

7 percent of all other documented anthropogenic emissions could be attributed to reservoirs. In addition, reservoirs are associated with increased malarial and other mosquito-induced diseases, because they provide a breeding ground for mosquitos (Rudd et al. 1993; Tremblay, Lambert, and Gagnon 2004; Keiser et al. 2005; Giles 2006; Kitchens 2013). Also, a major argument against large dams is that they are too big to solve any urgent energy or irrigation need, and planners are either susceptible to planning fallacies or intentionally deceptive when they push for big projects (Ansar et al. 2014; Flyvbjerg 2005, 2009). Ansar et al. (2014) compute that, on average, large dams take 8.6 years to be fully functional. Rangachari et al. (2000), in their report to the World Commission on Dams (WCD) on large dams in India, commented that in many cases, while costs were underestimated, benefits were consistently overestimated.

After losing some support in the 1980s–mid-1990s, there is again a renewed interest in dams, as indicated by ongoing construction of the mega 900 MW Dahuaqiao Dam in China. With growing demand for food and power, international funding agencies are more open to investing in dams that provide irrigation as well as electricity. Better design and better management of dams are essential to improved outcomes.

3 Economic design modeling implications of dams and water supply systems

The construction of dams requires collective action at both the regional and national levels. Throughout the world, water-user associations are involved in diverting water resources for agriculture, mining, and energy. Generally, smaller dams are funded at the local level, while larger dams are funded at the state or international level. Government is often involved in large water projects meant for flood protection, hydroelectric power, and agricultural production. For example, in the United States, the Tennessee Valley Authority (TVA), created in 1933, is a government-sanctioned initiative investing in hydroelectric dams, navigational canals, and road networks (Kline and Moretti 2014).

Since the 1970s, benefit–cost analysis has become a major economic method used by governments and international organizations to assess the benefit of water projects (Schoengold and Zilberman 2007). Criteria used in the benefit–cost analysis includes the expected net present value of market and non-market benefits. It requires the use of economic surplus measures to assess market benefits as well as non-market valuation techniques (National Research Council [NRC] 2004).

Chakravorty et al. (2009) consider water systems to include several components: a water extraction and divergence source (e.g., a dam), a water conveyance mechanism, and a distribution network that allocates water to farmers. The design of a system may include physical parameters as well as managerial parameters, such as the size of the dam (this affects the expected benefit from hydroelectric power, water storage and irrigation, flood protection, and recreation), the lining of the canals (relates to distributional losses from source to use), and incentives for allocation among water users (principles for water pricing). Social welfare maximization may lead to an optimal design by equating the expected discounted social marginal benefits of key dams and the parameters of their expected discounted social marginal costs. If, however, a water project aimed at providing irrigation water is controlled by a profit-maximizing monopoly, it may under-divert and undersupply water. Furthermore, a lack of attention to the environmental services provided by water at the source may result in excessively large diversions and dams. When the design of water projects ignores or underinvests in conveyance, it results in shorter canals and reduces the benefits derived from the projects (Chakravorty, Hochman, and Zilberman 1995). Similarly, when water trading is disallowed or water is underpriced, this may lead to underinvestment in modern irrigation technologies and reduce the benefits of a project (Schoengold and Zilberman 2007). Thus, the economic calculus that determines the scale of a dam or reservoir must take into account the components of the system associated with the project.

The use of irrigation water provided by dams may change over time due to the availability of new technologies. The emergence of new irrigation technologies that include sprinkler and drip irrigation, as well as the use of weather data, affect water use, crop selection, and the profitability of water projects. Modern irrigation technologies increase water use efficiency, that is, the percentage of applied water used by the crop (Caswell and Zilberman 1986) and the timing of irrigation. The adoption of modern irrigation technologies by farmers tends to increase yield per hectare, and when combined with proper chemical applications, leads to a reduction in both input use and the residue of inputs not utilized by the crops (Caswell, Lichtenberg, and Zilberman 1990). The adoption of modern irrigation technologies is likely to save water if the marginal productivity of effective water is declining significantly with water application (technically, the elasticity of marginal productivity [EMP] is greater than one). However, if EMP is smaller than one, this need not be the case. The empirical evidence generally supports that the adoption of modern technologies can increase water use (Ward and Pulido-Velazquez 2008; Pfeiffer and Lin 2014).

Xie and Zilberman (2016) develop a framework on the optimal size of catchment reservoirs (which capture water in the wet season and release it in the dry season), taking into account the uncertainty of precipitation and climate as well as the social benefits and costs of water use. The optimal size of reservoirs is determined by balancing, at the social margin, the cost of construction, the expected cost from flood damage, and the net benefits from reservoir outflow use. They find that dam size increases as potential damage from floods increases, and as the value of output produced by released water increases, which depends on water allocation institutions (e.g., water rights systems, pricing schemes, and the industrial organization of water supply), and as the distribution of inflow skews rightwards, which can be caused by climate change. Another finding is that the adoption of water conservation technologies is not likely to occur when dams are too small (without sufficient scale to cover the cost of conservation) or excessively large (so there is little gain from marginal water conservation). This feature will make the marginal benefit of dam capacity discontinuous, and it will make the impact of overlooking the potential adoption of irrigation technologies on dam capacity choice ambiguous.

Zhao and Zilberman (1999) suggest that there are increasing returns to scale in the construction of dams. Therefore, in some cases where there is an expected increase in future output demand, large irrigation dams may not be fully utilized for several years after construction. While large dam capacity may expand when there is technological uncertainty and increasing returns to scale in construction, it may contract when certain demands for conservation technology and environmental safeguards increase.

Xie and Zilberman (2018) consider the optimal design of dams used for both reallocating water within seasons and storing water over time. They show that increased storage capacity and conservation technologies are not necessarily substitutes. The introduction of water-conserving technologies may actually increase the optimal size of dams when the marginal productivity of water is slowly decelerating (EMP < 1) or when it does so quickly, but the rate of change is small (EMP > 1 and second-order EMP < 2).

Other important theoretical works on storage capacity for agricultural water use include, but are not limited to, Fisher and Rubio (1997) and Truong (2012). Fisher and Rubio (1997) investigate the optimal real-time renovation of storage capacities that manage annual variations in water supply. They show long-run storage capacity is positively correlated with variance in water supply, which can be increased by climate change if the marginal benefit of water release is convex. Truong (2012) builds a competitive storage model to investigate the impact of reducing storage capacities that manage seasonal and annual variations in water supply on the irrigation sector. Results show that capacity reduction will increase the share of dam capacity utilized, on average, and that the value of the irrigation sector will decrease, while the impact on the average water price is ambiguous.

Most of the economic literature on optimal dam size and water supply management takes a microeconomic perspective. However, given the scale of dams and their importance in the overall economy, Kline and Moretti (2014) suggest using general equilibrium, structural, multi-sector models of the economy so that some of the dynamic macroeconomic implications of dams can be captured by growth theory models. One example is given in Hornbeck and Keskin (2015), who apply a two-sector model of economic growth for analyzing the impact of a large aquifer. Firms use technology $f(A, L, K, T)$, where A is the productivity parameter; L is labor input; K is capital input; T is total land; and w, r, and q are the prices of labor, capital, and land, respectively. The farmer maximizes profit $\Pi = f(A, L, K, T) - wL - rK - qT$ with the appropriate choice of L, K, and T, which are all functions of the reservoir. Assuming the normalized output price of 1, the presence of a reservoir R is assumed to enhance the productivity of industry A (one can think of agro-based industries). Optimal values of L and T are functions of wages and rental land price, whereas optimal K is a function of wages, interest rate on capital, and rental land price. Wages and rental land price are also functions of the reservoir, as it attracts industries and individuals to migrate near the reservoir. This increases economic activities and labor productivity. With this model, profitability increases with the size of the reservoir through its impact on all inputs of production:

$$\frac{d\Pi}{dR} = \frac{\partial f}{\partial A}\frac{\partial A}{\partial R} + \frac{\partial w}{\partial R}\left\{\left[\frac{\partial L^*}{\partial w}\left(\frac{\partial f}{\partial L} - w\right) - L^*\right] + \left[\frac{\partial K^*}{\partial w}\left(\frac{\partial f}{\partial K} - r\right)\right] + \left[\frac{\partial T^*}{\partial w}\left(\frac{\partial f}{\partial T} - q\right)\right]\right\}$$
$$+ \frac{\partial q}{\partial R}\left\{\left[\frac{\partial L^*}{\partial q}\left(\frac{\partial f}{\partial L} - w\right)\right] + \left[\frac{\partial K^*}{\partial q}\left(\frac{\partial f}{\partial K} - r\right)\right] + \left[\frac{\partial T^*}{\partial q}\left(\frac{\partial f}{\partial T} - q\right) - T^*\right]\right\}$$

When firms in a country behave like a price taker in the world market, then $\frac{\partial f}{\partial L} = w$, $\frac{\partial f}{\partial K} = r$ and $\frac{\partial f}{\partial T} = q$. This leads to $\frac{d\Pi}{dR} = \frac{\partial f}{\partial A}\frac{\partial A}{\partial R} - \frac{\partial w}{\partial R}L^* - \frac{\partial q}{\partial R}T^*$. Hence, firms, and by extension the aggregate industrial sector, continue to increase profit with reservoir construction as long as the productivity effect exceeds the increase in cost due to the labor impact and the increased value of land. The analysis suggests that a country with both lax labor laws that restrain wages and large amounts of fallow land may continue to build dams profitably, whereas a country with severely limited land endowments and high wages may not be inclined to build more dams.

4 Farmers' response to dams

The socioeconomic status of farmers living in the catchment areas of dams is likely to be heterogeneous, as will their response to dam construction. Baboo (1991), for example, noted that when the construction of Hirakud Dam began in Odyssa, India, wealthy and well-educated farmers migrated to cities, whereas poor farmers remained in designated colonies nearby. Kline and Moretti (2014) also noted that in the United States relatively poorer people lived near the reservoirs built by the TVA because those lands were cheaper due to environmental risks. However, impact evaluation papers (Duflo and Pande 2007; Blanc and Strobl 2014) tend to give light treatment to migration issues in catchment areas. Xabadia, Goetz, and Zilberman (2004) noted that in the absence of corrective water pricing, policies to address heterogeneity in land quality and suboptimal water use behavior, such as the delayed adoption of modern technologies, will be observed among farmers.

Dams also affect a farmer's risk exposure through decreased variance in production, thus affecting a farmer's expected income, *ceteris paribus*. Reduced risk may decrease the urgency for adopting risk-mitigating measures, including new technologies. Conversely, stability in production may increase predictability regarding yield and provide an incentive to invest in productive technology. Farmers' adoption of improved seeds, fertilizers, and efficient water conservation technologies may be affected by their access to stable sources of water. Kovacs et al. (2015) find a substitution of rice with soybean crops when the depth of the aquifer increased in their simulation model. Shakya and Flinn (1985), in their study of the adoption behavior of farmers in Nepal, found that the use of improved seeds was highest in the areas where irrigation facilities existed. However, Koundouri, Nauges, and Tzouvelekas (2006) found that farmers' adoption of new technology increases when their need to hedge against production risk is higher. This was similar to Gerhart's (1975) finding that farmers in Kenya were not adopting hybrid maize as long as they had other means to cope with the risks addressed by these hybrids. Zilberman et al. (2011), in their study of droughts in California, noted that only after the supply of water from reservoirs started to decrease did California farmers begin to respond by increasingly adopting water conservation technologies. Bhaduri and Manna (2014) showed that farm-level storage capacities can encourage the adoption of efficient irrigation technologies. Emerick et al. (2016) showed through a randomized experiment that reduced flood risk tends to improve farmers' income by increasing the intensity of use of complementary inputs like labor and land.

The provision for a stable water source has been shown to affect farmers' behavior in several other analyses of groundwater management. Shah, Zilberman, and Chakravorty (1995) indicate that the adoption of water conserving technologies increases with groundwater depletion, and the optimal management of groundwater requires a tax on reducing the groundwater reservoir level with pumping, which may enhance adoption compared to an open access system. Carey and Zilberman (2002) take the literature further by analyzing a farmer's decision to adopt a new technology under irreversibility and uncertainty. They consider the farmer as an individual facing uncertain prices in the water market and making decisions to invest in water extraction technology. Their quasi-irreversibility setting was distinct from the usual analysis in finance literature where uncertainty was in output price and not in input price. Treating output price as fixed for farmers, they found that anything that stabilizes the price of water is likely to promote the adoption of efficient technologies. Dams therefore promote the use of more efficient water conserving technology insofar as they decrease the variance of the price of water in the market.

Many water systems rely on both ground and surface water. Conceptual optimal control models were used to analyze groundwater management problems (Gisser and Sanchez 1980; Tsur and Graham-Tomasi 1991) while identifying optimal rules for substitution decisions between groundwater and surface water. Many of the studies that consider the conjunctive use of groundwater emphasize the stochasticity of surface water supply and use numerical techniques to find solutions. Knapp and Olson (1995) find that groundwater pumping decreases with surface flows. This implies that farmers will substitute their groundwater use with water available from reservoirs whenever available. Tsur (1990) finds that variance in the surface water supply increases the benefit from the stabilization role of an alternative water source (in his case, groundwater, but as he implies, this could as well be a reservoir). Bredehoeft and Young (1983) also suggest that farmers sometimes should totally disregard the variability in surface water flow and install pumping facilities to extract groundwater resources. If the cost of groundwater pumping and getting water from a reservoir are similar, then farmers benefit from a stable source of water, such as large reservoirs.

5 Impact of dams: empirical evidence

The challenge for empirical analysis based on conceptual models is to deal with issues of dimensionality and multiple correlation. Because dams may be involved with other dynamic investments, it is difficult to separate the impact of dams from other correlated large-scale investments or an agglomeration effect (Kline and Moretti 2014). Murphy, Shleifer, and Vishny (1989) argue that big pushes, such as large dams, transform a society's population, income, and industry by operating through demand externalities. Empirical studies try to find instruments and other methods that attempt to isolate the impact of dams. The verdict in the empirical literature regarding dam construction is mixed. In the past, empirical enquiries generally reported the welfare of people with and without dams (Hussain 2007), often showing a very large difference in the poverty between those irrigated and non-irrigated areas. These results did not have any causal interpretation as the locations of large dams are not randomly selected, and without a robust econometric method that has causal interpretation, evaluating the impacts of dams can be very difficult (Janaiah, Bose, and Agarwal 2000; Ersado 2005).

Duflo and Pande (2007) illustrate the benefit of precisely aiming to disaggregate the impact of dams using a simple fixed effect model, such as $y_{ist} = \beta_0 + \beta_1 D_{ist} + v_s + \mu_t + \varepsilon_{ist}$, where y is the economic variable of interest and D is the number of dams in a district i in state s and year t, which does not have any causal interpretation. They made an influential contribution by suggesting river gradient as an instrumental variable (IV) for dam placement. Their study compares the welfare of two regions upstream and downstream and conducts a robustness check for the migration of mainly the upstream people and the rainfall shocks. They suggest including district-level information directly and using only district and state year fixed effects. If one assumes that the annual variation in dam construction in districts within a state is uncorrelated with other district-specific shocks, the following equation can be written as $y_{ist} = \beta_0 + \beta_1 D_{ist} + \beta_2 D_{ist}^u + v_i + \mu_{st} + \varepsilon_{ist}$, where D_{ist}^u indicates the total number of dams upstream of district i. It will provide a reasonable causal estimate for the impact of dams for a district where the dams are located and for the district that is in the command area of a dam or many dams.

Duflo and Pande (2007) find that dams marginally improve welfare in the command areas, whereas they decrease welfare in the catchment areas. They argue that there is no significant movement of population in catchment districts, undermining claims of endogenous selection by population (as mentioned in Attwood 2005). Districts upstream saw a modest (0.7 percent) increase in irrigated land, whereas districts downstream saw a 1 percent increase in irrigated land. They also found that farmers do not substitute crop production toward more water intensive crops. However, downstream districts had an increased adoption of water-intensive and high-yield variety seeds. The use of fertilizers also increased downstream.

Kitchens (2013) reinforces Duflo and Pande's skepticism about the efficacy of dams, arguing that their estimates included the cumulative effect of other activities associated with dams. Kitchens (2013) documents that reservoirs built by the TVA were likely to have increased the incidence of malaria, and there would have been many more victims of malaria near these dam areas had there been no intervention of DDT or other vector control activities. Kitchens argues that the TVA increased mortality by 3 to 4.4 per 100,000 people and morbidity by 7.1 to 13.9 per 10,000 people. The estimated loss of human health and life due to the TVA reservoirs ranged from US$508 million to US$1.06 billion. The TVA provides an important setting for studying the impacts of dams because of the relatively extensive availability of data. Kline and Moretti

(2014) find that the population increase in TVA counties was significant, while the impact on land price or wages was not.

Subsequent studies have highlighted the positive impacts of dams. For example, Severnini (2014) focuses on a panel of 154 U.S. counties with the hydropower potential of more than 100 megawatts and uses a combination of a synthetic control and event study methods, arguing that the synthetic control method facilitates estimating the county level heterogeneity of the impact. The regression model is:

$$Y_{ct} = \sum_{y} \beta_y D_{ct}^y + \alpha_c + \gamma_{rt} + Z_c' \phi_t + X_{ct}' \lambda + \varepsilon_{ct}$$

where Y_{ct} is the outcome of interest; D_{ct}^y is the dummy, indicating whether the dam was constructed in the county y years after (or before for negative y) year t; α_c, γ_{rt} represents county and region year dummy variables; Z denotes county-specific time invariant variables; and X represents other socioeconomic variables that varied across counties and years. Though the impact differed across counties, on average the population of counties with pre-1950 dams grew by 51 percent within 30 years, whereas there was no such effect for dams built after 1950. His findings of increased population in the catchment area of dams is compatible with the related finding by Hornbeck (2012), who indicated that the main area of adjustment after the American Dust Bowl, which resulted in the massive decline of productivity in the agricultural sector, was population decline. Severnini also investigates the long-term effects of these dams and found that population grew by 120 percent in 60 years. These results were, importantly, independent of dam size. Severnini also calculates the effect of dams for different sectors: agriculture and manufacturing sectors benefited even after 60 years, whereas construction and trade did not, and where aggregate impact on counties was driven by agriculture. Other sectors that benefited significantly included real estate, medical, and legal services. Severnini argues that, based on the evidence, large dams continue to affect the economy for such a long term that they act like an instrument of a big push policy, pushing counties toward a higher sustainable path of development.

Severnini's finding that the manufacturing sector benefited from large dams slightly differs from Kitchens (2014), who clarifies that comparatively lower electricity prices faced by manufacturing firms did not add value to the manufacturing sector. This suggests that firms' location choices at the time were motivated by their desire to benefit from the availability of electricity in a TVA region.

Hansen et al. (2011, 2014) estimate the impact of major storage facilities in the western United States on farming decisions. Dams increase total crop acreage, encourage farmers to choose higher-valued, more water-intensive crops, and increase crop yields, particularly during severe droughts. These findings are consistent with the results in Sarsons (2015) and Takeshima et al. (2016) that, in India and Nigeria, respectively, incomes in agriculture dependent areas that are downstream from irrigation dams are less sensitive to droughts. Hansen, Lowe, and Xu (2014) found little short-term impact on farmland values but found long-term impacts on agricultural development. They show that dams reduce the water available for ecosystem use and increase seasonal volatility in the water supply.

There has been some empirical evidence of failing water project management. For example, Attwood (2005) notes that in India many of the canals are in shambles and asserts that improvement in the canal system could greatly improve irrigation. He blames fiscal irresponsibility and clumsy handling by an expansive bureaucracy in India for the inefficient system and the negative net gain from these large-scale irrigation projects. Citing Rangachari et al. (2000), he asserts that a 10 percent increase in water-use efficiency would create 14 million hectares of additional irrigated land. Even in areas where dams have increased productivity, poverty may be due to an inequitable wealth redistribution system. This fact undermines the external validity of works based on Indian datasets.

In an early analysis, Tsur (1990) argues that many stable sources of water, such as groundwater or reservoirs, lose their value with decreasing rainfall variability. It is also possible that removing from the analysis dams built due to political interference or in areas where rainfall variability is minimal, results obtained are likely to differ from those found by Duflo and Pande (2007).

Lipscomb, Mobarak, and Barham (2013) studied the impact of dams in Brazil, which, while very rich in water resources, is grappling with issues regarding the equitable supply of electricity to its population. They found significant positive effects of dams in Brazil: each 10 percent increase in electrification led to a 3.8 percent increase in the mean housing value. They addressed the endogeneity of dam placement by using a simulation-based prediction of grid-level electricity availability. Furthermore, they also used an Amazon indicator interacted with each decade as their instruments. Their model was as follows:

$$Y_{ct} = \alpha_c^1 + \gamma_t^1 + \beta \hat{E}_{c,t-1} + \varepsilon_{ct}$$

where $E_{c,t-1} = \alpha_c^2 + \gamma_t^2 + \theta Z_{c,t-1} + \eta_{ct}$, where Y was the relevant development outcome, E was the electrification in county c in time t, $\hat{E}$ was the instrumental variable for E, and Z was the proportion of the grid model forecast to be electrified. Their study found a significant effect of dams on poverty reduction and formal employment generation, which differs from Duflo and Pande due to the focus on hydropower dams rather than irrigation dams.

Strobl and Strobl (2011) showed that in South Africa large dams reduce cropland productivity in their vicinity but augment the impact of small dams. They used 20 years of panel data on land cover, rivers, and productivity and developed the following model:

$$CP_{it} = \alpha + \beta_1 D(L)_{it} + \beta_2 D(S)_{it} + \beta_3 UD(L)_{it} + \beta_4 X_{it} + \varepsilon_{it}$$

where CP is cropland productivity, $D(L)_{it}$ is the number of dams in basin i in year t, $D(S)_{it}$ is the number of small dams in basin i, and $UD(L)_{it}$ is the number of large dams located upstream from district i in year t, and X represents other socioeconomic variables. Endogeneity of dam placement is accounted for by noting that dam placement is largely a function of politics. They specify the following first-stage equation:

$$D(sz)_{it} = \delta_1 + \sum_{k=2}^{4} \delta_2 \left(RGr(k)_i \star P_{it} \right) + \delta_3 \left(M_i \star P_{it} \right) + \sum_{k=2}^{4} \delta_4 \left(RGr(k)_i \star l_t \right)$$
$$+ \delta_5 ERLENGTH_i (sz = small) \star P_{it} + w_{it} + v_i + \mu_{it}$$

where $RGr(k)_i$, k = 2,3,4 and indicates the fraction of perennial river in basin i with 1.5–3 percent, 3–6 percent, and a 6–~percent river gradient, $ERLENTH_i$ is the length of the Ephemeral River in basin i, P is policy proxy, M is river basinspecific time invariant characteristics, and $sz \in \{large, small\}$ is the size of the dams.

Dillon (2011) also investigated the impact of the size of irrigation dams on poverty and production. He matched villages in Africa on their observable characteristics and found that small dams caused a larger effect on agricultural production and agricultural income, whereas large dams had a larger effect on consumption per capita.

Ansar et al. (2014) used an outside view method that operates by first identifying a reference class and establishing an empirical distribution for the reference class of the parameter being estimated. They then compared the specific case at hand with the reference class distribution.

They used information on 245 dams (with 26 major dams that are either more than 150 meters tall or can store more than 15 million cubic meters of water or have more than 25 square kilometers of reservoir storage) that were built between 1934 and 2007. Their study showed that Pakistan's Diamer-Bhasha Dam is likely to cost up to US$25.4 billion (in 2008 AD value), rather than the official estimate of about US$12.7 billion. When inflation is factored in, this estimate may exceed US$35.08 billion. Furthermore, instead of the planner's estimated completion date of 2021, this dam is likely to be under construction until 2028.

Kline and Moretti (2014), while evaluating the TVA's long-run effectiveness, compared identical pre-program counties (including those similar to the TVA counties but not chosen) to the TVA counties. Their method was to use the Oaxaca–Blinder regression model for all counties before the TVA started as follows:

$$y_{it} - y_{it-1} = \alpha + \beta X_i + (\varepsilon_{it} - \varepsilon_{it-1})$$

where y_{it} is the dependent variable of interest and X is the time independent vector of pre-program characteristics. They found that counties selected for TVA intervention saw a significant decline in agricultural employment and a significant increase in manufacturing employment. For example, the TVA's impact on agricultural employment was −7.1 percent and on manufacturing employment was 5.3 percent compared to non-TVA but characteristically identical counties. They found that by replacing agricultural jobs with manufacturing jobs, the median family income in TVA counties had also increased.

In assessing and explaining the overall effects of dams, Lipscomb, Mobarak, and Barham (2013) find that large dams in Brazil increased productivity and thus positively affected social welfare. Lipton, Litchfield, and Faures (2003) claim that much of the gains from dams were due to increased market integration in labor and input markets that were associated with the economies of scale that large dams made possible. Severnini (2014) argues that agglomeration impacts were behind the observed growth in those territories where large dams were built during the TVA era. Kline and Moretti (2014) investigate the general equilibrium effect of the TVA in a structural approach, showing that the TVA's direct investments yielded a significant increase in national manufacturing productivity that exceeded its costs, while the agglomeration gains in the TVA region were offset by losses in the rest of the country. In the case of India, Attwood (2005) lists the historical episodes of inflation and shocks to population growth, arguing that large reservoirs contributed to social welfare by preventing flooding in India.

6 Conclusions

Benefit–cost analysis on the benefits of dams has been inspired by the continuous debate on the value and design of dams and the challenge of the increased utilization of water resources for improved economic well-being while reducing the negative economic, social, and environmental side effects. This chapter reviews the impact of large dams. The results to date provide mixed evidence on the different topics. A dam's impact is likely to be heterogeneous in nature based on land quality, the qualifications of the individuals affected, and the purpose of the dam itself.

Further research should concentrate on identifying and disaggregating the direct and indirect impacts of dams. For example, it is possible that dams built in urban districts increase the welfare of the people living in the vicinity compared to dams built in rural districts, because the migration of people in response to a dam's construction in urban districts is systematically different from the migration of those living in rural districts. Both structural, general equilibrium approaches and reduced formed, partial equilibrium approaches should be encouraged.

New research can also examine the sometimes lackluster effects of dams. For example, dams may have failed to increase welfare because they caused too much resettlement, attracted too many poor people to catchment areas with marginal land quality, or the management of the dams failed to optimally allocate water or maintain a reservoir (due to a lack of effective sedimentation management). Furthermore, most of the empirical literature is localized and considers dams from countries such as Brazil, India, South Africa, and the United States. The ability to generalize from these countries' outcomes needs to be further investigated. Cross-country analyses, which are lacking, may provide some insight on the external validity of the results reported in previous literature.

There are also several other aspects of large dams that are not well understood. Given that reservoirs sites are exhaustible resources, how should a government faced with limited resources exploit them? There needs to be more investigation of temporal rules and optimal switching times from reservoirs to groundwater. Though thousands of dams have been constructed, there is very little research that analyzes the decision making process for governments before investing in the construction of dams.

To be effective, economic research on dams needs to be better integrated with knowledge and understanding from other disciplines. Some of the questions arising from economic research should influence the scientific research agenda on the performance and impact of dams. Furthermore, policy makers need more information as they approach the planning, construction, and management of new dams. As this survey shows, dams should take into account other activities, such as new technologies in water conveyance, conservation, and crop and hydroelectric production. New technologies that allow for the reuse of wastewater and for desalinization should affect dam and water management projects.

References

Ansar, A., B. Flyvbjerg, A. Budzier, and D. Lunn. 2014. "Should We Build More Large Dams? The Actual Costs of Hydropower Megaproject Development." *Energy Policy* 69:43–56.

Attwood, D.W. 2005. "Big Is Ugly? How Large-Scale Institutions Prevent Famines in Western India." *World Development* 33(12):2067–2083.

Baboo, B. 1991. "State Policies and People's Response: Lessons from Hirakud Dam." *Economic and Political Weekly* 26(41):2373–2379.

Bacon, R.W., and J.E. Besant-Jones. 1998. "Estimating Construction Costs and Schedules: Experience with Power Generation Projects in Developing Countries." *Energy Policy* 26(4):317–333.

Barros, N., J.J. Cole, L.J. Tranvik, Y.T. Prairie, D. Bastviken, V.L. Huszar, and F. Roland. 2011. "Carbon Emission from Hydroelectric Reservoirs Linked to Reservoir Age and Latitude." *Nature Geoscience* 4(9):593–596.

Bhaduri, A., and U. Manna. 2014. "Impacts of Water Supply Uncertainty and Storage on Efficient Irrigation Technology Adoption." *Natural Resource Modeling* 27:1–24.

Blanc, E., and E. Strobl. 2014. "Is Small Better? A Comparison of the Effect of Large and Small Dams on Cropland Productivity in South Africa." *The World Bank Economic Review* 28(3):545–576.

Bredehoeft, J.D., and R.A. Young. 1983. "Conjunctive Use of Groundwater and Surface Water for Irrigated Agriculture: Risk Aversion." *Water Resources Research* 19(5):1111–1121.

Brown, C., and U. Lall. 2006. "Water and Economic Development: The Role of Variability and a Framework for Resilience." *Natural Resources Forum* 30(4):306–317.

Carey, J.M., and D. Zilberman. 2002. "A Model of Investment Under Uncertainty: Modern Irrigation Technology and Emerging Markets in Water." *American Journal of Agricultural Economics* 84(1):171–183.

Caswell, M., E. Lichtenberg, and D. Zilberman. 1990. "The Effects of Pricing Policies on Water Conservation and Drainage." *American Journal of Agricultural Economics* 72(4):883–890.

Caswell, M., and D. Zilberman. 1985. "The Choices of Irrigation Technologies in California." *American Journal of Agricultural Economics* 67(2):224–234.

Chakravorty, U., E. Hochman, C. Umetsu, and D. Zilberman. 2009. "Water Allocation Under Distribution Losses: Comparing Alternative Institutions." *Journal of Economic Dynamics and Control* 33(2):463–476.

Chakravorty, U., E. Hochman, and D. Zilberman. 1995. "A Spatial Model of Optimal Water Conveyance." *Journal of Environmental Economics and Management* 29(1):25–41.

Dillon, A. 2011. "Do Differences in the Scale of Irrigation Projects Generate Different Impacts on Poverty and Production?" *Journal of Agricultural Economics* 62(2):474–492.

Duflo, E., and R. Pande. 2007. "Dams." *The Quarterly Journal of Economics* 122(2):601–646.

Emerick, K., A. de Janvry, E. Sadoulet, and M.H. Dar. 2016. "Technological Innovations, Downside Risk, and the Modernization of Agriculture." *American Economic Review* 106(6):1537–1561.

Ersado, L. 2005. "Small-Scale Irrigation Dams, Agricultural Production, and Health: Theory and Evidence from Ethiopia." WBPR 3494. The World Bank, New York.

Fisher, A.C., and S.J. Rubio. 1997. "Adjusting to Climate Change: Implications of Increased Variability and Asymmetric Adjustment Costs for Investment in Water Reserves." *Journal of Environmental Economics and Management* 34:207–227.

Flyvbjerg, B. 2005. "Machiavellian Megaprojects." *Antipode* 37(1):18–22.

———. 2009. "Survival of the Unfittest: Why the Worst Infrastructure Gets Built and What We Can Do about It." *Oxford Review of Economic Policy* 25(3):344–367.

Food and Agriculture Organization [FAO]. 2015. *World Agriculture Toward 2015–2030*. Rome: FAO. www.fao.org/docrep/005/y4252e/y4252e06a.htm. Accessed January 1, 2017.

———. 2017. *Water Resources by Country*. Rome: FAO. www.fao.org/docrep/005/y4473e/y4473e08.htm. Accessed April 11, 2018.

Gerhart, J. 1975. *The Diffusion of Hybrid Maize in Kenya*. Mexico City: CIMMYT.

Giles, J. 2006. "Methane Quashes Green Credentials of Hydropower." *Nature* 444(7119):524–525.

Gisser, M., and D.A. Sanchez. 1980. "Competition Versus Optimal Control in Groundwater Pumping." *Water Resources Research* 16(4):638–642.

Hansen, Z.K., G.D. Libecap, and S.E. Lowe. 2011. "Climate Variability and Water Infrastructure: Historical Experience in the Western United States." In G.D. Libecap and R.H. Steckel, eds., *The Economics of Climate Change: Adaptations Past and Present*. Chicago, IL: University of Chicago Press, pp. 253–280.

Hansen, Z.K., S.E. Lowe, and W. Xu. 2014. "Long-Term Impacts of Major Water Storage Facilities on Agriculture and the Natural Environment: Evidence from Idaho (U.S.)." *Ecological Economics* 100:106–118.

Hornbeck, R. 2012. "The Enduring Impact of the American Dust Bowl: Short- and Long-Run Adjustments to Environmental Catastrophe." *The American Economic Review* 102(4):1477–1507.

Hornbeck, R., and P. Keskin. 2015. "Does Agriculture Generate Local Economic Spillovers? Short-Run and Long-Run Evidence from the Ogallala Aquifer." *American Economic Journal: Economic Policy* 7(2):192–213.

Hussain, I. 2007. "Poverty-Reducing Impacts of Irrigation: Evidence and Lessons." *Irrigation and Drainage* 56(2/3):147–164.

International Energy Association [IEA]. 2011. *Policies and Scenario*. Paris: IEA.

Janaiah, A., M.L. Bose, and A.G. Agarwal. 2000. "Poverty and Income Distribution in Rainfed and Irrigated Ecosystems: Village Studies in Chhattisgarh." *Economic and Political Weekly* 35(52):4664–4669.

Keiser, J., M.C. de Castro, M.F. Maltese, R. Bos, M. Tanner, B.H. Singer, and J. Utzinger. 2005. "Effect of Irrigation and Large Dams on the Burden of Malaria on a Global and Regional Scale." *The American Journal of Tropical Medicine and Hygiene* 72(4):392–406.

Kitchens, C. 2013. "A Dam Problem: TVA's Fight Against Malaria, 1926–1951." *The Journal of Economic History* 73(3):694–724.

———. 2014. "The Role of Publicly Provided Electricity in Economic Development: The Experience of the Tennessee Valley Authority, 1929–1955." *The Journal of Economic History* 74(2):389–419.

Kline, P., and E. Moretti. 2014. "Local Economic Development, Agglomeration Economies, and the Big Push: 100 Years of Evidence from the Tennessee Valley Authority." *The Quarterly Journal of Economics* 129(1):275–331.

Knapp, K.C., and L.J. Olson. 1995. "The Economics of Conjunctive Groundwater Management with Stochastic Surface Supplies." *Journal of Environmental Economics and Management* 28(3):340–356.

Koundouri, P., C. Nauges, and V. Tzouvelekas. 2006. "Technology Adoption Under Production Uncertainty: Theory and Application to Irrigation Technology." *American Journal of Agricultural Economics* 88(3):657–670.

Kovacs, K., M. Popp, K. Brye, and G. West. 2015. "On-Farm Reservoir Adoption in the Presence of Spatially Explicit Groundwater Use and Recharge." *Journal of Agricultural and Resource Economics* 40(1):23–49.

Kraljevic, A., J-H. Meng, and P. Schelle. 2013. *Seven Sins of Dam Building*. Berlin: WWF. http://awsassets.panda.org/downloads/wwf_seven_sins_of_dam_building.pdf. Accessed September 1, 2017.

Lipscomb, M., M.A. Mobarak, and T. Barham. 2013. "Development Effects of Electrification: Evidence from the Topographic Placement of Hydropower Plants in Brazil." *American Economic Journal: Applied Economics* 5(2):200–231.

Lipton, M., J. Litchfield, and J.M. Faures. 2003. "The Effects of Irrigation on Poverty: A Framework for Analysis." *Water Policy* 5(5/6):413–427.

Louis, V.L.S., C.A. Kelly, É. Duchemin, J.W. Rudd, and D.M. Rosenberg. 2000. "Reservoir Surfaces as Sources of Greenhouse Gases to the Atmosphere: A Global Estimate Reservoirs Are Sources of Greenhouse Gases to the Atmosphere, and Their Surface Areas Have Increased to the Point Where They Should Be Included in Global Inventories of Anthropogenic Emissions of Greenhouse Gases." *BioScience* 50(9):766–775.

McCully, P., 1996. *Silenced River: The Ecology and Politics of Large Dams*. London: Zed Books.

Murphy, K.M., A. Shleifer, and R.W. Vishny. 1989. "Industrialization and the Big Push." *Journal of Political Economy* 97(5):1003–1026.

National Research Council [NRC]. 2004. *Analytical Methods and Approaches for Water Resources Project Planning*. Washington, DC: The National Academies Press.

Pfeiffer, L., and C.Y. Cynthia Lin. 2014. "Does Efficient Irrigation Technology Lead to Reduced Groundwater Extraction? Empirical Evidence." *Journal of Environmental Economics and Management* 67:189–208.

Poudel, B.N. 2010. *Three Essays in Environmental and Agricultural Economics*. PhD dissertation, University of California at Berkeley.

Rangachari, R., N. Sengupta, R.R. Iyer, P. Banergy, and S. Singh. 2000. *Large Dams: India's Experience*. Cape Town, South Africa: Secretariat of World Commission on Dams.

Reisner, M. 1993. *Cadillac Desert: The American West and Its Disappearing Water*. New York: Penguin.

Rudd, J.W., R.E. Hecky, R. Harris, and C.A. Kelly. 1993. "Are Hydroelectric Reservoirs Significant Sources of Greenhouse Gases." *Ambio* 22(4):246–248.

Sarsons, H. 2015. "Rainfall and Conflict: A Cautionary Tale." *Journal of Development Economics* 115:62–72.

Schoengold, K., and D. Zilberman. 2007. "The Economics of Water, Irrigation, and Development." *Handbook of Agricultural Economics* 3:2933–2977.

Scudder, T.T. 2012. *The Future of Large Dams: Dealing with Social, Environmental, Institutional and Political Costs*. London: Earthscan.

Severnini, E.R. 2014. "The Power of Hydroelectric Dams: Agglomeration Spillovers." Discussion Paper 8082. Carnegie Mellon University, Pittsburgh, PA.

Shah, F.A., D. Zilberman, and U. Chakravorty. 1995. "Technology Adoption in the Presence of an Exhaustible Resource: The Case of Groundwater Extraction." *American Journal of Agricultural Economics* 77(2):291–299.

Shakya, P.B., and J.C. Flinn. 1985. "Adoption of Modern Varieties and Fertilizer Use on Rice in the Eastern Tarai of Nepal." *Journal of Agricultural Economics* 36(3):409–419.

Stone, R. 2011. "The Legacy of the Three Gorges Dam." *Science* 333(6044):817–817.

Strobl, E., and R.O. Strobl. 2011. "The Distributional Impact of Large Dams: Evidence from Cropland Productivity in Africa." *Journal of Development Economics* 96(2):432–450.

Takeshima, H., A.I. Adeoti, and O.A. Popoola. 2016. "The Impact on Farm Household Welfare of Large Irrigation Dams and Their Distribution across Hydrological Basins: Insights from Northern Nigeria." International Food Policy Research Institute [IFPRI], Washington, DC.

Tremblay, A., M. Lambert, and L. Gagnon. 2004. "Do Hydroelectric Reservoirs Emit Greenhouse Gases?" *Environmental Management* 33(1):S509 – S517.

Truong, C.H. 2012. "An Analysis of Storage Capacity Reallocation Impacts on the Irrigation Sector." *Environmental and Resource Economics* 51:141–159.

Tsur, Y. 1990. "Stabilization Role of Groundwater When Surface Water Supplies Are Uncertain: The Implications for Groundwater Development." *Water Resources Research* 26(5):811–818.

Tsur, Y., and T. Graham-Tomasi. 1991. "The Buffer Value of Groundwater with Stochastic Surface Water Supplies." *Journal of Environmental Economics and Management* 21(3):201–224.

United States Department of Agriculture, National Agricultural Statistics Service [USDA-NASS]. 2012. *2012 California Land Values and Cash Rents*. Washington, DC: USDA-NASS.

van der Zaag, P., and J. Gupta. 2008. "Scale Issues in the Governance of Water Storage Projects. *Water Resources Research* 44(10):Article W10417.

Ward, F.A., and M. Pulido-Velazquez. 2008. "Water Conservation in Irrigation Can Increase Water Use." *Proceedings of the National Academy of Sciences* 105(47):18215–18220.

Water Technology Net. 2016. *The World's Oldest Dams Still in Use*. www.water-technology.net/features/feature-the-worlds-oldest-dams-still-in-use/. Accessed October 11, 2017.

World Bank. 1996. "Estimating Construction Costs and Schedules: Experience with Power Generation Projects in Developing Countries." World Bank, Washington, DC.

World Commission on Dams [WCD]. 2000. *Cross-Check Survey*. Cape Town: WCD.

Xabadia, A., R. Goetz, and D. Zilberman. 2004. "Optimal Dynamic Pricing of Water in the Presence of Waterlogging and Spatial Heterogeneity of Land." *Water Resources Research* 40(7):WR002215.

Xie, Y., and D. Zilberman. 2016. "Theoretical Implications of Institutional, Environmental, and Technological Changes for Capacity Choices of Water Projects." *Water Resources and Economics* 13:19–29.

———. 2018. "Water Storage Capacity Versus Water Use Efficiency: Substitutes or Complements?" *Journal of Agriculture and Environmental Resource Economics* 5:1–35.

Zhao, J., and D. Zilberman. 1999. "Irreversibility and Restoration in Natural Resource Development." *Oxford Economic Papers* 51(3):559–573.

Zilberman, D., A. Dinar, N. MacDougall, M. Khanna, C. Brown, and F. Castillo. 2011. "Individual and Institutional Responses to the Drought: The Case of California Agriculture." *Journal of Contemporary Water Research and Education* 121:17–23.

Part 3
Trade, policy, and development

18
A synopsis of trade theories and their applications

Stephen Devadoss and Jeff Luckstead

1 Introduction

This chapter presents an overview of the major trade theories and their application to international trade. First, it starts with the Ricardian theory of comparative advantage and the two major extensions of this theory via Dornbusch, Fischer, and Samuelson's (1977) continuum of goods model and Eaton and Kortum's (2002) multi-country model with stochastic productivity. Second, it describes the basic Heckscher–Ohlin model using the dual mathematical approach and the corresponding graphical analysis. Third, it covers New Trade Theory models of monopoly, oligopoly, and monopolistic competition. This section also includes applications of New Trade Theory models to agricultural trade. Fourth, it presents the recent developments in Firm-Level Trade Theory based on heterogeneous firms and recent work in this area, including agricultural trade. The final section concludes this chapter.

2 Ricardian theory

The theory of comparative advantage, put forth by English economist David Ricardo in 1821, is the staple among trade economists, used to explain that technological differences across countries are the major force for trade. In its simplest terms, the Ricardian theory states that a country exports a commodity in which it has a comparative advantage in labor productivity. Consider the trade model of two countries (A and B), two commodities (Food F and Manufacturing M), and one factor of production (Labor). Suppose the input requirement to produce one unit of F and M in country A is less than that in country B, implying that country A has an absolute advantage in the production of both goods. Even though country A has an absolute advantage in both goods, this does not mean it will export both goods. To determine the comparative advantage, one must compute the relative input requirement between the two goods in each country, and then compare these relative input requirements across countries. Thus, if country A has a relatively lower input requirement in M production than in F production compared to that of country B, then country A will export M and import F. This also means that country B has a relatively lower input requirement in F production than in M production compared to that of country A, and consequently country B will export F and import M.

To examine the welfare implications of free trade under the Ricardian model, consider the Edgeworth box analysis in Figure 18.1 (Houck 1986). In this figure, the origin for country B is in the southwest corner. With its production possibility frontier given by the dashed line, country B produces either 15 units of M or 30 units of F when it completely specializes. Furthermore, the opportunity cost of one unit of F is half a unit of M, and, in turn, the autarky price ratio of F over M is $\frac{1}{2}$. The origin for county A is in the northeast corner. Its production possibility frontier is given by the solid line, which indicates that country A produces either 20 units of M or 30 units of F when it fully specializes. In country A, the opportunity cost of one F is $\frac{2}{3}$ M, which implies the autarky price ratio of F over M is $\frac{2}{3}$. Suppose under autarky, with the price ratio of $\frac{1}{2}$ and autarky equilibrium at point L, country B produces and consumes 10 units of M and 10 units of F. In addition, with the price ratio of $\frac{2}{3}$ and autarky equilibrium at point C, country A produces and consumes 10 units of M and 15 units of F. Under autarky, F is relatively less expensive in country B and M is relatively less expensive in country A. If trade is allowed, to fetch a relatively higher price country B will export F and country A will export M. With constant opportunity costs, country B will completely specialize in F production and country A will completely specialize in M production, and the world production point will be at D with the world price ratio between $\frac{1}{2}$ and $\frac{2}{3}$. With the free trade price ratio $\left(\frac{P_M}{P_F}\right)_{FT}$, the consumption point could be at G, where each country consumes more of both goods and thus their utility unambiguously increases from their respective autarky level. Exports are given as the difference in production and consumption of good F for country B and good M for country A.

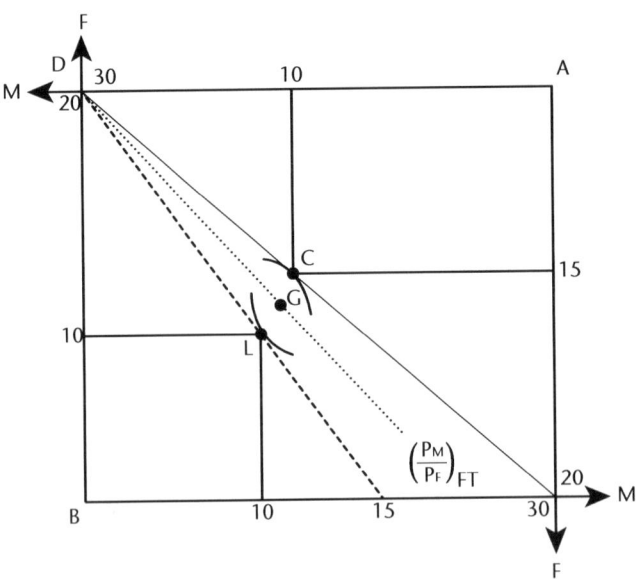

Figure 18.1 Edgeworth box

A synopsis of trade theories

2.1 Extensions of the basic Ricardian theory

Ricardian models that appear in the literature mostly deal with a discrete number of commodities. But a pioneering study by Dornbusch, Fischer, and Samuelson (DFS) (1977) revolutionized the Ricardian theory and also other trade theories by considering a continuum of goods. Furthermore, this model made it easier to analyze the Ricardian theory and to examine the effects of technological changes, economic growth, and changes in demand. The DFS model considers two countries ($i = 1,2$) and a technology that produces a continuum of goods indexed by $\theta \in [0,1]$ using labor. The production function for each good is $q_i(\theta) = \dfrac{l_i(\theta)}{a_i(\theta)}$, where $q_i(\theta)$ is the output of good θ in country i, $l_i(\theta)$ is the labor input, and $a_i(\theta)$ is the unit-labor requirement for producing good θ in country i. Each country is endowed with $\bar{l}$ units of labor where $\bar{l}_1 = \bar{l}_2 = \bar{l}$. Assume the unit labor requirement functions are $a_1(\theta) = e^{\mu\theta}$ for country 1 and $a_2(\theta) = e^{\mu(1-\theta)}$ for country 2. As θ increases, the unit-labor requirement rises and thus productivity $\left(\dfrac{1}{a_i(\theta)}\right)$ decreases in country 1, whereas the unit-labor requirement falls and productivity increases in country 2.

The representative consumer in country i maximizes utility over $c_i(\theta)$:

$$\int_0^1 \log(c_i(\theta))d\theta$$

The budget constraint implies expenditure on all goods consumed is equal to total labor income:

$$\int_0^1 p_i(\theta)c_i(\theta)d\theta = w_i \bar{l}_i.$$

Utility maximization yields:

$$c_i(\theta) = \frac{w_i \bar{l}_i}{p_i(\theta)}, \quad i = 1,2$$

Firms operate under perfect competition, and each representative firm in country i producing good θ maximizes profit over labor input $l_i(\theta)$:

$$\max p_i(\theta)\frac{l_i(\theta)}{a_i(\theta)} - w_i l_i(\theta), \quad i = 1,2$$

Profit maximization results in price equals marginal cost:

$$p_i(\theta) - a_i(\theta)w_i \leq 0, \quad i = 1,2, \forall \theta.$$

To complete this general equilibrium model, define the market clearing conditions for each good θ:

$$\sum_i q_i(\theta) = \sum_i c_i(\theta), \forall \theta \in [0,1],$$

the labor market clearing in each country:

$$\int_0^1 l_i(\theta)d\theta = \bar{l}_i, \quad i = 1,2$$

and the trade balance given by the value of exports equal to the value of imports:

$$\int_{\Theta_{1,2}} p_1(\theta) c_1(\theta) d\theta = \int_{\Theta_{2,1}} p_2(\theta) c_2(\theta) d\theta,$$

where $\Theta_{1,2}$ ($\Theta_{2,1}$) contains the set of goods that country 1 (2) exports to country 2 (1).

With both countries having the same labor endowment $\bar{l}_1 = \bar{l}_2$ and utility functions but different technologies, each country has a comparative advantage for half of the goods, as can be seen from $a_i(\theta)$. Under these conditions, the wage rates in both countries will be equal and can be normalized to 1.

From the profit maximization, the production pattern will be such that firms in country 1 will produce good θ if $a_1(\theta) w_1 < a_2(\theta) w_2 \Rightarrow \frac{a_1(\theta)}{a_2(\theta)} < \frac{w_2}{w_1} = 1$, and firms in country 2 will produce θ for which $a_1(\theta) w_1 > a_2(\theta) w_2 \Rightarrow \frac{a_1(\theta)}{a_2(\theta)} > \frac{w_2}{w_1} = 1$. With wage rates equal to 1, prices in country 1 are $p(\theta) = e^{\alpha\theta}$ for $\theta \in [0, \bar{\theta}]$ and in country 2 are $p(\theta) = e^{\alpha(1-\theta)}$ for $\theta \in (\bar{\theta}, 1]$. For these two countries with identical endowments and utility functions, $\bar{\theta} = \frac{1}{2}$. The consumer will choose domestic goods or imported goods by comparing prices: $p_i(\theta) = \min\{p_1(\theta), p_2(\theta)\}$. These results imply that country 1 has a comparative advantage over goods $[0, \bar{\theta}]$ and will produce and export these goods. Similarly, country 2 has a comparative advantage over goods $[\bar{\theta}, 1]$ and will produce and export these goods. Thus, as in the basic Ricardian model, the DFS model also finds that the pattern of trade depends on which country has the comparative advantage in the production of goods. These results are illustrated in Figure 18.2.

The extension of the two-country continuum-of-goods Ricardian model examined by DFS (1977) to many countries is arduous, but has been modeled by Wilson (1980). Eaton and Kortum (2002), by using stochastic productivity, were able to develop a many-country continuum Ricardian model, obtain an analytical solution, and derive a gravity equation of trade flows. They used a production function as in DFS (1977): $q_j(x) = x l_j(x)$ for commodity j, where the

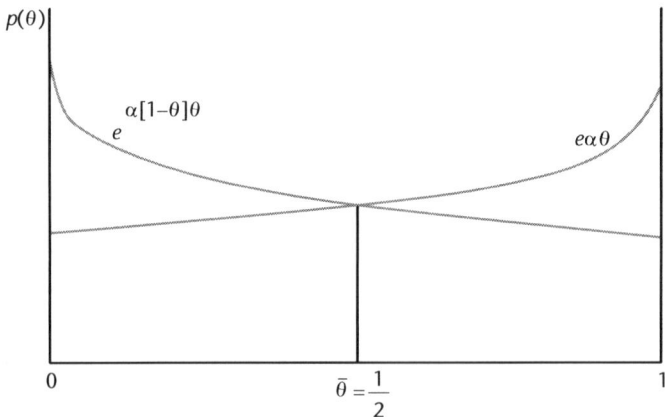

Figure 18.2 Pattern of production and trade

productivity parameter x is stochastic. The novel approach that Eaton and Kortum developed relies on the Frechet distribution for productivity in country $i = 1, \ldots, N$:

$$F_i(x) = e^{-T_i x^{-\varphi}}, \tag{1}$$

where $T_i > 0$ and $\varphi > 1$. Probability draws across x are independent and thus F is independent across countries. Parameter T_i is country-specific and determines the location of the distribution and thus governs the mean. The higher the T_i, the further to the right lies the probability density function, and country i is on average more efficient. Alternatively, a higher T_i implies a higher probability of a high efficiency for commodity j. T_i allows countries to have different average levels of productivity, and thus it refers to a country's state of technology; that is, it determines absolute advantage by each country over the continuum of goods. The shape parameter φ is related to the dispersion of productivity and plays a key role in determining the comparative advantage. A lower value of φ generates more heterogeneity (greater dispersion in productivities), implying that comparative advantage exerts a stronger force for trade.

Using the Frechet distribution for the productivity parameter, Eaton and Kortum derive the following gravity equation for trade flows:

$$Z_{in} = \frac{\left(\frac{\tau_{in}}{P_n}\right)^{-\varphi}}{\sum_{m=1}^{N} Z_m \left(\frac{\tau_{im}}{P_m}\right)^{-\varphi}} Z_n Q_i(T_i)$$

where Z_{in} is the value of exports of i to n, τ_{in} is the trade costs for exports from i to n, P_n is the price index in country n, Z_n is n's total spending on all goods, Q_i is total sales by country i, Z_m is m's total spending, τ_{im} is the trade costs between i and m, and P_m is the price index in m. In this gravity equation, Z_{in} depends on total expenditure in n, total sales of country i, bilateral trade costs between i and n, and all other countries' prices, outputs, and trade costs. The denominator captures the market size effect of all other countries: the larger the other countries, the smaller the bilateral sales from i to n. The price in country n also determines trade: a lower price implies imports from i will be smaller. The smaller the dispersion parameter φ, the larger the trade flows because of the greater heterogeneity and stronger comparative advantage. With $Q_i'(T_i) > 0$, improvements in T and smaller costs (input cost c and trade cost τ) increase the trade from i to n.

Eaton and Kortum (2012) also provide a succinct synopsis of Ricardian theory, continuum of goods model, and Eaton and Kortum's (2002) multi-country model with stochastic productivity. The most recent advances in Ricardian theory are to analyze the cause for the technological differences, such as institutional quality reflecting governance and corruption (Levchenko 2004), contract enforcement (Costinot 2009), and factor market rigidities (Cuñat and Melitz 2010, 2012).

2.2 Empirical tests

Empirical tests of the Ricardian theory are limited because of difficulties involved in measuring input prices, input requirements, and the number of commodities, and ascertaining whether trade involves two or many countries, along with a host of other issues. Yet, a few studies have empirically tested the Ricardian theory. The earliest studies include MacDougall (1951) and Balassa (1963). MacDougall (1951) utilized 1937 data on commodity prices and marginal/average cost for the United States and the United Kingdom to test the Ricardian theory. He observed that the U.S. wage rate was about double that in the United Kingdom. This would imply U.S. exports would be more than U.K. exports if the output per worker were more than double that of the United Kingdom. MacDougall found general support for the Ricardian theory based on this hypothesis. However, later studies by Balassa (1963) found no evidence of trade between the United States and the United Kingdom based on the theory of comparative advantage.[1]

More recent empirical work tested the Ricardian theory for agricultural commodities. In a series of two papers, Costinot and Donaldson (2012) and Costinot et al. (2016) use productivity estimates from agronomical studies to analyze comparative advantage in agricultural production and the impacts of climate change. Costinot and Donaldson (2012) conclude that Ricardian theory accurately predicts agricultural trade patterns. Costinot et al. (2016) develop a macro-level model of comparative advantage to analyze micro-level impacts of globalization on agriculture. They find that the impact of climate change on agricultural production will reduce global GDP by 0.23 percent, and the resulting changes in production pattern due to shifts in comparative advantage play an important role in reducing the negative effects of climate change.

Heerman et al. (2015) apply Eaton and Kortum's (2002) model to agriculture by characterizing comparative advantage using agro-ecological characteristics of each commodity. Their results show a distinct positive relationship in productivity for commodities between countries with comparable agro-ecological attributes. As expected from a comparative advantage–based model, bilateral trade between similar countries is highly sensitive to changes in trade costs.

Heerman (2017) builds on Heerman et al. (2015) to develop a general framework to analyze the impact of trade policy, where comparative advantage arises from agro-ecological characteristics on the trade patterns of agricultural commodities. This chapter draws on the discrete choice literature to develop a methodological estimate of the distributions for both agricultural productivity and trade costs.

3 Heckscher–Ohlin (H–O) theorem

While the Ricardian theory focuses on trade patterns due to comparative advantage resulting from differences in production technology between countries, H–O theorem[2] considers trade patterns due to comparative advantage arising from differences in factor abundance across countries.

This basic H–O model considers two countries (1 and 2), two commodities (X and Y), and two factors of production (K and L), and thus is known as the $2 \times 2 \times 2$ model. The H–O model focuses on trade patterns resulting from differences in factor endowments by making several assumptions. Here, we provide some of the key assumptions: (1) both countries use the same technology in the production of X and Y; (2) good X is labor intensive and good Y is capital intensive, and both goods are produced under constant returns to scale in both countries; (3) tastes and preference are identical in both countries; (4) factors are perfectly mobile across sectors within each country, but not across countries; and (5) country 2 is capital-abundant and country 1 is labor-abundant: $\left(\frac{K}{L}\right)_2 > \left(\frac{K}{L}\right)_1$.

Since country 2 is capital abundant and Y is capital intensive, country 2 will produce relatively more Y than X and export Y, and since country 1 is labor abundant and X is labor intensive, country 1 will produce relatively more X than Y and export X. In summary, the H–O Theorem states that a country will export a good that uses the abundant factor intensively. The rationale is that a relatively large supply of a particular factor of production will have a relatively low price. The good that utilizes the relatively low-priced factor intensively will have a relatively low price as well: $\left(\dfrac{P_X}{P_Y}\right)_1 < \left(\dfrac{P_X}{P_Y}\right)_2$. Thus, a nation will have a comparative advantage in the good that uses the nation's abundant factor of production intensively. Hence, the essence of the H–O theory is that trade is caused by the differences in factor availability across countries.

The H–O theorem can be also readily analyzed using Figure 18.3. The capital/labor ratio (k) is the proportion of capital to labor used to produce a given level of output. Given the technology and output level, factor prices will determine the capital/labor ratio. As the wage/rental ratio increases, firms in the X and Y industries tend to use more of K and less of L so k_X and k_Y increase, as depicted in the first quadrant of Figure 18.3. Assume no factor intensity reversal occurs, implying that k_Y is always greater than k_X.

Since output prices reflect production costs, an increase in the wage rate relative to the rental rate will increase the cost of X (labor-intensive good) relative to the cost of Y. It follows that an increase in the wage/rental ratio will increase the price of X (P_X) relative to the price of Y (P_Y), as shown in the lower portion of the diagram. Since the autarky price ratio $\dfrac{P_X}{P_Y}$ will be lower in country 1 than that in country 2, when trade is allowed, country 1 will export X and import Y. Similarly, country 2 will export Y and import X. Under complete free trade, the commodity price will be equal in both countries. Furthermore, the factor prices will also be equal in both countries; this is known as the Factor Price Equalization theorem. The important implication of this theorem is that if two countries engage in free trade, then there is no need for a factor to move from one country to the other. Thus, under the H–O theory, free trade eliminates the need for the immigration of workers seeking employment in the other country or capital moving across countries. In summary, H–O theory indicates factor abundance and factor intensities are the reasons for differences in comparative advantages and this results in autarky factor and commodity price differences across the countries.

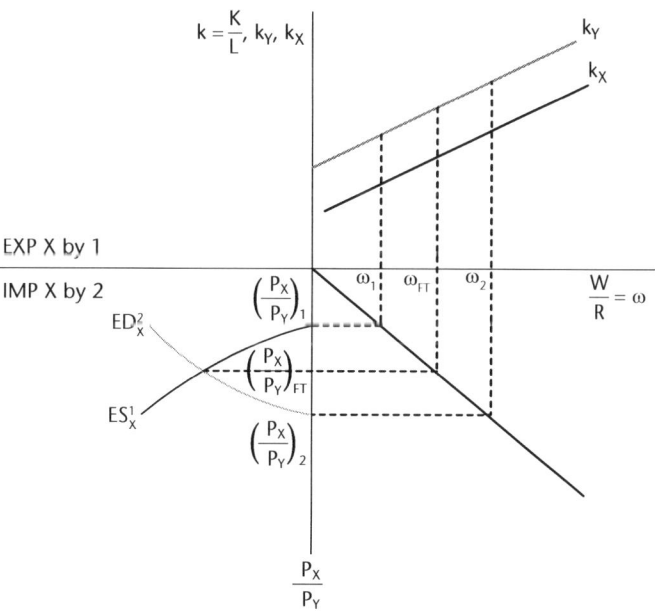

Figure 18.3 Illustration of H–O theory

3.1. Dual approach to H–O

The literature examines the H–O theory using both primal and dual mathematical formulations. See Bowen et al. (2012) for a detailed exposition of the primal approach. Here, we present the dual formulation, which was originally developed in a pioneering study by Jones (1965) and extended considerably by Batra (1973). For the 2 × 2 × 2 model, the full employment conditions are

$$C_{LY}Y + C_{LX}X = L$$
$$C_{KY}Y + C_{KX}X = K,$$

where C_{ij} is the quantity of the ith ($i = K, L$) factor used in the production of one unit of the jth ($j = X, Y$) commodity. Under perfect competition, the firms earn zero profits in the economy. This implies that the unit cost of production is exactly equal to the output prices (see Woodland (1982) and Wong (1995) for derivations of these equations):

$$wC_{LY} + rC_{KY} = p_Y$$
$$wC_{LX} + rC_{KX} = p_X.$$

The full employment and zero-profit conditions are converted with considerable additional work into equations of change (* denotes proportional change):

$$\lambda_{LY}Y^* + \lambda_{LX}X^* = L^* + \beta_L(w^* - r^*)$$
$$\lambda_{KY}Y^* + \lambda_{KX}X^* = K^* - \beta_K(w^* - r^*)$$
$$\theta_{LY}w^* + \theta_{KY}r^* = p_Y^*$$
$$\theta_{LX}w^* + \theta_{KX}r^* = p_X^*,$$

where λ_{ij} is the proportion of the ith factor used in the production of the jth good, θ_{ij} is the distributive share of the ith factor in the total earnings of the jth industry, and β_i represents the percentage changes in the employment of the ith input per unit of output in both commodities due to changes in the factor price ratio. Thus,

$$[\lambda] = \begin{bmatrix} \lambda_{LY} & \lambda_{LX} \\ \lambda_{KY} & \lambda_{KX} \end{bmatrix}$$

is the matrix of factor-allocation fractions where $\lambda_{iY} + \lambda_{iX} = 1$, and

$$[\theta] = \begin{bmatrix} \theta_{LY} & \theta_{KY} \\ \theta_{LX} & \theta_{KX} \end{bmatrix}$$

is the matrix of distributive shares where $\theta_{Lj} + \theta_{Kj} = 1$. Based on the assumption that good Y is capital intensive and good X is labor intensive, $|\lambda|$ and $|\theta|$ are negative and the product $|\lambda| \times |\theta|$ and β_i are positive. The equations of change can be combined to yield

$$X^* - Y^* = \frac{K^* - L^*}{|\lambda|} + \sigma_S(p_X^* - p_Y^*),$$

where

$$\sigma_S = \frac{X^* - Y^*}{p_X^* - p_Y^*} = \frac{\beta_K + \beta_L}{|\lambda||\theta|} \geq 0$$

is the elasticity of transformation between X and Y.

The above equation expresses relative supply changes in the economy as a function of endowment and output prices. For given output prices, the factor endowments determine the relative supply of goods. If the country is capital abundant and Y is capital intensive, then this country will export Y and import X. This completes the supply side of the economy.

The demand side consists of demand functions for the two commodities, which are functions of prices and income – (I): $D_Y = D_Y\left(\frac{p_X}{p_Y}, I\right)$ and $D_X = D_X\left(\frac{p_X}{p_Y}, I\right)$. National income, $I = wL + rK$, is equal to expenditure $p_Y D_Y + p_X D_X$. The two demand functions can be converted to proportional changes, and taking the difference yields

$$D_Y^* - D_X^* = -\sigma_D(p_X^* - p_Y^*) + (\eta_X - \eta_Y)I^*,$$

where σ_D is the elasticity of substitution in consumption

$$\sigma_D = \frac{\partial\left(\frac{D_Y}{D_X}\right)\left(\frac{p_X}{p_Y}\right)}{\partial\left(\frac{p_X}{p_Y}\right)\left(\frac{D_Y}{D_X}\right)},$$

and

$$\eta_Y = \frac{\partial D_Y}{\partial I}\frac{I}{D_Y} \text{ and } \eta_X = -\frac{\partial D_X}{\partial I}\frac{I}{D_X}, \eta_Y = \frac{\partial D_Y}{\partial I}\frac{I}{D_Y}$$

are income elasticities.

Combining the demand and supply functions yields the autarky equilibrium price ratio:

$$(p_1^* - p_2^*) = \frac{(K^* - L^*)}{|\lambda|(\sigma_D + \sigma_S)} + \frac{I^*(\eta_1 - \eta_2)}{(\sigma_D + \sigma_S)}. \tag{2}$$

Equation 2 indicates that the difference in the commodity price ratio depends on the factor endowments, elasticity of demand, elasticity of transformation between commodities, and the income changes. When income elasticities across goods are equal, then the change in commodity price ratio depends solely upon the change in the factor endowments, elasticity of demand, and the elasticity of transformation between the commodities.

If a country is capital abundant, then the numerator is positive and the denominator is negative ($|\lambda| < 0$), implying that the left-hand side of the equation is negative. Thus, the price of X is higher than the price of Y, and when trade is allowed this country will export Y and import X, which is the summary of the H–O theorem. However, it is worth noting that many of the H–O assumptions, particularly of identical technologies across countries, do not hold in the real world. As a result, some of the predictions of the H–O model, such as factor price equalization, do not fully materialize in the real world.

3.2 Extensions of the H–O model

The H–O theory has been extended in several facets. Bowen, Hollander, and Viaene (2012) present a detailed analysis of the H–O theory with many commodities and many factors. They conclude that, on average, a country will export those commodities that use abundant factors intensively and import those commodities that use the scarce factor intensively. As with

Ricardian theory, Dornbusch, Fischer, and Samuelson (1980) extend the basic H–O model to a continuum of goods and show that the geographic pattern of specialization is determinate, and the elasticity of substitution in production plays a crucial role for comparative static analysis.

Several studies have examined the role of imperfect competition for H–O and related theories. In particular, studies have analyzed the impacts of monopsony and oligopsony in input purchase (Magee 1971; Feenstra 1980; Markusen and Robson 1980; Devadoss and Song 2003a, 2003b, 2006). The general findings of these studies are that under factor market imperfection, the production possibility frontier is distorted and major trade theorems may not hold.

Studies have also analyzed the implications of dynamics for the H–O and related theories. Bajona and Kehoe (2010) have developed a dynamic H–O model by combining the static $2 \times 2 \times 2$ H–O model with a two-sector growth model. They found that factor price equalization may not hold in all periods and convergence or divergence may occur depending on the elasticity of substitution between traded goods. Caliendo (2010) studied the consequences of growth in factors of production for international trade between countries. His results showed that terms of trade have important implications for growth and production specialization, and a labor abundant country will remain poorer than a capital abundant country. In his model, factor price equalization may not hold in every period and countries' incomes do not converge to a steady-state level.

3.3 Empirical tests

A few studies have examined the application of H–O theory to agricultural trade. Gopinath and Kennedy (2000) have estimated that capital and land endowments are input factors that predict U.S. state level agricultural exports. Li (2012) applied Gopinath and Kennedy's (2000) approach to assess the importance of factor endowments in determining Chinese regional agricultural exports. The results showed that labor–land ratio in the eastern region, land–labor ratio in the central region, and capital–land ratio in the western region play a crucial role in influencing agricultural exports. Sheldon and Roberts (2008) have studied the conditions for the viability of U.S. biofuel exports under the framework of Heckscher–Ohlin and found that absolute advantage in cellulosic production is needed for biofuel exports. Reimer (2012) used a two-factor model (capital and water) to predict that a country that is abundant in water resources will export commodities that utilize water intensively.

4 New Trade Theory

The Ricardian and H–O theories postulate trade patterns based on comparative advantage arising from, respectively, technological and endowment differences across countries. Yet, a significant volume of trade occurs among developed countries with similar technologies and endowments. Furthermore, the traditional trade theories consider goods as homogeneous, but trade among industrial countries largely occurs in differentiated goods, leading to intra-industry trade. Grubel and Lloyd (1975) was the first comprehensive study that focused on intra-industry trade. Imperfect competition and economies of scale have contributed to the growth in trade among the developed countries. To explain these trade patterns, a large volume of literature mushroomed in the last four to five decades under the banner of New Trade Theory. This literature examines trade under a partial and general equilibrium framework of monopoly, oligopoly, and monopolistic competition. And trade policies under these frameworks – referred to as Strategic Trade Theory – have been studied extensively and generated results that lead to different conclusions depending upon the assumptions and market structure.

A synopsis of trade theories

4.1 Monopoly

This subsection presents the implications of monopoly to trade in a general equilibrium framework. In Figure 18.4, a small country's autarky equilibrium under perfect competition is at point A, at which the marginal rate of transformation equals the price ratio (p^A) and the marginal rate of substitution. Now suppose the X sector operates under monopoly and charges a price above its marginal cost. Then, the price ratio p^M is steeper because of the higher price for X, and the autarky equilibrium point is at B. Also suppose that the country opens to trade and the world market price ($p^\star$) happens to equal to the perfect competition price under autarky (p^A), which is depicted in Figure 18.5, where point B is the same monopoly equilibrium under autarky in Figure 18.4. Then, production and consumption under trade is point A, and movement from B to A is the pro-competitive gain from trade because the monopoly power is destroyed by the competition from abroad. However, it is likely that the world price ratio will be different from $p^\star$. If the world price ratio is equal to p^f, then production is at C and consumption is at D. The

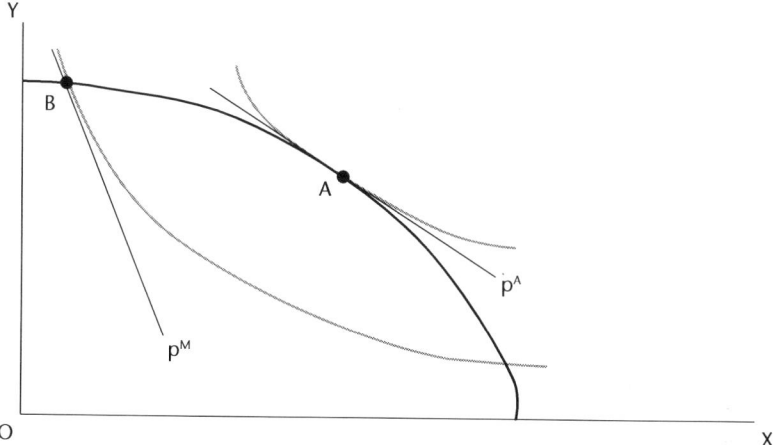

Figure 18.4 Monopoly in autarky

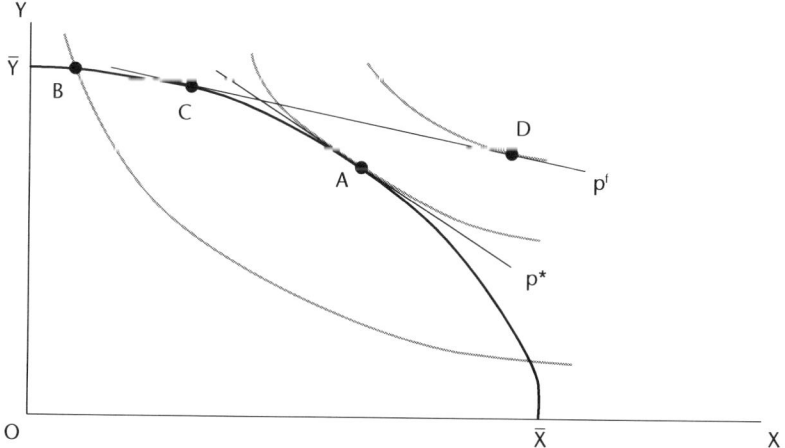

Figure 18.5 Monopoly and trade

movement from A to D is the gains from trade due to the usual comparative advantage. The presence of a monopoly will lead to larger gains from trade relative to perfect competition because the trade eliminates the monopoly pricing and induces additional gains due to pro-competition that will reinforce the normal comparative advantage gain (Melvin and Warne 1973).

4.2 Monopolistic competition

Following the work of Grubel and Lloyd (1975), Krugman (1979, 1980) provided a theoretical framework to explain trade patterns and intra-industry trade without Ricardian or H–O comparative advantage. His studies developed a theoretical model of monopolistic competition and explained that trade can lead to increased product variety and gains from trade. These models are founded on the Dixit–Stiglitz utility function and the increasing returns to scale of firms. Under such a market structure, opening trade increases the diversity of the consumption bundle, which leads to higher utility and gains from trade. For example, consumers prefer to own one car and one motorcycle rather than two cars or two motorcycles.

Krugman's model can be summarized briefly as follows. The economy consists of a representative consumer maximizing utility $U = \sum_{i=1}^{n} c_i^\theta$, with $0 < \theta < 1$, subject to the budget constraint $wL = \sum_{i=1}^{n} p_i c_i$, where c_i is consumption of good i, w is the economy-wide wage rate, p_i is the price of c_i, n is the number of goods produced, and L is the labor endowment. Producers of each good maximize their profit $\Pi_i = p_i x_i - (\alpha + \beta x_i)w$, where x_i is the output of firm i and labor requirement, $l_i = \alpha + \beta x_i$, implies increasing returns to scale. The model allows for free entry and exit. The output market clearing is $x_i = Lc_i$ and the input market clearing is $L = \sum_{i=1}^{n} l_i = \sum_{i=1}^{n} [\alpha + \beta x_i]$.

Once trade is allowed between two countries, consumers in both countries are able to consume twice the number of goods but half of each good as in autarky. This increase in variety in free trade augments the utility, and thus consumers benefit from greater diversity of goods.

Since the pioneering studies of Krugman, the literature on New Trade Theory burgeoned, leading to the oligopolistic competition and Strategic Trade Theory that are important to agricultural trade, which is the focus of the rest of this section.

4.3 Oligopolistic competition

Oligopoly models of New Trade Theory, except for a few studies, largely deal with partial equilibrium models. These models examine trade under quantity competition à la Cournot and price competition à la Bertrand. Under perfect competition, export subsidies are the most detrimental policy in terms of welfare loss as these subsidize the consumers in the importing country. However, studies have shown that in industries that operate under the Cournot competition, home government can subsidize export firms and improve the welfare of the home country at the expense of the foreign country and its firms by shifting rents; that is, profits are transferred from foreign firms to home firms (Spencer and Brander 1983; Brander and Spencer 1985; Spencer and Brander 2008). The reason for this result is that the firms that compete under Cournot tend to under produce, and subsidies by the home government encourage more production, which lowers the market share and profits of foreign firms. In contrast, when firms compete under Bertrand, they tend to over produce. In this case, the production tax on the home firms by the home government curtails production and increases price, which augments the profits of home firms at the expense of foreign firms, again shifting rents from foreign firms to home firms. This strand of literature is known as Strategic Trade Policy analysis.

Studies have also examined the free entry and exit of firms into the industry. Under free entry and exit, all the active firms earn zero profits because firms enter the market until the profit opportunities vanish. Once trade is allowed, because of the increased competition, some firms will exit in both the home and foreign markets. However, the total number of firms serving each market will be larger under trade than under autarky. Under this market structure, free trade lowers the average cost, both home and foreign consumers gain from trade, and total welfare equals consumer gains because profits are zero.

With a few exceptions, the studies that examined the oligopoly models largely neglected the general equilibrium implication for other sectors such as factor markets. Dixit and Grossman (1986) and Neary (1994) have incorporated factor markets into the general equilibrium oligopoly model. However, a more recent study by Neary (2016) has addressed this shortcoming extensively by modeling an economy with a continuum of sectors, with each sector comprising a small number of large oligopoly firms. These firms are large in their own industry but small at the national level, and thus do not exert any oligopsony power. In this model, a Ricardian factor market is included with factors moving across sectors, leading to an economy-wide determination of factor prices, and are thus exogenous to a particular firm, but not for the industry. The oligopoly firms compete strategically against their rivals in the domestic market and in the foreign markets. Neary (2016) uses this model to investigate gains from trade, income distribution between wages of workers and profits of the owners, production, and trade patterns. The results show the gains from trade arise from enhanced competition as more firms compete and from comparative advantage due to heterogeneity within each country.

4.4 Application of New Trade Theory to agriculture

Although the New Trade Theory literature is vast and extensive, application of this theory to agricultural trade is scant. This is because the agricultural economics literature generally considers agricultural commodity markets to operate under perfect competition (Sexton 2013). However, firms that trade agricultural goods – particularly when they engage in value added processing with an emphasis on quality and/or differentiation – can exert market power in both the domestic and world markets. When perfect competition breaks down in international markets, trade policy can reverse traditional welfare results, and government intervention can indeed be welfare improving. The welfare-improving channels typically included government policies that lead to a higher market share and profits for domestic firms, or policies that degrade the oligopoly power, resulting in lower prices to consumers. A few studies have analyzed the strategic trade in these markets.[3] For example, Anania, Bohman, and Carter (1992) examine the welfare effects of the U.S. Wheat Export Enhancement Program, which was designed to mitigate the effects of the export subsidies applied by the European Community. They find that, even in the context of Strategic Trade Theory, the wheat Export Enhancement Program was not welfare improving, largely due to its high cost; the European Community was hurt only slightly, and wheat exports increased only marginally.

Bagwell and Staiger (2001) find that many of the assertions of the strategic trade literature hold in perfectly competitive markets when the political motivations of governments are explicitly modeled. Consequently, as various agricultural markets operate under perfect or imperfect competition, agricultural trade disputes should be examined through, and are a key example of, strategic trade policy. Reimer and Stiegert (2006) review empirical literature, focusing on imperfect competition in agricultural and food markets. Their survey reveals that, even though many exporters of agricultural commodities engage in oligopoly competition, markup over marginal cost seems to be minimal, leaving little room for gains from government intervention. In these models, market power is determined by firm concentration and the elasticity of demand. If consumers have numerous substitutes, then, even with a high degree of firm concentration, markup will be diminished. Despite many cases of government intervention to make domestic firms more competitive internationally, few government policies are strategic from a New Trade Theory perspective. Sexton et al. (2007) show that over half of the intended benefits to farmers in developing countries resulting from government liberalization in international markets are captured by intermediaries, even those with only moderate levels of market power in the supply chain.

In a series of studies, Luckstead, Devadoss, and Mittelhammer (2014), Luckstead, Devadoss, and Dhamodhara, (2015); Luckstead, Devadoss, and Mittelhammer (2015), and Dhamodharan, Devadoss, and Luckstead (2016) develop strategic trade models to study the impact of trade policies on the oligopolistic competition in the Association of Southeast Asian Nations apple market, the U.S. and EU orange juice markets, and the U.S. apple juice market. Based on the theoretical models, they derive comparative static results for change in tariffs, antidumping duties, and productivity shocks. From the theoretical models, they formulate structural econometric models to estimate and statistically test the degree of oligopolistic competition and market power. The structural econometric model requires simultaneous estimation of a system of supply relations and demand functions. The models are then simulated to quantify the magnitude of the impacts of changes in trade policy parameters on quantities, prices, and welfare. This series of papers shows that modeling an oligopolistic industry with free entry and exit leads to results that are contrary to conventional wisdom. For instance, consider the orange juice market. Unilateral trade liberalization by the European Union increases the profits of Brazilian firms, leading to the entry of firms into the industry. As a result, the total Brazilian orange juice supply and exports expand to not only the European Union but also the United States. Consequently, despite a loss to the U.S. orange juice industry, U.S. consumption expands and welfare increases, even though the U.S. import tariff remains the same. However, if the number of firms is not endogenized, then EU trade liberalization will cause Brazilian firms to divert exports from the United States to the European Union and will reduce U.S. welfare.

5 Heterogeneous firms

The extent of heterogeneity in exporting firms was not prevalent until the boom in digital data in the late 1980s and early 1990s. Since then, the high degree of heterogeneity in exporting firms has been well documented. For example, using the 1992 U.S. Census of Manufacturers data, Bernard et al. (2003) observed that as few as 21 percent of all manufacturing plants engaged in exporting. Within these exporting firms, about two-thirds sold less than 10 percent of their products outside the United States and accounted for over half of all exports. Furthermore, less than 5 percent of all exporting plants exported more than 50 percent of their production. The fact that manufacturing firms that exported goods were very large and produced on average 5.6 times more than their non-exporting counterparts[4] explains three observations: exports account for about 14 percent of gross manufacturing output, a low level of export participation exists, and there is a higher export intensity by individual firms. This export heterogeneity is further emphasized by Bernard et al. (2007), who report that in the United States in 2000, only four percent of firms were exporters, of which the top 10 percent accounted for 96 percent of the value of exports. Thus, the firms that do export are more productive and tend to be more skill- and capital-intensive.

Two seminal papers are widely recognized in the heterogeneous firm literature: Eaton and Kortum (2002), which is discussed at length in the Ricardian theory section, and Melitz (2003), which is the focus of this section. Eaton and Kortum (2002) were the first to derive a gravity model for Ricardian theory that relied on perfect competition and productivity difference across sectors found in the trade data. Because of the perfect competition structure, the Eaton and Kortum (2002) model is an important framework for studying trade in agricultural commodities. Melitz (2003) employs heterogeneous firms in trade by bridging the monopolistic competition with firm entry developed by Krugman (1980) and the firm-level heterogeneity with endogenous entry and operating decisions advanced by Hopenhayn (1992). The Melitz model has been path-breaking and extensively used by trade economists because it captures the real-world observations in the firm-level trade data, is highly flexible, and is easily extended. These firm-level heterogeneous models also identified new gains from trade where resources are reallocated from low-productivity to high-productivity firms, leading to greater efficiency (Melitz and Ottaviano 2008). After presenting the foundation of Melitz's model, we discuss several extensions and examples in both general economics and agricultural economics.

5.1 Autarky

Consider a representative consumer, with Constant Elasticity of Substitution (CES) preference over a continuum of goods c indexed by ω. Then, utility maximization leads to the demand function $c(\omega) = C\left(\frac{p(\omega)}{P}\right)^{-\sigma}$ and expenditure function $r(\omega) = P \times C\left(\frac{p(\omega)}{P}\right)^{1-\sigma}$ for an individual commodity ω, where C is the aggregate consumption, $p(\omega)$ is the price of commodity $c(\omega)$, and $P = \left[\int_{\omega_i \in \Omega} p(\omega_i)^{1-\sigma} d\omega_i\right]^{\frac{1}{1-\sigma}}$ is the CES price index.

On the production side, there exists a continuum of firms with productivity indexed by θ that employ labor l and have access to production technology $l = \frac{q}{\theta} + f$, where q is the firm-level production and f is the fixed operating costs. Therefore, fixed operating costs are the same but productivity differs across firms. With Dixit–Stiglitz preferences and monopolistic competition, each consumer good indexed by ω is produced by only one firm indexed by θ, implying a one-to-one relationship between these two index variables. The firm that produces variety θ and faces a residual demand curve earns profits $\pi(\theta) = pq(p;\theta) - w\left(\frac{q(p;\theta)}{\theta} + f\right)$. Applying Bertrand competition yields the pricing rule:[5]

$$p(\theta) = \frac{w}{\theta \rho}, \qquad (3)$$

where the consumer's degree of product differentiation $\frac{\sigma}{\sigma-1} = \frac{1}{\rho}$ dictates the markup over marginal cost $\left(\frac{w}{\theta}\right)$. With the wage rate normalized to 1, profits can then be written as $\pi(\theta) = \frac{r(\theta)}{\sigma} - f$. The role of the productivity parameter θ is seen by taking the ratio of two firms' output $\frac{q(\theta_1)}{q(\theta_2)} = \left(\frac{\theta_1}{\theta_2}\right)^{\sigma}$ and revenue $\frac{r(\theta_1)}{r(\theta_2)} = \left(\frac{\theta_1}{\theta_2}\right)^{\sigma-1}$. With $\theta_1 > \theta_2$, these ratios imply that firm 1 is more productive and has a larger output and higher revenue than firm 2.

Fixed operating costs play a fundamental role in firms deciding to enter the industry and whether to operate. Firms that wish to enter into the industry pay a fixed entry fee, f_e, to draw a productivity level from a cumulative distribution function $G(\theta)$. Therefore, before paying the fixed entry fee, the firm knows only the likelihood of a particular efficiency draw. A firm incurs f_e only if expected profits are greater than or equal to the entry fee:

$$\int_{\bar{\theta}}^{\infty} \pi(\theta) dG(\theta) \geq f_e, \tag{4}$$

where $\bar{\theta}$ is the minimum productivity level for which a firm will operate.[6] After entering the market and realizing a productivity level, a firm makes an operating decision. A firm operates in the industry only if it earns non-negative profits ($\pi(\theta) \geq 0$); otherwise, it exits and never produces. Consequently, the minimum productivity level $\bar{\theta}$ satisfies

$$\pi(\bar{\theta}) = 0, \tag{5}$$

implying that the marginal firm that draws $\theta = \bar{\theta}$ earns zero profits, and any firm with $\theta > \bar{\theta}$ earns positive profits.

Since markets are defined by monopolistic competition, the demand is met by the production for each variety:

$$c(\theta) = q(\theta). \tag{6}$$

Furthermore, the labor market clears as the labor endowment L equals total labor demand:

$$L = M \int_{\bar{\theta}}^{\infty} l(\theta) dG(\theta) + M_e f_e,$$

where M is the total mass of operating firm and M_e is the total mass of firms that enter the market.[7]

Aggregation[8] of firm-level variables to the industry level is achieved by evaluating firm-level variables at the average productivity level of operating firms and multiplying by the total mass of firms. Consider a mass of firms M_e that enter the industry but have yet to make an operating decision. The number of firms that choose to operate is the percentage of firms with productivity levels equal to or above $\bar{\theta}$: $M = (1 - G(\bar{\theta})) M_e$.

Furthermore, the average productivity $(\tilde{\theta})$ given by a cutoff productivity level $\bar{\theta}$ is

$$\tilde{\theta}(\bar{\theta}) = \left(\int_{\bar{\theta}}^{\infty} \theta^{\sigma-1} \frac{g(\theta)}{1 - G(\bar{\theta})} d\theta \right)^{\frac{1}{\sigma-1}}, \tag{7}$$

where $\frac{g(\theta)}{1 - G(\bar{\theta})}$ is the distribution of operating firms. Then the aggregate (industry) price index is $P = M^{\frac{1}{1-\sigma}} p(\tilde{\theta})$, quantity is $Q = M^{\frac{1}{\rho}} q(\tilde{\theta})$, and profit is $\Pi = M_\pi(\tilde{\theta})$.

Per worker welfare ($W = P^{-1} = M^{\frac{1}{\sigma-1}} \rho \tilde{\theta}$) increases with the mass of firms (number of products) and average productivity. With $\bar{\theta}$ determined by the zero-profit condition 5, the aggregate results of this model are the same as a model with a representative firm with an average productivity of $\bar{\theta}$. However, the Melitz model reacts differently to shocks in the economy. For instance, even without a production shock, the average productivity, and thus welfare, can still change due to an exogenous shock that impacts the profitability of firms.

5.2 Free trade

If this economy opens to trade and firms incur no additional trade costs, then firm-level variables are not impacted, as each firm's output levels and profits are the same as with the closed economy, but production is now divided between sales in the home country and the foreign country. Nonetheless, the overall size of the economy (measured in terms of the mass of operating firms M) expands, which leads to an increase in welfare. As a result, without trade costs, the heterogeneous firms model does not provide any different result from that of Krugman's (1980) model with a representative firm. However, trade is not costless. Firms that sell their product abroad incur additional per-unit and fixed trade costs, which are critically important to the predictions of Melitz's model. Per-unit trade costs, which are measured in iceberg form $\tau > 1$, typically include transport costs and trade barriers, while one-time fixed export costs f_e arise from compliance with foreign regulations, foreign market research, marketing strategies, product adaptation to foreign standards and customs, and new distribution channels.

The revenues function ($r_d(\theta)$) for firms selling only in the domestic market is the same as in the autarky case, which leads to the pricing rule $p_d(\theta) = (\rho\theta)^{-1}$. Firms with a high enough productivity to trade do so with all n symmetric partners. Then, revenues for selling both domestically and exporting to all regions are $r_d(\theta) + r_x(\theta) = (1 + n\tau^{1-\sigma})r_d(\theta)$ (noting that $w_i = w_j = 1$ by symmetry). The pricing rule for exports includes the per-unit transport costs: $p_x(\theta) = \tau p_d(\theta)$. A firm that exports also necessarily sells in the domestic market because variable profit is strictly positive after paying the fixed production cost f. Consequently, a firm's profits from domestic sales and exports are separable:

$$\pi_d(\theta) = \frac{r_d(\theta)}{\sigma} - f$$

$$\pi_x(\theta) = \frac{r_x(\theta)}{\sigma} - f_x.$$

With two potential sources of revenue, firms make operating decisions for the domestic market with the cutoff level of $\bar{\theta}_d$ and for the export market with cutoff level of $\bar{\theta}_x$. A firm that sells domestically ($\pi_d(\theta) \geq 0$) also exports to the n export markets if export profits are positive: $\pi_x(\theta) \geq 0$. As a result, the cutoff productivity level for operating domestically $(\bar{\theta}_d)$ is given by $\pi_d(\bar{\theta}_d) = 0$, and the cutoff productivity for exporting $(\bar{\theta}_x)$ is $\pi_x(\bar{\theta}_x) = 0$. When $\bar{\theta}_d = \bar{\theta}_x$, all operating firms sell in both the domestic and export markets. We can write $\bar{\theta}_x$ as a function of $\bar{\theta}_d$ by taking the ratio of these two zero cutoff productivity conditions: $\bar{\theta}_x = \bar{\theta}_d \tau \left(\frac{f_x}{f}\right)^{\sigma-1}$. Therefore, if the fixed export cost is large enough, $\tau\sigma^{-1}f_x > f$, then a partitioning exists between firms that only operate domestically and those that also export $\bar{\theta}_d < \bar{\theta}_x$. Firms with productivity levels between $\bar{\theta}_d$ and $\bar{\theta}_x$ only sell in their domestic market as their profits would be lower if they were to export. This relationship between $\bar{\theta}_d$ and $\bar{\theta}_x$ leads to only one zero-cut off productivity condition:

$$\pi_d(\bar{\theta}_d) + \alpha n \pi_x(\bar{\theta}_d) = 0, \tag{8}$$

where $\alpha = \dfrac{1-G(\bar{\theta}_x)}{1-G(\bar{\theta}_d)}$ is the probability of exporting conditional on operating domestically. The free entry conditions is now:

$$\int_{\bar{\theta}_d}^{\infty} \left(\pi_d(\bar{\theta}_d) + \alpha n \pi_x(\bar{\theta}_d) \right) dG(\theta) \geq f_e. \tag{9}$$

To close the model, the commodity market clearing for each good is still given by Equation 6, and the labor market clearing now includes all labor for domestic production and exports:

$$L = M_d \int_{\bar{\theta}_d}^{\infty} l(\theta) dG(\theta) + n M_x \int_{\bar{\theta}_x}^{\infty} l(\theta) dG(\theta) + M_e f_e,$$

where M_d is the mass of firms operating in the domestic market and M_x is the mass of firms that export calculated as the probability of exporting conditional on operating domestically times the number of domestic operating firms (αM).

With the average productivity of domestic firms given by $\tilde{\theta}_d = \tilde{\theta}_d(\bar{\theta}_d)$ and the average productivity of exporting firms given by $\tilde{\theta}_x = \tilde{\theta}_x(\bar{\theta}_x)$, the weighted average productivity level is

$$\tilde{\theta}_t \left[\frac{M_d(\tilde{\theta}_d)^{\sigma-1} + n M_x (\tau^{-1} \tilde{\theta}_x)^{\sigma-1}}{M_d + n M_x} \right]^{\frac{1}{\sigma-1}}$$

Then, as before, the aggregate price index is $P = (M_d + n M_x)^{\frac{1}{1-\sigma}} p(\tilde{\theta}_t)$, quantity is $Q = (M_d + n M_x)^{\frac{1}{\rho}} q(\tilde{\theta}_t)$, and profits are $\Pi = (M_d + n M_x) \pi(\tilde{\theta}_t)$.

5.3 Autarky, free trade, and falling trade costs

The free trade cutoff productivity level for operating domestically is greater than the autarky cutoff level $(\bar{\theta}_d > \bar{\theta})$ because comparing the zero-cutoff profit Equations (5) and (8) reveals that total average profits increase as $\pi_d(\bar{\theta}_d) + \alpha n \pi_x(\bar{\theta}_d) > \pi(\bar{\theta})$. This implies that the increases in competition arising from trade increases the cutoff productivity level and average profits. With profits in the vertical axes and productivity in the horizontal axis, Figure 18.6 illustrates this result because profits in autarky (depicted by the line "Autarky π (θ)") crosses the horizontal axes closer to the origin than profits under free trade (given by the line "Free trade π (θ)"). This suggests that the least productive firms, with productivity levels between $\bar{\theta}$ and $\bar{\theta}_d$, no longer earn non-negative profits under free trade and exit production. However, firms with productivity above $\bar{\theta}_d$ but below $\bar{\theta}_x$ only operate in the domestic market. Moreover, firms with productivity greater than or equal to $\bar{\theta}_x$ operate both domestically and in the export market. In moving from autarky to free trade, the combined domestic market (with the exit of low productivity firms) and export market (where only the most productive firms export) effects cause market shares to shift from the least productive firms to the most productive firms, leading to a rise in aggregate productivity. The added competition results in a smaller mass of firms in the economy $M_d < M$. In contrast, there are a greater number of product varieties under free trade than under autarky for consumers, as $M_d + n M_x > M$. The increase in product varieties and higher productivity implies that trade always leads to positive welfare gains. While heterogeneity is not needed to model the segmenting of firms that only operate domestically and those that also export, it is necessary to model the rise in productivity, as resources are reallocated to highly efficient firms.

Next, we analyze the firm-level effects of opening to trade. Consider a firm that operates both in autarky and free trade: $\theta \geq \bar{\theta}_d$. Trade implies that the firm loses revenues and market share in the domestic market, as $r_d(\theta) < r(\theta)$, and, if the firm does not export, then it incurs a loss in total revenue. However, for a firm that sells both domestically and exports after free trade $\theta \geq \bar{\theta}_x$, export sales make up for any losses in the domestic market as $r(\theta) < r_d(\theta) + nr_x(\theta)$. Figure 18.6 depicts the impact on firm-level profits. Firms with productivity in the range $\theta \in [\bar{\theta}_d, \bar{\theta}_x)$ incur a net loss of profit because revenues and variable profit are lower after free trade. Furthermore, not all firms with $\theta > \bar{\theta}_x$ experience increase in profits. The marginal exporting firm $(\theta = \bar{\theta}_x)$ earns lower profits after trade because $\pi_x(\bar{\theta}_x) = 0$ and, from above, $r_d(\theta) < r(\theta)$. However, the change in profits increases in θ (represented by the line labeled "$\Delta \pi(\theta)$" in Figure 18.6), and for a high enough θ, profits are higher after the economy opens to trade than under autarky. These results underscore that in the heterogeneity model, all firms do not benefit from trade.

Melitz emphasized two channels to explain why free trade leads to the exit of the least productive firms. First, trade results in higher market competition because product variety increases and imports come from higher productivity firms. However, with monopolistic competition dictated by CES utility, the change in the number of varieties or their prices does not affect the

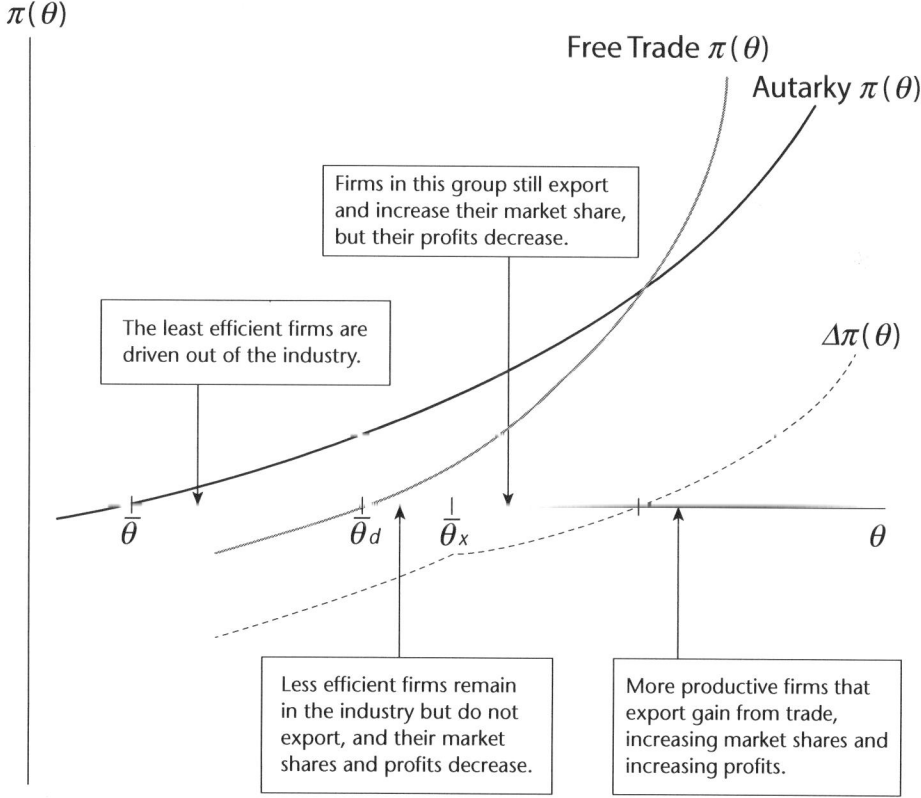

Figure 18.6 Reallocation of market shares and profits across firms as a result of free trade

price elasticity of demand for any good. Consequently, this channel does not occur in the current model (see Section 5.5 below for a model that relaxes the CES assumption). The second channel results from competition in the factor market for labor. Trade results in higher profits only for the most productive firms who increase production as they gain more market share and expand sales beyond the domestic market. Furthermore, the allure of higher potential profits with a high enough productivity draw leads to more entry, as seen in a rise in M_e. Combined, these two effects increase the demand for labor, which drives up the real wage rate. This hike in the real wage makes the least productive firms unprofitable and they exit.

With the World Trade Organization (WTO) making substantial progress in tariff reduction over the last seven decades, it is important to analyze the impact of a reduction in variable trade costs. A decline in the iceberg costs from τ to τ^0 shifts the zero profit curve to the right and profits from exports rise faster, causing the cutoff productivity of the firms that operate domestically to increase $\bar{\theta}_d' > \bar{\theta}_d$ and the cutoff productivity for exporting firms to decrease $\bar{\theta}_x' < \bar{\theta}_x$ (Figure 18.7). Therefore, falling trade costs expose domestic firms to further competition from highly productive exporting firms, and these firms increase the demand for labor, causing a higher real wage rate and further exit of the least productive firms. By contrast, a lower per-unit cost for exporting induces new entry into the export market. As with moving from autarky to free trade, all firms that continue to operate lose part of their domestic sales. Consequently, these firms lose both market share and profits. The more productive export firms make up for the loss of domestic

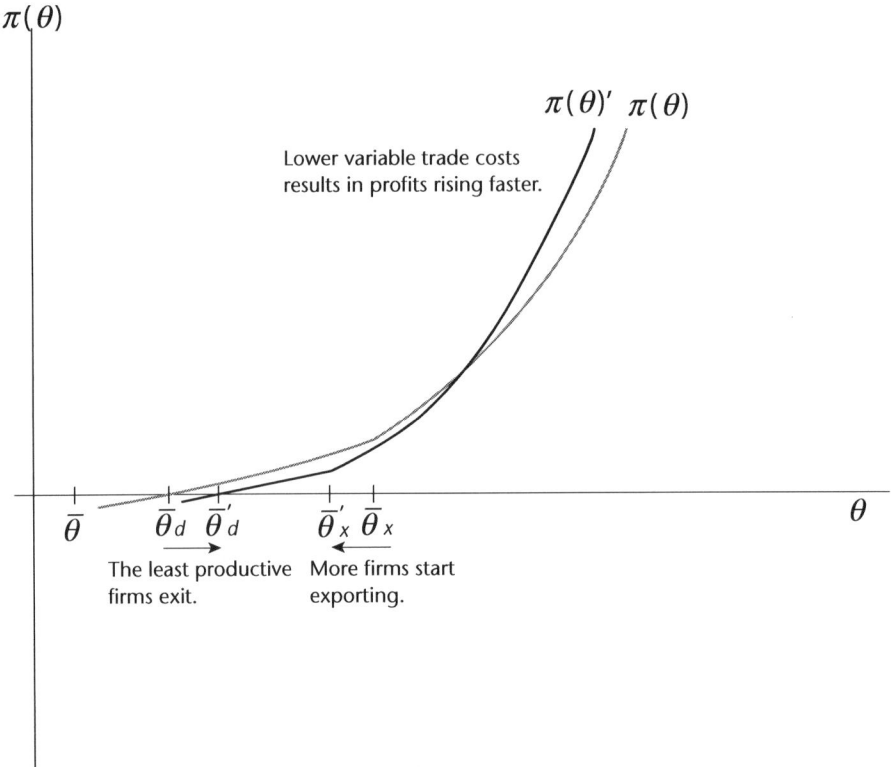

Figure 18.7 Reallocation of market shares and profits across firms due to a decline in per-unit trade costs

sales with exports, and the most productive firms further benefit with higher profits. Aggregate productivity and welfare increase as resources continue to shift to higher productivity firms.

5.4 Connecting firm-level heterogeneity to the data

Micro-level firm data analyzed by Axtell (2001) and Helpman, Melitz, and Yeaple (2004) provide strong evidence that U.S. firms are highly skewed, with a large number of small- and medium-sized firms and a small number of large firms. These studies also conclude that the Pareto distribution provides an excellent fit to this firm-level data, with cumulative distribution function for the random variable θ:

$$H(\theta) = 1 - \left(\frac{\lambda}{\theta}\right)^{\alpha}, \quad \theta \in (\lambda, \infty], \tag{10}$$

where $\lambda > 0$ is the scale parameter and $\alpha > 1$ is the shape parameter. Consequently, because of the tight connection to the data and allowing for analytical solutions, the Pareto distribution is commonly employed by firm heterogeneity studies where only a small proportion of firms are large and highly productive (Melitz and Ottaviano 2008).

5.5 Beyond Melitz and the agricultural supply chain

Since Melitz (2003), there has been a burgeoning literature surrounding firm-level heterogeneity.[9] Here, we provide an overview of notable extensions and then discuss implications of the Melitz model in the agricultural supply chain. With the explosion of foreign direct investment starting in the late 1980s, Helpman, Melitz, and Yeaple (2004) developed a model of a firm's decisions between exporting to, or establishing an affiliate in, the foreign market. With the CES demand, heightened competition due to trade does not affect the distribution of surviving firms. Therefore, employing a linear demand function, Melitz and Ottaviano (2008) specify a heterogeneous-firms model with endogenous markups to analyze the impact of market size and trade liberalization on firms' survivability as the real wage increases and they are exposed to greater competition. Helpman, Melitz, and Rubinstein (2008) connect the observation of zero trade flows in bilateral trade data to the theory by specifying a bounded distribution of productivity. They further discuss the implication for gravity model estimation. Helpman, Itskhoki, and Redding (2010) tackle the issues of wage inequality by incorporating labor market frictions in the monopolistic competition model with firm heterogeneity. Eaton, Kortum, and Kramarz (2011) specify a parsimonious model that includes uncertainty in both demand and fixed export costs to structurally estimate the behavior of heterogeneous French firms that export based on a simulated method of moments. Kugler and Verhoogen (2012) develop a model of heterogeneous firms that endogenously choose the quality of both inputs and output based on the empirical observation of the positive relationship between the firm's size and the price charged for output and the price paid for material inputs.

The value-added segment of the food supply is consistent with the Melitz framework because of the considerable variability in firm size and the high degree of product differentiation. Furthermore, as both consumers' income and tastes for varieties have expanded, the number of U.S. processed food and beverage-manufacturing establishments has also increased by 47 percent from 20,912 in 1992 to 30,659 in 2012 (U.S. Census Bureau 2015). Consequently, food and beverage processing exports have grown extensively by 178 percent, from $37 billion in 1998 to $104 billion in 2012 (BEA 2015). This increase in number of firms and their exports underscores the growth at the extensive margin as new firms enter and products are developed, which leads to new varieties in the market.

Several studies have focused on the impact of trade liberalization on agricultural value added within the Melitz framework. Zhai (2008) builds a 12 region and 14 sector Computable General Equilibrium (CGE) model based on a modified version of Melitz's model[10] of heterogeneity in all sectors except for primary agricultural products and the energy sector. Zhai implements the model to analyze the impact of hypothetical trade liberalization scenarios. The results show that a reduction in tariffs leads to welfare gains and an expansion of exports that are twice as large as those under the standard Armington assumption. Based on the evidence of the importance of the quality of inputs in food manufacturing, Tseng and Sheldon (2015) build a model of firm heterogeneity by taking into account that the quality of inputs are inversely related to the marginal cost of processed food, but complementary to intermediate inputs. In their model, firms optimally choose the quality of the final good. If the export destinations prefer higher quality goods, firms optimally choose higher-quality inputs. An increase in the preference for quality leads to more export firms operating in the market, while low-productivity domestic firms exit food processing. Luckstead and Devadoss (2016) develop a partial equilibrium version[11] of the Melitz model to analyze the impact of the now defunct Transatlantic Trade and Investment Partnership on prices, domestic production, bilateral trade, productivity, the measure of operating firms, and welfare in the processed food sector. This study develops a novel and unique method of calibrating heterogeneous firm models for policy analysis. Because non-tariff measures (including sanitary and phytosanitary measures, food labeling requirements, certification, traceability, etc.) distorting processed food trade more than traditional tariff measures, they analyzed the effects of both U.S.–EU bilateral tariff elimination and non-tariff barrier harmonization. Their results show that trade liberalization expands U.S.–EU cross hauling between about 87 to 95 percent and net welfare expands, not only in the United States and Europe, but also in the rest of the world.

Devadoss and Luckstead (2017) applied the heterogeneous firm model of Luckstead and Devadoss (2016) to study the recently completed Comprehensive Economic and Trade Agreement (CETA) between Canada and the European Union (EU) to liberalize bilateral trade between these two countries. They considered a four-region (Canada, the European Union, the United States, and rest of the world) model of the processed food industry. The firms in this industry are different in size and productivities, produce differentiated products, and engage in monopolistic competition. Their results show, under tariff elimination, bilateral trade flows between Canada and the European Union expand at both the intensive and extensive margins, and the number of firms operating in the export market rises. Net welfare in both of these countries increases even though tariff revenues fall. Although CETA does not liberalize non-tariff barriers (NTBs), they examine the impacts of a 40 percent cut in NTBs to highlight the benefits that would have accrued had CETA also covered NTBs. Under this scenario, the trade flows would have expanded significantly, and, more importantly, Canadian and EU welfare would have risen by *11.8-* and *39.4-fold*, respectively. Since CETA excludes the United States, the U.S. processed food industry loses due to the greater competition in the Canadian and EU markets, and the net U.S. welfare declines. U.S. firms' welfare losses are further exacerbated under the Canadian–EU NTB reduction scenario.

6 Conclusions

This chapter reviews major trade theories – the Ricardian Theory of comparative advantage, the Heckscher–Ohlin Theory of factor abundance, New Trade Theory of imperfect competition, and Firm-Level Heterogeneity Trade Theory – and presents more recent developments. The empirical applications of Ricardian and Heckscher–Ohlin theories are not frequently observed in agricultural economics. However, agricultural economics literature does apply New Trade

Theory and Firm-Level Heterogeneity Trade Theory to agricultural trade, which are covered in this chapter.

Notes

1. Also see Stern (1962) and MacDougall et al. (1962) for empirical work related to Ricardian theory.
2. This theory is also known as the factor proportion or Heckscher–Ohlin–Vanek or Heckscher–Ohlin–Samuelson theory.
3. While New Trade Theory has received minimal attention in the agricultural trade literature, a modest literature exists on imperfection competition and market power in domestic markets. Though market structure has been discussed, it is traditionally in the context of food manufacturing or food retailers (Marion 1986; Sexton and Lavoie 2001). However, mounting evidence shows that even markets that are commonly considered perfect in competition, such as wheat trade, are imperfectly competitive, as wheat is actually differentiated by protein content, and the now-defunct Canadian Wheat Board was able to influence market prices (Lavoie 2005). For further discussion in this area, see Kaiser and Suzuki (2006).
4. This stark size difference holds even when we consider only domestic sales; that is, exporting firms are larger and produce 4.8 times more than non-exporting firms.
5. Because monopolistic competition implies that each firm has a monopoly in their particular variety θ, Bertrand and Cournot competition yield the same pricing rule.
6. The model presented here is modified from the original Melitz model by abstracting from the dynamic environment.
7. The mass (or measure) is analogous to the number of firms or the number of product varieties. However, M or M_e cannot be a whole number in a continuous model, and thus is referred to as a measure or mass in the industrial organization and trade literature.
8. This aggregation method was initially modeled in Hopenhayn (1992) and was the first tractable way to aggregate firm-level variables to the industry level.
9. See Melitz and Trefler (2012) and Melitz and Redding (2014) for a detailed review of the importance of firm heterogeneity for trade in differentiated products.
10. Zhai (2008) abstracts from dynamics and endogenous entry, which implies that the total measure of firms that can potentially operate is fixed.
11. The partial equilibrium setting arises from (1) fixed income, (2) exogenous operating and export costs, and (3) positively sloped input supply function for intermediate inputs.

References

Anania, G., M. Bohman, and C.A. Carter. 1992. "United States Export Subsidies in Wheat: Strategic Trade Policy or Expensive Beggar-Thy-Neighbor Tactic?" *American Journal of Agricultural Economics* 74(3):534–545.
Axtell, R.L. 2001. "Zipf Distribution of U.S. Firm Sizes." *Science* 293(5536):1818–1820.
Bagwell, K., and R.W. Staiger. 2001. "Strategic Trade, Competitive Industries, and Agricultural Trade Disputes." *Economics & Politics* 13(2):113–128.
Bajona, C., and T.J. Kehoe. 2010. "Trade, Growth, and Convergence in a Dynamic Heckscher – Ohlin Model." *Review of Economic Dynamics* 13(3):487–513.
Balassa, B. 1963. "An Empirical Demonstration of Classical Comparative Cost Theory." *The Review of Economics and Statistics* 45(3):231–238.
Batra, R.N. 1973. *Studies in the Pure Theory of International Trade*. New York: Springer.
BEA. 2015. *International Economic Accounts*. Washington, DC: United States Department of Commerce. www.bea.gov/international/. Accessed on June 1, 2017.
Bernard, A.B., J. Eaton, J.B. Jensen, and S. Kortum. 2003. "Plants and Productivity in International Trade." *American Economic Review* 93(4):1268–1290.
Bernard, A.B., J.B. Jensen, S.J. Redding, and P.K. Schott. 2007. "Firms in International Trade." *Journal of Economic Perspectives* 21(3):105–130.
Bowen, H.P., A. Hollander, and J.-M. Viaene. 2012. *Applied International Trade*. New York: Palgrave Macmillan.
Brander, J.A., and B.J. Spencer. 1985. "Export Subsidies and International Market Share Rivalry." *Journal of International Economics* 18(1/2):83–100.

Caliendo, L. 2010. "On the Dynamics of the Hecksher–Ohlin Theory." Technical Report MFI, Working Paper Series No. 2010–2011, Milton Friedman Institute for Research in Economics, University of Chicago, Chicago, IL.

Costinot, A. 2009. "On the Origins of Comparative Advantage." *Journal of International Economics* 77(2):255–264.

Costinot, A., and D. Donaldson. 2012. "Ricardo's Theory of Comparative Advantage: Old Idea, New Evidence." *The American Economic Review* 102(3):453–458.

Costinot, A., D. Donaldson, and C. Smith. 2016. "Evolving Comparative Advantage and the Impact of Climate Change in Agricultural Markets: Evidence from 1.7 Million Fields Around the World." *Journal of Political Economy* 124(1):205–248.

Cuñat, A., and M.J. Melitz. 2010. "A Many-Country, Many-Good Model of Labor Market Rigidities as a Source of Comparative Advantage." *Journal of the European Economic Association* 8(2–3):434–441.

———. 2012. "Volatility, Labor Market Flexibility, and the Pattern of Comparative Advantage." *Journal of the European Economic Association* 10(2):225–254.

Devadoss, S., and J. Luckstead. 2017. "Implications of CETA for Canadian, E.U., and U.S. Processed Food Markets." *Canadian Journal of Agricultural Economics*. doi:10.1111/cjag.12162.

Devadoss, S., and W. Song. 2003a. "Factor Market Oligopsony and the Production Possibility Frontier." *Review of International Economics* 11(4):729–744.

———. 2003b. "Oligopsonistic Intermediate Input and Patterns of Trade." *International Economic Journal* 17(3):77–97.

———. 2006. "Oligopsony Distortions and Welfare Implications of Trade." *Review of International Economics* 14(3):452–465.

Dhamodharan, M., S. Devadoss, and J. Luckstead. 2016. "Imperfect Competition, Trade Policies, and Technological Changes in the Orange Juice Market." *Journal of Agricultural and Resource Economics* 41(2):189–203.

Dixit, A.K., and G.M. Grossman. 1986. "Targeted Export Promotion with Several Oligopolistic Industries." *Journal of International Economics* 21(3):233–249.

Dornbusch, R., S. Fischer, and P.A. Samuelson. 1977. "Comparative Advantage, Trade, and Payments in a Ricardian Model with a Continuum of Goods." *The American Economic Review* 67(5):823–839.

———. 1980. "Heckscher-Ohlin Trade Theory with a Continuum of Goods." *The Quarterly Journal of Economics* 95(2):203–224.

Eaton, J., and S. Kortum. 2002. "Technology, Geography, and Trade." *Econometrica* 70(5):1741–1779.

———. 2012. "Putting Ricardo to Work." *The Journal of Economic Perspectives* 26(2):65–89.

Eaton, J., S. Kortum, and F. Kramarz. 2011. "An Anatomy of International Trade: Evidence from French Firms." *Econometrica* 79(5):1453–1498.

Feenstra, R.C. 1980. "Monopsony Distortions in an Open Economy: A Theoretical Analysis." *Journal of International Economics* 10(2):213–235.

Grubel, H.G., and P. Lloyd. 1975. *International Trade in Differentiated Products*. London: Palgrave Macmillan.

Gopinath, M., and P.L. Kennedy. 2000. "Agricultural Trade and Productivity Growth: A State-Level Analysis." *American Journal of Agricultural Economics* 82(5):1213–1218.

Heerman, K.E.R., S. Arita, and M. Gopinath. 2015. "Asia–Pacific integration with China Versus the United States: Examining Trade Patterns Under Heterogeneous Agricultural Sectors." *American Journal of Agricultural Economics* 97(5):1324–1344.

Helpman, E., O. Itskhoki, and S. Redding. 2010. "Inequality and Unemployment in a Global Economy." *Econometrica* 78(4):1239–1283.

Helpman, E., M. Melitz, and Y. Rubinstein. 2008. "Estimating Trade Flows: Trading Partners and Trading Volumes." *The Quarterly Journal of Economics* 123(2):441–487.

Helpman, E., M.J. Melitz, and S.R. Yeaple. 2004. "Export Versus FDI with Heterogeneous Firms." *The American Economic Review* 94(1):300–316.

Hopenhayn, H.A. 1992. "Entry, Exit, and Firm Dynamics in Long-Run Equilibrium." *Econometrica* 60(5):1127–1150.

Houck, J.P. 1986. *Elements of Agricultural Trade Policies* (Vol. 191). New York: Palgrave Macmillan.

Jones, R.W. 1965. "The Structure of Simple General Equilibrium Models." *Journal of Political Economy* 73(6):557–572.

Kaiser, H.M., and N. Suzuki. 2006. *New Empirical Industrial Organization and the Food System.* New York: Peter Lang.

Krugman, P.R. 1979. "Increasing Returns, Monopolistic Competition, and International Trade." *Journal of International Economics* 9(4):469–479.

———. 1980. "Scale Economies, Product Differentiation, and the Pattern of Trade." *The American Economic Review* 70(5):950–959.

Kugler, M., and E. Verhoogen. 2012. "Prices, Plant Size, and Product Quality." *The Review of Economic Studies* 79(1):307–339.

Lavoie, N. 2005. "Price Discrimination in the Context of Vertical Differentiation: An Application to Canadian Wheat Exports." *American Journal of Agricultural Economics* 87(4):835–854.

Levchenko, A.A. 2004. *Institutional Quality and International Trade* (Vol. 4). Washington, DC: International Monetary Fund.

Li, X. 2012. "Technology, Factor Endowments, and China's Agricultural Foreign Trade: A Neoclassical Approach." *China Agricultural Economic Review* 4(1):05–123.

Luckstead, J., and S. Devadoss. 2016. "Impacts of the Transatlantic Trade and Investment Partnership on Processed Food Trade Under Monopolistic Competition and Firm Heterogeneity." *American Journal of Agricultural Economics* 98(5):1389–1402.

Luckstead, J., S. Devadoss, and M. Dhamodharan. 2015. "Strategic Trade Analysis of U.S. and Chinese Apple Juice Market." *Journal of Agricultural and Applied Economics* 47(2):175–191.

Luckstead, J., S. Devadoss, and R.C. Mittelhammer. 2014. "Apple Export Competition Between the United States and China in the Association of Southeast Asian Nations." *Journal of Agricultural and Applied Economics* 46(4):635–647.

———. 2015. "Imperfect Competition Between Florida and Sao Paulo (Brazil) Orange Juice Producers in the U.S. and European Markets." *Journal of Agricultural and Resource Economics* 40(1):164–178.

MacDougall, G.D. 1951. "British and American Exports: A Study Suggested By the Theory of Comparative Costs. Part I." *The Economic Journal* 61(244):697–724.

MacDougall, D., M. Dowley, P. Fox, and S. Pugh. 1962. "British and American Productivity, Prices, and Exports: An Addendum." *Oxford Economic Papers* 14(3):297–304.

Magee, S.P. 1971. "Factor Market Distortions, Production, Distribution, and the Pure Theory of International Trade." *The Quarterly Journal of Economics* 85(4):623–643.

Marion, B.W. 1986. *The Organization and Performance of the U.S. Food System.* Lexington, MA: DC Heath and Company.

Markusen, J.R., and A.J. Robson. 1980. "Simple General Equilibrium and Trade with a Monopsonized Sector." *Canadian Journal of Economics* 13(4):668–682.

Melitz, M.J. 2003. "The Impact of Trade on Intra-Industry Reallocations and Aggregate Industry Productivity." *Econometrica* 71(6):1695–1725.

Melitz, M.J., and G.I. Ottaviano. 2008. "Market Size, Trade, and Productivity." *The Review of Economic Studies* 75(1):295–316.

Melitz, M.J., and D. Trefler. 2012. "Gains from Trade When Firms Matter." *Journal of Economic Perspectives* 26(2):91–118.

Melvin, J.R., and R.D. Warne. 1973. "Monopoly and the Theory of International Trade." *Journal of International Economics* 3(2):117–134.

Neary, J.P. 1994. "Cost Asymmetries in International Subsidy Games: Should Governments Help Winners or Losers?" *Journal of International Economics* 37(3/4):197–118.

———. 2016. "International Trade in General Oligopolistic Equilibrium." *Review of International Economics* 24(4):669–698.

Reimer, J.J. 2012. "On the Economics of Virtual Water Trade." *Ecological Economics* 75:135–139.

Reimer, J.J., and K. Stiegert. 2006. "Imperfect Competition and Strategic Trade Theory: Evidence for International Food and Agricultural Markets." *Journal of Agricultural and Food Industrial Organization* 4:Article 6.

Sexton, R.J. 2013. "Market Power, Misconceptions, and Modern Agricultural Markets." *American Journal of Agricultural Economics* 95(2):209.

Sexton, R.J., and N. Lavoie. 2001. "Food Processing and Distribution: An Industrial Organization Approach." In B. Gardner and G. Rausser, eds., *Handbook of Agricultural Economics*. Amsterdam: North-Holland Publishing, pp. 863–932.

Sexton, R.J., I. Sheldon, S. McCorriston, and H. Wang. 2007. "Agricultural Trade Liberalization and Economic Development: The Role of Downstream Market Power." *Agricultural Economics* 36(2):253–270.

Sheldon, I., and M. Roberts. 2008. "U.S. Comparative Advantage in Bioenergy: A Heckscher-Ohlin-Ricardian Approach." *American Journal of Agricultural Economics* 90(5):1233–1238.

Spencer, B.J., and J.A. Brander. 1983. "International R&D Rivalry and Industrial Strategy." *The Review of Economic Studies* 50(4):707–722.

———. 2008. "Strategic Trade Policy." In S. Durlauf and L. Blume, eds., *The New Palgrave Dictionary of Economics*. 2nd ed. Basingstoke: Palgrave Macmillan. doi:10.1057/9780230226203.1632.

Stern, R.M. 1962. "British and American Productivity and Comparative Costs in International Trade." *Oxford Economic Papers* 14(3):275–296.

Tseng, E., and I. Sheldon. 2015. "Food Processing Firms, Input Quality Upgrading and Trade." In *2015 Allied Social Science Association (ASSA) Conference*, Boston, MA, January.

U.S. Census Bureau. 2015. *EC1200CADV1: All Sectors: Core Business Statistics Series: Advance Summary Statistics for the U.S. (2012 NAICS Basis): 2012*. Washington, DC: United States Census Bureau. www.census.gov/econ/census/index.html. Accessed January 20, 2017.

Wilson, C.A. 1980. "On the General Structure of Ricardian Models with a Continuum of Goods: Applications to Growth, Tariff Theory, and Technical Change." *Econometrica* 48(7):1675–1702.

Wong, K.-y. 1995. *International Trade in Goods and Factor Mobility*. Cambridge, MA: MIT Press.

Woodland, A.D. 1982. *International Trade and Resource Allocation*." Amsterdam: North-Holland Publishing.

Zhai, F. 2008. "Armington Meets Melitz: Introducing Firm Heterogeneity in a Global CGE Model of Trade." *Journal of Economic Integration* 23(3):575–604.

19
International agricultural trade
A road map

Andrew Schmitz and Ian Sheldon

1 Introduction

This chapter reviews several important aspects of international agricultural trade, including regional trade agreements, market access, non-tariff distortions, suspension agreements, trade embargos, trade barrier case studies, international trade agreements, agricultural trade in transitional economies, international movement of capital and labor, and gains from trade. Agricultural trade has expanded significantly over the past decade (Beckman, Dyck, and Heerman 2017). There is a continuing debate on the impact of agricultural trade. For example, President Trump contends that the North American Free Trade Agreement (NAFTA) needs to be reviewed because it is not in the best interest of the United States. Likewise, the United States withdrew from the Trans-Pacific Partnership Agreement (TPPA) for the same reason.

In earlier reviews of the literature on agricultural trade, including the Josling et al. (2010) paper on the AAEA 100th Anniversary Celebration, it is very apparent that the dimensions of agricultural trade are extremely complex. Because of these complexities, empirical analyses on the impact of trade policies are very difficult. These complexities include (1) agricultural trade is affected by both trade and agricultural policies (Schmitz and Schmitz 2012); (2) U.S. agricultural price supports have a major effect on the impact of U.S. agricultural tariffs and quotas; (3) policies outside of agriculture, such as the U.S. corn ethanol program (administrated through the U.S. Department of Energy), have major impacts on U.S. corn production and trade; (4) the role of the World Trade Organization (WTO) is not well understood; (5) multinational corporations are heavily engaged in international agricultural trade and influence the direction and magnitude of trade; (6) many large agricultural producer firms are involved in trade, in which firms domiciled in the United States produce, with their facilities, products abroad and then export these back to the U.S. market; and (7) international trade is affected by agricultural productivity, which in turn is affected by the cost of inputs, such as illegal immigrant farm labor – a debate that continues in the United States over the production of especially fruits and vegetables.

2 Earlier reviews

For international agricultural trade, the spatial equilibrium price models developed by Samuelson (1952) and Takayama and Judge (1964, 1971) were used to analyze the impact of trade policies. Applications of these models are found for wheat in Schmitz (1968), for fresh oranges in Zusman, Melamed, and Katzir (1969) and for ad valorem tariffs in Devadoss (2013). Variants of these models are still used widely in analyzing distortions in agricultural trade.

One of the early volumes on international trade, edited by Hillman and Schmitz (1979), examined the gains from trade (from both a theoretical and applied political perspective), the importance of exchange rates, and the methods and techniques for analyzing agricultural trade. Chapter authors included well-known trade economists such as Tim Josling, Alberto Valdez, Alex McCalla, and D. Gale Johnson.

In 2002, Karp and Perloff outlined several important ideas that underlie the empirical work in agricultural trade: the theory of comparative advantage, the theory of the second best, and the principle of targeting. They also discussed the extent to which uncertainty and missing insurance markets justify trade policy. In addition, they reviewed the attempts to measure trade elasticities, exchange rate effects, and market power. In the same year, Sumner and Tangermann presented background information on trade policies and agreements and conducted an economic analysis on trade agreements, highlighting the Uruguay Round of Agreements in 1994.

In 2005, Schmitz et al. published *International Agricultural Trade Disputes: Case Studies in North America*. Topics included NAFTA trade disputes, trade remedy laws, the Byrd Amendment, WTO rulings, and import controversies over shrimp and fresh garlic.

In 2010, Josling et al. reviewed the contribution of international agricultural economists over 100 years. Topics on changing trade issues over the past ten decades included understanding the behavior of international commodity prices, domestic policies and market instability, linkages between agricultural trade and exchange rate policies, market power and agricultural trade, the quantification of trade effects and agricultural policies, political economy of agricultural trade (including rent-seeking activities), and multinational trade negotiations.

3 In perspective

There are many topics that can be included in a handbook on trade. For example, Stephen Devadoss and Jeff Luckstead (Chapter 18, this volume) focus on many theoretical aspects of trade, and Michael Reed and Sayed Saghaian (Chapter 22, this volume) deal with the monetary aspects of international agricultural trade. Therefore, in this chapter, we do not deal with aspects of the purer theory of trade. Rather, we examine (1) the general equilibrium models of agricultural trade where a heavy emphasis is placed on estimating the impact of tariff and non-tariff barriers from multinational and international perspectives and (2) the international monetary aspects of trade. We focus on literature that would be considered a micro analysis of international trade, for example, individual commodity studies, dumping and countervailing duty cases, suspension agreements, and distortions caused by genetically modified organism (GMO) events.

From a broad perspective, in his 1973 work *Agriculture in Disarray*, D. Gale Johnson was the first to point out the many distortions that exist in international trade, and how they significantly impact individual countries. He also examined the distributional effects from tariff and non-tariff barriers.

More recently, a Global Trade Analysis Project (GTAP) was launched at Purdue University under Dr. Tom Hertel with the aim of convening a global network of researchers and policy makers to conduct quantitative analysis of international policy issues. The goal of GTAP is to improve the quality of quantitative analysis of global economic issues within an economy-wide framework. The standard GTAP model is a multiregional, multi-sector, computable general equilibrium model, with perfect competition and constant returns to scale.

Professor Kym Anderson (2009) and contributors provided an in-depth analysis of the many agricultural trade policies and distortions that exist worldwide. They concluded that the world moved three-fifths closer toward global free trade between 1980 and 2004. Without the agricultural price and trade policy reforms, global welfare would have been lower by US$233 billion annually. This analysis included 75 countries that together account for over 90 percent of the world's farmers.

4 Background to agricultural trade policies

To provide context for analysis of trade policy instruments applied to the agricultural sector, it is useful to start by describing the extent to which they have been subject to the rules of the General Agreement on Tariffs and Trade (GATT) and its successor, the World Trade Organization (WTO). Between 1947 and 1994, GATT held eight rounds of negotiations that substantially lowered developed-country tariffs in the manufacturing sector (Baldwin 2016). Based on the principle of reciprocity, developed countries exchanged access to each other's markets, extending such access to other members of GATT through the principle of non-discrimination. In the case of the agricultural sector, prior to the Uruguay Round of GATT, agriculture was never subject to disciplines on either border policy instruments, such as tariffs and non-tariff barriers, or domestic farm policies that had the potential to distort international trade in agricultural commodities.

The Uruguay Round Agreement on Agriculture (URAA), completed in 1994, was a major step forward in establishing a clear set of rules for agricultural trade (Josling et al. 1994). The structure of the URAA, and the negotiations that preceded it, focused on three key areas of policy as they relate to agricultural trade: *market access*, *export competition*, and *domestic support* for farmers. Focusing on developed countries, the URAA's key elements were as follows: (1) non-tariff barriers were to be translated into tariffs (*tariffication*), minimum market access was to be guaranteed through tariff rate quotas (TRQs), and all tariffs were to be bound and reduced by 36 percent over six years; (2) new export subsidies were banned, while existing export subsidies were to be reduced by 36 percent over six years; and (3) allowable farm subsidies were defined and the aggregate level of trade-distorting farm support was to be reduced by 20 percent over six years (key here was the recognition that domestic farm policies can stimulate production, leading to reduced imports or increased exports, thereby distorting the world market).

Given the URAA, what has been the progress toward liberalizing agricultural trade since 1994 and through the current Doha Round of trade negotiations in the WTO? With respect to market access, Gibson et al. (2001) reported that despite URAA commitments, the global average of agricultural tariffs was 62 percent, although this hides a considerable amount of detail both across and within countries. Jales et al. (2005) noted that developed countries typically have a number of very high agricultural tariffs and a large number of very low tariffs, which results in their having relatively low mean tariffs with a high degree of tariff dispersion. By contrast, developing countries tend to have higher mean agricultural tariffs and less tariff dispersion. In the case of TRQs, 1,400 have been introduced since 1995, with over-quota and in-quota tariffs averaging 123 and 63 percent, respectively, and there are also low fill rates for the quotas, the overall conclusion being that TRQs have not significantly increased market access (Meilke et al. 2001; Jales et al. 2005).

In the case of export subsidies, while the European Union (EU) accounted, on average, for 92 percent of their global use by value in the immediate five-year period after the URAA was implemented, the European Union did agree to its eventual elimination at the start of the Doha Round of negotiations in the WTO in 2001 (Young 2005). Finally, at the WTO Ministerial Meeting in Nairobi in 2015, the European Union, along with other developed countries such as the United States, agreed to immediately eliminate the use of export subsidies, while developing countries agreed to do the same by the end of 2018. Market intervention by the European Union at above world market prices had resulted historically in high stocks of agricultural commodities that could only be reduced through the use of export subsidies, which distorted world commodity prices. The European Union was able to commit to the reduction of subsidies because it has reformed its farm policies over the past 20 years.

With respect to trade-distorting farm policies, many developed countries, including the United States and the European Union, began to shift their focus toward so-called green box policies that are considered less trade distorting than the so-called amber box policies, although their overall levels of domestic farm support remained high in the late 1990s (Kennedy et al. 2001). In the Doha Round negotiations of the WTO, efforts have been made to place further constraints on the use of domestic farm policies by both developed and developing countries, but no progress has been made in completing these negotiations (Blandford 2005).

While the bias against trade in agricultural commodities due to both border and domestic farm policies has fallen in developed and developing countries since its peak in the mid-1980s (Anderson 2010), there is clear empirical evidence that they are traded considerably less than manufactured goods (Xu 2015). Reimer and Li (2010), using a new Ricardian trade model, have established that exiting trade costs are a major barrier to trade in agricultural commodities, and that, if removed, there would likely be a significant increase in the volume of agricultural trade. In other words, despite the URAA, and with little progress in the Doha Round of the WTO, agricultural trade remains less than liberalized. In this context, we now turn to analyzing the distorting effect of agricultural trade policies.

5 Regional trade agreements

As well as agricultural trade being brought under the disciplines of the WTO, it has also been affected by the increasing tendency toward countries signing bilateral and regional trade agreements (RTAs). RTAs have been part of the trade liberalization landscape since the formation of the European Economic Community (EEC) in 1958 and its subsequent expansion to 28 member countries. Importantly, due to the preferential treatment they offer, RTAs are inherently discriminatory compared to the multilateral GATT/WTO. As noted above, a key pillar of GATT/WTO is the principle of non-discrimination; that is, if some WTO member countries negotiate a reduction in tariffs, those tariff cuts have to be offered to all WTO member countries under Most Favored Nation (MFN) status. By contrast, countries that negotiate tariff reductions through an RTA do not offer them to other countries, a violation of the multilateral trading system that is allowed under GATT Article XXIV.

Prior to the early 1990s, there was only modest expansion in the number of RTAs, but post-1994 there has been rapid growth in the number being negotiated, resulting in an overlapping "spaghetti/noodle bowl" of RTAs (Baldwin 2006, 2016). There has been considerable debate among trade economists about the benefits of regional versus multilateral trade liberalization, Bhagwati (1993) in particular expressing concern that they are a "building block" to greater trade liberalization due to their complex, overlapping rules of origin (ROO), which cause inefficient trade in intermediate goods. In contrast, Baldwin (2006) argues that regionalism and multilateralism have been interdependent, making the argument that one occurring is an incentive for the other; for example, he argues that European regionalism stimulated the Kennedy Round of GATT in the 1960s. Baldwin (2016) concludes that trade diversion due to RTAs has not materialized, driven by the growth of international value chains and the associated push for "deep" integration. There is now an interesting debate about whether or not the WTO is still relevant as regards future trade liberalization efforts (Bagwell, Bown, and Staiger 2016; Baldwin 2016).

In the case of the United States, 14 RTAs have been signed since 1985, covering a total of 20 countries that include Australia, Canada, Chile, Israel, Korea, and Mexico, with tariffs either being reduced to either zero or phased-out over time. In terms of trade, these 20 countries account for 43 percent of U.S. agricultural exports, an increase from 29 percent in 1990 (USDA-FAS 2016). As a share of total U.S. agricultural exports, increases of 15 percent or more have occurred to RTA members for grains and feeds, dairy products, poultry and poultry products, beef and beef products, pork and pork products, and fruit and vegetables. Notably, corn exports to RTA members have increased to 60 percent of total U.S. corn exports, largely due to the reduction in trade barriers with Mexico under the North American Free Trade Agreement (NAFTA) (USDA-FAS 2016).

RTAs have been a focus of agricultural trade research. NAFTA was signed by the United States, Canada, and Mexico in 1994, creating what has become a highly integrated trade bloc that has significantly liberalized agricultural trade among the three countries (USDA-ERS 2015). Since 1993, U.S. agricultural exports to Mexico have grown from US$8.9 billion to US$38.6 billion in 2015; Canada and Mexico accounted for an average of 28 percent of U.S. agricultural exports over the 2012–2015 period, compared to an average of 19 percent over the 1990–1993 period (USDA-FAS 2016). Canada is now the top agricultural export market for the United States, while Mexico has become the top U.S. export market for corn, soymeal, and poultry (USDA-FAS 2016), and recently it became the largest market for U.S. pork exports. Pork production in NAFTA is an example of several complex cross-border agricultural supply chains that have evolved over the past 20 years (Hendrix 2017). Pigs born and weaned in Canada are exported to the United States at 8 to 12 weeks old, where they are fed out to slaughter weight and then slaughtered and processed in the United States, the pork products then being exported to Canada and Mexico.

The Trans-Pacific Partnership (TPP), signed in 2015, began as an expansion of the Trans-Pacific Strategic Economic Partnership Agreement (TPSEP, or P4) signed by Brunei, Chile, New Zealand, and Singapore in 2005. Beginning in 2008, additional countries joined the discussion for a broader agreement: Australia, Canada, Japan, Malaysia, Mexico, Peru, the United States, and Vietnam, bringing the total number of countries participating in the negotiations to 12. Agriculture was included in the agreement (Burfisher et al. 2014; USDA-FAS 2015a, 2015b). Current trade agreements between participating countries, such as NAFTA, would have been reduced to those provisions that did not conflict with the TPP or provide greater trade liberalization than the TPP. In January 2017, however, the United States withdrew from TPP (also, the Trump administration is challenging NAFTA). In terms of agriculture, there has been relatively little analysis done on whether reforms to NAFTA are needed. Likewise, little empirical analysis exists on the effect of the United States pulling out of the TPP agreement, although analysis prior to its signing indicated that U.S. agricultural exports would have increased US$2.8 billion by 2025, the U.S. gaining significant market access to Japan (USDA-FAS 2016). This lack of analysis of TPP and the impact of any NAFTA renegotiation on the agricultural sector should give encouragement to strengthen the empirical work on agricultural trade. As our survey illustrates, much of the work on NAFTA is narrow and deals with sugar and tomato suspension agreements between Mexico and the United States. There are, however, exceptions. Schmitz, Zhu, and Zilberman (2017) estimate the impact of TPP, but mostly with respect to the Japanese agricultural sector and Japanese farmer compensation under freer trade arrangements.

6 Market access[1]

As noted earlier, an important characteristic of the URAA was the focus on market access, be it through binding and cutting tariffs or through ensuring minimum market access via TRQs.

6.1 Import tariffs

There is a rich history on the use of tariffs (Schmitz, Furtan, and Baylis 2002). In the United States, for example, tariffs of 5 percent were established as early as 1789. In 1816, U.S. tariffs were increased to between 15 percent and 20 percent. After 1870, the sentiment for tariff reform in the United States grew rapidly. The Dawes Bill of 1872 lowered tariffs by 10 percent; however, the see-saw approach to tariffs gave rise to the McKinley tariff of 1890, which gave the United States the highest protection afforded by any of the previous tariff acts. After a brief move to lower tariffs in 1913, the United States returned to a protectionist policy in 1921. The most dramatic increase in U.S. tariffs came with the introduction of the Smoot–Hawley tariff in 1930. Canada immediately responded by placing its own higher tariffs on U.S. imports, which increased Canada's trade with Britain even more than after the well-known removal of the Corn Laws in 1820. Canada remained a significant agricultural exporter to Britain until the end of World War II (Marchildon 1998).

6.1.1 Small country tariffs

In Figure 19.1, the effects of a specific import tariff on different groups, including producers and consumers, are shown. In the small-country case, the domestic supply schedule is S, while the domestic demand schedule is D. The free trade price is P_w. If tariff T is introduced, the internal price in the small country will increase to P_1 and imports will be reduced to $(Q_2 - Q_1)$. Producers will gain $(P_1 P_w ca)$, consumers will lose $(P_1 P_w ed)$, and the government will collect $(abfd)$ in tariff revenue. The net loss from the tariff is $(acd + dfe)$.

6.1.2 Large country tariffs

Tariffs in the large country case impact international prices. There are many types of tariffs that fall within this category.

6.1.3 Optimal welfare tariffs

Optimal welfare tariffs have a long history in international economics. It is welfare improving for the importing country, but in aggregate there is a net loss when all countries are taken into account. In Figure 19.2, the excess supply curve of country A is given by ES_A, and the excess demand curve of country B is ED_B. The free trade price is P_w.

International agricultural trade

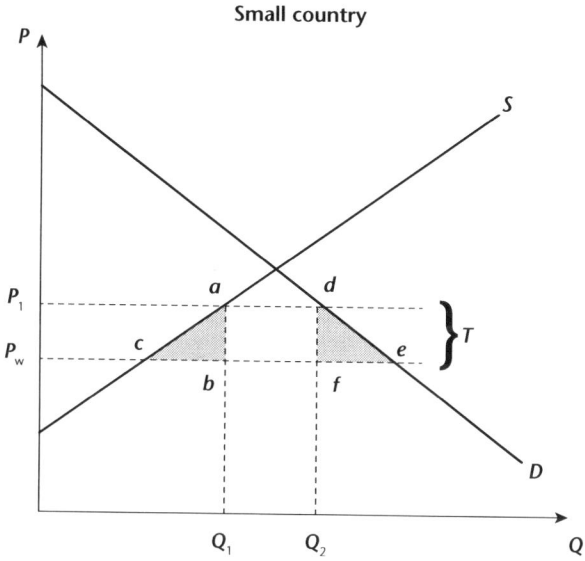

Figure 19.1 Import tariffs in a small country

Source: Reproduced from Schmitz, A., C.B. Moss, T.G. Schmitz, H.W. Furtan, and H.C. Schmitz. 2010. *Agricultural Policy, Agribusiness, and Rent-Seeking Behaviour*, 2nd ed. Toronto: University of Toronto Press.

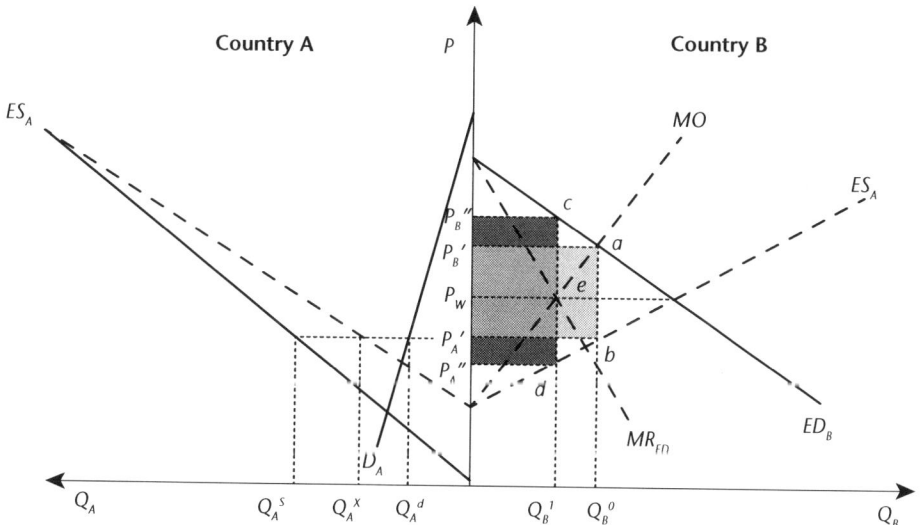

Figure 19.2 Optimal welfare and revenue tariffs

Source: Reproduced from Schmitz, A., C.B. Moss, T.G. Schmitz, H.W. Furtan, and H.C. Schmitz. 2010. *Agricultural Policy, Agribusiness, and Rent-Seeking Behaviour*, 2nd ed. Toronto: University of Toronto Press.

The optimal welfare tariff, $(P'_B - P'_A)$, is determined where the marginal outlay curve MO intersects the marginal revenue curve MR_{ED}.[2] If country B imposes an optimal welfare tariff $(P'_B - P'_A)$, it acts as a monopsonist on the buying of imports from country A. The total tariff collected will be $(P'_B P'_A ba)$. Consumers lose from the tariff while producers gain.

333

Andrew Schmitz and Ian Sheldon

6.1.4 Optimal revenue tariffs

Governments use revenue tariffs to collect revenue from imports. These types of tariffs have been used by the Japanese Food Agency and by the European Union (Carter and Schmitz 1979; Schmitz, Firch, and Hillman 1981). Conceptually, these tariffs are set so the government can exploit both exporters and domestic consumers. The optimal revenue tariff is determined where MO and MR_{ED} intersect (Figure 19.2). The optimal revenue tariff is $(P_B'' - P_A'')$ per unit of import. The tariff revenue collected is $(P_B'' P_A'' dc)$. Note that imports will be Q_B', which are below those under the optimal welfare tariff.

6.2 Tariff rate quotas

Since the URAA, TRQs are commonly applied to agricultural imports. Their original purpose was to ensure a minimum amount of market access after the process of tariffication. It is a combination of tariff and quota in which imports below a specified quantity enter at a low tariff or nonexistent level and imports above that quantity enter at a higher tariff. In Figure 19.3(1), P* is the foreign market price balancing total demand and total supply after Q_E' has been exported to the United States. $\tilde{Q}_F^S$ is the total supply in the foreign market and $\tilde{Q}_F^D$ is the amount consumed in the foreign market. Thus, the quantity exported to the United States is $\tilde{Q}_F^S - \tilde{Q}_F^D$. As a result of TRQs, the United States can set a higher domestic market price at $\tilde{P}$. Consider Figure 19.3(2): t_o is the lower tariff rate allowing a minimum access, $\tilde{Q}_E$, for foreign exports, and t_i is a higher tariff rate charged on imports above $\tilde{Q}_E$. S_F and D_F are the supply and demand of a commodity produced in the foreign market, respectively. Figure 19.3(3) depicts the excess supply from the foreign market ($ES_F = S_F - D_F$) and excess demand from the domestic market (e.g., ED_{US} in the United States). Adding the tariff to the excess supply curve yields the effective excess supply curve ES_F', which is the discontinuous line $abcd$. It is discontinuous at $\tilde{Q}_E$, resulting in the domestic market price of $\tilde{P}$.

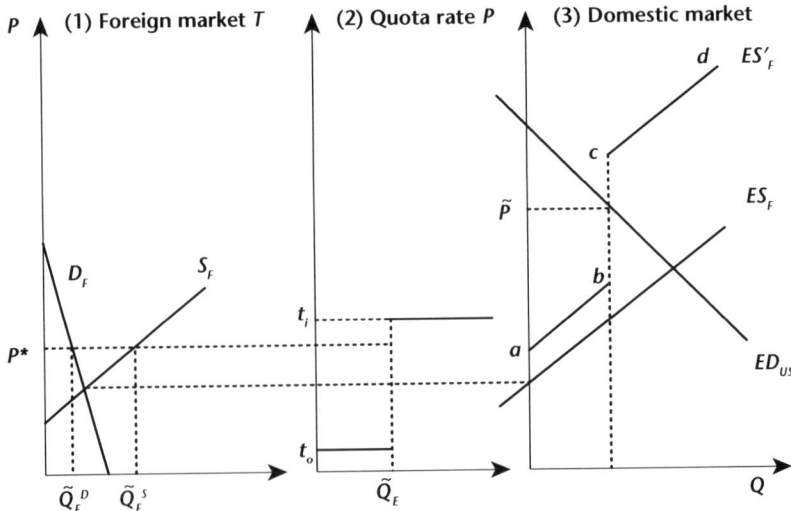

Figure 19.3 Tariff rate quota

Source: Reproduced from Schmitz, A., C.B. Moss, T.G. Schmitz, H.W. Furtan, and H.C. Schmitz. 2010. *Agricultural Policy, Agribusiness, and Rent-Seeking Behaviour*, 2nd ed. Toronto: University of Toronto Press.

Key to the political economy of TRQs is whether the quota is filled or not. In Figure 19.3(3), if ED_{US} intersects the excess supply curve over the range ab, the quota is not filled, import demand being too low. Here, the binding border policy is the lower in-quota tariff t_o, in which case the TRQ is equivalent to a simple import tariff that could be lowered in any future round of WTO negotiations. However, if ED_{US} intersects the excess supply curve over the range bc or cd, there is a potential for quota rents to be earned by firms that get import licenses. For example, in Figure 19.3(3), per-unit quota rents are measured as $(\tilde{P} - b)$ over and above what the United States receives from the in-quota tariff, rents increasing as ED_{US} shifts up to intersect cd. If the quota is auctioned to the highest bidder(s), all of the quota rents are captured by the importing country and the quota would likely be filled, but at least 50 percent of TRQs are administered on a license-on-demand basis which is inefficient if there is under-fill of the quota and also encourages wasteful rent-seeking behavior (Jales et al. 2005)

7 Non-tariff distortions in agricultural trade

Non-tariff distortions (NTDs) take many forms, including tariff rate quotas (TRQs), production quotas, import–export suspension agreements, export subsidies, countervailing duties, country of origin labeling (COOL), unreasonable/unjustified application of non-tariff measures (NTMs), such as sanitary and phytosanitary (SPS) measures, and other technical barriers to trade (TBT).

7.1 Production quotas

Agricultural production quotas are part of the landscape of countries, including the United States and Canada. In the beginning of 2000, production quotas were eliminated for both American and Canadian tobacco production. Also, the U.S. peanut program was ended (Schmitz, Haynes, and Schmitz 2016). Terminating production quota programs impact international trade. In Figure 19.4, total demand is Dt, domestic demand is Dd, and supply is S. Under a production quota, price is P_1, and Q_1 is produced. Total exports are $(Q_1 - Q^*)$. With the elimination of the quota, price falls to P_2, and quantity increases to Q_2. Exports increase from $(Q_1 - Q^*)$ to $(Q_1 - Q^{**})$.

The U.S. tobacco quota program was terminated in 2004. However, while tobacco and cigarette prices fell after the buyout, tobacco production decreased, as did exports. This was due to many factors, including the decline of consumer preference for tobacco products and the increase in cigarette taxes.

7.2 Export subsidies

As noted earlier, export subsidies introduced by large countries can also cause trade distortions to the importing countries. The United States and the European Union have extensively used export subsidies (Schmitz et al. 2010). The U.S. Export Enhancement Program (EEP) and the Dairy Export Incentive Program (DEIP), as direct export subsidy programs, were authorized originally in the 1985 Food Security Act to increase U.S. agricultural exports (Schmitz and Kennedy 2015). The 2008 Farm Bill repealed EEP and authorized only DEIP (Hanrahan 2013), but no DEIP subsidies have been provided since fiscal year (FY) 2010. Wheat ranked first among the commodities that benefited from EEP in terms of total value, with 73 percent of all EEP expenditures being spent on wheat from 1989 to 1993. The U.S. government spent US$1.15 billion to fund EEP in 1994, which was the highest annual EEP expenditure.

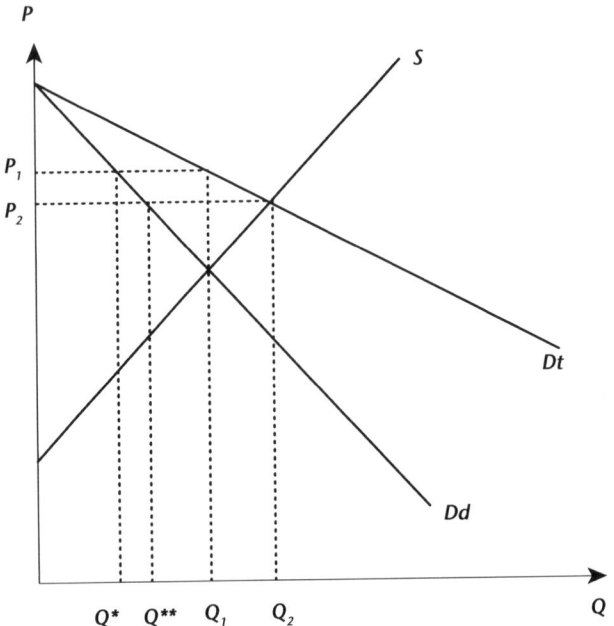

Figure 19.4 Trade and production quota
Source: Authors.

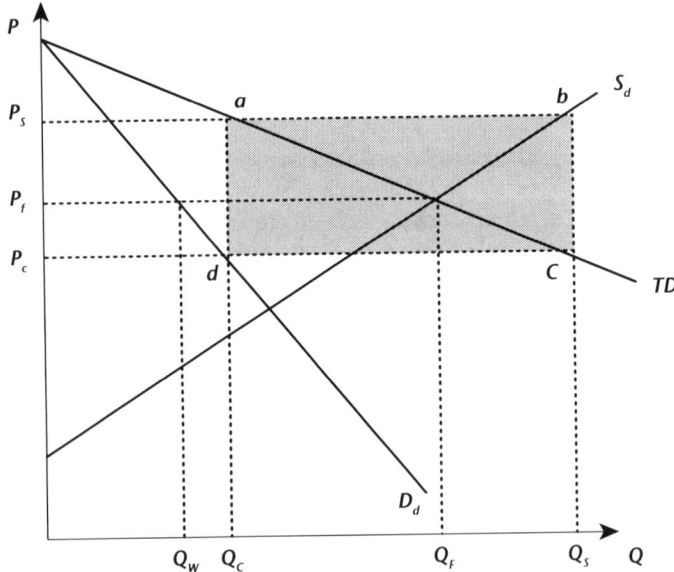

Figure 19.5 Price supports and export subsidies
Source: Reproduced from Schmitz, A., C.B. Moss, T.G. Schmitz, H.W. Furtan, and H.C. Schmitz. 2010. *Agricultural Policy, Agribusiness, and Rent-Seeking Behaviour*, 2nd ed. Toronto: University of Toronto Press.

The use of export subsidies can result from government domestic support programs. Consider Figure 19.5, where the government introduces a price support favorable to domestic producers at P_s. Domestic supply will increase to Q_S, while the export price will fall to P_c, which is lower than the free trade price P_f. Exports increase from $(Q_F - Q_w)$ to $(Q_S - Q_C)$. To clear the market, the export subsidies from the government are $(abcd)$. (Note that the total government deficiency payment is $(P_s bc P_c)$.)

7.3 Hidden subsidies in trade

In the analysis of international agricultural trade, one should be aware of hidden subsidies that are still globally prevalent. As illustrations, Babcock and Schmitz (1986) show with reference to the U.S. sugar policy that there are significant hidden subsidies as a result of quota import restrictions. Schmitz, Schmitz, and Dumas (1997) show the importance of combining both output and input subsidies. In the United States, water is highly subsidized. When this is combined with price supports, the trade impacts from tariffs and quotas are significantly greater than if only one instrument is accounted for. Schmitz, Schmitz, and Seale (2003) theoretically and empirically demonstrate that Brazil has provided significant subsidies to the sugar and ethanol industries through state trading enterprises.

7.4 Trade remedy actions

To deal with international subsidies, the WTO allows for trade remedy actions such as anti-dumping (AD) and countervailing duties (CVD). For example, in the United States AD actions can be imposed when two conditions are met: (1) the United States Department of Commerce (USDOC) determines that foreign goods are sold or are likely to be sold at less than fair value (LTFV), and (2) the United States International Trade Commission (USITC) determines that an industry in the United States is injured materially or threatened with injury from the sale of the goods. Dumping, or sales at LTFV, typically occurs when a company exports a product at a price lower than the price it normally charges in its own home market.

The WTO agreement allows governments to act against dumping when there is genuine material injury. A government must show that dumping is taking place, calculate the extent of dumping (the dumping margin), and show that dumping is causing injury. Typically, an anti-dumping action means charging extra import tariffs in order to bring the price closer to normal value. The dumping margin is the difference between the price or cost in the home market (the normal value) and the price in the export market.

The WTO agreement stipulates three methods for calculating the normal value. (1) a comparison of the export price with the price in the exporter's home market; (2) a comparison of the export price with the price charged in a third country market; or (3) a calculation based on the combination of the exporter's production costs, other expenses, and normal profit (constructed value).

CVDs can be imposed when (1) the USDOC determines that the government of a country, or any public entity within a country, provides a countervailable subsidy on the manufacture, production, or export of goods sold into the United States, and (2) the USITC determines that an industry in the United States is injured materially or threatened with injury by reason of these imports. The types of subsidies covered by CVD actions are specified in the WTO Agreement on Subsidies and Countervailing Measures (ASCM) (WTO 2003). Only specific subsidies that apply to a specific enterprise or industry group are usually actionable. The WTO agreement identifies three types of actionable subsidies: (1) subsidies that provide an advantage over rival exporters in a third-country market; (2) subsidies that benefit the domestic industry over exporters to that country; and (3) subsidies that damage the domestic industry in the importing country.

7.5 Genetically modified organisms

Discussion of NTDs or NTMs has focused on genetically modified organisms (GMOs). As summarized by the WTO, "trade problems arise when countries have different regulations regarding the testing and approval procedures necessary to place GMOs and their products on the market, or when they disagree about labeling and identification requirements."[3] The WTO has tried, and is trying, to resolve trade dispute cases involving GMOs. For instance, the United States, Canada, and Argentina challenged the European Union's *de facto moratorium* on GMOs, effective between June 1999 and August 2003. A WTO panel ruled on February 7, 2006, that the European Union's moratorium on GM products was illegal. Tremendous attention has also been focused on another form of NTM, which is the mandatory labeling requirement on GMOs currently enforced by 64 countries, including the 28 member nations of the European Union, Japan, Brazil, Australia, and China.

Due to concerns about losing the export market, especially in Japan and the European Union, GM wheat and GM rice (e.g., golden rice) have never been commercialized. Using GM wheat as an example, the welfare impact of commercializing GM wheat can be explained in Figure 19.6. It is assumed that the GM wheat yield in the exporting country increases due to the GM technology, and that both domestic and foreign consumers fully accept the new GM wheat variety. For simplicity, it is also assumed the domestic demand curve Dd and total demand curve D_t do not shift because of the adoption of GM wheat (note that due to this strong assumption,

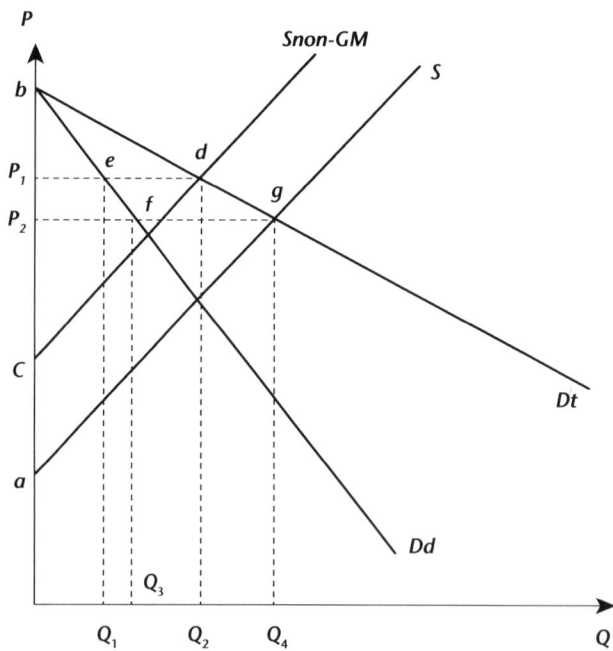

Figure 19.6 Gains from commercializing GM commodities
Source: Authors.

the welfare impact estimates of commercializing GMO wheat in the latter case study might be biased). Before the adoption of GM wheat, the free trade price of non-GM wheat is P_1 and the free trade export amount is (Q_2-Q_1). Due to the yield increase, the original supply curve of non-GM wheat, Snon-GM, shifts to S', which is the supply curve of aggregated wheat after adoption. The free trade price decreases to P_2 and the export amount becomes (Q_4-Q_3), and thus the export change is $[(Q_4-Q_3) - (Q_2-Q_1)]$. The consumer surplus of the exporting country increases by the area P_1P_2fe, and the consumer surplus of importing countries increases by the area $efgd$. The producer surplus change is $(P_2ga - P_1dc)$. The net welfare change of commercializing GM wheat is a gain represented by the area $cagd$.

7.6 Sanitary and phytosanitary measures

The Agreement on the Application of Sanitary and Phytosanitary (SPS) Measures, also known as the SPS agreement, entered into force with the establishment of the WTO in January 1995 (WTO 1998; Miano 2006). Sanitary requirements restrict imports that pose a risk of human or animal disease, and phytosanitary requirements control the risk of plant diseases from imported products (Dyck and Arita 2014).

According to the WTO SPS agreement (1998), countries are allowed to set their own standards. But it also says regulations must be based on science. Member countries are encouraged to use international standards, guidelines, and recommendations where they exist. However, members may use measures that result in higher standards if there is scientific justification. The agreement still allows countries to use different standards and different methods of inspecting products (WTO 1998).

SPS measures may often result in restrictions on trade, especially when the government is pressured to use SPS measures to protect domestic industries from international competition. Due to the technical complexity of SPS measures, they are difficult to challenge by those countries affected. In addition, it is difficult to get a quantitative handle on the impact of these types of measures.

One famous SPS case is Japan's beef imports from the United States. Japan suspended all imports of U.S. beef because of a disease called Bovine Spongiform Encephalopathy (BSE), also commonly referred to as mad cow disease, at the end of 2003. In December 2005, Japan lifted the two-year ban and resumed its imports of U.S. beef from cattle slaughtered younger than 21 months of age. However, in January 2006, Japan halted imports again because animal spines affected by BSE were found in a shipment of frozen beef from a firm in New York. In July 2006, Japan eased the ban on imports of beef only from cattle 20 months of age and younger. Japan's imports of U.S. beef have not recovered to pre-2003 levels (Dyck and Johnson 2013).

There is a growing literature on the impact of SPS regulations on agricultural trade. Ferrier (2014) investigated the impact of phytosanitary regulations on U.S. fresh fruit and vegetable (FFV) imports and found there is a high concentration of imports across countries: 18 of the 29 goods studied have one single country providing more than 80 percent of the total U.S. imports. Fresh olives, potatoes, and corn are the only permitted imports from Canada and Mexico (except for breeding or research purposes, or diplomatic reasons). Grant, Peterson, and Ramniceanu (2015) found that a new SPS measure reduces U.S. FFV exports by 44 percent to 81 percent. Arita, Mitchell, and Beckman (2015) estimated the effects of SPS measures and technical barriers to trade (TBT, a category of NTDs) on agricultural trade between the United States and the European Union. SPS and TBT measures have the most trade-impeding effects on U.S. poultry, pork, and corn, with estimated ad valorem tariff equivalent effects of 102 percent, 82 percent, and 79 percent, respectively. For U.S. beef, the EU SPS measures alone are equivalent to a 24 percent tariff.

Andrew Schmitz and Ian Sheldon

Unlike tariffs, SPS measures can bring about an improvement in overall welfare. This is the case if SPS regulations are put in place to remove a negative externality imposed by a trading exporter.

7.7 Export policies

As noted by Anderson (2010), developing countries have a history of taxing the exportable part of their agricultural sector compared to developed countries, who typically use export subsidies. In the past decade, though, both developed and developing countries have used export restrictions. Export restrictions can take many forms, including export taxes, export quotas, export bans, voluntary export restraints, and other policy instruments (e.g., minimum or reference prices, non-automatic licenses, and dual schemes).

7.8 Export taxes

The impact of export restrictions on food prices and agriculture trade can be illustrated by Figure 19.7 in the case of export taxes. The free trade price is P_F. The optimal export tax is c where supply in the exporting country S_E (for simplicity, assume there is no domestic demand in the exporting country) intersects the marginal revenue demand curve MR_{ED}. Producers' surplus in the exporting country decreases by $fcdp_F$ because of the lower domestic price at d, but producers in the importing country will gain due to a higher world price at a.

Generally, the value of agricultural exports is positively correlated with international commodity price indices. Export restrictions are mainly applied in raw materials like minerals and metals, agricultural commodities, and wood products. In agriculture, export restrictions are frequently used for food security purposes. As shown in Figure 19.8, when the number of export restriction measures increases, the international commodity price index often increases as well.

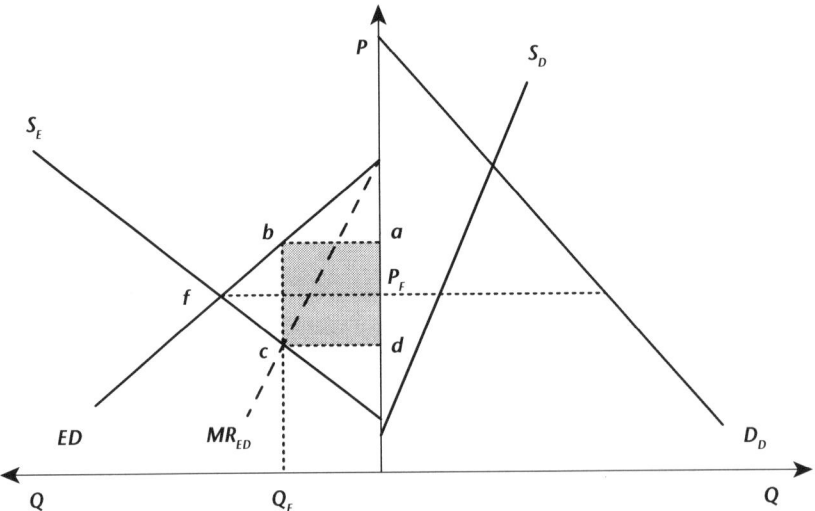

Figure 19.7 Export taxes

Source: Reproduced from Schmitz, A., C.B. Moss, T.G. Schmitz, H.W. Furtan, and H.C. Schmitz. 2010. *Agricultural Policy, Agribusiness, and Rent-Seeking Behaviour*, 2nd ed. Toronto: University of Toronto Press.

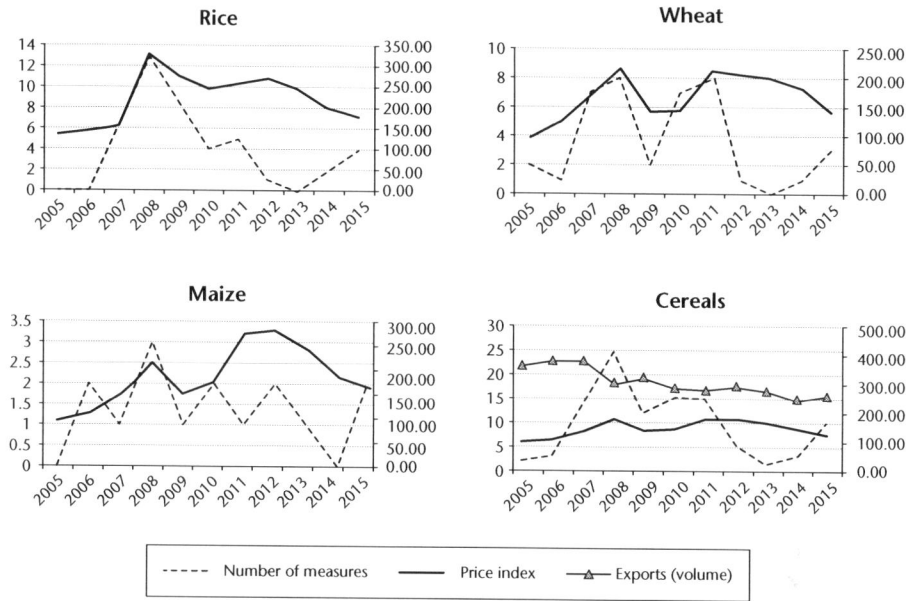

Figure 19.8 Commodity price index and number of export restriction measures
Source: ERA, FAO food price index, IMF commodity prices (Estrades 2015).

7.9 Voluntary export restraints

Producers in both import and export countries can also gain from voluntary export restrictions (VERs) (Figure 19.9). To increase price, producers in the export country will voluntarily restrict exports from Q_A^0 to Q_A^1 (where the marginal revenue curve MR_A corresponding to the excess demand ED_A from the import country equals the export country's supply curve S_A).

The higher price P_1 will encourage producers in the import country to increase their production from $Q_B^{S_0}$ to $Q_B^{S_1}$. Producers in the export country will gain $(P_1 P_0 dc-fde)$, and producers in the import country will gain $(P_1 P_0 ba)$.

8 Suspension agreements

Another illustration of non-tariff barriers is a suspension agreement between importers and exporters. For example, this type of agreement exists between the United States and Mexico for trade in tomatoes (Asci et al. 2016) and sugar (Schmitz and Lewis 2015). Consider Figure 19.10, where producers in both import and export countries form a cartel to maximize joint profits. The cartel quantity Q_C and price P_C are established where the marginal revenue curve MR intersects the joint supply curve $(S_E + S_I)$. Producers in both countries are made better off by forming a cartel. For instance, producers in the export country gain $(abP_f P_c-dcb)$.

9 Trade embargos

Trade embargos have been used over time by various countries and have been applied to various agricultural commodities. An often-cited embargo was the termination of U.S. soybeans to Japan in 1980. USDA undertook a study of the impact of embargos (USDA-ERS 1986). In

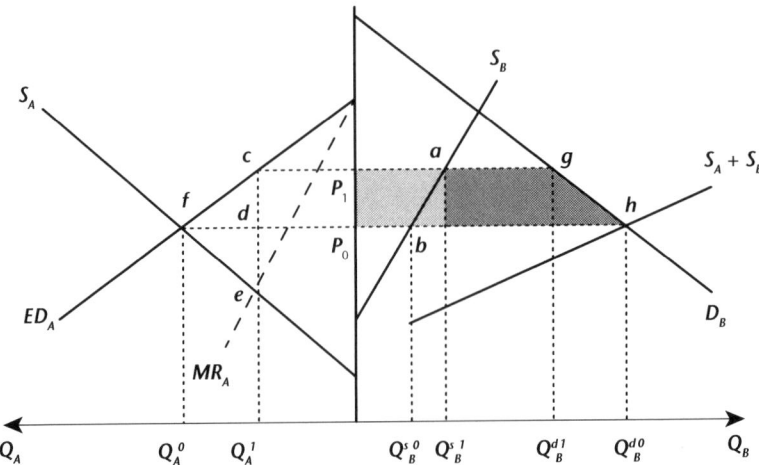

Figure 19.9 Voluntary export restraints
Source: Bredahl, M., A. Schmitz, and J.S. Hillman. 1987. "Rent Seeking in International Trade: The Great Tomato War." *American Journal of Agricultural Economics* 69(1):1–10.

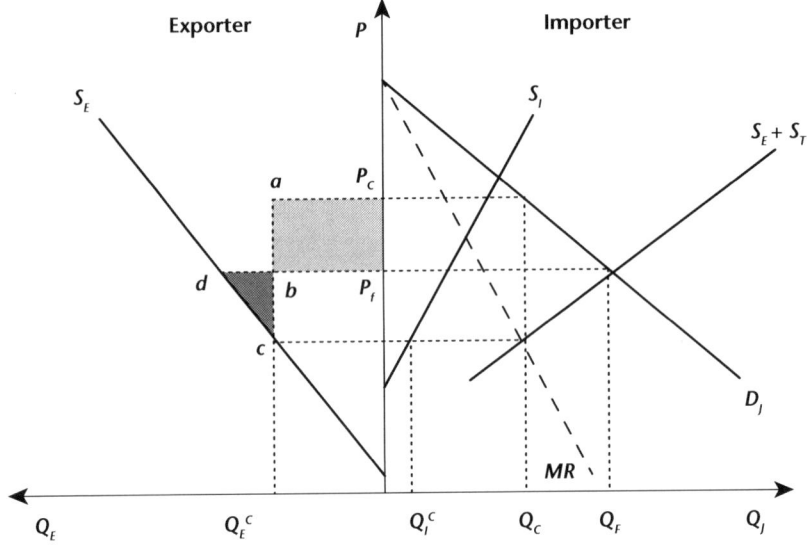

Figure 19.10 Producer import and export cartel
Source: Bredahl, M., A. Schmitz, and J.S. Hillman. 1987. "Rent Seeking in International Trade: The Great Tomato War." *American Journal of Agricultural Economics* 69(1):1–10.

early 2000, the Japanese blocked the importation of U.S. corn due to the U.S. shipment of non-licensed GM corn. This impact was analyzed by Schmitz, Schmitz, and Moss (2005) and Carter and Schmitz (2007). In 2013, China blocked corn exports from the United States because of GM corn (NGFA-NAEGA 2011). Kazakhstan, Russia, and Ukraine restricted wheat exports between 2007 and 2011 due to global commodity price peaks (Gotz, Djuric, and Glauben 2015).

Also, importing countries have used embargos to curtail imports. The topic of embargos was taken up by the USDA in the 1980s (USDA-ERS 1986). Their report shows that embargos historically have had a significant impact on trade volumes and changes in the sources of export goods. In this regard, there have been several interesting and major lawsuits that have evolved due to embargos. These lawsuits are outside the purview of antidumping and countervailing duty cases (discussed later). Some of these cases include (1) the StarLink® case, where Japan blocked the imports of corn from the United States because it contained the unapproved StarLink® gene (Schmitz, Schmitz, and Moss 2005; Carter and Schmitz 2007); (2) the rice controversy; and (3) the 2013 corn import embargo against the United States by China (NGFA-NAEGA 2011). In all of these cases, the termination of trade caused significant price drops and a reallocation of trade among trading nations. This literature contains interesting theoretical and empirical procedures on the appropriate methods to estimate damages from import embargos.

10 Selected case studies on trade barriers

10.1 Country of origin labeling

Country of origin labeling (COOL) for red meats was initially introduced as a requirement in the 2008 U.S. Farm Bill. According to USDA-AMS (2009),

> Country-of-origin-labeling (COOL) is a labeling law that requires retailers, such as full-line grocery stores, supermarkets and club warehouse stores, to notify their customers with information regarding the source of certain foods. Food products covered by the law include muscle cut and ground meats: beef, veal, pork, lamb, goat, and chicken; wild and farm-raised fish and shellfish; fresh and frozen fruits and vegetables; peanuts, pecans, and macadamia nuts; and ginseng.

COOL requirements increase the information available to consumers, and thus improve consumers' decisions on food quality and safety. However, given that COOL results in increased costs, the mandatory imposition of COOL may represent a NTD. The United States has COOL requirements on beef and pork imports from Canada and Mexico. The related legislation costs Canada US$1 billion annually. Canada and Mexico defeated this meat labeling rule through the WTO in May 2015. For the United States, failure to fix the COOL requirements could lead to tariffs on a wide range of U.S. products being exported to Canada and Mexico.

10.2 Cotton and sugar disputes[4]

Two widely discussed trade dispute cases (both involving Brazil) that dealt with cotton and sugar were reviewed by Powell and Schmitz (2005). In the first case, the WTO dispute settlement panel found that the United States was excessively subsidizing upland cotton and that these subsidies were causing serious prejudice to Brazil's cotton growers within the meaning of the Subsidies Agreement. In the second case, which involved the European Union's Common Agricultural Policy, a WTO panel held that the European Union had exceeded its agreed commitments on sugar in both the amount of exports and the level of subsidies. In response to the WTO ruling, the United States was forced to change the nature of its cotton policy, and the same was true for the European Union with respect to sugar. Certain elements of the U.S. cotton policy (domestic support to cotton under the marketing loan and countercyclical payment programs, and export credit guarantees under the GSM-102 program) were significantly reduced under the 2014 U.S. Farm Bill. In the European Union, the sugar beet quotas (a mainstay of the EU sugar policy) were eliminated in September 2017.

10.3 U.S. shrimp imports from China[5]

The warm-water shrimp-harvesting industry in the Gulf of Mexico and South Atlantic (GSA) region represents the most economically important component of all the domestic commercial seafood-harvesting sectors in the United States. Since 1975, shrimp harvesters in the GSA region have sought regulatory relief from imports on several occasions (Adams, Keithly, and Versaggi 2005). Examples include the following:

- In 1975, the USITC responded to the International Shrimp Congress, which led to adjustment assistance permitted under Title II of the Trade Act of 1974.
- U.S. Senator Breaux authored a bill to formulate a policy to provide for domestic shrimp industry protection. A temporary five-year import quota combined with a 30 percent ad valorem tariff was proposed. The bill failed.
- In 1985, the USITC responded to public hearings on imposing trade restrictions on shrimp imports but did not take any concrete action.

10.4 U.S. imports of garlic from China

China is one of the world's leading producers of garlic, a product used for cooking and medicinal purposes. In 1994, the Fresh Garlic Producers Association (FGPA) of the United States alleged that fresh garlic from China was being imported to the U.S. market at less than fair value. The USITC ruled in favor of the United States in the garlic antidumping case (USITC 1994; Yamazaki and Paggi 2005). Three five-year reviews have been initiated by the USITC since 1999 to examine whether the antidumping case should be revoked. They continue to conclude that its removal would likely lead to the continuation or recurrence of the material injury to the U.S. domestic garlic industry. In April 2012, the United States announced that it would maintain the existing AD duty on fresh garlic from China. The duty margin is set at 376.67 percent.

10.5 Beef dispute between the United States and Canada

In 1999, the Rancher–Cattlemen Action Legal Foundation (R-CALF) filed an AD petition with the USITC alleging that Canada was dumping live cattle onto the U.S. market. The USITC ruled in Canada's favor that live cattle from Canada were not being dumped on the U.S. market. In addition, R-CALF filed a CVD complaint against the Canadian cattle industry, contending that Canadian cattle were being fed cheap barley subsidized by the Canadian Wheat Board (CWB). Again, the USITC ruled in Canada's favor. The defense argued that neither case had an economic or empirical base (Wohlgenant and Schmitz 2005). However, the USDOC sided with R-CALF and imposed temporary duties on Canada (Schmitz et al. 2005).

Under the 2014 U.S. Farm Bill, the COOL provision was included. However, the WTO ruled against this provision, and thus the United States can no longer require that importation of commodities such as beef have country of origin labeling.

10.6 Canadian–U.S. lumber disputes[6]

The Canadian–U.S. lumber dispute centers on the allegation that the Canadian government has subsidized the lumber industry by setting stumpage fees as opposed to allowing the market to dictate prices (Berck 2005). The Canada–U.S. Softwood Lumber Agreement (SLA) was a measure to restrict Canadian exports of softwood lumber to the United States. Rather than a countervailing duty or export tax, the SLA utilized a quota-based system that provided a rent to Canadian lumber producers at the expense of U.S. consumers. Canadian lumber producers are better off under a voluntary quota than under free trade, with the entire sector gaining if quota are traded (van Kooten 2002, 2014). However, Canadian producers are also better off if countervailing duties or export taxes are transferred back to industry as a lump sum. In terms of log exports, both Canada and the United States restrict exports from public lands, as does Russia (Abbott, Stennes, and van Kooten 2009; Johnston and van Kooten 2017).

10.7 Apples

Luckstead, Devadoss, and Mittelhammer (2014) developed an apple trade model under imperfect competition to analyze the market power of U.S. and selected Chinese apple producers. They derived estimates of the effect of tariffs using the new empirical industrialization literature. The elimination of the Chinese tariff as per the Association of Southeast Asian Nations (ASEAN) China free trade area resulted in the contraction of U.S. exports to ASEAN.

Luckstead, Devadoss, and Dhamodharan (2015) considered U.S. antidumping duties on apple juice imports from China. While these duties benefited U.S. processors, they negatively impacted Chinese processors as well as U.S. consumers. Under oligopolistic competition, changes in U.S. tariff policy and Chinese productivity impact the market structure in both the United States and China along with prices and overall welfare.

10.8 Bananas

The long-lasting trade dispute between the United States and the European Union over the sale of bananas in the EU market ended in April 2001. This dispute came about because of a complex licensing system developed by the European Union. According to Josling and Taylor (2003), the mechanism used for allocating the licenses violated the WTO rules as they discriminated against the suppliers from Latin America. Josling and Taylor further emphasize that

> The complexity of the banana conflict . . . proved a test case for several aspects of trade policy, including the authority of the World Trade Organization (WTO), and its General Agreement on Trade in Services (GATS) and Dispute Settlement Body (DSB); it has challenged the development policies of the EU, as laid out in the Lomé Conventions as well as the policies of the producing regions; and it has exposed some sensitive issues of domestic political influence, not least in the US. Each of these aspects of the story would merit examination and reflection. But it is the interaction of these different policy issues that renders the banana case unique in the area of recent trade policy disputes. Solving such a complex puzzle involving a heady mix of political interests on both sides of the Atlantic was not easy.
>
> (Josling and Taylor 2003, pp. 1–2)

10.9 Mexico–U.S. sugar disputes

The mainstay of U.S. sugar policy is import quotas. Under NAFTA, Mexico agreed to limit its exports of sugar to the United States. This limit was supposed to be governed by certain conditions, including the sugar export surplus situation in Mexico. The United States has periodically contended that Mexico has oversupplied sugar to the United States. Interestingly, along the lines discussed earlier on tomato suspension agreements, the United States and Mexico agreed to a sugar suspension agreement in 2013. Who gained and who lost from this agreement has been open to discussion (Lewis and Schmitz 2015; Schmitz and Lewis 2015).

10.10 Ethanol from corn

The United States launched several controversial programs to promote the conversion of corn into ethanol. In the 1980s, a 54-cents-per-gallon tariff was imposed on imports. In 2004, U.S. blenders of transportation fuel received a tax credit of 46 cents per gallon. The import tariff and tax credit were the source of an ongoing ethanol dispute between the United States (a country that produces ethanol from corn) and Brazil (a country that produces ethanol from sugarcane). Even though both the tariff and the tax credit were removed in December 2011 (and these changes complemented the tariff elimination enacted by Brazil in 2010), the Energy Independence and Security Act in the United States established a new renewable fuel standard (RFS) that requires the use of 36 billion gallons of biofuels annually by 2022. Of this requirement, 15 billion gallons must be corn-based ethanol (de Gorter and Just 2009; Just and de Gorter 2009).

The use of corn to produce ethanol is promoted by the U.S. Department of Energy, not the USDA, even though it has a major impact on agricultural trade. As a result, often agricultural trade issues are brought about from non-agricultural governing bodies. One controversy surrounds using corn, a food crop, to produce fuel. In 2015, the United States used over 4 billion bushels of corn to produce ethanol (Figure 19.11). Schmitz, Moss, and Schmitz (2007) pointed out that under the ethanol program, there was no free lunch. While there are gainers from

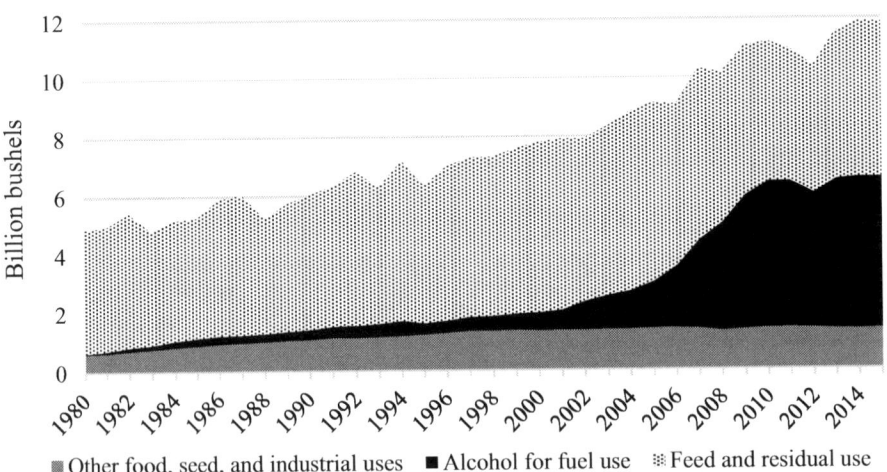

Figure 19.11 U.S. domestic corn use

Source: Calculated by USDA-ERS.

the program, there are also losers. Also, U.S. corn exports would be significantly higher under a zero-corn ethanol program. However, there is evidence that Brazil and the United States have begun collaborating in an effort to promote the global use of not just ethanol, but also biofuels in general.

11 International trade agreements

Over time, there have been many international trade agreements, such as NAFTA. Out of the Doha round, new agreements have emerged between Canada and the European Union, Japan and the European Union, and the Trans-Pacific Partnership (TPP) (Schmitz, Zhu, and Zilberman 2017). The first two Doha agreements have been ratified, but not the third. There is some debate regarding whether countries benefit more through bilateral, rather than multilateral, agreements.

In the case of NAFTA, it is generally accepted that this agreement has generated net gains despite several roadblocks. To highlight some of the issues involved under NAFTA, we focus on the tomato and sugar disputes between the United States and Mexico, and the lumber disputes between the United States and Canada, using the theoretical framework of Bredahl, Schmitz, and Hillman (1987). The theoretical argument put forward is that tariffs imposed by the United States against Mexico are a competitive trade instrument, while a cooperative agreement would involve quota allocations among countries. In the case of tomatoes, suspension agreements were created in 1996 and 2013 that regulated trade between the United States and Mexico.

In the case of sugar, Mexico agreed to limit exports of sugar to the United States. However, this agreement has often been challenged by the United States, arguing that Mexico has often exceeded allotted export amounts. These allegations have brought about the filing of antidumping and countervailing duty lawsuits by U.S. sugar producers. As in the case of tomatoes, from these disputes emerged a suspension agreement between the United States and Mexico for sugar. The first agreement was signed in December 2014 (Schmitz and Lewis 2015). This agreement was short-lived when the United States again brought legal action against Mexico, alleging that Mexico was dumping sugar into the U.S. market. This was resolved in May/June 2017 when a new suspension agreement was signed (Commerce.gov 2017).

12 Agricultural trade in transitional economies

Erokhin (2016) argues that foreign trade has had an immense impact upon modern economies. To succeed in the global marketplace, countries must strive for sustainable development in trade. For many transitional economies, food security is of vital concern. Moreover, uncertainties introduced by globalization can negatively impact food security. *Global Perspectives on Trade Integration and Economies in Transition* (Erokhin 2016) is an authoritative reference source for the latest research on the dynamics of transitional economies. It addresses specific roles of economies in transition on the global market. Topics include economic and trade policies in Central and Eastern Europe, Asia, Latin America, and Africa; inward and outward investment in transition economies; sustainable trade and investing on falling markets; regional trade integration as a response to threats of globalization; and expectations of the new global order due to growing involvement of economies in transition.

Furthermore, as detailed in Erokhin (2017), the process of food production and distribution has grown into a global corporate system, which has impacted sustainability on an international scale. Trade plays a vital role in stabilizing food supplies and food prices due to the co-existence of oversupply in some countries and shortages in others. Emerging economies are pursuing various policy options for ensuring food security. These policies include expanding investment in agriculture; encouraging climate-friendly technology; restoring degraded farmland; improving post-harvest storage and supply chains; and promoting niche products. Emerging economies face challenges as the growing middle class shifts from traditional staples to products with higher resource intensity, such as meat, fish, and dairy (Schmitz and Meyers 2015). This shift requires expansion of domestic agricultural capacity or greater reliance on imports. However, importing countries promote self-sufficiency in food through trade policy restrictions due to unreliability of world markets. This volume stresses the importance of balancing trade liberalization vis-à-vis protectionism in the context of food security.

We briefly examine the wheat market in Russia and Ukraine. Russia implemented an export tax in November 2007, which was increased to 40 percent in December 2007 and remained in effect until July 2008 (Götz, Djuric, and Glauben 2015). Later, in August 2010, Russia implemented an export ban on wheat through July 2011 due to record low harvests. In 2015, Burkitbayeva and Kerr (2015) estimated the impact of tariff reductions on wheat in transitional economies. Also, Leifert and Leifert (2015) discuss the rise of the Former Soviet Union region as a major grain exporter.

Figure 19.12 shows that the export restrictions in Russia were only partially successful when reducing domestic wheat prices; domestic prices in May and June 2008 were above

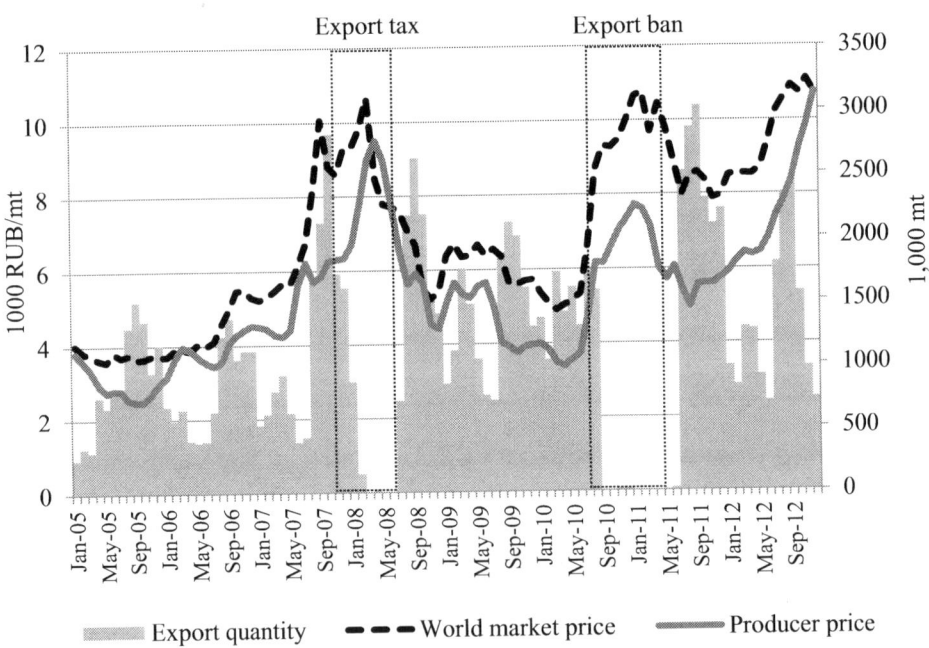

Figure 19.12 Development of world market price, producer price, and export quantities in Russia, 2005–2012

Note: Prices for Russia are average prices for the Central Federal District.

Source: Reproduced from Götz, L., I. Djuric, and T. Glauben, T. 2015. "Wheat Export Restrictions in Kazakhstan, Russia, and Ukraine: Impact on Prices along the Wheat-to-Bread Supply Chain." In A. Schmitz, and W.H. Meyers, eds., *Transition to Agricultural Market Economies: The Future of Kazakhstan, Russia and Ukraine*. Wallingford, UK: CABI.

world market price and the domestic wheat price has been rising steadily since May 2010. In response to rising food inflation, in December 2014, stringent quality monitoring was imposed in order to curb wheat exports. In February 2015, an additional tax on wheat exports was implemented to dampen high domestic wheat prices. Welton (2011) found that while the ban did not lower food prices in Russia, the ban did increase the international wheat price.

In 2007, Ukraine imposed an export duty on wheat. From 2006 to 2011, Ukraine imposed many separate export quotas for wheat (Kobuta, Sikachyna, and Zhygadlo 2012; Kobuta 2015) No wheat export duties were imposed for the 2000–2010 period. On November 15, 2012, Ukraine implemented the export ban on wheat, but lifted it at the end of April 2013. As a result, the gap between the world and the domestic wheat price increased substantially when the export-quota was implemented (Goychuk and Meyers 2013).

13 International movement of capital and labor and gains from trade

Unfortunately, standard economic trade impact analysis has limited coverage with respect to the international movement of capital and labor, multinational firms, and product trade. The early theoretical trade literature assumes that neither capital nor labor moves internationally. In a classic paper, Mundell (1957) relaxed this assumption and argued that tariffs caused an increase in international capital movements, which substitute for product trade. Later, Schmitz and Helmberger (1970) demonstrated that product trade, especially in agricultural products, could increase as a result of international capital movements. Within this context, where firms, including multinationals, that carry out trade, who gains and who loses from international trade becomes much more complex than under neoclassical arguments. For example, if a U.S. agricultural firm produces a product abroad, but then ships and sells this product in the United States, what is the effect of a tariff on this product? Likewise, for example, Smithfield Foods, Inc. is a large Chinese-owned major pork producer/processor located in the United States. What are the welfare impacts from trade-distorting measures imposed by the Chinese against their importation of pork? Duda and Sons, a large agricultural firm in Florida, produces agricultural products outside of the United States that find their way into the U.S. market. Imposing a tariff on vegetables shipped from Duda and Sons in the United States has a negative impact on Duda and Sons. This conclusion is the opposite of that reached under standard economic analysis where tariffs help domestic producers.

Likewise, studies on the impact of the movement of labor internationally are scanty. In reality, labor mobility is a major issue for U.S. agriculture, especially in the production of fruits and vegetables. The attempt to restrict labor from countries such as Mexico has a major negative impact on U.S. labor-intensive agriculture production due to increased U.S. imports of agricultural products produced abroad with cheap labor. Likewise, curtailing illegal immigration affects U.S. agricultural producers negatively and reduces the export of U.S. agricultural products. A discussion surrounding the impact of multinationals and trade is taken up in Schmitz et al. (2010, pp. 295–297), where they state that

> more and more of the importation of goods into a country, such as the United States, is done by multinationals. American capital (human and physical) moves from the United States to produce goods abroad. The goods, in turn, are shipped back to the United States and to third markets. The trend toward increased capital movements and product trade (where they are complements) has risen sharply.

14 Conclusions

From this overview, it should become apparent that agricultural trade contains many complex elements, some of which are not included in this chapter. The macro dimensions are taken up by Michael Reed and Sayed Saghaian (Chapter 22, this volume). Also, we do not deal with important topics such as (1) climate change and trade; (2) agricultural productivity and its impact on absolute and comparative advantage; (3) agricultural trade and its ties to food security (Schmitz, Kennedy, and Schmitz 2015, 2016, 2017); (4) trade in GMO food products and the link to malnourishment in less developed countries; and (5) how empirical analyses can be strengthened to better understand the forces that determine world agricultural trade.

Acknowledgments

The authors thank Professor Stephen Devadoss for technical assistance and contributions.

Notes

1. Part of this material is taken from "Agricultural Import Tariffs and Export Restrictions" (Schmitz, Zhu, and Schmitz 2016a) and "Nontariff Distortions in Agricultural Trade" (Schmitz, Zhu, and Schmitz 2016b). Also, there are tariffs not considered in this chapter (see, for example, Schmitz, Seale, and Schmitz 2006 on optimal Byrd tariffs).
2. The marginal outlay curve is the marginal cost of buying additional imports, while the marginal revenue curve is the revenue generated from the sale of the last unit of the product sold.
3. Source: World Trade Organization, SPS Training Module: Current Issues, Section 8.1 (Genetically Modified Organisms).
4. Additional work on cotton disputes and subsidies can be found in Ridley and Devadoss (2012, 2014).
5. International trade in fish and fish products, including shrimp, has increased significantly since 2000. Global production and trade in fish products are discussed in Anderson et al. (2017) and Asche, Anderson, and Garlock (Chapter 9, this volume).
6. Additional work on lumber disputes can be found in Devadoss (2006, 2008, 2013).

References

Abbott, B., B. Stennes, and G.C. van Kooten. 2009. "Mountain Pine Beetle, Global Markets, and the British Columbia Forest Economy." *Canadian Journal of Forest Research* 39(7):1313–1321.

Adams, C., W.J. Keithly, and S. Versaggi. 2005. "The Shrimp Import Controversy." In A. Schmitz, C. Moss, T. Schmitz, and W. Koo, eds., *International Agricultural Trade Disputes: Case Studies in North America*. Calgary: University of Calgary Press, pp. 224–235.

Anderson, J.L., F. Asche, T. Garlock, and J. Chu. 2017. "Aquaculture: Its Role in the Future of Food." In A. Schmtiz, P.L. Kennedy, and T.G. Schmitz, eds., *World Agricultural Resources and Food Security*. Bingley: Emerald, pp. 159–173.

Anderson, K. (editor). 2009. *Distortions to Agricultural Incentives: A Global Perspective*. Washington, DC: The World Bank.

———. 2010. "Krueger, Schiff, and Valdés Revisited: Agricultural and Trade Policy Reform in Developing Countries since 1960." *Applied Economic Perspectives and Policy* 32(2):195–231.

Arita, S., L. Mitchell, and J. Beckman. 2015. *Estimating the Effects of Selected Sanitary and Phytosanitary Measures and Technical Barriers to Trade on US-EU Agricultural Trade*. No. 212887. Washington, DC: United States Department of Agriculture, Economic Research Service.

Asci, S., J.L. Seale Jr., G. Onel, and J.J. Vansickle. 2016. "U.S. and Mexican Tomato Perception and Implications of the Renegotiated Suspension Agreement." *Journal of Agricultural and Resource Economics* 41(1):138–160.

Babcock, B.A., and A. Schmitz. 1986. "Look for Hidden Costs: Why Direct Subsidy Can Cost Us Less (And Benefit Us More) Than A "No Cost" Trade Barrier." *Choices* 2:18–21.

Bagwell, K., C.P. Bown, and R.W. Staiger. 2016. "Is the WTO Passé?" *Journal of Economic Literature* 54(4):1125–1231.

Baldwin, R. 2006. "Multilateralising Regionalism: Spaghetti Bowls as Building Blocs on the Path to Global Free Trade." *World Economy* 29(11):1451–1518.

———. 2016. "The World Trade Organization and the Future of Multilateralism." *Journal of Economic Perspectives* 30(1):95–116.

Beckman, J., J. Dyck, and K.E.R. Heerman. 2017. "The Global Landscape of Agricultural Trade, 1995–2004." Economic Information Bulletin 181, USDA, Washington, DC.

Berck, P. 2005. "Contested Trade in Logs and Lumber." In A. Schmitz, T. Schmitz, W. Koo, and C. Moss, eds., *International Trade Disputes: Case Studies in North America*. Tuscaloosa: University of Alberta Press, pp. 121–137.

Bhagwati, J. 1993. "Regionalism and Multilateralism: An Overview." In J. de Melo and A. Panagariya, eds., *New Dimensions in Regional Integration*. Cambridge: Cambridge University Press, pp. 22–51.

Blandford, D. 2005. "Disciplines on Domestic Support in the Doha Round." International Agricultural Trade Research Consortium (IATRC), St. Paul, MN.

Bredahl, M., A. Schmitz, and J.S. Hillman. 1987. "Rent Seeking in International Trade: The Great Tomato War." *American Journal of Agricultural Economics* 69(1):1–10.

Burfisher, M.E., J. Dyck, B. Meade, L. Mitchell, J. Wainio, S. Zahniser, S. Arita, and J. Beckman. 2014. "Agriculture in the Trans-Pacific Partnership." United States Department of Agriculture, Economic Research Service, Washington, DC.

Burkitbayeva, S., and W.A. Kerr. 2015. "Accession of KRU to the WTO: The Effect of Tariff Reductions on KRU and International Wheat Markets." In A. Schmitz and W.H. Meyers, eds., *Transition to Agricultural Market Economies: The Future of Kazakhstan, Russia, and Ukraine*. Boston, MA: CABI, pp. 183–190.

Carter, C.A., and A. Schmitz. 1979. "Import Tariffs and Price Formation in the World Wheat Market." *American Journal of Agricultural Economics* 61(3):517–522.

———. 2007. "Estimating the Market Effect of a Food Scare: The Case of Genetically Modified StarLink Corn." *The Review of Economics and Statistics* 89(3):522–533.

Commerce.gov. 2017. *U.S. and Mexico Strike Deal on Sugar to Protect U.S. Growers and Refiners, Ensure Supply to Consumers* [Press release, June 6]. Washington, DC: United States Department of Commerce, Office of Public Affairs (Commerce.gov).

de Gorter, H., and D.R. Just. 2009. "The Economics of the Blend Mandate for Biofuels." *American Journal of Agricultural Economics* 91(3):738–750.

Devadoss, S. 2006. "Is There an End to U.S.-Canadian Softwood Lumber Disputes?" *Journal of Agricultural and Applied Economics* 38(1):137–153.

———. 2008. "An Evaluation of Canadian and U.S. Policies of Log and Lumber Markets." *Journal of Agricultural and Applied Economics* 40(1):171–184.

———. 2013. "Ad Valorem Tariff and Spatial Equilibrium Models." *Applied Economics* 45(23):3378–3386.

Dyck, J., and S. Arita. 2014. "Japan's Agri-Food Sector and the Trans-Pacific Partnership." United States Department of Agriculture, Economic Research Service, Washington, DC.

Dyck, J., and R.J. Johnson. 2013. "Japan Announces New Rules for Imports of U.S. Beef." United States Department of Agriculture, Economic Research Service, Washington, DC.

Erokhin, V. (editor). 2016. *Global Perspectives on Trade Integration and Economies in Transition*. Hershey, PA: IGI Global.

———. 2017. *Establishing Food Security and Alternatives to International Trade in Emerging Economies*. Hershey, PA: IGI Global.

Estrades, C. 2015. "The Role of Export Restrictions in Agricultural Trade. International Agricultural Trade Research Consortium (IATRC), St. Paul, MN.

Ferrier, P. 2014. "The Effects of Phytosanitary Regulations on U.S. Imports of Fresh Fruits and Vegetables." United States Department of Agriculture, Economic Research Service, Washington, DC.

Gibson, P., J. Wainio, D. Whitley, and M. Bohman. 2001. "Profiles of Tariffs in Global Agricultural Markets." United States Department of Agriculture, Economic Research Service, Washington, DC.

Götz, L., I. Djuric, and T. Glauben. 2015. "Wheat Export Restrictions in Kazakhstan, Russia, and Ukraine: Impact on Prices along the Wheat-to-Bread Supply Chain." In A. Schmitz and W.H. Meyers, eds., *Transition to Agricultural Market Economies: The Future of Kazakhstan, Russia and Ukraine.*" Wallingford: CABI, pp. 191–203.

Goychuk, K., and W.H. Meyers. 2013. "Black Sea and World Wheat Market Price Integration Analysis. *Canadian Journal of Agricultural Economics* 62(2):245–261.

Grant, J., E. Peterson, and R. Ramniceanu. 2015. "Assessing the Impact of SPS Regulations on U.S. Fresh Fruit and Vegetable Exports." *Journal of Agricultural and Resource Economics* 40(1):144–163.

Hanrahan, C.E. 2013. "Agricultural Export Programs: Background and Issues. Congressional Research Service 7–5700. No. R41202. CRS, Washington, DC.

Hendrix, C.S. 2017. *Agriculture in the NAFTA Renegotiation. Trade and Investment Policy Watch*. Washington, DC: Peterson Institute of International Economics.

Hillman, J.S., and A. Schmitz (editors). 1979. *International Trade and Agriculture, Theory, and Policy*. Boulder, CO: Westview Press.

Jales, M., T. Josling, A. Nassar, and A. Tutweiler. 2005. "Options for Agriculture: From Framework to Modalities in Market Access/Domestic Support/Export Competition." International Agricultural Trade Research Consortium (IATRC), St. Paul, MN.

Johnson, D.G. 1973. *Agriculture in Disarray*. New York: Palgrave Macmillan.

Johnston, C.M.T., and G.C. van Kooten. 2017. "Impact of Inefficient Quota Allocation Under the Canada-U.S. Softwood Lumber Dispute: A Calibrated Mixed Complementarity Approach." *Forest Policy and Economics* 74:71–80.

Josling, T., K. Anderson, A. Schmitz, and S. Tangermann. 2010. "Understanding International Trade in Agricultural Products: One Hundred Years of Contributions by Agricultural Economics." *American Journal of Agricultural Economics* 92(2):242–246 (Special Issue Commemorating the Centennial of the AAEA).

Josling, T., M. Honma, J. Lee, D. MacLaren, B. Miner, D. Sumner, S. Tangermann, and A. Valdes. 1994. "The Uruguay Round Agreement on Agriculture: An Evaluation." International Agricultural Trade Research Consortium (IATRC), St Paul, MN.

Josling, T., and T. Taylor. 2003. *Banana Wars: The Anatomy of a Trade Dispute*. Wallingford: CABI.

Just, D.R., and H. de Gorter. 2009. "The Welfare Economics of a Biofuel Tax Credit and the Interaction Effects with Price Contingent Farm Subsidies." *American Journal of Agricultural Economics* 91(2):477–488.

Karp, L., and J. Perloff. 2002. "A Synthesis of Agricultural Trade Economics." In B. Gardner and G. Rausser, eds., *Handbook of Agricultural Economics*. Amsterdam: North-Holland, pp. 1946–1998.

Kennedy, P.L., L. Brink, J.H. Dyck, and D. MacLaren. 2001. "Domestic Support: Issues and Options in the Agricultural Negotiations." International Agricultural Trade Research Consortium (IATRC), St. Paul, MN.

Kobuta, I. 2015. "Wheat Export Development in Ukraine." In A. Schmitz and W.H. Meyers, eds., *Transition to Agricultural Market Economies: The Future of Kazakhstan, Russia and Ukraine*. Wallingford: CABI, pp. 51–60.

Kobuta, I., O. Sikachyna, and V. Zhygadlo. 2012. *Wheat Export Economy in Ukraine*. Paris: France.

Leifert, W.M., and O.L. Leifert. 2015. "The Rise of the Former Soviet Union Region as a Major Grain Exporter." In A. Schmitz and W.H. Meyers, eds., *Transition to Agricultural Market Economies: The Future of Kazakhstan, Russia, and Ukraine*. Wallingford: CABI, pp. 27–38.

Lewis, K.E., and T.G. Schmitz. 2015. "Measuring the Impact of Mexican Domestic Subsidies on the Trade of Sugar between the U.S. and Mexico." *Journal of Agribusiness* 33(1):17–38.

Luckstead, J., S. Devadoss, and M. Dhamodharan. 2015. "Strategic Trade Analysis of U.S. and Chinese Apple Juice Market." *Journal of Agricultural and Applied Economics* 47(2):175–191.

Luckstead, J., S, Devadoss, and R. Mittelhammer. 2014. "Apple Export Competition Between the United States and China in the Association of Southeast Asian Nations." *Journal of Agricultural and Applied Economics* 46(4):635–647.

Marchildon, G.P. 1998. "Canadian – American Agricultural Trade Relations: A Brief History." *American Review of Canadian Studies* 28(3):233–252.

Meilke, K., J. Rude, M. Burfisher, and M. Bredahl. 2001. "Market Access: Issues and Options in the Agricultural Negotiations." International Agricultural Trade Research Consortium (IATRC), St. Paul, MN.

Miano, T.J. 2006. "Understanding and Applying International Infectious Disease Law: UN Regulations during an H5N1 Avian Flu Epidemic." *Chicago-Kent Journal of International and Comparative Law* 6:26–252.

Mundell, R.A. 1957. "International Trade and Factor Mobility." *The American Economic Review* 47(3):321–335.

National Grain and Feed Association and North American Export Grain Association (NGFA/NAEGA). 2011. "Joint Statement on Media Reports of Lawsuit Involving Syngenta's Agrisure Viptera Biotech Corn." MIR 162. NGFA/NAEGA, Arlington, VA.

Powell, S.J., and A. Schmitz. 2005. "The Cotton and Sugar Subsidies Decisions: WTO's Dispute Settlement System Rebalances the Agreement on Agriculture." *Drake Journal of Agricultural Law* 10(2):287–330.

Reimer, J., and M. Li. 2010. "Trade Costs and the Gains from Trade in Crop Agriculture." *American Journal of Agricultural Economics* 92(4):1024–1039.

Ridley, W., and S. Devadoss. 2012. "Analysis of the Brazil-USA Cotton Dispute." *Journal of International Trade Law and Policy* 11(2):14–162.

———. 2014. "U.S.-Brazil Cotton Dispute and the World Cotton Market." *The World Economy* 37(8):1081–1100.

Samuelson, P. 1952. "Spatial Price Equilibrium and Linear Programming." *The American Economic Review* 42(3):283–303.

Schmitz, A. 1968. *An Economic Analysis of the World Wheat Economy in 1980*. Madison, WI: University of Wisconsin Press.

Schmitz, A., R.S. Firch, and J.S. Hillman. 1981. "Agricultural Export Dumping: The Case of Mexican Winter Vegetables in the U.S. Market." *American Journal of Agricultural Economics* 61(4):645–654.

Schmitz, A., H. Furtan, and K. Baylis. 2002. *Agricultural Policy, Agribusiness, and Rent-Seeking Behaviour* (1st ed.). Toronto: University of Toronto Press.

Schmitz, A., D. Haynes, and T.G. Schmitz. 2016. "The Not So Simple Economics of Production Quota Buyouts." *Journal of Agricultural and Applied Economics* 48(2):119–147.

Schmitz, A., and P. Helmberger. 1970. "Factor Mobility and International Trade: The Case of Complementarity." *The American Economic Review* 60(4):761–767.

Schmitz, A., and P.L. Kennedy. 2015. "Food Security and the Role of Food Storage." In A. Schmitz, P.L. Kennedy, and T.G. Schmitz, eds., *Food Security in a Food Abundant World: An Individual Country Perspective*. Bingley: Emerald, pp. 1–18.

Schmitz, A., P.L. Kennedy, and T.G. Schmitz, editors. 2015. *Food Security in a Food Abundant World: An Individual Country Perspective*. Bingley: Emerald.

———. 2016. *Food Security in an Uncertain World: An International Perspective*. Bingley: Emerald.

———. 2017. *World Agricultural Resources and Food Security*. Bingley: Emerald.

Schmitz, A., C.B. Moss, and T.G. Schmitz. 2007. "Ethanol: No Free Lunch." *Journal of Agricultural and Food Industrial Organization* 5(2):1–28.

Schmitz, A., C.B. Moss, T.G. Schmitz, H.W. Furtan, and H.C. Schmitz. 2010. *Agricultural Policy, Agribusiness, and Rent-Seeking Behaviour* (2nd ed.). Toronto: University of Toronto Press.

Schmitz, A., C.B. Moss, T.G. Schmitz, and W.W. Koo. 2005. *International Agricultural Trade Disputes: Case Studies in North America*. Calgary: University of Calgary Press.

Schmitz, A., and W.H. Meyers (editors). 2015. *Transition to Agricultural Market Economies: The Future of Kazakhstan, Russia, and Ukraine*. Wallingford: CABI.

Schmitz, A., J.L. Seale Jr., and T.G. Schmitz. 2006. "The Optimal Processor Tariff Under the Byrd Amendment." *International Journal of Applied Economics* 3(2):9–20.

Schmitz, A., M. Zhu, and D. Zilberman. 2017. "The Trans-Pacific Partnership and Japan's Agricultural Trade." *Journal of Agricultural and Food Industrial Organization* 15(1):Article 20170001.

Schmitz, T.G., and K.E. Lewis. 2015. "Impact of NAFTA on U.S. and Mexican Sugar Markets." *Journal of Agricultural and Resource Economics* 40(3):387–404.

Schmitz, T.G., and A. Schmitz. 2012. "The Complexities of the Interface Between Agricultural Policy and Trade." *The Estey Centre Journal of International Law and Trade Policy* 13(1):14–25.

Schmitz, T.G., A. Schmitz, and C. Dumas. 1997. "Gains from Trade, Inefficiency of Government Programs, and the Net Economic Effects of Trading. *Journal of Political Economy* 105(3):637–647.

Schmitz, T.G., A. Schmitz, and C.B. Moss. 2005. "The Economic Impact of StarLink Corn." *Agribusiness* 21(3):391–407.

Schmitz, T.G., A. Schmitz, and J.L. Seale. 2003. "Brazil's Ethanol Program: The Case of Hidden Sugar Subsidies." *International Sugar Journal* 105(1254):254–265.

Schmitz, T.G., M. Zhu, and A. Schmitz. 2016a. "Agricultural Import Tariffs and Export Restrictions." In *Reference Module in Food Science*. Amsterdam: Elsevier.

———. 2016b. "Nontariff Distortions in Agricultural Trade." In *Reference Module in Food Science*. Amsterdam: Elsevier.

Sumner, D., and S. Tangermann. 2002. "International Trade Policy and Negotiations." In B. Gardner and G. Rausser, eds., *Handbook of Agricultural Economics*. Amsterdam: North-Holland, pp. 1999–2055.

Takayama, T., and G. Judge. 1964. "Spatial Equilibrium and Quadratic Programming." *Journal of Farm Economics* 46(10):67–93.

———. 1971. *Spatial and Temporal Price and Allocation Models*. Amsterdam: North-Holland.

United States Department of Agriculture, Agricultural Marketing Service [USDA-AMS]. 2009. "Country of Origin Labeling (COOL)." United States Department of Agriculture, Agricultural Marketing Service, Washington, DC.

United States Department of Agriculture, Economic Research Service [USDA-ERS]. 1986. "Embargoes, Surplus Disposal, and U.S. Agriculture: A Summary." AIB 503. USDA-ERS, Washington, DC.

———. 2015. "NAFTA at 20: North America's Free-Trade Area and Its Impact on Agriculture." WRS-15-01. USDA-ERS, Washington, DC.

United States Department of Agriculture, Foreign Agricultural Service [USDA-FAS]. 2015a. "TPP Benefits for Specific Agricultural Commodities and Products." USDA-FAS, Washington, DC.

———. 2015b. "Agriculture-Related Provisions of the TPP." USDA-FAS, Washington, DC.

———. 2016. "Free Trade Agreements and U.S. Agriculture." USDA-FAS, Washington, DC.

United States International Trade Commission [USITC]. 1994. "Fresh Garlic from China." USITC, Washington, DC.

van Kooten, G.C. 2002. "Economic Analysis of the Canada-United States Softwood Lumber Dispute: Playing the Quota Game." *Forest Science* 48:712–721.

———. 2014. "The Benefits of Impeding Free Trade: Revisiting British Columbia's Restrictions on Log Exports." *Journal of Forest Economics* 20(4):333–347.

Welton, G. 2011. *The Impact of Russia's 2010 Grain Export Ban*. Oxford: Oxfam.

Wohlgenant, M., and A. Schmitz. 2005. "Canada – U.S. Beef Dumping and Countervailing Disputes." In A. Schmitz, C. Moss, T. Schmitz, and W. Koo, eds., *International Agricultural Trade Disputes: Case Studies in North America*. Calgary: University of Calgary Press, pp. 185–206.

World Trade Organization [WTO]. 1998. *Understanding the WTO Agreement on Sanitary and Phytosanitary Measures*. Paris: WTO.

———. 2003. *Subsidies and Countervailing Measures*. Paris: WTO.

Xu, K. 2015. "Why Are Agricultural Goods Not Traded More Intensively: High Trade Costs or Low Productivity Variation?" *World Economy* 38(11):1722–1743.

Yamazaki, F., and M.S. Paggi. 2005. "Fresh Garlic from the People's Republic of China and U.S. Trade Remedy Law." In A. Schmitz, C. Moss, T Schmitz, and W. Koo, eds., *International Agricultural Trade Disputes: Case Studies in North America*. Calgary: University of Calgary Press, pp. 245–268.

Young, L.M. 2005. "Export Competition Disciplines in the Doha Round. 2005." International Agricultural Trade Research Consortium (IATRC), St. Paul, MN.

Zusman, P., A. Melamed, and I. Katzir. 1969. *Possible Trade and Welfare Effects of EEC Tariff and "Reference Price" Policy on the European-Mediterranean Market for Winter Oranges*. Berkeley, CA: California Agricultural Experiment Station.

20
Use of subsidies and taxes and the reform of agricultural policy

G. Cornelis Van Kooten, David Orden, and Andrew Schmitz

1 Introduction

Governments have intervened in agriculture since the earliest of times. In ancient Egypt, a Hebrew slave named Joseph convinced the reigning Pharaoh to purchase grain at low prices during a period of good harvests and sell it during drought conditions. The good harvests lasted seven years, but when famine struck, the Egyptian treasury profited greatly; to save themselves from starvation, farmers sold everything they had to the Pharaoh, including their land, and subsequently they worked as tenants paying one-fifth of their harvest to the king (Gen 47: 13–26). In ancient Egypt, the success of such a buffer fund scheme was measured in terms of its benefit to the authority. Today, in contrast, buffer fund and other support schemes operate to the benefit of the agricultural producers who, in a democracy, are able to lobby for agricultural programs that create rents, which they then capture.

In this chapter, the various agricultural policies by which governments have supported agriculture are reviewed, in part, using tools (displayed graphically) of applied welfare economics. We discuss U.S., EU, and Canadian agricultural policies. Furthermore, since the policies and trade of major emerging economies have come to play important roles in world markets (Diaz-Bonilla 2015), we also briefly examine China's agricultural policies. Across these countries, government support policies have generally led to overproduction for major commodities, such as wheat and corn, but in the case of commodity marketing organizations and measures to control supply, government intervention has reduced output below free market levels, leading to higher prices and monopoly rents.

More broadly, governments worldwide have been willing to support farm prices and incomes for a variety of reasons, including protecting farmers from market variability. These include, *inter alia*, a desire for food security (Schmitz, Kennedy, and Schmitz 2017) and safety, to earn foreign exchange, and to protect an agrarian lifestyle and prevent the demise of rural communities. An important factor has been rent seeking on the part of producers (Rausser 1982, 1992; Rausser and Foster 1990; Anderson, Rausser, and Swinnen 2013). Farmers were better able to organize and lobby as their numbers declined, while consumers and taxpayers had little incentive to oppose the agricultural lobby as food expenditures declined relative to disposable income and concerns about food safety translated into support for agriculture. In contrast, developing

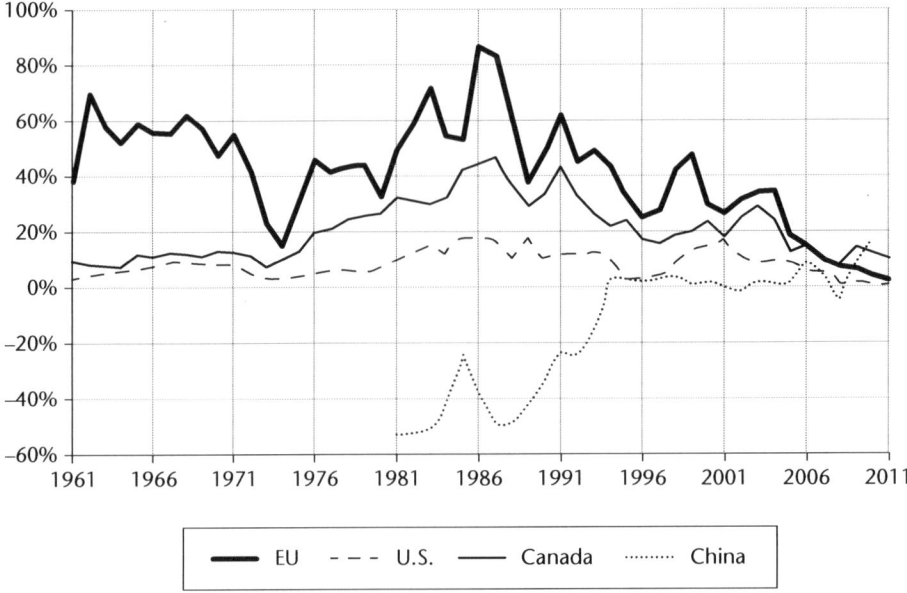

Figure 20.1 Nominal rates of assistance to agriculture, EU, U.S., Canada, and China, 1961–2011
Source: Anderson and Nelgen (2013).

countries have often discriminated against agriculture, particularly against exports, but their support levels have also tended to increase as incomes rose (Anderson 2009).

Agricultural support can be measured by the nominal rate of assistance (NRA), which is the percentage by which the domestic producer price is above (or below if negative) the border price of a similar product, net of transportation costs and trade margins. The NRA is an estimate of direct government policy intervention. The European Union has assisted its agricultural sector to a much greater extent than either the United States or Canada since 1960, at times raising prices by more than 80 percent above the border price (Figure 20.1).[1] Canada has also protected its agricultural sector, primarily via supply-restrictions in dairy, eggs, and poultry, while China has taxed agriculture and only recently begun to provide assistance (Figure 20.1).

Generally, the data show that policies have moved in the direction of reducing economic inefficiencies, but the levels of support remain substantial. As argued in the next sections, the direction of change in government agricultural programs, with backsliding on occasion, has been toward replacing crop-specific subsidies with ones that provide income support and target income volatility but are less distorting to production and trade – programs that decouple support payments from output.

2 Menu of agricultural support policies

An objective of government programs is to provide income support and reduce price instability. In principle, governments can seek to stabilize farm prices by purchasing farm commodities for later sale without raising the average price level. We illustrate this using a simplified buffer fund price stabilization model. We then discuss additional instruments used by governments to prop up and stabilize farm incomes, illustrating in each case the applied welfare implications of policies.

2.1 Stock-holding buffer fund stabilization

With buffer fund stabilization, the government purchases a commodity when prices are low, stores the commodity for a time, and then sells it when prices are high (Newberry and Stiglitz 1981; van Kooten and Schmitz 1985; van Kooten, Schmitz, and Furtan 1988). This simplified model, with later extensions, gained popularity because of the large amounts of grain and dairy products often stockpiled as a result of U.S. and EU policies – would such stockpiles arise in a pure buffer scheme? In Figure 20.2, assume stable demand but a stochastic supply function that varies symmetrically between S_0 and S_1, each with probability ½. With government intervention via a buffer fund program, producers no longer face prices that fluctuate between a low P_0 and high P_1, but they plan on a stable price of P_e. To stabilize price at P_e, the authority buys $(q_0 - q_e)$ when S_0 occurs and sells $(q_e - q_1) = (q_0 - q_e)$ when S_1 occurs, with government commodity purchase costs offset by government sales revenue. With stabilization, when S_0 occurs, consumers lose $\alpha + \beta + \delta + \gamma + \varphi$, while producers gain $\alpha + \beta + \delta + \gamma + \varphi + \sigma$, with a net gain to society of σ. When S_1 occurs, consumers gain $a + b + c$ with stabilization, while producers lose area a, with a net gain to society of $b + c$. The stock-holding buffer fund scheme leads to an average annual net gain of ½ $(\sigma + b + c)$ minus administrative and storage costs.

While producers are better off with the stabilization fund, their incomes are more variable than they would be in the absence of buffer fund stabilization; consumers lose with a buffer fund program. The framework in Figure 20.2 (the foundation of which is discussed in detail in Currie, Murphy, and Schmitz 1971) has been expanded to internationally traded commodities. As Hueth and Schmitz (1972) show, in an international trading environment, some countries can gain from instability while others lose. In practice, however, pure buffer stock price stabilization programs are often pushed into longer-term price support programs because periods of high prices are less frequent than theory suggests and because the authority is reluctant to drive down prices to the extent required to average them over the long term. The result is welfare losses to society rather than the gains anticipated in theory from pure buffer stabilization as shown in Figure 20.2.

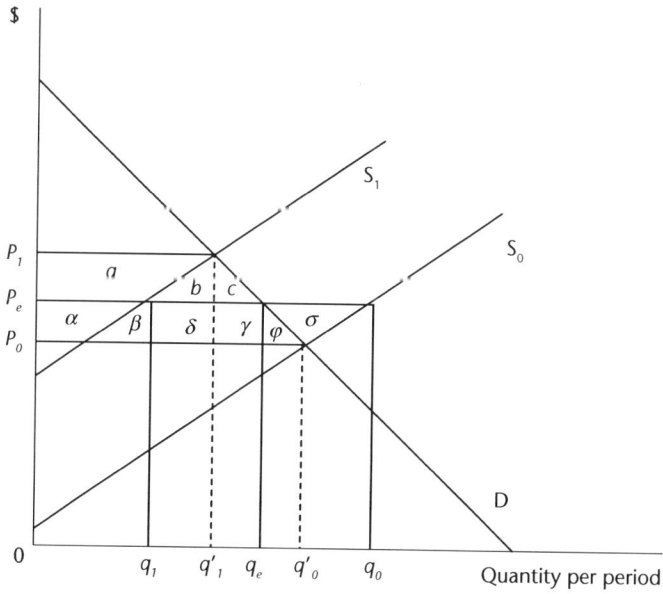

Figure 20.2 Buffer fund stabilization under price instability

2.2 Price and income support

Agricultural policy, especially in the United States and the European Union, has centered on intervention prices. Consider Figure 20.3, where intervention price (P_S) is set above the market-clearing price (P^*). This framework adds to the price stabilization model above (discussed in more detail in Just, Hueth, and Schmitz 2005) and is needed because it deals with policies in addition to buffer funds. Three generic options have characterized the price intervention policies. First, P_S is imposed on both producers and consumers. Supply q^s exceeds the market-clearing level q^*, while demand q^d is below the free market level (for now ignore the vertical supply curve S^R that is identified by the dark line). To maintain this market intervention, government must store the excess production ($q^s - q^d$) and block any imports the price support might otherwise attract. To keep prices above the market-clearing level, stock accumulation continues year-over-year (in contrast to the buffer fund price stabilization program in Figure 20.2).

Historically, to deal with growing commodity stocks, governments of exporting countries with price support programs resorted to subsidizing exports. These subsidies were constrained by commitments under the World Trade Organization's (WTO) Agreement on Agriculture that came into effect in 1995 and were eliminated as an allowable policy by agreement in 2015. Governments have also resorted to supply controls to limit production where intervention prices exceed market-clearing levels – in general with acreage or other restrictions that shift the supply curve leftward (not shown in Figure 20.3) or more draconically with fixed domestic production quotas (discussed later). For an importing country, an intervention price P_S would necessarily be set above the world market price, so import restrictions are required; indeed, the extent to which imports are regulated determines the eventual price. If no imports are allowed, the effective price is the domestic market-clearing (autarky) price; if some imports are permitted, P_S is below the autarkic price (the EU situation for some products in the 1960s and 1970s). Domestic producers gain while domestic consumers and foreign suppliers lose from the intervention.

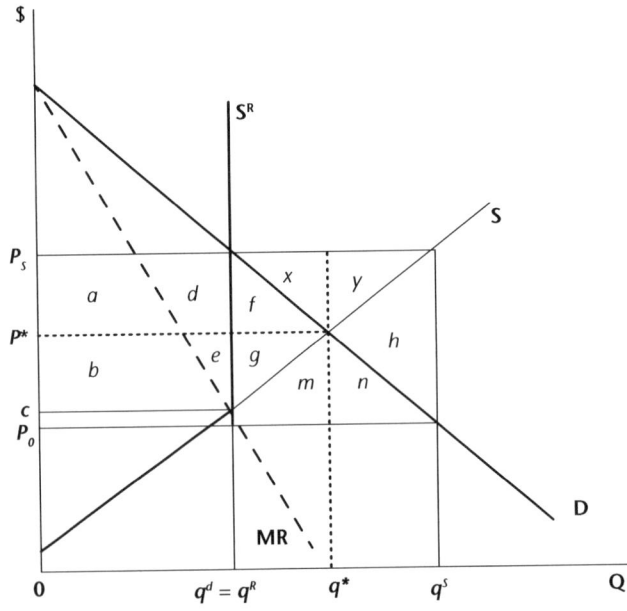

Figure 20.3 Price or income support

Governments of importing countries do not have to accumulate stocks, subsidize exports, or restrict supply, as is the case with an exporting country.

Consider a second option where producers are provided support through a "target" price P_S. Farmers still produce q^S, but now the market clears at P_0. Instead of having to acquire stocks, subsidize exports, or impose supply-shifting acreage reductions to balance supply and demand, the government makes a "deficiency payment" to farmers for the difference $(P_S - P_0)$ on each unit of output. The total cost to government is $(P_S - P_0) \times q^S$ and the net welfare loss is area h. To the extent that the lower consumer price stimulates exports, supply in the domestic market will be reduced and the price will rise above P_0. Import restrictions are required.

Finally, to reduce the fiscal costs and economic distortions associated with the intervention price, P_S can be applied to a fixed production level and not all output. For example, the level on which payments are made might be q^* (it could be lower). With a fixed output q^*, on which deficiency payments are made, producers and consumers are incentivized by the market-clearing price P^*. With this "decoupling" of payments from total production, the government makes total deficiency payments $(P_S - P^*) \times q^* = $ area $(a + d + f + x)$ without inducing economic inefficiencies. Taxpayers are worse off but producers are better off by this amount compared to no intervention. This is essentially a price-contingent income transfer that varies countercyclically as supply and demand shift around and P^* changes from year to year. Payments could also be made at a fixed level regardless of the market price, or regardless of whether the crop is produced. Support policy in the United States and the European Union has evolved over time in the direction of fixed payments based on some past (invariant) output level and set price, which may or may not vary over time. Middle-income countries, such as China, are now grappling with the same policy issues.

2.3 Marketing orders and supply quotas

Certain policy interventions involve marketing orders that facilitate monopoly. This was done for certain commodities in the Agricultural Marketing Agreement Act (1937) in the United States, via the Common Marketing Orders (1958) in the European Union, and with the Farm Products Agency Act (1972) in Canada. The economic implications of marketing orders that control or restrict supply (e.g., via quotas) are also demonstrated in Figure 20.3. A commodity marketing board acts as a monopolist and optimally restricts supply to q^R, where the industry marginal cost function (S) intersects the marginal revenue curve (MR).[2] The supply curve is kinked as indicated by the dark curve S^R – producers receive P_S and are allocated quota to prevent output from exceeding q^R.

From Figure 20.3, due to a restricted supply, consumers lose. Because the marginal cost to the producer is c, the wedge between price and marginal cost results in a policy induced scarcity (quota) rent (van Kooten and Taylor 1989). The total annual rent is $R_A = q^R \times (P_S - c)$ (area $a + d + b + e$); this is a gain to producers, but they experience a deadweight producer surplus loss of g. The value of the quota in perpetuity is $V = R_A / r$, where r is the rate used to discount future quota income. The rate r used to discount the annual rents will be high, although it will vary across producers, since the quota is not likely to continue into perpetuity as there is always a risk that a quota regime policy will be terminated. Not to be forgotten, consumers lose $a + d + f$, of which $a + d$ is transferred to producers and f is a deadweight loss.

Quota schemes have been used in sectors where countries are net importers or net exporters. For example, the European Union has employed a dairy quota to reduce the costs of export subsidies, while Canadian dairy supply management increases the domestic consumer price to the benefit of dairy producers (van Kooten 2017). In the latter situation, the quota regime imposes no cost on the public treasury, except for expenses related to its implementation and governance, but it requires import restrictions. In Figure 20.3, since the domestic price P_s exceeds the world price, trading partners will lobby to open up the supply-restricted market, which has resulted in at least minimal market access through the WTO's tariff-rate quota (TRQ) commitments, whereby a low tariff rate is applied to the quota-amount of imports and a much higher rate to imports above this amount.[3] In the case of the net exporter, the quota reduces subsidized sales of dairy products, thereby raising the world price toward its free trade level (leaving the domestic price unchanged) and reducing government expenditures; however, as noted earlier, export subsidies were no longer permitted after 2015. Any restriction on agricultural output essentially transfers income from consumers to producers – from domestic consumers in the case of the importer and from foreign as well as domestic consumers in the case of the net exporter.

2.4 Producer compensation

Over time, certain quota programs have been terminated, such as tobacco in the United States and Canada, and peanuts in the United States. To the extent that producers have a right to or an investment in quota, governments may need to compensate producers to get them to acquiesce to changes in a supply-restricted market. This might be required if the authority agrees to modify or eliminate a quota system as part of international trade negotiations. The authority might need to buy back quota, which requires a determination of a fair buy-back price.

There are several reasons why quota holders should not be compensated for the full amount of the quota rent. First, producers do not lose the entire quota rent when a quota is removed; they retain a large part of the quota rent area as producer surplus plus the producer component of the deadweight loss triangle (in the context of Figure 20.3, the removal of the quota would lead producers to lose $a + d$ but gain g). Second, many farmers have already recouped their investment in quota, although this depends on when they joined the marketing regime. Even dairy producers whose quota investment has already been paid off will engage in rent seeking to protect the quota windfall, although some will have squandered some of the windfall by failing to take advantage of new technology or markets. Third, while recent entrants into the sector are disadvantaged by less-than-full compensation, these producers are likely to be the most efficient producers, who might benefit if they could expand their operations without needing to purchase quota. In the absence of a quota regime, they might be able to access export markets.

Lastly, quota rents and any compensatory deficiency payments accrue over time. The rate used to discount the periodic rents determines the value of quota. If a market for quota exists, prices for quota can be used as a basis for setting compensation; they can also be used to estimate the discount rate purchasers use to value quota. Because there is uncertainty regarding the survival of the quota regime, future prices, and the size of the rent to which the quota buyer is entitled, buyers of quota generally employ a short payback period.

The issue of compensation has been addressed in the U.S. and Canadian tobacco buyout cases, and the termination of the U.S. peanut quota program. Orden (2007) shows that compensation was generous for the U.S. tobacco and peanut cases: the discounted value of the tobacco buyout was nearly double the prevailing market price of quota rights and for peanuts a costly new support program was initiated along with the buyout of quotas. Schmitz, Haynes, and Schmitz (2016) also empirically estimate that in the U.S. tobacco buyout case producers received compensation payments that far exceeded quota values. Like Orden's analysis, Schmitz and Schmitz (2010) find a similar outcome for the U.S. peanut quota buyout. In the Canadian tobacco buyout, Schmitz, Haynes, and Schmitz (2015) show that producers were compensated far beyond the value of tobacco quotas. The method employed by the European Union to dismantle milk quota differed from these cases, with the buyout quickly rolled into a single farm payment that was completely decoupled from production (van Kooten 2017).

3 Agricultural programs: an integrated analysis

The above discussion deals with specific programs, but little attention is paid to how several programs in place at any one time interact with each other. Multimarket and general equilibrium welfare economics deals with programs in the context of multiple commodities and inputs or the whole economy, extending the empirical analysis that deals with only a single commodity program. Even in a single program case, trade has to be taken into account because associated large rectangular losses far exceed closed model traditional deadweight loss triangular estimates. This can be exacerbated when multiple products and multiple programs are taken into consideration. Furthermore, agriculture is affected not only by programs such as price supports, quotas, and crop insurance, but also by input subsidies on water and fertilizer, for example, and more recently by energy programs. A single-commodity model combining price supports with an input subsidy in the context of international trade is given in Appendix B. As this model shows, the welfare costs of farm programs are much larger than the traditional welfare triangles discussed above.

The analysis of farm programs is complex, partly because of additional policy interventions that interact with the farm support programs, such as subsidized ethanol production that is made from corn in the United States. Programs such as U.S. ethanol subsidies are outside the purview of the USDA, even though they affect the agricultural sector. Significant work has been done on the impact of the U.S. ethanol program. One of the early models developed by Schmitz, Moss, and Schmitz (2007) is taken up in Appendix C. Generally, the benefits to U.S. grain and oilseed producers from the ethanol program appear to be positive, but the overall societal impact is generally negative.

4 Dismantling of market intervention

4.1 Reform in the United States

Consider the history of agricultural policy in the United States because it has been a key exporter and leader in implementing agricultural support policies.[4] Modern-era government intervention in agriculture began in the United States in 1862 with the Homestead Act, which allocated federal lands to private landowners who were willing to farm the land. The Morrill Act (1862 and 1890) created the land grant colleges. The Hatch Act (1887) created the agricultural experiment stations, while the Smith–Lever Act (1914) created the cooperative extension service. Farm support programs were implemented in the United States when overproduction of agricultural commodities during the Great Depression caused net farm income to fall from US$5.2 billion to US$1.4 billion (between 1929 and 1932), thereby affecting one-quarter of the U.S. labor force employed in agriculture. In response, as part of President Roosevelt's New Deal, the Agricultural Adjustment Act was passed in 1933, followed by the Agricultural Marketing Agreement Act (1937) and other permanent farm support legislation in 1938, 1948, and 1949. The 1933 Act created the Commodity Credit Corporation (CCC) and introduced non-recourse loans for cotton and corn, whereby the producer could take out a loan from the CCC when the harvest was realized. If the farmer defaulted, the CCC would take the commodity and store it, hopefully selling when prices rebounded. In conjunction with these market interventions, programs were initiated to focus on the environment – initially soil conservation, and later loss of wetlands and other environmental externalities.

The Depression-era legislation also created monopoly behavior that was prohibited under the Sherman Anti-Trust Act (1890) in other sectors of the economy. Price discrimination and supply restrictions using acreage allotments and marketing quotas were employed and taxes were levied on processors to fund expenditures; this was done to reduce production of basic commodities in an effort to increase farm-gate prices, and thereby farm incomes. In addition to the marketing arrangements, farmers were paid not to produce to further boost prices. These policies persisted into the 1960s and continue to this day; although they have been largely transformed by substantial reforms that have eased market interventions and reduced efficiency losses.

The broad objectives of initial U.S. farm legislation were to reduce rural poverty, promote soil conservation, provide crop yield insurance against local natural disasters, and make credit available to agricultural producers – objectives arising from Depression-era shocks to the sector. The main forms of support – loan rates above market-clearing price levels and imposition of supply controls – were redistributive through substantial intervention into commodity markets.[5] The problems associated with these programs became apparent once the war-related high commodity prices of the 1940s and early 1950s subsided, and as technological innovation and non-farm economic growth induced a massive labor migration out of agriculture – with fewer farmers better able to organize to lobby government for favorable income transfers.

With loan rates holding prices too high relative to market demand, CCC stocks increased throughout the late 1950s and 1960s, and then again in the late 1970s and early 1980s. To manage these situations, both long-term idling of cropland, rationalized on the basis of conservation, and short-term land idling, required for support payment eligibility, were used to address overproduction. Farmers were eventually mandated to meet certain conservation requirements to qualify for support payments, known as conservation cross-compliance, and additional conservation and environmental programs emerged. When prices and farm incomes were high, as during the 1970s and mid-1990s, overproduction and the accumulation of CCC stocks were not a problem and market outcomes tended to reign.

In the mid-1980s, a farm financial crisis arose for the United States as exports, prices, and farm incomes fell precipitously. With high loan rates, stocks accumulated and emergency short-term land set-asides were used to idle a record 78 million acres in 1983. There was a sharp cut in nominal loan rates (a move away from the first policy option described in Figure 20.3) while maintaining levels of the higher target prices that triggered deficiency payments (the second option described in Figure 20.3). Also, conservation-based, long-term land retirement was put in place in a new Conservation Reserve Program (CRP). The CRP enrolled 36 million acres by the early 1990s. The lowering of loan rates but not target prices pushed up deficiency payments, making U.S. exports more competitive, while farmers were paid to put land into the CRP, which allowed unpaid annual set-asides to decline. Payments to farmers rose to more than one-third of net farm income during the mid-1980s.

The 1985 Farm Bill also took a step toward reform by fixing the base acres and yields that determined deficiency payments, thereby reducing incentives to produce more (a move toward the third policy option described in Figure 20.3). Then to achieve budget savings, payments were limited to 85 percent of base acres in the early 1990s, with farmers given flexibility to plant various crops on the remaining 15 percent instead of being required to plant the base crop. In addition, rather than delivering crops to the CCC to repay their loans made at the loan rate, by 1990 producers could sell the crop at the lower market price and receive a "loan deficiency payment" for the difference. Thus, the loan rate no longer set a floor under the market price.

With growing competition in world markets as a result of the European Union's Common Agricultural Policy (CAP), the 1985 U.S. Farm Bill also created an Export Enhancement

Program designed to match EU export subsidies. International negotiations to reduce agricultural market distortions gained a momentum they had not had in the past. The result was the WTO Agreement on Agriculture negotiated between 1986 and 1994. Under this agreement, member countries committed to disciplines on market access (lower tariffs and minimum access through TRQs), export competition (primarily limits on export subsidies), and domestic support. The complex domestic support provisions set nominal limits on certain support that distorted production or trade as determined by WTO rules (so-called amber box support), while various other programs were exempted because they were deemed non-distorting ('green box' support). Finally, a 'blue box' was included to accommodate existing U.S. and EU support that was provided on fixed area and yield or livestock numbers, or on 85 percent or less of the base level of production and tied to programs that limited production.

In the 1996 Farm Bill, with the new WTO agreement in place, several additional reforms were implemented. Acreage set-aside authority was eliminated (while continuing the CRP and other conservation-based programs). Instead of deficiency payments based on target prices, farmers received fixed annual payments of around US$5 billion based on past production but not dependent on annual price levels. With payments no longer tied to prices, planting flexibility was extended to all base acres, with an exception that fresh fruits and vegetables, which are not supported with price or income interventions, could not be grown while retaining payment eligibility. Reform advocates argued that these new policies reflected a modern agriculture no longer in need of support programs.

The 1996 U.S. reforms were opportunistic from the farm sector's perspective, however. The move to fixed payments allowed farmers to capture support that otherwise would have disappeared when market prices temporarily rose above target price levels. Nor did the new WTO Agreement on Agriculture play much of a role. The U.S. payments simply shifted from one exempted category to another. The deficiency payments fell into the blue box until set-aside authority was ended; once this occurred, the replacement fixed payments fell into the green box.

When farm prices dropped again in the late 1990s, farm incomes fell and loan deficiency payments soared. The fixed payments were expanded and reauthorized. The 2002 Farm Bill also restored target prices and price-dependent deficiency payments on fixed base acreage and yields, with some options for base acreage and yield updating. However, the 1996 reforms that eliminated annual land set-asides and allowed planting flexibility on base acres were retained. Subsequently, the 2008 Farm Bill continued the support policies enacted in 2002, although it introduced an innovation: an optional revenue guarantee program. Instead of receiving deficiency payments, farmers would, under the Average Crop Revenue Election (ACRE), receive a payment on base acres planted to a program crop if revenue for that crop fell more than 10 percent below a moving average based on past national prices and state-level yields. While ACRE was a program available to farmers without paying an insurance premium, its introduction in 2008 reflected growing interest in crop insurance, particularly revenue insurance.

Modern crop insurance was introduced in 1980 with the Federal Crop Insurance Act, which mandated delivery by the private sector, thereby uniting financial intermediaries and the agricultural lobby. Under this Act, the federal government subsidized premiums at a rate of 30 percent and covered the administrative and operating (A&O) costs as well as the underwriting risks. The Act also mandated extension of insurance to more crops and areas than under previous insurance programs. The Federal Crop Insurance Reform Act of 1994 marked a further development as it enabled crop revenue insurance, as opposed to only yield insurance, while increasing the subsidy rate to 40 percent, although lowering A&O payments made directly to insurance companies and introducing a premium loading to cover long-term underwriting costs. Finally, the Agricultural Risk Protection Act (2000) raised the premium rate even further (so it averaged over 62 percent), increased A&O payments, and, most importantly, extended the premium subsidy to the Harvest Price Option introduced in the 2008 Farm Bill that permitted farmers to use either the price at planting or at harvest in risk programs.

Thus, by the early 2000s, under lobbying by both farmers and private insurers, whose interests had been aligned in the 1980 crop insurance legislation, crop insurance premiums had risen significantly and there was a move to near total reliance on revenue insurance (85 percent of insured acres by 2015) as opposed to yield insurance. Still, net indemnity payments for crop insurance had remained below US$2 billion annually until they doubled to more than US$4 billion in 2008. ACRE provided a supplement to the growing reliance on crop insurance, essentially providing support for revenue losses greater than 10 percent but less severe than those covered by insurance, which was generally losses greater than 25 percent.

Stronger market conditions after 2007 reduced deficiency payments and made crop revenue insurance the main pillar of U.S. farm support, with annual premium subsidies around US$7 billion. The fixed payments introduced in 1996 proved politically unpopular when farm prices and incomes were high. In the 2014 Farm Bill, this relatively non-distorting support program, once envisioned to herald an end to farm support, was ended instead. Target prices (now termed reference prices) were set at higher nominal levels, while a revenue guarantee program became more deeply entrenched as an option. Farmers again were given the choice of a deficiency payments program (Price Loss Coverage, or PLC) or a new revenue program (Agriculture Risk Coverage, or ARC). Enrollment in ARC was over 90 percent for corn and soybeans, largely because the reference revenue was a moving average that reflected the high prices during 2007–2013. Wheat enrollment in ARC was about half, while producers of several other crops, particularly rice and peanuts, chose PLC. Both programs made substantial payments as prices (and revenue) fell after the 2014 Farm Bill was enacted. For cotton, an earlier WTO ruling against the U.S. support programs in a case brought by Brazil resulted in replacement of the traditional support program by heavily subsidized insurance. Cotton farmers gained because they could receive support payments if they grew other crops on their old cotton base acres, and subsequent to the 2014 Farm Bill they lobbied for new support, mainly to categorize cotton as an oilseed crop.

4.2 Reform in the European Union

European price and income support for agriculture followed a similar trajectory to the United States, albeit several decades later.[6] Because many citizens faced possible starvation during World War II and much of the labor force was still employed in agriculture in the years following the war, there was widespread support within Europe for a Common Agricultural Policy (CAP). The Agricultural Guidance and Guarantee Fund (EAGF) was created at the Conference of Stresa (1958). A subsequent Council of Ministers in 1962 laid the groundwork for price support policy, border taxes and subsidies, production quotas, direct income support, and Common Market Organizations (CMOs). By 1970, payments to agriculture from the EAGF accounted for nearly 90 percent of the total budget of the European Community (now European Union, or EU). The proportion of the budget accounted for by agriculture has since then slowly declined so that in 2016 it amounted to only about 40 percent.

In Europe, as early as 1968, the EU Agricultural Commissioner, Sicco Mansholt, recognized that the CAP would lead to overproduction while doing little to enhance farm incomes. His plan to reform EU agriculture promoted the (1) consolidation of small farms to increase farm size to take advantage of economies of scale, (2) removal of more than two million hectares from crop production, and (3) reduced payments to smaller producers. These proposals were controversial and opposed by small farm holders.

Because of oversupply and growing support costs, the European Union began to implement, in the 1980s, Mansholt's ideas.[7] To reduce mounting stocks of butter and skim milk powder and export subsidies on these products, a milk production quota was adopted in 1984. At the same time, the increase in CAP spending was limited to the growth rate of GNP, and then to 74 percent of GNP growth in 1988. Subsequently, beginning in the 1990s, agricultural reforms came about for reasons that had as much to do with the evolution of the European Union – the politics of expansion and greater integration – as they did with agriculture per se. More specifically, agricultural reforms were driven by four factors: (1) the high and increasing costs of the CAP at a time when politicians wished to allocate more of the limited EU budget to other programs; (2) pressure for reform emanating from WTO negotiations, particularly for a reduction of export subsidies associated with high EU price supports; (3) the integration of new members as the European Union expanded from the EU-15 prior to 2004 to the EU-28 – more specifically, the implication for the CAP of adding several nations with large, underdeveloped agricultural sectors; and (4) increasing environmental concerns.

Over a period of 25 years, the agricultural reforms led to a much more market-oriented system, although one that retained high levels of farm income support. The European Union has not followed the path toward subsidized crop insurance that has become so pervasive in U.S. policy, in part because of the high level of income support. Since the early 1990s, priorities changed somewhat from market and income support to rural development.

The 1992 MacSharry Reform of the price support programs was fundamental: it essentially lowered guaranteed prices toward world prices and fully compensated farmers with deficiency payments made on fixed acreage and livestock numbers. This reform began to move the European Union away from its regime of heavy market intervention and allowed completion of the WTO negotiations, with the new payments falling in a blue box negotiated into the Agreement on Agriculture. Additional measures agreed to in 1999 (known as Agenda 2000) continued this reform movement, with a further lowering of support prices and partial compensation from additional blue box payments. The structure of CAP was also revised in Agenda 2000 into two pillars: "pillar 1" constituted the price and income support aspects, while "pillar 2" consolidated rural development and environmental programs, with the reallocation of program payments for farm support to pillar 2 termed "modulation."

The Fischler Mid-Term Review (2003) required that agricultural producers had to comply with food safety, animal welfare, and environmental standards to remain eligible for payments. The structure of support was once again altered to compensate farmers for lower support prices and to facilitate the entry of new members. Rather than continuing payments linked to production or prices, the European Union introduced a single payment scheme, also known as a single farm payment (SFP). Like in the United States in 1996, this represented a shift from payments linked to production that fell into the blue box to decoupled payments falling into the green box. Countries were able to choose among several approaches for determining the reference decoupled payment: (1) historic entitlements depending on farm-specific reference margins; (2) a regional basis for establishing a reference margin; or (3) a hybrid approach that combined the historic and regional approaches. The majority of countries opted for the historical approach, with some countries retaining elements of support that were not entirely decoupled from production. Nonetheless, a basic further shift to decoupling occurred; reforms were then continued, with agreements for sugar, fruits, vegetables, wine, and dairy between 2006 and 2010.

Thirteen new members entered the European Union between 2004 and 2013, and several had very large agricultural sectors. To prevent a dramatic increase in farm expenditures, eligibility for full payment was phased in over a period of ten years. Furthermore, the single farm payment was based on average historical commodity-based yields for the period 2000–2002 for the

EU-15, but for new entrants to the European Union, the equivalent period was 1995–1998 – a period characterized by historically low output following the demise of communism. Given differences between countries in the degree to which collectivization of farms had occurred in the countries of eastern and central Europe, the eligible SFP varied from €300 for small farms in Poland to €40,000 for large farms in Hungary and the Czech Republic. Meanwhile, agricultural producers in the EU-15 benefit from the CAP to a much greater extent than new entrants (with average payments of €7,995 per EU-15 farm and €3,653 per EU-27 farm). Although taxpayers in the EU-15 contribute more to farm support, the disparity in treatment between the latest entrants and the EU-15 could not continue because it would destabilize efforts at further political integration. Therefore, the CAP reforms of June 2013 attempted to address the convergence of farm payments between the EU-15 and other countries.

The 2013 reforms introduced a Basic Payment Scheme (BPS) that would eventually provide the same level of support to every hectare of agricultural land within a region. To prevent locking in cropping patterns, the direct payment would be independent of both the crop choice and the type of farm. In addition, producers could receive compensation for providing public goods in the form of environmentally friendly farming practices – a so-called greening component added to the basic payment if farmers were in compliance.

The dairy quota that the European Union imposed on countries in 1984 was slowly phased out as a result of the Mid-Term (Fischler) review. Initially, producers received a milk premium (deficiency payment) on the basis of their historic production levels. This was converted into an SFP and, subsequently, in 2015, into a direct payment under BPS that no longer depended on the amount of milk produced or even whether the farmer continued in the sector.

Under the reforms, as of 2015, the mandatory direct payments are the basic payment, the green payment, and payments for young farmers. Countries can determine how these payments are implemented, how to identify eligible agricultural practices, and whether to use more restrictive criteria to identify young farmer beneficiaries. The voluntary components left to the individual states are the redistributive payment, payments for areas with natural constraints, payments coupled to production, and the small farmers' scheme. While much of the production was thus decoupled from support payments, decisions left to the discretion of individual countries could have elements that continue to couple production with payments.

4.3 Reform in Canada

Canadian agricultural programs are characterized by a significant dichotomy between the crop and livestock sectors and the supply-managed dairy, egg, and poultry sectors. The major crops grown in and exported by Canada are grains, primarily wheat, durum wheat, and oilseeds, and these are also the crops that relatively receive the most government support (although fruits and horticultural crops are also eligible for various types of support).[8] Support for the supply-managed sectors is much higher.

4.3.1 Non-supply-managed sectors

Canada has traditionally had less crop price and income support than the United States or the European Union. By 2000, Canada had abandoned ad hoc subsidy programs, such as Western Grain Stabilization and the Special Grains Payment, which had been a response to low prices and U.S.–EU export subsidy programs. Transportation subsidies (e.g., feed freight assistance and the "Crow subsidy," which incentivized livestock production near population centers) had been abandoned, and single-desk selling of western grain was done away with in 2012 when the

Canadian Wheat Board was privatized. A five-year, cost-shared federal–provincial agricultural agreement, known as Growing Forward (GF), began April 1, 2008. It focused on competitiveness, innovation, the environment, and business risk management (BRM). In particular, previous agricultural BRM programs, whose development paralleled but also differed from the development of U.S. crop insurance and revenue guarantee programs, were overhauled and subsumed under four GF program components as follows:

1   *AgriInvest* is a government-matched savings account intended to help producers protect their gross margins from small declines. Each year, farmers could deposit up to 1.5 percent of their Allowable Net Sales (ANS) into the AgriInvest account, and this was matched by a government contribution not to exceed CA$22,500, as ANS was limited to CA$1.5 million annually.
2   *AgriStability* is a margin-based, whole-farm program that protects against larger income losses, with payments based on the difference between the realized gross margin (revenue minus certain production costs) in any year and a reference historical margin, with payments initially triggered when a producer's realized gross margin fell to 85 percent or lower than its reference margin. The reference margin was determined as a five-year Olympic average (lowest and highest margins removed) of realized gross margins. The coinsurance (what the farmer pays) was 30 percent when the realized margin fell between 70 percent and 85 percent of the reference margin, but only 20 percent when it fell below 70 percent.
3   *AgriRecovery* provides relief in the case of disasters such as disease outbreaks, permitting governments to fill risk gaps not covered by other government programs. This disaster-relief program was offered by the two levels of government to assist producers to recover the extraordinary costs of such disasters.
4   *AgriInsurance* provides protection to producers from production losses for specified perils, including economic losses arising from natural hazards, such as drought, insect infestations, et cetera – it is production insurance. AgriInsurance is an extension of the multi-peril crop insurance that has been available to Canadian farmers since 1959, although the range of products covered increased over time and especially so under Growing Forward.

The BRM suite of programs is cost shared between the federal and provincial governments and the producers who pay premiums or fees for AgriInsurance and AgriStability.

Because it requires knowledge of revenues and input costs, AgriStability indemnities are determined through the income tax system. This has led to producer uncertainty regarding their eligibility for payments and significant delays in financial support. The agricultural BRM programs were subsequently revised under Growing Forward 2 (GF2), which runs from 2013 to March 31, 2018 (with the federal government currently preparing legislation for a similar Growing Forward 3 suite of programs to follow GF2). GF2 made changes to AgriInvest and AgriStability while leaving the other programs unchanged.

With GF2, the trigger for AgriStability was reduced from 85 to 70 percent, while the amount for which farmers were responsible increased to 30 percent from 20 percent. Thereby, AgriStability met the requirements of the 1995 Agreement on Agriculture, but the changes required the government to modify the AgriInvest program. The producer contribution limit under AgriInvest was increased from 1.5 to 100 percent of ANS, but only 1 percent is now matched by the government. Furthermore, the government's annual matching contribution is limited to CA$15,000, although the account balance limit has increased from 25 percent of the historical average ANS under GF to 400 percent under GF2.

4.3.2 Supply-managed sectors

Supply management was introduced into Canada as a result of the 1972 Farm Products Agency Act. Although potato farmers, hog producers, and other producers attempted to establish supply management schemes in their sectors, supply management regimes only took hold in the dairy and the poultry (eggs, chicken, and turkey) sectors shortly after 1972. In the dairy sector, the federal government then abrogated its responsibility over trade and vested this responsibility with the provinces, which then suppressed interprovincial trade and essentially prevented exports of dairy products by quota holders and non-quota holders alike. Thus, the restrictive quota regime has prevented Canada from taking advantage of export opportunities in developing countries whose citizens desire safe food from developed-country suppliers, with the European Union, Australia, New Zealand, and the United States having taken advantage by expanding exports.

Furthermore, as a result of the WTO Agreement on Agriculture, import restrictions were replaced by TRQs so that in subsequent bilateral and multilateral trade negotiations Canada's trading partners have tried to increase the TRQ levels while lowering import duties. For example, under the 2016 bilateral Comprehensive Economic and Trade Agreement with the European Union, Canada agreed to small increases in its dairy TRQ, with the government indicating it would provide significant compensation to dairy producers in exchange. While the previous Conservative government had promised to compensate dairy producers CA$4.3 billion over 15 years to permit 17,700 metric tons of cheese imports, the Liberal government promised only CA$350 million over five years to help domestic producers compete with European imports.

4.3.3 Comparison with U.S. and EU dairy programs

Both the United States and the European Union employed support prices to protect dairy producers. The European Union adopted supply restricting quotas in 1984, as noted earlier, but subsequently abandoned them (as did Australia), while the United States continued to provide support prices with less stringent supply controls for a longer period. During the 1980s, the U.S. CCC bought 5 to 30 percent of milk production as butter and milk powder. Then, during the 1990s, the United States reduced the support price of milk and, along with a rise in the market price of industrial milk, markets improved to the extent that overproduction was no longer a problem. From 1961 to the mid-1990s, the European Union and Canada assisted milk producers at rates that generally exceeded border prices by 100 percent or much more, while those in the United States were generally much lower but still significant (Figure 20.4). With the exception of Canada and more recently China, NRAs for dairy producers had fallen to near zero by 2010.

Canada and the European Union established production quotas to reduce the costs of supporting farmers. While both jurisdictions prevented interregional trade, the main difference in the programs related to exports. The European Union was and continues to be the world's largest exporter of dairy products. Because it used subsidies to dump its surpluses, the European Union implemented a levy on member states if they exceeded their EU-determined quota (known as a "superlevy"). Canada may have had a robust dairy export sector in the early years of the quota, but it would now be considered a net importer. In both jurisdictions, the supra-authority essentially determined the level of quota available to each region. To maintain a quota regime, it is necessary for the supra-authority to prevent or at least control trade in dairy products between states or provinces.

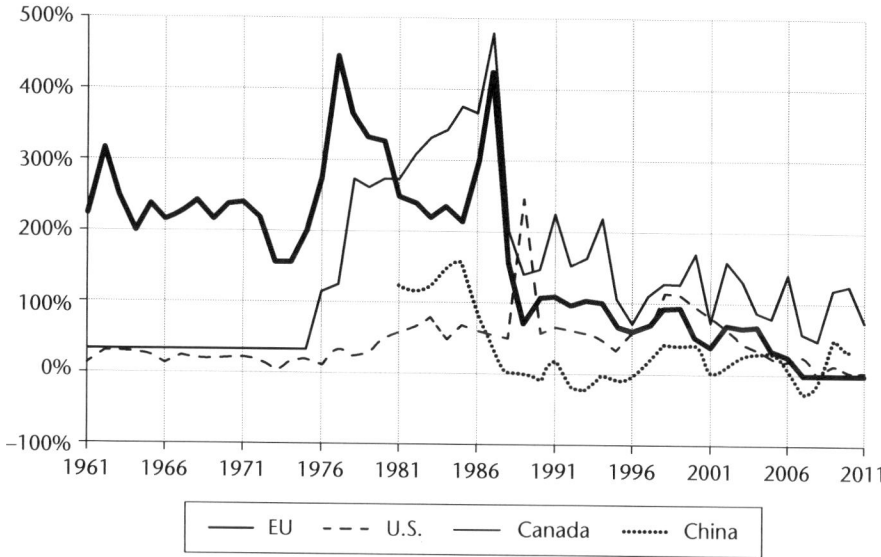

Figure 20.4 Nominal rates of assistance to milk producers, EU, U.S., Canada, and China
Source: Anderson and Nelgen (2013).

The inter-EU restriction of trade directly conflicts with the primary objective of its common agricultural market, which is to facilitate the movement of goods and factors of production among EU member states. Therefore, the overriding objective of greater market integration was an important reason why the European Union chose to disband the dairy quota, although the agreement within the WTO reached in 2015 to eliminate export subsidies for agricultural products was another contributing factor. These considerations should also be important in nudging Canada to abandon its quota regime as markets are increasingly globalized. This will undoubtedly require compensating dairy producers, but by how much remains an issue.

4.4 Recent policy challenges in China

Agricultural policy reforms have yet to play out in some of the major developing countries. Historically, these countries have often discriminated against agriculture, but support has come as the incomes from agriculture have increased (Anderson 2009; Anderson and Nelgen 2013). China is an important case in point.[9] China's exploitation of peasant agriculture was severe in the early stages of its post–World War II development. A subsequent privatizing of agriculture and a shift to market-based incentives, which began in the late 1970s, was followed by trade liberalization beginning in the 1990s and the more-recent provision of positive farm support – this is quite a remarkable turnaround. One consequence is that China now faces policy reform challenges similar to those faced by the United States, the European Union, and Canada over the past half century.

These challenges have come to light for China in the period since world agricultural prices increased for several years around 2008 and then have fallen back. For most years from 2000 to 2007, China was a net exporter of grains, primarily corn; then, after 2008, China's net imports of grains increased steadily and rapidly. Increased net grain imports accompanied increased domestic production of wheat, corn, and rice (OECD 2016). The increased grain output fits China's high-level policy strategy to attain food security with 95 percent self-reliance in wheat and rice, while complementing domestic production with livestock feed imports, particularly soybeans, which China imports in huge quantities under low tariffs.

In managing and adjusting to the changes in trade and production, China has resorted to various policy interventions. Rising support prices, increased input subsidies, and other support measures, including subsidies for crop insurance, have been among China's recent policies. The National Development and Reform Commission set support prices; prices for corn and wheat were raised by 50 to 60 percent between 2008 and 2015, while those of rice increased by 80 percent. As with the early U.S. and EU price support programs, rising support prices have been associated with increased government procurements and growing Chinese grain stocks as world prices declined. As estimated by the USDA, China's corn stocks increased the most, from an average of 52.8 million tons in crop years 2008/2009–2010/2011 to 110.7 million tons at the end of 2015/2016 (USDA-FAS 2016).

Once China acceded to the WTO in 2001, its price support programs have operated in the context of TRQs for wheat, corn, and rice that range between 4 and 9 percent of domestic consumption, with prohibitive over-quota tariff rates of 65 percent. Most of the TRQs are reserved for state-designated enterprises. The low fill rates of earlier years have risen in more recent years, despite support prices above world levels. Because it can limit competing imports, operation of China's support programs for wheat and rice is easier than for corn, for which rising administered prices in China have contributed to expanded imports of substitute feed grains, such as sorghum, barley, and DDGS (distillers' dried grain with solubles). These substitute grains are not subject to TRQs and have low bound tariffs under China's WTO commitments.

In 2016, the United States brought a complaint to the WTO about China's increased domestic agricultural support (asserted to exceed its WTO limit) and TRQ administration (asserted not to be on a transparent, predictable, or fair basis). Thus, China remains in the midst of reforms driven by its increased support levels. The corn price support program was reformed in 2016 to lessen market intervention.

5 A way forward through insurance and risk management?

While agricultural programs were initiated to stabilize prices and enhance the incomes of agricultural producers, they have proven problematic when support programs induce supply to exceed market demand, requiring further policies, such as export subsidies, public storage, and supply restrictions to deal with overproduction. Farm programs have also been a source of uncertainty because they are continually modified in response to actual commodity prices, budgetary pressures, and ongoing developments on the trade front, including the WTO's Agreement on Agriculture and bilateral and regional trade agreements. The effect of ongoing lobbying by various producer, consumer, and environmental groups is also uncertain. As a result, developed countries have moved toward support of the agricultural sector in ways that have minimized effects on crop choices and production. Two main options have emerged, as discussed earlier: decoupled direct income support or deficiency payments, and an increased focus on BRM and insurance.

Government intervention in the livestock sector (outside of the supply-managed industries) often comes via various forms of insurance and, particularly, protection against catastrophic losses. Crop revenue insurance (covering yield and price risk) has played an increasingly important role in recent government agricultural programs, as described above for the United States and Canada, and worldwide (Barnaby and Russell 2016; Woodard 2016; Zacharias and Paggi 2016).[10] Generous government subsidies in the United States have resulted in the extensive use of crop revenue insurance, while Canada has offered a somewhat different set of BRM programs that includes yield coverage and revenue protection. The European Union, on the other hand, has not developed such comprehensive risk-management assistance programs, while China now has the second largest crop insurance program.

Subsidized insurance raises questions of adverse selection and moral hazard. Requiring all farmers to participate in an insurance program eliminates adverse selection, at a cost to society. But no crop insurance program can eliminate the problem of moral hazard. To address both concerns and because adverse weather is the primary source of production risk, agricultural economists have proposed the use of weather-indexed insurance in lieu of crop insurance (Turvey 2001; Vedenov and Barnett 2004; Finger and Lehmann 2012). Weather-indexed insurance provides a payout to the farmer when weather is likely to be adverse for yields, and there has been some uptake of these programs, particularly in developing countries. The insurance provides an indemnity according to the performance of a weather index rather than on the basis of crop yield, revenue, or gross margin, which are influenced by the producer's agronomic choices where moral hazard arises.

Weather-indexed insurance differs from a financial weather derivative. With insurance, the concern is to discover actuarially fair premiums, even if the government might subsidize these. With financial derivatives, there is no search for an actuarially fair premium. Rather, there is a speculator on the other side of the market, and it is the speculator who willingly takes on some of the risk. Nonetheless, the entity offering the product needs to be concerned about its fair market value if it is to stay in business.

Financial weather products can be traded over the counter (OTC) or in existing markets. A farmer can contract with a financial intermediary to hedge against too little heat or rainfall on the basis of data from the nearest weather station or from a weather monitoring station placed on the farm by a company providing this service. This constitutes an OTC contract that blurs the line between privately provided weather insurance and a financial weather derivative, although there is nothing to prevent governments from also offering weather insurance. Cooling degree days (days multiplied by the number of degrees each day that temperature is above 18°C) and heating degree days (below 18°C) already trade on the Chicago Mercantile Exchange, purchased by companies to reduce their exposure to weather-related risks. Such financial weather derivatives can, in principle, complement crop insurance, or take the place of weather-based insurance. The advantage of both weather insurance and weather derivatives is that they eliminate problems of adverse selection and moral hazard because neither the behavior of farmers nor participation rates can influence weather outcomes. The major drawback relates to *basis risk*—the risk that payoffs of a hedging instrument do not correspond to the underlying exposures (Woodard and Garcia 2008; Musshoff, Odening, and Xu 2011).

Agricultural risks can be mitigated to varying degrees through market instruments such as futures contracts and options, including market-traded and OTC weather products (insurance and derivatives). However, financial products for mitigating weather risk alone result in many questions regarding the broader issue of agricultural risk and the role of government in risk mitigation. Grain farmers routinely make use of cash forward and minimum price contracts with grain handlers and processors, who can offer these contracts because they can backstop them with futures and options contracts. Farmers themselves do not generally participate in options and futures trading directly, nor do they purchase market-traded weather products or privately provided, index-based insurance products. With some exceptions, farmers are solely interested in subsidized crop yield and revenue insurance and protection against catastrophic loss. It appears that farmers are able to protect themselves in many cases by appropriate agronomic decisions that reduce risk, through forward contracts with handlers/processors, pooling with farm cooperatives, and, importantly, through off-farm sources of income. Consequently, few farmers purchase crop insurance unless it is subsidized, and often significantly so (Smith 2017).

Critics of U.S. crop insurance programs argue that it is the net income transfers associated with the premium, underwriting, and operating and administrative subsidies that drive the popularity of these programs, but that they are an inefficient means of making such transfers (Just, Calvin, and Quiggin 1999; Wright 2014; Smith 2015). However, given that the Federal Crop Insurance Act of 1980 mandated private delivery, in addition to agricultural producers, private crop insurance companies are major beneficiaries of government-sponsored crop insurance. This led to an alliance between private-sector insurance companies and agricultural stakeholders to lobby government for large income transfers via subsidies to crop insurance – to establish crop insurance as the principal means of supporting farmers. What all of this suggests is that there is still a lot to be sorted out in terms of future BRM and government's role, just as there has been in price and income support programs.

6 Conclusions

Despite their differing political structures and governing processes, there has been an evolution in U.S. and EU agricultural price and income support programs that have reduced economic inefficiencies, while transfers to farmers have remained substantial. The direction of government agricultural programs has shifted toward replacing crop-specific subsidies with ones that decouple support payments from output, providing income support and targeting income volatility with less distortion to production and trade. Conservation and other environmental concerns have gained prominence. Insurance programs have emerged as the main pillar of farm support in the United States and Canada, but not in the European Union. Middle-income countries such as China are now grappling with similar policy transformations.

Even in the presence of these changes, farm lobbies continue to demonstrate their political power in developed countries. In the United States, for example, high prices and farm incomes in the 1970s, mid-1990s, and from 2007 to 2013 did not bring an end to support programs; nor did Europe change its support levels to western European farmers when the EU expanded its membership into central and eastern Europe. International negotiations have failed to bring support levels down substantially. This is because of the effectiveness of worldwide rent-seeking behavior by farmers, who are now a smaller and wealthier interest group than in the past. From an international perspective, the form and direction of agricultural policies are influenced by challenges from outside a country's borders. Consider the case where Brazil, through the WTO, challenged the legitimacy of both U.S. cotton and EU sugar policies. Significant changes were made to both countries' policies to comply with the WTO ruling that favored Brazil (Powell and Schmitz 2005).

In analyzing the impact of agricultural policy, the framework introduced by Rausser (1982, 1990) is useful. He distinguishes between two types of agricultural policies. Policies that reduce transaction costs by correcting market failure, providing public goods or otherwise improving efficiency he refers to as political–economic resource transactions (PERTs) because of their neutral distributional impacts. In contrast, PESTs are political economic-seeking transfer policies that primarily redistribute wealth from one group to another without enhancing efficiency, and may even be harmful to efficiency. We have argued that many agricultural support programs are primarily of a PEST character, with reforms over time having reduced the economic inefficiencies associated with these income transfers, even though it is sometimes difficult to determine the PERT or PEST implications of a specific agricultural support program, since this involves knowing how the program affects and is impacted by a host of other government programs (see Appendices B and C).

Qualitative and quantitative assessment of why farm support policies survive and the impact they have requires further analysis. One approach has been the use of welfare economics as the basis for cost–benefit assessment. As Schmitz et al. (2010) point out,

> The theory of applied welfare economics . . . provides a powerful approach to empirical applications. Unfortunately, there are relatively few empirical applications of cost–benefit analysis to agricultural policy. . . . One of the most difficult aspects of formulating agricultural policy is the setting of goals and objectives for the agricultural sector. This is understandable when policy is analyzed within the context of rent-seeking behavior. The development of policy is usually muddled and confusing. . . . Governments have invested vast sums of money to stabilize agriculture, but many analysts disagree with the approach governments take.
>
> *(pp. 471, 469–470)*

In the context of this messy policy environment, we have addressed in this chapter primarily two questions: what are the distributional and net welfare effects of alternative farm programs, and how have the traditional commodity support programs evolved to increase efficiency and lessen deadweight welfare losses? As we approach the mid-twenty-first century, these support programs need to remain under scrutiny. Additional questions also arise. With the increased emphasis on business risk management and the increased prominence of subsidized crop insurance programs in several countries, new issues about efficient risk-management instruments and the appropriate role of the government are germane. Moreover, one can ask whether any farm programs are still needed at all since farm incomes and wealth have increased significantly after they were introduced during the Great Depression. There is work to be done to ensure the continued evolution and improvement of farm policies worldwide.

Notes

1 The OECD provides similar estimates (its Nominal Protection Coefficient or NPC) for more recent years, but these series only go back to 1986.
2 In Figure 20.3, the demand quantity q^d for the illustrated price support program is also shown equal to q^R, though this need not be the case; nor in practice might a government quota program necessarily mimic perfectly the optimal monopoly behavior.
3 If some imports are allowed, the analysis in Figure 20.3 changes somewhat since production and imports now need to be controlled. This case is illustrated in Appendix A.
4 For a classic recounting and analysis of the historical development of U.S. agriculture and agricultural legislation from 1790 to 1950, see Benedict (1953).
5 There are many excellent reviews of the development of U.S. farm policies since the 1930s. This synopsis draws upon Tweeten (1989); Orden, Paarlberg, and Roe (1999); Gardner (2002); Schmitz et al. (2010); Wright (2014); Novak, Pease, and Sanders (2015); Smith (2017), and Coppess (forthcoming). The National Agricultural Law Center provides the legislative texts of all U.S. farm bills, which are generally enacted at about five-year intervals. For discussion of the most recent 2014 legislation, see Orden and Zulauf (2015). Our coverage is not comprehensive, as we do not discuss, for example, the controversial U.S. sugar program.
6 The history and status of EU agricultural policy to 2010 is found in Oskam, Meester, and Silvis (2011).
7 For more information, see Sorrentino, Henke, and Severini (2011); Swinbank (2012); Henke et al. (2015); Josling and Tangermann (2015); and Matthews, Salvatici, and Scoppola (2017).
8 See Vercammen (2013) for an excellent overview of Canadian agricultural programs.

9 This section draws on Gale, Hansen, and Jewison (2015); Brink and Orden (2016);Yu (2017); Huang and Yang (forthcoming); and Zhong and Zhu (forthcoming).
10 In the United States, there has been a near complete shift from crop yield to crop revenue insurance in the past decade, partly because the premium subsidy rate averages more than 60 percent, while the government also covers underwriting and operating and administrative costs. This also raises concerns about distributional impacts (Lusk 2017).
11 In this model, the relative magnitude and distribution of the rents depends largely on the demand and supply elasticities, the amount of exports, and the per unit cost of the input subsidy. For example, the more elastic the supply, the greater the deadweight loss, and the higher the percentage of domestic production that is exported, the greater the net cost of the combined subsidies. Schmitz, Schmitz, and Dumas (1997) employ this model to demonstrate the existence of potential negative gains from trade.

References

Anderson, K. (editor). 2009. *Distortions to Agricultural Incentives: A Global Perspective, 1955–2007*. Washington, DC: The World Bank and Palgrave Macmillan.

Anderson, K., and S. Nelgen. 2013. *Updated National and Global Estimates of Distortions to Agricultural Incentives, 1955 to 2011*. Washington, DC: World Bank.

Anderson, K., G. Rausser, and J. Swinnen. 2013. "Political Economy of Public Policies: Insights from Distortions to Agricultural and Food Markets." *Journal of Economic Literature* 51:423–477.

Barnaby, G.A., and L.A. Russell. 2016. "Crop Insurance Will Be at the Center of the 2019 Farm Bill Debate." *Choices* 3rd Quarter.

Benedict, M.R. 1953. *Farm Policies of the United States, 1790–1950*. New York: Twentieth Century Fund.

Brink, L., and D. Orden. 2016. "The United States WTO Complaint on China's Agricultural Domestic Support: Preliminary Observations." *IATRC Conference*, Scottsdale, AZ, December.

Coppess, J. 2018. *The Fault Lines of Farm Policy*. Lincoln, NE: The University of Nebraska Press.

Currie, J.A., J.M. Murphy, and A. Schmitz. 1971. "The Concept of Economic Surplus and Its Use in Economic Analysis." *Economics Journal* 81:741–800.

De Gorter, H., and D.R. Just. 2010. "The Social Costs and Benefits of Biofuels: The Intersection of Environmental, Energy and Agricultural Policy." *Applied Economic Perspectives and Policy* 32:4–32.

———. 2009. "The Welfare Economics of a Biofuel Tax Credit and the Interaction Effects with Price Contingent Farm Subsidies." *American Journal of Agricultural Economics* 91:477–488.

Diaz-Bonilla, E. 2015. "Contextual Factors: Country Heterogeneity and Global Economic Conditions." In E. Diaz-Bonilla, ed., *Macroeconomics, Agriculture, and Food Security*. Washington, DC: IFPRI, pp. 53–94.

Finger, R., and N. Lehmann. 2012. "The Influence of Direct Payments on Farmers' Hail Insurance Decisions." *Agricultural Economics* 43:343–354.

Gale, F., J. Hansen, and M. Jewison. 2015. *China's Growing Demand for Agricultural Imports*. Washington, DC: USDA/ERS.

Gardner, B. 2002. *American Agriculture in the Twentieth Century: How It Flourished and What It Cost*. Cambridge, MA: Harvard University Press.

Henke, R., M.R.P. D'Andrea, T. Benos, T. Castellotti, F. Pierangeli, S.R. Lironcurti, F. De Filippis, M. Giua, L. Rosatelli, T. Resl, and K. Heinschink. 2015. *Implementation of the First Pillar of the CAP 2014–2020 in the EU Member States*. Brussels: European Parliament.

Huang, J., and G. Yang. 2017. "Understanding Recent Challenges and New Food Policy in China." *Global Food Security* 12:119–126.

Hueth, D.L., and A. Schmitz. 1972. "Trade in Intermediate and Final Goods." *Quarterly Journal of Economics* 86(3):351–365.

Josling, T., and S. Tangermann. 2015. *Transatlantic Food and Agricultural Policy: Fifty Years of Conflict and Convergence*. Cheltenham: Edward Elgar Publishing.

Just, R.E., L. Calvin, and J. Quiggin. 1999. "Adverse Selection in Crop Insurance: Actuarial and Asymmetric Information Incentives." *American Journal of Agricultural Economics* 81(4):834–849.

Just, R.E., D.L. Hueth, and A. Schmitz. 2005. *The Welfare Economics of Public Policy: A Practical Approach to Project and Policy Evaluation*. Cheltenham: Edward Elgar Publishing.

Lusk, J.L. 2017. "Distributional Effects of Crop Insurance Subsidies." *Applied Economic Perspectives and Policy* 39(1):1–15

Matthews, A., L. Salvatici, and M. Scoppola. 2017. "Trade Impacts of Agricultural Support in the EU." IATRC Commissioned Paper 19. International Agricultural Trade Research Consortium, St. Paul, MN.

Musshoff, O., M. Odening, and W. Xu. 2011. "Management of Climate Risks in Agriculture: Will Weather Derivatives Permeate?" *Applied Economics* 43:1067–1077.

Newberry, D.M.G, and Stiglitz, J.E. 1981. *The Theory of Commodity Price Stabilization: A Study in the Economies of Risk*. New York: Oxford University Press.

Novak, J.L., J.W. Pease, and L.D. Sanders. 2015. *Agricultural Policies in the United States*. New York: Routledge.

OECD (Organization for Economic Cooperation and Development). 2016. *Producer and Consumer Support Estimates Database. Data and Definitions and Sources*. Paris: OECD.

Orden, D. 2007. "Feasibility of U.S. Farm Program Buyouts: Is It a Possibility for U.S. Sugar." In K.M. Huff, K. Meilke, R. Knutson, R. Ochoa, and J. Rude, eds., *Achieving NAFTA Plus*. College Station, TX: Texas A&M University, pp. 147–162.

Orden, D., R. Paarlberg, and T. Roe. 1999. *Policy Reform in American Agriculture*. Chicago, IL: University of Chicago Press.

Orden, D., and C. Zulauf. 2015. "The Political Economy of the 2014 Farm Bill." *American Journal of Agricultural Economics* 97:1298–1311.

Oskam, A., G. Meester, and H. Silvis (editors). 2011. *EU Policy for Agriculture, Food and Rural Areas*. Wageningen, Germany: Wageningen Academic Publishers.

Powell, S.J., and A. Schmitz. 2005. "The Cotton and Sugar Subsidies Decisions: WTO's Dispute Settlement System Rebalances the Agreement on Agriculture." *Drake Journal of Agricultural Law* 10(2):287–330.

Rausser, G.C. 1982. "Political Economics Markets: PERTs and PESTs in Food and Agriculture." *American Journal of Agricultural Economics* 64:821–833.

———. 1992. "Predatory Versus Productive Government: The Case of U.S. Agricultural Policies." *Journal of Economic Perspectives* 6(3):133–157.

Rausser, G.C., and W.E. Foster. 1990. "Political Preference Functions and Public Policy Reform." *American Journal of Agricultural Economics* 71:641–652.

Schmitz, A., D. Haynes, and T.G. Schmitz. 2015. "Alternative Approaches to Compensation and Producer Rights." *Canadian Journal of Agricultural Economics* 64(3):439–454.

———. 2016. "The Not-So-Simple Economics of Production Quota Buyouts." *Journal of Agricultural and Applied Economics* 48(2):119–147.

Schmitz, A., P.L. Kennedy, and T.G. Schmitz. 2017. *World Agricultural Resources and Food Security*. Bingley: Emerald Publishing.

Schmitz, A., C. Moss, and T.G. Schmitz. 2007. "Ethanol: No Free Lunch." *Journal of Agricultural and Food Industrial Organization* 5(2):Article 3.

Schmitz, A., C. Moss, T. Schmitz, W.H. Furtan, and C. Schmitz. 2010. *Agricultural Policy, Agribusiness and Rent Seeking Behavior* (2nd ed.). Toronto: University of Toronto Press.

Schmitz, A., and T.G. Schmitz. 2010. "Benefit–Cost Analysis: Distributional Considerations under Producer Quota Buyouts." *Journal of Benefit–Cost Analysis* 1(1):Article 2.

———. 2012. "The Complexities of the Interface Between Agriculture and Trade." *The Estey Centre Journal of International Law and Trade Policy* 13(1):14–25.

Schmitz, T.G., A. Schmitz, and C. Dumas. 1997. "Gains from Trade, Inefficiency of Government Programs, and the Net Economic Effects of Trading." *Journal of Political Economy* 105(3):637–647.

Smith, V.H. (editor). 2015. *The Economic Welfare and Trade Implications of the 2014 Farm Bill*. Bingley: Emerald Publishing.

———. 2017. "The U.S. Federal Crop Insurance Program: A Case Study in Rent Seeking." Mercatus Working Paper, George Mason University, Arlington, VA.

Sorrentino, A., R. Henke, and S. Severini. 2011. *The Common Agricultural Policy After the Fischler Reform*. Farnham: Ashgate.

Swinbank, A. 2012. New Direct Payments Scheme: Targeting and Redistribution in the Future CAP, DG for Internal Policies. Policy Department B: Structural and Cohesion Policies. Agriculture and Rural Development. Note IP/B/AGRI/CEI/2011–097/E003/SC1.

Turvey, C.G. 2001. "Weather Derivatives for Specific Event Risks in Agriculture." *Review of Agricultural Economics* 23(2):333–351.

Tweeten, L. 1989. *Farm Policy Analysis*. Boulder, CO: Westview Press.

USDA-FAS (United States Department of Agriculture, Foreign Agriculture Service). 2016. *Production, Supply and Distribution Online*. Washington, DC: USDA/FAS.

van Kooten, G.C. 2017. "The Welfare Economics of Dismantling Dairy Quota in a Confederation of States." REPA Working Paper 2017–04, University of Victoria, Victoria, BC-Canada.

van Kooten, G.C., and A. Schmitz. 1985. "Commodity Price Stabilization: The Price Uncertainty Case." *Canadian Journal of Economics* 18:426–434.

van Kooten, G.C., A. Schmitz, and W.H. Furtan. 1988. "The Economics of Storing a Non-Storable Commodity." *Canadian Journal of Economics* 21:579–586.

van Kooten, G.C., and K.F. Taylor. 1989. "Measuring the Welfare Impacts of Government Regulation: The Case of Supply Management." *Canadian Journal of Economics* 22:902–913.

Vedenov, D.V., and B.J. Barnett. 2004. "Efficiency of Weather Derivatives as Primary Crop Insurance Instruments." *Journal of Agricultural and Resource Economics* 29(3):387–400.

Vercammen, J. 2013. "A Partial Adjustment Model of Federal Direct Payments in Canadian Agriculture." *Canadian Journal of Agricultural Economics* 61(3):465–485.

Vercammen, J., and A. Schmitz. 1992. "Supply Management and Input Concessions." *Canadian Journal of Economics* 25:957–971.

Woodard, J.D. 2016. "Crop Insurance Demand More Elastic Than Previously Thought." *Choices* 31(3).

Woodard, J.D., and P. Garcia. 2008. "Basis Risk and Weather Hedging Effectiveness." *Agricultural Finance Review* 68(1):99–117.

Wright, B.D. 2014. "Multiple Peril Crop Insurance." *Choices* 29(3):1–5.

Yu, W. 2017. *How China's Farm Policy Reforms Could Affect Trade and Markets: A Focus on Grains and Cotton*. Geneva, Switzerland: International Centre for Trade and Sustainable Development (ICTSD).

Zacharias, T.P., and M.S. Paggi. 2016. "Current Perspectives on the Crop Insurance Farm Safety Net." *Choices* 31(3):1–5.

Zhong, F., and J. Zhu. 2017. "Food Security in China from a Global Perspective." *Choices* 32(2):1–5.

Appendices

A. A further note on supply management

The key to supply management in the case of a net importing country is the use of import quotas and domestic-production controls (Vercammen and Schmitz 1992). Both of these policy instruments are modeled in Figure 20.A1. Domestic demand is given by the curve D_o and domestic supply by S. Under free trade, the domestic (border) price is P_b, domestic production is Q_1, and domestic consumption is Q_2, so that imports equal $Q_2 - Q_1$.

When an agreed-upon quantity of imports (a minimum access commitment) is given tariff-free access to the market, imports are restricted in the figure to $Q_2 - Q''_1 = Q'_1 - Q_m$. Now domestic producers face the demand curve D'. The domestic producers maximize profits when the production quota is set where the marginal revenue curve MR equals the supply curve S, which results in domestic production Q_m. Producers gain area $(P_e P_b ea - ehi)$, while the quota value is the discounted value of $(P_e - P_s)$ per unit of quota. The approximate quota value for the industry will be the discounted value of rectangle $P_e P_b ha$.

Consumers lose the entire area $P_e P_b db$. The availability of import quotas gives importers (who may be domestic food retailers) incentives for rent seeking, because import quotas have value equal to $(P_e - P_b)(Q'_1 - Q_m)$, or area $aecb$. This value arises because importers buy the product at P_b and sell it in the domestic market at P_e.

Triangle bcd in Figure 20.A1 is the deadweight loss (DWL) triangle $[(P_e P_b db)(P_e P_b ea - aecb)]$, and it is the net cost of the supply management program. The smaller the triangle, the greater is the efficiency with which income can be transferred from consumers to producers. The size of the income transfer can be large, while the DWL triangle can be small.

B. Input and output subsidies: multiplicative effects

Figure 20.A2 presents a combined input subsidy (such as for fertilizer or water) and price-support payment model, and also explicitly represents each policy program instrument separately (Schmitz, Schmitz, and Dumas 1997). Analyzing these instruments together and individually demonstrates that they operate multiplicatively rather than additively. S and S' represent the respective supply curve and the input-subsidized supply curve; D_d is the domestic demand curve and T_d is the total demand curve, including export demand, which is implicit and is not shown

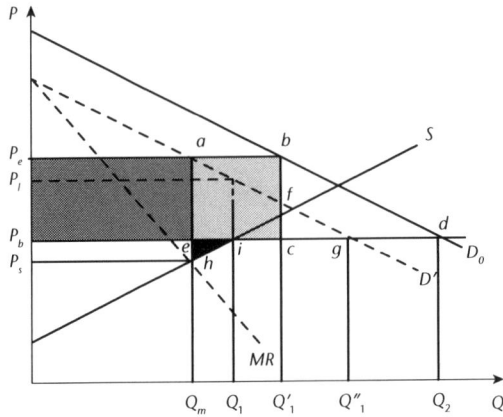

Figure 20.A1 Supply management with a TRQ

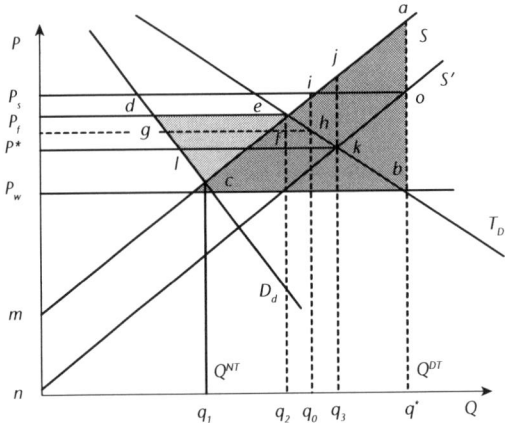

Figure 20.A2 Multiplicative effects due to multiple subsidy programs

directly in Figure 20.A2. (This model has been extended to incorporate international trade considerations where the results show that the welfare costs from distortions can be large due to the interface between domestic policy and trade [Schmitz and Schmitz 2012].)

Under the multiplicative effects (ME) scenario in Figure 20.A2, the intersection of the support price P_s and the subsidized supply curve S' establishes both the output quantity q^* and the world price P_w. The support price is affecting all production, but with deficiency payments so the consumer market clears at a lower price (the second support option as descried in Figure 20.3). Domestic producers receive the area $P_s P_f emno$ as a net gain, while domestic consumers gain the area $P_f P_w cd$. The area *dcbe* is referred to as slippage, because it represents the rents received by importing countries. Cost to the government for the input subsidy is the area *amno*, while the cost of the government price support payments equals the area $P_s P_w bo$. Therefore, the combined net domestic cost to society of the two subsidies applied together is the shaded (and mottled) area *aedcb*. The net cost comparison is made with reference to point *e*, where P_f and q_2 are free of distortions.[11]

For the ME model depicted in Figure 20.A2, domestic producers gain more rents from the input subsidy given by area *mnoi* than from the price support payments given by $P_s P_f ei$.

The majority of the price support payments from the government go to domestic consumers, namely area $P_f P_w cd$, and foreign countries ($dcbe$), rather than to producers. However, the actual distribution of these rents is an empirical matter, which illustrates how parameter changes affect the calculation and distribution of the subsidy rents and welfare losses.

A combination of the two subsidies distorts output more than when they act alone, causing the ME of the two instruments to be greater than a mere summation of the individual effects. For example, the production quantity q^* is established where the target price P_S intersects the input-subsidized supply curve S' at point o in Figure 20.A2 instead of at point i (associated with quantity q_0) where it would otherwise be given a price support only. Thus, adding the input subsidy to the price support increases production from q_0 to q^*. In addition to increased output, there is a significant decrease in the resulting price P_w needed to clear the world market. Both of these effects increase the size of the price support payments made by the government and, in conjunction with price supports, the aggregate size of the input subsidy is greater than without. This is why Figure 20.A2 is referred to as a "multiplicative effects" model.

One can also observe the individual effects of input subsidies and price supports. In Figure 20.A2, the net cost of the price support is given by area $dghei$. The net cost of the input subsidy is area $dlkje$.

C. Impact of the U.S. ethanol program on agriculture

The interrelated economics of ethanol production and agricultural policy in the United States is examined with the aid of Figure 20.A3, which is derived from Schmitz, Moss, and Schmitz (2007) and depicts the U.S. corn market; additional considerations are given by De Gorter and Just (2009, 2010). Under the loan rate provisions of U.S. farm bills, farmers are guaranteed a minimum loan rate price P_{LR} for each bushel and produce q_s bushels at P_{LR} if the supply curve is S, as shown in Figure 20.A3. The domestic demand curve is D_d, and with an export demand D_x, the total demand curve is D_T. These demand curves result in a market-clearing price of P_0 if there are loan deficiency payments (LDPs). With this market-clearing price, q_d is consumed domestically and q_x ($= q_s - q_d$) is exported. At this equilibrium, the LDPs paid to farmers based on the level of production are represented by the area $P_{LR}abP_0$, or the light-shaded area. In addition, farmers receive countercyclical payments (CCPs) based on their historical level of production q_h (typically 85 percent of historical base acreage and lagged historical yields) and the target price P_{TP}. Graphically, this payment is depicted by the area $P_{TP}dcP_{LR}$, or the darker shaded area. The net cost of the subsidy program from the U.S. perspective is area $keabg$, of which area $kebg$ is a gain to importers – the *slippage* effect.

In this original equilibrium, we assume that the market-clearing price P_0 is greater than the choke price for the derived demand curve for corn used to produce ethanol D_E, so that no ethanol is produced. Next, we assume that increases in the price of gasoline shift the derived demand for corn used to produce ethanol outward to D'_E (as indicated by the arrow). This changes the total demand curve to ($D_T + D'_E$). As drawn in Figure 20.A3, this rightward shift in the derived demand for corn from ethanol producers is sufficient to raise the equilibrium price of corn to the loan rate, eliminating the loan deficiency payments to farmers. Thus, there are no direct subsidies based on production, but there are indirect subsidies to corn producers via ethanol tax credits, and farmers continue to receive countercyclical payments that are not affected by production decisions.

Consider further the demand for corn derived from ethanol production. Starting from D'_E (which assumes a fixed oil price), a sufficiently large increase in corn prices above P_1, say, would choke off the demand for corn to produce ethanol. This point represents a corner solution in Figure 20.A3. However, if one assumes an increase in oil prices for a given price of corn, the derived demand curve for corn shifts to the right. From a theoretical perspective, the demand for corn for ethanol production could be positive without a tax credit. Thus, at least two factors affect ethanol production: a favorable oil to corn price ratio and a tax credit for ethanol production.

In the first shift of equilibrium shown in Figure 20.A3, it is assumed that the increased ethanol demand brings the market price up so that it equals the loan rate. Producers are not impacted by ethanol demand, even though corn prices rise. This is because the LDP no longer exists (the payment represented by the light-shaded area disappears), while the CCP (dark-shaded area) remains unchanged. Also, an important result is derived from the observation that market clearing prices rise from P_0 to P_{LR}, causing both domestic and export demand to fall for those components making up demand D_T. The demand for corn for ethanol is $q_s - q'_s$. Domestic consumers now pay a higher price for corn and related products, given demand D_d. Foreigners also pay a higher price for the corn they import.

To summarize, in this first shift of equilibrium in this model: (1) producers are unaffected by ethanol demand; (2) domestic consumers lose area $P_{LR}hgP_0$; (3) foreign importers lose area $hgbv$; (4) the government saves loan deficiency payments equal to area $P_{LR}abP_0$ – that is, there are government farm program cost reductions from the ethanol production; and (5) the consumers of ethanol fuel gain area wva. To calculate the *net effect* of ethanol, one needs to consider (1) the net welfare gain of area $aekgb$; (2) the consumer gain from the introduction of ethanol of area wva; and (3) the cost, if any, of the indirect ethanol subsidy. The first two components are positive while the last one is negative. Ethanol subsidies replace direct subsidies that affect consumers because the price of corn increases. Production costs are now covered, so direct subsidies are no longer binding.

To further show the interrelationship between ethanol production and government payments to corn farmers, assume that the derived demand for corn used to produce ethanol shifts further outward to D''_E. This increased derived demand causes the total demand for corn to shift outward to $(D_T + D''_E)$, increasing the market equilibrium price to P_2 (just below the target price) and the equilibrium quantity to q_t. Comparing this equilibrium with the equilibrium at the loan rate, producers gain area P_2yaP_{LR}, although part of this gain, given by area P_2zcP_{LR}, is offset by reductions in the countercyclical payments to farmers. Thus, the net producer gain is area $zcay$. This shift results in an economic loss to domestic consumers of area P_2mhP_{LR} and a loss to foreign consumers of area $mxvh$. Completing the model, the economic gain for ethanol producers is the area δmy. If the demand for ethanol shifts even farther to the right than D''_E, all government payments (including countercyclical payments) are eliminated. Thus, there is a direct link between tax credits to ethanol and farm program payments.

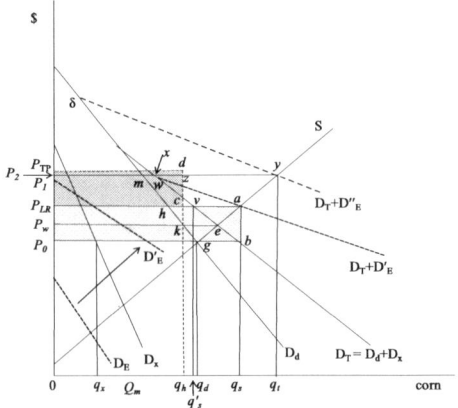

Figure 20.A3 The slippage effect in the corn ethanol case

21
The political economy of agricultural and food policies

Johan Swinnen

1 Introduction[1]

Food and agriculture have been subject to heavy-handed government interventions throughout much of history and across the globe, in both developing and developed countries. Political considerations are crucial to understand these policies, since almost all agricultural and food policies have redistributive effects and are therefore subject to lobbying and pressure from interest groups and are used by decision makers to influence society for both economic and political reasons (Schmitz et al. 2010). Some policies, such as import tariffs or export taxes, have clear distributional objectives and reduce total welfare by introducing distortions in the economy. Other policies, such as food standards, land reforms, or public investments in agricultural research, may increase total welfare, but, at the same time, also have distributional effects. These distributional effects will influence the preferences of different interest groups and thus trigger political action.

The inherent interlinkage between efficiency and equity issues in policy making is due to history, economics, and politics being closely related disciplines and often written about by the same authors, as reflected in the works of the original architects of the economics discipline, such as Adam Smith, John Stuart Mill, David Ricardo, etc. In the late nineteenth century, the economics discipline started separating itself from the "political economy" framework.

The revival (or return) of political economy started in the 1950s and 1960s and was referred to as "neoclassical political economy" or "new political economy," as economists started using their economic tools to analyze how the incentives of political agents and constraints of political institutions influenced political decision making.

The start of this field is often associated with certain publications, such as Downs' 1957 book, *An Economic Theory of Democracy*; Olson's 1965 book, *The Logic of Collective Action*; and Buchanan and Tullock's 1962 book, *The Calculus of Consent*, and classic papers by Krueger (1974) and Bhagwati (1982) on "rent-seeking." Stigler's (1971) "Theory of Economic Regulation" and contributions by Becker (1983) formed the basis of the (new) "Chicago school of political economy."

These theories and insights have been applied to analyze food and agricultural policies. Research was not only triggered by the emerging "new political economy," but also by the puzzling question: *why was agriculture subsidized in rich countries and taxed in poor countries?* New data,

Johan Swinnen

and in particular Krueger, Schiff, and Valdes (1991), showed that in countries where farmers were the majority of the population, farmers were taxed, while in countries where they were the minority, farmers received subsidies: the so-called development paradox (see further Wise 2004). The combination of an intriguing question, new theories to apply, and fascinating data induced a rich literature on the political economy of agricultural trade and distortions in the 1980s and the first part of the 1990s.[2]

Later, interest in the political economy of agricultural policies was sparked by a similar combination of factors as in the 1980s: new data, new theories, and new intriguing questions (Swinnen 2009, 2010). New political economy models focused on the role of institutions in economic policies (Persson and Tabellini 2000, 2003; Acemoglu 2003); better micro-foundations for analyzing political–economic decision making (Grossman and Helpman 1994);[3] and the role of (mass) media (Stromberg 2004; McCluskey and Swinnen 2010). New datasets improved indicators on institutional and political variables. The World Bank's project on distortions to agricultural incentives provided new and rich datasets on agricultural policies (Anderson 2009, 2016). Studies focused on how major institutional and political reforms had affected agricultural policy (Rozelle and Swinnen 2004, 2009, 2010). This includes the global shift from state-controlled to market-based governance of agricultural and food systems, not only in China and the former Soviet Union but also in Latin America and Africa (Swinnen, Vandeplas, and Maertens 2010). Other questions are related to the impact of changes in international organizations and trade agreements, such as the Uruguay Round Agreement on Agriculture (URAA), the World Trade Organization (WTO), the North American Free Trade Agreement (NAFTA), the enlargement of the European Union, and etc.

The turnaround in the global agricultural and food markets in the second half of the 2000s induced new economic and political debates on agricultural and food policies. Instead of export subsidies and import tariffs, export barriers and price ceilings were introduced to prevent food prices from rising. The food crisis also drew attention to the failure of agricultural policies to stimulate investment and productivity growth. Another emerging issue was whether there is a shift from traditional trade barriers (such as import tariffs) to so-called non-tariff measures. This was triggered by the rapid growth in public and private standards in global agri-food chains and a concern that, with binding WTO constraints on tariffs, governments were looking for other instruments to protect their markets.

This chapter starts with a discussion of political models and coalitions and then discusses the political economy of policies motivated primarily by redistributing income (or rents) between different groups in society, such as price and trade interventions in agricultural and food markets. The last part of the chapter focuses on policies that may stimulate growth, reduce externalities and volatility, redistribute income, or increase investments in public goods (research).

2 Political coalitions in agricultural and food policies[4]

Political economy models of agricultural and food policy often consider producers, consumers, and taxpayers as the main groups with vested interests in policy outcomes. The theoretical reason is its didactic use, that is, to avoid unnecessary complications in deriving policy effects and identify equilibria. The empirical reason is the absence of disaggregated information regarding the policy impacts on various agents within (or outside) the value chain.

It is well known that, in reality, many more agents lobby governments to introduce or remove certain policies (agents include landowners, seed and agro-chemical companies, rural banks, traders, food processors, retail companies, environmental groups, and food advocacy groups, among others.). These agents may be affected differently by policies, depending on the nature of the policy (e.g., whether the policy is targeted to the [raw] agricultural commodity or to a processed commodity) or the effect of farm subsidies on land or other production factors. As a consequence, these different agents have sometimes joined forces ("political coalitions") with farmers or with final consumers to influence policy makers in setting public policies. In other cases, they have opposed each other on policy issues. For example, sugar processors and farmers may jointly lobby for sugar import tariffs or quota but may oppose each other when governments consider regulating sugarcane or beet prices.

These coalitions are not static. There are several reasons why political coalitions may change: traditional power structures within value chains may change with some (sub)sectors growing and others declining with economic development, new technologies may bring new players into the value chains, new policy instruments may be introduced (or considered), etc. As an illustration, consider changed coalitions due to "new players" that have emerged, or because the same players are interested in "new things."

New players have emerged for a variety of reasons. Awareness of environmental issues and the lobbying of environmental organizations has increased in recent decades. In the United States, the Dust Bowl in the 1930s led to the introduction of a major conservation payment program. In the European Union, where environmental concerns have traditionally been less important in agricultural policy, lobbying by environmental Non-govermental organizations (NGOs) has had only limited influence on agricultural and food policy (Swinnen 2015). Technological advances, such as biotechnology and genetically modified (GM) crops, have created new vested interests and changed others. In the 1970s, there were no pro-GM or anti-GM lobbies since there were no GM crops. Biofuels have emerged as an important factor in agricultural markets and food policies due to rising oil prices and the search for renewable energy sources.

The emergence of new policies, such as crop insurance subsidies, has brought new sectors into the lobbying game for farm support programs. In the U.S., crop insurance companies have become an increasingly important interest group in agricultural policy discussions as crop insurance programs have become the largest expenditure item on recent farm bills (Coble and Barnett 2013; Glauber 2004).

The growing concentration in retail and the emergence of preferred supplier systems have made the retail sector a more powerful sector in the value chain. This may benefit consumers because for many agricultural policy issues consumer and retailer interests are aligned, and their political coalitions may be reinforced by growing retail concentration, such as insurance companies.

Vested interests may also change. In poor countries, consumers are most interested in having sufficient food at low prices. As incomes grow, consumers become more concerned with the quality of their food and with the environmental and ethical standards of their food. At the policy front, this has resulted in regulations on geographical indications (GI) – an issue that has created tensions in trade negotiations.

3 Price and trade interventions

In the second half of the twentieth century, there were major differences in agricultural and food policies between poor countries that taxed farmers and rich countries that subsidized farmers (and taxed consumers). This difference (known as the *development paradox*) was both huge and counterintuitive (Krueger, Schiff, and Valdes 1991). In poorer countries where farmers were the majority of the population and had most of the votes (generally countries that were not democracies, the political strength of numbers), farmers were losing out from agricultural policies that imposed significant taxes on them. In contrast, in richer countries where farmers were a small minority, farmers were subsidized despite the fact that their numbers in the political arena had declined.

Political economy studies have since explained that the differences in agricultural policies between rich and poor countries captured in the development paradox are due to differences in political economy equilibria caused by the combination of structural economic differences, information costs, changes in governance structures, etc.[5]

3.1 Structural change and political incentives

The structural changes that accompany economic development alter the costs and benefits of policies to various interest groups, and thus the incentives for political activities to be undertaken in order to influence governments. These, in turn, determine the government's political incentives and adjust the political–economic equilibrium (Gardner 1987; Swinnen 1994; Anderson 1995). First, economic growth typically coincides with a rise in urban–rural income disparities, as growth in industry and services outpaces growth in the agricultural sector, whose specific assets make it slow to adjust. This income gap creates incentives for farmers and agricultural companies to demand – and politicians to supply – policies that redistribute income in order to reduce that income gap. There are several mechanisms presented in the political economy literature that explain these countercyclical policies. One is the "relative income hypothesis" of Swinnen (1994) and de Gorter and Swinnen (2002), which when driven by changes in marginal utility determines political incentives for governments to respond to interest groups. Another is the "loss aversion" argument where political action is driven by interest groups that want to avoid losses coming from changing market conditions (Freund and Ozden 2008; Tovar 2009).

Second, in a poor economy, most workers spend a large share of their income on food. They will therefore strongly oppose an increase in food prices through government interventions such as import tariffs. Industrial capital will support worker opposition against food price increases because they are concerned about the inflationary effects on wages and their profits. In contrast, in a rich economy, most workers spend a (much) smaller share of their income on food, and only a relatively small part of this is the cost of raw materials (agricultural products). This effect is reinforced by declining opposition from industry because the inflationary pressure on wages from agricultural protection declines.

Third, for a given per-capita subsidy to farmers, it takes a much larger per-capita tax on consumers (or workers in other sectors) when there are many farmers and fewer consumers (as in poor countries) than when there are few farmers and many consumers (workers in other sectors), as in rich countries. In other words, even though the share of farmers in the voting population declines, less opposition to protecting farmers arises when there are fewer of them. Swinnen (1994) shows that, under plausible assumptions, the second of those two effects dominates. In summary, the combination of these factors causes a shift in the political economy equilibrium from taxing farmers to subsidizing farmers with economic growth.

3.2 Political organization

Improvements in rural infrastructure with economic development also enhance farmers' ability to organize for political action. Olson (1965) explained that collective action by relatively large groups is difficult because of free-riding incentives, implying that in poor countries it is costly to organize farmers politically. Consumers are often concentrated in cities, where coordination and collective action are easier than in the rural areas. However, as the number of farmers declines and rural infrastructure improves – the cost of political organization for farmers decreases.

In addition, the growth and concentration of agribusinesses and food-processing companies, which are often aligned with farm interests in lobbying for agricultural policies, strengthen pro-farm interests. In many countries, the growth of agricultural protection is often associated with the growth of cooperative agribusiness and food-processing companies.[6]

3.3 Information costs

Information plays a crucial role in political markets, organization, and policy design. Downs' (1957) "rationally ignorant voter" principle explains that it is rational for voters to be ignorant about certain policy issues if the costs of information are higher than the (potential) benefit of being informed. McCluskey and Swinnen (2004) argue that rational ignorance, be it in the political arena (voters) or in the economic arena (consumers), is still relevant today despite reductions of information costs with the growth of mass media and social media. People's opportunity costs and ideological differences between the information (media) source and the reader limit information consumption. The rationally ignorant voter argument implies that policies will be introduced that create concentrated benefits and dispersed costs (Stromberg 2004). This information effect reinforces the distributional effects caused by structural factors.

In addition, an enhanced rural communication infrastructure, through either public investments (as in many high-income countries earlier in the twentieth century) or technological innovations and commercial distributions (as in the spread of mobile phone use in developing countries), will reduce the relative costs of information in rural areas compared to urban areas. As a consequence, farmers will be better informed about policies, and they can use this enhanced information infrastructure to organize themselves better, improving the effectiveness of their lobbying activities. In summary, several information-related aspects of economic development cause a shift in the political economy equilibrium from supporting consumers to supporting farmers (Olper and Swinnen 2013).

3.4 Political reforms

There is a correlation between political regimes and economic development, with democratic regimes more prominent among richer countries than among poorer countries. Median-voter models predict that democracies tend to redistribute from the rich to the poor because the distribution of political power (measured by votes) is typically more equal than is the distribution of income and wealth (Alesina and Rodrik 1994; McGuire and Olson 1996). Because most farmers are typically poorer, this suggests that farmers may benefit from democratization. Moreover, the same factors that make it difficult for farmers to organize politically in poor countries (e.g., their large number and geographic dispersion) render farmers potentially more powerful in electoral settings (Varshney 1995; Bates and Block 2010).

It has been difficult to measure this empirically due to data and econometric constraints. Swinnen, Banerjee, and de Gorter (2001), using long-run data for Belgium, find that changes in electoral rules that have disproportionately benefited agriculture (e.g., extending voting rights to small farmers and tenants) have induced an increase in agricultural protectionism. Other electoral changes have not affected agricultural protection because they increase the voting power of both those for and against protection. Olper, Falkowski, and Swinnen (2014) analyze the impact of all democratic reforms since the 1960s (which were concentrated in developing and emerging countries) and find that, on average, democratization has benefit for farmers.

Johan Swinnen

4 Policy instrument choice

The distortionary effects of government interventions are as equally dependent on the choice of the instrument as on the level of the intervention. There is an extensive literature comparing the transfer efficiency and the distortions of various policy instruments in trade and agricultural policies (Gardner 1983; Alston and James 2002). The differences in distortionary effects of policies plays an important role in policy discussions and trade negotiations.

Another stylized fact of agricultural and food policies is the *anti-trade bias*, that is, the import-competing sectors are protected by taxing imports and exportable sectors are taxed more, thereby reducing exports (Anderson 2009, 2016). Trade-policy instruments are the most important agricultural and food policies to redistribute income between consumers and producers. In earlier history, they were often the only policies, but even today remain very important (Anderson, Rausser, and Swinnen 2013). In the years 2007 to 2012, the anti-trade bias took on a particular form, as many governments responded to rising food prices on world markets by restricting, sometimes outright banning, food exports, thereby exacerbating global price spikes (Alston, Ivanic, and Martin 2014; Pinstrup-Anderson 2014).

There are several reasons why trade policies are used. First, import-competing sectors have lower comparative advantage than do exporting sectors. Benefits from market returns are lower in sectors with a comparative disadvantage, while those sectors' incentives to seek income from government support are higher. In these (sub)sectors, returns to investment in lobbying activities dominate returns from market activities and thus indirectly support an anti-trade bias. A second factor is the so-called revenue motive of public policy. Tariff revenues and export taxes increase government revenues and improve their terms of trade, while export and import subsidies do the opposite. Third, distortions (deadweight costs) and budgetary costs of policy intervention typically increase with higher supply elasticities (Gardner 1983; de Gorter, Nielson, and Rausser 1992). Sectors with higher supply elasticities (e.g., exports) are subsidized less because it is more costly to do so and causes more distortions (Becker 1983; Gardner 1987).

Fourth, policy instruments differ not only in deadweight costs but also in implementation costs. The most obvious explanation for the broad use of trade taxes (either import tariffs or export taxes) is that they are the easiest and least costly to implement (Rodrik 1995; Dixit 1996). In many developing countries, tax-collection institutions are weakly developed and trade taxes (import tariffs or export taxes) are often an important – or the only substantive – source of tax revenue. In this case, governments have greater incentives to assist farmers.

Fifth, policy instruments also differ in their "transparency," that is, the information available concerning policies and their incidence. Politicians have an incentive to use policies that hide their costs or use policies that obfuscate the transfer itself (Magee, Brock, and Young 1989). This obfuscation perspective helps explain why non-budget methods of redistribution, such as tariffs, are politically preferable to direct subsidies. Sixth, governments may prefer distortionary policies, such as tariffs, when they have imperfect information on their target group or the amount of transfer needed (Foster and Rausser 1993; Mitchell and Moro 2006). The total transfers – even with deadweight costs – may be lower than would be the case with direct (lump-sum) transfers when governments need to secure a minimum amount of political support. In summary, the combination of these political economy forces causes anti-trade bias.

These political factors can be constrained if the counter-pressure is strong enough, for example, by integrating them into international trade agreements where (economic and political) costs and benefits can be weighed and compensated. In particular, the WTO has had a significant impact in the anti-trade bias and on policy instrument choice.

5 Policy reforms

Since the 1990s, there has been a change in the trend of agricultural protection and in policy instruments for several high-income countries. In Organization for Economics Co-Operation and Development (OECD) countries in the 1980s, the most important instruments were coupled policies – consistent with the strong anti-trade bias.[7] Their share in total support was 82 percent, whereas decoupled support made up only 10 percent. However, in the 1990s and 2000s, there was a dramatic change. By the late 2000s, the former had decreased to 49 percent and the latter had increased to 61 percent. The reduction of trade-distorting policies was significant in rich countries. Swinnen, Olper, and Vandemoortele (2012) find that the implementation of the General Agreements on Tarrifs and Trade (GATT) and WTO have reduced trade interventions, and thus the anti-trade bias. There was also a virtual abolition of all support measures in Australia and New Zealand (Anderson, Rausser, and Swinnen 2013).

The fall in agricultural protection was also strong in both Western and Eastern Europe, but for very different reasons. In Eastern Europe, economic and political liberalizations removed much of the heavy subsidies to farms that existed under the Communist regimes in the 1970s and 1980s (Liefert and Swinnen 2002; Anderson and Swinnen 2008). In the European Union, the Common Agricultural Policy (CAP) has been reformed significantly. Both the level of subsidies and the distortions caused by them have been significantly reduced since 1990. The change started with the integration of agricultural policies in the GATT/WTO as part of the Uruguay Round Agreement on Agriculture (URAA) in 1994.[8] Further changes in policy instruments, from market interventions to direct payments, increased the visibility of the transfers, as they occupied a large share of the EU budget. This increased transparency of the transfers may have been an additional cause of reforms in the European Union: over the past two decades, taxpayers have continuously pressured EU leaders to reduce agricultural subsidies.[9]

At the same time, developing countries have reduced taxation of agricultural exports. The dominance of trade policies remains, but the anti-trade bias has declined in recent decades, mainly owing to macroeconomic and trade policy reforms that reduced the taxation of farmers and the anti-trade bias in developing countries. These political economy changes are due to economic growth, structural adjustments, and changes in information costs and governance structures, as we explained earlier.

Anderson, Rausser, and Swinnen (2013, p. 7) conclude that "since the 1980s, both the anti-agricultural bias in developing countries and the pro-agricultural bias in high-income countries have diminished and the two groups' average rates of assistance to agriculture have converged toward zero." This means that – rather that the *divergence* observed in the 1950s to 1980s – there is now *convergence* in agricultural policies.

6 Price shocks and political economy of stabilization and development policy

With a brief exception in the early 1970s when prices moved up following the first oil crisis, global agriculture has been characterized by relatively stable and low prices for the past 50 years. Most of the global agricultural and food policy discussions focused on the reduction of taxes on farmers in developing countries and the removal of policies that subsidized farmers in rich countries. This changed with dramatic increases in food prices in the 2000s. Urban consumers across the world protested, and governments reacted rapidly to the price spikes. Many governments, in particular in developing and emerging countries, intervened to reduce the effect of the global food price spikes (Barrett 2014; Naylor 2014; Pinstrup-Andersen 2014). Governments used price and trade policies to counter global price movements and to insulate the domestic market from the international price spikes (Demeke, Pangrazio, and Maetz 2009).[10] At the same time, food price spikes triggered media and policy attention to the broader issues of hunger and rural poverty.

6.1 Trading-off volatility and distortions

A key argument in favor of policy interventions to stabilize prices is that price volatility causes inefficiencies in demand and supply (FAO 2011; Prakash 2011; World Bank 2012). Unexpected price changes make it difficult for consumers and producers to make optimal decisions, which reduces their confidence in the market and investment. Some criticize government interventions to insulate domestic markets from global price fluctuations for (1) being ineffective, (2) causing distortions in the economy, or (3) reinforcing price fluctuations when food exporters reduce supply and food importers increase demand (Anderson, Ivanic, and Martin 2014; Ivanic and Martin 2014).

From a political economy perspective, there is nothing surprising in the way that governments have reacted to changes in world market prices. Government interventions to counter market fluctuations are a key "stylized fact" of agricultural and food policies induced by the political economy mechanism caused by relative income hypothesis or loss aversions. Hence, even without taking into account possible additional benefits for consumers or producers from price stability, political mechanisms will induce governments to respond to international price increases by policy interventions that limit price rises on domestic markets by export constraints and vice versa through import tariffs when prices fall on the international markets.

To integrate benefits from price stability, Pieters and Swinnen (2016) derive a socially optimal distortions–volatility (DV) tradeoff[11] that takes into account both consumer and producer benefits from consumption distortions caused by deviations from the world market price. However, they find that many of the government policies are often far removed from socially optimal distortion–volatility (DV) combinations. Hence, political motives are very important.

6.2 Food prices and global development policy

Food crises push food security and agricultural development upward on policy makers' priority lists, especially when accompanied by mass media attention (Aksoy and Hoekman 2010; Headey 2013; Verpoorten et al. 2013; Guariso, Squicciarini, and Swinnen 2014).[12] The price spikes of 2007–2008, which created political instability in some countries, received heavy media coverage worldwide due to urban protests. As a result, the international mass media paid a disproportionate amount of attention to the short-run food insecurity of urban populations compared to the long-run food insecurity problems of rural populations (Cohen and Garret 2010; Maystadt, Tanb, and Breisinger 2014).

Thus, while for many years experts pointed at the low level of investment in developing country agriculture as a source of poverty and food security, it was only after the "food crisis" that media attention increased and that policy makers worldwide put rural poverty and underinvestment in agriculture on their priority list. Donor funding has since followed.[13]

The "food crisis" acted as a catalyst of attention on long-standing issues related to food security and agricultural production, which were made particularly salient by the fact that urban consumers – whose voice is typically heard the most by mass media and policy makers – were hit the hardest by the spikes in food prices. What is remarkable is that, despite the fact that food insecurity for poor farmers in developing countries has been a major problem for a long time, an "urban (consumer) crisis" helped put the poor farmers' situation on top of the international aid agenda. Hence, food price spikes may have succeeded where others have failed in the past: to put the food insecurity problems of poor farmers on the policy agenda and to induce development policies and donor strategies to help them.

7 The political economy of food standards

Increasingly, agricultural production and trade are regulated by stringent public and private standards based on quality, safety, nutritional, environmental, and ethical aspects. An important critique is that standards are (non-tariff) trade barriers. Because trade agreements such as those made by the WTO have reduced tariffs, countries may use standards to shield their domestic markets from foreign competition (Fischer and Serra 2000; Brenton and Manchin 2003; Anderson, Damania, and Jackson 2004; Swinnen 2017). Convergence (or not) of standards is at the heart of trade negotiations, such as the Transatlantic Trade and Investment Partnership (TTIP).

Standards affect trade.[14] However, the implicit comparison with tariffs in the trade debate is not entirely valid. In a small open economy, the socially optimal tariff level is zero. A positive tariff level constrains trade, is harmful to social welfare, and is protectionist. However, this is not necessarily the case for standards since this ignores the potential benefits of standards. If the standard reduces asymmetric information or externalities, there is no simple relationship between the trade effects of a standard and the social optimum (Van Tongeren, Beghin, and Marette 2009; Marette and Beghin 2010; Sheldon 2012; Beghin 2013; Marette 2014). This result, however, does not imply that there are no protectionist elements in standards setting.

Standards can enhance aggregate welfare by reducing asymmetric information or negative externalities; they can also create rents for specific interest groups. Because of the distributional effects of standards, interest groups have a vested interest in influencing governments' decisions on standards. When interest groups have differing lobbying strengths, the political equilibrium will generally differ from the social optimum (Anderson, Damania, and Jackson 2004; Swinnen and Vandemoortele 2008, 2011).

The political equilibrium standard may be either too high or too low from a social welfare point of view. Influential lobby groups may push for either more stringent or less stringent standards depending on the relative magnitude of the price (demand) effect compared to the implementation cost (for producers) or the efficiency gain (for consumers) (Beghin, Maertens, and Swinnen 2015; Swinnen 2016). For example, if producers are more influential than consumers, then over-standardization results when producers' profits increase with a higher standard, and results in under-standardization otherwise. Higher profits for producers are more likely when the standard's price (demand) effect is large and when the implementation cost is small.

7.1 Development and pro-standard and anti-standard coalitions

This political economy can explain the empirically observed positive relationship between standards and economic development. First and most obvious, higher income levels are typically associated with higher consumer preferences for quality and safety standards, as reflected in higher efficiency gains. Second, the quality of institutions for enforcement of contracts and public regulations is correlated positively with development. Better institutions imply better enforcement and control of standards. While poor countries may have a cost advantage in the production of raw materials, better institutions in rich countries lower the marginal increase in production costs caused by standards. Third, higher education and the skills of producers, better public infrastructure, easier access to finance, etc., also lower implementation costs. Fourth is the different organization and structure of the media in rich and poor countries. Mass media is the main source of information for many people (McCluskey and Swinnen 2004). The cost of media information is higher and government control of the media is stronger in poor countries. Therefore, the media structure and information provision is likely to induce a more pro-standard attitude in rich countries than in poor countries because increased access to media increases attention to risks and the negative implications of low standards (Curtis, McCluskey, and Swinnen 2008).

In combination, these factors are likely to induce a shift of the political equilibrium from low standards to high standards with development. A pro-standard coalition of consumers and producers in rich countries results if consumers derive large efficiency gains from a standard, while producers incur only moderate increases in costs. In contrast, an anti-standard coalition may be present in poor countries if consumers are more concerned with low prices than with high quality (leading to small efficiency gains from a higher standard), while the implementation costs for producers may be large. Structural differences in information and media may reinforce the positive relationship between standards and development.

7.2 The persistence of standards: dynamic political economics

Some of the most important political aspects of standards relate to their dynamic effects. Dynamic political economic aspects of standards can provide an explanation for why there are different food standards in countries with similar levels of development (e.g., the European Union and the United States) and why such differences may persist.[15]

Once adopted, countries will tend to stick to the status quo in standards because implementation costs depend on the existing standard because of past investments. Differences in standards between countries may persist because of this and trade may enforce it. The reason is that producer or consumer preferences may change in a dynamic way once the standard is introduced.[16] The standard will affect comparative advantages and will thus induce producers to support maintaining the standard to protect them from (cheaper) non-standard imports. Hence, although standards may have been introduced due to consumer demands, their persistence in the long run results from (a coalition of consumer and) producer demands. Hence, hysteresis in standards can be driven by protectionist motives even if the initial standards were not introduced for protectionist reasons.

Empirical studies document persistence of standards over time and that the protectionist effects of standards may increase over time.[17] Such regulatory differences may cause major conflicts later as the industries lobby to impose their own standards on foreign producers.

Other studies document important historical changes in standards (Vogel 2003). However, significant "shocks" (both internal and external) to the political economy system may be required for such changes, that is, to move the political economy equilibrium to another equilibrium given the dynamic political and institutional constraints to overcome (Rausser, Swinnen, and Zusman 2011). One source of shocks is internal. One example is the effect of domestic crises on food standards. Modern public food safety and quality regulations began in the late nineteenth century over the use of cheap and sometimes poisonous ingredients in food production (Meloni and Swinnen 2015, 2017). Since then, tightening public standards for food has continued, with consumers demanding better protection (McCluskey and Swinnen 2011).

Another source of shocks is external. One example is the integration of countries with different standards through international agreements. This may cause either the removal of inefficient standards or the extension of inefficient standards to other countries with international integration.

8 The political economy of public investment in agricultural research

Public investment in agricultural research is an important source of productivity growth (Alston 2017). Studies have documented both high social rates of return to public agricultural research investments and significant underinvestment in research in both poor and rich countries (Ruttan 1982; Huffman and Evenson 1992).

One political economy explanation of the underinvestment by governments is spillover effects (or externalities) in a policy environment where government research investments in one country affect other countries.[18] Research has both public and private good characteristics because some of the benefits of research expenditures can be captured by specific groups/countries, while other benefits spill over to other groups/countries (Cornes and Sandler 1986). This affects governments' incentives to invest in research. Spillover effects can induce free-riding behavior, whereby benefits are reaped from investments made by others (Huffman and Evenson 1992). Alternatively, interactions between private sector investments and government incentives may lead to suboptimal public research investment (Ulrich, Furtan, and Schmitz 1986).

A different political economy explanation draws on the distributional effects of public research investments (de Gorter and Zilberman 1990; de Gorter, Nielson, and Rausser 1992; Rausser 1992). While society as a whole will gain from research, consumers and producers will have different preferences dependent upon how research affects their income. Each side will negatively react to government research spending if they do not benefit from the research.

Public investment in research has contributed to the dramatic increase in the productivity of agriculture beginning in the twentieth century. In turn, the increase in agricultural productivity has contributed to the long-term decrease in agricultural prices, thus benefiting consumers and putting pressure on farm incomes (Ruttan 1982; Alston 2017). When there is opposition to research because of income distribution effects, governments will underinvest in public research to balance the political costs and benefits of diverging from the social optimum. As de Gorter and Swinnen (1998) show, with unequal income distributional effects, a government maximizing political support will underinvest in research, both in rich countries and in poor countries. Gardner (1989) and Oehmke and Yao (1990) find that underinvestment occurs if farmers gain relatively less from research.

9 Policy interactions

So far, we have analyzed the political economy of various policies in isolation, meaning that we analyzed them as if there were no other policies. However, in reality, many public policies exist simultaneously and may interact with each other. In the study by de Gorter and Swinnen (1998), they distinguish between "*economic* interaction effects" (EIEs), which arise if one policy affects the distributional and welfare effects of other policies, and "*political* interaction effects" (PIEs), which occur when one policy affects the political incentives of governments to introduce or change other policies.

One form of (positive) EIE is when combined reforms reinforce the (beneficial) impacts of separate policy reforms. For example, in the reform strategies in China and Eastern Europe in the 1990s, land reforms and privatization strategies provided new opportunities and better incentives for farmers, while, at the same time, distortionary price and market policies were reduced or removed. In these cases, both policy reforms combined improved efficiency. An example of (negative) EIEs is the interaction between public agricultural research and commodity policies that regulate agricultural prices or production. Agricultural research increases productivity and may cause an increase in the distortions of existing regulations. Under some conditions, the increased distortions may outweigh the research benefits (Murphy, Furtan, and Schmitz 1993; Alston, Edwards, and Freebairn 1988).

Probably the best-known example of PIEs is the use of distributional policies for compensation purposes. Compensation is an important element in the political economy of policy reform or public investment (Rausser 1982; Rausser, Swinnen, and Zusman 2011; Anderson, Rausser, and Swinnen 2013).[19] Reforms to a more efficient policy usually imply gains for some groups and losses for others. Similarly, building a road may lead to major gains in rural development but may hurt those displaced by the construction of the road. If the gains outweigh the losses, it is socially optimal to implement the reforms or make the investment, since the gains of those who win are more than sufficient to compensate the losers. There are numerous empirical examples of "policy packages" that include compensation for certain groups. They are a traditional part of multi-annual agricultural policy decision making in both the European Union and the United States.[20]

An important problem with compensation, however, is the credible implementation of such schemes. Those who lose from reforms may oppose the reforms if they expect that (full) compensation will not take place. This may occur when governments lack the credibility to effectively provide compensation for reforms (Swinnen and de Gorter 2002), when governments offer only partial compensation to mitigate political opposition to reforms (Foster and Rausser 1993), when local institutions prevent effective compensation schemes (Swinnen 1997), or when there is uncertainty regarding the effect of the reforms (Fernandez and Rodrik 1991).

The inability of governments to credibly commit to compensate groups that are adversely affected is a prime cause of failures to implement aggregate welfare-improving policies. An important question is how to design mechanisms that constrain policy makers to bring the discretionary political equilibrium closer to the social optimum. One way is the creation of institutions that make policy reversal more difficult to enhance the credibility of policy makers to commit to future compensation. These institutions include independent central banks for monetary policies or international trade agreements, which impose constraints on government policies in agriculture and food.

Acknowledgments

I thank Andy Schmitz and Keith Coble for useful comments on the chapter, and the editors for their editorial guidance. Much of this chapter has benefited from discussions and earlier collaborations with Kym Anderson, Harry de Gorter, Jill McCluskey, Alessandro Olper, Gordon Rausser, and many others. Writing this chapter was supported by the KU Leuven's Methusalem project.

Notes

1 For a more extensive discussion on all the issues discussed in this chapter, see Swinnen (2018).
2 A survey of this literature is in de Gorter and Swinnen (2002).
3 See Rausser, Swinnen, and Zusman (2012) for a review of contributions in these fields.
4 See Swinnen (2015) for more details on this.
5 See Anderson, Rausser, and Swinnen (2013) for a more elaborate review.
6 Econometric studies by Gawande and Hoekman (2006) and López (2008) also show the influence of agribusiness and food companies' political contributions on U.S. policies.

7 "Coupled PSE" includes all policy transfers (e.g., tariffs, price support, and subsidies) directly linked (coupled) to agricultural production (most distortive). "Decoupled PSE" includes decoupled agricultural payments (least distortive).
8 The URAA appears to have had less impact on U.S. agricultural policies. Nonetheless, the U.S. administration has attempted to introduce policy reforms with an eye toward insuring that many U.S. agricultural subsidies are classified as "green box" (i.e., non-trade distorting) under the WTO agreement (Orden, Blandford, and Josling 2010).
9 Interestingly, the combination of increased budgetary transparency and high agricultural prices in the late 2000s and early 2010s are argued to have caused a shift (back) from direct payments towards more coupled programs, such as crop insurance payments (Coble and Barnett 2013; de Gorter, Drabsik, and Just 2015; Goodwin and Smith 2013).
10 While food-exporting countries suspend or ban exports, food-importing countries use price regulations to restrict price increases (Naylor 2014; Pinstrup-Andersen 2014). Also, governments (sometimes erroneously) assume that interventions will have an immediate effect – resembling what Swinnen (1996) calls "fire brigade policy making" when governments are confronted with unfamiliar shocks in the external environment.
11 Their model is based on Barrett (1996); Bellemare, Barrett, and Just (2013); and Gouel and Jean (2015).
12 Swinnen, Squicciarini, and Vandemoortele (2011) develop a model of the market for communications, mass media, and donor funding. Media attention is typically concentrated around events or shocks, the so-called bad news hypothesis (Hawkins 2002; McCluskey and Swinnen 2004; Swinnen and Francken 2006).
13 After the food crisis, international aid and research funding increased (Global Policy Forum 2013; Guariso, Squicciarini, and Swinnen 2013).
14 Only in very special circumstances do standards not affect trade: this is when the effect on domestic production exactly offsets the effect on consumption (Swinnen and Vandemoortele 2009).
15 See Swinnen and Vandemoortele (2011); Swinnen (2016, 2017); and Swinnen, Olper, and Vandemoortele (2016) for more technical analysis and details.
16 The case that producers have different preferences and consumers have the same is analogous. For example, Paarlberg (2008) argues that consumers on both sides of the Atlantic Ocean tend to dislike GM technology, but agribusiness lobbying has been much more pro-GM in the United States. In the longer run, it may be that because consumers live in different GM-food environments in the United States and the European Union, they develop different preferences. Consumer attitudes with respect to biotechnology are likely to be endogenous. Consumer preferences may shift in favor of biotechnology in countries where GM products are available, as opposed to countries that ban GM food products.
17 For example, Meloni and Swinnen (2013) show how stringent standards in the wine industry, which were first set in France around 1900 in response to pressure on wine growers, further tightened over time, and later spread to the rest of Europe with integration of other wine-producing countries into the European Union. Meloni and Swinnen (2015, 2016) also document how the introduction of food standards in the mid-nineteenth century, in response to the discovery by new scientific means of massive fraud and adulterations in food production, led to different regulatory approaches in different countries. These regulations and standards persisted for a long time and influenced production processes and consumer preferences in the domestic industries, leading to trade conflicts. Similarly, van Tongeren (2011) shows how a 500-year-old food law was the reason for trade disputes in the late twentieth century.
18 Studies have also argued that benefits of public research are overestimated due to deadweight costs of taxation (Fox 1985), terms of trade effects (Edwards and Freebairn 1984), the effects on unemployment (Schmitz and Seckler 1970), increased deadweight costs of existing commodity policies (Alston, Edwards, and Freebairn 1988; Murphy, Furtan, and Schmitz 1993), or lags in the effects of research (Alston and Pardey 1996).
19 Trade policy reform and compensation have a long history in the economics literature, going back to the early analyses of Adam Smith and David Ricardo. A crucial element in the arguments on the optimality of free trade are that the gains of the winners of trade liberalization are more than sufficient to compensate the losers of reform, an issue which has clearly become highly relevant again in recent years with discussions on the gainers and losers from globalization.
20 Modeling such joint policy decisions is complex and the identification of equilibria may be difficult, in particular when decision making institutions are modeled explicitly (see e.g., Pokrivcak et al. 2006).

References

Acemoglu, D. 2003. "Why Not a Political Coase Theorem? Social Conflict, Commitment, and Politics." *Journal of Comparative Economics* 31(4):620–652.

Aksoy, M.A., and B. Hoekman. 2010. *Food Prices and Rural Poverty*. Washington, DC: World Bank.

Alesina, A., and D. Rodrik. 1994. "Distributive Politics and Economic Growth." *Quarterly Journal of Economics* 109(2):465–490.

Alston, J.M. 2017. Fellows Address. "Reflections of Agricultural R&D, Productivity, and the Data Constraint: Unfinished Business, Unsettled Issues." *Proceedings of the American Journal of Agricultural Economics Conference*, Chicago, IL (August).

Alston, J.M., G.W. Edwards, and J.W. Freebairn. 1988. "Market Distortions and Benefits from Research." *American Journal of Agricultural Economics* 70:281–288.

Alston, J.M., and J.S. James. 2002. "The Incidence of Agricultural Policy." In B. Gardner and G. Rausser, eds., *Handbook of Agricultural Economics*. Amsterdam: North Holland, pp. 2073–2123.

Alston, J.M., and W. Martin. 1995. "Reversal of Fortune: Immiserizing Technical Change in Agriculture." *American Journal of Agricultural Economics* 77(95):251–259.

Alston, J.M., and P. Pardey. 1996. *Making Science Pay. The Economics of Agricultural R&D Policy*. Washington, DC: AEI Press.

Anderson, K. 1995. "Lobbying Incentives and the Pattern of Protection in Rich and Poor Countries." *Economic Development and Cultural Change* 43(2):401–423.

———. (editor). 2009. *Distortions to Agricultural Incentives: A Global Perspective, 1955–2007*. Washington, DC: World Bank and Palgrave Macmillan.

———. 2016. *Agriculture Trade, Policy Reforms, and Global Food Security*. London: Palgrave Macmillan.

Anderson, K., R. Damania, and L.A. Jackson. 2004. "Trade, Standards, and the Political Economy of Genetically Modified Food." CEPR Discussion Paper 0410. University of Adelaide, Adelaide, Australia.

Anderson, K., M. Ivanic, and W. Martin. 2014. "Food Price Spikes, Price Insulation, and Poverty." In J.-P. Chavas, D. Hummels, and B. Wright, eds., *The Economics of Food Price Volatility*. Chicago, IL: University of Chicago Press, pp. 311–344.

Anderson, K., G.C. Rausser, and J. Swinnen. 2013. "Political Economy of Public Policies: Insights from Distortions to Agricultural and Food Markets." *Journal of Economic Literature* 51(2):423–477.

Anderson, K., and J. Swinnen (editors). 2008. *Distortions to Agricultural Incentives in Europe's Transition Economies*. Washington, DC: World Bank.

Barrett, C.B. 1996. "On Price Risk and the Inverse Farm Size-Productivity Relationship." *Journal of Development Economics* 51(2):193–215.

———. 2014. *Food Security and Sociopolitical Stability*. Oxford: Oxford University Press.

Bates, R.H., and S. Block. 2010. "Agricultural Trade Interventions in Africa." In K. Anderson, ed., *The Political Economy of Agricultural Price Distortions*. Cambridge: Cambridge University Press, pp. 304–331.

Becker, G.S. 1983. "A Theory of Competition among Pressure Groups for Political Influence." *Quarterly Journal of Economics* 98:371–400.

Beghin, J. 2013. *Nontariff Measures with Market Imperfections: Trade and Welfare Implications*. Bingley: Emerald.

Beghin, J., M. Maertens, and J. Swinnen. 2015. "Nontariff Measures and Standards in Trade and Global Value Chains." *Annual Review of Resource Economics* 7:425–450.

Bellemare, M.F., C.B. Barrett, and D.R. Just. 2013. "The Welfare Impacts of Commodity Price Volatility: Evidence from Rural Ethiopia." *American Journal of Agricultural Economics* 95(4):877–899.

Bhagwati, J.N. 1982. "Directly Unproductive Profit Seeking Activities: A Welfare Theoretic Synthesis and Generalization." *Journal of Political Economy* 90:988–1002.

Brenton, P., and M. Manchin. 2003. "Making EU Trade Agreements Work: The Role of Rules of Origin." *The World Economy* 26(5):755–769.

Buchanan, J.M., and G. Tullock. 1962. *The Calculus of Consent*. Ann Arbor, MI: University of Michigan Press.

Coble, K.H., and B.J. Barnett. 2013 "Why Do We Subsidize Crop Insurance?" *American Journal of Agricultural Economics* 95(2):498–504.

Cohen, M.J., and J.L. Garrett. 2010. "The Food Price Crisis and Urban Food (In)security." *Environment and Urbanization* 22:467–482.

Cornes, R., and T. Sandler. 1986. *The Theory of Externalities, Public Goods, and Club Goods*. Cambridge: Cambridge University Press.

Curtis, K.R., J.J. McCluskey, and J. Swinnen. 2008. "Differences in Global Risk Perceptions of Biotechnology and the Political Economy of the Media." *International Journal of Global Environmental Issues* 8(1/2):77–89.

de Gorter, H., D.J. Nielson, and G.C. Rausser. 1992. "Productive and Predatory Public Policies: Research Expenditures and Producer Subsidies in Agriculture." *American Journal of Agricultural Economics* 74(1):27–37.

de Gorter, H., and J. Swinnen. 1998. "The Impact of Economic Development on Public Research and Commodity Policies in Agriculture." *Review of Development Economics* 2(1):41–60.

———. 2002. "Political Economy of Agricultural Policies." In B. Gardner and G. Rausser, eds., *The Handbook of Agricultural Economics*. Amsterdam: Elsevier, pp. 2073–2123.

de Gorter, H., and D. Zilberman. 1990. "On the Political Economy of Public Good Inputs in Agriculture." *American Journal of Agricultural Economics* 72:131–137.

Demeke, M., G. Pangrazio, and M. Maetz. 2009. "Country Responses to the Food Security Crisis: Nature and Preliminary Implications of the Policies pursued." In *FAO Initiative on Soaring Food Prices*. Rome: FAO, pp. 1–32.

Dixit, A.K. 1996. *The Making of Economic Policy: A Transaction Cost Politics Perspective*. Cambridge, MA: MIT Press.

Downs, A. 1957. *An Economic Theory of Democracy*. New York: Harper and Row.

Edwards, G.W., and J.W. Freebairn. 1984. "The Gains from Research into Tradable Commodities." *American Journal of Agricultural Economics* 66(1984):41–49.

Fernandez, R., and D. Rodrik. 1991. "Resistance to Reform: Status Quo Bias and the Presence of Individual Specific Uncertainty." *American Economic Review* 81:1146–1155.

Fischer, R., and P. Serra. 2000. "Standards and Protection." *Journal of International Economics* 52(2):377–400.

Food and Agricultural Organization of the United Nations [FAO]. 2011. *Price Volatility in Food and Agricultural Markets: Policy Responses*. Rome: FAO.

Foster, W.E., and G.C. Rausser. 1993. "Price-Distorting Compensation Serving the Consumer and Taxpayer Interest." *Public Choice* 77(2):275–291.

Fox, G.C. 1985. "Is the United States Really Underinvesting in Agricultural Research?" *American Journal of Agricultural Economics* 67:806–812.

Freund, C., and C. Ozden. 2008. "Trade Policy and Loss Aversion." *American Economic Review* 98(4):1675–1691.

Gardner, B.L. 1983. "Efficient Redistribution Through Commodity Markets." *American Journal of Agricultural Economics* 65(2):225–234.

———. 1987. "Causes of U.S. Farm Commodity Programs." *Journal of Political Economy* 95(2):290–310.

Gawande, K., and B. Hoekman. 2006. "Lobbying and Agricultural Trade Policy in the United States." *International Organization* 60:527–561.

Glauber, J. 2004. "Crop Insurance Reconsidered." *American Journal of Agricultural Economics* 86(5), 1179–1195.

Global Policy Forum. 2013. *Global Policy Forum – Financing of the UN Programmes, Funds, and Specialized Agencies*. New York: Global Policy Forum.

Goodwin, B.K., and V. Smith. 2013. "What Harm Is Done by Subsidizing Crop Insurance?" *American Journal of Agricultural Economics* 95(2):489–497.

Gouel, C., and S. Jean. 2015. "Optimal Food Price Stabilization in a Small Open Developing Country." *The World Bank Economic Review* 29(1):72–101.

Grossman, G.M., and E. Helpman. 1994. "Protection for Sale." *American Economic Review* 84(4):833–850.

Guariso, A., M.P. Squicciarini, and J. Swinnen. 2014. "Food Price Shocks and the Political Economy of Global Agricultural and Development Policy." *Applied Economic Perspectives and Policy* 36(3):387–415.

Hawkins, V. 2002. "The Other Side of the CNN Factor: The Media and Conflict." *Journalism Studies* 3(2):225–240.

Headey, D. 2013. "The Impact of the Global Food Crisis on Self-Assessed Food Security." *World Bank Economic Review* 27(1):1–27.

Huffman, W.E., and R.E. Evenson. 1992. "Contributions of Public and Private Science and Technology to U.S. Agricultural Productivity." *American Journal of Agricultural Economics* 74:751–756.

Ivanic, M., and W. Martin. 2014. "Implications of Domestic Price Insulation for Global Food Price Behaviour." *Journal of International Money and Finance* 42:272–288.

Krueger, A.O. 1974. "The Political Economy of the Rent-Seeking Society." *American Economic Review* 64(3):291–303.

Krueger, A.O., M. Schiff, and A. Valdes. 1991. *The Political Economy of Agricultural Pricing Policy*. Baltimore, MD: Johns Hopkins University Press for the World Bank.

Liefert, W., and J. Swinnen. 2002. "Changes in Agricultural Markets in Transition Countries." USDA-ERS, Washington, DC.

López, R.A. 2008. "Does 'Protection for Sale' Apply to the US Food Industries?" *Journal of Agricultural Economics* 9(1):25–40.

Magee, S.P., W.A. Brock, and L. Young. 1989. *Black Hole Tariffs and Endogenous Policy Theory*. Cambridge: Cambridge University Press.

Marette, S. 2014. *Non-Tariff Measures When Alternative Regulatory Tools Can Be Chosen*. Grignon, France: INRA-AgroParisTech.

Marette, S., and J. Beghin. 2010. "Are Standards Always Protectionist?" *Review of International Economics* 18(1):179–192.

Maystadt, J.F., J.F.T. Tanb, and C. Breisinger. 2014. "Does Food Security Matter for Transition in Arab Countries?" *Food Policy* 46:106–115

McCluskey, J.J., and J.F.M. Swinnen. 2004. "Political Economy of the Media and Consumer Perceptions of Biotechnology." *American Journal of Agricultural Economics* 86:1230–1237.

———. 2010. "Media Economics and the Political Economy of Information." In D. Coen, W. Grant, and G. Wilson, eds., *The Oxford Handbook of Government and Business*. Oxford: Oxford University Press, pp. 643–662.

———. 2011. "Media and Food Risk Perceptions." *The EMBO Journal* 12(7):467–486.

McGuire, M.C., and M. Olson Jr. 1996. "The Economics of Autocracy and Majority Rule: The Invisible Hand and the Use of Force." *Journal of Economic Literature* 34(1):72–96.

Meloni, G., and J. Swinnen. 2013. "The Political Economy of European Wine Regulations." *Journal of Wine Economics* 8(3):244–284.

———. 2015. "Chocolate Regulations." In M. Squicciarini and J. Swinnen, eds., *The Economics of Chocolate*. Oxford: Oxford University Press, pp. 268–306.

———. 2017. "Standards, Tariffs, and Trade: The Rise and Fall of the Raisin Trade between Greece and France in the Late Nineteenth Century." *Journal of World Trade* 51(4):711–740.

Mitchell, M., and A. Moro. 2006. "Persistent Distortionary Policies with Asymmetric Information." *American Economic Review* 96(1):387–393.

Murphy, J.A., W.H. Furtan, and A. Schmitz. 1993. "The Gains from Agricultural Research under Distorted Trade." *Journal of Public Economics* 51:161–172.

Naylor, R.L. (editor). 2014. *The Evolving Sphere of Food Security*. Oxford: Oxford University Press.

Olper, A., J. Fałkowski, and J. Swinnen. 2014. "Political Reforms and Public Policies: Evidence from Agricultural and Food Policy." *World Bank Economic Review* 28(1):21–47.

Olper, A., and J. Swinnen. 2013. "Mass Media and Public Policy for Agriculture." *World Bank Research Digest* 7(3):Article 6.

Olson, M. 1965. *The Logic of Collective Action*. New Haven, CT: Yale University Press.

Orden, D., D. Blandford, and T. Josling. 2010. "Determinants of United States Farm Policies." In K. Anderson, ed., *The Political Economy of Agricultural Price Distortions*. Cambridge: Cambridge University Press, pp. 160–190.

Paarlberg, R. 2008. *Starved for Science. How Biotechnology Is Being Kept Out of Africa*. Cambridge, MA: Harvard University Press.

Persson, T., and G. Tabellini. 2000. *Political Economics: Explaining Policy*. Cambridge, MA: MIT Press.

———. 2003. *The Economic Effects of Constitutions: What Do the Data Say?* Cambridge, MA: MIT Press.

Pieters, H., and J. Swinnen. 2016. "Trading-Off Volatility and Distortions? Food Policy During Price Spikes." *Food Policy* (61):27–39.

Pinstrup-Andersen, P. (editor). 2014. *Food Price Policy in an Era of Market Instability: A Political Economy Analysis*. Oxford: Oxford University Press.

Pokrivcak, J., Crombez, C., and J. Swinnen. 2006. "The Status Quo Bias and Reform of the Common Agricultural Policy: Impact of Voting Rules, the European Commission, and External Changes." *European Review of Agricultural Economics* 33(4):562–590.

Prakash, A. 2011. "Why Volatility Matters." In A. Prakash, ed., *Safeguarding Food Security in Volatile Global Markets*. Rome: FAO, pp. 3–26.

Rausser, G.C. 1982. "Political Economic Markets: PERTs and PESTs in Food and Agriculture." *American Journal of Agricultural Economics* 64(5):821–833.

———. 1992. "Predatory Versus Productive Government: The Case of U.S. Agricultural Policy." *Journal of Economic Perspectives* 6(3):133–157.

Rausser, G., J. Swinnen, and P. Zusman. 2012. *Political Power and Economic Policy: Theory, Analysis, and Empirical Applications*. Cambridge, UK: Cambridge University Press.

Rodrik, D. 1995. "Political Economy of Trade Policy." In G. Grossman and K. Rogoff, eds., *Handbook of International Economics*. Amsterdam: North-Holland, pp. 1457–1494.

Rozelle, S., and J.F.M. Swinnen. 2004. "Success and Failure of Reforms: Insights from Transition Agriculture." *Journal of Economic Literature* 42:404–456.

———. 2009. "Why Did the Communist Party Reform in China, But Not in the Soviet Union? The Political Economy of Agriculture Transition." *China Economic Review* 20(2):275–287.

———. 2010. "Agricultural Distortions in the Transition Economies of Asia and Europe." In K. Anderson, ed., *The Political Economy of Agricultural Price Distortions*. Cambridge: Cambridge University Press, pp. 191–214.

Ruttan, V.W. 1982. *Agricultural Research Policy*, Minneapolis, MN: University of Minnesota Press.

Schmitz, A., C.B. Moss, T.G. Schmitz, H.W. Furtan, and H.C. Schmitz. 2010. *Agricultural Policy, Agribusiness, and Rent-Seeking Behaviour* (2nd ed.). Toronto: University of Toronto Press.

Sheldon, I. 2012. "North-South Trade and Standards: What Can General Equilibrium Theory Tell Us? *World Trade Review* 11(3):376–389.

Stigler, J.G. 1971. "Theory of Economic Regulation." *The Bell Journal of Economics and Management Science* 2(1):3–21.

Stromberg, D. 2001. "Mass Media and Public Policy." *European Economic Review* 45(4–6):652–663.

———. 2004. "Mass Media Competition, Political Competition, and Public Policy." *Review of Economic Studies* 71(1):265–284.

Swinnen, J.F.M. 1994. "A Positive Theory of Agricultural Protection." *American Journal of Agricultural Economics* 76(1):1–14.

———. 1996. "Endogenous Price and Trade Policy Developments in Central European Agriculture." *European Review of Agricultural Economics* 23(2):133–160.

———. 1997. "Does Compensation for Disruptions Stimulate Reforms? The Case of Agricultural Reform in Central Europe." *European Review of Agricultural Economics* 24(2):249–266.

———. 2009. "The Growth of Agricultural Protection in Europe in the 19th and 20th Centuries." *The World Economy* 32(11):1499–1537.

———. 2010. "Political Economy of Agricultural Distortions: The Literature to Date." In K. Anderson, ed., *The Political Economy of Agricultural Price Distortions*. Cambridge: Cambridge University Press, pp. 81–104.

———. 2015. "Changing Coalitions in Value Chains and The Political Economy of Agriculture and Food Policy." *Oxford Review of Economic Policy* 31(1):90–115.

———. 2016. "Economics and Politics of Food Standards, Trade, and Development." *Agricultural Economics* 47(S1):7–19.

———. 2017. "Some Dynamic Aspects of Food Standards." *American Journal of Agricultural Economics* 99(2):321–338.

———. 2018. *The Political Economy of Agricultural and Food Policies*. Palgrave McMillan.

Swinnen, J., A. Banerjee, and H. de Gorter. 2001. "Economic Development, Institutional Change, and the Political Economy of Agricultural Protection: An Econometric Study of Belgium since the 19th Century." *Agricultural Economics* 26(1):25–43.

Swinnen, J., A. Olper, and T. Vandemoortele. 2012. "Impact of the WTO on Agricultural and Food Policies." *The World Economy* 35(9):1089–1101.

———. 2016. "The Political Economy of Policy Instrument Choice: Theory and Evidence from Agricultural and Food Policies." *Theoretical Economics Letters* 6(1):106–117.

Swinnen, J., P. Squicciarini, and T. Vandemoortele. 2011. "The Food Crisis, Mass Media and the Political Economy of Policy Analysis and Communication." *European Review of Agricultural Economics* 38(3):409–426.

Swinnen, J., and T. Vandemoortele. 2008. "The Political Economy of Nutrition and Health Standards in Food Markets." *Applied Economic Perspectives and Policy* 30(3):460–468.

———. 2009. "Are Food Safety Standards Different from Other Food Standards? A Political Economy Perspective." *European Review of Agricultural Economics* 36(4):507–523.

———. 2011. "Trade and the Political Economy of Food Standards." *Journal of Agricultural Economics* 62(2):259–280.

Swinnen, J., A. Vandeplas, and M. Maertens. 2010. "Liberalization, Endogenous Institutions, and Growth. A Comparative Analysis of Agricultural Reforms in Africa, Asia, and Europe." *The World Bank Economic Review* 24(3):412–445.

Swinnen, J.F.M., and H. de Gorter. 2002. "On Government Credibility, Compensation, and Under-Investment in Public Research." *European Review of Agricultural Economics* 29(4):501–522.

Swinnen, J.F.M., and N. Francken. 2006. "Summits, Riots and Media Attention: The Political Economy of Information on Trade and Globalization." *World Economy* 29:637–654.

Tovar, P. 2009. "The Effects of Loss Aversion on Trade Policy: Theory and Evidence." *Journal of International Economics* 78:154–167.

Ulrich, A., Furtan, H., and A. Schmitz. 1986. "Public and Private Returns from Joint Venture Research: An Example from Agriculture." *Quarterly Journal of Economics* 101(1):103–130.

Van Tongeren, F. 2011. "Standards and International Trade Integration: A Historical Review of the German 'Reinheitsgebot." In J. Swinnen, ed., *The Economics of Beer*. Oxford: Oxford University Press, pp. 51–61.

Van Tongeren, F., J. Beghin, and S. Marette. 2009. *A Cost–Benefit Framework for the Assessment of Non-Tariff Measures in Agro-Food Trade*. OECD: Paris, France.

Varshney, A. 1995. *Democracy, Development, and the Countryside: Urban–Rural Struggles in India*. Cambridge: Cambridge University Press.

Verpoorten, M., A. Arora, N. Stoop, and J.F.M. Swinnen. 2013. "Self-Reported Food Insecurity in Africa during the Food Price Crisis." *Food Policy* 39:51–63.

Vogel, D. 2003. "The Hare and the Tortoise Revisited: The New Politics of Consumer and Environmental Regulation in Europe." *British Journal of Political Science* 33:557–580.

Wise, T.A. 2004. "The Paradox of Agricultural Subsidies: Measurement Issues, Agricultural Dumping, and Policy Reform." Tufts University.

World Bank. 2012. *Responding to Higher and More Volatile World Food Prices*. Washington, DC: World Bank.

22
Macroeconomic issues in agricultural economics

Michael Reed and Sayed Saghaian

1 Introduction

This chapter covers the agricultural economics and related literature on macroeconomic issues in agriculture with a focus on international trade and finance, and how such factors as the monetary policy, exchange rate, and energy prices affect domestic and international markets for agricultural products. We do not cover the effects of interest rate changes on agricultural finance and land values even though those relationships are important for U.S. agriculture. Rather, we synthesize the literature on the more direct impacts of monetary policies and agricultural prices.

2 The beginnings of the literature

Ed Schuh (1974) was not the first to write about the importance of the exchange rate and U.S. agriculture, but he did recognize the importance of the move to flexible exchange rates by the world after the collapse of the Bretton Woods system in 1973. He argued that the U.S. dollar was seriously overvalued until the system's collapse and that the new exchange rate regime was an important structural change for U.S. agriculture. He was correct about the structural change issue, but his contention that a massive disequilibrium in world agriculture was eliminated was less true. He stated that the long-run supply curve for agriculture would determine future prices (which is also correct), but that the U.S. "could experience a permanent rise in the relative price of agricultural products, and the day of the U.S. consumer spending only 16 percent of his budget on food will be a thing of the past" (Schuh 1974, p. 11). The latter prediction has not happened.

Flexible exchange rates did not eliminate a U.S. balance of trade deficit, and the term "overvalued" currency did not mean that the U.S. dollar would depreciate consistently over time to increase agricultural prices. The term is still misunderstood and misused, but overvalued seems to mean that the amount of U.S. dollars demanded for investment in U.S. dollars, U.S. treasury bonds, and other dollar-denominated assets continues to be high and to generate a net outflow of U.S. dollars. In other words, the United States experiences a net trade of its currency for goods (a persistent balance of trade deficit) because of the international demand for its assets. Yet Schuh's observations were spot on about the importance of the exchange rate and other macroeconomic factors on U.S. agriculture.

The U.S. dollar has been the key currency since World War II because of the confidence in the U.S. economy. Krueger (1983) argues that the accumulation of dollars outside the United States and the overvaluation of the U.S. dollar caused the Bretton Woods system to collapse. The move to flexible exchange rates meant that the world economy was experiencing "as efficient an allocation of world resources as has existed at any time." (Krueger 1983, p. 866). This interdependence among world economies was due to five factors that Krueger referenced, but the most enduring was the floating exchange rate. The exchange rate is particularly important for agriculture due to the sensitivity of agricultural production and exports to the real exchange rate. Krueger (1983) believed the dollar's depreciation had unanticipated consequences for U.S. agriculture and agricultural exports, even though Schuh (1974) had anticipated these changes. Yet, neither of them recognized that the U.S. dollar could appreciate so rapidly in the 1980s, and that those fluctuations would be so important in the decades to come.

Schuh (1984) was concerned with the export performance of U.S. agriculture in the early 1980s. He believed that while the shift to flexible exchange rates was significant, changes in monetary policy induce international capital flows, which then induce changes in the value of the dollar that influence exports and imports. The observation that the United States must run a current account deficit because U.S. dollars are needed for capital outflows (transactions demand, precautionary demand, etc.), is still important today. Both the commodity boom in the 1970s and the bust in the 1980s were associated with the exchange rate. Schuh (1984) argued that the federal budget deficit was the reason because it created a demand for U.S. dollars and kept U.S. interest rates relatively high.

Batten and Belongia (1986) did not entirely agree with Schuh's assessment that the federal budget deficit hurt U.S. agricultural exports. While they agreed that the real appreciation of the U.S. dollar had contributed to lower U.S. agricultural exports, they disputed the causes for the appreciation. They argued that the monetary policy was not tight in the early 1980s and the linkage between monetary policy and the real exchange rate had been overhyped. They contended that the magnitude and duration of money supply effects on exchange rates and the magnitude and duration of the exchange rate effects on agricultural exports was overemphasized. Money growth only affects nominal exchange rates in the long run, and nominal exchange rates only change when relative future inflation expectations (among countries) change. The long-run trend in money growth is most closely associated with inflation, so they argued there is little correlation between money growth and the real value of the U.S. dollar. Their empirical analysis supported their proposition that neither the monetary policy nor the federal budget deficit has any effect on the real value of the U.S. dollar.

The international finance literature has also dealt with this new exchange rate regime. Economists watched the U.S. monetary policy carefully and analyzed its effects in the 1980s because of the increasing openness of the world economy, especially as related to the relaxation of exchange controls and free flow of currency. Frankel (1984) stresses the difference between international trade and finance, arguing that agricultural economists had neglected the latter. He discusses seven concepts that are important for agriculture from the international finance literature: money neutrality; interest rate parity; rational expectations; the magnification effect (that the exchange rate reaction to money supply increases is large because it might signal a change in monetary policy); overshooting; reaction to news; and the risk premium. He addresses the relationships between the nominal interest rate and unexpected increases in the money supply, attributing the decline in commodity prices in the early 1980s to the rise in real interest rates and the appreciation of the U.S. dollar. However, he argues that the effects of the U.S. dollar's appreciation are more than simply its effects through the law of one price (that is, the U.S. price being equal to the foreign price times the exchange rate). Rather, he argues, from a monetary approach, that portfolio effects also move commodity prices.

3 Monetary policy and commodity prices

The agricultural economics literature embraced the idea that money supply changes could affect nominal agricultural prices. Some of the early work integrated monetary policy directly into structural supply–demand models for agricultural commodities. Chambers and Just (1979) introduced the idea of a total exchange rate elasticity that takes into account adjustments in other markets. A broader view of exchange rate effects related to other commodities could capture overlooked income effects if exchange rate impacts are restricted to the law of one price. Chambers and Just (1982) take a monetary approach to the balance of payments, so that the exchange rate is mostly determined through balance of payments. They use a structural model of crops markets to estimate effects of money supply changes on disappearance, inventories, exports, and production. Results show that the impact of changes in domestic credit on the agricultural sector can be quite significant. Their long-run elasticities of corn and wheat exports to changes in domestic credit are large in the long run. Grain prices are also very responsive to fluctuations in domestic credit. They conclude that the burden of restrictive monetary policy may be unusually great for agricultural producers because upward pressure on the exchange rate has serious effects on the competitive position of U.S. exports.

Lachaal and Womack (1998) perform similar research for Canada. They use a structural system of production and price equations to model exportables, importables, home goods, and the real exchange rate. In their model, macroeconomic policies affect the price of home goods relative to importables and exportables, and thus the real exchange rate. They find that macroeconomic developments are important to Canadian agriculture and have a large impact on the Canadian farm economy.

The macroeconomic literature in agricultural economics, however, deviates from the structural equation systems due to the short-run nature of monetary impacts. The importance of measuring short-run impacts of money supply and other monetary policy on commodity prices leads naturally to time series analysis. Barnett, Bessler, and Thompson (1983) were the first to use Granger–Sims causality in their monetary analysis of agricultural prices. Using a monetarist explanation of money supply effects, they argue that money supply changes cause real shocks that alter relative prices. Prices that are quite flexible, such as many agricultural prices, can experience real price changes from monetary shocks. They test whether there is a two-way causality from (agricultural) prices to money. Their empirical findings are that changes in the U.S. money supply lead to increases in food and agricultural prices rather than merely accommodating any increases.

Barnett, Bessler, and Thompson (1983) find no significant evidence to support the idea that the money supply adjusts to accommodate agricultural price changes. Bessler (1984), however, finds an accommodation effect for industrial prices in Brazil using a Vector Autoregressive Model (VAR), implying that higher prices induce the monetary authority to loosen the money supply. He uses monthly data from 1964 to 1981, a very volatile period for prices and money supply changes in Brazil. There is significant causality from money to agricultural prices and the lags are quite short, but there is no significant evidence that the monetary authority pays attention to agricultural prices in its decisions.

Devadoss and Meyers (1987) use Bessler's VAR technique to analyze the relationship between relative agriculture prices and the money supply in the United States. They find that agricultural product prices respond faster than do manufactured product prices to money supply changes. Orden and Fackler (1989) focus more on time series techniques (particularly ordering restrictions in VAR models) in their analysis of monetary impacts on agricultural prices. They argue that recursive ordering implies strong assumptions about which variables affect each behavioral relationship. They argue that sensible results come from a model with simultaneity. In this model, there is an increase in agricultural prices with a positive monetary supply shock (and the interest rate falls and the exchange rate depreciates).

4 Agricultural goods as flex-priced

The early research using time series techniques moved the agricultural economics literature toward the differential effects between agricultural and other goods. This ultimately led to the idea of fixed-priced goods versus flex-priced goods and the possibility of overshooting. The genesis of this literature was the path-breaking article by Dornbusch (1976), which showed that exchange rates could overshoot their long-run equilibrium due to an unanticipated change in money supply with fixed real output in the short run. This overshooting occurs because of the interest rate parity condition and rational expectations for exchange rate changes. That is, as the fixed output assumption is relaxed, the effects on the interest rate and exchange rate diminish.

Bordo (1980) and Van Duyne (1979) use another, although similar, line of thought from the new world of international finance for the analysis of commodity prices, where capital flows easily between countries to equate savings and investment. They each present the idea that commodity (undifferentiated goods with prices determined through exchanges) prices tend to be flexible-priced goods, whereas most other goods are fixed-priced in the short run.

Bordo (1980) appeals to the early work of Cairnes (1873) in distinguishing short-run from long-run impacts for commodity price dynamics. Cairnes (1873) predicted that there would be short-run impacts of new money on different commodities due to differing elasticities of supply and demand, and that the effects would be neutral in the long run but not in the short run. Bordo (1980) distinguishes supply and demand elasticities from price flexibility by integrating the role of long-run contracts into commodity price determination. Highly standardized commodities with auction markets have short-run price dynamics that are quite different from industrial commodities that are not highly standardized. Such non-standardized products often have implicit, long-run fixed-price contracts. In addition, Bordo finds that contract length varies inversely with relative price variability.

Van Duyne (1979) constructed a two-sector, fix-price/flex-price model of an open world economy under flexible exchange rates. His portfolio-style model includes financial assets and the exchange rate, where the flex-price equilibrium is determined via a stock equilibrium. The model is used to examine the effects of a production shortfall in the commodity market or commodity speculation (increase in stockholding) on all variables in the model (including the exchange rate and trade flows). Asset markets play a major role in transmitting these shocks throughout the world.

Van Duyne (1979) found that these commodity disturbances have a big impact on international capital flows and the exchange rate, and that these effects are persistent. The stockholding of commodities is the key way that commodity shocks are transmitted into the financial variables, and then throughout the world economy. The stockholding of commodities and their potential use as an asset in a monetary model led to the early work of Chambers (1984) and Frankel (1986), which looks at monetary policy and its effects on commodities as a storage asset.

5 Commodities as assets in price formulations

Chambers (1984) was among the first to integrate monetary policy into a model that includes fixed-price and flex-priced goods. He argues that monetary policy influences the interest rate, which affects agricultural storage. In the short run, a restrictive monetary policy appreciates the exchange rate and increases the interest rate. Both of these changes deteriorate the competitive position of U.S. agriculture as an export industry, which is similar to the findings of Chambers and Just (1979).

Frankel's (1984) approach is different. He applies Dornbusch's (1976) model to a closed economy with fixed-price goods (industrial) and flex-priced goods (agricultural commodities). In this closed economy model, commodities are stored and held as assets whose price changes with unanticipated money supply changes. An increase in the money growth rate causes investors to shift from money to commodities, while a decrease in money growth causes investors to shift from commodities to bonds. Note that the unanticipated change in money supply in Frankel's analysis is expected to be permanent. Spot commodity prices will overshoot because investors must receive the same return on commodities and bonds (with the increased interest rate). So, monetary policy has an effect on real commodity prices due to the stickiness of some goods. These factors do not affect industrial goods because their prices are sticky and they are not held as assets. If there is an expected increase in the long-run money growth rate with no change in the current money supply, the interest rate will not rise fully to capture the higher inflation rate (because of real money balances). In this case, commodity prices will increase above their long-run rate.

The natural next step in the progression of research on macroeconomic issues is to extend Frankel's (1986) closed economy model to an open economy as Stamoulis and Rausser (1988) did. Their model includes monetary effects on commodity prices within a small open economy, where any change in the currency's value is offset by changes in the domestic interest rate (since the foreign interest rate is fixed). They include fixed-price and flex-priced goods, and there is overshooting of the flex-priced good if output is fixed in the short run. Their most important finding is that monetary policy can affect relative prices of agricultural products in the short run.

Most of the early research measuring macroeconomic impacts on agricultural prices has focused on the United States. Yet, there is much literature applied to other countries, such as the research by Taylor and Spriggs (1989). Using a VAR model to trace the dynamic response paths of manufactured and agricultural prices to a shock in the money supply, Taylor and Spriggs (1989) find that agricultural prices overshoot their long-run equilibrium and respond more quickly to monetary shocks in the short run, while manufactured goods prices catch up later. Interestingly, the relative U.S. exchange rate is the most significant monetary factor influencing instability in Canadian agricultural prices. Another important source of unexpected variation in agricultural prices is domestic money supply. Thus, not only is domestic monetary policy important for countries, but also the monetary policy of their important trading partners because of integrated international capital markets. Taylor and Spriggs conclude that Canadian agricultural policy analysts and farm lobbyists should look at both U.S. agricultural policies and macroeconomic policies.

6 Models that incorporate short-run effects with long-run relationships

The literature has also progressed as econometric techniques have advanced. Robertson and Orden (1990) use a VAR and a Vector-Error Correction (VEC) model in their analysis of monetary impacts on agricultural prices in New Zealand. The VEC model recognizes that there may be a long-run relationship among variables in a time series analysis (in this case, money neutrality might hold in the long run). They use M1 as the measure of money supply and find that money is neutral in the long run. New Zealand is a special case because it manages the exchange rate (so the exchange rate is exogenous) and is a price taker in the world market. They do not find overshooting of agricultural prices, arguing that this is because the exchange rate does not overshoot. Agricultural prices do adjust faster than manufacturing prices, though. Their analysis indicates that shocks in manufacturing prices induce monetary expansions, but agricultural prices do not (which could put a cost-price squeeze on farmers at times)

Dorfman and Lastrapes (1996) use a Bayesian approach to long-run money neutrality, allowing the horizon to vary from 100 months to infinite months. They allow real agricultural prices to change in the short run with money supply changes. Their model includes the bond market; a money market; crops, livestock, and energy markets; and markets for domestic and foreign aggregate output. Their emphasis is on the dynamic interactions among the nominal money supply; the real money supply; the nominal interest rate; real exchange rate; domestic aggregate output; and real crop, livestock, and energy prices. They find that a positive money-supply shock increases nominal money supply and reduces the interest rate steeply. Real energy prices respond positively to the money supply increase and the real exchange rate depreciates as the interest rate falls. Real livestock prices increase and hit a peak, but then decline to their long-run equilibrium (they overshoot). Real crop prices fall back to long-run equilibrium more slowly (they undershoot). So livestock and crop prices benefit in the short run from a positive money shock. It takes nine months for long-run money neutrality to hold. Yet they conclude that money-supply shocks are not enormously important to the variance of agricultural prices and that monetary policy's impact on agricultural sector productivity may be somewhat exaggerated.

Using a variant of Frankel's (1986) model, Lai, Hu, and Wang (1996) investigate the effects of anticipated monetary shocks on commodity prices. They find that while agricultural product prices rise immediately as future monetary expansion is anticipated, prices may overshoot or undershoot depending on the lead-time for the announcement. If the lead-time is short, prices will overshoot; if the lead-time is long, they will undershoot.

Saghaian, Reed, and Marchant (2002) generalize the Dornbusch (1976) model to include an agricultural sector which is similar to relaxing the closed economy assumption of Frankel (1986). Thus, there are flexible agricultural prices and sticky industrial prices. Their model uses a monetary approach to exchange rate determination in a small open economy. Monetary expansion reduces interest rates and causes a depreciation in the exchange rate that overshoots. Because the interest rate falls and the money supply increases, agricultural prices must overshoot for the agricultural market to clear. The degree of overshooting of agricultural prices depends on the stickiness of industrial prices, but the existence of a flexible exchange rate (and its overshooting) diminishes overshooting in agriculture.

In their analysis, Saghaian, Reed, and Marchant (2002) examine monthly data from 1975 to 1999 to empirically test whether overshooting occurs. Using a VEC model to test for long-run money neutrality and overshooting of commodity prices, they find that money is not neutral in the long run for agricultural prices (i.e., inflation will reduce long-run agricultural prices because their nominal increases do not match overall inflation). They also find agricultural prices adjust faster than industrial prices in the short run and overshoot their long-run equilibrium.

Saghaian, Hasan, and Reed (2002) apply the model of Saghaian, Reed, and Marchant (2002) to four Asian countries and incorporate contemporaneous correlation among the corresponding innovations to construct the impulse response functions (and therefore the degree of overshooting for agricultural prices). They investigate whether a highly industrialized country like

Korea has a higher degree of overshooting. They find overshooting of agricultural prices in three countries (Korea, Philippines, and Thailand) and that money is not neutral in these economies, but the degree of overshooting in Korea is lower than in Philippines.

Kwon and Koo (2009) argue that a vector moving average model is preferred over a VEC because it avoids the long-run identification issue associated with the endogenous variables. They use a graphical causality model to the contemporaneous causal structure among equations (similar to Saghaian, Hasan, and Reed 2002). They discover three things. First, money-supply shocks are negatively related to agricultural exports and positively related to exchange-rate shock. Second, it takes time for export quantity to adjust due to overshooting of commodity prices and a J-curve effect on agricultural exports (where adjustments in price occur faster than adjustments in exports) (Carter and Pick 1989). Third, unexpected movements in the exchange rate and interest rate are the main macroeconomic shocks causing fluctuations in the agricultural sector.

7 New linkages between agricultural and energy prices

Energy sector linkages to agriculture are important determinants of farm prices and income, especially in the current corn-based ethanol production environment, oil market volatility, and global economic conditions. These factors are of paramount importance to farmers as well as to consumers. A large portion of the variable costs of agricultural products in the form of fuel and fertilizer directly depends on energy prices. However, since 2000, the emergence of large-scale biofuel production has led to additional linkages between the energy and agriculture sectors. The stronger connections have also led to the "food versus fuel" debate. The new linkages that have made some agricultural crops, such as corn, more energy crops will be discussed here.

The energy and agriculture sectors have always been interlinked because energy is an input into farm production, processing, and distribution. Since 2000, crude oil prices and environmental concerns have led policy makers to adopt alternative energy sources like ethanol (Vedenov, Duffield, and Wetzstein 2006), and therefore establish other linkages between energy and agricultural prices. The linkages between ethanol and agricultural prices in Brazil have been studied due to the important role that sugar processing plays in that country's energy production. Balcombe and Rapsomanikis (2008) developed a generalized bivariate error correction model with nonlinear dynamic adjustments to study the nexus among sugar, ethanol, and crude oil markets in Brazil. Their results indicate that in the long run, oil prices affect Brazil's sugar prices.

Serra, Zilberman, and Gil (2011) used a methodology that jointly estimates cointegration and a multivariate generalized autoregressive conditional heteroscedastic process so that not only overall energy and agricultural prices, but also their volatilities can be affected by unanticipated shocks. They found that crude oil prices not only influence ethanol prices in the Brazilian ethanol and energy markets, but also price volatility, and that this volatility is transmitted, although weakly, to the sugar market. They also found strong linkages between energy and food prices in Brazil.

The emergence of large-scale biofuel production in the United States began in the mid-2000s and progressed rapidly due to U.S. energy policies. These biofuel/energy policies have reinforced the linkages between the agriculture and energy sectors, further strengthening the relationship between prices of agricultural commodity and energy prices, specifically among corn, ethanol, and crude oil prices (Serra and Zilberman 2013). Ethanol is the major liquid biofuel produced in the United States, and 98 percent of ethanol production is from corn (Renewable Fuels Association 2017). The percentage of corn used for ethanol production grew from 12 percent of the crop in 2004 to 38 percent of the crop in 2016, with potential for more growth in the future (U. S. Department of Agriculture 2017). Condon, Klemick, and Wolverton (2015) argue that an increase in ethanol production by one billion gallons increases corn prices by 3 to 4 percent.

The literature on linkages between energy and agricultural prices under U.S. policies developed quickly. Suddenly, ethanol production and the use of cereals in biofuels became an important issue with research methodologies readily available to provide estimates of policy impacts (particularly the renewable fuel standards). A voluminous literature bloomed.

Many studies focused on price-level interdependencies between energy and agricultural commodity prices, but the results were mixed. For example, using monthly data from January 1996 to December 2008, Saghaian (2010) investigated the causal relationship between energy and agricultural prices using a VEC model with directed graphs to examine the relationship among crude oil, ethanol, corn, wheat, and soybean prices. Results show a strong correlation between energy and agricultural prices, but the evidence on causal links between these two sectors is mixed. Myers et al. (2014) studied comovement between feedstock and energy prices using a VEC model but found no long-run comovement between these prices. However, Serra (2013) linked high agricultural prices in the 2007–2010 period directly to changes in U.S. renewable fuel standards.

Mensi et al. (2014) used various flexible vector autoregressive–generalized autoregressive conditional heteroscedasticity (VAR-GARCH) models to identify linkages among eight major energy and cereal commodities. Although they focused on returns from these commodities in a properly diversified portfolio, they had to estimate the relationship between energy and agricultural commodity prices (and their volatilities) in order to assess an optimal portfolio. They found significant linkages between energy and agricultural commodity prices and their volatilities that are indicative of the move toward studies focusing on volatility spillovers between energy and agricultural prices.

8 Volatility spillover effects

Agricultural commodity prices are reaching higher levels and experiencing higher price volatility with negative economic and social consequences (Wright 2011). The increased correlation between food and energy prices in recent years has likely led to stronger volatility spillovers between these prices (Tyner 2010). An important concern has been whether corn–ethanol–oil linkages transfer instability and risk from energy markets to already volatile agricultural markets.

De Gorter, Drabik, and Just (2015) argue that knowing the linkages between agriculture and energy is important to understanding changes in food prices, especially between ethanol and corn. In 2000, the coefficient of variation for corn ranged from 0.05 to 0.1 and, by 2010, it ranged from 0.08 to 0.25 (Trujillo-Barrera, Mallory, and Garcia 2012). The literature points to several factors for this increased price variability, such as supply and demand shocks, storage policies, macroeconomic conditions and policies, economic growth, climate change, speculation, and increased biofuel production from agricultural commodities (Balcombe and Prakash 2011; Wright 2011). An increase in food-price volatility has profound economic implications, raising concerns for consumers, producers, and policy makers. The tremendous increase in ethanol production and potential future increases make it imperative that models focus on measuring its impacts on price volatility for food crops.

High, volatile agricultural commodity prices have stimulated the literature on volatility spillover effects between the energy and agriculture sectors. While this work began later than the work on price linkages, it has developed faster. Most papers make heavy use of GARCH techniques. Zhang et al. (2009) studied volatility spillovers among weekly U.S. ethanol, corn, soybean, gasoline, and crude oil prices using many different models to study these relationships. Their results showed no spillover effects from ethanol price volatility to corn and soybean prices, but they found volatility transmission from agricultural commodity prices to energy prices. Yet they argue that these market signals and shocks to the system have short-run impacts, but no long-run impacts.

Serra and Gil (2012b) integrate commodity stocks and macroeconomic conditions (such as the interest rate) in their analysis of price volatility spillovers. They use corn and ethanol prices between 1990 and 2010 to fit a multivariate GARCH (MGARCH) model using parametric and semiparametric methods to find that an increase in ethanol price volatility causes an increase in corn price volatility. Yet this relationship is muted if large carryover stocks are present. They argue that higher stocks and a more stable macroeconomic environment lead to more stable agricultural commodity prices.

Du and McPhail (2012) studied the dynamic relationships among U.S. ethanol, corn, and gasoline futures with daily data using a MGARCH model. Then they estimated a structural VAR model. They found that while there was no long-run relationship between corn and biofuel prices, gasoline and ethanol prices transmit volatility to corn prices. With recent stronger corn-ethanol ties, their variance decomposition showed that for each market, a significant and large share of price variation is explained by price changes in the other two markets.

Trujillo-Barrera, Mallory, and Garcia (2012) used mid-week closing futures prices of corn, ethanol, and crude oil from 2006 to 2011 to study volatility spillover effects in the United States and found volatility transmission from the corn market to the ethanol market, but not the reverse. This spillover between corn and ethanol reached its peak during the financial crisis (just after 2009) when the price of oil fell dramatically.

Gardebroek and Hernandez (2013) used weekly spot prices to test volatility spillovers among crude oil, ethanol, and corn prices in the United States between 1997 and 2011 using multivariate GARCH methods. Their results indicated significant spillovers from corn to ethanol prices, especially after 2006, but no spillovers from ethanol to corn. In addition, they did not find major cross-volatility effects from crude oil to corn markets, and their results provided no evidence of volatility that energy markets cause price volatility in the U.S. corn markets.

Serra and Gil (2012a) used copula models to study the relationship between crude oil and biodiesel blends, and crude oil and diesel prices in Spain. They found an asymmetric dependence between crude oil and biodiesel prices and a symmetric dependence between diesel and crude oil prices.

Some researchers have studied volatility effects only between oil and agricultural commodities market, leaving out ethanol. Nazlioglu, Erdem, and Soytas (2013) studied volatility transmission between crude oil and agricultural commodities using the causality in variance test and impulse response functions using daily prices. They divided the sample into pre-crisis (January 1986 to December 2005) and post-crisis (January 2006 to March 2011). They did not find any volatility transmission between crude oil and agricultural commodity markets in the pre-crisis period (before 2006), but they did find that oil market volatility spilled over to the agricultural markets in the post-crisis period.

Haixia and Shiping (2013), analyzing the price volatility spillovers among China's crude oil, corn, and fuel ethanol markets, observed a higher interaction among these three markets after September 2008. Their results showed spillover effects from the crude oil market to the corn and ethanol markets, and bidirectional spillover effects between corn and ethanol markets.

Recent empirical research has also identified volatility transmission relationships between the global energy and food markets. Abdelradi and Serra (2015) used an asymmetric MGARCH model to study the price volatility spillovers between biodiesel blend and refined-sunflower oil prices in Spain. Their results showed a bidirectional and asymmetric volatility spillover between these two commodity prices.

Cabrera and Schulz (2016) used GARCH and stochastic volatility models to study oil, natural gas, biodiesel, rapeseed, and other energy/agricultural product prices in Germany. Much of their focus was on the choice of model and its accuracy in predicting jumps, mean volatility, and other distributional characteristics. They found the stochastic volatility model with moving average innovations as the best one for their data. In relation to linkages between energy and agriculture, they found no evidence that biodiesel prices influenced the volatility in agricultural commodity prices.

9 Future work

The literature on the linkages between energy and agricultural prices will likely continue to blossom as the profession analyzes these linkages during an era of lower energy prices and more biofuel substitutes for ethanol. Future studies could examine asymmetric volatility spillovers between agricultural commodity and energy prices, investigating whether energy and commodity price volatility responds differently to price increases than to price decreases. It is also unclear whether ethanol price variability is higher during price increases or whether ethanol price increases have a stronger impact on corn price volatility than price decreases. There is little evidence that food and biofuel price increases have the same interactions as price decreases (Serra and Zilberman 2013).

Many studies on energy and agriculture linkages use different data frequencies (e.g., daily, weekly, or monthly) and the profession needs a better understanding of how conclusions about volatility spillovers vary among time frames (Elyasiani, Perera, and Puri 1998; Gardebroek, Hernandez and Robles 2015). More research could identify a robustness among the different data frequencies used.

Future work on the effects of monetary policy and exchange rates on agricultural prices will likely continue, but the direction is unclear. It is likely that the huge demand for U.S.-denominated assets by the world community will continue to play an important role in exchange rate determination and therefore the transaction of U.S. prices into foreign prices. This means that U.S. agriculture will continue to feel pressure from a relatively strong U.S. dollar, but could experience surges of prosperity when exchange-rate dynamics lead to a dollar depreciation. The development of another key world currency, such as the Chinese yuan, will have a large impact on this dynamic, but development will take time and is uncertain.

A new area of macroeconomics that might also have an impact on the literature in agriculture is agent-based modeling (ABM); yet the literature is very scant (Lengnick 2013; Dosi et al. 2013). With ABM, the analytical unit is the individual agent interacting with others in an environment that adds to and forms an overall system. This type of modeling often uses numerical computation to solve complex problems that may be chaotic or involve nonlinear equilibria that are not smooth. This bottom-up approach to macroeconomics could have implications for movements in agricultural commodity prices.

References

Abdelradi, F., and T. Serra. 2015. "Asymmetric Price Volatility Transmission Between Food and Energy Markets: The Case of Spain." *Agricultural Economics* 46:503–513.

Balcombe, K., and A. Prakash. 2011. "The Nature and Determinants of Volatility in Agricultural Prices: An Empirical Study." In *Safeguarding Food Security in Volatile Global Markets, Food and Agriculture Organization* (UN), pp. 89–110.

Balcombe, K., and G. Rapsomanikis. 2008. "Bayesian Estimation and Selection of Nonlinear Vector Error Correction Models: The Case of the Sugar-Ethanol-Oil Nexus in Brazil." *American Journal of Agricultural Economics* 90:658–668.

Barnett, R., D. Bessler, and R. Thompson. 1983. "The Money Supply and Nominal Agricultural Prices." *American Journal of Agricultural Economics* 65:303–307.

Batten, D., and M. Belongia. 1986. "Monetary Policy, Real Exchange Rates and U.S. Agricultural Exports." *American Journal of Agricultural Economics* 68:422–427.

Bessler, D. 1984. "Relative Prices and Money: A Vector Autoregression on Brazilian Data." *American Journal of Agricultural Economics* 66:25–30.

Bordo, M. 1980. "The Effects of Monetary Change on Relative Commodity Prices and the Role of Long-Term Contracts." *Journal of Political Economy* 88:1088–1109.

Cabrera, B., and F. Schulz. 2016. "Volatility Linkages between Energy and Agricultural Commodity Prices." *Energy Economics* 54:190–203.

Cairnes, J.E. 1873. *Essays on Political Economy: Theoretical and Applied*. New York: Kelley (Reprint).

Carter, C., and D. Pick. 1989. "The J-Curve Effect and the U.S. Agricultural Trade Balance." *American Journal of Agricultural Economics* 71:712–720.

Chambers, R. 1984. "Agricultural and Financial Market Interdependence in the Short Run." *American Journal of Agricultural Economics* 66:12–24.

Chambers, R., and R. Just. 1979. "A Critique of Exchange Rate Treatment in Agricultural Trade Models." *American Journal of Agricultural Economics* 49:249–257.

———. 1982. "An Investigation of the Effect of Monetary Factors on U.S. Agriculture." *Journal of Monetary Economics* 9:235–247.

Condon, N., H. Klemick, and A. Wolverton. 2015. "Impacts of Ethanol Policy on Corn Prices: A Review and Meta-Analysis of Recent Evidence." *Food Policy* 51:63–73.

De Gorter, H., D. Drabik, and D. Just. 2015. *The Economics of Biofuel Policies: Impacts on Price Volatility in Grain and Oilseed Markets*. New York: Palgrave Macmillan.

Devadoss, S., and W. Meyers. 1987. "Relative Prices and Money: Further Results for the United States." *American Journal of Agricultural Economics* 69:838–842.

Dorfman, J., and W. Lastrapes. 1996. "The Dynamic Responses of Crop and Livestock Prices to Money-Supply Shocks: A Bayesian Analysis Using Long-Run Identifying Restrictions." *American Journal of Agricultural Economics* 78:530–541.

Dornbusch, R. 1976. "Expectations and Exchange Rate Dynamics." *Journal of Political Economy* 84:1161–1176.

Dosi, G., G. Fagiolo, M. Napoletano, and A. Roventini. 2013. "Income Distribution, Credit, and Fiscal Policies in an Agent-Based Keynesian Model." *Journal of Economic Dynamics and Control* 37:1598–1625.

Du, X., and L. McPhail. 2012. "Inside the Black Box: The Price Linkage and Transmission Between Energy and Agricultural Markets." *Energy Journal* 33:171–194.

Elyasiani, E., P. Perera, and T. Puri. 1998. "Interdependence and Dynamic Linkages Between Stock Markets of Sri Lanka and Its Trading Partners. *Journal of Multinational Financial Management* 8:89–101.

Frankel, J. 1984. "Commodity Prices and Money: Lessons from International Finance" *American Journal of Agricultural Economics* 66:561–566.

———. 1986. "Expectations and Commodity Price Dynamics: The Overshooting Model." *American Journal of Agricultural Economics* 68:344–348.

Gardebroek, C., and M. Hernandez. 2013. "Do Energy Prices Stimulate Food Price Volatility? Examining Volatility Transmission Between U.S. Oil, Ethanol and Corn Markets." *Energy Economics* 40:119–129.

Gardebroek, C., M. Hernandez, and M. Robles. 2015. "Market Interdependence and Volatility Transmission among Major Crops." *Agricultural Economics* 47:141–155.

Haixia, Wu, and L. Shiping. 2013. "Volatility Spillovers in China's Crude Oil, Corn and Fuel Ethanol Markets." *Energy Policy* 62:878–886.

Krueger, A. 1983. "Protectionism, Exchange Rate Distortions, and Agricultural Trading Patterns." *American Journal of Agricultural Economics* 65:864–871.

Kwon, D-H., and W. Koo. 2009. "Interdependence of Macro and Agricultural Economics: How Sensitive is the Relationship?" *American Journal of Agricultural Economics* 91:1194–1200.

Lachaal, L., and A. Womack. 1998. "Impacts of Trade and Macroeconomic Linkages on Canadian Agriculture." *American Journal of Agricultural Economics* 80:534–542.

Lai, C-C., S-W. Hu, and V. Wang. 1996. "Commodity Price Dynamics and Anticipated Shocks." *American Journal of Agricultural Economics* 78:982–990.

Lengnick, M. 2013. "Agent-Based Macroeconomics: A Baseline Model." *Journal of Economic Behavior and Organization* 86:102–120.

Mensi, W., S. Hammoudeh, D. Nguyen, and S-M. Yoon. 2014. "Dynamic Spillovers Among Major Energy and Cereal Commodity Prices." *Energy Economics* 43:225–243.

Myers, R., S. Johnson, M. Helmar, and H. Baumes. 2014. "Long-Run and Short-Run Co-Movements in Energy Prices and the Prices of Agricultural Feedstocks for Biofuel." *American Journal of Agricultural Economics* 96:991–1008.

Nazlioglu, S., C. Erdem, and U. Soytas. 2013. "Volatility Spillover Between Oil and Agricultural Commodity Markets." *Energy Economics* 36:658–665.

Orden, D., and P. Fackler. 1989. "Identifying Monetary Impacts on Agricultural Prices in VAR Models." *American Journal of Agricultural Economics* 71:495–502.

Renewable Fuels Association. www.ethanolrfa.org/wp-content/uploads/2016/02/10823-RFA.pdf. Accessed June 1, 2017.

Robertson, J., and D. Orden. 1990. "Monetary Impacts on Prices in the Short and Long-Run: Some Evidence from New Zealand." *American Journal of Agricultural Economics* 72:160–171.

Saghaian, S. 2010. "The Impact of the Oil Sector on Commodity Prices: Correlation or Causation?" *Journal of Agricultural and Applied Economics* 42:477–485.

Saghaian, S., M. Hasan, and M. Reed. 2002. "Monetary Impacts and Overshooting of Agricultural Prices: The Case of Four Asian Economies." *Journal of Agricultural and Applied Economics* 34:95–109.

Saghaian, S., M. Reed, and M. Marchant. 2002. "Monetary Overshooting of Agricultural Prices in an Open Economy." *American Journal of Agricultural Economics* 84:90–103.

Schuh, E. 1974. "Exchange Rate and U.S. Agriculture." *American Journal of Agricultural Economics* 56:1–13.

———. 1984. "Future Directions for Food and Agricultural Trade Policy." *American Journal of Agricultural Economics* 66:242–247.

Serra, T. 2013. "Time-Series Econometric Analyses of Biofuel-Related Price Volatility." *Agricultural Economics* 44:53–62.

Serra, T., and J. Gil. 2012. "Biodiesel as a Motor Fuel Price Stabilization Mechanism." *Energy Policy* 50:689–698.

———. 2012. "Price Volatility in Food Markets: Can Stock Building Mitigate Price Fluctuations?" *European Review of Agricultural Economics* 40:507–528.

Serra, T., and D. Zilberman. 2013. "Biofuel-Related Price Transmission Literature: A Review." *Energy Economics* 37:141–151.

Serra, T., D. Zilberman, and J. Gil. 2011. "Price Volatility in Ethanol Markets." *European Review of Agricultural Economics* 38:259–280.

Stamoulis, K., and G. Rausser. 1988. "Overshooting of Agricultural Prices." In P. Paarlberg and R. Chambers, eds., *Macroeconomics, Agriculture and Exchange Rates*. Boulder, CO: Westview Press, pp. 163–189.

Taylor, J., and J. Spriggs. 1989. "Effects of the Monetary Macro-Economy on Canadian Agricultural Prices." *Canadian Journal of Economics* 22:278–289.

Trujillo-Barrera, A., M. Mallory, and P. Garcia. 2012. "Volatility Spillovers in U.S. Crude Oil, Ethanol, and Corn Futures Markets." *Journal of Agricultural and Resource Economics* 37:247–262.

Tyner, W. 2010. "The Integration of Energy and Agricultural Markets." *Agricultural Economics* 41:193–201.

U.S. Department of Agriculture. www.ers.usda.gov/data-products/us-bioenergy-statistics/us-bioenergy-statistics/#Feedstocks. Accessed June 1, 2017.

Van Duyne, C. 1979. "The Macroeconomic Effects of Commodity Market Disruptions in Open Economies." *Journal of International Economics* 9:559–582.

Vedenov, D., J. Duffield, and M. Wetzstein. 2006. "Entry of Alternative Fuels in a Volatile U.S. Gasoline Market." *Journal of Agricultural and Resource Economics* 31:1–13.

Wright, B. 2011. "The Economics of Grain Price Volatility." *Applied Economic Perspectives and Policy* 33:32–58.

Zhang, Z., L. Lohr, C. Escalante, and M. Wetzstein. 2009. "Ethanol, Corn, and Soybean Price Relations in a Volatile Vehicle-Fuels Market. *Energies* 2009(2):320–339.

23

Models of economic growth

Application to agriculture and structural transformation

Terry L. Roe and Munisamy Gopinath

1 Introduction

The post-Solow resurgence of interest in economic growth has been stimulated by contributions to economic theory; by evolving features of the world economy, such as the divergence in the value added per worker among nations; and by advancements in numerical methods. Major contributors to the analytical literature attribute the resurgence of interest to the rediscovery of Frank Ramsey's 1924 contribution to the mathematical theory of savings by Cass (1965) and Koopmans (1965). Koopmans' extension allows for richer transitional dynamics of a modeled economy, laying the foundation for much of the empirical applications of growth theory starting in the 1970s. Confronting the theory with data has been stimulated by huge advancements in numerical techniques and methods to solve the models' equations of motion. While no single data-fitting method seems to dominate, datasets such as the Global Trade Analysis Project database (http://www.gtap.agecon.purdue.edu/databases/v9/) and the World Input–Output Database (Timmer et al. 2015) facilitate the initial estimates of, for example, sectoral labor, capital and land shares in sector value added, the input–output coefficients of intermediate inputs, and household expenditure shares, all of which can later be buttressed with statistical methods. Methods for estimating sectoral capital stocks and sectoral total factor productivities (TFP) have also advanced. Applications of dynamic programming, based on the earlier work of Stokey and Lucas (1989), and algorithms that are easily coded to numerically solve systems of nonlinear and non-autonomous differential equations have further facilitated research in this area. The fitting of structural models to data has not precluded the application of statistical methods to test hypotheses derived from theory. Virtually all of the empirical literature reviewed here, whether dealing with structural model specifications or reduced forms, attempts to take into account the general equilibrium implications.

Space limitations force a choice between pursuing multiple topics or selecting a theme and following the growth literature's contribution to the chosen theme: transitional growth with emphasis on agriculture. This limitation prevents us from addressing multiple but closely related topics, such as Smith, Nelson, and Roe (2015), who extend a growth model to capture the hydrology of groundwater extraction over time; Gaitan and Roe (2012), who show that recourses to political economy arguments are not necessary for a natural resource-rich country, all else being equal, to grow more slowly than a resource-poor country; and

Tajibaeva (2012), who models endogenous property rights in the development of an open resource-based economy such as those in Africa where the poaching of wildlife is prominent.

We begin with a brief discussion of early interests in transitional growth with an emphasis on agriculture. This discussion suggests the early economists seemed to ask many of the right questions, but methodological challenges restricted their queries into questions posed. We then follow the literature in a mostly chronological fashion. We find advances in methodology and see complementarity between structural dynamic general equilibrium models and statistical models. We also briefly consider intra-industry dynamics when firms are heterogeneous in size or productivity and the resulting resource reallocation for faster growth of such industries (Melitz 2003). Nevertheless, much remains to be investigated if we are to better understand what Parente and Prescott (2000) referred to as barriers to riches. This discussion avoids the temptation to express any of the models discussed in mathematical form, leaving this challenge to the reader.

2 Background

Since the early 1940s, economists have noted the empirical observation that, as economies develop, a structural transition occurs where agriculture experiences a secular decline in its share of the economy-wide labor force and a decline in its share of the gross domestic product (GDP). Bruce Johnston's (1970) survey of the literature emphasizes the importance of earlier economists such as Lewis (1955), Kuznets (1957), and Chenery (1960). Neither models of the Harrod–Domar type nor Solow (1957) provide insight into the transformation of sectors in the process of growth. More recently, Herrendorf, Rogerson, and Valentinyi (2013) document these features for the rich countries of Belgium, Finland, France, Japan, Korea, Netherlands, Spain, Sweden, the United Kingdom, and the United States over the nineteenth and twentieth centuries. The transformation tends to exhibit a decline in the sectoral shares of employment in agriculture, a rise in the share of labor in manufacturing,[1] and an increase in the share of workers in services. Sectoral value-added shares in the total GDP tend to follow this same pattern.

These observations suggest the important elements of longer-run economic growth in real GDP per capita entail the transition of resources among and within sectors of the economy as capital deepening occurs. Agriculture receives particular attention for many reasons, two of which are the following. First, growth in rural population on a relatively fixed land base can lead to diminishing returns to labor, impoverishment, and food scarcity, with negative implications to household savings and investment, health, and the demand for non-agricultural goods and services. Second, most economies, prior to experiencing trend rates of growth in real income per capita over an extended period of time, begin this take-off process with a relatively large share of resources in agriculture, suggesting that agriculture may be of major importance in contributing to this transition process.

Agriculture can contribute to this process by helping to lessen the otherwise rise in the real cost of labor faced by the manufacturing and service sectors, while also increasing agriculture's demand for manufacturing and service goods (Galor 2005).[2]

The literature in this general topic area is seeking to identify and measure the effect of several factors contributing to structural transformation. A partial list of factors, some of which are prominent in the literature discussed here and others that remain to be studied, includes:

1 Production technology

- The effect of capital deepening on transition growth when sectors differ in relative capital, labor, and land intensities.

- The rate of sectoral technological change and whether it is sector- and factor-biased.
- Whether the costs of intermediate factors of production cause negative internal terms of trade to agriculture over time due, for example, to the high cost of transportation and the lack of rural electrical services relative to other sectors of the economy.
- Whether, in the process of capital deepening, the elasticities of factor substitution become more elastic so as to lower the cost of substituting the relatively more abundant factors (e.g., capital) for the scarcer (e.g., labor).

2   Policy and institutions

- The nature of policy that tends to tax agricultural exportables and subsidize importables while, at the same time, protecting the import-competing sector of the non-farm economy. Within industries, these are policies that affect entry, exit, and job creation.
- The role of institutions, the rule of law, and the lack of property rights to land.
- The failure of capital markets to arbitrage asset prices so as to undervalue agricultural land, lowering its collateralized value and providing disincentives to husband land resources.
- Failure to provide rural education to raise literacy rates to levels that facilitate rural–urban labor market linkages.

3   Households

- The nature of Engel's law, which, in a closed economy, can dampen the growth in agricultural productivity per worker relative to productivity in other sectors of the economy.
- The opportunity for urban employment of multiple household members helps to increase the opportunity cost of household time, thus increasing demand for food and services supplied by supermarkets relative to the traditional marketing chain providing food and services to traditional food outlets.
- Residency choice of rural-to-urban migration which tends to decrease demand for goods and services in rural areas and increase them in urban areas, while at the same time changing the urban–rural endowment of labor.

This list is not intended to be exhaustive of possible factors affecting growth, but rather to suggest the possibility that multiple effects on growth are possible, with their relative importance varying by country and stage of development.

3   Overview of literature from 1990 to 2006

Commonly cited contributors to the literature on structural general equilibrium models of agriculture and economic growth include Matsuyama (1992), King and Rebelo (1993), Echevarria (1995, 1997), Gollin, Parente, and Rogerson (2002, 2004), and Gollin and Rogerson (2014). A review of some of this literature appears in Gollin (2010) and is only highlighted here. The literature since 2006 is given more attention in the next section.

Matsuyama (1992) is the first to develop a two-sector growth model of the representative household that takes into account the welfare and resources of their prospective descendants. He uses the model to show analytically that the effect of agricultural productivity on growth depends crucially on an openness to trade. His analysis relies on the view that agriculture is backward, which he implements in the model by assuming that the growth process is driven by capital accumulation and learning-by-doing in manufacturing. Evidence since then has shown that agricultural total factor productivity (TFP) often exceeds that of other sectors of the economy (Martin and Mitra 2001).

Echevarria (1995) posits a three-sector model: an agriculture, manufacturing, and service sector in which households maximize the discounted present value of utility. Factors of production are labor and capital, where land is aggregated into capital. Echevarria (1995) compares transition results for the case where the modeled economy is closed to replicate an infant industry argument and for the case where the agricultural and the manufacturing sector face given world prices. The Cobb–Douglas production functions for each sector differ in their labor and capital intensities and in their rate of neutral technological change. Household utility is non-homothetic to capture the main features of Engel's law in the short run, but utility converges to being homothetic in the long-run steady state equilibrium. The model is calibrated to data so a numerical solution reproduces certain average values in the United States and the Organization for Economic Cooperation and Development (OECD) countries.[3] The Gauss–Seidel algorithm is used to find the values of the state variable (capital) that maximizes the discounted present value of utility subject to flow budget and resource constraints. Differences in sectoral composition and income elasticities for the three goods are shown to explain an important part of the variation in growth rates observed across countries. The main finding is that if the country produces only primary and service goods, the effects of trade on growth are mixed in the sense that trade helps growth at low levels of income, but at higher levels trade slows growth. Echevarria (1997) builds on the earlier paper and finds that changes in sectoral composition, with the share of workers in agriculture declining, are driven largely by different income elasticities for the consumption goods produced by each of the economy's three sectors.

Gollin and Rogerson (2014) focus on transport costs and productivity in a static model consisting of three regions, two of which are modeled as remote agriculture and near agriculture. A city region produces a manufactured good. The manufactured good is employed in agriculture and is associated with a transport cost. The structural transition question posed is: why do large factions of the population remain in agriculture in poor countries? Subsistence production is depicted as an environment where rural households consume most of what they produce. To capture the essence of Engel's law, preferences over two goods are posited to require a minimum (subsistence) level of consumption. Technology is Cobb–Douglas, and land appears as a fixed factor in each of the agricultural regions. Analytical results find three channels which can lead to a greater allocation of labor to the agricultural sector: low TFP in agriculture, low TFP in the production of the intermediate good employed in agriculture, and high transportation costs. Numerical results reinforce these findings and show that if food is also costly to transport, it will be efficient for many people in the economy to locate near the source of production because food is relatively cheap in the places where it is produced.

At about the same time as Gollin, Parente, and Rogerson (2004), Irz and Roe (2005) developed a two-sector growth model that allowed for either an open or a closed economy. Sectoral technology is Cobb–Douglas with differences in sectoral factor intensities for labor and capital and differences in sectoral TFP. Agricultural land is fixed but allowed to degrade at an exogenous rate. Preferences are of the Stone–Geary type. The representative household owns all of the economy's resources and behaves as though it maximizes, in continuous time, its discounted present value of utility. Irz and Roe (2005) attempt to analyze how agricultural productivity influences the rate and pattern of growth in this Ramsey framework. A modest innovation is separating land from capital, which leads to a food self-sufficiency condition, and empirically solving such a model using the Time Elimination method that we discuss later. They find that household demand considerations and agriculture factor productivity matter to growth. Low levels of agricultural productivity and high food prices are shown to constrain savings. Low agricultural productivity represents a major bottleneck, limiting growth and the transition of resources out of agriculture. The Tiffen and Irz (2006) time series analysis of 30 countries reinforced these results. Together, results are consistent with the view that agricultural productivity growth is necessary because it releases a surplus beyond rural household food consumption and agricultural labor, while simultaneously generating demand for industrial goods and services.

4 Overview of literature since 2006

More recent contributions providing insights into structural transformation include papers by Foster and Rosenzweig (2008), Verma (2012), Herrendorf, Rogerson, and Valentinyi (2013, 2014, Herrendorf, Herrington, and Valentinyi (2015), Spolador and Roe (2013), Bustos, Caprettini, and Ponticelli (2016), and Eberhardt and Vollrath (2016). The latter two papers are more closely linked to the Foster and Rosenzweig contribution, so we discuss these papers last. In general, these attempts look at the critical role of agricultural TFP growth relative to that of other segments of the economy as well as sectoral intensities of factor use.

Verma (2012) develops a three-sector dynamic general equilibrium model of the Indian economy in which an infinitely lived representative household owns land, labor, and capital. There are three sectors: agriculture, manufacturing, and services. Economy-wide factors of production are capital and labor, while agriculture also includes land. The model is set up in discrete time, and the representative household owns the economy's resources and maximizes the discounted present value of utility. The model is reduced to three equations: (1) a Euler equation from household optimization that determines whether households choose a pattern of per capita consumption that rises, remains unchanged, or falls over time; (2) an intra-temporal equation over consumption goods; and (3) a resource constraint equation. Parameters are chosen based on the literature, others are statistically estimated, and an extensive growth accounting exercise yields capital stock and TFP for each of the three sectors. The accounting exercise is needed to provide estimates of initial capital stock and TFP estimates for the model's production functions. Another feature is to validate the model by assessing its ability to reproduce the data over the period 1980 to 2005. These details are noted because they suggest the trend in this literature is to more carefully fit models to data and validate the results.

An increase in service sector TFP is found to result from the liberalization policies adopted by India in the 1990s. While the baseline model performs well in matching sectoral value-added shares and their growth rates over time, it could not quantitatively match the levels of sectoral employment seen in the data. Including public capital in the model's production functions helped to rectify this weakness, which highlights the role of government policy in influencing India's structural transformation. Service sector TFP, which measured higher than the other two sectors, was the largest source of service sector value-added growth. The relatively high service sector TFP is attributed to post-1991 liberalization policies. This chapter highlights the importance of a non-traded goods sector on economic growth. The service sector includes transport, communications, financial and business services, public administration, and retail services. This sector, being relatively non-traded, tends to distinguish one country from another, relative to the firms in manufacturing in one country relative to another. International competition helps to encourage efficiency in the production of traded goods. This incentive is mostly absent in the production of service goods. Because this sector is typically a large portion of a country's GDP, institutional rigidities caused by the rule of law, property rights, or provision of utilities can have a major influence on the sector's role in structural transformation.

Many of the analyses discussed thus far model sectoral output as value added by capital, labor, and land. Households in these models consume value added as a good, when in fact, households consume commodities embodying intermediate factors of production. Input–output tables of many countries show the value of intermediate inputs in the value of commodities consumed can range over 50 percent. The paper by Herrendorf, Rogerson, and Valentinyi (2013) asks the question: what utility function should one use in applied work on structural transformation? Data on household expenditure shares for the U.S. economy over the period from 1947 to 2010 are used to answer this question. The authors calculate a time series of consumption value added. They find that if one adopts the final-commodity consumption expenditure specification, then a Stone–Gary utility function provides a good fit to U.S. time series data. If instead one adopts a consumption value-added specification, then a Leontief utility provides a reasonable fit to the data. A key contribution is their derivation of a sufficient condition under which a Stone–Geary utility function over final consumption expenditure is consistent with a Leontief utility function defined over consumption of value added.

The study by Herrendorf, Rogerson, and Valentinyi (2014) synthesizes and evaluates recent advances in the research on structural transformation. They also develop a benchmark model of a three final good sector economy with CES–Stone–Geary preferences to capture Engel effects. The production sector is modeled by four Cobb–Douglas technologies, one each for agriculture, manufacturing, service, and capital production. The technologies aggregate land into capital, and the technologies share common factor elasticities. Their model accommodates the fact that capital is not produced by the manufacturing sector alone, as data clearly show. In the presence of capital deepening, the marginal value products of sectoral labor and capital change is driven by capital deepening and household preferences. In spite of the model's assumption of common production elasticities and the aggregation of land into capital, agriculture's labor share is shown to decline and rise in the service sector.

The study by Herrendorf, Herrington, and Valentinyi (2015) focusing on the technology-side of the agriculture, industry, and service sectors poses the question: how important in structural transformation are differences across sectors in technological progress and technology parameters, including capital share, and the elasticity of substitution between capital and labor? Again, land is aggregated into capital. Technology is of a two-factor CES form. However, differences in exogenous factor augmentation and factor elasticities are considered. Using U.S. data, they find that differences in technical progress are the predominant force behind structural transformation. Surprisingly, they conclude that sectoral Cobb–Douglas production functions with equal capital shares do a reasonable job of capturing the main trends of U.S. structural transformation. This result is unlikely to replicate across countries.

Spolador and Roe (2013) focus on the structural transformation of the Brazilian economy from roughly post-1990 policy reform, with projections to 2034. The question is: why has Brazil's structural transformation been slow relative to other emerging market economies? Data show that Brazil's agriculture is more capital intensive than other emerging market economies, exceeding that of U.S. agriculture. The authors construct a three-sector (agriculture, manufacturing, and service) growth model with labor and capital as economy-wide resources and land specific to agriculture.[4] The service sector good is treated as a home good so its price is endogenous. The representative household maximizes the discounted present value of homothetic utility over agricultural, manufacturing, and service commodities (hence, not value added). Technology over primary factors of capital, labor, and land is Cobb–Douglas, but Leontief in the employment of the intermediate factors of production produced by the other sectors of the economy. Levels of initial capital stock and TFP estimates are based on a growth accounting exercise. Factor productivity is Harrod neutral-labor augmenting in industry and service production. In the case of agriculture, TFP is both labor and land augmenting. Data suggest that land augmenting TFP in Brazil exceeds that of labor. As a major agricultural exporter, the price of agricultural exports relative to the price industrial imports is allowed to trend over the time period 1994 to 2013, as given by the data, but assumed to converge monotonically to a zero rate of growth in relative prices in the long run.[5] Account is taken of average annual growth rate in the number of workers of 2.4 percent.

Specification of the analytical model and its numerical solution differs from the rest of the literature. First, equilibrium is specified in the dual (i.e., cost and value added) functions as in Roe, Smith, and Saracoglu (2010). Specifying the model in the primal leads to more complicated solution procedures. Equilibrium at each point in continuous time can be viewed as having two components: intra-temporal and inter-temporal. The intra-temporal equilibrium is similar in structure to that found in static trade theory, as in Woodland (1982). The inter-temporal component includes two nonlinear and non-autonomous differential equations. The non-autonomous equations rule out the time elimination method (Mulligan and Sala-i-Martin 1991) to solve the system. Instead, the relaxation method developed by Trimborn, Koch, and Steger (2008) is used. Mathematica was chosen to code the system.

A validation exercise shows the model results match sectoral GDP data surprisingly well over the 1994 to 2010 period, although it tends to underestimate the service sector GDP during the 1994 to 2000 sub-period. The results show relatively slow transitional growth, with agriculture's GDP share of total GDP declining slowly from about 13 percent in 1994 to about 11.5 percent during the 2004 to 2009 period, then returning to about 13 percent in the longer run. This non-monotonic pattern also appears, but to a lesser degree, in agriculture's labor and capital shares in total labor and capital. Service sector share in total GDP increases monotonically, while the manufacturing sector tends to decline by about 3 percent in the long run compared to its level in 2009. To better explain the model results, a growth accounting-like exercise is conducted on the model's reduced form factor rental rate, supply, and factor demand equations. From these reduced forms, Stopler–Samuelson and Rybczynski-like results can be obtained.

The non-monotonic pattern of agricultural growth is explained by two main effects. First, the service sector supplies about 23 percent of the value of agriculture's total employment of intermediate factors. A rise in the price of the service good results in negative domestic terms of trade effects to agriculture. This is a Rybczynski-like effect on service sector production. The service sector supply function has as arguments capital and labor. The sector is relatively labor intensive, causing a negative supply effect from growth in capital stock and a positive effect from growth in labor supply. Because transitional growth in the country's capital stock exceeds the growth in number of effective workers, the net effect is negative. Consequently, the service price must rise to clear its market. The second effect is a tradeoff in cost of agricultural production caused by the rise in real wages and a negative effect on costs from the decline in the capital rental rate. Since agriculture is capital intensive, the net effect of these forces gives rise to the non-monotonic pattern over time. Underlying the non-monotonic pattern is TFP associated with land-augmenting technical change of about 4 percent per annum, which exceeds that of labor by a factor of almost 2.5. Agriculture's external terms of trade effects on transition were negligible.

Why did manufacturing value-added share in total value added tend to stagnate and then decline, albeit modestly? There are three major factors affecting this sector's share in total value added. The external terms of trade were positive but negligible. The domestic terms of trade were negative and of primary importance. This effect is relatively large because the value of service sector intermediate factors in total intermediate factors employed by the sector is about 30 percent of its gross output. The net Rybczynski positive effect from growth in capital only modestly dominated the negative effect from growth in effective labor. The net effect of these forces led to growth in industrial sector gross output of about 4 percent per annum, not sufficient to increase this sector's value-added share in total value added.

The Foster and Rosenzweig (2008) study is unique in that it takes a more microeconomic perspective and focuses on the linkage between agricultural productivity and rural non-farm employment. The rural sector engages in the production of services to agriculture, such as metal working, machinery repair shops, agricultural processing, and education. Locally then, agriculture may serve as an engine of growth by creating a demand for these services. Foster and Rosenzweig (2008) consider how this process is affected by local conditions such as local rates of agricultural productivity growth, the presence of physical infrastructure that links local and regional markets, and opportunities for investment in human capital. They review the many problems and deficiencies in the data on rural employment and internal migration, and then construct a three-sector static general equilibrium model. The three sectors are agriculture, factory goods, and a local non-traded sector. Agricultural goods are produced using land and labor, non-traded goods are produced using labor, and factory goods are produced using externally provided capital and labor. The comparative statics of this framework are used to statistically test predictions using data from India.

Growth in crop yield is found to be weakly related to the proportion of workers primarily in the rural non-agricultural wage sector in 1971. By 1999, the overall non-farm share of the rural workforce is particularly high in areas experiencing the lowest growth in crop yields, and increases in agricultural productivity reduce migration from households with greater land holdings. Furthermore, growth in non-agricultural wage work is closely associated with the proportion of villages with a factory. The number of workers in the production of local services on net increases more in the high than in low growth in crop yield areas over the 1982 to 1999 period. Education is associated with migration. Out-migrants from agriculture are more likely to have completed primary schooling, and migrants are more likely to come from households with primary-schooled heads. Importantly, higher agricultural productivity significantly increases schooling. Thus, the analysis suggests a circuitous route. Out-migration from agriculture is indirectly linked to agricultural productivity through education and to the presence of a factory in the local village that together help pull literate workers from agriculture. This mechanism for labor to depart from agriculture does not appear in the structural dynamic models discussed thus far, and it suggests a possible modification of the models to accommodate and then validate this feature.

Bustos, Caprettini, and Ponticelli (2016) take advantage of the rather rapid effects of technical change in Brazil's agriculture and linkages to the industrial sector. They analyze the adoption of genetically engineered soybean seeds as labor-augmenting technological change, and the second harvesting season for maize as land-augmenting technological change. The simultaneous expansion of these two crops allows them to assess the effect of agricultural productivity on structural transformation. Like Foster and Rosenzweig (2008), a theoretical model is developed to highlight analytical predictions and to guide the interpretation of statistical results. Predictions are obtained from a theoretical model that initially assumes two sectors, manufacturing and agriculture, both of which produce tradable goods. Manufacturing employs only labor while agriculture employs labor and land in a CES technology where each factor is augmented by exogenous technological change. Later, the model is extended to include a non-traded good. Preferences are Cobb–Douglas. The statistical analysis first considers association between the two different forms of technological change and transition growth, and the second attempts to obtain insights into causation.

Bustos, Caprettini, and Ponticelli (2016) find that municipalities where the area planted to soybeans experienced an increase in agricultural output per worker also experienced a reduction in labor intensity in soybean production and an expansion in industrial employment. A 1 percent increase in agricultural labor productivity in soybean production led to a 0.16 percentage point decrease in agricultural employment share, and an increase in the manufacturing employment share of a similar magnitude. Using these estimates, by 2010, the labor savings effects explain 24 percent of the observed differences in the reduction of agricultural employment share across municipalities and 31 percent of the corresponding differences in the growth of the manufacturing employment share. They find this result to be consistent with the theoretical prediction that the adoption of labor augmenting technologies reduces labor demand in agriculture and induces the reallocation of workers toward the industrial sector. They also check whether causality might run in the other direction, namely, whether an increase in productivity in the industrial sector could raise labor demand and wages, inducing agricultural firms to switch to less labor-intensive crops such as soybeans. In the case of maize, switching from traditional to double cropping technology is associated with an increase in the area planted, an increase in labor intensity, a reduction in industrial employment, and faster growth in wages. They conclude that different effects of technological change in agriculture indicate that factor bias of technological change is a key determinant in the relationship between agricultural productivity and structural transformation.

Bustos, Caprettini, and Ponticelli (2016) also investigate the effects of technological change in agriculture on municipalities' service sector. They find that labor-augmenting technological change in soybean production on employment in local services is not significantly different from zero. Turning to the conceptual model for possible interpretation of this finding, they note two opposing forces. The labor supply effect is generated by a reduction in the marginal product of labor in soybean production that reduces agricultural employment. The demand effect for service is generated by higher land rental income from higher land productivity in the production of soybeans. They suggest, and then confirm, that the absence of an effect of local technical change in agriculture on employment in the local service sector can be expected if the share of land rents that accrue to landlords consuming services in the same municipality where they own land is small. They also find that labor-augmenting technical change tends to induce outmigration from the municipality in which it occurred.

To investigate the effects of land-augmenting technological change, they draw on the expansion in new lands brought under cultivation by chemicals to lower soil acidity and improve nutrients. They conclude that the effect of a second harvesting season for maize had significant effects on increasing labor demand in maize production. Consequently, this demand lowered out-migration from agriculture.

Eberhardt and Vollrath (2016) emphasize that the role of agriculture in structural transformation depends crucially on the elasticity of agricultural output with respect to labor. Since agricultural technology is relatively location-specific, focusing on TFP alone is not sufficient. The rest of the economy production is expressed as a linear function of labor and preferences are Stone–Geary. For a given productivity increase in agriculture, theory predicts that an economy with a low labor elasticity will experience larger shifts of labor between sectors, greater increases in agricultural labor productivity, and greater increases in GDP per capita than an economy with a high labor elasticity. If the elasticity is low (i.e., the marginal physical product of labor is highly inelastic), then agricultural output is relatively insensitive to the number of workers it employs. A productivity increase makes possible the release of workers while still meeting food demand. This environment frees up workers to the other sector, thus raising GDP per capita. High elasticity economies are predicted to not shift as many workers out of agriculture, and thus tend to produce fewer additional non-agricultural goods in response. In this case, real GDP per capita rises, but not by as much as seen in economies with a low elasticity.

The statistical model by Eberhardt and Vollrath (2016) draws on a 1961 to 2002 panel dataset of 128 countries, based mostly on Food and Agriculture Organization's (FAO) FAOSTAT database. Technology is presumed to vary across countries, but to be constant over time. By adopting a common factor-over-country framework, they are able to account for country/year variation in total factor productivity. The main objective is to see if the average value of agriculture's labor elasticity for temperate countries differs from equatorial countries. Dividing countries into four groups (arid, temperate and cold, equatorial, and highland) they find labor elasticities in production to be low in both arid (0.183) and temperate countries (0.166), but significantly higher in equatorial (0.405) and highland (0.265) countries. The next empirical question is whether these differences in elasticities imply a significant quantitative effect on development.

Eberhardt and Vollrath (2016) use data from South Korea to calibrate the static two-sector closed economy model so that it closely reproduces the observed decline in agriculture's labor share from 63 percent in 1963 to 8 percent by 2005. They then construct three simulated economies with labor elasticities of 0.15, 0.35, and 0.55, respectively, and impose the assumption that the agricultural share of total labor for each simulation begins at 80 percent. An exogenous increase in TFP of 20 percent is assumed for each simulation. The simulation results are compared to a base solution that is common to all simulations. They find that the economy with the lowest elasticity results in a larger movement of labor out of agriculture and a smaller increase in the real price of the agricultural good than in economies with higher labor share relative to the base solution. Labor productivity, agricultural consumption, non-agricultural consumption, and real income are all higher in the lower agricultural labor elasticity countries than in the higher elasticity countries. Countries with a high elasticity are shown to benefit from a growth in population relative to the low elasticity countries. Performing the same simulation with no TFP but assuming a 5 percent growth in population reverses the direction of the TFP results. In this case, the low elasticity countries experience an increase in the share of the population in agriculture and a higher real price of the agricultural good. Labor productivity, agricultural consumption, non-agricultural consumption, and real income are all lower in the low labor elasticity simulations than the corresponding values, relative to the same base, in the higher labor elasticity simulations. They conclude that a pessimistic aspect is that the inherent agricultural technology will continue to cause equatorial regions to move through the process of structural change at a slower pace than seen in temperate countries.

Most of the above growth models examine the intersectoral reallocation of resources arising from the Engel effects and relative TFP growth of agriculture. An emerging literature, from seminal contributions in industrial organization and new trade theory, is considering intra-industry organization and reallocation of resources when farms or firms are heterogeneous, for example, in size or productivity (Melitz 2003).[6] Like growth models, the impetus for resource reallocation is primarily terms of trade changes arising from presumably Engel effects or trade liberalization. However, productivity is often assumed exogenous and a key distinguishing factor among entities within an industry. Here, for instance, falling output prices of a sector would force out some of the low-productivity firms, with their resources being reallocated to their high-productivity counterparts. Evidence exists from a large set of countries (e.g., the United States, France, Chile, Korea, and Indonesia, among others), showing such effects significantly contribute to increases in average industry growth, especially in the manufacturing industries (Pavncik 2002). While this phenomenon is very relevant to agriculture, few studies have explored mechanisms inducing resource reallocation and resulting increases in average industry productivity growth (Gopinath, Sheldon, and Echeverria 2007).

5 Concluding remarks

We provide a non-mathematical overview of some of the literature associated with the economic growth of nations. The introduction section rationalizes the narrowing of our topic to the time period 1990 through 2016. The choice of starting date is linked to Landes (1990), who a number of authors have credited with rekindling interest, post-Solow (1957), in the economics of growth. We limit further our focus to contributions that provide insight into one or more of the major features of structural transformation that countries, realizing relatively sustained economic growth over time, tend to experience. Transformation entails three main features: change in the sector value-added share of total value added, change in the sector labor share of total labor employed (the preferred measured being hours worked), and change in sector labor productivity, typically measured as sector value added per worker, or hours worked. Economists' concern with aspects of transition date from at least 1940, with others, such as Gollin (2010), citing remarks made by Adam Smith and the transition of labor from agriculture that accompanied the specialization and division of labor. In this context, economic growth in value added per capita can be viewed through forces causing or limiting economic transition.

It is our view that contributions to the theory of economic growth, the increased availability of data, and advances in numerical and statistical methods have sharpened and extended our insights to the forces and consequences of transformation. The emergence of these new tools suggests to us that this area of query will receive increased attention.

Of the several factors contributing to structural transformation mentioned in this chapter, the literature thus far provides only some insight into the effects of production technology and household behavior on economic growth and, even in this case, whether findings are country-specific, as Eberhardt and Vollrath (2016) suggest, remains to be assessed. The importance of the rural non-farm sector and its pulling of labor from agriculture and, in return, supplying intermediate inputs to agriculture, as in Foster and Rosenzweig (2008), seems to be a strong candidate for further investigation.

We also observe various specifications of models. For statistical analyses, the analytical models tend to be static and more parsimonious than structural dynamic models. The more parsimonious frameworks provide predictions that better serve statistical analyses. Of the papers reviewed, statistical analyses tends to focus on reduced forms or components of structure, such as that of Eberhardt and Vollrath (2016) and their estimation of labor elasticities. The structural dynamic models and the statistical models are clearly complimentary in providing insights into transformation.

In the literature reviewed, our emphasis on agriculture necessarily omits the importance of other sectors of the economy on transformation. Hausmann, Hwange, and Rodrik (2005), for example, observe that investment in product discovery in manufacturing determines the types of goods a country produces and exports, which in turn shapes structural change. In an accompanying paper, Hausmann and Klinger (2006) focus on the determinants of structural change as a process by which countries upgrade the products they produce and export. The technology, capital, institutions, and skills needed to make newer products are more easily adapted from industrial product lines than from primary resource-based products such as agriculture or mining. An economy based on the production of primary commodities is likely slower than non-primary good economies in investing in skills, assets, and institutions that facilitate transformation toward new products of different technological content and higher unit values. Badibanga et al. (2013) extend this type of data analysis to compare the differences in the dynamics of structural transformation between China, Malaysia, and Ghana. Their analysis suggests that Ghana is unlikely to alter its primary sector focus, while the evolution of the industrial sectors in China and Malaysia becomes somewhat self-perpetuating due to spillovers from adapting and modifying old product lines into new industrial product lines.[7] While this literature is data driven without a structural model from which to draw inferences, it provides grounds for additional research into the importance of a country's endowments and initial conditions that may give rise to development-path dependency. A strategy for countries hoping to exit path dependency is to engage transnational firms to employ those resources a country has in relative abundance, as in the case of global supply chains discussed by Baldwin (2012).

The service sector's role in transformation received special attention in Verma (2012) and Gollin and Rogerson (2014) in terms of transportation costs. This sector typically comprises the largest share of a country's GDP. In a three-sector economy of the type modeled by Roe, Smith, and Saracoglu (2010), the service sector is closed, and industry and agriculture face world prices. In the presence of capital deepening, the marginal value product of labor rises in the most capital-intensive sectors relative to the service sector, which tends to be labor intensive. All else being constant, labor at old factor prices is pulled into the more capital-intensive sectors. As incomes rise, households consume more of all goods. For the service sector market to clear, it must raise its price to compete for economy-wide resources. The more inelastic the service sector supply response, the fewer resources it can profitably employ for a percent increase in its price, thus decelerating transitional growth. A rise in the service sector price can decrease the value-added prices faced by the traded goods sector. This negative internal terms of trade effect is larger for the traded goods sector, which employs a larger value share of service sector inputs in gross sector output. It is in this way that an "inefficient" service sector tends to "tax" other sectors of the economy.

We noted the absence of incentives for service sector firms compared to the competitive forces from foreign trade that typically prevail for firms in the traded goods sectors. The service sector embodies many components of the political economy that affect the provision services, such as transport, communications, education, financial and business services, public administration, and retail services, and the provision of health care and social safety nets. To the extent that governance impacts the provision of these services, the political economy of this process must surely affect transformation, and thus be worthy of investigation. Acemoglu (2009) reviews some of the economic growth of political economy, and Rodrik (2008) provides a discussion of institutions.

Much work is also needed in recognizing the intra-industry heterogeneity and the spatial organization of farms/firms and industries. Heterogeneity within industries allows for differential responses to sectoral terms of trade changes and the resulting resource reorganization might hinder or enhance the transition to a higher growth path. Likewise, spatial and agglomeration forces have much to offer in the improved understanding of the transition dynamics with emphasis on agriculture. Progress made thus far in fitting inter-temporal models to data should also facilitate studying the effects of climate change and the harvesting of natural resource stocks on structural transformation.

Disclaimer

The views expressed here are solely the author's and do not necessarily reflect USDA or ERS.

Notes

1 Rodrick (2015) finds that the hump-shaped relationship between industrialization and incomes has shifted downwards, particularly for Latin American countries, and moved closer to the origin. This pattern suggests these countries are running out of industrialization opportunities sooner and at much lower levels of income per capita.

2   We omit here the review of papers on policy that discriminates against or in favor of agriculture and the effects on economic growth. Reference to this literature appears in the World Bank report (2008).
3   Echevaria calibrated the model in a manner similar to that of Kydland and Prescott (1982).
4   The model is an extension of the framework in Roe, Smith, and Saracoglu (2010).
5   The evolution of the border price of agriculture relative to manufactures is implemented using the Interpolation command in Mathematica. The command constructs a function, in this case of time, from data such that the function reproduces each data point.
6   Space limitations prevent us from examining the spatial reallocation of resources within or across sectors. The economic geography literature offers insights on the organization of firms and the associated economic welfare outcomes (Combes, Mayer, and Thisse 2008).
7   The endogenous growth models pioneered by Romer (1990) and Grossman and Helpman (1991) exhibit scale economies from the fixed costs of producing new ideas, the production of which are facilitated by the growth in technical knowledge from the production of past ideas. In this sense, it is possible for the industrial sector to become more self-perpetuating with incentives to increase the scale of the market than a primary goods producing economy.

References

Acemoglu, D. 2009. *Introduction to Modern Economic Growth*. Princeton, NJ: Princeton University Press.

Badibanga, T., X. Diao, T. Roe, and A. Somwaru. 2013. "Measuring Structural Change: The Case of China, Malaysia and Ghana." *The Journal of Developing Areas* 47:373–393.

Baldwin, R. 2012. *Global Supply Chains: Why They Emerged, Why They Matter, and Where They Are Going*. Hong Kong: Fung Global Institute.

Bustos, P., B. Caprettini, and J. Ponticelli. 2016. "Agricultural Productivity and Structural Transformation: Evidence from Brazil." *American Economic Review* 106:1320–1365.

Cass, D. 1965. "Optimum Growth in an Aggregate Model of Capital Accumulation." *Review of Economic Studies* 32:233–240.

Chenery, H.B. 1960. "Patterns of Industrial Growth." *American Economic Review* 50(4):624–654.

Combes, P-P., T. Mayer, and J-F. Thisse. 2008. *Economic Geography: The Integration of Regions and Nations*. Princeton, NJ: Princeton University Press.

Eberhardt, M., and D. Vollrath. 2016. "The Effect of Agricultural Technology on the Speed of Development." *World Development*. doi:10.1016/j.worlddev.2016.03.017.

Echevarria, C. 1995. "Agricultural Development vs. Industrialization: Effects of Trade." *Canadian Journal of Economics* 28(3):631–647.

———. 1997. "Changes in Sectoral Composition Associated with Economic Growth." *International Economic Review* 38(2):431–452

Foster, A.D., and M.R. Rosenzweig. 2008. "Economic Development and the Decline of Agricultural Employment." In J. Behrman and T.N. Srinivasan, eds., *The Handbook of Development Economics* (Vol. 4). Amsterdam: North Holland, pp. 3051–3083.

Gaitan, B., and T.L. Roe. 2012. "International Trade, Exhaustible-Resource Abundance and Economic Growth." *Review of Economic Dynamics* 15:72–93.

Galor, O. 2005. "From Stagnation to Growth: Unified Growth Theory." In P. Aghion and S. Durlauf, eds., *The Handbook of Economic Growth* (Vol. 1). Amsterdam: North Holland, pp. 172–293.

Gollin, D. 2010. "Agricultural Productivity and Economic Growth." In R. Eveson and P. Pengali, eds., *Handbook of Agricultural Economics* (Vol. 4). Amsterdam: North Holland, pp. 3825–3866.

Gollin, D., S.L. Parente, and R. Rogerson. 2002. "The Role of Agriculture in Development." *American Economic Review* 92(2):160–164.

———. 2004. "Farm Work, Home Work, and International Productivity Differences." *Review of Economic Dynamics* 7(4):827–850.

Gollin, D., and R. Rogerson. 2014. "Productivity, Transport Costs, and Subsistence Agriculture." *Journal of Development Economics* 107:38–48.

Gopinath, M., I. Sheldon, and R. Echeverria. 2007. *Firm Heterogeneity and International Trade: Implications for Agricultural and Food Industries*. St. Paul, MN: International Agricultural Trade Research Consortium (IATRC). http://purl.umn.edu/9349. Accessed November 1, 2017.

Grossman, F., and E. Helpman. 1991. *Innovation and Growth in the Global Economy*. Cambridge, MA: MIT Press.

Hausmann, R.J., J.J. Hwange, and D. Rodrik. 2005. "What You Export Matters." CDI Working Paper 123, Center for International Development, Harvard University, Cambridge, MA.

Hausmann, R.J., and B. Klinger. 2006. "Structural Transformation and Patterns of Comparative Advantage in the Product Space." CDI Working Paper 128, Center for International Development, Harvard University, Cambridge, MA.

Herrendorf, B., C. Herrington, and A. Valentinyi. 2015. "Sectoral Technology and Structural Transformation." *American Economic Journal: Macroeconomics* 7(4):104–133.

Herrendorf, B., R. Rogerson, and A. Valentinyi. 2013. "Two Perspectives on Preferences and Structural Transformation." *American Economic Review* 103(7):2752–2789.

———. 2014. "Growth and Structural Transformation." In P. Aghion and S. Durlauf, eds., *Handbook of Economic Growth* (Vol. 2). Amsterdam: North Holland, pp. 855–941.

Irz, X., and T. Roe. 2005. "Seeds of Growth? Agricultural Productivity and the Transitional Dynamics of the Ramsey Model." *European Review of Agricultural Economics* 32(2):143–165

Johnston, B.F. 1970. "Agriculture and Structural Transformation in Developing Countries: A Survey of Research." *Journal of Economic Literature* 8(2):369–404.

King, R.G., and S.T. Rebelo. 1993. "Transitional Dynamics and Economic Growth in the Neoclassical Model." *American Economic Review* 83(4):908–931

Koopmans, T. 1965. "On the Concept of Optimal Economic Growth." In J. Johnnsen, ed., *The Econometric Approach to Development Planning*. Amsterdam: North Holland, pp. 225–287.

Kuznets, S. 1957. "Quantitative Aspects of the Economic Growth of Nations: II. Industrial Distribution of National Product and Labor Force." *Economic Development and Cultural Change* 5(4):1–111.

Kydland, F., and E.C. Prescott. 1982. "Time to Build and Aggregate Fluctuations." *Econometrica* 50:1345–1370.

Landes, D.S. 1990. "Why Are We So Rich and They So Poor?" *American Economic Association Papers and Proceedings* 80:1–13.

Lewis, W.A. 1955. *The Theory of Economic Growth*. London: George Allen and Unwing.

Martin, W., and D. Mitra. 2001. "Productivity Growth and Convergence in Agriculture Versus Manufacturing." *Economic Development and Cultural Change* 49:403–422.

Matsuyama, K. 1992. "Agricultural Productivity, Comparative Advantage, and Economic Growth." *Journal of Economic Theory* 58(2):317–334.

Melitz, M.J. 2003. "The Impact of Trade on Intra-Industry Reallocations and Aggregate Industry Productivity." *Econometrica* 71(6):1695–1725.

Mulligan, C.B., and X. Sala-i-Martin. 1991. *A Note on the Time-Elimination Method for Solving Recursive Economic Models*. Cambridge, MA: National Bureau of Economic Research (NBER).

Parente, S.L., and E.C. Prescott. 2000. *Barriers to Riches*. Cambridge, MA: The MIT Press.

Pavncik, N. 2002. "Trade Liberalization, Exit, and Productivity Improvements: Evidence from Chilean Plants." *Review of Economic Studies* 69:245–276

Rodrik, D. 2008. "Second-Best Institutions." *American Economic Review* 98(2):100–104

———. 2015. *Premature Deindustrialization*. Cambridge, MA: National Bureau of Economic Research (NBER).

Roe, T.L., R.B.W. Smith, and D.S. Saracoglu. 2010. *Multisector Growth Models: Theory and Application*. New York: Springer.

Romer, P.M. 1990. "Endogenous Technological Change." *Journal of Political Economy* 98(5), Part 2:S71–S102.

Smith, R.B.W., H. Nelson, and T. Roe. 2015. "Groundwater and Economic Dynamics, Shadow Rents and Shadow Prices: The Punjab." *Water Economics and Policy* 1(3). doi:10.1142/S2382624X15500149.

Solow, R.M. 1957. "A Contribution to the Theory of Economic Growth." *Quarterly Journal of Economics* 70:65–94.

Spolador, H., and T. Roe. 2013. "The Role of Agriculture on Recent Brazilian Economic Growth: How Agriculture Competes for Resources." *The Developing Economies* 51(4):333–359.

Stokey, N.L., and R.E. Lucas, Jr., with E.C. Prescott. 1989. *Recursive Methods in Economic Dynamics*. Cambridge, MA: Harvard University Press.

Tajibaeva, L.S. 2012. "Property Rights, Renewable Resources, and Economic Growth." *Environmental and Resource Economics* 51:23–41

Tiffen, R., and X. Irz. 2006. "Is Agriculture the Engine of Growth?" *Agricultural Economics* 35:79–89.

Timmer, M.P., E. Dietzenbacher, B. Los, R. Stehrer, and G.J. de Vries. 2015. "An Illustrated User Guide to the World Input-Output Database: The Case of Global Automotive Production." *Review of International Economics* 23(3):575–605.

Trimborn, T., K. Koch, and T.M. Steger. 2008. "Multidimensional Transitional Dynamics: A Simple Numerical Procedure." *Macroeconomic Dynamics* 12:301–319

Verma, R. 2012. "Can Total Factor Productivity Explain Value Added Growth in Services?" *Journal of Development Economics* 99(1):163–177.

Woodland, A.D. 1982. *International Trade and Resource Allocation*. Amsterdam: North-Holland.

World Bank. 2008. *Economic Growth in the 1990s: Learning from a Decade of Reform*. Washington, DC: The World Bank.

24
Rural development – theory and practice

Thomas G. Johnson and J. Matthew Fannin

1 Introduction

This chapter describes the theory and practice of rural development. To many people, the term "rural development" implies efforts to accelerate the rate of development in developing nations. However, this chapter focuses on the theory and practice applied to lagging regions in any context, including in developed nations. To precisely address this topic requires attention to the terms "rural" and "development."

A great deal of attention has been given to the problem of defining rural (Miller 2010; USDA ERS 2016, 2017; USDA NAL 2016; USDA ERS 2017). The precise distinction between rural and non-rural has implications for policy because it affects which communities, people, and businesses qualify for programs. It affects the classification of data and thus the results of research. In reality, rural and non-rural are not distinct but rather represent a continuum from remote and sparsely populated areas to large and dense urban areas. For the purposes of this chapter, we will focus on theories and practices that have particular utility in places that are at least to some degree remote, sparsely populated, or both.

Next, we must consider the meaning of development and the distinction between growth and development. Kindleberger and Herrick distinguish growth and development as follows:

> The differences between growth models and development models can be inferred from the differences in the processes of growth and development themselves. Variegated composition of the economy, the differences in sectoral responses, and the reactions of individuals within a setting of poverty are neglected or ignored by growth models, but they are at the heart of models of economic development.
>
> *(Kindleberger and Herrick 1983, p. 39)*

Thus, economic development involves changes in the structure of the economy – increases in some activities and declines in others, faster growth in some regions than others, and changes in the distribution of economic benefits among people and places. Development is also about the process of change, rather than simply the outcomes of change.

2 Theories of economic development

Given the definition of economic development above, economic development theories can be viewed as exhibiting characteristic of four types or dimensions (Johnson 1994): spatial, sectoral, distributional, and temporal. The process of theoretical development has usually involved the extension of earlier, more basic concepts along one or more of these dimensions. In this chapter, we will describe theories and methods according to their treatment of these dimensions. Those theories that articulate several or all of the dimensions are necessarily more sophisticated. The major theories of relevance to rural development are displayed in Table 24.1.

2.1 The simplest theories: aggregate, static, asectoral, and aspatial

The foundation of most macroeconomic theories is a system of mathematical identities arranged into an accounting system. Aggregated theories are based on highly aggregated identities. Accounting systems are defined by a series of accounting conventions and rules and are subject to definitions, accounting identities, and disaggregation schemes. The first step in developing accounts is to decide on an accounting stance. An accounting stance is a definition of who (individuals, households, firms, institutions, or governments) should be included and

Table 24.1 Theory

		Aggregate		Distributional	
		Static	Dynamic	Static	Dynamic
Aspatial	Asectoral	Economic based; regional business cycle theory	Harrod–Domar neoclassical growth; endogenous growth; cumulative causation	Human capital theory; social capital theory	Piketty Stiglitz life cycle income theory
	Sectoral	Trade theory; staples hypothesis; resource curse			
	Intersectoral	Ind. org. theory; portfolio theory; input–output theory	Dynamic input–output	Static social accounting theory	Comprehensive wealth theory
Spatial	Asectoral	Tiebout hypothesis; Roback–SEM		Geography of intergenerational mobility	
	Sectoral	Losch; Weber	Krugman's core–periphery model		
	Intersectoral	Von Thunen; central place theory; agglomeration theory		Static multi-region social accounting theory	Multi-region comprehensive wealth theory

who should not when adding up jobs, income, sales, benefits, costs, and various measures of economic performance.

Accounting systems are not, in themselves, theories, but are often the basis of theories when behavioral relationships are attached to them. The hypothesized behavioral relationships give the resulting theories predictive power. A static account is essentially an anatomical description of an economy over a defined period, such as a particular year. The behavioral processes that give life to an economy are the economy's physiology. These behavioral relationships describe the products and/or the process of change in each component of the accounts, thus permitting the generation of alternate accounts under different assumed or observed circumstances.

Perhaps the simplest theory of growth is the economic base theory (North 1955; Harris et al. 1998). Economic base theory (sometimes called export base theory) is essentially at the origin of our four-dimensional theoretical space. Economic base theory is built on a simple accounting identity – total supply of goods and services must equal total disposition:

$$Y + M \equiv C + E,$$

where Y is regional product, M is regional imports, C is expenditures by all actors (consumers, governments, and firms, both current and capital account), and E is total regional exports.

This simple accounting system becomes economic base theory when it is assumed that the level of expenditures and imports are linear functions of total regional output. Then,

$$C = cY$$
$$M = mY$$

where c is the propensity to consume ($0.0 < c < 1.0$) and m is the propensity to import ($0.0 < m < c$). Substituting,

$$Y - cY + mY = E$$

and

$$Y = 1/(1 - c + m)E,$$

then future values of Y can be predicted from E, because

$$\partial Y/\partial E = Y/E = 1/(1 - c + m).$$

The term, $1/(1 - c + m)$ is the economic (or export) base multiplier and will take values between 1 (at $c = m$) and ∞ (at $c = 1.0$ and $m = 0.0$). Because of the linearity assumption, the marginal multiplier and the average multiplier are equal.

In this very simple theory, the level of the two types of leakage (net savings, $1 - c$, and imports, m) limit the size of the regional economy. The theory includes no references to time and thus does not predict when the economy will grow or how long it will take to achieve any new level of regional product.

This simple theory can be easily extended to include other aspects of a regional economy. One can, for example, introduce the concepts of value added, taxes, and the leakage of local value added to non-residents, each of which are implicit in the previous model (see Frey 1989, for example). Another useful extension distinguishes between total industrial output (TIO) and gross regional product (GRP).

429

Economic base models are strictly demand-oriented models and view savings and investments as unrelated activities. Savings is a leakage from the economy, while investment is part of the demand that drives the economy. In an open economy, this may be a reasonable assumption in the short run, since local savings can leak to other regions. However, savings are generally invested and become part of final demand. Furthermore, investments determine the productive capacity of an economy.

Rural economists have long been interested not only in economic growth but also in economic stability. The economic base theory suggests that instability, like growth, may be generated exogenously through changes in exports. Regional business cycle theory extends economic base theory to explain regional instability as a consequence of export instability (Malizia and Ke 1993; Siegel et al. 1995).

2.2 Adding sectors – multi-sectoral theories

An interest in sectors is a natural one for agricultural and rural economists. Both classical and neoclassical schools of economic thought are organized on a sectoral basis. The work of Walras, Stone, Leontief, and a host of more recent economists has contributed to a broad literature on intersectoral relations.

The economic base theory is sometimes extended to apply to specific sectors or classes of sectors. The staples thesis (Innis 1936), for example, argues that early development of rural regions is often based on the exploitation of staples – food, energy, timber, and minerals. The thesis also incorporates a simple spatial element (i.e., a core–periphery relationship) in which the staples-based periphery is inhibited from future development by the core. The resource curse theory (Auty 1993) comes to a similar conclusion, arguing that a resource-based economy is discouraged from the development of more advanced sectors for a variety of reasons.

2.3 Intersectoral theories

Intersectoral theories go beyond multi-sectoral theories by considering the relationships among the identified sectors. The best known of these theories are those built on Social Accounting Systems. The most common applications of social accounts are input–output (Miller and Blair 2009) and Social Accounting Matrices (SAMs) (Pyatt and Round 1985). These are discussed in greater detail in the method section of this chapter.

The Social Accounting Matrix is a fundamental part of most sectoral theories and is a powerful instrument in its own right. The sectoral dimension has also been fertile ground for extension into the other dimensions. The SAM itself may be equally viewed as a sectoral or a distributional framework. Interregional input–output and spatial equilibrium models are among some of the most sophisticated models along the sectoral–spatial dimensions. Single region models are used somewhat apologetically because of their well-known limitations.

The concept of sectoral interaction has proven over and over again to be very powerful (Basu and Johnson 1996). By choosing different bases for aggregating and disaggregating, some very diverse frameworks can be developed. In effect, sectors can be disaggregated so as to define regions, or space disaggregated to define sectors.

Theories, such as industrial organization theory and Portfolio Variance Theory, have been adopted from other branches of economics and applied to regional economies (Siegel et al. 1995). Theorists hypothesize that less concentrated and more diverse economies are more stable. These theories are strictly static and aspatial, but they view economies through the lens of the sectors represented in the economy. Portfolio Variance Theory, when applied to regional economies, explicitly considers the covariance among sectors to describe economy-wide variability as a consequence of the variability of an economy's constituent sectors.

2.4 Static spatial theories

The one dimension that seems to be a necessary condition of regional science concepts is the spatial dimension, but spatial theories vary greatly in their spatiality. Theories about regions are not necessarily very spatial. Their reference to a particular place (often point) gives them location but no reference to other places or points. To be truly spatial, theories must treat space in a more sophisticated way than by simply identifying a location. They should incorporate distances, area, or relative location, explicitly.

The similarities between time and space are more than coincidental. Time and space (distance) are inextricably interwoven in economic relationships since they are sometimes complementary inputs and sometimes substitute commodities. Emory Castle (1988) makes this point when he refers to the value of space and the cost of distance. Space injects temporal lags into economic relationships. Supply, demand, prices, and competition are defined by spatial and temporal coordinates.

Spatial theories are likely to be necessarily more complicated than temporal theories, if only because space is itself two- or three-dimensional and multidirectional, while time can be treated as a single, unidirectional phenomenon. Thus, serial correlation is simpler than spatial autocorrelation and temporal gradients are simpler than spatial gradients.

The spatial dimension is particularly important in theories employed by rural economists. These theories, often associated with urban economics today, have their roots in rural problems and phenomena. These theories include Alfred Weber's theory of market orientation, which distinguishes between factor-oriented, market-oriented, and footloose firms; Von Thünen's theory describing rural land use relative to urban markets; August Lösch's theory of industrial location; and Christaller's central place theory, which predicts the location and hierarchical structure of spatial services centers (Hoover and Giarratani 1984). Other theories describe the location decisions of households and businesses. The Tiebout hypothesis (Tiebout 1956) predicts that households will tend to migrate to locations that provide the mix of public services and amenities that match their preferences best. This process is referred to as "voting with one's feet." Roback (1982) describes a similar phenomenon in which individuals bid up the price of land and bid down wage rates in order to live and work in areas with desirable amenities. Meanwhile, employers bid up the price of land and wages in order to locate where their profits are maximized. More recent static theories have been developed to explain spatial economic organization, including agglomeration theory (Romer 1986) and the related cluster theory (Porter 1998).

2.5 The temporal dimension

All theories of growth and development are to some degree temporal since they deal with changes in economic systems through time. However, most of these theories, including those discussed above, are comparative static models and describe economic systems in equilibrium. The focus of this section is with theories that are dynamic in the sense that they deal explicitly with the process of change as well as the result of the change. Dynamic economic theories describe systems that are in disequilibrium in some way. A dynamic theory is concerned with the path that an economy follows as it responds to changing exogenous and endogenous forces. Dynamic models may describe systems that move toward, and possibly achieve, equilibrium or steady states. Other theories describe economic systems that may move away from equilibrium. Cumulative causation is a model of a disequilibrium economic process in which regions do not converge but rather continue to become more disparate. Most dynamic models deal with capital stocks, savings, investment, resource adjustment, innovation, and changing technology.

Development is not always a uniform and smooth process of marginal adjustments. More often, it appears to be punctuated by some watershed effects. The selection of a growth path at any one time forecloses alternative choices that might have become available at a later stage of development, so path dependency is a part of most dynamic theories. Rather than being gradual and non-disruptive, economic adjustment often involves jolts and shocks. Development involves changes not only in the magnitude of variables but also in the structure of relationships among them. This introduces a qualitative difference between the requisites for understanding both stationary underdevelopment and the more complex process of sustained growth.

Like static and comparative static theories, dynamic models may be asectoral, sectoral, or intersectoral, and they may or may not have distributive dimensions.

The Harrod–Domar model (Domar 1947; Harrod 1948) adds a simple dynamism to the otherwise static economic base model. In contrast to the demand-oriented economic base model, the Harrod–Domar model is supply-oriented and focuses on the production-enhancing effects of investment. Gross regional product is modeled, not as a function of demand, but as a production function with capital stocks as the sole factor of production. Investment is assumed to equal savings. Thus, in the Harrod–Domar model, the rate of savings becomes an exogenous determinant of growth, conditioned by the productivity of capital. If productivity increases, then the rate of growth increases for each level of savings. This growth occurs without bounds because leakages are assumed to be zero, and demand is assumed to equal supply. But this dynamism is largely incomplete without an understanding of the dynamics of savings, productivity, and demand.

Neither the economic base nor the Harrod–Domar theories are very satisfactory theories of growth. The former theory depends on its assumption of unconstrained production, while the latter depends on its assumption of a completely closed economy and unlimited demand. Both theories provide useful insights into the role of short-term changes in demand and the longer-term role of savings and investment.

The neoclassical growth model (Solo 1956; Swan 1956) extends the Harrod–Domar perspective by adding other factors of production in addition to capital. Any number of factors can be added, but in the neoclassical tradition, the factors are land, labor, and capital. Capital includes all kinds of human-made factors, land represents natural endowments (including environmental resources), and labor represents population-related resources.

The aggregate neoclassical function is:

$$Y = f(K, L, N, t),$$

where Y is regional product, K is capital, L is labor, and N is land. According to the neoclassical growth model, growth (the derivative of regional product with respect to time) is a simple sum of the elasticities of the output of land, labor, and capital, each multiplied by its proportional rate of growth, plus a term for the exogenous growth due to technological change.

The neoclassical growth model has largely been applied at the national level, but it can be applied at the sub-national, regional level as well. At the regional level, the exogenous technological change factor includes regional technological change and other factors, such as unmeasured capital inputs (human capital for example), local investments in new technology, entrepreneurial capacity, changes in government and institutions, and spatially fixed factors such as discovered natural resources.

According to the neoclassical theory, economic growth is due to increases in inputs and exogenous technological change. The endogenous growth theories (Arrow 1971; Lucas 1988; Romer 1986, 1990) explain growth more completely by linking technological change to other changes in the economic system. The theory is particularly useful for regional and rural research since the theory recognizes that technology is not equally available everywhere.

The Arrow model makes technology (knowledge), A, a function of the level of capital K. New investment increases K but also increases knowledge:

$$Y = f[K, L, A(K)]\ ^1$$

In Lucas' model, human capital is explicit and technological change is a function of the level of human capital:

$$Y = f[K, L, H, A(H)].$$

In Romer's spillover model, there are two sectors – the goods producing sector and the knowledge producing sector – both of which are functions of capital and labor. Knowledge, then, is a factor in overall productivity:

$$Y = f[K, L, A(K, L)].$$

In each of these models, the rate of growth in technology, and therefore the productivity of labor and capital, is determined by growth in the labor force and investment in capital (built, natural, and human). Since investments in these factors differ over space, growth will differ over space. Rural regions will lag in productivity growth and competitiveness if investment in any of these types of capital lags behind the rates in urban areas. But these models are not explicitly spatial; nor are they sectoral or intersectoral.

The popular Leontief input–output theory has spawned a number of dynamic input–output models (Leontief 1953; Johnson 1985; Duchin and Szyld 1985) that combine intersectoral and dynamic features and in some cases spatial interactions (DiFrancesco 1998). Dynamic input–output is a sophisticated sectoral–temporal theory with considerable promise. When time is added to the spatial sectoral models, the timing of interregional, intersectoral activities can be addressed.

2.6 Explicitly spatial dynamic theories

Recent theorists have introduced space explicitly into models of economic growth and development. Krugman (1991) introduced the "New Economic Geography" – an outgrowth of new trade theory and new growth theory (Krugman 1998). The result is the core–periphery model. As its name suggests, the core–periphery model, much like the models by Von Thünen, Lösch, and Christaller, describes the relationship between urban and rural components of an economy. Unlike previous models, the theory is also explicitly dynamic, describing the process that an economy employs to move toward (or away from, perhaps) a steady state. The theory is described in detail in Fujita et al. (1999). Kilkenny (1998) critiques and refines the core–periphery model and draws implications for rural development policy.

2.7 The distributional dimension

Social welfare depends not only on efficiency, but also on the distribution of income, wealth, power, and opportunity among individuals. The distribution of wealth, income, and power affect other, non-distributional aspects of the economy, such as demand for public and private goods and services, growth potential, and industrial location. For these reasons, rural development researchers, especially those concerned with policy assessment, are frequently concerned with the incidence, or distribution, of the costs and benefits of economic phenomena. Distributional theories help us understand and explain the patterns of income and wealth among socioeconomic groups, across space, and over time. Unfortunately, there is very little depth or breadth of distributional theories available.

While we are often only concerned with the aggregated, static distribution (referred to as personal distribution of income or wealth), or the distribution among sectors of the economy (functional distribution of income or wealth), there are other occasions when we are concerned with distribution over space (rural/urban, for example) and over time.

Rural economists and sociologists have not only employed distributional theories to explain rural income distribution and rural poverty but have also contributed significantly to the development and refinement of these theories. Among the most notable contributions are those in the areas of human capital (Schultz 1961; Becker 1994) and social capital (Rupasingha et al. 2000, 2006). Theories of human and social capital have implications for both efficiency and equity (distribution). Distributional theories predict the value of people's assets and their ability to accumulate assets over time. Another area of significant contribution has been in the understanding of rural poverty, its causes, and its consequences (Weber et al. 2005; Tickamyer and Duncan 1990).

Social Accounting Systems can be extended to conceptually link the functional distribution of income distribution to the personal distribution of income (Miyazawa 1976; Bernat and Johnson 1991). Together these theories and concepts have been used to explain rural-urban disparities in income and rural poverty (Rural Sociological Society 1993).

Several recent advances in distributional research have implications for rural economists. While these are perhaps not complete theories, they do have conceptual implications and will lead to better theories. The first of these research areas is the geography of intergenerational mobility (Chetty et al. 2014). This research area involves the correlations of individuals' socio-economic mobility according to the location where they lived as children. Analysis to date indicates that there are clear geographic patterns in mobility. This research has only vague indications of the cause for these patterns, but their mere identification will lead to theoretical developments.

Another conceptual area of interest is the increasingly unequal distribution of income and wealth (Piketty 2014; Stiglitz 2012). Again, a well-developed theory for this trend has not been articulated, but new sources of data and innovative ways of analyzing the data provide an opportunity and an incentive to develop better theories.

Finally, a relatively new area of conceptualization is the comprehensive rural wealth framework (Pender et al. 2014). This theory recognizes the inclusion of a wide array of capital assets – financial capital, built capital, natural capital, human capital, social capital, intellectual capital, cultural capital, and political capital – in an individual's real wealth. It also recognizes the role of publicly provided assets in an individual's comprehensive wealth. The theory links comprehensive wealth to societal sustainability. We discuss the potential for this emerging area in the concluding section.

3 Rural development methods

An overview of rural development theories would be incomplete without a complementary understanding of methods that are typically applied to test hypotheses from these conceptual frameworks. Similar to Table 24.1, these methods can be broken down along the four dimensions of sector, space, time, and distribution. The categorization of these methods is presented in Table 24.2.

3.1 Aggregate, static, aspatial methods

The most basic of the aggregate, static, asectoral methods to have become commonplace in rural community development practice is the simple economic base multiplier (previously shown). Another static method commonly used by rural development practitioners is the location quotient (Isard et al. 1998). Location quotients evaluate individual economic sectors within an individual region. They assist in identifying sectors that are basic, that is, sectors driven by export demand, versus nonbasic sectors where demand for a sector's products and/or services are driven by internal demand within a region. Further, they have been used in analysis of regional clusters in the Porter tradition (Woodward and Guimaraes 2009), as well as examples of more targeted food industry cluster analysis (Goetz, Shields, and Wang 2009).

While location quotients evaluate the export base level of a region at a given point in time, shift-share analysis places more of its focus on understanding economic growth/change between points in time (Loveridge and Selting 1998). Typically using such variables as employment or income, shift-share analysis serves as a variance decomposition technique by disentangling the

Table 24.2 Methods

		Aggregate		Distributional	
		Static	Dynamic	Static	Dynamic
Aspatial	Asectoral	Export base multiplier	Panel or time series regression; system dynamics	Inequality indices; regression	
	Sectoral	Location quotient; difference in difference; shift-share			
	Intersectoral	Single region static; IO/CGE/ conjoined; diversity indices		Single region static; SAM/CGE/ conjoined	Single region comprehensive wealth account; system dynamics
Spatial	Asectoral				
	Sectoral	Regression	System dynamics	Multi-region static; SAM/CGE/ conjoined	Multi-region comprehensive wealth account; system dynamics
	Intersectoral	Multi-region static IO/CGE/ conjoined; cluster analysis			

contributions of growth between overall aggregate economic growth of a larger region, the growth of a specific sector in that larger region, and the economic growth due to specific sectoral effects within a subregion. Basic applications of shift-share analysis have included forecasting, strategic planning, and policy evaluation.

A common policy analysis method applied in rural development circles is difference in difference (double difference) estimation in concert with quasi-experimental matching methods (Ravallion 2008). In a rural development context, an individual region's (e.g., states, counties, or other functional economic regions) economic performance is explained through evaluating that performance compared against a control region. In classical difference in difference analysis, control regions are determined through such quasi-experimental control methods as the Mahalanobis Metric Estimator applied by Isserman and Rephann (1995) to understand the impacts of the Appalachian Regional Commission (ARC) and Propensity Score Matching (Pender and Reeder 2011), in order to identify the economic contributions of the Delta Regional Authority. Newer techniques involve developing a control; not one based on a single identifiable region but rather on a synthetic region that combines the characteristics of several regions to serve as the control (Abadie and Gardeazabal 2003; Abadie et al. 2010). For example, Munasib and Rickman (2015) use Synthetic Control Analysis techniques to evaluate the economic effects of oil and gas shale energy activities on nonmetropolitan counties in Arkansas, North Dakota, and Pennsylvania.

3.2 Intersectoral, static methods

Rural and regional development scholars wishing to evaluate intersectoral relationships have commonly used input–output-based techniques. As mentioned previously, when applying assumptions about the transformation of inputs to outputs, the double entry accounting system of input–output transforms into a model that can be used for understanding intersectoral linkages within a regional economy (Miller and Blair 2009). Input–output methods apply a Leontief production technology combined with assumptions about perfect elasticity of supply, resulting in a basic general equilibrium model of a region's economy. In the original input–output models developed by Leontief, intersectoral relationships were the focus of the model. The input–output model could evaluate how an individual exogenous change in final demand in one sector of the economy would result in an overall change in output across other individual sectors of that same regional economy. By summing the individual output effects for each individual sector, Leontief's input–output model could generate economic multipliers specific to each individual sector in a given region.

Other scholars, such as Sir Richard Stone, expanded Leontief's original input–output models by endogenizing the institutions of a region's economy, such as households, government, and trade (Pyatt and Round 1985). These models became known as Social Accounting Matrices (SAMs). This extension allowed for incorporating some or all of these traditional final demand sectors with industry sectors, thus incorporating their output effects into the economic analysis. Further, the incorporation of households allowed for predictions of distributional effects. For example, household sectors can be subdivided into multiple household income groupings, allowing for an understanding of how final demand changes in an economic sector have distributional effects across households in a region (Miyazawa 1976).

In addition, input–output/SAM models have been extended beyond evaluating intersectoral effects within an individual region. Interregional and multiregional models have been further developed to understand intersectoral effects between regions. Such models allow for an understanding of both interregional spillover effects, from a change in final demand in one region's sector on the change in output of a second region's sector (Miller and Blair 2009).

Further extensions of the input–output/SAM models have occurred through development of computable general equilibrium (CGE) models (Kilkenny and Robinson 1990; Kraybill et al. 1992; Kilkenny 1993), as well as conjoined models (Rey 2000; Shields et al. 2003). Computable general equilibrium models leverage the intersectoral relationships that are present in the input–output framework but relax the more limiting production function and elasticity assumptions for factor inputs such as labor. Another set of models that come from the input–output/SAM framework leverage subsets of these models to estimate or forecast economic outcomes. For example, the community policy analysis modeling system (COMPAS) represents a combination of the use of input–output models to project changes in employment demand (Johnson et al. 2006). These changes in employment demand serve as the inputs into a set of econometric models that project changes in labor market outcomes that further feed a set of econometric equations that project changes in local government revenues and expenditures.

While many input–output methodological advancements were contributed by regional and rural development economists in the 1970s and 1980s (for examples, see Petkovich and Ching 1978; Martin and Henry 1982), the method matured and transitioned more to application by rural development practitioners in local community outreach as well as local, regional, and national policy. Most input–output models applied today in the United States are sourced from private companies that have developed secondary data-based input–output (SAM) models. These include such companies such as IMPLAN (www.implan.com), REMI (www.remi.com), and EMSI (www.economicmodeling.com). The Bureau of Economic Analysis also calculates sectoral multipliers for regions through its Regional Input Output Modeling System (RIMS II) (www.bea.gov/regional/rims/index.cfm). Most all of the private company models can create input–output/SAM models at the county (and in some cases zip code) level of geography.

3.3 Spatial, static models

While input–output/SAM models represent some of the most applied nonparametric models by rural development scholars and practitioners in the last 30 years, the most applied econometric modeling techniques applied by this same group of scholars represent models using spatial econometric techniques. These models have been particularly effective in helping scholars deal with situations where the assumption of independently and identically distributed errors are violated in traditional regression analysis (LeSage and Pace 2009) due to spatial dependence. Spatial lag models that incorporate a spatial lag of the dependent variable (similar to time lags in time series modeling) allow for an increased understanding of variations in the dependent variable that are not otherwise explained by the covariation in the spatial exogenous variables. For example, Henry et al. (1997) used spatial lags to understand spread and backwash effects between rural hinterlands and urban regions of South Carolina. Goetz and Rupasingha (2002) used a Spatial Tobit model to understand the determinants of high-tech industry clustering. Kim, Marcouiller, and Deller (2005) applied a Spatial Error Model to estimate the contribution of amenities to demographic and economic growth and inequality change in rural and urban counties of the Upper Midwest. More recent extensions of the classic spatial econometric framework have included spatiotemporal dimensions (LeSage et al. 2012) where employment impacts were evaluated on urban and rural counties from Hurricane Ike. Further, spatial econometric estimators, such as Geographically Weighted Regression (GWR), allowed for spatial heterogeneity in parameter estimates (Fotheringham,

Brunsdon, and Charlton 2002). For example, Partridge et al. (2008) used GWR to show how many of the same causal factors can significantly impact economic growth positively and negatively in local regions that would otherwise be insignificant, using traditional regression approaches that attempt to identify a single global parameter to explain the functional relationship for an entire nation.

4 Current frontiers in rural development theory and methods – the rural wealth framework

Some of the most recent efforts and extensions to rural development theory and methods have centered around the rural wealth framework (Pender et al. 2014). The underlying tenets of the rural wealth framework share a common basis with the green accounting framework (Bartelmus et al. 1991; United Nations 2014), the community capitals framework (Emery and Flora 2006), and the inclusive wealth framework (Arrow et al. 2012; Christian 2014). The rural wealth framework attempts to measure the "comprehensive wealth" of a place by measuring/valuing each of its individual forms of wealth. The framework distinguishes stocks and flows, people-based wealth, and placed-based wealth, as well as public wealth and private wealth.

The rural wealth framework is built around the concepts of the Fisherian income. Fisherian income assumes that a flow of services comes from an instrument (wealth asset) and that a community's income is the total flow of services from all of its instruments (wealth assets) (Fisher 1906). Further, Nordhaus (1995) expands on Fisher's definition and argues that Fisherian income is equivalent to sustainable income. Given a constant return on the flow of services from instruments, if Fisherian income is constant or growing, the instruments (wealth assets) are constant or growing and are equivalent to sustainability.

Agricultural economics scholars have been working on topics of wealth measurement and the linkages of place-based wealth to issues of economic prosperity in rural regions. For example, Rupasingha, Goetz, and Freshwater (2006) developed one of the more popular indices of social capital used by economists. Deller et al. (2001) developed alternative proxies for measuring amenity stocks (natural wealth and physical wealth) and evaluated their contributions to overall economic growth. McGranahan (1999) developed a natural amenities scale that provides an index of natural wealth assets in place.

Current efforts by regional and rural development scholars have focused on the issues of wealth measurement as well as the contribution of the wealth assets of a place on major economic quality of life metrics. For example, Pender et al. (2012) provides an overview of potential secondary data proxies for various wealth asset categories at the state and local levels. Further, efforts by Schmit et al. (2017) have focused on strategies to quantify traditional intangible wealth assets in the context of local and regional food systems. Further extending the work of Chetty et al. (2014), Weber et al. (2017) identify such wealth assets as social capital as being significant positive contributors to social mobility in rural places.

Further, the rural wealth framework can also build on modeling structures that incorporate spatial, intersectoral, temporal, and distributional elements. The system dynamics framework popularized by Forrester (1993) has great potential for both understanding and providing decision support to rural communities interested in economic development strategies built on a foundation of the rural wealth framework (Collins 2014). Unlike traditional microeconomic theory or modern spatial theories (Roback 1982; Carlino and Mills 1987) that are focused on achieving equilibrium, system dynamics does not assume equilibrium, but is more focused on understanding the dynamic process of change in community wealth and economic outcomes (for examples, see Costanza and Gottlieb 1998 and Voinov et al. 1999). Johnson et al. (2011)

develop a system dynamics model that incorporates multiple forms of wealth assets to understand the interactions of policy interventions on economic, social, and environmental outcomes. The models incorporate multiple sectors of the economy (including agriculture) and multiple household income groups to understand distributional consequences as well as variables that evaluate environmental outcomes. Hence, this example presents a template on how to incorporate multiple forms of wealth and address rural development outcomes along sectoral, distributional, spatial, and temporal dimensions.

5 Conclusion

In the past 40 years, rural development has forged a path of theory and methods that have complemented but have increasingly become less dependent on agriculture and its role in measuring rural prosperity, distribution, etc. Adoption of theories and methods from regional science has increased among agricultural economists focusing on rural development. Future research will see an increasing understanding of the relative contributions of place and space toward people-based development outcomes with a particular emphasis on distribution as wealth assets and will see their returns become less equally distributed over time.

Note

1 Here, land is included in capital.

References

Abadie, A., A. Diamond, and J. Hainmueller. 2010. "Synthetic Control Methods for Comparative Case Studies: Estimating the Effect of California's Tobacco Control Program." *Journal of the American Statistical Association* 105:493–505.

Abadie, A., and J. Gardeazabal. 2003. "The Economic Costs of Conflict: A Case-Control Study for the Basque Country." *American Economic Review* 93(1):113–132.

Arrow, K.J. 1971. "The Economic Implications of Learning By Doing." In *Readings in the Theory of Growth*. London: Palgrave Macmillan, pp. 131–149.

Arrow, K.J., P. Dasgupta, L.H. Goulder, K.J. Mumford, and K. Oleson. 2012. "Sustainability and the Measurement of Wealth." *Environment and Development Economics* 17(3):317–353.

Auty, R.M. 1993. *Sustaining Development in Mineral Economies: The Resource Curse Thesis*. Routledge.

Bartelmus, P., C. Stahmer, and J. van Tongeren. 1991. "Integrated Environmental and Economic Accounting: Framework for a SNA Satellite System." *Review of Income and Wealth* 37(2):111–148.

Basu, R., and T.G. Johnson. 1996. "The Development of a Measure of Intersectoral Connectedness Using Structural Path Analysis." *Environment and Planning* A28:709–730.

Becker, G. 1994. *Human Capital: A Theoretical and Empirical Analysis with Special Reference to Education*. Chicago: The University of Chicago Press.

Bernat Jr, G.A., and T.G. Johnson. 1991. "Distributional Effects of Household Linkages." *American Journal of Agricultural Economics*. 73(2):326–333.

Carlino, G.A., and E.S. Mills. 1987. "The Determinants of County Growth." *Journal of Regional Science* 27(1):39–54.

Castle, E.N. 1988. "Policy Options for Rural Development in a Restructured Rural Economy: An International Perspective." In Gene F. Summers, John Bryden, Kenneth Deavers, Howard Newly, and Susan Sechler, eds., *Agriculture and Beyond: Rural Economic Development*. Madison: Department of Rural Sociology, University of Wisconsin, pp. 11–27.

Chetty, R., N. Henderson, P. Kline, and E. Saez. 2014. "Where is the Land of Opportunity? The Geography of Intergenerational Mobility in the United States." *Quarterly Journal of Economics* 129(4):153–1623.

Christian, M.S. 2014. "Human Capital Accounting in the United States: Context, Measurement, and Application." In Jorgenson, Dale W., J. Steven Landefeld, and Paul Schreyer, eds., *Measuring Economic Sustainability and Progress*. Chicago: University of Chicago Press, pp. 461–491.

Collins, R. 2014. "Using Inclusive Wealth for Dynamic Analyses of Sustainable Development: Theory, Reflection and Application." Paper presented at the 32nd International Conference of the System Dynamics Society. Delft, Netherlands.

Costanza, R., and S. Gottlieb. 1998. "Modelling Ecological and Economic Systems with STELLA: Part II." *Ecological Modelling* 112(2):81–84.

Deller, S.C., T. Tsai, D. Marcouiller, and D. English. 2001. "The Role of Amenities and Quality of Life in Rural Economic Growth." *American Journal of Agricultural Economics* 83(2):352–365.

DiFrancesco, R. 1998. "Large Projects in Hinterland Regions: A Dynamic Multiregional Input-Output Model for Assessing the Economic Impacts." *Geographical Analysis* 30(1):15–34.

Domar, E.D. 1947. "Expansion and Employment." *A.E.R.* 37(March):34–55.

Duchin, F., and D.B. Szyld. 1985. "A Dynamic Input–Output Model with Assured Positive Output." *Metroeconomica* 37:269–282.

Emery, M., and C. Flora. 2006. "Spiraling-Up: Mapping Community Transformation with Community Capitals Framework." *Journal of the Community Development Society* 37(1):19–35.

Fisher, I. 1906. *The Nature of Capital and Income*. London: Palgrave Macmillan.

Forrester, J.W. 1993. "System Dynamics and the Lessons of 35 Years." In *A Systems-Based Approach to Policymaking*. Boston, MA: Springer, pp. 199–240.

Fotheringham, A.S., C. Brunsdon, and M. Charlton. 2002. *Geographically Weighted Regression: The Analysis of Spatially Varying Relationships*. Chichester: Wiley.

Frey, D.E. 1989. "A Structural Approach to the Economic Base Multiplier." *Land Economics* 65(4):352–358.

Fujita, M., P. Krugman, and A.J. Venables. 1999. *The Spatial Economy: Cities, Regions, and International Trade*. Cambridge, MA: MIT Press.

Goetz, S.J., and A. Rupasingha. 2002. "High-Tech Firm Clustering: Implications for Rural Areas." *American Journal of Agricultural Economics* 84(5):1229–1236.

Goetz, S.J., M. Shields, and Q. Wang. 2009. "Identifying Food Industry Clusters." In Goetz, S.J., S.C. Deller, and T.R. Harris, eds., *Targeting Regional Economic Development*. London: Routledge, pp. 281–310.

Harris, T.R., G.E. Ebai, and J.S. Shonkwiler. 1998. "A Multidimensional Estimation of Export Base." *Journal of Regional Analysis and Policy* 28:3–17.

Harrod, R.F. 1948. *Towards a Dynamic Economics: Some Recent Developments of Economic Theory and Their Application to Policy*. London: Palgrave Macmillan.

Henry, M., D.L. Barkley, and S. Bao. 1997. "The Hinterland's Stake in Metropolitan Growth: Evidence from Selected Southern Regions." *Journal of Regional Science* 37(3):479–501.

Hoover, E., and F. Giarratani. 1984. *An Introduction to Regional Economics*. New York: McGraw-Hill Ryerson, Limited.

Innis, H.A. 1936. "Approaches to Canadian Economic History." *Commerce Journal* 26:24–30.

Isard, W., I.J. Azis, M.P. Drennan, R.E. Miller, S. Saltzman, and E. Thorbecke. 1998. *Methods of Interregional and Regional Analysis*. Aldershot: Ashgate Publishing Company.

Isserman, A., and T. Rephann. 1995. "The Economic Effects of the Appalachian Regional Commission: An Empirical Assessment of 26 Years of Regional Development Planning." *Journal of the American Planning Association* 61(3):345–364.

Johnson, T.G. 1985. "A Continuous Leontief Dynamic Input Output Model." *Papers of the Regional Science Association* 56:177–188.

———. 1994. "The Dimensions of Regional Economic Development Theory." *Review of Regional Studies* 24(2):119–126.

Johnson, T.G., S. Alva-Lizarraga, K. Refsgaard, T. Kampas, D. Psaltopoulos, and G. Frances. 2011. "Developing and Adapting the POMMARD Model." In Bryden, J.M., S. Efstratoglou, T. Ferenczi, K. Knickel, T.G. Johnson, K. Refsgaard, and K.J. Thomson, eds., *Towards Sustainable Rural Regions in Europe: Exploring Inter-Relationships Between Rural Policies, Farming, Environment, Demographics, Regional Economies and Quality of Life Using System Dynamics*. New York: Routledge Publishers, Chapter 5.

Johnson, T.G., D.M. Otto, and S.C. Deller. 2006. "Introduction to Community Policy Analysis Modeling." In Johnson, T.G., D.M. Otto, and S.C. Deller, eds., *Community Policy Analysis Modeling*. Ames, IA: Blackwell Publishing, pp. 3–16.

Kilkenny, M. 1993. "Rural Urban Effects of Terminating Farm Subsidies." *American Journal of Agricultural Economics* 75(4):968–980.

———. 1998. "Transport Costs and Rural Development." *Journal of Regional Science* 38(2):293–312.

Kilkenny, M., and S. Robinson. 1990. "Computable General Equilibrium Analysis of Agricultural Liberalization: Factor Mobility and Macro Closure." *Journal of Policy Modeling* 12(3):527–556.

Kim, K.K., D.W. Marcouiller, and S.C. Deller. 2005. "Natural Amenities and Rural Development: Understanding Spatial and Distributional Attributes." *Growth and Change* 36(2):273–297.

Kindleberger, C.P., and B. Herrick. 1983. *Economic Development*. New York: McGraw-Hill.

Kraybill, D.S., T.G. Johnson, and D.R. Orden. 1992. "Macroeconomic Imbalances: A Multiregional General Equilibrium Analysis." *American Journal of Agricultural Economics* 74(3):726–736.

Krugman, P. 1991. "Increasing Returns and Economic Geography." *Journal of Political Economy* 99:483–99.

———. 1998. "What's New About the New Economic geography?" *Oxford Review of Economic Policy* 14(2):7–17.

Leontief, W. 1953. *Studies in the Structure of the American Economy: Theoretical and Empirical Explorations in Input–Output Analysis*. New York: Oxford University Press.

LeSage, J., and R.K. Pace. 2009. *Introduction to Spatial Econometrics*. Boca Raton, FL: CRC Press.

LeSage, J., R.K. Pace, N. Lam, and R. Campanella. 2012. "Space-time Modeling of Natural Disaster Impacts." *Journal of Economic and Social Measurement* 36(3):169–191.

Loveridge, S., and A.C. Selting. 1998. "A Review and Comparison of Shift-Share Identities." *International Regional Science Review* 21(1):37–58.

Lucas, R.E. 1988. "On the Mechanics of Economic Development." *Journal of Monetary Economics* 22(1):3–42.

Malizia, E.E., and S. Ke. 1993. "The Influence of Economic Diversity on Unemployment and Stability." *Journal of Regional Science* 33(2):221–235.

Martin, T.L., and M.S. Henry. 1982. "Rural Area Consumer Demand and Regional Input–Output Analysis." *American Journal of Agricultural Economics* 64(4):752–755.

McGranahan, D.A. 1999. *Natural Amenities Drive Population Change*. AER Report No. 781. Economics Research Service. United States Department of Agriculture.

Miller, K. 2010. "Why Definitions Matter: Rural Definitions and State Poverty Rankings." *Rural Policy Research Institute Policy Brief*. www.rupri.org/Forms/Poverty%20and%20Definition%20of%20Rural.pdf. Accessed April 4, 2018.

Miller, R.E., and P.D. Blair. 2009. *Input–Output Analysis: Foundations and Extensions* (2nd ed.). Cambridge: Cambridge University Press.

Miyazawa, K. 1976. *Input–output Analysis and the Structure of Income Distribution*. Lecture Notes in Economics and Mathematical Systems Book Series (LNE, volume 116). Berlin, Heidelberg: Springer.

Munasib, A., and D.S. Rickman. 2015. "Regional Economic Impacts of Shale Gas and Tight Oil Control Analysis." *Regional Science and Urban Economics* 50:1–17.

Nordhaus, W.D. 1995. "How Should We Measure Sustainable Income?" Cowles Foundation Discussion Paper.

North, D.C. 1955. "Location Theory and Regional Economic Growth." *Journal of Political Economy* 63(3):243–258.

Partridge, M.D., D.S. Rickman, K. Ali, and M.R. Olfert. 2008. "The Geographic Diversity of U.S. Nonmetropolitan Growth Dynamics: A Geographically Weighted Regression Approach." *Land Economics* 84(2):241–266.

Pender, J., and R. Reeder. 2011. *Impacts of Regional Approaches to Rural Development: Initial Evidence on the Delta Regional Authority* (ERR-119). United States Department of Agriculture, Economic Research Service, June.

Pender, J., A. Marre, and R. Reeder. 2012. *Rural Wealth Creation: Concepts, Strategies, and Measures. Economic Research Report* (ERR-131). Economic Research Service, United States Department of Agriculture, March.

Pender, J.L., T.G. Johnson, B.A. Weber, and J.M. Fannin. 2014. "Introduction and Overview." In Pender, J.L., T.G. Johnson, B.A. Weber, and J.M. Fannin, eds., *Rural Wealth Creation*. New York: Routledge, pp. 3–15.

Petkovich, M.D., and C.T. Ching. 1978. "Modifying a One Region Leontief Input–Output Model to Show Sector Capacity Constraints." *Western Journal of Agricultural Economics* 3:173–179.

Piketty, T. 2014. *Capital in the Twenty-First Century*. Cambridge: The Belknap Press of Harvard University Press.

Porter, M. 1998. "Clusters and the New Economics of Competition." *Harvard Business Review* November–December:77–90.

Pyatt, G., and J.I. Round. 1985. *Social Accounting Matrices: A Basis for Planning*. Washington, DC: The World Bank.

Ravallion, M. 2008. "Evaluating Anti-Poverty Programs." In R. Evenson and T. Schultz, eds., *Handbook of Development Economics* (Vol 4). Amsterdam, the Netherlands: North Holland, pp. 3787–3846.

Rey, S.J. 2000. "Integrated Regional Econometric+ Input–Output Modeling: Issues and Opportunities." *Papers in Regional Science* 79(3):271–292.

Roback, J. 1982. "Wages, Rents, and the Quality of Life." *Journal of Political Economy* 90(6):1257–1278.

Romer, P. 1986. "Increasing Returns and Long Run Growth." *Journal of Political Economy* 94:1002–1037.

———. 1990. "Endogenous Technological Change." *Journal of Political Economy* 98(5, Part 2):S71–S102.

Rupasingha, A., S.J. Goetz, and D. Freshwater. 2000. "Social Capital and Economic Growth: A County-Level Analysis." *Journal of Agricultural and Applied Economics* 32(3):565–572.

———. 2006. "The Production of Social Capital in US Counties." *The Journal of Socio-Economics* 35(1):83–101.

Rural Sociological Society. 1993. *Persistent Poverty in Rural America*. Rural Studies Series. Boulder, CO: Westview.

Schmit, T.M.B. Jablonski, J. Minner, D. Kay, and L. Christensen. 2017. "Rural Wealth Creation of Intellectual Capital from Urban Local Food System Initiatives: Developing Indicators to Assess Change." *Community Development* 48(5):639–656.

Schultz, T.W. 1961. "Investment in Human Capital." *The American Economic Review* 51(1):1–17.

Shields, M., J.I. Stallmann, and S.C. Deller. 2003. "The Economic and Fiscal Impacts of the Elderly on a Small Rural Region." *Community Development* 34(1):85–106.

Siegel, Paul B., Thomas G. Johnson, and Jeffrey Alwang. 1995. "Regional Economic Diversity and Diversification: Seeking a Framework for Analysis." *Growth and Change* 26(2):261–284.

Solow, R.M. 1956. "A Contribution to the Theory of Economic Growth." *The Quarterly Journal of Economics* 70(1):65–94.

Stiglitz, J.E. 2012. *The Price of Inequality: How Today's Divided Society Endangers Our Future*. WW Norton & Company.

Swan, T.W. 1956. "Economic Growth and Capital Accumulation." *Economic Record* 32(2):334–361.

Tickamyer, A.R., and C.M. Duncan. 1990. "Poverty and Opportunity Structure in Rural America." *Annual Review of Sociology* 16(1):67–86.

Tiebout, C.M. 1956. "A Pure Theory of Local Expenditures." *Journal of Political Economy* 64(5):416–424.

United Nations. 2014. *System of Environmental Economic Accounting 2012 – Central Framework*. New York: United Nations.

United States Department of Agriculture, Economic Research Service. 2016. "Rural Definitions." www.ers.usda.gov/data-products/rural-definitions/. Accessed December 10, 2017.

———. 2017. "What is Rural?" www.ers.usda.gov/topics/rural-economy-population/rural-classifications/what-is- rural.aspx. Accessed December 10, 2017.

Voinov, A., R. Costanza, L. Wainger, R. Boumans, F. Villa, T. Maxwell, and H. Voinov. 1999. "Patuxent Landscape Model: Integrated Ecological Economic Modeling of a Watershed." *Environmental Modelling & Software* 14(5):473–491.

Weber, B., L. Jensen, K. Miller, J. Mosley, and M. Fisher. 2005. "A Critical Review of Rural Poverty Literature: Is There Truly a Rural Effect? *International Regional Science Review* 28(4):381–414.

Weber, B.A., J.M. Fannin, S.M. Cordes, and T.G. Johnson. 2017. "Upward Mobility of Low-Income Youth in Metropolitan, Micropolitan, and Rural America." *The Annals of the American Academy of Political and Social Science* 672:103–122.

Woodward, D., and P. Guimaraes. 2009. "Porter's Cluster Strategy and Industrial Targeting." In Goetz, S.J., S.C. Deller, and T.R. Harris, eds., *Targeting Regional Economic Development*. London: Routledge, pp. 68–83.

Part 4
Methods

25
Econometrics for the future

Hector O. Zapata, R. Carter Hill, and Thomas B. Fomby

1 Introduction

In this short chapter, we provide some guidance for researchers, teachers, and graduate students who are interested in analyzing economic data. Because our page limit is so small, our frame of reference will be small, primarily literature since 2010. For all relevant material up to 1985, see Judge et al. (1985). For literature and methods up to about 2010 see Greene (2018), Wooldridge (2010), Hayashi (2000), and Cameron and Trivedi (2005). *The Handbook of Econometrics*, published by Elsevier, presently consists of six volumes spanning 1983 to 2007 and provides exhaustive references. Greene (2018) has now been published.

1.1 Keeping current on econometrics

Keeping current in econometrics requires some work.

1. The *Journal of Economic Literature* (*JEL*) contains review and survey articles, book reviews, and a bibliography of new books. Use the *JEL* Classification System (www.aeaweb.org/econlit/jelCodes.php) in conjunction with EconLit (www.aeaweb.org/econlit/) to target research
2. The *Journal of Economic Perspectives* (1987–present) is publicly available online at no charge (www.aeaweb.org/journals/jep). It contains essays that are between general reading and academic journals. For example, Volume 31, No. 2, Spring 2017 contains "Symposium: Recent Ideas in Econometrics," which we cite in this chapter.
3. *Advances in Econometrics* (www.emeraldinsight.com/series/aeco) is a series of research volumes that publish original scholarly econometrics papers with the intention of expanding the use of developed and emerging econometric techniques by disseminating ideas on the theory and practice of econometrics throughout the empirical economic, business, and social science literature. The volume content is usually devoted to a single topic, such as the recent Volume 38, "Regression Discontinuity Designs," but are sometimes festschrifts in honor of a famous econometrician, such as Volume 36, "Essays in Honor of Aman Ullah." This series is a terrific way to explore new areas.

4   The *Journal of Econometrics* (www.journals.elsevier.com/journal-of-econometrics/special-issues) has special issues that are focused on a single area, as does the *Journal of Applied Econometrics* (http://onlinelibrary.wiley.com/journal/10.1002/(ISSN)1099-1255).

5   The software giants Statistical Analysis System (SAS) (www.sas.com/sas/books.html) and Stata (www.stata.com/bookstore/books-on-stata/) have publishing companies with books specialized on their computing systems in general and on specific topics such as econometrics. Stata users should consider the *Stata Journal* (www.stata-journal.com/), which publishes articles explaining new techniques and providing user-written code, which then sometimes is incorporated into Stata. *The Econometrics Journal* offers a detailed list of possible econometric software products (www.feweb.vu.nl/econometriclinks/software.html).

The purpose of this chapter is to mainly introduce recent developments in econometrics, remaining mindful that a number of the papers and ideas we discuss have gained the interest of agricultural and applied economists in a plethora of fields and applications. The next section of the chapter focuses mostly on frontiers in microeconometrics and related estimation and inference methods; the third section is on the big-data revolution, of which econometrics is at the center; next, an information theoretic approach to econometrics is discussed and a new development highlighted. The chapter closes with a section on time series econometrics that focuses on current developments that may merit consideration in future research. We feel strongly that as readers think a bit more deeply about the material and ideas discussed in this chapter that their future research will add empirical evidence in favor of its title.

2 Microeconometrics

2.1 Causal effects and policy evaluation

Imbens and Rubin (2015) provide a very complete analysis of causal modeling within the framework of potential outcomes paradigm. It begins with randomized controlled experiments and carries through regression, matching, propensity scores, instrumental variables, and compliance issues. Athey and Imbens (2017) provide a survey pointing to current issues, including regression discontinuity designs, synthetic controls, differences-in-differences methods, methods designed for network settings, and methods combining experimental and observational data. They also discuss the various supplementary analyses that support and defend the primary analysis. They argue that the method of synthetic control, Abadie et al. (2014), is the most important innovation in the policy evaluation literature in the last 15 years. These methods and topics use reduced form models. Structural model approaches, while now less frequent, are discussed by Low and Meghir (2017).

2.2 Regression discontinuity (RD) designs

When the separation into treatment and control groups follows a deterministic rule, such as, "Students receiving 75 percent or higher on the midterm exam will receive an award," RD designs are useful. How the award affects future academic outcomes might be a question of interest. The literature is introduced and summarized by Lee and Lemieux (2010) and Cattaneo and Escanciano (2016). The original sharp design, in which someone crossing the threshold is definitely given treatment, and fuzzy design, in which the probability of treatment increases for someone crossing the threshold, have been extended to multiple cutoffs (Cattaneo et al. 2016) and geographical limits (Keele and Titiunik 2015), to mention two. *Advances in Econometrics,*

Volume 38, "Regression Discontinuity Designs: Theory and Applications" (2017), contains a host of recent papers and extensions. Applied researchers will be interested in software for implementing these methods. See, for example, Calonico et al. (2017). Recent extensions, such as Card et al. (2015), include examining discontinuities in slopes, creating kinked regressions.

2.3 Instrumental variables

Imbens (2014) gives a historical perspective on the use of instrumental variables. He also discusses how and why a statistician's views about the subject differ from an economist's views. In some problems, many instrumental variables (IVs) are potentially available. How many should we use and which ones should we use? Caner et al. (2016) offer some guidance based on simulations for IV and generalized method of moments (GMMs), concluding that an adaptive Lasso is a good solution. Bekker and Crudu (2015) propose a jackknife estimator in the case of unknown heteroscedasticity and many IVs. Bekker and Wansbeek (2016) offer an alternative to limited information maximum likelihood (LIML) using "concentrated IV," which can lead to "Bekker standard errors" in the usual IV form.

2.4 Measurement errors and missing data

Errors-in-variables models, missing data, and measurement error are persistent problems in analyzing economic data. De Nadai and Lewbel (2016) consider a nonparametric model with errors on both sides of the equation and provide a GMM sieve estimator. Chalak (2017) discusses the consequences of instruments measured with error in the estimation of models with heterogeneous effects. Related to both errors-in-variables (EIV) models and IV is the problem of missing data. McDonough and Millimet (2017) examine imputation methods designed to increase the strength of IV. Abrevaya and Donald (2017) use a GMM method for dealing with missing values of explanatory variables. Chaudhuri and Guilkey (2016) use GMM for missing-at-random variables.

2.5 Weak and/or many instruments

Sanderson and Windmeijer (2016) approach the problem of diagnosing weak instruments from the point of view of the rank of the reduced form parameter matrix rather than the Stock and Yogo (2005) approach of considering the rank of the IV matrix. They propose a simple-to-calculate conditional F-test. Olea and Pflueger (2013) develop a test for weak instruments that is robust to heteroscedasticity, autocorrelation, and clustering. Wang and Kaffo (2016) consider bootstrap inference in the many/weak IV case and propose a modified bootstrap that provides a valid approximation in the case of LIML or Fuller's modified-LIML. In another strand of literature, Andrews (2017) uses a two-step process to create confidence sets when instruments might be weak. In the first step the identification is assessed, and in the second step a confidence region is constructed with controlled coverage distortions. The estimated set, or interval, covers with probability $1-\alpha-\gamma$ and then estimates γ. Chernozhukov and Hansen (2008) suggest an easy and clever test procedure for hypotheses on the coefficients of the endogenous variables. Substituting in the null hypothesis to form the constrained model, and then substituting the reduced form equation for the exogenous variables, results in an ordinary least squares (OLS) equation, the testing of which is equivalent to testing the original null hypothesis. This equation can be used if the instruments are weak and can be made robust. Kline (2016) examines the endogenous treatment variable in a nonlinear semiparametric or nonparametric model. Rather than approaching the problem of estimating the shape of the relationship between the outcome and treatment, which may be only partially identified, Kline investigates whether a relationship exists and the direction of the relation, increasing or decreasing.

2.6 Consequences of invalid IV

The primary requirement for an IV is that it be uncorrelated with the regression error. If it is correlated, then it is invalid and the resulting IV estimator is inconsistent. The Sargan test, or the J-test, for the validity of surplus instruments, is routinely used to check for orthogonality, though these tests tend to have low power. Kolesár et al. (2015) offer an alternative. They use additional variables that are correlated with the endogenous variable, but which may also have a direct effect on the outcome. In this case, LIML is no longer consistent, but the authors propose a bias-corrected two-stage least squares (2SLSs), introduced by Anatolyev (2013), that is consistent. They also show that the standard tests for the validity of the over-identifying, surplus IV test for the presence of the direct effects.

2.7 Control functions

A control variable, or control function, has some features of IVs but, unlike IVs, are included into the estimation equation in order to consistently estimate the coefficient of an endogenous variable without using IV methods. Wooldridge (2015) summarizes how the control functions are used with both linear and nonlinear equations. Control functions are useful in models in which the response to treatment varies across the population in a way that may be related to the endogenous variable. The control function estimator is consistent, under some additional assumptions, and, if the treatment variable is endogenous, bootstrap standard errors can be used. The conventional test for endogeneity or robust standard errors can be used, because under the null hypothesis of exogeneity, the usual test procedures are valid. In the case that the endogenous treatment is binary, we have a so-called endogenous switching model. Wooldridge (2015, pp. 432–433) shows how the control function estimates lead to the estimate of the local average treatment effect (LATE). See also Murtazashvili and Wooldridge (2016).

2.8 Robust variance estimation (HCCME)

Since White (1980), there has been a quest for an automatic fix-up of ordinary least squares (OLS) standard errors in the presence of conditional heteroscedasticity or to account for data clusters. Perhaps because of the pervasiveness of their use, there has been a continual examination of the methods, in the practical sense of improving them, but also in the philosophical sense of "Is this what we should be doing?" It relieves the empirical researcher with a large sample of having to test for the presence of non-spherical regression errors, and/or from having to search for a more efficient generalized least squares (GLS) estimator. Good reading on both sides of the issue can be found in the Spring 2010 issue of the *Journal of Economic Perspectives*. See Angrist and Pischke (2010) and Leamer (2010). MacKinnon (2012) summarizes the many versions of the White heteroscedasticity consistent covariance matrix estimators (HCCME) that have evolved over the years.

A casualty of the HCCME progress has been the use of a GLS estimator or weighted least squares (WLS). When heteroscedasticity is present, not only are the conventional standard errors incorrect, but also the OLS estimator is inefficient relative to the WLS estimator. The problem is that in order to improve upon OLS using WLS, we must build and use a model of the conditional heteroscedasticity. Modern econometric practice stresses valid inference rather than inference based on the most efficient estimator, so that there is a school of thought that we should simply use OLS, with a sufficient number of control variables to satisfy conditional mean independence, and then use an HCCME. Romano and Wolf (2017) take a refreshing step back in time and argue that we can obtain asymptotically valid inference using WLS even if our model of the conditional heteroscedasticity is incorrect, if we combine it with a HCCME. The argument is that with just a little more work, we are able to have valid inference and narrower interval estimates with proper coverage. They offer a pre-test, two-step estimator.

While mentioning WLS, we should also mention the confusing topic of sample weights. The idea of sampling weights is this: if we have a random sample from a population, then sample statistics estimate population quantities consistently. If the sample either under represents or over represents portions of the population, and if the way the under and over sampling is known, then population quantities can be estimated by WLS with inverse probabilities of selection. See Solon et al. (2015).

2.9 Bootstrapping

Bootstrapping and the Jackknife are clearly some of the most powerful and useful numerical techniques to come along in recent memory. They are used to obtain valid standard errors for complex estimators and as a basis for the asymptotic refinement of distribution-critical values. It is impossible to assemble in a half page a list of useful references. Try using EconLit with a search for bootstrap, with a qualifier such as instrumental variables, panel data, spatial data, or time series data. Both Russell Davidson, at McGill University, and Joel Horowitz, at Northwestern, offer courses on bootstrapping, and with luck you might find syllabi and notes on their web pages. Jackknife IV estimators have already been mentioned in this chapter. Clustered data corrections are another flavor of robust estimation. Regression models with clustered data may suffer from heteroscedasticity across individual and cluster and serial correlation due to commonalities of the group, and perhaps time series correlation as well. The importance of correcting estimator variances for these combinations of problems cannot be understated. Conventional or even heteroscedasticity robust standard errors can understate the sampling error of the estimator by substantial amounts. Cameron and Miller (2015) survey cluster-robust inference methodology. However, Carter et al. (2017) point out that the original work of Halbert White assumed homogeneous clusters of equal sizes. They show that when clusters are of unequal size, the consistency of the variance estimator requires the number of clusters to approach infinity, not just the number of data points. In fact, they show that the appropriate measure of sample size is the number of clusters. An example they note is that when using cluster-robust inference with fixed effects, the effective sample size is one, so that the "estimator variance is undetermined" (Carter et al. 2017, p. 698). MacKinnon and Webb (2017) explore the question with simulations and determine that the wild cluster bootstrap performs better in situations such as state level clusters. See the time series section for other uses of bootstrapping.

2.10 Spatial methods

A great starting point is LeSage and Pace (2009). The authors introduce and motivate spatial models, discuss maximum likelihood and Bayesian estimation methods. Matlab users will be happy to know that there is a spatial econometrics toolbox. Stata users will be happy to know that Version 15 now has commands for spatial autoregressive models and enhanced Bayesian tools. A recent volume of *Advances in Econometrics*, edited by Baltagi et al. (2016), has papers dealing with both discrete and continuous dependent variables. All the usual econometric issues are relevant. See, for example, Qu et al. (2016), who use an IV estimator for a spatial dynamic panel. Theirs is another paper that is rich in citations to both theory and practice. Yang (2015) introduces an lagrange multiplier (LM) test for spatial dependence using a bootstrap-critical value. Bootstrapping in the spatial context, beyond simple blocking, is explored by Lahiri and Zhu (2006) and Lahiri (2013).

2.11 Panel data

Recent valuable textbooks are Baltagi (2013) and Pesaran (2015). Baltagi (2015) contains a collection of recent survey papers on a variety of panel data topics, including panel data models for discrete choice, multinomial choice, spatial models, nonparametric data models, measurement error models, and count data models. Heckman et al. (2016) consider modeling dynamic treatment effects, multi-stage decision problems, and show that the usual IV estimator does not estimate economically useful parameters, unless policy variables are IV. This work is loaded with references to the dynamic discrete choice literature. See the time series section for a discussion of panel unit-root tests.

2.12 Semiparametric and nonparametric methods

Horowitz (2009) introduces the concepts in an accessible way. *Advances in Econometrics*, Volume 36, "Essays in Honor of Aman Ullah" (2016; edited by Gonzalez-Rivera, Hill, and Lee), is a collection of original papers related to nonparametric and semiparametric methods, but also panel data and information theory. *Advances in Econometrics*, Volume 25, "Nonparametric Econometric Methods" (edited by Li and Racine), includes several original survey papers, as well as papers on copula and density estimation.

2.13 Shrinkage estimation

With high dimensional models, the use of shrinkage estimation is widespread. This includes the "big data" applications but also dynamic panel data models. Recent examples are Lu and Su (2016) and Qian and Su (2016). In the tradition of the James–Stein literature, Hansen (2015) shows that we can shrink maximum likelihood estimators towards subspaces defined by nonlinear restrictions. As in the traditional Stein-rule modeling, when the dimension reduction is greater than two, the new estimator has asymptotic risk less than the maximum likelihood estimator (MLE).

2.14 Bayesian methods

On August 2, 2017, a quick scan of Google Scholar with the keywords "Bayesian Approach Econometrics" since 2017 shows 7,350 articles. The titles indicate linkages to everything from spatial econometrics to cointegration to sports. It is impossible in a brief space to make too many useful comments. We have cited several sources in other parts of this chapter, e.g., Stock and Watson (2017) and Mullainathan and Spiess (2017). Koop (2017) also takes us towards the usefulness of the Bayesian methodology with "big data." It was suggested to us that variational encoders[1] will become widely used in the future. The important reference is by Kingma and Welling (2014). The "old" Bayesian methods, of combining sample and nonsample information to produce a posterior density and shrinkage estimators, are still viable. For example, the popular software Stata 15 offers an expanded package of options for carrying out Bayesian analysis. Griffiths and Hajargasht (2016) and Jacobi et al. (2016) illustrate Bayesian methods for dealing with endogeneity and treatment effects.

3 Big data

What are its major terms? Wide and long datasets; data preparation; supervised versus unsupervised learning; prediction/classification; variable selection and input space compression; machine learning methods developed in various fields of computer science; data partitioning; N-fold model building; model stabilization; trimmed ensemble modeling; and lift in profits or benefits offered by the application of big-data analytics. Unlike in economics, where hypothesis testing and causality issues are of primary interest, in the analysis of big data the focus is typically on building predictive models that provide out-of-sample predictions that are as accurate as possible. For an introduction to the area, which economists are just beginning to explore, see Varian (2014). The closely related concept of machine learning is introduced to economists particularly by Mullainathan and Spiess (2017). For introductory textbook presentations of big data analytic methods, see Shmueli et al. (2016), Tan et al. (2006), and Hastie et al. (2009).

To illustrate some big data concepts, let us consider a business metaphor.

3.1 Wide and long dataset

A very large appliance manufacturer has a warranty business that currently insures full repair services for 250,000 washing machines; these machines have remote sensors on them that generate 100 separate readings on each of their warranted machines every 15 minutes. For one year, that represents a data file of dimension 8,760,000,000 by 101 if we consider the current age of the machine in addition to the 100 readings. Thus, one aspect of big data analysis is large dataset management. For more discussion of additional areas of application of big data analytics and some of the vendor software packages used in big data analytics, see Nisbet et al. (2009) and Shmueli et al. (2016). For discussion of open source software packages R, Rattle, and Python for use in big data analytics, see Harrington (2012) and Williams (2011).

3.2 Supervised learning

In terms of the business metaphor, let us say the analysts decide to build a classification model to predict machine failure (see Hastie et al. 2009 for a discussion of machine learning and prediction), where a machine that fails within a month of measurements being taken is assigned a dependent variable outcome of $y = 1$. Assume that, during the period of observation, 6,000 machines failed. Then, to have a balanced sample, the analysts randomly select 6,000 machines that did not fail within a month of prior readings. These observations are assigned a dependent variable outcome of $y = 0$. The process of building a classification model for subsequent use in diagnosing potential machine failure is called supervised learning.

3.3 Variable selection and input space compression

A major concern, of course, is the very large number of predictors present in the classification problem. A logical next step in the model building process is for the analysts to identify the most important indicators of subsequent machine failure. There are several possible ways to reduce the dimension of the input space. The analysts could conduct a principal components analysis on the inputs and retain the components that explain, say, 90 percent of the variation in the data. See Pearson (1901) and Jolliffe (2002) for a discussion of principal component analysis. Alternatively, they could use the Lasso technique on a linear probability model, starting with 101 inputs (age being reserved) to pare down the number of "important" inputs. See Tibshirani (1996) for the development of the Lasso method. Finally, they could estimate 100 bivariate logit models, one input at a time, and select the input variables that produce the smallest p-values of the input coefficient estimates. In the spirit of big data analytics, the analysts could look at the three separate "important input variable" sets and take an intersection of the variables in the sets to move forward into the modeling phase of the study.

3.4 Machine learning, validation, and model stabilization

The next phase of big data model-building involves forming a partition of the 12,000 observations into three randomly drawn datasets – training, testing, and validation datasets of 4,000 observations each. The training dataset is used to make initial builds of a whole set of classification models – linear probability, logit/probit, K-Nearest Neighbors, Naïve Bayes, Artificial Neural Networks, Support Vector Machines, CART and CHAID Decision Trees, and Random Forests of Decision Trees, to name a few. See Hastie et al. (2009) for an explanation of these methods as well as shrinkage methods. Some models are "pruned" on the training dataset by using in-sample goodness-of-fitness criteria, while other methods use the test (second) dataset to obtain pruned models. This pruning process is intended to keep analysts from "overtraining" (overfitting) models that can arise from the extensive mining of the data that all methods (both econometric and machine learning) use when dealing with observational data.

To judge the predictive accuracy of these competing models and to select optimal cutoff (threshold) probability values for the choice models, an appropriate predictive accuracy measure must be chosen. One way this can be done is by considering the relative costs of making Type I (false positive) choice errors versus Type II (false negative) choice errors. Suppose that the cost of a Type I error is one-fifth the cost of a Type II error. This suggests using a weighted error rate measure for judging the goodness of the competing models. One aspect of big data analysis, and information theory, is that there is an essential reliance on a "loss function" related to modeling objectives.

3.5 Ensemble modeling

But in another spirit of predictive analysis of big data, it is assumed that no one model is destined to be the uniquely most accurate predictive model on independent datasets yet to be realized. In this spirit, ensemble model building has taken on a central role in big data analytics. Such an independent dataset is the third (validation) dataset that was reserved in the original partitioning of the data. This dataset helps validate "trimmed" ensemble models that are more accurate than any one single model previously determined by the training and test datasets. By a trimmed ensemble model, we mean the selection of a small number of the best classification models as determined by the smallest weighted error rates computed on the validation (second) dataset and then scoring the predictive accuracy of the trimmed ensemble on the validation (third) dataset. The predictions of the trimmed ensembles are obtained by using either weighted averages of the predicted probabilities of the individual models making up the ensemble or a simple majority rule of the predictions of the constituent models. If properly constructed, the best trimmed ensemble model will, with a high probability, provide more weighted predictive accuracy than any of the separate methods alone. For discussion on the building of ensemble models, see Diederich (2000), Seni and Elder (2010), and Zhou (2012).

Given the determination of an effective ensemble model, the analysts are prepared to apply it to new datasets on warranted machines to determine which machines should receive preemptive maintenance to reduce warranty costs over a current practice of, say, only being reactive in repairing failed machines and not conducting any preventive maintenance driven by analytics. It is the profit "lift" offered by predictive analytic techniques that demonstrates the benefits of carefully analyzing big data using a mixture of econometric and machine learning tools and scoring new data with them.

3.6 Causality, treatment effects, and big data

Finally, apart from the majority of big data problems being predictive in purpose, there is recent research involving the application of big data methods to policy analysis and causality and treatment effect problems. See Athey and Imbens (2016) and Wager and Athey (2017) for how random forests might be used for policy evaluation. Belloni et al. (2013), Belloni et al. (2014), and Chernozhukov et al. (2015) connect big data methods, with many IV and/or controls, to the estimation of treatment effects. Machine Learning methods have even been used for demand estimation; see Bajari et al. (2015).

4 The information theoretic approach

In this section, we argue that traditional econometric modeling approaches do not provide a reliable basis for making inferences about the causal effect of a supposed treatment of data in observational and quasi-experimental settings. Recognizing that one's knowledge regarding the underlying economic behavioral system and observed data process is complex, partial, and incomplete, Judge and Mittelhammer (2012) suggest an information theoretic basis for estimation, inference, model evaluation, and prediction.

In coping with quantitative economic information recovery problems, a natural solution is to use information theoretic estimation and inference methods that are designed to deal with the nature of economic–econometric models and data, as well as the resulting behavioral stochastic inverse problem. In this context, the family of distribution-entropic functions described by Cressie and Read (1984) and Read and Cressie (1988) provide a basis for linking the data and the unobserved and unobservable behavioral model parameters. This permits the researcher to exploit the statistical machinery of information theory to gain insights relative to the underlying adaptive–causal behavior of a dynamic economic system. In developing an information theoretic econometric approach to estimation and inference, the single parameter Cressie–Read (CR) family of entropic functions represents a way to link the likelihood-entropic behavior informational functions with the underlying sample of data to recover estimates of the unknown parameters.

In the context of the above paragraph, consider the Cressie–Read (CR) multi parametric entropy family of power divergence measures:

$$I(p,q,\gamma) = \frac{1}{\gamma(\gamma+1)} \sum_{i=1}^{n} p_i \left[\left(\frac{p_i}{q_i} \right)^{\gamma} - 1 \right]. \tag{1}$$

In Equation (1), the value of γ indexes members of the CR family, p_is represent the subject probability distribution, the q_is are reference probabilities, and p and q are $n \times 1$ vectors of p_is and q_is, respectively. The usual probability distribution characteristics of $p_i, q_i \in [0,1] \forall_i$, $\sum_{i=1}^{n} p_i = 1$, and $\sum_{i=1}^{n} q_i = 1$ are assumed to hold. In Equation (1), as γ varies, the family of estimators that minimize power divergence exhibit qualitatively different sampling behavior that includes Shannon's entropy, the Kullback–Leibler measure, and, in a binary context, the logistic distribution divergence (see Gorban et al. 2010; Judge and Mittelhammer 2011, 2012). The CR family of power divergences is defined through a class of additive convex functions that represents *a broad family of likelihood functional relationships* and test statistics within a moments-based estimation context. All well-known divergences–entropies belong to the class of CR functions. In addition, the CR measure exhibits proper convexity in p for all values of γ and q and embodies the required probability system characteristics, such as invariance with respect to a monotonic transformation of the divergence measures. In the context of extremum metrics, the general CR family of power divergence statistics represents a flexible family of pseudo-distance measures from which to derive empirical probabilities.

The CR statistic is a single index family of divergence measures that can be interpreted as encompassing a wide array of empirical goodness-of-fit and estimation criteria. As g varies, the resulting estimators that minimize power divergence exhibit qualitatively different sampling behavior. One possibility is to use data-consistent empirical sample moments–constraints, such as

$$\mathbf{h}(\mathbf{Y}, \mathbf{X}, \mathbf{Z}; \boldsymbol{\beta}) = n^{-1}[\mathbf{Z}'(\mathbf{Y} - \mathbf{X}\boldsymbol{\beta})]^p \to \mathbf{0} \tag{2}$$

where Y, X, and Z are respectively a $n \times 1$, $n \times k$, $n \times m$ vector/matrix of dependent variables, explanatory variables, and instruments, and β is an unknown and unobservable parameter vector. A solution to the stochastic inverse problem, based on the optimized value of $I(p, q, \gamma)$ is one basis for representing a range of data sampling processes, likelihood functions, and ensembles. In this linear model context, if we use (1) as the information recovery criterion, along with moment-estimating function information (2), the estimation problem based on the CR divergence measure (CRDM) may, for any given choice of γ, be formulated as the following extremum-type estimator for b:

$$\hat{\boldsymbol{\beta}}(\gamma) = \underset{\beta \in B}{\operatorname{argmin}} \left[\min_{p} \left\{ I(p, q, \gamma) \Big| \sum_{i=1}^{n} p_i Z_i'(Y_i - X_i b) = 0, \sum_{i=1}^{n} p_i = 1, p_i \geq 0 \forall i \right\} \right] \tag{3}$$

where q is often taken as a uniform distribution. This class of estimation procedures is referred to as Minimum Power Divergence (MPD) estimation, and additional details of the solution to this stochastic inverse problem are provided in Judge and Mittelhammer (2011). The MPD optimization problem may be represented as a two-step process. In particular, one can first optimize with respect to the choice of the sample probabilities p, and then optimize with respect to the structural parameters β for any choice of the CR family of divergence measures identified by the choice of γ, given q. It is important to note that the *family* of power divergence statistics defined by (2) is symmetric in the choice of which set of probabilities are considered as the subject and reference distribution arguments of the function (3). In particular, whether the statistic is designated as $I(p, q, \gamma)$ or $I(p, q, \gamma)$, *the same collection* of members of the family of divergence measures are ultimately spanned. Three discrete CR alternatives for $I(p, q, \gamma)$, where $\gamma \in \{-1, 0, 1\}$, have received the most attention in the literature.

The applicability of this type of nontraditional information theoretic approach is illustrated in a number of recent applications, including Cho and Judge (2014) and Judge (2015) on recovering the optimum solution for the unknown pathway probabilities of a general binary behavioral network. See Zanin et al. (2012), Judge (2013), and Judge (2015) on nonlinear ordinal basis for recovering patterns in time series data and Miller and Judge (2015) on hidden Markov processes.

5 Developments in time series analysis

"To attempt to survey recent developments in time series analysis is a daunting task" was the first sentence in the introduction to Newbold (1981). It appears that this statement is timeless. Box and Jenkins (1970), Granger and Newbold (1977), and Dickey and Fuller (1979) were cited. Dickey and Fuller's article was referenced in relation to the asymptotic distribution of maximum likelihood (ML) estimators of autoregressive processes with a unit root. Curiously, spurious regression (Granger and Newbold 1974) was excluded. The late 1970s brought along vector autoregressive models (VAR) for analyzing and forecasting macroeconomic time series (e.g., Sims 1980), and these models became popular in macroeconomic forecasting. Testing for unit roots and the specification of VARs opened the flood gates to a vast literature that brought the two together in search for adequate time series model specification. A key question in this early literature was whether to detrend or to difference a time series. At the time, detrending and differencing were being used almost interchangeably in empirical modeling. Seminal among major contributions of the 1980s were Nelson and Plosser (1982) on trends and random walks in macroeconomic data; Engle (1982) on volatility modeling; Phillips (1987) on the functional analysis of unit roots; the Nobel Prize winners Engle and Granger (1987) on cointegration analysis; and Johansen and Juselius (1990) on maximum likelihood estimation with cointegrated VARs.[2]

Of importance to this section is where developments in time series are having an impact in the empirical literature and what future time series econometrics may bring along! In relation to where we are today, a brief review of recent published empirical articles in agricultural and applied economics reveals interest in maximum likelihood estimation of the error-correction model (ECM) with codependence, causality testing, nonlinear modeling such as asymmetric ECMs, and other time-varying smooth transition autoregressive models (e.g., Douc et al. 2014), tests of stationarity and cointegration in nonstationary panel data (e.g., Hassler and Hosseinkouchack 2017), and modeling dependence using copulas (e.g., Joe 2015). Global VARs (e.g., Peseran et al. 2004 and Dees et al. 2007) have been used to analyze spikes in agricultural commodity markets, since food commodity price fluctuations impact poverty and food security worldwide. Our modest aim in what follows is to highlight some new promising work in testing for unit roots and time series modeling.

5.1 Testing for unit roots

5.1.1 Unit roots testing as of 2015

Almost all that needs to be known about unit roots of relevance to applied research is found in Choi (2015) and Newbold and Leybourne (2003). Choi's discussion of the functional central limit theorem and of the general limit theory for regressions with nonstationary time series is a good introduction to the subject. After Choi, a few contributions focus on aspects of approximating functional limit theory by bootstrapping, series boundedness vs I(0) behavior in mean-reversion, second-order stationarity, unit-roots in noncausal ARs with t-distributed errors, panel unit root tests and estimation, and the most recent research on nonstationarity in cross-sectional distributions. These works are at the frontier of developments and will significantly impact future work.

5.1.2 Bootstrap methods to approximate functional Brownian motion

Advances in computing technology have made more feasible the use of bootstrap methods, such as block, wild, and sieve, on residuals from autoregressive regressions and have been recently adopted to approximate the null distributions of the Dickey–Fuller tests (e.g., Phillips 2010). Hybrid bootstrap tests that combine residual-based bootstrap methods with resampling methods that perturb the minimand of the objective function (e.g., Li et al. 2014), for example, are adaptable to time series with uncorrelated but possibly dependent innovations, another example of which is series with time-varying conditional variance. The idea is to approximate the desired functional of Brownian motion rather than the normal distribution, as is the case in most standard unit-root tests. Functional analysis of unit roots (e.g., Horvath et al. 2015) sets stationarity as the null hypothesis, as in the well-known Kwiatkowski, Phillips, Schmidt, and Shin (KPSS) test, and the unit root as the alternative (a unit root is the hypothesis in the standard Dickey–Fuller test).

5.1.3 Mean reversion and unit roots

Mean reversion in unit roots has been studied in commodity futures prices and in financial time series. Whether mean reversion is driven by the boundedness of series or by the I(0)-ness is analyzed in Carrion et al. (2016). One implication might be that an I(1) bounded process could be characterized as a I(0) process because natural bounds on the time series can prevent the process from drifting away from the mean value, so the researcher may conclude mean reversion is due to the boundedness of the time series alone. Structural breaks in the series can complicate matters, and bounded unit root tests should capture such properties. New testing procedures address this data property and may lead to re-assessing works in commodity and financial futures analyses.

5.1.4 Second order stationarity in multivariate time series

This is the subject of Jentsch and Rao (2015). They use a discrete Fourier series transform and generate a Portmanteau-type statistic similar to the common Box–Ljung statistic for white noise on residuals of time series models. Again, bootstrap methods are used to estimate the variance and the asymptotic distribution of the covariance test, in stationary and nonstationary processes.

5.1.5 Unit roots in noncausal ARs (NCAR)

If the assumed unpredictability of the error term by past values of the time series, needed for the causal autoregressive (AR) model, breaks down, possibly due to omitted variables, then noncausal autoregressions (NCARs) may be an alternative. Nonstationarity often found in economic time series has escaped previous NCAR developments, but Saikkonen and Sandberg (2016) propose NCAR unit-root tests of the Wald-type based on local maximum likelihood and bootstrap methods to model skewed errors in order to relax assumptions related to symmetric distribution of the errors. Monte Carlo simulations are used to examine the small sample properties of the tests, and the results are satisfactorily supportive in terms of size and better power relative to stationary NCAR alternatives. An empirical application in the presence of leptokurtic errors lends support to the proposed tests.

5.1.6 Nonstationarity in cross-sectional distributions

A number of time series contain cross-sectional distributions, and examples include household income and expenditures, global temperature, individual earnings, etc. Chang et al. (2016) propose a new framework to analyze, using principal component analysis, nonstationarity in time series of state densities and apply it to analyze cross-sectional distributions of individual earnings and the intra-month distributions of stock returns. Given the number of recurrent climate change, CO_2, and crop yield analyses, for example, using functional time series models of this type may find wide applicability.

5.1.7 Panel unit roots

A comprehensive treatment of panel unit roots can be found in Choi (2015), Chapter 7. Tests for unit-roots of individual series in a time series panel or a large multivariate time series, adding cross-sectional correlations, appears in Smeekes (2015), who proposes a sequential testing method and uses block bootstrap to obtain asymptotically valid critical values. The literature on the subject and alternative estimators are discussed in some detail, as well as some bootstrap alternatives. One result in Smeeke's simulation study is that in cases of panel data where T is small compared to N, simulation results support the propose method; an empirical application is also included.

5.2 Econometric time series modeling developments

5.2.1 Granger causality

Economic policy analysis and the integration of Granger's time series ideas, such as causality, is the subject of White and Pettenuzzo (2014). They remind readers, as is also more comprehensively articulated in Judge et al.'s (1985) *Theory and Practice of Econometrics*, 2nd ed. (TPE2), that econometric theory and practice should be informative and useful to policy makers. This article and the references therein are a must read for all practitioners interested in Granger causality. One interplay to two of Granger's major ideas in time series, causality and cointegration, is found in Zapata and Rambaldi (1997), who evaluated alternative tests for Granger causality in cointegrated systems. An issue basically untouched, but also discussed in some of Granger's early work, is that of temporal aggregation. In practice, temporal aggregation can generate spurious effects in tests of Granger causality, an issue often discussed early on by Granger (Granger and Siklos 1995). Time series data may be aggregated via temporal aggregation or through selective sampling. For temporally aggregated time series, Ghysels et al. (2016) use a mixed-frequency VAR (MF–VAR) to introduce a new class of Granger causality tests (called *max*-test), with all series aggregated to a common low frequency and also at whatever sample frequency is available for all series. This work has implications for a large number of empirical works where weekly, monthly, quarterly, and annual data are available but perhaps not for all series of the model in question. An extension of this chapter, with application to business cycles and uncertainty in financial markets, is Götz et al. (2016). In a search for flexibility in capturing nonlinear economic dynamics in time series data, Diks and Wolski (2017)[3] introduces a full nonparametric generalized autoregressive conditional heteroskedasticity (GARCH) model of nonlinear Granger causality and illustrates its use in the study of the weather expectation effects on price relationships in grain markets in the U.S. This model may capture the interest of practitioners where similar effects tend to be excluded in bivariate and more general multivariate parametric models.

5.2.2 VARMA models

It has been common practice in applied work to approximate a vector autoregressive moving average (VARMA) model with an AR specification, which has practical appeal. However, VARMA and error correction model (ECM)-VARMA models provide a more parsimonious alternative. A new methodology for identifying the structure of reduced form VARMA models applicable to both asymptotically stationary and cointegrated time series is developed by Poskitt (2016), and also in Poskitt and Yao (2017). It is easy to visualize an application for which a VAR specification is short due to data restrictions but for which estimation and inferences may be suspect. Examples have been provided in the literature where a structural shock would require a sample size of several thousand observations and a number of lags that is proportionally very long to adequately capture the effects. A macroeconomic application of factor-augmented VARMA, including forecasting and impulse response analysis, is Bedock and Stevanović (2017). Athanalopoulus et al. (2016) emphasize that there is no need to restrict the structure of ECMs to VARs and that ECM-VARMA may provide more accurate forecasts of macroeconomic time series. This literature is growing, and ongoing developments are likely to provide a rich ground for future research. For researchers interested in VARMA modeling with R, Tsay (2014) is a good start. See also Lütkepohl (2005) and Box et al. (2016).

5.2.3 Nonlinear models

Extensive theoretical and empirical contributions have appeared on the use of nonlinear time series models and breaks (e.g., Enders 2016, Ch. 7; Diks and Wolski 2016; Douc, Moullines, and Stoffer 2014). Tong (2015), who introduced threshold models in the late 1970s, provides a historical perspective on applications, developments, and the remaining challenge of how to best model multivariate time series using threshold models. This literature is and will continue to be of much empirical interest because of the noisy, bouncy, breaky, and fat-tail nature of economic and financial data. An example of how to compress this literature in the context of financial econometrics is Ling et al. (2015), who highlight several articles that have had significant impact on this area of research. The discussion of the articles is excellent reading for students of financial econometrics; those interested in particular topics can proceed to read the full papers. Researchers interested in contributing to new developments of their own may want to spend time on the last section of the article, which covers nonlinear modeling with structural breaks, with examples such as jump resonance in measuring seasonal economic effects, threshold (TAR) models in panel data, conditionally heteroscedastic AR models with thresholds (T-CHARM), and threshold autoregressive moving average (ARMA) models. A financial econometrics computational enthusiast would benefit from using Boffelli and Urga (2016). A model not discussed in Ling et al. (2015) that has gained research interest, partly because it permits the estimation of the impact of news on specific financial assets, is the expansion of the GARCH process to asymmetric generalized dynamic conditional correlation (AG–DCC), originally introduced by Cappiello et al. (2006). Kuester et al. (2006), and related developments, assess Value-at-Risk prediction models (e.g., mixed normal-GARCH, asymmetric power autoregressive conditional heteroskedasticity (ARCH), etc.) and extreme value theory (e.g., skewed threshold GARCH (t-GARCH), skewed threshold extreme value theory (ST-EVT)), which has attracted a wide research audience.

5.2.4 Modeling with I(2) series

A 2017 issue of the *Econometrics Journal* highlights recent developments in cointegration and modeling with I(2) series. Applications to cointegrated VARs and testing restrictions is the subject in Boswick and Paroulo (2017), who focus on translating economic hypotheses from the factor loading matrix of the I(2) variables on the cointegrating relations, and Doornik (2017), who triangularizes the I(2) model to introduce a new estimation procedure of the unrestricted and restricted I(2) models.

5.2.5 Modeling mixed-root series

Most applied research has remained distant from mixed roots. When economic time series are nonstationary with unit roots, there may exist effects that are too complex to capture with standard estimation approaches. Fisher et al. (2016) develop a structural econometric model containing I(1) and I(0) variables in which the equations for the I(0) variables have shocks that can be either permanent or transitory. The specification of the structural VAR (SVAR) is at the heart of the proposed method, along with the appropriate specification of I(0) series. Fisher et al. (2016) may shed new light on a number of findings with SVARs when marginally I(1), which are I(0) in reality, are modeled as I(1) series. The findings and interpretation of results with this method, relative to standard methods, may surprise empirical researchers in terms of where transitory versus permanent shocks emerge and the direction of related shocks. For robust inference, using self-generated instruments (i.e., instrumental variables relying on regressor variables (IVX) instrumentation), for example, refer to Phillips and Lee (2016); they also make reference to forthcoming developments in this area of multivariate models.

5.2.6 Frequency domain analysis

The analysis of time series in the time domain has dominated the literature, but frequency domain approaches are gaining ground. Wilson (2017) applies spectral factorization to construct multivariate impulse responses and derive their statistical properties. Details on this modeling strategy are found in Wilson et al. (2016) for both frequency and time domain analyses. This is a good resource on data and computer code using R. In addition to VARs with exogenous variables, this book covers multivariate spectral analysis, structural VARs, continuous time models, and partial coherence in time series. The book is a good resource for teaching advanced time series analysis courses and is a quasi-quantum leap into the future.

5.2.7 Functional analysis for dependent time series

Analogous to its role in unit-root analysis, functional analysis is becoming prominent in the analysis of dependent data. Horvath and Rice (2015), and the references therein, is a compact account of this literature. It provides historical and new tests of independence in functional time series. In financial econometrics, autoregressive conditional heteroscedasticity (GARCH) models are now available (e.g., Aue et al. 2017) for modeling high-frequency volatility (functional generalized autoregressive conditional heteroskedasticity (fGARCH)). Kim (2017) uses quasi-maximum likelihood estimation on high frequency financial data with a finite observation period to estimate the parameters of a generalized autoregressive conditional heteroskedasticity-Itô (GARCH-Itô) model. Kim provides asymptotic results for the proposed estimator, simulation analysis, and an empirical application. These models are at an early stage of development. Given that market volatility, speculative behavior, financial turmoil, and financial news continuously impact financial assets, thus making risk management and prediction complex, this literature will continue to grow and so will applications from current and forthcoming developments.

5.2.8 New to time series econometrics?

If you are new to time series econometrics, pioneers in time series analysis like Box and Jenkins continue to impact the literature by bridging two centuries of analysis of time series (Box et al. 2016). This time around they "R having fun," which makes the book easy to follow. One approach to reading this book is to stop at about chapter 12, then go to multivariate time series in Tsay (2014), collect R cookies and over 300 examples and exercises there, then go back and forth between the two books doing R-mixing as needed. Enders (2016) has been popular in applied time series econometrics, so you may make this one handy for an easy-to-read reference on theory and empirical examples. Surely Shumway and Stoffer (2017) will be great additional company! All about nonlinear time series models is in Douc at al. (2014). One book that reminds me of the TPE2 (Judge et al.), but for time series econometrics, is Hunter et al. (2017). When the interest is on multivariate cointegration models for impulse response analysis and forecasting, this book offers great guidance. It also contains an appendix on easy-to-follow distribution theory using Brownian motion. Hunter et al. (2017) is so easy to handle that one can hold it in one hand and write an econometrics program with the other. If your interest is in wavelets and G-stationary processes, along with more classical time series in R, then an applied guide is Woodward et al. (2017). State density estimation with nonstationary data will intrigue many, so go back to Chang et al. (2016), and follow their ongoing work and recent citations of that paper to remain on the frontier of such developments.

5.2.9 Econometric software programs

Applied researchers are not left alone when it comes to software programs. SAS/econometric time series (ETS) and other SAS libraries, Matlab on time series, Stata's introduction to time series analysis and related books, regression analysis of time series (RATS) (and cointegration analysis of time series (CATS) in RATS on maximum likelihood estimation (MLE) for error correction models (ECMs)), EViews, R, and GAUSS deserve mentioning since these packages follow developments in time series econometrics.

5.2.10 The last two decades

This time series review did not intend to bypass some backbone textbooks and contributions in time series econometrics; we just wanted to limit the section to recently published works that are impacting and will continue to impact applied work. The same can be said of lots of articles that without doubt will impact the literature and are often omitted from reviews of this nature. We close this section by pointing the reader to Stock and Watson (2017), both influential to the time series literature, who tell us the story of the past 20 years of time series econometrics through ten pictures. Those ten pictures, and Stock and Watson's insights, serve as reminders, as done in Judge et al.'s (1985) TPE2 for all econometrics, that close connections between the theory and practice of econometrics is a driver of econometrics for the future.

Acknowledgments

The authors would like to acknowledge the input and advice of Juan Carlos Escanciano, Bill Griffiths, Eric Hillebrand, Ivan Jeliazkov, Daniel Millimet, Chris Skeels, and Mike Smith. A special thank you to George Judge for his advice on the information theoretic portion of the chapter. Any errors are those of the authors.

Notes

1 See https://jaan.io/what-is-variational-autoencoder-vae-tutorial/.
2 To illustrate the popularity of such articles, Google Scholar shows at the time of this writing that the number of citations were around 7,500 for Granger and Newbold (1974), 22,000 for Dickey and Fuller (1979), 11,800 for Sims (1980), 22,400 for Engle (1982), 6,200 for Nelson and Plosser (1982), 3,300 for Phillips (1987), 13,800 for Johansen and Juselius (1990), and 43,320 for Engle and Granger (1987). The dominance of these and many other related papers in the literature is clearly obvious. Arguably, contemporaneous developments were built upon these and a few other highly cited works. Box and Jenkins have passed the test of time, and this may be driven by the importance of forecasting and control problems, not just in economics and business, but also in the physical sciences. Box was a chemist first and then moved to statistics later; history tell us that he thought of time series as boring until Jenkins came along with practical uses of time series for forecasting and control problems (Mills 2013). They have a new edition with more co-authors (Box et al. 2016), which has been highly cited.
3 Literature on the estimation of parametric nonlinear regressions with nonstationary time series and new asymptotic theory for the cointegrating regression model is found in Chan and Wang (2015).

References

Abadie, A., A. Diamond, and J. Hainmueller. 2014. "Comparative Politics and the Synthetic Control Method." *American Journal of Political Science* 59:495–510.
Abrevaya, J., and S.G. Donald. 2017. "A GMM Approach for Dealing with Missing Data on Regressors." *Review of Economics and Statistics*, forthcoming.
Anatolyev, S. 2013. "Instrumental Variables Estimation and Inference in the Presence of Many Exogenous Regressors." *The Econometrics Journal* 16:27–72.
Andrews, I. 2017. "Valid Two-Step Identification–Robust Confidence Sets for GMM." *Review of Economics and Statistics*, forthcoming.
Angrist, J.D., and J. Pischke. 2010. "The Credibility Revolution in Empirical Economics: How Better Research Design is Taking the Con out of Econometrics." *Journal of Economic Perspectives* 24:3–30.
Athanasopoulos, G., D.S. Pokitt, F. Vahid, and W. Yao. 2016. "Determination of Long-run and Short-run Dynamics in EC-VARMA Models via Canonical Correlations." *Journal of Applied Econometrics* 31:1100–1119.
Athey, S., and G.W. Imbens. 2016. "Machine Learning Methods for Causal Effects." www.nasonline.org/programs/sackler-colloquia/documents/athey.pdf. Accessed on September 1, 2017.
———. 2017. "The State of Applied Econometrics: Causality and Policy Evaluation." *Journal of Economic Perspectives* 31.3–32.
Aue, A., L. Horvath, and D.F. Pellatt. 2017. "Functional Generalized Autoregressive Conditional Heteroscedasticity." *Journal of Time Series Analysis* 38:3–21.
Bajari, P., D. Nekipelov, S.P. Ryan, and M. Yang. 2015. "Machine Learning Methods for Demand Estimation." *American Economic Review: Papers & Proceedings* 105(5):481–485.
Baltagi, B. 2013. *Econometric Analysis of Panel Data* (5th ed.). Hoboken, NJ: Wiley.
——— (editor). 2015. *The Oxford Handbook of Panel Data*. Oxford: Oxford University Press.
Baltagi, B.H., J.P. LeSage, and R.K. Pace (editors). 2016. *Advances in Econometrics, Volume 37 – Spatial Econometrics: Qualitative and Limited Dependent Variables*. Bingley: Emerald Press.
Bedock, N., and D. Stevanović. 2017. "An Empirical Study of Credit Shock Transmission in a Small Open Economy." *Canadian Journal of Economics* 50(2):541–570.

Bekker, P.A., and F. Crudu. 2015. "Jackknife Instrumental Variable Estimation with Heteroskedasticity." *Journal of Econometrics* 185:332–342.

Bekker, P.A., and T. Wansbeek. 2016. "Simple Many-Instruments Robust Standard Errors Through Concentrated Instrumental Variables." *Economics Letters* 149:52–55.

Boffelli, S., and G. Urga. 2016. *Financial Econometrics Using Stata*. College Station: Stata Press.

Belloni, A., V. Chernozhukov, and C. Hansen. 2014. "High-dimensional Methods and Inference on Structural and Treatment Effects." *Journal of Economic Perspectives* 28(2):29–50.

Belloni, A., V. Chernozhukov, and W. Ying. 2013. "Honest Confidence Regions for a Regression Parameter in Logistic Regression with a Large Number of Controls." Cemmap Working Paper, Centre for Microdata Methods and Practice, No. CWP67/13. doi:10.1920/wp.cem.2013.6713.

Boswick, H.P., and P. Paroulo. 2017. "Likelihood Ratio Tests of Restrictions on Common Trends Loading Matrices in I(2) VAR Systems." *Econometrics* 5(3):28. doi:10.3390/econometrics5030028.

Box, G.E.P., and G.M. Jenkins. 1970. *Time Series Analysis, Forecasting and Control*. San Francisco: Holden-Day.

Box, G.E.P., G.M. Jenkins, G.G. Reinsel, and G.M. Ljung. 2016. *Time Series Analysis, Forecasting and Control* (5th ed.). Hoboken: Wiley.

Calonico, S., M.D. Cattaneo, M.H. Farrell, and R. Titiunik. 2017. "rdrobust: Software for Regression Discontinuity Designs." *Stata Journal* 17:372–404.

Cameron, A.C., and D.L. Miller. 2015. "A Practitioner's Guide to Cluster-Robust Inference." *Journal of Human Resources* 50:317–372.

Cameron, A.C., and P.K. Trivedi. 2005. *Microeconometrics: Methods and Applications*. Cambridge: Cambridge University Press.

Caner, M., E. Maasoumi, and J.A. Riquelme. 2016. "Moment and IV Selection Approaches: A Comparative Simulation Study." *Econometrics Review* 35:1562–1581.

Cappiello, L., R.F. Engle, and K. Sheppard. 2006. "Asymmetric Dynamics in the Correlations of Global Equity and Bond Returns." *Journal of Financial Econometrics* 4(4):537–572.

Card, D., D. Lee, Z. Pei, and A. Weber. 2015. "Inference on Causal Effects in a Generalized Regression Kink Design." *Econometrica* 83:2453–2483.

Carrion-I-Silvestre, J.L., and M.D. Gadea. 2016. "Bounds, Breaks and Unit Root Tests." *Journal of Time Series Analysis* 37:165–181.

Carter, A.V., K.T. Schnepel, and D.G. Steigerwald. 2017. "Asymptotic Behavior of a t Test Robust to Cluster Heterogeneity." *Review of Economics and Statistics* 99:698–709.

Cattaneo, M.D., and J.C. Escanciano. 2016. "Introduction." In M.D. Cattaneo and J.C. Escanciano, eds., *Advances in Econometrics: Volume 38, Regression Discontinuity Designs: Theory and Applications*. Bingley: Emerald Press, pp. ix–xxii.

Cattaneo, M.D., L. Keele, R. Titiunik, and G. Vazquez-Bare. 2016. "Interpreting Regression Discontinuity Designs with Multiple Cutoffs." *Journal of Politics* 78:1229–1248.

Chalak, K. 2017. "Instrumental Variables Methods with Heterogeneity and Mismeasured Instruments." *Econometric Theory* 33:69–104.

Chang, Y., C.S. Kim, and J.Y. Park. 2016. "Nonstationarity in Time Series of State Densities." *Journal of Econometrics* 192:152–167.

Chaudhuri, S., and D.K. Guilkey. 2016. "GMM with Multiple Missing Variables." *Journal of Applied Econometrics* 31:678–706.

Chernozhukov, V., and C. Hansen. 2008. "The Reduced Form: A Simple Approach to Inference with Weak Instruments." *Economics Letters* 100:68–71.

Chernozhukov, V., C. Hansen, and M. Spindler. 2015. "Post-selection and Post-Regularization Inference in Linear Models with Many Controls, and Instruments." *American Economic Review, Papers & Proceedings* 105(5):486–490.

Cho, W.T., and G. Judge. 2014. "An Information Theoretic Approach to Network Tomography." *Applied Economic Letters* 22:1–6.

Choi, I. 2015. *Almost All About Unit Roots: Foundations, Developments, and Applications*. Cambridge: Cambridge University Press.

Cressie, N., and T. Read. 1984. "Multinomial Goodness-of-Fit Tests." *Journal of the Royal Statistical Society, Series B* 46:440–464.

Dees, S., F.D. Mauro, M.H. Pesaran, and L.V. Smith. 2007. "Exploring the International Linkages of the Euro Area: A Global VAR Analysis." *Journal of Applied Econometrics* 22(1):1–38.

De Nadai, M., and A. Lewbel. 2016. "Nonparametric Errors in Variables Models with Measurement Errors on Both Sides of the Equation." *Journal of Econometrics* 191(1):19–32.

Dickey, D.A., and W.A. Fuller. 1979. "Distribution of the Estimators for an Autoregressive Time Series with a Unit Root." *Journal of the American Statistical Association* 74:427–431.

Diederich, T.G. 2000. "Ensemble Methods in Machine Learning." First International Workshop on Multiple Classifier Systems, Cagliari, Italy. http://web.engr.oregonstate.edu/~tgd/publications/mcs-ensembles.pdf. Accessed September 1, 2017.

Diks, C., and M. Wolski. 2016. "Nonlinear Granger Causality: Guidelines for Multivariate analysis." *Journal of Applied Econometrics* 31:1333–1351.

Doornik, J.A. 2017. "Maximum Likelihood Estimation of the I(2) Model Under Linear Restrictions." *Econometrics Journal* 5(2):19 doi:10.3390/econometrics5020019.

Douc, R., E. Moulines, and D. Stoffer. 2014. *Nonlinear Time Series, Theory, Methods, and Applications with R Examples*. New York: CRC Press.

Enders, W. 2016. *Applied Econometric Time Series* (4th ed.). Hoboken: John Wiley & Sons, Inc.

Engle, R.F. 1982. "Autoregressive Conditional Heteroscedasticity with Estimates of the Variance of United Kingdom Inflation." *Econometrica* 50(4):987–1007.

Engle, R.F., and C.W.J. Granger. 1987. "Co-integration and Error Correction: Representation, Estimation and Testing." *Econometrica* 55:251–276.

Fisher, L.A., H-S Huh, and A.R. Pagan. 2016. "Econometric Methods for Modeling Systems with Mixture of I(1) and I(0) Variables." *Journal of Applied Econometrics* 31:892–911.

Ghysels, E., J.B. Hill, and K. Motegi. 2016. "Testing for Granger Causality with Mixed Frequency Data." *Journal of Econometrics* 192:207–230.

Gonzalez-Rivera, G., R.C. Hill, and T-H. Lee (editors). 2016. *Advances in Econometrics, Volume 36, Essays in Honor of Aman Ullah*. Bingley: Emerald Press.

Gorban, A., P. Gorban, and G. Judge. 2010. "Entropy: The Markov Ordering Approach." *Entropy* 12(5):1145–1193.

Götz, T.B., A. Hecq, and S. Smeekes. 2016. "Testing for Granger Causality in Large Mixed-Frequency VARs." *Journal of Econometrics* 193:418–432.

Granger, C.W.J., and P. Newbold. 1974. "Spurious Regressions in Econometrics." *Journal of Econometrics* 2:111–120.

———. 1977. *Forecasting Economic Time Series*. New York: Academic Press.

Granger, C.W.J., and P.L. Siklos. 1995. "Systematic Sampling, Temporal Aggregation, Seasonal Adjustment, and Cointegration: Theory and Evidence." *Journal of Econometrics* 66:357–369.

Griffiths, W.E., and G. Hajargasht. 2016. "Some Models for Stochastic Frontiers with Endogeneity." *Journal of Econometrics* 190:341–348.

———. 2018. *Econometric Analysis*, (8th ed.). London: Pearson.

Hansen, B.E. 2015. "Efficient Shrinkage in Parametric Models." *Journal of Econometrics* 190:115–132.

Harrington, P. 2012. *Machine Learning in Action*. Greenwich: Manning Publications Company.

Hassler, U., and M. Hosseinkouchack. 2017. "Panel Cointegration Testing in the Presence of Linear Time Trends." *Econometrics* 4(4) doi:10.3390/econometrics4040045.

Hastie, T., R. Tibshirani, and J. Friedman. 2009. *The Elements of Statistical Learning: Data Mining, Inference, and Prediction* (2nd ed.). New York: Springer.

Hayashi, F. 2000. *Econometrics*. Princeton, NJ: Princeton.

Heckman, J.J., J.E. Humphries, and G. Veramendic. 2016. "Dynamic Treatment Effects." *Journal of Econometrics* 191:276–292.

Horowitz, J.L. 2009. *Semiparametric and Nonparametric Methods in Econometrics*. New York: Springer.

Horvath, L., and G. Rice. 2015. "Testing for Independence Between Functional Time Series." *Journal of Econometrics* 189:371–382.

Hunter, J., S.P. Burke, and A. Canepa. 2017. *Multivariate Modelling of Non-Stationary Economic Time Series*. London: Palgrave MacMillan.

Imbens, G. 2014. "Instrumental Variables: An Econometrician's Perspective." *Statistical Science* 29:323–358.

Imbens, G.W., and D.B. Rubin. 2015. *Causal Inference for Statistics, Social and Biomedical Sciences: An Introduction*. Cambridge: Cambridge University Press.

Jacobi, L., H. Wagner, and S. Frühwirth-Schnatter. 2016. "Bayesian Treatment Effects Models with Variable Selection for Panel Outcomes with an Application to Earnings Effects of Maternity Leave." *Journal of Econometrics* 193:234–250.

Jentsch, C., and S.S. Rao. 2015. "A Test for Second Order Stationarity of a Multivariate Time Series." *Journal of Econometrics* 185:124–161.

Joe, H. 2015. *Dependence Modeling with Copulas*. New York: CRC Press.

Johansen, S., and K. Juselius. 1990. "Maximum Likelihood Estimation and Inference on Cointegration: With Applications to the Demand for Money." *Oxford Bulletin of Economics and Statistics* 52(2):169–210.

Jolliffe, I.T. 2002. *Principal Components Analysis* (2nd ed.). New York: Springer.

Judge, G. 2013. "Fellow's Opinion Corner: Econometric Information Recovery." *Journal of Econometrics* 176:1–2.

———. 2015. "Entropy Maximization as a Basis for Information Recovery in Dynamic Economic Behavioral Systems." *Econometrics* 3:91–100.

Judge, G.G., W.E. Griffiths, R.C. Hill, H. Lütkepohl, and T. C Lee. 1985. *The Theory and Practice of Econometrics* (2nd ed.). John Wiley & Sons, Inc.

Judge, G., and R. Mittelhammer. 2011. *An Information Theoretic Approach to Econometrics* New York: Cambridge University Press.

———. 2012. "Implications of the Cressie-Read Family of Additive Divergences for Information Recovery." *Entropy* 14(12):2427–2438.

Keele, L.J., and R. Titiunik. 2015. "Geographic Boundaries as Regression Discontinuities." *Political Analysis* 23(1):127–155.

Kim, D. 2017. "Statistical Inference for Unified GARCH-Itô Models with High-Frequency Financial Data." *Journal of Time Series Analysis* 37:513–532.

Kingma, D.P., and M. Welling. 2014. "Auto-Encoding Variational Bayes." arXiv:1312.6114[stat.ML].

Kline, B. 2016. "Identification of the Direction of a Causal Effect by Instrumental Variables." *Journal of Business & Economic Statistics* 34:176–184.

Kolesár, M., R. Chetty, J. Friedman, E. Glaeser, and G.W. Imbens. 2015. "Identification, and Inference with Many Invalid Instruments." *Journal of Business & Economic Statistics* 33:474–484.

Koop, G. 2017. "Bayesian Methods for Empirical Macroeconomics with Big Data." *Review of Economic Analysis*, in press.

Kuester, K., S. Mitinik, and M.S. Paolella. 2006. "Value-at-Risk Prediction: A Comparison of Alternative Strategies." *Journal of Financial Econometrics* 4(1):53–89.

Lahiri, S.N. 2013. *Resampling Methods for Dependent Data*. New York: Springer Science & Business Media.

Lahiri, S.N., and Zhu, J. 2006. "Resampling Methods for Spatial Regression Models Under a Class of Stochastic Designs." *Annals of Statistics* 34:1774–1813.

Leamer, E.E. 2010. "Tantalus on the Road to Asymptopia." *Journal of Economic Perspectives* 24:31–46.

Lee, D.S., and T. Lemieux. 2010. "Regression Discontinuity Designs in Economics." *Journal of Economic Literature* 48:281–355.

LeSage, J.P., and R.K. Pace. 2009. *Introduction to Spatial Econometrics*. Boca Raton: CRC Press.

Li, G., C. Leng, and C-L Tsai. 2014. "A Hybrid Bootstrap Approach to Unit Root Tests." *Journal of Time Series Analysis* 35:299–321.

Li, Q., and J.S. Racine (editors). 2009. *Advances in Econometrics, Volume 25, Nonparametric Econometric Methods*. Bingley: Emerald Press.

Ling, S., M. McAleer, and H. Tong. 2015. "Frontiers in Time Series and Financial Econometrics: An Overview." *Journal of Econometrics* 189:245–250.

Low, H., and C. Meghir 2017. "The Use of Structural Models in Econometrics." *Journal of Economic Perspectives* 31:33–58.

Lu, X., and L. Su. 2016. "Shrinkage Estimation of Dynamic Panel Data Models with Interactive Fixed Effects." *Journal of Econometrics* 190:148–175.

Lütkepohl, H. 2005. *The New Introduction to Multiple Time Series Analysis*. Berlin: Springer-Verlag.MacKinnon, J.G. 2012. "Thirty Years of Heteroskedasticity-robust Inference." In X. Chen and N. Swanson, eds., *Recent Advances and Future Directions in Causality, Prediction, and Specification Analysis*. New York: Springer, pp. 437–461.

MacKinnon, J.G., and M.D. Webb. 2017. "Wild Bootstrap Inference for Wildly Different Cluster Sizes." *Journal of Applied Econometrics* 32:233–254.

McDonough, I.K., and D.L. Millimet. 2017. "Missing Data, Imputation, and Endogeneity." *Journal of Econometrics*, forthcoming.

Miller, D., and G. Judge. 2015. "Information Recovery in a Dynamic Statistical Markov Model." *Econometrics* 3:187–198.

Mills, T. 2013. *A Very British Affair: Six Britons and the Development of Time Series Analysis During the Twentieth Century*. London: Palgrave Macmillan.

Mullainathan, S., and J. Spiess. 2017. "Machine Learning: An Applied Econometric Approach." *Journal of Economic Perspectives* 31(2):87–106.

Murtazashvili, I., and J.M. Wooldridge. 2016. "A Control Function Approach to Estimating Switching Regression Models with Endogenous Explanatory Variables and Endogenous switching." *Journal of Econometrics* 190:252–266.

Nelson, C.R., and C.R. Plosser. 1982. "Trends and Random Walks in Macroeconomic Time Series: Some Evidence and Implications." *Journal of Monetary Economics* 10:139–162.

Newbold, P. 1981. "Some Recent Developments in Time Series Analysis." *International Statistical Review* 49:53–66.

Newbold, P., and S.J. Leybourne. 2003. *Recent Developments in Time Series, Vol. I & II*. Northampton: An Elgar Reference Collection.

Nisbet, R., J. Elder, and G. Miner. 2009. *Handbook of Statistical Analysis & Data Mining Applications*. Cambridge: Academic Press.

Olea, J.L.M., and C. Pflueger. 2013. "A Robust Test for Weak Instruments." *Journal of Business & Economic Statistics* 31:358–369.

Pearson, K. 1901. "On Lines and Planes of Closest Fit to Systems of Points in Space." *Philosophical Magazine* 2(11):559–572.

Pesaran, M.H. 2015. *Time Series and Panel Data Econometrics*. Oxford: Oxford University Press.

Pesaran, M.H., T. Schuermann, and S.M. Weiner. 2004. "Modeling Regional Interdependence Using Global Error-correcting Macroeconometric Model." *Journal of Business Economic Statistics* 22(2):129–162.

Phillips, P.C.B. 1987. "Time Series Regressions with a Unit Root." *Econometrica* 55:277–301.

———. 2010. "Bootstrapping I(1) Data." *Journal of Econometrics* 158:280–284.

Phillips, P.C.B., and J.H. Lee. 2016. "Robust Econometric Inference with Mixed Integrated and Mildly Explosive Regressors." *Journal of Econometrics* 192:433–450.

Poskitt, D.S. 2016. "Vector Autoregressive Moving Average Identification for Macroeconomic Modeling: A New Methodology." *Journal of Econometrics* 192:468–484.

Poskitt, D.S., and W. Yao. 2017. "Vector Autoregressions and Macroeconomic Modeling: An Error Taxonomy." *Journal of Business & Economic Statistics* 35(3):407–419 doi:10.1080/07350015.2015.1077139.

Qian, J., and L. Su. 2016. "Shrinkage Estimation of Regression Models with Multiple Structural Changes." *Econometric Theory* 32:1376–1433.

Qu, X., X. Wang, and L. Lee. 2016. "Instrumental Variable Estimation of a Spatial Dynamic Panel Model with Endogenous Spatial Weights When T Is Small." *Econometrics Journal* 19:261–290.

Read, T., and N. Cressie. 1988. *Goodness-of-Fit Statistics for Discrete Multivariate Data*. New York: Springer-Verlag.

Romano, J.P., and M. Wolf. 2017. "Resurrecting Weighted Least Squares." *Journal of Econometrics* 197:1–19.

Saikkonen, P., and R. Sandberg. 2016. "Testing for a Unit Root in Noncausal Autoregressive Models." *Journal of Time Series Analysis* 37:99–125.

Sanderson, E., and F. Windmeijer. 2016. "A Weak Instrument F-Test in Linear IV Models with Multiple Endogenous Variables." *Journal of Econometrics* 190:212–221.

Seni, G., and J. Elder. 2010. *Ensemble Methods in Data Mining: Improving Accuracy Through Combining Predictions*. Williston: Morgan & Claypool Publishers.

Shmueli, G., N.R. Patel, and P.C. Bruce. 2016. *Data Mining for Business Analytics: Concepts, Techniques, and Applications in Microsoft Office with XLMINER* (3rd ed.). Hoboken: Wiley.

Shumway, R.H., and D.S. Stoffer. 2017. *Time Series Analysis and Its Applications: With R Examples* (4th ed.). New York: Springer.

Sims, C.A. 1980. "Macroeconomics and Reality." *Econometrica* 48:1–49.

Smeekes, S. 2015. "Bootstrap Sequential Tests to Determine the Order of Integration of Individual Units in a Time Series Panel." *Journal of Time Series Analysis* 36:398–415.

Solon, G., S.J. Haider, and J.M. Wooldridge. 2015. "What Are We Weighting For?" *Journal of Human Resources* 50:301–316.

Stock, J.H., and M.W. Watson. 2017. "Twenty Years of Time Series Econometrics in Ten Pictures." *Journal of Economic Perspectives* 31(2):59–86.

Stock, J.H., and M. Yogo. 2005. Testing for Weak Instruments in Linear IV Regression. In D.W.K. Andrews and J.H. Stock, eds. *Identification and Inference for Econometric Models, Essays in Honor of Thomas Rothenberg*. New York: Cambridge University Press, pp. 80–108.

Tan, P-N., M. Steinbach, and V. Kumar. 2006. *Introduction to Data Mining*. Boston: Addison-Wesley.

Tibshirani, R. 1996. "Regression Shrinkage and Selection Via the Lasso." *Journal of the Royal Statistical Society, Series B* 58:267–288.

Tong, H. 2015. "Threshold Models in Time Series Analysis-Some Reflections." *Journal of Econometrics* 189:485–491.

Tsay, R.S. 2014. *Multivariate Time Series Analysis: With R and Financial Applications*. Hoboken, NJ: John Wiley & Sons.

Varian, H.R. 2014. Big Data: New Tricks for Econometrics." *Journal of Economic Perspectives* 28(2):3–28.

Wager, S., and S. Athey. 2017. "Estimation and Inference of Heterogeneous Treatment Effects Using Random Forests." *Journal of the American Statistical Association*, forthcoming.

Wang, W., and M. Kaffo. 2016. "Bootstrap Inference for Instrumental Variable Models with Many Weak Instruments." *Journal of Econometrics* 192:231–268.

White, H. 1980. "A Heteroscedasticity-consistent Covariance Matrix Estimator and a Direct Test for Heteroscedasticity." *Econometrica* 48(4):817–838.

White, H., and D. Pettenuzzo. 2014. "Granger Causality, Exogeneity, Cointegration, and Economic Policy Analysis." *Journal of Econometrics* 178:316–330.

Williams, G. 2011. *Data Mining with Rattle and R*. New York: Springer.

Wilson, G.T. 2017. "Spectral Estimation of the Multivariate Impulse Response." *Journal of Time Series Analysis* 38:381–391.

Wilson, G.T., R. Reale, and J. Haywood. 2016. *Models for Dependent Time Series*. New York: CRS Press.

Woodward, W.A., H.L. Gray, and A.C. Elliot. 2017. *Applied Time Series Analysis with R* (2nd ed.). Boca Raton: CRC Press.

Wooldridge, J.W. 2010. *Econometric Analysis of Cross Section and Panel Data* (2nd ed.). Cambridge, MA: MIT Press.

———. 2015. "Control Function Methods in Applied Econometrics." *The Journal of Human Resources* 50:420–445.

Yang, Z. 2015. "LM Tests of Spatial Dependence based on Bootstrap Critical Values." *Journal of Econometrics* 185(1):33–59.

Zanin, M., L. Zunino, O. Rosso, and D. Papo. 2012. "Permutation Entropy and its Main Biomedical and Econophysics Applications: A Review." *Entropy* 14:1553–1577.

Zapata, H.O., and A.N. Rambaldi. 1997. "Monte Carlo Evidence on Cointegration and Causation." *Oxford Bulletin of Economics and Statistics* 59(2):285–298.

Zhou, Zhi-Hua. 2012. *Ensemble Methods: Foundations and Algorithms*. Boca Raton: CRC Press.

26
On the evolution of agricultural econometrics

David A. Bessler and Marco A. Palma

1. Introduction

In William Stanley Jevons' 1871 classic, *The Theory of Political Economy*, economists were introduced to the idea of marginal utility ("final utility" from the subjective evaluation of an agent's utility in use $i(u_i)$): "When the distribution is completed, we ought to have $\Delta u_1 = \Delta u_2 \ldots$ we must have, in other words, the final degrees of utility in the two uses equal" (Jevons 1871). Earlier, in his introduction to the same book, Jevons laid down a challenge for econometrics: "I know not when we shall have a perfect system of statistics, but the want of it is the only insuperable obstacle in the way of making Political Economy an exact science" (Jevons 1871, p. 116).[1]

Early neoclassical economists seem to have little interest in meeting Jevons' challenge. Alfred Marshall, in a letter written to Henry Moore on his ordinary least squares regression analysis presented in *Law of Wages* (Moore 1911), wrote: "the *ceteris paribus* clause – though formally adequate – seems to me impracticable" (G. Stigler 1962)

Despite this cool reception, economists did pursue Jevons' insuperable (impossible or challenging) obstacle.[2] In this chapter, we explore agricultural economists' efforts in this pursuit.[3] We begin with the early years of the twentieth century and the work with non-experimental data (Henry Moore's work). We discuss our struggles with the identification issue in our study of both consumer and producer behavior with observational data. We follow with a discussion of the promise of instrumental variables, and the resulting disappointments, leading to our fallback to non-structural time series methods to capture forecasting opportunities, which leads to the subsequent abandonment of observational data in favor of experimental data to obtain a clearly unambiguous structure. While this struggle was carried out in agricultural economics, a similar path was followed in economics in general. In the early years, econometrics and agricultural econometrics were co-mingled, as the problems studied were oftentimes agricultural, even if those studying such problems were not trained as agriculturalists. Henry Moore is a case in point – his appointment was as Professor of Political Economy at Columbia University, having served earlier at Smith College and received his PhD at Johns Hopkins University (G. Stigler 1962). Nevertheless, Moore studied the demand for agricultural products (i.e., corn, oats, hay, and potatoes from the agricultural sector, and pig iron from the industrial sector). Below we consider the work of Trygve Haavelmo, University of Oslo and University of Aarhus. One of his early papers on the demand for food was co-authored with Meyer Girshick, at the time head of the Statistical Techniques Section of the Bureau of Agriculture (Fox 1989).[4]

Henry Moore's 1914 work focused on "demands" for several agricultural commodities and one industrial commodity. There he recognized Marshall's criticism (cited earlier) with respect to *ceteris paribus*. Moore argued that "demands' are "the average changes that society is actually undergoing" (Moore 1914). These "demands" are not the "classical theory of demand which is limited to the simple enunciation of this one characteristic, *ceteris paribus*" (Moore 1914). Moore (1914) did not equate his "statistical demand" curve with Marshall's *ceteris paribus* demand curve; nevertheless, he was criticized (see Stigler (1962)) for his results on pig iron – a "statistical demand" curve with a positive slope.

Elmer Working (1927) shows that with equilibrium observational data, failure to account for shifts in demands and supplies (due to omitted variables), when studied with ordinary least squares, can give us regularities that sometimes look like demands, supplies, or neither. Figure 26.1 has become a fairly standard undergraduate aide in helping to communicate problems in the interpretation of ordinary least squares with observational (equilibrium) data on price (P) and quantity (Q) generated by the intersection of supply curves ($S_i, i = 1,2,3$) and demand curves ($D_i, i = 1,2,3$) in years 1, 2, and 3. The equilibrium data (all one ever sees with observational data), when studied with ordinary least squares regression, yields a positively sloped line, labeled here as the ordinary least squares (OLS) line of best fit. The relative shifts (omitted variables causing the shifts) in D and S from year to year govern the slope of the OLS line (Hamilton 1994).

Working's conclusion (last paragraph) in his celebrated 1927 paper still haunts many of us nearly 90 years later:

> The matter of correlation between shifts of the demand and supply curves is a more difficult problem to deal with. Every effort should be made to discover whether there is a tendency for the shifting of these to be interdependent. In case it is impossible to determine this, it should be carefully noted that the demand curve which is obtained is quite likely not to hold true for periods other than the one studied, and cannot be treated as corresponding to the demand curve of economic theory.
>
> *(Working 1927, p. 235)*

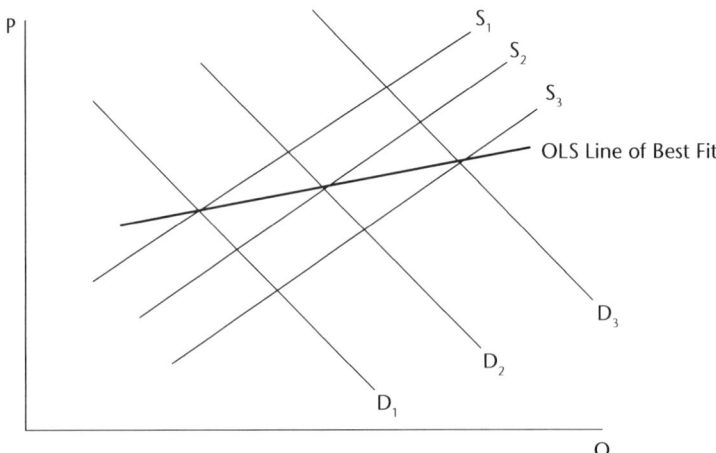

Figure 26.1 OLS line of best fit from supply (*i*) equals demand (*i*) equilibrium data

Moore's problem (and Working's apt description) became known as an early case of the identification problem, where demand and supply together determine both price and quantity. So, for example, if we have the two algebraic expressions given here on demand and supply:

Demand: $q_{dt} = \gamma_{10} + \gamma_{11} p_t + e_{dt}; \gamma_{11} < 0$,
Supply: $q_{st} = \gamma_{20} + \gamma_{21} p_t + e_{dt}; \gamma_{21} > 0$,
Equilibrium: $q_{dt} = q_{st}$

where q_{dt} and q_{st} are quantity demanded and supplied in period t, p_t is the observed price in period t, γ_{ij} are the unobserved parameters to be estimated from the data, and e_{dt} and e_{dt} are unobserved innovations (or errors) summarizing other variables, not included in the model, that moved the demand (d) or the supply (s) line through time.

The parameters of interest, γ_{11} and γ_{21}, in the equations given here are under-identified, meaning that with observational data on P and Q, there is no unique set of parameter values generating this system.[5]

Working's call for efforts to sort out the shifting demands and supplies was answered two years after it was published by Phillip Wright (who had help from his son, Sewell Wright, formerly of the Bureau of Agriculture). Philip Wright introduced us to Instrumental Variables (IV) as a possible solution to "Moore's Problem." Here P and Q are connected via latent variables, say, rainfall (R) and income (I). OLS regression of P on Q gives biased and inconsistent parameter estimates of $\partial P / \partial Q$ (as Moore's OLS regressions on pig iron data suggested). One may have to look for variables R and I, such that I causes P and Q, the latter only through P, and R causes Q and P, the latter only through Q. This new system (the IV augmented system) is identified. The counting rules (given in endnote 5) have been restated recently in graphical form as independent causal paths by Arefiev (2016).

In the early days of agricultural econometrics, *a priori* theory was used as our source of exogenous variables (for identification).[6] The early econometricians at the United States Department of Agriculture (USDA) issued regular forecasts for crop and livestock quantities and prices. They "solved" the identification problem using time delays (lags) in supply to define independent exogenous (predetermined) variables. Current supply depended on lagged prices (among other variables) and current demand depended on disposable income, own prices, and prices of substitute goods (Fox 1989). Ezekiel (1938) presented the general theory behind agricultural supply and demands that "solved" Moore's problem using lagged price to generate supply, and current price to generate demand, under the descriptive title "Cobweb Theorem."[7]

As suggested earlier, agricultural economists worked closely with more general economists in the first half of the twentieth century. A particularly productive cooperation was between Fredrick Waugh of the USDA and Ragnar Frisch (Fox 1989). Frisch's combining of economic theory with statistics made him the founder of "econometrics" and first editor of *Econometrica* (Blaug 1985). In considering Jevons' original contribution of marginal utility, Frisch suggested going directly to individuals with what he labels "the interview method" (Frisch 1932b). In his 1932 paper, "New Orientation of Economic Theory: Economics as an Experimental Science," Frisch suggests that

> it is this which I describe as the second advance since the classics [the first being subjective valuation] and which gives me the belief that [it] is right to state that economics are now in progress of development into an experimental science.
>
> *(Frisch 1932a, p. 100)*

Waugh appears not to have embraced Frisch's call for working directly with individual agents (people) but did cooperate with Frisch on work with observational data.[8]

Trygve Haavelmo, Frisch's student, made this link to the experiment clear in his 1944 *Econometrica* supplement, "Probability Approach to Econometrics." That is, he defined two types of experiments: "experiments that would verify hypotheses if we artificially isolate the phenomena of interest from 'other influences,'" and the experiments from the "stream of experiments that Nature is steadily turning out from her own enormous laboratory, and which we merely watch as passive observers" (Haavelmo 1944). He notes that in the first type of experiment, theory and facts can be brought into agreement by either changing the theory or by changing the facts we consider. In the second type of experiment, we can only try to adjust our theories to the facts. That is to say:

> In the second case we can only try to adjust our theories to reality as it appears before us... [W]e try to choose a theory and a design of experiments to go with it, in such a way that the resulting data would be those which we get by passive observation of reality. And to the extent that we succeed in doing so, we become master of reality – by passive agreement.
> *(Haavelmo 1944, pp. 14-15)*

Nearly three-quarters of a century after Jevons' challenge to make economics a science, Haavelmo defined two paths: the experimental path (what we label below as the random assignment path) and the observational path (his passive observations). Later, we discuss both paths and a third hybrid or merged path under the title of quasi-experiments.[9] The first (random assignment) and the second (observational or passive observational and not quasi-experimental) are what we interpret as Koopmans' (1947) characterization of Newton's (the former) and Kepler's (the latter) style of econometrics.[10] We begin our review of these with the Neyman–Rubin–Holland model of causal inference in experiments. We follow with quasi-experiments and use the basic experimental model as a guide to confront observational data where practitioners know (or pretend to know) something about the causes of their exogenous variables. We end with a discussion of econometric methods with observational data that do not have such knowledge.

Earlier, we briefly reviewed early work in agricultural econometrics. Virtually all of these attempts have been made with observational data. Moore ignored *ceteris paribus* and summarized historical regularities between the variables of interest. This approach gave us regularities that sometimes looked like demand and sometimes looked like supply. As suggested by others, we could either go into the laboratory and conduct controlled experiments or use observational (historical) data to obtain econometric structures. If economic reality is sufficiently simple and reversible,[11] historical data can be relied on to give us relevant observations on economic structure.

Econometricians at USDA, in their efforts to build practical decision making models (Fox 1989), proceeded from theory to quantify basic demand and supply elasticities. These were viewed as structural (in Haavelmo's sense) because they were thought to lie behind period-to-period changes in the data and (like a working bridge over a river) would hold up under policy changes. Or, using Haavelmo's characterization, they would be more like the interrelated systems of an automobile (the fuel system, braking system, electrical system, etc.) to represent a deeper and (in some cases) preferred level of knowledge relative to an understanding of the relationship between the pressure of one's foot applied to the accelerator and the speed of the car.

The solution to the unknown variables problem appears to be randomization. This requires that we take Frisch's suggestion seriously and gather data from individual subjects. That is, here the analyst must conduct experiments in which subjects are randomly assigned to different levels of the treatment variable. In the following section, we outline a description of the experimental problem and the role of randomization in providing internally valid results.

2 Neyman–Rubin–Holland model

Using the Neyman–Rubin–Holland Model,[12] we observe an individual consumer, u_i, from the population of consumers U.[13] Fundamental is the *potential value* of the dependent variable, a given value of the "independent" variable. For a particular unit, u_i (an individual consumer), $Q_p(u_i)_{t0}$ is the quantity that would be purchased by u_i if price is set at p at time t_0. $Q_{p+\Delta p}(u_i)_{t0}$ gives the quantity that would be purchased by the same unit i at time t_0 if price is set at $p+\Delta p$. Such variables are defined for each individual u_i, whether the individual receives treatment p or treatment $p+\Delta p$. We define the effect of a change in p, at time t_0, on Q that would be purchased by consumer u_i as $\Delta Q_{t0} = Q_{p+\Delta p}(u_i)_{t0} - Q_p(u_i)_{t0}$. This effect is not observable because as we cannot observe $Q_{p+\Delta p}(u_i)_{t0}$ and $Q_p(u_i)_{t0}$ on the same unit u_i and at the same time t_0. We cannot observe the economic unit's response to a change in p under *ceteris paribus*. Holland (1986) defines this as the "Fundamental Problem of Causal Inference": (in our notation) we cannot observe $Q_{p+\Delta p}(u_i)_{t0}$ and $Q_p(u_i)_{t0}$ on the same unit; in general, it is impossible to observe the effect of P on Q.

Holland outlines three alternative assumptions that may make it possible to observe (in our case) the "effect of P on Q": temporal stability, unit homogeneity, and random assignment. In Bessler and Dearmont (1996), we argue that the first two alternative assumptions are not likely to hold, and we dismiss them as practical solutions to Holland's Fundamental Problem. The third alternative assumption, for large numbers of subjects, is attractive, giving us an "equality by randomization" for large N (number of subjects) (Campbell and Stanley 1966).[14] Of course, we never observe the potential outcomes ($Q_{p+\Delta p}(u_i)_{t0}$ or $Q_p(u_i)_{t0}$). What we observe is the outcome an individual selects when receiving price p or $p + \Delta p$. By construction, the only difference from the treatment group, those receiving the price of $(p + \Delta p)$ at t_0, and the control group, those receiving the price of p at t_0, is Δp and the units, u_i (the units in the treatment group and the units in the control group). All other things are equal by chance. We have (representing the expectation operator applied at time t_0, as $E_{t0}\{\bullet\}$).[15] That is, $E_{t0}\{Q_{p+\Delta p}\} = E_{t0}\{Q|p+\Delta p\}$ and $E_{t0}\{Q_p\} = E_{t0}\{Q|p\}$, giving us a measure of the Average Causal Effect of P on Q. The observed ratio, $\Delta Q / \Delta P = [E_{t0}\{Q|p+\Delta p\} - E_{t0}\{Q|p\}] / \Delta_p$, approximates the derivative. Key here is the condition: individuals are assigned randomly to the treatment $(p + \Delta p)$ and control (p) groups.

3 Agricultural production and markets: attempts to find structure

The agricultural economics literature began its econometric study of agricultural production with production functions with observational, not experimental, data. Tintner and Brownlee (1944) studied data from a random sample of Iowa farms for 1939 (see the description in Mundlak 2001). Yet, there has been a non-trivial literature on production function estimation using experimental data. Here the works of Earl Heady and others were based on experimental trial data, giving us unbiased and consistent parameter estimates of the underlying production surface (Dean 1960; Heady and Dillon 1961). More recent work by Sri Ramaratnam et al. (1987) and Mitchell, Gray, and Steffey (2004), among others, offers a complement to the extensive work done in this area in the 1950s and 1960s. Random assignment of input ensures they are not dependent on output levels or other unknown omitted variables – random assignment assures no simultaneous equation bias because inputs to the production function are assigned via a random device (die) and not assigned according to innovations (errors) in the underlying production function (more on this later).

Similarly, experimental methods have been applied in consumer demand (Brunk and Federer 1953). Starting in the 1990s, the interest of agricultural economists in using experiments to assess potential markets of food and agricultural products increased substantially. Experimental auctions became the predominant tool for the valuation of market and non-market goods.[16] Experimental auctions have been applied to measure willingness to pay for a wide range of agricultural and food product valuations, including production methods, technology adoption, policy changes, and social and environmental programs, to name a few. Lusk and Shogren (2007) provide a historical perspective of the use of experimental auctions, going back to the early work in induced value experiments (Smith 1976) to the most recent empirical applications of food and environmental valuations.

In the Potential Outcome Average Causal Effect model outlined above, Holland (1986) concluded that it is only through random assignment experiments that we can conclude causation (structure). Yet, his colleague,[17] Donald Rubin (1974, p. 688), appeared more open to study in non-laboratory (observational) settings:

> [T]o ignore existing observational data may be counter-productive. It seems more reasonable to try to estimate the effects of the treatments from nonrandomized studies than to ignore these data and dream of the ideal experiment or make "armchair" decisions without the benefit of data analysis.

Rubin may be viewed as a bit controversial in his haste to extend the potential outcome and average causal effect model to observational studies. As quoted in Li and Mealli (2014, 445), Rubin described a lunch conversation with Jerzy Neyman as follows:

> Shortly after I arrived in Berkeley, I gave a talk on missing data and causal inference. The next day, I went to lunch with Neyman and I said something like "It seems to me that formulation of causal problems in terms of missing potential outcomes is an obvious thing to do, not just in random assignment experiments, but also in observational studies." Neyman answered to the effect – no, causality is far too speculative in non-randomized studies.

Below, we discuss quasi-experiments, recognizing that researchers may not know just how their data differ from data generated from a random assignment experiment. Our subjective judgement is to follow Rubin in this direction (as most of agricultural economics has, either knowingly or unknowingly), with a strong sense of humility toward any result found.

Quasi-experiments look for assumptions behind the data-generating process that allow us to isolate the causal (structural) influence of, say, P on Q. Possibly the most ubiquitous assumption in post-Haavelmo agricultural econometrics is that the right-hand side variables in either single-equation or multiple-equation models are not correlated with their error terms. That is to say, while not assuming "explanatory" variables were set by randomization, it is often (almost always) assumed that they do not respond to the omitted variables (which would show up in the error term) generating y (the dependent variable). We generally relied on prior theory of the consumer or producer to specify our "structural" models, but that theory (almost never) included a theory on how our right-hand side variables were generated. These were termed (deemed) exogenous to the model; however, they came to find their values at a particular data point was not questioned.[18]

The consumer and producer problem specification begins with a theory on how the consumer selects a bundle, Q, based on prices (of consumer goods) or input prices (of producer inputs) and output price (of producer outputs). These choices are based on utility maximization subject to a budget constraint for the consumer and profit maximization subject to a technology constraint for the producer. Prices, utility, technology, and money income are assumed to be exogenous (not determined in the model). The Qs, consumption bundles, and input levels and outputs, are all endogenous (determined in the model).

This was the model essentially given to us by Jevons (1871), reaching its zenith in Hicks (1946) and in Samuelson (1947). It is generally the case that we do not ask whether there are omitted variables that affect both P and Q. Data may be generated, as theory suggests, by a directed edge, P → Q, but, as well, both are moved by a third (or a fourth or more) variable L, so that something like Figure 26.2 holds in the data.

The random assignment experiment discussed above makes P exogenous (not just in our theory but in its data generating process as well). If P is set by randomization, it is orthogonal to L (there is no arrow from L to P). But in observational data, such an arrow may well exist (and we generally do not have a theory on what L actually consists of). Agricultural economists have recognized that such omitted variables may plague their work. We generally think about what may be in L and condition on those variables in our econometric models. We discuss these models under the heading of quasi-experimental, as the list of all possible "confounders" is potentially long (and possibly time varying). So one is never sure there is a sufficiently rich set L to condition on (indeed, we may never know exactly what is in set L) (Malivaud 1980; see, as well, Kuznets 1953).

A further complication arises as right-hand side variables may be related among themselves, giving rise to what has been labeled as multicollinearity. For example, causal patterns among prices are generally not treated in a systematic form. Might, for example, the price of one soft drink be set according to the price of another soft drink? Or might the manager of the meat case at a retail grocery store price chicken or pork by looking first at (dependent upon) the price of beef?[19] Of course, the researcher generally does not know the answer to these questions. Among the possible solutions is to go out into the field and find out by looking at what actually happens. Another possibility is to work with quasi-experimental methods, as outlined below.

To be fair, the "complications" described here are possibly of second-order effects and are probably small relative to first-order effects. But not always, if two prices are linked to quantity

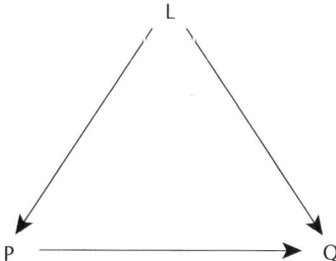

Figure 26.2 Causal diagram where the unobserved variable L confounds the relationship between P and Q

as a chain, such that P_1 affects Q_2 only through P_2, as in Figure 26.3. We have to have a model on how both Q_2 and P_2 are generated to capture the full effect of price change of P_1 on Q_2:

$$P_1 \to P_2 \to Q_2$$

A standard demand system specification will miss the cross-price effects of P_1 on Q_2. We need $\partial Q_2 / \partial P_1 = \partial Q_2 / \partial P_2 \times \partial P_2 / \partial P_1$. That is, how does P_2 behave as we change P_1? In a standard demand system, having both P_1 and P_2 on the right-hand side of an equation explaining Q_2 will result in P_2 screening off the influence of P_1 on Q_2 (which may be termed acceptable if we are modeling conditional demands). If the arrow graph in Figure 26.2 is amended to include another directed edge, that from P_1 directly to Q_2 (in addition to those already present above), then any cross-price elasticity will capture this latest directed edge but will again miss the chain of arrows from P_1 to P_2 to Q_2. These questions on how our exogenous variables are generated are not usually asked in agricultural economics.[20]

The "theoretically sound" estimates of the Almost Ideal Demand System (Deaton and Muellbauer 1980) are possibly so specified because prior theory assumed prices were determined exogenously, and, while the consumer (probably) does not care how prices are determined, the methods used to estimate our parameters certainly do care. Notice in the almost ideal demand system (AIDS) model given below we see no description of how prices (p_{ij}) are generated:

$$w_{it} = \alpha_i + \sum_{i=1}^{N} \gamma_{ij} \ln(p_{ij}) + \beta_i \ln(X_t / P_t) + u_t,$$

where w_{it} = expenditure share on product i in period $t = p_{it} \times q_{it}/X_t$, X_t = agent's total expenditures on N commodities, P_t = an aggregate price index, p_{it} = the price of product j in period t, and u_t = a disturbance (error) term (assumed to be) not correlated with $\ln(p_{it})$ or $\ln(X_t/P_t)$. AIDS models of food demand have been the standard in the agricultural economics literature for about 30 years. In particular, the theoretical properties of consumer demand (e.g., homogeneity in prices and income (X), and Slutsky symmetry on cross-price elasticities) are easily applied. Here there is no allowance for $p_{it} \to p_{kt}, j \neq k$. Furthermore, one rarely discusses what is in the disturbance term (only that whatever is in it, it is not correlated with prices or total expenditures). Similar structures are behind the Rotterdam model and Exact Affine Stone Index (EASI) models of consumer behavior.[21] A few important applications of such work in the agricultural economics literature are Blanciforti and Green (1983), Capps and Havlicek (1984), Chalfant (1987), Eales and Unnevehr (1988), Green and Alston (1990), Mittelhammer, Shi, and Wahl (1996), and Chen et al. (2014). A recent summary and comparison of elasticity estimates are given in Okrent and Alston (2011).

Other important work, classified under the heading of imperfect competition, allows for $Q \to P$ (see the summary paper by Sexton and Lavorie 2001, or price-dependent demand systems by Eales and Unnevear 1994 and Holt and Balagtas 2009). Here price responds to quantity, not unlike Marshall's (1896) fish market (without refrigeration, fresh-caught fish must be sold by the end of the day).

$$P_1 \longrightarrow P_2 \longrightarrow Q_2$$

Figure 26.3 Causal diagram where two prices and one quantity are linked via a causal chain

Just how one decides on whether one has a price-dependent or a quantity-dependent system is a bit problematic (see Thurman's [1986] use of instrumental variables). A survey of all interesting econometric applications of consumer theory to demand analysis is beyond the scope of the current chapter. We cite others' work, such as the study by Okrent and Alston (2011), summarizing the considerable work on elasticity estimation. Other quality work, now a bit dated, is discussed in Gardner and Rausser (2001).

A similar line of analysis can be applied to econometric studies of producer behavior in agricultural economics. Here one begins with a profit maximization problem, subject to a technology (production) constraint, where profit (π) is written as:

$$\pi = Pf(x_1, x_2, \ldots, x_n) - r_1 x_1 - \ldots - r_n x_n$$

where P is the output price r_i, $i = 1, \ldots n$ are n input prices, respectively, x_i, $i = 1, \ldots n$ are choice input variables, and $f(x_1, x_2, \ldots, x_n)$ represents a technology surface that varies with the x_i.

If one estimates the production surface ($y = f(x) + e$), where y and x are observational and e is an unobserved error term, the resulting parameter estimates from ordinary least squares are biased and inconsistent because input levels will almost surely be correlated with the error term (years of drought early in the year may have less harvest labor applied than years having abundant or optimal rainfall). As reviewed by Mundlak (2001), the early work by agricultural economists Tintner (1944) and Tintner and Brownlee (1944) estimated agricultural production functions on a random cross section sample of Iowa farms. The work that followed used similar observational data and thus fits our classification as quasi-experimental. Mundlak (2001) offers nine observations based on this early work on observational data. His first observation is stressed here: "the estimates are not robust." In getting us to this point, the original authors and Mundlak note that the firms studied used different technology, input quality was (generally) not addressed, and inputs were endogenous (unlike those that are found in the experimental studies cited above).

Here output (Q) depended on input levels (X), but unmeasured or perhaps unknown variables affected Q and could (under reasonable assumptions) affect levels of X in observational studies. A key proposed solution was to find an instrument that moved X, but not Q directly, only indirectly through X. In the graph below, R (rainfall) affects Q (quantity produced) and X (input levels, say labor). One needs an instrument (we label this P_x, say the price in input X in Figure 26.4) that affects X but does not link to R (or has no path to R going from the instrument in the direction of the arrows through X then to R (i.e., we have to go against the arrow at X to get to R). If we predict X using only the knowledge of P_x in a first stage regression, we could

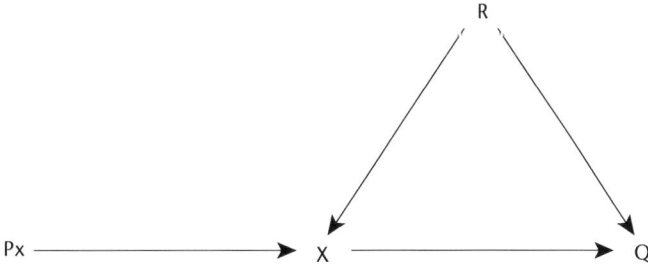

Figure 26.4 Causal diagram illustrating a proper instrumental variable, input price (P_x), solves the endogeneity issue between quantity produced (Q) and input levels (X), confounded by rainfall (R)

obtain an unbiased estimate of the beta connecting X and Q from a second-stage regression of Q on X^p, where X^p is the prediction of X from the first-stage regression.

Of course, the instrumental variable setup summarized in Figure 26.4 rests on the fact that there is no path linking P_X and Q that does not go through X, or R does not directly affect P_X (there is no arrow from R directly or indirectly to P_X).[22]

Panel data have been successfully employed in the study of the production problem, where the stochastic error term varies by individual farms but not through time (e_i), varies over time but not farms (u_t), and varies over both individuals and time (v_{it}). The above error structure is then imposed on a single production relation to explain output (y_{it}):

$$y_{it} = +\beta z_{it} + e_i + u_t + v_{it}$$

Hildreth proposes estimation via maximum likelihood with jointly Gaussian error components (Hildreth 1950; Bessler et al. 2010). Other contributions to the econometric analysis of the agricultural production problem with panel data include Hoch (1962) and Mundlak (1978).

The dual representation of the production problem was a major focus of applied econometric talent in the 1970s and 1980s, where profit and cost functions were expressed as a function of input and output prices. Conditioning information on technology (generally expressed as a function of time) or fixed inputs were also included (Mundlak 2001). Shumway (1983), for example, estimates the product supply parameters for six Texas field crops (cotton, sorghum, wheat, corn, rice, and hay) and their corresponding input demands for fertilizer and hired labor using seemingly unrelated regressions on a quadratic functional form in prices of each and in quantities of labor, acres of land, time, weather, and government payments. Important other applications in this "duality literature" are reviewed by Mundlak (2001), including Binswanger (1974), Kako (1978), Antle (1984), Lopez (1984), and Chambers and Just (1989). How prices interact with each other (if at all) is not considered in this literature (much like the applied demand literature discussed above).

Other quasi-experimental works that have occupied agricultural econometricians include modeling limited dependent variables. Here logit and probit modeling, which link back to Griliches (1957), use logistic function to help model the adoption of hybrid corn. Related extensions are given in the works of Hanemann (1984), Englin and Shonkwiler (1995), and Herriges and Kling (1999). Discussions of these latter cases are found in Bessler et al. (2010).

4 Non-structural modeling with observational data

The problems discussed above in agricultural econometricians' work induced several to concentrate on model building rather than finding structure in observational data to capture regularities that yielded good forecasts. The time series models were either univariate or small multivariate models (Leuthold et al. 1970), where little prior economic theory was used to guide specification; rather, patterns in the data governed specification. The univariate form of these models were motivated by the work of Box and Jenkins (1976). Earlier work in agriculture by Nerlove (1958) anticipated the flexibility of these models, where he shows that adaptive expectations representation had its origins in the earlier work of Hicks (1946). The study by Nerlove (1972) showed how more general models in the ARIMA (Autoregressive Integrated Moving Average) class could be used to represent quasi-rational expectations of agents in traditional structural models.

As we moved beyond univariate models to multivariate models to be able to say more about policy and to accommodate more sophisticated expectations schemes, we came to appreciate the work of Haavelmo (1944) and his admonition that for those studying observational data the task is to summarize the regularities in the observed data. The key here is that with observational data we have no control over the data-generating process on all of our variables (variables on the left-hand side and the right-hand side of the "=" sign in an equation). The Marshallian notion of *ceteris paribus* could not be assumed with any sense of legitimacy. Sims (1980) makes this point clear with an example from agriculture:

> However certain we are that the tastes of consumers in the U.S. are unaffected by temperature in Brazil, we must admit that it is possible that U.S. consumers, upon reading of a frost in Brazil in the newspapers, might attempt to stockpile coffee in anticipation of the frost's effect on price. Thus variables known to affect supply enter the demand equation, and vice versa through the expected price.
>
> *(Sims 1980, p. 8)*

Sims recognized what Milton Friedman (1953) advised more than a quarter century earlier, not to mix-up the elements of our (demand and supply) filing system:

> Viewed as a language, theory has no substantive content; it is a set of tautologies. Its function is to serve as a filing system organizing empirical material and facilitating our understanding of it; and the criteria by which it is to be judged are appropriate to a filing system. Are the categories clearly and precisely defined? Are they exhaustive? Do we know where to file each individual item, or is there considerable ambiguity? Is the system of headings and subheadings so designed that we can quickly find an item we want, or must we hunt from place to place? Are the items we shall want to consider jointly filed? Does the filing system avoid elaborate cross-references?
>
> *(Friedman 1953, p. 7)*

Haavelmo's recommendation, with observational data, is to summarize what the data say. Sims' Vector Autoregression (VAR) did just this. Here every variable in a small system is allowed to affect every other variable in that system at lags (time delays). These are reduced form models whose coefficients are combinations of the underlying "structural" demand and supply parameters. The system is estimated equation-by-equation by ordinary least squares, where each equation has the same right-hand side variables in its algebraic representation. Hsiao (1979) suggests a subset VAR where certain lags of certain variables have coefficient values set equal to zero (prior to estimation); here, seemingly unrelated regression is used for efficient estimation.

The VAR work is estimated in autoregressive form:

$$\Phi(B)X_t = e_t,$$

where X_t is an m × 1 vector of variables observed at period t, $\Phi(B)$ is an m × m matrix of autoregressive parameters written in terms of the backshift operator ($B^k X_t = X_{t-k}$), and e_t is a m × 1 vector of error or innovation terms (representing new information at time t). We generally communicate VAR results in what is known as its moving average (MA) form.

Multiply both sides of the autoregressive form by $\Phi(B)^{-1} = \Theta(B)$ to get the corresponding MA form:

$$X_t = \Theta(B)e_t.$$

Key to presentation of terms (or its finite approximation) is how VAR researchers treat contemporaneous innovations ($\hat{e}_t$, which is the observed innovation (error) and not the theoretical innovation ((error)e_t). In the derivation given above, we need to specify how the innovations in contemporaneous time affect one another, if at all. That is to say, we have an identification problem, just as the structural econometricians faced, as described earlier. If series i and j have innovations that are correlated in current time, then in providing users the moving average representation (in one of its popular forms), do we allow series i to respond to a shock in series j according to their innovation correlation? Or should we let series j respond to shocks in series i? Which comes first in contemporaneous time, i or j?

Sims and early VAR users used a Cholesky factorization of contemporaneous covariance to (apparently) solve this problem. Early users of this form in agricultural economics include Bessler (1984), Chambers (1984), and Orden (1986). Cooley and LeRoy (1985), writing in the general macroeconomic literature, gave Sims and his followers a harsh critique for the use of the (clearly arbitrary) Cholesky factorization. Indeed, the merits of Sims' 1980 offering was to allow the data to inform us about relationships among our variables; yet imposing the Cholesky factorization of contemporaneous covariance brought subjective knowledge into the system, knowledge that was perhaps no better than that brought to the structural econometrics of the pre-1980s from which Sims was attempting to remove us.

Two fixes to the VAR modeling strategy answered Cooley and LeRoy's critique. In the mid-1980s, Bernanke (1986) and Sims (1986) studied non-Cholesky factorizations on contemporaneous covariance, which Doan (2010) labeled the Bernanke factorization. Orden and Fackler (1989) studied agricultural systems with this factorization, using theory-based knowledge to orthogonalize innovations. The Cholesky VAR and the Bernanke VAR, while possibly correct in their applications, mixed both empirical regularities (among lagged relations) and *a priori* knowledge (in contemporaneous time). About a decade later, Swanson and Granger (1997) suggested a completion of what some saw as the 1980 promise of Sims (to purge our econometric models of "incredible identification" assumptions) by allowing the data to determine the form of both lagged relations and contemporaneous structure. Using the work of Judea Pearl and his students in computer science, Swanson and Granger (1997) recognized that screening off patterns in data can inform researchers on the underlying causal structure on VAR innovations. While they confined their 1997 study of innovations to causal chains $e_1 \rightarrow e_2 \rightarrow e_3$, their work opened up more general modeling of contemporaneous structures of forks $e_1 \leftarrow e_2 \rightarrow e_3$ and inverted forks $e_1 \rightarrow e_2 \leftarrow e_3$, by Bessler and Akleman (1998), who built on the works of Pearl (1995) and Spirtes, Glymour, and Scheines (1993).

Other time series work related to volatility over time and reduced rank (cointegration) has generated much of the work by agricultural econometricians. In the study by Baillie and Myers (1991) on Bivariate Generalized Autoregressive Conditional Heteroskedastic models of cash and futures price innovations for six commodities, the optimal hedge ratios are found to be nonstationary, counter to usual assumption. Cointegration applications by Ardeni (1989), Bessler and Covey (1991), and Schroder and Goodwin (1991), among others, consider the "law of one price" in markets separated by space, time, or form. Robertson and Orden (1990) study monetary impacts on prices in New Zealand using a cointegration framework. Monte Carlo work has been carried out by Bewley et al. (1994) and Zapata and Rambaldi (1997).

The models discussed here under the heading of time series models are not structural and thus ought to be used for policy analysis with caution. Their main strength is forecasting in both univariate and multivariate forms. A rich literature on forecasting, using both point and probability forecasts, has developed in post-1970s agricultural econometrics. The autoregressive representation allows researchers to project ahead any horizon into the future though sequential forecasting, using the parameters fit from known observed data, usually with OLS regression. As they go beyond one step ahead, forecasted values at time $t + 1, t + 2, \ldots t + k$ are used as if they are the "true" values to generate forecasts for horizon $t + k + 1, \ldots$ This recursion, studied as the "Law of Iterated Projections" (Sargent 1979), is frequently communicated through "the chain rule of forecasting."

In an unconstrained setting, the "chain rule" sets all future innovations (errors) to zero. When time series models are used for policy (which we recommend be done with considerable caution), the policy settings can be viewed as placing a set of (nonzero) restrictions on future innovations (errors). One can then assess the likely success of a policy by the size of the implied errors relative to observed errors from earlier time periods (Sims (1982). Agricultural economists have not adapted this last setting much, if at all. Bessler and Kling (1989) did something close, in terms of the chain rule, but did not study the implied errors for reaching the policy.

Forecasting research in agriculture has a long history. In fact, Henry Moore, the first econometrician to use ordinary least squares to address the estimation of the demand curve (discussed earlier), published a book on forecasting cotton prices and yields (Moore 1917). Stine (1929) followed with an appraisal of the early successes. Nielson (1953) made the case for forecasts as a basis of farm planning models in the 1950s. Ferris (2010) offered a discussion of results on forecasting through extension programs. A modestly rich literature in agriculture exists comparing econometric models with futures markets, including Just and Rausser (1981), Garcia, Hudson, and Waller (1988), Allen (1994), and Vercammen and Schmitz (2001). Since 2000, work with factor models (VARs with a very large number of series) to reduce dimensionality has had some applications in energy markets (Garcia-Martos, Rodriguez, and Sanchez 2011). While most forecasting work in agriculture is with point forecasts, studies by Kling and Bessler (1989), Fackler and King (1990), and Trujillo-Barrera, Garcia, and Mallory (2016) deal with probability forecasts using probability calibration. The idea here is that any probability issued, whether it be subjective or model-based, should be subjected to testing (in and out of sample sense). Do the probabilities issued come to bear, after the fact, the proper number of times (in a relative frequency sense)? For example, if a model of average cotton yields for several counties in Texas for the year 2017 offers a probability of 0.6 that average cotton yield will be in some interval (a different interval perhaps in each county), we should observe (after the fact) that 60 percent of these counties should have their 2017 average county yield in the interval for which 0.6 was offered by the model. The model is thus said to be well-calibrated at 0.6. The mathematics behind such work is laid out in Dawid (1984), invoking the probability integral transform.

Recent work in applied copula modeling has garnered interest in agricultural and energy markets. The idea here is to write the joint probability distribution on data of interest as a product of the copula, describing the dependence between or among variables and their individual marginal distributions (Sklar 1973). This gives us a type of simplicity in modeling, which has shown promise (Gao, Zhang, and Wu 2015; Goodwin and Hungerford 2015).

5 Summary

These and other econometric techniques and advances have been applied and developed in agricultural economics for almost 150 years. We apologize to those whose work we did not include in this review. We do, however, believe we have demonstrated that agricultural economists have begun to see the way over the "insuperable obstacle" Jevons described to us in making economics a science. Science does two things: it explains (through hypothesis generation and testing) and it predicts. Our stance as we approach 150 years of econometrics with the neoclassical model in agriculture is that perhaps different models are best used to perform the required tasks of science. We explain with structural models, we test in the laboratory, and we predict with reduced forms of stochastic difference equations (VARs, copulas, and the like). The way over Jevons' obstacle appears to be a recognition of how we obtain data and just how (or if) *ceteris paribus* is violated in any particular application. But with Leon Walras (1954) and general equilibrium looming in our background, just how we do that is problematic (and makes all the difference).

We saw in Henry Moore's work with observational agricultural prices and quantities that regression techniques could (at least) give the proper sign, but not always. Later, Elmer Working (1927) made it clear what Moore's problem was, and the father–son team of Phillip and Sewall Wright solved (at least on paper) the problem with their introduction of instrumental variables (S. Wright 1921; P. Wright 1928). While such solutions do hold on paper (their mathematics were correct), the fact that economic agents (people) are not content to limit their vision to either the demand or the supply files, where we knew our IV estimators would work, suggests that econometric identification of independent instruments is sometimes questionable with observational data. This led Sims (1980), from the macroeconomics literature, to suggest reduced form vector autoregressions, where every variable in a system is dependent on lags of every other variable in the system. There is nothing special here about macroeconomics; the same mixing of files in our economic filing system, which Friedman (1953) warns us about and Sims (1980) describes, lurks in microeconomics studies with observational data as well. It is not the macro versus micro that makes the difference; it is how the data are obtained. The VAR, cointegration, or copula models are not structural in Haavelmo's (or Frisch's) sense, but they can be used to generate regularities that (with some adjustments) yield good out-of-sample forecasts.

The path to solving the identification problem is still not settled with observational data. We have seen the creative work of Orden and Fackler (1989) making some progress with the identification issue on innovations of a vector autoregression in contemporaneous time. Some later contributions following Swanson and Granger (1997) have offered a fully data-based route to identification. Yet many edges (relationships among variables economists want to study) remain undirected.

The alternative path to dealing with the identification issue is to go into the laboratory and assign treatments via random assignment. Here agricultural economists were actively engaged in production function estimation in the 1950s and 1960s, but generally moved away from this line thereafter. A recent revival of experiments in agricultural marketing and demand analysis is testimony that a good method will not lie dormant forever, and so it has not. Such experiments can yield internally valid results (just identified results) but remain open for advances in external validity. One of the criticisms of laboratory experiments, even with random assignment, is how accurately they portray the real world. Field experiments may be a way to bridge the gap between randomization and external validity (Roe and Just 2009). Such questions take us back to Egon Brunswik's 1955 classic on representative design, where we keep in mind the population we are addressing our study toward in setting up our laboratory experiment. Pearl and Bareinboim (2014) offer a modern version of Brunswik's (1955) suggestion. A new wave of "field experiments" in economics has emerged since 2000 to study human behavior (Harrison and List 2004; Levitt and List 2009).[23] In general, field experiments use random assignment in naturally occurring settings. This connection to reality combined with randomization is a promising venue to improve the external validity of experimental work while preserving its internal validity. We seem to be addressing Brunswik's call in our move into field experiments. Recent work highlights the importance of field experiments to food choices and nutrition and health (Nayga 2008; Berning, Chouinard, and McCluskey 2010; List and Samek 2015; Li et al. 2016).

The structural models that we worked hard on during the decades between 1950 and 1990 are quite possibly just that, structural, but we have no guarantees. The random assignment experiment does have such a guarantee (at least probabilistically for large N), and the "summaries of regularities" approach (Sims' reduced forms) never claimed structure. We suggest out-of-sample forecasting competitions on all econometric offerings. Cross section work, cross section and time series work, and limited dependent variable models all can be studied with respect to evidence on forecasts of new data, data not used for construction of the initial model. These can be tested, much as Dyson, Eddington and Davidson (1920) suggest, through a strong future of forecasting competitions to be offered in our professional journals.

Notes

1. To our knowledge, Jevons' statistical work was, primarily, on prices and the value of gold (see S. Stigler 1986, p. 5) and did not attempt to measure marginal utilities, as, say Frisch (see the discussion and citation given below) did some 60 years after Jevons' *Theory*. Jevons, nonetheless, did call for the measurements of the purely deductive results presented in his theory: "if only commercial statistics were far more complete and accurate than they are at present."
2. Microsoft Word thesaurus offers two (among others) definitions of the word "insuperable": "impossible" and "challenging." Herein, we hope to convey the latter usage with respect to Jevons' obstacle in making economics a science.
3. Our description of the early years, say from 1871 to 1950, and the "potential outcomes" model of Rubin and others, have been presented earlier, in somewhat different settings, in Bessler and Dearmont (1996) and Bessler ((2013).
4. Of historical interest are photos and conference proceedings of the Cowles Commission (http://cowles.yale.edu/archives/1937-3rd-annual-conf). There one notes several agricultural economists in attendance, including L. Bean, S. Ciriacy-Wantrup, O. Stine, G. Tintner, and E. Working.
5. Usual counting rules on the rank conditions for identification are given in Griffiths, Hill, and Judge (1993) as follows:

    g^* = the number of endogenous variables present in the ith equation.
    k^* = the number of exogenous variables present in the ith equation (including the intercept).
    K = the total number of exogenous variables in the system.
    If: $(g^* + k^* - 1) < K$ or $(g^* - 1) < (K - k^*)$, then the equation is *over-identified*.
    If: $(g^* + k^* - 1) = K$ or $(g^* - 1) = (K - k^*)$, then the equation is *just-identified*.
    If: $(g^* + k^* - 1) > K$ or $(g^* - 1) > (K - k^*)$, then the equation is *under-identified*.
    Both equations are under-identified, meaning there is no unique set of parameters that can be found from the data on P and Q.

6. Actually, the first work that we are aware of that solved Moore's problem with what was later to be labeled "instrumental variables" was by Phillip Wright (1928); these were introduced in an appendix (written by Sewall Wright) using graphical representations (arrow graphs) (S. Wright 1921). Sewall Wright worked at USDA.
7. See Ezekiel (1938), as well as the more recent piece placing the Cobweb model in its historical setting by Nerlove (2010).
8. The celebrated Frisch–Waugh Theorem is perhaps the most lasting product of Waugh's cooperation with Frisch: running a regression on de-trended variables (Y and X) gives the same OLS parameter estimates as running the regression of Y on X and a time trend t (Frisch and Waugh 1933).
9. Unlike Stock and Watson (2007), who consider only a small subset of observational studies as quasi-experimental, our unconventional use of quasi-experimental refers to the mental state of the researcher and not some unknowable reality. Natural experiments would fall under Stock and Watson's definition, whereas we assume some knowledge about omitted variables. Perhaps it is better for us to label our quasi-experimental as pseudo-experimental.
10. Rausser and Bessler (2016) offer a more detailed discussion of Koopmans (1947) and the role of prior theory and data in economic modeling.

11 By reversible, Haavelmo means that if one selects q* when price is at p* and q** when price is at p* + Δp*, does one return to q* when price is changed back to p*? Or might one behave in a non-reversible manner? (see Haavelmo 1944).
12 This model is actually found in the work of Jerzy Neyman (1923) in his study of yields of crop varieties in agricultural test plots. Later, presumably independent versions of this model are given in Rubin (1974) and Holland (1986). An earlier version of this model for isolating the demand for a product is given in Bessler and Dearmont (1996). The details here are similar to that description.
13 Here I indexes individual agents from a population of U agents, I = 1,2, ... ,U. It is not the same as that used by Jevons, as discussed in our introduction. There, I indexes different uses (not individuals).
14 Campbell and Stanley (1966) give extensive discussion to factors that jeopardize both internal and external validity.
15 Here notation is important: $E_{t0}\{Q_{p+\Delta p}\}$ is the expected potential outcome at time t_0 for an individual facing prices $p + \Delta p$. The quantity $E_{t0}\{Q_{p+\Delta p}\}$ is never observed. $E_{t0}\{Q|p + \Delta p\}$ is the actual outcome (Q selected) at t_0 if the consumer faces price $p + \Delta p$. This quantity is observed. Because of random assignment, we can insert the "=" sign between these two (for large N) and infer the effect of P on Q.
16 Experimental auctions take advantage of random assignment in a real market institution where products and money exchanges take place. They are mostly designed to be incentive compatible, meaning there is a cost to participants associated with deviating from the true product valuations. While they are flexible for experimenters to control features within the auction (e.g., treatments and market price/auction winner selection), there are factors outside the control of the auction design (e.g., related substitutes/complements).
17 Holland and Rubin worked together at Harvard and Education Testing Service (Holland 1986).
18 Econometric methods used here will usually be either single equation models fit with Ordinary Least Squares or systems fit with Seemingly Unrelated Regressions or Maximum Likelihood methods. It is generally the case that either hypothesis testing or actual parameter estimation is based on the assumption of Gaussian innovations (errors). See any of several good introductory texts, including Stock and Watson (2007), Woodridge (2009), and Green (2015). More advanced treatments are found in Morgan and Winthrop (2007) and Angrist and Pischke (2009).
19 With aggregate U.S. data, Wang and Bessler (2006) find the beef price leads (causes) pork and poultry prices.
20 All science makes such "circumscription" assumptions; what's in and what's out of the model? (Pearl 2000). Of course, the data are not necessarily generated by such restricted systems. The reference on circumscription is McCarthy (1986). Economists have a longer history with a similar idea, that of separability (weak): the marginal rate of substitution between any two commodities (services) within a group is independent of the quantity of any commodity not in the group (Green 1971). Agricultural economists have an impressive history with this notion (LaFrance 1991, 1993; Moschini, Moro, Green 1994). These studies generally do not ask whether prices are related to each other.
21 The Rotterdam model is discussed in Theil (1965) and Barten (1966). The EASI model is developed in Lewbel and Pendakur (2009).
22 Margolis (2015) provides a nice discussion of instrumental variables and graphical models. Other comments on IV methods are found in Bessler (2013).
23 The term "field experiments" was used due to the nature of the early experiments conducted in agricultural fields. The connection of agriculture to subsequent related work is discussed in Herberich, Levitt, and List (2009) and Levitt and List (2009).

References

Allen, P.G. 1994. "Economic Forecasting in Agriculture." *International Journal of Forecasting* 10:81–135.
Angrist, J., and J-S. Pischke. 2009. *Mostly Harmless Econometrics: An Empiricist's Companion*. Princeton, NJ: Princeton University Press.
Antle, J.M. 1984. "The Structure of U.S. Agricultural Technology, 1910–1978." *American Journal of Agricultural Economics* 66:414–421.
Ardeni, P.G. 1989. "Does the Law of One Price Really Hold?" *American Journal of Agricultural Economics* 71:661–669.

Arefiev, N. 2016. *Graphical Interpretations of Rank Conditions for Identification of Linear Gaussian Models.* Moscow: National Research University Higher School of Economics.

Baillie, R.T., and R.J. Myers. 1991. "Bivariate GARCH Estimation of Optimal Commodity Futures Hedge." *Journal of Applied Econometrics* 16:109–124.

Barten, A.P. 1966. *Theorie en empirie van een volledig stelsel van vraegvergelijkingen.* PhD dissertation, University of Rotterdam, Rotterdam, Netherlands.

Bernanke, B. 1986. "Alternative Explanations of the Money–Income Correlation." *Carnegie–Rochester Conference Series on Public Policy* 25:49–99.

Berning, J.P., H.H. Chouinard, and J.J. McCluskey. 2010. "Do Positive Nutrition Shelf Labels Affect Consumer Behavior? Findings from a Field Experiment with Scanner Data." *American Journal of Agricultural Economics* 93:364–369.

Bessler, D.A. 1984. "Relative Prices and Money: A Vector Autoregression on Brazilian Data." *American Journal of Agricultural Economics* 66(1):25–30.

———. 2013. "On Agricultural Econometrics." *Journal of Agricultural and Applied Economics* 45(4):617–637.

Bessler, D.A., and D.G. Akleman. 1998. "Farm Prices, Retail Prices, and Directed Graphs: Results for Pork and Beef." *American Journal of Agricultural Economics* 80(5):1145–1150.

Bessler, D.A., and T. Covey. 1991. "Cointegration: Some Results on U.S. Cattle Prices." *The Journal of Futures Markets* 11(4):461–474.

Bessler, D.A., and D. Dearmont. 1996. "Ceteris Paribus: An Evolution in Agricultural Econometrics." *European Review of Agricultural Economics* 23:262–280.

Bessler, D.A., J.H. Dorfman, M.T. Holt, and J.T. LaFrance. 2010. "Econometric Developments in Agricultural and Resource Economics: The First 100 Years." *American Journal of Agricultural Economics* 92(2):571–589.

Bessler, D.A., and J.L. Kling. 1989. "The Forecast and Policy Analysis." *American Journal of Agricultural Economics* 79:503–507.

Bewley, R., D. Orden, M. Yang, and L.A. Fisher. 1994. "Comparison of Box-Tiao and Johansen Canonical Estimators of Cointegrating Vectors in VEC(1) Models," *Journal of Econometrics* 64:3–27.

Binswanger, H. 1974. "A Cost Function Approach to the Measurement of Elasticities of Factor Demand and Elasticities of Substitution." *American Journal of Agricultural Economics.* 56:377–386.

Blanciforti, L., and R. Green. 1983. "An Almost Ideal Demand System Incorporating Habits: An Analysis of Expenditures on Food and Aggregate Commodity Groups." *The Review of Economics and Statistics* 65:511–515.

Blaug, M. 1985. *Great Economists Since Keynes.* Cambridge: Cambridge University Press.

Box, G., and G. Jenkins. 1976. *Time Series Analysis: Forecasting and Control.* San Francisco, CA: Holden-Day.

Brunk, M.E., and W.T. Federer. 1953. "Experimental Designs and Probability Sampling in Marketing Research." *Journal of the American Statistical Association* 48:440–452.

Brunswik, E. 1955. *Systematic and Representative Design of Psychological Experiments.* Berkeley, CA: University of California.

Campbell, O.T., and J.C. Stanley. 1966. *Experimental and Quasi-Experimental Designs for Research.* Boston, MA: Houghton Mifflin Company.

Capps, O., and J.H. Havlicek. 1984. "National and Regional Household Demand for Meat, Poultry and Seafood: A Complete Systems Approach." *Canadian Journal of Agricultural Economics* 32:93–108.

Chalfant, J. 1987. "A Globally Flexible Almost Ideal Demand System." *Journal of Business and Economics Statistics* 5:233–242.

Chambers, R.G. 1984. "Agricultural and Financial Market Interdependence in the Short Run." *American Journal of Agricultural Economics* 66:12–24.

Chambers, R.G., and R.E. Just. 1989. "Estimating Multiproduct Technologies." *American Journal of Agricultural Economics* 71:980–995.

Chen, Z., E.A. Finkelstein, J.M. Nonnemaker, S.A. Karns, and J.E. Todd. 2014. "Predicting the Effects of Sugar-Sweetened Beverage Taxes on Food and Beverage Demand in a Large Demand System." *American Journal of Agricultural Economics* 96:1–25.

Cooley, T., and S. LeRoy. 1985. "Atheoretic Macroeconometrics: A Critique." *Journal of Monetary Economics* 16:283–308.

Dawid, P. 1984. "Statistical Theory: A Prequential Approach." *Journal of the Royal Statistical Society* 147:278–292.

Dean, G. 1960. "Consideration of the Time and Carryover Effects in Milk Production Functions." *Journal of Farm Economics* 42:1512–1514.

Deaton, A., and J. Muellbauer. 1980. "An Almost Ideal Demand System." *American Economic Review* 70:312–326.

Doan, T. 2010. *RATS Reference Manual*. Evanston, IL: ESTIMA.

Dyson, F.W., A.S. Eddington, and C. Davidson. 1920. "A Determination of the Deflection of Light by Sun's Gravitational Field, from Observations Made at the Total Eclipse of May 29, 1919." *Philosophical Transactions of the Royal Society of London* 1:1–44.

Eales, J.S., and L.J. Unnevehr. 1988. "Demand for Beef and Chicken Products: Separability and Structural Change." *American Journal of Agricultural Economics* 70:521–532.

———. 1994. "The Inverse Almost Ideal Demand System." *European Economic Review* 38:101–115.

Englin, J., and J.S. Shonkwiler. 1995. "Estimating Social Welfare Using Count Data Models: An Application to Long-Run Recreation Demand under Conditions of Endogenous Stratification and Truncation." *Review of Economics and Statistics* 77:104–112.

Ezekiel, M. 1938. "The Cobweb Theorem." *Quarterly Journal of Economics* 52:255–280.

Fackler, P.L., and R.P. King. 1990. "Calibration of Option-Based Probability Assessments in Agricultural Commodity Markets." *American Journal of Agricultural Economics* 72:73–83.

Ferris, J. 2010. "The USDA/Land Grant Extension Outlook Program – A History and Assessment." *AAEA/CAES/WAEA Joint Conference*, Denver, CO. http://ageconsearch.tind.io//bitstream/101723/2/USDALandGrantOutlookProgram.pdf. Accessed June 2017.

Fox, K.A. 1989. "Agricultural Economists in the Econometric Revolution: Institutional Background, Literature, and Leading Figures." *Oxford Economic Papers* 41:53–70.

Friedman, M. 1953. *The Methodology of Positive Economics*. Chicago, IL: The University of Chicago Press.

Frisch, R. 1932a. "New Orientation of Economic Theory. Economics as an Experimental Science." *Nordic Statistical Journal* 4:97–111.

———. 1932b. *New Methods of Measuring Marginal Utility*. Tubingen, Germany: Mohr.

Frisch, R., and F. Waugh. 1933. "Partial Time Trend Regression as Compared with Individual Trends." *Econometrica* 1:387–401.

Gao, Y., Y. Zhang, and X. Wu. 2015. "Penalized Exponential Series Estimation of Copula Densities with an Application to Intergenerational Dependence of Body Mass Index." *Empirical Economics* 48:61–81.

Garcia, P., M.A. Hudson, and M.L. Waller. 1988. "The Pricing Efficiency of Agricultural Futures Markets: An Analysis of Previous Research." *Southern Journal of Agricultural Economics* 20:119–130.

Garcia-Martos, C., J. Rodriguez, and M.J. Sanchez. 2011. "Forecasting Electricity Prices and Their Volatilities Using Unobserved Components." *Energy Economics* 33(6):1227–1239.

Gardner, B., and G. Rausser (editors). 2001. *Handbook of Agricultural Economics*. Amsterdam: North Holland.

Goodwin, B.K., and A. Hungerford. 2015. "Copula-Based Models of Systemic Risk in U.S. Agriculture: Implications for Crop Insurance and Reinsurance Contracts." *American Journal of Agricultural Economics* 97:879–896.

Green, H.A.J. 1971. *Consumer Theory*. London: Penguin Modern Economics.

Green, R., and J.M. Alston. 1990. "Elasticities in AIDS Models." *American Journal of Agricultural Economics* 72:442–445.

Green, W.H. 2015. *Econometric Analysis*. Upper Saddle River, NJ: Prentice Hall.

Griffiths, W.E., R.C. Hill, and G.G. Judge. 1993. *Learning and Practicing Econometrics*. New York: Wiley.

Griliches, Z. 1957. "Hybrid Corn: An Exploration in the Economics of Technological Change." *Econometrica* 25:501–522.

Haavelmo, T. 1944. "The Probability Approach in Econometrics." *Econometrica* 12:1–118.

Hamilton, J. 1994. *Time Series Analysis*. Princeton, NJ: Princeton University Press.

Hanemann, W.M. 1984. "Discrete/Continuous Models of Consumer Demand." *Econometrica* 52:541–561.

Harrison, G.W., and J.A. List. 2004. "Field Experiments." *Journal of Economic Literature* 42(4):1009–1055.

Heady, E.O., and J.L. Dillon. 1961. *Agricultural Production Functions*. Ames, IA: Iowa State University Press.

Herberich, D.H., S.D. Levitt, and J.A. List. 2009. "Can Field Experiments Return Agricultural Economics to the Glory Days?" *American Journal of Agricultural Economics* 91(5):1259–1265.

Herriges, J., and C. Kling. 1999. "Nonlinear Income Effects in Random Utility Models." *The Review of Economics and Statistics* 81:62–72

Hicks, J.R. 1946. *Value and Capital*. Oxford: Clarendon.

Hildreth, C. 1950. "Combining Cross Section and Time Series." Cowles Commission Discussion Paper No. 347, Cowles Foundation, Colorado Springs, CO.

Hoch, I. 1962. "Estimation of Production Function Parameters Combining Time-Series and Cross-Section Data." *Econometrica* 30:34–53.

Holland, P. 1986. "Statistics and Causal Inference." *Journal of the American Statistical Association* 81:945–960.

Holt, M.T., and J.V. Balagtas. 2009. "Estimating Structural Change with Smooth Transition Regression: An Application to Meat Demand." *American Journal of Agricultural Economics* 91:1424–1431.

Hsiao, C. 1979. "Autoregressive Modeling of Canadian Money and Income Data." *Journal of the American Statistical Association* 74:553–560.

Jevons, W.S. 1871. *The Theory of Political Economy*. London: Palgrave MacMillan and Company.

Just, R.E., and G.C. Rausser. 1981. "Commodity Price Forecasting with Large-Scale Econometric Models and the Futures Market." *American Journal of Agricultural Economics* 63(2):197–215.

Kako, T. 1978. "Decomposition Analysis of Derived Demand for Factor Inputs: The Case of Rice Production in Japan." *American Journal of Agricultural Economics* 60:628–635.

Kling, J., and D.A. Bessler. 1989. "Calibration-Based Predictive Distributions: An Application of Prequential Analysis to Interest Rates, Money, Prices and Output." *Journal of Business* 62(4):447–500.

Koopmans, T.J. 1947. "Measurement Without Theory." *Review Economics Statistics* 29(3):161–172.

Kuznets, G. 1953. "Measurement of Market Demand with Particular Reference to Consumer Demand for Food." *Journal of Farm Economics* 35:878–895.

LaFrance, J. 1991. "When Is Expenditure 'Exogenous' in Separable Demand Models?" *Western Journal of Agricultural Economics* 16:49–62.

———. 1993. "Weak Separability in Applied Welfare Analysis." *American Journal of Agricultural Economics* 75(3):770–775.

Leuthold, R.M., A.J.A. MacCormick, A. Schmitz, and D.G. Watts. 1970. "Forecasting Daily Hog Prices and Quantities: A Study of Alternative Forecasting Techniques." *Journal of the American Statistical Association* 65:90–107.

Levitt, S.D., and J.A. List. 2009. "Field Experiments in Economics: The Past, the Present, and the Future." *European Economic Review* 53(1):1–18.

Lewbel, A., and K. Pendakur. 2009. "Tricks with Hicks: The EASI Demand System." *American Economic Review* 99:827–863.

Li, F., and F. Mealli. 2014. "A Conversation with Donald B. Rubin." *Statistical Sciences* 3:439–457.

Li, Y., M.A. Palma, S.D. Towne, J.L. Warren, and M.G. Ory. 2016. "Peer Effects on Childhood Obesity from an Intervention Program." *Health Behavior and Policy Review* 3:323–335.

List, J.A., and A.S. Samek. 2015. "The Behavioralist as Nutritionist: Leveraging Behavioral Economics to Improve Child Food Choice and Consumption." *Journal of Health Economics* 39:135–146.

Lopez, R.E. 1984. "Estimating Substitution and Expansion Effects Using a Profit Function Framework." *American Journal of Agricultural Economics* 66:358–367.

Lusk, J.L., and J.F. Shogren. 2007. *Experimental Auctions*. Cambridge: Cambridge University Press.

Malinvaud, E. 1980. *Statistical Methods of Econometrics*. Amsterdam: North Holland.

Margolis, M. 2015. "Graphics as a Tool for the Close Reading of Econometrics (Settler Mortality Is Not a Valid Instrument for Institutions)." Unpublished Paper, University of Guanajuato, Mexico.

Marshall, A. 1896. *Principles of Economics*. London: Palgrave Macmillan and Company.

McCarthy, J. 1986 "Applications of Circumscription to Formalizing Common Sense Knowledge." *Artificial Intelligence* 26:89–116.

Mitchell, P.D., M.E. Gray, and K.L. Steffey. 2004. "A Composed-Error Model for Estimating Pest-Damage Functions and the Impact of the Western Corn Rootworm Soybean Variant in Illinois." *American Journal of Agricultural Economics* 85:1392–1396.

Mittelhammer, R.C., H. Shi, T.I. Wahl. 1996. "Accounting for Aggregation Bias in Almost Ideal Demand Systems." *Journal of Agricultural Resources and Economics* 21:247–262.

Moore, H.L. 1914. *Economic Cycles: Their Law and Cause*. New York: Palgrave MacMillan.

———. 1917. *Forecasting the Yield and Price of Cotton*. New York: Palgrave MacMillan.

———. 1911. *Laws of Wages: An Essay in Statistical Economics*. New York: MacMillan.

Morgan, S., and C. Winthrop. 2007. *Counterfactuals and Causal Inference: Methods and Principles for Social Research*. New York: Cambridge University Press.

Moschini, G., D. Moro, and R. Green. 1994. "Maintaining and Testing Separability in Demand Systems." *American Journal of Agricultural Economics* 76:61–73.

Mundlak, Y. 1978. "On Pooling of Time Series and Cross Section Data." *Econometrica* 46:69–85.

———. 2001. "Production and Supply." In B. Gardner and G. Rausser, eds., *Handbook of Agricultural Economics*. Amsterdam: North Holland, pp. 3–85.

Nayga, R.M. 2008. "Nutrition, Obesity, and Health: Policies and Economic Research Challenges." *European Review of Agricultural Economics* 35(3):281–302.

Nerlove, M. 1958. *The Dynamics of Supply: Estimation of Farmers' Response to Price*. Baltimore, MD: Johns Hopkins University Press.

———. 1972. "Lags in Economic Behavior." *Econometrica* 40:221–251.

———. 2010. "Cobwebs." In M. Blaug and P. Lloyd, eds., *Famous Figures and Diagrams in Economics*. Cheltenham: Edward Elgar Publishing, pp. 184–190.

Neyman, J. 1923. "On Application of Probability Theory to Agricultural Experiments. Essay on Principles." *Statistical Sciences* 5:465–480.

Nielson. 1953. "The Use of Long-Run Price Forecasts in Farm Planning." *Journal of Farm Economics* 35:1000–1007.

Okrent, A.M., and J.M. Alston. 2011. "Demand for Food in the United States: A Review of the Literature, Evaluation of Previous Estimates, and Presentation of New Estimates of Demand." Giannini Foundation Monograph 48, University of California, Berkeley.

Orden, D. 1986. "Agricultural Trade and Macroeconomics: The United States Case." *Journal of Policy Modeling* 8:27–51.

Orden, D., and P. Fackler. 1989. "Identifying Monetary Impacts on Agricultural Prices in VAR Models." *American Journal of Agricultural Economics* 71:495–502.

Pearl, J. 1995. "Causal Diagrams for Empirical Research." *Biometrica* 82:669–710.

———. 2000. *Causality: Models, Reasoning, and Inference*. New York: Cambridge University Press.

Pearl, J., and E. Barenboim. 2014. "External Validity: From Do-Calculus to Transportability across Populations." *Statistical Sciences* 29:579–595.

Rausser, G., and D.A. Bessler. 2016. "Information Recovery and Causality: A Tribute to George Judge." *Annual Review of Resource Economics* 8:7–23.

Robertson, J.C., and D. Orden. 1990. "Monetary Impacts on Prices in the Short and Long Run: Some Evidence from New Zealand." *American Journal of Agricultural Economics* 72:160–171.

Roe, B.E., and D.R. Just. 2009. "Internal and External Validity in Economics Research: Tradeoffs between Experiments, Field Experiments, Natural Experiments, and Field Data." *American Journal of Agricultural Economics* 91(5):1266–1271.

Rubin, D. 1974. "Estimating Causal Effects of Treatments in Randomized and Non-Randomized Studies." *Journal of Educational Psychology* 66:688–701

Samuelson, P.A. 1947. *Foundations of Economic Analysis*. Cambridge, MA: Harvard University Press.

Sargent, T.J. 1979. *Macroeconomic Theory*. New York: Academic Press.

Schroder, T.C., and B.K. Goodwin. 1991. "Price Discovery and Cointegration for Live Hogs." *The Journal of Futures Markets* 11:685–696.

Sexton, R.J., and N. Lavorie. 2001. "Food Processing and Distribution: An Industrial Organization Approach." In B. Gardner and G. Rausser, eds., *Handbook of Agricultural Economics*. Amsterdam: North Holland, pp. 863–932.

Shumway, C.R. 1983. "Supply, Demand, and Technology in a multiproduct industry: Texas Field Crops." *American Journal of Agricultural Economics* 65:748–760.

Sims, C. 1986. "Are Forecasting Models Useful for Policy Analysis." *Quarterly Review of Federal Reserve Bank* 10:2–16.

———. 1980. "Macroeconomics and Reality." *Econometrica* 48:1–48.

———. 1982. "Policy Analysis with Econometric Models." Brookings Papers on Economic Activity, Brookings Institution, Washington, DC.

Sklar, A. 1973. "Random Variables, Joint Distributions, and Copulas." *Kybernetica* 9:449–460.

Smith, V.L. 1976. "Experimental Economics: Induced Value Theory." *The American Economic Review* 66(2):274–279.

Spirtes, P., C. Glymour, and R. Scheines. 1993. *Causation, Prediction, and Search.* New York: Springer-Verlag.

Sri Ramaratnam, S., D.A. Bessler, M.E. Rister, J. Matocha, and J. Novak. 1987. "Fertilization under Uncertainty: An Analysis Based on Producer Yield Expectations." *American Journal of Agricultural Economics* 69(2):349–357.

Stigler, G.J. 1962. "Henry L. Moore and Statistical Economics." *Econometrica* 30:1–21.

Stigler, S. 1986. *The History of Statistics: The Measurement of Uncertainty Before 1900.* Cambridge, MA: Harvard University Press.

Stine, O.C. 1929. "Progress in Price Analysis and an Appraisal of Success in Price Forecasting." *Journal of Farm Economics* 11:128–140.

Stock, J., and M. Watson. 2007. *Introduction to Econometrics.* New York: Pearson, Addison-Wesley.

Swanson, N., and C.W.J. Granger. 1997. "Impulse Response Functions Based on a Causal Approach to Residual Orthogonalization in Vector Autoregressions." *Journal of the American Statistical Association* 92:357–367.

Theil, H. 1965. "The Information Approach to Demand Analysis." *Econometrica* 30:67–87.

Thurman, W.N. 1986. "Endogeneity Testing in a Supply and Demand Framework." *Review of Economics and Statistics* 68:638–646.

Tintner, G. 1944. "A Note on the Derivation of Production Functions from Farm Records." *Econometrica* 12:26–34.

Tintner, G., and O.H. Brownlee. 1944. "Production Functions Derived from Farm Records." *Journal of Farm Economics* 26:566–571.

Trujillo-Barrera, A., P. Garcia, and M. Mallory. 2016. "Price Density Forecasts in the U.S. Hog Market: Composite Procedures." *American Journal of Agricultural Economics* 98:1529–1544.

Vercammen, J., and A. Schmitz. 2001. "Marketing and Distribution: Theory and Statistical Measurement." In B. Gardner and G. Rausser, eds., *Handbook of Agricultural Economics.* Amsterdam: North Holland, pp. 1137–1181.

Walras, L. 1954. *Elements of Pure Economics.* London: George Allen and Unwin.

Wang, Z., and D.A. Bessler. 2006. "Price and Quantity Endogeneity in Demand Analysis: Evidence from Directed Acyclic Graphs." *Agricultural Economics* 34:87–95.

Wright, P. 1928. *The Tariff on Animal and Vegetable Oils.* New York: Palgrave MacMillan.

Wright, S. 1921. "Correlation and Causation." *Journal of Agricultural Research* 220:557–585.

Wooldridge, J.M. 2009. *Introductory Econometrics: A Modern Approach.* Mason, OH: South-Western Cengage Learning.

Working, E.J. 1927. "What Do Statistical 'Demand Curves' Show?" *Quarterly Journal of Economics* 39:212–235.

Zapata, H.O., and A.N. Rambaldi. 1997. "Monte Carlo Evidence on Cointegration and Causation." *Oxford Bulletin of Economics and Statistics* 59:285–298.

27
New empirical models in consumer demand

Timothy J. Richards and Celine Bonnet

1 Introduction

Advances in available data, econometric methods, and computing power have created a revolution in demand modeling over the past two decades. Highly granular data on household choices means that we can model very specific decisions regarding purchase choices for differentiated products at the retail level. In this chapter, we review the recent methods in modeling consumer demand that have proven useful for problems in industrial organization and strategic marketing.

Analyzing problems in the agricultural and food industries requires demand models that are able to address heterogeneity in consumer choice in differentiated-product markets. Discrete choice models, for example, are particularly adept at handling problems that concern potentially dozens of choices as they reduce the dimensionality of product space into the smaller space occupied by product attributes. Discrete choice models, however, suffer from the independence of irrelevant alternatives (IIA) problem, so improvements on the basic logit model – the nested logit, mixed logit, and Bayesian versions of each – have been developed that are more relevant for consumer demand analysis and a wide range of applied problems. Yet, the fundamental assumption that consumers make discrete choices among products remained unsatisfying for a large class of problems.

Beyond the well-understood problems with the logit model, there are many settings in which choices are not exactly discrete. Families that purchase several brands of soda, for example, make multiple discrete choices, as do consumers who purchase Sugar Pops (cereal) for their children and granola for themselves. Consumers who purchase a certain cut of beef make a discrete choice among the several they face, but then make a continuous choice as to how much to purchase. Often, our interest lies more in the structure of the continuous part than in the discrete part. Consumers also reveal a demand for variety when their purchase cycle is a week, anticipating three meals per day for the next seven days when purchasing food. This demand for variety is often manifest in multiple discrete–continuous decisions, each with a continuous quantity (Bhat 2005, 2008). In this chapter, we describe the evolution of demand models that describe more types of purchases and more accurately capture purchases observed in "the real world."

There have been substantial advances in the application of "traditional" demand systems to the study of differentiated-products markets in recent years (see Baggio and Chavas 2009 for a recent example). However, our focus here is on settings in which the assumption that consumers allocate all of their income (or budget) among competing goods is fundamentally unsettling.

Developments in the spatial econometrics literature opened up an entirely new way of thinking of demand models (Pinkse, Slade, and Brett 2002; Slade 2004). When we think about the demand for differentiated products, our notion of differentiation is all about distance, whether in geographic, attribute, demographic, or even temporal space. The differences between products can be expressed in terms of each definition of space. Most importantly for applied problems, writing demand models in terms of the spatial distance between products can potentially reduce a high-dimension problem to one that is simpler and empirically tractable. We briefly review the spatial econometrics literature and the "distance metric" approach to demand estimation.

Finally, we address the frontier of demand analysis. Researchers working in "big data" have realized the power of machine learning methods to understand data patterns in largely atheoretic but incredibly powerful ways (Bajari et al. 2015; Belloni, Chernozhukov, and Hansen 2014; Varian 2014). Once limited to only forecasting and prediction, machine learning models have become increasingly important in econometric inference, again driven by the availability of massive datasets, both in terms of their depth (number of observations) and breadth (number of predictors).

We complete the chapter by suggesting some useful applications for new consumer demand models, such as empirical industrial organization, behavioral inference, and determining causality in treatment-effect models.

2 Models of discrete choice

When products are highly differentiated, the fundamental assumption of representative consumer models, that consumers buy a small proportion of each item in the dataset, falls apart. Rather, with access to data on a highly disaggregate set of products, say at the UPC-level among ready-to-eat cereals or yogurt, it is more accurate to describe the decision process as choosing only one brand from potentially dozens on offer. Building on the conceptual framework for discrete choices of items from Luce (1959), researchers in economics (McFadden 1974) and marketing (Guadagni and Little 1983) began to build a family of demand models that could describe purchases as discrete choices among differentiated items. Based on the assumption that preferences are randomly distributed among individuals, discrete choice models grew to become a standard approach to demand analysis due to their tractability and their ability to reduce high-dimension problems to relatively simple estimation routines. In this section, we describe the general model, the mixed logit, and other specific cases.

2.1 Models of demand

Variation in choice among consumers is driven by the assumption that tastes are randomly distributed over individuals. Consider a consumer h who faces a set of J alternatives. For each alternative $j \in J$, he obtains a certain level of utility U_{hj}. The consumer i chooses the alternative j that gives him the highest utility: $\forall k \neq j, U_{hj} > U_{hk}$. Some attributes of the alternative j and some characteristics of the consumer h are observed and some others are not. Therefore, the indirect utility of the consumer i for the alternative j can be decomposed into two components: $U_{hj} = V_{hj} + \epsilon_{ij}$, where V_{hj} is a function of observed characteristics and ϵ_{ij} is a random term that captures unobserved factors. Different models are derived from the specification of the distribution of this error term.

The general model, namely the mixed logit model or the random coefficient logit model, which approximates any random utility model representing discrete choices (McFadden and Train 2000), can be specified as follows:

$$U_{hj} = \alpha_h p_j + \sum_k \gamma_{hk} b_{kj} + \xi_j + \epsilon_{hj}, \qquad (1)$$

where p_j is the price of the alternative j, α_h is the marginal utility of income for the consumer h, b_{kj} are the kth observed attribute of the alternative j and γ_{hk} is the parameter associated with each observed variable that captures a consumer's tastes, ϵ_j represents the unobserved time invariant characteristics of the alternative j, and ϵ_{hj} is the error term that is assumed to follow a Type I Extreme Value distribution. The coefficients α_h and $\gamma_h = (\gamma_{h1}, \gamma_{h2} \ldots \gamma_{hk})$ vary over consumers with density $f(\alpha_h, \gamma_h)$ which, in most applications, is specified as normal or lognormal. The analyst then estimates the mean vector and covariance matrix of the random distribution.[1] The distribution $f(\alpha_h, \gamma_h)$ can also depend on observed characteristics of consumers (Bhat 2000), in which case the random coefficients are then specified as

$$\begin{pmatrix} \alpha_h \\ \gamma_h \end{pmatrix} = \begin{pmatrix} \alpha \\ \gamma \end{pmatrix} + \Pi D_h + \sum v_h, \qquad (2)$$

where α and γ are the mean marginal utility of income and the mean taste for characteristics, respectively, Π is a matrix of coefficients that measure taste for consumers according to their observed characteristics and D_h, Σ is a matrix of coefficients that represent the variance of each additional unobserved characteristic v_h and possible correlations between them.

Consumers choose the alternative that maximizes their utility. The individual probability of choosing the alternative j for the consumer h is given by

$$S_{hj} = P\left[U_{hj} > U_{hi}, \forall_i = 1, \ldots, J, i \neq j | b_j, p_j, \xi_j, D_h, v_h \right]$$

$$= \frac{\exp\left(\alpha p_j + b_j \gamma_h + \xi_j + [p_j, b_j](\Pi D_h + \sum v_h)\right)}{\sum_{l=1}^{j} \exp\left(\alpha p_l + b_l \gamma_h + \xi_l + [p_l, b_l](\Pi D_h + \sum v_h)\right)}$$

and the aggregated probability, which is the market share of the alternative j, is

$$s_j = \int S_{hj} f(v_h) dv_h,$$

assuming the alternatives cover the entire market of interest. We discuss alternatives for introducing an "outside option" to expand the definition of market share. Estimating market shares, however, is typically only relevant when used to estimate price elasticities.

2.2 Demand elasticities

Price elasticities for the mixed logit reflect very general and flexible patterns of substitution among products (McFadden and Train 2000) and take the following form for the demand of the alternative j with respect to the alternative k:

$$\eta_{jk} = \frac{\partial s_j}{\partial p_k} \frac{p_k}{s_j} = \begin{cases} \dfrac{p_j}{s_j} \int \alpha_h s_{hj}(1 - s_{hj}) f(v_h) dv_h & \text{if } j = k \\[6pt] -\dfrac{p_k}{s_j} \int \alpha_h s_{hj} s_{hk} f(v_h) dv_h & \text{if } j \neq k \end{cases} \qquad (3)$$

When the price of the alternative *j* varies, the probability of choosing the other alternatives varies according to their attributes and the ones of the alternative *j*. Introducing consumer characteristics and unobserved individual components takes into account the heterogeneity of consumers' preferences, which, in turn, creates the flexibility we desire. However, one drawback of this method is that it lacks a closed form, so simulation methods are required to estimate all parameters and obtain price elasticities. Although the mixed logit is the most general form, McFadden and Train (2000) show that all other forms of the logit model are, in fact, special cases of the mixed logit.

2.3 Particular cases

Two common discrete choice models can be derived from the mixed logit by imposing restrictions on the random variables describing consumer preferences: the simple logit and the nested logit. Constraining the random variables that describe unobserved heterogeneity permits closed form expressions for choice probabilities and for price elasticities.

The simple logit model differs from the mixed logit in that the parameters are assumed to be fixed: $\alpha_h = \alpha$ and $\beta_h = \beta$. With this assumption, the aggregated choice probabilities are then written as the logit expression:

$$S_j = \frac{\exp\left(\alpha p_j + \sum_l \gamma_l b_{lj} + \xi_j\right)}{\sum_{l=1}^{j} \exp\left(\alpha p_l + \sum_l \gamma_l b_{lj} + \xi_k\right)} \tag{4}$$

and the price elasticities become more tractable:

$$\eta_{jl} = \begin{cases} \alpha p_j (1 - s_j) \text{ if } j = l \\ -\alpha p_l s_l \text{ if } j \neq l \end{cases} \tag{5}$$

The cross-price elasticities of the alternative *j* with respect to the alternative *l* only depend on the alternative *l* whatever the alternative *j* considered. Therefore, when the price of an alternative changes, the share of each other alternative is affected in exactly the same way. Moreover, this simple model exhibits the IIA property referred to earlier, as the ratio of the probabilities of two alternatives *j* and *l* is independent from changes in the price of other alternatives. Although restrictive, the IIA property provides a very convenient form for the choice probabilities, which also explains its popularity.

Some additional assumptions on the distribution of the error term generate another closed-form expression for choice probabilities and offer more flexibility in substitution patterns than the simple logit. In particular, when the set of alternatives can be decomposed into several subsets, and alternatives within each subset are correlated in demand, the nested logit results. In the nested logit, the IIA property holds for alternatives belonging to the same group but does not hold for alternatives in different subsets. Assuming each alternative belongs to a group $g \in \{1,\ldots,G\}$, the number of alternatives within each group g is J_g and the error term can be written as $\epsilon_{ij} = \zeta_{ig} + (1 - \sigma_g) v_{ij}$. While v_{ij} follows a Type I Extreme Value distribution, ζ_{ig} is common to all alternatives of the group g and has a cumulative distribution function that depends on σ_g, with $\sigma_g \in [0,1]$. Importantly, the parameter σ_g measures the degree of correlation between alternatives within the group g. When σ_g tends toward 1, preferences for alternatives of the group g are perfectly correlated, meaning that the alternatives are perceived as perfect substitutes. When σ_g tends toward 0 for all $g = 1,\ldots,G$; the nested logit model is equivalent to the simple logit model.

In the nested logit model, the analytical expression for the choice probabilities is:

$$s_j = s_{j|g} s_g \text{ where } s_{j|g} = \frac{\exp\left(\frac{\alpha p_j + \Sigma_l \gamma_l b_{lj} + \xi_j}{1-\sigma_g}\right)}{\frac{\exp(I_g)}{1-\sigma_g}} \text{ and } s_g = \frac{\exp(I_g)}{\exp(I)} \quad (6)$$

$$I_g = (1-\sigma_g) \ln\left(\sum_{j \in J_g} \exp\left(\frac{\alpha p_j + \Sigma_l \gamma_l b_{lj} + \xi_j}{1-\sigma_g}\right)\right) \text{ and } I = \ln \sum_{g=1}^{G} \exp(I_g).$$

The nested logit model is more general and substitution patterns are more flexible than in the multinomial logit model. When the price of the alternative k belonging to the group g varies, the cross-price elasticities η_{jk} are not identical, whether j belongs to the same group or not. The price elasticities of the demand of the alternative j with respect to the alternative k is

$$\eta_{jk} = \begin{cases} \frac{-\alpha p_j}{1-\sigma_g} p_j \left(1-(1-\sigma_g)s_j - \sigma_g s_{j|g}\right) \text{ if } j=k \text{ and } j,k \in g \\ \frac{\alpha p_j}{1-\sigma_g} p_j \left((1-\sigma_g)s_j - \sigma_g s_{j|g}\right) \text{ if } j \neq k \text{ and } j,k \in g \\ -\alpha p_k s_k \text{ if } j \neq k \text{ and } j \in g, \ k \in h \end{cases} \quad (7)$$

In this discussion, we considered the simple case of the nested logit models with two nests. In some situations, three or more nests may be appropriate, where the probability expression is a relatively straightforward generalization of the two-nest case. Goldberg (1995) considers a five-nest case, while Brenkers and Verboven (2006) use three levels. In general, the parameters of the multinomial and nested logit models can be estimated by maximum likelihood, whereas mixed logit models require the use of the simulated maximum likelihood (Train 2003). Most econometric software (Stata, Nlogit, R) contains algorithms that allow for the relatively simple, efficient estimation of any logit variant, whether with random coefficients or not.

3 Models of discrete–continuous choice

For many products – consumer non-durables such as food and beverages or environmental amenities such as parks or fisheries – the choice process is more appropriately described as discrete–continuous rather than either purely discrete or entirely continuous. There are many classes of goods for which people do not purchase a single-item, but rather a variable amount or variable weight of a specific product. For example, meat, fresh produce, or even bottled water can all be described as discrete–continuous. In this section, we consider two modeling approaches and develop one in more detail.

3.1 Discrete–continuous choice models

Why might choices be made in a discrete–continuous way, and what does this imply about the nature of the underlying utility functions? Discrete–continuous choices are typically characteristic of the product-class. Durable goods with either variable amounts of usage or inputs, non-durable goods that are purchased in varying quantities, or many services are good examples. The consumer may purchase one alternative out of many in the consideration set, but then purchase an amount that varies continuously, and in a way that differs from other consumers in the dataset.

Many choices involve a discrete and then a continuous choice in the same purchase that invalidates the underlying econometric assumptions of our traditional demand model: the choice of a brand or variety and the volume to buy is the most obvious in a food-demand context (Chintagunta 1993). In each case, the relevant data contains a large number of zeros for the alternatives that were not purchased, and continuous purchase amounts for those that were. There are two ways of dealing with this issue econometrically: (1) creating an *ad hoc* econometric model that accounts for the selection bias created by the discrete choice process within a continuous modeling framework, or (2) estimating a model of discrete–continuous choice that is grounded in a single, unifying utility-maximization framework (Wales and Woodland 1983).

The early models of Heckman (1979) and Lee, Maddala, and Trost (1980) are of the first form, based indirectly in the theory of utility-maximization theory, but dealing with the econometric issues associated with a censored dependent variable in a statistically correct way. That is, if the dependent variable is inherently "zero-positive," then there are clear statistical problems with applying standard ordinary least squares estimation methods. The most common method of estimating these models relies on the Heckman two-stage approach in which a probit model is applied to the buy/no-buy problem in the first stage, and then the inverse Mill's Ratio is used as a regressor in the second stage ordinary least squares (OLS) regression to correct for the sample selection bias. If there is a way to describe the data generating process directly, however, it will nearly always be preferred.

Corner solutions to utility-maximization problems can be modeled in an empirically tractable way. Hanemann (1984) describes an approach based on the Kuhn–Tucker conditions for utility maximization that formally introduces the empirical restrictions on a model of discrete–continuous choice implied by the theory. His approach is similar to that of Wales and Woodland (1983), who describe a method of estimating demand systems in the presence of corner solutions, or zeros in the dependent variable. Hanemann's (1984) model describes a particular setting in which only one choice is made and a continuous amount is purchased. Although this may be a simplification of many choice environments, his approach represented a substantial advance in structural demand modeling.

The intuition behind the approach is as follows: Assume a perfect-substitutes world in which the good with the lowest price-per-unit of quality is purchased (Deaton and Muellbauer 1980). The choice of a particular good is determined in a random utility framework, so it is governed by the distribution of the unobserved heterogeneity that drives the specification for perceived quality. Conditional on this choice, therefore, the expected purchase quantity is found by solving for the implied demand from a known indirect utility function and applying a change of variables from the random unobserved heterogeneity term to the quantity-demand term. The result is an expected expenditure amount that is a parametric function of the arguments of the implicit quality function of the good in question.

Formally, the model consists of two stages, with the first describing the discrete choice of goods and the second the distribution of the continuous volume purchased. The direct utility function for the perfect substitutes model is given by the general class of utility function written as:

$$u = (x_1, x_2, \psi_1, \psi_2, z) = u = (x_1\psi_1, x_2\psi_2, z), \tag{8}$$

for two goods, where x_j is the quantity of good j, ψ_j is the quality of good j, and z is all other goods such that income, y, is exhausted. This is a perfect substitutes model because maximizing utility subject to an income constraint implies that only one good is purchased, the one with the lowest ratio of price-to-quality, or $\frac{p_j}{\psi_j}$. Clearly, the specification for perceived, or expected, quality is key to the model, because it determines which good is purchased. Hanemann (1984) describes both a linear and multiplicative quality function, but we focus on the multiplicative function with attributes for good j given by b_j so that the quality function is written as $\psi_j(b_j, \varepsilon_j) = \exp\left(\alpha_j + \sum_k \gamma_k b_{jk} + \varepsilon_j\right)$, where ε_j is the random component of quality that is assumed to be independent and identically distributed (iid) extreme-value distributed in the model development to follow. By parameterizing the quality function this way, the choice probabilities are given by

$$\pi_j = \Pr\left[\varepsilon_j + \alpha_j + \sum_k \gamma_k b_{jk} - \ln p_j \geqslant \varepsilon_i + \alpha_i + \sum_k \gamma_k b_{ik} - \ln p_i\right], \forall i, \tag{9}$$

so with the extreme-value assumption, the probability of choosing item j becomes

$$\pi_j = \frac{\exp\left(\alpha_j + \sum_k \gamma_k b_{jk} - \frac{1}{\mu \ln p_j}\right)}{\sum_i \exp\left(\alpha_i + \sum_k \gamma_k b_{ik} - \frac{1}{\mu \ln p_i}\right)} \tag{10}$$

where μ is the logit scale parameter.

From the logit choice probability, we then find the distribution of demand for the commodity by applying a change of variable technique based on the conditional demand function for x_j. Assuming an indirect utility function from a simple bivariate utility model:

$$v(p_j, y) = \left(\frac{\theta}{(\rho-1)}\right) p_j^{1-\rho} - \frac{\exp(-\eta y)}{\eta}, > 0,$$

then the conditional demand function is found by applying Roy's theorem to find: $x_j(p_j, \psi_j, y) = \theta p_j^{-\rho} \psi_j^{\rho-1} \exp(\eta y)$, and, after substituting the expression for quality:

$$x_j(p_j, \psi_j, y) = \theta p_j^{-\rho} \exp(\eta y) \exp\left((\rho-1)\lambda_j\right) \exp\left((\rho-1)\varepsilon_j\right), \tag{11}$$

where $\lambda_j = \alpha_j + \sum_k \gamma_k b_{jk} - \ln p_j$, or the mean quality function less prices. We then apply a change of variables from ε_j to x_j and take the mean of the resulting conditional distribution to find

$$E\left[\ln p_j x_j | \varepsilon_j + \lambda_j > \varepsilon_i + \lambda_i\right] = \ln \theta + \eta y + (\rho-1)\left[\mu \ln\left(\sum_i \exp\left(\frac{\lambda_i}{\mu}\right)\right) + 0.5722\mu\right], \tag{12}$$

where 0.5572 comes from the expectation of an extreme value (EV) random variable.

The expected demand function can then be estimated using maximum likelihood or the two-stage estimator described in Hanemann (1984). The two-stage estimator involves estimating the values of $\frac{\lambda_i}{\mu}$ from a logit maximum likelihood routine, and then using the estimated values in the demand equation to estimate the remaining parameters with OLS. Although this was the recommended approach in 1984, it is more efficient to estimate everything together with maximum likelihood estimation (MLE). Because this discrete–continuous specification is derived from a single utility maximization problem, the choice to purchase and how much to purchase are internally consistent, but the primary drawback is that price elasticities are restricted to -1. Despite this fact, the model is relatively flexible, as Hanemann (1984) describes several other utility specifications that will work in this framework. Applications of the multiple–discrete model to estimating food demand include Chintagunta (1993) and Richards (2000).

4 Models of multiple–discrete and multiple–discrete continuous choice

In the last two decades, researchers in transportation (Bhat 2005, 2008), marketing (Dube 2004; Hendel 1999), and environmental economics (Phaneuf, Kling, and Herriges 2000) recognized that individuals in many settings not only make discrete–continuous choices, but often make multiple discrete choices, such as choosing several brands of soda on each trip to the store, or more than one variety of apple. In this section, we describe three models that are able to address the (1) multiple–discrete, (2) multiple–discrete continuous, and (3) multiple–discrete continuous with complementarity issues in flexible, tractable ways. We introduce the intuition underlying the first two specifications and develop the latter, most general model more formally.

4.1 Models of multiple–discrete choice

Like the models of discrete–continuous choice developed previously, models of multiple–discrete–continuous (MDC) choice are also grounded in the theory of utility-maximization. However, they tend to be comprised of sub-problems, each describing a different part of the decision process, that are solved together in one utility-maximization framework. MDC models written this way explain an important observation in quantitative marketing, namely if consumers make multiple, discrete purchases on each trip to the store, then there is a revealed demand for variety. For example, if consumers buy both Diet Coke and Coke on a trip to the store, they are clearly either anticipating a change in tastes from consumption occasion to consumption occasion (between purchase occasions) or are buying for others in the family. The structure of the model that accounts for this demand for variety is based on the general theme of identifying corner solutions from a single utility-maximization problem (Dube 2004).

The utility maximization process assumes consumers have a number of consumption occasions between purchases. The total utility from a purchase occasion therefore sums over the sub-utility functions that describe the utility from each consumption occasion. Consumers maximize the utility from a purchase occasion and not a consumption occasion. Therefore, the expected quantity purchased at each visit to the store is composed of the distribution of demands for each consumption occasion and the distribution that governs the number of consumption occasions (a count-distribution). The three components to the demand model are (1) the count-data model that governs the number of consumption occasions, (2) the sub-utility function that determines what is consumed on each consumption occasion, and (3) the total utility maximization process at each purchase occasion. Consumers are assumed to maximize utility subject to a budget constraint, and the Kuhn–Tucker conditions are used to derive estimable demand models for each purchase occasion. MDC models are able to produce elasticities that appear reasonable and have proven useful in applied industrial organization models, where accurately conditioning for consumer demand is critical.

4.2 Models of multiple–discrete continuous choice

The MDC model described above, however, assumes that each of the purchases is still only discrete and that consumers either purchase a constant amount or the discrete purchases themselves are for different quantities. In this section, we describe a model that synthesizes the corner-solution approach developed in Section 2, with the multiple discrete logic outlined above. Originally applied to problems in transportation (Bhat 2005, 2008), where individuals often choose multiple modes of transportation and use each for varying distances or amounts of time, the application to food demand is fairly obvious. Namely, for many categories of products, consumers purchase many different brands or varieties in the same category and purchase a continuous amount of each. For example, Richards, Gomez, and Pofahl (2012) describe a problem in the demand for fresh produce. Items within each sub-category (e.g., apples) are purchased by the variety, but the amounts are typically measured in pounds. A substantial proportion of the consumers in that data reported purchasing multiple varieties on each purchase occasion, whether due to varying tastes within the household or a desire to not have to eat the same kind of apple time after time.

As in the MDC case, the underlying model is consistent with utility maximization, and the Kuhn–Tucker conditions for constrained utility maximization are used to derive the demand model. Unlike the MDC model, however, the multiple–discrete continuous model of Kim, Allenby, and Rossi (2002) and Bhat (2005, 2008) generates demand equations that describe the joint probability distribution for continuous quantities of a discrete set of items chosen from a larger consideration set. By including the utility from a numeraire good that is always consumed, demands for each of the other "inside goods" are derived using an equilibrium argument: the utility from a good that is purchased must be at least as great as the utility from the always-consumed numeraire good. Assuming consumers make random errors in utility maximization and that these errors are Type I Extreme Value distributed, the resulting system of purchase probabilities is derived. Remarkably, the demand equations nest a simple logit model when only one item is purchased. Bhat (2005, 2008) shows how unobserved heterogeneity can be accommodated by allowing for random parameters in the usual way. Typically, the resulting multiple–discrete continuous extreme value (MDCEV) model is estimated using simulated maximum likelihood.

4.3 Generalized model of multiple–discrete continuous choice

The MDCEV model described previously has become a common method of estimating multiple–discrete continuous demand models. However, this class of model still retains the critical weakness that all products are restricted to be substitutes. When the problem involves items in multiple categories (e.g., milk, bread, and cereal), then any reasonable model would need to accommodate the possibility that some items may be complements. Vasquez-Lavin and Hanemann (2008) derive generalized versions of the described multiple–discrete continuous model that do just that. Pairs of items can be complements, depending on the sign of an interaction parameter. Formally, the utility function for this generalized multiple–discrete continuous extreme value (GMDCEV) model is written as:

$$u_j^h\left(q_{ij}^h, \Omega\right) = \frac{1}{\alpha_1}\left(q_{1j}^h\right)^{\alpha_1} \phi_{1j}^h + \sum_{i=2}^{I}\left[\frac{\gamma_i}{\alpha_i}((\frac{q_{ij}^h}{\gamma_i}+1)^{\alpha_i} - 1)(\phi_{ij}^h + 1/2 \sum_{i \neq k, i \neq 1}^{K} \theta_{ik} \frac{\gamma_k}{\alpha_k})((\frac{q_{kj}^h}{\gamma_k}+1)^{\alpha_k} - 1))\right] \quad (13)$$

$j = 1, 2, \ldots, J; h = 1, 2, \ldots, H$, where q_{ij}^h is the amount of good i purchased by household h on occasion j, Ω is a vector of parameters to be estimated,

$$\pi_{ij}^h = \Phi_{ij}^h + 1/2 \sum_{i \neq k, i \neq 1}^{K} \theta_{ik} \frac{\gamma_k}{\alpha_k} ((\frac{q_{kj}^h}{\gamma_k} + 1)^{\alpha_k} - 1) \tag{14}$$

is the baseline marginal utility for good i on occasion j by household $h\left(\pi_{ij}^h > 0\right)$, α_i are parameters that reflect the curvature of the utility function ($-\infty < \alpha_i \leq 1$), and γ_i is the product-specific utility translation parameter $\gamma_i > 0$. Note that because $\gamma_1 = 0$ by assumption, the numeraire good is not subject to satiation effects.

The parameters α_i and γ_i are largely what separate the MDCEV (GMDCEV in our case) model from others in the class of discrete, multiple–discrete, or discrete–continuous models (Richards, Gomez, and Pofahl 2012). In mathematical terms, γ_i is a translation parameter that determines where the indifference curve between q_{1j} and q_{2j} becomes asymptotic to the q_{1j} or q_{2j} axis, and thereby where the indifference curve intersects the axes. The parameter α_i, on the other hand, determines how the marginal utility of good i changes as q_{ij} rises. If $\alpha_i = 1$, then the marginal utility of i is constant, indifference curves are linear, and the consumer allocates all income to the good with the lowest quality-adjusted price (Deaton and Muellbauer 1980). As the value of α_i falls, satiation rises, the utility function in good i becomes more concave, and satiation occurs at a lower value of q_{ij}: importantly, if the values of Π_{ij}^h are approximately equal across all varieties, and if the individual has relatively low values of α_i, then that individual can be described as "variety seeking" and purchase some of all choices, while the opposite will be the case if α_i values are relatively high (close to 1.0) and the perceived qualities differ (Bhat 2005).

The GMDCEV incorporates additive separability in a form suggested by Vasquez-Lavin and Hanemann (2008) in that utility is quadratic in quantities. Both complementary ($\theta_{ik} > 0$) and substitute ($\theta_{ik} < 0$) relationships are permitted between pairs of products, so the GMDCEV represents a very general corner-solution model. Bhat, Castro, and Pinjari (2015) show that allowing unrestricted own-quadratic effects can lead to negative values for baseline utility, so restrict $\theta_{ii} = 0$. This restriction makes sense as the data are not likely to identify additional nonlinearities with respect to own-price effects but should reflect interactions between products that are not part of any additively separable demand system. Each of the constant terms, or ϕ_{ij}^h parameters, can also be written as functions of demographic or marketing mix variables to address concerns regarding the importance of observed heterogeneity.

As with the MDC and MDCEV models, the Kuhn–Tucker approach is used to solve for a discrete/continuous demand system implied by the utility function described above. By solving the Kuhn–Tucker conditions for the constrained utility maximization problem, the GMDCEV demand functions consist of a mixture of corner and interior solutions that are a product of the underlying utility structure and are not simply imposed during econometric estimation.

Assuming the optimization procedure is solved only up to a random error ϵ_{ij}^h, and assuming the errors are distributed Type I Extreme Value, the econometric model assumes a particularly straightforward form as the utility-maximizing solution collapses to

$$P\left(q_{1j}^h, q_{2j}^h, \ldots, q_{Mj}^h, 0, 0, \ldots, 0\right) = \frac{1}{\mu^{M-1}} |J| \left(\prod_{K=1}^{M} e^{\frac{V_{kj}^h}{\mu}} \right) \left(\sum_{i=1}^{I} e^{\frac{V_{ij}^h}{\mu}} \right)^{-M} (M-1)!, \tag{15}$$

where $|J|$ is the Jacobian of the transformation from the errors to the demand quantities, $V_{kj}^{h}\left(p_{1},p_{2},\ldots,p_{I},y^{h}\right)$ is the indirect utility function implied by the choice model above, and M varieties are chosen out of I available choices. In this estimating equation, μ is the logit scale parameter. In fact, when $M = 1$, or only one alternative is purchased, and there are no cross-category effects, the GMDCEV model becomes a simple logit. Although appearing complicated, the GMDCEV model is estimated in a straightforward way using MLE, or in a random-coefficient variant using the simulated maximum likelihood (SML) approach described in the final section below.

4.4 Shopping-basket models

Consumers typically purchase many items together, from dozens of categories and many brands. Typically, empirical models of consumer demand focus only on one category at a time, ignoring potentially important interactions with items in other categories. If consumers purchase groceries by the shopping basket, and not just one item at a time, then it is reasonable to estimate models that take into account the demand for many categories, and the potential for complementarity, on each shopping occasion (Ainslie and Rossi 1998; Kwak, Duvvuri, and Russell 2015; Manchanda, Ansari, and Gupta 1999; Russell and Petersen 2000). Complementarity matters because retailers set prices as if consumers purchase items together in the same shopping basket (Smith 2004). In this section, we present an alternative way of modeling consumers' shopping-basket choice process: the multivariate logit (MVL) model.

Like the GMDCEV model, the MVL model is derived from a single utility-maximization process. Unlike the GMDCEV, however, the choice process is based on the random-utility assumption. We begin by describing the nature of the MVL utility function and how it is used to describe consumers' choices among several items that may appear together in their shopping basket. Consumers $h = 1,2,3,\ldots,H$ in the MVL model select items from among $i = 1,2,3,\ldots,N$ categories, c_{iht}^{r}; in assembling a shopping basket, $\mathbf{b}_{ht}^{r} = (c_{1ht}^{r}, c_{2ht}^{r}, c_{3ht}^{r},\ldots,c_{Nht}^{r})$ on each trip, t, conditional on their choice of store, r. Define the set of all possible baskets in r as $b_{ht}^{r} \in \mathbf{B}^{r}$. Our focus is on purchase incidence, which is the probability of choosing an item from a particular category on a given shopping occasion, and we model demand at the category level by assuming consumers purchase one item per category across multiple categories.

Consumers choose among categories to maximize utility, U_{ht}^{r}, and we follow Song and Chintagunta (2006) in writing utility in terms of a discrete, second-order Taylor series approximation:

$$U_{ht}^{r}\left(b_{ht}^{r}\mid r\right)=V_{ht}^{r}\left(b_{ht}^{r}\mid r\right)+\varepsilon_{ht}^{r} \quad (16)$$
$$=\sum_{i=1}^{N}\pi_{iht}^{r}c_{iht}^{r}+\sum_{i=1}^{N}\sum_{j\neq 1}^{N}\theta_{ijh}^{r}c_{iht}^{r}c_{jht}^{r}+\varepsilon_{ht}^{r},$$

where π_{iht}^{r} is the baseline utility for category i earned by household h on shopping trip t in store r; c_{iht}^{r} is a discrete indicator that equals 1 when category i is purchased in store r, and 0 otherwise; ε_{ht}^{r} is a Gumbel-distributed error term that is iid across households and shopping trips; and θ_{ijh}^{r} is a household-specific parameter that captures the degree of interdependence in demand between categories i and j in store r: specifically, $\theta_{ijh}^{r} < 0$ if the categories are substitutes, $\theta_{ijh}^{r} > 0$ if the categories are complements, and $\theta_{ijh}^{r} = 0$ if the categories are independent in demand. To ensure identification, we restrict all $\theta_{ii}^{r} = 0$ and impose symmetry on the matrix of cross-purchase effects, $\theta_{ijh}^{r} = \theta_{jih}^{r}\;\forall i,j \in r$ (Besag 1974; Cressie 1993; Russell and Petersen 2000).

The probability that a household purchases a product from a given category on a given shopping occasion depends on both perceived need and marketing activities from the brands in the category (Russell and Petersen 2000). Therefore, the baseline utility for each category depends on a set of category ($\mathbf{X}_i$) and household ($\mathbf{Z}_h$) specific factors such that: $\pi'_{iht} = \alpha'_{ih} + \beta'_{ih}\mathbf{X}'_i + \gamma'_{ih}\mathbf{Z}_h$, where perceived need, in turn, is affected by the rate at which a household consumes products in the category, the frequency that they tend to purchase in the category, and any other household demographic measures. Category factors include marketing mix elements, such as prices, promotion, or featuring-activities. As with any other demand model, unobserved heterogeneity can be included by allowing any of these parameters to be randomly distributed over households.

With the error assumption in Equation (16), the conditional probability of purchasing in each category assumes a relatively simple logit form. Following Kwak, Duvvuri, and Russell (2015), we simplify the expression for the conditional incidence probability by writing the cross-category purchase effect in matrix form, suppressing the store index on the individual elements, where $\Theta'_h = [\Theta_{1h}, \Theta_{2h}, \ldots, \Theta_{Nh}]$ and each Θ_{ih} represents a column vector of a $N \times N$ cross-effect Θ'_h matrix with elements Θ'_{ijh}. With this matrix, the conditional utility of purchasing in category i is written as

$$U'_{ht}(c'_{iht}|c'_{jht}) = \pi'^{'}_{ht}b'_{ht} + \Theta'_{ih}b'_{ht} + \varepsilon_{ht} \tag{17}$$

for the items in the basket vector b'_{ht}. Conditional utility functions of this type potentially convey important information and are more empirically tractable than the full probability distribution of all potential assortments (Moon and Russell 2008); however, they are limited in that they cannot describe the entire matrix of substitute relationships in a consistent way and are not econometrically efficient in that they fail to exploit the cross-equation relationships implied by the utility maximization problem. Estimating all N of these equations together in a system is one option, but Besag (1974) describes how the full distribution of b'_{ht} choices are estimated together.

Assuming the Θ'_h matrix is fully symmetric and the main diagonal consists entirely of zeros, then Besag (1974) shows that the probability of choosing the entire vector b'_{ht} is written as:

$$\Pr(b'_{ht}|r) = \frac{\exp\left(\pi'^{'}_{ht}b'_{ht} + \frac{1}{2}b'^{'}_{ht}\Theta'_h b'_{ht}\right)}{\sum_{b'_{ht} \in B'}\left[\exp\left(\pi'^{'}_{ht}b'_{ht} + \frac{1}{2}b'^{'}_{ht}\Theta'_h b'_{ht}\right)\right]}, \tag{18}$$

where $\Pr(b'_{ht}|r)$ is interpreted as the joint probability of choosing the observed combination of categories from among the 2^N potentially available from N categories, still conditional on the choice of store r. Assuming the elements of the main diagonal of Θ is necessary for identification, while the symmetry assumption is required to ensure that (18) truly represents a joint distribution, a multivariate logistic (MVL) (Cox 1972) distribution of the category-purchase events. Essentially, the model in (18) represents the probability of observing the simultaneous occurrence of N discrete events in a shopping basket at one point in time.

Given the similarity of the choice probabilities to logit-choice probabilities, the elasticities are similar to those shown above for the logit model, but recognizing the fact that cross-price elasticities for items within the same basket will differ from those in different baskets (Kwak, Duvvuri, and Russell 2015). In the absence of unobserved heterogeneity, the MVL model is estimated using maximum likelihood in a relatively standard way, but when random parameters are used, the model is estimated using the SML method described below.

The MVL is powerful in its ability to estimate both substitute and complementary relationships in a relatively parsimonious way but suffers from the curse of dimensionality. That is, with N products, the number of baskets is N^2-1, so the problem quickly becomes intractable for anything more than a highly stylized description of the typical shopping basket.

5 Spatial econometrics and the distance metric model

There is a rich history of modeling the demand for differentiated products solely in terms of their attributes (Lancaster 1966). In fact, the mixed logit model relies on attribute variation among items in a category of products to identify differences in price elasticities, and to project the demand for differentiated products from a high-dimensional product space to a lower-dimensional attribute space. It is both convenient and intuitive to think of products not necessarily in terms of their brand or variety, but in terms of the attributes that comprise them. Slade (2004) exploits attribute variation among a large number of beer brands in developing the "distance metric" (DM) demand model as an alternative means of overcoming the curse of dimensionality in differentiated-products demand analysis and avoiding the IIA problem associated with logit models. In this section, we briefly review the power of spatial econometrics more generally and show how the DM model represents a fundamentally different way of estimating demand.

5.1 Spatial econometrics and demand estimation

In this model, attribute-variation is another way of circumventing the IIA characteristic of logit-based demand systems. Because the distance between products in attribute space is a primitive of the consumer choice process, the matrix of substitution elasticities is completely flexible, unlike a simple logit. Slade (2004) applies a similar notion of product differentiation to the discrete choice model by assuming the price-coefficient to be a function of attributes; however, a disadvantage of this approach is that a consumer's price-response in a discrete-choice model of demand is determined by the marginal utility of income, which is a characteristic of the individual that cannot logically vary over choices. Rather, the DM model described here includes attribute-distance as a direct argument of the utility function.

The DM approach to demand estimation is similar to the address model of Anderson, de Palma, and Thisse (1992) and Feenstra and Levinsohn (1995), in that the utility from each choice depends upon the distance between the attributes contained in that choice and the consumer's ideal set of product attributes, where the ideal product reduces to the product chosen by a representative consumer. The DM models account for the utility-loss associated with distance by introducing a spatial autoregression parameter to measure the extent to which differentiation from other products raises (or lowers) the utility from choosing product j according to the relative distances between products and the ideal attribute mix of a given consumer.

The distance metric–multinomial logit (DM–MNL) uses a nonlinear utility-loss function, where mean utility from product j falls (or rises) in the distance from all other products, measured by the distance matrix $\mathbf{W}$. Each element of $\mathbf{W}$ measures the Euclidean distance between each pair of products, so the element w_{jl} measures the distance between product j and product l in a multi-attribute space. The importance of differentiation is estimated through a spatial-autoregressive parameter. Formally, mean utility for product $j = 1, 2, \ldots, J$ in week $t = 1, 2, \ldots, T$, and is written in vector notation (with bold notation indicating a vector), as

$$\boldsymbol{\delta} = \boldsymbol{\beta}'\mathbf{x} + \lambda \mathbf{W}\boldsymbol{\delta} - \alpha \mathbf{p} + \boldsymbol{\xi}, \tag{19}$$

where $\boldsymbol{\delta}$ is a $JT \times 1$ vector of mean utility, $\mathbf{x}$ is a $JT \times K$ matrix of demand shifters, $\mathbf{p}$ is a $JT \times 1$ vector of prices, and $\boldsymbol{\xi}$ is a random error unobserved by the econometrician. The vector $\boldsymbol{\beta}$ and scalar parameters λ and α are all estimated from the data. The matrix $\mathbf{W\delta}$ measures the effect of product differentiation on utility according to attribute distance, which defines the λ as a spatial autoregression parameter (Anselin 2002).

As a spatial autoregression parameter, λ is interpreted as the extent to which utility is affected, positively or negatively, by the distance between the chosen product and all other products in the choice set. Autoregression reflects the notion that consumers evaluate the utility attainable from each product relative to the utility that can be attained from consuming other available products in the choice set. By convention, $\mathbf{W}$ is defined as a measure of inverse-distance, or proximity, so that greater product differentiation in the product category reduces utility when $\lambda > 0$ (i.e., utility rises with attribute proximity) and increases utility when $\lambda < 0$.

Solving Equation (19) for mean utility gives: $\boldsymbol{\delta} = (\mathbf{I} - \lambda \mathbf{W})(-1)(\boldsymbol{\beta}'\mathbf{x} - \alpha \mathbf{p} + \boldsymbol{\xi})$, where $(\mathbf{I} - \lambda \mathbf{W})^{-1}$ is the Leontief inverse, or spatial multiplier matrix (Anselin 2002). In spatial models, the concept of the multiplier is critical and powerful because it measures how changes to one observation ripple throughout the entire system. For example, if one price changes exogenously, the demand for all other products changes according to the spatial multiplier matrix. In a context where $\mathbf{W}$ measures the distance between individuals consuming the product in a social network, the multiplier measures how strong the peer- or bandwagon-effects for the product are.

Assuming utility varies among consumers in a random way, utility is written as $u_i = \boldsymbol{\delta} + \varepsilon_i$, where ε_i is an iid random error that accounts for unobserved consumer heterogeneity. Further assuming ε_{ij} is Type I Extreme Value distributed, and aggregating over consumers, the DM–MNL model yields a market share expression for item j given by $S_j = \exp(\delta_j) / \left(1 + \sum_{l=1}^{J} \exp(\delta_l)\right)$, where S_j is the volume-share of product j, which can be linearized using the approach in Berry (1994) and Cardell (1997) and estimated using MLE. However, because the $\mathbf{W}$ matrix must be inverted during estimation, an MLE routine may encounter computational issues. Kelejian and Prucha (1999) describe a generalized method of moments (GMM) routine that avoids these issues and accounts for the likely endogeneity of prices or any other marketing mix elements for that matter.

There are many other ways of applying the DM concept to demand modeling. The MNL model above is similar to Slade (2004) and Pinkse and Slade (2004), in that we explicitly incorporate a distance-metric component in the demand model; however, attribute distance enters in a structural way in Equation (19) through the utility function. Rojas and Peterson (2008) and Pofahl and Richards (2009) describe two other approaches using more traditional demand systems. The point is that including attribute space through the DM logic is very general – projecting demand into attribute space or even social space (Richards, Hamilton, and Allender 2014) not only reduces the dimensionality problem associated with differentiated-products analysis, but adds flexibility and the ability to study a wider range of applied problems.

6 Machine learning

Advances in computing power and in the creation of huge datasets generated by virtually any web-based activity have renewed interest in "big data" methods for analyzing consumer-demand problems (Bajari et al. 2015; Varian 2014). While the definition of what exactly constitutes big data remains elusive, it has come to be associated instead with a set of analytical methods rather than attributes of the data itself. When presented with virtually unlimited numbers of observations and possibly thousands of explanatory variables, researchers have turned to machine learning (ML) methods instead of traditional econometric techniques. Using ML methods to analyze demand data, however, is fundamentally different from any of the frameworks discussed above, in that the outputs are different and the objectives of the analysis differ accordingly.

6.1 Studying demand data with ML methods

ML, or statistical learning more generally, is typically used as a prediction tool. In fact, models are evaluated on the basis of their ability to fit out-of-sample instead of on some sort of in-sample metric, as is usually done in econometrics. The model that is able to produce the lowest root mean squared error (RMSE) on a cross-validation sample of the data is the winner. That said, recent advances in the literature on machine learning investigate how big data models can be used to study causal inference (Athey and Imbens 2015) or to generate marginal effects similar to econometric models of demand (Bajari et al. 2015; Varian 2014). In this section, we review six machine learning techniques and how they can be applied to demand data. Our discussion draws heavily on James et al. (2014), which is a valuable and standard reference in this area.

Many of the methods are actually variants on standard econometric approaches, using the concept of least squares in different ways to estimate large models. At the risk of oversimplification, these methods can be classified into either *regularization* approaches or *tree-based* methods. Regularization involves reducing a regression problem to a smaller one by restricting some coefficients that are close to zero, exactly to zero, focusing on the nonzero estimates. Tree-based methods, on the other hand, seek to order predictor variables according to their importance and determine critical breaks in regions of statistical support. In this section, we consider three of the former (forward stepwise regression, lasso, and support vector machines) and three of the latter (bagging, random forests, and boosting). We also provide a brief discussion of cross-validation as a method for model selection.

6.2 Regularization and penalized regression

Analysts in the ML literature generally have no qualms with using *forward stepwise regression* as a method for selecting the best linear model. Forward stepwise regression begins by estimating a null model and then adding variables in succession and choosing the predictor at each step that produces the lowest cross-validated prediction error, Akaike information criterion (AIC), Bayesian information criterion (BIC), or adjusted R^2. While econometricians may have conceptual issues with the data mining aspect of forward stepwise regression, that is the point of machine learning. With large datasets of very high dimension, forward stepwise regression is often a very pragmatic and effective tool for model selection, particularly given the power of cross-validation when the size of the dataset permits holding out a large number of observations for training purposes.

A second class of models is known as shrinkage, penalized regression, or regularization methods. Regularization means that the coefficients on some predictors are reduced to zero in estimation if their statistical effect is, for all practical purposes, zero. They are referred to as shrinkage methods because they effectively shrink the size of the predictor set according to the number of zero coefficients that are assigned. Principal among these methods is the lasso, which minimizes an objective function that includes a penalty for many, large regression coefficients:

$$LASSO = \min_{\beta} \sum_{i=1}^{n} \left(y_i - \beta_0 - \sum_{j=1}^{p} \beta_j x_{ij} \right)^2 + \lambda \sum_{j=1}^{p} |\beta_j|, \qquad (20)$$

where λ is referred to as a "tuning parameter" that controls the extent to which the choice of parameters is constrained by the penalty. When $\lambda = 0$, lasso estimates are clearly equal to the least squares estimates, and when λ is sufficiently large, all parameter estimates will be reduced to zero. Lasso estimates are particularly valuable in settings where p is large relative to n – that is, in high-dimensional datasets with relatively few observations. In this case, the approach has the effect of shrinking parameter estimates for non-important variables to zero, effectively becoming a means of selecting variables based on their values as predictors. With sufficient data, cross-validation methods over a grid that includes a wide range of possible parameter values is used to determine the value of λ that minimizes out-of-sample forecast error. As a shrinkage model, lasso is similar to ridge regression, but the latter, which uses a quadratic rather than absolute-value penalty, never reduces any coefficient estimates exactly to zero, but only shrinks them toward zero. If the problem is dimensionality, ruling some variables out is important.

Support vector machines (SVM) are designed for classification (i.e., assigning observations in the dataset to binary classes). They are unique in that they rely on the notion of a *maximal margin classifier* (MMC), which is an algorithm that chooses the parameters of a separating hyperplane, familiar to economists as the core construct in duality theory, in order to maximize the minimum distance between the hyperplane and data observations. However, the base MMC method suffers from the fact that the data are often not sufficiently well behaved to identify a unique hyperplane that cleanly separates all the observations into one class or another. That is, the MMC solution does not exist.

Consequently, the SVM approach is based on a *support vector classifier* (SVC) method that allows for some observations to lie on the wrong side of the margin, or even on the wrong side of the hyperplane. In the SVC optimization routine, however, only observations that either lie on or on the wrong side of the margin enter into the calculation, as the objective function values for the others are very small. Therefore, these vectors are known as *support vectors* because they determine the location of the margin alone. Despite the fact that the SVC method is more flexible than the MMC in the sense that it admits violations of the strict MMC principle, it still constrains the margin to be linear. In many, if not most, datasets, the classification margin is not linear. SVM were developed specifically to allow for nonlinear classification margins.

Support vector machines (SVM) are a special class of SVC that introduce a larger feature space created from polynomials of the original features. Although a margin defined with an SVM is still linear in the expanded set of features, it can be highly nonlinear in the original, un-transformed features. The SVM algorithm is the same as that developed for the SVC but relies on the recognition that only the support vectors matter. That is, others that do not enter the solution are formally excluded. And, the calculation used to find the location of the margin depends only on the inner product of all the vectors that matter, or the *kernel* of the data. When the kernel is linear, the inner product is simply the correlation between each pair of vectors. But different kernels can be used to allow for support vectors that describe highly nonlinear class boundaries. For example, a polynomial kernel of degree d can produce nonlinear boundaries, and a radial kernel even describes a circular region of support, separating observations into highly flexible patterns of association within the data. In essence, an SVM is an SVC with a nonlinear kernel.

6.3 Tree-based methods

Regression trees, on the other hand, are a means of determining the relative importance of a predictor variable in influencing an output variable. If the data are continuous, a regression tree algorithm searches for a split value of the most important predictor and then calculates predicted values for the output variable for values above and below the split value. Once all observations are assigned in one branch, the algorithm then seeks the predictor variable that best explains the next split for each of the new branches, and so on. Because this *recursive binary splitting* algorithm begins at the top and makes the error-minimizing decision for that split only, it is referred to as a greedy algorithm (James et al. 2014).

Predictive accuracy is evaluated out-of-sample through a *k*-fold cross validation method: Divide the training data into $k = 1,2,\ldots,K$ subsets, or folds (of equal size), train the model on the data in $k - 1$ folds, and calculate the mean-square-error (MSE) on the kth fold. Repeat for each of the other *k*-folds, estimating on each of the other *k* and finding MSE on the $k - 1$ fold so that there are *k* estimates of the MSE, and average the MSE that results. The result is a measure of the *k*-fold cross-validated MSE. A simpler alternative is leave-one-out cross-validation (LOOCV), which excludes one observation from training and then fits the model on the left-out data. However, the LOOCV measure has high variance, as the fitted value is averaged over only one observation per run.

Formally, the objective function for a standard regression tree approach minimizes the residual sum of squares (RSS) given by:

$$RT = \sum_{j=1}^{J} \sum_{i \in Rj} \left(y_i - \hat{y}_{R_j}\right)^2,$$

where y_i is the observed value of the variable of interest, and $\hat{y}R_j$ is the mean value of the variable in the region R_j. In other words, the tree structure divides the data into regions based on values of the predictor space $X_1, X_2, \ldots, X_p$ and then calculates mean values of the response variable for each realized value of the predictor variables and chooses the regions in order to minimize the residual sum of squares. James et al. (2014), however, argue that the base regression tree approach may not produce the best result. Other approaches that average predictions over many trees (i.e., a forest of them, in fact) can typically outperform classical methods of classification or prediction.

The three most common methods are bagging, random forests, and boosting. Bagging, or bootstrap aggregation (Breiman 1996), draws a large number of random samples from the data (bootstrap samples) and fits regression trees using cross-validation to determine the optimal structure of each tree. By averaging the predictions from all the bagged predictions, a sum-of-squares minimizing prediction set is derived. Bagging often represents a substantial improvement in predictive ability relative to a basic regression tree because averaging over a large number of samples provides much more information than a simple, single sample. Intuitively, when the metric for refining the fit of the tree is cross-validation, averaging across different slices of the same dataset is far more likely to produce results that are representative of the data generating process as a whole. Bagging suffers, however, when one or two predictors dominate so that each random sample produces a tree that looks like all the others.

Random forests represents a variant on the bagging approach in which a random sample of *m* predictors out of the total set of *p* predictors is considered at each split in the tree. Only one of the *m* predictors can be used at each split, and a new sample is drawn each time a split decision is to be made. In this way, some models will contain entirely different predictors than others as not every model can simply draw on the most important predictor every time. When bagged regression-tree models are not constrained in this way, their predictions will be highly correlated, so averaging the predictions does not produce much benefit, as each separate run does not add much new information. In fact, bagging is a special case of the random forests method as bagging and random forests are exactly equivalent when the number of predictors in the random forest algorithm (*m*) is set equal to the total number of predictors (*p*). "De-correlating" the predictions from the models and then averaging the result typically produces more accurate predictions because each new sub-sample brings independent information for finding which variable is most important in predicting values of the variable of interest (James et al. 2014). In general, the number of predictors in each sample is set at a fraction (1/3) of the total number of predictors. Comparing a number of alternative regression tree methods, Bajari et al. (2015) and Varian (2014) find that the random forests approach is the most effective in minimizing MSE in out-of-bag (OOB) samples.

In a regression tree context, boosting uses the notion of fitting several trees to the same data in a fundamentally different way. Boosting uses a process of "slow learning" in which the tree is not built on many independent bootstrapped samples as in bagging, but in sequence, building on the tree fit before it. Each tree is relatively small, with potentially only a few terminal nodes. Once the initial tree is fit to the training dataset, the residuals are saved and a new tree is fit to the residuals. In this way, the boosting algorithm proceeds in a manner that is similar to stepwise regression, considering new predictors in sequence until the remaining residuals are minimized. At each iteration, or new tree, the updated predictions are only allowed to be influenced by the new predictions up to a "shrinkage parameter" that causes the evolution of the tree to move more slowly. The parameter is often set at 0.01 or 0.001. Boosted regression trees that evolve slowly are typically the best performing.

7. Practical considerations

7.1 Data sources

Historically, econometricians began studying markets for differentiated products using aggregate datasets. The data consisted of markets shares or volume sold, average prices, and primary product attributes for each product over several time periods and/or geographical areas (Berry, Levinsohn, and Pakes 1995). Econometricians interested in food demand are relatively lucky because firms such as Nielsen and IRI Marketing Research began collecting "syndicated" scanner data on a highly disaggregated basis in the late 1990s.[2] Scanner data provide price and movement data on individual items called Stock Keeping Units (SKU) or Universal Product Codes (UPC). IRI, Nielsen, and Kantar in Europe also maintain consumer panel datasets. Consumer panel data are collected by individual households with hand-held scanning devices. They also contain detailed information on the product, the place of the purchase, and attributes of the household. However, household panel datasets do not provide any information about the alternatives that the consumer faces on each shopping occasion.

7.2 Choice sets

When the set of the alternatives that the consumer faces is unknown by the researcher, additional assumptions are needed. Most traditional demand models are estimated under the assumption that consumers are aware of all available alternatives or the models use information at the aggregate-level data to infer the set of available alternatives for consumers (Berry, Levisohn and Pakes 1995; Nevo 2001). However, in markets with rapidly changing product lines or stockouts, it seems unlikely that consumers have full information on all alternatives. Researchers in marketing and economics highlight how the limited cognitive abilities of consumers restrict their attention to some alternatives (Mehta, Rajiv, and Srinivasan 2003). Hence, the choice set is reasonably assumed to be heterogeneous across consumers, limited in size, and endogenously determined. For example, Bruno and Vilcassim (2008) extend traditional discrete choice models using a random distribution of choice sets and find that not accounting for varying product availability on the UK chocolate confectionery market leads to biased demand estimates. Further, Goeree (2008) estimates a discrete choice model with limited consumer information using advertising data and consumer characteristics and finds that full information models predict upward biased price elasticities that imply greater competition among firms than is realistically the case.

7.3 Outside good

To predict changes in total demand in response to a price change, researchers need to include a measure of how much demand can change, regardless of the set of goods in the choice set. This is accomplished through the outside option. The outside option represents either an aggregate of other alternatives that are considered as further substitutes, or non-purchasing behavior. If the outside option is not included, then the model can be used to predict changes in market shares among consumers who already chose the alternatives, or conditional demand, but not in total demand because the model essentially does not contain any room to expand. In general, for discrete, discrete–continuous, or multiple discrete–continuous models, the mean baseline utility for one option is typically set to zero. This definition of the outside option, which amounts to delimiting the relevant market when competitive analysis is the goal, is a key issue because it could affect the level of utility and subsequent price-elasticity estimates (Foncel and Ivaldi 2005). In the literature, different approaches have been taken depending on the dataset used. For example, Besanko, Gupta, and Jain (1998) use the number of all household shopping trips to compute the share of non-purchase behavior. Villas-Boas (2007) restricts her analysis to primary brands and retailers, and then defines the other small brands and retailers as the outside option. Bonnet and Réquillart (2013) use observed purchases of other product categories that are more or less substitutes for their focal soft-drink categories to define the outside option. More formally, the relevant market and the outside option could be deduced from a test based on household budget allocation decisions akin to a test of separability in a traditional demand system setting (Allais, Etilé, and Lecocq 2015).

7.4 Estimation methods

When the choice probabilities have a closed form expression, we can easily use the maximum likelihood method to estimate the parameters θ. Define $P_{ht}(\theta)$ as the probability that the consumer h chooses any alternative or a bundle of alternatives on purchase occasion t. The probability of the sequence of observed choices of consumer h is then

$$S_h(\theta) = \prod_{t=1}^{T} P_{ht}(\theta),$$

And, assuming that each consumer's choice is independent of that of other consumers, the log likelihood function could be written as $LL(\theta) = \sum_{h=1}^{H} \ln S_h(\theta)$.

When unobserved heterogeneity in consumer preferences are introduced via random parameters, the choice probabilities no longer have a closed-form expression. The log likelihood function is then a multiple integral that cannot be solved analytically. In this case, SML is necessary (Train 2003). SML approximates choice probabilities for any given value using the following algorithm: first, we take R random draws from the chosen distributions and compute the simulated probability $SP_{ht}(\theta) = \frac{1}{R}\sum_{r=1}^{R} P_{ht}(\theta^r)$. Second, the simulated likelihood function is then calculated as

$$SLL(\theta) = \sum_{h=1}^{H} \ln\left(\prod_{t=1}^{T} SP_{ht}(\theta)\right),$$

and can be optimized in a third step. If R rises faster than $\sqrt{HT}$, the maximum likelihood estimator is consistent, asymptotically normal and efficient, and equivalent to maximum likelihood. In practice, a large number of random draws are needed, and a large number of simulations is typically very computationally expensive. To reduce the number of simulations, randomized and scrambled Halton sequences are often used (Bhat 2003), where the simulation error falls with the number of Halton draws.

Random utility models such as those presented in this chapter are consistently estimated if the observed characteristics of alternatives b_j are independent from the error term ε_{hj} in each baseline utility function. If we assume $\varepsilon_{hj} = \xi_j + e_{hj}$, where ξ_j is the unobserved term that captures all unobserved product characteristics and e_{hjt} is an individual-specific error term, the independence assumption cannot hold if unobserved factors included in ξ_j (and then included in the error term e_{hj}) are correlated with observed factors (included in b_j). In this case, the estimated impact of the observed factor captures not only that factor's effect, but also the effect of the correlated, unobserved factor. Unobserved product characteristics could include attributes that are not measured, or marketing efforts such as advertising, sales promotions, and shelf position that are observed by the retailer but not by the econometrician. The resulting endogeneity means that all parameter estimates will be biased and inconsistent. For example, if the unobserved factor is advertising, we know that firms maximize profits with respect to both price and advertising, so, in general, these decisions cannot be independent. Firms might raise the price of their products when they advertise if they believe that doing so stimulates demand. Alternatively, firms may lower price when they advertise (e.g., as a part of a sale), so the possibility of either case makes the sign of the bias ambiguous.

Endogeneity in discrete choice models is typically addressed through the control function approach (Petrin and Train 2010). Define the vector of observed product attributes as: $b_j = (b_j^0; y_{jh})$, where b_j^0 is the vector of exogenous product attributes and y_{jh} the endogenous variable. The control function method is a two-step approach in which the endogenous variable y_{jh} is regressed on the exogenous product attributes x_{jht} and instrumental variables Z_j in the first-stage. If the first-stage model is written as: $y_j = Z_j \gamma + b_j^0 \tau + \bar{\omega}_j$, then $\bar{\omega}_{jh}$ is the error term. Assuming a joint normal distribution between $\bar{\omega}$ and ξ_j, we can rewrite the indirect utility function as $U_{jh} = V(b_j, \theta) + \lambda \widehat{\bar{\omega}}_j + \sigma \eta_j + \vartheta_{jh}$, where η_j is a standard normal distributed variable and σ is the associated standard deviation. The estimated error term $\widehat{\omega}_j$ includes some omitted variables that are correlated with the endogenous variable y_j and not captured by the other exogenous variables of the demand equation b_j^0 or by the instrumental variables Z_j. Introducing this term in the indirect utility function captures unobserved product characteristics that vary across time, and essentially purges the equation of bias, as the endogenous variable y_j is now uncorrelated with the new error term $\vartheta_{jh} = \xi_j + \varepsilon_{hj} - \lambda \widehat{\bar{\omega}}_j$. However, because the demand model contains variables that are themselves estimated, the standard errors of the estimated demand parameters must be adjusted accordingly (Karaca-Mandic and Train 2003).

The choice of instrumental variables Z_j is crucial. Good instruments must be independent of the error term ξ_j, make economic sense, and be sufficiently correlated with the endogenous regressors, but must not be correlated between themselves. To control for price endogeneity, three kinds of instruments are generally used. First, input prices are generally uncorrelated with customer choices, but are correlated with prices from the theory of the firm (Bonnet and Dubois 2010). Assuming no spatial correlation between markets, prices in other markets can also be valid proxies for the cost of production (Hausman, Leonard, and Zona 1994; Nevo 2000). Finally, attributes of other products are not correlated with the demand for the product in question, but are likely to be correlated with its price (Berry, Levinsohn, and Pakes 1995). If other variables are thought to be endogenous, then similar instruments must be found. For example, Richards and Hamilton (2015) instrument for endogenous variety, while Allais, Etilé, and Lecocq (2015) instrument for label choices.

8 Conclusions and implications

In this chapter, we review a broad selection of methods that have been used to study problems in consumer demand over the last 20 years and provide a hint as to the types of models likely to be used in the near future. In each case, the form of the model is driven by both the type of data that are available and the question at hand. While most practical applications of these models involve demand elasticities, they are equally adept at producing demand forecasts, or for inference and drawing conclusions regarding the causal effect of a policy treatment. As computing power and data gathering capabilities advance, our methods will surely keep pace.

Notes

1. Triangular, uniform Rayleigh or truncated normal distributions are also used in the literature.
2. Syndication means that cooperating stores send their data to IRI or Nielsen, who then combine the chain-specific data to produce standardized datasets of the entire market; they share the data with retailers and manufacturers.

References

Ainslie, A., and P.E. Rossi. 1998. "Similarities in Choice Behavior Across Product Categories." *Marketing Science* 17:91–106.

Allais, O., F. Etilé, and S. Lecocq. 2015. "Mandatory Labels, Taxes, and Market Forces: An Empirical Evaluation of Fat Policies." *Journal of Health Economics* 43:27–44.

Anderson, S.P., A. de Palma, and J.F. Thisse. 1992. *Discrete Choice Theory of Product Differentiation*. Cambridge, MA: MIT Press.

Anselin, L. 2002. "Under the Hood Issues in the Specification and Interpretation of Spatial Regression Models." *Agricultural Economics* 27(3):247–267.

Athey, S., and G.W. Imbens. 2015. "Machine Learning Methods for Estimating Heterogeneous Causal effects." Working Paper, Graduate School of Business, Stanford University, Stanford, CA.

Baggio, M., and J.P. Chavas. 2009. "On the Consumer Value of Complementarity: A Benefit Function Approach." *American Journal of Agricultural Economics* 91:489–502.

Bajari, P., D. Nekipelov, S.P. Ryan, and M. Yang. 2015. "Machine Learning Methods for Demand Estimation." *American Economic Review* 105:481–485.

Belloni, A., V. Chernozhukov, and C. Hansen. 2014. "Inference on Treatment Effects After Selection Among High-Dimensional Controls." *The Review of Economic Studies* 81:608–650.

Berry, S.T. 1994. "Estimating Discrete-Choice Models of Product Differentiation." *The RAND Journal of Economics* 25(2):242–262.

Berry, S.T., J. Levinsohn, and A. Pakes. 1995. "Automobile Prices in Market Equilibrium." *Econometrica* 63(4):841–890.

Besag, J. 1974. "Spatial Interaction and the Statistical Analysis of Lattice Systems." *Journal of the Royal Statistical Society. Series B (Methodological)* 36:92–236.

Besanko, D., S. Gupta, and D. Jain. 1998. "Logit Demand Estimation Under Competitive Pricing Behavior: An Equilibrium Framework." *Management Science* 44(11):1533–1547.

Bhat, C.R. 2000. "Incorporating Observed and Unobserved Heterogeneity in Urban Work Mode Choice Modeling." *Transportation Science* 34:228 238.

———. 2003. "Simulation Estimation of Mixed Discrete Choice Models Using Randomized and Scrambled Halton Sequences." *Transportation Research Part B: Methodological* 37(9):837–855.

———. 2005. "A Multiple Discrete Continuous Extreme Value Model: Formulation and Application to Discretionary Time-Use Decisions." *Transportation Research Part B: Methodological* 39(8):679–707.

———. 2008. "The Multiple Discrete–continuous Extreme Value (MDCEV) Model: Role of Utility Function Parameters, Identification Considerations, and Model Extensions." *Transportation Research Part B: Methodological* 42(3):274–303.

Bhat, C.R., M. Castro, and A.R. Pinjari. 2015. "Allowing for Complementarity and Rich Substitution Patterns in Multiple Discrete Continuous Models." *Transportation Research Part B: Methodological* 81:59–77.

Bonnet, C., and P. Dubois 2010. "Inference on Vertical Contracts Between Manufacturers and Retailers Allowing for Non-Linear Pricing and Resale Price Maintenance." *The RAND Journal of Economics* 41(1):139–164.

Bonnet, C., and V. Réquillart. 2013. "Tax Incidence with Strategic Firms in the Soft Drink Market." *Journal of Public Economics* 106:77–88.

Breiman, L. 1996. "Bagging Predictors." *Machine Learning* 24(2):123–140.

Brenkers, R., and F. Verboven. 2006. "Market Definition with Differentiated Products: Lessons from the Car Market." In J.P. Choi, ed., *Recent Developments in Antitrust: Theory and Evidence*. Cambridge, MA: MIT Press, pp. 153–186.

Bruno, H.A., and N.J. Vilcassim. 2008. "Research Note-Structural Demand Estimation with Varying Product Availability." *Marketing Science* 27(6):1126–1131.

Cardell, N.S. 1997. "Variance Components Structures for the Extreme-Value and Logistic Distributions with Application to Models of Heterogeneity." *Econometric Theory* 13(2):185–213.

Chintagunta, P.K. 1993. "Investigating Purchase Incidence, Brand Choice, and Purchase Quantity Decisions of Households." *Marketing Science* 12(2):184–208.

Cox, D.R. 1972. "The Analysis of Multivariate Binary Data." *Journal of the Royal Statistical Society Series C* 21:113–120.

Cressie, N.A.C. 1993. *Statistics for Spatial Data*. New York: John Wiley and Sons.

Deaton, A., and J. Muellbauer. 1980. *Economics and Consumer Behavior*. New York: Cambridge University Press.

Dubé, J.P. 2004. "Multiple Discreteness and Product Differentiation: Demand for Carbonated Soft Drinks." *Marketing Science* 23(1):66–81.

Feenstra, R.C., and J.A. Levinsohn. 1995. "Estimating Markups and Market Conduct with Multidimensional Product Attributes." *Review of Economic Studies* 62(1):19–52.

Foncel, J., and M. Ivaldi. 2005. "Operating System Prices in the Home PC Market." *Journal of Industrial Economics* 53(2):265 297.

Goeree, M.S. 2008. "Limited Information and Advertising in the US Personal Computer Industry." *Econometrica* 76(5):1017–1074.

Goldberg, P.K. 1995. "Product Differentiation and Oligopoly in International Markets: The Case of the US Automobile Industry." *Econometrica* 63:891–951.

Guadagni, P.M., J.D. Little. 1983. "A Logit Model of Brand Choice Calibrated on Scanner Data." *Marketing Science* 2:203–238.

Hanemann, W.M. 1984. "Discrete/Continuous Models of Consumer Demand." *Econometrica* 53:541–561.

Hausman, J., G. Leonard, and J. Zona. 1994. "Competitive Analysis with Differentiated Products." *Annales d Economie et de Statistique* 34:159–180.

Heckman, J.J. 1979. "Sample Selection Bias as a Specification Error." *Econometrica* 48:153–161.

Hendel, I. 1999. "Estimating Multiple–discrete Choice Models: An Application to Computerization Returns." *Review of Economic Studies* 66(2):423–446.

James, G., D. Witten, T. Hastie, and R. Tibshirani. 2014. *An Introduction to Statistical Learning: With Applications in R*. New York: Springer.

Karaca-Mandic, P., and K. Train. 2003. "Standard Error Correction in Two-Stage Estimation with Nested Samples." *Econometrics Journal* 6(2):401–407.

Kelejian, H.H., and I.R. Prucha. 1999. "A Generalized Moments Estimator for the Autoregressive Parameter in a Spatial Model." *International Economic Review* 40(2):509–533.

Kim, J., G.M. Allenby, and P.E. Rossi. 2002. "Modeling Consumer Demand for Variety." *Marketing Science* 21(3):229–250.

Kwak, K., S.D. Duvvuri, and G.J. Russell. 2015. "An Analysis of Assortment Choice in Grocery Retailing." *Journal of Retailing* 91:19–33.

Lancaster, K.J. 1966. "A New Approach to Consumer Theory." *Journal of Political Economy* 74:132–156.

Lee, L.F., G.S. Maddala, and R.P. Trost. 1980. "Asymptotic Covariance Matrices of Two-Stage Probit and Two-Stage Tobit Methods for Simultaneous Equation Models with Selectivity." *Econometrica* 42:491–503.

Luce, R.D. 1959. *Individual Choice Behavior*. New York: John Wiley & Sons.

Manchanda, P., A. Ansari, S. Gupta. 1999. "The Shopping Basket: A Model for Multicategory Purchase Incidence Decisions." *Marketing Science* 18:95–114.

McFadden, D. 1974. "Conditional Logit Analysis of Qualitative Choice Behavior." In P. Zarembka, ed., *Frontiers in Econometrics*. New York: Academic Press, pp. 105–142.

McFadden, D., and K. Train. 2000. "Mixed MNL Models for Discrete Response." *Journal of Applied Econometrics* 15(5):447–470.

Mehta, N., S. Rajiv, and K. Srinivasan. 2003. "Price Uncertainty and Consumer Search: A Structural Model of Consideration Set Formation." *Marketing Science* 22(1):58–84.

Moon, S., and G.J. Russell. 2008. "Predicting Product Purchase from Inferred Customer Similarity: An Autologistic Model Approach." *Management Science* 54:71–82.

Nevo, A. 2000. "Measuring Market Power in the Ready-to-Eat Cereal Industry." *Econometrica* 69(2):307–342.

Petrin, A., and K. Train. 2010. "A Control Function Approach to Endogeneity in Consumer Choice Models." *Journal of Marketing Research* 47(1):3–13.

Phaneuf, D.J., C.L. Kling, and J.A. Herriges. 2000. "Estimation and Welfare Calculations in a Generalized Corner Solution Model with an Application to Recreation Demand." *Review of Economics and Statistics* 82(1):83–92.

Pinkse, J., and M.E. Slade. 2004. "Mergers, Brand Competition, and the Price of a Pint." *European Economic Review* 48(3):617–643.

Pinkse, J., M.E. Slade, and C. Brett. 2002. "Spatial Price Competition: A Semiparametric Approach." *Econometrica* 70:1111–1153.

Pofahl, G.M., and T.J. Richards. 2009. "Valuation of New Products in Attribute Space." *American Journal of Agricultural Economics* 91(2):402–415.

Richards, T.J. 2000. A Discrete/Continuous Model of Fruit Promotion, Advertising, and Response Segmentation." *Agribusiness* 16(2):179–196.

Richards, T.J., M.I. Gomez, and G.F. Pofahl. 2012. "A Multiple–discrete/Continuous Model of Price Promotion." *Journal of Retailing* 88(2):206–225.

Richards, T.J., and S.F. Hamilton. 2015. "Variety Pass-Through: An Examination of the Ready-to-Eat Cereal Market." *Review of Economics and Statistics* 91(1):166–179.

Richards, T.J., S.F. Hamilton, and W.J. Allender. 2014. "Social Networks and New Product Choice." *American Journal of Agricultural Economics* 96(2):489–516.

Rojas, C., and E.B. Peterson. 2008. "Demand for Differentiated Products: Price and Advertising Evidence from the US Beer Market." *International Journal of Industrial Organization* 26(1):288–307.

Russell, G.J., and A. Petersen. 2000. "Analysis of Cross Category Dependence in Market Basket Selection." *Journal of Retailing* 76:367–392.

Slade, M.E. 2004. "Market Power and Joint Dominance in UK Brewing." *Journal of Industrial Economics* 52:133–163.

Smith, H. 2004. "Supermarket Choice and Supermarket Competition in Market Equilibrium." *The Review of Economic Studies* 71:235–263.

Song, I., and P.K. Chintagunta. 2006. "Measuring Cross-Category Price Effects with Aggregate Store Data." *Management Science* 52:1594–1609.

Train, K.E. 2003. *Discrete Choice Methods with Simulation*. New York: Cambridge University Press.

Varian, H.R. 2014. "Big Data: New Tricks for Econometrics." *Journal of Economic Perspectives* 28(2):3–27.

Vásquez-Lavin, F., and M. Hanemann. 2008. "Functional Forms in Discrete/Continuous Choice Models with General Corner Solution." CUDARE Working Paper 1078, Department of Agricultural and Resource Economics, University of California-Berkeley, Berkeley.

Villas-Boas, S.B. 2007. "Vertical Relationships Between Manufacturers and Retailers: Inference with Limited Data. *Review of Economic Studies* 74:625–652.

Wales, T.J., and A.D. Woodland. 1983. "Estimation of Consumer Demand Systems with Binding Non-Negativity Constraints." *Journal of Econometrics* 21(3):263–285.

28
A survey of semiparametric regression methods used in the environmental Kuznets curve analyses

Krishna P. Paudel and Mahesh Pandit

1 Introduction

The environmental Kuznets curve (EKC) hypothesis states that the relationship between pollution and per capita income generally appears as an inverted U-shaped curve, as shown in Figure 28.1. The notion presented by the EKC hypothesis is that pollution grows rapidly in the early stages of a country's industrialization because high priority is given to increased production, and people are more interested in income than environmental qualities (i.e., green production practices, reducing pollutants in industry, etc.). Additionally, at the height of industrialization, environmental quality is considered a luxury good. As a country advances beyond the industrialization phase into an economy that is primarily service dominated, people's demand for environmental quality increases. Further, at that stage, people are willing to pay for better water or air quality. EKC studies have been conducted on air pollution, water pollution, deforestation, toxic substances, waste, and energy-related variables. Empirical studies have either refuted or failed to reject the EKC hypothesis.

The environmental Kuznets curve (EKC) is an empirical phenomenon showing how some pollutants increase and then decrease with rising per capita income. The original thought behind the EKC comes from Grossman and Krueger (1991) (and the subsequent publication of their influential paper in *QJE* in 1995 (Grossman and Krueger 1995)) and Shafik and Bandyopadhyay (1992), from their study of economic growth and environmental quality during the North American Free Trade Agreement debate of the 1990s. They connected their findings with the production economy to show the existence of the EKC hypothesis. They stated that the EKC is the result of scale, technique, and composition effects. An increase in current production leads to an increase in pollution, which is referred to as the *scale effect*. Increased adoption of more *efficient* technologies decreases pollution. An increase in economic growth shifts the economy from a manufacturing base to one that is more service oriented in its scope. This is commonly referred to as the *composition effect*. Hence, if technique

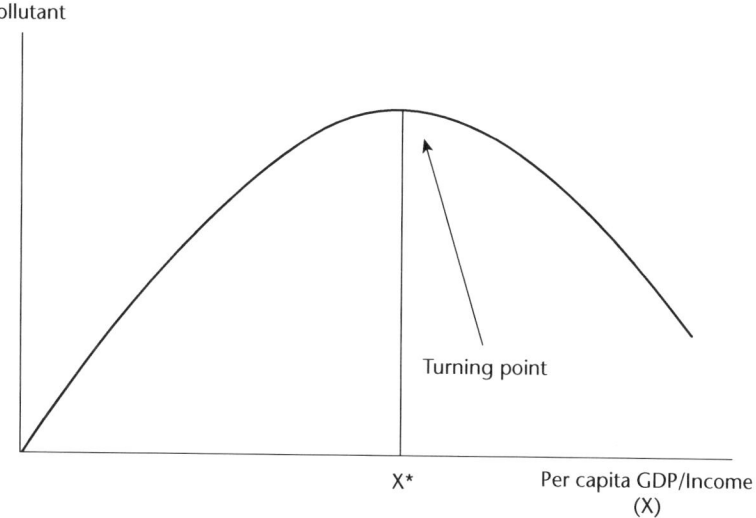

Figure 28.1 Environmental Kuznets curve

and composition effects are higher than the scale effect over a particular period, the EKC relationship is established.

The shape of the curve, however, is very sensitive to the data period, location, and pollutant considered in the analysis (Harbaugh et al. 2002). Since the early 1990s, a number of empirical studies have been conducted, but many of these studies have refuted the inverted U-shape of the curve, indicating that a more flexible functional form is required to examine the EKC hypothesis. The debate over the EKC hypothesis has settled to some extent, but the research direction is moving towards developing a theoretical model to better understand the EKC hypothesis and estimating the empirical model using a flexible model specification.

This chapter reviews the literature on the effects of economic growth on environmental quality that have used semiparametric and nonparametric methods. Since the mid-1990s, research has been conducted to examine the existence of the EKC on different types of environmental quality indicators such as air, water, forest, and energy consumption. The literature on the topic is continuously growing with various findings that are inconsistent with the traditional belief that an inverted U-shaped relationship holds for all pollutants.

One of the current and important debates on EKC research is the use of a functional form implemented to examine the environmental quality and economic growth relationship. During the 1990s, parametric models with polynomial specifications were generally used (example: Grossman and Krueger 1991). A parametric model requires distributional assumptions in order to estimate relevant model parameters. If the distributional assumption is not valid, the inferences drawn from the wrong model are inconsistent, biased, and inefficient. Generally speaking, the true relationship between variables is unknown. From an econometric perspective, a complex model or flexible model is required to extract more information from data, so the use of nonparametric or semiparametric models has begun to emerge in the EKC literature (see Millimet et al. 2003; Paudel et al. 2005).

Many researchers have focused only on the effects of economic growth on environmental pollution to examine the EKC hypothesis (e.g., Grossman and Krueger 1991). Other factors, such as population density, political freedom, and farmland, also play important roles in determining the concentration of pollutants. If these important variables are omitted in model specification, the results obtained will not be consistent due to omitted variable bias. Several authors (Van 2003; Roy et al. 2004; Paudel et al. 2005; Van and Azomahou 2007; Lawell et al. 2018) used additional variables other than income in their models to partial out the effects of these variables so that they could establish a more accurate relationship between economic growth and environmental quality.

1.1 Theoretical EKC model

There are two strands of theoretical literature in EKC – one draws its theoretical underpinnings from growth theory, while the other bases its rationale on ideas drawn from static utility-maximization theory. We provide a summary of representative papers covering both strands of literature and demonstrate the essence of the two approaches.

Many researchers have proposed different theories behind the EKC hypothesis. Lopez (1994) describes the inverted U-shaped relation as a production function. He shows that as the substitution elasticity between conventional input and pollution falls, then the relative curvature of income in the utility function falls and the inverted U-shaped relationship gets established. This suggests that firms pay an increasing price for pollution while it is less costly to reduce pollution by changing the production technology to an environmentally friendly one. On the other hand, non-homothetic[1] preference implies that consumers are willing to give up additional consumption to receive a better environmental quality. McConnell (1997) studies the role of income elasticity of demand for environmental quality and concludes that it is not the income elasticity of demand for environmental quality that shapes EKC. It is, rather, due to lower values placed on pollution reduction as income rises.

John and Pecchenino (1994) use an overlapping generation model and provide a theoretical explanation for the inverted U-shaped correlation between environmental quality and income. They conclude that "the relationship between growth and the quality of the environment is complex." Andreoni et al. (2001) use a Cobb–Douglas utility function to explain the income and pollution relationship. They propose that utility depends on consumption and pollution, and pollution depends on consumption levels and pollution control efforts. They suggested that an inverted U-shaped EKC relationship occurs if there are increasing returns to scale in terms of the pollution control effort. This case is likely due to many factors, such as population growth and technological changes.

Kelly (2003) develops an EKC model from a stock externalities perspective. According to him, the marginal benefit and the marginal cost of pollution control rise with income over the economic growth path. If the marginal benefit rises faster than the marginal costs, the emission-income relationship has a negative slope for a given level of income and vice-versa. Recently, Brock and Taylor (2010) extend the Solow growth model in the EKC framework, also known as the Green Solow Model. Due to diminishing returns, development begins with rapid economic growth, and emissions rise with the output growth, but fall with ongoing technological progress. At first, emission of pollutants overwhelms the impact of technological progress, and emission levels rise. As countries mature and approach a balanced growth path, the impact of slower growth on emission is overwhelmed by the impact of technological progress, and emission levels decline. So, diminishing returns and technological progress are responsible for generating the inverted U-shaped EKC.

1.2 Static models

Following Andreoni and Levinson (2001), let us consider that an individual maximizes utility from consumption of private good (C) and pollution (P). The utility function is given as:

$$U = U(C, P) \tag{1}$$

where $\frac{\partial U}{\partial C} = U_C > 0$ and $\frac{\partial U}{\partial P} = U_P < 0$. Since consumption C designates normal goods and P designates non-normal goods, U is quasi-concave in C and $-P$. Pollution enters in as a byproduct from the consumption of goods, and individuals allocate resources to reduce pollution or prevent it from happening. Let us denote the resources spent cleaning the environment by E. Hence, pollution is a function of consumption and environmental effort:

$$P = P(C, E) \tag{2}$$

where $P_c > 0$ and $P_E < 0$. Further, suppose that M is total income available to spend on either C or E. Hence, the resource constraint is given by:

$$C + E = M \tag{3}$$

To illustrate, let us consider simple utility and pollution functions as given:

$$U = C - zP \qquad z > 0 \tag{4}$$
$$P = C - C^\alpha E^\beta \qquad \alpha, \beta > 0 \tag{5}$$

where z represents the marginal disutility from pollution. The second term ($C^\alpha E^\beta$) in Equation (5) represents "abatement." Maximizing the utility function (4), subject to the constraint (5), yields the optimal solution for C and P as follows:

$$C^\star = \frac{\alpha}{\alpha + \beta} M \text{ and } C^\star = \frac{\beta}{\alpha + \beta} M. \tag{6}$$

Substituting the optimal value of $C^\star$ and $E^\star$ from Equation (6) into Equation (5), the optimal pollution is given as:

$$P^\star(M) = \frac{\alpha}{\alpha + \beta} M - \left(\frac{\alpha}{\alpha + \beta}\right)^\alpha \left(\frac{\beta}{\alpha + \beta}\right)^\beta M^{\alpha + \beta} \tag{7}$$

Differentiating this equation with respect to M yields

$$\frac{\partial P^\star}{\partial M} = \frac{\alpha}{\alpha + \beta} - (\alpha + \beta)\left(\frac{\alpha}{\alpha + \beta}\right)^\alpha \left(\frac{\beta}{\alpha + \beta}\right)^\beta M^{\alpha + \beta - 1} \tag{8}$$

Equation (8) indicates that when $\alpha + \beta > 1$, the pollution level P^* follows an inverted U-shape curve with respect to income. This is the condition for increasing returns to scale. Kijima et al. (2010, p. 1193) explain this relation as:

> For low income (M) the consumption level is also low, and the increasing return of abatement indicates that the effect from the abatement effort has little impact on environmental quality. At this condition, the representative agent does not want to spend much money on abatement, and so the pollution level rises with an increase in income. In contrast, for a sufficiently high level of income, a high level of consumption causes the agent much disutility from pollution. In fact, the impact of abatement on utility value is higher due to the increasing return, and the agent optimally spends more resource on abatement. Thus pollution levels decrease with higher level of income. Hence combining these two conditions implies the existence of EKC.

1.3 Dynamic model

John and Pecchenino (1994) develop an overlapping generation model with two time periods. According to these authors, a person allocates his income between consumption and abatement efforts for two time periods. Let w_t represent the wage an individual is generating at time t. Utility at current period t is a function of consumption and environmental quality at the later period and is given as

$$U_t = U(c_{t+1}, E_{t+1}) \tag{9}$$

where, c_t = consumption at period t and E_t = environmental quality at period t. A higher value of E represents better environmental quality. The environmental quality can be pressed using the following dynamics equation:

$$E_{t+1} - E_t = -bE_t - \beta c_t + \gamma m_t, \tag{10}$$

where m_t is the investment in environmental maintenance and improvement, and $b, \beta,$ and γ are positive constant. Let the production function be given as $Y_t = F(K_t, N_t)$, where Y is output, K is capital stock, and N is the labor. Assuming first-order homogenous production function, the output per capita can be expressed as $y = f(k)$. Here, $k_t = K_t/N_t$.

John and Pecchenino have derived the equations for dynamic equilibrium path using the following equations:

$$r_t = f'(k_t) - \delta = r(k_t) \tag{11}$$
$$w_t = f(k_t) - k_t f'(k_t) = w_t(k_t), \tag{12}$$
$$U_1(c_{t+1}, E_{t+1})(1 + r_{t+1}) - \gamma U_2(C_{t+1}, E_{t+1}) = 0 \tag{13}$$
$$k_{t+1} = s_t \tag{14}$$

where U_i represents the partial derivative with respect to the ith argument, δ is depreciation rate, r_t is interest rate at period t, and s_t is the saving amount of generation t.

Assume that economy starts with a little capital only. In that scenario, firms do not have enough capital to spend on environmental pollution abatement, i.e., $m_t = 0$. Hence, environmental quality deteriorates initially. After a certain period of time, as capital stock accumulates and the income rises, firms are more willing to pay for enhancing environmental quality and investing in more environmentally friendly production processes. Due to this phenomenon, the income and pollution relationship exhibits an inverted U-shape curve.

1.4 EKC policy

The inverted U-shaped relationship between economic growth and environmental quality reveals that sufficient economic growth is one possible solution in the abatement of environmental pollution. This is an important motivation that leads us to examine the EKC hypothesis. If this is true, we are led to ask the question, do environmental problems reduce automatically with the rise in per capita income? Alternatively, do people start caring about environment once they become richer? Empirical studies have shown that there is no unique answer to this question, because the results are susceptible to the pollutant type, the time period, and geographical location, to name a few.

Thus, economic growth does not control environmental quality itself automatically (see more details in Vincent (1997) and Criado (2008)). This answer leads to another question, and that is whether or not environmental policies are needed in order to improve environmental quality? Grossman et al. (1995) and Dasgupta et al. (2002) suggest that improvement in environmental quality comes through environmental regulations. Effective policy significantly reduces environmental degradation and the environmental cost of growth (Panayotou 1997).

To illustrate, an increase in economic growth changes the preference and environmental regulation that leads to change in production (Tsurumi et al. 2010). Environmentally friendly regulations play a significant role in improving environmental quality. Tsurumi et al. (2010) suggest a tradeoff between economic growth and environmental quality depends on the technique effects. The magnitude of the technique effect is important to implement environmental policy, and stringent environmental regulation leads to an improvement in environmental quality. If the technique effect is not sufficient to reduce environmental degradation, environmental regulations are required to reduce pollution.

Developing countries ignore their environmental problem until they are further along their industrial development path and have become wealthier; however, these countries should consider formulating regulations at a less stringent level in the beginning and then ratchet up those regulations as their economy matures (Carson 2010). Acemoglu et al. (2012) and Carson (2010) conclude that since environmental regulation and abatement efforts are required to control environmental degradation, an optimal time for abatement and policy should be determined. Further, Stern (2004) suggests that a new technology needs to be adopted in high-income countries before it is adopted in low-income countries to improve environmental quality. However, a "one size fits all" approach for finding a solution for all countries would not work, so heterogeneous technologies that are country-specific would be needed.

Empirical studies have shown that there exists a cubic-shaped relation for some pollutants, implying a chance of further degradation of environmental quality after improvement. We believe that this might be due to the following consequence: as a country becomes wealthier, the demand for industrial products rises. This higher demand subsequently raises the production of goods and, as a result, increases with it the emission of pollutants as a byproduct of the production process. Unless alternative technology is invented, depending upon the extant environmental regulations and the condition of the environment, as per capita consumption of resources increases so do pollution levels. This is evident with electricity consumption in developed countries and the related by-product of the generation of that electricity, i.e., air pollutants. Returning to the motivation of EKC studies, economic growth might be the solution for environmental degradation, but it might not be true for stock pollutants because of the irreversibility and catastrophic impact on the environment.

The objective of this chapter is to survey recent developments in nonparametric and semiparametric methods and their use in EKC literature. The remainder of this chapter is organized as follows. First, we describe how EKC hypotheses are examined in empirical studies. Second, we provide a detailed review of the recent developments in semiparametric econometric methods and how these advances are implemented in the empirical EKC literature. We then discuss existing model specification test statistics and the additional variables used in the EKC literature. The details provided here should be beneficial in shaping the direction of future studies on the EKC. We will show potential improvement that can be achieved in EKC estimation using the most recently developed techniques.

2 Existing EKC model

The most general parametric panel model specification used in the EKC literature is a polynomial form equation with two or three degrees for income. According to Stern (2004), the polynomial model in the EKC is specified as follows:

$$P_{it} = \gamma_i + \phi_t + \beta_1 y_{it} + \beta_2 y_{it}^2 + \beta_3 y_{it}^3 + x_{it}\alpha + \epsilon_{it} \qquad i=1,...,n\ ;\ t=1,...,T \qquad (15)$$

where the first two terms are the intercept parameters of two-way fixed effects for individuals (such as county or state or countries) and times. The intercept parameters control location- and time-specific factors in the panel data model, respectively. In some cases, researchers have used only a one-way effect model, arguing that country or geographic effects are constant. In that case, either $\gamma_i = 0$ or $\epsilon_t = 0$. P_{it} represents pollution level for the individual county or watershed i at time t. Pollutants are usually measured in concentration. y_{it} is the measure of economic growth and is usually measured in per capita income or per capita gross domestic product (GDP). β_1, β_2, and β_3 are associated coefficients for y_{it}, y_{it}^2, and y_{it}^3, respectively. If a quadratic model is used, then β_3 is restricted to zero (i.e., $\beta_3 = 0$). Variable x_{it} includes other factors that affect the pollution emission, such as population density, farm crop land, and political freedom; and ϵ_{it} is a contemporaneous error term that can take different structures according to model specification.

By definition, the EKC hypothesis implies that the relationship between income and pollution emissions is nonlinear. Sometimes, it is difficult to parameterize a nonlinear relationship with a parametric specification. In this case, a nonparametric or semiparametric model may be more useful than a parametric model, as the former does not require any distributional assumptions. In the EKC literature, many researchers have employed nonparametric or semiparametric model specifications with an economic growth variable entered as a nonparametric component and other variables entered as parametric components. The model, which contains both parametric and nonparametric components, is a semiparametric model. A semiparametric, partially linear regression model (Bertinelli et al. 2005; Millimet et al. 2003; Robinson 1988) is specified as

$$P_{it} = \gamma_i + \phi_t + g(y_{it}) + x_{it}\alpha + \epsilon_{it}\quad i=1,...,n\ and\ t=1,...,T \qquad (16)$$

where $g(.)$ is some unknown smooth function. Other parameters are defined as in Equation (15). The nonparametric component can extract more information from the data about the curvature of the regression at any specific value of y.

Table 28.1 Smoothing approach and literature

Smoothing approach	Literature
Kernel smoothing	Millimet et al. (2003), Stern (2004), Paudel et al. (2005), Paudel and Poudel (2012), and Lawell et al. (2018)
Spline smoothing	Van (2003), Criado (2008), Luzzati et al. (2009), Zanin et al. (2012), and Kim (2013)

In the EKC literature, we find two different approaches to estimate the smoothness of a function. The two approaches are kernel smoothing and spline smoothing, both of which have been extensively used by researchers. Table 28.1 gives examples of EKC literature differentiated by smoothing technique used in semiparametric models.

A more flexible smoothing technique is also used in the EKC literature. Van and Azomahou (2007) use the smooth coefficient model as proposed by Li et al. (2002). The smooth coefficient model is specified as follows:

$$y_{it} = g(y_{it}) + x_i' \alpha(y_{it}) + \epsilon_{it}.$$

where all the representations are the same as above. The semiparametric model (16) is nested in this model and can be obtained from a restriction $\alpha(y_{it}) = \alpha$.

3 Recent advances in semiparametric model

Given the debate on the functional form used to examine the EKC hypothesis, we are interested in searching for the latest developments in nonparametric and semiparametric methods that can be used to examine the EKC hypothesis. Semiparametric regression combines parametric and nonparametric regressions, which are found to be better than running only parametric or nonparametric regressions (Pandit et al. 2013; Paudel et al. 2005). Semiparametric regression relaxes the distributional assumption of a parametric model and reduces the curse of dimensionality associated with a nonparametric method. The semiparametric regression method is used in various subject areas. The two approaches used to smooth variables using nonparametric and semiparametric regression methods are kernel smoothing and spline smoothing of a variable entered as a nonparametric component. A spline model is a parametric approach of fitting a nonlinear model, whereas kernel smoothing is a locally weighted average regression method.

The nonparametric and semiparametric statistical methods have been used in economic research since the 1960s but have only gained widespread use since the early 1990s. Since that time, a new development of estimation procedures has been continually evolving. One of the most used semiparametric models is a semiparametric partially linear model that is developed by Robinson (1988). Li et al. (2002) generalized the model proposed by Robinson (1988) and generated a semiparametric smooth coefficient model using local least squares with a kernel function. This model is more flexible than the partially linear model.

Many variables (such as gender, location, etc.) in economic models are also categorical or binary variables. It is easy to perform statistical analysis if all variables are continuous; however, mixed data containing continuous and categorical variables are tedious to manipulate in a semiparametric regression model compared to a parametric regression model. Many authors have proposed new methodologies to account for mixed variables in the semiparametric model. For example, Racine et al. (2004) propose a new method for nonparametric regression estimation to include both categorical and continuous variables in a semiparametric model. Using kernels along with the cross-validation method for smoothing parameters, they show that the proposed estimator performs much better than the conventional nonparametric estimators in the presence of mixed data. Furthermore, multivariate-based distributions used in economic research are another difficulty in the semiparametric estimation procedure. To account for this phenomenon, Chen and Fan (2006) suggest a copula-based semiparametric stationary Markov model characterized by a parametric copula and a nonparametric marginal distribution. A copula serves as a heuristic in constructing a multivariate regression and represents general types of dependence.

In addition to a mixed model developed by Racine and Li (2004), Li et al. (2009) develop a nonparametric estimation procedure for treatment effects models, which can include categorical and continuous variables. They show that their method is capable of performing better than the conventional nonparametric method. Details on the kernel-based estimation procedure for categorical variables can be found in Li et al. (2011). Ma et al. (2015) and Nie and Racine (2012) also develop a spline-based nonparametric regression model that includes both continuous and categorical variables. In addition to handling a mixed model in the spline-based semiparametric regression model, Ma and Racine (2013) propose estimating using the model as an additive regression spline model.

All of these models can handle categorical variables and require at least one continuous variable. However, Chen and Fan (2011) develop a categorical semiparametric coefficient model that can handle all categorical variables in a nonparametric component in a semiparametric model. We also observe a rapid growth in the literature that uses nonparametric and semiparametric models using panel data. Detailed discussion on a semiparametric model using panel data is found in Ullah et al. (1998) and Ai et al. (2008). Griffin et al. (2010) propose a Bayesian fully nonparametric regression estimation procedure from a combination of Bayesian nonparametric density estimation and a nonparametric regression model. Copulas are usually used to fit the multivariate distribution. Qian et al. (2012) have developed a semiparametric panel data model using a first differencing method based on the marginal integration of a locally linear smoothed higher-dimensional function.

4 Model consistent specification test

Appropriate nonparametric model specification test statistics are necessary to compare nonparametric and semiparametric models. We review model specification tests used to compare parametric, nonparametric, and semiparametric models in the EKC literature in this section. In the 1980s, the nonparametric technique for model specification is first suggested by Ullah (1985) and Robinson (1988). Many studies have proposed test statistics to compare nonparametric or semiparametric versus parametric models (Delgado and Stengos 1994; Fan et al. 1996; Hong et al. 1995; Zheng 1996). All of these test statistics are used in the EKC study. For example, the test statistic developed by Hong and White (1995) was used by Paudel et al. (2005). This test statistic is based on the covariance between the residual from the parametric and the discrepancy between the parametric and nonparametric fitted values. The decision is made based on the asymptotic normal distribution, so it does not address the nonlinearity of the data. Li et al. (1998) develop a specification test to identify better models between the parametric partial linear model and the semiparametric partial linear model. Because this test is based on the wild bootstrap technique, it performs better than the test statistics that depend on the assumption of an asymptotic normal distribution. We observe that this method is fairly common in the EKC literature to compare parametric and nonparametric or semiparametric models (e.g., Azomahou et al. 2006; Millimet et al. 2003; Van 2010; Poudel et al. 2009; Roy et al. 2004).

Semiparametric model estimation techniques such as the kernel method have been used to construct consistent model specification tests. Robinson (1988) tests the suitability of parametric vs. semiparametric regression models using such a process. Similarly, Hardle et al. (1993) suggest the use of the wild bootstrap procedure. Further, semiparametric test statistics are also used to check the endogeneity of variables by some researchers. Blundell and Powell (2004) introduce a testing procedure for determining the endogeneity of variables by implementing semiparametric methods in an income–consumption relationship using British family expenditure survey data.

Li et al. (2002) introduce a more flexible semiparametric model as well as test statistics to check model specification. The test statistics developed by Li et al. (2002) are used by Vangeneugden et al. (2011). All of these test statistics mentioned above have a drawback in that they do not work when there are categorical variables that entered as a nonparametric component. Hsiao et al. (2007) developed a new test to overcome this drawback. Using simulation results, they found that the proposed test has a significant advantage over other conventional frequency-based kernel tests. The test statistics developed by Hsiao et al. (2007) are used by Paudel and Poudel (2012) in their EKC paper.

5 Semiparametric estimation of the EKC

In this section, we discuss how semiparametric models have been used in the EKC literature. A summary of journal articles that have used semiparametric models in the EKC study is provided in Table 28.2. The table provides author, year of publication, type of additional variables included in the model other than income, type of parametric and semiparametric model and model specification test used, and their major finding, including turning points (TP) if they found the existence of an EKC in their research. Table 28.2 shows that the use of the semiparametric method in the EKC literature is increasing. Generally, the parametric models estimated in the EKC are of quadratic and cubic forms.

Millimet et al. (2003) use a flexible semiparametric model to study the existence of the EKC. They test the existence of the EKC for sulfur dioxide (SO_2) and nitrogen oxide (NO_x) emissions from 1929–1994 using a panel dataset at the U.S. state-level. They consider a fixed effects cubic model as a parametric model. Spline smoothing[2] and Robinson (1988) partial linear models are used as a semiparametric model. Income is entered as a nonparametric variable in the semiparametric model. As expected, they find the existence of the EKC for sulfur dioxide and nitrogen oxide. They use Zheng (1996) and Li and Wang (1998) model specification tests to compare the results from parametric and semiparametric models. The model specification tests show the semiparametric model performs better than the parametric model. This suggests that a semiparametric model is a more flexible model compared to the parametric model.

Van (2003) uses an additive partial linear model developed by Hastie et al. (1990) that is a spline-based semiparametric model. He uses data on protected areas in 89 countries to examine the EKC hypothesis. In addition to per-capita GDP, he considers other factors such as trade, population density, education, and political institutions. These variables are parametrically entered in the semiparametric model. Van (2003) finds that EKC does not exist for the protected area. To compare parametric and nonparametric model specifications, he computes gain statistics developed by Hastie and Tibshirani (1990). Test results show that the semiparametric model performs better than a parametric model.

Roy and van Kooten (2004) also examine the existence of the EKC for three nonpoint source air pollutants: carbon monoxide (CO), nitrogen oxide (NO_x), and ozone (O_3), using adjusted partial linear models allowing heteroscedasticity. Li and Wang's (1998) specification

Table 28.2 Existing published studies that have used semiparametric techniques in environmental Kuznets curve estimation

Literature	Types of models used	Additional variables used	Model specification test	Finding and Turing points
Millimet et al. (2003)	*Parametric*: two-way fixed effects, cubic *Semiparametric*: Robinson (1988)	No	Zheng (1996) and Li and Wang (1998)	EKC existed for SO_2 and NO_x PS-SO_2: $16,417, FS-$NO_x$: $8,657, PS-$NO_x$ $10,570
Van (2003)	*Parametric*: OLS *Semiparametric*: Hastie and Tibshirani (1990)	Trade, population density, education, and political institution	Gain statistics developed by Hastie and Tibshirani (1990)	No EKC for protected areas
Roy and van Kooten (2004)	*Parametric*: Linear and cubic models *Semiparametric*: Wand and Jones (1994), Linton and Nielsen (1995)	Population density, % minorities, % unemployed, % labor in manufacturing, % with high school etc.	Li and Wang (1998)	EKC exists for NO_x, does not exists for CO and O_3
Bertinelli and Strobl (2005)	*Parametric*: quadratic fixed effects *Semiparametric*: Robinson (1988)	No	Ullah (1985)	Linear relationship between pollutant and income, i.e., no EKC existed for CO_2, SO_2
Paudel, Zapata, and Susanto (2005)	*Parametric*: fixed and random effects panel *Semiparametric*: Robinson (1988)	Weighted income, population density	Hong and White (1995)	EKC existed for nitrogen and dissolved oxygen but not for phosphorous TP-N: $12,993
Azomahou, Laisney, and Van (2006)	*Parametric*: within cubic panel estimation *Semiparametric*: Wand and Jones (1994), Linton and Nielsen (1995)	No	Li and Wang (1998)	EKC yes for CO_2 in a parametric model, no for CO_2 in a nonparametric model TP CO_2: $13,258
Van and Azomahou (2007)	*Parametric*: fixed and random effect panel *Semiparametric*: smooth coefficient model by Li et al. (2002)	Trade, population growth rate, population density, literacy rate, political institution	Li et al. (2002)	EKC does not exist for deforestation

Study	Method	Control variables	Testing	Findings
Criado (2008)	*Parametric:* cubic panel fixed effects *Semiparametric:* Wood (2006) approach	No	V-test, Yatchew's (2003) pooling test	EKC existed for CH_4, CO, CO_2, NMVOC TP CH_4: $17,300; CO: $16,800; CO_2: $16,400; NMVOC: $17,200
Luzzati and Orsini (2009)	*Parametric:* fixed effects panel *Semiparametric:* generalized additive			TP for energy consumption: low income countries: none; middle income countries: $57,500; high income countries: $18500; other countries: $9,000
Poudel, Paudel, and Bhattarai (2009)	*Parametric:* fixed effect *Semiparametric:* Robinson (1988)	Forestry, population density, illiteracy, income weight	Li and Wang (1998)	EKC exists with N Shaped
Li (2011)	*Parametric:* quadratic fixed effects *Semiparametric:* B-spline		Average mean square error, bootstrap confidence band	OECED-countries support for EKC
Zanin and Marra (2012)	*Parametric:* OLS *Semiparametric:* additive mixed model, penalized regression spline		Restricted likelihood ratio test (RLRT) by Crainiceanu and Ruppert (2004)	Existence of EKC for CO_2, France and Switzerland (U shape), Austria-N, Denmark, M Shaped
Chiu (2012)	*Parametric:* OLS *Semiparametric:* panel smooth transition regression (PSTR) of Gonzalez et al. (2005)	Population density, trade openness, political freedom	F-version LM, and pseudo LR test	$3,021 and $3,103
Kim (2013)	*Parametric:* quadratic and cubic *Semiparametric:* kernel based semiparametric model		Upper confidence bands (UCB)	EKC exists for SO_2 and CO_2

test is used to compare a quadratic model against the semiparametric model. Compared to the previous literature, they use a log of income in their model. They use linear, quadratic, and cubic models and find that income is very sensitive to model specification. They do not find the existence of EKC for these pollutants, which is also consistent with the findings of Millimet et al. (2003) for NO_x. As with the previous research, they use Li and Wang's (1998) model specification test and find that the semiparametric model is better compared to the quadratic model.

Bertinelli and Strobl (2005) estimate the relationship between pollutants (sulfur oxide (SO_2) and carbon dioxide (CO_2)) using Robinson's (1988) partial linear regression using 108 and 122 cross country observations for SO_2 and CO_2, respectively. In contrast to previous literature, Bertinelli and Strobl (2005) find interesting results that there exists a linear relationship between these pollutants and income. This implies that no EKC exist for these pollutants. The linear hypothesis is tested against the semiparametric model using a method suggested by Ullah (1985). The bootstrap procedure suggested by Lee et al. (2001) is used to obtain the standard error, and the standard error is used to find the significance of the test statistic. They fail to reject the null of a linear relationship between income and pollution.

EKC has been also tested at the local level for water pollutants by Paudel et al. (2005). They estimate an EKC for nitrogen (N), phosphorus (P), and dissolved oxygen (DO) at the watershed level for 53 parishes for the period of 1985–1998 using the data collected by the Department of Environmental Quality. One-way and two-way fixed and random effects with quadratic and cubic models are estimated as parametric models. Using the Hausman (1978) test, they find that the fixed effect model was better than the random effects model. Like the previous literature, they also use the Robinson (1988) partial linear model as a semiparametric model. A method suggested by Hong and White (1995) is used to compare the parametric model against a semiparametric model. As expected, they find that the semiparametric model captured nonlinearity better than the quadratic and cubic models. They observe mixed results on the existence of EKC, i.e., the EKC exists for nitrogen but not for phosphorus and dissolved oxygen.

Azomahou et al. (2006) study the empirical relationship between CO_2 emission and economic development using panel data from 100 countries over the period 1960–1996. They investigate the relationship using a cubic parametric model and a nonparametric model and find that the parametric model shows an inverted U-shape relation but the nonparametric model does not support this shape. They use test statistics suggested by Li and Wang (1998) to compare results obtained from parametric and semiparametric models and observe that the null of the correct parametric model is rejected in favor of the nonparametric model.

Land cover with forest is an indicator of environmental quality because it helps to sequester CO_2 from air. It also provides other benefits such as recreation and soil erosion protection. Deforestation can cause severe environmental damage. Van and Azomahou (2007) investigate the relationship between deforestation and economic growth with a panel dataset of 59 developing countries over the period 1972–1994, using parametric and semiparametric models. They estimate quadratic and cubic fixed and random effects models. They compare a fixed effects versus a random effects model using the Hausman test. The test favored a random effects model contradictory to the finding from Paudel et al. (2005). They use a smooth coefficient model suggested by Li et al. (2002), which is a more flexible model than the models used by previous researchers (e.g., Robinson's model). Using this model, they found that there is no EKC for deforestation. However, they find that the other variables (e.g., population density, political institutions) considered in the model have significant effects on deforestation. They test the robustness between parametric and semiparametric models using test statistics proposed by Li et al. (2002) and find that a parametric model is preferred against the semiparametric model.

In general, many researchers have used panel data to study the pollution–GDP relationship. However, they assume the temporal (stability of the cross-sectional regressions over time) and spatial (stability of the cross-sectional regressions over individual units) homogeneity assumption of the panel data. Criado (2008) questioned these assumptions on model estimation and proposed a nonparametric poolability test of Yatchew (2003) in the EKC to avoid functional misspecification. Criado (2008) used a balanced panel of 48 Spanish provinces over the 1990–2002 time period to examine an EKC for air pollutant emission: methane (CH_4), carbon-monoxide (CO), carbon-dioxide (CO_2), and non-methanic volatile organic compounds (NMVOC). His findings indicate that the temporal poolability assumption holds in the Spanish provinces for three pollutants (CH_4, CO, and CO_2), but spatial homogeneity does not hold for all four pollutants. The pooled nonparametric regression suggests existence of an EKC.

A semiparametric model is not only used in the analysis of air and water pollutants. It is also tested over all types of EKC hypotheses. Luzzati and Orsini (2009) used a semiparametric model suggested by Wood (2006) to examine an EKC hypothesis on absolute energy consumption and gross domestic product (GDP) per capita for 113 countries over the period 1971–2004. They used both parametric fixed and random effect models as parametric models. They found the existence of an EKC for energy consumption.

Likewise, the previous research of Poudel et al. (2009) used a semiparametric model to examine an EKC for CO_2 using data from 15 Latin American countries. They used quadratic and cubic one way fixed and random effects models as parametric models and Robinson's partial linear model. They used a test statistic suggested by Li and Wang (1998) for model specification and found that parametric model specification is rejected in favor of semiparametric specification. Their main finding was that they observed "N"-shaped income-CO_2 relation shape for Latin American countries.

Van (2010) examines the existence of EKC on per capita energy consumption using data from the Energy Information Administration (EIA) that includes a balanced panel of 158 countries and territories for the period of 1980–2004. He estimates both parametric and semiparametric models to study the energy pollution relationship. The model used by him is more general than the model used by Luzzati and Orsini (2009). He does not find the existence of EKC on energy consumption. This finding is contradictory to the finding from Luzzati and Orsini (2009). The test statistic suggested by Li and Wang (1998) is used to compare parametric versus semiparametric models with the result that a semiparametric model is suitable for their data.

Wang (2011) propose a flexible nonparametric approach to study the existence of EKC on sulfur emissions from hard coal, brown coal, petroleum, and mining activities from most of the countries of the world over the 1850–1990 period. She also uses B-spline smoothing on the semiparametric model and found mixed results between OECD and non-OECD countries. The results show that an EKC exists in OECD countries, but not in non-OECD countries. A correctly specified model produces the least average mean squared error (AMSE), and is usually used to test for goodness-of-fit statistics. Wang (2011) use average mean squared error to compare parametric versus nonparametric specifications. The smallest AMSE value for a semiparametric model implies that the semiparametric model is better for this data.

In a recent study, Omay (2013) uses a penalized spline regression method to examine the existence of an EKC for carbon dioxide (CO_2) using data from ten developed countries. The results were mixed. The penalized spline method is more general than the spline regression used in the previous literature. In a penalized spline smoothing method, the smoothing parameter is selected automatically, so it is more reliable than spline or B-spline regression. They observed an EKC with an inverted U-shape for France and Switzerland, an "N"-shaped for Austria, an inverted "L"-shaped for Finland and Canada, and an "M"-shaped relation for Denmark. They use a restricted likelihood ratio test (RLRT) suggested by Ruppert et al. (2003) to compare robustness between semiparametric and parametric models. This test statistic is equivalent to testing the presence of random effects for spline regression coefficients. The random effect parameterizes the deviations of a smooth function from a given linear term (Zanin and Marra 2012). The results suggest that the parametric (quadratic or cubic) model is not adequate to capture nonlinearity between pollution and income.

Chiu (2012) also studies an EKC hypothesis in deforestation using data from 52 developing countries over the 1972–2003 period. Chiu uses a panel smooth transition regression (PSTR) model. Results support the EKC hypothesis that, with an increase in real income, deforestation increases initially, and after reaching a certain income level, declines. He uses an F-version of the likelihood ratio test and a pseudo-likelihood ratio test to check model specification. Chiu (2012) finds the existence of an EKC hypothesis for deforestation.

In a recent study, Kim (2013) studies the relationship between air pollution (NO_x and SO_2) emissions and per capita income from 1929–1994 to estimate an EKC model. These data are the same used by Millimet et al. (2003). Kim (2013) uses a kernel-based semiparametric model. He proposes a Uniform Confidence Band (UCB) for the nonparametric component g(.) to test parametric model specifications against the nonparametric model. According to this test statistic, if the nonparametric 95 percent upper confidence band contains a parametric estimate, then we fail to reject the null hypothesis that the parametric specification is correct. He uses that the null of parametric model specification is rejected in favor of a semiparametric model. He also observe the existence of an EKC for sulfur dioxide and carbon dioxide.

6 Future directions

As we mentioned in the previous section, identification of parametric and nonparametric components is a major step in a semiparametric model. Hahn and Ridder (2013) develop a three-step estimation procedure to estimate a semiparametric model. The first step is to select a parametric or a nonparametric component. The second step consists of estimating a nonparametric regression and the third step of estimating a finite-dimensional parametric component. Pandit et al. (2013) selected nonparametric variables using a method suggested by Ruppert et al. (2003) in the off-farm labor allocation. Many EKC papers use a *priori* information to identify a variable that should enter as a nonparametrically or a parametrically, except some articles such as Pandit and Paudel (2016). In this paper, the authors test each variable to find its suitability to enter parametrically or nonparametrically in the model.

Another important aspect in the EKC semiparametric literature is the estimation method used. The new development in the semiparametric literature has provided many alternative methods. Ai et al. (2014) investigate a fixed effects partial panel data model and suggest polynomial spline series approximation and the use of a profiled least-square procedure. They develop the least-square dummy variables estimator for the parametric component and a series estimator for the nonparametric component. Gao et al. (2015) relax the independence for time series data to use both categorical and continuous data in kernel method and extend this so that a mix of continuous and/or categorical variables can appear in the nonparametric part of a partially linear time series model. Feng et al. (2017) develop a semiparametric varying-coefficient categorical panel data model with purely categorical covariates. They propose a two-stage local polynomial estimation for the nonparametric component by applying the additive structure and the series estimator. This method can be applied in the EKC literature with mixed type (continuous and categorical) independent variables.

Many papers, such as those by Lin and Liscow (2012), have brought our attention to the issue of endogeneity in the EKC model, primarily caused by the income variable. Chen et al. (2014) provide an instrumental variable model for a semiparametric consumption-based asset pricing model. This paper contributes toward more conditions needed for the nonlinear, nonparametric models than outlined by Fisher (1966) and Rothenberg (1971) to avoid nonlinearities overwhelming linear effects. Also, Rodriguez-Poo and Soberon (2014) develop higher-dimensional kernel weight and time-varying coefficient models of unknown form with individual effects arbitrarily correlated with the explanatory variables in an unknown way. Yao and Zhang (2015) propose two kernel-based semiparametric IV estimators that relax the tight functional form assumption on the covariates and the reduced form. They have used explicit algebraic structures and are easily implemented without numerical optimizations. If we use their method in the semiparametric EKC model, we may able to remove the bias and endogeneity problem.

Rodriguez-Poo and Soberon (2017) provide recent developments in the functional form of semiparametric and fully nonparametric panel data models that are linearly separable in the innovation and the individual-specific term. They provide an effective way to avoid the curse of dimensionality problem in a nonparametric regression model by using semiparametric additive models. These types of models can be used in EKC studies when there are many explanatory variables other than income.

Although the EKC hypothesis has been tested and proved theoretically and empirically in many cases, the universal existence of the EKC for all pollutants for all time and all locations cannot be established. Even if the presence of EKC can be established in some cases, there is a need to identify and decompose the effects of income and the effects of other confounding factors such as policy variables. We hope attempts will be made to tease out these effects in future research using some of the advanced methods used in nonparametric and semiparametric econometrics literature.

7 Conclusions

This chapter surveyed recent developments in the semiparametric econometric method and the recent use of these developments on EKC studies. From these studies, we found that the partial linear model developed by Robinson (1988) and its extensions are mostly used to test the EKC hypothesis. We observed that many researchers used kernel-based partial linear models in EKC (e. g., Azomahou et al. 2006; Millimet et al. 2003; Paudel et al. 2005; Poudel et al. 2009). Others have used an alternative of kernel regression known as spline smoothing. Further, we found various forms of spline smoothing–based semiparametric models are used in the EKC literature. For example, Millimet et al. (2003) used spline smoothing. Van (2003) used an additive partial linear model suggested by Hastie and Tibshirani (1990), and Luzzati and Orsini (2009) used a spline additive model suggested by Wood (2006). These types of models provide the best mean squared fit as well as prevent overfitting, an important concern in nonparametric smoothing. There are different flexible types of splines used in nonparametric regression. B-spline and P-spline smoothing are more flexible than a simple spline. P-spline is the most flexible method, where an optimum smoothing is determined by the data itself. B-spline and P-spline models are used by Wang (2011) and Zanin and Marra (2012) in EKC studies, respectively.

Various advances in econometrics that capture nonlinearity are still absent in the EKC literature. Although many authors have used additional variables in addition to income, these additional variables are mainly included in a parametric form (Paudel et al. 2005; Van and Théophile 2007; Van 2003). It is likely that these variables may have nonlinear effects, too. We need to investigate whether these variables should enter parametrically or nonparametrically in a model. This type of approach is used by Pandit et al. (2013) in off-farm labor supply decisions by farm operators and their spouses. Pandit and Paudel (2016) used a semiparametric seemingly unrelated model to understand the relationship between water pollutants and income. They indicated that pollution and income can be related to each other so those should not be treated independently. Lawell et al. (2018) used a nonparametric endogenous EKC model for water pollution. They used an instrumental variable approach to address the endogeneity concern of a nonparametric income variable in the semiparametric model.

Racine and Li (2004) suggested a nonparametric estimation procedure which allows both continuous and categorical variables, heretofore absent in previous EKC literature. If a researcher wants to include multiple explanatory variables in the EKC and if some of these are categorical in nature (variables such as political liberties, civil rights, etc.), then Racine and Li's approach comes in handy. Further, Hsiao et al.'s (2007) relaxed model specification tests by Li and Wang (1998) also admit both continuous and categorical variables. Small samples are commonly used in EKC literature, so the usual model specification tests are not valid for the small sample size. The specification test suggested by Hsiao et al. (2007) uses the bootstrap method to derive significance level and therefore works well with a finite-sample.

We observed that spline-based semiparametric models are frequently used in recent EKC literature, but the authors have not included categorical variables as nonparametric components. Nie and Racine (2012) proposed nonparametric spline regression for mixed data, which can be used in future EKC research. Lin and Liscow (2012) observed that the reduced form model used to examine the EKC hypothesis has an endogeneity problem; a semiparametric instrumental regression model developed by Darolles et al. (2011), Horowitz (2011), Santos (2012), and Delgado and Parmeter (2014) can be used in the EKC studies. Another development that can be used in the EKC studies is a dynamic panel semiparametric model, which has been missing so far.

To identify an appropriate functional form between environmental quality and economic growth, we reviewed advanced literature in econometrics specifically related to nonparametric and semiparametric models. Then, we explained how the new developments have been used in EKC literature. We observed that there is still an ongoing debate about the use of econometric specification in EKC analyses. We found that recent studies have focused on relaxing distributional assumptions using nonparametric and semiparametric models. Existing studies have indicated a semiparametric model is better compared to a parametric model. Hence, the EKC hypothesis can be analyzed more accurately using recent econometrics advances in nonparametric/semiparametric models. Future research should consider using a more flexible form of econometric modeling in EKC studies.

Notes

1 A monotone preference relation $\succeq$ on $X = R_+^L$ is homothetic if all indifference sets are related by proportional expansion along rays; that is, if $x \sim y$, then $\alpha x \sim \alpha y$ for any $\alpha \geq 0$.
2 Although Millimet et al. (2003) used spline as a parametric model, spline smoothing is a parametric approach of estimating a nonparametric model (Ruppert et al. 2003).

References

Ai, C., and Q. Li. 2008. "Semi-parametric and Non-parametric Methods in Panel Data Models." In P.S. László Mátyás, ed., *The Econometrics of Panel Data: Fundamentals and Recent Developments in Theory and Practice*. Berlin Heidelberg: Springer-Verlag, pp. 451–478.

Ai, C., J. You, and Y. Zhou. 2014. "Estimation of Fixed Effects Panel Data Partially Linear Additive Regression Models." *The Econometrics Journal* 17:83–106.

Andreoni, J., and A. Levinson. 2001. "The Simple Analytics of the Environmental Kuznets Curve." *Journal of Public Economics* 80:269–286.

Azomahou, T., F. Laisney, and P. Nguyen Van. 2006. "Economic Development and CO2 Emissions: A Nonparametric Panel Approach." *Journal of Public Economics* 90:1347–1363.

Bertinelli, L., and E. Strobl. 2005. "The Environmental Kuznets Curve Semi-parametrically Revisited." *Economics Letters* 88:350–357.

Blundell, R.W., and J.L. Powell. 2004. "Endogeneity in Semiparametric Binary Response Models." *Review of Economic Studies* 71:655–679.

Brock, W.A., and M.S. Taylor. 2010. "The Green Solow Model." *Journal of Economic Growth* 15(2):127–153.

Carson, R.T. 2010. "The Environmental Kuznets Curve: Seeking Empirical Regularity and Theoretical Structure." *Review of Environmental Economics and Policy* 4:3–23.

Chen, X., V. Chernozhukov, S. Lee, and W.K. Newey. 2014. "Local Identification of Nonparametric and Semiparametric Models." *Econometrica* 82:785–809.

Chen, X., and Y. Fan. 2006. "Estimation of Copula-based Semiparametric Time Series Models." *Journal of Econometrics* 130:307–335.

Chiu, Y.-B. 2012. "Deforestation and the Environmental Kuznets Curve in Developing Countries: A Panel Smooth Transition Regression Approach." *Canadian Journal of Agricultural Economics/Revue Canadienne d'Agroeconomie* 60:177–194.

Crainiceanu, C. M., & D. Ruppert. 2004. "Likelihood Ratio Tests for Goodness-of-Fit of a Nonlinear Regression Model." *Journal of Multivariate Analysis* 91(1):35–52.

Criado, C.O. 2008. "Temporal and Spatial Homogeneity in Air Pollutants Panel EKC Estimations." *Environmental and Resource Economics* 40:265–283.

Darolles, S., Y. Fan, J. Florens, and E. Renault. 2011. "Nonparametric Instrumental Regression." *Econometrica* 79:1541–1565.

Dasgupta, S., B. Laplante, H. Wang, and D. Wheeler. 2002. "Confronting the Environmental Kuznets Curve." *Journal of Economic Perspectives* 16(1):147–168.

Delgado, M.S., and C.F. Parmeter. 2014. "A Simple Estimator for Partial Linear Regression with Endogenous Nonparametric Variables." *Economics Letters* 124(1):100–103.

Delgado, M.A., and T. Stengos. 1994. "Semiparametric Specification Testing of Non-nested Econometric Models." *The Review of Economic Studies* 61:291–303.

Fan, Y., and Q. Li. 1996. "Consistent Model Specification Tests: Omitted Variables and Semiparametric Functional Forms." *Econometrica* 64:865–890.

Feng, G., J. Gao, B. Peng, and X. Zhang. 2017. "A Varying-Coefficient Panel Data Model with Fixed Effects: Theory and an Application to US Commercial Banks." *Journal of Econometrics* 196:68–82.

Fisher, F. 1966. *The Identification Problem in Econometrics*. New York: McGraw-Hill.

Gao, Q., Long, L., and J.S. Racine. 2015. "A Partially Linear Kernel Estimator for Categorical Data." *Econometric Reviews* 34:959–978.

González, A., T. Teräsvirta, and D. Van Dijk. 2005. "Panel Smooth Transition Model and an Application to Investment under Credit Constraints." Working Paper, Stockholm School of Economics, Stockholm, Sweden.

Griffin, J.E., and M.F. Steel. 2010. "Bayesian Nonparametric Modelling with the Dirichlet Process Regression Smoother." *Statistica Sinica* 20:1507–1527.

Grossman, G.M., and A.B. Krueger. 1991. *Environmental Impacts of a North American Free Trade Agreement (No. w3914)*. Cambridge, MA: National Bureau of Economic Research.

Grossman, G.M., and A.B. Krueger. 1995. "Economic Growth and the Environment." *Quarterly Journal of Economics* 110(2):353–377.

Hahn, J., and G. Ridder. 2013. "Asymptotic Variance of Semiparametric Estimators with Generated Regressors." *Econometrica* 81:315–340.

Harbaugh, W.T., A. Levinson, and D.M. Wilson. 2002. "Reexamining the Empirical Evidence for an Environmental Kuznets Curve." *Review of Economics and Statistics* 84(3):541–551.

Hardle, W., and E. Mammen. 1993. "Comparing Nonparametric versus Parametric Regression Fits." *The Annals of Statistics* 21:1926–1947.

Hastie, T., and R. Tibshirani. 1990. *Generalized Additive Models*. New York: Chapman and Hall.

Hausman, J.A. 1978. "Specification Tests in Econometrics." *Econometrica* 46(6):1251–1271.

Hong, Y., and H. White. 1995. "Consistent Specification Testing via Nonparametric Series Regression." *Econometrica* 63:1133–1159.

Horowitz, J.L. 2011. "Applied Nonparametric Instrumental Variables Estimation." *Econometrica* 79:347–394.

Hsiao, C., Q. Li, and J.S. Racine. 2007. "A Consistent Model Specification Test with Mixed Discrete and Continuous Data." *Journal of Econometrics* 140:802–826.

John, A., and R. Pecchenino. 1994. "An Overlapping Generations Model of Growth and the Environment." *Economic Journal* 104(427):1393–1410.

Kelly, D.L. 2003. "On Environmental Kuznets Curves Arising from Stock Externalities." *Journal of Economic Dynamics and Control* 27:1367–1390.

Kijima, M., K. Nishide, and A. Ohyama. 2010. "Economic Models for the Environmental Kuznets Curve: A Survey." *Journal of Economic Dynamics and Control* 34(7):1187–1201.

Kim, K.H. 2013. "Inference of the Environmental Kuznets Curve." *Applied Economics Letters* 20:119–122.

Lee, T.-H., and A. Ullah. 2001. "Nonparametric Bootstrap Tests for Neglected Nonlinearity in Time Series Regression Models." *Journal of Nonparametric Statistics* 13:425–451.

Lawell, C.Y.C.L., K. P. Paudel, & M. Pandit. 2018. "One Shape Does Not Fit All: A Nonparametric Instrumental Variable Approach to Estimating the Income-Pollution Relationship at the Global Level." *Water Resources and Economics* 21:3–16.

Li, Q., C. J. Huang, D. Li, and T. Fu. 2000. "Semiparametric Smooth Coefficient Models." *Journal of Business & Economic Statistics* 20(3):412–422.

Li, Q., J.S. Racine, and J.M. Wooldridge. 2009. "Efficient Estimation of Average Treatment Effects with Mixed Categorical and Continuous Data." *Journal of Business & Economic Statistics* 27:206–223.

Li, Q., and S. Wang. 1998. "A Simple Consistent Bootstrap Test for a Parametric Regression Function." *Journal of Econometrics* 87:145–165.

Li, Q.I., D. Ouyang, and J.S. Racine. 2011. "Categorical Semiparametric Varying-Coefficient Models." *Journal of Applied Econometrics*: n.p.

Lin, C.-Y.C., and Z.D. Liscow. 2012. "Endogeneity in the Environmental Kuznets Curve: An Instrumental Variables Approach." *American Journal of Agricultural Economics* 95:268–274.

Linton, O., and J.P. Nielsen. 1995. "A Kernel Method of Estimating Structured Nonparametric Regression Based on Marginal Integration." *Biometrika* 82(1):93–100.

Lopez, R. 1994. "The Environment as a Factor of Production: The Effects of Economic Growth and Trade Liberalization." *Journal of Environmental Economics and Management* 27:163–184.

Luzzati, T., and M. Orsini. 2009. "Investigating the Energy-environmental Kuznets Curve." *Energy* 34:291–300.

Ma, S., and J.S. Racine. 2013. "Additive Regression Splines with Irrelevant Categorical and Continuous Regressors." *Statistica Sinica* 23:515–541.

Ma, S., J.S. Racine, and L. Yang. 2015. "Spline Regression in the Presence of Categorical Predictors." *Journal of Applied Econometrics* 30:705–717.

McConnell, K.E. 1997. "Income and the Demand for Environmental Quality." *Environment and Development Economics* 2:383–399.

Millimet, D.L., J.A. List, and T. Stengos. 2003. "The Environmental Kuznets Curve: Real Progress or Misspecified Models?" *Review of Economics and Statistics* 85:1038–1047.

Nguyen Van, P., and T. Azomahou. 2007. "Nonlinearities and Heterogeneity in Environmental Quality: An Empirical Analysis of Deforestation." *Journal of Development Economics* 84:291–309.

Nie, Z., and J.S. Racine. 2012. "The CRS Package: Nonparametric Regression Splines for Continuous and Categorical Predictors." *The R Journal* 4:48–56.

Omay, R.E. 2013. "The Relationship between Environment and Income: Regression Spline Approach." *International Journal of Energy Economics and Policy* 3:52–61.

Panayotou, T. 1997. "Demystifying the Environmental Kuznets Curve: Turning a Black Box into a Policy Tool." *Environment and Development Economics* 2:465–484.

Pandit, M., and K.P. Paudel. 2016. "Water Pollution and Income Relationships: A Seemingly Unrelated Partially Linear Analysis." *Water Resources Research* 52:7668–7689.

Pandit, M., K.P. Paudel, and A.K. Mishra. 2013. "Do Agricultural Subsidies Affect the Labor Allocation Decision? Comparing Parametric and Semiparametric Methods." *Journal of Agricultural and Resource Economics* 38:1–18.

Paudel, K.P., and B.N. Poudel. 2012. "Functional Form of Water Pollutants-income Relationship Under the Environmental Kuznets Curve Framework." *American Journal of Agricultural Economics* 95:261–267.

Paudel, K.P., and M.J. Schafer. 2009. "The Environmental Kuznets Curve under a New Framework: The Role of Social Capital in Water Pollution." *Environmental and Resource Economics* 42:265–278.

Paudel, K.P., H. Zapata, and D. Susanto. 2005. "An Empirical Test of Environmental Kuznets Curve for Water Pollution." *Environmental and Resource Economics* 31:325–348.

Poudel, B.N., K.P. Paudel, and K. Bhattarai. 2009. "Searching for an Environmental Kuznets Curve in Carbon Dioxide Pollutant in Latin American Countries." *Journal of Agricultural and Applied Economics* 41:13–27.

Qian, J., and L. Wang. 2012. "Estimating Semiparametric Panel Data Models by Marginal Integration." *Journal of Econometrics* 167:483–493.

Racine, J., and Q. Li. 2004. "Nonparametric Estimation of Regression Functions with Both Categorical and Continuous Data." *Journal of Econometrics* 119:99–130.

Robinson, P.M. 1988. "Root-n-Consistent Semiparametric Regression." *Econometrica* 56:931–954.

Rodriguez-Poo, J.M., and A Soberon. 2014. "Direct Semi-Parametric Estimation of Fixed Effects Panel Data Varying Coefficient Models." *The Econometrics Journal* 17:107–138.

Rodriguez-Poo, Juan, M., and A. Soberon. 2017. "Nonparametric and Semiparametric Panel Data Models: Recent Developments." *Journal of Economic Surveys* 31:923–960.

Rothenberg, T.J. 1971. "Identification in Parametric Models." *Econometrica* 39(3):577–591.

Roy, N., and G. Cornelis van Kooten. 2004. "Another Look at the Income Elasticity of Non-Point Source Air Pollutants: A Semiparametric Approach." *Economics Letters* 85:17–22.

Ruppert, D., P. Wand, and R.J. Carroll. 2003. *Semiparametric Regression*. Cambridge: Cambridge University Press.

Santos, A. 2012. "Inference in Nonparametric Instrumental Variables with Partial Identification." *Econometrica* 80:213–275.

Shafik, N., and S. Bandyopadhyay. 1992. *Economic Growth and Environmental Quality: Time-Series and Cross-Country Evidence* (Vol. 904). Washington, D.C.: World Bank Publications.

Stern, D.I. 2004. "The Rise and Fall of the Environmental Kuznets Curve." *World Development* 32:1419–1439.

Tsurumi, T., & S. Managi. 2010. "Decomposition of the Environmental Kuznets Curve: Scale, Technique, and Composition Effects." *Environmental Economics and Policy Studies* 11(1–4):19–36.

Ullah, A. 1985. "Specification Analysis of Econometric Models." *Journal of Quantitative Economics* 2:187–209.

Ullah, A., and N. Roy. 1998. "Nonparametric and Semiparametric Econometrics of Panel Data." In A. Ullah, ed., *Handbook of Applied Economic Statistics*. New York: Marcel Dekke, CRC Press, pp. 579–604.

Van, P.N. 2003. "A Semiparametric Analysis of Determinants of a Protected Area." *Applied Economics Letters* 10:661–665.

Van, P.N. 2010. "Energy Consumption and Income: A Semiparametric Panel Data Analysis." *Energy Economics* 32:557–563.

Van, P.N., and A. Théophile. 2007. "Nonlinearities and Heterogeneity in Environmental Quality: An Empirical Analysis of Deforestation." *Journal of Development Economics* 84:291–309.

Vangeneugden, T., et al. 2011. "Marginal Correlation from an Extended Random-Effects Model for Repeated and Overdispersed Counts." *Journal of Applied Statistics* 38:215–232.

Vincent, J.R. 1997. "Testing for Environmental Kuznets Curves Within a Developing Country." *Environment and Development Economics* 2:417–431.

Wand, M.P., and M.C. Jones. 1994. *Kernel Smoothing*. Boca Raton, FL: CRC Press.

Wang, L. 2011. "A Nonparametric Analysis on the Environmental Kuznets Curve." *Environmetrics* 22:420.

Wood, S.N. 2006. *Generalized Additive Models: An Introduction with R*. New York: Chapman & Hall/CRC.

Yao, F., and J. Zhang. 2015. "Efficient Kernel-Based Semiparametric IV Estimation with an Application to Resolving a Puzzle on the Estimates of the Return to Schooling." *Empirical Economics* 48:253–281.

Yatchew, A. 2003. *Semiparametric Regression for the Applied Econometrician*. Cambridge: Cambridge University Press.

Zanin, L., and G. Marra. 2012. "Assessing the Functional Relationship Between CO2 Emissions and Economic Development Using an Additive Mixed Model Approach." *Economic Modelling* 29:1328–1337.

Zheng, J.X. 1996. "A Consistent Test of Functional Form Via Nonparametric Estimation Techniques." *Journal of Econometrics* 75:263–289.

29

Reconstructing deterministic economic dynamics from volatile time series data

Ray Huffaker, Ernst Berg, and Maurizio Canavari

1 Introduction

Price series data taken from real-world markets are often volatile and seemingly random in appearance. Economists questioned how volatility persists when rational agents could realize above-normal profits over time by engaging in countercyclical strategies that would smooth out price cycles (Hayes and Schmitz 1987). They reasoned that either economic agents behave naïvely with linear cobweb price adjustments (Ezekiel 1938; Harlow 1960; Waugh 1964) or agents cannot predict economic cycles to the extent required to formulate countercyclical strategies (Chavas and Holt 1991).

Why would economic cycles be unpredictable? The dominant "efficient-markets hypothesis" blames exogenous random shocks that continually bombard otherwise intrinsically stable markets (Fama 1970). Persistent volatility reflects the market's natural self-corrective mechanism, as rational economic agents re-adjust supply to demand in response to shocks.

There is another possibility: markets may be intrinsically unstable (Chavas and Holt 1991; Galtier 2013; Huffaker et al. 2017; Minsky 1992). Economic studies have demonstrated that persistent volatility might be due to destabilizing endogenous market factors, including highly inelastic demands (Chavas and Holt 1993; Chavas and Holt 1991); nonlinear cobweb price expectations (Jensen and Urban 1984); and financial, institutional, and biophysical constraints frustrating supply from matching demand (Berg and Huffaker 2015). In general, nonlinear dynamics teaches us that complexity can emerge deterministically from relatively simple low-dimensional nonlinear ("chaotic") systems, and, despite being deterministic, these systems can produce unpredictable output (Glendinning 1994; Gregson and Guastello 2011; Kaplan and Glass 1995).

Accurately distinguishing between exogenous and endogenous volatility has important implications for economic policy. Stable markets are self-correcting and thus support the adoption of non-interfering laissez-faire policies. Alternatively, intrinsically unstable markets do not provide a self-correcting mechanism and thus make a case for government intervention designed to reduce volatility and/or mitigate its negative impacts. An instructive case study is provided by the recent failure of macroeconomic models to reproduce the temporal patterns of booms and busts observed in the 2008 financial crisis (Huffaker 2015). In a special hearing entitled "Building a Science of Economics for the Real World," the U.S. Congress concluded that "if major crises are a recurrent feature of the economy then our models should incorporate this possibility," and expressed frustration that "because our experts' way of looking at the economy left them blind to the crisis that was building, we were unprepared to deal with the crisis" (U.S. House of Representatives 2010). In a series of articles entitled "Big Economic Ideas" (*The Economist* 2016a), the *Economist* recommended that "like physicists, [economists] should study instability instead of assuming that economies naturally self-correct" (*The Economist* 2016b) and re-awaken to Minsky's "financial instability hypothesis" (Minsky 1992) explaining how financial booms systematically breed their own busts (*The Economist* 2016c).

The Congressional hearing clearly demonstrates that economic models used for public policy are "live ammunition." Models corresponding poorly to the real-world facing policy makers "leave the real problem unaddressed, waste resources, and impede learning" (Saltelli and Funtowitz 2014). An economist empaneled at the hearing recommended that the National Science Foundation formally audit economic models used in public policy. This follows an earlier recommendation by Oreskes et al. (1994) that modelers have the burden "to demonstrate the degree of correspondence between the model and the material world it seeks to represent when public policy and public safety are at stake" (Oreskes et al. 1994). The European Commission's Joint Research Centre formally audits models used to assess the impacts of EU initiatives, legislation, and policy.

We cannot reliably diagnose whether real-world economic dynamics are stochastic or deterministic simply from observing volatility in time series data. The theory of randomness from the philosophy of science literature teaches that mathematically random output can be generated by both indeterministic and deterministic processes, and we cannot detect the generating process simply by observing output volatility (Horan 1994). Nor can we rely on conventional linear methods (e.g., autocorrelation plots), since they fail to detect nonlinear deterministic structure in time series data (Huffaker 2015; Kaplan and Glass 1995). Finally, we cannot validate that real-world economic dynamics are, for example, stochastic simply by demonstrating that predictions of a stable linear–stochastic market model provide a "good fit" to observed values. This conventional model-centric approach commits the logical fallacy of "affirming the consequent": if P, then Q; Q, therefore P. Or, in this context: if the model is true, it provides a good fit; the stochastic model provides a good fit, therefore it is true (Oreskes et al. 1994). The problem is that other models structured very differently can be parameterized to also provide good fits (Hornberger and Spear 1981; Rykiel 1991).

We propose a pre-modeling data diagnostics framework – based on nonlinear time series analysis (NLTS) methods developed in the mathematical physics literature – designed to detect whether observed time series data are most likely generated by linear–stochastic or nonlinear deterministic dynamics. The centerpiece of NLTS is "phase space reconstruction," which reconstructs system dynamics from a single observed time series variable.

Phase space representations of system dynamics record the level of system (state) variables at each point in time. Phase space "trajectories" connecting these points depict the evolution of the system through time from given initial levels. In a "dissipative" dynamic system, long-run system dynamics are bounded within a subset of phase space wherein trajectories evolve along an "attractor" – a geometric structure with noticeable regularity (Brown 1996). Early practitioners assumed that they needed data on all system variables to construct phase space dynamics. This severely limited phase space as an empirical tool because we cannot reasonably identify or measure all variables interacting in real-world dynamic systems. A major breakthrough occurred when mathematicians discovered that phase space dynamics could be reconstructed from a single time series variable using delay coordinates (Breeden and Hubler 1990).

We illustrate phase space reconstruction with data generated by a dynamic Walrasian price and quantity adjustment model (Appendix). The model is parameterized to generate periodic time series solutions (Figure 29.1a). Phase space trajectories are attracted toward a "limit cycle" attractor along which they oscillate periodically thereafter (Figure 29.1b). Thanks to Takens' theorem (Takens 1980), we can use either the price or the quantity series – we do not require both – to reconstruct system dynamics in a "shadow" phase space.

Takens' theorem allows us to use a lagged copy of an observed variable to serve as a surrogate for an unobserved variable. This procedure, "time-delay embedding," ensures one-to-one correspondence between the original and the reconstructed attractors so long as a shadow phase space has sufficient dimensions to contain the original attractor. This means that the phase spaces are "topologically" equivalent, so that dynamic properties in the original phase space are preserved in a shadow phase space. In this illustration, we know that the dynamic system generating the data is two dimensional, so a shadow phase space must also be at least two dimensional for topological equivalence. However, in practice, we cannot directly observe the real-world dynamic system that generated observed data; consequently, we must rely on statistical tests (discussed later) to estimate the required dimensionality of a shadow phase space. In sum, NLTS potentially allows us to investigate real-world system dynamics from a single observed time series variable and diagnose whether complex long-run dynamic behavior can be adequately captured in a nonlinear deterministic model with a small number of state variables.

In Figure 29.1c, shadow phase space is reconstructed from both time series to demonstrate the dynamic resemblance between the original and shadow attractors. The one-to-one correspondence between the original and shadow phase spaces implies that the shadow phase spaces also correspond one-to-one. This is the foundation of the "convergent cross mapping" method

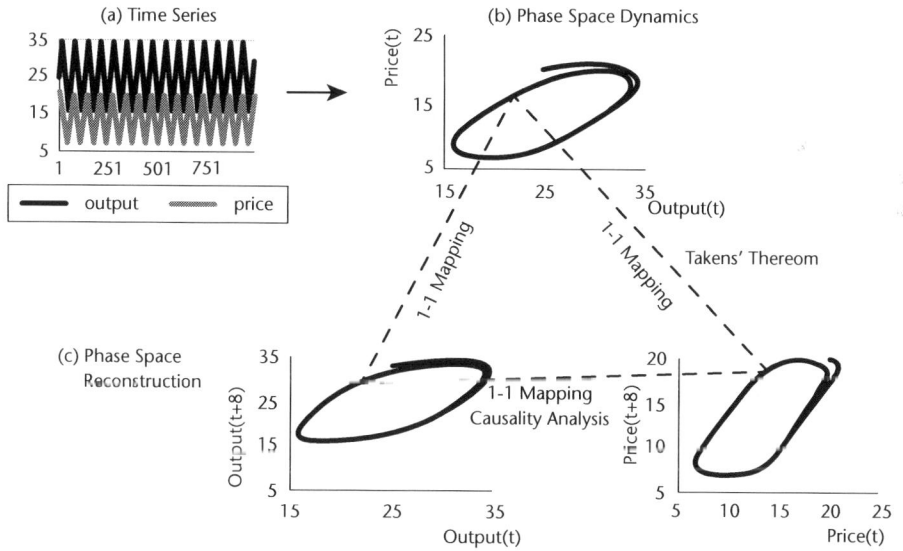

Figure 29.1 Phase space reconstruction. (a) Time series solutions to a dynamic Walrasian output and price adjustment model (Appendix) are periodic. (b) Phase space trajectories gravitate toward, and evolve along, a limit cycle attractor. (c) Reconstruction of shadow phase spaces using an eight-period delayed copy of the output series as a surrogate for the price series (left plot) and an eight-period delayed copy of the price series as a surrogate for the output series (right plot). The shadow phase spaces are topologically equivalent to the original phase space in (b), and thus either can be used to investigate system dynamics

(discussed later) used to detect causal interactions among multiple observed system variables in nonlinear deterministic dynamic systems (Sugihara et al. 2012).

NLTS, like any other tool, may not work all the time. In practice, NLTS can reveal deterministic structure only to the extent that it is embedded into observed time series data, which are often noisy, short, and nonstationary. Noisy data obscure systematic real-world dynamics (Kot 1988). Time series that are short relative to the lengths of the system's characteristic behavioral cycles may be nonstationary, causing modeling problems discussed below. NLTS methods are designed to mitigate the limitations of noisy, short, and nonstationary time series data in empirically reconstructing real-world dynamics. However, we can only reasonably hope to reconstruct a "sampling" or "skeleton" of the true real-world attractor (Ghil et al. 2002). Finally, some problems cannot be resolved: observed time series data may not rest on the real-world attractor, or a low-dimensional real-world attractor may not exist in the first place (Williams 1997).

The literature strongly advises against the conventional early practice of relying on topological measures estimated from observed time series data to positively demonstrate that real-world system dynamics are chaotic (Kantz and Schreiber 1997). Topological measures of deterministic chaos (discussed later) are based on asymptotic properties best approximated with high-quality and massive datasets, such as those generated by laboratory experiments designed expressly to detect chaos. However, these measure are unreliable when estimated from short and noisy real-world data (McSharry 2011; Schreiber 1999). They can be reliably used in NLTS diagnostics, which relaxes inquiry from positively proving deterministic chaotic dynamics in observed data to providing empirical evidence supporting low-dimensional, nonlinear, and deterministic dynamics as a viable explanation for volatility (Schreiber 1999).

Our objectives for this chapter are to propose a strategy for reliably applying NLTS diagnostics, introduce key components (while providing sources of more detailed explanations), and illustrate the strategy by summarizing some recent applications.

2 A strategy for NLTS diagnostics

Based on an exhaustive review of NLTS methods in the mathematical physics literature, we propose a strategy for implementing NLTS in applied empirical economic research designed to withstand expert scrutiny. We emphasize from the start that we see NLTS as an aid to – not a replacement for – theoretical modeling explaining observed phenomena. We join the chorus of voices recommending "getting to know your data" as an essential part of evidentiary scientific inquiry (Haack 1999). NLTS diagnostics can inform model specification by providing empirical evidence of key variables interacting in a real-world dynamic system and of the number of state variables required to adequately reproduce detected dynamics. We also emphasize that we do not intend the strategy to serve as a "treasure map" to find chaos buried in the universe. NLTS is capable of reconstructing real-world dynamics detected in the data, whether they be linear, nonlinear, stochastic, or deterministic. The important consideration is to make an accurate diagnosis.

In a nutshell, we first apply "signal processing" methods to separate the observed time series into a "signal' component (isolating structured variation in the data) and a "noise" component (isolating unstructured variation) (Figure 29.2).

We test an isolated signal for the "nonlinear stationarity" required by NLTS diagnostics. We next diagnose a stationary signal for underlying system dynamics with "phase space reconstruction' (to detect the presence of a low-dimensional shadow attractor) and "surrogate data testing" (to test the hypothesis that visual regularity observed in a reconstructed attractor was not mimicked by linear–stochastic dynamics). Empirical evidence of low-dimensional, nonlinear, deterministic dynamics opens the door to causality testing among multiple-observed time series

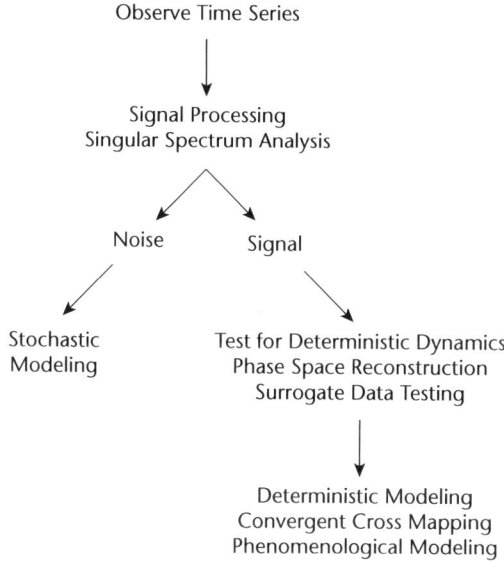

Figure 29.2 A strategy for NLTS diagnostics

variables ("convergent cross mapping") and data-driven ("phenomenological") modeling capable of extracting equations of motion reproducing a shadow phase-space attractor. Finally, the noise component can be investigated with a broad array of familiar stochastic approaches. For example, extreme value statistics (EVS) provide a useful noise diagnostic: "return-level plots" showing expected waiting ("return") times before particular extreme noise levels are realized (Katz 2010). For example, EVS has been applied in economic research to model extreme fluctuations in food prices (Huffaker et al. 2016) and in the growth rate of U.S.gross domestic product (GDP) (Huffaker 2015).

We now consider each of the components in Figure 29.2 in more detail. More in-depth presentation is available in Kantz and Schreiber (1997), Gregson and Guastello (2011), and Huffaker et al. (2017).

2.1 Signal processing

We apply the "singular spectrum analysis" (SSA) method of signal processing (Elsner and Tsonsis 2010; Ghil et al. 2002; Golyandina et al. 2001). SSA proceeds in three stages to separate signal (including trend and oscillatory components) from noise in observed data: decomposition, grouping, and reconstruction.

In the decomposition stage, we construct a "trajectory" matrix whose columns are the observed time series and its delayed copies. SSA requires only one parameter: the "window length" L, which sets the number of rows in the trajectory matrix. A rule of thumb is to set window length proportional to the dominant cycle length in the Fourier power spectrum and below one-half the length of the time series (Golyandina et al. 2001; Hassani 2007). "Singular value decomposition" methods are used to decompose the trajectory matrix into a sum of matrices. The eigenvalues from the decomposition measure the proportion of variance in the time series explained by each component. The cumulative proportion of variance explained by signal components provides a measure of signal "strength."

In the grouping stage, we consult a series of diagnostics provided by the decomposition to divide the decomposed matrices into groups corresponding to various signal and noise components of the time series.

In the reconstruction stage, we convert the grouped signal and noise matrix components from the decomposition stage back into time series, isolating these components with "diagonal averaging." In the end, we have time series vectors isolating signal components (trend and oscillations) from the noise component.

2.2 Stationarity testing

We test a strong signal – one that explains a majority of the variance in the time series – for the nonlinear stationarity required by NLTS. If the signal is stationary, we can reasonably expect that it is the realization of a single dynamic system. We do not expect reconstructed phase space trajectories to jump from one shadow attractor to another. The "weak" stationarity required by linear time series methods does not suffice for NLTS because an attractor can shift through time even if mean and variance do not vary significantly across segments of the corresponding time series (Schreiber 1997). Consequently, Schreiber (1997) devised "nonlinear cross prediction" to test for similar nonlinear behavior (rather than similar statistical parameters) across non-overlapping segments. The time series is stationary for purposes of nonlinear analysis if each segment can be used to skillfully cross predict values on the others with a "nonlinear prediction" algorithm (Kaplan and Glass 1995).

In general, the nonlinear prediction algorithm reconstructs a shadow attractor from the time series using time-delay embedding. Points on the attractor are divided into "learning" and "validation" bases. With one-step-ahead prediction, the final point in the learning base is taken as the reference point on the attractor, and its nearest neighboring points are identified and advanced one period. The prediction is calculated as a weighted average of the advanced neighboring points, where weights depend on the distance of a neighbor from the reference point. Next, the learning base is augmented by first point in the validation base, and the process continues until all points in the validation base have been predicted. This not only predicts points on the attractor, but also actual time series values, since time-delay embedding ensures that the first element of a multidimensional point on an attractor at a given time is the time series observation itself at that time. Prediction skill is assessed with a goodness-of-fit measure (such as the Pearson correlation coefficient) comparing actual and predicted time series values. In the context of nonlinear cross prediction, each segment of the time series serves in turn as the learning base to cross predict values on the others.

From this point on, when we refer to time series data, we mean a stationary signal that has been isolated from the observed time series.

2.3 Phase space reconstruction

The time-delay embedding method of phase space reconstruction requires that we select three parameters: the "embedding delay," the "Theiler window," and the "embedding dimension."

The embedding delay is the time needed to separate delayed copies of the time series for an accurate reconstruction. For the Walrasian model, we reconstructed shadow limit cycle attractors accurately portraying the original by selecting an embedding delay of eight periods (Figure 29.1b,c). In general, the embedding delay is selected to create statistical independence between a time series and its successive delayed copies. If the embedding delay is too short, reconstructed system dynamics may not have sufficient time to evolve. Alternatively, if it is too long, reconstructed system dynamics may skip over important structure. A popular rule-of-thumb is to select the embedding delay as the first minimum of the "average mutual information" (AMI) function – a probabilistic measure of the extent to which a time series is related to delayed copies of itself (Williams 1997).

When attempting to visualize deterministic structure in a shadow attractor, we need to know that neighboring points are close because of the attractor's geometry and not simply because they are proximate in time. The Theiler window aids us in this effort by excluding a window of serially correlated points in reconstructing the attractor (Kantz and Schreiber 1997). The Theiler window is conventionally selected as the first minimum of the autocorrelation function measuring the linear correlation of a time series with successive delayed copies (Williams 1997).

The embedding dimension identifies the minimum number of delayed copies required to accurately reconstruct the shadow attractor. In practice, the embedding dimension provides empirical evidence for whether observed complexity conceals long-term dynamics that are reducible to a low-dimensional shadow attractor. In the Walrasian illustration, we reconstructed shadow attractors from data generated by a two-dimensional model (Appendix). So, we knew that the embedded dimension needed to be at least two (including one of the time series and one of its delayed copies) to ensure one-to-one correspondence between the original and shadow attractors (Figures 29.1b,c).

In general, the embedding dimension is conventionally estimated using the "false nearest neighbors" (FNN) test. Given the embedding delay and the Theiler window, the FFN test first reconstructs a shadow attractor for two embedding dimensions and measures the distances between points in phase space. The test omits points whose time indices fall within the Theiler window. The test repeats this exercise for three embedding dimensions and computes the percentage of false nearest neighbors, that is, the percentage of points that grow apart as the embedding dimension increases from two to three. This is repeated for successively higher embedding dimensions, and the selected value is the minimum dimension for which the percentage of false nearest neighbors falls below a given threshold (Williams 1997).

We can illustrate the concept of false nearest neighbors with a real-world example. Ticket prices for soccer matches tend to be higher at mid-field because spectators have a better vantage point to accurately assess the true distances between players. Alternatively, spectators behind goal posts mistakenly perceive that players are closer than they really are – the players are false nearest neighbors. Due to this distorted geometric perception, spectators at the end of the field have more difficulty watching attacks on the other goal unfold.

We use the estimated embedding parameters to construct an "embedded data matrix." The columns of this matrix are composed of the time series and its successive delayed copies, so that the number of columns equals the embedding dimension. The rows are multidimensional points on a shadow attractor. We can visualize the attractor geometrically by scatterplotting these points. The row dimension of the embedded data matrix (and thus the number of points on the attractor) equals $n - (m - 1)d$, where n is the length of the time series, m is the embedding dimension, and d is the embedding delay.

2.4 Surrogate data testing

The systematic geometric appearance of a reconstructed shadow attractor provides evidence that deterministic system dynamics may be concealed in volatile observed data. However, the possibility remains that apparent structure was generated fortuitously by mimicking stochastic dynamics. "Surrogating data testing" was developed to statistically test the null hypothesis that observed data were generated by a stochastic real-world dynamic process (Schreiber and Schmitz 2000; Theiler et al. 1992).

We first generate an ensemble of randomized "surrogate" time series vectors that destroy structured variation in the observed time series while maintaining selected statistical properties. For example, "independently and identically distributed" (IID) surrogates are constructed as random draws (without replacement) from the data. They shuffle the time order of the data while maintaining the same probability distribution. Consequently, IID surrogates would be used to test the null hypothesis that all structure in the data is given by the probability distribution. In comparison, "amplitude adjusted Fourier transform" (AAFT) surrogates – the most popular in applied work – test whether structure in the data is limited to the probability distribution and the Fourier power spectrum. If this turns out to be the case, we reject the alternative hypothesis of deterministic dynamics.

We next reconstruct a shadow attractor for each surrogate data vector and take measurements of hallmark characteristics of low-dimensional, nonlinear, deterministic dynamics ("discriminating statistics"). Discriminating statistics include the "correlation dimension" (measuring the fractal (non-integer) dimensionality of an attractor); the "maximum Lyapunov exponent" (measuring sensitivity to initial conditions by virtue of the average exponential divergence of trajectories from neighboring initial conditions); and "nonlinear prediction skill" (providing evidence of the attractor's deterministic structure via iterative one-step-ahead prediction as discussed earlier).

Finally, we apply nonparametric rank-order statistics to determine whether the discriminating statistics taken from the empirically reconstructed attractor (i.e., the shadow attractor reconstructed from the observed data) fall within the extreme ranges of values taken from surrogate attractors. If so, we accept the alternative hypothesis that geometric structure visualized in the empirically reconstructed attractor is generated by low-dimensional, nonlinear deterministic real-world dynamics.

2.5 Convergent cross mapping

"Convergent cross mapping" (CCM) offers a mathematically rigorous method for testing whether multiple observed time series variables causally interact in the same real-world dynamic system (Sugihara et al. 2012). CCM determines that a "forcing" variable drives a "response" variable by measuring whether there is a one-to-one correspondence between the shadow attractors reconstructed from each variable (Figure 29.1c). In particular, CCM measures how skillfully the shadow attractor reconstructed from the response variable can cross predict values of the forcing variable. The rationale is that forcing variables embed their dynamics into response variables.

CCM may mistakenly detect causal interactions (*false positives*) in the case of *generalized synchrony* (Sugihara et al. 2012; Ye et al. 2015). This can occur when two non-interactive time series variables are synchronized because they are both driven by a third variable. Ye et al. (2015) developed "extended (delayed) CCM" as a second test to screen out false positives in the face of generalized synchrony.

2.6 Phenomenological (data-driven) modeling

In general, phenomenological models mathematically describe relationships among empirically observed phenomena without attempting to explain underlying mechanisms (Hilborn and Mangel 1997). In the context of NLTS, we can use phenomenological modeling to extract equations of motion governing a real-world dynamic system from a single observed time series variable (Baker et al. 1996; Brunton et al. 2016).

Reconstructing economic dynamics from data

We summarize how this works for three variables ($X(t)$, $Y(t)$, $Z(t)$) in Figure 29.3. If we are limited to observing only a single time series variable, $X(t)$, we can use its time-delayed copies as surrogates for the unobserved variables: ($X(t)$, $Y(t) = X(t + d)$, $Z(t) = X(t + 2d)$) and reconstruct a shadow phase space in the delayed coordinates. Alternatively, if we observe all three variables, we reconstruct a shadow space for each and apply CCM to test whether they causally interact in the same real-world dynamic system. From phase space reconstruction, we learn the minimum model dimensionality required to capture system dynamics and have empirical evidence of causally interactive variables to include in the specification.

We next fit a system of ordinary differential equations (ODE) to the observed variables ($X(t)$, $Y(t)$, $Z(t)$). We can numerically estimate the time derivatives corresponding to these variables with "finite differencing" (Gouesbet and Macquet 1992; Wei-Dong et al. 2003). If we do not wish to impose theoretical restrictions on the model, we can formulate a purely phenomenological (data-driven) specification by generating a polynomial expansion for each ODE (Wei-Dong et al. 2003) or by including trigonometric terms (Brunton et al. 2016). We illustrate this with a third-order polynomial expansion in Figure 29.3. Alternatively, we can take a hybridized modeling approach by incorporating mechanistic (theory-driven) restrictions into the specification. If the specification is linear in the parameters (such as the polynomial expansion), we can use ordinary least squares regression or "least absolute shrinkage and selection operator" (LASSO) regression that simultaneously estimates coefficients and selects essential regressors (Tibshirani 1996).

Finally, we numerically integrate the model to test whether scatterplotted model-simulates are capable of reproducing the shadow attractor empirically reconstructed from the observed

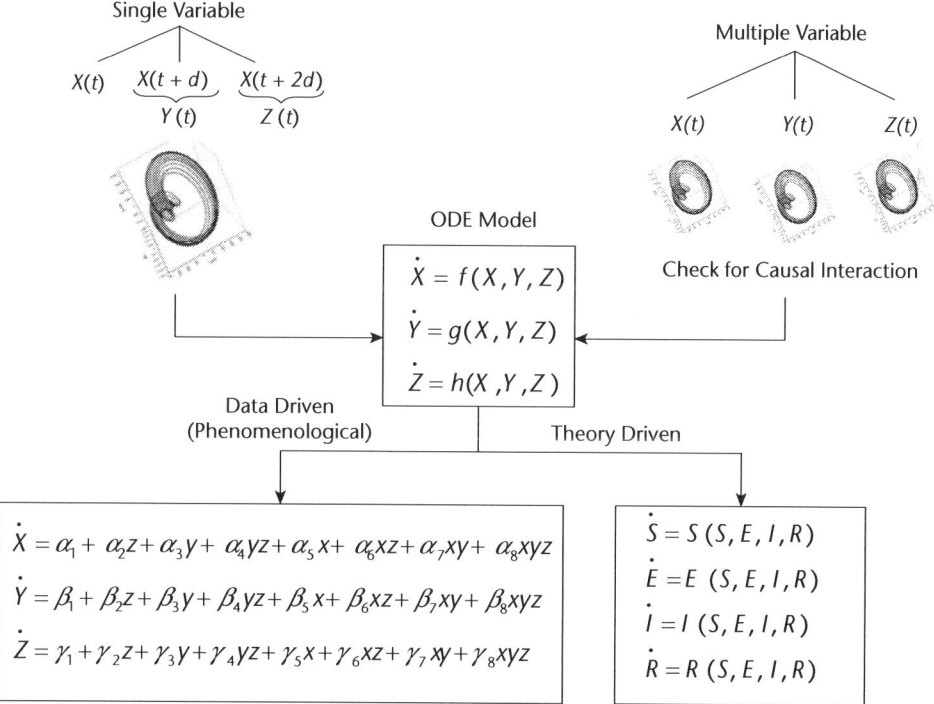

Figure 29.3 Phenomenological modeling

data. If not, we need to test an alternative specification that corresponds better to real-world dynamics.

3 NLTS with R

The following R packages run NLTS methods: Singular Spectrum Analysis (RSSA (Golyandina and Korobeynikov 2014)); Phase Space Reconstruction and Surrogate Data Analysis (tseriesChaos (Di Narzo and Di Narzo 2013), and nonlinearTseries (Garcia 2015)); fractal (Constantine and Percival 2014)); Convergent Cross Mapping (multispatialCCM (Clark 2014)); OLS regression (stats); LASSO regression (glmnet (Friedman et al. 2016)); and ODE solver (deSolve (Soetaert et al. 2015)). Huffaker et al. (2017) provide R code that facilitates, automates, extends, and explains the use of these packages in NLTS.

4 Case study illustrations

Food-price volatility is widely viewed as a serious threat to food security (G20 2011; HLPE 2011). The effectiveness of laissez-faire versus public-interventionist policies to reduce volatility itself, or to buffer its negative impacts, depends critically on whether agricultural markets are naturally stable and self-correcting or inherently unstable and persistently volatile (Galtier 2013). We present two case studies illustrating application of NLTS to diagnose and model market dynamics concealed in volatile food prices.

4.1 German hog-price cycle

We analyzed a weekly record of average producer prices (€ per kg carcass weight) for slaughtered pigs of quality E to P for the state of North Rhine-Westphalia, Germany from January 1990 to December 2011 (1144 observations) (Berg and Huffaker 2015). German hog prices oscillated aperiodically (non-repeatedly) over time (Figure 29.4, black curves). Early work attributed these oscillations to naïve price expectations (Buchholz 1982; Hanau 1928). Recent work attributed it to purely stochastic drivers and modeled it as a randomly shifting sinusoidal oscillation with time varying amplitudes (Parker and Shonkwiler 2014). We applied NLTS to empirically test the hypothesis that the German hog market is intrinsically unstable, and, if so, to see whether we could accurately simulate endogenous instability with a theory-based, low-dimensional, nonlinear, deterministic dynamic market model.

Our empirical results support the hypothesis of endogenous market instability. We isolated a strong signal in hog prices composed of a dominant annual oscillation (explained by the hog-gestation cycle), and a longer-term five-year oscillation (Figure 29.4, gray curve). We succeeded in reconstructing a low (three) dimensional shadow attractor with an embedding delay of 20 weeks. It is a torus-type attractor exhibiting the annual and five-year oscillations (Figure 29.4, lower left corner). Surrogate data testing soundly rejected the hypothesis that the systematic appearance of the attractor is randomly generated.

We also succeeded in reproducing the empirical attractor with a deterministic, nonlinear, dynamic-systems model specified as a fifth order system of ordinary differential equations (Figure 29.4, left side). The model consists of three state variables (hog prices, supply, and production capacity), and three dynamic processes (price adjustment, supply adjustment, and investment adjustment) (Figure 29.4, right side). The model provided valuable insight into endogenous economic drivers of volatile hog-price dynamics, including low price elasticity of demand and an investment lag resulting from the need for producers to

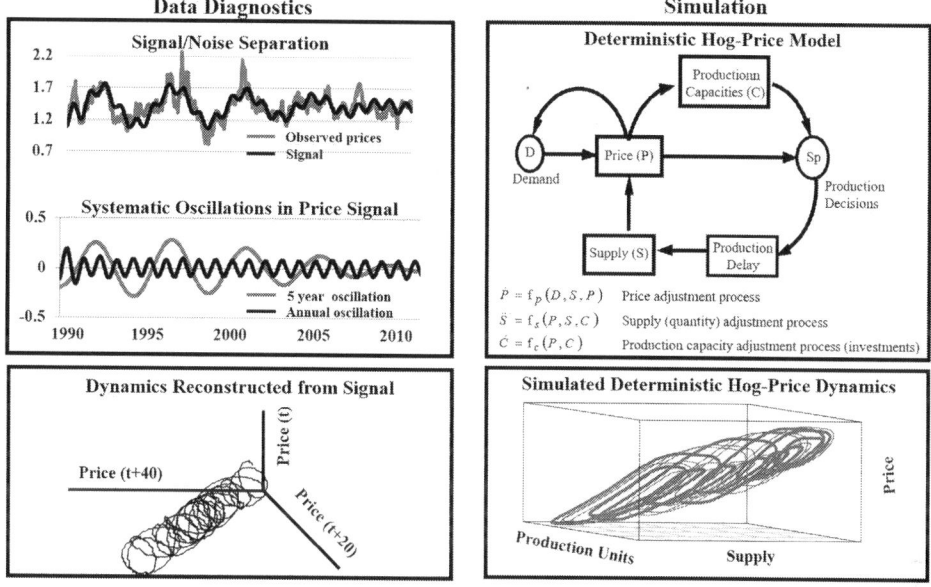

Figure 29.4 NLTS applied to analyze economic dynamics of German hog-price cycle

consolidate finances after a big investment (likely explaining the detected five-year oscillatory component).

4.2 Italian organic fruit price volatility

We applied NLTS to empirically reconstruct and characterize the economic dynamics of an organic fruit market in Italy (Huffaker et al. 2016). Our dataset included organic apple, pear, orange, and lemon prices (€/kg) recorded weekly at the Milano Ipercoop over eight years (2003–2010, 421 weeks). Over this period, the market was evolving from a niche organic fruit market serving a limited number of customers to a large retail market with widespread distribution (Canavari and Olson 2007).

Similar to the German hog-price series, Italian organic fruit prices (mean removed) were volatile (Figure 29.5a, gray curves). Signal processing isolated strong signals in each price series composed of trend and dominant annual and semiannual oscillatory components (Figure. 29.5a, black curves). We successfully reconstructed a low (three) dimensional shadow attractor from each price series. For example, the shadow attractor reconstructed from the apple price signal strongly depicts the combined semiannual and lower-frequency annual cycling isolated in signal processing (Figure. 29.5b). Surrogate data testing provided strong statistical evidence that geometric regularity of the shadow attractors is not randomly generated. In sum, NLTS diagnostics indicate that observed price volatility in this Italian organic fruit market is inherently unstable and driven by low-dimensional, nonlinear, deterministic dynamics.

We applied convergent cross mapping to find that the price signals for the four price signals strongly interact in a single market. Armed with this information, we turned to the challenge of formulating a deterministic phenomenological market model capable of reproducing the diagnosed market dynamics with the apple, pear, orange, and lemon price signals as state variables. The model is a four-dimensional system of ordinary differential equations composed of

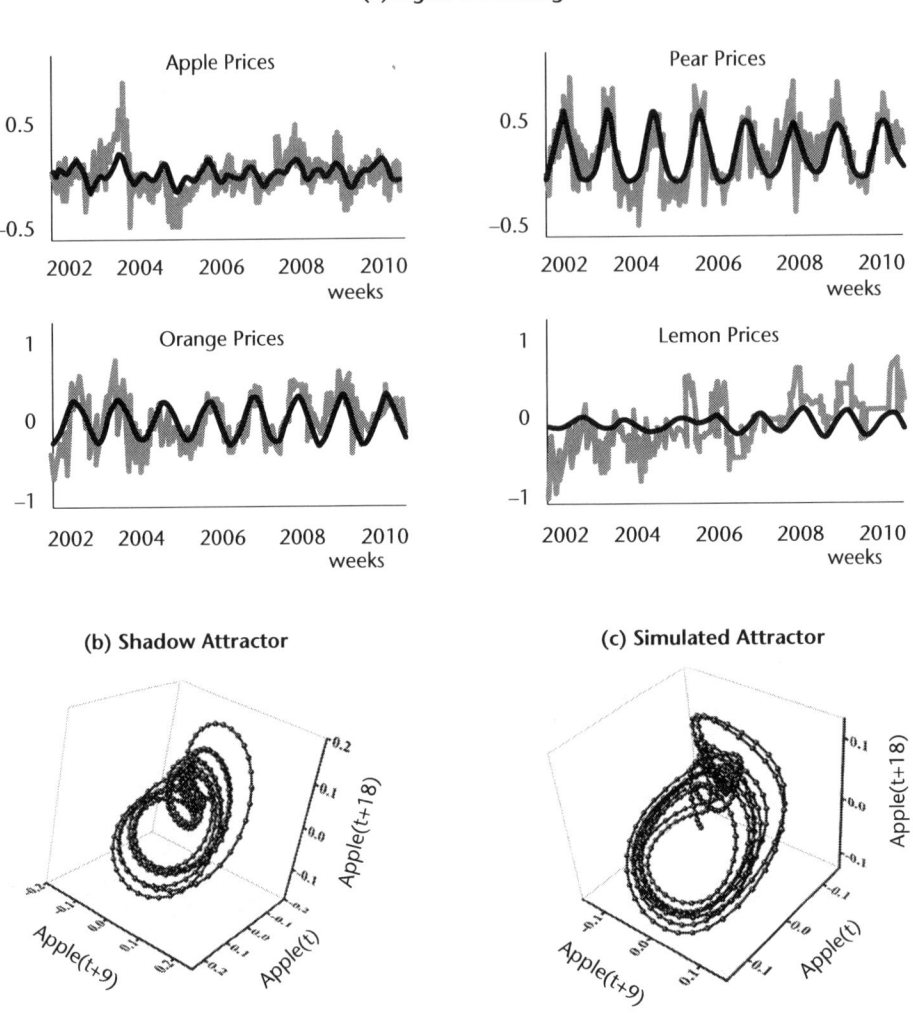

Figure 29.5 NLTS applied to analyze economic dynamics of Italian organic fruit prices (mean removed). Observed prices are plotted in gray and isolated signals in black

polynomial interaction terms. The model succeeded in reproducing empirically detected market dynamics deterministically since the attractor reconstructed from model-simulated apple prices (Figure 29.5c) bears striking correspondence to the empirically reconstructed shadow attractor (Figure 29.5b). Demonstrating the utility of phenomenological economic modeling, we used the model to investigate how the prices interact in a market "ecosystem" over time. Following ecosystem modeling methods (Hastings 1978), we computed cross partial derivatives to determine the marginal impact that an incremental increase in one price has on the growth rate of another and used this information to categorize price interactions in terms of predator–prey, competitive, and symbiotic relationships. We found strong (and surprising) evidence that price interactions shifted systematically over the course of each year.

5 Conclusion

This chapter has introduced nonlinear time series analysis (NLTS) – a collection of methods that allow economists to empirically diagnose whether observed volatility in economic time series data is due to stochastic shocks to otherwise stable markets or intrinsically unstable market dynamics. If data are sufficiently informative, economists can potentially use NLTS to reconstruct phase-space market dynamics and extract equations of motion from a single price series. The ability to diagnose market dynamics from a single price series allows economists to replace the conventional presumption that stochastic dynamics are required to model real-world volatility with hard empirical evidence of whether this is really the case and to recommend effective market stabilization policies corresponding to the reality faced by policy makers. We propose that economists use NLTS methods to test for systematic behavior in time series data before resorting to conventional stochastic exploratory methods unable to detect this valuable information.

Appendix

We used R package deSolve (Soetaert et al. 2015) to solve the Walrasian quantity (y) and price (p) adjustment model formulated by Shone (2002) in Example 4.17:

$$\dot{y} = \beta(p - 0.87 - 0.5y)$$
$$\dot{p} = \alpha(-0.02p^3 + 0.8p^2 - 9p + 50 - y)$$

Parameter values are set at: $\alpha = 1$ and $\beta = 2$ (Shone 2002).

Acknowledgment

RH acknowledges support from NIFA. We are grateful to Dr. Gerhard Schiefer (Center for Food Chain and Network Research, University of Bonn, Germany) for providing numerous opportunities to present this material and develop productive international collaborations at the annual IGLS-FORUM "System Dynamics and Innovation in Food Networks."

References

Baker, G., J. Gollub, and J. Blackburn. 1996. "Inverting Chaos: Extracting System Parameters from Experimental Data." *Chaos* 6:528–533.
Berg, E., and R. Huffaker. 2015. "Economic Dynamics of the German Hog-Price Cycle." *International Journal on Food System Dynamics* 6.64–80.
Breeden, J., and A. Hubler. 1990. "Reconstructing Equations of Motion from Experimental Data with Unobserved Variables." *Physica Review A* 42:5817–5826.
Brown, T. 1996. "Measuring Chaos Using the Lyapunov Exponent." In E. Kiel and E. Elliott, eds., *Chaos Theory in the Social Sciences: Foundations and Applications*. Michigan: University of Michigan, pp. 53–66.
Brunton, S., J. Proctor, and J. Kurtz. 2016. "Discovering Goerning Equations from Data By Sparse Identification of Nonlinear Dynamic Systems." *PNAS* 113:3932–3937.
Buchholz, H. 1982. "Zyklische Preis-und Mengenschwankungen auf Agrarmarkten." In H. Buchholz, G. Schmitt, and E. Woehlken, eds., *Landwirtschaft und Markt. Arthur Hanau zum 80*. Hannover: Geburtstag, pp. 87–112.
Canavari, M., and K. Olson. 2007. "Current Issues in Organic Food: Italy." In M. Canavari and K. Olson, eds., *Organic Food: Consumers' Choices and Farmers' Opportunities*. New York: Springer, pp. 171–183.

Chavas, J., and M. Holt. 1991. "On Nonlinear Dynamics: The Case of the Pork Cycle." *American Journal of Agricultural Economics* 73:819–828.

———. 1993. "Market Instability and Nonlinear Dynamics." *American Journal of Agricultural Economics* 75:113–120.

Clark, A. 2014. "MultispatialCCM: Multispatial Convergent Cross Mapping." https://cran.r-project.org/package=multispatialCCM. Accessed October 7, 2017.

Constantine, W., and D. Percival. 2014. "Fractal: Fractal Time Series Modeling and Analysis." https://cran.r-project.org/package=fractal. Accessed October 7, 2017.

Di Narzo, A., and F. Di Narzo. 2013. "tseriesChaos: Analysis of Nonlinear Time Series." https://cran.r-project.org/package=tseriesChaos. Accessed October 7, 2017.

The Economist. 2016a. "Big Economic Ideas." July 23.

———. 2016b. "If Economists Reformed Themselves." May 16.

———. 2016c. "Minsky's Moment." July 30.

Elsner, J., and A. Tsonsis. 2010. *Singular Spectrum Analysis*. New York: Plenum Press.

Ezekiel, M. 1938. "The Cobweb Theorem." *The Quarterly Journal of Economics* 52:255–280.

Fama, E. 1970. "Efficient Capital Markets: A Review of Theory and Empirical Work." *Journal of Finance* 25:383–417.

Friedman, J., et al. 2016. "Lasso and Elastic-Net Regularized Generalized Linear Models." https://cran.r-project.org/web/packages/glmnet/index.html. Accessed October 7, 2017.

G20. Ministerial Declaration: Action Plan on Food Price Volatility and Agriculture. www.oecd.org/g20/topics/agriculture/2011-06-23_-_Action_Plan_-_VFinale.pdf. Accessed November 1, 2017.

Galtier, F. 2013. "Managing Food Price Instability: Critical Assessment of the Dominant Doctrine." *Global Food Security* 2:72–81.

Garcia, C. 2015. "NonlinearTseries: Nonlinear Time Series Analysis." https://cran.r-project.org/package=nonlinearTseries. Accessed October 7, 2017.

Ghil, M., et al. 2002. "Advanced Spectral Methods for Climatic Time Series." *Reviews of Geophysics* 40:1–41.

Glendinning, P. 1994. *Stability, Instability and Chaos: An Introduction to the Theory of Nonlinear Differential Equations*. Cambridge: Cambridge University Press.

Golyandina, N., and A. Korobeynikov. 2014. "Basic Singular Spectrum Analysis and Forecasting with R." *Computational Statistics and Data Analysis* 71:934–954.

Golyandina, N., V. Nekrutkin, and A. Zhigljavsky. 2001. *Analysis of Time Series Structure*. New York: Chapman & Hall/CRC.

Gouesbet, G., and J. Macquet. 1992. "Construction of Phenomenological Models from Numerical Scalar Time series." *Physica D* 58:202–215.

Gregson, R., and S. Guastello. 2011. *Introduction to Nonlinear Dynamic Systems Analysis*. New York: CRC.

Haack, S. 1999. "Defending Science – Within Reason." *Principia* 3:187–211.

Hanau, A. 1928. "Die Prognose der Schweinepreise. 2." In erw. und erg. Aufl. des Sonderh. 2. Berlin: Hobbing (Vierteljahrshefte zur Konjunkturforschung: Sonderh ed.).

Harlow, A. 1960. "The Hog Cycle and the Cobweb Theorem." *Journal of Farm Economics* 42:842–853.

Hassani, H. 2007. "Singular Spectrum Analysis: Methodology and Comparison." *Journal of Data Science* 5:239–257.

Hastings, A. 1978. "Global Stability of Two Species Systems." *Journal of Mathematical Biology* 5:399–403.

Hayes, D., and A. Schmitz. 1987. "Hog Cycles and Countercyclical Production Response." *American Journal of Agricultural Economics* 69:762–770.

Hilborn, R., and M. Mangel. 1997. *The Ecological Detective: Confronting Models with Data*. Princeton, NJ: Princeton University Press.

HLPE. "Price Volatility and Food Security." www.fao.org/fileadmin/user_upload/hlpe/hlpe_documents/HLPE-price-volatility-and-food-security-report-July-2011.pdf. Accessed November 1, 2017.

Horan, B. 1994. "The Statistical Character of Evolutionary Theory." *Philosophy of Science* 61:76–95.

Hornberger, G., and R. Spear. 1981. "An Approach to the Preliminary Analysis of Environmental Systems." *Journal of Environmental Management* 12:7–18.

Huffaker, R. 2015. "Building Economic Models Corresponding to the Real World." *Applied Economic Perspectives and Policy* 37:537–552.

Huffaker, R., M. Bittelli, and R. Rosa. 2017. *Nonlinear Time Series Analysis with R*. Oxford: Oxford University Press.

Huffaker, R., M. Canavari, and R. Munoz-Carpena. 2016. "Distinguishing Between Endogenous and Exogenous Price Volatility in Food Security Assessment: An Empirical Nonlinear Dynamics Approach." *Agricultural Systems*. doi:10.1016/j.agsy.2016.09.019. Accessed October 7, 2017.

Jensen, R., and R. Urban. 1984. "Chaotic Price Behavior in a Non-Linear Cobweb Model." *Economics Letters* 15:235–240.

Kantz, H., and T. Schreiber. 1997. *Nonlinear Time Series Analysis*. Cambridge: Cambridge University Press.

Kaplan, D., and L. Glass. 1995. *Understanding Nonlinear Dynamics*. New York: Springer.

Katz, R. 2010. "Statistics of Extremes in Climate Change." *Climatic Change* 100:71–76.

Kot, M., W. Schaffer, G. Truty, D. Graser, and L. Olsen. 1988. "Changing Criteria for Imposing Order." *Ecological Modeling* 43:75–110.

McSharry, P. 2011. "The Danger of Wishing for Chaos." In S. Guastello and R. Gregson, eds., *Nonlinear Dynamical Systems Analysis for the Behavioral Sciences Using Real Data*. New York: CRC Press, pp. 539–558.

Minsky, H. 1992. "The Financial Instability Hypothesis." The Jerome Levy Institute of Bard College, Working Paper No. 74. www.levyinstitute.org/pubs/wp74.pdf. Accessed on November 1, 2017.

Oreskes, N., K. Shrader-Frechette, and K. Belitz. 1994. "Verification, Validation, and Confirmation of Numerical Models in the Earth Sciences." *Science* 263:641–646.

Parker, P., and S. Shonkwiler. 2014. "On the Centenary of the German Hog Cycle: New Findings." *European Review of Agricultural Economics* 41:47–61.

Rykiel, E. 1991. "Testing Ecological Models: The Meaning of Validation." *Ecological Modeling* 90:229–244.

Saltelli, A., and S. Funtowitz. 2014. "When All Models Are Wrong." *Computer Modeling* Winter 2014:79–85.

Schreiber, T. 1997. "Detecting and Analyzing Nonstationarity in a Time Series with Nonlinear Cross Predictions." *Physical Review Letters* 78:843–846.

———. 1999. "Interdisciplinary Application of Nonlinear Time Series Methods." *Physics Reports* 308:1–64.

Schreiber, T., and A. Schmitz. 2000. "Surrogate Time Series." *Physica D* 142:346–382.

Shone, R. 2002. *Economic Dynamics*. Cambridge: Cambridge University Press.

Soetaert, K., T. Petzoldt, and R. Setzer. 2015. "Solving Differential Equations in R: Package deSolve." *Journal of Statistical Software* 39:1–25.

Sugihara, G., et al. 2012. "Detecting Causality in Complex Ecosystems." *Science* 338:496–500.

Takens, F. 1980. "Detecting Strange Attractors in Turbulence." In D. Rand, and Young, L. ed., *Dynamical Systems and Turbulence*. New York: Springer, pp. 366–381.

Theiler, J., et al. 1992. "Testing for Nonlinearity in Time Series: The Method of Surrogate Data." *Physica D* 58:77–94.

Tibshirani, R. 1996. "Regression Shrinkage and Selection Via the LASSO." *Journal of the Royal Statistical Society: Series B* 58:267–288.

U.S. House of Representatives. 2010. "Building a Science of Economics for the Real World." *U.S. Government Printing Office*, Serial No. 111–106.

Waugh, F. 1964. "Cobweb models." *Journal of Farm Economics* 46:732–750.

Wei-Dong, L., et al. 2003. "Global Vector-Field Reconstruction of Nonlinear Dynamical Systems from a Time Series with SVD Method and Validation with Lyapunov Exponents." *Chinese Physics* 12:1366–1373.

Williams, G. 1997. *Chaos Theory Tamed*. Washington, DC: John Henry Press.

Ye, H., et al. 2015. "Distinguishing Time-Delayed Causal Interactions Using Convergent Cross Mapping." *Scientific Reports* doi:10.1038/srep14750.

30
Agricultural development impact evaluation

J. Edward Taylor

1 Introduction

Beginning with Adam Smith, economists have tried to understand why some people and countries are rich while others are poor – typically in the hope of alleviating poverty. Most poor countries are agricultural, and most of the world's poor live in rural areas. Agricultural development research focuses on the role of agriculture in the development process, the design of policies to promote agricultural growth, the evaluation of agricultural development interventions on poverty, and the impacts of poverty interventions on agriculture. Agricultural and development economists have been at the forefront of discovering new methods to identify impacts, including for the preponderance of situations in which the "gold standard" of experiments using randomized controlled trials (RCT) breaks down or is not politically, ethically, or logistically feasible. This chapter offers an overview of these methods and a selective review of some of the agricultural development studies that use them.

A vast literature in development economics uses experimental, econometric, and simulation methods to evaluate impacts of a diversity of interventions and other exogenous shocks – which we can regard broadly as "treatments" – on a wide range of outcomes, from school attendance to crop productivity, mortality, poverty, and people's subjective well-being. This review focuses on the impacts of agricultural interventions as well as the impacts of other treatments on agricultural outcomes. It follows a progression from the most astructural, reduced-form experimental methods to the most structural econometric and simulation models. Four inescapable conclusions emerge: (1) no single tool is sufficient to evaluate impacts; (2) different methods are more or less appropriate in different situations; (3) all have strengths and weaknesses, even under the most ideal circumstances; and (4) combinations of approaches often can provide a deeper and more reliable understanding of impacts than stand-alone approaches.

The key subjects of agricultural evaluations in developing countries include the impacts of new products of agricultural research (technologies, ideas, etc.); the effects of specific agricultural development projects and programs, including institutions such as agricultural extension; and the design of policies affecting agriculture, including sector-specific (e.g., agricultural subsidy) and trade policies. These subjects, by their nature, demand an assortment of evaluation tools. For some, such as impacts of new technologies and cost-efficiency analysis of new projects, evaluation can be conducted *ex ante*. For others, like evaluating the benefits from agricultural technologies long-since rolled out or counterfactual analysis of existing policies, evaluation has to be conducted *ex post*. RCTs are valuable for some types of impact evaluation (e.g., see Voors et al. 2016; De Janvry, Dustan, and Sadoulet 2011), but they are far from being the only tool in the shed. RCTs are not feasible or logically possible to address most of the critical agricultural development questions touched upon in this review.

The most convincing impact evaluation studies have a basis in economic theory, regardless of the empirical approaches they employ. The theoretical development of agricultural household models (Singh, Squire, and Strauss 1986; Kuroda and Yotopoulos 1978; Barnum and Squire 1979; Nakajima 1969) provided a basis for estimating response elasticities and demonstrating complex interactions between production and the consumption behavior of micro agents. Agricultural household models are a useful foundation for generating hypotheses about microeconomic outcomes and behavior that are testable using experimental or non-experimental approaches. They are also a basic building block for constructing models to evaluate the impacts of agricultural policies, development projects, or other shocks using simulation techniques. Many of the evaluations considered in this chapter are grounded explicitly or implicitly in agricultural household theory.

2 Agricultural development and the experimental revolution

In 1997, Mexico launched the most ambitious poverty program ever to be carried out in a developing country. The PROGRESA (now PROSPERA, a.k.a. *Oportunidades*) program's objective was to combat rural poverty by giving cash to poor women, conditional upon their children being enrolled in school and in the local medical clinic. Due to logistical and budgetary constraints, the government could not roll out the program to all eligible households in the first year, so it opted to implement the program randomly across localities, expanding it to the full population of eligible rural households two years later. This effectively created a randomized control group, ideal to evaluate short-run impacts of this ambitious public assistance program on a wide range of outcomes.

The extensive PROGRESA literature includes almost no mention of agricultural impacts. Nevertheless, PROGRESA left an indelible mark on development economics research that may be unmatched by any other government program. It inspired the use of randomized experiments and no doubt laid the foundation for an increasing use of RCTs for agricultural development impact evaluation.

In PROGRESA, randomization ensured that the treatment and control villages and households, on average, were identical except for the treatment. This provided an answer to the selection problem, the greatest challenge to identifying impacts of development programs and policies. Most of the time, PROGRESA-style randomization is not politically feasible or logically possible. Nevertheless, the idea behind randomization can be a helpful benchmark for thinking about identifying program and policy impacts related to agriculture.

2.1 The selection problem

Following Rubin (1974), consider a treatment (say, an intervention aimed at increasing agricultural productivity) and some outcome of interest (for example, per-acre crop productivity, farmer incomes, or farm-household nutritional status). Denote farm i's outcome by Y_i. If the farm gets treated directly by the policy or program, the outcome is Y_{1i}, and if it does not, it is Y_{0i}. Each farm has both a Y_{1i} and a Y_{0i}, but obviously we can only observe one of them.

Let D_i equal 1 if farm i gets treated and 0 otherwise. A concise (and conventional) way to represent the outcomes is:

$$Y_i = \begin{cases} Y_{1i} \text{ if } D_i = 1 \\ Y_{0i} \text{ if } D_i = 0 \end{cases} \quad (1)$$

The observed outcome for farm i is what it would be without the treatment, Y_{0i}, plus the effect of the treatment, $(Y_{1i} - Y_{0i})D_i$. Define the actual treatment effect as rho; then, we can express the outcomes as:

$$Y_i = Y_{0i} + \underbrace{(Y_{1i} - Y_{0i})}_{\rho} D_i \quad (2)$$

The treatment effect, ρ, is what we typically wish to estimate in studies of impact evaluation. It is the change in the outcome that is *caused* by the treatment, or the average treatment effect (ATE). (If the person does not get treated, $D_i = 0$, so this second term is zero.)

Now suppose we simply compare expected or average outcomes for those who do and do not get the treatment, like maize output per hectare on farms that have and have not adopted a new seed variety. The expected or average outcome given that a farm adopts the new technology is $E[Y_i|D_i = 1]$, and the expected outcome for farms that did not receive or choose the technology treatment is $E[Y_i|D_i = 0]$. The average difference between the two is:

$$E[Y_i|D_i = 1] - E[Y_i|D_i = 0] \quad (3)$$

This difference is not the average effect of the technology on the adopters because it includes selection bias. The average effect of the treatment on the treated (ATT) is the difference between (1) the expected outcome for farms with the treatment, given that they got it ($E[Y_{1i}|D_i = 1]$), and (2) the expected outcome for these same farms *if they had not been treated* ($E[Y_{0i}|D_i = 1]$). In other words, ATT = E[] − E[]:

$$E[Y_{1i}|D_i = 1] - E[Y_{0i}|D_i = 1] \quad (4)$$

Imagine the farmers who adopt the new technology (the first term). If, after they get the treatment, we could send them back in time and then not treat them, we would have the second term. If that person did not change in any other way, the difference between the two terms would be the true ATT.

Obviously, we cannot both treat and not treat the same farmers with a new technology. We have to compare those who adopted the technology with those who did not. This leaves us with selection bias. Selection bias is the difference between (1) expected yield for those who adopted the new technology, if they had not adopted it (the second term in expression (4): ($E[Y_{0i}|D_i = 1]$), and (2) expected yield without the treatment for the farmers who did not adopt ($E[Y_{0i}|D_i = 0]$). In other words:

$$\text{Selection Bias } = E[Y_{0i}|D_i = 1] - E[Y_{0i}|D_i = 0] \quad (5)$$

A common concern in studies of technology impacts is that those who adopt are different from non-adopters in ways that influence outcomes with or without the new technology. Adopters may be better educated, less cash- and credit-constrained, have better access to markets for inputs and outputs, and have a greater willingness to take on the risks inherent in adopting a new technology. Early literature on the determinants of agricultural technology adoption provides theoretical and empirical support for this (e.g., Feder, Just, and Zilberman 1985; Lipton and Longhurst 1989; Perrin and Winkelmann 1976; Burke 1979; Byerlee and Harrington 1983). Feder (1980) and Just and Zilberman (1983) show theoretically that a farmer's likelihood of adopting a new crop variety is likely to depend on differences in expected yields between the new and old varieties, the expected difference in variance of harvest, and the farmer's risk aversion. All three are likely to vary from one potential adopter to another. Partial rather than complete adoption is likely, except perhaps in the case of the most risk-neutral farmers or farms on which yield impacts are large and impacts on yield variance are minimal. This means that differences in impacts of a new technology on crop yields and other outcomes are likely to depend on the extent of adoption, not simply on the decision of whether or not to adopt, and that selection bias is likely to be a significant problem when measuring the impacts of technology change.

Randomization would solve the selection problem. If the technology treatment were truly random, then on average the farmers who adopt would be identical to those who do not except for the treatment. Their outcomes without the treatment, on average, would be the same: $E[Y_{0i}|D_i = 1] = E[Y_{0i}|D_i = 0]$. Selection bias would vanish, leaving only the ATT, and the ATT would be the same as the ATT on the population (ATE). This is why, in a well-designed RCT, we can estimate the ATE simply by comparing average outcomes between treatment and control groups, like in a drug trial.

One might imagine an idealized thought experiment that induces a randomly chosen group of farmers (the treatment group) to adopt a new technology and a randomly chosen control group not to adopt. The researcher could compare per-hectare yields at the end of the crop season to yields in the pre-treatment baseline and calculate the difference in yields for each farmer. For both farmer groups, this difference would reflect impacts of weather and other exogenous variables, but for the treatment group it would include the impact of the "technology treatment" as well. If we found that the difference between the average difference in yield between treatment and control groups (difference in difference) were positive and significant, we could conclude that the technology treatment significantly increased yields. Alternatively, we could estimate an ordinary least-squares (OLS) regression corresponding to (2):

$$Y_i = \alpha + \rho D_i + \varepsilon_i \tag{6}$$

In this equation, α is expected yield without the treatment, and as before ρ is the ATE. The stochastic error ε_i is independent of D_i as long as the treatment is random. If the treatment is random, OLS gives a linear unbiased estimate of the ATE – unless there are general equilibrium (GE) effects of the program on nonparticipants. We consider GE effects later in this chapter.

2.2 RCT studies of impacts of agricultural interventions

This is essentially what Beaman et al. (2013) did in a study of the impacts of fertilizer use on rice yields in Mali. Production economists often treat fertilizer as simply an input in the production function, but due to a variety of constraints (liquidity, risk, market, etc.), fertilizer application is dismally low in sub-Saharan Africa. "Nudging" farmers to use some fertilizer could be transformative; by adding a new input to the production function, it is akin to technology change.

In this evaluation, a simple field experiment provided free fertilizer to female rice farmers in an effort to measure how they choose to use the fertilizer when they have it, what changes they make to their agricultural practices, and the impact on profits. Because the fertilizer was given free of charge, this was essentially a transfer experiment, in which the transfer was a productive input. Uptake was not an issue. Because the fertilizer treatment was random, a difference-in-difference analysis comparing treatment and control farmers gave the causal impact of the treatment. The study found that rice output increased significantly (by 31 percent on average); however, there was no statistically significant impact on profits. Beaman et al. (2013) attributed this non-result with regard to profit to farm heterogeneity and stochastic yield variability: "It may simply be difficult to learn about the returns to fertilizer for a particular farmer, given their land quality and other inputs, when the signal is weak and there are large fluctuations in profits due to other shocks" (p. 382).

This experiment may be useful for understanding the impacts of free input transfers on production and profits. However, impacts of input subsidies, like Malawi's famed Farm Input Subsidy Program (FISP), or of other productive interventions, like promoting the adoption of new seed varieties or offering access to credit, are more complex. One cannot (and would not want to) force farmers in a treatment group to plant a new seed, purchase a subsidized input, or take out a loan. Often, some (even most) of the individuals offered the treatment do not take it (in experimenter's parlance, they are non-compliers). Farmers in a treatment group might choose not to adopt new technologies, like higher-yielding or lower-risk seeds, unless they are convinced that they will be better off as a result – no matter what experimental trials at research stations might find. As a result, it is generally not feasible to implement an RCT that resembles the randomized thought experiment above for most agricultural interventions, and the paucity of RCTs in the agricultural development literature reflects this.

One could use an encouragement design that provides information about a new technology or some other inducement to a random treatment group of farmers but not to a control group, and hope there is sufficient uptake in the treatment group to measure program impacts. This approach, commonly referred to as "intention to treat (ITT)," is based on the initial treatment assignment and not on the treatment actually received; it ignores uptake (noncompliance) and anything else that happens after randomization takes place. Using ITT instead of actual treatment in Equation (6) does not give the ATE – it gives the average ITT effect, which typically is more conservative due to the dilution effect of noncompliance on the treated sample. Nevertheless, it is often argued that the ITT effect is more relevant than the ATE from a real-world policy perspective, given the compliance question and the high costs of monitoring uptake. This concern is not limited to economic treatments: the U.S. Food and Drug Administration guidelines state that an ITT analysis should be included as part of clinical drug trials (Gupta 2011; USFDA 1988).

The randomized information treatment can be an instrument to predict adoption of the new technology in an instrumental variable (IV) approach designed to estimate the impact of adoption on crop yield in Equation (6). If the experimentally constructed information instrument is strong, this might be a feasible way to estimate the expected effects of the technology treatment on compliers (the local average treatment effect, or LATE; Imbens and Angrist 1994). If the only ones who are treated are those who were randomly assigned to the treatment group, this IV approach can give the ATT.

Duflo, Kremer, and Robinson (2011) performed an ITT analysis in their study of fertilizer use in Kenya. The authors hypothesized that farmers procrastinate and postpone the purchase of fertilizer until later periods, when they may be too impatient to purchase it. Their RCT offered a treatment group of farmers inducements in the form of small, time-limited discounts on the cost of acquiring fertilizer (free delivery) just after harvest. This increased fertilizer use by 47 to 70 percent. Offering free delivery later in the season had a smaller impact, even with a 50 percent subsidy on fertilizer.

Elabed and Carter (2015) offered the possibility of purchasing index insurance with multiple encouragements to a random sample of cotton cooperatives in Mali. The results reveal evidence of *ex ante* impacts of insurance on farmers' behavior. Farmers in cooperatives who were offered insurance felt more insured on average than farmers in the control group of cooperatives not offered insurance. They increased their area planted in cotton, and they also increased their expenditure on seeds per hectare.

An improved understanding of the potential benefits from new agricultural technologies prior to their release is an area in which economists' insights are likely to have a lasting impact on national systems. A growing number of studies use RCTs for *ex ante* evaluations of new crop technologies under farmers' real-world conditions. Dar et al. (2013) conducted an RCT to test impacts of a flood-tolerant rice variety, randomly selecting farmers in Orissa India to receive a 5-kg packet of improved (Swarna-Sub1) seed. This experiment produced treatment and control groups that were similar with respect to observed characteristics. The test found a 45 percent increase in yields compared to an existing variety when fields were submerged for ten days, offering hope that flood-tolerant varieties can produce efficiency gains and highlighting benefits from agricultural research on stress-tolerant varieties. People with lower incomes tend to populate flood-prone areas; thus, this technology can achieve equity gains as well. A similar RCT by Emerick et al. (2016) found that farmers who adopted new flood-tolerant varieties increased their labor and fertilizer use and credit demand. This "crowding in" of other inputs explains much of the productivity gain from the new variety and opens up the possibility that risk-reducing technologies might increase agricultural productivity even in normal years. Bernard et al. (2016) gave subsidies in the form of vouchers for improved seeds to a random sample of farmers in the Democratic Republic of Congo. The subsidies encouraged high adoption of improved seeds; however, they also had environmental implications by increasing the conversion of forest and savanna to cropland.

Unlike in medical experiments, farmers generally know whether or not they receive the actual treatment; there is no placebo. An exception was Bulte et al.'s (2014) double-blind RCT, in which farmers did not know whether they received an improved cowpea variety or a placebo traditional variety. The study found that harvests were the same for people who knew they received the modern seeds and for people who did not know what type of seed they got; however, people who knew they received the traditional seeds did much worse. Farmers who did not know which seed they received and those who knew they received the modern seed planted more than farmers who knew they received the traditional seed. This article offers some insight into how farmers' behavioral responses, particularly with regard to complementary input choices, might compromise the validity of RCTs involving agricultural technologies.

2.3 RCT studies of agricultural impacts of non-agricultural interventions

Developing-country governments face severe budget constraints and tradeoffs between investing in social programs, like social cash transfers (SCTs) that target poor households, and programs aimed at raising agricultural productivity. Gertler et al. (2012) found evidence that beneficiaries of Mexico's PROGRESA program increased their investment in productive assets, which in turn raised their agricultural income and consumption. These findings suggest that SCTs can have sustainable long-term impacts on poor households.

A series of experimental studies in sub-Saharan Africa tested whether SCTs can have productive impacts in beneficiary households (Davis et al. 2016). Researchers randomly assigned villages or village clusters to the SCT treatment, similar to what was done in Mexico's PROGRESA program but without conditionality. Baseline and follow-on surveys gathered information on crop and livestock production, input use, and sales in these as well as a set of randomly chosen control sites. Eligible households at the treatment sites received periodic cash payments that varied across countries and, in some countries, across eligible households. Follow-on surveys were conducted one to two years after payments were initiated. These studies found that, even though SCTs usually targeted asset- and labor-poor households, cash transfers had productive impacts on beneficiary households' agricultural production, input use, asset accumulation, and/or sales in most of the SCT programs.

SCTs and other development programs, while benefiting treated eligible households, also can create production and income spillovers to ineligible households. For example, SCT beneficiaries might share their cash with other households. Increased food demand by SCT beneficiaries might stimulate production in ineligible households that have sufficient land and other assets to respond to the demand, raising their income, which in turn might stimulate other production and incomes. Estimating spillovers experimentally requires baseline and follow-on data on both SCT-eligible and SCT-ineligible households at both treatment and control sites.[1] Such data are extremely rare, and it appears that only two studies have used experiments to estimate SCT spillovers (Angelucci and Di Giorgi 2009 and Taylor et al. 2017).

Imperfections in credit and insurance markets are likely to influence agricultural production, particularly given weather-related yield uncertainty and long time lags between planting and harvests. A rich literature uses experimental methods to identify the impacts of alternative risk and credit interventions on agricultural outcomes. Karlan et al. (2014) conducted experiments of various designs in northern Ghana to study impacts of increased liquidity and insurance on agricultural investments. Researchers offered randomly selected groups of farmers cash grants, grants of rainfall index insurance or the opportunity to purchase this insurance, or a combination of the two. They found evidence of a strong demand for weather index insurance. Insurance led to increased agricultural investment as well as to riskier agricultural production choices. By eliminating self-selection bias, randomized treatments made it possible to discover that uninsured risk is a binding constraint, and when farmers have access to insurance, they increase expenditures on their farms. In subsequent years, farmers' demand for insurance increased with their own receipt of insurance payouts as well as with payouts to others in the farmers' social networks and recent poor rain outcomes.

2.4 Experimental challenges

Randomization might appear straightforward, but researchers seeking to use RCTs and other experimental methods to identify the impacts of agricultural policies (and other things) inevitably confront multiple challenges. To begin with, RCTs generally are not useful for *ex post* analysis, for example, of policies or technologies long since rolled out. This problem plagues the many evaluations of impacts of Consultative Group on International Agricultural Research (CGIAR) and agricultural research in general. Most government programs are captive to political clocks and resource constraints that do not allow for pilot evaluations or randomized roll-outs. Once a project is rolled out, governments and donors can be reluctant to invest resources in follow-on data collection and evaluation, and they may have a vested interest in not knowing the answer.

Often, the challenges in implementing experiments relate to ethical considerations that (quite understandably) cause actual implementation to stray from the experimental ideal (Barrett and Carter 2010). These challenges include avoiding the adverse consequences of experiments (the principal of "do no harm"), informed consent (subjects of RCTs usually are unaware that they are in an experiment), blindedness (subjects know whether or not they got a treatment), and targeting (development organizations and governments have scarce resources to carry out development projects, so it seems illogical and possibly unethical to target them not randomly but where they are likely to have the greatest impact).

RCTs are infeasible in most situations, wherein it is not logically possible to define treatment and control groups or governments are unwilling to do so. It is generally not possible or politically feasible to implement large-scale policies in which some farmers but not others receive input subsidies, output price supports, infrastructure investments, access to new technologies, or export-crop promotion. In Malawi's famed Farm Input Subsidy Program (FISP), Mexico's Programa de Apoyos Directos al Campo (PROCAMPO) decoupled income payment program, and Ethiopia's Productive Safety Net Program (PSNP) – three of the world's largest agricultural subsidy schemes – initial randomization would have been theoretically possible, but there never was a control group. Macro policies and events defy analysis via RCTs. It is logically impossible to treat some farmers but not others with global price shocks, macro policies, or climate change.

RCT challenges can be particularly acute in the case of agricultural interventions, wherein observed outcomes are the result of many choices by heterogeneous individuals acting under diverse conditions. Often it is difficult to discern exactly what treatments farmers are getting (or think they are getting). Barrett and Carter (2010, p. 522) write:

> Unobservable perceptions of a new product, contract, institutional arrangement, technology or other intervention vary among participants and in ways that are almost surely correlated with other relevant attributes and expected returns from the treatment.

This results in what Barrett and Carter call "faux exogeneity."

Randomization should result in treatment and control groups having characteristics that are identically distributed prior to the treatment, a property called "balance" – though this is only true asymptotically, not in small samples. The best RCT studies carefully test whether treatment and control groups are balanced. Significant differences in characteristics could reflect departures from randomization or possibly just bad luck. If significant differences are found, those baseline variables should be included as controls in Equation (6). Researchers add other terms to the equation to increase efficiency or for other ends, including interactions between the treatment and baseline characteristics to uncover differences in the ATT across subgroups of treated farms. In practice, the test equation can become quite complex, and as the equation grows, it becomes clear that the real contribution of RCTs is to experimentally generate instruments for econometric modeling. Issues that arise in any econometric model (and fill the pages of econometrics textbooks) are potentially relevant to these RCT evaluations.

Researchers also need to ensure that the control group is viable, including being isolated from the treatment group. Control-group contamination can arise if the two groups interact in some way, directly or indirectly. Treated farmers might share information with farmers in the control group, creating a situation similar to Miguel and Kremer's (2004) celebrated worms study. If both treated and control farmers' yields increase as a result, the intervention ironically might be so successful the impact cannot be identified using an RCT!

Interventions that raise incomes in the treatment group can create general equilibrium effects in village or regional economies that transmit impacts to non-treated groups via local markets. Control group contamination is very likely to arise when both treatment and control groups are in the same localities – for example, villages or village-clusters. It is often difficult to randomize at the village/cluster level because there are too few arguably similar sites to perform a proper statistical analysis, subject to budgetary and other constraints.[2]

Human behavior, heterogeneity among subjects and across environments, and other factors raise serious questions about both internal validity (solving the selection problem) and external validity (the ability to generalize results). RCTs might be considered analogous to *in vivo* experimentation in the sciences: a treatment is administered to a group of subjects who are part of a larger organism (i.e., an economy). Internal validity is always a concern in biological experiments, but the questions discussed above can make social experiments particularly challenging, especially in the context of large-scale government programs in which researchers face difficulties in maintaining control over design and implementation.

External validity – the ability to generalize from an RCT to a larger population – also poses unique challenges in social experiments. Findings from RCTs carried out in one population (e.g., region) may not apply to other populations in different agro-climatic, cultural, and market settings – including ones into which a program might be expanded. As Teele (2014) points out, when one moves from small-scale experiments to large-scale (e.g., countrywide) policy implementation, general equilibrium feedback effects are likely to influence outcomes, raising questions about the external validity of RCTs. It is not surprising, therefore, that most impact evaluations of agricultural interventions do not employ RCTs (IEG 2011). Researchers have had to grapple with other methods to evaluate the impacts of policy, market, environmental, and other shocks on agricultural outcomes in developing countries.

3 Laboratory experimental studies of farmer behavior

The challenges inherent in carrying out RCTs to evaluate new technologies have inspired an alternative, laboratory experimental approach to elicit farmers' preferences and the behavioral parameters likely to influence technology adoption and agricultural outcomes. Laboratories in the field can provide an alternative to RCTs if they generate instruments effective at explaining farmers' behavior outside the lab.

Given yield uncertainty and the inherent subjective risks associated with adopting new technologies, farmers' risk aversion can play a critical role in technology adoption decisions. Binswanger (1980) used a pioneering experimental approach to elicit risk aversion by observing farmer reactions to gambles involving real payoffs. He found that, at high payoff levels, nearly all farmers are moderately risk-averse. Risk aversion decreases with wealth but not significantly. Binswanger attempted, unsuccessfully, to use his experimentally generated risk aversion index to explain farmers' technology choices.

A considerable literature extends and refines experimental approaches to elicit farmers' risk aversion, discount rates, and other behavioral parameters. Examples include Harrison et al. (2010), Mosley and Verschoor (2005), and Cardenas and Carpenter (2008). Bauer and Chytilová (2010) used field experiments to estimate subjective discount rates in Ugandan villages, finding a negative correlation with education. Tanaka et al. (2016) found that people in higher-income villages were less loss-averse and more patient.

In a particularly interesting application of behavioral experiments, Mani et al. (2013) found evidence that poverty saps attention and reduces effort: sugarcane farmers from Tamil Nadu, India, showed lower cognitive performance on tests before harvest, when poor, than after harvest, when relatively rich. Liu (2013) used a field experiment to elicit the risk preferences of Chinese farmers. She found that farmers who are more risk or loss averse adopt Bt cotton later, while farmers who overweight small probabilities adopt Bt cotton earlier.

Governments and donors spend substantial sums to develop new seed varieties that promise higher yields, greater yield stability (e.g., resistance to weather and other environmental shocks), or other characteristics. Such investments are for naught if farmers do not value the characteristics embodied in the new technology. Lybbert (2006) used a novel experimental approach to elicit how farmers value mean yield growth versus yield stability. This study was inspired by Binswanger but reflects some important advances in field test methods. Experimental games in south India found that farmers valued higher mean yield growth far more than reduced downside yield risk or yield stability, a finding unaffected by household wealth or risk exposure. This stands in contrast with prevailing beliefs among crop breeders and raises questions about the agricultural research focus on stabilizing yields and/or how to influence farmers' valuation and willingness to adopt new yield-stabilizing technologies.

When it comes to providing reliable information about the impacts of agricultural development interventions, laboratory experiments raise questions similar to those from *in vitro* experiments in the biological sciences. It is unclear how well insights from laboratory experiments reflect behavior in real-world situations (internal validity) or how reliably they may be generalizable to larger populations (external validity).

4 Identification through econometrics

Given the practical and logical barriers to implementing experiments, econometrics has been an indispensable tool to study impacts of projects and policies on agriculture as well as impacts of agriculture on other outcomes of interest. In practice, there can be a fine line between experimental and econometric approaches, inasmuch as most RCTs end up generating econometric models. One might think of a randomized treatment (or ITT) as an ideal (or almost ideal) instrument to include in an econometric impact evaluation model.

Unless treatments are random, OLS estimators are likely to run up against the selection problem when evaluating project impacts or estimating the impact of agriculture on other outcomes, like the effect of agricultural productivity on household incomes. Omitted (including unobserved) variables correlated with the treatment may be correlated with the outcome, confounding results. Because of this, OLS has generated few contributions to the agricultural development impact evaluation literature in recent years, unless accompanied by RCTs. Matching methods and IV techniques are more promising; however, their viability depends on having access to convincing instruments. IV methods using cross-section data are subject to the influences of unobserved variables on the modeled outcome. Matched panel data make it possible to sweep away the influences of invariant unobserved variables, a step towards isolating program impacts. This explains their popularity for agricultural impact evaluation. Nevertheless, influences of varying unobserved variables remain, potentially confounding panel estimates.

An extensive literature uses econometric methods to identify agricultural policy impacts and uncover processes that shape agricultural development. It is not possible to do justice to this literature here. A selection of econometric methods and papers employing them to evaluate agricultural impacts follows.

4.1 Matching

When individuals are not randomly assigned to treatment and control groups, selection bias is likely to confound estimates of treatment effects. If data are not available from RCTs, sometimes researchers can use matching methods to emulate RCTs. Propensity score matching (PSM) is the most common matching method employed in agricultural development impact evaluations. First proposed by Rosenbaum and Rubin (1983) as a means to reduce selection bias when estimating treatment effects with observational datasets, PSM is widely used in the economic analysis of policy interventions as well as in the biological sciences, including medical trials.

The basic idea behind PSM is to reduce selection bias by comparing treated and non-treated individuals who appear as similar as possible except for the treatment. For example, one might imagine reducing selection bias by comparing impacts of new seed varieties on farms of similar sizes – some observed to adopt the new technology and others not. Selection bias might be reduced further by comparing impacts on farms that are similar on an n-dimensional vector of characteristics; however, this usually becomes infeasible if n is large.

PSM summarizes the n-dimensional vector of individual characteristics into a single index, the propensity score (PS). Any standard probability model can be used to estimate the PS, but probit and logit models are most commonly used. The PS is simply the predicted $PR(D_i = 1 \mid X_i)$, where X_i is a set of individual characteristics. Once each individual's PS has been estimated, the sample is divided up into intervals within which the PS is not significantly different between treated and non-treated individuals, and outcomes of interest (e.g., crop yields) are compared between the two groups.

The validity of PSM depends, in part, on treated and control groups with the same PS being observationally identical on average. Conditional independence (also called "selection on observables" or "unconfoundedness") assumes that for a given set of covariates participation must be independent of potential outcomes. Smith and Todd (2005) rightly note that there may be systematic differences between the outcomes of adopters and non-adopters, even after conditioning on observables. Such differences may arise because of selection into treatment based on unmeasured characteristics. If unobserved variables influence selection into the treatment, they may undermine matching as an alternative to randomized experiments. In the past decade or two, a number of studies have explored the implications of unobserved variables for confidence intervals, including the construction of "Rosenbaum bounds" (Rosenbaum 2002; Caliendo and Kopeinig 2008).

In practice, the conditional independence assumption raises identification challenges similar to those in the econometric estimation of program impacts.[3] The possibility that treatment and control groups are not observationally identical on average, or balanced, also crops up in RCTs. The best RCT studies carefully check for balance with respect to observed variables and attempt to control for it econometrically, by including the unbalanced variables along with the randomized treatment dummy in a regression equation to test for impacts. With regard to balance, the advantage of having a randomized treatment is clear; a lack of balance raises the specter of a treatment not actually being random.

PSM may be useful if a treatment is not random but baseline data were collected on treated and non-treated individuals before the treatment. If *ex post* data are used (say, in the absence of a baseline survey), one must assume that the characteristics used to estimate the PS did not change because of the treatment. If farmers' self-selection into new technologies depends on both observables and unobservables, methods that assume selection on observables, including PSM, are likely to be biased. The bias is likely to be more severe if a study relies on cross-sectional surveys with limited data on the situation of the farm prior to adopting.

Despite these challenges, PSM can offer some insights into the likely impacts of agricultural interventions and other treatments when randomization is not possible or logically feasible and when researchers have access to a convincing set of variables that can be used to construct the PS. The intuition of comparing individuals who appear very similar except for the treatment underlies PSM as well as RCTs.

Becerril and Abdulai (2010) used PSM to evaluate the impact of adopting improved maize germplasm on farm household income and poverty in Oaxaca and Chiapas, Mexico. PSM made it possible to form matches between pairs of households with the same PS, in which one adopted improved seeds and the other did not. This study found that adopting improved maize varieties significantly increased households' income (proxied by per-capita expenditure) and reduced their probability of falling below the poverty line by 19–31 percent. Kassie, Shiferaw and Muricho (2011) found that adopting improved groundnut varieties significantly increased crop income and reduced poverty in a PSM evaluation involving Ugandan households.

Berhane et al. (2014) extended PSM to evaluate the impacts of exposure to Ethiopia's Productive Safety Net Program (PSNP) on agricultural asset accumulation and other outcomes. The PSNP provides food-insecure households with cash assistance linked to interventions designed to increase agricultural productivity. Unlike Mexico's PROGRESA, PSNP was not rolled out randomly; thus, RCT evaluations were not possible. This study compared outcomes for households after different periods receiving PSNP cash transfers, a PSM "dose response" approach. Pre-treatment characteristics were used to predict households' years of exposure to PSNP treatments. The study found that exposure to the PSNP increased beneficiaries' livestock holdings. The impact was larger when beneficiaries also received encouragement and advice aimed at increasing their income from agricultural activities and accumulating assets. Keswell and Carter (2014) used a variant of PSM to study the impacts of land redistribution on poverty in South Africa. Approval and sales delays in transferring land varied across households in an arguably exogenous way. These were used along with other covariates to predict the land transfer treatment and duration of treatment. The study found that land transfers significantly increased beneficiaries' living standards after an initial negative effect.

4.2 Instrumental variables

Econometric impact evaluation invariably faces the challenge of finding instruments correlated with the treatment but not the outcome in question. In RCTs, randomization creates an ideal instrument. Unanticipated exogenous shocks can create a "natural experiment" opportunity to test for impacts. Used as a proxy, comparisons of outcomes before and after natural disasters, droughts, and unanticipated economic events can sometimes offer insights into the impacts of treatments that defy randomization. Weather shocks are the most natural "natural experiment." Undeniably exogenous, they can be an ideal instrument to evaluate a wide range of impacts. However, they are not likely to be a good instrument if the outcome of interest is related to agricultural production, because the weather obviously affects harvests directly.

A prolific literature uses weather shocks as instruments to evaluate impacts of variables that arguably are linked to the weather or simply as an exogenous control when modeling agricultural and other outcomes, often in conjunction with panel methods (discussed below). Miguel (2005) used extreme rainfall, which is associated with poor harvests, as an instrument for income in his study of the impacts of poverty on violent crime. Jessoe, Manning, and Taylor (2016) used weather shocks as a proxy for climate change in a study of impacts on labor allocations to agriculture and other activities in rural Mexico. Impacts depended upon when in the maize crop cycle the shocks occur. Jayachandran (2006) used rainfall as a proxy for agricultural productivity shocks in India. She found that productivity shocks decreased wages more for rural workers who were poorer, less able to migrate, and more credit-constrained. However, lower wages acted as insurance for landowners. Rozelle, Taylor and de Brauw (1999) used village migration histories and weather shocks as instruments to study the impacts of migration and remittances on crop production in China. They discovered that the loss of labor to migration decreased yields while migrant remittances increased them.

There are many examples of studies using unexpected economic or political events as a strategy to identify impacts, mostly outside of agricultural development. Examples include Card's (1990) celebrated study of the impacts of the Mariel Boatlift on the Miami labor market; Yang's (2008) study of the impacts of the Asian financial crisis on welfare in migrant-sending Philippine households; Halliday's (2006) estimate of the impact of an earthquake on migration in El Salvador; and Hornbeck's (2012) study of effects of the American Dust Bowl on land values. The biggest challenge in these "before and after" studies is making certain that other variables affecting the outcome of interest did not change at around the same time the shock hit. For example, if large aid programs rushed in after the disaster but before the "after" data could be collected, the impacts of the disaster may be confounded with the impacts of the aid; it will be difficult or impossible to separate the two. A novel twist on quasi-experiments was Duflo's (2001) study of impacts of school construction in Indonesia, which recognized that new schools would not affect schooling or employment outcomes for people who were too old to go to school, but they would have an effect on children who were the right age to benefit. Randomness in birth year was used to identify impacts of school construction on schooling attainment and employment. This study did not consider agricultural outcomes, but Charlton and Taylor (2017) found a significant negative impact of schooling on agricultural employment in Mexico by using a similar approach.

Gibson, McKenzie, and Stillman (2011), in a novel study of impacts of migration on Tonga Islander incomes, used a quasi-experimental approach exploiting New Zealand's random allocation of entry visas through the Pacific Access Category (PAC). By randomly denying Tongan applicants the right to move to New Zealand, PAC created a control group of individuals with presumably the same income outcomes as the migrants would have had if they had not migrated. Really, this is akin to an "intent to treat" experiment, because not every Tongan who won the PAC migration lottery had migrated at the time of the study. The study tested for migration impacts on a variety of outcomes, including two related to agriculture: migration negatively affected livestock holdings and total agricultural income (though not agricultural income per household member).

One of the most novel quasi-experimental studies related to agriculture in a developing country is Aker's (2010) evaluation of the impact of cell phones on grain prices in Niger. The efficient functioning of food markets depends critically on traders' ability to arbitrage, which in turn depends on access to information. In this study, traders' access to mobile phones was the treatment, which clearly is endogenous: it depends in part on traders' decision to invest in mobile phones. However, the construction of cell phone towers between 2001 and 2006 resulted in traders having access to cell services in some parts of the country but not others. Using cell phone towers as an instrument, the study found that mobile phones explained a 10 to 16 percent reduction in grain price dispersion.

4.3 Switching regression

When RCTs are not possible or logically feasible, an alternative to PSM is to use endogenous switching regression (ESR). A first-stage dichotomous choice model (usually a probit or logit) is used to predict which individuals (e.g., farmers) are treated (e.g., with a new technology) and which are not. Predicted treatment status can be used as an IV replacing the randomized treatment in Equation (6). Alternatively, it can be used to construct an inverse-Mills ratio (IMR) to correct for sample selection bias in separate regressions modeling the outcome with and without the treatment (Heckman 1979). An essential but often overlooked ingredient of switching models is to provide a plausible story about why the change in question would lead to a completely new regime. This might be straightforward in some cases (e.g., a change in the production function associated with adoption of a new seed technology) but less so in others (e.g., a change in consumption regime due to the same seed technology change). Under general conditions, a significant coefficient on the IMR term signals the presence of sample selection bias, and the IMR controls for this bias while estimating the outcome equation. In ESR models, like IV and PSM, the validity of the approach turns critically on having access to one or more statistically and intuitively convincing instruments (Z) that determine whether the individual received the treatment but do not affect the outcome of interest directly.

A number of studies use ESR to test the impacts of agricultural technologies on household welfare. Asfaw et al. (2012a) found that adopting improved legume technologies had a significant positive impact on consumption expenditure in Tanzania. Khonje et al. (2015) found that adoption of improved maize led to significant gains in crop incomes, consumption expenditure, and food security while reducing poverty in eastern Zambia.

Studies that compare results from PSM and ESR typically find that results are robust to the estimation method used. Asfaw et al. (2012b) found evidence that the adoption of improved pigeonpea in Tanzania significantly increased consumption expenditures and reduced poverty in both PSM and ESR models. Shiferaw et al. (2014) found consistent evidence from PSM and ESR that the adoption of new wheat varieties increased food security in Ethiopia. By controlling for *ex ante* differences between adopters and non-adopters, these studies' findings suggest that households that did not adopt would have benefited had they adopted the new seed varieties.

ESR, like PSM, must confront the challenge of finding suitable instruments to estimate *ex ante* likelihoods of treatment. The most convincing instruments make theoretical sense while passing statistical tests for constituting good instruments. In Asfaw et al.'s (2012b) study, the authors hypothesized that the likelihood of adopting the new technology depended on farmers' prior exposure and access to improved seeds. They used four instruments to reflect this prior exposure: access to information from extension workers, access to information from radio/television, experience in participatory variety selection (PVS) in the previous year, and constrained access to improved seeds. They established the acceptability of these instruments by conducting a simple rejection test based on Di Falco et al. (2011): that the instruments are significantly correlated with technology adoption but do not affect the welfare outcome among households that did not adopt improved varieties. Three of the four instruments passed this test.

4.4 Panel methods

The key contribution of panel data in impact evaluation is to enable researchers to "sweep away" the influences of unobserved, time-invariant variables on measured outcomes. As longitudinal data have become increasingly available, panel econometrics have become a method of choice for agricultural development impact evaluation.

An excellent example is a thread of research on the role of social learning in the diffusion of new agricultural technologies, which research and extension services have long recognized and attempted to exploit (Bindlish and Evenson 1997; Bandiera and Rasul 2006; Munshi 2004). The way in which new evaluation strategies build upon existing ones in an effort to econometrically identify impacts is evident by comparing the initial and seminal study in this area (Foster and Rosenzweig 1995) with a more recent one (Conley and Udry 2010).

The Green Revolution brought new high-yielding seeds to India's farmers, whose production technology had changed little for decades. Many did not adopt the new seeds, which agronomic field tests had shown to be significantly more productive than traditional seed varieties. Foster and Rosenzweig (1995) used matched panel data that tracked Indian farmers over time and found that the profitability of the new seeds for farmers increased as their neighbors gained experience growing them. Because of this, farmers whose neighbors had experience growing the new seeds planted more of their own land with high-yielding varieties (HYVs). Thus, the authors concluded, farmers who adopted the new technology created benefits not only for themselves but also for others through "knowledge spillovers."

Conley and Udry (2010) pointed out two major challenges to identifying the impacts of social learning: defining the set of neighbors from whom an individual can learn, and distinguishing learning from other explanations, including interdependent preferences, technologies, or shared unobservable shocks. Lacking data on actual information connections among farmers, Foster and Rosenzweig (1995) used a village aggregate of neighbors' cumulative cultivated area in HYVs, lagged one survey round, as their measure of neighbors' HYV experience. They made careful use of IV fixed effects methods in an effort to address the second challenge and eliminate spurious correlations between individual and village variables. Conley and Udry (2010) addressed the question of identifying neighbors more directly by gathering detailed information on whom individuals knew and talked to about farming. This offered a way to define information links among pineapple farmers in Ghana. Many farmers began growing pineapple in the 1990s and learned through trial and error how much fertilizer to use to maximize their profit. By analyzing data on social networks and this trial-and-error process, the study found that farmers do indeed learn from their "information neighbors," but only when neighbors get a surprisingly good or bad result from trying a particular fertilizer dosage rate.

These panel studies document how the spread of information about new technologies can make farmers in poor countries more productive. The insights they provide have influenced the design of agricultural extension programs and sparked other studies seeking an even deeper understanding of the information and cognitive constraints that impede technology adoption. They offer some insights into why many farmers do not adopt new technologies or use inputs demonstrated to increase yields, but this mystery persists.

Suri (2011) used panel data to estimate a correlated random coefficient model and found evidence that farmer heterogeneity explains decisions to adopt hybrid maize. This study found that the group of Kenyan farmers with the highest estimated net returns from hybrids faced high costs of acquiring the new technology due to poor infrastructure and did not adopt, while farmers with lower returns did adopt.

In many African countries, farmers grow crops on plots controlled by different members of the household, including male and female plots. Pareto efficiency implies that these plots are managed to maximize income – the "size of the pie" – regardless of how the pie might get divvied up within the household. Udry (1996), using panel data from Burkina Faso, found that plots controlled by women were farmed much less intensively than male-controlled plots; yields were 30 percent lower on female plots within the same household. This finding suggests that heterogeneous preferences and access to resources within households affect not only equity, but also agricultural efficiency.

Udry's study illustrates both the strengths and weaknesses of fixed effects methods. On one hand, panel data made it possible to control for household fixed effects while testing for differences in input allocations between male- and female-controlled plots. On the other hand, unobserved differences across plots could easily confound estimates. The gender of the plot manager is a time-invariant plot characteristic, so it is not possible to control for plot fixed effects while testing for differences in management between male and female plots. Recognizing this, Udry controlled for a large number of indicators of plot characteristics (size deciles, toposequence, soil types, location) in his regression.

An extensive literature uses panel econometric methods to evaluate the impacts of agricultural subsidy programs in developing countries. It includes a suite of studies on Malawi's Farm

Input Subsidy Program (FISP), inspired by international controversy over the wisdom of providing input subsidies to farmers. A challenge to identifying impacts of this, like other interventions, is the possibly non-random character of the subsidy treatment. Ricker-Gilbert, Jayne, and Chirwa (2011) found that subsidized fertilizer crowded out purchases of commercial fertilizer more for rich than poor households, and Jayne et al. (2013) concluded that, partly because of this, the value of maize harvests increased by less than the cost of fertilizer subsidies. Chibwana, Fisher, and Shively (2012) found evidence that FISP increased land area in maize and tobacco while decreasing areas in other crops. Ricker-Gilbert et al. (2013) found that fertilizer subsidy programs in Malawi had a minimal effect on retail maize prices. Ricker-Gilbert (2013) found that FISP had a small negative impact on households' supply of informal wage labor to agriculture (*ganyu* labor), a small but not quite significant impact on the probability that a household demanded agricultural labor, and a small positive impact on the agricultural wage. Sadoulet, De Janvry, and Davis (2001) used a two-year household panel to show that cash payments from an agricultural subsidy program in Mexico had a multiplier effect on incomes within beneficiary households.

Studies of the impacts of trade on agricultural development make extensive use of panel methods that control for country fixed effects. For example, Grant and Lambert (2008) used a 21-year country panel analysis to test whether regional trade agreements (RTAs) are effective at promoting agricultural trade by dismantling member countries' agricultural tariff and non-tariff barriers. They found that RTAs have a disproportionately large impact on agricultural trade compared with non-agricultural trade. Cross-section trade studies are unable to control for the confounding influences of unobserved country characteristics while evaluating agricultural trade impacts.

Agricultural systems around the world are experiencing a marketing revolution in which supermarkets are transforming the food supply chain (Reardon et al. 2003). Michelson (2013) used panel methods to test whether the expansion of Walmart's procurement channels created economic opportunity or economic hardship for small farmers in Nicaragua. This study really lies at the nexus of panel and quasi-experimental methods, because it took advantage of the expansion of a super-marketing structure over time to test its hypothesis. It found that geographic location and transport options are decisive in enabling farmers to participate in new market systems; many farmers lack the endowments to enter new supply chains as the supermarket sector grows in the developing world. Panel methods supported the identification of impacts while providing this study with a dynamic focus: Michelson found that participation in supermarket supply chains directly affected the accumulation of productive assets by farmers positioned to take advantage of it.

An apparently inverse relationship between farm size and productivity has been a longstanding puzzle in agricultural and development economics. Do unobserved characteristics of large and small farms, correlated with productivity, explain the puzzle? Heltberg (1998) used panel econometric methods to test the impact of farm size on productivity. Including farm fixed effects permitted him to remove the confounding influences of unobserved farm attributes on productivity, which previously had plagued studies using cross-section data. The study found convincing evidence that productivity decreases with farm size. Kagin, Taylor, and Yúnez-Naude (2016), using panel data from Mexico, found an inverse relationship between farm size and both productivity and technical efficiency. Their findings offer a guardedly optimistic view of small farms' capacity to produce efficiently and potentially to adapt to changing market conditions. Using state-level panel data, Besley and Burgess (2000) found evidence that land reform had a significant impact on poverty reduction and agricultural wages, though possibly at the expense of lower per-capita income.

Panel data have enabled researchers to test key hypotheses related to farmers' *ex ante* and *ex post* responses to weather risk. Rosenzweig and Binswanger (1993), using panel data from rural India, discovered that exposure to weather risk caused farmers to invest in less risky portfolios. This reduced farmers' exposure to risk *ex ante*, but it also resulted in lower average farm profits. Lybbert et al. (2004) used an unusually long panel of herd histories from Ethiopia to model herd dynamics in a population that held almost all of its wealth in livestock. They found evidence that both covariate and idiosyncratic shocks were a key driver of herd dynamics. Their finding of multiple dynamic wealth equilibria points to an important role for policies to buffer pastoralists from weather shocks and prevent them from falling into poverty traps.

Fafchamps et al. (1998) used panel data to test the hypothesis that livestock sales buffered agricultural households from the impacts of weather shocks on consumption *ex post*. By tracking Burkina Faso households over a period of four years, which included a severe drought, they found that livestock sales played less of a consumption-smoothing role than expected. On average, sales of animals compensated for 15 to 30 percent of income shortfalls. Carter and Lybbert (2012), using threshold estimation with the same panel data, found that the average impact reported by Fafchamps et al. masked a more complex and troubling story. A small group of households at the upper part of the asset distribution almost completely smoothed their consumption by selling livestock when weather shocks hit. However, a large group at the lower end of the distribution destabilized their consumption in order to preserve livestock assets, in effect not putting food on the table in order to save their small herds. This had the short-run effect of avoiding falling into a poverty trap but the long-run implication of irreversibly harming the physical and cognitive development of the youngest and most vulnerable household members. Hoddinott (2006) found comparable results using panel data from Zimbabwe.

One of the most challenging identification issues in agricultural development concerns estimating the economic returns to agricultural research. The pathway by which R&D creates value – from research to dissemination and adoption to impacts of yields – makes benefits difficult to evaluate. The time lags from initial investment to yield outcome typically are long; for example, Byerlee and Traxler (1995) reported a lag of ten years between the start of research and the beginning of reaping benefits from Centro Internacional de Mejoramiento de Maíz y Trigo (CYMMIT)-developed new seed varieties. Costs have to be weighed against (discounted) future benefits. Lower food prices shift benefits from farmers to consumers. Farmers' willingness to adopt new technologies, considered previously, plays a key role in realizing benefits from agricultural research. These and other considerations make evaluating the returns to agricultural research difficult in developed countries and daunting in less-developed countries.

Alene and Coulibaly (2009), using country-level panel data with a distributed-lag model, found evidence that investment in agricultural research contributes significantly to productivity growth in sub-Saharan Africa, with aggregate returns to research estimated at 55 percent. This increases income and reduces poverty. A simultaneous equation causal chain model by Thirtle, Lin, and Piese (2003) found a powerful impact of agricultural productivity growth on poverty in both rural and urban areas of Africa, Asia, and Latin America. Changes in food prices were key to transmitting benefits to the poor. A review of various studies concluded that estimated rates of return to agricultural research were high, often exceeding 40 percent, though the variance of estimates also was high (Evenson 2001). Most estimates were consistent with countries' actual growth experience. Comparing benefits and costs to investments, there is evidence of significant underinvestment in agricultural research in developing countries, particularly research related to genetic improvement (Pardey, Alston, and Piggott 2006; Renkow and Byerlee 2010). Moyo et al. (2007) estimated the change in economic surplus from the adoption of virus-resistant peanut varieties in Uganda. They explore the implications of this technology for rural poverty by constructing a technology adoption profile from probit estimates of adoption probabilities and using a Foster, Greer, and Thorbeck (1984) poverty index. The study finds large research benefits but modest poverty impacts.

The returns to agricultural research literature highlights an important point usually overlooked by impact evaluations in development economics: Demonstrating benefits, as challenging as that may be, is not the same thing as demonstrating cost-effectiveness. Governments – particularly in poor countries – confront severe budget constraints and an imperative to channel resources into programs that yield large net economic benefits while attempting to achieve other goals. They can ill afford to invest in programs that are not the most cost-effective on the policy menu, even if RCTs and other methods demonstrate that they have significant positive impacts. Most impact evaluations address one side of the cost–benefit equation but not the other.

Absent panel data, researchers may be able to create retrospective panels from micro surveys, following Besley and Case (1993). Moser and Barrett (2006) used farmer recall data together with extension records to study technology adoption dynamics in Madagascar. They found that farmer education, liquidity, and labor availability significantly shape farmers' willingness to try new, labor-intensive technologies. Zaal and Oostendorp (2002) compiled retrospective data on population density, rainfall, crop prices, and terrace construction to identify the determinants of agricultural intensification on small farms in Kenya. Questioning Boserup (1965), they discovered that market access and windfall profits from the coffee boom in the late 1970s were at least as important as population pressure in explaining investments in terracing. Recall bias and its possible correlation with the outcome of interest are a major concern when using retrospective panels. Charlton and Taylor (2016) and Jessoe et al. (2016) used overlapping panels constructed from two or more survey rounds to test and control for recall bias.

Usually, panel methods control for unobservables that do not vary over time. Barrett et al. (2004) employ a different approach. They used fixed effects methods with cross-section data on households with multiple plots to disentangle the output effects of a new rice technology from effects attributable to farmer characteristics. Multiple plot observations per farm, combined with the fact that all farms in the sample used both an improved and traditional technology to grow rice, made it possible to compare yields between technologies on individual farms. Controlling for farmer fixed effects ensured that any endogeneity bias was plot-specific. Like Udry (1996), Barrett et al. (2004) included indicators of observable plot quality in their regression. The study found that the improved technology was associated with an 84 percent gain in productivity, of which roughly half was attributable to the technology and half to farmer characteristics. A small percentage was attributable to (observable) plot characteristics. The new technology also increased yield risk, limiting its attractiveness to risk-averse farmers. This study illustrates how to carry out fixed-effects estimation without panel data when there are multiple observations per household.

PSM, IV, and panel methods sometimes find larger higher order effects (e.g., on household income) than can be explained by lower (e.g., plot-level) impacts. If a study finds only marginal impacts on yields or productivity but large impacts on household income, this may be evidence of improper identification. Recognizing and seeking explanations for surprising findings, as well as testing for the robustness of the identification strategy – including the use of "placebo tests" – are an essential but often overlooked part of impact evaluations (Athey and Imbens 2017).

4.5 Experiments vs. non-experiments: weighing the evidence

The co-existence of experimental and non-experimental methods naturally raises the question of whether and to what extent the latter are biased. Some researchers have empirically attempted to compare methods. It appears that Lalonde (1986) was the first to attempt to rigorously test for biases in non-experimental versus experimental approaches. He compared experimental estimates from a randomized labor-market treatment (the National Supported Work [NSW] Demonstration) to non-experimental estimates using control groups constructed from household survey data. The non-experimental estimates failed to replicate the experimental results.

Lalonde did not consider PSM, but a number of subsequent studies did, with more mixed conclusions. Dehejia and Wahba (2002) used the same data as Lalonde and found that PSM results were similar to the experimental estimates. Smith and Todd (2005) replicated this analysis and concluded that the results were sensitive to a number of researcher decisions, including the choice of samples, the specification of the equation to estimate the PS, and the rules used to form matches of treated and otherwise similar individuals. McKenzie et al. (2010), using the New Zealand migration lottery as an instrument, found that OLS estimation of the impact of migration (lottery success) on weekly income (with and without other controls) yielded a significant positive impact. The authors then repeated the exercise, imagining that they did not have the random lottery success variable, using five non-experimental methods: first differences, OLS, difference-in-differences, matching, and IV. All were found to overstate the impact of migration on income. The overstatement of impacts ranged from 20 to 82 percent. Similar comparisons between experimental and non-experimental estimators have not been carried out to evaluate the impacts of agricultural programs.

5 Simulation approaches

Agricultural economics has a rich history of using structural simulations to model impacts of agricultural policies and other shocks. Simulation methods sometimes are considered anathema by experimental economists wedded to reduced-form impact evaluation methods. However, they are widely used in science, generally when the goal is to understand and evaluate the workings of complex systems. Indeed, there has been an overall shift in scientific research from *in vivo/vitro* to *in silico* methods.[4] Historically, simulations have played an important role in agricultural policy evaluation at the micro level using agricultural household models, as well as at the macro and economy-wide levels using econometric and economy-wide modeling techniques. Recently, they have become a new tool for project impact evaluation, as well.

5.1 Agricultural household simulations

Agricultural household (AH) models are a staple of agricultural development analysis at the micro level, including evaluation via simulation. In most applied AH analysis, commodity demands and either output supply and input demands or a production function are estimated econometrically with cross-section data, though estimates can be considerably more reliable with panel data. Estimated models are used to obtain response elasticities to output prices, wages, and other shocks, as well as to simulate impacts of agricultural and other development policies. A number of examples of AH models appear in Singh, Squire, and Strauss (1986).

The results of policy simulations and the parameterization of AH models depend critically on market closure assumptions at the household level. If agricultural households are perfectly integrated with outside markets, prices are fixed, and therefore the consumption and production sides of the AH model are separable or recursive. Missing markets at the household level create endogenous "shadow prices" that make the production and consumption sides of the model simultaneous rather than recursive and transmit impacts between the two. There is convincing evidence that small farmers face high transaction costs isolating them from outside markets (Barrett 2008). De Janvry, Fafchamps, and Sadoulet (1991) and Taylor and Adelman (2003) construct AH models to simulate the impacts of cash crop prices on both staple and cash crop production under alternative model closure assumptions.

The importance of market closure spawned a literature that econometrically tests for separability in agricultural households – something that does not lend itself to experimental methods. Most econometric tests for separability in developing countries focus on labor, specifically, the substitutability of family and hired labor (Benjamin 1992; Jacoby 1993; Skoufias 1994). Researchers use AH models as a basis to econometrically study a wide variety of questions ranging from crop supply response to production impacts of migration (Rozelle et al. 1999), crop genetic resource conservation (Arslan and Taylor 2009), valuation of non-marketed crop by-products (Magnan et al. 2012), and other outcomes. These studies often uncover evidence that farms are imperfectly integrated with markets and thus of nonseparability.

5.2 General equilibrium (GE) impacts

Development interventions and other exogenous shocks have impacts on the directly affected actors (referred to as the "treatment group" in this chapter). They also are likely to generate general equilibrium spillovers as their influences ripple through local, regional, national, and international economies. GE impacts create two challenges in agricultural development impact evaluation. First, if they transmit impacts to a control group, RCTs break down. This is why it is usually unwise to draw treatment and control groups from the same localities. Second, if they transmit impacts to individuals who are not eligible for the treatment, the ATT provides only a partial representation of impacts.

Researchers, when thinking about GE effects of treatments, need to consider not only the eligible treatment and control groups but also the non-treated ineligible groups. In project impact evaluations, both treatment and control groups are eligible for the treatment. To be viable, a control group must be limited to individuals who on average are identical to the treated except for the treatment. With very few exceptions, impact evaluations consider only the households that are eligible for the program.

Treatment and control sites also contain individuals who are not eligible for the treatment but nevertheless may be affected by GE spillovers. Examples include non-poor households that do not qualify for a poverty program, non-farm households that do not qualify for a crop subsidy or intervention to raise crop productivity, childless households that do not qualify for a program to increase school enrollment, etc. Very few impact evaluations using RCTs or econometric methods consider spillover effects on ineligible individuals. In an idealized RCT to evaluate the ATT as well as the average effect of the treatment on the ineligible (ATI), researchers would draw random samples of both eligible and ineligible groups at both treated and non-treated sites. The ATI would be estimated in the same way as the ATT, but by comparing differences in outcomes for the ineligible groups at treatment and control sites.

Randomized control trials almost never include ineligible groups in baseline or follow-on surveys. Doing so adds to evaluation costs. Identifying spillover effects usually requires large samples of ineligibles, because like ripples in a pond, spillovers tend to become diffused. Although a dearth of data impedes evaluation of GE spillovers in experimental or econometric evaluations, a rich literature uses structural GE models to simulate total impacts, including spillovers, in local, regional, national, and even international economies.

5.3 Computable general equilibrium (CGE) models

The advent of computable general equilibrium models in the wake of Arrow and DeBreu's (1954) theoretical demonstration of the existence of competitive equilibria opened up the possibility of simulating impacts on agricultural sectors as well as spillovers to other sectors of the economy. While AH models are structural representations of individual household groups, CGE models are structural representations of whole economies, constructed from economic data and used to simulate how economies might react to changes in policies or other external shocks. In most cases, they are aggregated multi-sectoral models of national economies, calibrated using national social accounting matrices (SAMs) (Stone 1986 and Burfisher 2011), although there are instances of multi-national CGE models as well as CGE models of regions within countries – even villages (Taylor and Adelman 1996). Production activities and households within CGEs may be highly aggregated or disaggregated into many different sectors and /or household groups. Most CGEs assume that all actors are price takers within the economy. Prices may be determined within the model (for nontradables) or outside the model (for tradables).

An advantage of CGE models is their ability to capture spillovers across production sectors as well as households. For example, agricultural policies and technology change can have positive impacts on urban as well as rural activities and households by influencing the supply of agricultural commodities and thus food prices, and they might influence rural–urban migration via wage effects. CGEs shift the focus of agricultural development impact analysis from micro actors to whole economies. A rich literature includes the use of country CGE models to evaluate economy-wide impacts of agricultural policies, trade policies, and other interventions in developing countries. Much of it uses the Global Trade Analysis Project (GTAP) platform developed by Hertel (1997). Only a few examples are given below to illustrate this approach and the insights it can provide about spillovers at an economy-wide level.

Arndt, Pauw, and Thurlow (2015) used an aggregate country CGE model to evaluate the impacts of Malawi's FISP, including the production and income spillovers it creates. Their simulations indicated that FISP benefit–cost ratios are 60 percent higher when indirect impacts are taken into account. Elbehri and Macdonald (2004) found that the adoption of Bt cotton increases producer surplus, returns to land, farm incomes, and exports in West and Central Africa. By raising farm incomes and reducing labor demand for pest control in cotton production, it creates positive spillovers for other sectors, including food production via commodity and labor markets. Arndt, Pauw, and Thurlow (2012) evaluate the impacts of biofuels production on growth and poverty in Tanzania, using a recursive dynamic computable general equilibrium model. Their simulations reveal that maximizing the poverty-reducing effects of biofuels production requires engaging and improving the productivity of smallholder farmers.

De Janvry and Sadoulet (2002) constructed prototypical regional CGE models to simulate the direct and spillover effects of technological change in food production. They found that direct effects of technological change on farm incomes are most important in Africa, which is largely agrarian and rural. Indirect effects on non-farm incomes are most pronounced in Asia, where there are many rural landless households, and in Latin America, where the poor are largely urban and poor rural household incomes are highly diversified, with a large non-agricultural income share. Simulations using this model suggest that technological change that focuses on tradable crops (to avoid depressing prices) maximize the direct impacts to farmers. This in turn suggests that technology policy may need to focus on high value-added export crops instead of food crops, and complimentary interventions to mitigate price declines may be required to ensure that technological change benefits small farmers. Dorosh and Haggblade (2003) used both CGE and fixed-price multiplier models to measure economic linkages from public investments in eight African countries. They found that indirect spillover effects are nearly as large as direct effects, and the impacts of agricultural investments favor the poor more than investments in other sectors. Huang et al. (2004) combined data from field trials with a modified GTAP model to study impacts of biotechnology on China's production, trade, and welfare. They found that welfare gains far outweigh the public biotechnology research expenditures. Economic linkages captured by the CGE model add to prospective welfare gains.

Srinivasan, Whalley, and Wooton (1993) used multi-country CGE models to examine impacts of regional trade agreements on agricultural performance. Robinson et al. (1993) and Levy and van Wijnbergen (1992) linked together CGEs of Mexico and the United States to simulate impacts of economic integration under a North American Free Trade Agreement (NAFTA) on agricultural production, rural households, and migration. NAFTA CGE models were useful in revealing complex interactions across markets, space, and economic actors in Mexico and the U.S. Anderson, Valenzuela, and Jackson (2008) used the GTAP global CGE model (version 6.1) to compare the impacts of Bt cotton adoption with the impacts of cotton subsidies and trade policies on economic welfare in developing countries. They concluded that welfare in developing countries could be increased more by allowing Bt cotton adoption than by removing all cotton subsidies and tariffs.

Hertel et al. (2010) integrated CGE and climate modeling to study the effects of climate change on poor households via food production in 15 different developing countries. Projected temperature and rainfall from climate simulations were input into CGE models that simulated impacts of the resulting crop production changes on the incomes of different household groups as well as prices for food and related products. Under one climate-change and food production scenario, prices for major staples rise 10 to 60 percent by 2030. In some non-agricultural household groups, the poverty rate rises by 20 to 50 percent in parts of Africa and Asia. Meanwhile, in other parts of Asia and in Latin America, some households specializing in agriculture gain from climate change. This ambitious study is an excellent example of using a simulation model to evaluate impacts where other modeling approaches are not feasible. It was also unique in bringing together experts in CGE modeling and climate science.

National CGEs offer useful insights into impacts of policies and other shocks on aggregate economies, including spillovers; however, they have been criticized for having weak econometric foundations. Hertel et al. (2007) took a step towards integrating CGE modeling with econometrics, using country data to estimate the elasticity of substitution among imports to evaluate the likely impacts of the Free Trade Agreement of the Americas (FTAA). By econometrically estimating this key parameter, they generated a distribution of model results with confidence intervals. The findings suggest that imports would increase as a result of the FTAA.

Aggregate CGEs are less useful to understand heterogeneous impacts within countries, particularly when high transaction costs result in impacts that are heterogeneous across space. Löfgren and Robinson (1999) illustrated this with a stylized country CGE model with nonseparable households. GE models for small economies emerged to fill the lacuna between aggregate CGE and micro AH models. The first of these models were for individual villages and employed linear SAM multiplier analyses (Adelman, Taylor, and Vogel 1988; Lewis and Thorbecke 1992; Subramanian and Sadoulet 1990). Subramanian and Qaim (2009) used this approach and found that Bt cotton adoption reduces labor requirements for pest control but substantially increases overall rural employment. Hired female workers benefit from higher harvest labor demand. The new technology stimulates growth in other agricultural and rural non-agricultural activities as incomes rise, increasing employment there. Faße, Winter, and Grote (2014) constructed a village SAM multiplier model to study the impacts of agro-forestry on a Tanzanian village economy. A novelty of their model is that it includes environmental accounts for changes in tree stocks.

The authors found that agro-forestry has a higher impact on incomes of the poorest households than do other bioenergy crops, including *J. curcas*, sugarcane, and cassava. It also reduces pressure on public forest reserves. Households use agro-forestry as a source of firewood as well as fruits for home consumption or sale.

Village SAM-based models were a first step towards building more flexible disaggregated CGE models that take into account nonlinearities and prices (Taylor and Adelman 1996). Often referred to as "village models," local GE models have been constructed for a wide range of economic spaces, from villages to small regions, island archipelagos, and entire rural economies. Local CGEs share a common modeling approach with national CGE models, but there are important differences between the two. Besides tending to focus on smaller economies, they typically are constructed from the bottom-up, using microsurvey data, instead of from the top-down, using a national SAM.

Taylor, Yúnez-Naude, and Dyer (1999) used a village–town GE model to simulate impacts of Mexican agricultural policies on production, employment, and migration. They found that liberalization of staple prices after NAFTA have a minimal impact on rural wages and migration and a positive effect on rural real incomes, contrary to results from aggregate CGE models. Taylor, Dyer, and Yúnez-Naude (2005) extended this analysis to all of rural Mexico, discovering that there are striking differences in behavioral responses to the same policy changes across regions and household groups. Dyer and Taylor (2011) concluded that subsistence activities limited the impacts of a sharp increase in global maize prices on rural welfare in Mexico, while keeping deforestation pressures in check.

Behrer, Manning, and Seidl (2017) constructed a local general equilibrium model to study the impacts of privatizing common-property grazing land in Patagonia, Chile. They showed that privatization leading to land use change has the potential to increase local wages. This contradicts a widely held view, based on Samuelson (1974) and Weitzman (1974), that privatization of common-property resources reduces economy-wide labor demand and wages. However, consistent with Weitzman (1974), economy-wide wages fall if privatization does not lead to land use change.

Some village models, inspired by developments in micro AH models, permit transaction costs to isolate subsistence households from agricultural and/or labor markets. Jonasson et al. (2014) included endogenous market participation for farmers facing transaction costs in their rural economy-wide (DEVPEM) models, extending the OECD Policy Evaluation Model (PEM; Dewbre et al. 2001) to six developing countries. Their results highlight the importance of market imperfections as well as income and production spillovers in shaping the impacts of agricultural support policies on production, incomes, and welfare. DEVPEM findings on price supports and input subsidy programs contrast with PEM findings from high-income countries. Dyer, Boucher, and Taylor (2006) integrated micro models of many individual agricultural households within a GE model of a Mexican village. They showed that a decrease in the market price of maize depresses commercial maize production while stimulating maize production on subsistence farms; however, poverty among subsistence households increases. The production impacts are contrary to what Robinson et al. (1993) and Levy and van Wijnbergen (1992) found but consistent with the increases in maize production actually observed in Mexico after NAFTA.

5.4 Local economy-wide impact evaluation (LEWIE)

LEWIE models nest micromodels of heterogeneous households and producers within GE models of larger economies – villages, regions, island archipelagos, or whole rural or national economies. Braverman and Hammer (1986) was a first step towards embedding agricultural household models into models of larger economies. Their partial equilibrium model imposed a labor market constraint that equated aggregate household labor demand and supply to generate an endogenous rural wage. This effort was partial and aggregate in the sense of assuming n identical agricultural households (or a single representative one), but it hinted at the future directions that disaggregated economy-wide modeling would take.

The LEWIE models described in Taylor and Filipski (2014) are essentially multi-agent models: econometrically estimated models of heterogeneous agricultural households, integrated within GE models of larger economies and used to simulate impacts of policies, projects, and other exogenous shocks on both a micro and economy-wide level. Econometric estimation of model parameters from micro-data makes it possible to take repeated random draws from all parameter distributions and construct multiple base models on which to simulate impacts and construct confidence intervals around simulated impacts, extending upon Hertel et al. (2007).

Filipski et al. (2017) simulate local economy-wide impacts of global saffron price shocks in Morocco. Local general equilibrium effects buffer income effects while augmenting impacts on some production sectors and employment, particularly of women. Filipski et al. (2015) and Thome et al. (2013) find that poverty programs in Lesotho and Kenya, respectively, significantly impact agricultural production and food consumption while creating income spillovers to non-treated households. Gupta et al. (2017) examine the impacts of productivity-increasing technological change in cotton production in Tanzania's Lake Zone. Lower cotton prices transmit benefits from small farmers to downstream gins, limiting positive income spillovers in the zone, unless there is excess gin capacity or a complementary intervention to expand markets for increased cotton output.

6 Conclusions

Agricultural development impact evaluation entails the use of multiple methods to assess the effects of diverse interventions on a variety of outcomes. There is no "gold standard" for agricultural impact evaluation. Economists, like other scientists, inevitably find themselves in the situation of using the best technologies available to evaluate impacts given the research question and data at hand. In theory, there is a very broad spectrum of identification possibilities, ranging from perfect and unequivocal identification of impacts that are generalizable to entire economies (which never happens) to situations in which the available data (or strategies to obtain data) are not sufficient to support studies that can improve our understanding of agricultural development impacts. All impact evaluation studies fall somewhere in between these two extremes. In the final analysis, audiences and journal reviewers decide whether an identification strategy is convincing or not – and rarely does everyone in the jury agree on the verdict.

In light of this, a promising approach is to combine approaches, as several of the studies in this chapter do. I conclude by highlighting a few examples wherein researchers creatively employed combinations of methods to identify impacts where stand-alone methods were unlikely to generate credible results.

An extensive literature combines weather instruments and panel methods to test hypotheses related to agricultural development; examples in this review include Fafchamps, Udry and Czukas (1998), Miguel (2005), and Carter and Lybbert (2012).

Estimating how credit affects production entails confronting serious selection problems related to differences between farmers who have credit and those who do not. Variables influencing credit access and credit demand are likely to also influence agricultural production directly. Guirkinger and Boucher (2008) combined switching regression with panel methods to evaluate impacts of credit on agricultural output in northern Peru. By tracking the same farmers over time, some of whom shifted between credit states, they found that credit constraints lowered the value of agricultural output in the study area by 26 percent.

Jessoe, Manning, and Taylor (2016) combined IV, panel, and simulation methods to study potential impacts of climate change in Mexico. They used 30 years of nationally representative household survey data as well as daily readings from weather stations to identify impacts of weather shocks on individuals' labor allocations to agriculture and other activities. Panel data at the individual level, combined with state–year fixed effects, made it possible to control for unobservable variables that could confound the study's results. The study found that temperatures harmful to maize production reduce the probability of local work in agriculture. Negative impacts reverberate into non-agricultural sectors of the rural economy such as retail, services, and construction. The contraction in local employment stimulates migration out of rural Mexico, either to other parts of Mexico or to the United States. Linking these econometric results to two major climate simulation models, the authors were able to project changes in rural employment and migration in 2075 under a variety of CO_2 emissions scenarios.

Aker (2010), in her quasi-experimental analysis of the impacts of cell phones on price dispersion across grain markets in Niger, carried out a robustness check using PSM. One might think that the "cell phone tower treatment" is exogenous to individual traders. Critical audiences of economists (and no doubt journal reviewers) noted that the rollout of cell towers might not be random, and variables influencing the rollout might correlate in some way with food prices. To test the robustness of her results, Aker combined the econometric estimation described earlier with PSM matching cell phone tower treated and untreated market pairs. As customary, she estimated the propensity score with a probit model that regressed the dichotomous cell-phone-tower-treatment variable on pre-treatment observables. She then included the propensity score as an additional control in the equation that tested for an effect of cell phone access on grain prices using a weighted least squares regression. Her findings were robust to these estimation methods. The PSM revealed that the effect of mobile phones was stronger for pairs of markets with higher transport costs.

Filipski et al. (2015) used local economy-wide impact evaluation (LEWIE) simulations to perform an *ex ante* evaluation of the effects of a SCT program on agricultural and livestock production as well as other outcomes in Lesotho. They found significant and positive production impacts and income spillovers to non-beneficiary households, with most production impacts concentrated in livestock. The results of an *ex post* experimental evaluation (Taylor et al. 2017) confirmed these results while suggesting that the *ex ante* simulations gave somewhat conservative estimates of local income and production multipliers. Production and income spillovers favored non-beneficiaries, consistent with the simulation findings. A quantile analysis revealed that the positive income spillovers primarily favored non-poor non-beneficiaries.

The juxtaposition of simulation and quasi-experimental methods yielded similarly fruitful results in two studies of income and production spillovers from refugee settlements. LEWIE simulations found positive spillovers on host country agricultural production and income within a 10-km radius around three Congolese refugee camps in Rwanda (Taylor et al. 2016). A novel quasi-experimental analysis, using nighttime lights data as a proxy for economic activity, found evidence of positive spillovers within a 10-km radius around refugee camps in northern Kenya (Alix-Garcia et al. 2017). Combining satellite luminosity readings with microsurvey data, the latter found evidence that employment and agricultural and livestock prices were the key mechanisms driving the spillovers – consistent with simulation findings from Rwanda.

Methodological diversity is essential for successful impact evaluation. Researchers who limit the scope of what they do to a "gold standard" RCT or a specific econometric method constrain the set of agricultural development questions they can address. Economic researchers and the developing countries they study cannot afford for the evaluation to be slave to the tool. New strategies that refine and exploit synergies among identification methods no doubt will emerge as agricultural impact evaluation evolves to address critical development issues in the future.

Acknowledgment

I am grateful to Jeffrey Alwang, Julian Alston, Michael Carter, Travis Lybbert, Bekele Shiferaw, Anubhab Gupta, Heng Zhu, and Miki Doan for their helpful suggestions and comments and to Andrew Schmitz for convincing me to take on this project. Taylor is a member of the Giannini Foundation of Agricultural Economics. This chapter covers a very prolific literature, and I apologize to the authors of many outstanding studies that could not be included in it.

Notes

1 Angelucci and Di Maro (2016) provide a framework to understand spillover effects in program evaluations experimentally.
2 Mexico's PROGRESA program evaluation was an exception, because it included a very large number of randomly chosen treatment and control villages.
3 Becker and Ichino (2002) and Caliendo and Kopeinig (2008) offer practical explanations of using PSM to estimate treatment effects.
4 Witness a sharp increase in usage of "*in silico*" contrasted with the declining use of "*in vivo*" and "*in vitro*" on the Google books Ngram viewer. *In silico* approaches include simulations, with structural models that might be calibrated with data from microsurveys, RCTs, or laboratory experiments.

References

Adelman, I., J.E. Taylor, and S. Vogel. 1988. "Life in a Mexican Village: a SAM Perspective." *The Journal of Development Studies* 25(1):5–24.
Aker, J.C. 2010. "Information from Markets Near and Far: Mobile Phones and Agricultural Markets in Niger." *American Economics Journal: Applied Economics* 2(3):46–59.
Alene, A.D., and O. Coulibaly. 2009. "The Impact of Agricultural Research on Productivity and Poverty in Sub-Saharan Africa." *Food Policy* 34(2):198–209.
Alix-Garcia, J., S. Walker, A. Bartlett, H. Onder, and A. Sanghi. 2017. "Do Refugee Camps Help or Hurt Hosts? The Case of Kakuma, Kenya." *Journal of Development Economics* 130(C):66–83. doi:10.1016/j.jdeveco.2017.09.005.
Anderson, K., E. Valenzuela, and L.A. Jackson. 2008. "Recent and Prospective Adoption of Genetically Modified Cotton: A Global Computable General Equilibrium Analysis of Economic Impacts." *Economic Development and Cultural Change* 56(2):265–296.
Angelucci, M., and G. De Giorgi. 2009. "Indirect Effects of an Aid Program: How do Cash Transfers Affect Ineligibles' Consumption?" *The American Economic Review* 99(1):486–508.
Angelucci, M., and V. Di Maro. 2016. "Programme Evaluation and Spillover Effects." *Journal of Development Effectiveness* 8(1):22–43.
Arndt, C., K. Pauw, and J. Thurlow. 2012. "Biofuels and Economic Development: A Computable General Equilibrium Analysis." *Energy Econ* 34:1922–1930.
———. 2015. "The Economy-wide Impacts and Risks of Malawi's Farm Input Subsidy Program." *American Journal of Agricultural Economics* 98(3):962–980. doi:10.1093/ajae/aav048.
Arrow, K.J., and G. Debreu. 1954. "Existence of an Equilibrium for a Competitive Economy." *Econometrica: Journal of the Econometric Society* 22:265–290.
Arslan, A., and J.E. Taylor. 2009. "Farmers' Subjective Valuation of Subsistence Crops: The Case of Traditional Maize in Mexico." *American Journal of Agricultural Economics* 91(4):956–972.

Asfaw, S., M. Kassie, F. Simtowe, and L. Lipper. 2012a. "Poverty Reduction Effects of Agricultural Technology Adoption: A Micro-Evidence from Rural Tanzania." *Journal of Development Studies* 48(9):1288–1305.

———. 2012b. "Impact of Modern Agricultural Technologies on Smallholder Welfare: Evidence from Tanzania and Ethiopia." *Food Policy* 37(3):283–295.

Athey, S., and G.W. Imbens. 2017. "The State of Applied Econometrics: Causality and Policy Evaluation." *Journal of Economic Perspectives* 31(2):3–32.

Bandiera, Oriana, and Imran Rasul. 2006. "Social Networks and Technology Adoption in Northern Mozambique." *Economic Journal* 116(514):869–902.

Barnum, H.N., and L. Squire. 1979. "An Econometric Application of the Theory of the Farm-Household." *Journal of Development Economics* 6:79–102.

Barrett, C.B. 2008. "Smallholder Market Participation: Concepts and Evidence from Eastern and Southern Africa." *Food Policy* 33(4):299–317.

Barrett, C.B., and M.R. Carter. 2010. "The Power and Pitfalls of Experiments in Development Economics: Some Non-Random Reflections." *Applied Economic Perspectives and Policy* 32(4):515–548.

Barrett, C.B., C.M. Moser, O.V. McHugh, and J. Barison. 2004. "Better Technology, Better Plots, or Better Farmers? Identifying Changes in Productivity and Risk among Malagasy Rice Farmers." *American Journal of Agricultural Economics* 86(4):869–888.

Bauer, M., and J. Chytilová. 2010. "The Impact of Education on Subjective Discount Rate in Ugandan Villages." *Economic Development and Cultural Change* 58(4):643–669.

Beaman, L., D. Karlan, B. Thuysbaert, and C. Udry. 2013. "Profitability of Fertilizer: Experimental Evidence from Female Rice Farmers in Mali." *The American Economic Review* 103(3):381–386.

Becerril, J., and A. Abdulai. 2010. "The Impact of Improved Maize Varieties on Poverty in Mexico: A Propensity Score-Matching Approach." *World Development* 38(7):1024–1035.

Becker, S.O., and A. Ichino. 2002. "Estimation of Average Treatment Effects Based on Propensity Scores." *The Stata Journal* 2(4):358–377.

Behrer, A.P., D.T. Manning, and A. Seidl. 2017. "The Impact of Institutional and Land Use Change on Local Incomes in Chilean Patagonia." *The Journal of Development Studies* 1–18.

Benjamin, D. 1992. "Household Composition, Labor Markets, and Labor Demand: Testing for Separation in Agricultural Household Models." *Econometrica* 60(2):287–322.

Berhane, G., D.O. Gilligan, J. Hoddinott, N. Kumar, and A.S. Taffesse. 2014. "Can Social Protection Work in Africa? The Impact of Ethiopia's Productive Safety Net Programme." *Economic Development and Cultural Change* 63(1):1–26.

Bernard, T., S. Lambert, K. Macours, and M. Vinez. 2016. *Adoption of Improved Seeds and Land Allocation, Evidence from DRC.* Paris: Paris School of Economics. http://lacer.lacea.org/handle/123456789/61240. Accessed April 13, 2018.

Besley, T., and R. Burgess. 2000. "Land Reform, Poverty Reduction, and Growth: Evidence from India." *The Quarterly Journal of Economics* 115(2):389–430.

Besley, T., and A. Case. 1993. "Modeling Technology Adoption in Developing Countries." *The American Economic Review* 83(2):396–402.

Bindlish, Vishva, and Robert E. Evenson. 1997. "The Impact of T&V Extension in Africa: of Kenya and Burkina Faso." *World Bank Research Observer* 12(2):183–201.

Binswanger, H. 1980. "Attitudes toward Risk: Experimental Measurement in Rural India." *American Journal of Agricultural Economics* 62(3):395–407.

Boserup, E. 1965. *The Conditions of Agricultural Growth: The Economics of Agrarian Change Under Population.* New York: Aldine.

Braverman, A., and J.S. Hammer. 1986. "Multimarket Analysis of Agricultural Pricing Policies in Senegal." In I. Singh, L. Squire, and J. Strauss, eds., *Agricultural Household Models, Extensions, Applications and Policy.* Baltimore: The Johns Hopkins University Press, pp. 233–254.

Bulte, Erwin, Gonne Beekman, Salvatore de Falco, Joseph Hella, and Pan Lei. 2014. "Behavioral Responses and the Impact of New Agricultural Technologies: Evidence from a Double-blind Field Experiment in Tanzania." *American Journal of Agricultural Economics* 96(3):813–830.

Burfisher, M.E. 2011. *Introduction to Computable General Equilibrium Models*. Cambridge: Cambridge University Press.

Burke, R.V. 1979. "Green Revolution Technologies and Farm Class in Mexico." *Economic Development and Cultural Change* 28(1):135–154.

Byerlee, D., and L.W. Harrington. 1983. New Wheat Varieties and Small Farmers. No 197286, Occasional Paper Series No. 3, International Association of Agricultural Economists, https://EconPapers.repec.org/RePEc:ags:iaaeo3:197286. Accessed April 13, 2018.

Byerlee, D., and G. Traxler. 1995. "National and International Wheat Improvement Research in the Post-Green Revolution Period: Evolution and Impacts." *American Journal of Agricultural Economics* 77(2):268–278.

Caliendo, M., and S. Kopeinig. 2008. "Some Practical Guidance for the Implementation of Propensity Score Matching." *Journal of Economic Surveys* 22(1):31–72.

Card, D. 1990. "The Impact of the Mariel Boatlift on the Miami Labor Market." *ILR Review* 43(2):245–257.

Cardenas, J.C., and J.P. Carpenter. 2008. "Behavioural Development Economics: Lessons from Field Labs in the Developing World." *Journal of Development Studies* 44(3):311–338.

Carter, M.R., and T.J. Lybbert. 2012. "Consumption versus Asset Smoothing: Testing the Implications of Poverty Trap Theory in Burkina Faso." *Journal of Development Economics* 99(2):255–264.

Charlton, D., and J.E. Taylor. 2016. "A Declining Farm Workforce: Analysis of Panel Data from Rural Mexico." *American Journal of Agricultural Economics* 98(4):1158–1180.

———. 2017. *Access to Secondary Schools Accelerating the Agricultural Transition: Analysis of Panel Data from Rural Mexico*. Montana State University, Department of Agricultural Economics and Economics.

Chibwana, C., M. Fisher, and G. Shively. 2012. "Cropland Allocation Effects of Agricultural Input Subsidies in Malawi." *World Development* 40(1):124–133.

Conley, Timothy G., and Christopher R. Udry. 2010. "Learning About a New Technology: Pineapple in Ghana." *American Economic Review* 100(1):35–69.

Dar, M.H., A. De Janvry, K. Emerick, D. Raitzer, and E. Sadoulet. 2013. "Flood-Tolerant Rice Reduces Yield Variability and Raises Expected Yield, Differentially Benefitting Socially Disadvantaged Groups." *Scientific Reports* 3.

Davis, B., S. Handa, N. Hypher, N.W. Rossi, P. Winters, and J. Yablonski (editors). 2016. *From Evidence to Action: The Story of Cash Transfers and Impact Evaluation in Sub Saharan Africa*. Oxford: Oxford University Press.

De Janvry, A., A. Dustan, and E. Sadoulet. 2011. *Recent Advances in Impact Analysis Methods for Ex-Post Impact Assessments of Agricultural Technology: Options for the CGIAR*. Consultative Group on International Agricultural Research, Independent Science and Partnership Council, April. https://gspp.berkeley.edu/assets/uploads/research/pdf/deJanvryetal2011.pdf. Accessed 4/13/2018.

De Janvry, A., M. Fafchamps, and E. Sadoulet. 1991. "Peasant Household Behavior with Missing Markets: Some Paradoxes Explained." *The Economic Journal* 101:1400–1417.

De Janvry, A., and E. Sadoulet. 2002. "World Poverty and the Role of Agricultural Technology: Direct and Indirect Effects." *Journal of Development Studies* 38(4):1–26.

Dehejia, R.H., and S. Wahba. 2002. "Propensity Score-Matching Methods for Nonexperimental Causal Studies." *The Review of Economics and Statistics* 84(1):151–161.

Dewbre, J., J. Antón, and W. Thompson. 2001. "The Transfer Efficiency and Trade Effects of Direct Payments." *American Journal of Agricultural Economics* 83(5):1204–1214.

Di Falco, S., M. Veronesi, and M. Yesuf. 2011. "Does Adaptation to Climate Change Provide Food Security? A Microperspective from Ethiopia." *American Journal of Agricultural Economics* 93(3):829–846.

Dorosh, P., and S. Haggblade. 2003. "Growth Linkages, Price Effects and Income Distribution in Sub Saharan Africa." *Journal of African Economies* 12(2):207–235.

Duflo, Esther. 2001. "Schooling and Labor Market Consequences of School Construction in Indonesia: Evidence from An Unusual Policy Experiment." *American Economic Review* 91(4):795–813.

Duflo, E., M. Kremer, and J. Robinson. 2011. "Nudging Farmers to Use Fertilizer: Theory and Experimental Evidence from Kenya." *The American Economic Review* 101(6):2350–2390.

Dyer, G.A., S. Boucher, and J.E. Taylor. 2006. "Subsistence Response to Market Shocks." *American Journal of Agricultural Economics* 88(2):279–291.

Dyer, G.A., and J.E. Taylor. 2011. "The Corn Price Surge: Impacts on Rural Mexico." *World Development* 39(10):1878–1887.

Elabed, Ghada, and Michael R. Carte. 2015. "*Ex-ante* Impacts of Agricultural Insurance: Evidence from a Field Experiment in Mali." University of California, Davis. https://basis.ucdavis.edu/sites/g/files/dgvnsk466/files/2017-05/impact_evaluation_0714_vdraft.pdf. Accessed April 13, 2018.

Elbehri, A., and S. Macdonald. 2004. "Estimating the Impact of Transgenic Bt Cotton on West and Central Africa: A General Equilibrium Approach." *World Development* 32(12):2049–2064.

Emerick, K., A. de Janvry, E. Sadoulet, and M.H. Dar. 2016. "Technological Innovations, Downside Risk, and the Modernization of Agriculture." *American Economic Review* 106(6):1537–1561.

Evenson, R.E. 2001. "Economic Impacts of Agricultural Research and Extension." *Handbook of Agricultural Economics* 1:573–628.

Faße, A., E. Winter, and U. Grote. 2014. "Bioenergy and Rural Development: The Role of Agroforestry in a Tanzanian Village Economy." *Ecological Economics* 106:155–166.

Fafchamps, Marcel, Christopher Udry, and Katherine Czukas. 1998. "Drought and Saving in West Africa: Are Livestock a Buffer Stock?" *Journal of Development Economics* 55(2):273–305.

Feder, G. 1980. "Farm Size, Risk Aversion and the Adoption of New Technology Under Uncertainty." *Oxford Economic Papers* 32(2):263–283.

Feder, G., R.E. Just, and D. Zilberman. 1985. "Adoption of Agricultural Innovations in Developing Countries: A Survey." *Economic Development and Cultural Change* 33(2):255–298.

Filipski, M., A. Aboudrare, T.J. Lybbert, and J.E. Taylor. 2017. "Spice Price Spikes: Simulating Impacts of Saffron Price Volatility in a Gendered Local Economy-Wide Model." *World Development* 91:84–99.

Filipski, M.J., J.E. Taylor, K.E. Thome, and B. Davis. 2015. "Effects of Treatment Beyond the Treated: A General Equilibrium Impact Evaluation of Lesotho's Cash Grants Program." *Agricultural Economics* 46(2):227–243.

Foster, Andrew D., and Mark R. Rosenzweig. 1995. "Learning by Doing and Learning from Others: Human Capital and Technical Change in Agriculture." *Journal of Political Economy* 103(6):1176–1209.

Foster, J., J. Greer, and E. Thorbecke. 1984. "A Class of Decomposable Poverty Measures." *Econometrica: Journal of the Econometric Society* 52(3):761–766.

Gertler, Paul J., Sebastian W. Martinez, and Marta Rubio Codina. 2012. "Investing Cash Transfers to Raise Long-Term Living Standards." *American Economic Journal: Applied Economics* 4(1):164–192.

Gibson, J., D. McKenzie, and S. Stillman. 2011. "The Impacts of International Migration on Remaining Household Members: Omnibus Results from a Migration Lottery Program." *Review of Economics and Statistics* 93(4):1297–1318.

Grant, J.H., and D.M. Lambert. 2008. "Do Regional Trade Agreements Increase Members' Agricultural Trade?" *American Journal of Agricultural Economics* 90(3):765–782.

Guirkinger, Catherine, and Stephen R. Boucher. 2008. "Credit Constraints and Productivity in Peruvian Agriculture." *Agricultural Economics* 39(3):295–308.

Gupta, A., J. Kagin, J.E. Taylor, M. Filipski, L. Hlanze, and J. Foster. 2017. "Is Technology Change Good for Cotton Farmers? A Local-Economy Analysis from the Tanzania Lake Zone." *European Review of Agricultural Economics* 1–30.

Gupta, Sandeep K. 2011. "Intention-to-Treat Concept: A Review." *Perspectives in Clinical Research* 2(3):109–112.

Halliday, T. 2006. "Migration, Risk, and Liquidity Constraints in El Salvador." *Economic Development and Cultural Change* 54(4):893–925.

Harrison, G.W., S.J. Humphrey, and A. Verschoor. 2010. "Choice Under Uncertainty: Evidence from Ethiopia, India and Uganda." *The Economic Journal* 120(543):80–104.

Heckman, J. 1979. "Sample Selection Bias as a Specification Error." *Econometrica* 47(1):153–161.

Heltberg, R. 1998. "Rural Market Imperfections and the Farm Size – Productivity Relationship: Evidence from Pakistan. *World Development* 26(10):1807–1826.

Hertel, T. (editor). 1997. *Global Trade Analysis: Modeling and Applications*. Cambridge: Cambridge University Press.

Hertel, T., D. Hummels, M. Ivanic, and R. Keeney. 2007. "How Confident Can We Be of CGE-Based Assessments of Free Trade Agreements?" *Economic Modelling* 24(4):611–635.

Hertel, Thomas W., Marshall B. Burke, and David B. Lobell. 2010. "The Poverty Implications of Climate-Induced Crop Yield Changes By 2030." *Global Environmental Change* 20(4):577–585.

Hoddinott, J. 2006. "Shocks and Their Consequences Across and Within Households in Rural Zimbabwe." *The Journal of Development Studies* 42(2):301–321.

Hornbeck, R. 2012. "The Enduring Impact of the American Dust Bowl: Short-and Long-Run Adjustments to Environmental Catastrophe." *The American Economic Review* 102(4):1477–1507.

Huang, J., R. Hu, H. van Meijl, and F. van Tongeren. 2004. "Biotechnology Boosts to Crop Productivity in China: Trade and Welfare Implications." *Journal of Development Economics* 75(1):27–54.

IEG (Independent Evaluation Group). 2011. *Impact Evaluations in Agriculture: An Assessment of the Evidence*. Washington, DC: World Bank. https://openknowledge.worldbank.org/handle/10986/27794. Accessed April 13, 2018.

Imbens, Guido W., and Joshua D. Angrist. 1994. "Identification and Estimation of Local Average Treatment Effects." *Econometrica* 62(2):467–475.

Jacoby, H.G. 1993. 'Shadow Wages and Peasant Family Labour Supply: An Econometric Application to the Peruvian Sierra. *The Review of Economic Studies* 60(4):903–921.

Jayachandran, S. 2006. "Selling Labor Low: Wage Responses to Productivity Shocks in Developing Countries." *Journal of Political Economy* 114(3):538–575.

Jayne, T.S., D. Mather, N. Mason, and J. Ricker-Gilbert. 2013. "How do Fertilizer Subsidy Programs Affect Total Fertilizer Use in Sub-Saharan Africa? Crowding Out, Diversion, and Benefit/Cost Assessments." *Agricultural Economics* 44(6):687–703.

Jessoe, K., D. Manning, and E.J. Taylor. 2016. "Climate Change and Labour Allocation in Rural Mexico: Evidence from Annual Fluctuations in Weather." *The Economic Journal* 128(609):230–261. doi:10.1111/ecoj.12448/full.

Jonasson, E., M. Filipski, J. Brooks, and J.E. Taylor. 2014. "Modeling the Welfare Impacts of Agricultural Policies in Developing Countries." *Journal of Policy Modeling* 36(1):63–82.

Just, R.E., and D. Zilberman. 1983. "Stochastic Structure, Farm Size and Technology Adoption in Developing Agriculture." *Oxford Economic Papers* 35(2):307–328.

Kagin, J., J.E. Taylor, and A. Yúnez-Naude. 2016. "Inverse Productivity or Inverse Efficiency? Evidence from Mexico." *The Journal of Development Studies* 52(3):396–411.

Karlan, D., R. Osei, I. Osei-Akoto, and C. Udry. 2014. "Agricultural Decisions After Relaxing Credit and Risk Constraints." *The Quarterly Journal of Economics* 129(2):597–652.

Kassie, M., B. Shiferaw, and G. Muricho. 2011. "Agricultural Technology, Crop Income, and Poverty Alleviation in Uganda." *World Development* 39(10):1784–1795.

Keswell, M., and M.R. Carter. 2014. "Poverty and Land Redistribution." *Journal of Development Economics* 110:250–261.

Khonje, M., J. Manda, A.D. Alene, and M. Kassie. 2015. "Analysis of Adoption and Impacts of Improved Maize Varieties in Eastern Zambia." *World Development* 66:695–706.

Kuroda, Y., and P.A. Yotopoulos. 1978. "A Microeconomic Analysis of Production Behavior of the Farm Household in Japan – a Profit Function Approach." 経済研究 29(2):116–129.

LaLonde, R.J. 1986. "Evaluating the Econometric Evaluations of Training Programs with Experimental Data." *The American Economic Review* 76(4):604–620.

Levy, S., and S. Van Wijnbergen. 1992. "Maize and the Free Trade Agreement Between Mexico and the United States." *The World Bank Economic Review* 6(3):481–502.

Lewis, B.D., and E. Thorbecke. 1992. "District-level Economic Linkages in Kenya: Evidence Based on a Small Regional Social Accounting Matrix." *World Development* 20(6):881–897.

Lipton, M., and R. Longhurst. 1989. *New Seeds and Poor People*. Baltimore: Johns Hopkins.

Liu, E.M. 2013. "Time to Change What to Sow: Risk Preferences and Technology Adoption Decisions of Cotton Farmers in China." *Review of Economics and Statistics* 95(4):1386–1403.

Löfgren, H., and S. Robinson. 1999. "Nonseparable Farm Household Decisions in a Computable General Equilibrium Model." *American Journal of Agricultural Economics* 81(3):663–670.

Lybbert, Travis J. 2006. "Indian Farmers' Valuation of Yield Distributions: Will Poor Farmers Value 'Pro-poor' Seeds?" *Food Policy* 31(5):415–441.

Lybbert, T.J., C.B. Barrett, S. Desta, and D. Layne Coppock. 2004. "Stochastic Wealth Dynamics and Risk Management Among a Poor Population." *The Economic Journal* 114(498):750–777.

Magnan, N., D.M. Larson, and J.E. Taylor. 2012. "Stuck on Stubble? The Non-Market Value of Agricultural Byproducts for Diversified Farmers in Morocco." *American Journal of Agricultural Economics* 94(5):1055–1069.

Mani, A., S. Mullainathan, E. Shafir, and J. Zhao. 2013. "Poverty Impedes Cognitive Function." *Science* 341(6149):976–980.

McKenzie, D., S. Stillman, and J. Gibson. 2010. "How Important is Selection? Experimental vs. Non-experimental Measures of the Income Gains from Migration." *Journal of the European Economic Association* 8(4):913–945.

Michelson, H.C. 2013. "Small Farmers, NGOs, and a Walmart World: Welfare Effects of Supermarkets Operating in Nicaragua." *American Journal of Agricultural Economics* 95(3):628–649.

Miguel, E. 2005. "Poverty and Witch Killing." *The Review of Economic Studies* 72(4):1153–1172.

Miguel, E., and M. Kremer. 2004. "Worms: Identifying Impacts on Education and Health in the Presence of Treatment Externalities." *Econometrica* 72(1):159–217.

Moser, C.M., and C.B. Barrett. 2006. "The Complex Dynamics of Smallholder Technology Adoption: The Case of SRI in Madagascar." *Agricultural Economics* 35(3):373–388.

Mosley, P., and A. Verschoor. 2005. "Risk Attitudes and the 'Vicious Circle of Poverty'." *The European Journal of Development Research* 17(1):59–88.

Moyo, S., G.W. Norton, J. Alwang, I. Rhinehart, and C.M. Deom. 2007. "Peanut Research and Poverty Reduction: Impacts of Variety Improvement to Control Peanut Viruses in Uganda." *American Journal of Agricultural Economics* 89(2):448–460.

Munshi, K. 2004. "Social Learning in a Heterogeneous Population: Technology Diffusion in the Indian Green Revolution." *Journal of Development Economics* 73(1):185–213.

Nakajima, C. 1969. "Subsistence and Commercial Family Farms: Some Theoretical Models of Subjective Equilibrium." In C. Wharton, ed., *Subsistence Agriculture and Economic Development*. Chicago: Aldine, pp. 165–185.

Pardey, P.G., J.M. Alston, and R. Piggott (editors). 2006. *Agricultural R&D in the Developing World: Too Little, Too Late?* Washington, DC: International Food Policy Research Institute.

Perrin, R., and D. Winkelmann. 1976. "Impediments to Technical Progress on Small Versus Large Farms." *American Journal of Agricultural Economics* 58(5):888–894.

Reardon, T., C.P. Timmer, C.B. Barrett, and J. Berdegué. 2003. "The Rise of Supermarkets in Africa, Asia, and Latin America." *American Journal of Agricultural Economics* 85(5):1140–1146.

Renkow, M., and D. Byerlee. 2010. "The Impacts of CGIAR Research: A Review of Recent Evidence." *Food Policy* 35(5):391–402.

Ricker-Gilbert, J. 2013. "Wage and Employment Effects of Malawi's Fertilizer Subsidy Program." *Agricultural Economics* 45:1–17.

Ricker-Gilbert, J., T.S. Jayne, and E. Chirwa. 2011. "Subsidies and Crowding Out: A Double-Hurdle Model of Fertilizer Demand in Malawi." *American Journal of Agricultural Economics* 93(1):26–42.

Ricker-Gilbert, J., N.M. Mason, F.A. Darko, and S.T. Tembo. 2013. "What Are the Effects of Input Subsidy Programs on Maize Prices? Evidence from Malawi and Zambia." *Agricultural Economics* 44(6):671–686.

Robinson, S., M.E. Burfisher, R. Hinojosa-Ojeda, and K.E. Thierfelder. 1993. "Agricultural Policies and Migration in a US-Mexico Free Trade Area: A Computable General Equilibrium Analysis." *Journal of Policy Modeling* 15(5–6):673–701.

Rosenbaum, P.R. 2002. *Observational Studies*. New York: Springer.

Rosenbaum, P.R., and D.B. Rubin. 1983. "The Central Role of the Propensity Score in Observational Studies for Causal Effects." *Biometrika* 70(1):41–55.

Rosenzweig, M.R., and H. Binswanger. 1993. "Wealth, Weather Risk and the Composition and Profitability of Agricultural Investments." *Economic Journal* 103:56–78.

Rozelle, S., J.E. Taylor, and A. DeBrauw. 1999. "Migration, remittances, and agricultural productivity in China." *The American Economic Review* 89(2):287–291.

Rubin, D.B. 1974. "Estimating Causal Effects of Treatments in Randomized and Nonrandomized Studies." *Journal of Educational Psychology* 66(5):688.

Sadoulet, E., A. De Janvry, and B. Davis. 2001. "Cash Transfer Programs with Income Multipliers: PROCAMPO in Mexico." *World Development* 29(6):1043–1056.

Samuelson, P.A. 1974. "Is the Rent-Collector Worthy of His Full Hire?" *Eastern Economic Journal* 1(1):7–10.

Shiferaw, B., M. Kassie, M. Jaleta, and C. Yirga. 2014. "Adoption of Improved Wheat Varieties and Impacts on Household Food Security in Ethiopia." *Food Policy* 44:272–284.

Singh, I., L. Squire, and J. Strauss. 1986. *An Overview of Agricultural Household Models-The Basic Model: Theory, Empirical Results, and Policy Conclusions, in Agricultural Household Models, Extensions, Applications and Policy*. I. Singh, L. Squire, and J. Strauss, eds. Baltimore: The World Bank and the Johns Hopkins University Press, pp. 17–47.

Skoufias, E. 1994. "Using Shadow Wages to Estimate Labor Supply of Agricultural Households." *American Journal of Agricultural Economics* 76(2):215–227.

Smith, J.A., and P.E. Todd. 2005. "Does Matching Overcome LaLonde's Critique of Nonexperimental Estimators?" *Journal of Econometrics* 125(1):305–353.

Srinivasan, T.N., J. Whalley, and I. Wooton. 1993. "Measuring the Effects of Regionalism on Trade and Welfare." In K. Anderson and R. Blackhurst, eds., *Regional Integration and the Global Trading System*. London: Harvester Wheatsheaf, pp. 52–79.

Stone, R. 1986. "Nobel Memorial Lecture 1984: The Accounts of Society." *Journal of Applied Econometrics* 1:5–28.

Subramanian, A., and M. Qaim. 2009. "Village-wide Effects of Agricultural Biotechnology: The Case of Bt Cotton in India." *World Development* 37(1):256–267.

Subramanian, S., and E. Sadoulet. 1990. "The Transmission of Production Fluctuations and Technical Change in a Village Economy: A Social Accounting Matrix Approach." *Economic Development and Cultural Change* 39(1):131–173.

Suri, T. 2011. "Selection and Comparative Advantage in Technology Adoption." *Econometrica* 79(1):159–209.

Tanaka, T., C.F. Camerer, and Q. Nguyen. 2016. "Risk and Time Preferences: Linking Experimental and Household Survey Data from Vietnam." In *Behavioral Economics of Preferences, Choices, and Happiness*. Japan: Springer, pp. 3–25.

Taylor, J.E., and I. Adelman. 1996. *Village Economies: The Design, Estimation and Application of Village-Wide Economic Models*. Cambridge: Cambridge University Press.

———. 2003. "Agricultural Household Models: Genesis, Evolution, and Extensions." *Review of Economics of the Household* 1(1):33–58.

Taylor, J.E., G.A. Dyer, and A. Yunez-Naude. 2005. "Disaggregated Rural Economywide Models for Policy Analysis." *World Development* 33(10):1671–1688.

Taylor, J.E., and M.J. Filipski. 2014. *Beyond Experiments in Development Economics: Local Economy-Wide Impact Evaluation*. Oxford University Press.

Taylor, J.E., M.J. Filipski, M. Alloush, A. Gupta, R.I.R. Valdes, and E. Gonzalez-Estrada. 2016. "Economic Impact of Refugees." *Proceedings of the National Academies of Sciences* 113(27):7449–7453. www.pnas.org/content/113/27/7449.abstract. Accessed 4/13/2018.

Taylor, J.E., A. Gupta, M. Filipski, K. Thome, B. Davis, L. Pellerano, and O. Niang. 2017. "Heterogeneous Spillovers from SCTs: Evidence from Lesotho." Working Paper. www.wider.unu.edu/sites/default/files/Events/PDF/Papers/PubEconConf2017-Gupta.pdf.

Taylor, J.E., A. Yúnez-Naude, and G. Dyer. 1999. "Agricultural Price Policy, Employment, and Migration in a Diversified Rural Economy: A Village-Town CGE Analysis from Mexico." *American Journal of Agricultural Economics* 81(3):653–662.

Teele, D.L. (editor). 2014. *Field Experiments and Their Critics: Essays on the Uses and Abuses of Experimentation in the Social Sciences*. New Haven, CT: Yale University Press.

Thirtle, C., L. Lin, and J. Piese. 2003. "The Impact of Research-Led Agricultural Productivity Growth on Poverty Reduction in Africa, Asia and Latin America." *World Development* 31(12):1959–1975.

Thome, K., M. Filipski, J. Kagin, J.E. Taylor, and B. Davis. 2013. "Agricultural Spillover Effects of Cash Transfers: What Does LEWIE Have to Say?" *American Journal of Agricultural Economics* 95(5):1338–1344.

Udry, Christopher. 1996. "Gender, Agricultural Production, and the Theory of the Household." *The Journal of Political Economy* 104(5):1010–1046.

US Food and Drug Administration (USFDA). 1988. *Guideline for the Format and Content of the Clinical and Statistical Sections of Applications*. Center for Drug Evaluation and Research, Food and Drug Administration, Department of Health and Human Services, 5600 Fishers Lane Rockville, MD 20857(301) 443–4330. www.fda.gov/downloads/Drugs/ . . . /Guidances/UCM071665.pdf.

Voors, M., M. Demont, and E. Bulte. 2016. "New Experiments in Agriculture." *African Journal of Agricultural and Resource Economics* 11(1):1–7.

Weitzman, M.L. 1974. "Free Access vs. Private Ownership as Alternative Systems for Managing Common Property." *Journal of Economic Theory* 8(2):225–234.

Yang, D. 2008. "International Migration, Remittances and Household Investment: Evidence from Philippine Migrants' Exchange Rate Shocks." *The Economic Journal* 118(528):591–630.

Zaal, F., and R.H. Oostendorp. 2002. "Explaining a Miracle: Intensification and the Transition Towards Sustainable Small-Scale Agriculture in Dryland Machakos and Kitui Districts, Kenya." *World Development* 30(7):1271–1287.

31
Estimating production functions

Daniela Puggioni and Spiro E. Stefanou

1 Overview

In his outstanding review chapter, Mundlak (2001, p. 5) boldly states that "hardly a subject in economics can be discussed with production sitting in the balcony rather than playing center stage." Indeed, there are many excellent overviews on the emergence and evolution of production analysis. Chambers (1988) presents a succinct history of the concept of the production function as a theoretical and empirical mechanism for analysis. More recently, Chavas, Chambers, and Pope (2010) have reflected on the emergence of production economics and farm management over the twentieth century. This contribution is highly recommended to the reader interested in the historical evolution of applied production analysis and concepts geared toward an application to agriculture.

Production behavior is grounded in two core elements. The first is the prospect of how input choices are transformed into output realizations and involves the technical specification of input–output responses, and input–input and output–output substitutions, and how input–output relationships interact with the production environment. The second addresses the forces driving input and output choices, which necessarily involves an assumption regarding the producer's objective given the relationship describing how inputs are transformed into outputs.

There are tens of inputs used on a farm; some are unique to the output to be produced and others are unique to the farm's resources. As we seek to engage in a broader analysis of production decision making, the analyst summarizes (essentially aggregates) inputs into broad categories. For example, an aggregate labor input category will include a variety of skilled hired laborers and even family labor; an intermediate inputs category will often include various agricultural chemicals (e.g., fertilizers, herbicides, pesticides), veterinary services, hired machinery services, fuel, and irrigation operation expenses, among others; and capital can represent various types of equipment and structures. Similarly, a farm may produce several outputs, and those may be summarized into broader categories such as crops, livestock, and permanent plantings. There is an extensive literature, both theoretical and applied, on creating these aggregate categories (Chambers 1988; Balk 2012). The point is that the first steps include identifying the key factors relating inputs to outputs and creating aggregates that reflect the transformation of inputs into outputs.

Modeling the production process involves making decisions on how to summarize inputs and outputs into variables that still embody the robust interplay between input–output and input–input relationships, while still being tractable in application. There are many frameworks to empirically address production frameworks. This chapter focuses on recent innovations in the econometric modeling of production functions. Section 2 reviews the notion of a production technology and how this technology is characterized historically. Section 3 addresses the revival of the econometrics estimation of the production function in its primal form. The revival is motivated to take advantage of the rise in firm/farm-level datasets over time, with only physical input data available. Various types of Census databases have led to renewed interest in estimating production technologies directly. The driving force behind this reemergence is the need to accommodate the endogeneity of inputs as well as output. Section 4 addresses some emerging directions, followed by Section 5 offering concluding comments.

2 Characterizing production technologies

The production technology is a statement that describes how a bundle of inputs (x) and fixed factors (z) can produce a bundle of outputs (y). The statement of the production technology can take on a set-theoretic form (Shepherd 1970) or a function-based form. For example, the case of the single output, multiple input production function is the most basic form where we specify:

$$y = f(x,z), \tag{1}$$

where $f(x,z)$ embody a set of desirable properties that come from the character of the technology T (Chambers 1988). When there are multiple outputs and multiple inputs, we tend to move toward a transformation function where we specify:

$$F(y,x,z) = 0. \tag{2}$$

There are restricted versions of the transformation function that admit separability in input and outputs, $F(y,x,z) = G(y) - H(x) = 0$, which can be interpreted as an output function (i.e., an output aggregate defined as $Y = G(y)$ is a function in inputs $H(x)$) (Powell and Gruen 1968; Hasenkamp 1976; Chambers 1988). In application, both the production function and transformation function are implicitly implying parametric specifications.

Our focus is the operationalization of the production technology with the specification of a functional form and the employment of econometric approaches to identify the structure of production. The functional form should relate to the character of the decision environment, which includes the economic objectives of the agents generating the data and, more broadly, the economic decision environment that can influence the relationships between factors of production (both variable and fixed) (Fuss, McFadden, and Mundlak 2014). Several key issues emerge in taking this path. The first issue is the choice of functional forms, which Fuss, McFadden, and Mundlak (2014) suggest should take into account the parsimony of parameters, the ease of interpretation and computation, and the ability to interpolate and extrapolate robustly. These are matters addressing the core functional specification of the technology. The second issue relates to how the error structure of the overall specification reflects on the relationship between variables. Our empirical investigations almost never enjoy the clean features of data based on experimental trials where the production environment is under the full control of the researcher. In typical settings, the data are generated by actors who are motivated by economic-based objectives operating in a market environment that may pose challenges of imperfect information, market structures outside of perfect competition, and external shocks arising from weather. The analyst does not have the same level of information and insight as the decision making agent when making decisions.

3 Production function estimation

3.1 Basic endogeneity issues

Econometric production functions, as we know them today, essentially relate productive inputs (e.g., capital and labor) to outputs and have their roots in the work of Cobb and Douglas (1928), who proposed production function estimation as a tool for testing hypotheses on marginal productivity and competitiveness in labor markets. Some researchers were quick to criticize their approach. For example, Mendershausen (1938) argued that the data used by Cobb and Douglas were too multicollinear to allow for a credible determination of the production function coefficients. Marschak and Andrews (1944) raised concerns that production function estimation is problematic due to the potential endogeneity of inputs.

To illustrate the issue, consider the Cobb–Douglas production function technology $Y_j = A_j K_j^{\beta_k} L_j^{\beta_l}$, with one output, Y_j, and two inputs: capital, K_j, and labor, L_j. A_j is the Hicks-neutral efficiency level of firm j, which is unobservable by the econometrician. Taking natural logs, the previous relation becomes linear

$$y_j = \beta_0 + \beta_k k_j + \beta_l l_j + \varepsilon_j \tag{3}$$

with lowercase letters expressing natural logarithms of the variables, for example, $\ln(K_j) = k_j$) and $\ln(A_j) = \beta_0 + \varepsilon_j$. The constant β_0 can be viewed as the mean efficiency level across firms, while ε_j is the deviation from that mean for each firm j. The term ε_j represents all residual disturbances left out of the production factors, such as firm-specific technology, efficiency, or management differences, functional form discrepancies, measurement errors in output, or unobserved sources of variation in output. The observation made by Marschack and Andrews is that, since the right-hand-side variables are chosen by the firm in some optimal or behavioral fashion, they cannot really be treated as independent. In fact, if the firm knows its ε_j, or some part of it, when making input choices, these choices will likely be correlated with ε_j. One could argue that capital can be considered a fixed input, as it is usually predetermined for the duration of the relevant observation period, and it is therefore orthogonal with respect to the disturbance term. The same argument, however, cannot apply to labor, even if we are willing to make the quite strong assumption that firms operate in perfectly competitive input and output markets and treat capital as a fixed input. If firms perfectly or imperfectly observe ε_j before choosing the optimal amount of labor to utilize in production, their choice will necessarily depend on ε_j and the usual exogeneity assumptions that are required for unbiasedness and consistency of ordinary least squares (OLS) are unlikely to hold. Empirical results have actually shown that both capital and labor are usually correlated with the error term, but most often the bias in the labor coefficient is larger than the bias on the capital coefficient. This is consistent with the view that labor is more easily adjustable than capital, is thus more variable, and therefore is more highly correlated with ε_j.

There is a second problem, perhaps less emphasized and documented in the literature, embedded in the OLS estimation of (3). Firm-level datasets are usually characterized by a significant level of attrition, that is, firms entering and exiting, but, obviously, researchers have only data on firms prior to exiting. Assume that firms can observe ε_j, then decide whether or not to exit, and choose labor and level of production optimally if they decide not to exit. Abstracting from the implications associated with firm dynamics, assume also that firms deciding to exit receive a non-negative remuneration equal to their sell-off value; thus, firms will exit if the variable profits are lower than the sell-off value. The problem here is that this exit condition will generate correlation between ε_j and K_j, conditional on continuing to be in the dataset (i.e., continuing to produce). This is because if firms know their ε_j when they have to decide whether or not to exit, firms continuing to produce will have ε_j drawn from a selected sample, and the selection will be partially dependent on the fixed input K_j. That is, as firms with higher fixed capital are able to afford lower ε_j without having to exit, the sample selection of the firms remaining in business will generate negative correlation between ε_j and K_j. Once again, the orthogonality conditions for OLS estimation would be violated.

3.2 Traditional solutions: fixed effects and instrumental variables

The earliest responses to the concerns about the necessity of considering the endogeneity issues in production functions estimation came through the increasing availability of panel data and developed, traditionally, along two main directions: fixed effects and instrumental variables.

3.2.1 Fixed effects

Hoch and Mundlak offer several contributions addressing the fixed effects methodology in economics in the context of production function estimation. To understand the essence of this approach, consider a modified formulation of (3):

$$y_{jt} = \beta_0 + \beta_k k_{jt} + \beta_l l_{jt} + \omega_j + \eta_{jt}, \qquad (4)$$

where η_{jt} is not observed by the firm before any production decision (input choice or exit), so that this term is not correlated with the firm's optimal choices. Conversely, firms have knowledge of ω_j when they make input and exit choices. Intuitively, ω_j can represent entrepreneurial ability, labor quality, or any other factor affecting the production that a firm can observe or predict, and it is usually defined as the firm's unobserved (by the econometrician) productivity. On the other hand, η_{jt} represents deviations from the expected values of these factors and can also be thought of as the conventional measurement error in y_{jt} that is uncorrelated with input and exit decisions. Clearly the endogeneity issues concern only ω_j and not η_{jt}. The fact that ω_j is assumed to be constant over time, or at least over the length of the available panel, is the basic premise behind fixed effects estimation and allows for consistent estimation of production function coefficients using differencing or least squares dummy variables estimation techniques. In general, this implies that (4) can be consistently estimated via OLS, specifying:

$$\left(y_{jt} - y_{j\bar{t}}\right) = \beta_k \left(k_{jt} - k_{j\bar{t}}\right) + \beta_l \left(l_{jt} - l_{j\bar{t}}\right) + \left(\eta_{jt} - \eta_{j\bar{t}}\right), \qquad (5)$$

where the notation $\left(x_{jt} - x_{j\bar{t}}\right)$ represents averaging over the time dimension for each individual firm.[1]

This approach is first stated briefly in Hoch (1955) and fully developed in Hoch (1962). In this latter contribution, Hoch makes use of combined time series and cross-section data in the estimation of production function parameters for a sample of 63 Minnesota farms over a six-year period from 1946 to 1951. The main goal of the study is to estimate the elasticity of output with respect to inputs in order to draw inferences regarding the allocation of resources by the economic units of the sample. A Cobb–Douglas specification is used to derive a condition stating that firms equate the value of the marginal product of each input to its price multiplied by a constant. This constant represents the elasticity of output with respect to that input and can be interpreted as returns to scale. If firms are in fact profit maximizers, the value of the constant should be one, as optimality requires the value of the marginal product to be the same as the price of the factor. Hoch argues that rationalizing the use of single equation estimates of the production function parameters is possible if one is willing to assume that firms maximize by differentiating anticipated output with respect to current input so that the observed input choices are not correlated with the disturbance term. The extent to which this assumption can be supported depends on the characteristics of the industry where the firm operates. In the case of agriculture, for example, it seems reasonable to believe that the term ε_j includes the effects of weather variability, which do not affect the optimal choice of inputs. In this context, a single equation estimation is justifiable. There are, however, other differences between firms, such as differences in technical efficiency that will influence both output and inputs. Hoch points out that if there are differences in technical efficiency between firms, that is, ω_j in (4) varies substantially across firms, firms that are more efficient will be able to produce more output for a given level of inputs and, by profit maximization, will tend to have higher levels on inputs; thus, the optimal choice of factors will depend on ω_j. A similar problem arises if productivity increases over time. As a way out of this difficulty, Hoch uses the analysis of covariance, exploiting the time series and cross-sectional dimensions of his data and estimating a system of equations, including firm-specific and time-specific fixed effects in the production function equation. Since differences between firms and periods of time affecting both output and input choices are accounted for in these fixed effects, he argues that his model does not suffer from simultaneous equations bias. Despite the innovative approach, Hoch's results are not very encouraging. Moving from time correction estimates, where only time effects are included, to analysis of covariance estimates, where farm effects are also considered, there is a significant drop in the estimated sum of elasticities, from almost 1 to approximately 0.75, which, in turn, generates unreasonably low estimated marginal returns (around 0.20) to labor. These figures force him to (questionably) interpret the shortfall as reflecting the fact that efficiency may increase with scale and that there may be returns to the unmeasured, fixed entrepreneurial factor.

Mundlak (1961) further exploits the fixed effect approach by focusing on obtaining unbiased production function estimates in the presence of unobserved managerial ability. He notes that, instead of trying to rationalize the concept and the meaning of managerial capacity to include some index of management in the production function, one should assume that, whatever management is, it does not change substantially over time and, for at least a two-year period, it can be assumed to remain constant. Mundlak assumes a Cobb–Douglas specification very similar to (1), apart from the fact that the management variable is included among the inputs and has its own (constant) coefficient to be estimated. However, since management is not directly observable, the specification taken to the data is exactly the same as (4), with $\omega_j = cm_j$ being firm j's fixed effect, m_j being management, and c being the constant multiplicative term associated with it. If the production function is fully specified and the assumptions of the classical regression model hold, unbiased and efficient estimates can be obtained using the analysis of covariance. Using a sample of 66 family farms in Israel from 1954 to 1958, Mundlak finds that management is positively correlated with most of the inputs and that the firm fixed effects are significantly different from zero, suggesting that the estimates obtained adopting a specification that does not include them are likely to be biased. However, the elasticity of output is fairly close to 1 when only time effects are considered. When only firm fixed effects are present, this elasticity drops to

0.87, and to 0.79 when both year and firm effects are included. Moreover, the model, including both year and firm effects, which are found to be significant, delivers, again, an unrealistically low elasticity labor of 0.11, demonstrating that controlling explicitly for management bias does not necessarily improve the credibility of the results.

The unsatisfactory results – low and often insignificant capital coefficients and unreasonably low returns to scale – obtained in the literature prove that the fixed effects framework, valid in theory, is not particularly successful in solving the endogeneity problem in practice. There are a number of reasons why this is the case. For the fixed effects methodology to be applicable, one needs to rely on the rather strong assumption that the unobserved productivity term ω_j is constant over time. This assumption is becoming less justifiable now that longer panel datasets are more easily available. Moreover, researchers are usually interested in studying major changes in the economic environment and, since significant changes are likely to affect different firms' productivity differently, firms are likely to adjust their optimal decision accordingly. If this is the case, ω_j will obviously not be constant over time anymore. Furthermore, the transformation of the data through differencing may actually aggravate this problem in the present measurement error in inputs, and the estimates obtained with fixed effects are actually even less reliable than the OLS estimates. Nonetheless, the fixed effects approach can be seen as a useful and simple reduced form way of exploring the data by decomposing a firm's heterogeneity into within and between effects.

3.2.2 Instrumental variables

The second classical solution to the endogeneity issue proposed in the literature is the use of instrumental variables. Consider a slight modification of (4):

$$y_{jt} = \beta_0 + \beta_k k_{jt} + \beta_l l_{jt} + \omega_{jt} + \eta_{jt}, \tag{6}$$

where now the term ω_{jt} is allowed to change by firm and over time. Valid instruments would be variables that are correlated with the endogenous explanatory variables, in this case inputs, but do not enter the production function explicitly and are not correlated with the production function residuals. The theory of production provides some indication regarding natural candidates to be valid instruments: input prices. Input prices certainly influence input choices, as they are part of the input demand functions, but do not directly enter the production function. Moreover, input prices need to be uncorrelated with ω_{jt}, and this will depend on the nature of competition in the input market. Specifically, if input markets are perfectly competitive, firms take input prices as given, thus input prices are appropriate instruments. Other possible instruments would be output prices, once again under the condition that output markets are competitive, or any other variable that shifts either the demand for output or the supply of inputs. Nevertheless, input prices are usually the most popular instruments because perfect competition is a more plausible assumption in input markets than in output markets.

In the wake of the "duality" revolution in production function theory, Nerlove (1963) is one of the very few successful contributions in the literature making use of input prices as instruments. His investigation on the returns to scale in electricity supply relies on several characteristics that render the U.S. electric power industry unique, because the output of a firm and the prices it pays for the production factors can be regarded as exogenous, even if the industry does not operate in perfectly competitive markets. Thus, the problem of the individual firm appears to be that of minimizing the total cost of production of output, subject to the given production function technology and factor prices, taking into account variations in efficiency among firms.

If the efficiency among firms varies neutrally and the factor prices vary across firms, the input choices are not independent, but determined jointly by the firm's efficiency, level of output, and factor. However, if factor price data are available and factor prices do not move proportionally, the unbiased estimation of a reduced form cost function is possible. The fundamental duality between cost and production function, demonstrated by Shepherd (1953), guarantees that the relation between the cost function empirically estimated and the underlying production function is unique.

Despite being a remarkable and innovative contribution, the peculiar environment and data used in Nerlove's study demonstrate why the instrumental variables approach, even if theoretically sound, may be challenging to apply in practice. First, firms do not usually report input prices, and when they do, especially in the case of labor costs, they tend to report average wage per worker or per hour of labor. Ideally, the cost of labor should measure exogenous differences in labor market conditions, but it often captures also some component of unmeasured worker quality. It is very possible, for example, that firms employing higher quality workers will pay higher average wages. In this case, the cost of labor will be correlated with the production function residuals, and its validity as an instrument will be compromised. Second, the use of input prices and instruments requires that these variables have sufficient variation to identify production function coefficients. While input prices clearly change over time, they usually do not vary significantly across firms, as inputs market conditions tend to be fairly national in scope. If input prices do not differ enough across firms in the data, or if the observed differences reflect unobserved input quality and not exogenous input market characteristics, the instrumental variable approach is not applicable. A third issue arises because the instrumental variables framework relies on the strong assumption that the term ω_{jt} in (6) evolves independently from input choices over time, and thus firms cannot affect the evolution of ω_{jt} through input decisions. If the evolution in ω_{jt} is correlated with some inputs, finding valid instruments would require identifying variables that affect only those input choices without simultaneously affecting other input choices. Since individual input choices most likely depend on the prices of all inputs of production, the task of selecting valid instruments in such a context can be extremely challenging. Finally, the instrumental variables approach only addresses the endogeneity of input choice, not the endogeneity of a firm's exit. If exit is endogenous, it will possibly depend, in part, on input prices, so that firms facing higher input prices will be more likely to exit. This generates correlation between input prices, used as instruments, and the residuals in the production function rendering the instruments invalid.

3.3 *Structural solutions*

In the last 20 years, the increasing availability of firm-level data opened the door to more structural approaches to identifying production function coefficients controlling for simultaneity and selection problems.

3.3.1 *The Olley and Pakes approach*

Olley and Pakes in their 1996 contribution propose an innovative empirical framework with the goal of quantifying the impact of deregulation on measures of plant-level productivity in the U.S. telecommunication equipment industry between 1974 and 1987. Considering firms operating through discrete time, making production decisions to maximize the present discounted value of current and future profits, Olley and Pakes make use of the following assumptions. First, the production function is given by:

$$y_{jt} = \beta_0 + \beta_k k_{jt} + \beta_a a_{jt} + \beta_l l_{jt} + \omega_{jt} + \eta_{jt}, \tag{7}$$

where a_{jt} is the age (in years) of the plant expressed in natural logarithms. The motivation for introducing a plant's age as an additional input is to analyze the impact of age on productivity. Second, the unobserved productivity ω_{jt} evolves exogenously following a first-order Markov process of the form:

$$p\left(\omega_{jt+1} \mid \{\omega_{j\tau}\}_{\tau=0}^{t}, I_{jt}\right) = p\left(\omega_{jt+1} \mid \omega_{jt}\right), \tag{8}$$

where I_{jt} is firm j's information set at time t, and current and past realization of ω, that is, $\{\omega_{jt}, \ldots, \omega_{j0}\}$ are assumed to be part of I_{jt}. This is simultaneously an econometric assumption on the statistical properties of the unobservable term ω_{jt} and an economic assumption on the way firms form their expectations on the evolution of their productivity over time. Specifically, at time $t + 1$ these expectations depend only on the realization that occurred at time t. Moreover, the first-order Markov process is assumed to be stochastically increasing over time; that is, a firm with a higher ω_{jt} today expects to have a better distribution of ω_{jt+1} tomorrow.

Third, capital is accumulated by firms through a deterministic dynamic investment process, specified as:

$$k_{jt} = (1-\delta) k_{jt-1} + i_{jt-1}. \tag{9}$$

This formulation implies that the firm's capital stock at period t was actually decided, through investment, at period $t - 1$. Finally, the per-period profit function is given by

$$\pi\left(k_{jt}, a_{jt}, \omega_{jt}, \Delta_t\right) - c\left(i_{jt}, \Delta_t\right). \tag{10}$$

Note that labor l_{jt} does not explicitly enter the profit function, as it is considered a variable, non-dynamic input. Labor is variable in the sense that it is chosen and utilized in production in the same period and it is non-dynamic, unlike capital, because current labor decisions do not impact future profits (i.e., labor is not a state variable). Therefore, $\pi(k_{jt}, a_{jt}, \omega_{jt}, \Delta_t)$ can be thought as a conditional profit function, where the conditioning is on the optimal static choice of the labor input. Note also that both $\pi(\cdot)$ and $c(\cdot)$ depend on Δ_t, which represents the economic environment where firms operate in a specific period. Δ is allowed to change over time but, in a given time period, is considered to be constant across firms.

The firm maximization problem can be described by the following Bellman equation:

$$V_t\left(k_{jt}, a_{jt}, \omega_{jt}, \Delta_t\right) = \max \left\{ \begin{array}{l} \Phi, \sup_{i_{jt} \geq 0} \pi\left(k_{jt}, a_{jt}, \omega_{jt}, \Delta_t\right) - c\left(i_{jt}, \Delta_t\right) \\ + \beta E\left[V_{t+1}\left(k_{jt+1}, a_{jt+1}, \omega_{jt+1}, \Delta_{t+1}\right) \mid I_{jt}\right] \end{array} \right\}, \tag{11}$$

where Φ represents the sell-off value of the firm, and I_{jt} is, once again, the information available to the firm at time t, that is, $(k_{jt}, a_{jt}, \omega_{jt}, \Delta_t, i_{jt})$. The Bellman equation specifies that each firm compares its sell-off value and the expected discounted returns of staying in business. If the current state variables ($k_{jt}, a_{jt}, \omega_{jt}, \Delta_t$) indicate that continuing in operation is not profitable, the firm will exit, while, in the opposite case, it will choose an optimal, positive investment level. Under the appropriate assumptions that an equilibrium exists and the difference in profits between continuing and exiting is increasing in ω_{jt} (i.e., firms with higher ω_{jt} are more likely to realize higher profits and thus decide to stay in business), the solution to the control problem in (11) generates an exit rule and an investment demand function. Defining χ_{jt} as the indicator function that takes the value of zero when the firm decides to exit, the exit decision rule and the investment demand function are written, respectively, as:

$$\chi_{jt} = \begin{cases} 1 \text{ if } \omega_{jt} \geq \bar{\omega}_{jt}\left(k_{jt}, a_{jt}, \Delta_t\right) = \bar{\omega}_t\left(k_{jt}, a_{jt}\right) \\ 0 \text{ otherwise} \end{cases} \quad (12)$$

and

$$i_{jt} = i\left(k_{jt}, a_{jt}, \omega_{jt}, \Delta_t\right) = i_t\left(k_{jt}, a_{jt}, \omega_{jt}\right). \quad (13)$$

Investment is assumed to be strictly monotonic in ω as, conditional on k_{jt} and a_{jt}, firms with higher ω_{jt} will optimally invest more.

As long as investment is positive, since (13) is strictly monotonic in ω_{jt}, it is possible to invert it and generate

$$\omega_{jt} = h_t\left(k_{jt}, a_{jt}, i_{jt}\right), \quad (14)$$

which implies that, given a firm's levels of k_{jt} and a_{jt}, the investment demand i_{jt} provides sufficient information about ω_{jt}. This is because Olley and Pakes assume that ω_{jt} is the only unobservable in the investment demand and there are no other unobservable (by the econometrician) variables that affect investment but not production.

Substituting (14) into (7) yields

$$y_{jt} = \beta_l l_{jt} + \phi_t\left(k_{jt}, a_{jt}, i_{jt}\right) + \eta_{jt}, \quad (15)$$

where $\phi_t\left(k_{jt}, a_{jt}, i_{jt}\right) = \beta_0 + \beta_k k_{jt} + \beta_a a_{jt} + h_t\left(k_{jt}, a_{jt}, i_{jt}\right)$. Equation (15) is taken to the data in a first-stage regression to recover an estimate of the labor coefficient. This is possible because the monotonicity and scalar unobservable assumptions allow for "observing" the unobservable ω through investment, eliminating the endogeneity problem for the labor coefficient. In this first stage of the estimation, however, the coefficients on capital and age are not identified, because in (14) it is not possible to separate the effect of capital and age on the investment decision from their effect on output.

Rewriting (7), taking the term $\beta_l l_{jt}$ to the left-hand-side and taking expectation of both sides, results in:

$$E\left[y_{jt} - \beta_l l_{jt} \mid I_{jt}, \chi_{jt} = 1\right] = E\left[\beta_0 + \beta_k k_{jt} + \beta_a a_{jt} + \omega_{jt} + \eta_{jt}\right]$$
$$= \beta_0 + \beta_k k_{jt} + \beta_a a_{jt} + E\left[\omega_{jt} \mid I_{jt-1}, \chi_{jt} = 1\right]. \quad (16)$$

The second line comes from the fact that k_{jt} and a_{jt} are known at time $t-1$ and η_{jt} is, by definition, uncorrelated with I_{jt-1} and exit. The last term of (16) can be expanded as

$$E\left[\omega_{jt} \mid I_{jt-1}, \chi_{jt}=1\right] = E\left[\omega_{jt} \mid I_{jt-1}, \omega_{jt} \geq \bar{\omega}_t\left(k_{jt}, a_{jt}\right)\right]$$
$$= \int_{\bar{\omega}_t(k_{jt},a_{jt})}^{\infty} \omega_{jt} \frac{p(\omega_{jt} \mid \omega_{jt-1})}{\int_{\bar{\omega}_t(k_{jt},a_{jt})}^{\infty} p(\omega_{jt} \mid \omega_{jt-1})} d\omega_{jt}, \quad (17)$$
$$= g\left(\omega_{jt-1}, \bar{\omega}_t\left(k_{jt}, a_{jt}\right)\right),$$

where the first equality depends on the exit rule expressed in (12) and the last two lines from the exogenous first-order Markov process assumption on ω_{jt}.

While it is possible to estimate ω_{jt-1}, since, for a given set of parameters $(\beta_0, \beta_k, \beta_a)$, $\omega_{jt-1}(\beta_0, \beta_k, \beta_a) = \phi_{jt-1} - \beta_0 - \beta_k k_{jt-1} - \beta_a a_{jt-1}$, there is not direct knowledge of $\bar{\omega}_t(k_{jt}, a_{jt})$, Olley and Pakes try to control for $\bar{\omega}$ using data on observed exit. In fact, the probability of continuing operating at period t, conditional on the information available in the previous period, is

$$\Pr\left(\chi_{jt}=1 \mid I_{t-1}\right) = \Pr\left(\omega_{jt} \geq \bar{\omega}_t\left(k_{jt}, a_{jt}\right) \mid I_{t-1}\right)$$
$$= \Pr\left(\chi_{jt}=1 \mid \omega_{jt}, \bar{\omega}_t\left(k_{jt}, a_{jt}\right)\right) = \hat{\varphi}_t\left(\omega_{jt-1}, k_{jt}, a_{jt}\right), \quad (18)$$
$$= \varphi_t\left(i_{jt-1}, k_{jt-1}, a_{jt-1}\right) = P_{jt}$$

where the first equality comes from the exit rule (12), and the remaining equalities come from (14) and the fact that k_{jt} and a_{jt} are deterministic functions of i_{jt-1}, k_{jt-1}, and a_{jt-1}. Olley and Pakes obtain an estimate of $\hat{P}_{jt}$, that is, the probability of firm j surviving to period t, through nonparametric methods. Equation (18) also implies that $\bar{\omega}_t(k_{jt}, a_{jt})$ is a function of ω_{jt-1} and P_{jt}. Thus, (16) becomes:

$$E\left[y_{jt} - \beta_l l_{jt} \mid I_{jt}, \chi_{jt}=1\right]$$
$$= \beta_0 + \beta_k k_{jt} + \beta_a a_{jt} + g\left(\omega_{jt-1}, f(\omega_{jt-1}, P_{jt})\right)$$
$$= \beta_0 + \beta_k k_{jt} + \beta_a a_{jt} + g'\left(\omega_{jt-1}, P_{jt}\right) \quad (19)$$
$$= \beta_0 + \beta_k k_{jt} + \beta_a a_{jt} + g'\left(\phi_{jt-1} - \beta_0 - \beta_k k_{jt-1} - \beta_a a_{jt-1}, P_{jt}\right).$$

The second stage in the Olley and Pakes estimation requires one to take to the data the following expression:

$$y_{jt} - \beta_l l_{jt} = \beta_k k_{jt} + \tilde{g}'\left(\phi_{jt-1} - \beta_0 - \beta_k k_{jt-1} - \beta_a a_{jt-1}, P_{jt}\right) + \xi_{jt} + \eta_{jt}, \quad (20)$$

where the function $\tilde{g}'$ includes the constant term β_0, and ξ_{jt} represents the innovation in productivity with $\xi_{jt} = \omega_{jt} - E\left[\omega_{jt} \mid \omega_{jt-1}, \chi_{jt}=1\right]$. Substituting $\hat{P}_{jt}$, $\hat{\phi}_{jt}$, and $\hat{\beta}_l$, (20) can be estimated approximating the $\tilde{g}'$ function with polynomial or kernel methods. The coefficients associated with capital and age, β_k and β_a, can be identified in (20) because, given the information structure, the innovation in productivity is uncorrelated with k_{jt} and a_{jt}, since these two variables are only a function of the information at $t-1$ so that the orthogonality condition $E\left[\xi_{jt}+\eta_{jt}\mid I_{jt-1},\chi_{jt}=1\right]=0$ holds. Conversely, the labor input at time t is plausibly correlated with η_{jt}, since it is free to adjust to shocks in productivity; thus, the first stage of the estimation is needed to identify β_l. Finally, in (20), β_k and β_a are identified, making use of the information provided by the cross-sectional variation in k_{jt} and a_{jt} and the time variation in input usage across firms that have the same ω_{jt-1} and P_{jt}.

The findings of Olley and Pakes demonstrate how important the bias created by not controlling for productivity and endogenous exit can be. Comparing the results obtained with their alternative method and the more classical OLS and fixed effect approaches, for both a balanced panel and the full sample (constructed by including exiting and entering firms), they find remarkable differences. Specifically, under OLS and fixed effects, the coefficient on labor is overestimated and the coefficient on capital is heavily underestimated with respect to the Olley and Pakes approach, and the differences are even larger when considering the balanced panel instead of the full sample.

3.3.2 The Levinsohn and Petrin approach

Levinsohn and Petrin (2003) take a similar approach to Olley and Pakes for conditioning out serially correlated unobserved shocks in production function estimation. The key difference in their contribution is that they use an intermediate input demand function as a proxy for productivity instead of the investment demand function. The rationale behind this choice is that the Olley and Pakes' procedure requires the investment function to be strictly monotonic in ω_{jt} to be inverted. Formally, the inversion can be done also in the presence of zero or lumpy investment levels, but zero or lumpy investment levels cast doubt on the strict monotonicity assumption on investment. Conversely, restricting the sample to the sole observations for which $i_{jt>0}$ could create a significant loss in efficiency. Specifically, Levinsohn and Petrin observe that in the Chilean manufacturing dataset from 1979 to 1986 they use in their study, more than 50 percent of the plant-year observations have zero investment level. Discarding these observations would imply losing more than half of the sample with an obvious efficiency loss. In addition, Levinsohn and Petrin note that investment is a control on a state variable that, by definition, may be costly to adjust. If investment is subject to non-convex adjustment costs, the investment function may present kinks that affect the reaction of investment to the transmitted productivity shock. In this case, the error term η_{jt} in (15) will be correlated with l_{jt}, and the identification assumption on β_l would not hold.

To avoid the issues related to potentially large efficiency loss and adjustment cost non-convexities while using investment, Levinsohn and Petrin suggest using intermediate inputs choices (energy and/or materials) to proxy for ω_{jt}, as these variables are rarely zero and do not suffer from significant adjustment cost; thus, the strict monotonicity assumption is more easily satisfied. They consider the production function

$$y_{jt}=\beta_0+\beta_k k_{jt}+\beta_l l_{jt}+\beta_m m_{jt}+\omega_{jt}+\eta_{jt}, \qquad(21)$$

where m_{jt} (intermediate input) is an additional input in production that, like labor, is assumed to be variable and non-dynamic. The intermediate input demand equation is specified as

$$m_{jt}=m_t\left(k_{jt},\omega_{jt}\right), \qquad(22)$$

with the subscript t indicating that factors like input prices, market structure, or demand condition that can influence the demand for materials are allowed to vary across time but not across firms. Note that (22) implies specific timing assumptions regarding the choice of m_{jt}. First, the intermediate input in period t is a function of ω_{jt}; that is, it is chosen at the time production takes place. Second, labor does not enter (22), meaning that labor is chosen at the same time as intermediate inputs and, therefore, l_{jt} has no impact on the optimal choice of m_{jt}.

Assuming monotonicity of the intermediate input demand in ω_{jt}, analogously to Olley and Pakes, (22) is inverted to generate

$$\omega_{jt} = h_t(k_{jt}, m_{jt}). \tag{23}$$

Then, substituting (23) into (21) yields

$$y_{jt} = \beta_l l_{jt} + \phi_t(k_{jt}, m_{jt}) + h_t(k_{jt}, m_{jt}) + \eta_{jt}. \tag{24}$$

The first stage of the Levinsohn and Petrin procedure involves obtaining an estimate for β_l and $\phi_t(k_{jt}, m_{jt}) = \beta_0 + \beta_k k_{jt} + \beta_m m_{jt} + h_t(k_{jt}, m_{jt})$, treating h_t nonparametrically. Note that β_k and β_m are not separately identified from the nonparametric function in this first stage. The second stage of Levinsohn and Petrin consists of estimating

$$y_{jt} - \beta_l l_{jt} = \beta_k k_{jt} + \beta_m m_{jt} + \tilde{g}'(\phi_{jt-1} - \beta_k k_{jt-1} - \beta_m m_{jt-1}) + \xi_{jt} + \eta_{jt}. \tag{25}$$

Since k_{jt} is assumed to be decided at period $t - 1$, it is orthogonal to the residual $\xi_{jt} + \eta_{jt}$. However, since m_{jt} is a variable input, it is certainly not orthogonal to the innovation component of productivity, ξ_{jt}, as ω_{jt} is observed at the time the intermediate input is chosen. Thus, Levinsohn and Petrin use its lag, m_{jt-1}, as an instrument for m_{jt}, such that the orthogonality condition $E[\xi_{jt} + \eta_{jt} | I_{jt-1}] = 0$ is satisfied for both k_{jt-1} and m_{jt}. Levinsohn and Petrin find that using materials or electricity as a proxy for the unobserved productivity yields statistically significant estimates of the production function parameters for the Chilean manufacturing industry. Moreover, in line with Olley and Pakes, they also observe that, comparing their estimates with estimates obtained using OLS, the coefficient on labor is consistently upward biased, and the opposite is true for the coefficient on capital. A final comparison between estimates obtained using the investment as proxy (Olley and Pakes method) and estimates using the intermediate input demand (Levinsohn and Petrin method) also delivers higher coefficients on labor and lower coefficients on capital under the Olley and Pakes method, suggesting that, at least in the case of the Chilean plants, the intermediate input seems to respond more fully to the productivity shock than investment.

3.3.3 The Ackerberg, Caves, and Frazer approach

Both the Olley and Pakes and Levinsohn and Petrin procedures rely on a crucial assumption regarding the nature of the labor decision; that is, labor is not a state variable and, therefore, does not have an impact in the firm's dynamic optimization problem. Nonetheless, Ackerberg, Caves, and Frazer (2015) note that if there are significant hiring or firing costs, or if a firm is highly unionized so that labor contracts are long term, current labor choices have dynamic implications, and labor becomes a state variable. In this case, (14) and (23) become $\omega_{jt} = h_t(k_{jt},$

l_{jt}, a_{jt}, i_{jt}) and $\omega_{jt} = h_t(k_{jt}, l_{jt}, m_{jt})$ respectively, and the labor coefficient β_l cannot be identified from the first stage in either the Olley and Pakes estimation or the Levinsohn and Petrin estimation, because it is not possible to separate the impact of labor on the production from its impact on the nonparametric $h_t(\cdot)$ function. In other words, if the optimal labor choice is determined according to $l_{jt} = f_t(\omega_{jt}, k_{jt}) = f_t(h_t(\cdot), k_{jt})$, it is not feasible to simultaneously estimate a fully nonparametric, time-varying function of (k_{jt}, ω_{jt}), along with a coefficient associated with a variable, l_{jt}, that is merely a time-varying function of those same variables (k_{jt}, ω_{jt}). Ackerberg, Caves, and Frazer further argue that the collinearity problem that prevents identification of β_l in the first stage is more serious and less easy to overcome in the LP approach than in the Olley and Pakes approach. This is because the former method uses a proxy for productivity, the intermediate input demand, that is chosen in period t simultaneously with labor and production level after observing ω_{jt}, while the latter method relies on investment as a proxy that, by definition, is not directly linked to period t outcomes. Ackerberg, Caves, and Frazer suggest an alternative estimation procedure that avoids the collinearity problems arising in the estimation of the labor coefficient, adopting a mild modification on the timing assumption for input choices. The main difference between their approach and those of Olley and Pakes and Levinsohn and Petrin is that, in the first stage, no coefficient is estimated; rather, the first stage serves the purpose of netting out the error from the production function η_{jt}. More specifically, in the case of the intermediate input demand, with a production function specified as in (6), Ackerberg, Caves, and Frazer assume that labor l_{jt} is chosen by firms at time $t - b$, with $(0 < b < 1)$, after capital k_{jt} was chosen at or before $t - 1$, but prior to m_{jt} being chosen at t. The productivity process is assumed to evolve according to a first-order Markov process between the sub-periods $t - 1$ and $t - b$; that is,

$$p(\omega_{jt} | I_{jt-b}) = p(\omega_{jt} | \omega_{jt-b}) \text{ and } p(\omega_{jt-b} | I_{jt-1}) = p(\omega_{jt-b} | \omega_{jt-1}), \tag{26}$$

which implies that labor and intermediates are both variable inputs with labor being "less variable" than intermediates. Also, labor is not a function of ω_{jt}, but of ω_{jt-b}. With these timing assumptions, the demand for intermediate input and labor, respectively, are given by

$$m_{jt} = m_t(k_{jt}, l_{jt}, \omega_{jt}) \text{ and} \tag{27}$$

$$l_{jt} = f_t(k_{jt}, \omega_{jt-b}), \tag{28}$$

and the collinearity problem has been solved as m_{jt} and is now a function of $(k_{jt}, \omega_{jt-b}, \omega_{jt})$. Substituting into the production function, the first stage estimating equation in the Ackerberg, Caves, and Frazer is given by

$$y_{jt} = \phi_t(k_{jt}, l_{jt}, m_{jt}) + \eta_{jt}, \tag{29}$$

where $h_t(\cdot)$ is, once again, the inverse of (27) used as proxy for ω_{jt}. As previously mentioned, Ackerberg, Caves, and Frazer run the first stage just to obtain an estimate of $\phi_t(k_{jt}, l_{jt}, m_{jt}) = \beta_0 + \beta_k k_{jt} + \beta_l l_{jt} + h_t(k_{jt}, l_{jt}, m_{jt})$, in order to isolate η_{jt} and proceed to the second stage to estimate both β_k and β_l. This now requires two independent moment conditions for identification. The Markov process assumption on ω_{jt} implies that $\omega_{jt} = E[\omega_{jt} | I_{jt-1}] + \xi_{jt} = E[\omega_{jt} | \omega_{jt-1}] + \xi_{jt}$, so that ξ_{jt} is mean-independent of all the information known at

$t-1$. Since k_{jt} is decided at $t-1$ and l_{jt-1} is decided at $t-b-1$, both k_{jt} and l_{t-1} are included in the information set I_{t-1}, so the orthogonality conditions required to identify β_k and β_l are

$$E\left[\xi_{jt}\begin{vmatrix}k_{jt}\\l_{jt-1}\end{vmatrix}\right]. \tag{30}$$

For a given set of parameters (β_k,β_l), Ackerberg, Caves, and Frazer compute the implied value of $\hat{\omega}_{jt}(\beta_k,\beta_l) = \hat{\phi}_{jt} - \beta_k k_{jt} - \beta_l l_{jt}$, then nonparametrically regress $\hat{\omega}_{jt}$ on $\hat{\omega}_{jt-1}$ to obtain $\hat{\xi}_{jt}$, and finally form the sample analogue

$$\frac{1}{T}\frac{1}{N}\sum_t\sum_n \xi_{jt}(\beta_k,\beta_l)\begin{pmatrix}k_{jt}\\l_{jt-1}\end{pmatrix}, \tag{31}$$

to estimate (β_k,β_l) by minimizing (31).

In the case of the investment demand used as a proxy for ω_{jt}, the Ackerberg, Caves, and Frazer procedure is analogous to the procedure just described, with the exception of (27) becoming

$$i_{jt} = i_t\left(k_{jt},l_{jt},\omega_{jt}\right) \tag{32}$$

and (29) becoming

$$y_{jt} = \phi_t\left(k_{jt},l_{jt},i_{jt}\right) + \eta_{jt}. \tag{33}$$

Once again, $h_t(\cdot)$ is the inverse of (32) and $\phi_t(k_{jt},l_{jt},i_{jt}) = \beta_0 + \beta_k k_{jt} + \beta_l l_{jt} + h_t(k_{jt},l_{jt},i_{jt})$. Ackerberg, Caves, and Frazer assert that their framework is completely consistent with labor choices having dynamic implications and with other unobservables (e.g., input prices shocks or dynamic adjustment costs) impacting a firm's choices of capital and labor. The results, obtained using the same Chilean manufacturing data as Levinsohn and Petrin, show how the estimates, obtained with their alternative procedure, differ significantly from both the classical OLS and Levinsohn and Petrin methods. Specifically, the returns to scale estimated under OLS are higher than under Ackerberg, Caves, and Frazer, and this is mainly due to the fact that, as expected, the coefficient on labor is upward biased when not controlling for productivity shocks. Comparing the Ackerberg, Caves, and Frazer estimates with the Levinsohn and Petrin estimates, they find that the coefficients are generally different in magnitude, with the Levinsohn and Petrin estimates of the labor coefficient being more often smaller than their Ackerberg, Caves, and Frazer counterparts, suggesting that $\hat{\beta}_l$, which comes from the first stage in Levinsohn and Petrin, may be downward biased because of the discussed collinearity issue.

4 Extensions

This section addresses two classic concerns arising from econometric production function estimation. The first is an extension of the simultaneity accommodating frameworks presented by the Olley and Pakes; Levinsohn and Petrin; and Ackerberg, Caves, and Frazer approaches, which initially take the productivity component in estimation, ω_{jt}, as predetermined. The interest in accommodating the potential endogeneity of productivity lends itself to a range of policy considerations. These considerations include addressing the role of research and development (R&D) in promoting productivity growth and eventually addressing the demand for R&D.

The second expands on the abstraction that the firm is producing one, undifferentiated output. The reality is that firms produce multiple outputs and often these outputs are produced for multiple markets. A framework that can address the multiple-output/multiple-input production technology estimation is the distance function approach.

4.1 Estimating endogenous productivity

An alternative response to the simultaneity issues in production function estimation came from the dynamic panel data literature, starting with Chamberlain (1982). Dynamic panel methods essentially extend the fixed effect framework, allowing for a more sophisticated error structure, and combine it with instrumental variables to control for collinearity.

This approach is developed by considering the production function

$$y_{jt} = \beta_k k_{jt} + \beta_l l_{jt} + \left(\alpha_j + \omega_{jt} + \eta_{jt}\right) \tag{34}$$

$$= \beta_k k_{jt} + \beta_l l_{jt} + \psi_{jt}, \tag{35}$$

where the composite error term ψ_{jt} is the sum of all three error components, that is, the unobserved, time-invariant, firm-specific effect αj, the productivity shock ω_{jt}, and the serially uncorrelated residual term η_{jt}. The dynamic panel methodology relies on specific assumptions regarding the evolution of the error components α_j, ω_{jt}, and η_{jt}, and the correlation structure between these error components and the explanatory variables k_{jt} and l_{jt}. Given these assumptions, the estimation procedure requires finding functions of the aggregated error term ψ_{jt} that are uncorrelated with past, present, or future values of the explanatory variables. Commonly the assumptions imposed are as follows. First, the time invariant error component α_j may be correlated with capital and labor. Second, the term η_{jt} is independent and identically distributed over time and uncorrelated with capital and labor in every period. Third, the productivity process is usually modeled as a first-order linear autoregressive process of the form $\omega_{jt} = \rho \omega_{jt-1} + \xi_{jt}$. Third, while ω_{jt} is likely to be correlated with k_{jt} and l_{jt}, the innovation on ω_{jt} between $t-1$ and t, ξ_{jt}, is uncorrelated with all the input choices prior to period t. This is because the innovation in ω_{jt} is observed by the firm after period $t-1$, so that ξ_{jt} is uncorrelated with input chosen at $t-1$ or earlier. Note that the rationale behind this last assumption is similar to that behind the second stage identification conditions in Olley and Pakes; Levinsohn and Petrin; and Ackerberg, Caves, and Frazer.

Generalized method of moments (GMM) estimators are employed, which take first differences to eliminate firm-specific effects, α_j, and use lagged instruments to correct for simultaneity in the first-differenced equations. These, in turn, can be applied to estimate production function coefficients. Blundell and Bond (2000), however, comment that the methodology described above tends to produce unsatisfactory results in the context of production function estimation. They mainly attribute this poor performance to the weak correlations that exist between the current growth rates of capital and labor and the lagged levels of these variables, which result in weak instruments in the first-differenced GMM estimation procedure.

In all the frameworks presented so far, the productivity process has been considered constant over a given period of time (fixed effects) or exogenous (structural approaches and dynamic panel methods), in the sense that firms' optimal decisions do not affect the evolution in productivity. This modeling choice is mainly driven by the fact that endogenizing this process is problematic in the context of standard estimation procedures. Nonetheless, it seems very reasonable to assume that firms can optimally choose to undertake activities to increment their productivity.

A straightforward example is investment in R&D, which generates knowledge-based assets accumulation, just like investment in physical capital, changing the firm's relative position with respect to other firms. Doraszelski and Jaumandreu (2013) develop a dynamic model of endogenous productivity change where firms carry out two types of investment: one in physical capital and another in knowledge through R&D expenditure. The estimation of an endogenous productivity process is challenging when data on R&D are unavailable, because the estimation procedures analyzed so far do not apply for the following reasons. First, input prices are invalid instruments, because when the productivity process is not exogenous, the transitions from current to future productivity are affected by the choice of the additional unobserved R&D input, whose optimal choice depends on the prices of all the other inputs. Second, the scalar unobservable assumption necessary for a structural estimation like Olley and Pakes; Levinsohn and Petrin; or Ackerberg, Caves, and Frazer is violated in this context because R&D for firm j at time t and ω_{jt} are both unobservable, and recovering the productivity process using capital, investment, or intermediate input demand may not be possible. Furthermore, even when data on R&D are available, there may still be problems with identification, as noted by Buettner (2004). The Doraszelski and Jaumandreu estimation procedure relies on R&D data and builds on the Levinsohn and Petrin insight that, since static inputs like labor and material are chosen once the current productivity realization is known, they contain useful information about it.

4.2 A nonparametric approach: the distance function

The emergence of the distance function as a representation of the production technology, introduced conceptually by Shepherd (1953, 1970), serves as an alternative to the production function (addressing one output and many inputs) or the transformation function (addressing multiple outputs and inputs) as a statement of the technology. The production function or transformation function frameworks define the production relations on the frontier, or boundary, of the production set that is equivalent to the maximal potential production. The distance function framework explicitly admits the prospect of firms operating at non-boundary points in the production set. As such, it allows for the prospect of firms operating inefficiently and measures the degree of resource waste.

In the presence of multiple outputs and multiple inputs, the distance function provides a functional characterization of the structure of the production technology; in fact, the distance function is a value function resulting from an optimization. As such, we can articulate a set of properties embodied by this function. Not only do distance functions characterize the structure of the production technology, they are also related directly to measures of technical efficiency. The duality relationships also exist. The most basic form is the radial distance function, which can take on one of two orientations: input or output.

The distance function is defined for a technology, T, and a given input–output bundle (x, y). The input-oriented distance function is defined as

$$D_I(x,y) = \left\{ \max_{\theta} \theta : \left(\frac{x}{\theta}, y\right) \in T, \theta \geq 1 \right\}. \tag{36}$$

This reflects an input-saving perspective where all inputs are contracted proportionally. The input distance function reflects a measure of the firm's deviation from the boundary of the technology set; hence, when $\theta = 1$, the firm is operating on the boundary of the technology set and is deemed technically efficient; otherwise, the firm can still contract its input use by $\frac{1}{\theta}$

to achieve the boundary bundle. Given the properties of the technology, T, the input distance function satisfies a corresponding set of properties:

I.1 $D_I(x,y)$ is increasing in x

I.2 $D_I(x,y)$ is decreasing in y

I.3 $D_I(x,y)$ is homogeneous of degree one in x

I.4 $D_I(x,y)$ is concave in x

We can also define an output distance function, which is oriented to scaling the outputs. Again, for a given input–output bundle (x, y), the output distance function is defined as $D_O(x,y) = \left\{ \min_\delta \theta : \left(x, \frac{y}{\theta}\right) \in T, \delta \leq 1 \right\}$. Here, the parameter δ serves as a scaling factor over all outputs. When $\delta < 1$, all outputs can be scaled up by the factor $\frac{1}{\delta} > 1$ and still be part of the feasible technology; conversely, when $\delta = 1$, we cannot produce more output and still be a feasible technology. Given the properties of the technology, T, the output distance function satisfies a corresponding set of properties:

I.1 $D_I(x,y)$ is increasing in x

I.2 $D_I(x,y)$ is decreasing in y

I.3 $D_I(x,y)$ is homogeneous of degree one in x

I.4 $D_I(x,y)$ is concave in x

Econometric estimation of distance functions to address productivity and efficiency change has recently been revisited in O'Donnell (2014). For the econometric estimation of the input distance function, a functional form is specified for either the input-oriented or the output-oriented distance function. Working with the input-oriented distance function, a functional form is specified, $D_I(x,y) = g(y_1, y_2 \ldots y_N; x_1, x_2 \ldots x_M)$, and then the standard transformation will first append an error term, $e\varepsilon$, where ε is a two-sided error, typically assumed to be distributed $N(0,\Omega)$ with a logarithmic transformation:

$$\ln D_I(x,y) = \ln g(y_1, y_2 \ldots y_N; x_1, x_2 \ldots x_M) + \varepsilon. \tag{37}$$

Recognizing that the input distance function is measuring the scaling factor needed to move a particular firm to the boundary, we can define $D_I(x,y) = \theta = e^u, u \geq 0$. Taken together,

$$0 = \ln g(y_1, y_2 \ldots y_N; x_1, x_2 \ldots x_M) + \varepsilon - u \tag{38}$$

and solving for y_1,

$$\ln y_1 = \ln(y_2 \ldots y_N; x_1, x_2 \ldots x_M) + \varepsilon - u, \tag{39}$$

the endogeneity aspects are clearly apparent. Clearly, all (or nearly all) outputs can be endogenous, leading to the need to use the approaches of Olley and Pakes; Levinsohn and Petrin; and Ackerberg, Caves, and Frazer.

5 Concluding remarks

This survey has demonstrated that even if production function estimation is challenging because of the possibility of simultaneity and selection bias, obtaining realistic and reliable estimates of production function coefficients is the first step to answering more complex and interesting economic questions.

Since firms' responses to changes in the operating environment typically depend on how these changes affect their productivity, to separate the evolution in productivity from the variation in input choices, which also react to changes in the environment, requires an explicit model describing how firms' optimal choices are made. The appropriateness of different models and assumptions remains an empirical issue that needs to be addressed in each specific case, given the environment and the available data. The literature has suggested a considerable variety of alternatives. Nonetheless, the common message emphasized in all the proposed approaches is that productivity studies must explicitly take into account the fact that changes in productivity are the main determinant of firms' response to the changes being analyzed. Therefore, changes in productivity cannot be ignored in any estimation procedure.

Note

1 In the case of first-differencing, (5) would be $(y_{jt} - y_{jt-1}) = \beta_k (k_{jt} - k_{jt-1}) + \beta_l (l_{jt} - l_{jt-1}) + (\eta_{jt} - \eta_{jt-1})$.

References

Ackerberg, D.A., K. Caves, and G. Frazer. 2015. "Identification Properties of Recent Production Function Estimators." *Econometrica* 83(6):2411–2451.

Balk, B.M. 2012. *Price and Quantity Index Numbers: Models for Measuring Aggregate Change and Difference*. Cambridge: Cambridge University Press.

Blundell, R., and S. Bond. 2000. "GMM Estimation with Persistent Panel Data: An Application to Production Functions." *Econometric Reviews* 19(3):321–340.

Buettner, T. 2004. *R&D and the Dynamics of Productivity*. PhD thesis, London School of Economics.

Chamberlain, G. 1982. "Multivariate Regression Models for Panel Data." *Journal of Econometrics* 18(1):5–46.

Chambers, R.G. 1988. *Applied Production Analysis: A Dual Approach*. Cambridge: Cambridge University Press.

Chavas, J-P., R.G. Chambers, and R.D. Pope. 2010. "Production Economics and Farm Management: A Century of Contributions." *American Journal of Agricultural Economics* 92(2):356–375.

Cobb, C.W., and P.H. Douglas. 1928. "A Theory of Production." *The American Economic Review* 18(1):139–165.

Doraszelski, U., and J. Jaumandreu. 2013. "R&D and Productivity: Estimating Endogenous Productivity." *Review of Economic Studies* 80(4):1338–1383.

Fuss, M., D. McFadden, and Y. Mundlak. 2014. "A Survey of Functional Forms in the Economic Analysis of Production." In M. Fuss and D. McFadden, eds., *Production Economics: A Dual Approach to Theory and Applications: Applications of the Theory of Production* (Vol. 2). Amsterdam, Netherlands: North Holland, pp. 219–268.

Hasenkamp, G. 1976. *Specification and Estimation of Multiple-Output Production Functions* (Vol. 120). New York: Springer.

Hoch, I. 1955. "Estimation of Production Function Parameters and Testing for Efficiency." *Econometrica* 23(3):325–326.

———. 1962. "Estimation of Production Function Parameters Combining Time-Series and Cross-Section Data." *Econometrica* 30(1):34–53.

Levinsohn, J., and A. Petrin. 2003. "Estimating Production Functions Using Inputs to Control for Unobservables." *The Review of Economic Studies* 70(2):317–341.

Marschak, J., and W.H. Andrews. 1944. "Random Simultaneous Equations and the Theory of Production." *Econometrica* 12(3):143–205.

Mendershausen, H. 1938. "On the Significance of Professor Douglas' Production Function." *Econometrica* 6(2):143–153.

Mundlak, Y. 1961. "Empirical Production Function Free of Management Bias." *Journal of Farm Economics* 43(1):44–56.

———. 2001. "Production and Supply." In B. Gardner and G. Rausser, eds., *Handbook of Agricultural Economics* (Vol 1). Amsterdam, Netherlands: North Holland, pp. 3–85.

Nerlove, M. 1963. "Returns to Scale in Electricity Supply." In C.F. Christ, ed., *Measurement in Economics – Studies in Mathematical Economics and Econometrics in Memory of Yehuda Grunfeld*. Stanford, CA: Stanford University. Press, pp. 167–200.

O'Donnell, C.J. 2014. "Econometric Estimation of Distance Functions and Associated Measures of Productivity and Efficiency Change." *Journal of Productivity Analysis* 41(2):187–200.

Olley, G.S., and A. Pakes. 1996. "The Dynamics of Productivity in the Telecommunications Equipment Industry." *Econometrica* 64(6):1263–1297.

Powell, A.A., and F. Gruen. 1968. "The Constant Elasticity of Transformation Production Frontier and Linear Supply System." *International Economic Review* 9(3):315–328.

Shepherd, R.W. 1953. *Cost and Production Functions*. Princeton, NJ: Princeton University Press.

———. 1970. *Theory of Cost and Production Functions*. Princeton, NJ: Princeton University Press.

Part 5
Production

32
Role of risk and uncertainty in agriculture

Jean-Paul Chavas

1 Introduction

Uncertainty is an important characteristic of agriculture (e.g., Just and Pope 2002; Chavas 2011). Agricultural production is always risky. For crops, unpredictable insect damage or weather shocks (including droughts and floods) can have large effects on yields. For livestock, diseases and their negative impact on animal production also matter. In addition, agricultural markets have been volatile for two reasons: (1) production shocks have generated unpredictable shifts in farm supply; and (2) the demand for food is very price-inelastic. As a result, agricultural prices can vary a lot over time, making them difficult to predict. The recent increase in food price volatility has raised renewed concerns about the adverse effects of high food prices on food insecurity among poor consumers (Chavas et al. 2014, 2017). Finally, over the last few decades, the agricultural sector has seen rapid technological progress, but the process of adoption of new technologies is slow, and there is always some uncertainty about what a new technology offers (e.g., Feder et al. 1985). Thus, uncertainty plays a role in almost every economic aspect of the agricultural sector.

The importance of uncertainty in agriculture has stimulated research in three general directions: (1) assessing exposure to risk and uncertainty; (2) assessing preferences toward risk and uncertainty; and (3) evaluating the economics of management and policy options used in risk allocation. Much progress has been made to advance academic knowledge in these areas. This article provides a broad overview of the progress made.

Investigating the economics of risk and uncertainty in agriculture raises several issues. The first issue is: how to define and measure risk? The measurement of risk exposure has typically relied on probability assessments. Savage (1954) has argued that the odds of facing any unpredictable event can be characterized in general by "subjective probabilities." Note that risk management can take place without probability assessments. This is illustrated in the well-known advice: "Do not put all your eggs in the same basket." This advice supports the idea that diversification strategies can help reduce risk exposure. Importantly, it is stated without mentioning probabilities. Yet, probabilities are very useful in empirical risk assessment. They have been found so useful that almost all empirical analyses of risk have relied on probability assessments. Some concerns remain, however, about whether imposing a probability structure on unpredictable events is always appropriate (e.g., Knight 1921). This has generated a distinction between risk (when probability assessments are appropriate) and uncertainty (when they are not). Section 2 reviews the evolving literature on this topic.

The second issue is the assessment of risk preferences. Following von Neumann and Morgenstern (1944), the expected utility model gives a framework to conduct economic analyses under risk, where a von Neumann–Morgenstern utility function represents risk preferences. In this context, Arrow (1965) and Pratt (1964) have established the linkages between risk aversion and the shape of the von Neumann–Morgenstern utility function. Such linkages have provided the conceptual basis for most of the empirical analyses of risk preferences. The evidence indicates that risk preferences can vary a lot across individuals (e.g., Halek 2001; Dohmen et al. 2011). In the context of agriculture, as shown by Lin et al. (1974), Dillon and Scandizzo (1978), Binswanger (1981), Chavas and Holt (1996), and others, most farmers have been found to be averse to risk and to "downside risk" (where downside risk means exposure to unfavorable events). This is the main reason why the economics of risk is important: exposure to risk (and downside risk) makes most decision makers worse off, implying that management and policy options that can reduce risk exposure also generate economic gains.

But does the expected utility model provide an accurate representation of behavior under risk? Psychologists have pointed out some discrepancies between the expected utility model and actual behavior. Such discrepancies have stimulated the development of prospect theory by Kahneman and Tversky (1979) and Tversky and Kahneman (1992). By allowing the "overweighing" of small probabilities, prospect theory provides more accurate representations of behavior under risk (compared to the expected utility model). As showed by Liu (2013) and Bocqueho et al. (2014), such overweighing of small probabilities is relevant in the analysis of agricultural risk. Section 3 reviews these issues.

The third topic concerns management and policy decisions under risk. As discussed in Section 4, there are many options to manage agricultural risk. These options vary with the type of risk and with the institutional environment. At the farm level, farmers have developed risk management strategies to cope with unforeseen events, including farm diversification and the adoption of risk-reducing technologies. At the multi-individual level, insurance schemes, forward contracting, and hedging on futures markets can also help reduce risk exposure. At the aggregate level, national and international government policies can be implemented to redistribute risk (including disaster assistance and food aid). Current challenges in the economic evaluation of agricultural risk are also discussed, along with needs for further research.

2 Assessing risk exposure

As noted in the introduction, the empirical analyses of risk exposure have relied extensively on probability assessments. A probability is a positive weight associated with the likelihood of occurrence of an uncertain event, with weights normalized to sum up to 1 across all possible events. For repeatable events, this is intuitive: an event is more probable if it occurs more frequently. This intuition is typically extended to non-repeatable events, in which case probabilities of specific events are treated as subjective and personal. Imposing a probability structure on risky events also provides a framework to analyze human learning: Bayes' theorem specifies how prior probabilities turn into posterior probabilities as an agent obtains new information.

When risky events are represented by numbers (e.g., mm of rainfall, bushel of corn, or dollar of income), the outcomes are random variables with a given (subjective) cumulative distribution function or its associated probability function. In this context, empirical risk assessment involves estimating probability functions of the relevant random variables. This can be done in two ways: (1) estimating moments of the probability function; (2) estimating the probability function itself.

Estimating the moments of the relevant probability function has been commonly used in applied risk analysis. These moments include the mean (a measure of central tendency), the variance (a measure of dispersion), the skewness (a measure of asymmetry of the probability function around the mean), etc. The simplest approach is to estimate two moments: the mean and the variance (the variance then measuring risk exposure). This approach has been very popular because of its simplicity (e.g., Harwood et al. 1999). A good example is the analysis of production risk proposed by Just and Pope (1979), who investigated the effects of fertilizer use on crop yield. They found that fertilizer tends to increase the variance of yield, thus providing evidence that fertilizer is "risk-increasing."

While the mean–variance approach is very convenient in applied risk analysis, questions have been raised about its general validity. A mean–variance approach is always valid when the probability function is "normal" (because mean and variance are sufficient statistics for a normal distribution). A mean–variance approach is also valid when risk management options affect only the mean and the standard deviation of the random variable (Meyer 1987). But beyond these two cases, a mean–variance analysis can be very restrictive. Indeed, the variance does not distinguish between "upside risk" and "downside risk"; thus, it cannot reflect exposure to downside risk alone. But introducing skewness in risk analysis can (as increasing skewness means reducing downside risk). This suggests a need to go beyond the first two moments (mean and variance) in risk analysis. Empirical estimates of risk exposure can be extended to include any number of moments of the probability function (e.g., Antle 1987). As will be further discussed, including at least three moments (mean, variance, and skewness) is important when the focus is on managing downside risk. Other extensions include the use of partial moments (Antle 2010) and quantile moments (Kim et al. 2014).

While the use of moments is convenient in empirical risk assessment, it can also be restrictive. For example, risk transfer mechanisms (such as disaster payments or insurance schemes) typically involve risk redistributions that cannot be easily summarized by moments. In this context, it is useful to examine the whole probability function. But estimating a function is empirically more challenging than estimating a few moments. One possibility is to assume that the probability function belongs to a parametric family (e.g., the beta distribution or the gamma distribution) and then estimate the associated parameters. For an application to crop yield risk, see Gallagher (1987). A more flexible approach is to estimate the probability/distribution function using non-parametric methods. This can be done using kernel estimation of the probability density function (e.g., Goodwin and Ker 1998) or using quantile regression applied to the estimation of the distribution function across all quantiles (e.g., Chavas and Shi 2015).

Finally, as noted in the introduction, there are uncertain situations where probability assessments can be difficult. As argued by Knight (1921), this indicates that imposing a probability structure on unpredictable events may not always be appropriate. Such arguments have stimulated research in two directions. One direction is to distinguish between risk and ambiguity (where ambiguity arises when precise probability assessments are not possible), leading to new inquiries about the role of ambiguity in economic decisions (Ellsberg 1961; Klibanoff et al. 2005; Halevy 2007). Another direction is to perform economic analyses without imposing a probability structure. As argued by Chambers and Quiggin (2000), this can be done using a "state-contingent" approach. But, as discussed by Chavas (2008), conducting applied risk analysis without probability assessment is not without its challenges.

3 Risk preferences and the cost of risk

Once risk assessment is done, the next issue is to evaluate how risk exposure affects behavior and welfare. Economists are typically interested in both aspects: the impact of risk on behavior is relevant when we want to understand how decisions are made under uncertainty, and the impact of risk and welfare is relevant when we try to evaluate the efficiency of risk allocation. In both cases, risk preferences matter. As a result, much interest has focused on the evaluation of risk preferences.

We start with the expected utility (EU) model. The EU model has provided the framework for many empirical analyses of risk preferences (e.g., Chavas 2004). Consider a decision maker making decisions x under risk. The risk is represented by the random variable e with a given subjective probability distribution $F(c) = prob(e \leq c)$. Denote the stochastic monetary payoff of choosing x under state e by $\pi(x,e)$. Following Savage (1954) and von Neumann and Morgenstern (1944), the expected utility (EU) model assumes that the decision maker chooses x so as to maximize

$$EU(\pi(x,e)) = \int_e U(\pi(x,e)) dF(e), \tag{1}$$

where E is the expectation operator and $U(\pi)$ is a utility function representing risk preferences. The utility function $U(\pi)$ is assumed to be strictly increasing. Conditional on x, let $M(x) = E(\pi(x,e))$ be the mean payoff. Following Pratt (1964) and Arrow (1965), for a given x, define the risk premium as the sure amount $R_{EU}(x)$ satisfying

$$EU(\pi(x,e)) = U(M(x) - R_{EU}(x)) \tag{2}$$

Under the expected utility (EU) model, the risk premium $R_{EU}(x)$ in (2) is the decision maker's willingness to pay to replace the random payoff $\pi(x,e)$ by its mean $M(x)$. For a given x, the risk premium $R_{EU}(x)$ is a measure of the implicit cost of risk. From Equation (2), we can define the certainty equivalent $CE_{EU}(x)$ as

$$CE_{EU}(x) = M(x) - R_{EU}(x) \tag{3}$$

From Equation (3), the certainty equivalent $CE_{EU}(x)$ is the expected payoff $M(x)$ net of the cost of risk $R_{EU}(x)$. Since the utility function $U(\pi)$ is strictly increasing, substituting Equations (2) and (3) into (1) implies that the decision maker chooses x so as to maximize the certainty equivalent $CE_{EU}(x)$. Thus, under the expected utility (EU) model, the certainty equivalent $CE_{EU}(x)$ in (3) provides a valid welfare measure under risk.

The sign of $R_{EU}(x)$ can be used to characterize the risk preferences of the decision maker. For a given x, the decision maker is said to be $\begin{Bmatrix} \text{risk averse} \\ \text{risk neutral} \\ \text{risk lover} \end{Bmatrix}$ when $R_{EU}(x) \begin{Bmatrix} > \\ = \\ < \end{Bmatrix} 0$. Thus, risk aversion corresponds to situations where the cost of risk is positive. As will be further discussed, risk aversion is a common characteristic of risk preferences among most decision makers. In this context, (3) shows that the cost of risk $R_{EU}(x)$ is an integral part of economic analysis. It makes it clear how management (as represented by the decision x) affects behavior and welfare. Indeed, x can be identified as risk-increasing (risk-decreasing) if a rise in x contributes to increasing (decreasing) the cost of risk $R_{EU}(x)$.

Under the expected utility (EU) model, the risk premium $R_{EU}(x)$ also provides useful information on the nature of risk preferences. Denote the variance of payoff by $Var(x) = E[(\pi(x,e) - M(x))^2]$ and its skewness by $Skew(x) = E[(\pi(x,e) - M(x))^3]$. Under differentiability and for a given x, Pratt (1964) and Antle (1987) have shown that the risk premium $R_{EU}(x)$ in (2) can be approximated as

$$R_{EU}(x) \cong -\frac{1}{2}\frac{U''(M(x))}{U'(M(x))}Var(\pi(x,e)) - \frac{1}{6}\frac{U'''(M(x))}{U'(M(x))}Skew(\pi(x,e)), \qquad (4)$$

Where $U'(M) = \frac{\partial U(M)}{\partial M} > 0$, $U''(M) = \frac{\partial^2 U(M)}{\partial M^2}$ and $U'''(M) = \frac{\partial^3 U(M)}{\partial M^3}$.

Equation (4) includes the Arrow–Pratt absolute risk aversion coefficient $-\frac{U''(M(x))}{U'(M(x))}$ and the skewness coefficient $-\frac{U'''(M(x))}{U'(M(x))}$. As discussed in Pratt (1964) and Arrow (1965), given $U'(M) > 0$, Equation (4) means that risk aversion corresponds to $U''(M) < 0$: any rise in risk (as measured by a higher variance Var in (4)) increases the cost of risk R_{EU}. Thus, risk aversion corresponds to a concave utility function $U(\pi)$. And, as argued by Antle (1987), $U'''(M) > 0$ corresponds to downside risk aversion: any increase in downside risk (as measured by a decline in skewness $Skew$ in (4)) implies a rise in the cost of risk R_{EU}.

Given $\pi > 0$, a common specification for the utility function $U(\pi)$ is

$$U(\pi) = \frac{\pi^{1-r} - 1}{1-r}, \qquad (5)$$

where $r = -\frac{U''(\pi)}{U'(\pi)}\pi$ is the Arrow–Pratt relative risk aversion coefficient, with $r > 0$ corresponding to risk aversion. Treating r as a constant, Equation (5) corresponds to risk preferences exhibiting constant relative risk aversion (CRRA) (Pratt 1964; Arrow 1965). Noting that (5) implies that $\frac{U'''(\pi)}{U'(\pi)} = \frac{r(r+1)}{\pi^2}$, it follows that when $r > 0$, the utility function in (5) corresponds to both risk aversion and downside risk aversion. In addition, under the specification (5), note that Equation (4) becomes

$$R_{EU}(x) \cong \frac{1}{2}\frac{r}{M(x)}Var(\pi(x,e)) - \frac{1}{6}\frac{r(r+1)}{M(x)^2}Skew(\pi(x,e)), \qquad (4')$$

which provides useful linkages between risk exposure (as measured by variance and skewness) and the cost of risk.

Under the expected utility (EU) model, the analysis of risk preferences reduces to estimating the utility function $U(\pi)$ in (1). Three broad characteristics have been uncovered: (1) most decisions makers are risk averse; (2) most decision makers are averse to downside risk; and (3) there is much heterogeneity in risk preferences across individuals (e.g., Halek 2001; Dohmen et al. 2011). This last characteristic (the presence of individual heterogeneity in risk aversion) makes the empirical analysis of risk management rather challenging. But the first two characteristics are consistent with the utility specification (5). On that basis, the CRRA specification given in (5) has been commonly used in the analysis of risk aversion. In this context, Gollier

(2004) has argued that, under CRRA, the coefficient of relative risk aversion r typically varies in range from 0 (corresponding to risk neutrality) to 5 (corresponding to a high level of risk aversion). Similar results apply to farmers. Most farmers have been found to be averse to risk and to downside risk (e.g., Lin et al. 1974; Dillon and Scandizzo 1978; Binswanger 1981; Chavas and Holt 1996). On that basis, under general conditions, risk matters: exposure to agricultural risk affects economic decisions and welfare (e.g., Chavas and Holt 1990, 1996; Just and Pope 2002).

The above discussion was based on the expected utility (EU) model in (1). But does the EU model provide an accurate representation of economic behavior and welfare? Note that the objective function in (1) is linear in the probabilities. As mentioned in the introduction, this is a strong assumption. Indeed, starting with Allais (1953), there is empirical evidence that decision makers do not behave according to the EU model (e.g., Kahneman and Tversky 1979). These arguments have led to more general (non-EU) models of risk preferences. This includes cumulative prospect theory (CPT), as proposed by Tversky and Kahneman (1992). A CPT model assumes that the decision maker chooses x so as to maximize

$$\int_e U(\pi(x,e))\, dG(F(e)), \tag{6}$$

where $U(\pi)$ is a strictly increasing utility function, and $G(F)$ is a strictly increasing function satisfying $G(0) = 0$ and $G(1) = 1$. An example for the function $G(F)$ is

$$G(F) = exp\left[-(-log(F))^\alpha\right], \tag{7}$$

where $\alpha \in (0,1)$ is a probability weight (Prelec 1998). When $\alpha = 1$, equations (6)–(7) reduce to the expected utility (EU) model (1). But when $0 < a < 1$, (7) introduces a non-EU preference function that is nonlinear in the probabilities. In this context, the function $G(F)$ has an inverted S-shape that reflects overweighting the probability of rare events. There is much evidence that many decisions makers "overweigh small probabilities" (e.g., Kahneman and Tversky 1979; Tversky and Kahneman 1992; Prelec 1998; Neilson 2003; Fehr-Dula and Hepper 2012; Barberis 2013). Such evidence also holds for farmers (e.g., Liu 2013; Bocquého et al. 2014). Compared to the EU model, the overweighing of small probabilities under CPT implies greater behavioral sensitivity to rare events.

Kahneman and Tversky (1979) have argued that the CPT model (6) can capture two common properties of risk preferences: (1) oversensitivity to rare events (as previously discussed); and (2) "loss aversion" associated with kinks in the utility function $U(\pi)$, losses being perceived as having greater welfare effects than equivalent gains. This second characteristic (loss aversion) provides another way to express aversion to downside risk.

Note that the EU characterization of the cost of risk discussed above can be readily extended to the CPT model. Indeed, under CPT, Equation (2) can be modified to define the cost of risk as the sure amount $R_{CPT}(x)$ that satisfies

$$\int_e U(\pi(x,e))\, dG(F(e)) = U(M(x) - R_{CPT}(x)). \tag{2'}$$

In the presence of rare events, the cost of risk $R_{CPT}(x)$ may be greater (compared to $R_{EU}(x)$ under EU). And, under CPT, Equation (3) can be modified to define the certainty equivalent $CE_{CPT}(x)$ as

$$CE_{CPT}(x) = M(x) - R_{CPT}(x). \tag{3'}$$

Again, the certainty equivalent CE_{CPT} in (3) provides a valid welfare measure under CPT. It shows that cost of risk $R_{CPT}(x)$ is an integral part of economic analysis, as it depends on the decision x. Implications for risk management and policy are discussed next.

4 Risk management and policy

The previous section presented two models of decision making under risk: the expected utility (EU) model given in (1) and the cumulative prospect theory (CPT) given in (6). Such models can be used to investigate how individuals manage risk and/or to evaluate the efficiency of risk allocation decisions. We review some useful insights obtained from these models when applied to agriculture.

A good overview of the economics of risk in agriculture is given in Harwood et al. (1999). Typically, farmers face price risk as well as production risk. Price risk and production risk tend to be negatively correlated. This is good for risk-averse farmers: the negative correlation means that production shocks are partially absorbed by opposite price adjustments, leading to reduced exposure to revenue risk (Harwood et al. 1999). This negative correlation between price risk and production risk tends to be stronger in dominant production regions (e.g., in the Corn Belt for corn). It is weaker in more marginal production regions.

Market globalization has several effects on revenue risk (Newbery and Stiglitz 1981). On the one hand, trade can reduce output price risk. For example, allowing for imports into a region facing crop failure reduces the effects of production shortfalls on domestic prices. This can generate substantial benefits from global markets. But market globalization can also increase revenue risk in two ways. First, it can increase price risk by exposing a region to price shocks induced by supply/demand shocks in different regions (Newbery and Stiglitz 1981). Second, it can reduce the negative correlation between price and production. Indeed, increasing the number of producing regions tends to reduce the effects of local production shocks on induced price adjustments. Thus, under globalization, revenue risk may increase as local production shocks become less absorbed by opposite price adjustments.

4.1 Diversification

Both price risk and production risk are subject to management. A management option affecting revenue risk is diversification. Under fairly general conditions, enterprise diversification reduces revenue risk (e.g., Chavas 2004). This is particularly relevant for agriculture, where both price risk and production risk are prevalent (Just and Pope 2002). Most farms produce more than one output. Lin et al. (1974) showed that most California farms are multi-output enterprises because of risk and risk aversion. Chavas and Di Falco (2012) obtained similar results for Ethiopian farms. In general, farm diversification reduces the cost of risk, thus providing incentive for farms to produce multiple outputs. But the incentive to diversify can vary across agroecosystems. For example, diversification incentives appear to be weaker on Asian irrigated farms. At least on irrigated paddy land, Kim et al. (2014) showed that such farms tend to specialize in rice production for two reasons: (1) irrigation reduces weather risk; and (2) rice is particularly well adapted to the irrigated paddy. Yet, on non-irrigated land, most farms tend to diversify to reduce revenue risk (Lin et al. 1974; Chavas and Di Falco 2012). Note that the risk reduction can involve both a lower variance and a higher skewness (i.e., a lower exposure to downside risk). Chavas and Di Falco (2012) and Kim et al. (2014) found that, when evaluating the effects of diversification on the cost of risk, the skewness effect can be greater than the variance effect. This indicates that a large part of the risk benefit from farm diversification comes from reducing exposure to downside risk.

4.2 Farm management

Farmers have many management options to reduce production risk. Farm production decisions can have direct impacts on production risk (Harwood et al. 1999). For example, Just and Pope (1979) showed that fertilizer tends to be "risk increasing": fertilizer use increases the variance of crop yield (as crop yield rises but only if it rains). Thus, risk averse farmers have an incentive to reduce the use of risk-increasing inputs. Other inputs are "risk reducing." An example is irrigation, which lessens the adverse effects of rainfall shocks on crop yield (Harwood et al. 1999). Another example is veterinary care, which contributes to reducing the incidence of animal diseases and its adverse impact on livestock productivity. Also, some outputs are riskier than others. In general, livestock production is less risky than crop production (as livestock output tends to be less sensitive to weather shocks). This explains why regions facing greater weather risk tends to specialize in livestock (e.g., in the Sahel). Also, drought sensitivity varies across crops. For example, wheat is more drought-resistant than corn. In the U.S., wheat is grown in the Western Great Plains (where drought risk is higher) while corn is grown in the Corn Belt (where rainfall risk is lower). Finally, as documented by Chavas and Shi (2015), crop rotations can reduce production risk. Crop rotations help control weed and pest populations, thus lessening the unpredictable adverse effects of weed and pest damage on crop yield. Chavas and Shi (2015) found that crop rotations can be particularly effective at reducing exposure to downside risk. In general, risk averse farmers have an incentive to implement management strategies that reduce their exposure to risk and downside risk.

4.3 Technology

Technology also affects production risk in agriculture. Genetic selection has played a major role. Over the last 70 years, geneticists have developed new crop varieties that are more pest-resistant and more drought-resistant. Besides increasing yields, such innovations have contributed to reducing the unpredictable effects of pest damage and drought on agricultural production. Starting in 1996, newly developed genetically modified (GM) seeds for corn, soybean, and cotton have been quickly adopted by many U.S. farmers. Using genetic material from other species, GM crops are better able to control pest damages. As a result, GM crops exhibit reduced exposure to both risk and downside risk. In this context, Chavas and Shi (2015) showed that a significant benefit from GM corn is a lower cost of risk.

While technology can help reduce risk exposure, the outcome of a new technology is typically not perfectly known before it is adopted. Thus, technology adoption always involves some uncertainty (Feder et al. 1985). For a farmer, two types of uncertainty are relevant: (1) is the new technology adapted to local agro-climatic conditions? and (2) does the farmer know how to manage it? Such uncertainties would be seen negatively by risk averse farmers. In this context, risk aversion would affect technology adoption (Feder et al. 1985). To the extent that risk aversion varies across farmers, this generates the prediction that early (late) adopters would be among farmers exhibiting lower (higher) risk aversion. But two other aspects of technology adoption are also relevant. First, information about a new technology can vary across individuals. Barham et al. (2014a) documented how the ability to learn varies significantly among U.S. farmers. In this context, we expect that the early (late) adopters would be among the quick (slow) learners. Second, is it appropriate to impose a probability structure on the uncertainty associated with a new technology? Barham et al. (2014b) presented evidence that, for U.S. farmers, the decision to adopt GM technology depends on ambiguity and ambiguity aversion (and not on risk aversion). More research is needed to explore these issues.

4.4 Dynamics of risk

Risk management also occurs in a dynamic context. This raises the question: how can an agroecosystem respond to stochastic shocks over time? A desirable property is its "resilience," where resilience is defined as the ability to recover quickly from an adverse shock. Alternatively, a lack of resilience means it is difficult for a system to recover from adverse shocks. Are agroecosystems resilient? In general, this depends on both management and technology. Chavas and Di Falco (2017) found evidence that English wheat yields are resilient: they recover fairly quickly from adverse shocks. This is important given current concerns about climate change. Will technological progress and improved management be sufficient to deal with the increased climate-induced uncertainty in agriculture? Additional research is needed to examine this issue.

4.5 Hedging

As previously noted, price risk is also important in agriculture. This is particularly relevant in agriculture for two reasons: (1) agricultural markets exhibit large price volatility; and (2) the production process takes time, implying that farm input decisions are made before output prices are known. Again, farmers have a number of options to manage price risk (Harwood et al. 1999). One option is hedging on futures markets. Hedging consists of taking opposite positions in the cash market and the futures market. For a producer, it means selling a futures contract at planting time (with delivery at harvest time) and buying it back at harvest time. When the futures price and cash price converge at harvest time, hedging reduces price risk. In situations where there is no price uncertainty, hedging effectively eliminates the effects of price risk (Feder et al. 1980). As a result, hedging can be a powerful tool to manage price risk.

But there is a puzzle: while agriculture involves significant price risk, only a small proportion (less than 10 percent) of farmers hedge (Garcia and Leuthold 2004). Why are so few farmers hedging? Several explanations have been proposed. First, the ability to reduce price risk through hedging deteriorates under production risk. For example, in the event of crop failure, a hedger would face future price risk by becoming a speculator on the futures market. Second, spatial price differences mean that the local cash price may not converge with the futures price as one approaches the contact delivery time, generating "basis risk." Again, this would make hedging less effective in reducing price risk. Third, the contract unit on futures markets may be "too large" for many farmers. Finally, farmers may have other options to manage price risk. An alternative option is using forward contracts. For example, a grain farmer may sign a contract at planting time selling his crop to a trader, the price paid for output being specified in the contract. This would have the same impact as hedging: in the absence of production risk, the farmer's exposure to price risk would be eliminated.

4.6 Risk transfers

Risk transfers are another means of managing risk. They involve resource transfers across individuals under particular risky events. How can we evaluate the efficiency of these risk transfers? Standard efficiency arguments apply: risk transfer schemes are efficient if they generate the largest possible aggregate benefit to society (Chavas 2011). Since most individuals are risk averse, this translates into schemes that minimize the aggregate cost of risk. Here the aggregate cost of risk is the sum of individual costs of risk, as discussed earlier. If all individuals were identical and faced the same risk, there would be no efficiency gain from risk redistribution. But we have seen that risk preferences can vary a lot across individuals. In addition, exposure to risk (and downside risk) can differ both over time and across space. For example, the prospects of facing hail damages are both location-specific and time-specific. This creates opportunities to improve the efficiency of risk allocation. Since most individuals are averse to downside risk, the focus has been on risk transfer schemes that reduce exposure to downside risk. Such schemes can improve efficiency if they contribute to reducing the aggregate cost of risk.

These inter-individual risk transfers can occur at all levels of society. First, they can occur at the household level. An example is the case of remittances. Members of a family working in different locations typically face different income shocks. This is a case where remittances can redistribute income across family members toward those adversely affected by shocks. To the extent that not all family members face the same shock, this can reduce exposure to risk and downside risk. Such schemes are commonly found in developing countries where they contribute to reducing the aggregate cost of risk. They also play a role in developed countries where off-farm income can help stabilize household income in rural areas (Harwood et al. 1999).

Second, risk transfers can be implemented by insurance contracts. Insurance markets can provide effective redistribution of risk. In agriculture, an example is the case of hail insurance. Hail risk has three notable characteristics: (1) hail occurs rarely and is widely distributed across time and place; (2) hail events cannot be easily prevented; and (3) hail damages are fairly easy to observe. These characteristics have been conducive to the development of insurance contracts implemented by insurance firms that can redistribute resources toward the individuals adversely affected by hail damages.

Third, risk transfers can be implemented at the national level. In this context, they include economic policies attempting to reduce exposure to downside risk in society. Examples are disaster assistance and various social safety nets (e.g., unemployment insurance transferring income toward individuals who have lost their job). This also includes agricultural policies that attempt to reduce price risk, production risk, or both. In the U.S., price support programs were in place before 1996. By establishing price floors, price support programs reduced exposure to low farm prices (e.g., Newbery and Stiglitz 1981). Since 1996, U.S. agricultural policy has moved away from price interventions and toward reducing revenue risk. The move has been associated with the subsidization of agricultural insurance (e.g., Smith and Glauber 2012). This requires a new look into the functioning of insurance markets. One puzzle is that, while most farmers have been found to be risk averse (e.g., Lin et al. 1974; Dillon and Scandizzo 1978; Binswanger 1981), they are typically not willing to pay much for crop insurance (e.g., Hazell et al. 1986; Smith and Glauber 2012). When most farmers face large risk, why are they are so unwilling to participate in crop insurance? Several possible explanations have been proposed. One explanation is that farmers see federal disaster assistance as providing sufficient protection against downside risk (with disaster relief paid by the taxpayers and not by the farmers). Another possibility is that farmers have other risk management options available (as previously discussed), and they see these options as being good substitutes for formal insurance. A third possibility is that farmers see ambiguity in current insurance contracts, with ambiguity aversion providing them a disincentive to buy insurance. Further research is needed to explore these issues.

Fourth, risk transfers can be implemented at the international level. These include famine relief and food aid, typically targeted to individuals facing dire situations of poverty and malnutrition. They also include buffer stock policies at the world level. Such policies would have the objective of increasing food availability in the event of major adverse shocks to the food supply.

While there are many policy options to manage agricultural risk, there are also important challenges. A first challenge arises from some empirical complexities in assessing the nature of risk aversion and risk preferences (as discussed in Section 3). A second challenge comes from the presence of much heterogeneity across individuals in both risk exposure and risk preferences. This heterogeneity makes it difficult to assess the cost of risk and the aggregate efficiency of alternative risk management schemes. A third challenge is that risk markets are typically incomplete. So far, without subsidies, it has proven problematical to rely on insurance markets to improve the efficiency of risk allocation in agriculture. This may be due in part to the difficulties of insurance firms in obtaining precise assessments of the spatial and temporal distribution of agricultural risk. Fourth, this last feature also applies to government policy, making it hard to design programs that can improve the allocation of risk at the national or international level. Current concerns about climate change and its effects on agricultural risk add to this challenge. Designing policies to implement efficient risk allocations requires good information about the evolving distribution of agricultural risk over both time and space. More research is needed to address these issues.

References

Allais, M. 1953. "Le Comportement de l'Homme Rationnel Devant le Risque: Critique des Postulats et Axiomes de l'École Américaine." *Econometrica* 21:503–546.

Antle, J.M. 1987. "Econometric Estimation of Producers' Risk Attitudes." *American Journal of Agricultural Economics* 69:509–522.

———. 2010. "Asymmetry, Partial Moments, and Production Risk." *American Journal of Agricultural Economics* 92:1294–1309.

Arrow, K.J. 1965. *Aspects of the Theory of Risk Bearing*. Helsinki: Yrjo Jahnssonin Saatio.

Barberis, N.C. 2013. "Thirty Years of Prospect Theory in Economics: A Review and Assessment." *Journal of Economic Perspectives* 27:173–196.

Barham, B.L., J.P. Chavas, D. Fitz, V. Rios Salas, and L. Schechter. 2014a. "Risk, Learning, and Technology Adoption." *Agricultural Economics* 45:1–14.

———. 2014b. "The Roles of Risk and Ambiguity in Technology Adoption." *Journal of Economic Behavior and Organization* 97:204–218.

Binswanger, H.P. 1981. "Attitudes toward Risk: Theoretical Implications of an Experiment in Rural India." *Economic Journal* 91:867–890.

Bocquého, G., F. Jacquet, and A. Reynaud. 2014. "Expected Utility or Prospect Theory Maximizers? Assessing Farmers' Risk Behaviour from Field-Experiment Data." *European Review of Agricultural Economics* 41:135–172.

Chambers, R.G., and J. Quiggin. 2000. *Uncertainty, Production, Choice and Agency: The State-Contingent Approach*. New York: Cambridge University Press.

Chavas, J.P. 2004. *Risk Analysis in Theory and Practice*. London: Elsevier.

———. 2008. "A Cost Approach to Economic Analysis under State-Contingent Production Uncertainty." *American Journal of Agricultural Economics* 90:435–446.

———. 2011. "Agricultural Policy in an Uncertain World." *European Review of Agricultural Economics* 38:383–407.

———. 2017. "On Food Security and the Economic Valuation of Food." *Food Policy* 69:58–67.

Chavas, J.P., and S. Di Falco. 2012. "On the Role of Risk versus Economies of Scope in Farm Diversification with an Application to Ethiopian Farms." *Journal of Agricultural Economics* 63:25–55.

———. 2017. "Resilience, Weather and Dynamic Adjustments in Agroecosystems: The Case of Wheat Yield in England." *Environmental and Resource Economics* 67:297–320.

Chavas, J.P., and M.T. Holt. 1990. "Acreage Decisions Under Risk: The Case of Corn and Soybeans." *American Journal of Agricultural Economics* 72:529–538.

Chavas, J.P., and M.T. Holt. 1996. "Economic Behavior Under Uncertainty: A Joint Analysis of Risk Preferences and Technology." *Review of Economics and Statistics* 78:329–335.

Chavas, J.P., D. Hummels, and B. Wright (editors). 2014. *The Economics of Food Price Volatility*. NBER Volume, Chicago: University of Chicago Press.

Chavas, J.P., and G. Shi. 2015. "An Economic Analysis of Risk, Management and Agricultural Technology." *Journal of Agricultural and Resource Economics* 40:63–79.

Dillon, J., and P. Scandizzo. 1978. "Risk Attitudes of Subsistence Farmers in Northeast Brazil: A Sampling Approach." *American Journal of Agricultural Economics* 60:425–435.

Dohmen, T., A. Falk, D. Huffman, U. Sunde, J. Schupp, and G.G. Wagner. 2011. "Individual Risk Attitudes: Measurement, Determinants and Behavioral Consequences." *Journal of the European Economic Association* 9:522–550.

Ellsberg, D. 1961. "Risk, Ambiguity and the Savage Axioms." *Quarterly Journal of Economics* 75:643–669.

Feder, G., R.E. Just, and A. Schmitz. 1980. "Futures Markets and the Theory of the Firm Under Price Uncertainty." *Quarterly Journal of Economics* 94:317–328.

Feder, G., R.E. Just, and D. Zilberman. 1985. "Adoption of Agricultural Innovations in Developing Countries: A Survey." *Economic Development and Cultural Change* 30:59–76.

Fehr-Duda, H., and T. Hepper. 2012. "Probability and Risk: Foundations and Economic Implications of Probability-Dependent Risk Preferences." *Annual Review of Economics* 4:567–593.

Gallagher, P. 1987. "U.S. Soybean Yields: Estimation and Forecasting with Nonsymmetric Disturbances." *American Journal of Agricultural Economics* 69:796–803.

Garcia, P., and R.M. Leuthold. 2004. "A Selected Review of Agricultural Commodity Futures and Options Markets." *European Review of Agricultural Economics* 31:235–272.

Gollier, C. 2004. *The Economics of Risk and Time*. Cambridge: MIT Press.

Goodwin, B.K., and A.P. Ker. 1998. "Nonparametric Estimation of Crop Yield Distributions: Implications for Rating Group-Risk Crop Insurance Contracts." *American Journal of Agricultural Economics* 80:139–153.

Halek, M., and J.G. Eisenhauer. 2001. "Demography of Risk Aversion." *Journal of Risk and Insurance* 68:1–24.

Halevy, Y. 2007. "Ellsberg Revisited: An Experimental Study." *Econometrica* 75:503–536.

Harwood, J., R. Heifner, K. Coble, J. Perry, and A. Somwaru. 1999. "Managing Risk in Farming: Concepts, Research, and Analysis." Agricultural Economic Report No. 774, Economic Research Service, U.S. Department of Agriculture, Washington, D.C.

Hazell, P., C. Pomareda, and A. Valdez. 1986. *Crop Insurance for Agricultural Development: Issues and Experience*. Baltimore: Johns Hopkins University Press.

Just, R.E., and R.D. Pope. 1979. "Production Function Estimation and Related Risk Considerations." *American Journal of Agricultural Economics* 61:276–284.

———. 2002. *A Comprehensive Assessment of the Role of Risk in U.S. Agriculture*. Boston: Kluwer Academic Publishers.

Kahneman, D., and A. Tversky. 1979. "Prospect Theory: An Analysis of Decision under Risk." *Econometrica* 47:263–291.

Kim, K., J.P. Chavas, B. Barham, and J. Foltz. 2014. "Rice, Irrigation and Downside Risk: A Quantile Analysis of Risk Exposure and Mitigation on Korean Farms." *European Review of Agricultural Economics* 41:775–815.

Klibanoff, P., M. Marinacci, and S. Mukerji. 2005. "A Smooth Model of Decision Making under Ambiguity." *Econometrica* 73:1849–1892.

Knight, F.H. 1921. *Risk, Uncertainty and Profit*. Boston: Houghton Mifflin Company. Lin, W., G. Dean, and C. Moore. 1974. "An Empirical Test of Utility vs. Profit Maximization in Agricultural Production." *American Journal of Agricultural Economics* 56:497–508.

Liu, E.M. 2013. "Time to Change What to Sow: Risk Preferences and Technology Adoption Decisions of Cotton Farmers in China." *Review of Economics and Statistics* 95:1386–1403.

Meyer, J. 1987. "Two-Moment Decision Models and Expected Utility." *American Economic Review* 77:326–336.

Neilson, W. 2003. "Probability Transformations in the Study of Behavior Toward Risk." *Synthese* 135:171–192.

Newbery, D.M., and J.E. Stiglitz. 1981. *The Theory of Commodity Price Stabilization: A Study in the Economics of Risk*. Oxford: Clarendon Press.

Pratt, J.W. 1964. "Risk Aversion in the Small and in the Large." *Econometrica* 32:122–136.
Prelec, D. 1998. "The Probability Weighting Function." *Econometrica* 66(3):497–527.
Savage, L. 1954. *The Foundations of Statistics*. New York: John Wiley and Sons.
Smith, V.H., and J.W. Glauber. 2012. "Agricultural Insurance in Developed Countries: Where Have We Been and Where Are We Going?" *Applied Economic Perspectives and Policy* 34:363–390.
Tversky, A., and D. Kahneman. 1992. "Advances in Prospect Theory: Cumulative Representation of Uncertainty." *Journal of Risk and Uncertainty* 5:297–323.
Von Neumann, J., and O. Morgenstern. 1944. *Theory of Games and Economic Behavior*. Princeton, NJ: Princeton University Press.

33

Investing in people in the twenty-first century

Education, health, and migration

Wallace E. Huffman

1 Introduction

T.W. Schultz became famous for developing a new concept called human capital, that is, the idea that investing in people improves their health, skills and competencies, knowledge or information base, and geographical location relative to consumption and work opportunities. Because investing in people is costly, he argued that the investment decisions should be deliberate in that the benefits are weighed against the costs. Moreover, it is not a story about how innate ability (e.g., IQ) predetermines your economic lot in life.

Furthermore, he argued that we can apply the principles of investment theory developed for decision making on physical capital to human capital investments. That is, an individual who decides to obtain additional years of schooling bears the costs of forgone earnings and direct outlays for tuition, books, and supplies, and after completing the schooling, that individual expects to obtain high earnings for future employment. With a little work, the rate of return on this investment can be computed and compared to rates of returns on other available investments.[1] Good investment decision making requires that the rate of return on education is greater than or equal to that of the best alternative uses of the funds (Schultz 1961a). This ambitious application of economic thinking to investments made in people shocked some economists, social scientists, and others when he presented it in his Presidential Address to the American Economics Association in St. Louis, Missouri (USA) in 1960. Shocked individuals had difficulty in separating the direct consumption value and investment value of education (Schultz 1961b).

Before this, Schultz had wrestled with the new idea of human capital for almost two decades. As a result of his post–World War II travels to Germany and Japan, he saw firsthand their miraculous speed of recovery from the immediate widespread devastation caused by the war. In contrast, the United Kingdom took a long time to recover. He concluded that the rapid recovery was due to the healthy and highly educated population in Germany and Japan relative to the United Kingdom. Education helps people be productive, and good health care keeps educated individuals able to engage in productive work longer and more intensely. These insights were useful new ideas about the primary source of economic growth of countries and regions.

Human capital is now a well-established part of economics. The widely used *Journal of Economic Literature* classification of topics in economics places human capital under the broad field of labor and demographic economics. However, its fruits also spill over to the fields of health, education, and welfare; economic development, technical change, and growth; agricultural, natural resource, and environmental economics; and urban, rural, and regional economics.

When Schultz was born in 1902, the distribution of the U.S. labor force across employment sectors was 36 percent in agriculture, 28 percent in industry, and 36 percent in services. In 1960 (the year Schultz gave his investing in human capital address to the AEA), the shares had shifted dramatically – only 9 percent in agriculture, 34 percent in industry, and 57 percent in services. Hence, over this slightly more than half a century, the United States became a service economy by having a majority of employment in the services sector. In 2016, the shares were 1.4 percent in agriculture, 15.5 percent in industry, and 83.1 percent in services. Rapidly growing service sectors have been in the health, education, professional, and business sectors. These structural changes suggest that the demand for the ability to do physical labor has largely disappeared and the opportunity to perform services, some of which are quite highly skilled, has been growing rapidly.

Invented in the late 1930s, the first electronic digital computer used vacuum tubes for processing data and was very slow. The invention of the transistor in 1947 furthered the development of information technology. A transistor is a semiconductor device that acts as an electrical switch and encodes information in a 0–1 form associated with the off-and-on positions of a switch. Invented in 1958, integrated circuits and memory chips that consist of many transistors were first developed for data processing. Starting in about 1970, the capacity of memory chips has increased at a continuous compound rate of 35 to 45 percent (Jorgenson 2009), while the price of memory chips has decreased by a factor of 27,270 times (or 41 percent per year compound).[2] This resulted in a staggering rate of decline in the price of information technology used for the storage of information and computing and provided the incentive for a rapid diffusion of information technology (IT) through new hardware and software (Jorgenson, Ho, and Samuels 2011).[3] Information technology and software have resulted in the computerization of routine tasks and the rapid displacement of labor in repetitive production and monitoring tasks (e.g., bookkeeping, clerical work, and robotics in the manufacturing of durable goods). Digital technical change has been a major force bringing rapid change to the labor market.

A broader wage comparison by education is instructive here. Over the period from 1979 to 2007, the constant dollar median weekly earning of full-time wage and salary for those with less than a high school diploma decreased by 28.2 percent for men and by 8.7 percent for women (Figure 33.1). Men who were high school graduates (but without any college) experienced a 16 percent decline, while the weekly earnings for women rose 4 percent. Men who had some college or an associate degree also experienced a small decline of 7 percent, while the weekly earnings for women rose 8.6 percent. Men who had a bachelor's degree or higher education experienced an 18 percent rise in earnings over this period, while women experienced a much larger rise of 33 percent. Overall, women did better financially than men during this period.

Job growth has favored high-skilled goods and service workers and, to a lesser extent, low-skilled service workers. Low- to medium-skilled goods-producing and monitoring jobs have been most adversely affected. For example, from 2002 to 2014, the share of high-skilled jobs (managers, professionals) increased by seven percentage points, and low-skilled jobs (e.g., local food and personal services) increased by three percentage points, while the share of middle-skilled routine (e.g., manufacturing, operatives–assemblers, secretarial, clerical) jobs decreased by nearly ten percentage points (OECD 2016).[4] The main factors underlying these developments are technological changes that automated medium-skill routine tasks (e.g., manufacturing, farming, secretarial, and administrative jobs) and the offshoring of standardized goods and services facilitated by globalization and international trade. Hence, new jobs being created frequently

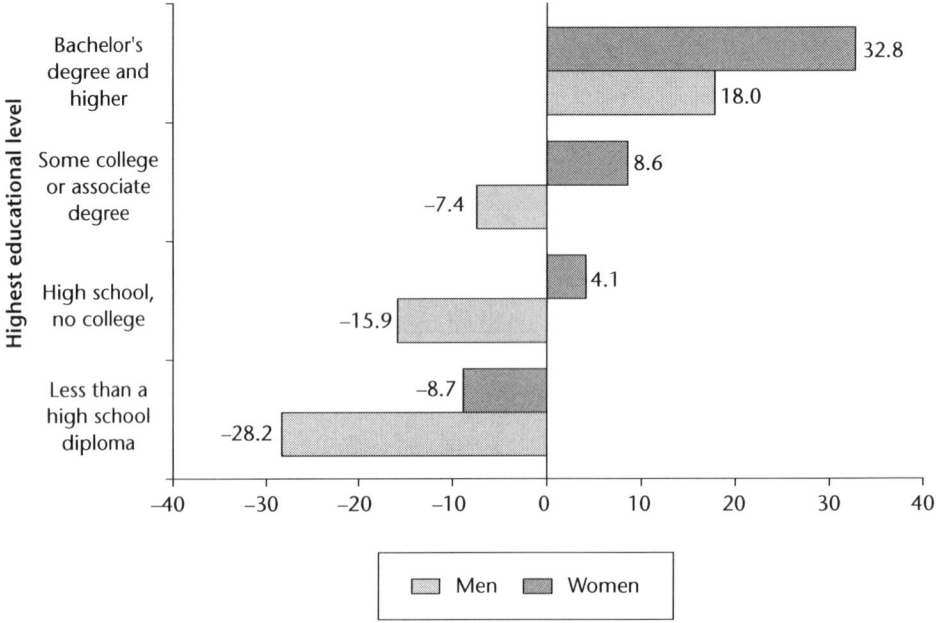

Figure 33.1 Change in constant dollar median usual weekly earnings, by educational attainment and gender, 1979–2007

require different skills, likely in a different industry, than those that were lost. Moreover, many of those losing their jobs are males with many years of job experience and significant accumulated firm- and industry-specific human capital (Becker 1993; Laing 2011).[5] With job losses and the structural change in skills needed for new jobs, these workers lose significant human capital and earning power. This means the immediate prospect of a significant wage decline in moving to a new job. Otherwise, middle-skilled displaced workers need to be upskilled to access higher-skilled and better wage jobs. If the individual becomes part of the force of the long-term unemployed or drops out of the labor force, society and families face major economic loses (OECD 2016).[6]

With the automation of work using computers and information technology, which displaced goods-producing workers doing repetitive activities, and the outsourcing of manufacturing and business service jobs to areas where labor is cheaper, the U.S. labor force is short on high-skilled workers and over supplied with middle-skilled workers. Furthermore, a large share of new entrants to the U.S. labor force do not have the education needed to compete well in the labor market of the twenty-first century. For example, in 2015, the U.S. ranked tenth (47 percent) in the world for the share of adult population ages 25 to 34 with at least two years of college (Figure 33.2).

In 2015, of U.S. adults ages 25 to 34, 10 percent had less than a high school diploma, 36 percent had a bachelor's degree or more, and 11 percent had an advanced degree (U.S. Census Bureau 2009; Ryan and Bauman 2016). A slightly different picture emerges when one looks at the education distribution of all U.S. adults ages 25 and older who were in the labor force in 2016, where 39 percent had a four-year college degree or higher, 28 percent had some college or a two-year college degree, 26 percent had a high school diploma, and 8 percent had less than a high school diploma (U.S. Department of Labor 2016).

Investing in people

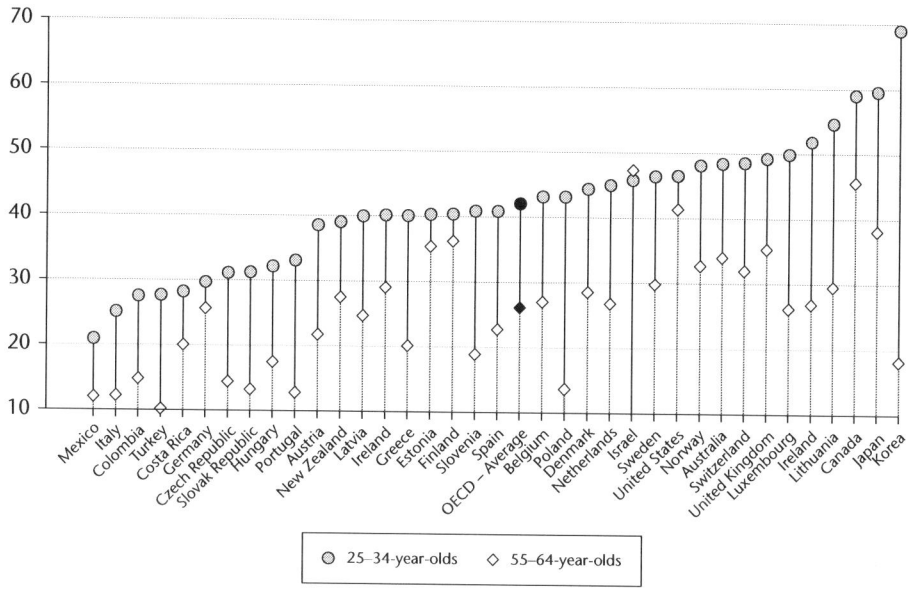

Figure 33.2 Population that has attained tertiary education, percentage by age group (2015)*

*Tertiary education is education beyond high school leading to a degree, which might be a two-year vocational degree.

U.S. public universities supply a large share of college-trained individuals. Although public universities struggled financially during the Great Recession of 2008–2009, they are now doing considerably better; however, a decline in the price of oil, metals, and agricultural commodities during 2015 and 2016 has inflicted a new source of hardship on some of them.

The purpose of this chapter is to show that human capital theory and labor market adjustments have important implications for investing in people for the twenty-first century. Section 2 identifies major types of human capital, Section 3 presents a model of the human capital investment decision, Section 4 reviews globalization and the changing world labor market, and Section 5 takes up the issue of who pays for a college education and how much. The final section presents some conclusions.

2 Types of human capital

The field of human capital has taken on a strong acquisition of skills and information flavor. Overall, K–12 education is a time when students learn cognitive skills such as reading, writing, completing mathematical operations, and understanding biological processes. These are valuable skills if one takes a job or continues to higher levels of education. However, the list of human capital topics has been expanded to include pre-school activities, informal education or information acquisition (e.g., short courses and seminars), and learning while working on the job (on-the-job training). For example, research by Heckman (2006) and Heckman, Pinto, and Savelyev (2013) report that pre-school activities of disadvantaged children help children develop social skills, sometimes called "personality" skills, that have long-term payoff: reduced externalizing behavior (aggressive, antisocial, and rule-breaking behavior), reduced drug use, better grades at school, and the likelihood of being employed as an adult.

619

Early discussions of investing in health focused on public health investments (mandatory vaccinations for contagious diseases and treatment of water and waste materials) and the inputs of medical services and pharmaceuticals. However, economists were among the first to hypothesize that good health is a product of how we choose to live our daily lives, that is, healthy lifestyle choices through better diet, more exercise, moderate alcohol consumption, and weight and stress management. In addition, other factors for pregnant women are access to prenatal medical care; an improved diet rich in vitamins, folic acid, and calcium; and behavioral modification (not smoking, drinking alcohol, or doing drugs). This set of events, which also has costs, has been shown to improve the health and birthweight of newborn babies (Rosenzweig and Schultz 1983). In addition, it is widely accepted that a poorly developed human organ system that arises from a poor fetal environment is one of the main causes of early onset of chronic diseases of old age (Fogel 1994). Moreover, research by Behrman and Rosenzweig (2004) has shown that increasing birthweight increases adult height, schooling attainment, and earnings for babies at most levels of birthweight, but does not affect adult body mass index (BMI) (Keng and Huffman 2007). Hence, a long-term payoff results from an investment in human capital as reflected in babies' birthweight.

Human migration creates another type of human capital. It is costly for individuals and families to relocate or migrate from one place to another, and it takes time to find employment and adjust to life in a new location. The study by Huffman and Feridhanusetyawan (2007) shows the response of migrating working-age males to wage opportunities and amenities at both the point-of-origin and destination. In addition, they show that additional education of adult males increases the probability of interstate migration, holding constant their expected wage differential between a new destination and current location. Hence, an individual's education level is a factor in whether a person is more geographically mobile. However, it is widely recognized that human migration changes and frequently provides benefits through improved education, employment, or consumption opportunities. For example, with the major shifts in the industrial (and occupational) distribution of the U.S. labor force, individuals first moved from farm to nonfarm jobs, and then to urban goods-producing jobs. With pressure from technical changes and international competition, individuals are having difficulty transitioning to low-skilled service jobs or high-skilled managerial jobs.

Immigration or international migration to the United States has had a major impact on U.S. history since at least the sixteenth century (Laing 2011, pp. 717–753). In particular, the United States had relatively unrestricted immigration until 1924, when the National Origins Immigration Act was passed. The 1924 Act limited new immigrant numbers to the countries that had the largest share of immigrants as determined in the U.S. census of population. Hence, Western and Northern Europeans were favored for a time. In 1965, this legislation was replaced by a family unification and refugee immigration policy. Potential immigrants who had family members who were U.S. citizens were given priority. This has had the effect of attracting a relatively large number of low-skilled individuals and parents of citizens. The Immigration Reform and Control Act of 1986 contained an amnesty provision that allowed illegal workers who could document their earlier U.S. work history to apply for a green card and later citizenship. The Illegal Immigration Reform and Immigrant Responsibility Act in 1996 established a means test for those seeking to bring family members from abroad.[7] Since 1970, the largest group of U.S. immigrants has been from Mexico, Central America, and Asia. Legal and illegal farm workers from Mexico have been a major part of the workforce in U.S. agricultural fruit and vegetable production, especially the harvesting of fresh produce, which cannot always be mechanized (Huffman 2014).

3 The human capital investment decision

Most human capital investment decisions take a similar structure – costs up front and expected payoff later. However, to be precise, consider the decision of whether to invest in a four-year college degree rather than stopping with a high school diploma or completing secondary school. Let Y_t^I = earnings of a high school graduate in year t (in constant dollars), Y_t^{II} = earnings while individual is enrolled in college, net of costs of completing a four-year or BA/BS college degree, and earnings after graduating from college in constant dollars (Figure 33.3). However, we simplify and assume that students while enrolled in a four-year college degree program have zero earnings (i.e., they forgo the earnings of a high school graduate). This is an important part of the cost of obtaining a four-year degree (see Figure 33.3, area C_1).

Also, college students incur the direct costs of attending college: expenditures for tuition, books, fees, and any net increase in housing, food, and clothing as a result of being a college student relative to being an earner with a high school diploma (see Figure 33.3, area C_2). These two types of costs are combined to obtain the total cost of obtaining a four-year college degree. In Figure 33.3, the earnings of a new college graduate are higher than for a high school graduate, and this difference is the net annual benefit of obtaining a college degree (area B). Since a dollar next year (or five or 40 years from now) is worth less to most individuals than a dollar today, we convert the education capital investment project into equivalent units by discounting. Let r_t be the discount rate (in constant dollars) appropriate for this education capital investment decision, for example, the real rate of interest encountered if the individual were to obtain a

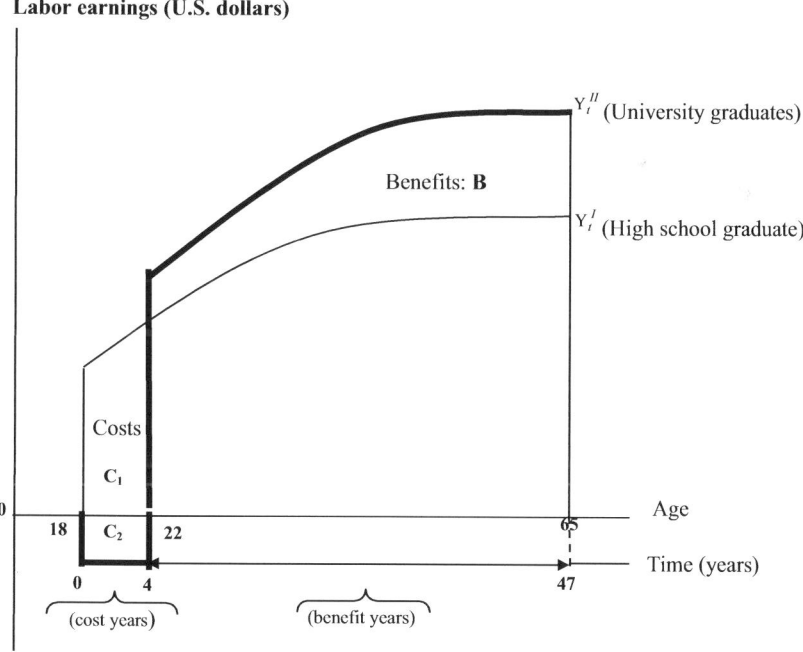

Figure 33.3 The economics of investing in a four-year college degree, given a high school diploma

Source: Adapted from (1981), p. 322.

college education loan, the real rate of return forgone on a passbook savings account or stocks or bonds if returns from these assets are used to self-finance the investment. Consider the net present value of investing at age 18 (NPV^{18}), the age at high school graduation, and assume retirement occurs at age 65:

$$NPV^{18} = \sum_{t=18}^{65}(Y_t^{II} - Y_t^{I})/(1+r_t)^t$$
$$= \sum_{t=18}^{22}(Y_t^{II} - Y_t^{I})/(1+r_t)^t + \sum_{t=22}^{65}(Y_t^{II} - Y_t^{I})/(1+r_t)^t = C + B, \quad (1)$$

where the first term on the right of the above expression equals (C) and the second term equals (B). This investment in a four-year college degree is a worthy investment if NPV^{18} is greater than or equal to zero. If the investment in a four-year college degree were delayed by two years, the forgone earning (and tuition costs) would increase a little, but more significant is the loss of two years of benefits at the end of the work life. More generally, investing in a four-year college degree at mid-life will not be a good investment because of shortening the benefits period and increasingly the forgone earnings cost while in college.

Huffman and Orazem (2007) present an agricultural household model where the marginal cost of human capital production is increasing each period but delaying results in a one-period loss of benefits. They conclude that large investments in human capital should occur early in the lifecycle. However, smaller increments in training may be a good investment later in an individual's lifecycle, but finite life is a drag on larger investments.

In many cases, it is useful to consider the internal rate of return on investing in a college degree: set the NPV^{18} = 0 and solve for the uniform discount rate $r^\star$ that makes NVP at age 18 equal to zero. A larger $r^\star$ implies a more attractive education investment project.

How has the payoff to a college degree relative to a high school degree varied over time? Figure 33.4 shows the ratio of the weekly wage rate for a four-year college graduate relative to a high school graduate from 1963 to 2008. The ratio was slightly less than 1.5 in 1963 and rose until 1971, reaching about 1.6. The ratio then declined to its starting value between 1978

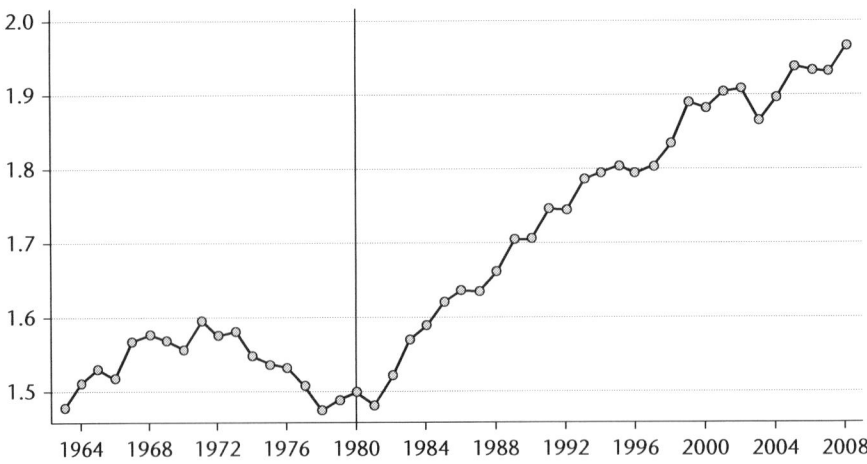

Figure 33.4 College–high school weekly wage ratio, 1963–2008

and 1981. However, starting in 1981, the ratio began trending strongly upward to about 1.95 in 2008. Later data show a slight increase between 2008 and 2012. This weekly wage ratio starting in 1980 shows the rapidly rising payoff to a college degree (Autor 2014).

Hence, with an increase since 1981 in the earnings of college graduates relative to high school graduates, the NPV[18] and $r\star$ have increased, implying that investing in a four-year college degree is an even better investment proposition than before. Autor (2014) documents that between 1981 and 2008, the net present value of a college versus a high school degree increased by a factor of 2.2 for men and by 2.7 for women.[8] If, in addition, individuals who complete a four-year college degree on average work more years than high school graduates, this further increases the NPV[18] and $r\star$.[9]

Moreover, Equation (1) and its internal rate of return alternative are powerful tools in addressing any human capital investment decision. As previously indicated, in almost all cases, the costs are upfront or at the beginning of the project and the benefits are in the future, generally ending at retirement (or at the end of life).

To provide information about the relative attractiveness of various types of human capital investments, a review of a wide variety of literature is used to assess the likely rate of return to human capital investments of various types (Card 1999; Heckman, LaLonde, and Smith 1999; Psacharopoulos and Patrinos 2004; Dougherty 2005; Heckman 2006; Heckman and LaFontaine 2006; Heckman, Lochner, and Todd 2006; Huffman and Evenson 2006a; Henderson, Polachek, and Wang 2011). In Table 33.1, this information is grouped into four categories: (1) extremely high, (2) high, (3) medium to low, and (4) other highly variable. The extremely high category includes the improved gestation environment of babies, pre-school programs of disadvantaged

Table 33.1 Likely inflation-adjusted returns to human capital investments

	Type of human capital investment
Extremely high (>25%)	
	Improved prenatal environment for babies
	Pre-school programs for disadvantaged children (especially development of social skills)
	Elementary school diploma
High (10–25%)	
	High school diploma
	Four-year college degree (BA/BS)
	Advanced degree I (MA/MS)
	Advanced degree II (PhD, MD, JD, DD)
Medium to Low (0–9%)	
	Some high school
	Some college or two-year college degree (AA/AS)
	General equivalency diploma (GED certificate)
	Job training (O-J-T)
	Job training (Job Training Partnership Act)
Other: Highly variable	
	Migration
	Information, including agricultural extension, which is actually quite high

children, and an elementary school diploma. The high category includes a high school diploma, four-year college degree, and advanced degrees I (master's level) and advanced degrees II (PhD, MD, JD, DD). Investing in migration and information seem likely to have highly variable returns.

4 Globalization and the changing world labor market

The occupational distribution of the U.S. labor force changed between 1950 and 2005, as did wage rate growth. Table 33.2, adapted from Autor and Dorn (2013), shows (Panel A) the distribution in 1950, 1980, and 2005 by six major occupational groups, arrayed by skill level, as represented by the mean wage in 1980. Over this 55-year period, relatively steady

Table 33.2 Levels and changes in employment shares and mean real log hourly wage rates by major occupation groups, 1950–2005: occupations ordered by average wage level in 1980

Job type	Level			Percent growth (growth per 10 years)	
	1950	1980	2005	1950–1980	1980–2005
Panel A. Share of employment					
Managers/professional/ technicians/ finance/ public security	22.3	31.6	40.9	41 (13.8)	30 (11.9)
Production/craft	5.1	4.8	3.0	−5 (−1.8)	−38 (−15.1)
Transportation/construction/ mechanics/mining/farm	29.2	21.6	18.2	−26	−15
Machine operators/ assemblers	12.6	9.9	4.6	−21 (−7.0)	−54 (−21.5)
Clerical/retail sales	22.2	22.2	20.4	10 (3.4)	−8 (−3.3)
Service occupations	10.7	9.9	12.9	−7 (−2.3)	30 (11.9)
Panel B: Mean hourly wage ($/2004)					
Managers/professional/ technicians/ finance/ public security	11.21	16.95	21.10	61 (20.48)	31 (12.5)
Production/craft	9.39	15.64	15.18	50 (16.8)	−3 (−1.2)
Transportation/construction/ mechanics/mining/farm	7.69	13.60	13.87	57 (18.9)	2 (0.9)
Machine operators/ assemblers	7.69	11.94	12.70	44 (14.7)	6 (2.3)
Clerical/retail sales	7.39	11.25	13.46	42 (14.1)	18 (7.3)
Service occupations	4.39	8.17	9.58	62 (20.7)	16 (6/4)

Source: Adapted from Autor and Dorn (2013).

growth in the share of employment occurred in occupations at the top of the skill distribution (i.e., managers/professionals/ technicians/finance/public safety) and at the bottom of the skill distribution (i.e., local personal service occupations). However, a steady decline (or no change) in the employment shares of the four intermediate skilled occupations occurred. The decline is most pronounced for the sub-period from 1980 to 2005, when the employment-share in all four of the intermediate-skilled occupations showed a decline, for example, 21.5 percent per decade for machine operators/assemblers and 15.1 percent for the production/craft occupation. At the highest and lowest skilled occupation groups, the employment-share increased by 11.9 percent per decade over this 35-year period.

Between 1950 and 1980, real wage rate growth occurred across all six occupation groups, ranging from 14 to 21 percent, being highest at the top and bottom of the skill distribution. However, over the sub-period from 1980 to 2005, wage growth was most pronounced at the top and two bottom occupations in the skill distribution (i.e., services occupations and clerical/retail sales). The real wage actually declined by 1.2 percent per decade for workers in the production/craft occupational group.

Autor and Dorn (2013) argue that service occupations at the bottom of the occupational skill distribution comprise the new "manual labor." This type of work requires physical dexterity and flexible interpersonal communication skills. These occupations include food service workers, security guards, janitors, gardeners, cleaners, home health aides, childcare workers, hairdressers and beauticians, and recreation occupations. The services associated with these occupations are demanded locally and, hence, must be supplied locally. These service jobs are not subject to international competition and are slightly affected by new computer and information technologies.

At the top of the occupational distribution are high-education goods and service production occupations. In these jobs, new computer and information technologies have been complementary with high-educated labor where data analysis is key to good decision making, such as for professionals and managers. Losers during the past 25 years have been those who were employed in low-education, goods-producing occupations such as production and craft occupations; operative and assembler occupations; and transportation, construction, mechanical, mining, and farming occupations. Autor and Dorn (2013) argue that these are occupations where computerization of routine tasks has occurred, such as in repetitive production and monitoring activities, and bookkeeping and clerical work. Workers in production occupations have also faced growing international competition from low-wage production workers in developing countries. In some cases, assembling durable goods requires several cross-border transfers of parts or partially assembled durables, as is the case for automobile assembling in Mexico using U.S. or Japanese technology.[10]

Globalization of the markets for many goods and financial assets has implications for the U.S. labor market of the future as well as for future human capital investment opportunities. Manufactured goods are highly transportable; hence, jobs in manufacturing are especially vulnerable to intra-country and international relocation with competition. New evidence, however, suggests that both manufacturing jobs and some service jobs may be at risk to international competition (Jensen and Kletzer 2005).

Imports from low-income countries were the fastest growing components of U.S. trade from 1972 to 1997, increasing more rapidly than aggregate imports. As more U.S. trade barriers are eliminated, Bernard, Jensen, and Schott (2006) show that low-wage countries like China and India will export to the United States many of the more labor-intensive products formerly produced domestically. After China joined the World Trade Organization in 2000, the United States conferred permanent normal trade relations on them. The policy change is notable for it seemed to eliminate the possibility of future tariff increases and the uncertainty with which they were associated. With this reduction in uncertainty, China rapidly accelerated its exports of manufactured goods to the United States (Pierce and Schott 2016). However, in April 2018 the U.S. has attempted to levy new tariffs against China, which then brought retaliation. The outcomes from these new actions remain to be determined.

This so-called product cycling – where the U.S. firms and workers move out of labor-intensive products like clothing and shoes as lower-cost developing counties move in – is a key feature of endowment-driven trade theory. Given the higher relative wages in the United States, it is virtually impossible for U.S. firms to earn profits producing labor-intensive goods. As a result, industries like apparel, footwear, and leather goods more generally have all but disappeared, while more skill-intensive and capital-intensive sectors, such as instruments and software creation, thrive here.

However, there are multiple margins of adjustment to low-wage country imports (e.g., exit and product upgrading). Labor-intense plants are relatively more susceptible to low-wage country imports than are capital-intensive and skill-intensive plants in the same industry. As a result, within-industry activity should shift toward relatively capital-intensive and skill-intensive plants. It is important to focus on low-wage country import penetrations, such as from countries that have per capita GDP that is less than 5 percent of the U.S. level. This attention to where imports originate is motivated by the factor proportions framework (capital–labor, capital–skilled labor, or capital–unskilled labor ratios) and allows for a cleaner test of the influence of comparative advantage than aggregate import penetration, which treats imports from high-wage and low-wage countries symmetrically (Findlay 1995).

Figure 33.5 displays the association between the U.S. low-wage country import shares and the U.S. industry average annual wage for disaggregated U.S. industries. Low-wage U.S.

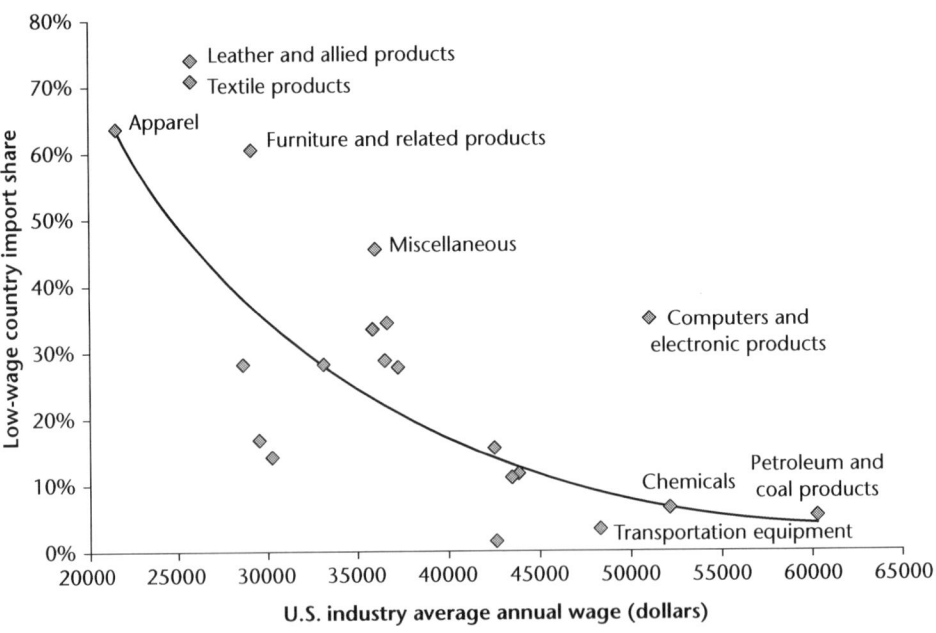

Figure 33.5 Low-wage U.S. industries face low-wage country import competition by U.S. industry average annual wage ($s) (NAICS classes 31, 32, and 33)

Source: Jensen and Kletzer (2008).

manufacturing jobs can be classified as those that pay an annual wage of less than $40,000 (or roughly a wage of $20 per hour in 2006) and high-wage jobs pay $40,000 or more. Figure 33.5 shows low-skill, low-wage, labor-intensive activities in the manufacturing sector face high levels of low-wage country import competition for apparel, leather and allied products, textile products, furniture and related products, and miscellaneous products (which include toys). In contrast, high-wage, high-skill industries face low competition from low-wage countries, as reflected in the low-wage country import share for transportation equipment, chemicals, and petroleum and coal products. An outlier to this trend is the computer and electronic equipment industry, which has a high average wage and relatively high low-wage country import competition. This exception is most likely due to the increased fragmentation of consumer electronics production, where the underlying components, like semiconductors, are high-wage, high-skill activities produced in the United States and shipped to China for low-wage, labor-intensive assembly (Jensen and Kletzer 2008).

U.S. manufacturing plants seem to adjust to international competition from low-skilled, low-wage countries in three dimensions. At the industry level, exposure to low-wage country imports is negatively associated with plant survival and employment growth (Bernard, Jensen, and Schott 2006). Within industries, the higher the exposure of the industry to low-wage country imports, the larger is the relative performance difference between capital-intensive and labor-intensive plants. Moreover, a positive association exists between exposure to low-wage country imports and industry switching. Plants that switch industries shift into industries that have less exposure to low-wage country imports and greater capital-intensity and skill-intensity than the industries they left behind. In manufacturing, it is the low-wage, labor-intensive industries like apparel that are most vulnerable to low-wage import competition. The United States continues to have a strong export presence in high-wage, skill-intensive manufacturing industries. These results support the view that U.S. manufacturing is moving away from comparative disadvantage activities and toward comparative advantage industries via exit, growth, and industry switching.

Some U.S. industries remain quite competitive, having large exports per worker. Figure 33.6 shows the association between U.S. exports per worker in manufacturing and U.S. industry average annual wage. It confirms the story from the import competition graph – low-wage, low-skilled U.S. industries export little per worker (e.g., apparel, textile products, leather and allied products, and furniture and related products). However, it also shows that in high-wage, high-skilled U.S. industries, exports per worker are high (e.g., transportation equipment, computer and electronics, and petroleum and coal products). Summing up, while lower-paying, labor-intensive U.S. industries face intense international competition from low-wage, labor-abundant countries, the United States continues to have a comparative advantage in high-wage, capital- and technology-intensive manufacturing, and workers are doing well there.

A new direction in potential international competition in services includes both U.S. imports and U.S. exports. Some services require face-to-face interactions (e.g., haircuts, legal counseling, and medical treatments), while others do not (e.g., accounting, architectural services, software publishing, securities and commodity trading, and research and development [R&D]). Occupational groups with low employment shares in tradable activities require a physical presence to deliver them (e.g., education, health care practitioners, health care support workers, food preparers, and janitorial workers).

Jensen and Kletzer (2008) present new evidence on the potential tradability of services. They argue that good information on whether a service has the potential to be internationally tradable can be gleaned from evidence on intra-country tradability of services. For example, many service activities (e.g., movie and music recording production, securities and commodities

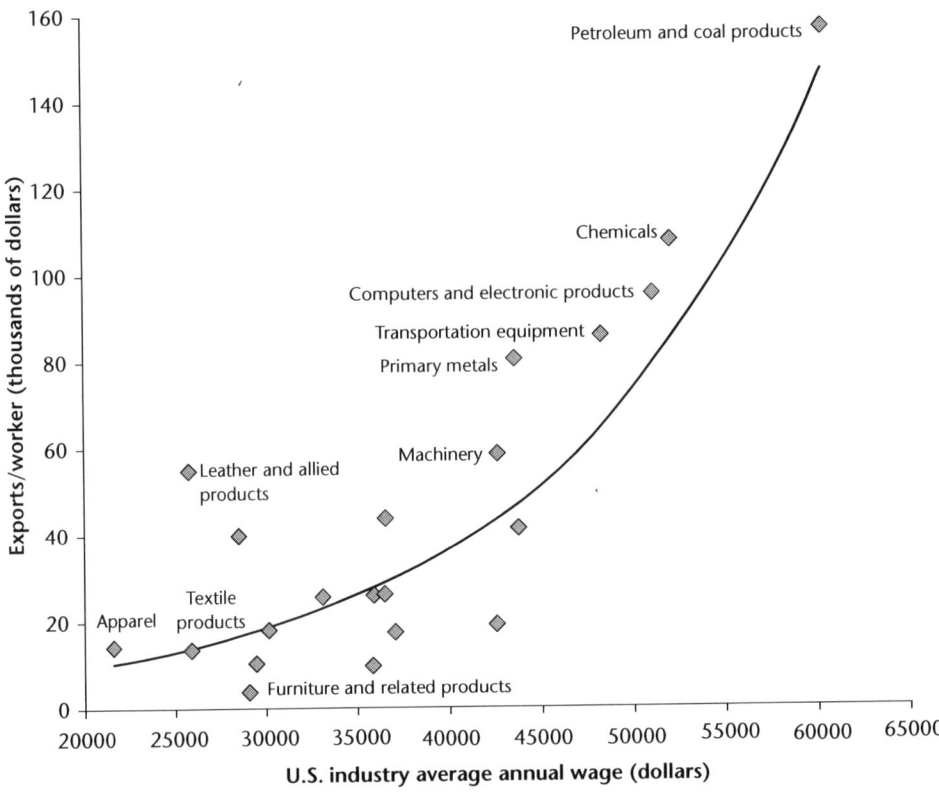

Figure 33.6 Exports per worker in U.S. manufacturing rise with wages (NAICS Classes 31, 32, 33)
Source: Jensen and Kletzer (2008).

trading, software and engineering services, and air-travel plan reservations) appear to be tradable both within the United States and internationally. Those service activities that require face-to-face interactions are far less likely to be domestically or internationally tradable. The share of U.S. employment that is in tradable professional services is 13.7 percent, which is larger than the share of employment in tradable manufacturing industries of 12.4 percent. Some big service sectors, such as education, health care, personal services, and public administration, do have low shares of employment in tradable industries. Also, a relatively small share of employment is in tradable retail, wholesale trades, and agriculture. When workers in tradable occupations (e.g., computer programmers, the retail banking industry, or medical transcriptionists in the health care industry) in nontradable industries are included, the share of the U.S. workforce in tradable service activities is even higher (Jensen and Kletzer 2008).

While many services appear tradable, Jensen and Kletzer (2008) suggest that only about one-third of U.S. jobs in these activities will face meaningful competition from low-wage countries or risk being offshored in the next decade. Tradable service jobs, such as those in engineering or research and development firms, are good jobs. Workers in tradable service activities have higher than average earnings. Part of this premium is due to these workers having high educational attainment, but even controlling for differences in education and other personal characteristics, workers in tradable service activities have 10 percent higher earnings.

Within the set of professional service industries, a worker in a tradable industry and a tradable occupation has earnings almost 20 percent higher than similar professional service workers in a nontradable industry or occupation.

High earnings in tradable service activities do not mean that these jobs will be "lost" to low-wage countries. High-wage, high-skill activities are consistent with U.S. comparative advantage. The United States continues to export high-wage, high-skill business services like computer software publishing, satellite telecommunications services, and integrated record production and distribution (Figure 33.7). Most issues about offshoring focus on the jobs that might be lost but neglect to emphasize that the United States has comparative advantage in many service activities. Jensen and Kletzer (2008) suggest that increased exports of services are likely to benefit many U.S. firms and high-skilled workers in the future. They suggest that at least two-thirds of tradable business service jobs are skilled enough to be consistent with U.S. comparative advantage. Many U.S. service workers and firms are likely to be beneficiaries of increased trade in services through increased export opportunities.

However, it is important to emphasize that many impediments exist to trade in services, ranging from language and cultural differences to regulatory and technical barriers. These impediments are likely to protect U.S. firms and service workers from import competition but

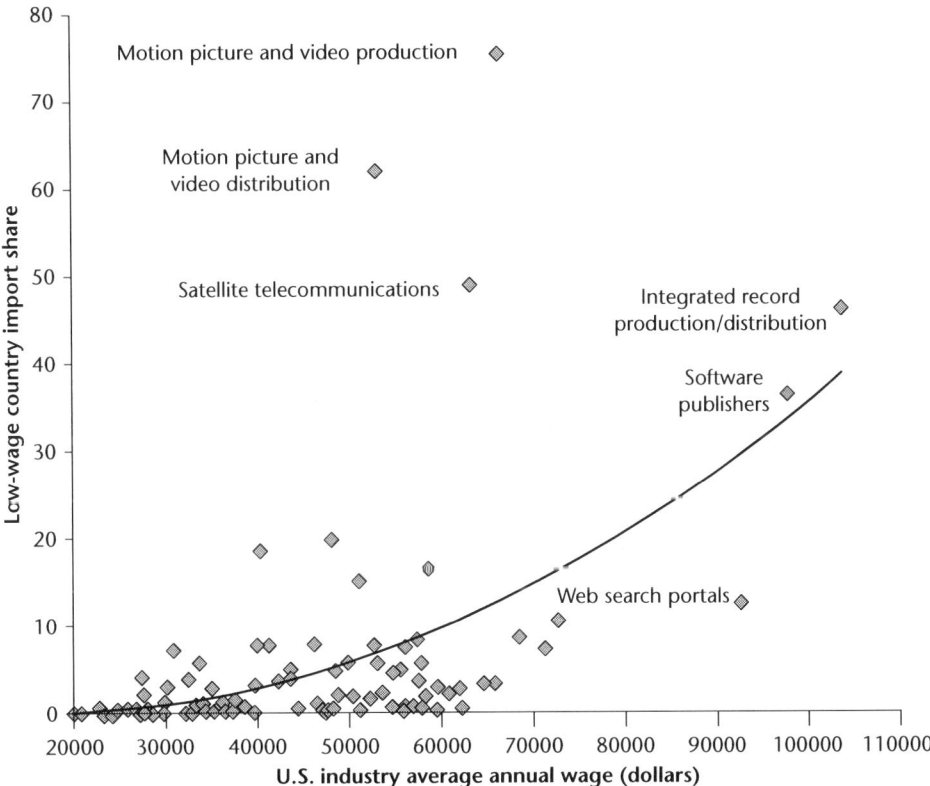

Figure 33.7 U.S. exports per worker in business services rise with industry wages (NAICS Classes 51, 54, 56)

Source: Jensen and Kletzer (2008).

are also likely to impede U.S. firms and service workers from rapidly growing exports. These impediments reduce the gains to the United States from trade in services and the increased living standards that could result. If harmonization of regulations and expanding mutual recognition of professional standards and accreditation could occur, the future potential of increased benefits of trade in services could develop over the next decade.

In summary, between 1998 and 2004, all U.S. industries experienced a major downturn in employment by 13 percent in nontradable agriculture, mining, and manufacturing, and 23 percent in tradable agriculture, mining, and manufacturing. In contrast, service sector employment increased by 10 percent in nontradable and 13 percent in tradable services. Thus, tradable manufacturing industries experienced large losses relative to nontradable manufacturing, but tradable service industries had employment growth similar to nontradable service industries in the United States.

Trade-affected displaced workers are legally entitled to largely uncapped support, including training, but they are a small minority among all displaced workers in the United States. Most displaced workers are from manufacturing (and a few from agriculture). However, displaced workers who qualify can receive both training and long-term unemployment benefits; in addition, workers over age 50 can opt for a wage supplement to cover 50 percent of the wage loss in a new job over a two-year period (OECD 2016).

5 Who pays for a college education in the United States?

Who pays for a college education in the United States? Sallie Mae (the nation's leading provider of paid-for-college programs) and Gallop conducted one of the first surveys of college students and parents of students 18 to 24 years of age (Sallie Mae 2016). The reference was the 2015–2016 academic school year. The survey asked about the total cost of college tuition and related expenses (tuition, books, fees, and room and board) and the method by which parents and students pay for college. The estimated average cost of a year at college was $23,289 for four-year state universities and $41,762 for four-year private colleges and universities (Figure 33.8). At four-year public universities, parents' income and savings account for 32 percent, scholarships and grants for 29 percent, student borrowing for 17 percent, student income and savings for 12 percent, parent borrowing for 7 percent, and money from other relatives and friends for 4 percent of the annual cost for education. Hence, for students attending four-year public schools, the student and their parents paid on average $16,603, or 71 percent of the cost. For students attending four-year private schools, the student and parents paid on average $24,580, or 59 percent of the cost.

In contrast, at four-year private universities, scholarships and grants account for 41 percent, parents' income and savings for 26 percent, students' borrowing for 11 percent, parent borrowing for 7 percent, and money from other relatives and friends for 5 percent of the annual cost for education. The study shows that in addition to both parents and students sharing the costs of a year of college, scholarships and grants have become more important over time.

Public universities receive a subsidy from state government appropriations to support undergraduate education and other teaching, research, and outreach activities. Using a Delta Project Report (Desrochers and Hurlburt 2016), the estimate of state government subsidies to public four-year universities for undergraduate education was $6,757 per full-time equivalent student in the 2013 academic year, or the average total cost of a year of college education at a four-year public institution was roughly $30,046 ($23,706 + $6,757). This is an average state government subsidy rate to public universities of roughly 22 percent.[11]

Investing in people

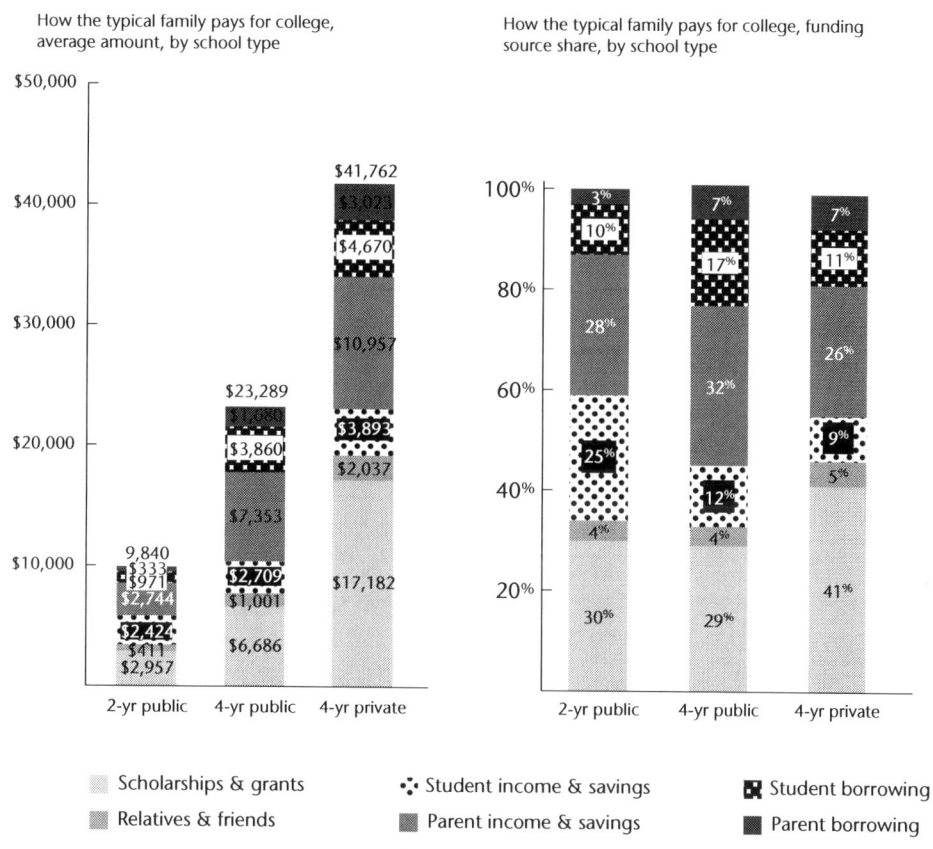

Figure 33.8 How families pay for college: funding source shares by school type, 2015–2016

The size of this state government subsidy as a percent of total costs has been approximately unchanged since 2007.

6 Land-grant universities

In the United States, land-grant universities are special among public universities. They were established by the Morrill Act of 1862 for the teaching of agricultural and mechanical arts to common people. The needs of farmers (and families) across the United States for new scientific knowledge to help them compete and prosper led to the Hatch Act of 1887, which provided federal funding for state agricultural experiment stations to undertake agriculture and home economics research. Although early financial support was mainly federal, state governments later assumed the majority funding role, which was natural given their emphasis on applied and basic research to assist local agriculture (and families). The establishment of a federal–state extension service was aided by the Smith–Lever Act of 1914. Hence, the primary structure of land-grant universities was established by the early twentieth century (Goldin and Katz 1999; Huffman and Evenson 2006b).

In the first half of the twentieth century, a growing supply of high school graduates from families with modest means were produced by the high school movement, and some of them chose to attend college (Goldin and Katz 1997). During the post–World War II period, a major transformation of the land-grant system resulted in an expansion in scale (size) and scope (number of specialized departments and professional schools). Land-grant universities became major centers of complementary research and teaching activities, where faculty trained for research and advancing the state of knowledge also engaged in undergraduate and graduate education. The mass exodus of people out of agriculture between 1948 and 1970, most with high school diplomas, provided a growing demand for college education. Public universities responded by offering an increasingly diverse range of undergraduate majors and degrees. Throughout this era, undergraduate students were primarily in-state students, many of whom were seeking a college education so they could continue to live and work in their home state. Abundant evidence exists that state governments were willing to provide large subsidies to public institutions of higher education during these years because it directly benefited local agriculture, business, and natural resource development and the citizens of the state.[12]

Jin and Huffman (2016) show that investments in U.S. public agricultural research and extension were a major factor explaining the increase in state agricultural productivity (and output) between 1970 and 2004 in the United States. Moreover, they show that the real internal rate of return from within-state and spillover effects of public agricultural research was 67 percent and for within-state extension was 100 percent during this period. However, the stock of public agricultural research peaked in 2005, and then started a slow decline (Huffman 2016). With a total research lag of 33 years, it is impossible to quickly make up for past underfunding of public agricultural research.

Just and Huffman (2009) summarize other changes facing public universities. College students have become more mobile, looking across state borders for the best educational deal. Even if they attend the local land-grant university, a large share of them expect to take jobs in other states. The U.S. Department of Agriculture (USDA) has dramatically reduced its block grant funding of agricultural research and extension, and scientists have been increasingly encouraged to seek competitive grant funding at the federal level (National Institute of Health, National Science Foundation, and U.S. Department of Agriculture.). This means that U.S. states can expect to capture a smaller share of the future benefits from the education of their graduates and the discoveries of their scientists. Hence, as state governments reduce their appropriations for land-grant and other public universities, public universities are raising tuition rates at a relatively rapid rate to cover a larger share of these lost funds for student education.

Where are we headed? One consequence of budget realities is that public land-grant universities are transformed into mixed public–private universities. The components of the university that provide relatively large within-state benefits, such as colleges of agriculture with their experiment stations and extension services or colleges of education may justify large state subsidies, including low tuition rates. Other units that provide training for degrees that are similar to those offered by non-land-grant public and private universities would charge higher tuition rates – rates comparable to private universities with similar quality degree offerings. Research faculty are being asked to seek grant funds from outside of state sources.

Given that the benefits of college education and research discoveries undertaken in any given state extend increasingly beyond its boundaries, regional groups of states or perhaps the federal government might take new responsibility for raising resources and allocating them to instruction and research in public universities. This could be structured so that the cost burden of instruction and research are more closely tied to the area(s) receiving the benefits, including positive externalities. This is the principle of "fiscal equivalence" from public economics which has been proposed by Mancur Olson (Olson 1969, 1986).

There is the option of land-grant and other public universities continuing on their path of slowly being converted into private universities and eventually receiving insignificant state financial support. A short menu of options for land-grant universities in the future is provided in this chapter — some which seem better than others for the long term. However, the route followed will significantly impact investments in people in the twenty-first century.

7 Conclusions and recommendations

T.W. Schultz was an incredibly insightful man — judging ideas and people with unusual expertise. His idea of human capital, investing in people, has steadily expanded to an increasing array of activities, ranging from modern economic growth and development to the economics of households and other non-market activities. However, the United States ranks tenth among 30+ OECD countries in the share of its population ages 25 to 34 years old who have completed at least two years of college education (Figure x.2). It also ranks fourth among countries having the smallest increase in the share of this age group over the past 25 years. With 66 percent of the U.S. labor force having at least a two-year college degree, and 83 percent of the labor force employed in the service sector, many individuals in the United States are working in low-skilled or medium-skilled services with less than a college degree (e.g., those in local service jobs).

Land-grant universities have a very special place in the training of undergraduate and graduate college students and in undertaking the work of advancing the frontiers of knowledge in many areas, including agriculture and life sciences, human sciences, and engineering. These are high-skill, education-intensive activities and cannot be successfully undertaken by part-time faculty or faculty that spend 100 percent of their time teaching. At the start of the twentieth century, major universities started to make scientific discoveries or research an important part of university jobs for professors. This feature distinguishes U.S. universities from those in many other parts of the world. New institutional funding mechanisms for instruction, research, and extension are needed for U.S. land-grant universities.

With technological change eliminating routine jobs and the globalization of standardized goods production eliminating manufacturing jobs, the United States is faced with reduced employment in middle-skilled, goods-producing industries but growth in high-skilled service occupations. Hence, the skills of those losing jobs do not match the skills needed for workers where rapid growth is occurring, i.e., a mismatch exists. Workers losing their jobs face firm- and industry-specific human capital losses, and the stakes for them and their families are high. They are at risk from an unhealthy environment — long-term unemployment, malnutrition, drug use, family troubles, divorce, and suicides. The big challenge is how to save these individuals and redirect them into useful training to enable re-entry into a productive life. Some type of carefully structured re-training and nutrition education programs with counseling might be successful. Otherwise, these displaced unemployed individuals need some type of extended unemployment benefits, possibly with diet and family counseling.[13]

Understanding where the future comparative advantage of the U.S. industries, occupations, and workers lies is important in planning for future educational investments. For example, the U.S. population working in low-skilled, low-wage U.S. manufacturing jobs will face ever-increasing competition from low-wage, unskilled workers in other large countries. Job loss and falling real wage rates are expected to continue in the United States. Young individuals should be advised to pursue college degrees and skills that are complementary with the new technologies and not subject to elimination by its application. Displaced workers need early re-employment support, which is possible only where there are vacant jobs that require skills that are similar to the lost jobs. They also need job-search assistance and low-cost job-search training.

Unfortunately, re-training programs for displaced workers have a bad reputation of being a poor investment of resources (Heckman 2006; OECD 2016), but changes can be made to improve them. First, it is necessary to promote demand-driven training at the local level by matching workers' pre-assessed skills and abilities closely with employers' needs. Second, it is important to expand professional career guidance and stackable training (complimentary training) and skill credentials for displaced workers. Third, it is useful to pinpoint future training needs early on and target training to those most in need. Fourth, a rigorous evaluation of training programs and training innovations is important to eliminate ineffective training and interventions.

New computers and information technology are being included in farm tractors, equipment, and buildings. Auto steer (assist) technology is now available on tractors and combines, where they work well on long straight rows but still require manual turning at the ends of fields. This technology has displaced much of the farm labor previously allocated to routine field operating tasks. Attentive supervision by alert operators is still required. Otherwise, large, expensive farm machinery is at risk of damage. New programs are needed to maintain the mental alertness of these machine operators and supervisors.

Investing in people for the twenty-first century is a very important activity, with important decisions at many levels. The goal of U.S. higher education should be to significantly increase the share of high school graduates who obtain four-year college or advanced degrees that prepare them with skills to undertake non-routine jobs and be complementary with new information technologies.

Notes

1 If the investment can be undertaken at different points in time, or in size of initial investment, or with complementary human capital (e.g., schooling and migration), the investment should be undertaken when the net present value is largest, provided it is positive.
2 Over a 40-year period, starting in 1974, the decline in the price of memory chips is by a factor of about 8,900,000.
3 From 1947 to 2010, information technology–using U.S. industries accounted for 55 percent of value added (Jorgenson, Ho, and Samuels 2014).
4 There was a 0.5 percentage point decline in medium-skilled workers in middle-skilled non-routine jobs.
5 Firm-specific on-the-job training increases labor productivity only in the firm where the training occurs. To reduce labor turnover, it is optimal for the firm and worker to share the benefits and costs of this type of training. However, when the demand for labor is sufficiently large, firms may still lay off their workers with firm-specific training. This results in human capital loss to both the worker and the firm.
6 On average, just over 3 percent of U.S. adult workers ages 20 to 64 with tenure of one year or more are displaced from their job every year (1999–2013). One-half of these displaced workers return to work within one year of job loss (OECD 2016, pp. 32 and 36).
7 It also excluded welfare payments, including food stamps, for illegal aliens.
8 The calculations include the cost of tuition at public university, a work–life after college of 42 years and a real discount rate of 3 percent.
9 Petter, Lyttkens, and Nystedt (2016) report new results from a large study of twins showing that individuals with 13 or more years of schooling have about three years' longer life expectancy at age 60 than those with only ten years of schooling.
10 Pierce and Schott (2016) liken the sharp drop in U.S. manufacturing employment after 2000 to a change in U.S. trade policy that eliminated potential tariff increases on Chinese imports. At the plant level, shifts toward less labor-intensive production and exposure to the policy via input–output linkages also contributed to the decline in manufacturing employment over this time period.
11 Full educational costs are identified as direct instructional costs plus spending for student services and the instructional share of central academic and administrative support. These costs do not include the implicit rental on capital in classrooms or laboratories and the equipment used for teaching.
12 Goldin and Katz (1999) emphasize that a large share of the early engineering graduates was employed by the government sector, and graduates of two-year and four-year colleges were employed primarily

by local school districts. Also, early graduates of colleges of agriculture, veterinary medicine, and the sciences were largely employed in local agriculture.
13 An encouraging report found unemployed individuals enrolling in an intensive 13-week program in computer coding were immediately hired after completing the program.

References

Autor, D. 2014. "Skills, Education, and the Rising Earnings Inequality Among the 'Other 99 Percent'." *Science* 344:843–850.

Autor, D.H., and D. Dorn. 2013. "The Growth of Low-Skill Service Jobs and the Polarization of the U.S. Labor Market." *American Economic Review* 103:1553–1597.

Becker, G.S. 1993. *Human Capital: A Theory and Empirical Analysis* (3rd ed.). New York: National Bureau of Economic Research and The University of Chicago.

Behrman, J.R., and M.R. Rosenzweig. 2004. "Returns to Birthweight." *The Review of Economics and Statistics* 86:586–601.

Bernard, A.B., J.B. Jensen, and P.K. Schott. 2006. "Survival of the Best Fit: Exposure to Low-Wage Countries and the (Uneven) Growth of U.S. Manufacturing Plants." *Journal of International Economics* 68:219–237.

Card, D. 1999. "The Casual Effect of Education on Earnings." In O.C. Ashenfelter and D. Card, eds., *Handbook of Labor Economics* (Vol. 3A). New York: Elsevier, pp. 1802–1864.

Desrochers, D.M., and S. Hurlburt. 2016. *Trends in College Spending: 2003–2013*. Washington, DC: American Institute for Research (Delta Cost Project).

Dougherty, C. 2005. "Why Are the Returns to Schooling Higher for Women Than for Men?" *Journal of Human Resources* 40:969–988.

Findlay, R. 1995. *Factor Proportions, Trade, and Growth*. Cambridge, MA: MIT Press.

Fogel, R.W. 1994. "Economic Growth, Population Theory, and Physiology: The Bearing of Long-Term Processes on the Making of Economic Policy." *American Economic Review* 84:369–395.

Goldin, C., and L.F. Katz. 1997. *Why the United States Led in Education: Lessons from Secondary School Expansion, 1910 to 1940*. Cambridge, MA: National Bureau of Economic Research (NBER).

———. 1999. "The Shaping of Higher Education: The Formative Years in the United States, 1890 to 1940." *Journal of Economic Perspectives* 13:37–62.

Heckman, J.J. 2006. "Skill Formation and the Economics of Investing in Disadvantaged Children." *Science* 312:1900–1902.

Heckman, J.J., and P. LaFontaine. 2006. "Bias-Corrected Estimates of GED Returns." *Journal of Labor Economics* 24:661–700.

Heckman, J.J., R.J. LaLonde, and J.A. Smith. 1999. "The Economics of Econometric of Active Labor Market Programs." In O. Ashenfelter and D. Card, eds., *Handbook of Labor Economics* (Vol. 3A). New York: North-Holland, pp. 1865–2097.

Heckman, J., J. Lochner, and P. Todd. 2006. "Earnings Functions, Rates of Return and Treatment Effects: The Mincer Equation and Beyond." In E.A. Hanushek and F. Welch, eds., *Handbook of the Economics of Education*. Amsterdam. North-Holland, pp. 307–458.

Heckman, J.J., R. Pinto, and P. Savelyev. 2013 "Understanding the Mechanisms Through Which an Influential Early Childhood Program Boosted Adult Outcomes." *American Economic Review* 103:2052–2085.

Henderson, D.J., S.W. Polachek, and L. Wang. 2011. "Heterogeneity in Schooling Rates of Return." *Economics of Education Review* 30:1202–1214.

Huffman, W.E. 2014. "Agricultural Labor: Demand for Labor." In N. Van Alfen, ed., *Encyclopedia of Agriculture and Food Systems* (Vol. 1). San Diego, CA: Elsevier, pp. 105–122.

———. 2016. "New Insights on the Impacts of Public Agricultural Research and Extension." *Choices* 31:1–6.

Huffman, W.E., and R.E. Evenson. 2006a. "Do Formula or Competitive Grant Funds Have Greater Impacts on State Agricultural Productivity?" *American Journal of Agricultural Economics* 88:783–798.

———. 2006b. *Science for Agriculture: A Long-Term Perspective* (2nd ed.). Ames, IA: Blackwell Publishing.

Huffman, W.E., and T. Feridhanusetyawan. 2007. "Migration, Fixed Costs and Location-Specific Amenities: A Hazard Rate Analysis for a Panel of Males." *American Journal of Agricultural Economics* 89:368–382.

Huffman, W.E., and P.F. Orazem. 2007. "Agriculture and Human Capital in Economic Growth: Farmers, Schooling, and Nutrition." In R. Evenson and P. Pingali, eds., *Handbook of Agricultural Economics: Agricultural Development: Farmers, Farm Production, and Farm Markets.* New York: Elsevier Science, pp. 2281–2342.

Jensen, J.B., and L.G. Kletzer. 2005. *Tradable Services: Understanding the Scope and Impact of Service Outsourcing.* Washington, DC: Peterson Institute for International Economics.

———. 2008. *"Fear" and Offshoring: The Scope and Potential Impact of Imports and Exports of Services.* Washington, DC: Peterson Institute for International Economics.

Jin, Y., and W.E. Huffman. 2016. "Measuring Public Agricultural Research and Extension and Estimating Their Impacts on Agricultural Productivity: New Insights from U.S. Evidence." *Agricultural Economics* 47:15–31.

Jorgenson, J.W. 2009. "Introduction." In D.W. Jorgenson, ed., *The Economics of Productivity.* Northampton, MA: Elgar, pp. ix–xxviii.

Jorgenson, D.W., M.S. Ho, and J.D. Samuels. 2011. "Information Technology and U.S. Productivity Growth: Evidence from a Prototype Industry Production Account." *Journal of Productivity Analysis* 36:159–175.

———. 2014. "What will Revive U.S. Economic Growth? Lessons from a Prototype Industry-Level Production Account for the United States." *Journal of Policy Modeling* 36:674–691.

Just, R.E., and W.E. Huffman. 2009. "The Economics of Universities in a New Age of Funding Options." *Research Policy* 38:1102–1116.

Keng, S.H., and W.E. Huffman. 2007. "Binge Drinking and Labor Market Success: A Longitudinal Study on Young People." *Journal of Population Economics* 20:35–54.

Laing, K. 2011. *Labor Economics.* New York: W.W. Norton.

OECD. 2016. *Back to Work: United States: Improving the Re-employment Prospects of Displaced Workers.* Paris: Organisation for Economic Co-operation and Development (OECD).

Olson, M. 1969. "The Principle of Fiscal Equivalence: The Division of Responsibilities among Different Levels of Government." *American Economic Review* 59:479–487.

———. 1986. "Toward a More General Theory of Government Structure." *American Economic Review* 76:120–125.

Petter, L., D.H. Lyttkens, and P. Nystedt. 2016. "The Effect of Schooling on Mortality: New Evidence from 50,000 Swedish Twins." *Demography* 53:1135–1168.

Pierce, J.R., and P.K. Schott. 2016. "The Surprising Swift Decline of US Manufacturing Employment." *American Economic Review* 106:1632–1662.

Psacharopoulos, G. 1994. "Returns to Investments in Education: A Global Update." *World Development* 22:1325–1343.

Psacharopoulos, G., and H.A. Patrinos. 2004. "Returns to Investment in Education: A Further Update." *Education Economics* 12:111–134.

Rosenzweig, M.R., and T.P. Schultz. 1983. "Estimating a Household Production Function: Heterogeneity, the Demand for Health Inputs, and Their Effects on Birth Weight." *Journal of Political Economy* 91:723–746.

Ryan, C.L., and K. Bauman. 2016. *Educational Attainment in the United States: 2015.* Washington, DC: U.S. Census Bureau.

Sallie Mae. 2016. *How America Pays for College: 2016.* Newark, DE: Sallie Mae.

Schultz, T.W. 1961a. "Investing in Human Capital." *American Economic Review* 51:1–17.

———. 1961b. "Investing in Human Capital: Reply." *American Economic Review* 51:1035–1039.

U.S. Census Bureau. 2009. *Educational Attainment in the United States: 2007.* P20–560. Washington, DC: U.S. Department of Commerce.

U.S. Department of Labor. 2016. *Employment Status of the Civilian Population 25 Years of Age of Over by Education Attainment, 2016.* Washington, DC: U.S. Department of Labor.

34

The economics of biofuels

Farzad Taheripour, Hao Cui, and Wallace E. Tyner

1 Introduction

Biofuels are liquid fuels primarily used to power transportation vehicles. Unlike conventional transportation fuels produced from fossil fuels, biofuels are derived directly from biogenic materials. In light of the distinction in the sources of biomass and adopted technologies, biofuels are generally classified into conventional and advanced biofuels.

Conventional biofuels, also referred to as first generation biofuels, have well-established technologies and processes that produce biofuels on a commercial scale. These biofuels typically include ethanol produced from starch-bearing grains such as corn and wheat; ethanol produced from sugarcane and sugar beets; and biodiesel produced from rapeseed, soybeans, palm oil, and other oilseeds and animal fats.

While easily produced from accessible feedstock, first generation biofuels compete directly with food production since these biofuels need crops as their main feedstock. The production of conventional biofuels from food-feed crops, among other factors, raised crop prices in the second part of the 2000s, and that triggered a major debate on the "food versus fuel" tradeoff. Production of conventional biofuels has caused environmental concerns due to the potential negative effects on water availability and quality, biodiversity, land degradation, and greenhouse gas (GHG) emissions. Also, producing biofuel feedstocks may require more cropland to be converted from pasture and forest, which have a greater capacity for carbon sequestration than cropland.

Unlike conventional biofuels, which mostly come from crops, advanced biofuels, including second and third generation biofuels, use different feedstocks. Second generation biofuels are made from cellulosic feedstocks, such as crop and forest residues and dedicated energy crops. Third generation biofuels are derived from algae. Advanced biofuels can be produced using land that is not usually used for producing traditional crops, but given that the technologies for producing advanced biofuels are still immature and production costs are very high, advanced biofuels are produced in very small quantities compared with conventional biofuels.

Global biofuel production has experienced fast growth since 2000. In 2014, global biofuel production reached 33 billion gallons, about four times the production in 2004, accounting for 4 percent of the world's road transport fuel in 2014 (International Energy Agency [IEA] 2017). While the production of ethanol and biodiesel both increased over this period, the share of biodiesel rose from about 8 percent in 2004 to about 25 percent in 2014.

Between 2004 and 2014, the global production of biofuels was dominated by Brazil, the European Union, and the United States (Table 34.1). The three regions jointly accounted for 84 percent of global biofuel production in 2014, with the United States alone contributing to nearly half of the global total. In particular, the United States was the largest fuel ethanol producer in the world. Out of the nearly 25 billion gallons of fuel ethanol produced globally in 2014, the United States produced about 14 billion gallons, followed by Brazil at nearly 7 billion gallons of ethanol that year.

The European Union was the largest biodiesel producing region. In 2014, it produced about 3 billion gallons of biodiesel, with the global biodiesel production totaling 8 billion gallons the same year. The surge of biofuels production across the world was largely driven by government policies. The world's dominant biofuel producers (i.e., Brazil, the European Union, and the United States) have biofuel policies in place supporting the development of domestic biofuel industries.

In the United States, the earliest policy that provided subsidies to the biofuel industry was the Energy Policy Act of 1978. Since then, there have been a number of major changes and amendments to U.S. federal government policies in support of the biofuel industry, including varying forms of subsidies such as excise tax exemptions, tax credits, etc. Especially during the mid-2000s, the steep rise in oil prices, growing concerns over energy security and GHG emissions, and the desire to support domestic farm and rural economies combined to reinvigorate support for biofuels (National Research Council [NRC] 2012). Tyner (2008) provided a detailed description of the history of ethanol subsidy policy in the United States. Today, the major federal policy that regulates the biofuel industry in the United States is the Renewable Fuel Standard (RFS). The RFS, established under the Energy Policy Act of 2005, stipulates the minimum amount of each type of biofuel to be consumed in the United States for the years ahead. The RFS was later amended in the Energy Independence Act of 2007, which substantially expanded the biofuel mandates through 2022, when total biofuel production in the United States is required to reach 36 billion gallons ethanol equivalent (Figure 34.1). This mandate also specified the quantitative requirement for the following four groups of biofuels:

1. *Conventional biofuels*: the only category that permits corn ethanol. Legally, there is no mandate for corn ethanol, but this category effectively constitutes the mandate for corn ethanol. This category mandates that new ethanol plants and facilities commencing construction after December 19, 2007, must achieve a reduction of lifecycle GHG emissions of at least

Table 34.1 Biofuel production in 2014 (billion gallons)

Ethanol		Biodiesel	
United States	14.35	European Union	3.13
Brazil	6.59	United States	1.27
European Union	1.35	Brazil	0.91
China	0.70	Indonesia	0.90
Canada	0.47	Germany	0.77
Thailand	0.29	Argentina	0.31
France	0.17	France	0.19
Germany	0.14	Thailand	0.15
Argentina	0.12	Spain	0.15
India	0.11	Poland	0.08
Others	0.46	Others	0.22
World	24.75	World	8.10

Source: EIA (2017).

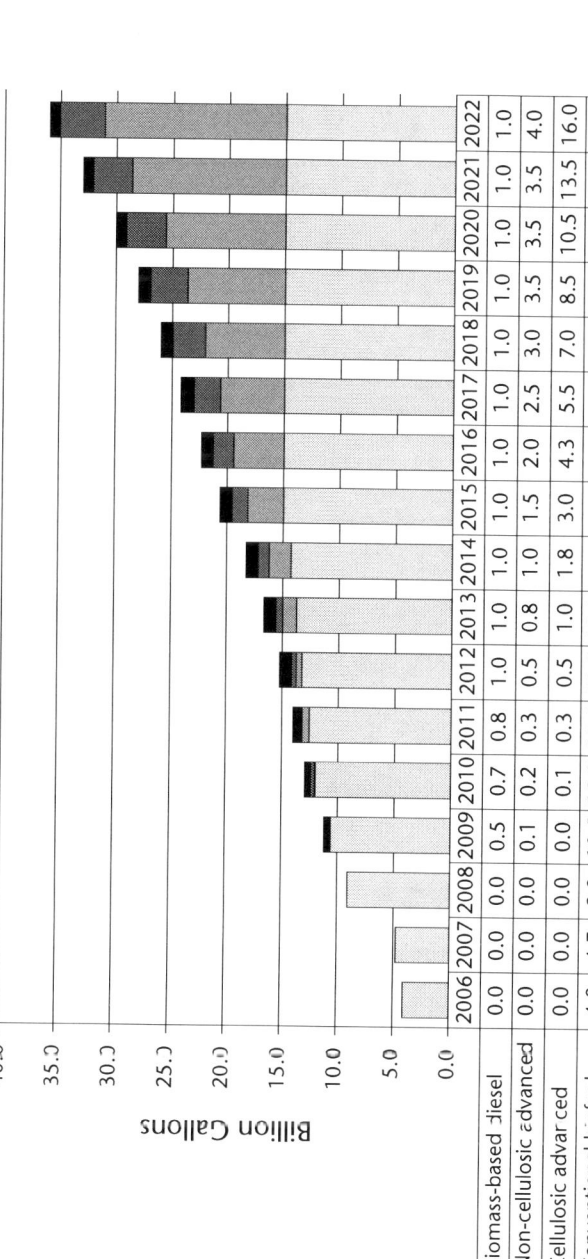

Figure 34.1 Renewable fuel standard 2007–2022 (Tyner 2015)

20 percent compared to petroleum-based gasoline and diesel (2005 baseline) to qualify as a renewable fuel. The RFS also requires this category to reach 15 billion gallons by 2015 and remain at that level.

2 *Cellulosic biofuels*: biofuels derived from cellulosic feedstocks such as corn stover, miscanthus, switchgrass, and some other woody chips and crops that achieve at least a 60 percent reduction of GHG emissions relative to their petroleum-based counterparts; 16 billion gallons ethanol equivalent of cellulosic biofuels are required by 2022.

3 *Biodiesel*: biomass-based diesel uses soybeans and other vegetable oils, waste cooking oil, and animal fats as feedstock. This category is required to reach a 50 percent GHG reduction threshold. The legislated mandate was 1 billion gallons, but the United States Environmetal Protection Agency (EPA) has increased the level over time; in 2017, it was 2 billion gallons.

4 *Other advanced biofuels*: this category encompasses any biofuels other than conventional biofuels. This category may include sugarcane ethanol, cellulosic biofuels, and biodiesel. Biofuels under this category must reach at least a 50 percent reduction in GHG emissions relative to petroleum-based fuels. The overall target for this category is 4 billion gallons ethanol equivalent by 2022.

While the RFS sets up annual targets for different categories of biofuels, the U.S. Environmental Protection Agency may by law review and revise the requirements for each individual year. So the actual volume requirements for a particular year may be different from what was stated in the initial RFS (Tyner 2015).

Brazil used to be the world's leader in ethanol production until it was surpassed by the United States in the mid-2000s. In Brazil, an early policy support for the domestic biofuel industry was provided in the National Alcohol Program (PROALCOOL) in 1975 as a response to the 1973 oil crisis (Goldemberg 2006). This government-run program regulated the use of hydrous ethanol as a stand-alone fuel and anhydrous ethanol as a fuel additive to be blended with gasoline. This program was largely implemented through research and development (R&D) funding, volume mandates, subsidies, guaranteed loans, etc. (Treesilvattanakul 2013). The price control policies were phased out during the 1990s (Moschini, Cui, and Lapan 2012). Some mandates over the production and trade of ethanol were removed between 1996 and 2000, but other policy supports, including tax incentives and other incentives supporting demand, still remained in effect (Janda, Kristoufek, and Zilberman 2012). Today the main policy in Brazil is the blending mandate, which is currently 27 percent ethanol for gasoline and 8 percent biodiesel for diesel (United States Department of Agriculture [USDA] 2016).

The biofuels-related policy in the European Union was designed primarily to meet its obligations under the commitment to the Kyoto targets for GHG emission reductions (Janda, Kristoufek, and Zilberman 2012). The EU biofuel support policies are included in several regulations, with the most notable one being the Energy and Climate Change Package, enacted in legislation in 2009. The package established the so-called 20/20/20 targets. It stipulates that by 2020: a 20 percent reduction in GHG emissions compared to the 1990 level; a 20 percent improvement in energy efficiency relative to the "business as usual" scenario; and a 20 percent share of renewable energy consumption in the European Union's total energy consumption (Dixson-Declève 2012). As part of this legislation, the Renewable Energy Directive (RED) also requires a 10 percent minimum target for renewable energy to be consumed in the transport sector (European Union [EU] 2009). In 2016, a revised RED increased the 20 percent renewable energy share target to 27 percent to be met by 2030 (European Commission [EC] 2016).

The remainder of this chapter proceeds as follows. Section 2 provides an overview of the techno-economic analysis of biofuels. Section 3 discusses the implications of biofuels for indirect land use change and GHG emissions. Section 4 examines the relationship between biofuel prices, fossil fuel prices, and food prices. Finally, Section 5 reviews the welfare implications of biofuels-related policies.

2 Techno-economic analysis of biofuels

The profitability and energy efficiency[1] of converting food-feed crops to biofuels, in particular for corn ethanol, has been a major concern since the 1990s. The early papers in this area showed that producing corn ethanol is not profitable and its net energy contribution is negative (Giampietro, Ulgiati, and Pimentel 1997; Pimentel 1991, 1998). Also, some countries, such as Brazil, expanded their ethanol industries through hidden subsidies (Schmitz, Seale, and Schmitz 2003). Over time, this perception changed with improvements in technology and increases in crude oil prices. Shapouri, Duffield, and Wang (2002) concluded that the energy efficiency of corn ethanol has increased due to technological improvements in the production processes of corn and ethanol. These authors showed that corn ethanol generates 34 percent more energy than it consumes. Others, such as Tyner and Taheripour (2008), showed that the U.S. ethanol industry gained a positive margin during the biofuel boom and when the crude oil prices were growing fast. The profit margin of the ethanol industry has changed frequently. As shown in Figure 34.2, the return over operating costs (a measure that represents profitability) for the ethanol industry followed a downward trend between 2007 and 2009. Then it fluctuated up and down several times between 2009 and 2011 and passed the level of US$1 per gallon at the end of 2011. After that, it dropped to around zero in 2012, and then followed an upward trend to US$2 per gallon in 2104 before dropping to around US$0.50 in 2015.

More recently, this line of research has shifted to a study of feasibility and the profitability of producing second generation biofuels using cellulosic material. Many papers have developed techno-economic analyses for various potential biofuel pathways that can be used to convert forest and crop residues and other types of biomass to biofuels. Some papers (Carriquiry, Du, and Timilsina 2011; Gubicza et al. 2016; Naik et al. 2010; Nigam and Singh 2011; Sims et al.

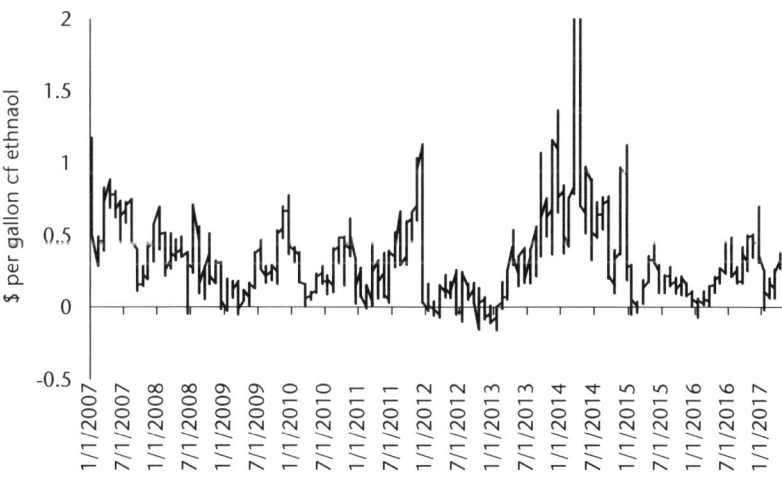

Figure 34.2 Return over operating costs of corn ethanol industry: daily data from January 1, 2007 to April 21, 2017 generated by the Center for Agricultural and Rural Development (CARD), www.card.iastate.edu, Iowa State University, Ames, Iowa

2010) reviewed these studies, compared their results, and concluded that it is still too early to commercially produce biofuels from cellulosic materials at a large scale, even when taking into account the existing Renewable Identification Number (RIN) credits defined in the U.S. RFS. More recently, several papers have developed stochastic techno-economic analysis for several biofuel pathways to take into account the probability of changes in economic and engineering variables (Bittner, Zhao, and Tyner 2015; Zhao, Brown, and Tyner 2015; Zhao, Yao, and Tyner 2016; Bann et al. 2017; Yao et al. 2017).

3 Induced land use change and associated greenhouse gas emissions

Induced land use changes due to biofuel production and policy is an important and controversial issue. There has been growing literature attempting to estimate Induced Land Use Changes (ILUC) and their corresponding emissions for alternative biofuel pathways. In general, two approaches have been followed to accomplish this task. The first approach uses economic models to determine the conversion of natural land (forest and pasture) to cropland due to an expansion in the demand for biofuels (Khanna and Crago 2012). The economic models used in this field usually are combined with emission factor models to convert ILUC to land use emissions. The emission factor models basically trace carbon fluxes per hectare by land type (Plevin et al. 2014; Dunn et al. 2016). The second approach uses biophysical data and estimates the land requirement for biofuel production (Tilman, Hill, and Lehman 2006; Fargione et al. 2008). Regardless of the details, this approach relies on three main variables to calculate the land requirement for each biofuel pathway: magnitude of biofuel production, feedstock yield per unit of land, and conversion rate of feedstock to fuel. This approach assumes that new cropland is needed for each and every drop of biofuel and simply uses these three variables to determine how much land is needed to produce a particular type of biofuel. Similar to the first approach, the second approach also uses the emissions factor models to convert the land requirement to land use emissions.

The economic approach to estimating ILUC typically takes into account what has been called "market mediated responses." In the context of biofuels expansion, market mediated responses may include the following:

- reductions in consumption for biofuel feedstock (e.g., corn) in non-biofuel uses (e.g., in food or feed) and in consumption of other crops due to higher crop prices;
- crop switching to produce more of the biofuel feedstock (e.g., corn) at the expense of producing less of the other crops (e.g., producing more corn instead of wheat, sorghum, barley, oats, etc.);
- conversion of forest or pasture to cropland to meet the need for additional cropland (known as change on the extensive margin);
- intensification in crop production such as yield increase, multiple cropping, and use of existing idle land;
- changes in crop trade reflecting changes in crop production and consumption in the rest of the world.

The economic models incorporating "market-mediated responses" are in contrast with the biophysical approach, which ignore these responses. Given the distinction between the two types of approaches (economics versus biophysical), it can be deduced without difficulty that the projections of biophysical analyses would generally exaggerate the induced land use change due simply to the lack of "market-mediated responses." While most economic models are designed to take into account market mediated responses, the various models that have been used handle these responses differently. Therefore, the extent to which the market-mediated responses contribute to the estimated ILUC for a typical biofuel pathway varies across the economic models used in this field.

In what follows, we review some important papers that used economic models to estimate ILUC and emissions for several biofuel pathways. Among all alternative biofuel pathways, U.S. corn ethanol has been studied most frequently. Hence, we first concentrate on this biofuel, and then we review the existing estimates for other biofuel pathways.

The study by Searchinger et al. (2008) was the first peer-reviewed study introducing the economic mechanism, that is, market-mediated responses, to estimate land use change due to increased U.S. ethanol production. Using a partial equilibrium model,[2] they projected that about 0.73 hectares of new cropland are needed to produce 1,000 gallons of U.S. corn ethanol, and that generates 100 g CO_2e/MJ emissions. However, by highlighting several biophysical variables and conditions implicitly embedded in this chapter, it can easily be shown that the same amount of new cropland is needed to produce 1,000 gallons of corn ethanol if we simply assume the following: (1) 2.7 gallons of ethanol per bushel of corn, (2) corn yield of 146.8 bushels per acre (average of U.S. corn yield in 2000–2013), (3) no switching across crops, (4) no change in U.S. exports of corn or other crops, and (5) returning one-third of corn used for ethanol as DDGS (Distiller's Dried Grains with Solubles) to livestock industry (Taheripour, Cui, and Tyner 2017). Therefore, the induced land use change calculated from this simple biophysical way, which ignores market-mediated responses, does not differ much from what was estimated by Searchinger et al. (2008).

This implies that while being equipped with an economic mechanism, Searchinger et al. (2008) implicitly assumed no market-mediated responses. Among all missing market-mediated responses, as Gohin (2014) pointed out, Searchinger et al. (2008) assumed no crop yield effects. Different yield assumptions may lead to huge differences in ILUC and associated emissions across modeling results. To clarify, this does not amount to claiming that a partial equilibrium framework cannot explain yield effects. Mosnier et al. (2013), using a global partial equilibrium (PE) model – The Global Biosphere Management Mode (GLOBIOM) – demonstrated that ILUC and GHG emissions due to U.S. RFS could change significantly if the assumed yield growth rate has been changed. Several studies using either a PE or a computable general equilibrium (CGE) framework have also shown that when yield response was introduced, ILUC and GHG emissions could reduce substantially (Hertel et al. 2010; Dumortier et al. 2011; Golub and Hertel 2012; Broch, Hoekman, and Unnasch 2013).

To show the extent to which market-mediated responses can affect the estimated ILUC, here we briefly review the estimated ILUC obtained from a CGE model, Global Trade Analysis Project-Biofuels (GTAP-BIO), which has been developed and frequently modified to take into account market-mediated responses according to historical observations. We begin with Hertel et al. (2010), which showed that if we take into account reductions in crop demands (including corn) in non-ethanol uses, switching among crops, improvements in crop yields, expansion in supply of DDGS (by-products of ethanol), and competition among economic sectors for limited resources all induced by market forces, then the land requirement for producing 1,000 gallons of U.S. corn ethanol drops to 0.29 hectares, and that generates 27 g CO_2e/MJ emissions. These numbers are 60 percent and 73 percent smaller, respectively, than the corresponding numbers calculated by Searchinger et al. (2008). While Hertel et al. (2010) showed that market-mediated responses could alter the estimated ILUC for corn ethanol, the CGE framework employed by these authors assumed that there is no idled marginal land and that multiple cropping does not exist. In addition, due to the lack of data, they made some simplifying assumptions to determine demand for new cropland land and to allocate land among its alternative uses. In particular, they assumed that: the productivity of new cropland is two-thirds the productivity of existing cropland across the world; land can move across alternative uses (forest, pasture, and cropland) with uniform land transformation elasticities of 0.2 across the world; and cropland can move across alternative crops with uniform land transformation elasticities of 0.5 across the world.

Taheripour and Tyner (2013a, b) and Taheripour et al. (2012) altered these assumptions according to actual observations by (1) estimating the productivity of new cropland versus new existing cropland using a terrestrial ecosystem model (TEM) for each region and each agro-ecological zone (AEZ); (2) including additional nests in the land module of the model to take into account the fact that the conversion of pasture land to cropland is easier and less expensive than the conversion of forest; and (3) tuning the land transformation parameters of the model for each region based on observed changes in land use changes across the world. In addition, they obtained data on available cropland pasture[3] for Brazil and the United States and included that in their pool of available cropland for these two countries. With these modifications, the demand for new cropland for producing 1,000 gallons of U.S. corn dropped to 0.11 hectares, with 13.3 g CO_2e/MJ emissions.

Several studies (Cassman 1999; Brady and Sohngen 2008; Alston, Babcock, and Pardey 2010; Alexandratos and Bruinsma 2012; Ausubel, Wernick, and Waggoner 2013; Ray and Foley 2013; Borchers et al. 2014; Byerlee, Stevenson, and Villoria 2014; Lewis and Kelly 2014; Hertel and Baldos 2015) have shown that some of the increase in agricultural output in recent decades can be attributed to intensification, including yield improvement and utilizing existing land more efficiently with multiple cropping and the cultivation of unused cropland. However, intensification activities varied significantly across regions (Siebert, Portmann, and Döll 2010; Babcock and Iqbal 2014). In response to the findings of these papers, Taheripour, Cui, and Tyner (2016) modified the GTAP-BIO model to incorporate multiple cropping; take into account the fact that, across the world, idled land could return to crop production in response to biofuel production; and consider heterogeneity in crop yields with respect to changes in crop prices by region. They showed that with the observed intensification in crop production, the land requirement for U.S. corn ethanol is about 0.05 hectares per 1,000 gallons, and this will generate 8.7 g CO_2e/MJ GHG emissions. These values are significantly lower than all the corresponding values provided in the earlier studies mentioned above.

The above analyses indicate that over time the estimated ILUC emissions for corn ethanol obtained from the GTAP-BIO model have followed a downward[4] trend over time due to (1) the model improvements, which bring more market-mediated responses into account and (2) tuning the model parameters to actual observations.

The model used by Searchinger et al. (2008) and the GTAP-BIO model are not the only models used in this field. Many PE and CGE models were used to estimate emissions due to U.S. corn ethanol and other biofuels. The first column of Table 34.2 shows the results obtained from various models for corn ethanol. This column shows that the most recent estimates for this biofuel pathway are considerably lower than the older ones. The last three columns of Table 34.2 show estimates for land use emissions for several other biofuel pathways. These figures demonstrate a major disparity among modeling results.

While many papers have estimated ILUCs for the first generation of biofuels, only a few papers have provided these estimates for the second generation biofuels (Wu, Wu, and Wang 2006; Hill et al. 2009; Huang et al. 2013; Taheripour, Zhao, and Tyner 2017). According to these papers, converting crop and forest residues to biofuels is expected to generate negligible ILUC. Producing biofuels from dedicated energy crops such as miscanthus and switchgrass on marginal lands needs lots of land. However, these energy crops generate net soil carbon sequestration. This

Table 34.2 Estimated induced land use emissions for several biofuel pathways (g CO_2e/MJ)

Studies	Implemented model	Corn ethanol	Sugarcane ethanol	Soybean biodiesel	Rapeseed biodiesel
Searchinger et al. (2008)	FAPRI	107	–	–	–
Al-Riffai, Dimaranan, and Laborde (2010)	MIRAGE	54.1	17.8	74.5	53
EPA (2010)	FASOM-FAPRI	30.1	3.8	40.7	–
Hertel et al. (2010)	GTAP-BIO	27	–	–	–
Dumortier et al. (2011)	FAPRI	14	–	–	–
Laborde (2011)	MIRAGE	10.3	13.4	55.8	53.8
Taheripour and Tyner (2013b)	GTAP-BIO	13.3	–	–	–
CARB (2015)	GTAP-BIO	19.8	11.8	29.1	14.5
Valin et al. (2015)	GLOBIOM	14	17	150	65
Taheripour, Cui, and Tyner (2016): 2004 database and intensification	GTAP-BIO	8.7	4.7	16.9	15.7
Taheripour, Cui, and Tyner (2016): 2011 database and intensification	GTAP-BIO	12	3.2	18.3	13.7

means that the ILUC emissions of these crops are smaller than their contributions to soil carbon sequestration. Khanna and Crago (2012) and Wicke et al. (2012) have reviewed, examined, and compared a large number of papers that assessed ILUCs and their corresponding emissions for a wide range of biofuel pathways.

4 Biofuels and their price impacts

A great body of literature has addressed the price impacts of biofuel production and policy, providing both theoretical foundations and empirical evidence. This literature covers several different but related subjects. A few papers developed analytical models to explain how biofuel production and policy could affect food and fuel prices over short-run and long-run time horizons. In line with these attempts, many papers used numerical PE and CGE models to project the magnitudes of these price impacts. Some papers examined the extent to which biofuel production and policy could change the nature of price volatility in commodity markets. Many papers conducted research to empirically estimate the price impacts of biofuels using econometric techniques and actual observations. In parallel to these efforts, a few analyses were developed to explain observed historical changes in crop prices and their links to the major events that occurred over time around the world. These studies reviewed the existing estimates in this area, studied the historical changes in crop prices, and used basic economic theory. In turn, we review the major findings of these studies in two subsections. The first one examines the impacts of biofuel production and policy on crop and food prices. The second subsection discusses the effects of biofuel production on fuel prices.

4.1 Impacts of biofuel production and policy on commodity markets

Analytical and numerical analyses developed in this area commonly indicated that converting food-feed crops to biofuels (due to policy intervention or market forces) will increase crop prices, which eventually leads to higher food prices (McPhail and Babcock 2008; Tyner and Taheripour 2008; de Gorter and Just 2009; Gohin and Tréguer 2010; Hertel, Tyner, and Birur 2010; Hochman, Rajagopal, and Zilberman 2010; Beckman et al. 2011; Hertel and Beckman 2011; Taheripour, Hertel, and Tyner 2011; Babcock 2012; Wu and Langpap 2015). However, there is no agreement on the extent to which either biofuel production or policy increases crop and food prices. Hochman et al. (2011) have summarized the estimated price impact of biofuels across the literature and shown that it varies from 3 percent to 75 percent. However, it is extremely difficult to compare the results of the existing projections for several reasons. The existing projections refer to various time horizons of different lengths. Alternative models represent different crop aggregation schemes. Some represent individual crops (e.g., corn, wheat, soybeans, etc.), and some show aggregated crop categories (e.g., coarse grains, oilseeds, fruits, etc.). Several models have evaluated the price impacts of individual biofuels, while others have evaluated the joint impacts of several biofuels produced in one region or across the world. Some models have reported the regional price impacts, while others have reported their global price consequences. Finally, the existing results cover a wide range of biofuel targets and policies. For these reasons, in what follows we highlight several examples to cover a wide range of results.

We begin with the projection made by Babcock (2012), who, using several backcasting simulations obtained from the FAPRI model, estimated the impacts of produced corn ethanol in the United States between 2005 and 2009 on the prices of major crops and several livestock products. Babcock's results showed that the additional ethanol produced in the United States in 2009 compared with 2004 increased the prices of corn, soybeans, wheat, and rice by 27 percent, 5 percent, 10 percent, and 1.2 percent, respectively. The additional ethanol produced in 2009 compared to 2004 was about 8.5 billion gallons. The results also showed that the corresponding price impacts for livestock products, including eggs, broilers, pork, and beef, were about 5 percent, 1.7 percent, 0.7 percent, and 0.5 percent, respectively.

Taheripour, Hertel, and Tyner (2011), who evaluated the joint impacts of the U.S. and EU biofuel mandates on crop and livestock prices using the GTAP-BIO model, projected that these mandates could increases the price of coarse grain, oilseeds, and other crops in the United States by 12.6 percent, 11.2 percent, and 6 percent, respectively. Their corresponding projections for the European Union were 6.9 percent, 19.1 percent, and 5.8 percent, respectively. The price impacts on the livestock products were projected to be around 1 percent for both regions. The imposed mandates in this work were defined for the 2006 to 2015 period.

While these results showed relatively moderate price impacts for biofuel production mandates, some modeling practices that use stylized models projected larger price impacts. For instance, Wu and Langpap (2015), using a very simple stylized CGE model (covering only three final goods) that ignores many long-run market-mediated responses, projected that a 10 percent mandated blend with no ethanol subsidy increases the price of corn by 30 percent, leading to a 1.8 percent increase in the food price. Their analyses indicated that adding ethanol subsidies barely increases these price impacts, but doubling the blend levels (from 10 percent to 20 percent) doubles the price impacts. This shows that the mandate is the binding policy.

In another modeling effort, Tyner and Taheripour (2008) showed that the short-run price impacts of biofuel policies are tied to the price of crude oil. That is, they showed that, *ceteris paribus*, for a given mandate (e.g., 15 billion gallons of corn ethanol), the impacts on crop prices decrease as the crude oil price increases. At higher crude oil prices, the market equilibrium represents higher crop prices, because the higher the crude oil price, the higher the ethanol production. Hence, at higher crude oil prices, a biofuel mandate generates a lower impact on crop prices compared with market outcomes. For example, from the simulation results provided by Tyner and Taheripour (2008), it is clear that when the price of crude oil is US$60 per barrel, mandating 15 billion gallons of corn ethanol increases the corn price by 83 percent. This figure drops to 43 percent and 22 percent, respectively, as the crude oil price increases to US$80 and US$100 per barrel. That is, at higher crude oil prices, market forces encourage ethanol producers to produce more ethanol, which reduces the price impacts of biofuel mandates. These figures indicate that at a crude oil price of US$50 per barrel, any drop in U.S. biofuel mandate could cause major reductions in commodity prices with major adverse consequences for crop producers.

Hochman, Rajagopal, and Zilberman (2010) also used a PE model to evaluate the impacts of biofuel production on crop prices. They concluded that with some exception, in general, about 10–15 percent of the observed increases in crop prices in 2007 were due to biofuel production. Unlike other PE models used in this area, which assume perfect competition in the crude oil market, the PE model used in this work assumes that the Organization of Petroleum Exporting Countries (OPEC), as a cartel-of-nations (CON), reduces its oil production to compete with biofuels, and that sets a higher crude oil price compared with a competitive market price. They argued that with this setup, biofuel production leads to a larger reduction in crude oil production compared with a competitive market, and that increases the environmental benefits of biofuel production. They also claimed that because OPEC acts as a CON, the price impacts of biofuel production on crop prices were reduced. In addition, their PE model considered the fact that biofuel production reduces crude oil price, which eventually reduces the production costs of food items.

In this line of research, several papers have developed analyses to better understand the nature of energy–agricultural price relationships using economic models or econometric approaches. For example, McPhail and Babcock (2008), Gohin and Tréguer (2010), Hertel, Tyner, and Birur (2010), Beckman et al. (2011), and Hertel and Beckman (2011) have all examined the price linkages using the PE and CGE models. In general, these efforts conclude that (1) biofuel production from agricultural resources ties commodity prices to the crude oil price and that energy policies have the potential to affect the links between energy and agricultural markets; (2) biofuel production/mandate increases commodity prices; (3) biofuel policies and agricultural policies interact; and (4) binding biofuel mandates usually reduce the vulnerability of agricultural markets to the instability of energy prices. However, biofuel production at a large scale, in the absence of a binding mandate, will make the transmissions of energy price volatility into agricultural markets stronger.

On the other hand, numerous papers have developed various econometric analyses using time series data to examine the energy and agricultural linkages. After a thorough review, Serra and Zilberman (2013) concluded that the findings of the biofuel-related price transmission econometric analyses are in line with the findings of the PE and CGE models. They showed that (1) energy prices affect long-run agricultural price levels, and (2) instability in energy markets has been transferred to commodity markets, particularly after 2000. Serra and Zilberman (2013) also highlighted some time series work that deviates from economic theory. For example, they referred to Zhang et al. (2009) and Wu, Guan, and Myers (2011). The first article found that corn determines crude oil prices, and the second article assumed that the spillover impact of variations in crude oil on corn price is a function of the ratio of ethanol consumption to gasoline consumption.

Finally, in parallel to the modeling and econometric methods, several attempts have been made to determine the main drivers of changes in commodity prices in the biofuel era after 2005. In this line of research, the first main attempt has been made by Abbott, Hurt, and Tyner

(2008). To explain the rapid growth of crop and food prices in 2007, these authors reviewed the existing literature and developed their own analyses. They concluded that expansion in global demand for food due to rapid growth in developing countries (basically China and India), depreciation in the value of the U.S. dollar, and expansion in biofuel production were the major sources of the price hike in 2007. In response to the observed price shocks of 2011, Abbott, Hurt, and Tyner (2011) revised their work to highlight the role of the U.S. ethanol mandate; the income growth in developing countries; the holding of large stocks of corn in China and the massive imports of soybeans by this country; the limits on land expansion and switching among crops that jointly transfer the price impact of shortage from one commodity to other commodities; and weather conditions as the major drivers of the 2011 crop prices. Using a numerical model, Hochman et al. (2014) made similar conclusions and in particular highlighted the role of changes in crop inventories. In line with these efforts, Abbott (2014) examined additional factors that may have affected the links between energy and agricultural prices since 2005. For instance, he analyzed the contributions of the structure of the U.S. biofuel mandate and its implementation mechanism, the roles of short-run and long-run price volatilities, the importance of "blend wall" and "binding RFS," banning the use of methyl tertiary butyl ether (MTBE), and ethanol industry capacity. From these analyses, Abbott (2014) argued that about 33 percent of the observed increases in corn prices between 2005 and 2009 were due to ethanol production.

In conclusion, regardless of the variability in results, the main body of the existing literature confirms that biofuel production has had significant impacts on food prices. However, biofuels have not been the only drivers of higher crop and food prices since 2005. The extent to which biofuels contributed to the observed price hikes during the biofuel era is extremely hard to determine and should be subject to more investigation.

4.2 Impacts of biofuel production and policy on gasoline and crude oil prices

In general, the main body of literature confirms that producing biofuels lowers the prices of crude oil and gasoline mildly. Gallagher et al. (2003) showed that an ethanol mandate of 5 billion gallons reduces the price of gasoline by 2 percent. Rajagopal et al. (2007) concluded that the U.S. ethanol tax credit, which increases biofuel production, leads to a 3 percent reduction in gasoline price. Ando, Khanna, and Taheripour (2010) estimated that without tax credits, the U.S. RFS reduces the producer price of gasoline by 6 percent compared with a status quo simulation, but with biofuel tax credits, the U.S. RFS reduces the price of gasoline by 5.5 percent. Du and Hayes (2009), using an econometric approach, estimated that ethanol production lowered the average U.S. wholesale gasoline price by 8 percent over the sample period from January 1995 to March 2008.

Hochman, Rajagopal, and Zilberman (2010) estimated that biofuels produced in 2007 lowered the price of crude oil by about 1.1 percent. Taheripour and Tyner (2014), using a global CGE model (GTAP), showed that the U.S. ethanol mandate could reduce the price of crude oil at the global scale by 1.2 percent to 2.9 percent, depending on the implemented biofuel and agricultural policies. Wu and Langpap (2015), using a stylized CGE model, estimated that a 10 percent mandated ethanol blend with no tax credit reduces the price of gasoline by 5 percent. Their results showed that this level of mandated blend with a US$1 tax credit on ethanol production drops the gasoline price by 6 percent. Moschini, Lapan, and Kim (2016), using a PE model, showed that the U.S. ethanol mandates designed for 2022 lower the price of crude oil by only 2.5 percent. Figure 34.3 illustrates these price impacts. Because the policies being modeled were targeted at U.S. ethanol, which is a tiny fraction of the global crude oil market, the larger

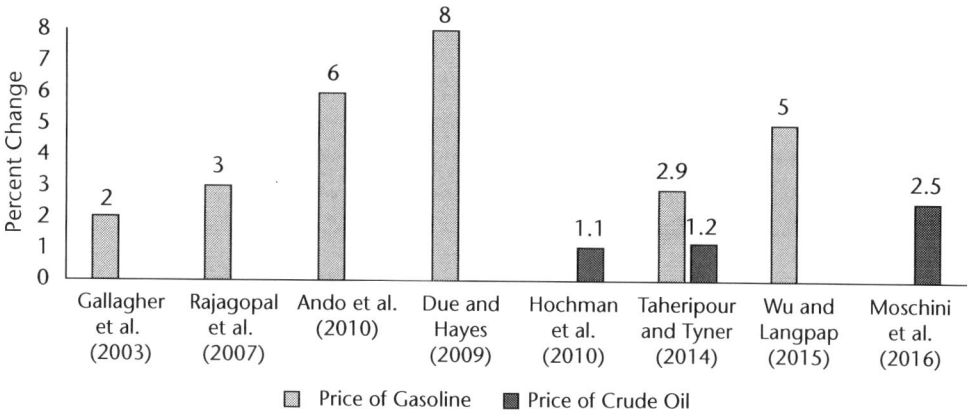

Figure 34.3 Estimated impacts of biofuel production on gasoline and crude oil prices

crude oil price impact estimates seem not to be plausible. Also, the studies generally ignored the difference in energy content between ethanol and gasoline, which would reduce further the price impacts in energy terms.

5 Welfare impacts

The existing biofuel policies were defined and designed to expand biofuels production and consumption based on three main arguments: (1) production and consumption of biofuels, compared with traditional fossil fuels, generate less GHG emissions; (2) biofuel production increases national security and reduces dependency on fossil fuels, and (3) biofuel production helps to improve rural areas and increases farmers' incomes. While these arguments have been repeated frequently in the literature, with many papers trying to determine the extent to which biofuel production helps to reduce GHG emissions, no major attempt has been made to assess the monetary values of changes in social welfare due to these factors. However, numerous efforts have been made to determine the extent to which biofuel policies affect economic welfare due to changes in the consumer and producer prices of goods and services and also changes in government net revenues/expenditures.

While studies investigating the welfare implications of biofuel policies generally suggest that the past and present biofuel policies essentially transfer wealth from fuel consumers (motorists) to biofuel and feedstock producers, the findings from these studies on overall welfare vary significantly. Some studies concluded that biofuel policies would result in an overall reduction in social welfare in the United States (Gardner 2007; Khanna, Ando, and Taheripour 2008; de Gorter and Just 2009; Hertel, Tyner, and Birur 2010; Pouliot and Babcock 2016), while other studies showed welfare improvements (Elobeid and Tokgoz 2008; Cui et al. 2011; Huang et al. 2013; Moschini, Lapan, and Kim 2016) or a negligible welfare change (Wu and Langpap 2015). Some of these studies, along with their estimates on U.S. welfare, are listed in Table 34.3. Given that the policies essentially force into the market a product that would not be there under market conditions, it is hard to accept the welfare increasing results.

In comparing the results of these studies, one should take into account several factors. The economic models used to calculate the welfare impacts of biofuels are different in their capabilities to capture the domain of commodities affected by biofuel policies and also the way

Table 34.3 Changes in U.S. social welfare due to biofuel policies

Description	Modeling framework	Reference scenario	Policy scenario	Welfare change (million US$)
Gardner (2007)	Closed, partial equilibrium	No ethanol subsidy and farm subsidy	Ethanol subsidy or farm subsidy	−91, −18, −665, −37, −689, −41
Schmitz, Moss, and Schmitz (2007)	Open, partial equilibrium	No interaction between ethanol and agricultural subsidy	With interaction between ethanol and agricultural subsidy	−1744, 2274
Khanna, Ando, and Taheripour (2008)	Closed, partial equilibrium	Fuel tax without ethanol subsidy	Ethanol subsidy	−500[a]
Elobeid and Tokgoz (2008)	Open, partial equilibrium	No subsidy and trade barrier	Ethanol subsidy and trade barrier	2812[b]
de Gorter and Just (2009)	Open, partial equilibrium	No ethanol subsidy and farm subsidy	Ethanol subsidy and/or farm subsidy	−1291, −913, −613
Ando, Khanna, and Taheripour (2010)	Closed, partial equilibrium	Fuel tax and ethanol subsidy, no mandate	Ethanol mandate	−8500, 5000[c]
Pouliot and Babcock (2016)	Open, partial equilibrium	Non-binding mandate	Binding mandate	−85, −842, 21, −642
Moschini, Lapan, and Kim (2016)	Open, partial equilibrium	No biofuel mandate	Current, optimal, or 2022 mandates	4430, 5431, 296
Cui et al. (2011)	Open, general equilibrium	Fuel tax, no ethanol subsidy and mandate	Current/optimal ethanol subsidy and/or mandate	6200, 7000, 7700[d]
Hertel, Tyner, and Birur (2010)	Open, general equilibrium	Ethanol subsidy, no mandate	U.S. and EU mandates	−8000
Lapan and Moschini (2012)	Open, general equilibrium	Ethanol subsidy, no mandate	Ethanol mandate, no subsidy	positive
Wu and Langpap (2015)	Open, general equilibrium	No biofuel subsidy or mandate	Biofuel subsidy and mandate	very small (consumer utility)

Notes:

a This number is calculated as the difference between two scenarios: Status Quo Fuel Taxes with Ethanol Subsidy and Status Quo Fuel Taxes without Ethanol Subsidy, as presented in Table 34.1 of the cited paper.

b This number is the sum of net welfare change for U.S. ethanol market and U.S. corn market presented in Table 34.5 of the cited paper.

c These two numbers are the differences between cases A and D, respectively, for the low and high gas supply elasticities in the cited paper.

d These numbers are the differences between several policy scenarios (Status Quo, Optimal Subsidy, and Optimal Mandate) and the No Ethanol Policy scenario presented in Table 3B of the cited paper.

that these policies interact with the existing energy, agricultural, regulatory, and trade policies. Some models, in particular the stylized PE and CGE models, capture only the impacts on the main markets that are directly affected by biofuel policies, including crude oil, gasoline, ethanol biodiesel, corn, and oilseeds. These models ignore the implications of biofuel policies on the markets of other goods and services and on the markets of primary inputs such as labor, land, and capital. They do not capture the interactions between biofuel policies and other existing energy and agricultural policies. Incorporating these interactions could alter the welfare impacts, as explained by Schmitz, Moss, and Schmitz (2007), de Gorter and Just (2009), and Taheripour and Tyner (2014). The large global CGE models used in this area – which cover markets for a wide range of good and services, take into account changes in demand for and supplies of primary inputs, and trace interactions between the existing taxes and biofuel policies – may provide more comprehensive welfare assessments for biofuel policies.

In addition to the differences previously mentioned, in comparing the results of welfare assessment practices, one should take into account that the economic models used in this area use different databases and various parameters and follow different assumptions in modeling the behaviors of consumers and producers. Finally, it is important to note that the welfare assessment practices cover a wide range of biofuel targets and policies.

Using several examples, we further review some of these results. We begin with Ando, Khanna, and Taheripour (2010), who showed that the U.S. ethanol mandate (including 15 billion gallons of corn ethanol and 5.5 billion gallons of cellulosic ethanol, plus tax credits of US$0.45/gallon for corn ethanol and US$1.01/gallon for cellulosic ethanol) reduces U.S. welfare by −US$8,500 million, when the gasoline supply is relatively inelastic. They calculated that the policy generates US$5,000 million welfare gains if the supply of gasoline is very elastic. This indicates the sensitivity of the results with respect to the elasticity of supply of gasoline. Their model is a typical PE model that only considers a few markets and ignores interactions between biofuel and agricultural policies and represents a closed economy.

In contrast, Schmitz, Moss, and Schmitz (2007), using a PE model that takes into account interactions between biofuel and agricultural policies, reported a different story for the welfare implication of ethanol production in the United States. Based on the 2006 economic conditions, these authors estimated that an expansion in U.S. corn ethanol by about 5.88 billion gallons improves U.S. welfare by US$2.27 billion when accounting for the fact that ethanol production reduces the need for a corn subsidy. If we ignore the saving in treasury costs, the U.S. welfare drops by −US$1.74. These calculations were made using an elasticity of 0.5 for corn supply. As concluded by Schmitz, Moss, and Schmitz (2007), these figures clearly confirm that including tax interaction effects could alter the welfare impacts significantly.

Taheripour and Tyner (2014) examined the welfare impacts of an expansion in U.S. ethanol from its 2004 level (3.41 billion gallons) to meet the 15 billion gallons of the U.S. RFS under three alternative cases to finance the policy using a large global CGE model (GTAP-BIO). This model covers production, consumption, and the trade of goods and services across the world and takes into account markets for primary inputs, including land, labor, capital, and natural resources. In this example, the first policy uses a tax on gasoline consumption to finance the required subsidies to boost ethanol production to 15 billion gallons. In the second policy, the required subsidy for ethanol production is financed by a tax on gasoline consumption and by cutting agricultural production subsidies to zero. In the third policy, the required subsidy for ethanol production is covered by an income tax and by cutting agricultural production subsidies to zero. The results show that the first, second, and third policy options reduce U.S. welfare by −US$16.8 billion, −US$15.3 billion, and −US$14.9 billion. These results confirm that cutting agricultural production subsides has a positive welfare impact, but it does not eliminate the distortionary impacts of the ethanol mandate.

Finally, Cui et al. (2011), unlike the previous typical biofuel welfare analyses, used a stylized small CGE model to calculate the monetary values of reductions in emissions due to ethanol production. Their results showed that (1) an expansion in U.S. corn ethanol by about 10 billion gallons induced by a mandate supported by an ethanol tax credit improves U.S. welfare by about US$6.2 billion; (2) the combination of a mandate and an ethanol subsidy increases U.S. emissions by 19.2 million tCO_2; and (3) the expansion in emissions is due to increases in total fuel consumption, assuming that the emissions intensity of ethanol per unit of energy is 75 percent of the emissions intensity of gasoline. The model used by Cui et al. considered only changes in consumer and producer surpluses for crude oil, gasoline, ethanol, and corn. Hence, like many other stylized models, the welfare impacts obtained from this model do not reflect many changes that biofuel production may induce.

It is important to note that using PE models, some papers (e.g., Huang et al. 2013 and Moschini et al. 2016) have showed that the RFS generates substantial welfare gains for the U.S. economy due to changes in the terms of trade in favor of the U.S., which is a large country in the world markets for energy and agriculture. Taheripour and Tyner (2014), using a global CGE model, confirmed that the biofuel mandate generates gains for the U.S. economy through the trade channel. However, they showed that the trade gains are not large enough to offset the inefficiency welfare losses of biofuel policies.

6 Conclusions

Biofuel economics is a fascinating topic with many different aspects and angles to be examined. We have tried to cover some of the important dimensions here. The reality is that biofuels are government-created industries in all three of the major producing regions: Brazil, the European Union, and the United States. Corn and sugarcane ethanol are cost competitive due to government mandates and incentives. When governments mandate a non-market solution, they do so because they perceive that there are non-market benefits that exceed the losses in mandating a non-market solution. In this case, the benefits are reduced GHG emissions, increased energy security, and improved rural development. While the literature covers the very important topics noted here, such as food versus fuel and welfare impacts, there is little on the "value" of the reasons most commonly cited for the government intervention. Thus, despite all the good research that has been done, we still do not have a good comparison of the value of the perceived benefits of biofuel interventions and the estimated costs.

Notes

1 Energy efficiency is measured in terms of energy returned on energy invested (EROEI), ignoring solar energy capture in the corn production process. The energy efficiency varies across alternative biofuel pathways.
2 The model was developed by the Center for Agriculture and Rural Development at Iowa State University based on the models developed by the Food and Agricultural Policy Research Institute (FAPRI) at Iowa State and the University of Missouri.
3 This type of land represents marginal cropland that has been cultivated for crop production in the past, but currently is used as pastureland. This type of land can return to crop production if needed.
4 The study by Taheripour et al. (2013) showed that water scarcity, which limits irrigation, may increase ILUCs and their corresponding emissions.

References

Abbott, P. 2014. "Biofuels, Binding Constraints, and Agricultural Commodity Price Volatility." In J.P. Chavas, D. Hummels, and B.D. Wright, eds., *The Economics of Food Price Volatility*. Chicago, IL: University of Chicago Press, pp. 91–131.

Abbott, P., C. Hurt, and W. Tyner. 2008. *What Is Driving Food Prices*. Oak Brook, IL: Farm Foundation.

———. 2011. *What Is Driving Food Prices in 2011*. Oak Brook, IL: Farm Foundation.

Al-Riffai, P., B. Dimaranan, and L. Laborde. 2010. *Global Trade and Environmental Impact Study of the EU Biofuels Mandate*. Washington, DC: IFPRI.

Alexandratos, N., and J. Bruinsma. 2012. "World Agriculture Toward 2030/2050: The 2012 Revision." ESA Working Paper, FAO, Rome.

Alston, J.M., B.A. Babcock, and P.G. Pardey. 2010. "The Shifting Patterns of Agricultural Production and Productivity Worldwide." MATRIC, University of Iowa, Ames, IA.

Ando, A.W., M. Khanna, and F. Taheripour. 2010. "Market and Social Welfare Effects of the Renewable Fuels Standard." In *Handbook of Bioenergy Economics and Policy*. New York: Springer, pp. 233–250.

Ausubel, J.H., I.K. Wernick, and P.E. Waggoner. 2013. "Peak Farmland and the Prospect for Land Sparing." *Population and Development Review* 38(s1):221–242.

Babcock, B.A. 2012. "The Impact of U.S. Biofuel Policies on Agricultural Price Levels and Volatility." *China Agricultural Economic Review* 4(4):407–426.

Babcock, B.A., and Z. Iqbal. 2014. "Using Recent Land Use Changes to Validate Land Use Change Models." 14-ST 109. I. S. U. Center for Agricultural and Rural Development, Iowa State University, Ames, IA.

Bann, S.J., R. Malina, P. Suresh, M. Pearlson, W.E. Tyner, J.I. Hileman, and S. Barrett. 2017. "The Costs of Production of Alternative Jet Fuel: A Harmonized Stochastic Assessment." *Bioresource Technology* 227:179–187.

Beckman, J., T. Hertel, F. Taheripour, and W. Tyner. 2011. "Structural Change in the Biofuels Era." *European Review of Agricultural Economics* 39(1):137–156.

Bittner, A., X. Zhao, and W.E. Tyner. 2015. "Field to Flight: A Techno-Economic Analysis of Corn Stover to Aviation Biofuels Supply Chain." *Biofuels, Bioproducts & Biorefining* 9:201–210.

Borchers, A., E. Truex-Powell, S. Wallander, and C. Nickerson. 2014. *Multi-Cropping Practices: Recent Trends in Double Cropping*. Washington, DC: USDA.

Brady, M.P., and B. Sohngen. 2008. "Agricultural Productivity, Technological Change, and Deforestation: A Global Analysis." *Proceedings of AAEA Conference*. Orlando, FL, July.

Broch, A., S.K. Hoekman, and S. Unnasch. 2013. "A Review of Variability in Indirect Land Use Change Assessment and Modeling in Biofuel Policy." *Environmental Science and Policy* 29:147–157.

Byerlee, D., J. Stevenson, and N. Villoria. 2014. "Does Intensification Slow Crop Land Expansion or Encourage Deforestation?" *Global Food Security* 3(2):92–98.

California Air Resources Board (CARB). 2015. *Calculating Carbon Intensity Values from Indirect Land Use Change of Crop-Based Biofuels*. Sacramento, CA: CARB.

Carriquiry, M.A., X. Du, and G.R. Timilsina. 2011. "Second Generation Biofuels: Economics and Policies." *Energy Policy* 39(7):4222–4234.

Cassman, K.G. 1999. "Ecological Intensification of Cereal Production Systems: Yield Potential, Soil Quality, and Precision Agriculture." *Proceedings of the National Academy of Sciences* 96(11):5952–5959.

Cui, J., H. Lapan, G. Moschini, and J. Cooper. 2011. "Welfare Impacts of Alternative Biofuel and Energy Policies." *American Journal of Agricultural Economics* 93(5):1235–1256.

de Gorter, H., and D.R. Just. 2009. "The Welfare Economics of a Biofuel Tax Credit and the Interaction Effects with Price Contingent Farm Subsidies." *American Journal of Agricultural Economics* 91(2):477–488.

Dixson-Declève, S. 2012. "Fuel Policies in the EU: Lessons Learned from the Past and Outlook for the Future." In T. Zachariadis, ed., *Cars and Carbon*. New York, Springer, pp. 97–126.

Du, X., and D.J. Hayes. 2009. "The Impact of Ethanol Production on U.S. and Regional Gasoline Markets." *Energy Policy* 37(8):3227–3234.

Dumortier, J., D.J. Hayes, M. Carriquiry, F. Dong, X. Du, A. Elobeid, J.F. Fabiosa, and S. Tokgoz. 2011. "Sensitivity of Carbon Emission Estimates from Indirect Land-Use Change." *Applied Economic Perspectives and Policy* 33(3):428–448.

Dunn, J.B., Z. Qin, S. Mueller, H-Y. Kwon, M.M. Wander, and M. Wang. 2016. *Carbon Calculator for Land Use Change from Biofuels Production (CCLUB) Users' Manual and Technical Documentation.* Lemont, IL: Argonne National Laboratory (ANL).

Elobeid, A., and S. Tokgoz. 2008. "Removing Distortions in the U.S. Ethanol Market: What Does It Imply for the United States and Brazil?" *American Journal of Agricultural Economics* 90(4):918–932.

Energy Information Administration (EIA). 2017. *Biofuels International Data.* Washington, DC: EIA.

Environmental Protection Agency (EPA). 2010. *Renewable Fuel Standard Program (RFS2) Regulatory Impact Analysis.* EPA-420-R-10–006. Washington, DC: EPA.

European Commission (EC). 2016. *Renewable Energy Directive.* Brussels: EC.

European Union (EU). 2009. "Directive 2009/28/EC of the European Parliament and of the Council of 23 April 2009 on the Promotion of the Use of Energy from Renewable Sources and Amending and Subsequently Repealing Directives 2001/77/EC and 2003/30/EC." *Official Journal of the European Union* 5.

Fargione, J., J. Hill, D. Tilman, S. Polasky, and P. Hawthorne. 2008. "Land Clearing and the Biofuel Carbon Debt." *Science* 319(5867):1235–1238.

Gallagher, P.W., H. Shapouri, J. Price, G. Schamel, and H. Brubaker. 2003. "Some Long-Run Effects of Growing Markets and Renewable Fuel Standards on Additives Markets and the U.S. Ethanol Industry." *Journal of Policy Modeling* 25(6):585–608.

Gardner, B. 2007. "Fuel Ethanol Subsidies and Farm Price Support." *Journal of Agricultural and Food Industrial Organization* 5(2):Article 4.

Giampietro, M., S. Ulgiati, and D. Pimentel. 1997. "Feasibility of Large-Scale Biofuel Production." *BioScience* 47(9):587–600.

Gohin, A. 2014. "Assessing the Land Use Changes and Greenhouse Gas Emissions of Biofuels: Elucidating the Crop Yield Effects." *Land Economics* 90(4):575–586.

Gohin, A., and D. Tréguer. 2010. "On the (De) Stabilization Effects of Biofuels: Relative Contributions of Policy Instruments and Market Forces." *Journal of Agricultural and Resource Economics* 35(1):72–86.

Goldemberg, J. 2006. "The Ethanol Program in Brazil." *Environmental Research Letters* 1(1):Article 014008.

Golub, A.A., and T.W. Hertel. 2012. "Modeling Land-Use Change Impacts of Biofuels in the GTAP-BIO Framework." *Climate Change Economics* 3(3):Article 1250015.

Gubicza, K., I.U. Nieves, W.J. Sagues, B. Barta, K. Shanmugam, and L.O. Ingram. 2016. "Techno-Economic Analysis of Ethanol Production from Sugarcane Bagasse Using a Liquefaction Plus Simultaneous Saccharification and Co-Fermentation Process." *Bioresource Technology* 208:42–48.

Hertel, T.W., and U.L.C. Baldos. 2015. *Global Change and the Challenges of Sustainably Feeding a Growing Planet.* New York: Springer.

Hertel, T.W., and J. Beckman. 2011. "Commodity Price Volatility in the Biofuel Era: An Examination of the Linkage Between Energy and Agricultural Markets." In J. Graff Zivin and J. Perloff, eds., *The Intended and Unintended Effects of U.S. Agricultural and Biotechnology Policies.* Chicago, IL: University of Chicago Press, pp. 189–221.

Hertel, T.W., A.A. Golub, A.D. Jones, M. O'Hare, R.J. Plevin, and D.M. Kammen. 2010. "Effects of U.S. Maize Ethanol on Global Land Use and Greenhouse Gas Emissions: Estimating Market-Mediated Responses." *BioScience* 60(3):223–231.

Hertel, T.W., W.E. Tyner, and D.K. Birur. 2010. "The Global Impacts of Biofuel Mandates." *The Energy Journal* 31(1):75–100.

Hill, J., S. Polasky, E. Nelson, D. Tilman, H. Huo, L. Ludwig, J. Neumann, H. Zheng, and D. Bonta. 2009. "Climate Change and Health Costs of Air Emissions from Biofuels and Gasoline." *Proceedings of the National Academy of Sciences* 106(6):2077–2082.

Hochman, G., D. Rajagopal, G. Timilsina, and D. Zilberman. 2011. "The Role of Inventory Adjustments in Quantifying Factors Causing Food Price Inflation." World Bank Policy Research Working Paper 5744.

———. 2014. "Quantifying the Causes of the Global Food Commodity Price Crisis." *Biomass and Bioenergy* 68:106–114.

Hochman, G., D. Rajagopal, and D. Zilberman. 2010. "Are Biofuels the Culprit? OPEC, Food, and Fuel." *The American Economic Review* 100(2):183–187.

Huang, H., M. Khanna, H. Önal, and X. Chen. 2013. "Stacking Low Carbon Policies on the Renewable Fuels Standard: Economic and Greenhouse Gas Implications." *Energy Policy* 56:5–15.

International Energy Agency (IEA). 2017. *Bioenergy*. Paris: IEA.

Janda, K., L. Kristoufek, and D. Zilberman. 2012. "Biofuels: Policies and Impacts." *Agricultural Economics* 58(8):372–386.

Khanna, M., A.W. Ando, and F. Taheripour. 2008. "Welfare Effects and Unintended Consequences of Ethanol Subsidies." *Applied Economic Perspectives and Policy* 30(3):411–421.

Khanna, M., and C.L. Crago. 2012. "Measuring Indirect Land Use Change with Biofuels: Implications for Policy." *Annual Review of Resource Economics* 4(1):161–184.

Laborde, D. 2011. *Assessing the Land Use Change Consequences of European Biofuel Policies*. Washington, DC: IFPRI.

Lapan, H., and G. Moschini. 2012. "Second-Best Biofuel Policies and the Welfare Effects of Quantity Mandates and Subsidies." *Journal of Environmental Economics and Management* 63(2):224–241.

Lewis, S.M., and M. Kelly. 2014. "Mapping the Potential for Biofuel Production on Marginal Lands: Differences in Definitions, Data, and Models Across Scales." *ISPRS International Journal of Geo-Information* 3(2):430–459.

McPhail, L.L., and B.A. Babcock. 2008. "Ethanol, Mandates, and Drought: Insights from a Stochastic Equilibrium Model of the U.S. Corn Market." CARD WP 3–2008. Iowa State University, Ames, IA.

Moschini, G., J. Cui, and H. Lapan. 2012. "Economics of Biofuels: An Overview of Policies, Impacts, and Prospects." *Bio-Based and Applied Economics* 1(3):269–296.

Moschini, G., H. Lapan, and H. Kim. 2016. "The Renewable Fuel Standard: Market and Welfare Effects of Alternative Policy Scenarios." *Proceedings of the AAEA Conference*. Boston, MA.

Mosnier, A., P. Havlík, H. Valin, J. Baker, B. Murray, S. Feng, M. Obersteiner, B. McCarl, S. Rose, and U. Schneider. 2013. "Alternative U.S. Biofuel Mandates and Global GHG Emissions: The Role of Land Use Change, Crop Management, and Yield Growth." *Energy Policy* 57:602–614.

Naik, S.N., V.V. Goud, P.K. Rout, and A.K. Dalai. 2010. "Production of First and Second Generation Biofuels: A Comprehensive Review." *Renewable and Sustainable Energy Reviews* 14(2):578–597.

National Research Council (NRC). 2012. *Renewable Fuel Standard: Potential Economic and Environmental Effects of U.S. Biofuel Policy*. Washington, DC: National Academies Press.

Nigam, P.S., and A. Singh. 2011. "Production of Liquid Biofuels from Renewable Resources." *Progress in Energy and Combustion Science* 37(1):52–68.

Pimentel, D. 1991. "Ethanol Fuels: Energy Security, Economics, and the Environment." *Journal of Agricultural and Environmental Ethics* 4(1):1–13.

———. 1998. "Energy and Dollar Costs of Ethanol Production with Corn." *Hubbert Center Newsletter* 98/2.

Plevin, R., H. Gibbs, J. Duffy, S. Yui, and S. Yeh. 2014. "Agro-Ecological Zone Emission Factor (AEZ-EF) Model (v47)." Center for Global Trade Analysis, Department of Agricultural Economics, Purdue University, West Lafayette, IN.

Pouliot, S., and B.A. Babcock. 2016. "Compliance Path and Impact of Ethanol Mandates on Retail Fuel Market in the Short Run." *American Journal of Agricultural Economics* 98(3):744–764.

Rajagopal, D., S.E. Sexton, D. Roland-Holst, and D. Zilberman. 2007. "Challenge of Biofuel: Filling the Tank Without Emptying the Stomach?" *Environmental Research Letters* 2(4):044004.

Ray, D.K., and J.A. Foley. 2013. "Increasing Global Crop Harvest Frequency: Recent Trends and Future Directions." *Environmental Research Letters* 8(4):044041.

Schmitz, A., C.B. Moss, and T.G. Schmitz. 2007. "Ethanol: No Free Lunch." *Journal of Agricultural & Food Industrial Organization* 5(2):Article 3.

Schmitz, A., J.L. Seale, and T.G. Schmitz. 2003. "Sweetener-Ethanol Complex in Brazil, the United States, and Mexico." *International Sugar Journal* 105(1259):505–513.

Searchinger, T., R. Heimlich, R. Houghton, F. Dong, A. Elobeid, J. Fabiosa, S. Tokgoz, D. Hayes, and T-H. Yu. (2008). "Use of U.S. Croplands for Biofuels Increases Greenhouse Gases Through Emissions from Land Use Change." *Science* 319(5867):1238–1240.

Serra, T., and D. Zilberman. 2013. "Biofuel-Related Price Transmission Literature: A Review." *Energy Economics* 37:141–151.

Shapouri, H., J.A. Duffield, and M.Q. Wang. 2002. *The Energy Balance of Corn Ethanol: An Update*. Washington, DC: USDA-ERS.

Siebert, S., F.T. Portmann, and P. Döll. 2010. "Global Patterns of Cropland Use Intensity." *Remote Sensing* 2(7):1625–1643.

Sims, R.E., W. Mabee, J.N. Saddler, and M. Taylor. 2010. "An Overview of Second Generation Biofuel Technologies." *Bioresource Technology* 101(6):1570–1580.

Taheripour, F., H. Cui, and W.E. Tyner. 2017. "An Exploration of Agricultural Land Use Change at the Intensive and Extensive Margins: Implications for Biofuels Induced Land Use Change." In Z. Qin, U. Mishra, and A. Hastings, eds., *Bioenergy and Land Use Change*. Washington, DC: American Geophysical Union (Wiley), pp. 19–37.

Taheripour, F., T.W. Hertel, and J. Liu. 2013. "The Role of Irrigation in Determining the Global Land Use Impacts of Biofuels." *Energy, Sustainability, and Society* 3(1):Article 4.

Taheripour, F., T.W. Hertel, and W.E. Tyner. 2011. "Implications of Biofuels Mandates for the Global Livestock Industry: A Computable General Equilibrium Analysis." *Agricultural Economics* 42(3):325–342.

Taheripour, F., and W.E. Tyner. 2013a. "Biofuels and Land Use Change: Applying Recent Evidence to Model Estimates." *Applied Sciences* 3:14–38.

———. 2013b. "Induced Land Use Emissions Due to First and Second Generation Biofuels and Uncertainty in Land Use Emission Factors." *Economics Research International* (Article ID 315787).

———. 2014. "Welfare Assessment of the Renewable Fuel Standard: Economic Efficiency, Rebound Effect, and Policy Interactions in a General Equilibrium Framework." In A. Pinto and D. Zilberman, eds., *Modeling, Dynamics, Optimization and Bioeconomics I*. New York: Springer, pp. 613–632.

Taheripour, F., X. Zhao, and W. Tyner. 2017. "The Impact of Considering Land Intensification and Updated Data on Biofuels Land Use Change and Emissions Estimates." *Biotechnology for Biofuels* 10:191.

Taheripour, F., Q. Zhuang, W.E. Tyner, and X. Lu. 2012. "Biofuels, Cropland Expansion, and the Extensive Margin." *Energy, Sustainability, and Society* 2:25.

Tilman, D., J. Hill, and C. Lehman. 2006. "Carbon-Negative Biofuels from Low-Input High-Diversity Grassland Biomass." *Science* 314:1598–1600.

Treesilvattanakul, K. 2013. *Impacts and Feasibility of the U.S. and the EU Sustainability Criteria on Existing Land-Use Practices*. PhD dissertation, Purdue University, West Lafayette, IN.

Tyner, W.E. 2008. "The U.S. Ethanol and Biofuels Boom: Its Origins, Current Status, and Future Prospects." *BioScience* 58(7):646–653.

———. 2015. "Biofuel Economics and Policy: The Renewable Fuel Standard, the Blend Wall, and Future Uncertainties." In A. Dahiya, ed., *Bioenergy: Biomass to Biofuels*. Amsterdam: Elsevier, pp. 511–521.

Tyner, W., and F. Taheripour. 2008. "Biofuels, Policy Options, and Their Implications: Analyses Using Partial and General Equilibrium Approaches." *Journal of Agricultural and Food Industrial Organization* 6(2):9.

United States Department of Agriculture (USDA). 2016. *Brazil Biofuels Annual Report 2016*. Washington, DC: USDA.

Valin, H., D. Peters, M. van den Berg, S. Frank, P. Havlik, N. Forsell, C. Hamelinck, J. Pirker, A. Mosnier, and J. Balkovic. 2015. *The Land Use Change Impact of Biofuels Consumed in the EU Quantification of Area and Greenhouse Gas Impacts*. Utrecht, Netherlands: ECOFYS.

Wicke, B., P. Verweij, H. van Meijl, D.P. van Vuuren, and A.P. Faaij. 2012. "Indirect Land Use Change: Review of Existing Models and Strategies for Mitigation." *Biofuels* 3(1):87–100.

Wu, F., Z. Guan, and R.J. Myers. 2011. "Volatility Spillover Effects and Cross Hedging in Corn and Crude Oil Futures." *Journal of Futures Markets* 31(11):1052–1075.

Wu, J., and C. Langpap. 2015. "The Price and Welfare Effects of Biofuel Mandates and Subsidies." *Environmental and Resource Economics* 62(1):35–57.

Wu, M., Y. Wu, and M. Wang. 2006. "Energy and Emission Benefits of Alternative Transportation Liquid Fuels Derived from Switchgrass: A Fuel Life Cycle Assessment." *Biotechnology Progress* 22(4):1012–1024.

Yao, G., M.D. Staples, R. Malina, and W.E. Tyner. 2017. "Stochastic Techno-Economic Analysis of Alcohol-to-Jet Fuel Production." *Biotechnology for Biofuels* 10:18, 13 pages.

Zhang, Z., L. Lohr, C. Escalante, and M. Wetzstein. 2009. "Ethanol, Corn, and Soybean Price Relations in a Volatile Vehicle-Fuels Market." *Energies* 2(2):320–339.

Zhao, X., T.R. Brown, and W.E. Tyner. 2015 "Stochastic techno-economic evaluation of cellulosic biofuel pathways." *Bioresource Technology* 198, pp. 755–763.

Zhao, X., G. Yao, and W.E. Tyner. 2016. "Quantifying Breakeven Price Distributions in Stochastic Techno-Economic Analysis." *Applied Energy* 183:318–326.

35
Farm management
Recent applications

Patricia Duffy and Sam Funk

1 Introduction

> Essentially, all models are wrong, but some are useful.
>
> Box and Draper (1987, p. 424)

Farm management has a number of different meanings, depending on the context. At a basic and practical level, farm management refers to the act of running or managing a farm, which involves making long-term strategic decisions and implementing those decisions via short-run tactics. Farm management also refers to a field in agricultural economics primarily concerned with the analysis of the profits and profitability of the farm firm. As with its practical counterpart, the academic field of farm management emphasizes decision making at the level of the individual farm. Broadly, these decisions involve the acquisition and use of limited resources for achieving the firm's goals.

Farm-level decisions involve day-to-day operations, such as when and how to apply pesticides, and also include decisions with long-run consequences such as choices concerning farm size and legal organization. Other decisions with longer run impacts include the acquisition or control of depreciable assets, such as buildings, machinery and equipment; investing in perennial enterprises such as fruit trees; or converting cropland to pasture or vice versa. Insurance strategies, financial investments, and participation in farm programs are other choices that impact farm profits. Typically, decisions involving marketing or advertising farm products have not been considered part of the scope of farm management, but rather as falling under the marketing field. However, such choices have important implications for farm profits, and marketing decisions have become especially important as increasing numbers of farms participate in direct marketing through local food systems.

Farm management is one of the more thoroughly analyzed areas in agricultural economics, with some perhaps viewing it as akin to Latin in terms of being a dead language. Ironically, several advances in management and economic theory have recently become possible with the availability of vast amounts of data and the increasing depth of research-based knowledge of farm operations. The fundamental need for food by all humanity makes farm management a critical concern whenever scarcity of food is feared, as is the case in many parts of the world. Further, in developed economies, desires for "designer" foods and a focus on aesthetics rather than availability also result in research questions that impact farm management. The diversity of questions that surround farm management also include a host of traditional agricultural economics topics such as production, marketing, finance, and environmental and natural resources management – as well as interdisciplinary topics from agronomy and crop science, soil science, animal science, climatology, and other disciplines.

As an academic field, farm management is closely linked to production economics. Earl Heady (1948) noted that while some see a division between these two fields, with farm management concerning individual farms and production economics addressing the industry level, he himself believed that "there is little academic justification in drawing this fine line between the study of the tree and the forest"(Heady 1948, p. 205). Heady went on to discuss the complementarity of firm and industry-level analyses and gave as an example the problem that would arise if firm-level analyses ignored the potential impact of industry-level adoption. He also noted the converse that "applied phases of certain broad problem [sic] can best be studied through the individual farm" (Heady 1948, p. 206).

Unlike Heady, most applied economists see a distinction between farm management and production economics in terms of individual versus aggregate analyses but recognize the deep connection between the two fields. Research on efficiency, productivity change, and their measurement is often illustrated with farm data. Management insights where farm (firm) level data is compared to aggregate data can be found, for example, in Mugera, Langemeier, and Ojede (2016). Separate chapters in this Handbook discuss production economics, and readers interested in farm management are encouraged to review them.

When agricultural policy impacts farm-level decisions, there is often an overlap between policy analysis and farm management. Similarly, problems with farm management implications can be found in the intersection between conservation and resource issues. An example of a policy application can be found in an extension publication by Schnitkey et al. (2016), who provided estimates of 2015 payments from Agricultural Risk Coverage at the county level for corn and soybeans across the U.S., showing their significant impact on farm profitability. Examples from the research literature will be discussed in a subsequent section of this chapter.

2 Farm management and economic theory

The foundational discipline of farm management is economics. Although this may seem obvious given that farm management is typically taught in departments of agricultural economics, it is nevertheless a point worth making, as farm management also requires a broad knowledge of production agriculture and also of principles drawn from accounting, management, finance, marketing, and, at the more advanced level, statistics and operations research. As Malcolm (2004, p. 401) noted,

> In the context of farm management analysis, the core discipline means the discipline that organises the practically obtainable relevant information about a question or series of questions into a framework and form which enables an informed, reasoned, rational choice to be made between alternative actions faced by management.

This distinction is important because farming systems work also frequently focuses on decision making at the level of a farm firm, but unlike farm management, farming systems work is thoroughly interdisciplinary, and the economic perspective may not be dominant or may even be absent.

Chavas, Chambers, and Pope (2010) noted that one of the most important contributions of production economics and farm management was the integration of economic theory with farm-level decision making. They cited papers by Heady (1948), Waugh (1951), Mundlak (1961), and others to support this assertion. Neoclassical theory has been especially relevant in farm management (and in production economics as well) because farm firms generally fit the base assumptions of this model. The individual farms are typically small enough that no one firm can influence price (e.g., perfect competition), and the products are often homogeneous.

Some of the basic principles from the neoclassical theory of the firm that are used at every level of farm management, from foundational courses taught to undergraduates through advanced applications, include diminishing marginal products (also called the law of variable proportions), marginal analysis, the least-cost principle, and opportunity costs. An opportunity cost is defined as the income that could be received by using a resource in its most profitable alternative use (Kay, Edwards, and Duffy 2016). Typical examples are the opportunity costs for unpaid operator labor and management and for the equity capital invested in the business.

The phenomenon of diminishing marginal products occurs when additional units of a variable input are used in combination with one or more fixed inputs. In this situation, eventually the amount of extra output gained per extra input added will decline. Because of diminishing marginal products, the economically optimal amount of a resource to use will not be the amount that corresponds to maximum yield, unless that input is free.

Marginal analysis can be used to determine the optimum level of input used (or output produced) by looking at changes in costs and benefits associated with changes in input level. The benefit of using an additional unit of input is typically expressed as the revenue obtained from selling the extra output produced by the use of the additional unit of input. If this benefit is more than the costs of acquiring the input, more input should be used, and vice versa. A good example of how agricultural economists can apply basic marginal analysis to obtain highly useful results can be found in Traxler and Byerlee (1993), where optimal levels of nitrogen fertilizer were found in a joint product framework in the production of wheat grain and straw. Further applications of simple marginal analysis involve finding the least-cost combination of inputs to produce a given output or allocating a limited input among competing uses.

Other applications involve a whole-farm approach, in which the producer maximizes profit for the entire operation, considering several potential inputs and outputs. Mathematically, we optimize profit (Π) subject to "quasi-fixed" factors of production such as land, and possibly machinery and labor.

$$\max \Pi = P^* Y(X, Z) - W^* X \qquad (1)$$
$$\text{Subject to: } Z \leq Z^0 \qquad (2)$$

Where P is a (1 × N) vector of output prices, Y is a (N × 1) vector of outputs, W is a (1 × M) vector of variable input costs, X is a (M × 1) vector of variable inputs, Z is a (K × 1) vector of quasi-fixed factors, Z^0 is the (K × 1) vector of the endowment levels of the quasi-fixed factors, and N, M, and K are the numbers of products, variable inputs, and quasi-fixed factors, respectively. In the case of imperfect competition, P and W may in turn be functions of output and input levels. In production economics, this behavioral model would typically be used as the basis for deriving the form of a function that will be estimated for the industry (see for example Mishra et al. 2004), while in farm management the application would be to an individual firm.

Duality is a concept that links a primal problem to a dual problem, for example, profit maximization and cost minimization. It is useful in studies of profit, cost, input demand, and supply or production functions. Varian (1992) outlines several uses of duality and provides proofs and useful references. Duality also provides a means of solving for imputed resource valuation (shadow prices) and provides insight into linear programming solutions, as noted in McCarl and Spreen (2007). Coelli et al. (2005) present several expansions from duality through total factor productivity and efficiency, linking production economics and farm management.

While profit-maximization (or the related problem of cost-minimization) is useful in a number of applications, both in farm management and agricultural production economics, current work often involves a more complex decision structure that incorporates risk preferences and other factors, such as conservation goals, into a utility framework. (See for example, Patten, Hardaker, and Pannell 1988 or Chouinard et al. 2008). In this utility framework, the decision maker is assumed to maximize a function in which utility is increasing in income and other factors, such as leisure and conservation, and typically decreasing in income variability. In contrast to early static models, time preference for money may be included in analyses, often modeled as multi-stage optimization (e.g., Rae 1971).

Whether based on the theory of the firm or on utility theory, a theoretical model is by nature an abstraction from reality. No model can fully account for the complicated reality it attempts to represent. However, a well-constructed model is nevertheless useful in providing insights into complex problems. As an applied field, farm management ideally should combine the theoretical model with institutional knowledge, either to explain or predict real-world phenomenon or to recommend a course of action. Thus, to be useful, the theoretical model can't be so simple as to ignore key relationships, but at the same time the model can't be so complicated that the questions of interest cannot be answered with available quantitative techniques. A pair of articles in the *Australian Journal of Agricultural Research* provides a critique of the growing irrelevance of farm management research to farm management problems as the theoretical underpinnings became increasingly complex (McCown, Brennan, and Parton 2006; McCown and Parton 2006). The tension between sophisticated theory and on-farm reality in farm management research is not a new problem but traces its history to the early days of the field, as will be discussed in a subsequent section.

3 Quantitative methods in modern farm management

To support better decision making, farm management professionals use a variety of quantitative techniques, including budgeting, statistical analyses and econometrics, and operations research methods. For these tools to be useful, good data are essential. Consequently, farm management professionals often collect information on costs of production through farm surveys or rely on such information collected by others.

An important tool for many farm management applications is budgeting. Budgets are forward planning tools reflecting the best estimate of what will happen in the future. Enterprise budgets provide an estimate of the expected revenue, expenses, and profit from producing a unit of an individual enterprise. Reliable and realistic budgets require a great deal of knowledge of expected prices and yields, as well as detailed information on production practices. Budgets are assumed to be based on the most profitable level of variable inputs and to reflect accurately the level of fixed costs – or sometimes the most common (or recommended) practices for an area. Sources of information may include published historical data, outcomes of agricultural experiments, expert knowledge, and farm surveys. Enterprise budgets are often prepared by farm management professionals, such as those who work for state extension systems, as a planning guide for producers in a region. Producers who use third-party budgets are cautioned to adjust these budgets to fit the individual farm situation.

Enterprise budgets should typically reflect opportunity costs, meaning that the profit is an economic profit, not an accounting profit. If certain opportunity costs are excluded, the "profit" from the enterprise is considered to be a return to the inputs that didn't have associated costs in the budget. Because of the difficulty often found in determining an opportunity cost of management, third-party enterprise budgets may omit this factor, meaning that the budget's bottom line is a return to management. Care should be taken when using a budget from outside of the geographic region for which it was developed, but often these out-of-area budgets can nevertheless be adopted with input from those familiar with the enterprise and area in question.

Other types of budgeting used in farm management include whole-farm budgets, which provide an estimate of revenue, expenses, and profit for the enterprise as a whole, as well as cash flow budgets, which track cash receipts and cash expenditures. A whole-farm budget is based on a whole-farm plan stating the types and amounts of enterprises to produce and the combinations of inputs that will be used to produce them. Because farm plans are highly dependent on individual circumstances, these are rarely produced by third parties, except in case studies or in order to illustrate the feasibility of introducing a new product or making other such changes for a "typical" farm in a region. Similarly, cash flow budgets are usually developed by farm operators, rather than farm management researchers or extension professionals, with exceptions occurring when the financial feasibility of adopting a new technology might be assessed. Budgets or summary farm information of various types or for certain activities or investment categories are often used for benchmarking farms – frequently via comparisons to averages or tiers of peer groups. Langemeier (2015) presents an example of benchmarking machinery investment and costs for a sample farm. Those interested in learning more about basic budgeting techniques in farm management work should consult a general farm management textbook, e.g., Kay, Edwards, and Duffy (2016) or Boehlje and Eidman (1984).

Another quantitative tool used by farm management professionals is statistics. Statistics is an essential tool of all agricultural economists, and farm management is no exception. Farm management researchers frequently work in interdisciplinary teams with production scientists; as such, in addition to a thorough grounding in statistical regression techniques, they also need a good understanding of experimental design and the appropriate analyses of data from experiments, which may include multivariate methods.

Other important tools come from operations research, including mathematical programming and simulation. While programming models have been used in a number of applications in agricultural economics, they lend themselves particularly well to applications in farm management where limited resources must be allocated to meet the firm's overall objective. Such models are especially useful when constant returns to scale and fixed proportions production reasonably describe the production processes; however, more complicated models employing separable programming and/or nonlinear techniques can often be used in cases where these assumptions do not hold. A text by Kaiser and Messer (2011) provides a good introduction to the use of programming models in agricultural economics generally, including farm management applications, and numerous examples of their use in farm management related research can be found in the literature (see for example, Bhaskar and Beghin 2010; Ritten et al. 2010; and Mérel et al. 2014). Simulation has also been used in a variety of different ways in farm management applications, typically to compare outcomes from different strategies in a risky environment (see for example, Canchi et al. 2010; Woodard et al. 2012; and Atallah et al. 2015). Separate chapters in this Handbook address operations research techniques in detail.

4 A brief history of farm management

To understand the current state of farm management work, it is important to first understand its origins. The history of farm management is deeply entwined with the history of agricultural economics itself. The organization we know today as the Agricultural and Applied Economics Association (AAEA) began its life as the American Farm Economic Association in 1919, following a merger of the National Association of Agricultural Economists, founded in 1915, and the American Farm Management Association, founded in 1910 (Runge 2008).

At the time of the founding of our profession, about 30 percent of the United States' population lived on farms. Through the early decades of the twentieth century, there were over six million farms in the United States. In addition, as G.A. Billings of the U.S. Office of Farm Management pointed out in in an address published in the inaugural issue of the *Journal of Farm Economics*, the high demand for agricultural products and the scarcity of farm labor caused by World War I gave "rise to conditions which demand greater concentration of effort on farm management problems" (Billings 1919, p. 3).

One of the earliest long works in the then emerging field of agricultural economics was a text by Henry C. Taylor (1905), a Cornell University professor with a bachelor and master's degree in agriculture and a PhD in Economics. This textbook, titled *An Introduction to the Study of Agricultural Economics*, devoted much of its content to issues in farm management. As Runge (2008) notes, this book applied concepts from Marshall's *Principles of Economics* (which was published in 1890) to farm production. The book also dealt at length with the acquisition and control of land in agriculture.

Although Taylor's text was firmly based in the then relatively new principles of the neoclassical theory of the firm, early farm management texts and papers often took a different approach, based on empiricism largely unguided by theory. The general lack of economic theory in the early farm management work is understandable, given the nature of its early practitioners. Taylor (1928) noted that only 21 percent of researchers in farm economics at that time held PhDs, and many had little training of any kind in economics, but instead had their training in agronomy, horticulture, or other production sciences. Johnson (1955) spoke of two traditions in farm management. Practitioners of one group he named the "heirs" because of their reliance on classical economic theory, and the other he called the "unendowed." This latter group consisted of the agricultural scientists who had not studied economics, but instead came to their interest in farm management through their work with farmers.

The contribution of the "heirs" to our field is perhaps obvious to modern-day agricultural economists, but Johnson also pointed out that the empirical line of farm management provided significant contributions, including an important focus on real-world problems. In addition, Johnson said, the empirically oriented group provided production economists with incentives to revise theoretical concepts when theoretical results were at odds with reality or even to develop new theoretical concepts as needed. Finally, Johnson noted that the emphasis on budgeting prepared agricultural economists to use linear programming. Along these lines, Warren (1932) had earlier stated that he believed it fortunate that foundational work in farm management was typically done by agronomists, as these early professionals brought what he called "the immediate adoption of the scientific rather than the philosophical method of procedure" (p. 7).

The evolution of the field brought with it an increased reliance on economic theory. The history of farm management and production economics from 1946 to 1970 is nicely laid out in a book chapter by Jensen (1977). Jensen noted the importance of Black, Heady, and Johnson in advancing the use of theoretical economics in farm management studies. In particular, he pointed out that Heady's seminal 1948 article emphasized the need for neoclassical theory and statistical analysis in farm management research.

In the 1950s, a new analytical technique was put into use by researchers in farm management. Linear programming, developed during World War II by Danzig and others, began to appear in the farm management literature. In 1951, Waugh applied linear programming to the problem of determining the least-cost ration for dairy cows. In the same year, Hildreth and Reiter (1951) applied what they called a "linear production model" to a crop rotation problem. Applications of linear programming became increasingly frequent in agricultural economics in general and farm management in particular through the rest of the decade. In 1958, Heady and Candler wrote an influential textbook specifically designed to show how linear programming could be used in agricultural economics applications.

With the development of computers and the continuing increase in computational power, both statistical analyses and mathematical programming applications became progressively more sophisticated. Integer and mixed integer programming applications began appearing in the 1960s (e.g., Musgrave 1962). Quadratic programming was developed by Freund in the mid-1950s (Freund 1956) and provided a method for the analysis of decision making under risk, but computational constraints at that time greatly limited its practical application. Consequently, Hazell (1971) developed MOTAD (minimization of the total absolute deviations) as a linear approximation, and Thomas et al. (1972) proposed the use of separable programming as a way of incorporating risk into farm decision models. Agrawal and Heady's influential textbook (1972) described a number of variations on linear programming, including variable resource programming, quadratic programming, integer programming, and dynamic programming. In 1983, Tauer developed target MOTAD, a variation on the MOTAD model, consistent with second degree stochastic dominance.

While the concept of dynamic programming was developed in the early 1950s, its initial applications were limited by computational power. Burt and Allison (1963) published one of the early applications of this technique in farm management in a study of the decision of leaving land fallow versus planting. In the 1980s and 1990s, with the advent of "supercomputing," applications became more commonplace. Zacharias and Grube (1986), for example, used dynamic programming to find the most profitable pest management strategies for corn and soybeans produced in rotation. Burt (1981) used a dynamic optimization model to study farm-level soil conservation decisions.

From the 1980s onward, farm management research tended to veer in two directions. In our profession's major disciplinary journal, the *American Journal of Agricultural Economics*, "pure" farm management work became less common. More often, research involving decision making at the firm level was linked to either farm policy (especially crop risk management) or issues in resource economics. Research involving costs, budgets, or analyses of optimal levels of a given input more typically appeared in interdisciplinary journals or in other outlets specifically geared toward such applications. Examples of both types of work are provided in the next section.

5 Farm management outlets and information in the twenty-first century

Today, only around 2 percent of the U.S. population lives on farms, and farm numbers have dropped to a little over two million. The demographic shifts explain, to a great degree, why the field of agricultural economics in the United States has broadened over time, so that farm management, once the dominant field, is now one of many. Graduate programs may have a course in farm management, but typically it would not be required for doctoral students. Instead, graduate students interested in farm management are expected to take courses in operations research, microeconomic theory, and statistical methods. Their dissertation or thesis work would then apply these skills to farm management topics.

Cooperative extension programs continue to provide a great deal of useful tools and information for farm owners and operators, much of it freely available on the web. Enterprise budgets and price forecasts are included on some sites, as is information on farmland rental rates and information on labor management. While a full accounting of all the extension farm management programs and decision support tools is beyond the scope of this chapter, a few examples are provided. The Department of Agricultural Economics at Kansas State University sponsors the AgManager.info website as "a comprehensive source of information, analysis, and decision-making tools for agricultural producers, agribusinesses, and others" (Kansas State University 2016). Iowa State University's website has a "crop and livestock land use analyzer," a spreadsheet application that allows producers to enter their own information to compare estimated profits from alternative farm plans (Iowa State University 2016).

The farmdoc website, maintained by the Department of Agricultural and Consumer Economics at the University of Illinois with contributors from (and links to) other institutions, provides "crop and livestock producers in the U.S. Cornbelt with round-the-clock access to economic information and analysis to better manage their farm businesses" (University of Illinois 2016). Texas A&M University's website has, among many other things, a decision aid to help with crop insurance choices and a decision aid that assesses the impact on profit of techniques to combat aflatoxin in corn (Texas A&M University 2016). A number of states also have farm business analysis associations, which provide individual assistance to producers in financial analysis, tax management, and estate planning.

The current state of research in farm management reflects a division between highly disciplinary work on the one hand and work that is of a more applied nature on the other. Our primary disciplinary journal, the *American Journal of Agricultural Economics* (*AJAE*), continues to regularly publish articles that involve "farm management" in its broadest sense, meaning work that involves analyses concerning decision making at the level of the individual farm firm. However, unlike much of the earlier disciplinary work, current research in the *AJAE* typically has little direct application to agricultural producers but is instead aimed at the scholarly academic community.

The general focus of material found in the *AJAE* involves either novel approaches in the application of economics to high-priority issues in agriculture or expanding economic theory relevant to our field. An example of an applied *AJAE* article germane to farm management can be found in Färe, Grosskopf, and Lee (1990), a first-time application of a nonparametric approach to expenditure-constrained profit maximization. Some other articles published in the last two decades in the *AJAE* illustrate how on-farm decisions are tied to policy and resource conservation issues. Just and Kropp (2013) used a framework for profit-maximization at the farm level to analyze land-use distortions associated with agricultural support payments from the government. Coble et al. (1996) depicted the demand for crop insurance at the farm-level using panel data from a set of Kansas farms. Schoengold, Ding, and Headlee (2015) estimated the impact of risk-reducing government programs on the use of conservation tillage. Other such examples can be identified by scanning the abstracts of the articles in the journal.

Applications relevant to farm management can similarly be found in the *Journal of Agricultural and Applied Economics*, the *Journal of Agricultural and Resource Economics*, and other outlets for disciplinary work in agricultural economics, including a host of international journals such as *Agricultural Economics*, the *Journal of Agricultural Economics*, the *Canadian Journal of Agricultural Economics*, and the *Australian Journal of Agricultural Economics*. *Land Economics*, another well-known journal, also occasionally publishes research that relates to farm management if the work connects to the journal's subject matter (e.g., Chouinard et al. 2008; Bryan, Deaton, and Weersink 2015). The *Choices* online publication sponsored by the AAEA focuses primarily on public policy and general interest applications of economics, with an audience ranging from the profession to the general public – not an audience specifically of farm decision makers.

Farm management work with more direct on-farm applications can be found in the *Journal of the American Society of Farm Managers and Rural Appraisers*, which is published annually by the society. Almost every article in this journal involves farm management issues of direct concern to farm operators. A recent volume contains an article that uses marginal analysis to assess the profitability of adopting individual-nozzle control for sprayers (Smith and Dhuyvetter 2016). Another article compares rates for custom farm work with machinery costs (Edwards 2016). And yet another involves a cost and return analysis for meat goats (Qushim, Gillespie, and McMillin 2016). Some work in this journal clearly reflects the heritage of early farm management professionals who collected farm survey data to determine costs of production (Ringelberg, Gunderson, and Widmar 2016).

Several farm management topics have been developed from the general management literature and also from the somewhat more closely related agribusiness management literature. Gray et al. (2004) examined how management literature intersects with and enhances the contributions of economics in innovation assessment. This work may not seem to have a direct application to farm management, but it is certainly not precluded, and this area holds many useful applications, as do other examples from the management and business literature.

Farm management research may also appear in journals with an interdisciplinary focus. The *Agronomy Journal*, for example, has a section devoted to "crop economics, production, and management," where agricultural economists frequently publish research related to the economics of crop production. *Hort Technology*, the *Journal of Dairy Science*, and other production journals also provide space for farm management research. *Agricultural Systems* provides an outlet for the related research field of farming systems analysis. Finally, it needs to be mentioned that although agricultural production directly employs only a small percentage of the U.S. labor force, agriculture remains the largest employer in many developing nations, and trade in agricultural products is an important engine for economic growth. The International Food and Agribusiness Management Association (IFAMA) publishes the *International Food and Agribusiness Management Review*, which frequently includes research relevant to farm management.

6 Farm management: future applications

Big data is more than a catch phrase. It represents the potential to utilize volumes of data that were until recently unable to be captured, stored, analyzed, and converted into practical applications. O'Donoghue et al. (2016) demonstrate the application of big data at the farm level to integrate production and financial data so as to enhance financial and environmental performance with benchmarking and also with improved information for decision making. Rapid growth in technology and information science continues to expand opportunities for differentiated actions with targeted results based on data from multiple platforms. Data and analysis can now link to real-time applications in systems to promote activities that target optimized outcomes based on more refined expectations. These predictive systems continue to develop as they benefit from expanded data insights.

The data analytics from firms such as Google or Amazon create buzz about their ability to target consumer decisions. Agriculture may find the prescriptive power of big data guiding decisions regarding seed genetics and variable input levels in the field – or with the farm financial and production management schemes as proposed in O'Donoghue et al. (2016). The Monte Carlo simulations used previously to generate expectations for yields in analyses of crop insurance or other program participation may be enhanced to degrees not deemed possible a decade before this writing. Rather than probabilistic distributions defining model outcomes, results may be guided by correlated factors leading to expected outcomes. The predictive capability of models (or machine learning) based on associated (correlated) variables can be strong. An even greater contribution of economics to this exciting field may be in explaining the potential changes and likely outcomes of future drivers as we use big data to look behind the curtain of probabilistic outcomes.

As the ownership, control, and structure of farms change, so may the financial tools and information that farm decision makers find most useful. Owner–operators are fewer in number, with absentee owners, specialized managers, and owner–general managers increasing on U.S. farms. These changing demographics may require new ways to communicate on-farm activities, structure equitable rents or shares, or implement new skills such as employee hiring and retention. Basic financial pro-forma documents may need supplementation with more complex economic analytics as farm managers seek to leverage data and extract more profits through superior decisions.

Farm management may have seemed as if it were one of the most thoroughly covered topics in science and thus a place where new discoveries were unlikely. However, as a critical underpinning of society with extensive history and a broad-based knowledge, farm management may provide a treasure chest that is just being opened through new theoretical and empirical applications for science as we move forward into the age of big data.

References

Agrawal, R.C., and E.O. Heady. 1972. *Operations Research Methods for Agricultural Decisions*. Ames, IA: Iowa State University.

Atallah, S.S., M.I. Gómez, J.M. Conrad, and J.P. Nyrop. 2015. "A Plant-Level, Spatial, Bioeconomic Model of Plant Disease Diffusion and Control: Grapevine Leafroll Disease." *American Journal of Agricultural Economics* 97:199–218.

Bhaskar, A., and J.C. Beghin. 2010. "Decoupled Farm Payments and the Role of Base Acreage and Yield Updating Under Uncertainty." *American Journal of Agricultural Economics* 92:849–858.

Billings, G.A. 1919. "President's Address, American Farm Management Association, Baltimore, MD, January 8, 1919." *Journal of Farm Economics* 1:3–7.

Boehlje, M.D., and V.R. Eidman. 1984. *Farm Management*. New York: John Wiley & Sons.

Box, G.E.P., and N.R. Draper. 1987. *Empirical Model-Building and Response Surfaces*. New York: John Wiley & Sons.

Bryan, J., B.J. Deaton, and A. Weersink. 2015. "Do Landlord-Tenant Relationships Influence Rental Contracts for Farmland or the Cash Rental Rate?" *Land Economics* 91:650–663.

Burt, O.R. 1981. "Farm Level Economics of Soil Conservation in the Palouse Area of the Northwest." *American Journal of Agricultural Economics* 63:83–92.

Burt, O.R., and J.R. Allison. 1963. "Farm Management Decisions with Dynamic Programming." *Journal of Farm Economics* 45:121–136.

Canchi, D., N. Li, K. Foster, P.V. Preckel, A. Schinckel, and B. Richert. 2010. "Optimal Control of Desensitizing Inputs: The Case of Paylean." *American Journal of Agricultural Economics* 92:56–69.

Chavas, J.-P., R.G. Chambers, and R.D. Pope. 2010. "Production Economics and Farm Management: A Century of Contributions." *American Journal of Agricultural Economics* 92:356–375.

Chouinard, H.H., T. Paterson, P.R. Wandschneider, and A.M. Ohler. 2008. "Will Farmers Trade Profits for Stewardship? Heterogeneous Motivations for Farm Practice Selection." *Land Economics* 84(1):66–82.

Coble, K.H., T.O. Knight, R.D. Pope, and J.R. Williams. 1996. "Modeling Farm-Level Crop Insurance Demand with Panel Data." *American Journal of Agricultural Economics* 78:439–447.

Coelli, T.J., D.S. Prasada Rao, C.J. O'Donnell, and G.E. Battese. 2005. *An Introduction to Efficiency and Productivity Analysis* (2nd ed.). New York: Springer Science + Business Media, LLC.

Edwards, W.M. 2016. "Have Custom Rates Kept Pace with Machinery Costs?" *Journal of ASFMRA*. www.asfmra.org/2016-journal-of-asfmra/. Accessed July 15, 2016.

Färe, R., S. Grosskopf, and H. Lee. 1990. "A Nonparametric Approach to Expenditure-Constrained Profit Maximization." *American Journal of Agricultural Economics* 72:574–581.

Freund, R.J. 1956. "The Introduction of Risk into a Programming Model." *Econometrica* 24:253–256.

Gray, A., M. Boehlje, V. Amanor-Boadu, and J. Fulton. 2004. "Agricultural Innovation and New Ventures: Assessing the Commercial Potential." *American Journal of Agricultural Economics* 86:1322–1329.

Hazell, P.B.R. 1971. "A Linear Alternative to Quadratic and Semivariance Programming for Farm Planning Under Uncertainty." *American Journal of Agricultural Economics* 53:664–665.

Heady, E.O. 1948. "Elementary Models in Farm Production Economics Research." *Journal of Farm Economics* 30:201–225.

Heady, E.O., and W. Candler. 1958. *Linear Programming Methods*. Ames, IA: Iowa State University Press.

Hildreth, C., and S. Reiter. 1951. "On the Choice of a Crop Rotation Plan." In T.C. Koopmans, ed., *Activity Analysis of Production and Allocation*. New York: John Wiley & Sons, pp. 177–188.

Iowa State University, Extension and Outreach. 2016. "Ag Decision Maker: An Agricultural Economics and Business Website." www.extension.iastate.edu/agdm/. Accessed August 10, 2016.

Jensen, H.R. 1977. "Farm Management and Production Economics, 1946–1970." In L.R. Martin, ed., *A Survey of Agricultural Economics Literature* (Vol. 1). St. Paul: North Central Publishing Company, pp. 3–89.

Johnson, G.L. 1955. "Results from Production Economic Analysis." *Journal of Farm Economics* 37:206–222.

Just, D.R., and J.D. Kropp. 2013. "Production Decisions from Static Decoupling: Land Use Exclusion Restrictions." *American Journal of Agricultural Economics* 95:1049–1067.

Kaiser, H.M., and K.D. Messer. 2011. *Mathematical Programming for Agricultural, Environmental, and Resource Economics*. New York: Wiley.

Kansas State University, Department of Agricultural Economics. 2016. "About AgManager.info." http://AgManager.info. Accessed August 24, 2016.

Kay, R., W. Edwards, and P. Duffy. 2016. *Farm Management*. New York: McGraw-Hill.

Langemeier, M. 2015. "Benchmarking Machinery Investment and Cost." *farmdoc daily (5)*: 170, Department of Agricultural and Consumer Economics, University of Illinois at Urbana-Champaign. http://farmdocdaily.illinois.edu/2015/09/benchmarking-machinery-investment-and-cost.html. Accessed August 27, 2016.

Malcolm, B. 2004. "Where's the Economics? The Core Discipline of Farm Management Has Gone Missing!" *The Australian Journal of Agricultural and Resource Economics* 48:395–417.

McCarl, B.A., and T.H. Spreen. 2007. *Applied Mathematical Programming Using Algebraic Systems*. http://agecon2.tamu.edu/people/faculty/mccarl-bruce/mccspr/thebook.pdf. Accessed August 27, 2016.

McCown, R.L., L.E. Brennan, and K.A. Parton. 2006. "Learning from the Historical Failure of Farm Management Models to Aid Management Practice. Part 1. The Rise and Demise of Theoretical Models of Farm Economics." *Australian Journal of Agricultural Research* 57:143–156.

McCown, R.L., and K.A. Parton. 2006. "Learning from the Historical Failure of Farm Management Models to Aid Management Practice. Part 2. Three Systems Approaches." *Australian Journal of Agricultural Research* 57:157–172.

Mérel, P., F. Yi, J. Lee, and J. Six. 2014. "A Regional Bio-economic Model of Nitrogen Use in Cropping." *American Journal of Agricultural Economics* 96:67–91.

Mishra, A.K., C.B. Moss, and K. Erickson. 2004. "Valuing Farmland with Multiple Quasi-Fixed Inputs." *Applied Economics* 36:1669–1675.

Mugera, A.W., M.R. Langemeier, and A. Ojede. 2016. "Contributions of Productivity and Relative Price Changes to Farm-Level Profitability Change." *American Journal of Agricultural Economics* 98:1210–1229.

Mundlak, Y. 1961. "An Empirical Production Function Free of Management Bias." *Journal of Farm Economics* 43:44–66.

Musgrave, W.F. 1962. "A Note on Integer Programming and the Problem of Increasing Returns." *Journal of Farm Economics* 44:1068–1076.

O'Donoghue, C., A. McKinstry, S. Green, R. Fealy, K. Heanue, M. Ryan, K. Connolly, J.C. Desplat, B. Horan, and P. Crosson. 2016. "A Blueprint for a Big Data Analytical Solution to Low Farmer Engagement with Financial Management." *International Food and Agribusiness Management Review* Special Issue 19(A):131–154.

Patten, L.H., J.B. Hardaker, and D.J. Pannell. 1988. "Utility Efficient Programming for Whole-Farm Planning." *Australian Journal of Agricultural Economics* 32:88–97.

Qushim, B., J.M. Gillespie, and K. McMillin. 2016. "Analyzing the Costs and Returns of U.S. Meat Goat Farms." *Journal of ASFMRA*. www.asfmra.org/2016-journal-of-asfmra/. Accessed July 15, 2016.

Rae, A.N. 1971. "Stochastic Programming, Utility, and Sequential Decision Problems in Farm Management." *American Journal of Agricultural Economics* 53:448–460.

Ringelberg, J., M. Gunderson, and D. Widmar. 2016. "Strategies and Time Allocation of Large Commercial Agricultural Producers." *Journal of ASFMRA*. www.asfmra.org/2016-journal-of-asfmra/. Accessed July 15, 2016.

Ritten, J.P., W.M. Frasier, C.T. Bastian, and S.T. Gray. 2010. "Optimal Rangeland Stocking Decisions Under Stochastic and Climate-Impacted Weather." *American Journal of Agricultural Economics* 92:1242–1255.

Runge, C.F. 2008. "Agricultural Economics." In S.N. Durlauf and L.E. Blume, eds., *The New Palgrave Dictionary of Economics* (2nd ed.), Vol. 1. New York: Palgrave Macmillan, pp. 78–88.

Schnitkey, G., N. Paulson, J. Coppess, and C. Zulauf. 2016. "2015 Estimated ARC-CO Payments." *farmdoc daily (6):157*, Department of Agricultural and Consumer Economics, University of Illinois at Urbana-Champaign. http://farmdocdaily.illinois.edu/2016/08/2015-estimated-arc-co-payments.html. Accessed August 24, 2016.

Schoengold, K., Y. Ding, and R. Headlee. 2015. "The Impact of AD HOC Disaster and Crop Insurance Programs on the Use of Risk-Reducing Conservation Tillage Practices." *American Journal of Agricultural Economics* 97:897–919.

Smith, C.M., and K.C. Dhuyvetter. 2016. "Determining the Optimal Level of Control on Sprayers and Planters." *Journal of ASFMRA*. www.asfmra.org/2016-journal-of-asfmra/. Accessed July 15, 2016.

Tauer, L.W. 1983. "Target Motad." *American Journal of Agricultural Economics* 65:606–610.

Taylor, H.C. 1905. *An Introduction to the Study of Agricultural Economics*. New York: Macmillan.

Taylor, H.C. 1928. "Research in Agricultural Economics." *Journal of Farm Economics* 10:33–41.

Texas A&M University, Department of Agricultural Economics. 2016. "Extension and Related Links." http://agecon.tamu.edu/extension/. Accessed August 22, 2016.

Thomas, W., L. Blakeslee, L. Rogers, and N. Whittlesey. 1972. "Separable Programming for Considering Risk in Farm Planning." *American Journal of Agricultural Economics* 54:260–266.

Traxler, G., and D. Byerlee. 1993. "A Joint Product Analysis of the Adoption of Modern Cereal Varieties in Developing Countries." *American Journal of Agricultural Economics* 75:981–989.

University of Illinois, Department of Agricultural and Consumer Economics. 2016. "About farmdoc." www.farmdoc.illinois.edu/about/. Accessed August 24, 2016.

Varian, H.R. 1992. *Microeconomic Analysis* (3rd ed.). New York: W.W. Norton & Company, Inc.

Warren, G.F. 1932. "The Origin and Development of Farm Economics in the United States." *Journal of Farm Economics* 14:2–9.

Waugh, F.V. 1951. "The Minimum-Cost Dairy Feed: An Application of Linear Programming." *Journal of Farm Economics* 33:299–310.

Woodard, J.D., A, D. Pavlista, G.D. Schnitkey, P.A. Burgener, and K.A. Ward. 2012. "Government Insurance Program Design, Incentive Effects, and Technology Adoption: The Case of Skip-Row Crop Insurance." *American Journal of Agricultural Economics* 94:823–837.

Zacharias, T.P., and A.H. Grube. 1986. "Integrated Pest Management Strategies for Approximately Optimal Control of Corn Rootworm and Soybean Cyst Nematode." *American Journal of Agricultural Economics* 68:704–715.

36
Economics of agricultural biotechnology

David Zilberman, Justus Wesseler, Andrew Schmitz, and Ben Gordon

1 Introduction

A key element to increase agricultural productivity has been the continuous improvement of crops and livestock through selective breeding, transgenesis, and mutagenesis. The discovery of DNA has provided a new way to enhance human health and the capacity to improve animals and crops for higher production. The growing understanding of the basic mechanism governing living organisms has immense potential for providing new forms of medical treatment and crop varieties. The Cohen–Boyer patent that enabled splicing of genes to make recombinant proteins led to the foundation of the medical biotechnology industry in the 1970s (Hughes 2001). There have been major efforts to apply the new molecular and cell biology knowledge to agriculture. The first application of genetically modified organisms (GMOs) in crop production was the spraying of strawberry fields with the Ice-minus bacteria. There were also attempts to make genetically modified varieties that would enable non-leguminous crops to fix nitrogen. However, for example, even though the Flavr Savr tomato variety had a longer shelf life, it did not last long on the commercial market. The commercially successful application of GMOs for pest control was introduced in the mid-1990s and included GMO varieties that are either resistant to insects (in particular, Bt varieties) or tolerant of herbicides (in particular, Roundup Ready varieties), as well as virus-resistant papaya. Most of the early development leading to the present situation of GMO and biotechnology can be found in Cook-Deegan and Heaney (2010), Wieczorek and Wright (2012), and Rangel and Maurer (2015).

Early, Monsanto held a dominant position in controlling many of the crucial patents and intellectual property rights (IPR). Within the United States, regulation of genetic engineering[1] (GE) has been the subject of much debate. Today, agricultural biotechnology is widely used in fiber (cotton) and feed (soybean, corn), but not much in food for human consumption. The United States, Brazil, and Argentina are the major producers of transgenic corn and soybeans. While GE cotton is produced in several countries, such as China, India, and the United States, the production of GE cotton is banned in most of Europe and many African countries. Debates about the future of biotechnology continue. While there is some optimism for the increased use of biotechnology in agriculture, there may be negative environmental consequences associated with GMOs. Now, gene editing, which is another form of biotechnology, poses new challenges.

This chapter reviews the economic literature on the impact of agricultural biotechnology on crop productivity/yield/cost and adoption patterns and on the environment. The gainers and losers from agricultural biotechnology are documented. Despite the gains that can be derived from biotechnology, there is strong resistance, especially among consumers for GM products. Opposition comes from many groups, including Greenpeace. For major commodities such as wheat and rice, GM varieties have not been commercialized. This has resulted in significant economic losses, especially for consumers in less developed countries.

2 Agricultural biotechnology: impacts of agricultural production, markets, and welfare

Bennett et al. (2013) distinguish between three types of agricultural biotechnology: first-generation, second-generation, and third-generation traits. The majority of traits in the pre-market stage or on the commercial market are first-generation traits. First-generation traits are agronomic qualities, which primarily contribute to pest resistance and herbicide tolerance of crops and are widely used commercially. More than 50 percent of the traits in field trials are second-generation traits, which enhance product quality and stress tolerance and alter crop growth in some way. There are very few field trials of third-generation traits, which are qualities that allow for inherent plant creation of industrial products and pharmaceutical drugs.

The use of GE technologies in agriculture has increased since 1996 (Figure 36.1). Approximately 90 percent of the maize, soybean, and cotton produced in the United States, Brazil, and Argentina are GE crops. Close to 100 percent of the sugar beets produced in the United States and canola produced in Canada are GE crops, and roughly 90 percent of cotton produced in India are GE crop varieties. Both herbicide-tolerant and pest-resistant varieties occupy more

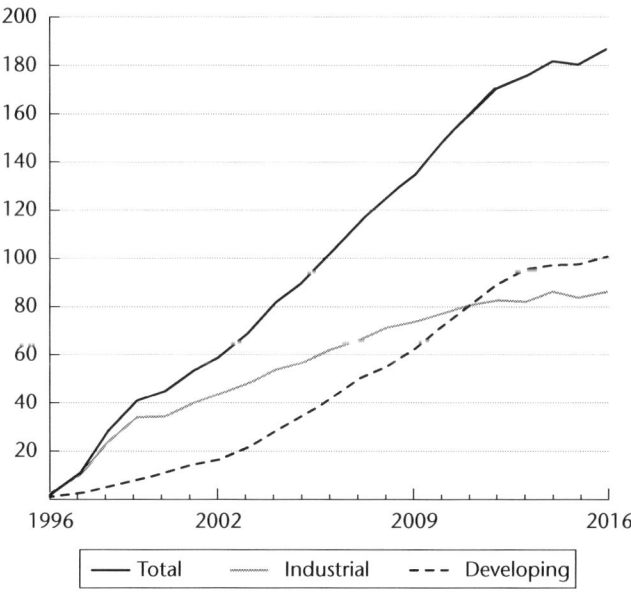

Figure 36.1 Global area of biotech crops, 1996–2016 (in million hectares)
Source: ISAAA (2016).

than 50 percent of the global GE area. Across significant crop areas, farmers plant "stacked" varieties (i.e., containing multiple GE traits).

Much of the literature focuses on the impact of pest-control traits in maize, soybean, and cotton. Lichtenberg and Zilberman (1986) use a damage control framework to examine output where consideration is given to pest damage, which depends both on the initial pest infestation and the amount and effectiveness of control. Although the adoption of GE varieties may tend to increase yield by reducing pest damage, it also may result in a reduction of yield if the GE trait is inserted into an inferior variety compared to the variety it replaces (Qaim and Zilberman 2003). Analysis by Qaim and Zilberman (2003) suggests that GE varieties are more likely to increase yield in locations that are more vulnerable to significant pest infestations, where alternative pest control mechanisms are not effective, and when the traits are inserted in superior varieties. Their analysis further suggests that GE varieties are likely to have much more significant yield effects in developing countries where alternative pest controls are less available. While in developed countries, GE varieties may have a stronger effect on the reduction of pesticide use.

Klümper and Qaim (2014) examine the effects of GE soybeans, maize, and cotton based on a meta-analysis of 147 original studies on yield, pesticide use, and farmer profits (Figure 36.2). They find that, on average, adoption of GE crops increases yield by more than 20 percent and reduces pesticide use and cost by more than 36 percent. Additionally, although these crops slightly increase production costs (~3 percent), farmer profit is increased by close to 70 percent. Furthermore, they find that the effects are more pronounced in developing countries (e.g., the yield effect in developing countries is 13 percent greater than the average, and pesticide quantity reduction is 10 percent lower than the average). The National Research Council (NRC 2010) finds that the major benefits from adoption of GE in the United States have been reduction in pesticide use as measured by toxicity, non-pecuniary benefits like reduced vulnerabilities to toxins and reduced effort by farmers, and greater control of farming production. Krishna, Qaim, and Zilberman (2015) find that in the case of Bt cotton, the adoption of Bt technology increased the mean yield and decreased yield uncertainty and yield risk; the combined positive effect has been a major contributor to the fast adoption of Bt cotton in India. Fernandez-Cornejo et al. (2014) find that in the United States, adoption of GE varieties benefits farmers through increases in yield, reductions in pest management costs (especially insecticides), and reductions in overall management costs. The yield effect of GE varieties has increased over time as the number of traits in "stacked" varieties has increased. Benefits also increase as traits are introduced to new crops, like alfalfa and sugar beets. In the United States, GMO sugar beet yields have increased (Kennedy, Lewis, and Schmitz 2017), whereas non-GMO sugarcane yields have not increased since 2000 (Schmitz and Zhu 2017). In addition, future work should consider the potential for the use of GMO varieties for crops such as cassava, where non-GMO cassava yields have remained flat over many years (Moss and Schmitz, forthcoming).

Research by Schmitz and Moss (2016) suggests that benefits from the adoption of GMOs may be understated, as these fit well with modern farm mechanization that includes no-till air drills, which are widely used in continuous cropping practices in the United States. Recent studies have used disaggregated agricultural production in different regions to assess the impact of the adoption of GE varieties given heterogeneity in terms of land quality and crop variety, as well as uncertainty in terms of weather. Huffman, Jin, and Xu (2016) find that GE varieties have provided a significant hedge against crop damage due to excess summer heat and loss of soil moisture. Shi, Chavas, and Lauer (2013) demonstrate that the impact of GE traits depends on the variety into which they are introduced. For example, genes that carry transgenic traits can interact with already existing genes or other transgenic genes. This interaction could be favorable and lead to higher crop yields, or negative, leading to a reduction in yield. However,

Economics of agricultural biotechnology

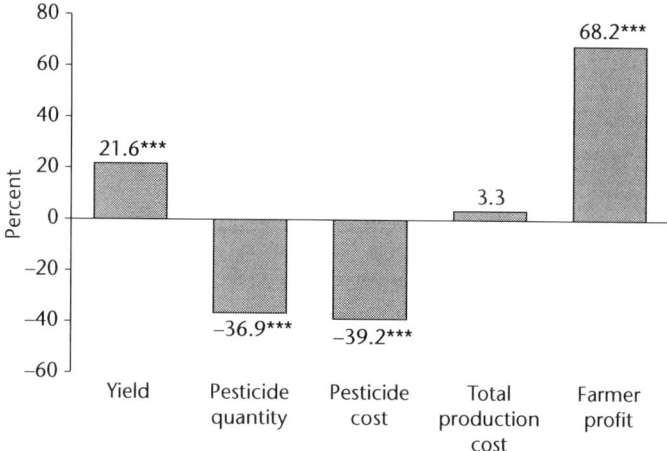

Figure 36.2 Impact of GM crop adoption
Source: Klümper and Qaim (2014).

positive gene interactions can be selected for by geneticists and plant breeders by replacing low-yielding varieties with superior ones. Furthermore, Shi et al. (2013) argue that some of the yield effect associated with GE traits may be the result of the introduction of GE traits into already improved varieties. Lusk, Tack, and Hendricks (2017) use panel data of U.S. corn yields between 1980 and 2015 combined with weather and soil characteristics data to show that the yield effects of GE varieties vary across location and that estimating the impact requires accounting for changing weather, soil, and other sources of variability to explain adoption and to assess yield effects. As theory suggests, GE varieties tend to counter some of the negative effects arising from pest and weather variability and are more likely to be adopted in areas with greater vulnerability to pest infestation. Assuming a homogenous GE effect, the analysis finds that, on average, adoption of GE varieties increases corn yield by 17 percent relative to the five-year average yield prior to GE use.

Unlike the early literature that consists of individual case studies for fields or regions, Barrows, Sexton, and Zilberman (2014b) use aggregate data to study overall agricultural output and land allocation between GE and non-GE varieties in the production of corn, soybean, and cotton for 26 countries between 1996 and 2010. In their regression analysis, they separate the contribution of country, time, and technology, with different specifications, for yield per acre for each of the three crops. They find that, on average, adoption of GE varieties increased cotton yield between 17 percent and 25 percent, for corn between 8 percent and 25 percent, and for soybeans between 1 percent and 3 percent. These impacts represent an intensive margin effect, where farmers switch from one technology to another. In addition, the adoption of GE varieties has an extensive margin effect, where overall acreage of a crop increases or decreases. Barrows, Sexton, and Zilberman (2014b) also develop a method to estimate the extensive margin effect. For example, with adoption of GE soybean varieties, acreage increased significantly, and, in many cases, the use of Roundup Ready varieties allowed for double cropping. Barrows, Sexton, and Zilberman (2014a) combine their estimates of yield effects with those from the literature to assess the supply effect of the introduction of GE varieties. They estimate that GE varieties increased the supply of corn in the range of 2 percent to 19 percent, of cotton 0 percent to 29 percent, and of soybeans 2 percent to 39 percent. Combining these estimates with several other demand price elasticities, they assess the impact of the introduction of GE varieties on prices and find that the adoption of GE corn varieties reduced the expected price of corn by 13 percent, cotton by 18 percent, and soybeans by 2 percent to 65 percent. Thus, the beneficiaries of GE crops include consumers and feedlot operators. The price reduction on consumer price (e.g., meat) due to GE was much smaller than the commodity price because it (e.g., corn as feedstock) makes up only a fraction of food prices.

Brookes and Barfoot (2014) assess the impact of the introduction of GE varieties on farmers' welfare in developed and developing countries. They find that the net farm-level income gain was US$18.8 billion in 2012 and more than US$100 billion, in nominal terms, for 1995 to 2012. The gains are equal for farmers in developing and developed countries. GE technology has added 122 million tons of global soybeans and 230 million tons of maize since their introduction in the mid-1990s. Insect-resistant traits have benefited farmers in developing countries through lower cost of production and those in both developed and developing countries through increased yield. Herbicide-tolerant varieties have increased incomes mostly through reduced cost of production, enabling the adoption of no-tillage production systems and, importantly, double cropping of soybean after wheat in Latin America. The NRC (2010) considers several studies that assess the welfare impact of adoption of GE varieties among various groups. It finds benefit sharing among technology providers, farmers, livestock producers, and consumers. As the adoption of the technology increases, the share of the consumer gain increases, but the distribution of benefits varies across regions.

Mahaffey, Taheripour, and Tyner (2016) use computable general equilibrium (CGE) models to assess the impact of a global GMO ban. The CGE approach has the advantage of being able to consider market interaction and adjustments, but this approach depends on the elasticities that are used, as well as assumptions of the impact of alternative technologies. Their analysis recognizes heterogeneity among regions but ignores, for example, the extensive margin effect and tends to underestimate the effects of GM varieties on soybean production. They estimate that a global ban on GMOs would reduce the acreage of cotton by 1.4 percent and increase the acreage of sorghum. The prices of coarse grains, in this scenario, would increase by 3.5 percent and of soybean by 4.5 percent; the overall price of food would increase by 0.8 percent. A key result is that agricultural acreage would increase significantly. Global cropland would increase by 3.1 hectares, including 0.6 million hectares of deforestation. The global welfare loss would be US$8.5 billion, and much of it would fall on consumers. The United States would be more affected than the global average as agricultural supply would decrease by up to 3.35 percent and food prices would increase by 0.81 percent, which is US$13 billion in additional food cost. Banning GMOs would be costly to the United States and China, as their capacity to increase productivity is limited, as well as to India and other South Asian countries due to greater food cost loss. At the same time, Brazil and Argentina would gain from higher food prices and the ability to expand production.

While there are obvious gains from the adoption of GE technologies, the effectiveness of both the insecticide-resistant and herbicide-resistant technologies have decreased due to insect and weed resistance to the technologies (NRC 2010). One approach to address insect resistance is to establish refuges and allocate some area of the field to non-GE varieties. There is a growing literature on the design of optimal refuge, taking into account population dynamics and balancing the benefits of resistance buildup reduction with the cost of reduced yield (Devos et al. 2013). However, insect resistance may occur and development of new GE treatment with multiple modes of pest control will provide new avenues to overcome resistance. The emergence of weed resistance can be reduced by weed management (Frisvold and Reeves 2010) and by the use of selective herbicides (Smyth et al. 2011). This may reduce the environmental benefits of herbicide-resistant technologies. Nevertheless, there are environmental benefits for a long period until resistance becomes significant, and sequential development of new traits and more environmentally benign pesticides can sustain these benefits further (Wesseler, Scatasta, and Fall 2011).

3 GE crops: the impact on health and the environment

Environmental groups, like Greenpeace, have played a major role in the resistance to GE technology in agriculture, and concerns about health and environmental effects have been major reasons for delaying, or even banning, the technology (Paarlberg 2009; Apel 2010). However, the impact of GE technologies on health and the environment has been under intense inquiry. While there have been several studies purporting to show that GE technologies may imperil human health, after scientific scrutiny, all these studies have been retracted or discredited (Xia et al. 2015). This does not imply that GE technologies are without flaws or risks. GE in agriculture consists of many traits, and either the traits themselves or their implementation may result in negative outcomes. This is the reason that the screening and assessment of the technology both before and after its introduction is essential. Prominent science academies around the world have not found any evidence that GE technology poses new risks to human health or the environment (Herring and Paarlberg 2016). E.O. Wilson, the eminent biologist who developed the concept of biodiversity, stated (see Douglas 2001) "I'll probably get it in the neck from my conservationist colleagues, but we've got to go all out on genetically modified crops. There doesn't seem to be any other way of creating the next green revolution without GMOs."

Several studies have found that the adoption of GE crops has had positive effects, mostly through reduced farmer exposure to dangerous insecticides (Qaim 2009). From surveys, it has been found that millions of smallholder cotton growers in China have reduced their exposure to dangerous pesticides and improved their health by adopting Bt cotton (Pray et al. 2002). In addition, there is a large potential to improve health by adoption of GE rice in China (Tan, Zhan, and Chen 2011). GE varieties can address severe mycotoxin problems. For example, the pest protection capabilities of Bt corn have been shown to decrease the level of aflatoxin and carcinogenic chemicals and provide significant health and economic benefits throughout the world, especially in developing countries (Wu 2006). China and other Asian countries have experienced severe aflatoxin contamination (Wang and Liu 2007), which can be partially addressed by the use of GE varieties (Wu and Butz 2004).

Because of the yield effect of GE technologies and the inelastic nature of the demand for food, the adoption of agricultural biotechnology tends to increase the agricultural footprint in terms of land, chemical, water, and energy use. Since adoption of herbicide-tolerant varieties has enabled the expansion of double-cropping and no-tillage practices, it has also contributed to the reduction of greenhouse gas (GHG) emissions (Smyth et al. 2011; Bennett et al. 2013). Barrows, Sexton, and Zilberman (2014b) find that, at minimum, the GHG emissions savings (CO_2 equivalent) in 2010 was 0.15 gigatons, which is roughly one-eighth the emissions of all passenger cars in the United States. Mahaffey, Taheripour, and Tyner (2016) find that if GE technology were banned in agriculture, it would increase CO_2 equivalent by approximately 0.9 billion tons annually. If the intensification of adoption within the countries that already allow GE technology increased to at least the U.S. level, Mahaffey, Taheripour, and Tyner (2016) find that global cropland would further decline by 0.8 million hectares and GHG emissions would decline by 0.2 billion tons annually. Zilberman (2015) argues that adoption of GE technology in developing countries in Africa would have a significant impact on GHG emissions reduction and allow for greater adaptation to climate change and that ignoring this technology by the International Panel on Climate Change (IPCC) is a mistake.

The impact of GE varieties on agrobiodiversity has also been a concern, especially since high-yield Green Revolution varieties have replaced local varieties. Transgenic traits can be introduced to local varieties. The main advantage of transgenic varieties is that they allow precise modification of existing varieties. Transgenic varieties may preserve agrobiodiversity if they are introduced to all local varieties and may actually enhance biodiversity if transgenic traits are permitted to restore some varieties highly vulnerable to diseases or pests, given such traits are available.

Krishna, Qaim, and Zilberman's (2015) study on the impact of the introduction of Bt cotton on crop biodiversity in India finds that in the early stage of adoption of Bt cotton, when only a few varieties were approved, agrobiodiversity of cotton declined. In later stages, when the number of approved varieties increased, so did crop biodiversity, and Bt adoption had a significant effect on product diversity, mostly due to partial adoption. Farmers have adopted a large number of modified varieties and, at the same time, have maintained some traditional varieties. The potential of GE varieties to restore an extinct variety has been illustrated in the case of the American chestnut. It once flourished in the eastern U.S. forests, but after three billion trees died, mainly due to invasive species, the species has become endangered. If a new transgenic variety is developed and approved, it could be the first endangered tree species to be restored (Powell 2014). The cases of Bt cotton in India and the American chestnut illustrate that the restrictive regulation of GE varieties actually leads to reduced crop biodiversity.

This discussion has focused mainly on corn, soybeans, and cotton, but there are many other GE commodities, including tomatoes, squash, salmon, and canola (NewHealthGuide.org 2017). The first GM food to reach American markets was the tomato, which allowed for longer shelf-life and transport time. GM squashes eliminated some of the susceptibility to virus diseases. GM salmon may be available soon but will only be available as farm-raised fish. Canola is a major crop grown on the Canadian prairies and is mainly produced from GMO varieties. This crop has added greatly to farm diversification and profitability.

4 Intellectual property rights (IPR), regulation

Traditionally, the public sector provided and distributed genetic materials to farmers at low cost. Over time, the role of the private sector in developed crop genetic materials has increased and firms have been incentivized by plant variety rights to develop improved and locally appropriate varieties (Kolady and Lesser 2009). In the case of agricultural biotechnology, its development has been the outcome of the "educational and industrial complex" of research and regulation (Graff et al. 2003), whereby after the initial innovations introduced by universities, the patent rights to develop these innovations were transferred to the private sector. Private companies have invested significantly in further developing the innovations and going through the regulatory process. Despite reductions in public research investments (Alston, Beddow, and Pardey 2009) and liberalization of the input market, private sector research and development does not imply a diminished role and need for public sector investments because public and private sector research efforts are generally complementary (Pray and Fuglie 2015), but there are exceptions (Ulrich, Furtan, and Schmitz 1986).

GE technologies with high regulatory cost and investment requirements have been one area emphasized by the public sector. The introduction of patent rights for GE traits has been a source of concern, both in terms of distribution of revenue and access to technologies. The development of GE technologies requires both private sector revenue investment and public sector research spillover (Pray and Fuglie 2015). Concerns that the private sector may not invest in technologies that benefit the poor in "orphan crops" (small crops that do not generate enough revenue) are reasonable. Thus, the public sector needs to invest in development of GE traits for this neglected sector. To overcome some of the IPR constraints in these cases, Graff et al. (2003) suggest the establishment of an IPR clearinghouse to assist in the development of GE traits for neglected crops. This clearinghouse could obtain access to the large body of IPR already owned by the public sector, negotiate favorable terms with the private sector, and reduce transaction costs to product development. Graff et al. (2003) also suggest that offices that transfer technologies to universities will engage in "precise" licensing policies, selling the right to develop patents in developed countries to companies but preserving the rights for development in neglected markets.

5 Regulations and their costs

Policy makers and scientists have agreed that the safety of using new GE technologies requires regulation. Common standards have been developed under the guidance of the OECD and summarized in the "Blue Book" and have been adopted by many OECD and non-OECD countries, but their implementation varies by country (Schiemann 2006). Differences in interpretation of these standards have resulted in asynchronicity in approval between countries and might hamper international trade.

Regulation of biotechnology is a special case of safety regulation, which according to Arrow et al. (1996) is optimized when the marginal benefit of safety is equal to its marginal cost. Coase (1960) argues that a well-defined liability system with full information can reach optimal outcomes. In recognizing implementation challenges, Coase (2006) suggests that society will select the most efficient regulatory system, taking into account implementation costs and existing regulations. For example, relying on tort law to ensure optimal behavior may be constrained due to imperfections of the legal system and ability to avoid liability through bankruptcy (Shleifer 2010). Rothbard (1982) recognizes individual heterogeneity and argues for an individual rights–based approach in establishing safety standards, where tort law applies for actual damage against an individual. However, Kolstad, Ulen, and Johnson (1990) suggest that asymmetric information about possible safety risks necessitates a combination of *ex ante* regulations and *ex post* liability rules rather than only *ex post* liability.

In the context of regulating GE technology, important differences exist among countries. For example, the regulatory framework of the United States is very different from that of the European Union, due largely to differences in regulatory principles (Schmitz et al. 2010, pp. 406–408). For example, in the United States, it is argued that GM and non-GMO crops are similar based on substantial equivalence. Conversely, the European Union rejects the substantial equivalence rule and instead uses a Precautionary Principle that limits the adoption of GMO crops. Also, unlike the United States, the European Union requires mandatory labeling of GMO crops.

A major challenge is designing *ex ante* regulations that take into account benefits and costs. One important issue, in the context of GE technology, is irreversible benefits and costs invoked to justify the use of the Precautionary Principle. Lack of sufficient knowledge of the downside effects of GE technologies may be used to justify the delay of approval for the use of the technology. In particular, the findings in a Cornell laboratory that exposure to Bt maize could kill the larvae of monarch butterflies motivated EU regulators to freeze the approval process for all Bt maize varieties (Van den Belt 2003). While similar findings were not found in the field, regulators, in the name of precaution, decided to delay decisions until more information was available. Van den Belt argues that the use of the Precautionary Principle is flawed because *not* taking action may also be very costly. Furthermore, it is akin to assuming "guilty until proven innocent," which is against a fundamental legal principle. It is also impossible to prove a negative (i.e., something does not exist). Finally, he argues that strict application of the Precautionary Principle would prevent any innovation whatsoever.

Nevertheless, regulations are needed, and a benefit–cost analysis that differentiates between reversible and irreversible benefits and costs can provide information for decisions regarding the most appropriate regulatory approach. When designing the regulatory process, it is important to remember that the approval process is not costless. Frequently, the regulatory process includes many redundancies. For example, the same test may be required by several neighboring authorities and significantly delay introduction of innovations in both developed and developing countries (Just, Alston, and Zilberman 2006). The approval costs include both the direct cost of testing and the cost of delayed benefits. Kalaitzandonakes, Alston, and Bradford (2007) show that the regulatory compliance costs are in the range of between US$6 million and US$15 million for the approval of insect- and herbicide-resistant maize depending on the specific regulatory requirements. Smyth, Kerr, and Phillips (2017), in a meta-analysis covering 17 countries, estimate the average regulatory compliance costs to be about US$7.8 million per genetically engineered event, while calculations by the industry estimates the cost at about US$35.1 million per trait.

The delays caused by the approval processes can be even higher. In several countries, the regulatory environment has prevented the approval of new technologies or caused several years of delay. Smyth, Kerr, and Phillips (2017) estimate the average approval length for genetically modified crops in the Unites States to be on average 1,327 days (3.6 years) for the period 1988 to 1997 and 2,467 days (6.7 years) for the period 1998 to 2015, an increase in time length by more than 80 percent. Demont, Wesseler, and Tollens (2004) and Wesseler et al. (2017) use a real option model to assess the maximum incremental social tolerable irreversible costs (MISTICs) that would justify a delay in approval. For sugar beets (Demont, Wesseler, and Tollens 2004), this is between €102 million and €163 million per year in the European Union, and for Bt corn and HR corn (Wesseler et al. 2017) about €1,285 million and €637 million, respectively.

Wesseler and Zilberman (2014) find that the delay of approval of golden rice in India may have resulted in hundreds of thousands of lost eyesight cases, which translates to billions of dollars, and thus the value of environmental gain and risk reduction that justifies the delay would have to be substantial. Similarly, Wesseler et al. (2017) find that the cost of approval delay of various GE technologies in Africa resulted in hundreds of thousands of malnourishment cases, which translates into hundreds of millions of dollars. Zilberman, Kaplan, and Wesseler (2015) find that the cost of banning the use of GE technology for wheat and rice, and restricting its use for maize, is in the hundreds of billions of dollars. Thus, the option value literature provides a good mechanism to assess the cost of regulatory delay.

The regulation of GE technologies has multiple venues. The strictest form of regulation is a complete ban on the production or consumption of GE products. Agricultural GE products are banned outright in some countries, while in others there exists a de facto ban through the delay of decisions to approve them (Smyth, Phillips, and Kerr 2015). The approval requirements also pose strict constraints. *Ex ante* safety regulations are another set of obstacles, because only products that pass risk assessment are allowed to be commercialized. Smyth, Phillips, and Kerr (2015) demonstrate that for GE crops, increased severity of these regulations reduces the probability of successful completion, is likely to reduce investment in agricultural research and development (R&D), and may hamper achieving food security in the future. In particular, complying with biosafety requirements of the Cartagena Protocol has served as an obstacle in developing countries for introducing GE products, and the delays associated with it are especially challenging (Falck-Zepeda and Zambrano 2011).

Another set of regulations are those that aim to allow for co-existence of GE and non-GE crops. These regulations may include segregation storage and restriction of land use to avoid drift. Co-existence regulations are also dependent on purity standards that set an upper bound on the share of GE substances in non-GE products. Tougher regulations and penalties for violations may make it unfeasible or very costly for farmers to adopt GE technology. Beckman, Soregaroli, and Wesseler (2006) develop a framework to assess the impact of *ex ante* and *ex post* co-existence regulations on adoption of GE technologies in Europe. They find the impact varies across location due to heterogeneity of conditions and regulations. In some locations, co-existence regulations have practically banned adoption, while in other countries it has increased the cost of adoption. Beckman, Soregaroli, and Wesseler (2010) find that co-existence regulations in Europe may provide incentives to regional agglomeration of GE and non-GE farming practices. Tougher regulations may reduce the share of GE production. Huffman and McCluskey (2014) suggest that segregation and identity preservation have been achieved in specialty crops where this cost is small relative to the crop price, but when it comes to major commodities, the cost of segregation is significant and there are gains from establishing unified standards.

Labeling requirements are another set of regulations. On the surface, labeling is desirable because it provides additional information to consumers. However, the performance of labeling depends on whether the labeling is mandatory or voluntary, what the label implies, and what should be the parameters of mandatory labeling. Huffman and McCluskey (2014) suggest that labeling costs are lower when segregation and tolerance levels are agreeable among nations but higher when they need to conform to different standards. When GE labeling is not mandatory, non-GE manufacturers who use labeling must cover the cost of certification and monitoring, which raises the cost and reduces the appeal of labels. When GE labeling is mandatory, there are additional costs. If labels serve as a warning, such as with cigarettes, then it has negative impacts on adoption of GE products (Costanigro and Lusk 2014). Mandatory GE labeling is strictly enforced in Europe, but not in the United States. Although U.S. regulators have not found enough health concerns to justify mandatory labeling, several states, such as California, have tried to require labeling, but have failed, most likely due to the additional cost of regulations (Zilberman et al. 2014). Recently, the U.S. government introduced disclosure standards as a compromise between groups supporting strict labeling and those opposing it (Bovay and Alston 2018).

Zilberman et al. (2014) suggest there is a difference between willingness to pay and willingness to vote in support of mandatory labeling. There is a large share of the population that has a small willingness to pay for GE-free products, and their support of labeling is relatively weak – so when the cost of labeling is high, they are swayed against it.

6 Genetically modified foods

In terms of consumer resistance to GMOs, Schmitz and Zhu (2017) develop a gain theoretical approach to illustrate why many groups, such as Greenpeace, have been successful in slowing down consumer adoption of GMO products. They show that anti-GMO groups can spin the facts against GMOs, while it is virtually impossible for GMO supporters and producers of the products, such as Monsanto, to counteract these spins.

However, in support of GMOs, Huffman (2015) finds that consumers are willing to spend more for biotech potato products because of reduced levels of a chemical compound (acrylamide) that is naturally present in starchy food and has been linked to cancer. Biotechnology and genetic modification have yielded more promising results than conventional plant breeding techniques in reducing levels of acrylamide. Huffman underscores the importance of educating consumers about biotechnology in the production of healthy foods. In this regard, legal issues arise over GMOs. For example, Kershen (2000) argues that firms open themselves up to legal issues by not adopting GMOs.

Nevertheless, confusion arises over the debate on health impacts of GMOs because the significant difference between GMOs and non-GMOs is not well understood (Virginia Farm Bureau 2017). The following example on GMO sugar beets and non-GMO sugarcane sheds light on what is a GMO and what is not. As discussed earlier, the EU, for example, generally blocks imports of GMO products. Refined sugar derived from GM sugar beets is identical to refined sugar derived from non-GM sugarcane (Klein, Altenbuchner, and Mattes 1998). In fact, no traces of GM DNA can be found in refined sugar originating from GM sugar beets, which means that sugar consumed from non-GM sugarcane and sugar consumed from GM sugar beets are identical. Consumable sugar (e.g., bagged sugar, sugar used as an ingredient in soft drinks) is refined and, by definition, refined sugar does not have any detectable DNA or protein; therefore, refined sugar derived from GM seeds is molecularly identical to refined sugar derived from non-GM seeds (Klein, Altenbuchner, and Mattes 1998; SIBC 2014). Thus, bagged sugar and sugar in soft drinks originating from GM sugar beets and non-GM sugarcane are nutritionally identical, as are the tastes of the sugars (Klein et al. 1998; SIBC 2014).

In the context of food labeling, many non-GMO foods contain a non-GMO label. However, this is not the case for GMO foods; rarely does one find a GMO product labeled as GMO. GMO labeling is not required as such because the term "GMO" has a negative connotation. Including a voluntary GMO label would decrease product demand and make GMO products less desirable.

From an agricultural perspective, there is a continuous effort through investment in GMO varietal development to prevent the destruction of certain agricultural industries. For example, citrus greening and citrus canker diseases have caused millions of dollars in damages to the Florida citrus industry. Despite the attempts at finding a cure for these diseases, a major production problem still remains. Part of solving this problem may be developing a GMO orange variety that is resistant to citrus greening (Grosser and Dutt 2017). If the industry were to fail, Florida's largest orange competitor, Brazil, is set to monopolize the market. According to Dr. Jude Grosser (personal communication, December 2017), a plant geneticist from IFAS at UF:

> HLB (Citrus Greening Disease or Huanglongbing) [can be solved] by both biotechnology-facilitated conventional breeding and by a transgenic solution... [However], there remains a significant problem with consumer acceptance of a GMO solution. Tropicana is owned by PepsiCo and Minute Maid/Simply Orange by Coca Cola – both international companies that worry a lot about their international reputation. Although both companies support GMO research, I'm still skeptical that they will pursue a GMO solution if other alternatives are available. They still want to retain consumer confidence in the European Union – thus I'm not sure how eager they would be to commercialize a GMO solution. Moreover, they both have anti-biotechnology labels currently on their orange juice products.
>
> We have tested many constructs, and . . . have some beautiful transgenic clones (mostly sweet oranges) that have been in the field for 6 years now, with little or no infection, and no significant symptoms. However, these trees are just coming through juvenility.
>
> In my view, a transgenic solution to HLB will require trees to have at least two transgenes that provide resistance by two different mechanisms . . . We are now producing transgenic plants with stacked genes in an effort to achieve this "long-term, stable" resistance.

GMOs are often discussed within the context of international trade, and trade of GMO commodities remains controversial. As discussed earlier, trade was a major factor in the importation by Africa of GMO products; the EU exerts pressure on Africa not to import GMO products.

Furthermore, GMOs have also created legal cases in the context of international marketing. For example, in the StarLink case (Schmitz, Schmitz, and Moss 2005), Japan sued the United States for shipping them a product that contained the StarLink gene (a product that was never licensed).

Additionally, two of the major commodities consumed and traded worldwide are wheat and rice, but GMO varieties have never been commercialized. The work by Lakkakula, Haynes, and Schmitz (2015) examines the effect in the marketplace if GMOs were to be introduced. They study the world rice economy and demonstrate that adoption of GMO rice would generate significant societal gains, especially for less developed countries. Furthermore, Lakkakula, Haynes, and Schmitz (2015) also consider genetic engineering from a food security perspective. Focusing on the effect of the introduction of a GM rice variety, they argue that GM rice would increase global yield by 5 percent and would give rise to a consumer gain of US$23.4 million. They also argue that the adoption of GM rice varieties would have a far greater impact on rice prices in developing and transitional countries, thus helping to ensure food security throughout the world. It is interesting to note that a Vitamin A-fortified rice variety (golden rice) is available but has never been commercialized.

In an international trade context, the likelihood that GMO wheat will be adopted and traded internationally is remote. A major buyer is Japan, which strongly resists GMO wheat even though it would be possible to separate GMO from non-GMO wheat in international markets. However, significant improvements have been made in non-GMO wheat varieties that have resulted in significant consumer and producer gains (Ulrich, Furtan, and Schmitz 1987).

Finally, it is important to keep in mind, in discussion of GMOs, that wealthy people have the choice and ability to purchase non-GMO products. The empirical evidence shows they are willing to pay a premium for non-GMO over GMO products. In addition, what consumers do not realize is that all the gains from GMOs discussed above will generally benefit consumers much more than producers. This statement is well grounded in the theory of rates of return to research.

7 Political economy and attitudes

The determination of GE technology regulations reflects political and social realities. Herring and Paarlberg (2016) suggest that GE crops are diverse and the decision to treat them as one entity (i.e., GMOs) is justified on political and social attitudes more than a reflection of scientific reality. Regulations are reflections of the interests and relative power of different political groups. Zilberman et al. (2013) argue that one reason the United States has a more favorable approach toward GE technology is because it was developed by U.S. companies, which, in turn, would hurt European-based chemical companies. Thus, the latter have sought to suppress the technology. Farmers' attitudes toward GE technology have been mixed – while the technology reduces costs, it also may reduce output prices. Wesseler, Ihle, and Zilberman (2016) show that, in retrospect, since the introduction of GE technology, the profitability of U.S. farmers has increased relative to European farmers. Some environmental groups have taken a negative attitude toward GE technology either due to the Precautionary Principle or because the technology is associated with major companies. Graff, Hochman, and Zilberman (2009) model the political debate as a two-stage game, where different parties shape their opinion and then aim to affect public opinion to achieve their desired policy. Consumer resistance to GMOs can also be modeled within a game theoretical framework (Zhu and Schmitz 2017). Environmental groups may have swayed public opinion in Europe and "won" the political economy debate, in part due to the parliamentary system where small environmental parties have strong political say, but policy may change as new information changes perception. The literature emphasizes the importance of timing and random shocks in affecting attitudes toward biotechnology and policies. Gaskell et al. (1999) emphasize that the major differences in regulations of GE technology between Europe and the United States reflect differences in trust in regulatory agencies. GE technology in agriculture was introduced in the 1990s during the period of concerns about mad cow disease and hoof-and-mouth disease, where government actions were highly criticized.

Paarlberg (2009) emphasizes that the global debate on GE technology has reached a stable equilibrium for fiber and feed, but not for food. In the Western Hemisphere, the United States is the hegemon with regulations that favor the use of GE technologies in the production of maize, soybeans, and other crops, while in Europe and Africa, where Europe is the hegemon, GE products have minimal adoption rates.

8 Conclusions

Agricultural biotechnology has established itself as a major technology, especially in providing feed and fiber. Yet, at the same time, its utilization has been heavily constrained and much of its potential in major food crops and the developing world has been underutilized. There is evidence that the technology has significant effects on yield and welfare with no obvious environmental costs. Yet political economic considerations are decelerating the adoption of GE technologies. The current regulatory delay associated with the approval process is costlier than the direct approval costs. Substantial economic gains can be expected from harmonization of approval policies that reduce the time length for adoption of GE technologies. While the benefits of GE technologies are quite significant, they still pose some risks. Therefore, benefit–cost analysis should be applied in an efficient and timely manner for decision making on regulations.

The delay associated with regulation not only affects the utilization of the technology, but also the speed of discovery. Due to regulatory delays, humanity has been able to utilize only a small fraction of the potential of GE technologies. The current experience with GE technologies suggests that the poor stand to gain significantly from these technologies through increased yield, fewer resource constraints, and adaptations to climate change. Mechanisms that provide access to GE technologies, especially for the disadvantaged, should be introduced.

Even so, science is advancing. In 2010, a new form of biotechnology – gene editing – was introduced. Many applications of this biotechnology have emerged. Furthermore, gene editing can contribute to addressing biological conservation challenges (NASEM 2016). The ownership of intellectual property governing the technology is being contested, but already there are large numbers of developments that may result in applications in both plant and animal agriculture (Egelie et al. 2016). Gene editing allows the elimination or transfer of genes. For example, the United States Department of Agriculture (USDA) has decided to deregulate a mushroom engineered to resist browning by deleting some genes (Waltz 2016), but it is unclear how European authorities will regulate it and whether gene editing that inserts external genes will be treated as a transgenic (Spring et al. 2016). This is not the last development – biological discoveries will continue to provide new tools that may change the agricultural land and agricultural economy. Policy and regulation will determine to what extent these technologies are utilized and who benefits from them.

Note

1 Genetic engineering refers to all forms of plant breeding using biotechnology, including, most commonly, genetically modified organisms (GMOs).

References

Alston, J.M., J.M. Beddow, and P.G. Pardey. 2009. "Agricultural Research, Productivity, and Food Prices in the Long Run." *Science* 325(5945):1209–1210.

Apel, A. 2010. "The Costly Benefits of Opposing Agricultural Biotechnology." *New Biotechnology* 27(5):635–640.

Arrow, K.J., M.L. Cropper, G.C. Eads, R.W. Hahn, L.B. Lave, R.G. Noll, and R. Stavins. 1996. *Benefit–Cost Analysis in Environmental, Health, and Safety Regulation*. Washington, DC: American Enterprise Institute.

Barrows, G., S. Sexton, and D. Zilberman. 2014a. "Agricultural Biotechnology: The Promise and Prospects of Genetically Modified Crops." *The Journal of Economic Perspectives* 28(1):99–119.

———. 2014b. "The Impact of Agricultural Biotechnology on Supply and Land-Use.: *Environment and Development Economics* 19(6):676–703.

Beckmann, V., C. Soregaroli, and J. Wesseler. 2010. "Ex-Ante Regulation and Ex-Post Liability under Uncertainty and Irreversibility: Governing the Coexistence of GM Crops." *Economics* 4:2010–2009.

———. 2006. "Co-Existence Rules and Regulations in the European Union." *American Journal of Agricultural Economics* 88(5):1193–1199.

Bennett, A.B., C. Chi-Ham, G. Barrows, S. Sexton, and D. Zilberman. 2013. "Agricultural Biotechnology: Economics, Environment, Ethics, and the Future." *Annual Review of Environment and Resources* 38:249–279.

Bovay, J., and J. Alston. 2018. "GMO Food Labels in the United States: Economic Implications of the New Law." *Food Policy* (forthcoming).

Brookes, G., and P. Barfoot. 2014. "Economic Impact of GM Crops: The Global Income and Production Effects 1996–2012." *GM Crops & Food* 5(1):65–75.

Coase, R. 1960. "The Problem of Social Cost." *Journal of Law and Economics* 3:1–44.

———. 2006. "The Conduct of Economics: The Example of Fisher Body and General Motors." *Journal of Economics & Management Strategy* 15(2):255–278.

Cook-Deegan, R., and C. Heaney. 2010. "Patents in Genomics and Human Genetics." *Annual Review of Genomics and Human Genetics* 11:383–425.

Costanigro, M., and J.L. Lusk. 2014. "The Signaling Effect of Mandatory Labels on Genetically Engineered Food." *Food Policy* 49:259–267.

Demont, M., J. Wesseler, and E. Tollens. 2004. "Biodiversity versus Transgenic Sugar Beets – the One Euro Question." *European Review of Agricultural Economics* 31(1):1–18.

Devos, Y., L.N. Meihls, J. Kiss, and B.E. Hibbard. 2013. "Resistance Evolution to the First Generation of Genetically Modified Diabrotica-Active Bt-Maize Events By Western Corn Rootworm: Management and Monitoring Considerations." *Transgenic Research* 22(2):269–299.

Douglas, E. 2001. "Darwin's Natural Heir." *The Guardian*, February 17.

Egelie, K.J., G.D. Graff, S.P. Strand, and B. Johansen. 2016. "The Emerging Patent Landscape of CRISPR-Cas Gene Editing Technology." *Nature Biotechnology* 34(10):1025–1031.

Falck-Zepeda, J.B., and P. Zambrano. 2011. "Socio-Economic Considerations in Biosafety and Biotechnology Decision Making: The Cartagena Protocol and National Biosafety Frameworks." *Review of Policy Research* 28(2):171–195.

Fernandez-Cornejo, J., S. Wechsler, M. Livingston, and L. Mitchell. 2014. *Genetically Engineered Crops in the United States*. Washington, DC: USDA.

Frisvold, G., and J. Reeves. 2010. "Resistance Management and Sustainable Use of Agricultural Biotechnology." *AgBioForum* 13(4):343–359.

Gaskell, G., M.W. Bauer, J. Durant, and N.C. Allum. 1999. "Worlds Apart? The Reception of Genetically Modified Foods in Europe and the United States." *Science* 285(5426):384–387.

Graff, G.D., S.E. Cullen, K.J. Bradford, D. Zilberman, and A.B. Bennett. 2003. "The Public – Private Structure of Intellectual Property Ownership in Agricultural Biotechnology." *Nature Biotechnology* 21(9):989–995.

Graff, G.D., G. Hochman, and D. Zilberman. 2009. "The Political Economy of Agricultural Biotechnology Policies." *AgBioForum* 12(1):4.

Grosser, J., and M. Dutt. 2017. *Genetic Engineering 'Fastest Method' to Save Florida Citrus Industry from Greening Disease.* https://geneticliteracyproject.org/2017/11/07/genetic-engineering-fastest-method-save-floridas-citrus-industry-greening-disease/. Accessed January 29, 2018.

Herring, R., and R. Paarlberg. 2016. "The Political Economy of Biotechnology." *Annual Review of Resource Economics* 8:397–416.

Huffman, W. 2015. "New Research from Iowa State University Economist Finds Consumers Willing to Spend More for Biotech Potato Products." Iowa State University News Service, Ames, IA.

Huffman, W., Y Jin, and Z Xu. 2016. "The Economic Impacts of Technology and Climate Change: New Evidence from U.S. Corn Yields." *International Consortium on Applied Bioeconomy Research* (ICABR), Ravello, Italy.

Huffman, W., and J.J. McCluskey. 2014. "The Economics of Labeling GM Foods." *AgBioForum* 17(2):156–160.

Hughes, S.S. 2001. "Making Dollars Out of DNA: The First Major Patent in Biotechnology and The Commercialization of Molecular Biology 1974–1980." *Isis* 92(3):541–575.

ISAAA. 2016. "Global Status of Commercialized Biotech/GM Crops: 2016." ISAAA Brief No. 52. ISAAA, Ithaca, NY.

Just, R.E., J.M. Alston, and D. Zilberman (editors). 2006. *Regulating Agricultural Biotechnology: Economics and Policy* (Vol. 30). Amsterdam: Springer.

Kalaitzandonakes, N., J.M. Alston, and K.J. Bradford. 2007. "Compliance Costs for Regulatory Approval of New Biotech Crops." *Nature Biotechnology* 25(5):509–511.

Kennedy, P.L., K.E. Lewis, and A. Schmitz. 2017. "Food Security Through Biotechnology: The Case of Genetically Modified Sugar Beets in the United States." In A. Schmitz, P.L. Kennedy, and T.G. Schmitz, eds., *World Agricultural Resources and Food Security.* Bingley: Emerald, pp. 35–52.

Kershen, D.L. 2000. The risks of going non-GMO. *Oklahoma Law Review* 53(4):631–652.

Klein, J., J. Altenbuchner, and R. Mattes. 1998. "Nucleic Acid and Protein Elimination during the Sugar Manufacturing Process of Conventional and Transgenic Sugar Beets." *Journal of Biotechnology* 60:145–153.

Klümper, W., and M. Qaim. 2014. "A Meta-Analysis of the Impacts of Genetically Modified Crops." *PLOS ONE* 9(11):e111629.

Kolady, D.E., and W. Lesser. 2009. "But Are They Meritorious? Genetic Productivity Gains Under Plant Intellectual Property Rights." *Journal of Agricultural Economics* 60(1):62–79.

Kolstad, C.D., T.S. Ulen, and G.V. Johnson. 1990. "Ex Post Liability for Harm vs. Ex Ante Safety Regulation: Substitutes or Complements?" *American Economic Review* 80(4):888–901.

Krishna, V., M. Qaim, and D. Zilberman. 2015. "Transgenic Crops, Production Risk, and Agrobiodiversity." *European Review of Agricultural Economics* 43(1):137–164.

Lakkakula, P., D.J. Haynes, and T.G. Schmitz. 2015. "Genetic Engineering and Food Security: A Welfare Economics Perspective." In A. Schmitz, P.L. Kennedy, and T.G. Schmitz, eds., *Food Security in an Uncertain World: An International Perspective.* Bingley: Emeral, pp. 179–193.

Lichtenberg, E., and D. Zilberman. 1986. "The Econometrics of Damage Control: Why Specification Matters." *American Journal of Agricultural Economics* 68(2):261–273.

Lusk, J.L., J. Tack, and N. Hendricks. 2017. "Heterogeneous Yield Impacts from Adoption of Genetically Engineered Corn and the Importance of Controlling for Weather." NBER Working Paper No. 23519, NBER, Washington, DC.

Mahaffey, H., F. Taheripour, and W.E. Tyner. 2016. "Evaluating the Economic and Environmental Impacts of a Global GMO Ban." *Journal of Environmental Protection* 7:1522–1546.

Moss, C.B., and A. Schmitz. Forthcoming. "Distribution of Agricultural Productivity Gains in Selected Feed the Future African Countries." *Journal of Agribusiness in Developing and Emerging Economies.*

National Academies of Sciences, Engineering, and Medicine. 2016. *Gene Drives on the Horizon: Advancing Science, Navigating Uncertainty, and Aligning Research with Public Values.* Washington, DC: National Academies Press doi:10.17226/23405.

National Research Council [NRC]. 2010. *The Impact of Genetically Engineered Crops on Farm Sustainability in the United States.* Washington, DC: National Academies Press.

NewHealthGuide.org. 2017. *List of Foods That Are Genetically Modified.* NewHealthGuide.org.

Paarlberg, R. 2009. *Starved for Science: How Biotechnology Is Being Kept Out of Africa.* Cambridge, MA: Harvard University Press.

Powell, W. 2014. "The American Chestnut's Genetic Rebirth." *Scientific American* 310:68–73.

Pray, C.E., and K.O. Fuglie. 2015. "Agricultural Research by the Private Sector." *Annual Review of Resource Economics* 7(1):399–424.

Pray, C.E., J. Huang, R. Hu, and S. Rozelle. 2002. "Five Years of Bt Cotton in China – the Benefits Continue." *The Plant Journal* 31(4):423–430.

Qaim, M. 2009. "The Economics of Genetically Modified Crops." *Annual Review of Resource Economics* 1:665–693.

Qaim, M., and D. Zilberman. 2003. "Yield Effects of Genetically Modified Crops in Developing Countries." *Science* 299(5608):900–902.

Rangel, G., and A. Maurer. 2015. "From Corgis to Corn: A Brief Look at the Long History of GMO Technology. *Science in the News.* http://sitn.hms.harvard.edu/flash/2015/from-corgis-to-corn-a-brief-look-at-the-long-history-of-gmo-technology/. Accessed January 29, 2018.

Rothbard, M. 1982. "Law, Property Rights, and Air Pollution." *Cato Journal* 2(1):55–99.

Schiemann, J. 2006. "The OECD Blue Book on Recombinant DNA Safety Considerations: Its Influence on ISBR and EFSA Activities." *Environmental Biosafety Research* 5:233–235.

Schmitz, A., and C.B. Moss. 2016. "Mechanized Agriculture: Machine Adoption, Farm Size, and Labor Displacement." *AgBioForum* 18(3):Article 6.

Schmitz, A., C. Moss, T.G. Schmitz, H.W. Furtan, and H.C. Schmitz. 2010. *Agricultural Policy, Agribusiness, and Rent-Seeking Behaviour.* Toronto: University of Toronto Press.

Schmitz, A., and M. Zhu. 2017. "The Economics of Yield Maintenance: An Example from Florida Sugarcane." *Crop Science* 57(6):2959–2971.

Schmitz, T.G., A. Schmitz, and C.B. Moss. 2005. "The Economic Impact of StarLink Corn." *Agribusiness: An International Journal* 21(3):391–407.

Shi, G., J.P. Chavas, and J. Lauer. 2013. "Commercialized Transgenic Traits, Maize Productivity, and Yield Risk." *Nature Biotechnology* 31(2):111–114.

Shleifer, A. 2010. "Efficient Regulation." NBER Working Paper No. w15651. NBER, Washington, DC.

Smyth, S.J., M. Gusta, K. Belcher, P.W.B. Phillips, and D. Castle. 2011. "Changes in Herbicide Use After Adoption of HR Canola in Western Canada." *Weed Technology* 25:492–500.

Smyth, S.J., W.A. Kerr, and P.W.B. Phillips. 2017. *Biotechnology Regulation and Trade.* Amsterdam: Springer.

Smyth, S.J., P.W. Phillips, and W.A. Kerr. 2015. "Food Security and the Evaluation of Risk." *Global Food Security* 1:16–23.

Spring, T., D. Eriksson, J. Schiemann, and F. Hartung. 2016. "Regulatory Hurdles for Genome Editing: Process- vs. Product-Based Approaches in Different Regulatory Contexts. *Plant Cell Reports* 35(7):149.

Sugar Industry Biotech Council (SIBC). 2014. *Frequently Asked Questions.* Washington, DC: SIBC.

Tan, T., J. Zhan, and C. Chen. 2011. "The Impact of Commercialization of GM Rice in China." *American Eurasian Journal of Agricultural and Environmental Science* 10:296–299.

Ulrich, A., W.H. Furtan, and A. Schmitz. 1986. "Public and Private Returns from Joint Venture Research: An Example from Agriculture." *Quarterly Journal of Economics* 37:103–129.

———. 1987. "The Cost of a Licensing System Regulation: An Example from Canadian Prairie Agriculture." *Journal of Political Economy* 95(1):160–178.

Van den Belt, H. 2003. "Debating the Precautionary Principle: "Guilty until Proven Innocent" or "Innocent until Proven Guilty?" *Plant Physiology* 132(3):1122–1126.

Virginia Farm Bureau. 2017. *New Surveys Reveal Consumers Still Confused About GMOs*. www.vafb.com/membership-at-work/news-resources/ArticleId/3006/new-surveys-reveal-consumers-still-confused-about-gmos. Accessed January 29, 2018.

Waltz, E. 2016. "Gene-Edited CRISPR Mushroom Escapes U.S. Regulation." *Nature* 532(7599):293.

Wang, J., and X.M. Liu. 2007. "Contamination of Aflatoxins in Different Kinds of Foods in China." *Biomedical Environmental Science* 20(6):483–487.

Wesseler, J., R. Ihle, and D. Zilberman. 2016. "Does Biotechnology Make a Difference?" *International Consortium on Applied Bioeconomy Research* (ICABR), Ravello, Italy.

Wesseler, J., S. Scatasta, and E.H. Fall. 2011. "Environmental Benefits and Costs of GM Crops." In C. Carter, G. Moschini, and I. Sheldon, eds., *Genetically Modified Food and Global Welfare*. Bingley: Emerald, pp. 173–199.

Wesseler, J., R. Smart, J. Thomson, and D. Zilberman. 2017. "Foregone Benefits of Important Food Crop Improvements in Sub-Saharan Africa." *PLOS ONE* 12(7):e0181353.

Wesseler, J., and D. Zilberman. 2014. "The Economic Power of the Golden Rice Opposition." *Environment and Development Economics* 19(6):724–742.

Wieczorek, A.M., and M.G. Wright. 2012. "History of Agricultural Biotechnology: How Crop Development Has Evolved." *Nature Education Knowledge* 3(10):9.

Wu, F. 2006. "Mycotoxin Reduction in Bt Corn: Potential Economic, Health, and Regulatory Impacts." *Transgenic Research* 15(3):277–289.

Wu, F., and W. Butz. 2004. *The Future of Genetically Modified Crops: Lessons from the Green Revolution*. Santa Monica, CA: Rand.

Xia, J., P. Song, L. Xu, and W. Tang. 2015. "Retraction of a Study on Genetically Modified Corn: Expert Investigations Should Speak Louder During Controversies Over Safety." *Bioscience Trends* 9(2):134–137.

Zhu, M., and T.G. Schmitz. 2017. "A Signaling Game in the Controversy Over Genetically Engineered Foods." *Applied Economic Policies and Perspective* (in review).

Zilberman, D. 2015. "IPCC AR5 Overlooked the Potential of Unleashing Agricultural Biotechnology to Combat Climate Change and Poverty." *Global Change Biology* 21(2):501–503.

Zilberman, D., S. Kaplan, E. Kim, G. Hochman, and G. Graff. 2013. "Continents Divided: Understanding Differences Between Europe and North America in Acceptance of GM Crops." *GM Crops & Food* 4(3):202–208.

Zilberman, D., S. Kaplan, E. Kim, and G. Waterfield. 2014. "Lessons from the California GM Labeling Proposition on the State of Crop Biotechnology." In S. Smyth, P. Phillips, and D. Castle, eds., *Handbook on Agriculture, Biotechnology and Development*. Cheltenham: Edward Elgar, pp. 538–549.

Zilberman, D., S. Kaplan, and J. Wesseler. 2015. "The Loss from Underutilizing GM Technologies." *AgBioForum* 18(3):312–319.

Part 6
Marketing

37
The financial economics of agriculture and farm management

Charles B. Moss, Jaclyn D. Kropp, and Maria Bampasidou

1 Introduction

Economics is the study of the allocation of scarce resources. A subfield of economics, agricultural economics, focuses on the analysis of policies in the agriculture and food and fiber sectors of the economy. The area of financial economics of agriculture and farm management is the study of capital allocation in the agricultural production process. By capital, we are referring to a stock of pre-produced goods, including machinery and other productive inputs used in production, natural resources such as land, and financial capital.

This overall definition of agricultural finance is somewhat different than standard undergraduate textbook definitions. One approach to agricultural finance is from the standpoint of financial management, which studies an individual farm operator's financial decisions. For example, should a farmer buy a tractor or tract of farmland? If the farmer purchases the tractor, how should it be financed? Alternatively, more general economics courses may frame the problem in the context of financial economics, which studies the supply and demand for capital, including the effects of monetary and fiscal policy. The approach taken here is the "middle ground," discussing the general market for assets and ownership.

In agricultural finance, a farmer must obtain use rights to the inputs used in production. A farmer must rent or purchase land, obtain the services of machinery such as tractors, purchase inputs that will be consumed in the production process such as fertilizer and hired labor, and access financial capital. To obtain these resources, the farmer could use his own savings or obtain financial resources from other individuals through contracts (e.g., either debt contracts that specify a fixed return to a lender or equity agreements that specify an arrangement for a share of the profits from the production activity). Hence, financial analysis involves two markets: (1) the market for physical or real capital and (2) the market for financial capital. Financial decisions are defined by the acquisition of inputs in both the real capital and financial capital markets.

For a variety of reasons that we will develop through this chapter, the farmer can choose to finance production through equity or debt. In agriculture, the most common practice is to finance through debt. However, the debt market is not homogeneous. We discuss three segments of the debt market based on the time covered by the contract and what the proceeds of the loan are to be used for:

- Operating credit (operating capital) is the short-term credit market used to purchase inputs that are used up in a single production period (i.e., fertilizer, fuel, labor).
- Intermediate credit is associated with the purchase of factors of production that have a life greater than one year but less than ten years (i.e., combines, tractors, and buildings).
- Long-term credit (trade credit) is associated with the purchase of long-term assets such as farmland (real estate debt).

These lines of credit are not set in stone, because cash is fungible. There may be farmers who use cash generated through long-term borrowing to meet short-term capital needs.

In this chapter, we will focus on the agricultural debt market to discuss capital market theory, the market for farmland, the Capital Asset Pricing Model, DuPont Analysis, risk balancing/optimal debt, and credit rationing. This discussion marks a difference between agricultural finance and the more general discipline of finance, which focuses primarily on the valuation of traded instruments such as stocks and bonds. The list of topics presented in this chapter is not exhaustive. In the sections that follow, we also discuss extensions of the theories presented and present areas for future research in agricultural finance.

2 Austrian and neoclassical capital market theories

The discussion regarding the value of production assets and ownership has evolved through different schools of thought. There is a tendency in most modern capital theory to abstract away from the assets to focus on the "aggregate value of capital." When we do so, we are no longer concerned with tractors, but with the dollar value of tractors plus the dollar value of combines plus the dollar value of other assets. One of the primary contributions of the Austrian theory of capital was the focus on real capital. The Austrian school contends that the focus on dollars of capital confuses or hides the value of the real capital item. For example, consider Roscher's development of capital:

> Suppose a nation of fisher-folk, with no private ownership in land or capital, dwelling naked in caves and living on fish caught by hands in pools left by the ebbing tide. All the workers here may be supposed equal, and each man catches and eats three fish per day. But now one prudent man limits his consumption to two fish per day for 100 days, lays up in this way a stock of 100 fish, and makes use of this stock to enable him to apply his whole labour-power to making a boat and a net. By the aid of this capital he catches from the first perhaps thirty fish a day.
>
> Here the Physical Productivity of capital is manifest in the fact that the fisher, by aid of capital, catches more fish than he would otherwise have caught – thirty instead of three. Or to put it quite correctly, a number somewhat under thirty. For the thirty fish which are now caught in a day are the result of more than one day's work. . . . In this surplus is manifested the physical productivity of capital.

(Bawerk 2007)

The question asked by Roscher and the other Austrians was how much value proceeds from the physical item (i.e., the fish net and boat) versus from the investment activity (i.e., the fact that

Financial economics

the fisherman lives on less fish for a time to construct the boat and net). There is also an implicit capital market. Suppose that one person is willing to forgo consumption while another builds the net and boat (e.g., the first person invests in the net or boat by paying for the second person's consumption while building the boat or net). What is the return to the physical asset versus the capital investment? The real asset has a claim on the return; in essence, a portion of the return is the value of the marginal product of the boat. Based on the Austrian theory, the physical asset is separate from the purchasing power.

In the modern context, there is a supply and demand for physical assets such as tractors and other equipment. A tractor has a marginal value product (MVP) over time (a technique that accounts for the time value of money, discussed later). In an efficient market, the present value of the MVP should equal the cost of purchasing the tractor. This relationship should hold no matter the source of financing (i.e., through either equity or debt). The source of financing, either by forgoing consumption or by investment, implies the price of acquiring capital in the capital market. This price of postponed consumption dictates the present value of capital. We should stress at this point that macroeconomic policies could affect the capital market through the effect of these policies on interest rates.

To develop the neoclassical capital market, we start with a slight reformation of the consumer's optimization problem:

$$\max_{C_t, C_{t+1}} U(C_t, C_{t+1})$$
$$s.t. C_t + \frac{1}{1+i} C_{t+1} \leq W \quad , \tag{1}$$

where C_t is the consumption at time t, C_{t+1} is the consumption at time $t+1$, i is the market interest rate, and W is the consumer's initial wealth, which is often referred to as the endowment. To understand the implied investment problem, consider the implications of the budget constraint. Rearranging the budget inequality in Equation (1) slightly yields $C_{t+1} \leq (W - C_t)(1 + i)$. The consumption in period $t + 1$ is the principal invested plus the interest rate earned by investing in the capital market. Hence, the consumer's problem involves finding the point where the marginal rate of substitution of consumption between periods equals the return in the capital market. This point is depicted as the combination $\{C_t^*, C_{t+1}^*\}$ in Figure 37.1a. In Figure 37.1a, the quantity $W - C_t^*$ is invested at time t so that $(W - C_t^*)(1 + i)$ is available for consumption at time $t + 1$.

Next, consider what happens if the interest rate increases from i to $\tilde{i}$. As depicted in Figure 37.1a, an increase in the interest rate from i to $\tilde{i}$ causes consumption at time t to decline (e.g., because consumption at time t has become relatively more expensive in terms of consumption at time $t + 1$). Hence, the optimal decision involves investing more today (e.g., $W - \tilde{C}_t > W - C_t^*$), and as a result the amount consumed in period $t + 1$ ($\tilde{C}_{t+1}$) increases. Note that this increase in consumption at time $t + 1$ is due to two factors: (1) the increased interest rate implies that the return on investment increases, so the same level of investment at time t will yield a higher rate of consumption at time $t + 1$, and (2) the consumer chooses to invest more, thus decreasing consumption at time t to shift consumption to time $t + 1$.

Building on the basic consumer model presented in Figure 37.1a, we add the concept that the consumer is also a producer in Figure 37.1b. Specifically, we replace the standard budget line that we have used to represent the capital market with a production possibilities frontier. Thus, we assume that the individual makes a decision to produce combinations of goods at time t and $t + 1$, which the individual then consumes. In this case, the consumer maximizes

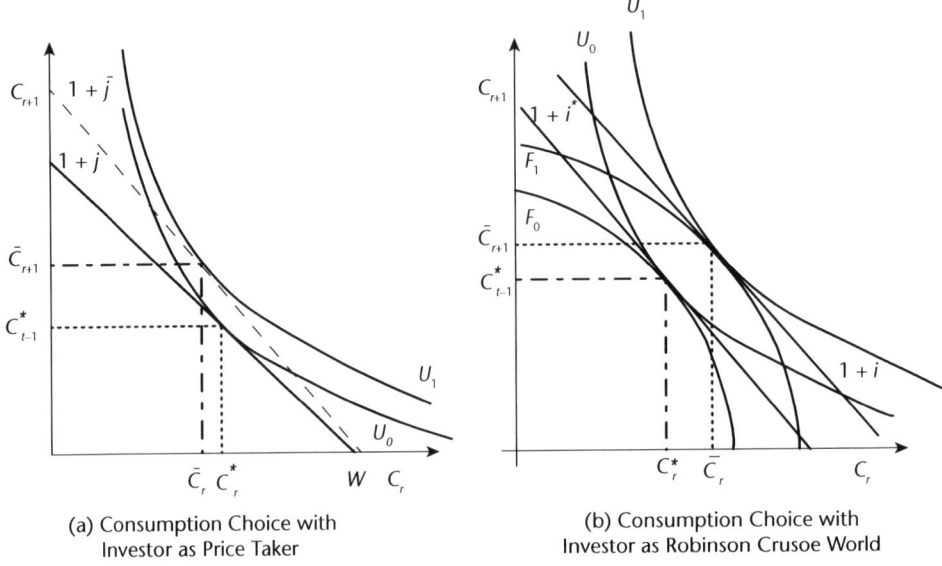

Figure 37.1 Consumption decisions between two periods in time

utility by equating the marginal rate of substitution for the utility function (e.g., the ratio of the marginal utility of consumption at time t to the marginal utility of consumption at time $t + 1$) with the marginal rate of transformation for the production possibility frontier (e.g., the ratio of the marginal cost of production at time t to the marginal cost of production at time $t + 1$). This tangency yields an implicit interest rate (i.e., relative price of consumption at time t in terms of consumption at time $t + 1$) comparable with the preceding formulation. The formulation in Figure 37.1b is often referred to as the Robinson Crusoe or no-trade economy. For example, we assume that F_0 is the production possibilities frontier of some initial endowment (i.e., ten goats and 20 bushels of corn). With this endowment, Robinson Crusoe chooses to produce and consume $\{C_t^*, C_{t+1}^*\}$. At this point, the tangency implies an interest rate of i^*. Next, we assume that the production possibilities frontier shifts out from F_0 to F_1. Several factors could cause such a shift. First, Robinson Crusoe's endowment could increase (i.e., two goats could wander into his camp). Second, a technological change could occur (i.e., Crusoe could discover something about cultivating corn). Either shift would cause an increase in consumption in both time periods. In addition, the shift could result in a change in the implied interest rate.

The development of the consumption/production model moves us closer to a model of investment. We could envision a family of production possibilities frontiers that are functions of the investor's initial wealth $F_0^j(W), j = 1, \ldots J$. One of these frontiers would be the budget constraint in Figure 37.1a. Under this scenario, $F_0(W)$ is the union of all such production possibility sets. Alternatively, we could envision F_0 as a set of investment opportunities facing an investor such as a farmer. In either case, Figure 37.1 is typically assumed to be a no-trade scenario. There is no market for trading consumption today for consumption tomorrow.

Figure 37.2 introduces the possibility of trade in the capital market. Starting with Figure 37.2a, F_0 is the production possibility frontier from Figure 37.1b and $\tilde{U}$ is the investor's utility function. In this case, we assume that the investor has access to a capital market, represented by the line M, where the investor can borrow or lend at an interest rate of i. The existence of

Financial economics

this capital market separates the investor's decision into two different decisions. First, the investor determines what combination $\{C_t, C_{t+1}\}$ to produce based on the tangency between the production possibilities frontier and the capital market line M. In Figure 37.2a, the investor chooses to produce $\{C_t^*, C_{t+1}^*\}$. This production decision then determines the feasible set for consumption (much like the decision depicted in Equation (1) and Figure 37.1a). Mathematically, the constraint for consumption is based on the investor's implicit wealth (W^*):

$$W^* = C_t^* + \frac{1}{1+i}C_{t+1}^* \tag{2}$$

In Figure 37.2a, the investor chooses to consume the combination $\{\tilde{C}_t, \tilde{C}_{t+1}\}$. Again, the resources from production determines the budget line for consumption $\tilde{C}_t + \frac{1}{1+i}\tilde{C}_{t+1} \leq W^* = C_t^* + \frac{1}{1+i}C_{t+1}^*$. Given the choice of production and consumption in Figure 37.2a, the investor is a lender. Specifically, the investor lends $C_t^* - \tilde{C}_t$ to the capital market at time t and receives a payment of $\tilde{C}_{t+1} - C_{t+1}^* = (C_t^* - \tilde{C}_t) \times (1+i)$ at time $t + 1$.

Figure 37.2b depicts a borrower under the same scenario. In Figure 37.2b, the investor borrows $\hat{C}_t - C_t^*$ at time t and pays $C_{t+1}^* - \hat{C}_{t+1} = (\hat{C}_t - C_t^*) \times (1+i)$ at time $t + 1$. Given that the production possibilities frontiers and interest rates are the same in each panel of Figure 37.2, the differences in the utility functions determines which individuals are lenders and which individuals are borrowers. However, it is possible that two investors with similar indifference functions may face different production possibilities frontiers. In this case, it is the slopes of the production possibilities frontiers that determine which investor is the borrower and which is the lender.

It is worthwhile to briefly introduce the concept of an equilibrium in the capital market. Figure 37.2 implicitly takes the interest rate as a given. However, if we limit the market to the two individuals depicted, the interest rate would be the price in the capital market – the equilibrium interest rate would be determined by that interest rate such that borrowing equaled lending (e.g., $C_t^*(i) - \tilde{C}_t(i) = \hat{C}_t(i) - C_t^{**}(i)$).

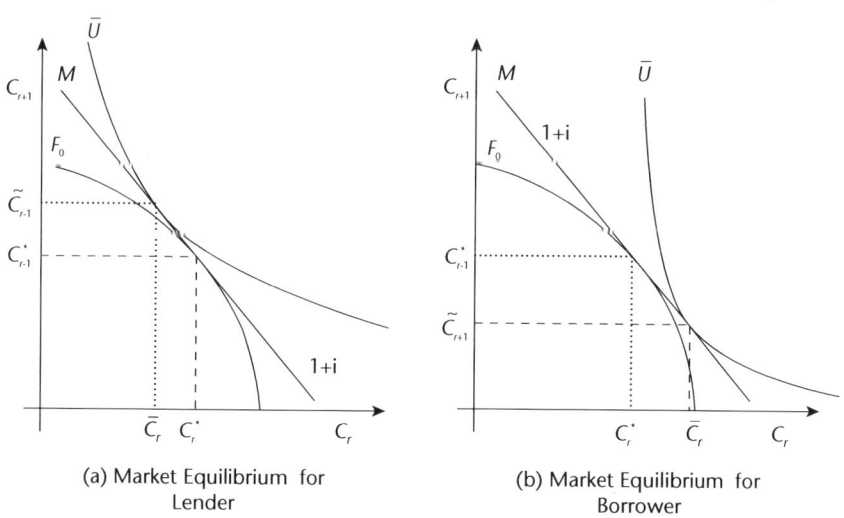

(a) Market Equilibrium for Lender

(b) Market Equilibrium for Borrower

Figure 37.2 Consumption decisions with a capital market

The important implication of the capital market is that the market is always preferred to the no-trade equilibrium. Intuitively, we could rotate the market line for the lender scenario until the market solution equals the no-trade point (i.e., as we decrease the interest rate, the lender offers less money to the capital market until the lender no longer lends any money). At the interest rate where the lender offers no money to the capital market, the solution is identical to the no-trade solution in Figure 37.1b. Thus, the no-trade equilibrium forms a lower bound to the utility generated by the capital market. Furthermore, the capital market separates the production and consumption decisions. This is referred to as Fisher separation.

3 Markets for agricultural assets

Expanding on the general model of the capital market equilibrium in the preceding section, we address the question of how long-term agricultural assets are valued. Specifically, a farmer would purchase an asset if its Net Present Value (NPV) is non-negative:

$$NPV = -I_0 + \sum_{t=1}^{N} \frac{E[CF_t \mid \Omega_0]}{(1+i)^t} \geq 0, \tag{3}$$

where I_0 is the initial investment, $E[CF_t \mid \Omega_0]$ is the expected cash flow from the investment in period t based on information available at time 0 where this information is denoted Ω_0, and i is the discount rate for the firm. This discount rate is usually the firm's weighted average cost of capital, which accounts for the cost of debt financing, the cost of equity financing, and the firm's capital structure (mix of debt and equity). The implicit "market clearing condition" from Equation (3) is a binary sum across farmers. Specifically, let $Q_m^*(I_0, \Omega_0, i) = 1$ be individual m's decision to purchase the investment because its $NPV > 0$, and let $Q_m^*(I_0, \Omega_0, i) = 0$ be the decision not to purchase the asset because its $NPV < 0$. For demonstration purposes, let us consider a tractor for our investment. The total demand for a tractor could then be derived as

$$x = \sum_{m=1}^{M} Q_m^*(I_0, \Omega_0, i), \tag{4}$$

where $m = 1, \ldots M$ is the set of all farmers that may be interested in the tractor in question. Intuitively, $Q_m^*(I_0, \Omega_0, i)$ is a decreasing function of the cost of the tractor (I_0) and the interest rate (i). $Q_m^*(I_0, \Omega_0, i)$ is also an increasing function of factors that increase the expected cash flow such as future commodity prices.

The demand function in Equation (4) has several "moving parts." First, it is based on expectations of the future based on current information. Inherent in this expectation process is some level of risk or uncertainty of returns associated with purchasing the investment (Moss 2010). Several approaches have been suggested to deal with risk. For our purposes, we will consider the Capital Asset Pricing Model (CAPM) that is typically used to value stocks (Sharpe 1954; Lintner 1965a, b; Mossin 1966). The CAPM explains how the price of a security/bond depicts differences in the risk/return relationship in a well-operating securities market. In general, the CAPM for security j can be written as

$$\bar{r}_j = \alpha_j + \beta_j \bar{r}_m, \tag{5}$$

where $\bar{r}_j$ is the average observed return on security j, $\bar{r}_m$ is the average observed return on the market (a portfolio of investments that have a return that represents the market as a whole), and β_j is the measure of the relative cost of risk for security j.

Backing the analysis, the value bid-price of a security can be defined by

$$\bar{r}_j = \frac{P_{ej} - P_{0j}}{P_{0j}} = r_f + \beta_j\left(\bar{r}_m - r_f\right), \tag{6}$$

where P_{ej} is the expected future price of investment j and P_{0j} is the current period price of investment j. The idea is that P_{0j} changes according to information available to investors in such a way that the return on investment equals the risk equilibrium. Following this intuition, we can substitute the Risk Adjusted Discount Rate (RADR) in Equation (6) into the present value formulation in Equation (3) for the interest rate. From this expression, we can conjecture a market for long-term agricultural assets such that if $NPV_t > 0$ farmers increase their bid for the long-term asset (e.g., I_0 increases) until an equilibrium is reached. Alternatively, if $NPV_t < 0$, the bid for the long-term asset declines. Given these arguments, we conjecture that the market price for long-term assets is the present value of the expected cash flows over time (i.e., the results of Equation (3) such that $I_0 = 0$).

One could still raise the question: how do we estimate β_j? As indicated in our development of the risk equilibrium, most of the early work on β_j involved the estimation of risk/return relationships in security prices – the equilibrium rate of return was determined by investors bidding on stocks in the capital market. Furthermore, these stocks imply ownership of a large portfolio of productive assets, not a particular asset. One concept would be to use the overall β_j to construct an overall marginal cost of capital for a firm. By extension, to the degree that firms in a particular industry have similar β_js, we may conjecture that firms in that industry have a common discount rate. Note that the firm's beta (β_j) is also a function of that firm's capital structure.

While the CAPM formulation is typically used to examine differences in returns on securities arising from differences in risk, a similar formulation called a single index model can be used to adjust the discount rate for a particular investment (for a discussion of the single index model for risk, see Collins (1986)). Using the empirical results from Equation (5) as the risk-coefficient for a particular investment, the present value formula in Equation (3) could be specified as

$$NPV = -I_0 + \sum_{t=1}^{N} \frac{E[CF_t \mid \Omega_0]}{(1+r^\star)^t} \geq 0, \tag{7}$$

where $r^\star$ is the weighted average cost of capital for a particular farm.

3.1 Markets for farmland

In much of our discussion on asset valuation, we worked with a generic asset – using the case of a tractor. As such, we have focused primarily on the demand side – what farmers are willing to pay for tractors. Implicitly, the supply of tractors lies outside the farm sector. In this subsection, we turn our attention to the valuation of farmland, the sector's biggest asset. The total quantity of farmland available to the farm sector is largely fixed. Land may be removed from the farm sector for a variety of reasons, such as urban, recreational, or environmental uses. However, the exit of farmland tends to be irreversible. That is, once farmland has been converted to other uses, particularly urban use, it seldom returns to farming. Hence, assessment of farmland utilization and farmland valuation is key in agricultural finance and farm management.

A strand of the agricultural finance literature examines the relation between farmland values and current and expected returns on farmland (e.g., Phipps 1984; Featherstone and Baker 1987; Falk 1991; Falk and Lee 1998; Lence and Miller 1999; Moss and Schmitz 2003; Moss and Katchova 2005; Mishra, Moss, and Erickson 2008). Most studies on farmland valuation follow a Ricardian rent approach (e.g., Barry 1980; Featherstone and Baker 1987; Moss et al. 1990).

Specifically, the contention is that the value of farmland is the present value of excess rents to farmland (i.e., the return on farmland after all variable inputs have been paid). Furthermore, farmland is assumed to yield these returns into the infinite future (e.g., in Equation (3), N → ∞). Taken together, the value of farmland is sometimes written as

$$V = \frac{p'y - w'x}{r^\star}, \tag{8}$$

where p is the vector of output prices, y is the vector of output levels, w is the vector of input prices, and x is the vector of input levels. Barry (1980) estimated a risk-adjusted discount rate for farmland using a variant of the CAPM model. He found that "farm real estate values at the national or regional levels contributes little systematic risk to a well-diversified portfolio" (Barry 1980, p. 552). Barry (1980) was one of the first to use the CAPM model to analyze the returns of farmland. Several studies extended his model, including Irwin, Forster, and Sherrick (1988), Collins (1988), Bjornson and Innes (1992), and Bjornson (1994). For a review of the studies, please refer to Moss and Katchova (2005).

Apart from questions of valuation, changes in the value of farmland have significant implications for the farm sector. As farmland values have increased over time, farmland has become a larger share of the agricultural balance sheet. In 1960, farmland accounted for 70.7 percent of agricultural assets. By 2016, the share of the balance sheet in farmland increased to 82.6 percent (computed using data from USDA/ERS 2017). This change in concentration raises two questions: (1) what are the implications for this convergence on the financial well-being of the farm sector, and (2) what is driving this concentration and will it continue over time? To answer the first question, we must consider ownership of the assets. A standard starting point for the discussion of ownership of assets is the accounting identity:

$$A = L + E, \tag{9}$$

where A is the total dollar value of assets controlled by the firm, L is the firm's liabilities, and E is the dollar value of ownership interest or equity (developed more extensively in a following section). Thus, the change in asset levels equals the change in equity plus the change in liabilities (debt) $[dA = dE + dL]$. Assuming the firm's asset values from liabilities or debt is fixed by contract, the differential of the accounting identity yields:

$$dA = dE \Rightarrow dA_L p_L + dp_L A_L + dA_O p_O + dp_O A = dE, \tag{10}$$

where A_L is a quantity index for farmland, p_L is the price of farmland, A_0 is a quantity index for other agricultural assets, and p_0 is the price of those assets. Our standard assumption is that the level of farmland has either remained constant or declined slightly; hence, most of the concentration of the agricultural balance sheet is due to increases in farmland prices, disinvestment in other agricultural assets (i.e., $dA_0 < 0$), or declines in the relative price of these non-land assets. Assuming that $dA_L = dA_0 = dp_o = 0$, and dividing by E yields:

$$\begin{aligned}\frac{dE}{E} &= \frac{A_L p_L}{E} \frac{dp_L}{p_L} \\ &= \frac{A_L p_L}{A} \frac{A}{E} \frac{dp_L}{p_L}, \\ \frac{dE}{E} &= s_L \frac{1}{1-\delta} \frac{dp_L}{p_L}\end{aligned} \tag{11}$$

where S_L is the share of farmland in total agricultural asset values and δ is the debt-to-asset ratio (measuring the firm's capital structure). Hence, the volatility in agricultural equity is an increasing function of the share of farmland in the overall farm balance sheet and the firm's leverage position (see Collins' DuPont expansion).

It is also important to note that while increases in farmland values positively affect asset values and farmers' equity positions, increases in farmland values typically do not improve farmers' liquidity positions. Considering the structure of the balance sheet, farmland is typically regarded as the least liquid of all assets, meaning that it cannot be converted to cash quickly. Markets for farmland tend to be thin; generally, a particular track of land is only sold once a generation or perhaps less frequently. Thus, it can be difficult to value farmland due to the lack of transactional data. Furthermore, the majority of farmland is inherited, passed down from one generation to the next, rather than purchased on an "open call market"; this has implications for the continuance of family farms and presents barriers of entry to new and beginning farmers who must purchase farmland to enter into agriculture. For many of these beginning farmers, leasing land from retired farmers or absentee landowners (heirs of farmers) may provide a foothold in agriculture.

Equation (11) is consistent with Schmitz's discussion of the boom/bust cycle for agricultural assets (Schmitz 1995). Specifically, Schmitz details the historical episode for farmland values beginning in 1972 and ending in 1993. In 1973, wheat prices experienced a dramatic rise as a consequence of a significant wheat purchase by the Soviet Union. This purchase provided the initial impetus for a dramatic increase in farmland values that occurred from 1972 through 1981. The later stages of this boom benefited from an expansionist monetary policy that attempted to reduce the impact of the oil crises of the mid-1970s. As farmland values increased, farmer wealth (equity) rose. Furthermore, the boom undoubtedly benefited from increased farm debt (e.g., as farmland prices rose, farmers used the increased value to support higher debt levels that contributed to additional upward pressures on farmland values – again as supported in Equation (11)). The good times for agriculture started to slow down with the radical change in monetary policy in 1979 as the Federal Reserve shifted from a policy that focused on unemployment to one that focused on price stability (e.g., reducing inflation). As the interest rate increased, the downward pressure on farmland values was amplified by reductions in agricultural exports. The gains to farmer wealth from the 1970s were quickly reversed as the sector slid into the farm financial crisis of the 1980s.

Given this tendency of boom/bust cycles in farmland values to contribute to financial crises in agriculture, the question is whether the rise in farmland values starting in 2008 portends similar financial difficulties as those experienced in the 1980s. During the boom/bust cycle from 1972 through 1993, farmland values accounted for a maximum of 78 percent of agricultural assets. In 2015, farmland values have increased to 82 percent of agricultural asset values. Partially offsetting this increased share of farmland values, however, is the current level of agricultural debt, which is much lower than at the beginning of the last bust cycle. In addition, the most recent rise in farmland prices has been in part supported by historically low interest rates growing from the Federal Reserve's attempt to offset the onset of the Great Recession in 2008, as well as policies promoting ethanol (Henderson and Gloy 2009; Kropp and Peckham 2015).

There has also been a great deal of research investigating the impacts of agricultural support policies on farmland values and rental rates (Weersink et al. 1999; Lence and Mishra 2003; Mishra, Moss, and Erickson 2008; Kirwan 2009; Goodwin, Mishra, and Ortalo-Magne 2011). While these studies find different capitalization rates, in general, it is believed that governmental subsidies affect farmland values and are capitalized into farmland values and rental rates.

Apart from the impacts of subsidies, other non-farm factors affect farmland values. One such factor is technological change. In general, improvements in technology have increased the yields across the agricultural landscape. Early advances in agronomics such as the development of hybrid corn, as well as more recent advances such as genetically modified crops, have increased the yield per acre of most field crops, contributing to higher returns per acre and ultimately higher farmland prices. However, technical progress may have contributed to changes in the distribution of farmland prices. Historically, differences in farmland values were explained in part by differences in soil quality and irrigation practices (e.g., Xu, Mittelhammer, and Barkley 1993; Craig, Palmquist and Weiss 1998; Huang et al. 2006), as well as proximity to other agricultural operations, for example livestock industries (see Wing and Wolf 2000; Vukina and Wossink 2000). More productive soils were priced higher than less productive soils. Since the mid-1990s, several innovations in technologies such as variable rate technologies and precision agriculture may have reduced some of this regional variation as farmers increase their rates of return to farmland by tailoring management to specific soil attributes. Additional questions about farmland pricing involve the effect of urban pressure on farmland values (e.g., Huang et al. 2006; Livanis et al. 2006).

4 Farmer debt and debt–equity choice

Studies of debt–equity choice are almost always embedded in the framework of Collins (1985) and Barry, Baker, and Sanint (1981), hereafter called the Collins–Barry model. The debt–equity choice for the farm assumes that new capital in agriculture will come from debt or changes in asset values (either profits or capital appreciation, typically from farmland). Collins (1985) proposed a structural model taking into consideration business risk, expected return from farm operations, expected capital gains from land, and interest rates. Using a DuPont formulation, the rate of return on equity in a given period is a function of the rate of return on assets and a leverage multiplier:

$$\frac{r_{PO}}{E} = \frac{r_{PO}}{A} \times \frac{A}{E}, \qquad (12)$$

where r_{po} is the return on the portfolio of assets owned by the firm, E is equity, and A is assets.

The simple representation of Equation (12) allows for formulations of the debt–equity choice. Considering leverage to be the ratio of debt to assets, $\delta = D/A$ yields the following representation:

$$\frac{r_{PO}}{E} = \frac{r_{PO}}{A} \frac{1}{1-\delta}. \qquad (13)$$

Next, we adjust the rate of return on equity (r_E) by subtracting out the interest expense (δi) and adding the rate of appreciation for assets held by the firm (a):

$$r_E = \left[\frac{r_{PO}}{A} + a - \delta i\right]\frac{1}{1-\delta} = \left[r_A - \delta i\right]\frac{1}{1-\delta}, \qquad (14)$$

where r_E is the rate of return on equity, r_A is the rate of return on assets (the sum of the operating return on assets r_{po}/A and the rate of appreciation), and i is the cost of capital (interest rate). The above formulation introduced by Collins (1985) allows for the interest rate and the anticipated changes in asset values to be considered in the debt–equity decision.

There have been several studies that extended the Collins–Barry expected utility model of debt–equity choice and risk balancing. Featherstone et al. (1988) integrated the effect of farm program payments. Moss, Ford, and Boggess (1989) introduced income tax considerations to examine the impact of eliminating the 60 percent capital gains tax deduction on the firm's optimal leverage position; based on the optimal leverage position, the impact on the probability of equity loss is examined. Turvey and Baker (1989) employ the Collins–Barry model to examine how optimal hedging decisions may be impacted by debt decisions, by incorporating gains from hedging in the rate of return on assets. In addition, they consider how the optimal debt-to-asset ratio δ^* adjusts to hedging. Collins and Karp (1993) introduce a stochastic optimal control model of farm debt–equity choice that models risk attitudes and leverage choices. In their study, they consider failure risk (a scenario of a potential bankruptcy) rather than wealth variability, and they control for age, wealth, and the opportunity cost of farming. In a recent study, Moss (2014) decomposes the asset portfolio into operating assets and real estate and presents a model where appreciation to agricultural assets accrues to farmland.

Now let us turn our attention to the impact of additional borrowing on the rate of return on equity:

$$\frac{\partial}{\partial \delta}\left[(r_A - \delta i)\frac{1}{1-\delta}\right] = \frac{r_A - i}{(1-\delta)^2}. \tag{15}$$

Given that $0 \leq \delta < 1$, an additional unit of debt increases the rate of return on equity as long as $r_A > i$ (e.g., the rate of return on assets is greater than the cost of capital). This result is obvious. However, the result in Equation (15) implicitly ignores the impact of risk and uncertainty. In practice, the effect of leverage on risk has been the topic of much of the research into the choice of debt by farmers.

One of the first models focusing on the effect of leverage on firm risk was Gabriel and Baker (1980), which decomposed risk into business and financial risk. Specifically, Gabriel and Baker start by defining the business risk (e.g., risk of profitability of the firm) as the normal risk related to the risk associated with random output prices and output levels. Hence, farmers choose a set of inputs x based on a vector of input prices w in anticipation of producing a set of outputs $y = y(x)$ that will produce a revenue of. $p'y$. The insight of Gabriel and Baker is that the risk implied by this set of choices is endemic to the agricultural enterprise – largely independent of financial decisions made by the firm. Based on this endemic business risk, financial decisions expand this risk exponentially, where "financial risk is defined to be the added variability of the net cash flows of the owner's equity that results from the fixed financial obligation associated with debt financing and cash leasing" (Gabriel and Baker 1980, p. 560).

Gabriel and Baker develop financial risk (FR) as

$$FR = \frac{\sigma_2}{\bar{c}'x - \bar{D}} - \frac{\sigma_1}{\bar{c}'x} \tag{16}$$

where σ_1 is the standard deviation of the operating cash flows (i.e., the cash flows of the farm without debt or leasing obligations), σ_2 is the standard deviation of the returns with the debt or leasing obligations, $\bar{c}'x$ is the netput profit function (i.e., a function where $x_k > 0$ denotes an output and $x_k < 0$ denotes an input), and $\bar{D}$ is the fixed level of debt obligation. Given this definition of financial risk, Gabriel and Baker define the total risk (TR) facing the firm:

$$TR = \frac{\sigma_1}{\bar{c}'x - \bar{D}} = \frac{\sigma_1}{\bar{c}'x}\frac{\bar{c}'x}{\bar{c}'x - \bar{D}} \tag{17}$$

(see Appendix A.1). Given this total risk, Gabriel and Baker formulate a risk constraint:

$$\frac{\sigma_1}{\bar{c}'x} \frac{\bar{c}'x}{\bar{c}'x - \bar{D}} \leq \gamma. \tag{18}$$

Hence, the farmer is hypothesized to choose a level of leverage or business risk such that total risk is less than or equal to some risk index (γ). Suppose that there is an exogenous change (i.e., an increase in the level of price support) so that the level of business risk declines (e.g., σ_1 declines). It is also possible that the farm operator could engage in activities such as enterprise or crop diversification, purchasing crop insurance, hedging, or forward contracting that reduce the level of business risk. In either case, business risk is reduced and hence the farmer could increase his borrowing by decreasing $\bar{c}'x - \bar{D}$. As the level of borrowing increases, the level of total risk increases. In principle, some increase in borrowing would return the overall financial risk to the original constraint in Equation (18).

In addition to the overall risk-balancing model, Gabriel and Baker suggest a formulation that incorporates the concept of liquidity. The linkage between leverage and liquidity is especially important for agriculture. The return to agricultural equity is typically from two sources: operating profits and capital gains, primarily due to the appreciation of farmland. While operating returns provide liquidity to make loan payments, capital gains can only be accessed by selling the asset (farmland) or by additional borrowing against the increased value of the asset.

Collins (1985) provides an alternative formulation of risk-balancing based on expected utility. Specifically, Collins assumes the farmers choose the debt level that maximizes utility,

$$\max_{\delta} - \exp\left(-\rho W_0 (1 + r_E)\right)$$
$$\text{s.t. } r_E \sim N\left(\frac{\mu_A - \delta i}{1 - \delta}, \frac{\sigma_A^2}{(1 - \delta)^2}\right), \tag{19}$$

where δ is the debt-to-asset ratio, ρ is the Arrow–Pratt absolute risk aversion coefficient, W_0 is the initial level of wealth, μ_A is the expected return on agricultural assets (including both operating returns and asset appreciation), and σ_A^2 is the variance of the rate of return on assets. Equation (19) follows from the linkage between the rate of return on equity and the rate of return on assets presented in Equation (15). Given this formulation, the optimal level of debt becomes

$$\delta^* = 1 - \frac{\rho \sigma_A^2}{\mu_A - i} \tag{20}$$

Assuming that $\mu_A - i \gg 0$, these results indicate that the optimal level of debt is an increasing function of the expected return on agricultural assets and a decreasing function of the variance of the rate of return on agricultural assets (e.g., the riskiness of agriculture), the cost of capital, and the farmer's risk aversion.

Featherstone et al. (1988) use Collins' model to demonstrate how agricultural policies that reduced risk may actually increase the probability of financial difficulties in agriculture. However, other studies, such as Kropp and Katchova (2011) and Kropp and Whitaker (2011), suggest that agricultural support policies can reduce the recipient's cost of borrowing, and improve liquidity and repayment ability. Specifically, agricultural programs that provide a price floor may actually cause an increase in leverage sufficient to increase the probability of financial difficulty.

Empirical models of the Collins–Barry debt–choice model have not always provided support for the risk-balancing hypothesis. This can be attributed to some strong assumptions in the Collins–Barry formulation, such as the constant interest rate, borrower risk profile homogeneity, full credit access, and non-stochastic borrowing costs (Cheng and Gloy 2008; Wu, Guan, and Myers 2014), in addition to the proper estimation of the variance that changes either across time or across individuals. Related studies include Moss, Shonkwiler, and Ford (1990), Jensen and Langemeier (1996), Ramirez, Moss, and Boggess (1997), Escalante and Barry (2003), Turvey and Kong (2009), de Mey et al. (2014), Uzea et al. (2014), Ifft, Kuethe, and Morehart (2015), and Bampasidou, Mishra, and Moss (2017).

The optimal leverage relationship in Equation (20) raises several questions for additional research. First, the variance in Equation (20) is actually the result of a variety of production decisions, such as the mixture of enterprises, marketing decisions, and the decision to purchase crop insurance. To the extent that these decisions reduce the risk of the rate of return on agricultural assets, they may increase the level of debt used by the farm firm. Second, the derivation of the optimal debt level using the risk balancing relationship (even in the guise of stochastic optimal control; e.g., Ramirez, Moss, and Boggess 1997) fails to recognize certain facets of borrowing. Specifically, the risk-balancing model really does not consider cash flow constraints. Many lenders, particularly in periods of economic uncertainty such as the farm financial crisis of the mid 1980s, consider the cash flow requirements implied by additional debt. These models may lead to a more formal stress-test of the farm's financial condition (i.e., requiring that farmers be able to meet their cash flow obligations under certain adverse events). Furthermore, many agricultural lenders also rely on credit scoring models that utilize information pertaining to the borrower's solvency, profitability, liquidity, and quality of collateral to determine the borrower's creditworthiness.

Finally, other theories of capital structure such as Pecking Order theory focus on internal and external sources of capital and the cost associated with each of these sources. The Pecking Order is loosely related to the decision of firms to pay out dividends (i.e., make owner withdrawals) versus funding investment alternatives using retained earnings (or cash on hand). The theory suggests that retained earnings are the preferred source of financing due to the lower transaction costs associated with using retained earnings for investment purposes versus securing additional funds through debt or outside equity. In the corporate finance world, the Pecking Order theory justifies an increase in the prevalence of growth stocks. However, in the case of farm firms, Pecking Order decisions mean balancing the marginal utility of income (e.g., owner withdrawals) against the marginal rate of return on investment.

5 Asset ownership: choice of debt and equity

We now turn our attention to the ownership of assets and capital. We return to the differentiated accounting identity $dA = dE + dL$ and abstract a little for the point of discussion. The claim on firm asset values from liabilities or debt is fixed by contract. Hence, dL represents new borrowing or repayment of principle. Assuming $dL = 0$, the change in the asset valuation is "owned" or "claimed" by the equity holders $dA = dE$.

The change in the ownership value of the firm can be decomposed as presented in Figure 37.3. The horizontal axis in Figure 37.3 is the return on the assets controlled by the firm (r_A) Following the Collins formulation described above, the firm controls A assets with $\delta A = L$ (where $\delta A = L$) assets financed using debt and $(1 - \delta)A$ assets financed from equity (where $(1 - \delta)A = A - L$). The vertical axis is the return to either the owner (R_E) or the lender (R_L). We begin by assuming a zero debt level. The relationship between changes in the asset values and returns to the owner are given by curve e. Mathematically, any change in asset values (either from income or changes in the market value of assets) accrues to the owner since debt is held constant.

Next, we assume some nonzero debt level ($0 < \delta \leq 1$). Under normal operations, we would assume that the operator will make enough from operations to pay the loan off at the end of the production period and have money left over:

$$R_A \geq -A + (1+i)L \Rightarrow \begin{cases} R_E = R_A - Li \geq 0 \\ R_L = Li \end{cases}, \tag{21}$$

where i is the interest rate. Notice that this is a slight change over the way one typically thinks about the return to the farmer. This result represents the region to the right of the axis in Figure 37.3.

If, on the other hand, the return is less than what is required to pay off the loan,

$$R_A < -A + (1+i)L \Rightarrow \begin{cases} R_E = -A + L \\ R_L = R_A + [A - L] \end{cases}, \tag{22}$$

then the decision maker is better off defaulting on the loan, losing equity of $-A + L$ (see Appendix B). The lender receives the firm's original equity ($A - L$) plus the rate of return on assets. The minimum return to firm is $-A + (1 + i)L$, while the maximum return is positive infinity (again the likelihood of this event is bounded to zero).

Seeking an answer to the question of whether there is an alternative to the debt market, the answer is yes, firms can sell equity – typically stocks or shares in limited partnerships. In these typical forms of equity investment, the investor shares proportionally in the gains. In addition, there are scenarios where the investor participates in losses. In the case of stocks, the loss is typically limited to the price of the stock (e.g., the stock value could fall to zero). In agricultural finance, we are typically interested in three ownership forms:

1   Sole proprietorship: under this organization, the individual directly owns assets that he employs in entrepreneurial activities. The liabilities and obligations of the firm are also the liabilities and obligations of the individual. The proprietor or owner/operator has a claim on any residual (excess of the value of the assets of the firm over its obligations).

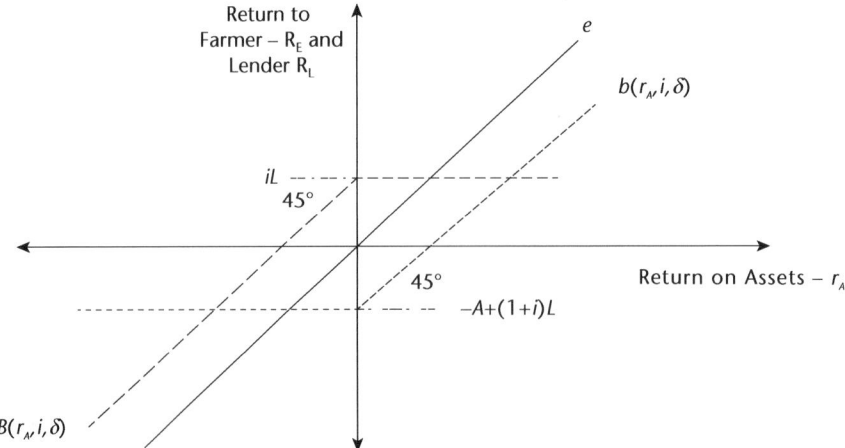

Figure 37.3 Equity and ownership claims

2   Partnership: partnerships may exist under a variety of legal frameworks, from fairly informal partnerships based on "handshake agreements" to more elaborate limited liability partnerships. At the most basic level, a partnership is an agreement linking the interests of two individuals. The agreement specifies the intent of the collaboration, the expectations of each party, and the claim that each party will have on the proceeds from the collaboration. Under some types of partnerships, the liabilities and obligations of the firm are also the liabilities and obligations of the individual.

3   Corporation: a corporation is a legal entity by which a group of individuals collaborates by placing some of their assets at risk (i.e., an investment of a portion of their assets in common stock). As a legal entity, the corporation may enter into contracts that do not bind the individual owners. Importantly, the liabilities and debt of the corporation do not pass to the individuals. However, the owners of the corporation only have a claim to the assets of the corporation embodied in the terms of their stock.

Historically, most farms have been organized as sole proprietorships and depend on debt for additional capital.

In general, the various ownership forms can be formulated as financial options. An option is a contingent asset, that is, an asset that has value contingent on the value of another asset. Options give the holder the opportunity but not the obligation to buy or sell an asset. For example, a call option gives the holder the right but not the obligation to purchase an asset (such as a stock) at a given (strike) price. If the stock price is higher than the strike price, the owner of the call option will exercise his option, purchase the stock at the strike price, and then sell the stock at the market price; his return will be the difference between the market price and the strike price. A put option is a similar contract that gives the holder the right to sell an asset at a specified price.

Figure 37.4 presents the payoff or profit function for call and put options, along with the probability density function for a stock price at the expiration date. Comparing Figure 37.4

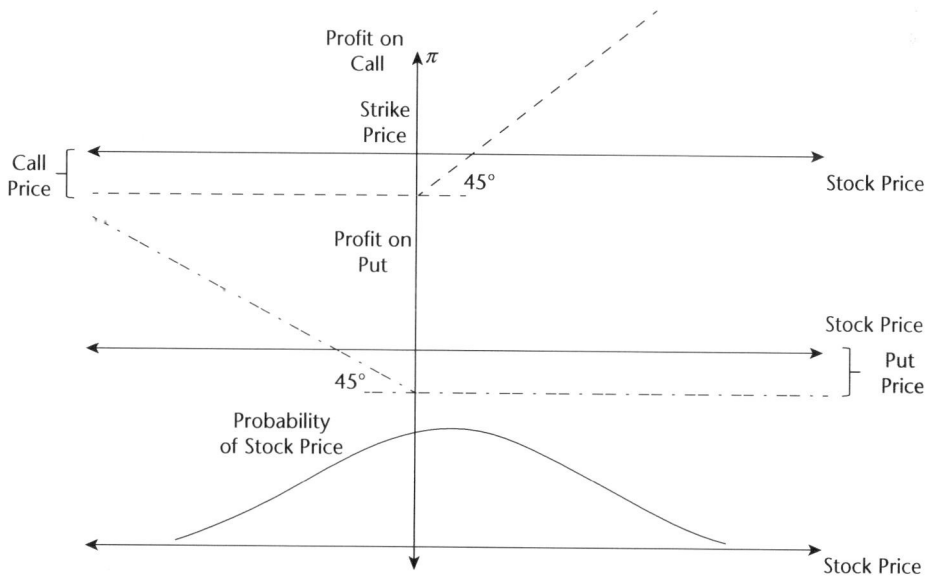

Figure 37.4 Payoff functions for call and put options

with Figure 37.3, the payoff function for the owner–operator resembles the call option where the strike price is a return on equity of zero. The lender's position in Figures 37.3 and 37.4 resembles "selling a put."

6 Credit market equilibrium

The Gabriel–Baker and the Collins models focus on the producer's choice. However, there are two players in the credit market: the borrower and the lender. In equilibrium, the interest rate charged by the lender depends on the opportunity rate of return to the bank (i.e., the return on loans with similar risk that the bank could make) as well as the demand for credit. The return to the bank increases as the interest rate charged increases. However, the interest rate that banks are able to charge is limited by the demand for credit (e.g., what producers are willing to pay). Hence, the interest rate in the credit market is determined by lending risk and the opportunity returns available to banks.

The market for agricultural debt is complicated because of information difficulties: asymmetric information and agency problems. Specifically, there are different types of borrowers. Some borrowers are good credit risks – they are profitable farmers who pay their bills. Other farmers may be poor credit risks for a variety of reasons; the farmer may be less profitable due to managerial skills or the farmer may have accumulated past debt because of bad market outcomes (i.e., low prices). In either case, acquiring a loan is a contracting process where the farmer provides information to the banker, who attempts to determine the likelihood that the loan will be repaid with interest.

We assume that the farmer could be one of two types of farmers: high-risk or low-risk. The high-risk farmer has a lower expected return on assets and a higher standard deviation for those returns, while the low-risk farmer has a higher expected rate of return and a lower standard deviation for those returns. The payoff for the borrower follows the payoff function for the call presented in Figure 37.4. Assume that the payoff function is $B(r_A, i, \delta)$, where i is the stated interest rate and δ is the debt-to-asset ratio (or share of assets borrowed). Essentially, the borrower only retains ownership of the asset (i.e., pays off the loan) when the return is greater than the alternative. Essentially, the lowest possible return is the loss of the down payment or collateral (e.g., $1 = \delta$). If the return is lower than the loss of the down payment, the borrower forfeits the down payment or collateral and gives the lender the return on assets (see Appendix C).

A standard construct in finance is that participants have to be paid to accept higher risk. Hence, riskier stocks earn higher returns on average. Put slightly differently, the common assumption is that the market return separates investments in the market. We discussed this axiom earlier in our discussion of CAPM. Hence, one would expect that riskier borrowers would pay higher interest rates. However, two factors conspire to make this "separation by interest rate" difficult in the loan market. First, the lender's ability to differentiate between the two borrower types is imperfect. Second, the "kinked" return function for the borrower may make it profitable for the borrower to default on the loan.

To develop the implications for the "kinked" payoff functions, we transform the results in Equations (21) and (22) from returns to the firm and lender into the rate of return to the firm and lender, starting with the implications for Equation (21) (where the firm pays off the loan), $R_A \geq -A + (1+i)L$. Note that the value to the high-risk borrower is declining throughout the range of possible interest rates regardless of the down payment requirement. Hence, there is an incentive for the high-risk borrower to mimic – always report to be less risky than the borrower actually is. However, note that the return to the low-risk borrower decreases in the interest rate because of the change in the down payment requirement.

Essentially, at the time of the loan application, the borrower knows more about the project's risk and the financial risks of the firm than the lender. However, the interest rate alone will fail to separate the high-risk and low-risk borrowers. Stiglitz and Weiss (1981) show that these informational asymmetries can lead to credit-rationing – the demand for credit will exceed the supply of credit at the current market interest rate. Increasing the market interest rate to eliminate the excess demand for credit is not feasible because this would drive out the low-risk borrower and leave only the high-risk borrower in the market. As a result, lenders must take additional actions, such as requiring down payments or collateral or expending resources to monitor the borrower during the loan period (Stiglitz and Weiss 1981). Similar work in agricultural economics includes Innes (1990). Furthermore, lenders have developed various methods of assessing the riskiness of potential borrowers, including credit scoring and credit risk appraisal practices.

7 Summary and suggestions for further study

This chapter developed financial economic models useful in understanding the financial market for agriculture. It began by sketching out the properties of the neoclassical capital market, and then extended the neoclassical model to include the effect of risk on capital structure. Specifically, the chapter demonstrated the effect of the sector's dominant asset – farmland – on farmers' wealth. In addition, we demonstrated how the optimal debt model follows the risk-balancing model, which shows how financial risk magnifies the business risk associated with agricultural production. Next, we sketched a general market model based on insights from the Stiglitz and Weiss (1981) model of credit rationing.

Historically, agricultural finance has been "farmer-centric." Most of the discussion of farm financial issues at the beginning of the twentieth century involved meeting the capital needs of the farm sector given the increased demand for borrowing due to increased mechanization. This period saw the establishment of the Farm Credit System and the Federal Reserve System. However, at the beginning of the twenty-first century, U.S. agriculture credit needs are being met by a variety of lenders, including savings and loans, Farm Credit, insurance companies, suppliers, and guaranteed loans through the United States Department of Agriculture (USDA). Current issues tend to focus around the effects of the distribution of farmland. Specifically, the agenda for current studies in agricultural finance are bounded by two events: (1) the boom/bust cycle in agriculture in the 1970s and 1980s that ended in the farm financial crisis of the mid-1980s and (2) the significant increase in farmer wealth from 2008 through 2013. During the boom cycles, farmer wealth significantly increased because of increased farmland values and higher grain prices. While both booms cycles were due to increased global demand for grains, the boom from 2008 through 2013 was driven primarily by higher corn prices to promote corn ethanol. The bust cycle caused some farmers to become insolvent as farmland values fell. Furthermore, a large share of agricultural output is increasingly concentrated on a few wealthy farm firms. This concentration is due in part to the impact of capital gains on returns to agricultural assets. As discussed in this chapter, an increasing share of agricultural income is a "holding return" from the appreciation of farmland. The interaction of the returns from operation and the returns to agricultural assets and increases in farmland prices is still cloudy. Changes in farmland prices appear to be consistent in the long run, but the level of farmland prices is often higher than can be justified using present value analysis. Regardless of the reasons for this anomaly, the returns from holding farmland provide an impetus for growth for larger farmers and additional pressure toward concentration of farmland ownership. Thus, understanding the factors driving farmland values and the effect of the higher farmland values on farmer wealth, wealth transfer, and access of credit remain an important avenue of research. Furthermore, the transition of farmland holdings to the next generation of farmers continues to be an active avenue of research.

Another area of future research in agricultural finance involves the potential arbitrage between returns on farmland and returns on agricultural assets. From the late 1990s through 2008, arbitrage of risky assets became an important component of the financial marketplace. This arbitrage typically involved the purchase of one asset, such as a mortgage-backed security, using money raised by a short-sale of another asset, such as a U.S. Treasury bond. The assumption of this transaction was that the mortgage-backed security was underpriced (i.e., the interest rate was too high) compared to its relative risk. The concept was that when enough investors recognized that the mortgage-backed security was underpriced, the market price of these instruments would increase, thus reducing the spread between mortgage-backed securities. The arbitrageur would then reverse the position: sell the mortgage-backed securities and buy back U.S. Treasury bonds at a profit. As the scenario sketched out suggests, this transaction (and the returns to arbitrage) may imply a substantial risk. In fact, arbitrage of mortgage-backed instruments contributed to the financial crisis of 2008. However, the shift from a traditional buy-and-hold implicit in the farm mortgage market to a more arbitrage-oriented market for agricultural assets may provide insights into the farm financial market. A key component in understanding arbitrage in financial assets is the risk inherent in each asset (i.e., the one purchased and the one sold). If this variance (risk) is known or at least bounded from above, the arbitrage position is typically profitable, but unforeseen variances may occur, such as the effect of arbitrage on agricultural asset values (primarily farmland). Furthermore, as with the questions raised by other financial assets, the primary effect of arbitrage on asset pricing is dependent on changes in this variance over time. These questions are compounded by the fact that farmland is infrequently traded, coupled with the fact that most of what we know about farmland values is based on annual farmer surveys.

References

Bampasidou, M., A.K. Mishra, and C.B. Moss. 2017. "Modeling Debt Choice in Agriculture: The Effect of Endogenous Asset Values." *Agricultural Finance Review* 77(1):1–16.

Barry, P.J. 1980. "Capital Asset Pricing and Farm Real Estate." *American Journal of Agricultural Economics* 62(3):549–553.

Barry, P.J., C.B. Baker, and L.R. Sanint. 1981. "Farmers' Credit Risks and Liquidity Management." *American Journal of Agricultural Economics* 63(2):216–227.

Bawerk, E.B. 2007. *Capital and Interest – A Critical History of Economic Theory* (Kindle ed.). Evergreen Review, Inc.

Bjornson, B. 1994. "Asset Pricing Theory and the Predictable Variation in Agricultural Asset Returns." *American Journal of Agricultural Economics* 76:454–464.

Bjornson, B., and R. Innes. 1992. "Risk and Return in Agriculture: Evidence from an Explicit-Factor Arbitrage Pricing Model." *Journal of Agricultural and Resource Economics* 17:232–252.

Cheng, M.L., and B.A. Gloy. 2008. "The Paradox of Risk Balancing: Do Risk Reducing Policies Lead to More Risk for Farmers? American Association of Agricultural Economics Conference, Orlando FL, July.

Collins, R.A. 1985. "Expected Utility, Debt–Equity Structure, and Risk Balancing." *American Journal of Agricultural Economics* 67(3):627–629.

———. 1986. "Risk Analysis with Single-Index Portfolio Models: An Application to Farm Planning." *American Journal of Agricultural Economics* 68(1):152–161.

———. 1988. "The Required Rate of Return for Publicly Held Agricultural Equity: An Arbitrage Pricing Theory Approach." *Western Journal of Agricultural Economics* 13:163–168.

Collins, R.A., and L.S. Karp. 1993. "Lifetime Leverage Choice for Proprietary Farmers in a Dynamic Stochastic Environment." *Journal of Agricultural and Resource Economics* 18(2):225–238.

Craig, L.A., R.B. Palmquist, and T. Weiss. 1998. "Transportation Improvements and Land Values in the Antebellum United States: A Hedonic Approach." *Journal of Real Estate Finance and Economics* 16(2):173–189.

de Mey, Y., F. Van Winsen, E. Wauters, M. Vancauteren, L. Lauwers, and S. Van Passel. 2014. "Farm-Level Evidence on Risk Balancing Behavior in the EU-15." *Agricultural Finance Review* 74(1):17–37.

Escalante, C.L., and P.J. Barry. 2003. "Determinants of the Strength of Strategic Adjustments in Farm Capital Structure." *Journal of Agricultural and Applied Economics* 35(1):67–78.

Falk, B. 1991. "Formally Testing the Present Value Model of Farmland Prices." *American Journal of Agricultural Economics* 73:1–10.

Falk, B., and B.S. Lee. 1998. "Fads Versus Fundamentals in Farmland Prices." *American Journal of Agricultural Economics* 80:696–707.

Featherstone, A.M. and T.G. Baker. 1987. "An Examination of Farm Sector Real Asset Dyanmics: 1910–85." *American Journal of Agricultural Economics* 69(3):532–546.

Featherstone, A.M., C.B. Moss, T.G. Baker, and P.V. Preckel. 1988. "The Theoretical Effects of Farm Policies on Optimal Leverage and the Probability of Equity Loss." *American Journal of Agricultural Economics* 70(3):572–579.

Gabriel, S.C., and C.B. Baker. 1980. "Concepts of Business and Financial Risk." *American Journal of Agricultural Economics* 62(3):560–564.

Goodwin, B.K., A.K. Mishra, and F. Ortalo-Magne. 2011. "The Buck Stops Where? The Distribution of Agricultural Subsidies." NBER Working Paper. No. 16693. National Bureau of Economic Research, Cambridge, MA.

Henderson, J., and B. Gloy. 2009. "The Impact of Ethanol Plants on Cropland Values in the Great Plains." *Agricultural Finance Review* 69(1):36–48.

Huang, H., G. Miller, B. Sherrick, and M. Gomez. 2006. "Factors Influencing Illinois Farmland Values." *American Journal of Agricultural Economics* 88(2):458–470.

Ifft, J.E., T. Kuethe, and M. Morehart. 2015. "Does Federal Crop Insurance Lead to Higher Farm Debt Use? Evidence from the Agricultural Resource Management Survey (ARMS)." *Agricultural Finance Review* 75(3):349–367.

Innes, R.D. 1990. "Limited Liability and Incentive Contracting with Ex-Ante Action Choices." *Journal of Economic Theory* 52(1):45–67.

Irwin, S.H., D.L. Forster, and B.J. Sherrick. 1988. "Returns to Farm Real Estate Revisited." *American Journal of Agricultural Economics* 70:580–587.

Jensen, F.E., and L.N. Langemeier. 1996. "Optimal Leverage with Risk Aversion: Empirical Evidence." *Agricultural Finance Review* 56(1):85–97.

Kirwan, B.E. 2009. "The Incidence of US Agricultural Subsidies on Farmland Rental Rates." *Journal of Political Economy* 117(1):138–164.

Kropp, J.D., and A.L. Katchova. 2011. "The Effects of Direct Payments on Liquidity and Repayment Capacity of Beginning Farmers." *Agricultural Finance Review* 71(3):347–365.

Kropp, J.D., and J.G. Peckham. 2015. "U.S. Agricultural Support Programs and Ethanol Policies Effects on Farmland Values and Rental Rates." *Agricultural Finance Review* 72(2):169–193.

Kropp, J.D., and J.B. Whitaker. 2011. "The Impact of Decoupled Payments on the Cost of Operating Capital." *Agricultural Finance Review* 71(1):25–40.

Lence, S.H. and D.J. Miller. 1999. "Transaction Costs and the Present Value Model of Farmland: Iowa, 1900–1994." *American Journal of Agricultural Economics* 81(2):257–272.

Lence, S.H., and A.K. Mishra. 2003. "The Impacts of Different Farm Programs on Cash Rents." *American Journal Agricultural Economics* 85(3):753–761.

Lintner, J. 1965a. "Security Prices, Risk, and Maximal Gain from Diversification." *Journal of Finance* 20:587–615.

———. 1965b. "The Valuation of Risk Assets and the Selection of Risky Investments in Stock Portfolios and Capital Budgets." *Review of Economics and Statistics* 47:13–37.

Livanis, G., C.B. Moss, V.E. Breneman, and R.F. Nehring. 2006. "Urban Sprawl and Farmland Prices." *American Journal Agricultural Economics* 88(4):915–929.

Mishra, A.K., C.B. Moss, and K.W. Erickson. 2008. "The Role of Credit Constraints and Government Subsidies in Farmland Valuations in the U.S.: An Options Pricing Model Approach." *Empirical Economics* 34:285–297.

Moss, C.B. 2010. *Risk, Uncertainty and the Agricultural Firm.* Hackensack, NJ: World Scientific Publishing Company.

———. 2014. *Farmland Values, Farmer Equity and Equilibrium in the Agricultural Capital Market.* Jersey City, NJ: Farm Credit System Funding Corporation.

Moss, C.B., S.A. Ford, and W.G. Boggess. 1989. "Capital Gains, Optimal Leverage, and the Probability of Equity Loss." *Agricultural Finance Review* 49:27–34.

Moss, C.B., and A. Katchova. 2005. "Farmland Valuation and Asset Performance." *Agricultural Finance Review* 65(2):119–130.

Moss, C.B., and A. Schmitz (eds). 2003. *Government Policy and Farmland Markets: The Maintenance of Farmer Wealth.* Ames, IA: Iowa State University Press.

Moss, C.B., J.S. Shonkwiler, and S.A. Ford. 1990. "A Risk Endogenous Model of Aggregate Agricultural Debt." *Agricultural Finance Review* 50:73–79.

Mossin, J. 1966. "Equilibrium in a Capital Asset Market." *Econometrica* 34:768–783.

Phipps, T. 1984. "Land Prices and Farm-Based Returns." *American Journal of Agricultural Economics* 66(4):422–429.

Ramirez, O.A., C.B. Moss, and W.G. Boggess. 1997. "A Stochastic Optimal Control Formulation of the Consumption/Debt Decision." *Agricultural Finance Review* 57:29–38.

Schmitz, A. 1995. "Boom/Bust Cycles and Ricardian Rent." *American Journal of Agricultural Economics* 77(5):1110–1125.

Sharpe, W.F. 1954. "Capital Asset Prices: A Theory of Market Equilibrium Under Conditions of Risk." *Journal of Finance* 19:425–442.

Stiglitz, J.E., and A. Weiss. 1981. "Credit Rationing in Markets with Imperfect Information." *American Economic Review* 71(3):393–410.

Turvey, C.G., and T.G. Baker. 1989. "Optimal Hedging under Alternative Capital Structures and Risk Aversion." *Canadian Journal of Agricultural Economics* 37(1):135–143.

Turvey, C.G., and R. Kong. 2009. "Business and Financial Risks of Small Farm Households in China." *China Agricultural Economic Review* 1(2):155–172.

Uzea, N., K. Poon, D. Sparling, and A. Weersink. 2014. "Farm Support Payments and Risk Balancing: Implications for Financial Riskiness of Canadian Farms." *Canadian Journal of Agricultural Economics* 62(4):595–618.

Vukina, T., and A. Wossink. 2000. "Environmental Policies and Agricultural Land Values: Evidence from the Dutch Nutrient Quota System." *Land Economics* 76(3):413–429.

Weersink, A., S. Clark, C.G. Turvey, and R. Sarker. 1999. "The Effect of Agricultural Policy on Farmland Values." *Land Economics* 75(3):425–439.

Wing, S., and S. Wolf. 2000. "Intensive Livestock Operations, Health, and Quality of Life among Eastern North Carolina Residents." *Environmental Health Perspectives* 108(3):233–238.

Wu, F., Z. Guan, and R. Myers. 2014. "Farm Capital Structure Choice: Theory and an Empirical Test." *Agricultural Finance Review* 74(1):115–132.

Xu, F., R.C. Mittelhammer, and P.W. Barkley. 1993. "Measuring the Contributions of Site Characteristics to the Value of Agricultural Land." *Land Economics* 69(4):356–369.

Appendix A
Gabriel and Baker formulation of financial and total risk

Gabriel and Baker reformulate Equation 16 to focus on the financial risk component:

$$FR = \frac{\sigma_2 - \sigma_1}{\bar{c}'x - \bar{D}} + \frac{\sigma_1}{\bar{c}'x}\frac{\bar{D}}{\bar{c}'x - \bar{D}}. \tag{A.1}$$

As a starting point, we assume that leverage decisions do not change the variability of cash flows (i.e., $\sigma_2 = \sigma_1$). This scenario is consistent with the assumption that debt payments are fixed. Subtracting a constant from a sequence does not change the variance. Hence, the first term in Equation A.1 drops out, yielding

$$FR = \frac{\sigma_1}{\bar{c}'x}\frac{\bar{D}}{\bar{c}'x - \bar{D}}. \tag{A.2}$$

In Equation A.2, financial risk is determined by the business risk (e.g., $\sigma_1 / \bar{c}'x$) and the share of cash flows that go to paying fixed debt obligations (e.g., $\bar{D} / (\bar{c}'x - \bar{D})$).

Appendix B
Distribution of income for different rates of return on assets

As a starting point, we assume that the rate of return on assets is sufficient to pay the interest on the loan (i.e., Equation 21). This result implies

$$R_A \geq -A + L + iL$$
$$\geq -E + iL \quad , \tag{B.1}$$

given that $A = E + L \Rightarrow E = A - L$, Thus, the result actually implies that the farmer has a positive level of equity for some return on assets.

Also note that as the rate of return on assets approaches the point where the firm forfeits its assets from the left:

$$\lim_{R_A \to \left[-A+(1+K)L\right]} R_L = -A + (1+i)L + [A - L] = iL \tag{B.2}$$

Hence, the minimum return to the lender is negative infinity (although the likelihood of this event is bounded to zero) and the maximum return is iL. However, in practical terms, a lender would never seize the borrower's asset if the asset had a negative return. In fact, a lender would only seize an asset if the market value of the asset less the transactional cost of seizing the asset through foreclosure was positive, and hence the minimum loss to the lender in practice is the original value of the loan.

The rate of return to the firm becomes

$$r_E = \frac{R_E}{E} = \frac{R_A - iL}{E}$$
$$= \frac{R_A}{A}\frac{A}{E} - \frac{L}{A}\frac{A}{E}i$$
$$= \frac{r_A - \delta i}{(1-\delta)} \quad , \tag{B.3}$$

which is consistent with Equation 14. Notice for this region $r_L = i$-, the borrower pays off the loan, so the return to the lender is simply the stated interest rate. Turning to Equation 22 (where the firm defaults on the loan $R_A < -A + (1 + i)$, the rate of return to the firm becomes:

$$r_E = \frac{R_E}{E} = \frac{-A+L}{E} = -\frac{E}{E} = -1. \tag{B.4}$$

The rate of return to the lender becomes:

$$\begin{aligned} r_L &= \frac{R_L}{L} = \frac{R_A + A - L}{L} \\ &= \frac{R_A}{A}\frac{A}{L} + \frac{A}{L} - 1 \\ &= (r_A + 1)\frac{1}{\delta} - 1. \end{aligned} \tag{B.5}$$

As a final point, the "kink" point in the rate of return space can be derived as $r_A^\star = -1 + (1+K)\delta$.

Appendix C
Credit market equilibrium

The expected value of the loan to the borrower is then:

$$\tilde{b}(K,\delta,\mu,\sigma^2) = \int_{-\infty}^{\infty} b(r_A,i,\delta) f(r_A;\mu,\sigma^2) dr_A .\tag{C.1}$$

The rate of return to the borrower can be written as:

$$\tilde{B}(K,\delta,\mu,\sigma^2) = \int_{-\infty}^{r_A^*} \left[\frac{(r_A+1)}{\delta}-1\right] f(r_A;\mu,\sigma^2) dr_A + \int_{r_A^*}^{\infty} K f(r_A;\mu,\sigma^2) dr_A ,\tag{C.2}$$

while the rate of return to the firm can be written as:

$$\tilde{b}(K,\delta,\mu,\sigma^2) = -\int_{-\infty}^{r_A^*} f(r_A;\mu,\sigma^2) dr_A + \int_{r_A^*}^{\infty} \left[\frac{r_A-\delta K}{1-\delta}\right] f(r_A;\mu,\sigma^2) dr_A .\tag{C.3}$$

38
Futures markets and hedging

T. Randall Fortenbery

1. Introduction

The concept of hedging against risky outcomes has likely been around since humans first began trading goods and services. According to Hull (2014), a hedge is defined as a "trade to reduce risk." As such, it may take many forms. For example, a consumer who purchases car insurance is hedging against the potential loss associated with being in an accident. The consumer is trading an insurance premium for protection against a larger potential loss if he or she is involved in an accident.

In the agricultural and applied economics literature, a hedge generally refers to a market agent taking a position in a futures derivative (futures contract or futures contract option) traded on an organized, formal exchange, to protect against any future adverse price movement in an underlying physical commodity. We will refer to the physical commodity as a cash asset. The cash asset may be physical inventory already held, future production not yet realized, or an expected future purchase of the asset.

The futures and options contracts used to initiate hedged positions are standardized in all aspects except price: each contract represents a specific, non-negotiable quantity of the underlying asset; has fixed quality parameters; and designates either a specific location where delivery takes place at contract expiration, or a reference against which the contract will be settled by cash payment at expiration (in other words, if physical delivery does not take place, a payment is exchanged between the futures contract buyer and seller that represents the difference between the initially traded price and its final value at contract expiration). As a result, all futures traders are trading identical assets, and the only variation across trades is the price at which transactions take place. The trader of a futures contract faces a legal obligation to deliver (in the case of a futures contract seller) or take delivery (in the case of a futures contract buyer) of the underlying asset at the initial price (the price at which the contract was bought or sold), at some specified future time period.

Options on futures contracts provide the holder the right to buy or sell a futures contract (call options provide the right to buy a contract and put options the right to sell) at a specific price for a specific delivery period, but not the obligation to do so. A particular option provides the right to buy or sell a single futures contract. Thus, options on futures contracts allow holders to hedge at a predetermined futures price if it turns out, over the life of the option, that the potential hedge price is more attractive than the realized cash price. In return for this right, the option buyer pays a premium to the option seller, who then has an obligation to facilitate the hedge if the option holder so desires.

According to Purcell and Koontz (1999), trade in organized futures contracts goes back at least several centuries, with evidence of futures type trading in both China and Japan. Contributors to Wikipedia suggest even earlier futures type transactions and argue futures trading existed in Mesopotamia as early as 1750 BC. Regardless of when futures trading actually started, modern futures trading and contract design can be traced to the founding of the Chicago Board of Trade in the mid-1800s.

2 Social value of futures trading

The academic literature often points to two social benefits that arise from futures market trade activity: price discovery and risk management. The price discovery function refers to futures market prices accurately reflecting, given all information currently available, what the current fair market value is of the commodity being traded for future delivery, i.e., when the futures contract expires. The risk management function refers to the ability of market participants to directly use the futures contract to hedge the future value of a cash asset, and thus avoid future price changes that might occur as the current information set changes. For the risk management function to work, hedgers must be able to enter the futures market at or near prices currently being traded for later delivery, and they must be able to exit the futures market when risk protection is no longer necessary. This, in turn, requires a liquid market – one in which there is sufficient trading volume to allow entry and exit by individual traders without an individual trader's action resulting in significant price changes.

Despite the focus on price discovery and risk management as broad social benefits in justifying the existence of future markets, however, the Chicago Board of Trade was originally founded to provide an orderly process for large and active speculators and merchandizers to facilitate trade among themselves. It was not initially founded to provide a platform for hedging beyond the initial members, nor to provide the larger community with information on the likely future value of various commodities.

2.1 Modern futures market history

Chicago emerged as the nation's grain trading center as early as the 1830s, and by 1845 over 1 million bushels of wheat per year were being exported to other U.S. cities and overseas through Chicago grain facilities. According to Ferris (1988), this level of activity attracted a significant amount of speculation. Merchants would buy grain directly from farmers in hopes of earning substantial returns through both the storage of grain over time and the sale of grain to distant consumers. Thus, they were speculating on both spatial and temporal price differences for grain.

To facilitate grain movement, the Illinois and Michigan canal was constructed, and by 1848 the canal connected the Chicago and Illinois rivers. This allowed more efficient movement of grain from the central Illinois production regions to the Chicago trade center. As trade expanded, a group of Chicago businessmen decided that their speculative interests could be furthered through a formal trade organization, and in 1848 they founded the Chicago Board of Trade (CBOT). According to Ferris' (1988) citation of the original CBOT founding announcement, its purpose was to "promote uniformity and equity in trade, facilitate a speedy adjustment to business suits, to acquire and disseminate valuable information, and generally to secure its members the benefits of cooperation."

By the time the Civil War started, Chicago had become the nation's granary center. Chicago's grain traders provided access to almost all the corn and wheat produced in the U.S. The premier wheat growing region of the country was along the Wisconsin–Illinois border, west of Chicago, and Illinois had become the number one corn growing state (Ferris 1988). Almost all this grain production was marketed through Chicago trade houses, and the corn trade soon launched Chicago into the dominant global futures market.

In 1898, the Chicago Mercantile Exchange (CME) was founded as the Chicago Butter and Egg Board, and this soon became the world's primary livestock futures exchange. The CBOT and CME were the world's two largest futures exchanges until their merger in 2007. They absorbed other U.S. exchanges through 2008 and today dominate the world's trading in both commodity and financial futures under the name CME Group.

2.2 Understanding price relationships

Hedging in a futures market involves using a futures contract traded today as a proxy for a cash transaction expected to occur on some future date. For this to be effective, there must be a predictable relationship between the futures contract price at or near contract expiration, and the cash price at which the hedged transaction will occur. In other words, cash and futures must converge towards the end of the hedge period, as illustrated in Figure 38.1.

In a perfect hedge, the cash transaction happens concurrently with the expiration of the futures contract, and there is no difference in the cash price and the futures price at futures contract expiration. In this case, a futures hedge completely covers all cash price risk. However, this situation is very rare.

For most commodities, the cash (or spot) transaction will not occur in the same market as the delivery location specified for an expiring futures contract. There can also be variations in the quality of the cash commodity relative to the futures contract's specification. Further, since most commodity futures markets (crude oil is an exception) do not have an expiration and delivery every month, there is often a temporal difference between the cash transaction date and the

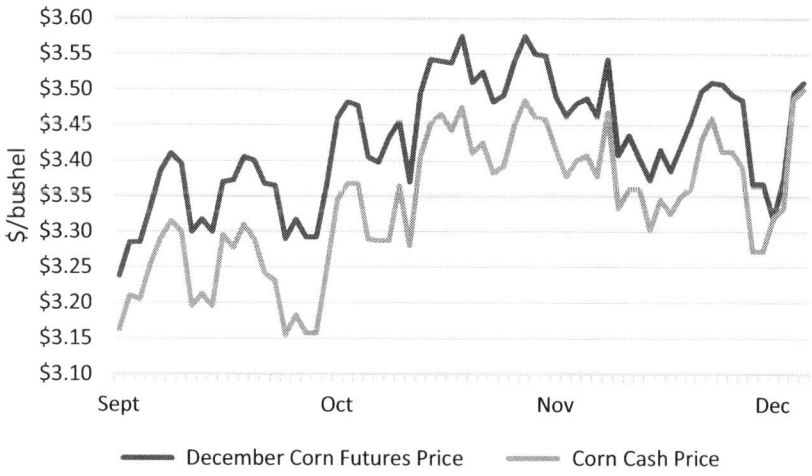

Figure 38.1 U.S. corn prices 2016

futures expiration. All these can lead to differences between the final price of the cash asset and the futures price used to hedge the cash price.

Soybean futures, for example, are available for delivery in January, March, May, July, August, September, and November. Delivery against a soybean futures contract requires the seller to deliver to the buyer a shipping certificate for 5,000 bushels of soybeans (the quantity of soybeans in a single futures contract) to be shipped from one of five delivery areas along the Illinois River or the single delivery area on the Upper Mississippi River. Locally grown soybeans cannot be delivered against a futures contract. Further, the delivery certificate must be for delivery from specific elevators in each delivery area that have been pre-certified for futures market delivery. One of the delivery areas specifies delivery at a price equal to the expiring futures contract price, while delivery from the other five areas requires an additional transportation premium added to the expiring futures contract price. Thus, even for soybeans certified for delivery against a futures contract, the actual delivery price is affected by the geographic location of the soybeans being delivered through the delivery certificate.

An Iowa soybean farmer that expects to harvest and sell soybeans in October cannot hedge his or her final sales price on a futures contract for October delivery in Central Iowa; instead, the farmer must use a futures contract for delivery in November at one of the predetermined delivery areas. If there is a difference between soybean prices in October versus November, or between Central Iowa and the futures market delivery areas, then the November futures contract may not provide a perfect hedge for an October cash sale.

While a perfect hedge seldom exists in commodity markets, a successful hedge can still be placed if the futures and cash prices converge to a predictable difference at the conclusion of the hedge period. If this is not the case, then the hedger does not have a useful prediction of the "expected" cash price realized through the hedge.

Because of the differences in quality characteristics, geographic location, and/or transaction dates for cash versus futures markets, the prices in futures and cash markets will usually differ. This difference is called the basis, and it is generally calculated by commercial traders as cash minus futures price. Thus, an effective hedge trades away cash price risk for basis risk – as long as cash and futures are expected to converge to a predictable difference, then basis risk will generally be less than cash price risk, and thus total market risk is reduced through hedging. If that is not the case, then there is no assurance that hedging is a price risk reducing activity.

Despite an expectation of convergence, futures and spot prices may not change by the same amount over time; thus, the basis itself experiences some variation. The more the cash commodity to be traded deviates from the standardized futures contract specifications, the more basis can vary. In fact, in some cases the futures contracts used to hedge and the cash commodity traded at the end of the hedge are not even for the same commodity. For example, there is no futures contract for soft white winter wheat. However, there is a Chicago traded futures contract for soft red winter wheat. While they both have very different final demand centers, their prices are sufficiently correlated that the soft red winter wheat futures contract is often used to hedge the final cash price of soft white winter wheat. A hedge where the futures and cash prices are for different commodities is called a cross-hedge. Cross-hedges generally experience greater basis risk compared to hedges where the futures and cash commodities are identical, but as long as the futures and cash prices in a cross-hedge move in the same direction, and their final relationship (the final basis) at the end of the hedge is somewhat predictable, cross-hedges can still be net price risk reducing (i.e., the basis risk is less than the risk faced by the unhedged cash position).

There has been a significant amount of empirical research over the last several decades measuring, either directly or indirectly, basis risk for a whole host of futures/cash commodity pairs.

Much of the early work on futures/cash price relationships was conducted using static regression models. They were usually formulated as:

$$S_t = \sigma + \beta F_t + e_t \tag{1}$$

where S is the cash price, F is futures price, and e measures the stochastic difference between the two. This specification assumes that new information affects both cash and futures prices instantaneously and identically (Zapata and Fortenbery 1996). However, Dewbre (1981) has shown that these assumptions are overly restrictive. Under some conditions, the two markets may not move to their long-run equilibrium immediately following a market shock. This realization led to the use of cointegration theory to test market relationships and basis behavior.

The application of cointegration tests in measuring price relationships allows for at least two types of increased realism: (1) that prices might be nonstationary (i.e., their means and/or variances might change with time), and (2) that two prices (say a futures and cash price for similar assets) should respond to outside shocks in a similar way over the long run (assuming markets are efficient), but in the short run might deviate from their equilibrium relationship when the shock is first realized. This could be because of asymmetric timing in realizing the shock between participants in the different markets or differences in initial valuations of the shock. However, assuming market efficiency, participants in both markets should come to value the shock in a similar way once all information is known and evaluated.

The seminal paper applying cointegration measures to commodity markets was published by Garbade and Silbur in 1985. This then led to a large literature focused on cash/ futures price relationships in a myriad of commodity markets. Most of the early work involved models employing a bivariate regression between cash and futures prices (Bessler and Covey 1991; Fortenbery and Zapata 1993).

The theory underlying cointegration is that even if prices for similar assets in different markets deviate from equilibrium for short periods of time, they should exhibit a long-run equilibrium to which they return. As such, cointegration has been used to infer price efficiency and price leadership (determining whether one price in a series leads others and thus serves as the center for price discovery) and, following a shock, the amount of time it takes for a series of prices to return to their long-run equilibrium. The longer the time necessary for the two series to return to long-run equilibrium, the greater the basis risk in a hedged position.

The basic cointegration model used in commodity markets is of the form:

$$F_t - S_t = \alpha + \delta(F_{t-1} - S_{t-1}) + \epsilon_t \tag{2}$$

The larger δ, the greater the allowed disparity before futures and cash prices are brought back to equilibrium. If δ is close to 1, cash and futures do not converge quickly, and then basis risk is magnified. If δ is close to zero, futures and cash converge quickly, and basis risk is reduced compared to a larger δ value.

While not always tested in the context of hedging, Zapata and Fortenbery argue that failure to reject cointegration implies basis risk from hedging cash with futures is less than cash price risk but increases as δ increases. However, they also show that failure to detect bivariate cointegration between a futures and cash price series does not necessarily suggest a hedge is not overall risk reducing. If two markets are either spatially or temporally linked by another nonstationary variable (say interest rates when the cash transaction and futures expiration dates are not the same), then not finding a cointegrating relationship might be a function of a misspecified model (the correct specification would be a trivariate model that includes interest rates, for example).

In this case, depending on the behavior of interest rates, a hedge may still be overall price risk reducing even though the bivariate cointegration model between futures and cash failed to identify a single, unique cointegrating vector.

2.3 Hedging

As noted earlier, hedging involves substituting a futures market position today for a transaction you intend to make in the cash market later. For a farmer who expects to sell grain at harvest, a hedge at planting time would involve selling a futures contract that will expire as close to, but after, the actual harvest sale of the grain being produced. For example, if a corn farmer wanted to protect against the possibility that corn prices will fall between planting and harvest, he or she would sell a corn futures contract at planting. If harvest is to occur in October, then the nearest corn futures contract delivery month following harvest is December; thus, the farmer sells a December corn futures contract.

The futures market for corn requires corn to be delivered at an elevator along the Illinois or Upper Mississippi rivers during the delivery period (similar to the soybean futures delivery described earlier).[1] Further, the grain to be delivered against a futures contract must already be in storage at an approved elevator – it must be graded and certified that it meets all delivery specifications related to the futures contract. Also, because it is a December futures contract, actual delivery of the corn against the contract cannot happen until on or about December 1. As a result, the farmer cannot feasibly deliver the corn harvested from his or her field at harvest against a futures contract. To offset the hedge, the farmer simply buys the futures contract back at harvest, and then sells the harvested grain to a local grain elevator or coop.

A simple example is illustrated in Example 1 in Table 38.1a. In this example, a farmer expects to harvest 90,000 bushels of corn in October. To protect against adverse price changes during the production season, the farmer sells 18 December corn futures contracts (each contract is for 5,000 bushels of corn). The December contract is sold because that is the one that has a delivery date closest to, but not before, the expected cash sale date.

Table 38.1a Hedging a sales price

Assume it is April 1, and a corn farmer wants to protect the price received for October corn production. December corn futures contracts are trading for $4.50 per bushel. The farmer hedges October corn production by selling December futures on April1. The expected cash price in October is derived by adjusting the April 1 futures price for the expected basis in October, and the futures broker's commission.

Date	Futures market	Cash market	Basis
Sept 1	Corn farmer sells 18 December corn futures contracts for $4.50 per bushel	Establishes an expected October selling price of $4.50 − $0.25 − $0.005 (Futures + Basis − Comm.) $4.245 cents per bushel	Expected to be −$0.25

Scenario 1. Assume prices fall over the summer, and December corn futures are trading for $3.50 per bushel on October 15. The producer buys back the December corn futures contract he/she sold on April 1, and sells the harvested corn to the local co-op. The basis was accurately forecast and cash prices are $0.25 per bushel below the December futures price in October.

Date	Futures market	Cash market	Basis
October 15	Corn farmer buys 18 December corn futures contracts for $3.50 that he sold on April 1 for $4.50	Sells 90,000 bushels of corn to the local co-op cash price $3.25 futures profit +$0.995	
	futures profit $1.00 broker's comm. -$0.005 ---------- Net Futures Profit $0.995	---------- **Net Selling Price $4.245 per bushel**	-$0.25

The combination of a cash price of $3.25 per bushel plus a futures profit of $0.995 nets the corn farmer an effective corn price of $4.245 per bushel, which is what was expected.

Scenario 2. Assume prices rise over the summer, and the December futures are trading for $5.50 per bushel on October 15. The producer buys the December futures contracts back at a loss, but is able to sell corn to the local buyer at a higher price. The basis was accurately forecast and cash prices are $0.25 below futures in October.

Date	Futures market	Cash market	Basis
October 15	corn farmer buys 18 December futures contracts for $5.50 that he sold on April 1 for $4.50.	Sells 90,000 bushels of corn to the local coop cash price $5.25 futures loss -$1.005	
	futures loss $1.00 broker's comm. -$0.005 ---------- Net Futures Loss $1.005	---------- **Net Selling Price $4.245 per bushel**	-$25

The combination of a cash price of $$5.25 minus a futures loss of $1.005 nets the corn producer an effective corn price of $4.245 per bushel, which is what was expected.

Scenario 3. This is the same as scenario 1, except in this case the basis turns out to be weaker than expected by $0.10 Per bushel. A weaker basis means that the cash price is lower relative to futures than had been expected. If the basis is weaker by 10 cents, the cash price is 10 cents lower than it would have been if the basis forecast had been correct.

Date	Futures market	Cash market	Basis
October 15	corn farmer buys 18 December futures contracts for $3.50 that he sold on April 1 for $4.50	Sells 90,000 bushels of corn to the local coop	
	futures profit $1.00 broker's comm. -$0.005 ---------- Net Futures Profit $0.995	cash price $3.15 futures profit +$0.995 ---------- **Net Selling Price $4.145**	-$0.35 (10 cents weaker than expected)

While the futures hedge did protect the cash position from most of the price decline, the unexpected weakening of the basis did result in a lower net sales price than originally anticipated.

Scenario 4. This is similar to scenario 3, except the basis ends up stronger than expected. A stronger than expected basis means the cash price is higher relative to the futures contract than originally anticipated.

(*Continued*)

Table 38.1a Continued

Date	Futures market	Cash market	Basis
October 15	Corn farmer buys 18 December futures contracts for $3.50 that he sold on April 1 at $4.50.	Sells 90,000 bushels of corn to the local coop cash price $3.35 futures profit +$0.995	−$0.15
	futures profit $1.00 broker's comm. −$0.005	----------	(10 cents stronger than expected)
	Net Futures Profit $0.995	**Net Selling Price $4.345**	

This time the producer ends up better off than expected as the result of a stronger basis. Futures prices over the hedge period fell by more than cash prices resulting in a stronger than expected cash market. In the case of rising prices (such as in scenario 2), cash price would have to rise more than futures prices for the basis to strengthen.

Conclusions from Table 38.1a
The net selling price of a hedged cash commodity will not change regardless of whether prices rise or fall over the hedge period as long as the basis is accurately forecast. Once a commodity is hedged, price risk has been traded for basis risk. It is critical to have as accurate an expectation of basis as possible in order to minimize the possibility of receiving a net selling price which is less than the expected selling price.

Table 38.1b Hedging a purchase price

Assume it is April 1 and a grain elevator wants to hedge price for 90,000 bushels of corn they expect to buy at harvest. Their risk is prices go up, and corn is more expensive in October. Assume December corn futures are trading for $4.50 per bushel on April 1. The elevator would need to purchase 18 corn contracts to hedge October corn purchases. The expected October purchase price is derived by adjusting the April 1 futures price for the expected basis in October and the broker's commission. The commission increases the net cost of buying corn to the elevator, so the commission is added to the expected hedge price.

Date	Futures market	Cash market	Basis
April 1	Grain elevator buys 18 December corn futures contracts $4.50	Establishes an expected October purchase price of $4.50 − $0.25 + $.005 (Futures + Basis + Comm.) $4.255 per bushel	Expected to be −$0.25

Scenario 1. Assume prices fall over the summer, and the December corn futures contract is trading for $3.50 per bushel on October 15. The grain elevator sells the 18 December futures contracts they bought on April 1, and purchases 90,000 bushels of corn from a local farmer. The basis was accurately forecast and cash prices On October 15 are $0.25 per bushel below the December corn futures contract.

Date	Futures market		Cash market		Basis
October 15	Grain elevator sells 18 December corn futures contracts at $3.50 that it bought on April 1 for $4.50		Buys 90,000 bushels of corn from a local farmer		
			cash price	$3.25	−$0.25
			futures loss	+$1.005	
	futures loss	$1.00			
	broker's comm.	+ $0.005			

	Net Futures Loss	$1.005	**Net Purchase Price $4.255 per bushel**		

Note that the futures loss and the broker's commission are added to the net buying price of the grain elevator. These items increase the elevator's net corn purchase price. The combination of a cash price of $3.25 per bushel plus a futures loss of $1.005 nets the grain elevator an effective corn price of $$4.255 per bushel, which is what was expected.

Scenario 2. Assume prices rise over the summer, and the December n futures are trading for $5.50 per bushel on October 15. The grain elevator sells 18 December futures contracts for $5.50 per bushel, and purchases corn form a local farmer. The basis was accurately forecast and cash prices in October are $0.25 per bushel below the December corn futures contract.

Date	Futures market		Cash market		Basis
October 15	Grain elevator sells 18 December corn futures contracts for $5.50 that it bought on April 1 for $4.50		Buys 90,000 bushels of corn from local farmer		
			cash price	$5.25	−$0.25
			futures profit	−$0.995	
	futures profit	$1.00			
	broker's comm.	−$0.005			

	Net Futures profit	$ 0.995	**Net Purchase Price $4.255**		

The combination of a cash price of $5.50 per bushel minus a futures profit of $0.995 nets the grain elevator an effective corn purchase price of $4.255 per bushel, which is what was expected.

Scenario 3. This is the same as scenario 1, except the basis turns out weaker than expected by $0.10 per bushel. A weaker basis means the cash price is lower relative to futures than had been expected. If the basis is weaker by 10 cents, the cash price is 10 cents lower than if the basis forecast had been correct.

Date	Futures market		Cash market		Basis
October 15	Grain elevator sells 18 December corn futures contracts for $3.50 that it bought on April 1 for $4.50		Buys 90,000 bushels of corn from a local farmer		−$0.35
			cash price	$3.15	
			futures loss	+$1.005	(10 cents weaker than expected)
	futures loss	$1.00			
	broker's comm.	+ $0.005			

	Net Futures Loss	$1.005	**Net Purchase Price $4.155 per bushel**		

While the futures hedge offset the improvement in cash prices from most of the price decline, the unexpected weakening of the basis did result in a lower net purchase price than originally anticipated.

Scenario 4. This is similar to scenario 3, except that the basis ends up stronger than expected. A stronger than expected basis means that the cash price is higher relative to the futures contract than had been originally anticipated.

(*Continued*)

Table 38.1b Continued

Date	Futures market		Cash market		Basis
October 15	Grain elevator sells 18 December corn futures contracts for $3.50 that it sold on April 1 for $4.50		Sells 90,000 bushels of corn to the local coop		
			cash price $3.35		−$0.15
			futures profit +$1.005		
	futures loss	$1.00	---------		(10 cents stronger
	broker's comm.	+$0.005			than expected)
	Net Futures Profit	$1.005	**Net Selling Price $4.355**		

This time the buyer ends up worse off than expected as the result of a stronger basis. Futures prices over the hedge period fell by more than cash prices resulting in a stronger than expected cash market. In the case of rising prices (such as in scenario 2), cash price would have to rise more than futures prices for the basis to strengthen.

Conclusions from Table 38.1b
The net purchase price of a hedged cash commodity will not change regardless of whether prices rise or fall over the hedge period as long as the basis is accurately forecast. Once a commodity is hedged, price risk has been traded for basis risk. It is critical to have as accurate an expectation of basis as possible in order to minimize the possibility of incurring a net purchase price which is greater than the expected purchase price.

In order to sell December corn futures contracts, the farmer must hire the services of a futures trading broker, someone who is licensed to trade futures and has access to the futures market. In this example, assume the broker charges the farmer $0.005 per bushel ($25 per contract) to facilitate the hedge.

The farmer expects the October basis (the difference between his or her local cash price and the October futures price for the December corn contract) to be −$0.25 per bushel; in other words, the farmer expects the October cash price to be $0.25 per bushel below the December futures price in October. The forecast basis may be based on historical experience over the last three to five years, or it may be generated from an empirical model estimated and disseminated by university or industry grain marketing specialists (see Sanders and Manfredo (2006) for a discussion of basis forecasting models that can be used in practice).

On October 15, the farmer harvests the corn crop and sells it in the local cash market. At that time, he or she simultaneously buys back the December futures contracts previously sold in April. Because the farmer forecast the October basis perfectly, he or she gets exactly what was calculated as the expected cash price back in April. Note that this happens whether prices rise or fall over the hedge period (Scenarios 1 and 2, respectively). The producer insured against the risk of lower prices by hedging, but also gave up the opportunity for higher prices in return for eliminating the price risk.[2]

In general, however, basis cannot be perfectly forecast, and hence hedging does not eliminate all risk. Hedgers essentially trade price risk for basis risk. Scenarios 3 and 4 illustrate the implications of prediction errors in forecasting the final basis at the time the hedge is placed.

The prediction error occurs because the farmer sells the harvested grain in a month different from the month the futures contract expires and in a location different from the delivery location specified by the futures contract. Basis risk is represented by the deviation between the forecast and the realized basis levels. If a hedger could predict basis perfectly and faced no other risk, then the risk minimizing hedge would be one that covered 100 percent of expected cash market exposure. However, because most hedges involve basis risk, the optimal risk reducing hedge is usually not one that covers 100 percent of the cash asset position. As with research examining price dynamics between futures and cash prices, there have been dozens of studies focused on empirical calculations of optimal hedge ratios – the percentage of cash market risk that should be hedged given basis and other risks and the decision maker's risk reducing objective.

In practice, the hedging programs for many firms are much more complicated than the examples from Table 38.1a. Consider, for example, a gasoline refinery that is both buying crude oil and selling gasoline after some time lag (the time it takes to refine crude into gasoline and other products). In the refiner's case, there are futures contracts for both the input (crude oil) and the output (gasoline). However, the specific formulation of gasoline being produced and delivered to consumers can change based on geographic location and season (winter blends in the northern climates differ from summer blends). Just like corn and soybeans, the futures contract specifications are fixed and do not change with delivery location or season. In addition, there is clearly a relationship between the cost of the input and the price of the output for a refiner. As such, the price dynamics for a gasoline refinery (and many other types of manufacturers) present complications in addition to basis risk. This has led to a large literature looking at optimal hedge strategies in more challenging environments.

Much of the early work focused on static type models, while more recent work has introduced dynamics that explicitly account for risky outcomes and transactions costs associated with multiple hedging horizons (for example, a refinery may be hedging both purchases and sales across different time periods simultaneously and thus may have futures positions for multiple expiration dates).

The early static models were generally of two types: (1) regression models where cash price changes are regressed on futures price changes, with the coefficient estimated representing the percentage of the cash position to be hedged in futures, and (2) optimization strategies using quadratic programming (QP) models based on a mean variance (E–V) framework. The objective of early work was often to identify a hedge ratio (the combination of cash and futures positions) that minimized cash price variance. This is called the minimum variance hedge ratio (MVHR). MVHRs are based on the relationship between changes in futures and cash asset prices over the life of the hedge. According to Hull (2015), the minimum hedge ratio (h^*) can be estimated from a regression of cash price changes on futures price changes. h^* is calculated as:

$$h^* = \rho \frac{\sigma_C}{\sigma_F} \tag{3}$$

where σ_C is the standard deviation of cash price changes, σ_F is the standard deviation of futures price changes, and ρ is the coefficient of correlation between cash and futures price changes.

For a perfect hedge, $\rho = 1$, and $\frac{\sigma_C}{\sigma_F} = 1$, thus $h^* = 1$. In this case, changes in the futures price identically match changes in the cash price – i.e., there is no basis risk. Conversely, if the standard deviation of futures price changes is twice the standard deviation of cash price changes, then the minimum variance hedge ratio will be 0.5. The proportion of the cash price variation that is reduced through hedging is a measure of hedging effectiveness and is represented by the R^2 of the regression.

The QP models employed have generally been formulated with an objective of minimizing the variance of hedge returns for some minimum acceptable return level. They are solved iteratively for various levels of minimum acceptable return, and an efficient frontier is traced showing the optimal futures positions associated with each level of minimum acceptable return (see Fortenbery and Hauser 1990 for an example). In both the regression and optimization cases, it is generally assumed that historical relationships remain stable over time.

While calculating MVHRs using either regression or optimization strategies has become quite common in practice, the academic literature has attempted to increase the understanding of cash/futures price relationships, and thus optimal hedging strategies, by introducing both dynamics (the possibility that the price relationships between cash and futures are not static across time) and transactions costs associated with both margin management of a futures position (i.e., margin requirements)[3] and costs of entering/exiting the market.

An early paper introducing dynamics to measure hedge performance in agricultural markets (i.e., attempting to quantify the riskiness of different potential outcomes) was Tronstad and Taylor (1991). They employed a stochastic dynamic programming (SDP) model that considered the impacts of pre-tax income, storage decisions, and basis risk on the optimal hedge positions of wheat farmers. Essentially, they argued that farmers' hedge decisions would change as prices changed because price levels impact income, and with progressive tax rates there would be larger percentage impacts on after tax income in high price environments compared to low price environments. This, in turn, might affect whether and how much farmers would sell at harvest and how much they would store into the next tax season. This would obviously affect their level of hedging and the appropriate futures delivery months to use for hedging. Tronstad and Taylor found that producers who adjusted their hedge positions based on expected after tax income received both higher returns to wheat production, as well as a lower standard deviation in wheat returns.

Park (2008) used a SDP model similar to Tronstad and Taylor's to evaluate potential hedging strategies for ethanol plants. He considered the simultaneous hedging of two inputs, corn and natural gas, and the primary output, ethanol. In 2008, there were a series of high profile bankruptcies among ethanol plants. One of the causes cited was the plants' practice of only hedging one side of their operating margin. Essentially, plants locked in relatively high corn prices, either by hedging or forward cash contracting, in the summer of 2008, but did not hedge the price for ethanol. When both commodity and energy markets collapsed in November 2008, they were stuck with high price corn and an ethanol price below breakeven. Park investigated whether a more comprehensive hedge strategy that included managing both sides of the margin equation, as well as accounting for the interactions between energy and corn markets, would have resulted in superior risk management opportunities. He found that if ethanol producers had utilized his dynamic hedge strategy they would have been able to withstand the market downturn.

Over the last several years, hedging models have become increasingly rigorous and have incorporated several measures of risk beyond just the price risk of the cash assets being hedged. They have built on the early dynamic models and have considered price dynamics and volatility spillovers between several markets simultaneously, as well as various time series representations of both price behavior and hedge returns.

Chang, Lai, and Chuang (2010) compared eight different empirical models ranging from regression models, several different specifications of Error Correction models, and state-space models to look at hedging effectiveness under different price regimes in energy markets. They discovered that hedging effectiveness in gasoline and crude oil markets tended to be higher in markets with rising prices than markets with prices in a downward trend. In addition, the performance of the individual models differed based on underlying market conditions. These results lead one to conclude that the percentage of the cash price exposure hedged should be greater in rising markets than declining markets. Further, hedgers should be prepared to calculate their optimal hedge positions using different methods or models, depending on underlying market conditions.

Tejedaq and Goodwin (2014) looked at dynamic multi-product hedging in the soybean complex. Specifically, they were interested in whether accounting for time-varying correlations across the various soy complex commodities improved overall hedging effectiveness. They employed a regime-switching dynamic correlation model and compared the results to a single commodity time-varying hedge model and the performance of naïve hedge ratios. They found that the multi-product hedge ratios out-performed the hedge ratios from the other models. They also discovered that the optimal hedge ratios in the multi-product model were significantly lower for soybean meal and soybeans compared to other models. This, in turn, led to the conclusion that accounting for multi-product correlations would allow for effective hedging programs to be implemented at lower cost compared to hedging programs that did not account for cross-product price correlations, because the size of the futures positions would be reduced.

The papers referenced here comprise a very small subset of the research focused on optimal hedging over the last several years. However, they do point to the increased complexity of both the models used and the resulting hedge strategies identified by the modeling efforts. Much of the research quantifies potential returns associated with increased sophistication in designing and implementing a hedge program. However, in many cases both the modeling strategies and the hedge program results may require skill sets not held by a potential hedger. In Park's case, for example, it is unlikely that the manager of a small ethanol plant would be sufficiently trained to develop, code, and then estimate a SDP model to identify the simultaneous optimal hedge ratios for the plant's various inputs and outputs. Even if he or she did have that skill set, the manager may face a serious time constraint in conducting that activity. When Park's work was done, the average ethanol plant in the U.S. produced about 40 million gallons of ethanol a year, often with very few full-time employees. As a result, to take advantage of the opportunities identified by Park, the plant manager would likely need to outsource the price risk management activities of the plant.

What seems to be missing from the current literature is an actual benefit–cost analysis of adopting more sophisticated hedging models, along with the implementation of the models by industry. While recent work has identified the potential benefits of increased sophistication in analyzing hedging opportunities, there has not been a major focus on understanding whether the increased benefits are offset by the costs associated with adopting more rigorous models, especially when that activity would likely need to be outsourced. This may be particularly important in the case of farmers, cooperatives, and small bioenergy firms that do not have professional, full time risk managers.[4] Further, many of the brokerage firms these entities use to access futures markets may not have needed modeling capacity in house.

3 Conclusions

There has been tremendous progress over the last couple of decades measuring market relationships between cash and futures prices and using that information to evaluate hedge strategies for a myriad of unique situations. In many cases, however, the research addresses improvements in hedge performance with increased model sophistication but does not consider the potential costs firms lacking rigorous training in model development would face in order to benefit from the new insights. There has also been a tendency to assume that all businesses of similar industries, regardless of ownership structure, face the same objectives in managing price risks associated with cash assets. This may be an overly restrictive assumption.

Because of the cash flow implications of managing margin accounts, privately owned firms and investor owned firms may have a different *ex post* reaction to a hedge where the cash price improved over the life of the hedge, but losses in the futures market offset the cash price improvement. In an investor owned firm, the Board of Directors and individual stock holders may not recognize that a futures market loss still represents a risk reducing strategy given *ex ante* uncertainty as to future price direction. They may view the futures market loss as an unacceptable hit on the company's bottom line, especially if it occurs several times over a quarter or year. Risk managers in these types of companies may face significant pressure to continually adjust their hedge position as prices change, depending on whether the changes are in favor of or against the futures side of the hedge. A better understanding of both the incentives and constraints faced by risk managers in different type ownership structures and the applicability of dynamic hedge programs to accommodate these differences in incentives would add substantially to the current literature.

Notes

1 While corn and soybeans are similar in terms of delivery requirements, in general futures contract delivery requirements and timing vary significantly by commodity.
2 An option on a futures contract would have allowed the producer to get rid of some of the downside price risk without eliminating all of the upside potential should prices have risen over the hedge period. A good description of using futures options versus futures contracts for price risk protection can be found in Purcell and Koontz (1999).
3 Futures contracts are traded on margin. Both sellers and buyers of futures post a performance bond, called a margin, to insure they will continue to honor their initial commitment to either deliver or take delivery as prices change. In addition to the initial margin, futures markets are marked-to-market daily based on price changes between one day and the next. This requires all traders to deposit additional funds to their margin account if prices go against their initial position (prices rise for a futures seller or fall for a futures buyer) or allows them to remove money from their margin account if prices move in favor of their initial position. While initial margin requirements typically represent a very small percentage of the total value of the future contract traded, the daily mark-to-market activity does imply, even for a hedger, that they can experience negative cash flows over the life of the hedge if prices move in favor of their cash position and against their futures position. This negative cash flow risk requires hedgers to be well capitalized over the life of the hedge in order to insure they can maintain their hedged position. Thus, even though the hedge works, as shown in Table 38.1, regardless of whether prices rise or fall, the direction of price change does have important cash flow implications over the life of the hedge.
4 There is a significant literature that suggests few farmers use futures markets under any scenario, thus more sophisticated/complicated models of hedging behavior may not generate any benefit for this agribusiness sector. Most forward pricing by producers is done in the cash market through forward contracts offered by the businesses that buy directly from farmers. A short review of the literature, and a discussion of reasons farmers tend to avoid futures markets, can be found in Mark, Brorsen, Anderson, and Small (2008).

References

Bessler, David, and Ted Covey. 1991. "Cointegration: Some Results on U.S. Cattle Prices." *The Journal of Futures Markets* 11(4):461–474.

Chang, Chiao-Yi, Jing-Yi Lai, and i-Juan Chuang. 2010. "Futures Hedging Effectiveness Under the Segmentation of Bear/Bull Energy Markets." *Energy Economics* 32(2):442–449.

Dewbre, Joe H. 1981. "Interrelationships Between Spot and Futures Markets: Some Implications of Rationale Expectations." *American Journal of Agricultural Economics* 63(5):926–933.

Ferris, William G. 1988. *The Grain Traders: The Story of the Chicago Board of Trade*. Michigan State University Press.

Fortenbery, T. Randall, and Robert Hauser. 1990. "Investment Potential of Agricultural Futures Contracts." *American Journal of Agricultural Economics* 72(3):721–726.

Fortenbery, T. Randall, and Hector O. Zapata. 1993. "An Examination of Cointegration Between Futures and Local Grain Markets." *The Journal of Futures Markets* 73:921–932.

Garbade, Kenneth D., and William L. Silbur. 1985. "Price Movements and Price Discovery in Futures and Cash Markets." *The Review of Economics and Statistics* 65(2):289–297.

Hull, John C. 2015. *Options, Futures, and Other Derivatives* (9th ed.). New York: Pearson Education Inc.

Mark, Darrell R., B. Wade Brorsen, Kim B. Anderson, and Rebecca M. Small. 2008. "Price Risk Management Alternatives for Farmers in the Absence of Forward Contracts with Grain Merchants." *Choices* 23(2):22–25.

Park, Hwan Il. 2008. *Dynamic Risk Management Strategies for an Ethanol Producer with Impacts of Ethanol Production on the U.S. Corn Price*. Unpublished PhD dissertation, Madison: University of Wisconsin.

Purcell, Wayne D., and Stephen R. Koontz. 1999. *Agricultural Futures and Options: Principles and Strategies* (2nd ed.). Upper Saddle River, NJ: Prentice Hall.

Sanders, Dwight R., and Mark R. Manfredo. 2006. "Forecasting Basis Levels in the Soybean Complex, a Comparison of Time Series Methods." *Journal of Agricultural and Applied Economics* 38(3):513–523.

Silbur, William L. 1985. "The Economic Role of Futures Markets." In Anne E. Peck, ed., *Futures Markets: Regulatory Issues*. Washington, DC: American Enterprise Institute for Public Policy Research, pp. 83–114.

Tejeda, Herman A., and Barry K. Goodwin. 2014. "Dynamic Multiproduct Optimal Hedging in the Soybean Complex – Do Time-Varying Correlations Provide Hedging Improvements?" *Applied Economics* 46(27):3312–3322.

Tronstad, Russell, and C. Robert Taylor. 1991. "Dynamically Optimal After-Tax Grain Storage, Cash Grain Sale, and Hedging Strategies." *American Journal of Agricultural Economics* 73(1):75–88.

Wikipedia. "Futures Exchange." https://en.wikipedia.org/wiki/Futures_exchange. Accessed February 27, 2017.

Zapata, Hector O., and T. Randall Fortenbery. 1996. "Stochastic Interest Rates and Price Discovery in Selected Commodity Markets." *Review of Agricultural Economics* 18(4):643–654.

39
Cooperative extension system
Value to society

Russell Tronstad and Mike Woods

1 Cooperative extension system: value to society

The 1862 Morrill Act provided land for each state to establish and maintain at least one college to teach courses related to agriculture and mechanical arts. The Hatch Act of 1887 established State Agricultural Experiment Stations (SAES), and land-grant colleges were charged with overseeing these stations by conducting original research and verifying experiments regarding agriculture (Huffman and Evenson 2006; Smith and Smith 2007). To aid in the dissemination of information available from SAES and land grants to the public, the Smith–Lever Act of 1914 was instituted to create the Cooperative Extension System (CES). A traditional activity of the CES has been to reduce the lag time between the development of new information and its application to increase agricultural productivity (Ahearn,Yee, and Bottom 2003). Federal monies were allocated to support CES through formula funds with the requirement that these funds be matched by state funds, thus formalizing the structure of CES as being jointly funded by the federal government (i.e., United States Department of Agriculture, USDA) and states through land-grant colleges (Thomson 1984). County governments soon became financial partners so that CES has a very unique three-pronged partnership of financial support and accountability.

The Second Morrill Act provided funding for 1890 institutions so that equal educational opportunities could be provided for African-American students who were denied admission to 1862 land-grant universities. Seventeen institutions are included in the 1890 system today, and they are predominantly located in the South. These 1890 institutions have Teaching, Research, and Extension programs that almost exclusively serve minority rural communities, focusing on small-scale and limited-resource producers (Tegene et al. 2002). For many years, 1890 institutions only received Extension funds through 1862 institutions in limited amounts, but today there is more funding (Comer et al. 2006). In 2016, capacity grant or statutory formula funding to support agricultural and forestry extension activities at 1890 institutions totaled over US$43 million, or about US$2.5 million per institution (USDA-NIFA 2017).

In 1992, Congress recognized campuses with at least 25 percent Hispanic enrollment among undergraduates and other criteria as Hispanic-Serving Institutions (HSIs) (Laden 2004). The National Institute of Food and Agriculture (NIFA) provides competitive grants to help HSIs conduct higher education programs for food and agricultural sciences. New Mexico State University and the University of Arizona (recently designated in 2018) are currently the only land-grant universities recognized as a HSI on their main campuses. More recently, the 1994 Elementary and Secondary Education Act conferred land grant status on 29 Native American colleges and authorized funding for agriculture and natural resource Extension programs (APLGU 2012).

As with any entity that produces a public good, its value to society is not readily available using market prices. To further complicate matters, the lag length associated with many public good benefits that CES provides may be immediate or ongoing for years (Anderson and Feder 2007). For example, information on rangeland improvements may yield watershed and wildlife benefits to a region for many years in the future (Howery 2000). Likewise, effective training provided to at-risk youth on how to contribute to their communities as upright citizens should have non-excludable and non-rivalrous benefits to society for many years to come (McKee, Talbert, and Barkman 2002). Thus, this chapter is not intended to provide a single rate of return or metric for CES's value to society, but rather to provide an overview of CES's recent valuation literature, trends in funding by region and general program area, and future direction.

Wang (2014) describes how the financial support for CES has shifted over time from federal formula funding to state support funding. For example, between 1935 and 1945, almost 60 percent of CES's support was from formula funds, whereas between 2005 and 2012, about 80 percent of CES's base support has been from state funds. Not only has funding shifted, but the demographics of agriculture have also greatly changed.

Daly (1981) notes that in 1870, shortly after the Morrill Act, almost 50 percent of employed individuals worked in agriculture, and one farmworker could feed five individuals. By the time of the Smith–Lever Act, 30 percent of the workforce was engaged in farming, and more than 50 percent of the population lived in rural areas. By 1980, only 4 percent of workers were employed in production agriculture, and each farmworker could support the food needs for 70 other individuals. This trend has continued, with the number of people currently fed annually by one U.S. farmer reaching around 155 individuals, as noted by Secretary of Agriculture Tom Vilsack (USDA 2010). Clearly, productivity gains associated with U.S. agriculture have been very impressive since the land-grant system was implemented with mandates involving Teaching, Research, and Extension (Huffman and Evenson 2006). But untangling the role that CES has played in these productivity gains is undeniably complicated (Wang et al. 2012). Since CES was initially created to address rural and agricultural issues, one may question why CES still exists with such a small percentage of the population involved in production agriculture or living in rural areas.

2 Recent literature valuing CES

CES is a rather unique institution with investments in human capital and strong linkages to applied research, communities, and agriculture. The three-way funding structure of CES has likely provided a level of credibility and accountability such that its conduct and performance should be different from other federal or state government agencies and non-profit organizations. CES represents a significant public investment, and as with any public resource, measures of the impact or value of CES are important (Lutz and Swoboda 1972). Agricultural economists and others have reported several studies quantifying the economic value of CES or agricultural research. One challenge in studies like these is isolating the relative impacts of CES versus Research, although it can be argued that the significance of the land-grant system is best viewed by considering the synergistic impacts of Research, Teaching, and CES operating together.

Table 39.1 provides a summary of selected recent studies that have valued CES for the United States. Studies vary from estimating the rate of return to the number of farmers and jobs tied to the CES. Jin and Huffman (2016) estimate an internal rate of return (IRR) to public

Table 39.1 Selected studies and measures that quantify the economic value of CES and/or agricultural Research

Study	Area of study	Type of analysis/ geographic area	Lag period	Time period	Measures for Cooperative Extension System (CES), Research (R), or both (CES&R)
Jin and Huffman (2016)	Return to public agricultural Research	Total factor productivity, net measures / 48 contiguous states	35 yrs.	1935 to 2002	R: 67 percent internal rate of return
	Return to ag. and natural resource Extension		5 yrs.	1966 to 2004	CES: > 100 percent internal rate of return
Kelsey and Goetz (2016)	Jobs and economic activity tied to all of their state's CES	Economic model of IMPLAN/ State of Pennsylvania	N/A	2015	CES: loss of US$50.5 million in state funds would lead to forfeiture of another US$91 million in USDA and county funds plus another US$134 million of indirect and induced losses
Goetz and Davlasheridze (2016)	Impact of Extension on farm numbers	Two-stage recursive net farmer model / 50 states	3 yrs.	1983 to 2010	CES: 28 percent more net farmer exits would have occurred without extension
Alston, Andersen, James, and Pardey (2011)	Economic returns to U.S. public agricultural Research and Extension	Econometric model, marginal benefit–cost ratio / 48 states	50 yrs.	1949 to 2002	CES&R: 22.7 percent internal rate of return and 9.9 percent modified internal rate of return (benefits reinvested at real rate of 3 percent per annum)
Akobundu, Alwang, Essel, Norton, and Tegene (2004)	Farm visits by Extension to limited-resource farmers	Two-stage farm participation model/ state of Virginia	~ 3 to 10 yrs.	~ 1990 to 2000	CES: US$3,300 larger net farm income per farm from each additional farm extension agent visit at the mean

agricultural Research and Extension of 67 percent and 100 percent, respectively. They find public agricultural Research and Extension to be substitutes rather than complements. Extension benefits are modeled to decline exponentially over five years (i.e., weights of 0.51, 0.26, 0.13, 0.07, and 0.03 for t through $t - 4$). To the extent that CES provides unique education through personal communication with specific farmers and individuals, it is argued to have some private good attributes that will rapidly become obsolete (Cornes and Sandler 1996; Evenson 2001). Research benefits are modeled to accrue with trapezoidal weights, starting and ending at zero, over a 35-year period. Under their lag assumptions, 90 percent of the return to agricultural Extension is realized before benefits begin accruing to Research and the benefits from Extension exceed their initial investment in the first year.

Alston et al. (2011) argue that agricultural productivity gains associated with Extension and Research activities are intertwined and somewhat inseparable. Indeed, many if not most institutions require faculty with Extension appointments to give evidence of applied research with original scholarship that is validated through peer review or widespread adoption. They also argue that the high rates of return reported for public agricultural Research and Extension are methodologically flawed since the government does not have the ability to reinvest the flow of benefits at the same rate of return assumed in an IRR calculation. Assuming that benefit flows are reinvested at a real rate of 3 percent per annum, they calculate a modified IRR for both public agricultural Research and Extension at an average 9.9 percent per annum across all states versus a 22.7 percent IRR without this modification. Own-state benefit–cost ratios averaged 21:1 or 32:1 whether excluding or accounting, respectively, for spillover benefits to other states.

Using Impact Analysis for Planning (IMPLAN), Kelsey and Goetz (2016) estimate the jobs and economic activity that would be lost to Pennsylvania if the state removed all support to their CES. They conclude that eliminating US$50.5 million in state funding to CES would result in the loss of US$78 million in federal funds and another US$134 million from indirect and induced economic losses. In addition, losses would occur from CES programs not disseminating science-based knowledge and other public goods to their state. Alston et al. (2011) find that the lag length associated with agricultural Research and Extension benefits can last for up to 50 years, suggesting that lost opportunities or benefits associated with current funding decisions will accrue over a long time span.

Using state-level data from 1983 to 2010, Goetz and Davlasheridze (2016) estimate that a US$100 reduction in federal program payments to every farmer would result in 8 to 11 fewer farmers exiting per state, while transferring the same dollars to CES would increase the average number of farmers per state by 41 to 67 annually. They conclude that the impact of a dollar directed to CES yields relatively large job gains compared to other economic job stimulus and development programs, even though they note that having more farmers is not necessarily beneficial in itself. Because farmers tend to share the knowledge they learn from CES with other farmers, this is believed to highly leverage the dollars spent on CES (Schnitkey et al. 1992; Diekmann, Loibl, and Batte 2009).

Akobundu et al. (2004) analyzed the impact of CES on limited-resource farmers in Virginia by an 1890 institution grant that targeted small minority farmers. Net farm incomes of producers participating in the grant were evaluated against other equivalent non-participating producers, as identified through Virginia Agricultural Statistics Service, in race, assets, and farm income. They found significant increases in net farm incomes of producers participating with CES, but only after multiple visits from an Extension agent. Provided there is sufficient intensity of participation, they conclude the program significantly increases net farm income. Overall, total producer benefits of up to US$5 million were realized from CES investments that were under US$0.5 million.

Hoag (2005) asks the question of whether CES is on the brink of extinction. He notes that budget threats may imply CES's time has come and gone. Has CES been "captured" by agricultural interests? Is there a need to broaden services offered or focus on Extension's core and original mission? Hoag worries that the appropriateness of the original public CES model is weakened because people are more educated and information is easy to gather. He provides economic principles to follow for revitalizing CES that include a greater focus on public goods where CES has a competitive advantage. That is, focusing on agriculture and rural areas may be more politically appealing than diffusing resources across urban clientele who do not recognize or speak up for CES, even though rural and agriculture populations are shrinking. This seems to be one of the key dilemmas facing CES: how does CES expand its support base and still enjoy the relatively strong support derived from traditional clientele?

Kalambokidis (2004) argues that CES needs to be able to persuade citizens and policy makers of the value of CES to those who are not directly served by CES (i.e., public goods that CES provides). For example, CES programs that induce farmers to adopt more environmentally sound practices or nutrition education that lead families to sound public health provide external benefits to individuals not in direct contact with CES. Where CES programs have strong public value through external benefits, this message must be articulated to government officials and the citizens who elect them (Kalambokidis 2011; Franz 2013). In cases where public value is not strong, Kalambokidis argues that if grants, user fees, or other sources of funding are unable to fund the program, then it should be dropped. Revenue raised through fees could be used to support programs with substantial public value that are unable to generate sufficient revenue on their own or to reinvest in the same program that generated the revenue. The next section considers how federal formula full-time equivalent (FTE) funding has shifted by program area and region in recent years.

3 Recent data on funding trends for CES

CES has a long history of utilizing formula funds provided to each state to build capacity and target programs to locally identified needs. Some discussion has focused on moving toward competitive funding (Hanson and Just 2001). Each approach has favorable arguments, according to Huffman and Evenson (2006). They note that formula funds allow for the development of a core or base for Research and CES efforts. Formula funds carry no general university overhead or indirect costs. Huffman and Evenson further note that the competitive approach tends to favor universities that already have the infrastructure required to compete for nationally focused research awards. Finally, a nationally competitive grant effort may tend to reallocate efforts away from state-level research identified as important by individual land-grant universities. Conversely, competitive grants may allow for more direct accountability. Focusing on targeted national priorities may utilize scarce resources in a more effective manner, implying there is no need for as many land-grant university colleges of agriculture as we have now. But a local focus and visibility with state legislators that enhances the status of a land-grant university may also lead to more positive impacts for local clientele, the ultimate provider of county, state, and federal funding. Next, we describe the value society places on CES through its legislators by describing recent trends and funding sources of CES.

3.1 Federal formula FTEs by program area and region

Table 39.2 through Table 39.6 present data on federal formula funded FTEs by general program area for the United States and by region. Data were provided by the National Institute of Food and Agriculture: Planning, Accountability, and Recording Staff (USDA-NIFA-PARS 2016) and only include federal capacity or formula CES funding. Each state decides every year how many programs they want to report and what names are associated with these CES programs (USDA-ERS 2017). Thus, even though each state does not directly control the FTEs or funds that are

Table 39.2 Capacity or federal formula funded CES FTEs by program area for the United States, 2007–2015

Year	Agriculture	Natural resources & environment	Community development	4-H youth	Nutrition/obesity/healthy lifestyles and food safety	Home economics/family and child development	Administration & misc.	Total FTEs
2007	3,847.4	1,244.9	831.2	3,008.0	1,906.5	1,046.8	62.3	11,947.1
2008	3,784.1	1,242.6	776.7	2,675.5	1,891.1	954.2	87.7	11,411.9
2009	4,195.0	1,432.3	725.3	3,408.4	2,628.3	962.8	84.3	13,436.4
2010	3,842.7	1,440.0	615.1	2,656.6	2,344.9	710.1	29.8	11,639.2
2011	4,022.0	1,631.5	552.1	2,931.1	2,349.6	697.5	15.8	12,199.6
2012	3,620.7	1,666.1	487.4	2,641.0	2,398.6	653.9	15.0	11,482.7
2013	3,452.6	1,552.9	647.4	2,452.4	2,301.3	456.2	12.4	10,875.2
2014	3,417.6	1,269.5	610.9	2,383.5	1,913.7	717.4	10.0	10,322.6
2015	3,460.2	1,316.8	597.0	2,511.6	1,875.3	811.3	6.0	10,578.2
Compounded annual % change	−1.3%	0.7%	−4.1%	−2.2%	−0.2%	−3.1%	−25.4%	−1.5%

Table 39.3 Federal formula funded CES FTEs by program area for the Northeast, 2007–2015

Year	Agriculture	Natural resources & environment	Community development	4-H youth	Nutrition/obesity/healthy lifestyles and food safety	Home economics/family and child development	Administration & misc.	Total FTEs
2007	555.5	413.9	197.8	764.1	639.9	276.8	36.8	2,884.8
2008	457.7	313.0	193.0	634.4	467.4	250.0	20.2	2,335.7
2009	885.1	455.4	225.0	1,281.6	1,157.0	245.1	18.3	4,307.5
2010	630.0	507.7	78.8	832.8	738.8	76.9	18.6	2,883.6
2011	798.4	459.9	84.8	916.9	628.2	107.7	14.6	3,010.5
2012	768.1	423.5	74.3	844.0	574.3	81.0	13.8	2,779.0
2013	595.3	322.3	120.8	585.0	498.4	52.3	11.2	2,185.3
2014	503.2	271.9	181.8	514.0	310.1	73.5	5.0	1,859.5
2015	492.5	248.9	187.1	588.5	378.2	53.0	6.0	1,954.2
Compounded annual % change	−1.5%	−6.2%	−0.7%	−3.2%	−6.4%	−18.7%	−20.3%	−4.8%

allocated through the formula, they do control the general categories under which these FTEs are reported. In looking through the array of program names and groupings provided by all states, FTEs were categorized into seven general program areas: (1) agriculture; (2) natural resources and environment; (3) community development; (4) 4-H youth; (5) nutrition/obesity/healthy lifestyles and food safety; (6) home economics/family/child development; and (7) administration and miscellaneous. This is a somewhat finer breakdown than the four categories used in Ahearn, Yee, and Bottom (2003) of agriculture and natural resources; community resource development; 4-H and youth development; and home economics and human nutrition.

These data provide a picture of national CES support for the nine years of 2007 through 2015. Table 39.2 shows that total federal formula funded FTEs for the United States have moved up and down but cumulatively have declined by 1.5 percent annually over this period. Significant declines occurred in community development (−4.1 percent), 4-H (−2.2 percent), and home economics (−3.1 percent). It is interesting to note that agriculture declined by −1.3 percent, slightly less than the overall (−1.5 percent), and natural resources actually increased (0.7 percent). Agriculture accounts for the largest portion of FTEs (32.4 percent) over this period, followed by 4-H (23.7 percent), nutrition/food safety (18.9 percent), natural resources (12.3 percent), home economics (6.7 percent), community development (5.6 percent), and administration/miscellaneous (0.3 percent). CES programs related to natural resources and the environment increased on a relative basis by more than any other area at 0.7 percent annually and increased from 10.4 to 12.4 percent of all FTEs. Aside from administration/miscellaneous, FTEs for community development declined more than any other category at −4.1 percent.

As shown in Table 39.3, overall annual FTEs in the Northeast changed by −4.8 percent.[1] Unlike at the national level, where community development declined the most, community development held its ground by more than any other area by declining only −0.7 percent annually, although it is a relatively small allocation of total FTEs (5.6 percent). Sharp declines in FTEs were experienced in home economics (−18.7 percent), nutrition/food safety (−6.4 percent), and natural resources (−6.2 percent), even though FTEs surged in 2009 for most areas. FTEs surged in 2009 for the Northeast; much of this increase is tied to New York, which is by far the largest recipient of FTEs for the Northeast (69.2 percent for 2009). Some programs remained the same with increases, while new program categories were reported for global food security and hunger; positive youth development; and childhood obesity and food safety. The largest program area for the Northeast was 4-H (28.8 percent) followed by agriculture (23.5 percent), nutrition/food safety (22.3 percent), natural resources (14.3 percent), community development (5.6 percent), home economics (5.0 percent), and administration/miscellaneous (0.6 percent).

Going against the trend for all other regions, total FTEs actually increased marginally in the North Central region (Table 39.4), although it was a minimal 0.2 percent. Significant annual FTE declines occurred for community development (−12.1 percent) and 4-H (−3.9 percent), while other program areas, except for administration/miscellaneous, had an increase. Increases were fairly substantial for home economics (5.1 percent), nutrition/food safety (4.7 percent), and natural resources (4.7 percent), while agriculture increased at just 0.1 percent. Agriculture in the North Central region accounted for 33.8 percent of the FTEs, similar to that for all the United States (32.4 percent), but by 2015, 4-H FTEs (11.9 percent) were about half of what they were for all the United States (23.7 percent); offsetting the relative smaller 4-H are community development, nutrition/food safety, and home economics.

In the South, total FTEs declined slightly (−0.8 percent annually), but FTEs were much more stable across program areas and in total than for any of the other regions (Table 39.5). Some of this stability may be tied to the number of 1890 institutions located in this region. Community development grew by a respectable 3.0 percent and natural resources grew a modest 0.7 percent.

Table 39.4 Federal formula funded CES FTEs by program area for the North Central region, 2007–2015

Year	Agriculture	Natural resources & environment	Community development	4-H youth	Nutrition/ obesity/ healthy lifestyles and food safety	Home economics/family and child development	Administration & misc.	Total FTEs
2007	777.1	197.3	299.7	362.0	287.9	241.7	16.0	2,181.7
2008	840.3	223.4	250.7	369.1	536.8	213.0	13.3	2,446.6
2009	834.9	218.4	225.5	310.6	590.0	225.1	20.2	2,424.7
2010	736.6	262.1	276.0	280.3	606.4	205.6	10.0	2,377.0
2011	733.8	338.8	163.2	286.3	567.8	195.3		2,285.2
2012	658.0	303.7	92.2	285.6	542.0	100.9		1,982.4
2013	676.0	288.9	81.9	266.4	596.0	97.8		2,007.0
2014	710.3	237.5	88.4	298.0	490.9	199.3		2,024.4
2015	783.0	284.2	106.7	264.0	416.1	358.7		2,212.7
Compounded annual % change	0.1%	4.7%	−12.1%	−3.9%	4.7%	5.1%	−100.0%	0.2%

Table 39.5 Federal formula funded CES FTEs by program area for the South, 2007–2015

Year	Agriculture	Natural resources & environment	Community development	4-H youth	Nutrition/ obesity/ healthy lifestyles and food safety	Home economics/family and child development	Administration & misc.	Total FTEs
2007	1,932.0	437.7	170.7	1,526.7	838.4	479.4	9.5	5,394.4
2008	1,891.6	439.4	195.9	1,378.5	712.7	441.8	34.5	5,094.4
2009	1,817.9	470.3	199.8	1,408.6	741.6	452.6	33.5	5,124.3
2010	1,837.1	470.1	208.2	1,316.0	726.2	371.1	1.2	4,929.9
2011	1,787.9	465.6	255.7	1,333.1	835.1	370.7	1.2	5,049.3
2012	1,765.3	462.6	286.0	1,221.4	906.0	454.2	1.2	5,096.7
2013	1,711.6	512.5	374.2	1,365.8	876.4	265.3	1.2	5,107.0
2014	1,755.3	455.6	259.5	1,342.8	811.8	427.1	5.0	5,058.1
2015	1,785.0	461.4	215.9	1,453.7	769.0	375.4		5,060.4
Compounded annual % change	−1.0%	0.7%	3.0%	−0.6%	−1.1%	−3.0%	−100.0%	−0.8%

Table 39.6 Federal formula funded CES FTEs by program area for the West, 2007–2015

Year	Agriculture	Natural resource & environment	Community development	4-H youth	Nutrition/obesity/healthy lifestyles and food safety	Home economics/family and child development	Administration & misc.	Total FTEs
2007	582.8	196.0	163.0	355.2	140.3	48.9		1,486.2
2008	594.5	266.8	137.1	293.5	174.2	49.4	19.7	1,535.2
2009	657.1	248.2	75.0	407.6	139.7	40.0	12.3	1,579.9
2010	639.0	200.1	52.1	227.5	273.5	56.5		1,448.7
2011	701.9	367.2	48.4	394.8	318.5	23.8		1,854.6
2012	429.3	476.3	34.9	290.0	376.3	17.8		1,624.6
2013	469.7	429.2	70.5	235.2	330.5	40.8		1,575.9
2014	448.8	303.5	81.2	228.7	300.9	17.5		1,380.6
2015	399.7	322.3	87.3	205.4	312.0	24.2		1,350.9
Compounded annual % change	−4.6%	6.4%	−7.5%	−6.6%	10.5%	−8.4%		−1.2%

Again, agriculture (−1.0 percent) moved at about the same rate as the total for the region, in part because agriculture is the largest component, accounting for 35.5 percent of all FTEs. The South directs relatively more FTEs to 4-H and home economics than the rest of the United States, while natural resources, community development, and nutrition/food safety are marginally under the levels for all the United States.

The West experienced the largest shift in FTEs across program areas (Table 39.6). Agriculture declined at a noticeably greater rate (−4.6 percent) than other program areas (−1.2 percent), yet agriculture still changed relatively less than the other program areas. It is still the largest program area, although it was reduced to 29.6 percent of all FTEs by 2015. Relative to other regions, large FTE reductions occurred for home economics (−8.4 percent), community development (−7.5 percent), and 4-H (−6.6 percent). By 2015, home economics only accounted for 1.8 percent of total FTEs, substantially less than the 7.7 percent for all the United States. Also, 4-H has fewer FTEs in the West (19.1 percent) than in other regions of the United States (23.7 percent), while natural resource FTEs are noticeably larger (20.3 percent vs. 12.3 percent). Interestingly, the share of FTEs dedicated to agriculture and natural resources for the West increased from 48 percent in 1992 (Ahearn, Yee, and Bottom 2003) to 55.9 percent for the 2007–2015 period. The West contains relatively large tracts of public land and appears to have experienced an increasing demand for resource policy issues in managing these public lands. Large produce production and urban population compared to other regions may explain why nutrition/food safety is relatively larger for this region.

Below are some general observations from these regional data. First, agriculture is the largest program area in terms of FTE allocations, except marginally for the Northeast, and it has been the most stable program area. Agriculture accounts for 32.4 percent of all FTEs for the United States. Second, FTE allocations to 4-H make it the largest program area in the Northeast at 28.8 percent of all FTEs, while 4-H ranks second in the South (26.9 percent), and third for the West (19.1 percent) and the North Central region (13.7 percent). Third, natural resource FTEs rank second in the West (20.3 percent), while this ranks fourth for the other regions. Fourth, all the regions show some variation in priorities and stability, ranging from the South showing very little change between program areas to the West showing the most significant reallocations.

3.2 Federal formula FTEs by metric

Tables 39.7 through 38.10 utilize the same data as above but report FTEs on the basis of population, farm numbers, farm acres, and gross agricultural sales. As shown in Table 39.7, all regions demonstrate a declining FTE level per population over the 2007–2015 period. The South accounts for the largest percentage of the U.S. population (33.6 percent) and a relatively larger percentage of total FTEs (44.2 percent). Thus, the South has the highest overall FTEs per million people of 48.1, whereas the West is less than half this level at 21.1. Relative to the U.S. average of 36.7, the Northeast is above this level by about 4 FTEs (41.6), while the North Central region is about 5 FTEs lower (31.0). The Northeast has experienced the largest annual FTE decline per person at −5.2 percent, while the North Central region has experienced the smallest decline of just −0.1 percent. These differences are primarily due to changes in FTEs, since the population growth rates for both the Northeast region (0.4 percent) and the North Central region (0.3 percent) were below the 0.8 percent average of all the United States. Population growth rates for the South (1.1 percent) and West (1.0 percent) were above those of all the United States, and their declines in FTEs per person were similar, at −2.0 and −2.2 percent.

Table 39.8 presents FTE data by region on a per farm basis. The Northeast averages 15.1 FTEs per thousand farms, more than double the next highest level from the South at 6.6 FTEs

Table 39.7 Total federal CES FTEs funded per million individuals by region and for the United States, 2007–2015

Year	Northeast	North Central	South	West	United States
2007	45.8	30.9	55.4	21.2	39.1
2008	36.9	34.5	51.7	21.7	37.1
2009	67.6	34.1	49.3	22.1	43.2
2010	45.1	33.3	48.6	20.1	37.2
2011	47.0	31.9	49.3	25.5	38.7
2012	43.1	27.6	49.2	22.1	36.2
2013	33.8	27.9	48.8	21.2	34.0
2014	28.6	28.1	47.7	18.4	32.0
2015	30.0	30.6	47.2	17.8	32.6
Compounded annual % change	−5.1%	−0.1%	−2.0%	−2.2%	−2.3%

Table 39.8 Total federal CES FTEs funded per thousand farms by region and for the United States, 2007–2015

Year	Northeast	North Central	South	West	United States
2007	15.9	2.4	6.7	4.6	5.4
2008	12.9	2.8	6.4	4.7	5.2
2009	24.0	2.8	6.5	4.8	6.2
2010	16.2	2.8	6.3	4.4	5.4
2011	17.1	2.7	6.5	5.7	5.7
2012	15.9	2.3	6.7	5.0	5.4
2013	12.5	2.4	6.7	4.9	5.2
2014	10.7	2.4	6.7	4.3	5.0
2015	11.3	2.7	6.8	4.2	5.1
Compounded annual % change	−4.2%	1.1%	0.2%	−0.9%	−0.7%

Table 39.9 Total federal CES FTEs funded per million farm acres by region and for the United States, 2007–2015

Year	Northeast	North Central	South	West	United States
2007	110.7	6.1	20.8	5.3	13.0
2008	89.9	6.9	19.6	5.5	12.4
2009	166.4	6.8	19.7	5.7	14.6
2010	112.1	6.7	19.0	5.2	12.7
2011	117.3	6.5	19.5	6.7	13.3
2012	107.1	5.6	19.8	5.8	12.6
2013	84.3	5.7	19.8	5.7	11.9
2014	71.7	5.8	19.7	5.0	11.3
2015	75.3	6.3	19.7	4.9	11.6
Compounded annual % change	−4.7%	0.4%	−0.6%	−1.2%	−1.4%

Table 39.10 Total federal CES FTEs funded per billion dollars (2015 USD) in agricultural sales, 2007–2015

Year	Northeast	North Central	South	West	United States
2007	148.2	15.3	63.2	17.9	36.2
2008	115.2	15.5	59.0	18.7	32.9
2009	243.6	16.3	63.5	20.8	41.6
2010	148.4	14.8	55.8	17.7	33.3
2011	145.0	12.5	55.6	20.1	31.6
2012	128.9	10.0	52.3	16.8	27.7
2013	99.1	10.6	50.9	15.8	26.5
2014	79.0	10.5	49.4	12.9	24.3
2015	93.4	13.0	54.2	14.6	28.1
Compounded annual % change	−5.6%	−2.0%	−1.9%	−2.5%	−3.1%

per thousand farms. The North Central has the smallest average of 2.6 FTEs per thousand farms, and the West at 4.7 FTEs per thousand farms is also below the U.S. average of 5.4. Again, the Northeast experienced the largest decline at −4.2 percent, primarily due to fewer FTEs rather than fewer farms, which declined at a rate of −0.6 percent. The North Central, South, and West saw their farm numbers decline at a compounded annual rate of −0.9, −1.0, and −0.2 percent, respectively, over the 2007–2015 period.

Another way to view the data in Table 39.8 is to calculate the number of farms per agriculture and natural resource FTE. Ahearn, Yee, and Bottom (2003) report that in 1992 there were 434 farms per agriculture and natural resource FTE for the North Central region. This number has increased to 848 farms per FTE for the years of 2007–2015, almost doubling in a 20-year period. As in 1992, the Northeast continues to have the highest relative CES coverage level with just 175 farms per agriculture and natural resource FTE. The West and South also have a bit higher coverage than the U.S. (414), with 378 and 340 farms per FTE.

FTEs per million farm acres operated are described in Table 39.9. Again, the Northeast has by far the highest level at 103.9. But the magnitude is even much greater than FTEs per farm, since the next highest level is the South at 19.7. To put these numbers in perspective, the number of acres covered by one agricultural and natural resource FTE equals 25,483; 113,688; 323,547; and 349,387 for the Northeast, South, West, and North Central regions, respectively. The Northeast has the highest gross sales per acre at US$797, but this is less than double the U.S. average of US$408 and hardly enough to justify from 4.5 to 13.7 times fewer crop acres per agricultural and natural resource FTE than the other regions. This metric may also be why the Northeast experienced a much higher decline in FTEs per farm acre (−4.7 percent) than all of the other regions. The North Central region actually grew slightly at 0.4 percent, in part because it experienced the largest percentage decline in farm acres (−0.19 percent) of all regions.

Table 39.10 describes FTEs per billion dollars of agricultural sales or cash receipts by region. Again, the Northeast has by far the highest FTE level at 133.4, well above the coverage levels for the South (56.0), West (17.2), and North Central (13.2) regions. The North Central region accounts for 45.8 percent of U.S. agricultural sales, well above that contributed by the South (24.5 percent), West (24.1 percent), and Northeast (5.5 percent). When considering just agriculture and natural resource FTEs, one FTE is funded for every US$20.3 million of farm sales in the Northeast, a much lower level than for the North Central region (US$169.3), West (US$105.0), and South ($US40.3).

What follows are general observations from these regional metrics. First, total FTEs per million people are quite similar across regions relative to the agricultural metrics, as the largest for the South (48.1) is about double the lowest level in the West (21.1). Second, the Northeast has by far the highest level of FTEs (agricultural and natural resource or total) per farm, farm acres, and agricultural sales. The Northeast exceeds other regions by a magnitude of anywhere from 5 to 19 times the level of the lowest region. Third, even though farm numbers have continued to decline, each agricultural and natural resource FTE for the North Central region currently covers about twice as many farmers as in 1992 (Ahearn, Yee, and Bottom 2003). Fourth, the West has the lowest total FTEs per farm acre, in part because they are relatively more urban and have the lowest average gross sales per farm acre at US$324.

3.3 Recent funding for two 1862 land-grant institutions

One lesson learned from the national FTE data just described is that individual states do make decisions and resource allocation choices that vary depending on local needs, institutional structure, and political/budget realities. To gain insight into the dollar composition and recent changes in funding metrics, we compare CES funding data available for two 1862 institutions: the University of Arizona (UA) and Oklahoma State University (OSU). Of the two states, Arizona is much more urban, with 89.8 percent of the population urban and ranking as the tenth most urban state, including Puerto Rico, while Oklahoma is the thirty-sixth most urban state, with about three times the percentage of its population rural (33.8 percent) compared to Arizona (11.2 percent; United States Bureau of the Census [USBC] 2010).

Recent funding trends and metrics for both institutions for the years of 2010 and 2015 are shown in Tables 39.11 and 39.12. Constant dollar state appropriations were 21.4 percent less for OSU in 1995 than in 2015,[2] even though Oklahoma's state share of their total CES budget in 1995 (66.3 percent) was similar to 2015 (69.4 percent). In 2010, state appropriations for OSU were slightly over 75 percent of all CES revenue sources. However, OSU's total CES real budget increased 21.5 percent between 1995 and 2015. Oklahoma's economic reliance on the energy industry no doubt has contributed to the ups and downs in OSU's CES funding, and federal formula funds have helped minimize these fluctuations, even though their capacity to do so has been eroding. However, grants and contracts have increased to the point that these are now about 85 percent of federal formula funds.

State appropriations as a percent of total funds declined at a similar annual rate for both OSU (−5.7 percent) and UA (−6.5 percent). From 2010 to 2015, state support for UA declined by 9.7 percent, and its funding per person declined by 15.4 percent as Arizona's population grew by 6.8 percent. Grants and contracts increased at a robust annual rate for both the UA (5.1 percent) and OSU (10.0 percent), so that their percentage of total funds grew to 37.5 and 14.1 percent of their 2015 budgets, respectively. County dollars have declined for UA so that they are only 2.2 percent of the total funds for 2015. Total sources of CES funds have grown slightly for the UA (0.6 percent annually), while they have declined at OSU (−1.3 percent). In spite of their gradual decline, federal formula funds have remained relatively stable compared to other sources.

Total funding per person for OSU in 2015 at US$10.73 per person is 4.3 times higher than for UA, in large part due to Oklahoma being a much more rural state. Both institutions showed a decline in this metric by securing less total funding per person (US$−1.19, or −10.0 percent, for OSU; US$−0.09, or −3.5 percent, for UA) from 2010 to 2015. Funding per farm is around 60 to 70 percent more for UA than OSU, probably because average sales per farm for Arizona (US$213K) are more than double those of Oklahoma (US$92K). Also, average farm size for Arizona (1,333 acres) is about three times that of Oklahoma (439 acres). Data for funding indicate

Table 39.11 Funding changes and metrics for Oklahoma State CES, 2010 and 2015

Funding source	2010 to 2015 annual % change	Funding ($2015)/ person 2010	2015	Funding ($2015)/ farm 2010	2015	Funding ($2015)/ farm acre 2010	2015	Cents of funding/$ gross ag sales 2010	2015	2015 funding level	% of Total (2015)
State Appropriations Fed. formula funds	−2.8%	$8.95	$7.45	$403.35	$373.63	$0.97	$0.85	0.501¢	0.408¢	$29,142,844	69.4%
Smith–Lever 3(b)3c	−1.5%	$1.67	$1.48	$75.09	$74.43	$0.18	$0.17	0.093¢	0.081¢	$5,805,331	13.8%
EFNEP 3(d)	−1.9%	$0.33	$0.28	$14.66	$14.24	$0.04	$0.03	0.018¢	0.016¢	$1,110,726	2.6%
Grants & contracts	10.0%	$0.98	$1.51	$44.16	$75.74	$0.11	$0.17	0.055¢	0.083¢	$5,907,816	14.1%
TOTAL	−1.3%	$11.92	$10.73	$537.26	$538.03	$1.29	$1.23	0.668¢	0.588¢	$41,966,717	100.0%

Table 39.12 Funding changes and metrics for the University of Arizona CES, 2010 and 2015

Funding source	2010 to 2015 annual % change	Funding ($2015)/ person 2010	2015	Funding ($2015)/ farm 2010	2015	Funding ($2015)/ farm acre 2010	2015	Cents of funding/ $ gross ag sales 2010	2015	2015 Funding level	% of Total (2015)
State appropriations Fed. appropriations	−2.0%	$1.36	$1.15	$483.52	$403.06	$0.36	$0.30	0.232¢	0.190¢	$7,859,750	46.3%
Formula (S-L, EFNEP)	−2.3%	$0.23	$0.19	$82.29	$67.67	$0.06	$0.05	0.040¢	0.032¢	$1,319,651	7.8%
Grants & contracts	3.6%	$0.56	$0.63	$200.53	$220.77	$0.15	$0.17	0.096¢	0.104¢	$4,305,016	25.4%
County funds	−4.4%	$0.07	$0.06	$26.31	$19.36	$0.02	$0.01	0.013¢	0.009¢	$377,428	2.2%
Non-Fed. grants & contracts	8.8%	$0.21	$0.30	$74.90	$105.60	$0.06	$0.08	0.036¢	0.050¢	$2,059,234	12.1%
Gifts, patents, and other funds	4.1%	$0.14	$0.16	$48.20	$54.34	$0.04	$0.04	0.023¢	0.026¢	$1,059,556	6.2%
TOTAL	0.6%	$2.58	$2.49	$915.75	$870.80	$0.69	$0.65	0.440¢	0.410¢	$16,980,635	100.0%

state demographics and agricultural characteristics likely dictate not only the magnitude of dollars received, but also the ability of their respective 1862 institutions to provide CES programs.

4 Future directions for CES

CES celebrated 100 years of existence in 2014. Centennial celebrations at the national level and in many states noted significant contributions made by the CES arm of the land-grant university system. CES, arguably, has had a major part in the growth of agricultural productivity and enhanced the quality of life in rural America during its proud history. The "structure" of CES has notable attributes, including its unique partnership across county, state, and federal government levels. CES is grounded in reliance on sound science- and research-based information. Finally, CES is engaged with the public it serves, such as farmers, ranchers, and youth. One good example would be the partnership with U.S. farmers and ranchers to provide on-farm "demonstrations" for farmers and ranchers in other countries for what research-based practices or technology can do to enhance productivity. How will the conduct and performance of CES be impacted over the next 100 years if the "structure" of CES is significantly altered? Here are a few questions that might be considered as CES looks forward in anticipation of serving the public for the next 100 years:

1. The United States has become more urbanized with fewer farmers. How will this impact the mission of and potential political support for CES? One of the principal focus areas for CES has continued to be agriculture. Some critics and observers feel CES should expand its support base to provide services directly beneficial to urban residents, which may translate into more support for the organization. Others argue that now is the time to focus on our core strengths, such as agriculture, and focus on communicating to our traditional clientele, as well as urban residents, how beneficial CES at providing a safe and reliable (and relatively inexpensive) food supply. It appears the answer to this set of questions varies depending on the geographic location of state CES programs and the prevailing local political climate. For example, the U.S. Northeast region has shifted away from agriculture relatively more than the other U.S. regions.
2. U.S. consumers enjoy a relatively safe and reliable food supply. Consumer preferences and expectations have evolved as incomes rise and more consumers are further removed from actual farming operations. Consumers are concerned and asking questions regarding everything from health and environmental implications of GMOs and local foods to urban–rural interface issues on dust particulates and open space. Is CES prepared to interact with the newly emerging consumer concerns of this century? This is really an expansion of the points contained in #1 above. Many traditional CES programs are geared to working directly with producers, and this has certainly had positive and tangible impacts. However, the growing urban population in many states brings to bear many new questions about food and related social and economic topics. Some CES organizations may be somewhat challenged to adjust to these evolving demographic and social trends with declining FTEs.
3. The public is generally less supportive of taxes and investments supporting public goods. There are financial pressures at all levels of CES funding, including county, state, and federal government levels. If formula-funding and state-funding support continues to erode, will CES be able to maintain its original historic structure of having an FTE residing in all counties in every state?
4. Related to declining support for public good investment by the public – will CES be able to articulate a message to the voting public that allows for survival and the

effective delivery of services? Epplin (2012) presents a good summary of the impact the public land-grant system (including CES) has had on productivity and the quality of life. He argues our Founding Fathers intuitively understood social benefits would accrue from public investments in Teaching, Research, and Extension. Epplin notes a critical role for agricultural economists is to help students understand the economic theory that explains why taxing citizens to pay for public education is good for students and society.

5   What role do agricultural economists play as CES faces these sweeping trends in our national and global economies?

CES is undergoing a restructuring in many states. Martin (2016) noted that the catalyst for restructuring CES in Ohio was the budget. The process moved forward through a series of meetings and input solicitation from stakeholders. One lesson learned was that policy makers responded positively with budget allocations that followed this process. Openness and reporting of positive impacts can have a fruitful outcome. Zublena (2016) indicated that budget pressures led to reorganization and changes in North Carolina. Key philosophies included transparency, honesty/integrity, system-wide thinking, a future focus, and a commitment to action. Core programs were a focus, as was the appropriate use of technology. Self-assessment and a focus on future structure are typical of discussions occurring across the country as CES administrators grapple with tighter budgets and evolving demographic and economic trends.

Barkley and Rohrer (1962) advocate that CES should be evaluated and compared against organizations similar in structure and function. They suggest that programs of public education, health and welfare, and soil conservation share common features with CES. Cook (1986) provides multiple conditions required for CES to make a positive difference: (1) CES must develop a dynamic, public-good oriented mission; (2) CES must develop a well-defined strategic directive; (3) CES must improve management of its human resources; (4) CES must structure itself to carry out its strategic mission; and (5) the urgency of the "future of cooperative extension" issue must be addressed aggressively.

Technology is seen as one viable alternative as CES adjusts to new organizational approaches. The CES system has invested in a more visible web presence through an initiative called eXtension (2017). The website for eXtension states that knowledge resources are provided from land-grant universities and CES. Teams of CES professionals across the country come together to provide research-based information, educational resources, webinars, ask-an-expert opportunities, and access to various social media sites. Topics selected are driven by perceived needs and interests and include many familiar CES focus areas: community, disaster issues, energy, environment, family, farm, health/nutrition, lawn/garden, pest management, and youth. Certainly this effort places CES in the position of providing information via social media as many other information providers do as well. This effort has potential but does have many competitors in the "marketplace" of public information.

While CES is likely to continue to see a gradual erosion in public support through reduced state and federal formula funds, more funding through grants, contracts, and gifts, along with greater efficiencies and a more targeted focus, should keep CES viable. Greater efficiencies are going to come at the expense of having a smaller or no resident FTE presence in small rural counties. More regionalization of agents and specialists by location and specialty is likely to occur. While technologies can remove many of the barriers associated with distance, the recognition and value of having CES expertise at the local level will be important for many rural communities despite improvements in communications technologies.

5 Concluding comments

Many argue that the land-grant system is one of the most significant institutions created in the United States. The CES is a critical part of the system. Important values and attributes of CES include a local presence (having agents in every county of each state) and a research-based outreach program providing unbiased information to clientele and the public. The CES is also unique in the funding partnership between federal, state, and local levels of government. Finally, CES offers a national network through the various land-grant universities where local agents/educators have access to a vast network of researchers and educators to address locally identified problems. The institution of CES has over 100 years of existence with a proud history and legacy. Recent changes in demographics and shifts in public support for public investments have challenged CES entities in many states. Reorganization and reallocation of resources has occurred. Discussions about priorities and the need to demonstrate the value of CES to the voting public occur at all levels of government.

While CES will likely survive, shifts in funding such that grants and contracts make up a larger percentage of CES's budget will likely shift the focus and priorities of CES to where these funds are most easily attained at the margin compared to local needs assessment. While many grants and contracts can be complementary in focus to local needs, CES should cautiously evaluate external funding sources to ensure that the needs and political advocacy of local stakeholders is not being missed or jeopardized when taking on time commitments embodied with external funds. Agriculture and rural communities have been the focus of CES since its beginning, and we see this support base as crucial for CES's future, even though the population of farmers and rural communities continues to decline. That is, even highly urbanized states still have legislators and advocates with agricultural interests who influence state policy and budget allocation decisions. After all, everyone is still receiving essentially all of their food and a good portion of their fiber from production agriculture, although some see a future where food may not be sourced from farms (Griggs 2014).

In addition to bringing science-based information to every stakeholder and community issue, CES needs to be ready to serve the role of facilitator or mediator for bringing together communities that have rural–urban interface and conflict issues. CES will need to provide education to all interested parties, not just those willing to pay for services. Articulating the public goods benefits of CES programs, particularly to those that are not direct users or recipients, will be critical for keeping public support. Fees are likely to occur for direct users of CES programs that have private good attributes, at least partially for recovering costs. But in order to produce CES public goods at an optimal level, CES needs to be able to convince voters and legislators of the value of their public goods in both difficult and good budget times. Youth becoming less likely to commit crimes on the public from attending 4-H or producers learning practices that minimize the spread of invasive weeds beyond their property boundaries are examples needed to sell legislators on the public goods value of CES. Explaining how individuals in society who are not direct recipients of CES's programs nevertheless benefit from is a key dimension that everyone in CES needs to better articulate and market to society (Kalambokidis 2011). While collecting and tracking numbers for CES is not all bad, these efforts have been around for a long time (Lutz et al. 1972) and are not sufficient by themselves for making an optimal funding case for CES.

Society and the agricultural industry will face many challenges in the coming years. The agricultural economics profession recently conducted a national effort led by the Agricultural and Applied Economics Association (AAEA) and The Council on Food, Agricultural and Resource Economics (CFARE) to identify areas where agricultural economists can assist

(Nelson 2016). Priorities identified include climate change, big data, development/trade, food security, evolving consumer preferences related to food and food production, natural resource sustainability, rural economic growth, and agricultural innovation. Certainly, these are important issues and one need only observe recent national elections to find discussions related to many of these ideas. If agricultural economists are to be of assistance, a strong and viable land-grant university system is critical for success. Strong Teaching and Research programs linked to a CES that delivers sound science-based information where needed will ensure that the public is best served. CES needs effective communications and connections with communities and stakeholders to be an effective conduit for science-based research and career opportunities for students attending land-grant institutions. While CES programs involve outreach, they need to be much more to ensure CES remains vibrant in the future. That is, CES programs need to provide a living resource and presence within communities so they are at the forefront of preserving and enhancing the well-being of current and future community members. To be successful, CES needs to deliver relevant science-based knowledge to communities so that they see the value in sending their children to land-grant institutions to obtain valuable knowledge and skills and to become productive citizens and leaders that advocate for the land-grant system, including CES.

Notes

1 Regions were defined as in Ahearn, Yee, and Bottom (2003), except for adding PR to the South and AK, GU, and HI to the West. States in each region are: Northeast – CT, DE, MA, ME, MD, NH, NJ, NY, PA, RI, VT, and WV; North Central – IA, IL, IN, KS, KY, MI, MN, MO, ND, NE, OH, SD, and WI; South – AL, AR, FL, GA, LA, MS, NC, OK, PR, SC, TN, TX, and VA; and West – AZ, AK, CA, CO, GU, HI, ID, MT, NM, NV, OR, UT, WA, and WY.
2 County dollars put into OSU's CES were not available. Retrieving comparable CES funding numbers before 2010 at UA was difficult.

References

Ahearn, M., J. Yee, and J. Bottom. 2003. "Regional Trends in Extension System Resources." ERS Information Bulletin No. 781, USDA-ERS, Washington, DC.
Akobundu, E., J. Alwang, A. Essel, G.W. Norton, and A. Tegene. 2004. "Does Extension Work? Impacts of a Program to Assist Limited-Resource Farmers in Virginia." *Review of Agricultural Economics* 26(3):361–372.
Alston, J.M., M.A. Andersen, J.S. James, and P.G. Pardey. 2011. "The Economic Returns to U.S. Public Agricultural Research." *American Journal of Agricultural Economics* 93(5):1257–1277.
Anderson, J.R., and G. Feder. 2007. "Agricultural Extension." In R. Evenson and P. Pingali, eds., *Handbook of Agricultural Economics*. Amsterdam, Netherlands: North Holland, pp. 2345–2378.
Association of Public and Land Grant Universities [APLGU]. 2012. *The Land-Grant Tradition*. Washington, DC: APLGU.
Barkley, P.W., and W.C. Rohrer. 1962. "An Interpretive View of an Institutional Process: Measuring Effectiveness and Changeability of the Cooperative Extension Service." *Journal of Farm Economics* 44(5):1740–1744.
Comer, M.M., T. Campbell, K. Edwards, and J. Hillison. 2006. "Cooperative Extension and the 1890 Land-Grant Institution: The Real Story." *Journal of Extension* 44(3):A4.
Cook, M.L. 1986. "Restructuring Agricultural Economics Extension to Meet Changing Needs: Discussion." *American Journal of Agricultural Economics* 68(5):1307–1309.
Cornes, R., and T. Sandler. 1996. *The Theory of Externalities, Public Goods and Club Goods*. New York: Cambridge University Press.
Daly, P.A. 1981. "Agricultural Employment: Has the Decline Ended?" *Monthly Labor Review* 104(11):11–17.

Diekmann, F., C. Loibl, and M. Batte. 2009. "The Economics of Agricultural Information: Factors Affecting Commercial Farmers' Information Strategies in Ohio." *Applied Economic Perspectives and Policy* 31(4):853–872.

Epplin, F.M. 2012. "Market Failures and Land Grant Universities." *Journal of Agricultural and Applied Economics* 44(3):281–289.

Evenson, R.E. 2001. "Economic Impacts of Agricultural Research and Extension." In B. Gardner and G. Rausser, eds., *Handbook of Agricultural Economics*. Amsterdam, Netherlands: North Holland, pp. 574–628.

eXtension. 2017. https://extension.org/. Accessed February 13, 2017.

Franz, N.K. 2013. "Improving Extension Programs: Putting Public Value Stories and Statements to Work." *Journal of Extension* 51(3):TT1.

Goetz, S.J., and M. Davlasheridze. 2016. "State-Level Cooperative Extension Spending and Farmer Exits." *Applied Economic Perspectives and Policy* 39(1):65–86.

Griggs, B. 2014. "How Test-Tube Meat Could be the Future of Food." *Cable News Network*, April 30.

Hanson, J.C., and R.E. Just. 2001. "The Potential for Transition to Paid Extension: Some Guiding Economic Principles." *American Journal of Agricultural Economics* 83(3):777–784.

Hoag, D.L. 2005. "Economic Principles for Saving the Cooperative Extension Service." *Journal of Agricultural and Resource Economics* 30(3):397–410.

Howery, L.D. 2000. "A Summary of Livestock Grazing Systems Used on Rangelands in the Western United States and Canada." Coop. Ext. Pub. No. AZ1184, University of Arizona, Tucson, AZ.

Huffman, W.E., and R.E. Evenson. 2006. *Science for Agriculture: A Long-Term Perspective* (2nd ed.). Ames, IA: Blackwell Publishing.

Jin, Y., and W.E. Huffman. 2016. "Measuring Public Agricultural Research and Extension and Estimating Their Impacts on Agricultural Productivity: New Insights from U.S. Evidence." *Agricultural Economics* 47:15–31.

Kalambokidis, L. 2004. "Identifying the Public Value in Extension Programs." *Journal of Extension* 42(2):A1.

———. 2011. "Spreading the Word about Extension's Public Value." *Journal of Extension* 29(2):A1.

Kelsey, T.W., and S.J. Goetz. 2016. "Potential Economic Impacts of Zero Funding for Agricultural Research and Extension Programs Extend Far Beyond the University." Penn. State Extension, University Park, PA.

Laden, B.V. 2004. "Hispanic-Serving Institutions: What are They? Where are They?" *Community College Journal of Research and Practice* 28(3):181–198. doi:10.1080/10668920490256381.

Lutz, A.E., and D.W. Swoboda. 1972. "Accountability in Extension." *Journal of Extension* 10(4):45–48.

Martin, K. 2016. *Lessons Learned Through Restructuring of Ohio State University Extension*. Stillwater, OK: Oklahoma State University.

McKee, R.K., B.A. Talbert, and S.J. Barkman. 2002. "The Challenges Associated with Change in 4-H/Youth Development." *Journal of Extension* 40(2):A5.

Nelson, G. 2016. *What I Learned About Priorities and What They Might Mean for You*. Stillwater, OK: Oklahoma State University.

Schnitkey, G., M. Batte, E. Jones, and J. Botomogno. 1992. "Information Preferences of Ohio Commercial Farmers: Implications for Extension." *American Journal of Agricultural Economics* 74(2):486–496.

Smith, E.G., and R.D. Smith. 2007. "Will Extension be Relevant in the 21st Century?" In R.D. Knutson, S.D. Knutson, and D.P. Ernstes, eds., *Perspectives on 21st Century Agriculture: A Tribute to Walter J. Armbruster*. Ann Arbor, MI: Sheridan Books, pp. 56–61.

Tegene, A., A. Effland, N. Ballenger, G. Norton, A. Essel, G. Larson, and W. Clarke. 2002. "Investing in the People: Assessing the Economic Benefits of 1890 Institutions." ERS Publication No. 1583. USDA-ERS, Washington, DC.

Thomson, J.S. 1984. "Extension's Federal Funding: Who Is Entitled?" *Journal of Extension* 22(6):F2.

United States Bureau of the Census [USBC]. 2010. "Census of Population." USBC, Washington, DC.

United States Department of Agriculture [USDA]. 2010. "Secretary of Agriculture Tom Vilsack's 2010 Briefing on the Status of Rural America." USDA, Washington, DC.

United States Department of Agriculture, Economic Research Service [USDA-ERS]. 2017. "Data Files: U.S. and State-Level Farm Income and Wealth Statistics." USDA-ERS, Washington, DC.

United States Department of Agriculture, National Institute of Food and Agriculture [USDA-NIFA]. 2017. "Agricultural Extension Programs at 1890 Institutions." USDA-NIFA, Washington, DC.

United States Department of Agriculture, National Institute of Food and Agriculture, Planning, Accountability and Recording Staff [USDA-NIFA-PARS]. 2016. "Personal communication." USDA-NIFA-PARS, Washington, DC.

Wang, S.L. 2014. "Cooperative Extension System: Trends and Economic Impacts on U.S. Agriculture." *Choices* 29(1):1–8.

Wang, S.L., E. Ball, L. Fulginiti, and A. Plastina. 2012. "Accounting for the Impacts of Public Research, R&D Spill-ins, Extension, and Roads in U.S. Regional Agricultural Productivity Growth, 1980–2004." In K. Fuglie, S. Wang, and V. Ball, eds., *Productivity Growth in Agriculture: An International Perspective*. Wallingford: CABI, pp. 13–31.

Zublena, J. 2016. *Our Strategic Vision for the Next Century*. Stillwater, OK: Oklahoma State University.

40

Theory of cooperatives
Recent developments

Michael L. Cook and Jasper Grashuis

1 Introduction

This update responds to the call by King et al. (2010) to expand and extend our understanding of theories and frameworks that explore the complexities of the agricultural cooperative, an organizational form which continues to be successful and even dominants in many sectors of the European and North American agri-food industries. The specific objective of our chapter is to highlight the cooperative theoretical work produced by scholars since the last surveys by Sexton (1984), Staatz (1989), and Cook et al. (2004). In illustrating the current state of the art, we intend to compare and contrast the main findings and conclusions while identifying new challenges and opportunities for future research directions.

Using the search term "agricultural cooperative", we searched publications in the following databases: Scopus, ScienceDirect, EBSCOhost, Web of Science, Google Scholar, and ProQuest. We then conducted a focused search by using the following criteria: (1) the article is published in 2005 or later, (2) the article is published in a peer-reviewed book or journal, (3) the article is theoretical and not empirical or conceptual in its orientation, and (4) the article relates to organizations owned and controlled by farm producers. In total, we identified 29 articles as appropriate and relevant for our literature review (see Table 40.1). Overall, we find theoreticians have made recent advances to our collective understanding of cooperative (1) performance, (2) governance, (3) management, and (4) finance. As before, formal analyses almost always find a positive impact of supply and marketing cooperatives on the welfare of farm producers and food consumers, yet there is increasing recognition of numerous constraints and inefficiencies. For example, theoreticians have been modeling the negative consequences of organizational growth in terms of heterogeneous utility functions of members, directors, and managers to inform agency decisions. When the principal-agent relationship is given, management behavior is often analyzed in relation to risk or vertical integration. As cooperatives face pressure to grow further, there is also growing attention to the debt or equity decision to inform solutions to the tension between patronizing and capitalizing the cooperative. Altogether, the theoretical literature is thus evolving in its conceptualization of cooperatives as complex and diverse business organizations which nonetheless remain able to positively impact both farm producers and food consumers.

Table 40.1 Overview of cooperative theory publications from 2005 to 2016

Year	Author(s)	Title	Book/Journal
2005	Bogetoft	An information economic rationale for cooperatives	*European Review of Agricultural Economics*
2005	Giannakas and Fulton	Process innovation activity in a mixed oligopoly: the role of cooperatives	*American Journal of Agricultural Economics*
2006	Evans and Guthrie	A dynamic theory of cooperatives: the link between efficiency and valuation	*Journal of Institutional and Theoretical Economics*
2006	Hueth and Marcoul	Information sharing and oligopoly in agricultural markets: the role of the cooperative bargaining association	*American Journal of Agricultural Economics*
2007	Fulton and Giannakas	Agency and leadership in cooperatives	*Vertical Markets and Cooperative Hierarchies*
2007	Olesen	The horizon problem reconsidered	*Vertical Markets and Cooperative Hierarchies*
2007	Rey and Tirole	Financing and access in cooperatives	*International Journal of Industrial Organization*
2009	Bontemps and Fulton	Organizational structure, redistribution and the endogeneity of cost: cooperatives, investor-owned firms and the cost of procurement	*Journal of Economic Behavior and Organization*
2009	Hovelaque et al.	Effects of constrained supply and price contracts on agricultural cooperatives	*European Journal of Operational Research*
2009	Ligon	Risk management in the cooperative contract	*American Journal of Agricultural Economics*
2009	Mérel et al.	Cooperatives and quality-differentiated markets: strengths, weaknesses, and modeling approaches	*Journal of Rural Cooperation*
2009	Saitone and Sexton	Optimal cooperative pooling in a quality-differentiated market.	*American Journal of Agricultural Economics*
2010	Drivas and Giannakas	The effect of cooperatives on quality-enhancing innovation	*Journal of Agricultural Economics*
2010	Fatas et al.	Blind fines in cooperatives	*Applied Economic Perspectives and Policy*
2012	Feng and Hendrikse	Chain interdependencies, measurement problems and efficient governance structure: cooperatives versus publicly listed firms	*European Review of Agricultural Economics*
2012	Fulton and Giannakas	The value of a norm: open membership and the horizon problem in cooperatives	*Journal of Rural Cooperation*
2013	Deng and Hendrikse	Uncertainties and governance structure in incentives provision for product quality	*Governance of Alliances, Cooperatives and Franchise Chains*
2013	Dietl et al.	Explaining cooperative enterprises through knowledge acquisition outcomes	*Managerial and Decision Economics*
2013	Fulton and Giannakas	The future of agricultural cooperatives	*Annual Review of Resource Economics*
2013	Liang and Hendrikse	Cooperative CEO identity and efficient governance: member or outside CEO?	*Agribusiness*
2014	Hueth and Moschini	Endogenous market structure and the cooperative firm	*Economics Letters*

(Continued)

Table 40.1 (Continued)

Year	Author(s)	Title	Book/Journal
2014	Kopel and Marini	Strategic delegation in consumer cooperatives under mixed oligopoly	Journal of Economics
2015	Agbo et al.	Agricultural marketing cooperatives with direct selling: a cooperative–non- cooperative game	Journal of Economic Behavior and Organization
2015	Deng and Hendrikse	Managerial vision bias and cooperative governance	European Review of Agricultural Economics
2015	Fulton and Pohler	Governance and managerial effort in consumer-owned enterprises	European Review of Agricultural Economics
2015	Hueth and Marcoul	Agents monitoring their manager: a hard-times theory of producer cooperation	Journal of Economics and Management Strategy
2015	Mérel et al.	Cooperative stability under stochastic quality and farmer heterogeneity	European Review of Agricultural Economics
2016	Giannakas et al.	Horizon and free-rider problems in cooperative organizations	Journal of Agricultural and Resource Economics
2016	Liang and Hendrikse	Pooling and the yardstick effect of cooperatives	Agricultural Systems

2 Performance and market structure

The primary intention of the theoretical work reviewed in this section is to inform dynamics of economic efficiency, producer and consumer welfare, or market structure. There is limited overlap, however, as most models and frameworks place emphasis on various variables to explain changes in supply, demand, and price. Invariably, the conclusion of the formal analyses is that supply and marketing cooperatives have a positive impact on the welfare of farm producers and food consumers, respectively, although the degree of success is often dependent on solving its inherent constraints and inefficiencies.

First, by comparing pure and mixed duopolies, Giannakas and Fulton (2005) inform process innovation activity by input supply cooperatives as compared to firms. Because of its objective to maximize member patron welfare, the supply cooperative is assumed to have greater incentive to invest in process innovation to decrease the cost of input production. The model addresses the difficulty of member equity acquisition in the presence of heterogeneous member patron objectives and preferences, in particular with regard to capitalizing long-term growth opportunities. Even so, the supply cooperative is demonstrated to have a positive impact on total process innovation activity. Subsequently, the decrease in the welfare of the input suppliers is exceeded by the increase in the welfare of the member patrons. The model indicates the heavy reliance on retained income is not necessarily fatal, as input supply cooperatives engage in process innovation to drive competitiveness.

While Giannakas and Fulton (2005) studied process innovation, Drivas and Giannakas (2010) instead emphasized product and service innovation by consumer cooperatives. Like Giannakas and Fulton (2005), Drivas and Giannakas (2010) concluded the presence of the cooperative has a positive impact on innovation activity as well as welfare. The total effect, however, is dependent on the degree of consumer heterogeneity, which implies elasticity of demand to product quality differentiation. Generally, the greater the responsiveness to product differentiation, the greater the likelihood of innovation activity by the cooperative. The formal findings by Giannakas and Fulton (2005) and Drivas and Giannakas (2010) suggest cooperatives should earmark future income for investment in research and development, in particular as product differentiation is of rising importance in the agri-food industry.

Hueth and Marcoul (2006) provided a formal analysis of the welfare effect of bargaining associations, which are prominent in the Californian fruit and vegetable sector. While its primary purpose is to affect the market structure by improving the bargaining power of its member patrons, Hueth and Marcoul (2006) also envisioned an independent impact of information sharing on the individual and the collective ability to meet demand. As each association receives an imperfect signal of future demand, sharing information is argued to reduce the variance of the signal error. However, while sharing information facilitates an increase in net welfare, the model indicates the dominant first-stage strategy is to withhold information. To avoid the Prisoner's Dilemma, Hueth and Marcoul (2006) recommended a contractual obligation to report information for collective price discovery.

Saitone and Sexton (2009) analyzed member patron heterogeneity in terms of product quality in relation to revenue pooling, which attenuates risk to farm producers from stochastic production of low- and high-quality products. In addition, revenue pooling decreases the incentive to overproduce high-quality products. In the first stage of the sequential game, the cooperative announces the pooling rate. Each farm producer decides to sell output to the cooperative or another company in the next stage. As indicated by the model, defection is most likely by producers of high-quality products, as the premium is in part shared with producers of low-quality products. Finding the optimal pooling rate is complicated by the degree of member patron heterogeneity in cost functions and risk preferences. Not all pooling arrangements are implementable, which implies cooperatives do not have a large margin for error.

Similarly, Mérel et al. (2009) applied the Hotelling model in a mixed duopoly to compare the performance of open and closed membership cooperatives in terms of quality-based competition. While open membership cooperatives have a yardstick effect on the industry by forcing competitive honesty, member patrons do not have incentive to make investments in value-added ventures, as part of the benefit is misappropriated to external free riders. In addition, the inability to dissuade low-quality producers or to attract high-quality producers is suggestive of the low competitiveness of open membership cooperatives in industries where demand is more responsive to quality as opposed to price. By comparison, closed membership cooperatives have a greater capacity to start value-added operations, but the yardstick effect on the industry is not as strong. The model has implications for policy makers who contemplate the tradeoff between producer and consumer welfare.

Fulton and Giannakas (2013) analyzed the impact of spatial dispersion on the price received by farm producers from processors. The analysis thus considers the fact that farm producers face variable costs of transportation. In the pure duopoly, the best-response functions and the Nash equilibrium prices depend on whether monopsonist behavior is local or regional. When introducing the cooperative to the mixed market, Fulton and Giannakas (2013) considered the impact of agency problems and membership access barriers, which directly and indirectly impact the price received by member patrons. The model in part informs the pricing strategies of large regional or even national cooperatives with member patrons in many states.

Hueth and Moschini (2014) developed a three-stage entry-deterrence model with a monopolist and a consumer cooperative. In the first stage, entry by the monopolist is dependent on the fixed entry cost and the likelihood of future competition from the consumer cooperative. When considering entry in the second stage, the consumer cooperative must incur the fixed entry cost as well as the cost of coordination, which increases with membership size. If the coordination cost is not high enough to prevent formation of the consumer cooperative, the incumbent market leader may deter entry by lowering its price and allowing the consumer coalition to reap the benefits. However, deterrence may decrease firm profit to the point where initial entry in the first stage is no longer viable. As such, their study demonstrated that the first-mover advantage of a profit-maximizing firm is at times negated by the entry threat of consumer cooperatives.

Agbo et al. (2015) employed a theoretical model to analyze a dual market structure in which individual farm producers simultaneously compete and cooperate. Homogeneous farm output is either sold to the marketing cooperative, which is active on a competitive non-local market, or to end consumers on the local market. Given price on the national market, each member patron decides (1) the optimal quantity to be produced and (2) the optimal quantity to be sold on the local market and to be supplied to the cooperative for sale on the non-local market. According to the model, the local market assumes an oligopsonistic nature as the existence of the marketing cooperative induces tacit collusion by its member patrons to lower local supply. With emphasis on price discrepancies in local and non-local markets, the model by Agbo et al. (2015) informs the decision by marketing cooperatives to allow direct selling or to bind member patrons to exclusive supply agreements.

Liang and Hendrikse (2016) analyzed the yardstick effect of cooperatives on firms in a non-competitive market. In the mixed market, the firm discriminates to secure supply of heterogeneous quality. The firm offers a reservation wage which equals the marginal cost of production for each farm producer. Meanwhile, the open membership cooperative uses price pooling, which induces adverse selection in terms of attracting low-quality producers. While full price pooling forces the firm to increase the reservation wage offered to high-quality producers, the yardstick effect is even stronger in case of partial price pooling, which implies the price is in part based on the heterogeneous product quality. With partial price pooling, the equilibrium market structure is a mixed market with two cooperatives. Like Saitone and Sexton (2009) and Mérel et al. (2009), Liang and Hendrikse (2016) thus inform pricing and pooling strategies by cooperatives that market fruit, vegetables, nuts, and other products of heterogeneous quality.

3 Governance

Assuming a microeconomic perspective, recent cooperative theory has analyzed the complex interrelationships of member patrons, board directors, and managers. Most commonly, a multi-stage model is developed from an agency theory perspective to formally study (1) the relationship of member patrons to other member patrons or (2) the relationship of member patrons to managers. Recent advances in cooperative theory thus address the existence of multiple utility functions with many parameters and constraints. As such, the analyses for the most part apply to cooperatives in which control is delegated to one or more non-member managers who have resource allocation authority.

3.1 Heterogeneous member patron preferences

Bogetoft (2005) developed a model similar to Karantininis and Zago (2001) but did not include open or closed membership as a constraint. The model features producers of a homogeneous good with differential cost functions that are not known to the cooperative. There is consequently an adverse selection problem as the cooperative cannot identify the low- and high-cost producers. In order to maximize net benefit, the cooperative must exclude high-cost producers and attract low-cost producers, which is accomplished by means of particular combinations of individual rationality, incentive compatibility, and budget balancing constraints. Because of expected profit at the production stage as well as the processing stage, the cooperative is believed to produce and process the optimal quantity, which is higher as compared to the firm processor.

Like Bogetoft (2005), Fatas et al. (2010) also analyzed the free-rider problem in terms of heterogeneous product quality in an experimental model. Interestingly, the developed model excludes monitoring as the primary solution to the free-rider problem. Instead, the cooperative uses the success ratio (R), given as the ratio of the observed quality to the maximum quality, as an indicator at the aggregate level. For each individual member patron, the exclusion or punishment probability is $1 - R$, which implies individual dependence on the collective. If full benefit exclusion is the punishment, quality is expected to increase by 75 percent. However, as indicated by Fatas et al. (2010), such a blind mechanism is rather unfair and has yet to be implemented in practice.

Another type of problem, namely the control and influence problem, is addressed by Bontemps and Fulton (2009), who explicitly modeled the impact of agency cost and democratic cost on the optimal contract. As compared to the monopsonist firm, the cooperative is characterized by higher output, which implies consumer welfare is superior, all else being equal. However, as output increases, the model anticipates the benefit distribution to skew toward the relatively efficient member patrons. If the efficient member patron is not representative of the average member patron, then individuals or groups of individuals will engage in influence activities. Subsequently, a control and influence problem may arise and cause agency cost and democratic cost, which may facilitate relative inefficiency. The main result informs the member governance system, which traditionally is characterized by the one member, one vote approach. However, the model indicates an efficiency-based system is expected to be superior.

Deng and Hendrikse (2013) advanced a principal–agent model to analyze the traditional relationship of many farmers at the upstream stage and one processor at the downstream stage. The model assumes yield uncertainty, risk aversion, and quality differentiation on the farm, as well as demand uncertainty in the market. In the open membership cooperative, a free-rider problem emerges as the marginal cost of product quality improvement is exceeded by its marginal benefit. In fact, product quality is decreasing in free riding, which itself is increasing in membership size. When in competition with a firm processor, the optimal income rights structure of the cooperative is given by a certain combination of the pooling ratio, the product quality incentive, and the base payment. Generally, because of the dual relationship to production risk and free riding, a low (high) pooling ratio is compatible with a low (high) quality incentive and a high (low) base payment, but product quality is never expected to be as high as compared to the firm processor.

Mérel et al. (2015) addressed the same problem of adverse selection, free riding, and heterogeneous quality. Again, distinction is made between low- and high-quality producers, who may have limited incentive to join the cooperative at any rate of pooling if no countermeasure is taken. In any situation, low-quality producers prefer full pooling, as risk sharing is optimized. For the high-quality producers, detection is only prevented if the benefit of risk sharing surpasses the decreased payoff. As demonstrated by Mérel et al. (2015), there is a stable pooling arrangement if producers are not too risk neutral, producer heterogeneity is not too great, and the price discount for low quality is not too low. Such fragile conditions imply member patron heterogeneity is difficult to address. Furthermore, any plan must likely be dynamic as heterogeneity is not a static concept.

3.2 CEO identity

The first example is offered by Fulton and Giannakas (2007), who placed emphasis on member commitment as a function of agent behavior and performance. The principal–agent model comprises three periods. In the first period, the principal screens two types of leaders in the employment market: member welfare maximizers and profit maximizers. With proper incentives, the hired agent signals her identity or objective in the second period. Subsequently, the third period is characterized by a mixed oligopoly market in which competition with a firm is based on price and quality. For the cooperative, its market share is determined by its product quality, which in turn is determined by member commitment. If the leader represents member objectives, member commitment and product quality will be relatively high.

Thus, using backward induction, the cooperative is encouraged to make an investment in screening leadership candidates who will represent member objectives to ensure member commitment.

Similarly, Liang and Hendrikse (2013) formulated a principal–agent model to analyze the identity of the cooperative CEO as a member or non-member. Of course, the member CEO is also an input supplier to the cooperative, which implies a fundamental difference in utility functions. The model addresses the impact of CEO payoff on upstream and downstream activities. Generally, the incentive must be higher for the member CEO so as to divert attention from the upstream to the downstream activity. However, CEO optimality is also dependent on the marginal productivities at the two stages. For example, when marginal productivity is equal across the two stages, a member CEO will always be more efficient. Thus, the model implies that the common decision by large cooperatives to hire non-member CEOs is in part motivated by the upstream bias in the utility function of the member CEO as well as low complementarities between value chain segments.

Deng and Hendrikse (2015) further analyzed the position of the CEO in a three-stage model comparing a cooperative with a member CEO, a cooperative with a non-member CEO, and a firm. The model considers the process of project evaluation and acceptance, where the project is first presented to the CEO and then to the board of directors. Judgment of the expected payoff of the project is in part determined by the positive and negative bias toward upstream and downstream activities, respectively, by the member CEO and vice versa by the non-member CEO. Bias implies an error in the internal valuation of projects. Efficiency of each governance structure, as given by expected payoff, is dependent on the magnitude of managerial bias, the difference in managerial bias (between the CEO and the board directors), and the upstream or downstream nature of the project. According to the main result, a member CEO is most appropriate if the majority of the growth potential is in the upstream segment of the value chain, while a non-member CEO is appropriate if the cooperative will invest in downstream activities.

3.3 CEO payment

Kopel and Marini (2014) contributed to the discussion on CEO payment, but from the perspective of a consumer cooperative that is not engaging in forward or backward integration. Kopel and Marini (2014) demonstrated that a variable pay contract for the non-member CEO has a detrimental impact on the cooperative, as the explicit emphasis on financial performance is in direct opposition to member patron utility parameters. Instead, it is in the best interest of the consumer cooperative to offer a fixed wage to an internal CEO whose objective is to set price equal to marginal cost. Thus, using backward induction, the consumer cooperative is never expected to hire a non-member CEO. Comparatively, the firm charges a higher price and sells a lower output in the final stage of the game as compared to the cooperative, for which profit is relatively low.

Fulton and Pohler (2015) also applied emphasis on the manager in their three-stage model, where the manager bonus is set in stage one, managerial effort is chosen in stage two, and member patron utility is determined in stage three by the price and quality of the product. As compared to firm shareholders, member patrons and board directors have greater incentive to monitor management as risk bearing is much higher, but the quality and quantity of monitoring is also impacted by off-farm income and age. From a managerial perspective, the combined impact of governance and remuneration is dependent on the utility and sensitivity of the manager. Furthermore, considering the ambiguous nature of performance, remuneration tied to performance is unlikely to fully align principal and agent interests, which implies monitoring is critical to the economic viability of the cooperative.

The importance of monitoring is also illustrated by Hueth and Marcoul (2015), who built a multi-task, five-stage model to find parameters for the optimal alignment of interests in the principal–agent relationship. Unlike the previous three publications, however, there is no explicit discussion of CEO wage or CEO bonus. In addition to a monitor, each organization is modeled to have an entrepreneur and an input supplier, which for the cooperative is the same individual. Because risk bearing is relatively high, each member patron has strong incentive to monitor the behavior of the entrepreneur. In fact, the quantity of monitoring can offset any deficiency in its quality by the board of directors. Because board directors are also member patrons, director–manager collusion is less likely as compared to the firm. By extension, the model indicates agency cost is relatively low for the cooperative, thus explaining why some transactions (projects) are governed by the cooperative and other transactions are governed by the firm.

4 Management

Related to governance, recent theory is also developed to inform the management and deployment of joint assets by managers and executives for the benefit of member patrons. As compared to the theory discussed in the previous section, the next publications do not place managerial action or behavior within parameters of monitoring or principal–agent interests. CEO identity, CEO payment, and member patron heterogeneity are exogenous to the formal analyses. Instead, management behavior is often analyzed in relation to risk or vertical coordination.

4.1 Risk

As compared to the risk of input supply or market access, Ligon (2009) argued production risk management is typically suboptimal in the cooperative, which is especially problematic when the quantity and quality supplied by its member patrons is susceptible to great variability and uncertainty. The formal solution to the problem is defined by proportionality of income to average patronage, not current patronage. Full risk sharing implies below expected yield in period t is buffered by mean past yield in periods $t - k$, and member patron i will be subsidized by member patron j, which intensifies the concept of group action. Of course, such risk sharing by the collective is likely to inspire several problems, including the free-rider problem and the influence problem. Consequently, if production risk sharing is to be at all feasible, the cooperative must also consider exclusive long-term supply agreements so as to dissuade member patrons with above expected yield from exiting.

Hovelaque et al. (2009) also analyzed risk management by means of contracting with member patrons who produce a differentiated good. The constrained supply chain model contains three elements: (1) the objective function of the cooperative, (3) the consumer–cooperative relationship, and (3) the member–cooperative relationship. The cooperative must determine how much of the basic product and how much of the differentiated product to produce dependent on stochastic demand. As member supply is unconstrained, cooperative profit is only superior to firm profit in case of a price increase of the non-differentiated good. The solution to farm risk management is the extension of individualized spot price contracts, which allow member patrons to align risk preferences to expected risk in the stochastic market environment. As compared to the basic contract, the individualized contract is estimated to increase the mean price as well as its standard deviation.

4.2 Vertical coordination

Feng and Hendrikse (2012) developed a multi-task principal–agent model to address differences in corporate and cooperative governance. The model consists of a two-stage non-cooperative game, where the principal chooses the optimal incentive in the first stage and the agent chooses the optimal action in the second stage. As usual, the agent is assumed to maximize expected utility, while farm profit maximization is the objective of the principal. Vertical integration is the key variable, and optimality of the organizational mode is determined in part by the complementarity of the upstream and downstream stages. If the downstream stage is not complimentary to the upstream stage, its value is not obvious to member patrons and the CEO will have limited incentive to invest. If the CEO does invest, the production and cost functions must be complimentary or the cooperative will risk relative inefficiency. The model implies management should not pursue non-member business if farm profit maximization or member return optimization is the true objective.

Dietl et al. (2013) developed a four-stage model to explain the cooperative mode of organization in terms of knowledge. Two variations of the model are considered: (1) two producers who collectively own the processing plant, and (2) two producers who independently supply a firm. A distinction is made between generalizable and non-generalizable knowledge, where the latter implies human asset–specific investment, which is often necessary for vertical expansion. Overall, the model concludes that the cooperative acquires less non-generalizable knowledge than the market, but more generalizable knowledge than the market if there is sufficient incentive for large member patrons. If so, the generated net welfare surplus is optimal if the impact of the knowledge on production cost is sufficiently large. The model thus explains cooperative investment in non-member business activities.

5 Finance

While the literature on corporate finance is in an advanced stage of development, the same is not true of cooperative finance. Yet cooperative finance is distinct from corporate finance, in part because of the dual function of organized farm producers as both patrons and capitalists. Thus, unlike the firm, the cooperative is not characterized by a clear separation of control and finance, which has severe implications for its capital structure. In order to better understand the cooperative debt or equity decision, recent theoretical contributions have placed emphasis on the tension between the desire to patronize and the obligation to capitalize the cooperative.

The lone exception is by Evans and Guthrie (2006), who advanced a dynamic theory of the cooperative by placing emphasis on the equity problem inherent to the ownership structure of traditional cooperatives. According to the authors, most cooperatives face three sources of inefficiency: (1) overproduction as marginal cost is equated to average revenue and not marginal revenue, (2) underproduction as the cost of owned capital is subsidized by other member supplies, and (3) overproduction as the return on owned capital is determined by current and not past patronage. According to the theoretical model, inefficiency is solved by fair value share pricing, which implies ownership is valued at the current value of future earnings. Evans and Guthrie (2006) thus advocate the implementation of ownership transferability and equity appreciability, which are both deviations from the capital structure of the traditional cooperative.

Rey and Tirole (2007) first provided a theoretical contribution to the analysis of free-rider and horizon problems in open and closed cooperatives by developing a two-period framework. Consistent with property rights theory, growth investment in period $t-1$ is suboptimal in the open cooperative if less than 100 percent of the benefit is appropriable in period t. In fact, the cooperative may not even be formed as the new generation of member patrons in period t appropriate some of the rent generated by the previous generation of member patrons. In case of member patron discrimination, a large membership fee for new member patrons in period t is necessary to incentivize new and existing member patrons in period $t-1$ to make necessary investments in long-term growth.

In contrast to Rey and Tirole (2007), Olesen (2007) challenged the common assumption of underinvestment by member patrons with short horizons. In fact, Olesen (2007) argued the horizon problem is more likely to cause overinvestment as opposed to underinvestment. However, the alternative hypothesis is dependent on the availability of an exit payment, which is determined in the period before investment. If the exit payment is at least as large as the original investment, member patrons with some probability of exit have incentive to invest redeemable equity. However, as indicated by the model, the exit payment may facilitate liquidation of the cooperative if too many member patrons exit, suggesting a large reserve of unallocated equity is necessary to provide stability.

Fulton and Giannakas (2012) extended the formal discussion of the horizon problem by placing emphasis on the objectives of the member patrons of consumer cooperatives. Similar to Rey and Tirole (2007), Fulton and Giannakas (2012) built a two-period framework to model the interactions of two generations of member patrons. According to the model, investment in each period is impacted by the horizon problem, which increases the cost of equity and thus also increases the necessary return on equity to incentivize member patron investment. However, because the formation of cooperatives is often motivated by a lack of market alternatives for goods with inelastic demand, the model indicates the negative impact of the horizon problem may not be as severe as long as the consumer surplus generated by the cooperative is large enough.

Finally, Giannakas et al. (2016) also concluded the formation of the cooperative is dependent on the length of the time horizons of the first member patrons. If the expected payoff is too far in the future, *ex ante* investment in joint assets is not an optimal strategy for the individual farm producers. If the horizon problem is solved, a free-rider problem emerges as new member patrons appropriate part of the rent generated by existing member patrons. According to the model, the best response is in part determined by the impact of organizational size on income. If the operation is defined by size economies, it is in the best interest of existing member patrons to not impose any entry barriers. By contrast, the enforcement of membership fees or base capital structures is optimal if an increase in membership size is detrimental to operational efficiency. The authors thus explain why many dairy, fruit, vegetable, and nut marketing and processing cooperatives, for which returns to scale are rarely increasing, often implement some degree of closed membership.

6 Summary and conclusion

In general, the 29 articles we reviewed in this chapter employ the research approach King (2012) characterizes as "economic analysis," where a minimal set of assumptions and rigorous analytical reasoning result in an efficiency-oriented set of policy or strategic implications. There is, however, a trend toward a complementary "economic design" process motivated by seeking solutions to problems identified in the "economic analysis" approach. Rather than solely focusing on what is, such articles begin with a purpose to identify what outcomes might yield satisfactory results. Such a method opens pathways to bridge the gap between outreach–engagement and research. Advances in utilizing more behavioral and institutional branches of applied economics temper the risks of pursuing the "what could be" or "what ought to be" types of academic output. Our review highlights a number of these economic design advances.

As indicated by the reviewed publications, another general development in the theoretical literature is the flexible or multidimensional conceptualization of the agricultural cooperative. Previous work usually approached the cooperative as (1) an extension of the farm, (2) an independent firm, or (3) a coalition of farms. Recent theory has departed from such rigid conceptualizations and instead approached the cooperative as a complex organization with multiple and competing objectives that may or may not allow a stable solution. Many studies have emphasized a single specific parameter or constraint, either by itself or in relation to some objective of the cooperative, while price and quantity no longer serve as the de facto outcome variables. Instead, theoretical work is often advanced to find solutions to problems of product quality, supply commitment, or member equity investment.

Theorists thus increasingly consider the real multidimensional nature of agricultural cooperatives, suggesting the gap between theory and practice is perhaps closing. Altogether, the primary purpose of theoretical work is arguably to help inform or explain the various challenges and opportunities faced by agricultural cooperatives in the increasingly global and complex marketplace. Although the general ability to test theories and frameworks in practice is hampered by the limited availability of sophisticated data, the reviewed publications in our chapter should provide inspiration for future empirical as well as theoretical research on agricultural cooperatives.

References

Agbo, M., D. Rousseliere, and J. Salanié. 2015. "Agricultural Marketing Cooperatives with Direct Selling: A Cooperative – Non-Cooperative Game." *Journal of Economic Behavior & Organization* 109:56–71.

Bogetoft, P. 2005. "An Information Economic Rationale for Cooperatives." *European Review of Agricultural Economics* 32(2):191–217.

Bontemps, P., and M. Fulton. 2009. "Organizational Structure, Redistribution and the Endogeneity of Cost: Cooperatives, Investor-owned Firms and the Cost of Procurement." *Journal of Economic Behavior and Organization* 72:322–343.

Cook, M.L., F.R. Chaddad, and C. Iliopoulos. 2004. "Advances in Cooperative Theory since 1990: A Review of Agricultural Economics Literature." In G.W.J. Hendrikse, ed., *Restructuring Agricultural Cooperatives*. Rotterdam, The Netherlands: Erasmus University, pp. 65–90.

Deng, W., and G.W. Hendrikse. 2013. "Uncertainties and Governance Structure in Incentives Provision for Product Quality." In *Governance of Alliances, Cooperatives and Franchise Chains*. New York: Springer, pp. 179–203.

———. 2015. "Managerial Vision Bias and Cooperative Governance." *European Review of Agricultural Economics* 42(5):797–828.

Dietl, H.M., T. Duschl, M. Grossmann, and M. Lang. 2013. "Explaining Cooperative Enterprises through Knowledge Acquisition Outcomes." *Managerial and Decision Economics* 34(3–5):258–271.

Drivas, K., and K. Giannakas. 2010. "The Effect of Cooperatives on Quality-Enhancing Innovation." *Journal of Agricultural Economics* 61(2):295–317.

Evans, L., and G. Guthrie. 2006. "A Dynamic Theory of Cooperatives: The Link Between Efficiency and Valuation." *Journal of Institutional and Theoretical Economics* 162(2):364–383.

Fatas, E., F. Jimenez-Jimenez, and A. Morales. 2010. Blind Fines in Cooperatives. *Applied Economic Perspectives and Policy* 32(4):564–587.

Feng, L., and G.W. Hendrikse. 2012. "Chain Interdependencies, Measurement Problems and Efficient Governance Structure: Cooperatives Versus Publicly Listed Firms." *European Review of Agricultural Economics* 39(2):241–255.

Fulton, M., and K. Giannakas. 2007. "Agency and Leadership in Cooperatives: Endogenizing Organizational Commitment." In *Vertical Markets and Cooperative Hierarchies*. Dordrecht, Netherlands: Springer Netherlands, pp. 93–113.

———. 2012. "The Value of a Norm: Open Membership and the Horizon Problem in Cooperatives." *Journal of Rural Cooperation* 40(2):145–161.

———. 2013. "The Future of Agricultural Cooperatives." *Annual Review of Resource Economics* 5(1):61–91.

Fulton, M., and D. Pohler. 2015. "Governance and Managerial Effort in Consumer-owned Enterprises." *European Review of Agricultural Economics* 42(5):713–737.

Giannakas, K., and M. Fulton. 2005. "Process Innovation Activity in a Mixed Oligopoly: The Role of Cooperatives." *American Journal of Agricultural Economics* 87(2):406–422.

Giannakas, K., M. Fulton, and J. Sesmero. 2016. "Horizon and Free-Rider Problems in Cooperative Organizations." *Journal of Agricultural and Resource Economics* 41(3):372–392.

Hovelaque, V., S. Duvaleix-Tréguer, and J. Cordier. 2009. "Effects of Constrained Supply and Price Contracts on Agricultural Cooperatives." *European Journal of Operational Research* 199(3):769–780.

Hueth, B., and P. Marcoul. 2006. "Information Sharing and Oligopoly in Agricultural Markets: The Role of the Cooperative Bargaining Association." *American Journal of Agricultural Economics* 88(4):866–881.

———. 2015. "Agents Monitoring their Manager: A Hard-Times Theory of Producer Cooperatives." *Journal of Economics & Management Strategy* 24(1):92–109.

Hueth, B., and G. Moschini. 2014. "Endogenous Market Structure and the Cooperative Firm." *Economics Letters* 124:283–285.

Karantininis, K., and A. Zago. 2001. "Endogenous Membership in Mixed Duopsonies." *American Journal of Agricultural Economics* 83(5):1266–1272.

King, R.P. 2012. "The Science of Design." *American Journal of Agricultural Economics* 94(2):275–284.

King, R.P., M. Boehlje, M.L. Cook, and S.T. Sonka. 2010. "Agribusiness Economics and Management." *American Journal of Agricultural Economics* 92(2):554–570.

Kopel, M., and M.A. Marini. 2014. "Strategic Delegation in Consumer Cooperatives Under Mixed Oligopoly." *Journal of Economics* 113(3):275–296.

Liang, Q., and G.W. Hendrikse. 2013. "Cooperative CEO Identity and Efficient Governance: Member or Outside CEO?" *Agribusiness* 29(1):23–38.

———. 2016. "Pooling and the Yardstick Effect of Cooperatives." *Agricultural Systems* 143:97–105.

Ligon, E. 2009. "Risk Management in the Cooperative Contract." *American Journal of Agricultural Economics* 91(5):1211–1217.

Mérel, P.R., T.L. Saitone, and R.J. Sexton. 2009. "Cooperatives and Quality-Differentiated Markets: Strengths, Weaknesses, and Modeling Approaches." *Journal of Rural Cooperation* 37(2):201.

———. 2015. "Cooperative Stability Under Stochastic Quality and Farmer Heterogeneity." *European Review of Agricultural Economics* 42(5):765–795.

Olesen, H.B. 2007. "The Horizon Problem Reconsidered." In Kostas Karantininis, and Jerker Nilsson (eds.). *Vertical Markets and Cooperative Hierarchies*. Netherlands: Springer, pp. 245–253.

Rey, P., and J. Tirole. 2007. "Financing and Access in Cooperatives." *International Journal of Industrial Organization* 25(5):1061–1088.

Saitone, T.L., and R.J. Sexton. 2009. "Optimal Cooperative Pooling in a Quality-Differentiated Market." *American Journal of Agricultural Economics* 91(5).1224–1232.

Sexton, R.J. 1984. "Perspectives on the Development of the Economic Theory of Co-operatives." *Canadian Journal of Agricultural Economics* 32(2):423–436.

Staatz, J.M. 1989. Farmer Cooperative Theory: Recent Developments. ACS Research Report No. 84. Washington, DC: U.S. Department of Agriculture.

41
Agribusiness economics and management

Michael Boland and Metin Çakır

1 Introduction

Since the Applied and Agricultural Economics Association (AAEA) created sections in 2001, the Agribusiness Economics and Management (AEM) section has consistently been the first or second most popular section as measured by numbers of members. The AAEA gave agribusiness its own topical heading in 1987. King et al. (2010) provided an excellent summary of AEM scholarship over the first 100 years of the AAEA. In this review, we build upon that work by providing a deeper synthesis of the notable studies on demand analysis and agribusiness management, with the aim of informing future research. An important conclusion drawn from this discussion is that agribusiness economics and industrial organization, once considered separate research tracks, no longer having any meaningful distinction from each other. Thus, our objective is to provide a summary history of AEM and discuss current and future issues in this field, with a particular emphasis on doctoral training and the needs of faculty conducting research in AEM.

King et al. (2010) discussed AEM scholarship in nine topical areas: economics of cooperative marketing and management; design and development of credit market institutions; organizational design; market structure and performance analysis; supply chain management and design; optimization of operational efficiency; development of data and analysis for financial management; strategic management; and agribusiness education. Agribusiness is included in the *Journal of Economic Literature* (*JEL*) codes under Code Q and, in particular, Q13, Agricultural Markets and Marketing, Cooperatives, and Agribusiness. However, AEM research can also be found under Code L, Industrial Organization, including (1) L1 Market Structure, Firm Strategy, and Market Performance; (2) L2 Firm Objectives, Organization, and Behavior; and (3) L6 Industry Studies: Manufacturing; and in particular Code L66, Food, Beverages, Cosmetics, Tobacco, Wine, and Spirits. Other occasional *JEL* codes might be D23, Organizational Behavior, Transaction Costs, or Property Rights, or D85, Network Formation and Analysis: Theory.

The word "agribusiness" has been widely debated since the creation of the word in 1957 by John Davis and Ray Goldberg (p. 2), who defined agribusiness as

> the sum total of all operations involved in the manufacture and distribution of farm supplies; production operations on the farm; and the storage, processing, and distribution of

farm commodities and items made from them. Thus, agribusiness essentially encompasses today the functions which the term agriculture denoted 150 years ago.

This definition includes our discussion on AEM and industrial organization. Goldberg (2018) has written extensive case studies on agriculture and food issues and has been nominated for the World Food Prize. Joe Bain's (1950, 1951) classic book on industrial organization and its concepts were influential for many AEM faculty, and these were reinforced by Michael Porter's (1980, 1985) work on strategy, which was based on industrial organization concepts. The next two sections review two aspects of AEM research that were not widely discussed in the King et al. (2010) work.

2 The importance of understanding demand in AEM research

Changing consumer trends have been the primary force of transformation in the contemporary food economy in the United States and worldwide.[1] As a result, understanding food demand and food retailing have become central topics of AEM research. Two decades ago, Kinsey and Senauer (1996) documented key consumer trends and the dramatic changes in the food retailing landscape and argued that

> the food system has shifted 180 degrees from being producer driven to being consumer driven. The power in the system is at the retail end because retailers receive the information about consumers' preferences first. This information gives them the power to compete with other retailers, to negotiate with vendors and to respond to consumers. (p. 1190)

Kinsey and Senauer further argued that the transmission of information from downstream to upstream would facilitate vertical coordination in the food system.

More recent reviews written by other authors share the same view that food and agricultural markets have become more consumer-oriented and consumer expectations of food quality are increasingly higher and diverse (Costanigro and McCluskey 2011; Unnevehr et al. 2010). Sexton (2013), in his presidential address to the Agricultural and Applied Economics Association, AAEA, identifies three key characteristics of modern agricultural markets in the United States. Two of the features are consumers' increasing demand for quality and differentiation, and, in response, firms' increasing vertical coordination and control via use of production and marketing contracts at different stages of the food system. In addition, Sexton highlights increasing concentration and consolidation in the vertical food supply chain, particularly at the retailing and processing stages. Consequently, these trends, which have been in the works for a few decades, impacted agribusiness economics research to increasingly focus on understanding consumer preferences and food retailer pricing, using the empirical tools of consumer theory and industrial organization.

The grocery retailing industry first witnessed the replacement of small grocery stores by larger grocery chains, which was referred as the "supermarket revolution," and then, starting from 1970s, the industry witnessed a proliferation of food retailer outlets with emerging supercenters, warehouse clubs, and mass merchandizers (Connor 1999; Marion 1986). In today's markets, around 70 percent of all food-at-home sales occur in traditional food stores such as supermarkets, convenience stores, and specialty stores.[2] Warehouse clubs and supercenters have around 16 percent and drugstores about 6 percent of all food-at-home sales (Economic Research Service, U.S. Department of Agriculture 2017). Most recently, the share of online grocery sales has grown substantially.

The food retailing industry has been constantly evolving, driven by the forces of demand and rivalry. The industry is being designed to respond to consumers' needs and wants directly and quickly while keeping low prices by eliminating inefficiencies in procurement, inventory management, and distribution. In her presidential address to the AAEA, Kinsey (2001, p. 1120) described the evolution of the food retailing landscape thusly: "At each juncture, retail products and selling formats that better served consumers' needs and save them time came to dominate the landscape."

2.1 Consumer trends and evolving markets

Retailers face monumental challenges to stay competitive as the rivalry is intense, consumer preferences are increasingly diverse, and the contemporary food shopping environments are complex. There is voluminous literature on the impacts of retailer strategies and factors affecting consumer choices. For example, studies that focus on why retailers own store brands and how they succeed found that retailers use store brands to generate store differentiation and store loyalty (Corstjens and Lal 2000), improve efficiency in the vertical chain by eliminating double marginalization[3] (Mills 1995), and improve bargaining position for purchases of national brands (Narasimhan and Wilcox 1998). The success of store brands varies greatly by retailer and by product category. Store brands tend to be more successful in categories that are large, not widely advertised, have fewer manufacturers and high margins (Hoch and Banerji 1993). Similarly, the closely related literature on understanding purchase behavior documents a large amount of evidence on how socio-demographic, economic, and environmental factors affect people's food choices. For example, studies that focus on factors impacting healthful food choices show that price, income, and other socio-demographics such as race and education are among significant factors that explain consumer choices (Andreyeva et al. 2010; Beatty et al. 2014; Cullen et al. 2007; Darmon and Drewnowski 2008). In addition, studies find that other factors, such as store format and shopping frequency, have significant effects on the healthfulness of food purchases (Volpe et al. 2013; Rudi and Çakır 2017).

The primary driver of the complexity of modern food markets is that consumers' perception of food quality is multidimensional – one that has expanded from conventional attributes such as taste and appearance to healthfulness, genetics, convenience, and consistency of food, as well as to characteristics that define where and how food is produced (Kinsey 1993, Sexton 2013). In response, in order to meet consumers' diverse preferences and capture quality premiums, firms introduce new products and redesign existing ones. This creates a dynamic market environment in which effective communication of product information to consumers via marketing activities, such as advertising, branding, labeling, and packaging, becomes critical to firms' ability to capture quality premiums. Unnevehr et al. (2010) provide an excellent review of the literature, focusing on the role of information and quality in affecting consumer choices.[4]

2.2 Current research topics in industrial organization

Some of the key questions that AEM economists are interested in answering are: what types of consumers buy what types of products from which types of stores? How do firms compete horizontally and vertically? How important are consumer search costs? How do consumers respond to non-price marketing strategies? How important is brand loyalty (or how high are consumer-switching costs)? What are the impacts of mergers and acquisitions on prices and markup? How to measure gains from a new product? How to assess effects of a new tax? There is a long history of the application of industrial organization and demand models to answer these and other related questions. Early literature used the structure conduct performance, or SCP, paradigm that focuses on cross-section study of many industries to analyze firms' strategies and the competitiveness of markets (see Caswell 1992 and Connor 1999 for reviews). The SCP studies are critiqued for inherent limitations of cross-section approach in estimating structural parameters of market conduct and their use of accounting data to measure performance. Starting from the 1990s, the profession saw an increasing number of studies evaluating the competitiveness of agricultural markets using New Empirical Industrial Organization (NEIO) models. A strand of this literature uses numerical simulation models to evaluate the implications of market power exertion at different levels of the food supply chain for welfare and public and business policies (Alston et al. 1997; Çakır and Nolan 2015; Sexton et al. 2007; Sexton and Zhang 2001), while another strand uses structural econometric models to estimate the degree of firms' market power exertion (Çakır and Balagtas 2012; Mérel 2009; Muth and Wohlgenant 1999). The NEIO models are typically applied to homogenous goods markets using market level data. Thus, they are severely limited in providing any micro-level inferences on firms' competitive strategies, such as branding, advertising, and the introduction of new products. In fact, the limitations of SCP and NEIO models have been considered among the primary reasons for the historically perceived separation of industrial organization and agribusiness research. Caswell (1992) notes that

> differences in traditional focus, as well as in modeling approaches, contribute to the idea that industrial organization and agribusiness research are separate strains. Industrial organization research focuses primarily on market performance and in the recent past has emphasized cross-sectional research that evaluates performance across industries. Case study work has been a part of this research tradition but has only recently regained some prominent ground. Agribusiness research, on the other hand, primarily focuses on firm strategy and firm performance with the term competitiveness coming into widespread use to summarize these issues. Being based upon firm experience, agribusiness research necessarily focuses more heavily on case studies with cross-sectional work being secondary.
>
> (p. 539)

In her review, Caswell argues that industrial organization and agribusiness research have much in common by identifying the primary areas in which they intersect and discussing the contributions of studies that fall in those areas. After 25 years, we have seen that the historical distinction has almost completely disappeared since firm level analysis has become central to the industrial organization field due to improved techniques in estimation of structural demand and supply models, increased availability of highly granular micro-level data, and improved computing power. In particular, advancements in the estimation of demand systems in differentiated product markets opened up vast opportunities for research in agribusiness economics.

There are two general approaches to estimating demand in differentiated product markets: a product space approach and a characteristics space approach. The product space approach assumes that consumers have preferences over products and is widely employed to estimate market-level demand using models such as the Almost Ideal Demand System (Deaton and Muellbauer 1980). This approach has two important limitations in estimating demand for differentiated products. The first is the large number of parameters to be estimated in order to capture substitution patterns if the product space is large, known as the dimensionality problem. This problem can be alleviated to a certain degree by adopting a multi-stage budgeting approach to modeling demand (Gorman 1971). Hausman et al. (1994) provide a classic application of the approach to estimate demand for differentiated products in the U.S. beer market. The second limitation is that the consumer heterogeneity can only be incorporated at the aggregate level, measuring the influence of average demographics on demand instead of explicitly reflecting the individual heterogeneity.

The characteristics space approach to estimating demand in differentiated product markets overcomes the dimensionality problem and is able to address heterogeneity in preferences. In this approach, the consumer utility is represented as a function of product characteristics, a composite commodity that represents the outside option and an individual-specific random taste shock. The probability distribution assigned to the random taste shock gives rise to different discrete choice models. Based on the assumed distribution, the researcher estimates the probability of an individual choosing the product that gives the highest utility. The market demand for each product is then obtained by numerical aggregation of the estimated probabilities over individuals. The approach offers a solution to the dimensionality problem by projecting products onto characteristics space. This allows recovery of substitution patterns across large product categories by estimating parameters only on product characteristics and prices. In addition, the approach allows incorporating individual heterogeneity by expressing model parameters as functions of household characteristics.

The logit model and its more flexible forms, such as the nested logit and the mixed logit models (McFadden 1974; McFadden and Train 2000; Train 2003), have become workhorse models of demand in differentiated product markets primarily due to advancements in estimation procedures made by Berry (1994), Berry et al. (1995, 2004), Nevo (2001), and Petrin and Train (2010). These models are used to study a large number of topics, such as (1) firms' pricing strategies and collusive behavior (Nevo 2001, *ready-to-eat breakfast cereal*; Kim and Cotterill 2008, *processed cheese*); (2) vertical relationships (Villas-Boas 2007, *supermarket*; Bonnet et al. 2013, *ground coffee*); (3) price transmission (Hellerstein 2008, *beer*; Nakamura and Zerom 2010, *coffee*); (4) competition in multiple strategic tools (Draganska and Jain 2005, *yogurt*; Richards and Hamilton 2015, *ready-to-eat breakfast cereal*); (5) consumer response to labeling, branding, and packaging (Allender and Richards 2012, *ice cream and soda*; Çakır and Balagtas 2014, *bulk ice cream*; Hainmueller et al. 2015, *fair trade*; Zhu et al. 2015, *ready-to-eat breakfast cereal*); (6) promotion and advertising (Dubois et al. 2017, *potato chips*); (7) impacts of public policy on prices and consumption (Bonnet and Réquillart 2013, *soda*; Villas-Boas 2009, *coffee*); (8) mergers and acquisitions (Nevo 2000, *ready-to-eat breakfast cereal*; Dubé 2005, *soda*); (9) introduction of new products (Nevo 2003, *ready-to-eat breakfast cereal*; Wang and Çakır 2017, *apples*); (10) entry (Hausmann and Leibtag 2007, *supermarket*); and (11) food store choice (Turolla 2016; Taylor and Villas-Boas 2016).

Discrete choice models are limited by the assumption that consumers are assumed to purchase one unit of the product that gives the highest utility. This assumption may be tenuous for a class of problems, such as purchases of random weight products (e.g., fruits, meats, cheese, etc.), where consumers first make a discrete choice of which alternative to buy and then decide how much to buy. This behavior is modeled in a discrete–continuous choice modeling framework (Dubin and McFadden 1984; Haneman 1984; Richards 2000). Alternatively, a multiple-discrete continuous choice modeling is used if consumers are assumed to choose a number of different alternatives and purchase a continuous amount of each (Bhat 2005, 2008; Richards et al. 2012).[5]

Our discussion has primarily focused on the increasing importance of the demand analysis in differentiated product markets for agribusiness research. However, it should be noted that the wide availability of panel datasets also facilitated adoption of reduced form econometric models such as the fixed effects and difference-in-differences. For example, recent notable studies use the difference-in-differences approach to investigate issues pertaining to grocery retailer mergers

(Allain et al. 2017; Çakır and Secor 2018), advertising (Dhar and Baylis 2011), food product recalls (Villas-Boas and Toledo 2016), and ecolabeling (Hallstein and Villas-Boas 2013).

3 Understanding the impact of management in firm performance

The inclusion of management in "agribusiness economics and management" as opposed to just "agribusiness economics" is important. Management education is a crucial part of the uniqueness of the food economy. The ability to attempt to isolate the effect of management on profitability continues to be needed but neglected in research due to lack of data. Boyd et al. (2007) have the most current literature review of understanding the role of management and its impact on profitability. This research has used farm-level data and agribusiness-level data. One stream of research has focused on identifying financial and management variables that affect profitability. A second stream of research has focused on isolating the impact of management variables on profitability. A third stream of research has focused on identifying industry, diversification, corporate, and firm-specific variables and measuring their impact on profitability. Many of these studies have provided recommendations for producers who manage farming operations and managers of food businesses and agribusinesses. Virtually all of the studies have used cross-sectional, time series data to determine the impact on performance.

Many studies have been conducted examining relationships between financial ratios and management factors and various performance measures. These studies used a variety of statistical (e.g., equality of means testing, stochastic dominance) and econometric and statistical procedures (e.g., regression, discriminant analysis) to answer questions related to firm growth (Baab and Keen 1982; Schrader et al. 1985; Van Dyne and Rhodes 1987; Lerman and Parliament 1990; Barton et al. 1993; Holmes 1994; Ginder and Henningsen 1994; Harris and Fulton 1996; Forster 1996; Parcell et al. 1998; and Ariyaratne et al. 2000). In general, these studies have found that firms with greater profitability were less leveraged, less diversified, and had better liquidity management. Only Barton et al. found a significant relationship between firm size and performance.

Cooperatives have been the most frequently studied type of business form by AEM faculty. There is a long history of public–private partnerships on extension, research, and teaching within departments of agricultural economics. Boards of directors set policy on the amount of equity a cooperative maintains on its balance sheet. Directors, through decisions on asset investment and equity management, decide the amount of interest expense, patronage payable, debt repayment, and similar variables. Factors such as equity and total assets are determined by the decisions of directors. Variables such as net margin (e.g., return on sales), asset turnover, and current ratio are more under the control of a manager. There has been a steady stream of doctoral research on cooperatives, as noted by Boland and Crespi (2010). More recently, this has included institutional design (Chaddad and Cook 2004).

None of these studies measured specific management factors as variables (i.e., prices paid for specific inputs or received for outputs, a productivity measure such as employee productivity, etc.). Agribusinesses have multiple plants and locations and often buy in bulk for all locations, making it more difficult to isolate productivity measures. It is difficult to measure management, and thus management is often an omitted variable and part of the unexplained variation in these models. Thus, there have been fewer studies that have estimated the statistical relationship between performance and various explanatory variables.

Given these difficulties, the management literature has used regression analysis to measure the statistical relationship between performance and various independent variables. Performance as measured by return on assets (e.g., McGahan and Porter 2002; Schumacher and Boland 2005) or Tobin's q (McGahan 1999) is divided into variables representing average profits that accrue to all firms in a given industry (industry effects) and average profits that accrue to firms that are diversified (corporate effects), and the residual profits are assumed to accrue to firms with better (or worse) management of resources. These studies have found that greater performance comes from the industry a firm operates in (e.g., industry membership) than whether a firm is diversified. The residual returns are important but not as important as industry membership in determining profitability.

4 The use of decision cases

A number of decision cases written by AEM faculty, which are used by teachers, have been published in *Review of Agricultural Economics* or *RAE* (now the *Applied Economics Perspectives and Policy*), *American Journal of Agricultural Economics* or *AJAE*, *Journal of Natural Resources and Life Sciences Education* (*JNRLSE*), *International Food and Agribusiness Management Review* (*IFAMR*), and *Case Research Journal* (*CRJ*). The first three journals have AAEA sponsorship. The *RAE* published case studies from 1996 to 2010 and the *AJAE* from 2011 to present. The *JNRLSE* and *IFAMR* have continually published cases over time, and the *CRJ* is the highest ranked decision case journal. However, the largest set of cases written by AEM faculty were published in the *RAE*. We did not review the cases published by Harvard Business School Publishing since their form for cases is not similar to that academic journals publish. We examined cases where teaching notes were available to learn more about what topics AEM faculty were interested in over the past 20 years.

The *RAE* was the largest publisher of case studies, with 72 decision case studies published over the 1996 to 2010 time period, with all but 15 published before 2005. The lack of published cases after 2005, coupled with the elimination of cases in *AEPP* in 2010, suggests that there were far fewer submissions or acceptable teaching cases during this period. The largest topical area was production agriculture, with 26 cases written with the farmer as a manager facing (1) an investment decision such as a robotic milker (Hyde et al. 2007), (2) a change in strategic orientation (McFadden et al. 2009; Ehmke et al. 2004), and (3) portfolio selection of enterprises (Pritchett 2004; Gould and Carlson 1998). However, the largest set of farm decisions were vertical coordination, including (1) production contracts for dry beans (Chambers and King 2002) or marketing contracts including hard white winter wheat (Taylor et al. 2005), malt barley (Boland and Brester 2006), pork production (Buhr 2004; Swinton and Martin 1997), and asparagus (Worley et al. 2000) and (2) vertical integration including lamb processing (Boland et al. 2007), dry bean processing (Boland et al. 1998), sugar beet processing (Brester and Boland 2004), and pineapple (Piana et al. 2005). Cooperatives were the largest type of agribusiness firm considered in these cases, with 9 of the 14 cases written from the firm perspective, including Sun-Maid Growers (Sanchez et al. 2008), Sunkist Growers (Pozo et al. 2009), and Coopuxe (Chaddad and Boland 2009).

Beginning in 2011, the *AJAE* began an annual invited case competition with a selection process as part of their annual meeting, with several papers having the possibility of being published and with a broader mission of case topics. AEM faculty have published cases on farm succession planning (Yeager and Stutzman 2014), horizontal integration (Secor and Boland 2017), and business development (Darby et al. 2013). The *IFAMR* has accepted cases since its inception in 1991, with a wide variety of case formats. It is much harder to categorize what is a case in these journals since they are not readily identifiable, but we identified 68 articles that might be categorized as cases by AEM faculty on a wide variety of topics, with many similar to such as Neves et al. (2016).

5 Role of institutions in understanding AEM scholarship

An understanding of AEM scholarship requires understanding the framework of its institutions. Many of the institutions developed during two historical periods. The first was the late 1950s and early 1960s and then the second during the mid-1980s and early 1990s. Boland and Akridge (2004) summarized these developments. In 1955, the Board of the Foundation for American Agriculture appointed a committee entitled "Agribusiness Assembly, Teaching, and Research." Following the Davis and Goldberg book, various symposia followed, including the "National Study on Agribusiness Education" and a series of publications in the *Journal of Farm Economics* in the early 1960s. These helped spur the development of the early undergraduate programs in AEM and graduate theses and doctoral research. For example, one of the first such programs was at Montana State in 1961. As King et al. (2010) note, this research followed the Davis and Goldberg definition of agribusiness in its broadest sense with applications of linear programming and optimization techniques being widely used as techniques. The National Agribusiness Education Commission in the late 1980s sparked the development of further institutions, including graduate degree programs or specializations within existing graduate degree programs and executive educations programs such as the Center for Food and Agricultural Business at Purdue University, which was founded in 1986 (Downey 1989).

It can be argued that AEM faculty have more professional societies to choose from than any other branch of agricultural and applied economics. In addition to the AAEA, AEM faculty can choose to participate in the activities of the International Food and Agribusiness Management Association (founded in 1991) or Food Distribution Research Society (founded in 1968); and the U.S. Department of Agriculture multi-state research coordinating committee and information exchange group WERA-72, Agribusiness Scholarship Emphasizing Competitiveness. The addition of the AEM section to AAEA in 2001 provided further leadership with an annual section meeting. Many AEM faculty also participate in industry–academic events in the National Grocers Association, Produce Marketing Association, and National Council on Farmer Cooperatives.

The working papers commissioned by the National Food and Agribusiness Management Education Commission (NFAMEC) documented the growth in undergraduate AEM degrees and majors (Boland and Akridge 2004). While many departments in applied and agricultural economics have faculty who teach in AEM subjects, it is important to note that certain departments have significant depth in AEM subjects, which are important when analyzing the importance of academic institutions. John Davis and Ray Goldberg at Harvard University created the word agribusiness and began the long-standing Agribusiness Seminar in 1957. In 1993, the Harvard Kennedy School, led by Ray Goldberg, created the Private and Public Science, Academic, and Consumer Food Policy Group (PAPSAC) program in food policy.

Boland and Crespi (2010) conducted a census of every dissertation written in departments of agricultural economics through 2005. The authors analyzed agribusiness management, food business, and industrial organization topics and reported that these constituted 6.6 percent of all agricultural economics dissertations over the 1951 to 2005 time period. The authors identified 493 dissertations that were in the Agribusiness Management (172), Food Business (72), and Industrial Organization (249) subject categories. Agribusiness Management and Food Business were important subject categories in the 1950s, 1960s, and 1970s, but by the mid-1980s had decreased significantly, with only a handful of dissertations in each of these categories from 1987 to 2002; Purdue led the category with three times as many doctoral students as any other program. The majority of these dissertations, if published, were found in *Agribusiness: An*

International Journal, Journal of Agribusiness, International Food and Agribusiness Management Review, and *Journal of Food Distribution Research.*

In a report to the U.S. Department of Agriculture on the Higher Education programs national needs doctoral program from 1984 to 2002, Boland and Thielen (2004) identified 192 fellowships funded in the national need of marketing and management. Completed dissertations on 109 fellows were found, with Purdue, Kansas State, and Illinois ranking the highest, with 23 fellows in academic or government positions. Since 2001 (until 2017), the United States Department of Agriculture (USDA) changed the topical areas for the national needs such that management and marketing are no longer a subject category, but a strong argument can be made that this continues to be a strong national need. We analyzed the doctoral data reported by the NFAMEC reports, Boland and Crespi (2010), and National Needs Fellowship reports. Since 2005, there has been greater diffusion of fellowships and AEM dissertations among a far greater number of academic units.

Boland and Akridge (2004) found that Purdue faculty or faculty who had received their dissertation from Purdue wrote 83 percent of all undergraduate teaching materials, such as textbooks and cases. Purdue University operates the internationally recognized Center for Food and Agricultural Business, which conducts a full year of executive education programs for agribusiness firms. Faculty with extension appointments teach in the center. Center faculty teach in the joint MS/MBA degree program in agribusiness, taught executive-style with an annual cohort of 25 students or so. Only Kansas State offers a similar master's degree taught executive-style in agribusiness. Other programs, such as Texas A&M, Florida, Georgia, and Mississippi State, offer an on-campus agribusiness degree program. The ability to offer such programs suggests having a core group of faculty teaching and carrying out creative activities in AEM topics, especially in industrial organization. In 2017, virtually every academic unit whose roots are in agricultural economics have faculty doing research in AEM topics. This is true globally, with faculty from Wagenein and many other universities having faculty doing work on AEM issues.

6 Current and future issues

The need for faculty to teach AEM undergraduate courses is a major issue identified by professional society leaders and academic unit leaders and recognized as a national need by the USDA National Institute of Food and Agriculture. Currently, the number of academic jobs with an emphasis on undergraduate teaching or applied research, which may or may not have an Agricultural Experiment Station appointment, or some type of outreach or a formal extension appointment in some aspect of agribusiness extension such as business development, remain strong. Given the need for terminal degrees for accreditation purposes and the growing number of degrees, this need is going to remain strong across the world.

6.1 The food economy continues to be unique

The growing number of international students in graduate programs, which has been the case for several decades, as noted in Boland and Crespi's (2010) census of dissertations, suggests a lack of knowledge on the uniqueness of agriculture. Sporleder and Boland (2011) note that the unique factors that are historically associated with the agribusiness sector include: risk emanating from the biological nature of supply chains; the role of buffer stocks within the supply chain; the scientific foundation of innovation in production agriculture having shifted from capital to chemistry to biology; information technology influences on supply chains; that the prevalent market structure at the farm gate remains oligopsony; relative market power shifts in supply chains away from food manufacturers downstream to food retailers who operate as chain captains; and the globalization of agriculture supply chains.

These also include concepts such as marketing orders and the concept of a marketing year, which is not the same as a calendar year; the price discovery process with different market structure in different steps; the role of government policy programs; the prevalence of institutions and closely held firms such as cooperatives and family owned businesses; these and other concepts are critical for AEM teachers and researchers. Yet, these are typically taught, if at all, in undergraduate courses with the word agriculture as an adjective in front of the functional roles of policy, finance, marketing, management, and strategy. Ideally, these concepts would be integrated into topics taught in graduate programs to help faculty jumpstart their teaching careers.

6.2 Training of doctoral students for AEM academic positions

Arizona State and Texas A&M have doctoral programs in agribusiness but with very small student numbers, despite strong doctoral graduate students overall in other fields. Missouri has a doctoral program built around institutional economics but low numbers in the doctoral program. Traditional agricultural and applied economics programs have strong fields, with graduate students in development and natural resource or environmental economics, but many lack critical mass in applied production economics, industrial organization, agricultural marketing, or similar fields, which is where many AEM faculty had their roots. A typical doctoral program in agricultural and applied economics includes courses in microeconomics, macroeconomics, econometrics, and field courses. In addition, prerequisites in mathematical statistics and math are required.

These are useful and needed courses but do not answer the need for a deeper knowledge of agribusiness institutions, which is necessary for the overwhelming number of doctoral students who do not have a production agriculture background. Are there alternative programs to training doctoral students for careers in AEM? Most AEM faculty will enter academic positions with undergraduate teaching, and most major doctoral granting institutions have a Graduate Student Teaching certificate, which is distinct from its Graduate Teaching Assistant training programs. Certainly, two graduate courses in collegiate teaching would be useful for AEM faculty. Harvard's doctorate in business economics, which is distinct from its doctorates in management or economics, requires field courses in business history, industrial organization, and similar concepts, in addition to the usual microeconomic and macroeconomic courses. These field courses teach students to do scholarly research, often with case study methodology, in industry analyses and other industrial organization topics, whose publication outlets might be *The Business History Review*, *Economic Geography*, or similar academic journals. There is no such requirement in any existing doctoral program in agricultural and applied economics.

6.3 Reinvestment in teaching materials

Many textbooks and teaching materials are offered digitally. Yet, with an examination of copyright dates and content, it is obvious that there is a great need to update textbooks and teaching materials, in general, in AEM classes. In particular, teaching materials in the form of detailed lesson outlines and teaching notes are needed for instructors who lack knowledge of agriculture. In addition, because more and more students do not come from production agriculture backgrounds, knowledge about food topics such as aquaculture, nontraditional livestock such as goats, and perennial crops such as tree nut, pome, citrus, and stone fruits is critical for textbooks.

In addition, knowledge about industrial organization topics such as a changing retail grocery and supermarket systems and changing consumer demands should be a part of updated teaching materials. More than 20 courses in business strategy are taught compared with almost none in 1985. Business strategy courses typically play an integrative role in agribusiness programs, and yet, while these courses use business strategy textbooks, there are few teaching materials on food topics to complement these books. There is a clear need for a strategy textbook with an agribusiness and food focus. The same situation exists for food marketing, where monopolistic competition widely exists, as opposed to agricultural commodities, where perfect competition is an appropriate model. While many decision cases are readily available, they require updates. In addition, our observation is that the quality of teaching notes as measured by the ability of a non-case author to use the case successfully in the classroom is inadequate. Many appear to be written as an afterthought once the case was written.

6.4 Case research as a methodology

Case research is an important methodological tool in social sciences but is not taught in doctoral programs where AEM faculty are trained. The AAEA and International Food and Agribusiness Management Association (IFAMA) have an annual Graduate Student Case Study Competition, and the Food Distribution Research Society (FDRS) has a similar competition for undergraduates. In addition, the National Grocers Association sponsors a case study competition in which a number of AEM faculty participate. By any measure, case research has become an important measure of scholarship but is not taught as a methodology.

With the exception of Penrose (1960) and Wysocki (1998), we could not identify a formal case study research topic despite a number of examples that we believe are worth pursuing. There are a number of corporate records for food firms in public libraries that could serve as a starting point for such a research case. The recognition of Coase (1937), Ostrom (1990), Williamson (2005), Hart (2017), and Holmström (1979) with Nobel Memorial Prizes in Economics has provided a theoretical background for AEM faculty for conducting research and dissertation topics. All of these would likely employ case research methodology. However, little empirical work has been done, with notable exceptions being Balbach (1998), Goodhue (2000), and Knoeber (1989). The food economy has a wide variety of contextual variables based in cultural, economic, historical, organizational, political, and social factors that could be used in analyzing a firm's decision making across competitors and within an industry over time.

6.5 New datasets are becoming available for research

A number of new datasets are becoming available to select AEM faculty whose academic units have the ability to handle confidentiality concerns, privacy, ability to work with Institutional Resource Boards, and faculty who can work with very large data. In addition, financial resources to fund these protocols is critical. Examples include Compustat, EuroStat, IRI, Nielsen, and the University of Chicago Kilts Center market research data. These have become important sources of data for faculty and doctoral students conducting empirical research.

6.6 The role of cooperative extension service and engagement

Educational programs, whether curricula taught through a traditional extension program or one-off outreach programs on various topics, continue to be important. Graduate students working in the area of AEM should be exposed to these efforts. Increasingly, many of these programs are done through public–private partnerships. The two most widespread traditional extension programs by AEM faculty are cooperative director leadership programs and business development programs. Cooperative director leadership programs include annual training in topics such as finance, governance, human resources, leadership, and strategy. More than $30 million have been invested by cooperative stakeholders in endowments in at least 14 academic departments to carry out these programs. There has been a reinvestment of faculty in these positions in recent years. Business development is a second area where there has traditionally been strong extension programming by AEM faculty. A number of endowments exist in a number of academic departments to carry out this work. Much of it is centered on rural entrepreneurship, including the nine steps of business developments.

The on-campus master's degree programs and two master's programs taught executive-style were previously discussed. Increasingly, AEM faculty are being asked to teach food policy concepts to graduate students in other colleges, such as public health, public affairs, veterinary medicine, law, business, and agricultural science. The rationale for using AEM faculty is because they possess the knowledge across the food system to place the policy discussion within the context of the food system and how it impacts agribusinesses in the food system.

Summary

We provide a summary history of agribusiness economics and management and discuss recent advancements and current and future issues in the field. A synthesis of the recent notable studies in the field concludes that the boundaries between the fields of agribusiness economics and management and industrial organization have become widely blurred, and we make the argument that they no longer have any meaningful distinction from each other. Continued work on seeking to find ways to quantify the impact of management and understand consumer trends are important topics for AEM faculty and doctoral students doing empirical work. Opportunities to use the concepts from recent Nobel Prize winners and apply them empirically to the food economy exist but have not been widely adopted yet in dissertations. There continues to be a global unmet need for AEM scholars who can successfully teach undergraduate students, advise graduate students, and communicate to stakeholders.

Notes

1 Kinsey (2001, p. 1113) defines the food economy as "the entire food chain from the laboratories that slice, dice and splice genes in everything from our crop seeds, pharmaceuticals, and animals, to the cream cheese we spread on our bagels."
2 Around 90 percent of the sales of traditional food stores were at supermarkets, and the remaining sales were at convenience stores and specialty stores (U.S. Department of Agriculture, Economic Research Service 2017).
3 Double marginalization occurs when successive firms that have their respective market powers in a vertical supply chain charge their own markups.
4 Unnevehr et al. (2010) review studies on information and quality in three broadly related areas: i) how food markets are affected by product attributes, quality, and heterogeneous consumer preferences and concerns, ii) how information affects food markets and food consumers, and iii) how to measure consumer demand and willingness to pay for food product attributes using hedonic price models and methods of nonmarket valuation.
5 For a detailed review of consumer demand models applied to the analysis of differentiated products markets, see Richards and Bonnet (2016).

References

Allain, M-L., C. Chambolle, S. Turolla, and S.B. Villas-Boas. 2017. "The Impact of Retail Mergers on Food Prices: Evidence from France." *Journal of Industrial Economics* 65(3):469–509.

Allender, W.J., and T.J. Richards. 2012. "Brand Loyalty and Price Promotion Strategies: An Empirical Analysis." *Journal of Retailing* 88(3):323–342.

Alston, J.M., R.J. Sexton, and M. Zhang. 1997. "The Effects of Imperfect Competition on the Size and Distribution of Research Benefits." *American Journal of Agricultural Economics* 79(4):1252–1265.

Andreyeva, T., M.W. Long, and K.D. Brownell. 2010. "The Impact of Food Prices on Consumption: A Systematic Review of Research on the Price Elasticity of Demand for Food." *American Journal of Public Health* 100(2):216–222.

Ariyaratne, C.B., A.M. Featherstone, M.R. Langemeier, and D.G. Barton. 2000. "Measuring X-Efficiency and Scale Efficiency for a Sample of Agricultural Cooperatives." *Agricultural and Resource Economics Review* 29(2):198–207.

Babb, E.M., and R.C. Keen. 1982. "Performance of Midwest Cooperative and Proprietary Grain and Farm Supply Firms." Agricultural Experiment Station Bulletin No. 366, Purdue University, West Lafayette, IN.

Balbach, J. 1998. "Chapter 7: The Effect of Ownership on Contract Structure, Costs, and Quality: The Case of the U.S. Beet Sugar Industry." In J.S. Royer and R.T. Rogers, eds., *The Industrialization of Agriculture: Vertical Coordination in the U.S. Food System*, London, UK, pp. 155–184.

Barton, D.G., T.C. Schroeder, and A.M. Featherstone. 1993. "Evaluating the Feasibility of Local Cooperative Consolidations: A Case Study." *Agribusiness* 9(3):281–294.

Beatty, T.K., B.H. Lin, and T.A. Smith. 2014. "Is Diet Quality Improving? Distributional Changes in the United States, 1989–2008." *American Journal of Agricultural Economics* 96(3):769–789.

Berry, S., J. Levinsohn, and A. Pakes. 1995. "Automobile Prices in Market Equilibrium." *Econometrica: Journal of the Econometric Society* 63(4):841–890.

———. 2004. "Differentiated Products Demand Systems from a Combination of Micro and Macro Data: The New Car Market." *Journal of Political Economy* 112(1):68–105.

Berry, S.T. 1994. "Estimating Discrete-Choice Models of Product Differentiation." *The RAND Journal of Economics* 25(2):242–262.

Bhat, C.R. 2005. "A Multiple Discrete-Continuous Extreme Value Model: Formulation and Application to Discretionary Time-Use Decisions." *Transportation Research Part B: Methodological* 39(8):679–707.

———. 2008. "The Multiple Discrete-Continuous Extreme Value (MDCEV) Model: Role of Utility Function Parameters, Identification Considerations, and Model Extensions." *Transportation Research Part B: Methodological* 42(3):274–303.

Boland, M.A., and J.T. Akridge. 2004. "Undergraduate Agribusiness Programs: Focus or Falter." *Review of Agricultural Economics* 26(4):564–578.

Boland, M.A., A.M. Bosse, and G.W. Brester. 2007. "Mountain States Lamb Cooperative." *Review of Agricultural Economics* 29(1):157–169.

Boland, M.A., and G. Brester. 2006. "Coors' Malt Barley Contracting Program." *Review of Agricultural Economics* 28(1):272–283.

Boland, M.A., and J. Crespi. 2010. "From Farm Economics to Applied Economics. The Evolution of a Profession as Seen through a Census of its Dissertations from 1951 to 2005." *Applied Economic Perspectives and Policy*.

Boland, M.A., S. Daniel, J. Katz, J. Parcell, and I.R. de Aristizabal. 1998. "The 21st Century Alliance: The Case of the Dry Edible Bean Cooperative." *Review of Agricultural Economics* 20(2):654–665.

Boland, M.A., and L. Thielen. 2004. "Industry Note: The USDA CSREES HEP: Doctoral Fellowships in the National Need of Management and Marketing." *International Food and Agribusiness Management Review* 7(1):67–69.

Bonnet, C., P. Dubois, S.B. Villas-Boas, and D. Klapper. 2013. "Empirical Evidence on the Role of Nonlinear Wholesale Pricing and Vertical Restraints on Cost Pass-Through." *Review of Economics and Statistics* 95(2):500–515.

Bonnet, C., and V. Réquillart. 2013. "Impact of Cost Shocks on Consumer Prices in Vertically-Related Markets: The Case of the French Soft Drink Market." *American Journal of Agricultural Economics* 95(5):1088–1108.

Boyd, S., M.A. Boland, K. Dhuyvetter, and D. Barton. 2007. "The Persistence of Profitability in Local Farm Supply and Grain Marketing Cooperatives." *Journal of Agricultural and Applied Economics* 59(1):201–210.

Brester, G., and M.A. Boland. 2004. "The Rocky Mountain Sugar Growers Cooperative: Sweet or Sugar-Coated Visions of the Future." *Review of Agricultural Economics* 26(2):287–302.

Buhr, B.L. 2004. "Case Studies of Direct Marketing Value-Added Pork Products in a Commodity Market." *Review of Agricultural Economics* 26(2):266–279.

Çakır, M., and J.V. Balagtas. 2012. "Estimating Market Power of US Dairy Cooperatives in the Fluid Milk Market." *American Journal of Agricultural Economics* 94(3):647–658.

———. 2014. "Consumer Response to Package Downsizing: Evidence from the Chicago Ice Cream Market." *Journal of Retailing* 90(1):1–12.

Çakır, M., and J. Nolan. 2015. "Revisiting Concentration in Food and Agricultural Supply Chains: The Welfare Implications of Market Power in a Complementary Input Sector." *Journal of Agricultural and Resource Economics* 40(2):203–210.

Çakır, M., and W.G. Secor. 2018. "Heterogeneous Impacts from a Retail Grocery Acquisition: Do National and Store Brand Prices Respond Differently?" *Agribusiness*. https://doi.org/10.1002/agr.21545.

Caswell, J.A. 1992. "Using Industrial Organization and Demand Models for Agribusiness Research." *Agribusiness* 8(6):537–548.

Chaddad, F., and M.A. Boland. 2009. "Strategy-Structure Alignment in the World Coffee Industry: The Case of Coopuxe." *Review of Agricultural Economics* 31(3):653–665.

Chaddad, F.R., and M.L. Cook. 2004. "Understanding New Cooperative Models: An Ownership-Control Rights Typography." *Review of Agricultural Economics* 26(2):348–360.

Chambers, W., and R.P. King. 2002. "Changing Agricultural Markets: Industrialization and Vertical Coordination in the Dry Edible Bean Industry." *Review of Agricultural Economics* 24(2):495–511.

Coase, R. 1937. "The Nature of the Firm." *Economica* 4(16):386–405.

Connor, J.M. 1999. "Evolving Research on Price Competition in the Grocery Retailing Industry: An Appraisal." *Agricultural and Resource Economics Review* 28(2):119–127.

Corstjens, M., and R. Lal. 2000. "Building Store Loyalty Through Store Brands." *Journal of Marketing Research* 37(3):281–291.

Costanigro, M, and J.J. McCluskey. 2011. "Hedonic Analysis and Product Characteristic Models." In J.L. Lusk, J. Roosen, and J. Shogren, eds., *Oxford Handbook on the Economics of Food Consumption and Policy*. Oxford: Oxford University Press, pp. 152–180.

Cullen, K., T. Baranowski, K. Watson, T. Nicklas, J. Fisher, S. O'Donnell, J. Baranowski, N. Islam, and M. Missaghian. 2007. "Food Category Purchases Vary by Household Education and Race/Ethnicity: Results from Grocery Receipts." *Journal of the American Dietetic Association* 107(10):1747–1752.

Darby, P.M., J.D. Detre, T.B. Mark, and M.E. Salassi. 2013. "A Sugar Crossroad: Is Biomass an Opportunity or a Problem?" *American Journal of Agricultural Economics* 95(2):519–526.

Darmon, N., and A. Drewnowski. 2008. "Does Social Class Predict Diet Quality?" *The American Journal of Clinical Nutrition* 87(5):1107–1117.

Davis, J.H., and R.A. Goldberg. 1957. *A Concept of Agribusiness*. Division of Research, Graduate School of Business. Boston: Harvard University.

Deaton, A., and J. Muellbauer. 1980. "An Almost Ideal Demand System." *The American Economic Review* 70(3):312–326.

Dhar, T., and K. Baylis. 2011. "Fast-Food Consumption and the Ban on Advertising Targeting Children: The Quebec Experience." *Journal of Marketing Research* 48(5):799–813.

Downey, W.D. (editor). 1989. "Agribusiness Education in Transition: Strategies for Change." Report of the National Agribusiness Education Commission Lincoln Institute of Land Policy, Cambridge, MA.

Draganska, M., and D.C. Jain. 2005. "Product-Line Length as a Competitive Tool." *Journal of Economics & Management Strategy* 14(1):1–28.

Dubé, J.P. 2005. "Product Differentiation and Mergers in the Carbonated Soft Drink Industry." *Journal of Economics & Management Strategy* 14(4):879–904.

Dubin, J.A., and D.L. McFadden. 1984. "An Econometric Analysis of Residential Electric Appliance Holdings and Consumption." *Econometrica* 52:345–362.

Dubois, P., R. Griffith, and M. O'Connell. 2017. "The Effects of Banning Advertising in Junk Food Markets." *The Review of Economic Studies* 85(1):396–436. doi:10.1093/restud/rdx025.

Ehmke, C., C. Dobbins, A. Gray, A. Miller, and M. Boehlje. 2004. "Which Way to Grow at MBC Farms?" *Review of Agricultural Economics* 26(4):589–602.

Forster, L. 1996. "Capital Structure, Business Risk, and Investor Returns for Agribusinesses." *Agribusiness* 12(5):429–442.

Ginder, R.G., and K.R. Henningsen. 1994. *Financial Standards for Iowa Agribusiness Firms, 1990–1993*. Ames, IA: Department of Economics, Iowa State University.

Goldberg, R.A. 2018. *Food Citizenship*. Oxford: Oxford Press.

Goodhue, R.E. 2000. "Broiler Production Contracts as a Multi-Agent Problem: Common Risk, Incentives, and Heterogeneity." *American Journal of Agricultural Economics* 82(3):606–622.

Gorman, W. 1971. "Two Step Budgeting." Mimeo.

Gould, B.W., and K.A. Carlson. 1998. "Strategic Management Objectives of Small Manufacturers: A Case Study of the Cheese Industry." *Review of Agricultural Economics* 20(2):612–630.

Hainmueller, J., M.J. Hiscox, and S. Sequeira. 2015. "Consumer Demand for Fair Trade: Evidence from a Multistore Field Experiment." *Review of Economics and Statistics* 97(2):242–256.

Hallstein, E., and S.B. Villas-Boas. 2013. "Can Household Consumers Save the Wild Fish? Lessons from a Sustainable Seafood Advisory." *Journal of Environmental Economics and Management* 66(1):52–71.

Hanemann, W.M. 1984. "Discrete/Continuous Models of Consumer Demand." *Econometrica* 53:541–561.

Hart, O. 2017. "Incomplete Contracts and Control." *American Economic Review* 107(7):1731–1752.

Harris, A., and M. Fulton. 1996. "Comparative Financial Performance Analysis of Canadian Co-operatives, Investor-Owned Firms, and Industry Norms." Centre for the Study of Co-operatives, University of Saskatchewan, Saskatoon, Saskatchewan.

Hausman, J., and E. Leibtag. 2007. "Consumer Benefits from Increased Competition in Shopping Outlets: Measuring the Effect of Wal-Mart." *Journal of Applied Econometrics* 22(7):1157–1177.

Hausman, J., G. Leonard, and J.D. Zona. 1994. "Competitive Analysis with Differentiated Products." *Annales d'Economie et de Statistique* 34:159–180.

Hellerstein, R. 2008. "Who Bears the Cost of a Change in the Exchange Rate? Pass-Through Accounting for the Case of Beer." *Journal of International Economics* 76(1):14–32.

Holmes, J. 1994. "Financial Benchmarking: An Application to the Retail Fertilizer and Chemical Industry." Unpublished M.S. thesis. Department of Agricultural Economics, Purdue University, West Lafayette, IN.

Holmström, B. 1979. "Moral Hazard and Observability." *The Bell Journal of Economics* 10(1):74–91.

Hyde, J., J.W. Dunn, A. Steward, and E.R. Hollabaugh. 2007. "Robots Don't Get Sick or Get Paid Overtime, But Are They a Profitable Option for Milking Cows?" *Review of Agricultural Economics* 29(2):366–380.

Kim, D., and R.W. Cotterill. 2008. "Cost Pass-Through in Differentiated Product Markets: The Case of Processed Cheese." *The Journal of Industrial Economics* 56(1):32–48.

King, R.P., M. Boehlje, M.L. Cook, and S.T. Sonka. 2010. "Commemorating the Centennial of the AAEA." *American Journal of Agricultural Economics* 92(2):554–570.

Kinsey, J. 1993. "Changing Societal Demands." In Daniel Padberg, ed., *Food and Agricultural Marketing Issues for 21st Century*. College Station, TX. The Food and Agricultural Marketing Consortium, Texas A&M University, pp. 54–67.

Kinsey, J., and B. Senauer. 1996. "Consumer Trends and Changing Food Retailing Formats." *American Journal of Agricultural Economics* 78(5):1187–1191.

Kinsey, J.D. 2001. "The New Food Economy: Consumers, Farms, Pharms, and Science." *American Journal of Agricultural Economics* 83(5):1113–1130.

Knoeber, C. 1989. "A Real Game of Chicken: Contracts, Tournaments, and the Production of Broilers." *Journal of Law, Economics and Organization* 5:271–292.

Lerman, Z., and C. Parliament. 1990. "Comparative Performance of Cooperatives and Investor-Owned Firms in US Food Industries." *Agribusiness* 6(6):527–540.

Marion, B.W., and NC-117 Committee. 1986. *The Organization and Performance of the U.S. Food System*. Lexington, MA: Lexington Books.

McFadden, D. 1974. "Conditional Logit Analysis of Qualitative Choice Behavior." In P. Zarembka, ed., *Frontiers in Econometrics*. New York: Academic Press, pp. 105–142.

McFadden, D., and K. Train. 2000. "Mixed MNL Models for Discrete Response." *Journal of Applied Econometrics* 15(5):447–470.

McFadden, D., W. Umberger, and J. Wilson. 2009. "Growing a Niche Beef Market: A Targeted Marketing Plan for Colorado Homestead Ranches." *Review of Agricultural Economics* 31(4):984–998.

McGahan, A. 1999. "The Performance of US Corporations: 1981–1994." *The Journal of Industrial Economics* 47:373–398.

McGahan, A., and M. Porter. 2002. "What Do We Know About Variance in Accounting Profitability?" *Management Science* 48:834–851.

Mérel, P.R. 2009. "Measuring Market Power in the French Comté Cheese Market." *European Review of Agricultural Economics* 36(1):31–51.

Mills, D.E. 1995. "Why Retailers Sell Private Labels." *Journal of Economics & Management Strategy* 4(3):509–528.

Muth, M.K., and M.K. Wohlgenant. 1999. "Measuring the Degree of Oligopsony Power in the Beef Packing Industry in the Absence of Marketing Input Quantity Data." *American Journal of Agricultural Economics* 81(3):638–643.

Nakamura, E., and D. Zerom. 2010. "Accounting for Incomplete Pass-Through." *The Review of Economic Studies* 77(3):1192–1230.

Narasimhan, C., and R.T. Wilcox. 1998. "Private Labels and the Channel Relationship: A Cross-Category Analysis." *The Journal of Business* 71(4):573–600.

Neves, M.F., A.W. Gray, and B.A. Bourquard. 2016. "Copersucar: A World Leader in Sugar and Ethanol." *International Food and Agribusiness Management Review* 19(2):207–240.

Nevo, A. 2000. "Mergers with Differentiated Products: The Case of the Ready-To-Eat Cereal Industry." *The RAND Journal of Economics* 31(3):395–421.

———. 2001. "Measuring Market Power in the Ready-To-Eat Cereal Industry." *Econometrica* 69(2):307–342.

———. 2003. "New Products, Quality Changes, and Welfare Measures Computed from Estimated Demand Systems." *The Review of Economics and Statistics* 85(2):266–275.

Ostrom, E. 1990. *Governing the Commons: The Evolution of Institutions for Collective Action*. Cambridge: Cambridge University Press.

Parcell, J.L., A.M. Featherstone, and D.G. Barton. 1998. "Capital Structure Under Stochastic Interest Rates: An Empirical Investigation of Midwestern Agricultural Cooperatives." *Agricultural Finance Review* 58:49–61.

Penrose, E.T. 1960. "The Growth of the Firm: A Case Study of the Hercules Powder Company." *Business History Review* 34(1):1–23.

Petrin, A., and K. Train. 2010. "A Control Function Approach to Endogeneity in Consumer Choice Models." *Journal of Marketing Research* 47(1):3–13.

Piana, C., A. Featherstone, and M.A. Boland. 2005. "Vertical Integration in Ecuador: The Case of Fresh Cut Fruit." *Review of Agricultural Economics* 27(4):593–604.

Pozo, V., M.A. Boland, and D. Sumner. 2009. "Sunkist Growers: Refreshing the Brand." *Review of Agricultural Economics* 31(3):628–639.

Pritchett, J. 2004. "Risk Decision Analysis: MBC Farms' Horse Hay Enterprise." *Review of Agricultural Economics* 26(4):579–588.

Richards, T.J. 2000. "A Discrete/Continuous Model of Fruit Promotion, Advertising, and Response Segmentation." *Agribusiness* 16(2):179–196.

Richards, T.J., and C. Bonnet. 2016. "Models of Consumer Demand for Differentiated Products." Working Paper 16–741. Toulouse School of Economics, Toulouse, France.

Richards, T.J., M.I. Gomez, and G.F. Pofahl. 2012. "A Multiple-Discrete/Continuous Model of Price Promotion." *Journal of Retailing* 88(2):206–225.

———. 2015. "Variety Pass-Through: An Examination of the Ready-To-Eat Breakfast Cereal Market." *Review of Economics and Statistics* 97(1):166–180.
Rudi, J., and M. Çakır. 2017. "Vice or Virtue: How Shopping Frequency Affects Healthfulness of Food Choices." *Food Policy* 69:207–217.
Sanchez, D., M.A. Boland, and D. Sumner. 2008. "Sun-Maid Growers." *Review of Agricultural Economics* 30(2):360–369.
Schrader, L.F., E.M. Babb, R.D. Boynton, and M.G. Lang. 1985. "Cooperative and Proprietary Agribusinesses: Comparison of Performance." Research Bulletin 982, Purdue University, Agricultural Experiment Station, West Lafayette, IN.
Schumacher, S., and M.A. Boland. 2005. "The Persistence of Profitability Among Firms in the Food Economy." *American Journal of Agricultural Economics* 87:103–115.
Secor, W., and M.A. Boland. 2017. "Entry and Exit Patterns in Corn-Ethanol Plants." *American Journal of Agricultural Economics* 99(2):524–531.
Sexton, R.J. 2013. "Market Power, Misconceptions, and Modern Agricultural Markets." *American Journal of Agricultural Economics* 95(2):209–219.
Sexton, R.J., I. Sheldon, S. McCorriston, and H. Wang. 2007. "Agricultural Trade Liberalization and Economic Development: The Role of Downstream Market Power." *Agricultural Economics* 36(2):253–270.
Sexton, R.J., and M. Zhang. 2001. "An Assessment of the Impact of Food Industry Market Power on U.S. Consumers." *Agribusiness* 17(1):59–79.
Sporleder, T., and M.A. Boland. 2011. "Exclusivity of Agrifood Supply Chains: Nine Fundamental Economic Characteristics." *International Food and Agribusiness Management Review* 11(5).
Swinton, S.M., and L.L. Martin. 1997. "A Contract on Hogs: A Decision Case." *Review of Agricultural Economics* 19(1):207–218.
Taylor, M., G. Brester, and M.A. Boland. 2005. "General Mills and Its Hard White Wheat Contracting Program." *Review of Agricultural Economics* 27(1):117–129.
Taylor, R., and S.B. Villas-Boas. 2016. "Food Store Choices of Poor Households: A Discrete Choice Analysis of the National Household Food Acquisition and Purchase Survey (FoodAPS)." *American Journal of Agricultural Economics* 98(2):513–532.
Train, K.E. 2003. *Discrete Choice Methods with Simulation*. Cambridge: Cambridge University Press.
Turolla, S. 2016. "Spatial Competition in the French Supermarket Industry." *Annals of Economics and Statistics/Annales d'Économie et de Statistique* 121/122:213–259.
Unnevehr, L., J. Eales, H. Jensen, J. Lusk, J. McCluskey, and J. Kinsey. 2010. "Food and Consumer Economics." *American Journal of Agricultural Economics* 92(2):506–521.
U.S. Department of Agriculture, Economic Research Service. 2017. Retail Trends. www.ers.usda.gov/topics/food-markets-prices/retailing-wholesaling/retail-trends/. Accessed December 5, 2017.
Villas-Boas, S.B. 2007. "Vertical Relationships between Manufacturers and Retailers: Inference with Limited Data." *The Review of Economic Studies* 74(2):625–652.
———. 2009. "An Empirical Investigation of the Welfare Effects of Banning Wholesale Price Discrimination." *The RAND Journal of Economics* 40(1):20–46.
Villas-Boas, S.B., and C. Toledo. 2016. "Safe or Not? Consumer Responses to Recalls with Traceability." *Cudare Working Papers*. UC Berkeley.
Van Dyne, D.L., and V.J. Rhodes. 1987. "Departmental Savings and Loss Characteristics for 12 Locally Owned Farmer Cooperatives." Agricultural Experiment Station, SR 359, University of Missouri, Columbia.
Volpe, R., A. Okrent, and E. Leibtag. 2013. "The Effect of Supercenter-Format Stores on the Healthfulness of Consumers' Grocery Purchases." *American Journal of Agricultural Economics* 95(3):568–589.
Wang, Y., and M. Çakır. 2017. "The Welfare Impacts of New Demand-Enhancing Agricultural Products: The Case of Honeycrisp Apples." Working Paper. https://ageconsearch.umn.edu/record/258360?ln=en. Accessed December 9, 2017.
Williamson, O.E. 2005. "The Economics of Governance." *The American Economic Review* 95(2):1–18.

Worley, T., R. Folwell, J. Foltz, and A. Jaqua. 2000. "Management of a Cooperative Bargaining Association: A Case in the Pacific Northwest Asparagus Industry." *Review of Agricultural Economics* 22(2):548–565.

Wysocki, A.F. 1998. *Determinants of Firm-Level Coordination Strategy in a Changing Agri-Food System*. PhD thesis, Michigan State University Department of Agricultural Economics.

Yeager, E.A., and S.A. Stutzman. 2014. "Deer Creek Farms: Tradition Into the Future." *American Journal of Agricultural Economics* 96(2):598–605.

Zhu, C., R.A. Lopez, and X. Liu. 2015. "Information Cost and Consumer Choices of Healthy Foods." *American Journal of Agricultural Economics* 98(1):41–53.

Reviewers[1]

Jeff Alwang
Virginia Tech

Keith Coble
Mississippi State University

Stephen Devadoss
Texas Tech University

Jeff Dorfman
University of Georgia

Cesar Escalante
University of Georgia

Valentina Haterska
Auburn University

Dermot Hayes
Iowa State University

Curtis Jolly
Auburn University

Ani Katchova
The Ohio State University

Henry Kinnucan
Auburn University

Will Martin
World Bank

Andy McKenzie
University of Arkansas

Chuck Moss
University of Florida

Jim Novak
Auburn University

Rulon Pope
Brigham Young University

Deepak Rajgopalan
University of California–Los Angeles

C. Richard Shumway
Washington State University

Stacy Sneeringer
ERS, USDA

Kurt Stephenson
Virginia Tech and VPI University

Henry Thompson
Auburn University

Sheldon Wu
University of Wisconsin–Madison

Note

1 In addition to reviewers listed above, the three co-editors reviewed most of the included papers in the Handbook.

Index

Page numbers in italics indicate figures in bold indicate tables on the corresponding pages

Abadie, A. 446
Abbitt, R. J. 217
Abbott, P. 647–648
Abdelradi, F. 408
Abdulai, A. 558–559
Abrevaya, J. 447
access, food 178–180
access, market 332–335, *333–334*
Acemoglu, D. 423, 517
Ackerberg, D. A. 592–594, 596, 597
Adamowicz, W. L. 212
Adams, R. M. 195
adaptation and climate change 198–201, 205
additionality in habitat conservation programs 221–222
Adelman, I. 567
Adhikari, R. 178
Advances in Econometrics 445, 449, 450
Agbo, M. 752
agent-based modeling (ABM) 408
aggregation under heterogeneity 16, 18–19
Agreement on the Application of Sanitary and Phytosanitary (SPS) Measures 339–340
Agribusiness: An International Journal 767–768
agribusiness economics and management (AEM): consumer trends and evolving markets and 762; current and future issues 768–771; current research topics in industrial organization and 762–765; decision cases 766; impact of management in firm performance and 765–766; importance of understanding demand in 761–765; introduction to 760–761; role of institutions in understanding scholarship in 767–768
Agricultural Adjustment Act 361
Agricultural and Applied Economics Association (AAEA) 2, 744, 761, 770
agricultural development: conclusions on evaluating 571–573; experimental revolution and 549–556; identification through econometrics 557–566; introduction to 548–549; laboratory experimental studies of farmer behavior 556–557; simulation approaches 566–571
agricultural economics: applied research in 4–5; biofuel (*see* biofuels); climate change mitigation actions and 203–204; economic growth in (*see* economic growth models); education in 1–2, 5–6, 769–770; farm management and 659–661; finance in (*see* finance, agricultural); futures markets in (*see* futures markets and hedging); Heckscher-Ohlin (H-O) theorem and 306–310, *307*; heterogeneous firms and 314–322, *319–320*; macroeconomic issues in (*see* macroeconomic issues); natural capital and ecosystems services in (*see* ecosystem services; natural capital); New Trade Theory and 310–314, *311*, 323n3; political economy and (*see* political economy); as a profession 1; Ricardian theory and 301–306, *302*, *304*; water management and (*see* water management and economics)
Agricultural Economics 665
Agricultural Guidance and Guarantee Fund (EAGF) 364
agricultural household (AH) models 566–567
Agricultural Marketing Agreement Act 361
agricultural policy: agricultural support 356–361, *357–358*; applications of behavioral economics to 91; climate change mitigation 202–203; food security and 183–184; product quality and reputation and 104–105; water management perverse incentives from 274–277; *see also* decision making; trade theories
agricultural production and behavioral economics 86–91
Agricultural Resource Management Survey (ARMS) 58, 164
Agricultural Risk Protection Act 363–364

Index

agricultural support: conclusions on 372–373; dairy programs 368–369, *369*; dismantling of market intervention in 361–370; for ethanol 379–380, *380*; integrated analysis of programs in 361; introduction to 355–356, *356*; marketing orders and supply quotas 359–360; menu of policies in 356–361, *357–358*; price and income support *358*, 358–359; producer compensation 360–361; reform in Canada 366–369, *369*; reform in China 369–370; reform in the European Union 364–366; reform in the United States 361–364; stock-holding buffer fund stabilization 357, *357*; supply management and 377–380, *378*, *380*; way forward through insurance and risk management 370–372; *see also* finance, agricultural; subsidies; taxes, export
Agricultural Systems 666
Agriculture in Disarray 328
Agronomy Journal 666
Ahearn, M. 117
Ahmed, S. A. 198
Ai, C. 526
aid, food 179, 180
Aisabokhae, R. 200
AJAE (American Journal of Agricultural Economics) 2
Aker, J. C. 560, 572
Akerlof, G. A. 98
Akerman, M. 254
Akleman, D. G. 478
Akobundu, E. 731
Akridge, J. T. 768
Albers, H. J. 224
Alene, A. D. 564
Alix-Garcia, J. M. 212, 222
Allais, O. 508
Allen, P. G. 479
Allenby, G. M. 496
Allison, J. R. 664
Almost Ideal Demand System 474, 763–764
Alston, J. M. 75, 474, 475, 678, 731
ambiguity in agricultural production 90–91
American Culinary Federation 114
American Journal of Agricultural Economics 2, 664, 665
Anania, G. 313
Anastasiadis, S. 240
Anatolyev, S. 448
Anderson, J. L. 143, 147, 148, 340
Anderson, K. 103, 329, 387, 569
Anderson, P. M. 71
Anderson, S. 177
Anderson, S. P. 500
Ando, A. W. 216–218, 224–225, 648, 651
Andreoni, J. 514, 515
Andrews, I. 447
Andrews, W. H. 583
Anfinson, C. 180–181

Angrist, J. D. 448
animal slaughtering 52
Ansar, A. 287, 293
antibiotics: bans on use of 169–170; common property resource perspective on antibiotic resistance management and 161–163; conclusions on 170–171; farm level regulatory approaches for 166–167, **167**; introduction to 159–161; as production inputs in 163–167, **167**; simple model, with implications for production structure 164–166; studies on consumer valuations and industry initiatives to limit use of 167–170
anti-trade bias 386
Antle, J. 178, 476, 607
Appalachian Regional Commission (ARC) 436
Appelbaum, E. 55, 56
apples 345
Applied and Agricultural Economics Association (AAEA) 760
aquaculture 145–149, *146*, *148*; *see also* seafood
Ardeni, P. G. 478
Ariely, D. 19
ARIMA (Autoregressive Integrated Moving Average) 476
Arkansas Global Rice Model 176–177
Armbruster, Walter 2
Arndt, C. 568
Arrow, K. J. 193, 568, 604
Asche, F. 139
Asfaw, S. 561
Association of Southeast Asian Nations (ASEAN) 345
asymmetric paternalism 76
Athey, S. 446, 453
Atkin, R. 103
Attavanich, W. 195, 199
Attwood, D. W. 292
Australian Journal of Agricultural Economics 665
Australian Journal of Agricultural Research 661
Austrian capital market theories 690–694
autarky 315–317; free trade, and falling trade costs 318–321, *319–320*
Autor, D. 623, 624–625
availability, food 176–178
Axtell, R. L. 321
Azomahou, T. 524

Babcock, B. A. 88, 215, 646, 647
Baboo, B. 289
Badibanga, T. 422
Bagwell, K. 313
Baillie, R. T. 478
Bain, J. S. 55, 761
Bajari, P. 453, 505
Bajona, C. 310
Baker, C. B. 698, 699–700, 709

Balassa, B. 305–306
Balcombe, K. 405
Baldos, U. L. C. 198
Baldwin, R. 331, 423
Baltagi, B. 449, 450
bananas 345
Bandyopadhyay, S. 512
Banerjee, A. 385
Banerjee, S. 219
Banff National Park, Canada 213
bans on antibiotic use in agriculture 169–170
Barenboim, E. 480
Barfoot, P. 674
Barham, B. L. 91, 610
Barham, T. 293, 294
Barkley, A. P. 170
Barkley, P. W. 743
Barnett, B. J. 87
Barnett, R. 401
Barrett, C. B. 555, 565
Barros, N. 286
Barrows, G. 673
Barry, P. J. 696, 698–699
Basic Payment Scheme (BPS), European Union 366
Batka, M. 180
Batra, R. N. 308
Batten, D. 400
Bauer, M. 556
Bayesian methods 450
Baylis, K. 183
Beaman, L. 551–552
Becerril, J. 558–559
Becker, G. S. 20, 381
Beckman, J. 647
Beckmann, V. 679
Bedock, N. 458
beef dispute between the U. S. and Canada 344
Beghin, J. C. 75, 178
behavioral economics: agricultural production and 86–91; ambiguity in agricultural production and 90–91; conclusions on 91–92; general consumption behaviors in 85–86; introduction to 84–85; models 88–90; obesity and 75–77; potential for application to agricultural policy 91
behavioral exploitation 77
Behavioral Risk Factor Surveillance System 70
behavioral welfare economics 13–19; econometric support for 22–23, 30n16
Behrer, A. P. 570
Behrman, J. R. 620
Bekker, P. A. 447
Bellemare, M. F. 90
Belloni, A. 453
Belongia, M. 400
Bennett, A. B. 671

Berck, P. 217
Bergès-Sennou, F. 63
Berhane, G. 559
Berkes, F. 256
Bernanke, B. 478
Bernard, A. B. 625
Bernard, D. J. 169
Bernard, J. C. 168, 169
Bernard, T. 553
Bernell, S. 72
Bernheim, B. D. 17, 20, 25
Berning, J. 101
Berry, S. T. 56, 64–66, 501
Bertinelli, L. 524
Bertone Oehninger, E. 279
Bertrand-Nash game 56
Besag, J. 499
Besanko, D. 506
Besley, T. 563, 565
Bessler, D. A. 75, 401, 471, 478, 479
beverage and food processing 122–129, **123–127, 129**
Bewley, R. 478
Bhaduri, A. 290
Bhagwati, J. N. 381
Bhargava, S. 77–78
Bhat, C. R. 496
big data 451–453; causality, treatment effects, and 453; information theoretic approach to 453–455
Binswanger, H. 476, 556, 564, 604
biodiesel 640
biofuels: biodiesel 640; cellulosic 640; conclusions on 652; conventional 639–640; induced land use change and associated greenhouse gas emissions and 642–645, **645**; introduction to 637–641, **638**, *639*; price impacts of 645–649, *649*; Renewable Fuels Standard program and 183, 203, 638, *639*; techno-economic analysis of *641*, 641–642; welfare impacts of 649–652, **650**
biotechnology, agricultural: adoption of genetically modified foods and 679–681; conclusions on 682; impact on health and the environment 675–676; impacts on agricultural production, markets, and welfare *671*, 671–675, *673*; intellectual property rights (IPR) regulation and 676–677; introduction to 670–671; political economy and 681–682; regulations and their costs in 677–679; *see also* genetically modified organisms (GMOs)
Birur, D. K. 647
Bivariate Generalized Autoregressive Conditional Heteroskedastic models 478
Bjornson, B. 119, 696
Blanc, E. 286
Blanciforti, L. 474
Blank, S. C. 115

Index

Blundell, R. W. 521, 595
Bocqueho, G. 89, 604
body mass index (BMI) 70–71, 72, 181, 620
Boffelli, S. 458
Bogetoft, P. 752–753
Boggess, W. G. 216, 699
Bohman, M. 313
Boland, M. A. 767, 768
Bond, C. A. 102
Bond, S. 595
Bonnano, G. 61
Bonnet, C. 63, 65, 506
Bontemps, C. 65
Bontemps, P. 753
Bontems, P. 63
boom/bust cycles 697
bootstrapping 449, 456
Bordo, M. 402
Boserup, E. 565
Boswick, H. P. 459
Bottom Billion 180
Bouamra-Mechemache, Z. 63, 65
Boucher, S. R. 571
Bougherara, D. 90
Bowen, H. P. 308, 309
Box, G. E. P. 455, 458, 476, 658
Boxall, P. C. 218
Box-Ljung statistic 456
Boys, K. A. 115
Braat, L. C. 254
Bradford, K. J. 678
Braverman, A. 570–571
Bredahl, M. 347
Bredehoeft, J. D. 290
Brenkers, R. 492
Bresnahan, T. F. 55, 56
Bretton Woods system 399–400
Brock, W. A. 514
Brookes, G. 674
Brower, C. H. 163
Brown, A. 115
Brown, G. 161, 162
Brown, J. 269
Brownian motion 456
Brownlee, O. H. 471, 475
Brozovic, N. 274
Bruin, K. C. de 205
Bruno, H. A. 506
Brunswik, E. 480
Buceviciute, L. 53
Buchanan, J. M. 381
budgeting, farm 662
Buettner, T. 596
buffer fund stabilization 357, 357
built capital 256
Bulte, E. 553
Bureau of the Biological Survey, Canada 213–214

Burgess, R. 563
Burt, O. R. 664
Buschena, D. E. 89
Bushak, L. 101
Bustos, P. 416, 419–420
Butcher, K. F. 71
Butt, T. A. 200
Byerlee, D. 564, 660

Cabrera, B. 408
cage-free eggs 114
Cairnes, J. E. 402
Calculus of Consent, The 381
Caliendo, L. 310
California Air Resources Board (CARB) 203
California water resources management *see* water management and economics
Calonico, S. 447
Calvin, L. 20
Cameron, A. C. 445, 449
Camm, J. D. 216
Canada, policy reforms in 366–369, *369*
Canada–U.S. Softwood Lumber Agreement (SLA) 345
Canadian Journal of Agricultural Economics 665
Canadian Wheat Board (CWB) 344, 367
Caner, M. 447
cap-and-trade strategy 202, 233–236, 243–246, 250n6
capital: gains from trade and international movement of labor and 349; Heckscher-Ohlin (H-O) theorem and 306–310, *307*; types of 256–257, *257*; *see also* human capital; natural capital
Capital Asset Pricing Model (CAPM) 694–695, 696
capital markets: for agricultural assets 694–698; asset ownership and 701–704, *702–703*; Austrian and neoclassical theories of 690–694, *692–693*; for farmland 695–698
Cappiello, L. 458
Capps, O. Jr. 73, 75, 474
Caprettini, B. 416, 419–420
carbon dioxide and crop yields 195–196
Card, D. 447, 560
Cardell, N. S. 501
Cardenas, J. C. 556
Carey, J. M. 290
Carpenter, J. P. 556
Carpenter, S. 254
Carrion-I-Silvestre, J. L. 456
Carroll, R. J. 526
Carson, R. T. 517
Cartagena Protocol 678–679
Carte, M. R. 553
Carter, A. V. 449
Carter, C. A. 313, 342

Carter, M. R. 555, 559, 564, 571
Case, A. 565
Casey, P. 75
Cass, D. 412
Castriota, S. 103
Caswell, J. A. 104, 763
catch shares 135
Cattaneo, M. D. 446
Caves, K. 592–594, 596, 597
Cawley, J. 72–73, 85
celluosic biofuels 640
Centennial Issue of the American Journal of Agricultural Economics, The 2
Center for Behavioral and Experimental Agri-Environmental Research (CBEAR) 91
Centers for Disease Control and Prevention (CDC) 70, 71, 160
Centro Internacional de Mejoramiento de Maiz y Trigo (CYMMIT) 564
Chakravorty, U. 287, 290
Chalak, K. 447
Chalfant, J. 474
Chamberlain, G. 595
Chambers, R. G. 90, 401–403, 476, 478, 581, 605, 659
Chamblee, J. F. 223
Chandler, A. D. 166
Chandon, P. 77
Chang, C.-Y. 724
Chang, Y. 457, 460
Charlton, D. 565
Chaudhuri, S. 447
Chavas, J.-P. 581, 604, 605, 609–611, 659, 672
Chee, E. 73
Chen, C. C. 196
Chen, J. 182–183
Chen, X. 218, 224, 520, 527
Chen, Z. 60, 474
Chenery, H. B. 413
Chern, W. S. 50, 55
Chernozhukov, V. 447, 453
Chesapeake Bay Watershed Nutrient Credit Exchange 246
Chetty, R. 24, 438
Chiang, S.-C. 102
Chibwana, C. 563
Chicago Board of Trade (CBOT) 714–715
Chicago Mercantile Exchange (CME) 715
China, policy reforms in 369–370
Chintagunta, P. K. 495, 498
Chirwa, E. 563
Chiu, Y.-B. 526
Cho, W. T. 455
Choi, I. 455, 457
Choices 666
Cholesky factorization 478
Chou, S. Y. 72

Chuang, i-Juan 724
Chytilova, J. 556
Clark, B. 168
Clark, C. W. 145
Clark, L. F. 180
Clarke, J. A. 221
Clarke, R. 55
Clean Power Plan 202
climate change 41–42; adaptation and 198–201, 205; concluding comments on 205; crop yields and 195–196, *196*, 199; effects on agriculture 194–198, *196–197*; effects on regions and land values *197*, 197–198; food security and 182–183, 198; GE crops and 675–676; habitat conservation and 217; introduction to 191–194, *192*; livestock and 196–197, 199; mitigation 201–205; projected *192*, 192–194; recent history of 191–192; societal action on 204–205; water management and 279
Clodius, R. L. 50
coalitions, political 382–383
Coase, R. 677
Cobb-Douglas production functions 415, 417, 583–584
Coble, K. H. 665
Cochrane, N. 182
Coelli, T. J. 661
Cohen-Boyer patent 670
collective reputation 101–102
college education: cooperative extension system (CES) and (*see* cooperative extension system (CES)); by land-grant universities 631–633; who pays for 630–631, *631*; *see also* human capital
Collins, A. 89
Collins, R. A. 695, 696, 698–700
Collins-Barry model 698–701
Committee on World Food Security (CFS) 45–46
Commodity Credit Corporation (CCC) 361, 362
Common Agricultural Policy (CAP), European Union 364–368, 387
Common International Classification of Ecosystem Services (CICES) 258, **259–260**
Common Market Organizations (CMOs) 364
competition: in food retailing 110–114, **111**; globalization and 627–629; monopolistic 312; oligopolistic 312–313
composition effect 512–513
Comprehensive Economic and Trade Agreement (CETA) 322
computable general equilibrium (CGE) models 28–29, 322, 437, 568–570, 651–652
Condon, N. 406
conflict and world food 42–43
Conley, T. G. 562
Connecticut Nitrogen Credit Exchange (CNCE) 246, 247

Conservation Reserve Enhancement Program (CREP) 218
Conservation Reserve Program (CRP) 213, 220–225, 275, 362
Constant Elasticity of Substitution (CES) preference 315, 320, 321
Consultative Group on International Agricultural Research (CGIAR) 554
consumer cooperatives agriculture (CSA) 115
consumer demand models: conclusions and implications 508; discrete choice 489–492; discrete-continuous choice 492–495; introduction to 488–489; machine learning 501–505; multiple-discrete and multiple-discrete continuous choice 495–501; practical considerations with 505–508
consumers: agribusiness economics and management (AEM) and trends and evolving markets with 762; behavioral welfare economics for 16–18; consumption behaviors of 85–86; in economic growth models 416–417; GM technology preferences of 393n16, 681–682; and industry initiatives to limit antibiotic use in animals 167–170
consumption behaviors 85–86
convenience stores 112
conventional biofuels 639–640
convergent cross mapping (CCM) 535–536, 540
Cook, M. L. 748
Cook-Deegan, R. 670
Cooley, T. 478
cooperative extension system (CES) 770–771; concluding comments on 744–745; future directions for 742–743; recent data on funding trends for 732–740, **733**, **735–736**, **738–739**, **741**; recent literature valuing 729–732, **730**; value to society of 728–729
cooperatives: CEO identity in 753–754; CEO payment in 754–755; finance of 756–757; governance of 752–755; heterogeneous member patron preferences in 752–753; introduction to 748, **749–750**; management of 755–756, 765; performance and market structure of 750–752; summary and conclusion on 757–758; vertical coordination of 756
corn: ethanol from *346*, 346–347, 379–380, *380*, 405–406, 651–652 (*see also* biofuels); futures markets and hedging **718–719**, 718–720
Costanigro, M. 101, 102, 103
Costanza, R. 254, 256, 257, 258, **259–260**, 266
Costco stores 52–53
Costello, C. 217, 223
Costinot, A. 306
Cotterill, R. W. 52, 65
cotton and sugar disputes 343–344
Coulibaly, O. 564

Council on Food, Agricultural and Resource Economics (CFARE) 744
country of origin labeling (COOL) 343
Covey, T. 478
Cowling, K. 55
Crago, C. L. 645
Cramer, Gail. L. 2
credit market equilibrium 704–705, 712
Crespi, J. M. 51, 52, 58–59, 767, 768
Cressie, N. 453
Cressie-Read (CR) multi parametric entropy family 453–454
Criado, C. O. 517, 525
Crocker, T. 232
cropping adaptations 199
crop yields: climate change and 195–196, *196*; impact of biotechnology on 671, 671–675, *673*; risk management and 610–611
crude oil prices 648–649, *649*
Cui, J. 648
cumulative prospect theory (CPT) 608
Currie, J. A. 357
Currie, J. M. 12
Currie, J. S. 72
Cutler, D. M. 71
Czukas, K. 571

Dahlhausen, J. L. 168
Daily, G. 254
dairy programs 368–369, *369*
Dales, J. 232
Daly, H. E. 254, 255, 257, 266
Daly, P. A. 729
dams *see* water supply and dams
Danzig, G. 664
Dar, M. H. 553
Darby, M. R. 101
Darko, F. A. 563
Darolles, S. 528
Dasgupta, S. 517
Davidson, C. 481
Davidson, R. 449
Davies, S. W. 55
Davis, B. 563
Davis, D. E. 120
Davis, G. C. 73
Davis, J. 760–761, 767
Davlasheridze, M. 731
Dawid, P. 479
Dearmont, D. 471
Deb, P. 116
DeBrauw, A. 559
Debreu, G. 568
debt and debt-equity: asset ownership and 701–704, *702–703*; farmer 698–701
De Castro, J. M. 86

decision making: climate change and 200–201; econometrics and 470; regarding human capital investment *621–622*, 621–624, **623**; in welfare economics 19–20; *see also* agricultural policy
de Gorter, H. 384, 385, 391, 406, 651
de Groot, R. 254
Dehejia, R. H. 566
De Janvry, A. 563, 567, 568
Delgado, M. S. 528
Deller, S. C. 437, 438
Delmastro, M. 103
demand: agribusiness economics and management (AEM) and 761–765; discrete models of 489–490; E. J. Working on 468–469; elasticities of 490–491; experiments on consumer 472; Heckscher-Ohlin (H-O) theorem and 306–310, *307*; IMPACT model supply and demand projections 37–39, *38–39*; "new" industrial organization analysis of the food industry and 53–54
demand models, consumer: conclusions and implications 508; discrete choice 489–492; discrete-continuous choice 492–495; introduction to 488–489; machine learning 501–505; multiple-discrete and multiple-discrete continuous choice 495–501; practical considerations with 505–508
Demont, M. J. 678
Demsetz, H. 55
De Nadai, M. 447
Deng, W. 753, 754
de Palma, A. 500
Detre, J. D. 117
Devadoss, S. 314, 322, 328, 345, 401–402
development paradox 383
DEVEPEM models 570
Dewbre, J. H. 717
Dhamodharan, M. 314, 345
Dhar, R. 65
Dharmasena, S. 73, 75
Diagne, M. 176
Dickey, D. A. 455
Diederich, T. G. 453
Dierx, A. 53
Dietary Guidelines for Americans (DGA) 75
Dietl, H. M. 756
Di Falco, S. 561, 609, 611
Diffenbaugh, N. S. 198
Diks, C. 457
Dillon, A. 286, 293
Dillon, J. 604
Dinar, A. 200
Direct Marketing Act of 1976 115
direct policy interventions 76
direct sales/local foods 114–116
discrete choices, models of 489–492, 764

discrete-continuous choice, models of 492–495
dismantling of market intervention 361–370
distance function 596–597
distance metric–multinomial logit (DM-MNL) 500–501
distortions-volatility tradeoff 388
distribution and processing, structural changes in 117–129, **118–127**, **129**
diversification for risk management 609
Dixit, A. K. 313
Doan, T. 478
Dobson, P. W. 60
Donald, S. G. 447
Donaldson, D. 306
Donnenfeld, S. 102
Doornik, J. A. 459
Dor, A. 73
Doraszelski, U. 596
Dorfman, J. 404
Dorfman, R. 99
Dorn, D. 624–625
Dornbusch, R. 301, 303, 310, 402, 403
Dorosh, P. 568
Douc, R. 460
Downs, A. 381, 385
Drabik, D. 406
Draganska, M. 65
Draper, N. R. 658
Drivas, K. 750
Du, X. 87, 88, 407, 648
duality 661
dual reputation 102
Dubois, L. 75
Dubois, P. 63
Ducks Unlimited 214, 219, 222
Duffield, J. A. 641
Duffy, P. 2
Duflo, E. 291, 293, 552, 560
Dumas, C. 337
Dumler, T. J. 278
Durand-Morat, A. 176
Duvaleix-Tréguer, A. 98
Dyer, G. A. 570
dynamic management of water resources 270–271
Dyson, F. W. 481

Eales, J. S. 474
Eaton, J. S. 301, 303, 305–306, 314–315, 321
Eberhardt, M. 416, 420–422
Ebert, A. W. 183
Echevarria, C. 414, 415
Econometrica 469
econometrics: agricultural development impact identification through 557–566; attempts to find structure in agricultural production and markets using 471–476; big data 451–453; introduction

to 445–446, 467–471; machine learning 452, 501–505; micro- 446–451; Neyman-Rubin-Holland model 471; non-structural modeling with observational data 476–479; in production functions estimation (*see* production functions); software programs 460; spatial 500–501; summary of 479–481; support for behavioral welfare economics in 22–23, 30n16; time series analysis in 455–460

Econometrics Journal 446, 459

economic growth models: background on 413–414; concluding remarks on 422–423; intra-temporal equilibrium 418; introduction to 412–413; labor and 419–420; liberalization policies and 416; in open *vs.* closed economies 415–416, 421; overview of literature, 1990–2006 414–416; overview of literature since 2006 416–421; service sector and 423; structural transformation and 417–418, 420; technology and 417, 419; three-sector 415, 416; transport costs and productivity 415; two-sector 414–415

economic interaction effects (EIEs) 391–392

Economics of Ecosystems and Biodiversity (TEEB) 258, **259–260**

Economic Theory of Democracy, An 381

ecosystem services 257–258; short history of 254–255; types of 258, **259–260**; valuation of 261–265, **263, 264**; *see also* natural capital

Eddington, A. S. 481

Edgeworth box 302, *302*

education *see* cooperative extension system (CES); human capital

Edwards, E. C. 273

Ehrlich, P. 254

EKC *see* environmental Kuznets curve (EKC) analyses

Ekins, P. 265

Elabed, G. 553

elasticities, demand 490–491

elasticity of marginal productivity [EMP] 288

Elbehri, A. 568

Elder, J. 453

Ellickson, P. B. 53

embargos, trade 341–343, *342*

Emerick, K. 290, 553

emissions reduction credit (ERC) trading mechanisms 243–246, 250n5

Endangered Species Act 212

endogeneity 583–584

endogenous productivity 595–596

endogenous product quality 99

endogenous switching regression (ESR) 560–561

energy prices 405–406

Engle, R. F. 455

Englin, J. 476

ensemble modeling 452–453

environmental impacts of GE crops 675–676

environmental Kuznets curve (EKC) analyses: conclusions on 527–528; dynamic model 516–517; existing model 518–519, **519**; future directions in 526–527; introduction to 512–518, *513*; model consistent specification test 520–521; policy 517–518; recent advances in semiparametric model 519–520; semiparametric estimation of 521–526, **522–523**; smoothing approach and literature 518–519, **519**; static models 515–516; theoretical model 514–515

Epplin, F. M. 743

Erokhin, V. 347

error-correction model (ECM) 455, 724

error-in-variables (EIV) models 447

Erutku, C. 60

Escanciano, J. C. 446

ethanol *346*, 346–347, 379–380, *380*, 405–406, 637–638, 651–652; *see also* biofuels

Etilé, F. 508

European Union, policy reforms in the 364–368

Evans, L. 756

Exact Affine Stone Index (EASI) 474

Exclusive Economic Zone (EEZ) 134

exogenous product quality 98–99

expected utility (EU) model 606–609

experiments *vs.* non-experiments, evidence from 565–566

export policies 340

export subsidies 335–337, *336*

export taxes 340, *340*

eXtension 743

extreme value statistics (EVS) 537

Ezekiel, M. 469

facing allowances 61

Fackler, P. L. 478, 479, 480

Fafchamps, M. 564, 567, 571

Falkowski, J. 385

false nearest neighbors (FNN) test 539

Färe, R. 665

Farley, J. 255

Farm Bill of 1985 362–363

Farm Bill of 2008 364

Farm Bill of 2014 364

farmers: behavior, laboratory experimental studies of 556–557; debt and debt-equity choice of 698–701; response to dams by 289–291; supports for (*see* agricultural support); weather-indexed insurance for 371–372

Farm Input Subsidy Program (FISP) 552, 555, 562–563

farmland, capital markets for 695–698

farm management: 21st century outlets and information on 664–666; brief history of 663–664; economic theory and 659–661; finance (*see* finance, agricultural); future

applications in 666–667; introduction to 658–659; quantitative methods in modern 661–662; risk management and 610; *see also* cooperatives
Farm Products Agency Act 368
Faße, A. 569
Fatas, E. 753
Faures, J. M. 294
Featherstone, A. M. 700
Feder, G. 551
Federal Crop Insurance Act 363, 372
Feenstra, R. C. 500
Fenech, A. 266
Feng, G. 526
Feng, H. 87, 88
Feng, L. 756
Feridhanusetyawan, T. 620
Fernandez-Cornejo, J. 672
Ferris, J. 479
Ferris, William G. 714
Filipski, M. A. 572
Filipski, M. J. 571
finance, agricultural: asset ownership and 701–704, *702–703*; Austrian and neoclassical capital market theories in 690–694, *692–693*; cooperative extension system (CES) funding and 732–740, **733**, **735–736**, **738–739**, **741**; cooperatives 756–757; credit market equilibrium and 704–705, 712; distribution of income for different rates of return on assets in 710–711; farmer debt and debt-equity in 698–701; introduction to 689–690; markets for agricultural assets and 694–698; markets for farmland 695–698; summary and suggestions for further study in 705–706
Finkelstein, E. A. 71, 73, 74
firm reputation 100–101
Fischer, C. 163
Fischer, S. 301, 303, 310
Fischler Mid-Term Review 365
Fisher, A. C. 194, 195, 197, 288
Fisher, F. 527
Fisher, F. M. 55
Fisher, L. A. 459
Fisher, M. 563
fisheries 140–145, *141–142*, 152n7, 153n10, 153n16; *see also* seafood
fixed effects in estimating production functions 584–586
Fleckinger, P. 101
Fleming, A. 159
Fleming-Dutra, K. E. 163
Flinn, J. C. 290
Folke, C. 256
Fon, V. 25
Food, Agriculture, Conservation, and Trade Act of 1990 213

food aid 179, 180
Food and Agricultural Organization (FAO) 139, 175–176, 421
Food Consumption Score (FCS) 180
food industry: introduction to organization of 50–51, *51*; market structure 51–53; "new" industrial organization analysis of 53–56, *55*; processor buyer power in 57–59; retailer buyer power and in 60–65; vertical market control 57
food processing *see* processing, food
food retailing *see* retailing, food
Food Safety Modernization Act (FSMA) 183–184
food security 42–43; access and 178–180; climate change and 182–183, 198; conclusions on 184; food availability and 176–178; global and local governance and 45–46; introduction to 175–176; stability and 181–184; utilization and 180–181
foodservice outlets 112
food standards, political economy of 389–390
Fooks, J. R. 219
Ford, S. A. 699
Forrester, J. W. 438
Forster, D. I. 696
Fortenbery, T. R. 717
fossil fuels 637
Foster, A. D. 416, 419, 422, 561–562
Foster, J. 565
Frankel, J. 400, 402, 403, 404
Frazer, G. 592–594, 596, 597
Free Trade Agreement of the Americas (FTAA) 569
free trade and heterogeneous firms 317–318
frequency domain analysis 459
Fresh Garlic Producers Association (FGPA) 344
Freshwater, D. 438
Friedman, M. 477, 480
Frisch, R. 469–470, 481n8
Frisvold, G. 212
Froot, K. A. 164
fuel taxes 202, 250n5; *see also* biofuels
Fukase, E. 179
Fuller, W. A. 455
Fulton, M. 750, 751, 753, 754, 757
Fulton, M. E. 183
functional analysis for dependent time series 459–460
Fuss, M. 582
futures markets and hedging 611; conclusions on 725–726; examples of **718–719**, 718–720; introduction to 713–714; margin trading in 726n3; short history of 714–715; social value of futures trading in 714–720, *715*, **718–719**; summary of findings from examples of **720–722**, 720–725; understanding price relationships in *715*, 715–718

Index

Gabriel, S. C. 699–700, 709
Gabriel and Baker equation formulation 699–700, 709
Gabszewicz, J. J. 97, 98
Gaitan, B. 412
Galbraith, J. K. 60
Gallagher, P. 605
Gallagher, P. W. 648
ganyu labor 179
Gao, Q. 526
Garbade, K. D. 717
Garber-Yonts, B. 216
Garcia, P. 88, 407, 479
Gardebroek, C. 407
Gardner, B. L. 2, 391, 475
garlic imports 344
Gaskell, G. 682
gasoline prices 648–649, *649*
Gates, D. M. 73
Gaughan, J. 199
Gauss–Seidel algorithm 415
GE crops *see* biotechnology, agricultural
General Agreement on Tariffs and Trade (GATT) 150, 329, 330
general equilibrium (GE) impacts simulations 567
general equilibrium welfare analysis 24–28, *25, 27*
generalized method of moments (GMMs) 447, 595
generalized multiple–discrete continuous extreme value (GMDCEV) model 496–498
generalized synchrony 540
genetically modified organisms (GMOs) 91, 101, 670–671; adoption of foods from 679–681; consumer preferences and 393n16, 681–682; food availability and 177; non-tariff distortions *338*, 338–339; risk management and 610; *see also* biotechnology, agricultural
GeoPlatform 225
Gerhart, J. 290
German Climate Computing Center 193
German hog-price cycle 542–543, *543*
Gertler, P. J. 553
Ghysels, E. 457
Giannakas, K. 750, 751, 753, 757
Gibson, J. 560
Gibson, P. 329
Gil, J. 405, 407
Girshick, M. 468
Glaeser, E. L. 71
Glauber, J. W. 183
Global Biosphere Management Mode (GLOBIOM) 643
Global Fishing Watch 226
global governance 45–46
globalization and the labor market **624**, 624–630, *626, 628–629*
global mean temperature (GMT) 193

Global Perspectives on Trade Integration and Economies in Transition 347
Global Trade Analysis Project (GTAP) 329
Global Trade Analysis Project-Biofuels (GTAP-BIO) 643–644
Glymour, C. 478
Goemans, C. 103
Goeree, M. S. 506
Goetz, R. 290
Goetz, S. J. 437, 438, 731
Goetzel, R. Z. 73
Gohin, A. 643, 647
Golan, A. 56
Goldberg, P. K. 492
Goldberg, R. 760–761, 767
Gollier, C. 607–608
Gollin, D. 414, 415, 423
Gomez, M. I. 496
Gomez-Baggethun, E. 254
Goodhue, R. E. 58
Goodwin, B. K. 87, 478, 725
Google Earth Engine 226
Google Scholar 450, 461n2
Gopinath, M. 310
Gordon, D. V. 140
Gotz, T. B. 457
governance reforms 45–46
government role in addressing obesity 73–75, **74**
Govindasamy, R. 116
Graff, G. D. 677, 681
Graham-Tomasi, T. 270, 290
Grain Inspection, Packers, and Stockyards Administration (GIPSA) 52
Granger, C. W. J. 455, 478
Granger causality 457–458
Grant, J. H. 563
Gray, A. 666
Gray, M. E. 471
Great Depression 361–362, 383
Greater Miami River Watershed Trading Pilot (GMRWTP) 247
Great Recession of 2007–2009 121, 619
Green, R. 474
Greene, K. N. 86, 445
greenhouse gas (GHG) emissions 42, 44, 192, 194, 639–641; GE crops and 675–676; induced land use change and associated 642–645, **645**; mitigation 201–204; *see also* biofuels; climate change
Greenpeace 675
Green Revolution technologies 37, 561–562
Green Solow Model 514
Greer, J. 565
Gregson, R. 537
Griffin, J. E. 520
Griffiths, W. E. 451
Griliches, Z. 476

Grosser, J. 680
Grosskopf, S. 665
Grossman, G. M. 313, 512, 517
Grossman, M. 72
Grote, U. 569
groundwater-energy nexus 278–279
groundwater management districts (GMDs) 277–278
Grube, A. H. 664
Grubel, H. G. 310, 312
Guan, Z. 647
Guastello, S. 537
guided sampling 99
Guilkey, D. K. 447
Guirkinger, C. 571
Gupta, A. 571
Gupta, S. 506
Guthman, J. 116
Guthrie, G. 756
Guttormsen, A. 147

Haavelmo, T. 467–468, 470, 477
Haber-Bosch process 176
habitat conservation: auctions and agglomerations bonuses 218–219; brief history of 213–214; conclusions on 226; contract length and 218; dynamic targeting in 217; evaluation of programs for 220–225; funding mechanisms 219–220; future directions in 225–226; introduction to 211–212; issues in design and implementation of programs for 214–220; targeting strategies in 215–217
Haggblade, S. 568
Hahn, J. 526
Haixia, W. 407
Hajargasht, G. 451
Halliday, T. 560
Hallstein, E. 151
Hamilton, S. F. 61–62, 64–65, 508
Hammer, J. S. 570–571
Handbook of Econometrics, The 445
Hanemann, W. M. 195, 197, 476, 493–496
Hannah, L. 225
Hannesson, R. 140
Hansen, B. E. 450
Hansen, C. 447
Hansen, M. C. 226
Hansen, Z. K. 292
Harberger, A. C. 27
Hardle, W. 521
Harrington, P. 451
Harrison, G. W. 556
Harvest Price Option 364
Harwood, J. 609
Hasan, M. 404
Hastie, T. 451, 452, 521, 527
Hatch Act 361, 631, 728

Hausman, J. A. 524, 763–764
Hausmann, R. J. 422
Havlicek, J. H. 474
Hayashi, F. 445
Hayes, D. J. 648
Haynes, D. J. 177, 360, 681
Hazell, P. B. R. 664
Heady, E. 471, 659, 660
health impact of GE crops 675–676
Heaney, C. 670
Heckman, J. J. 23, 24, 450, 493, 619
Heckscher-Ohlin (H-O) theorem 306–310, *307*
hedging *see* futures markets and hedging
hedonic analysis 102–103
Heerman, K. E. R. 306
Hellerstein, D. M. 218
Hellerstein, R. 64
Helpman, E. 321
Heltberg, R. 563
Hendricks, N. 673
Hendrikse, G. W. 752, 753, 754, 756
Hennessy, D. A. 87, 88, 163, 166
Henry, M. 437
Herkert, J. R. 220
Hernandez, M. 407
Herrendorf, B. 413, 416, 417
Herrick, B. 427
Herriges, J. 476
Herrington, C. 416, 417
Hertel, T. 569, 571, 643
Hertel, T. W. 198, 329, 646, 647
heterogeneous firms 314–315; autarky, free trade, and falling trade costs in 318–321, *319–320*; autarky and 315–317; beyond Melitz and the agricultural supply chain 321–322; connecting firm-level heterogeneity to the data from 321; free trade and 317–318
Hicks, J. R. 16, 473, 476
hidden subsidies 337
Hildreth, C. 664
Hill, M. R. 218
Hillman, J. S. 328, 347
Hispanic-Serving Institutions (HSIs) 728–729
Hoag, D. L. 731–732
Hobbs, J. E. 180
Hoch, I. 584–585
Hochman, G. 646, 647, 648, 681
Hoddinott, J. 564
Holland, P. 471, 472
Hollander, A. 309
Holt, M. T. 604
Homans, F. R. 135, 144
Hong, Y. 520, 524
Hopenhayn, H. A. 315
horizontal product differentiation 97–98
Hornbeck, R. 289, 560
Horowitz, J. L. 449, 450, 528

Index

Hort Technology 666
Horvath, L. 459
Hotelling, H. 97
Hotelling model 97, 751
Household Hunger Scale (HHS) 180
Hovelaque, V. 755
Howden, S. M. 199
Hsiao, C. 477, 521, 528
Hu, S.-W. 404
Huang, J. 569
Hudson, M. A. 479
Hueth, B. 750–751, 755
Hueth, D. L. 12, 25, 26, 28, 357
Huffaker, R. 537, 542
Huffman, W. E. 620, 622, 632, 672, 679, 729–730
Hull, John C. 713, 723
human capital 256; conclusions and recommendations on 633–634; globalization and changing world labor market and **624**, 624–630, *626*, *628–629*; introduction to 616–619, *618–619*; investment decisions regarding *621–622*, 621–624, **623**; land-grant universities and 631–633; types of 619–620; and who pay for college education in the U. S. 630–631, *631*
Hung, M. F. 239
Hunter, J. 460
Hunter River Salinity Trading Program (HRST) 247
Hurt, C. 647–648
Hussain, I. 286
Huxley, T. 140
Hwang, Y.-J. 168

Illegal Immigration Reform and Immigrant Responsibility Act 620
Ilzkovitz, F. 53
Imbens, G. W. 446, 447, 453
immigration 620
Impact Analysis for Planning (IMPLAN) 731
import tariffs 332–334, *333*
income: climate change and 198; distribution for different rates of return on assets 710–711; food access and 179–180; and price supports *358*, 358–359
Income-Based Food Security (IBFS) indicator 178
Inderst, R. 60
Individual Fishing Quotas (IFQs) 135
induced land use change and associated GHG emissions 642–645, **645**
information costs 385
information theoretic approach to big data 453–455
Innes, R. 61–62, 212, 696, 705
innovation: food availability and 176–177; in institutions 44–45; in policies 44; technological 43–44

input space, big data 451–452
institutional innovations 44–45
instrumental variables (IV) 447, 448, 481n6, 507–508; in estimating production functions 586–587; RCT 559–560
insurance and risk management 370–372, 612–613
intellectual property rights (IPR) regulation 676–677
Intergovernmental Panel on Climate Change (IPCC) 191–193, 200, 201; trade embargos 341–343, *342*
international agricultural trade: agreements on 347; apples 345; background to policies on 329–330; bananas 345; beef dispute between the U. S. and Canada 344; conclusions on 350; cotton and sugar disputes 343–344; country of origin labeling in 343; earlier reviews of 328; ethanol from corn *346*, 346–347; export policies 340; export subsidies 335–337, *336*; export taxes 340, *341*; genetically modified organisms in *338*, 338–339; hidden subsidies 337; import tariffs 332–334, *333*; international movement of capital and labor and gains from 349; introduction to 327; lumber disputes between the U. S. and Canada 345; market access and 332–335, *333–334*; non-tariff distortions (NTDs) 335–341, *336*, *338*, *340–341*; in perspective 328–329; production quotas 335; regional trade agreements 330–332; sanitary and phytosanitary measures 339–340; sugar disputes between the U. S. and Mexico 346; suspension agreements 341; trade embargos 341–343, *342*; trade remedy actions 337–338; in transitional economies 347–349, *348*; U. S. garlic imports from China 344; U. S. shrimp imports from China 344; voluntary export restraints 341, *342*
International Agricultural Trade Disputes: Case Studies in North America 328
International Council for the Exploration of the Sea (ICES) 140
International Food and Agribusiness Management Association (IFAMA) 666
International Food and Agribusiness Management Review 666, 768
International Food Policy Research Institute (IFPRI) 37, 41
International Model for Policy Analysis of Agricultural Commodities and Trade (IMPACT) model supply and demand projections 37–39, *38–39*, 41, 43, 47–48n2, 48n4
intra-temporal equilibrium 418
investment, human capital *621–622*, 621–624, **623**
Irwin, S. H. 696
Irz, X. 415, 416
Issacharoff, S. 17

Isserman, A. 436
Italian organic fruit price volatility 543–544, *544*
Itskhoki, O. 321
Ivanic, M. 182

Jackson, L. A. 569
Jacobi, L. 451
Jacobs, K. L. 224
Jacquet, F. 89
Jaffe, A. B. 216
Jain, D. 506
Jales, M. 329
James, G. 502, 504
Jarosz, L. 116
Jaumandreu, J. 596
Jayachandran, S. 559
Jayne, T. S. 563
Jenkins, G. 455, 476
Jensen, H. H. 75
Jensen, J. B. 625, 627–629
Jentsch, C. 456
Jessoe, K. 559, 565, 572
Jevons, W. S. 276, 467, 470, 473, 479, 481n1
Jin, Y. 182, 632, 672, 729–730
John, A. 516
Johnson, D. G. 328
Johnson, G. V. 677
Johnson, R. 103
Johnson, T. G. 438
Johnston, B. 413
Jolliffe, I. T. 452
Jonasson, E. 570
Jones, R. W. 308
Josling, Tim 327, 328, 345
Journal of Agribusiness 768
Journal of Agricultural and Applied Economics 665
Journal of Agricultural and Resource Economics 665
Journal of Agricultural Economics 665
Journal of Applied Econometrics 446
Journal of Dairy Science 666
Journal of Econometrics 446
Journal of Economic Literature 445, 616, 760
Journal of Economic Perspectives 445, 448
Journal of Farm Economics 767
Journal of Food Distribution Research 768
Journal of the American Society of Farm Managers and Rural Appraisers 666
J-test 448
Judge, G. 328, 445, 453–455, 457, 460
Juroszek, P. 196
Just, D. R. 76, 85–90, 181, 406, 651, 665
Just, R. E. 12, 20, 25–26, 28, 50, 55, 90, 401, 551; on econometrics 476, 479; on risk management 605, 610
Jyoti, D. 75

Kaffo, M. 447
Kagin, J. 563
Kahneman, D. 11, 16, 75, 84, 604, 608
Kaiser, H. M. 56, 662
Kako, T. 476
Kalaitzandonakes, N. 176, 678
Kalambokidis, L. 732
Kantz, H. 537
Kaplan, S. 678
Karantininis, K. 752
Karlan, D. 554
Karni, E. 101
Karp, L. S. 56, 328, 699
Kassie, M. 559
Katchova, A. L. 696, 700
Katzir, I. 328
Katz v. Walkinshaw 278
Kayombo, A. 182
Keane, M. 22, 23
Kehoe, T. J. 310
Kelejian, H. H. 501
Kelly, D. L. 514
Kelsey, T. W. 731
Kennedy, P. L. 177, 182, 310
Kergna, A. O. 200
Kerr, S. 240
Kerr, W. A. 180, 183, 678
Kershen, D. L. 680
Keskin, P. 289
Keswell, M. 559
Key, N. 164
Khanal, A. R. 179
Khanna, M. 645, 648, 651
Kijima, M. 516
Kim, D. 459
Kim, H. 648
Kim, J. 496
Kim, K. 609
Kim, K. H. 526
Kim, K. K. 437
Kindleberger, C. P. 427
King 760, 761
King, R. G. 414
King, R. P. 479, 748, 757
Kingma, D. P. 450
Kinsey, J. 85, 761
Kitchens, C. 291, 292
Klapper, D. 65
Klein, B. 100–101
Klemick, H. 406
Kletzer, L. G. 627–629
Kline, B. 447–448
Kline, P. 289, 291–292, 294
Kling, C. 476
Kling, J. L. 479
Klinger, B. 422
Klümper, W. 672

Knapp, K. C. 290
Knight, F. H. 605
Knutti, R. 192
Koch, K. 418
Koester, U. 181
Kolesar, M. 448
Kolstad, C. D. 677
Koo, W. W. 176, 405
Koontz, S. R. 714
Koop, G. 450
Koopmans, T. J. 470
Kopel, M. 754
Kortum, S. 301, 303, 305–306, 314–315, 321
Kosa, K. A. 71
Köszegi, B. 17, 20
Koundouri, P. 290
Kovacs, K. 290
Kramarz, F. 321
Kranton, R. E. 99
Kremer, M. 552, 555
Krishna, V. 672, 676
Kristkova, Z. S. 178
Kroger stores 52
Kropp, J. D. 665, 700
Krueger, A. 400
Krueger, A. B. 512
Krueger, A. O. 381, 382
Krugman, P. R. 312, 315, 317
Krupnick, A. J. 237, 238
Kucharik, C. J. 199
Kuester, K. 458
Kugler, M. 321
Kuhn-Tucker conditions 493
Kuosmanen, T. 241
Kuwayama, Y. 274
Kuznets, S. 413
Kwiatkowski, Phillips, Schmidt, and Shin (KPSS) test 456
Kwon, D.-H. 405

labeling 104–105; country of origin 343; GE products 679, 680
labor *see* human capital
laboratory experimental studies of farmer behavior 556–557
Lachaal, L. 401
Lahiri, S. N. 450
Lai, C.-C. 404
Lai, Jing-Yi 724
Laibson, D. 19
Lake Taupo nitrogen trading program 247
Lakkakula, P. 177, 681
LaLonde, R. J. 566
Lambert, D. M. 563
Land Economics 665
Landes, D. S. 422

land-grant universities 631–633; cooperative extension system (CES) and (*see* cooperative extension system (CES))
land market spillovers in habitat conservation programs 223–224
Landon, W. 103
Langemeier, M. 659, 662
Langpap, C. 646, 648
Lapan, H. 648
large country tariffs 332
Lasso technique 452
Lastrapes, W. 404
Lauer, J. 672
Laukkanen, M. 241
Lavoie, N. 50, 53, 55–56, 66
Lawley, C. 221–222, 224–225
Law of Wages 467
Laxminarayan, R. 161, 162, 163
Layton, D. F. 161
Le, P. V. 194
Leamer, E. E. 448
Lecocq, S. 508
Lee, D. S. 446
Lee, H. 665
Lee, J. H. 459
Lee, L. F. 493
Lee, T.-H. 524
Lee, Y. N. 90
Leffler, K. B. 100–101
Leland, H. E. 98, 104
Lemieux, T. 446
Leonard, B. 277
LeRoy, S. 478
LeSage, J. P. 449
Letson, D. 183
Levine, P. B. 71
Levinsohn, J. 56, 64–66, 500, 591–592, 594, 596, 597
Levy, S. 569
Lewbel, A. 447
Lewis, D. J. 216
Lewis, K. E. 177
Lewis, W. A. 413
Leybourne, S. J. 455
Li, M. 330
Li, Q. 519, 521, 524, 525, 527, 528
Li, X. 310
Liang, Q. 752, 754
Libecap, G. D. 277, 292
libertarian paternalism 76
Lichtenberg, E. 216, 222, 672
Ligon, E. 755
Lin, C.-Y. C. 272, 275, 279, 527, 528
Lin, H. 223
Lin, L. 564
linear programming 664

794

Ling, S. 458
Lin Lawel, C.-Y. C. 272, 273, 279
Lipscomb, M. 293, 294
Lipton, M. 294
Liscow, Z. D. 527, 528
List, J. A. 86
Litchfield, J. 294
Liu, E. M. 556, 604
Liu, X. 223
livestock and climate change 196–197, 199
Lloyd, P. 310, 312
Lobell, D. B. 200
local economy-wide impact evaluation (LEWIE) 570–571
local governance and world food 45–46
Lochner, L. J. 24
Loewenstein, G. 19, 77–78
Löfgren, H. 569
Logic of Collective Action, The 381
logit models 764
Lomas, P. L. 254
Lopez, R. 514
Lopez, R. E. 476
Louis, V. L. S. 286–287
Loureiro, M. L. 71
Low, H. 446
Lowe, S. E. 292
Lowenstein, G. 17
Lu, X. 450
Lubchenco, J. 254
Lubowski, R. N. 220–221, 225
Lucas, R. E. Jr. 412
Luce, R. D. 489
Luckstead, J. 314, 322, 328, 345
lumber disputes between the U. S. and Canada 345
Lusk, J. L. 168, 673
Lütkepohl, H. 458
Luzzati, T. 525, 527
Lyambabaje, A. 179
Lybbert, T. J. 90, 557, 564, 571

Ma, S. 520
MacDonald, J. M. 58
MacDonald, S. 568
MacDougall, G. D. 305–306
machine learning 452, 501–505
MacKinnon, J. G. 448, 449
macroeconomic issues: agricultural goods as flex-priced 402; beginnings of literature on 399–401; commodities as assets in price formulations 403; future work 408; introduction to 399; models that incorporate short-run effects with long-run relationships 403–405; monetary policy and commodity prices 401–402; new linkages between agricultural and energy prices 405–406; volatility spillover effects 406–408
MacSharry Reform 365
Maddala, G. S. 493
Mahaffey, H. 674, 675
Majumdsar, B. 170
Malcolm, B. 659
Mäler, K.-G. 19
Mallory, M. L. 217, 225, 407, 479
malnutrition, triple burden of 39–40
Malthus, T. 175, 184
Mancino, L. 85
Mani, A. 270, 556
Manna, U. 290
Manning, D. 559, 570, 572
Mansholt, S. 364–365
Marchant, M. 404
Marcouiller, D. W. 437
Marcoul, P. 750–751, 755
Marine Protected Areas 145
Marini, M. A. 754
market access 332–335, *333–334*
marketing, food: competition in food retailing and 110–114, **111**; conclusions on 129–130; direct sales/local foods 114–116; introduction to 108–110, **109**, *109*; modeling approach to study direct 116–117; orders and supply quotas 359–360; seafood 149–151; structural changes in food distribution and processing and 117–129, **118–127**, **129**; vertical coordination and 128–129, **129**
marketing orders and supply quotas 359–360
market intervention, dismantling of 361–370
market structure, food industry 51–53; vertical control in 57
Marra, G. 527
Marra, M. C. 224
Marschak, J. 583
Marschak's Maxim 24
Marshall, A. 10
Martin, K. 743
Martin, L. 2
Martin, W. 179, 182, 183
Mason, N. M. 563
Mason, R. 89
Masson, R. T. 102
matching methods, RCT 557–559
Matlab 449
Matsuyama, K. 414
Mattos, F. 88
Maurer, A. 670
Mayer, W. 102
McBride, W. D. 164
McCalla, A. 328
McCarl, B. A. 182–183, 192, 195, 196, 199–201, 203, 205, 661

795

Index

McCluskey, J. J. 99, 101–104, 385, 679
McConnell, K. E. 514
McConnell, V. 212
McCorriston, S. 50, 52, 63, 65
McCullough, M. P. 101
McDonough, I. K. 447
McFadden, B. 181
McFadden, D. 491, 582
McGowan, J. J. 55
McGranahan, D. A. 438
McKenzie, D. 560, 566
McNamara, P. E. 161, 170
McPhail, L. 407, 647
McQuade, T. 102
mean reversion and unit roots 456
measurement of food access 178–179
Meghir, C. 446
Melamed, A. 328
Melitz, M. J. 315, 319, 321–322
Menapace, L. 102, 104
Mendelsohn, R. 197, 199, 200
Mendershausen, H. 583
Mensi, W. 406
Mérel, P. R. 751, 752, 753
Merenlender, A. 217
Messer, K. D. 662
Metcalf, J. 116
metrics, obesity 70–71
Meyerhoefer, C. D. 72–73
Meyers, W. 401–402
Meyers, W. H. 176, 177
Michelson, H. C. 563
microeconometrics 446–451
micronutrient malnutrition 39–40
migration 620
Miguel, E. 555, 571
Miljkovic, D. 179
Mill, J. S. 381
Millennium Ecosystem Assessment 258, **259–260**
Miller, D. 455
Miller, D. L. 449
Miller, G. Y. 161, 170
Millimet, D. L. 447, 521, 524, 526, 527
Mills, D. E. 63
Mill's Ratio 493, 560
Minimum Power Divergence (MPD) estimation 454
minimum variance hedge ratio (MVHR) 723–724
Minnesota Pollution Control Agency 246
Minnesota River Basin Trading 246
Minten, B. 181
Mishra, A. K. 115, 117, 179
Mitchell, D. 182
Mitchell, P. D. 471
mitigation, climate change 201–205
Mittelhammer, R. C. 314, 345, 453, 454, 474
mixed-root series modeling 459

Mobarak, M. A. 293, 294
Mojduszka, E. M. 104
monetary policy 401–402
Monier-Dilhan, S. 63
monopoly *311*, 311–312
Monsanto 670
Monson, J. 115
Montes, C. 254
Montgomery, W. D. 238–240
Mooney, H. 254
Moore, H. 467–469, 479, 480
Moretti, E. 289, 291–292, 294
Morgenstern, O. 604, 606
Morrill Act 361, 631, 728, 729
Moschini, G. 102, 104, 648
Moser, C. M. 565
Mosley, P. 556
Mosnier, A. 643
Moss, C. B. 179, 342, 346, 361, 379, 651, 672, 696, 699
MOTAD (minimization of the total absolute deviations) model 664
Moyo, S. 564
Msangi, S. 180
Mu, J. E. 199
Mueller, W. F. 50
Mugera, A. W. 659
Mullainathan, S. 450, 451
multiple-discrete and multiple-discrete continuous choice models 495–501
Munasib, A. 436
Mundell, R. A. 349
Mundlak, Y. 475, 476, 581, 582, 584, 585, 660
Munro, G. R. 145
Muricho, G. 559
Murphy, J. A. 12
Murphy, J. M. 357
Murphy, K. M. 20, 291
Murtazashvili, I. 448
Mussa, M. 97
Musser, W. N. 89
Myers, R. 406
Myers, R. J. 478, 647

National Animal Health Monitoring System (NAHMS) 171
National Center for Ecological Analysis and Synthesis (NCEAS) 254–255
National Development and Reform Commission, China 370
National Food and Agribusiness Management Education Commission (NFAMEC) 767
National Health Interview Survey 85
National Health Wellness Survey 73
National Origins Immigration Act 620
National Pollution Discharge Elimination System (NPDES) 244, 245

National Restaurant Association 114
National School Lunch Program (NSLP) 75
National Science Foundation (NSF) 254, 534
Natural Areas Conservation Program (NACP), Canada 214
natural capital: basic principles 255, *255*; concepts and definitions of 256–257, *257*; short history of 254–255; sustainability and 265–266; *see also* ecosystem services
Nature 255
Nature Conservancy (TNC) 219
Nature's Services: Societal Dependence on Natural Ecosystems 254
Nauges, C. 290
Nayga, R. M. Jr. 71
Neary, J. P. 313
Neilson, W. S. 89
Nelson, C. R. 22, 455
Nelson, H. 412
Nelson, P. 99–100
neoclassical theory 660, 690–694, *692–693*
Nerlove, M. 476, 586–587
Neuse River Basin Total Nitrogen Trading 247
Neven, D. 98
Nevo, A. 56, 64, 65
Newbold, P. 455
Newburn, D. S. 217
new empirical industrial organization (NEIO) 50–51, 55–56, 763
"new" industrial organization analysis of the food industry 53–56, *55*
New Trade Theory 310–314, *311*, 323n3
Neyman-Rubin-Holland model 471
Nguyen, N. P. 249
Nguyen Van, P. 524
Nie, Z. 520, 528
Nielson, R. 479
Nisbet, R. 451
Noleppa, S. 177
noncausal autoregressions (NCARs) 456–457
nonlinear models 458
nonlinear time series analysis (NLTS) methods: case study illustrations 542–544, *543–544*; conclusion on 545; diagnostics strategy 536–542, *537*, *541*; introduction to 533–536, *535*; phase space reconstruction 534–535, *535*, 538–539; phenomenological (data-driven) modeling 540–542, *541*; with R 542; signal processing 537–538; stationarity testing 538; surrogate data testing 539–540
nonstationarity in cross-sectional distributions 457
non-structural modeling with observational data 476–479
non-tariff distortions (NTDs) 335–341, *336*, *338*, *340–341*
nontraditional grocery retailers 110–111, **111**
Nord, M. 75

Nordhaus, W. D. 197, 205, 438
North American Free Trade Agreement (NAFTA) 327, 331–332, 347, 382, 512, 569
North American Industrial Classification System (NAICS) 51–52, 118, 125
North American Waterfowl Management Plan (NAWMP) 214, 219–220, 221, 227n3
Norwood, F. B. 168
nutrition, adequate 180–181

Oates, W. E. 237, 238
Oaxaca-Blinder regression model 294
obesity 39–40; behavioral economics and 75–77; causes of 71–72; concluding remarks on 77–78; consequences of 72–73; food utilization and 181; government interventions for 73–75, **74**; introduction to 70; metrics of 70–71
ocean ecosystem 151–152
O'Donnell, C. J. 597
O'Donoghue, C. 666–667
O'Donoghue, T. 19
Oehmke, J. F. 179, 391
Ojede, A. 659
Okrent, A. M. 75, 475
Olea, J. L. M. 447
Olesen, H. B. 757
oligopolistic competition 312–313
Olley, G. S. 587–591, 592–593, 594, 596, 597
Olmstead, A. L. 101
Olper, A. 385
Olson, L. J. 290
Olson, M. 381, 384, 632
Olynk Widmar, N. J. 169
Omay, R. E. 525
Onel, G. 179
Oostendorp, R. H. 565
optimal revenue tariffs 334
optimal welfare tariffs 332–333, *333*
Orazem, P. F. 622
Orden, D. 360, 403, 478, 480
Organization for Economic Cooperation and Development (OECD) 415, 677
Organization of Petroleum Exporting Countries (OPEC) 647
Orozco, V. 65
Orsini, M. 525, 527
Ortega, D. L. 166, 169
Otani, Y. 25
Otsuki, T. 170
Ottaviano, G. I. 321
ownership, asset 701–704, *702–703*

Paarlberg, R. 682
Pace, R. K. 449
Padberg, D. I. 104
Painter, J. E. 76–77
Pakes, A. 56, 64–66, 587–594, 596–597

Palmer Drought Severity Index (PDSI) 195
Pan, X. 168
Pande, R. 291, 293
Pandit, M. 526, 528
panel data 450, 564–565
panel methods, RCT 561–565
panel unit roots 457
Parente, S. L. 413, 414, 415
Paris Climate Agreement 202
Park, H. I. 724, 725
Park, J. 199
Park, T. 116, 117
Parker, D. P. 212, 224
Parkhurst, G. M. 219
Parmeter, C. F. 528
Paroulo, P. 459
partial equilibrium welfare analysis 24–25, *25*, 27
Partridge, M. D. 438
Patterson, R. 104
Paudel, K. P. 116, 180, 270, 520, 521, 524, 526, 528
Pauw, K. 568
Pearce, D. 254, 258
Pearl, J. 478, 480
Pearson, K. 452
Pecchenino, R. 516
penalized regression 502–503
Pender, J. L. 438
Pennings, J. M. 88
Pennsylvania Nutrient Credit Trading Program (PNTP) 247
Perdue Farms 169
Perloff, J. M. 56, 328
Permanent Cover Program (PCP), Canada 213
Pesaran, M. H. 450
Pesek, J. D. 168
Petersen, A. 501
Peterson, H. H. 87–89
Petrin, A. 591–592, 594, 596, 597
Pettenuzzo, D. 457
Pfeiffer, L. 272, 275, 279
Pflueger, C. 447
phase space reconstruction 534–535, *535*, 538–539
phenomenological (data-driven) modeling 540–542, *541*
Phillips, P. C. B. 455, 459
Phillips, P. W. B. 678
Phu, N. V. 521, 525, 527
Phu Nguyen, V. 519
Piese, J. 564
Pieters, H. 178, 388
Pinkse, J. 501
Pinto, R. 619
Pischke, J. 448
Planet Networks 226
Plantinga, A. J. 72, 216, 220–221
Plosser, C. R. 455

Pofahl, G. 52, 496, 501
Pohler, D. 754
Polasky, S. 216, 217, 223
policy instrument choice 386
political economy: of food standards 389–390; GE technology and 681–682; introduction to 381–382; policy instrument choice in 386; policy interactions and 391–392; political coalitions in 382–383; price and trade interventions 383–385; price shocks and 387–389; of public investment in agricultural research 391
political instability 43
political interaction effects (PIEs) 391–392
political reforms 385
pollution *see* water quality trading (WQT)
pollution effect system (POS) 238–239
Ponticelli, J. 416, 419–420
Pope, R. D. 581, 605, 610, 659
population growth, global 177
Porter, M. 761
Poskitt, D. S. 458
Potential Outcome Average Causal Effect model 472
Poudel, B. N. 521
poultry slaughtering 52
Powell, J. L. 521
Powell, S. J. 343
Prairie Habitat Joint Venture (PHJV), Canada 214, 221
Pratt, J. W. 604, 607
Precautionary Principle 678, 681
Prelec, D. 19
Prescott, E. C. 413
Price, J. 86
Price, S. 178
prices: commodities as assets in formulations of 403; flexibility of 402; in futures market hedging *715*, 715–718; global development policy and 388–389; impacts of biofuels on 645–649, *649*; income support and *358*, 358–359; monetary policy and commodity 401–402; new linkages between agricultural and energy 405–406; shocks to, political economy and 387–389; time series data on (*see* nonlinear time series analysis (NLTS) methods); and trade interventions in political economy 383–385; volatility and production instability of 182
private labels 63
processing, food 51–52; beverage and 122–129, **123–127, 129**; buyer power 57–59; structural changes in food distribution and 117–129, **118–127, 129**; vertical market control in 57
producers: behavioral welfare economics for 13–16; compensation supports for 360–361
product cycling 626

production functions: characterizing production technologies and 582–583; Cobb-Douglas 415, 417, 583–584; concluding remarks on 598; endogeneity and 583–584; estimation of 583–594; extensions 594–597; fixed effects and instrumental variables in estimating 584–587; overview of 581–582; structural solutions in estimating 587–594

production quotas 335

Productive Safety Net Program (PSNP) 555, 559

Programa de Apoyos Directos al Campo (PROCAMPO) 555

PROGRESA program 549, 553–554, 559

project and policy evaluation in welfare economics 14–15

propensity score matching (PSM) 557–559

property rights 277–278

Prucha, I. R. 501

Prugh, T. 257

Pruitt, J. R. 168

Purcelll, W. D. 714

Putsis, W. P. 65

Qaim, M. 569, 672, 676

Qian, J. 450, 520

QJE 512

Qu, X. 450

quality, product: endogenous 99; exogenous 98–99; future research on 105; horizontal or vertical product differentiation and 97–98; introduction to 96–97; policy implications of 104–105

quantitative methods in modern farm management 661–662

quasi-experiments 472–473

Quetelet index 70–71

Quiggin, J. 20, 605

Rabin, M. 17, 19, 20, 89

Racine, J. 519–520

Racine, J. S. 520, 528

Raiser, G. 2

Rajagopal, D. 647, 648

Ramaratnam, S. 471

Rambaldi, A. N. 457, 478

Ramsey equation 193

Rancher–Cattlemen Action Legal Foundation (R-CALF) 344

random behavior in welfare economics 19–20

randomized control trials (RCTs) 549; experimental challenges with 554–556; instrumental variables 559–560; laboratory experimental studies of farmer behavior 556–557; matching methods 557–559; panel methods 561–565, 564–565; studies of agricultural impacts of non-agricultural interventions using 553–554; studies of impacts of agricultural interventions using 551–553; switching regression 560–561

Rangachari, R. 292

Rangel, A. 17, 20, 25

Rangel, G. 670

Rao, S. S. 456

Rapsomanikis, G. 405

Rashad, I. 72

Rausser, G. C. 387, 403, 475, 479

Read, T. 453

Rebelo, S. T. 414

recursive binary splitting algorithm 504

Redding, S. 321

Reed, M. 328, 350, 404

reform, policy: in Canada 366–369, *369*; in the European Union 364–368; in the United States 361–364

reforms, political 385

regional trade agreements (RTAs) 330–332, 563

Regmi, M. 180

Regnier, E. 149

regression discontinuity (RD) designs 446–447

regularization 502–503

regulation: costs of 677–679; intellectual property rights (IPR) 676–677

Reilly, J. M. 195, 199

Reimer, J. J. 310, 313, 330

Reiter, S. 664

relative income hypothesis 384

Renewable Fuels Standard program 183, 203, 638, *639*

Rennhoff, A. 63, 64

Rephann, T. 436

reputation 99–100; collective 101–102; dual 102; empirical work on 102–104; firm 100–101; future research on 105; policy implications of 104–105

Réquillart, V. 63, 65, 506

research and development (R&D): estimating endogenous productivity and 595–596; food availability and 177–178; nonparametric approach to production functions in 596–597; public investment in 391

resource constraints and food availability 178

retailing, food 52–53; agribusiness economics and management (AEM) and 761–765; competition in 110–114, **111**; emergence of direct sales/local foods in 114–116; evaluating vertical coordination in 63–65; growth of nontraditional 110–111, **111**; private labels 63; retailer buyer power and 60–65; strategic responses by traditional grocery outlets 113–114; vertical market control in 57

Rey, P. 756–757

Reynaud, A. 89

Reynolds, T. 183

Rhode, P. W. 101

Ricardian theory 301–306, *302, 304*
Ricardo, D. 301, 381, 393n19
Ricci, J. A. 73
Rice, G. 459
Richards, T. J. 52, 64–65, 495, 496, 501, 508
Rickard, B. 75, 104
Ricker-Gilbert, J. 563
Rickman, D. S. 436
Ridder, G. 526
Risk Adjusted Discount Rate (RADR) 695
risk and uncertainty: assessing exposure to 604–605; in cooperatives 755; diversification for 609; dynamics of 611; farm management for 610; hedging and 611; insurance and 370–372, 612–613; introduction to 603–604; management and policy 609–613; risk preferences and cost of 606–609; risk transfers for 611–613; technology and 610
risk transfers 370–372, 611–613
Roberts, M. J. 195, 225, 310
Robertson, J. 403
Robertson, J. C. 478
Robinson, J. 552
Robinson, P. M. 519, 521, 524, 527
Robinson, S. 569
robust variance estimation (HCCME) 448–449
Rodriguez-Poo, J. M. 527
Roe, B. 168
Roe, T. L. 412, 415, 416, 417, 418, 423
Rogerson, R. 413, 414, 415, 416, 417, 423
Rogerson, W. P. 101
Rohrer, W. C. 743
Rojas, C. 501
Rojo-Gimeno, C. 164
Romano, J. P. 449
Roosen, J. 168
Roosevelt, F. D. 361
Rosen, S. 97, 103
Rosenbaum, P. R. 557–558
Rosenzweig, M. R. 416, 419, 422, 561–562, 564, 620
Rosin, O. 76
Rossi, P. E. 496
Rothbard, M. 677
Rothenberg, T. J. 527
Roy, N. 521
Rozelle, S. 559
R package 542
Rpuasingha, A. 437
Rubin, D. B. 446, 472, 549, 557–558
Rubio, S. J. 288
Ruhm, C. J. 71
rules of origin (ROO) 331
Rungie, C. 168
Rupasingha, A. 438
Ruppert, D. 526

rural development: conclusion on 439; introduction to 427; methods of **435**, 435–438; rural wealth framework and 438–439; theories of economic development and 428–434
rural wealth framework 438–439
Rutherford, T. F. 28
Rybczynski effects 418

Sacks, W. J. 199
Sadoulet, E. 563, 567, 568
Safeway stores 52
Saffer, H. 72
Sage, J. L. 116
Saghaian, S. 328, 350, 404, 406
Saikkonen, P. 456
Saitone, T. L. 51, 52, 58–59, 169, 170, 751, 752
Salant, S. W. 102
Salassi, M. 177
Salop, S. C. 97
Samek, A. S. 86
Sampson, G. S. 151
Samuelson, P. A. 301, 303, 310, 328, 473, 570
Sandberg, R. 456
Sanderson, E. 447
Sandmo, A. 87
Sanint, L. R. 698
sanitary and phytosanitary measures 339–340
Santos, A. 528
Saracoglu, D. S. 418, 423
Sargan test 448
Sarsons, H. 292
Savage, L. 603, 606
Savelyev, P. 619
Scandizzo, P. 604
Scarf, H. 28
Schaefer, M. B. 140
Schamel, G. 103
Scharfstein, D. S. 164
Scheines, R. 478
Schiff, M. 382
Schlenker, W. 195, 197
Schmit, T. M. 438
Schmitz, A. 12, 25, 26, 177, 182, 221, 479; on agricultural support 343, 357, 360, 361, 373, 379; on biotechnology 672; on ethanol 651; on international agricultural trade 328, 332, 337, 342, 347, 349
Schmitz, T. G. 177, 181, 332, 337; on agricultural support 342, 346, 360, 361, 379; on biotechnology 681; on ethanol 651
Schneider, U. A. 196, 201, 203
Schnitkey, G. 659
Schott, P. K. 625
Schreiber, T. 537, 538
Schroeder, K. G. 177
Schroeder, T. C. 87, 478
Schroeter, J. R. 55

Schubert, K. 149
Schuh, E. 399, 400
Schultz, T. W. 616, 617, 633
Schulz, F. 408
Schumacher, S. 254
Science 254
Scitovsky, T. 12
Scott, J. M. 217
seafood: aquaculture and 145–149, *146*, *148*; concluding remarks on 152; defined 152n1; fisheries 140–145, *141–142*; introduction to 134–135; markets and trade in 149–151; ocean ecosystem and 151–152; production of *136–137*, 136–139
Seale, J. Jr. 183
Seale, J. L. 337
Searchinger, T. 643, 644
Sears, L. 272, 273
Secchi, S. 161, 162–163
Sedlacek, J. 192
Seidl, A. 570
Seidu, A. 179
semiparametric and nonparametric methods in econometrics 450; environmental Kuznets curve (EKC) analyses 512–528
Sen, A. 179
Senauer, B. 761
Sene, S. O. 116
Seni, G. 453
Seo, S. N. 199
Serra, T. 405, 406, 407, 408, 647
service sector 423
Severnini, E. R. 292, 294
Severson, K. 101
Sexton, R. J. 50–53, 55–56, 58–59, 66, 169, 170, 314, 761; on cooperatives 748, 751, 752
Sexton, S. 673
Shaffer, G. 61–62
Shafik, N. 512
Shah, F. A. 290
Shah, P. 217, 218
Shaked, A. 97
Shakhramanyan, N. G. 196
Shakya, P. B. 290
Shapiro, C. 100, 102, 104, 105
Shapiro, E. N. 222
Shapiro, J. M. 71
Shapouri, H. 641
Shaw, D. 197, 239
Sheldon, I. M. 55–56, 310
Shepherd, R. W. 596
Sherman Anti-Trust Act 362
Sherrick, B. J. 696
Shi, G. 610, 672, 673
Shi, H. 474
Shiferaw, B. 559, 561
Shiping, L. 407

Shively, G. 563
Shleifer, A. 291
Shmueli, G. 451
Shonkwiler, J. S. 476
shopping-basket models 498–500
Shortle, J. 240
shrimp imports 344
shrinkage estimation in econometrics 450
Shumway, R. H. 460, 476
signal processing 537–538
Silbur, W. L. 717
Simon, H. A. 75
Sims, C. 477, 478, 480
Sims, K. R. 222
simulations 566–571; agricultural household (AH) 566–567; computable general equilibrium (CGE) 28–29, 322, 437, 568–570; general equilibrium (GE) impacts 567; local economy-wide impact evaluation (LEWIE) 570–571
Singh, I. 566
singular spectrum analysis (SSA) method 537–538
Sitienei, I. 179
Skees, J. R. 87
Slade, M. E. 500, 501
slaughtering, animal 52
slippage in habitat conservation programs 222
small country tariffs 332, *333*
Small Is Beautiful: A Study of Economics As If People Mattered 254
Smarter Lunchrooms program 86
Smeekes, S. 457
Smith, A. 261, 381, 393n19, 422, 548
Smith, C. E. 103
Smith, J. A. 558, 566
Smith, M. D. 144, 147
Smith, R. B. W. 412, 418, 423
Smith, V. H. 183
Smith-Lever Act 361, 631, 728
Smith-Ramirez, R. 222
Smyth, S. J. 183, 678
Sneeringer, S. 164, 166
Soberon, A. 527
Social Accounting Matrices (SAMs) 436
social capital 256
social cash transfers (SCTs) 553–554
social value of futures trading 714–720, *715*, **718–719**
Soil Conservation and Domestic Allotment Act 213
Solon, G. 449
Solow, R. M. 413, 422
Song, I. 498
Soregaroli, C. 679
South Nation River Watershed Trading 247
spatial econometrics 500–501

spatial interactions and temporal spillovers in habitat conservation programs 224–225
spatial management of water resources 271–274
spatial methods in econometrics 449–450
Species at Risk Act, Canada 212
Sperling, R. 55–56
Spiess, J. 450, 451
Spirtes, P. 478
Spolador, H. 416, 417
Sporleder, T. 768
Spreen, T. H. 661
Spriggs, J. 403
Springborn, M. R. 279
Squire, L. 566
Srinivasan, T. N. 569
Staatz, J. M. 748
stability and food security 181–184
Staiger, R. W. 313
Stamoulis, K. 403
standards, food 389–390
Startz, R. 22
Stata Journal 446
State Water Efficiency and Enhancement Program (SWEEP), California 274, 275–276
stationarity testing 538
Statistical Analysis System (SAS) 446
Stavins, R. N. 216, 220–221
Steffey, K. L. 471
Steger, T. M. 418
Stein, J. C. 164
Steiner, P. O. 99
Stern, D. I. 517
Stern, N. 193, 194
Stern Review 205
Sterns, J. 117
Stevanovic, D. 458
Stiegert, K. 313
Stigler, J. G. 381
Stiglitz, J. E. 705
Stillman, S. 560
Stine, O. C. 479
Stock, J. H. 447, 450, 460
stock-holding buffer fund stabilization 357, *357*
Stoffer, D. S. 460
Stokey, N. L. 412
Stone, R. 436
Stone-Geary type preferences 415, 417, 420
St. Petersburg paradox 87
Strauss, J. 566
Strobl, E. 286, 293, 524
Strobl, R. O. 293
structural modeling in welfare analysis 24
structure-conduct-performance (SCP) approach 50, 55
Su, L. 450
Subramanian, A. 569

subsidies 202; export 335–337, *336*; input and output 377–379, *380*; *see also* agricultural development
Sudhir, K. 64
sugar disputes 343–344, 346
sugar-sweetened beverages (SSBs), taxes on 73–74, **74**
Sugden, R. 21
Sumaila, U. R. 145
Sumner, D. A. 75, 169, 170
supercenters, grocery 110–111
supervised learning 451
Supplemental Nutrition Assistance Program (SNAP) 75, 86
supply: E. J. Working on 468–469; Heckscher-Ohlin (H-O) theorem and 306–310, *307*; IMPACT model supply and demand projections 37–39, *38–39*; price volatility and production instability 182; quotas and marketing orders 359–360; supply chain efficiency 180
supply, water *see* water supply and dams
supply chain efficiency 180
supply management 377–380, *378*, *380*
support vector machines (SVM) 503
surface water management 270
surrogate data testing 539–540
suspension agreements 341
sustainability and natural capital 265–266
Sustainable Development Goals (SDGs) 45
Sutton, J. 97
Suzuki, N. 56
Swallow, S. K. 220
Swanson, N. 478
Swigert, J. M. 181
Swinnen, J. 178, 384, 385, 387, 388, 391
switching regression 560–561
Sykuta, M. E. 119

Taber, C. 24
Tack, J. 673
Taheripour, F. 644, 646–648, 651–652, 674, 675
Tajibaeva, L. S. 413
Takayama, T. 328
Takeshima, H. 292
Tan, P.-N. 451
Tanaka, T. 556
tariff rate quotas (TRQs) *334*, 334–335, 360, 370
tariffs, import 332–334, *333*
taxes, export 340, *340*
Taylor, C. R. 724
Taylor, E. J. 559, 572
Taylor, J. 403
Taylor, J. E. 559, 563, 565, 567, 570, 571
Taylor, M. S. 514
Taylor, R. 176
Taylor, T. 345

technological innovation 43–44, 419; cooperative extension system (CES) and 743; food availability and 176–177; labor-augmenting 420; risk management and 610
Teele, D. L. 556
Teillant, A. 163
Teisl, M. F. 151, 168
Tejeda, Herman A. 725
Tembo, S. T. 563
temperature-humidity index (THI) 199
Tennessee Valley Authority (TVA) 287, 289, 291–292, 294
Territorial Use Rights Fisheries (TURF) systems 144
Teshome, Y. 178
Thaler, R. 75, 84, 88
Thayer, A. 182–183
Theory and Practice of Econometrics 457
Theory of Political Economy, The 467
Thirtle, C. 564
Thisse, J.-F. 97, 98, 500
Thomas, W. 664
Thome, K. 178, 571
Thompson, R. 401
Thorbecke, E. 565
Thurlow, J. 568
Thurman, W. N. 224
Tibshirani, R. 452, 527
Tiffen, R. 416
time series analysis 455–460; convergent cross mapping 535–536, 540; reconstructing deterministic economic dynamics from volatile (*see* nonlinear time series analysis (NLTS) methods)
Tintner, G. 471, 475
Tirole, J. 101, 756–757
tobacco quotas 335
Tobol, Y. 76
Todd, P. E. 558, 566
Tollens, E. 678
Tong, H. 458
total allowable catch (TAC) 134–135
Total Economic Valuation (TEV) 262
total factor productivity (TFP) 415, 416, 418, 421
Towe, C. 221–222
trade, international *see* international agricultural trade
trade remedy actions 337–338
trade theories: conclusions on 322–323; Heckscher-Ohlin (H-O) theorem 306–310, *307*; heterogeneous firms and 314–322, *319–320*; introduction to 301; New Trade Theory 310–314, *311*, 323n3; Ricardian 301–306, *302, 304*; *see also* agricultural policy
trading-ratio system (TRS) 239
Train, K. 491

transitional economies, agricultural trade in 347–349, *348*
Trans-Pacific Partnership (TPP) 331–332, 347
Traxler, G. 564, 660
tree-based methods 503–505
Tréguer, D. 647
Trimborn, T. 418
Trivedi, P. K. 116, 445
Tronstad, R. 724
Trost, R. P. 493
Trujillo-Barrera, A. 407, 479
Truong, C. H. 288–289
Tsai, F. T.-C. 270
Tsay, R. 458, 460
Tsur, Y. 270, 279, 290, 293
Tsurumi, T. 517
Tullock, G. 381
Turner, R. K. 265
Turnovsky, S. J. 87
Tversky, A. 11, 16, 75, 604, 608
Twyman, J. 181
Tyner, W. 644, 646–648, 651–652, 674, 675
Type I errors 452
Type II errors 452
Tyson Foods 169
Tzouvelekas, V. 290

Uchida, S. 222
Udry, C. R. 562, 565, 571
Uematsu, H. 115, 117
Ulen, T. S. 677
Ullah, A. 520
uncertainty *see* risk and uncertainty
undernourishment 39–40
UNFCCC 200
United States Department of Agriculture (USDA) 52, 58, 75; behavioral tools used by 91; block grant funding of agricultural research and education 632; Environmental Quality Incentives Program 246; food quality and 99; Food Safety Inspection Service 169; National Organic Standard (NOS) 116
United States Department of Health and Human Services (HHS) 75
United States Geological Survey (USGS) 226
unit roots: mean reversion and 456; in noncausal ARs 456–457; panel 457; testing 455–456
Unnevehr, L. J. 474
urbanization *40*, 40–41; institutional innovation and 44–45
Urga, G. 458
Uruguay Round Agreement on Agriculture (URAA) 329, 382
Urzúa, S. 23, 24
Useche, P. 181
utilization and food security 180–181

Index

Valderrama, D. 143, 148
Valdez, A. 328, 382
Valentinyi, A. 413, 416, 417
Valenzuela, E. 569
Valletti, T. M. 60
valuation, ecosystem services 261–265, **263**, **264**
Van den Belt, H. 678
Van De Verg, E. 237, 238
Van Duyne, C. 402
Vangeneugden, T. 521
van Ittersum, K. 76–77
van Kijk, M. 178
van Kooten, G. C. 221, 521
van Meijl, H. 178
Van Wijnbergen, S. 569
variable selection, big data 451–452
Varian, H. R. 451, 505, 661
Variyam, J. N. 85
Vásquez-Lavin, F. 496
Vector Autoregressive Model (VAR) 401–403, 455, 458–459, 477–479
Vector-Error Correction (VEC) model 403–404
Verboven, F. 492
Vercammen, J. 479
Verhoogen, E. 321
Verma, R. 416, 423
Verschoor, A. 556
vertical coordination 63–65, 128–129, **129**; of cooperatives 756
vertical market control 57
vertical product differentiation 97–98
vertical restraints 61–63
Viaene, J.-M. 309
Vickers, J. 61
Viju, C. 183
Vilcassim, N. J. 506
Villas-Boas, S. B. 63–65, 151, 506
Villavicencio, X. 195
Vilsack, T. 729
Vincent, J. R. 517
Vishny, R. W. 291
volatility, price 388; spillover effects of 406–408
Vollrath, D. 416, 420–422
voluntary export restraints 341, *342*
Von Neumann, J. 604, 606
Von Tiedemann, A. 196
von Ungern-Sternberg, T. 60
von Witzke, H. 177
Vosti, S. A. 75
Vytlacil, E. J. 23, 24

Wade, M. A. 170
Wager, S. 453
Wahba, S. 566
Wahl, T. I. 474
Wailes, E. J. 176
Wales, T. J. 493
Waller, M. L. 479
Walls, M. A. 212
Walmart stores 52–53
Walras, L. 480
Wand, P. 526
Wang, H. 117
Wang, H. H. 169
Wang, M. Q. 641
Wang, S. 521, 524, 525, 528
Wang, S. L. 729
Wang, T. 166
Wang, V. 404
Wang, W. 205, 447
Wansbeek, T. 447
Wansink, B. 76–77
Ward, M. B. 65
waste, food 181
water management and economics: climate change and 279; conclusion on 279–280; dynamic 270–271; groundwater-energy nexus 278–279; introduction to 269–270; perverse incentives from policy 274–277; property rights and 277–278; spatial 271–274; surface 270
water quality trading (WQT): basic cap-and-trade model in 233–236; concluding comments on 248–249; cost-minimization and net benefit maximization 249n2; dynamics of 239–241; emissions reduction credit (ERC) trading mechanisms in 243–246; introduction to 232; multiple pollutants and 241; nonuniform mixing and multiple receptors in 236–239; programs for 246–248; theoretical background of 233–246; trade ratios 250n3; unobservable emissions and stochastic processes in 241–243
Waterson, M. 55, 60
water supply and dams: background on 285–287; conclusions on 294–295; economic design modeling implications of 287–289; farmers' response to 289–291; impact of 291–294; introduction to 285
Watson, M. W. 450, 460
Waugh, F. 469–470, 481n8, 660, 664
Wauthy, X. Y. 98
weather-indexed insurance 371–372
Webb, M. D. 449
Weber, B. A. 438
Weiss, A. 705
Weitzman, M. L. 570
welfare economics: alternative methodological approaches in 28–29; behavioral 13–19, 22; conceptual generalizations in welfare measurement in 11–13; conclusions on 29; general equilibrium welfare analysis in 24–28, *25*, *27*; introduction to 10–11; mistakes in decision making and random behavior in 19–20; sufficient statistics for welfare analysis

in 24; when should policy makers disregard revealed preferences in 21
welfare impacts of biofuels 649–652, **650**
Welling, M. 450
Welton, G. 349
Wesseler, J. 678, 679, 681
Western Grain Stabilization and the Special Grains Payment 366
Westman, W. 254
Wetland Reserve Program (WRP) 213
Whalley, J. 569
Whitaker, J. B. 700
White, H. 448, 457, 520, 524
Wicke, B. 645
Wieczorek, A. M. 670
Wilcove, D. S. 217
Wilde, P. 75
Wildlife Habitat Incentive Program (WHIP) 213
Wilen, J. E. 135, 144
Williams, G. 451
Williamson, O. E. 58
Willig, R. D. 12, 16
willingness to accept (WTA) 261
willingness to pay (WTP) 11–13, 261; aggregation and discrete choice under heterogeneity and 18–19; aggregation under heterogeneity and 16; antibiotic use in animals and 159, 167–170; behavioral welfare economics for producers and 13–14; change in revenue and 30n8; ex post compensation and 30n7; measurement for price changes 30n6; project and policy evaluation and 14–15; when should policy makers disregard revealed preferences and 21
Wilson, C. A. 303
Wilson, G. T. 459
Wilson, J. S. 170
Windmeijer, F. 447
wine industry 393n17
Winfree, J. 101, 102
Winter, E. 569
Wolf, C. A. 168
Wolf, M. 449
Wolff, H. 212
Wolski, M. 457
Wolverton, A. 406
Womack, A. 401
Women's, Infants, and Children Program (WIC) 75, 86
Wong, L. 221
Woodland, A. D. 418, 493
Woods, A. 525
Woodward, W. A. 460
Wooldridge, J. M. 445, 448
Wooton, I. 569
Working, E. J. 22, 468–469, 480
World Bank 182

World Commission on Dams 286, 287
world food: climate change and 41–42; conclusions on 46; conflict and 42–43; current and emerging challenges in 39–43; future strategies to fill the gaps in 43–46; IMPACT model supply and demand projections 37–39, *38–39*; introduction to 37; malnutrition and 39–40; reforms in global and local governance and 45–46; uncertainty in global political landscape and 43; urbanization and *40*, 40–41
World Food Program 180
World Food Summit 175
World Health Organization (WHO) 70, 160
World Organisation for Animal Health 160–161
World Trade Organization (WTO) 138, 150, 320, 327, 329–330, 368, 370, 382; 1985 Farm Bill and 362–363; Agreement on the Application of Sanitary and Phytosanitary (SPS) Measures 339–340; price and income support and 358; trade remedy actions 337–338
World Wildlife Fund 286
Wozniak, S. J. 117
Wright, M. G. 670
Wright, P. 469, 480, 481n6
Wright, S. 480
Wu, F. 647
Wu, J. 215, 216, 222, 223, 646, 648
Wu, S. Y. 58–59
Wu, X. 195, 199

Xabadia, A. 290
Xie, Y. 288
Xu, Z. 672

Yang, D. 450, 560
Yang, W. 224–225
Yaniv, G. 76
Yao, F. 527
Yao, W. 458
Yatchew, A. 525
Yeaple, S. R. 321
Yellowstone National Park 213
Young, R. A. 290
Yúnez-Naude, A. 563, 570

Zaal, F. 565
Zacharias, T. P. 664
Zago, A. 752
Zanin, L. 527
Zanin, M. 455
Zapata, H. O. 457, 478, 717
Zemel, A. 279
Zhai, F. 322
Zhang, J. 527
Zhang, L. 183
Zhang, M. 53

Index

Zhang, Y. W. 200
Zhang, Z. 407, 647
Zhao, J. 288
Zhao, Y. 64
Zheng, J. X. 521
Zhou, Z.-H. 453
Zhu, J. 450
Zhu, M. 332

Zilberman, D. 89, 405, 551, 681; on biofuels 647, 648; on biotechnology 672, 673, 675, 676, 678, 681; on food security 182; on habitat conservation 215; on water supply and dams 288, 290
Zink, C. F. 17
Zublena, J. 743
Zusman, P. 328